ALUMINIUM–LITHIUM ALLOYS III

Proceedings of the
Third International
Aluminium-Lithium Conference
sponsored and organized by
The Institute of Metals

University of Oxford
8-11 July 1985

EDITED BY

C Baker
Alcan International Ltd, Banbury, Oxfordshire

P J Gregson
The University of Southampton

S J Harris
The University of Nottingham

C J Peel
Royal Aircraft Establishment, Farnborough, Hampshire

The Institute of Metals
LONDON
1986

Book 358
ISBN 0 904357 80 5

published in January 1986 by
The Institute of Metals
1 Carlton House Terrace
London SW1Y 5DB

British Library Cataloguing in Publication Data

International Aluminium–Lithium Conference
(3rd : 1985 : University of Oxford)
Aluminium–Lithium Alloys III : proceedings of the
third International Aluminium–Lithium Conference
sponsored and organised by The Institute of Metals,
University of Oxford, 8–11 July 1985.

1. Aluminum–Lithium alloys

I. Title II. Baker, C. III. Institute of Metals
669'.722 TN775

ISBN 0-904357-80-5

Compiled and produced by the Institute's CRC Unit
from original typescripts and illustrations
provided by the authors.

Printed and made in England by
Dotesios Printers Ltd, Bradford-on-Avon, Wiltshire

Aluminium-Lithium Alloys III

VOLUME CONTENTS (Books I and II)

Paper Number at Conference shown in brackets

BOOK II

APPENDIX

SESSION CHAIRMEN

C BAKER

Alcan International Ltd., Banbury (UK)

R GRIMES

Alcan International Ltd., Banbury (UK)

P J GREGSON

University of Southampton (UK)

V GEROLD

Max-Planck-Institut für Metallforschung, Stuttgart (FRG)

L SABETAY

Cegedur Pechiney, Paris (France)

H M FLOWER

Imperial College, London (UK)

S T ERICSSON

Linköping University (Sweden)

R R SAWTELL

Alcan Laboratories, Pennsylvania (USA)

H JONES

University of Sheffield (UK)

S J HARRIS

University of Nottingham (UK)

P J WINKLER

Messerschmitt-Bölkow-Blohm GmbH, München (FRG)

N E PATON

Rockwell International, Canoga Park, California (USA)

M V HYATT

Boeing Commercial Airplane Company, Seattle, Washington (USA)

E A STARKE

University of Virginia, Charlottesville (USA)

FOREWORD by the Chairman of the Conference Organising Committee

C BAKER

THE THIRD International Aluminium–Lithium Conference, held in Oxford, England from the 8th to the 11th July 1985, was organised by the Metals Technology Committee of The Institute of Metals. It was attended by 268 delegates from 16 countries predominantly representing the aerospace and aluminium industries. A total of 66 papers were presented and 16 poster contributions exhibited.

The topics included in these proceedings have been rearranged from the programme order so as to group compatible subjects together. Book I covers the subjects of production, products and joining, fatigue behaviour, corrosion and oxidation, stress corrosion, hot forming and RSR, powder and mechanically alloyed materials. Book II includes physical metallurgy and structure-property relationships. Each poster paper has been included in the section appropriate to its subject matter. Since the order of the papers presented here differs from the conference programme, the original paper numbers are shown (bracketed) in the table of contents.

The number of papers and delegates reflects the increasing interest in Al–Li alloys particularly for aircraft structural applications. The papers from the major aluminium producers Alcan, Alcoa, Cegedur-Pechiney and Reynolds showed that commercial production on a limited scale is now underway. Three alloys (2090, 8090 and 8091) have been registered with the Aluminium Association.

Some alloys are achieving production and property targets in sheet, plate, extrusions and forgings. The fatigue properties appear to show improvements over conventional alloys but fracture toughness and stress corrosion, particularly in the short transverse direction, are often below the specified values. The loss of lithium on heat treatment is also a problem but fears concerning salt bath heat treatment of Al–Li alloys seem to be unfounded.

Al–Li alloys have also been shown to be superplastic and this property could extend the market. Studies of RSR and powder metallurgy of Al–Li alloys have suggested that it may be possible to obtain higher performance alloys by these routes.

The foreword to Book II deals with the physical metallurgy of these alloys, while the Concluding Summary by Dr Peel covers the conclusions from the conference in more detail.

Overall the conference was highly successful, the session audience being maintained to the end. The increase in attendance and in the number of presented papers above what was on offer at the 1980 and 1983 conferences is a reflection of the growing importance of Al–Li alloys. The imminent availability of the wrought products on a larger scale will add further impetus to whole development.

I should like to thank the members of the organising committee, Dr C J Peel, Dr S J Harris and Dr P J Gregson, for their work and Dr E A Starke Jr and Dr T H Sanders Jr for their help in ensuring that the Third Conference formed part of a series. Also I wish to thank the authors, session chairmen, Geoffrey Barbour and the Institute staff for their part in making this such a successful meeting.

COLIN BAKER

Alcan International,
Banbury, Oxfordshire

BOOK I

Overview

D LITTLE

The author is with
Airbus Industrie,
Blagnac, France

The aircraft is unique among transportation vehicles in that it depends on forward momentum for the lift force needed to sustain it in equilibrium in its operating medium, namely air.

The drag component of this lift force accounts for some 30% of the total drag of a typical subsonic airliner in cruising flight.

It follows therefore that reductions in empty weight are a powerful means of reducing lift induced drag and so saving fuel.

For a typical 220 seat twin-engined airliner a reduction of one tonne in empty weight produces a fuel saving of 80 kg on a 1,000 n.m. flight. If we consider that this same aircraft will make approximately 1,000 such trips each year, the annual saving will amount to 80 tonnes of fuel; worth almost $24,000 with fuel at 90c/US gallon.

This means that a weight reduction of 1 kg is worth $24/year in fuel saving and if we want to see a return on capital over five years or less it is worth spending up to $120 for every kilo just for the fuel saving.

This, however, is only part of the story. Air-line income is essentially passenger fares and freight revenue. On many routes the number of passengers that an aircraft can carry or its ability to carry freight will be limited by allowable take-off weight. This may be either the maximum certificated take-off weight (MTOW) or the limiting take-off weight as determined by the aircraft's performance and the airfield conditions.

Fig. 1 shows the components of take-off weight for a modern airliner for a 4,400 mile flight. It can be seen that weight saving can exercise a very powerful leverage in terms of increasing payload capabilities and hence the earning power of an aircraft.

The extent to which this added earning power actually becomes hard cash for the airline will depend very much on the nature of traffic on his particular route network and the corresponding yields. In many cases, however, it is worth paying a significantly higher price for weight saving that can be justified by the reduction in fuel burn alone.

For a new aircraft at the initial design stage, weight saving takes on even greater significance. Fig. 2 shows what happens in the case of a long range aircraft if we can cut weight early enough. For every tonne of weight we remove we save also the fuel required to carry it, which means that the aircraft can be scaled down, which again brings fuel saving.

If we can reduce by five tonnes the MWE of an initial design for an aircraft of 200 tonnes MTOW, then we can reduce the MTOW needed by almost nine tonnes and the engine thrust required will also go down by almost 4.5%.

If this iterative process can actually reduce the engine size as well, the overall saving could be even greater.

The designer of a new aircraft or a new derivative of an existing design must make a judgement concerning the extent to which weight can be reduced and the ways in which it can be achieved. He must decide which new technologies can be developed to the point where they can be included in the design definition.

For an entirely new aircraft all aspects of technology will have to be assessed. Advances

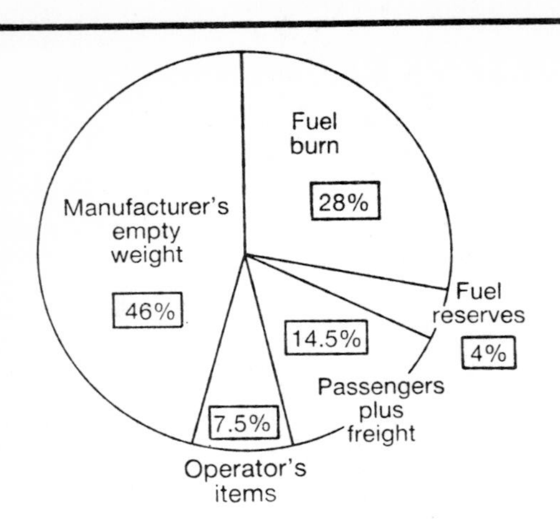

Aircraft takes-off at max. weight

A 1% saving in
aircraft weight empty plus
operators items will
increase available revenue payload
by almost 4%

1 Twin-engined medium-range aircraft, 4400 n.m. stage

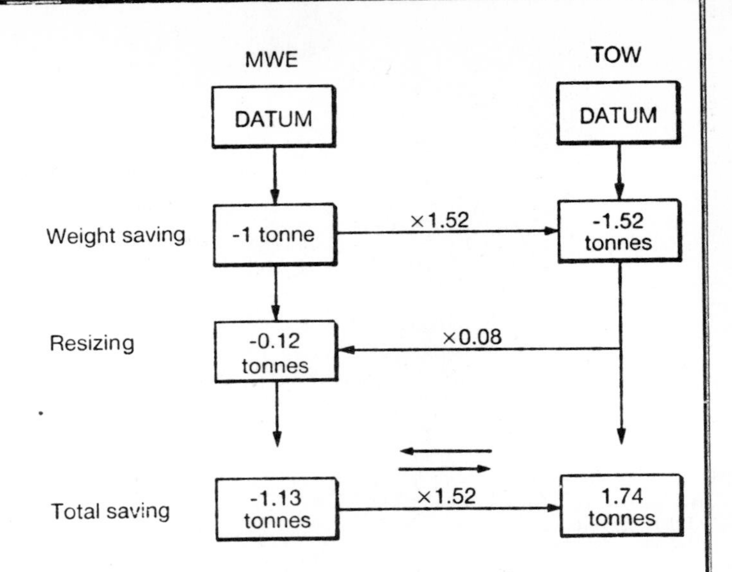

2 Weight reduction - new design; example of long range aircraft, 5000 n.m. stage

3 Superplastic forming - diffusion bonding SPFDB. Underwing access panels for A310 41% lighter than machined aluminium alloy panels fitted to previous aircraft: 31 kg weight saving per aircraft and no increase in manufacturing cost

4 A310-300 fin box in CFRP:

2 integral shell segments (skin, stringers and chord-wise stiffeners)
4 spar segments
22 rib segments
8 bearing supports
60 joining pieces

96 parts total (excluding fasteners)

This compares with 2072 separate parts (excluding fasteners) for the metal fin box. The CFRP box is also 22% lighter.

5 Production costs with advanced materials

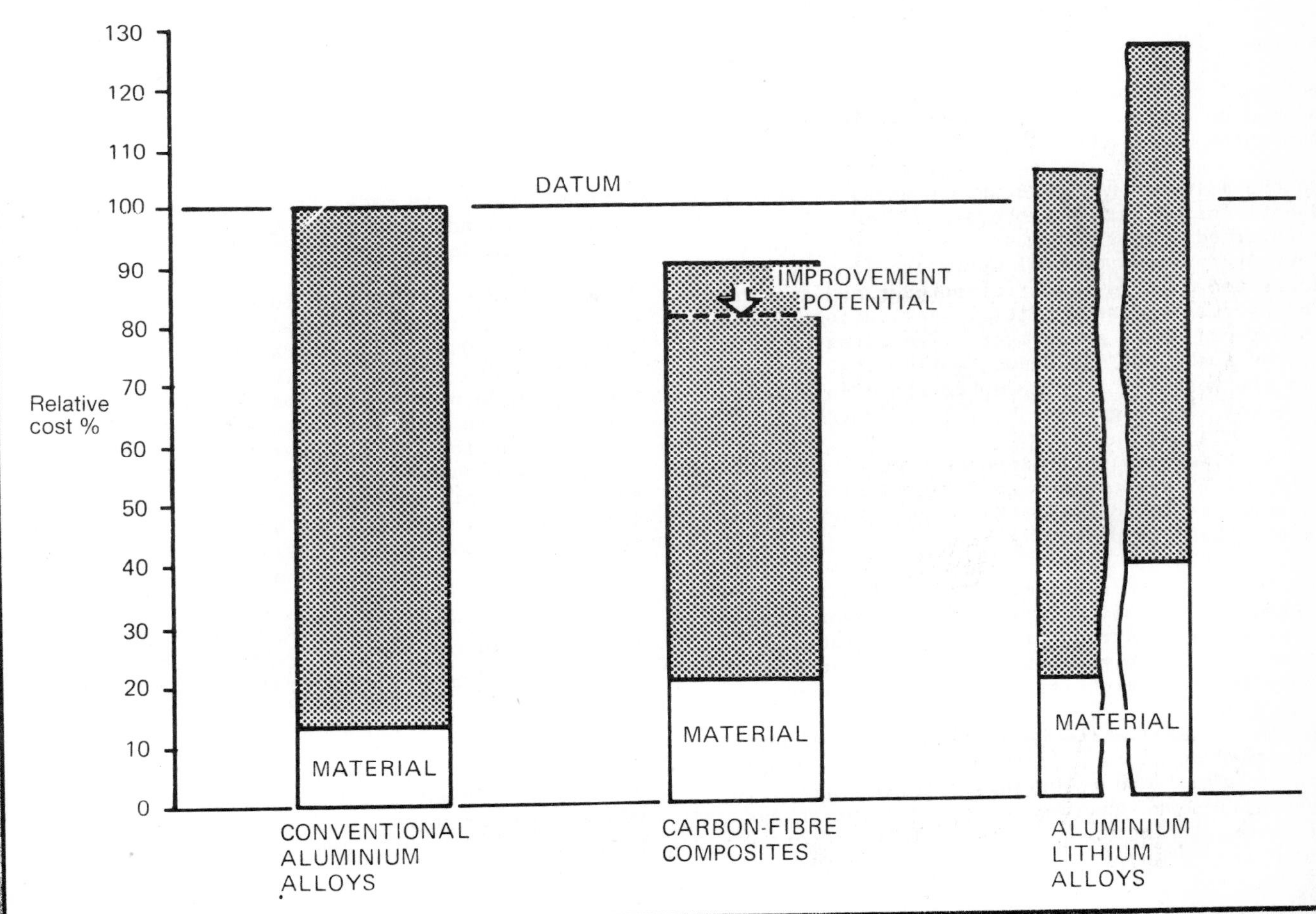

in aerodynamics could lead to thicker wing sections and hence lighter wing structures or greater aspect ratio. New types of powerplant could reduce fuel consumption to the point that the aircraft could be resized for the same range and payload. New systems concepts such as "fly-by-wire" can save weight in themselves and in addition can be used to reduce structure weight by operating in an "active" mode to reduce aerodynamic loads.

In addition to these aspects, the designer will be looking to new materials, both metallic and non-metallic. For derivative designs he will be looking mainly towards new materials.

As far as metallic materials are concerned we in Airbus Industrie have used new high strength aluminium alloys of the 7010-7050-7150 type to increase the payload/range capability of the A310 without increase in weight. Had we not used these alloys the aircraft would have been approximately 240 kg heavier.

We have also used titanium alloys in place of steel in a large number of applications since the beginning of the A300 program.

More recently we have begun to exploit the excellent superplasticity and diffusion bonding properties of titanium alloys to produce components which can be lighter in some instances than those made from aluminium alloy. An example of this is the underwing access panel on the A310, shown in Fig. 3.

Before I come back to metallic materials, and aluminium lithium in particular, let me say something about advanced composite materials.

On the A310 and A300-600 we are using substantial quantities of carbon fibre and aramid fibre reinforced composite materials. They are used notably for some control surfaces, landing gear doors and a large number of shrouds and fairings. The success of these initial applications has encouraged us to go further. The latest version of the A310, the -300, due to fly later this month, has the vertical fin in carbon fibre composite material (see Fig. 4).

The A320, due to enter into service in Spring 1988, will have the horizontal tail also in this material as well as all the control surfaces, except the leading edge slats.

Now to aluminium-lithium. The first and obvious attraction of this material is its lower density compared with current alloys of comparable strength. Its higher stiffness can also be exploited for weight saving in new designs.

As regards derivative versions of current designs a further attraction is that existing tooling can be used. Non-recurring costs are therefore generally lower than for a more radical change of material; for example, to carbon reinforced composite.

On the other hand the recurring or manufacturing costs associated with aluminium-lithium will almost certainly be higher, and this is an important consideration in the choice of material for a specific component.

I have already cited the example of the A310-300 fin box where the change from aluminium alloy construction to carbon fibre composite led to a 95% reduction in the number of individual parts to be assembled.

The associated reduction in the number of man-hours required to produce this assembly brought about an overall saving in manufacturing costs despite the higher material cost. It is difficult to see how increased cost can be avoided with aluminium-lithium alloys (see Fig. 5).

Nevertheless, as I said at the beginning, we are willing to pay for weight saving. How much we can pay is another matter and the cost of alternative construction methods must obviously be considered.

Another factor which has to be considered when making a choice of material for an aircraft component is the potential for development of the aircraft. A wing box, for example, is something which clearly favours aluminium alloys.

Fig. 6 shows the detail of a wing panel which is typical of Airbus design. With numerically controlled machine tools a change in design weights for the aircraft can be handled very simply by reprogramming the machines to remove more or less material as required. In contrast, a wing box made from carbon composite material would be much less easily adaptable to changes arising out of design development, and if modifications have to be made, they are likely to be both complex and costly and to require longer for their implementation.

If we consider an aircraft fuselage, there are other factors favouring metallic materials over composites. One of the attractions of composites is that we can orient the fibres according to the direction of the principal loading of the structure. In a typical fuselage with a complex loading distribution (pressurization, bending, torsion) and multiple cutouts for doors, windows, servicing points and so on, this possibility no longer exists, although the convenience of filament winding may be considered where the pressure shell loading is less complex.

We also have to consider the possible hazard to passengers in the event of a survivable accident, and the tendency of carbon composites to splinter like wood does not recommend them for fuselage structure.

Generally speaking, composite structures are more difficult to repair, and this is one of the reasons why we have up to now used them for components, such as the fin, which are less exposed to accidental damage (from, for example, ground

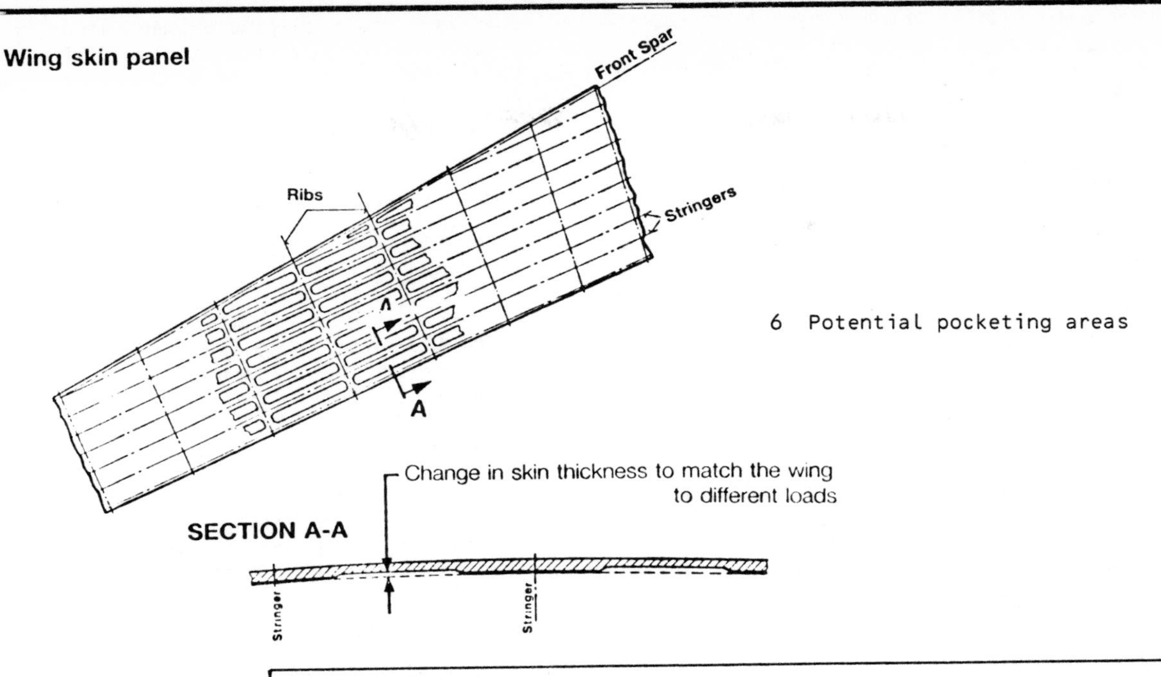

6 Potential pocketing areas

Airbus Industrie program

- Materials working group of Airbus Industrie and partners.
- 4 step program :
  - 1. Substitution for current alloys in non-critical parts ; examples : equipment racks, mounting brackets
  - 2. Substitution for current alloys in non-fatigue-critical structure
  - 3. Substitution for current alloys in fatigue-critical structure
  - 4. New design to make optimum use of Al-Li alloy properties

Applications

- Initial application to A320 structure foreseen as being limited to wing fixed leading edge and some intermediate ribs
- Application to A310 / A300-600 non-critical airframe parts foreseen on aircraft to be delivered in 1989
- Application to primary structure foreseen for 1990 provided :
  - satisfactory development and availability of required alloys
  - positive benefits to operators can be demonstrated.

7 Aluminium–lithium alloys - Airbus Industrie program

Alloy form	
Forging & Extrusions	Aerospatiale
Plate	British Aerospace, CASA
Sheet	MBB, CASA, Fokker (Associate)

8 Al–Li qualification program: partners' participation

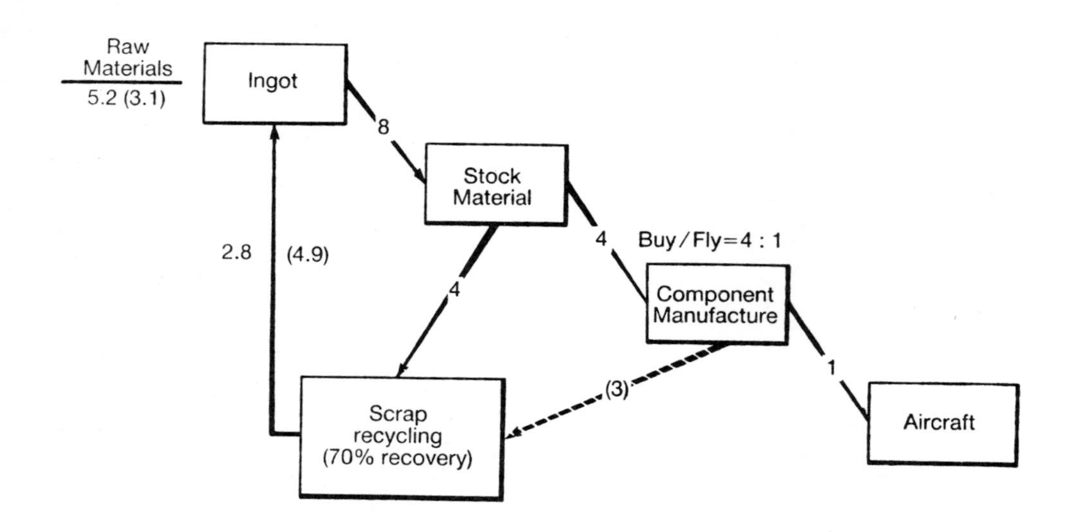

9 Al–Li cost/scrap recovery. Recycling of scrap from aircraft manufacturer can reduce raw materials input by 40%

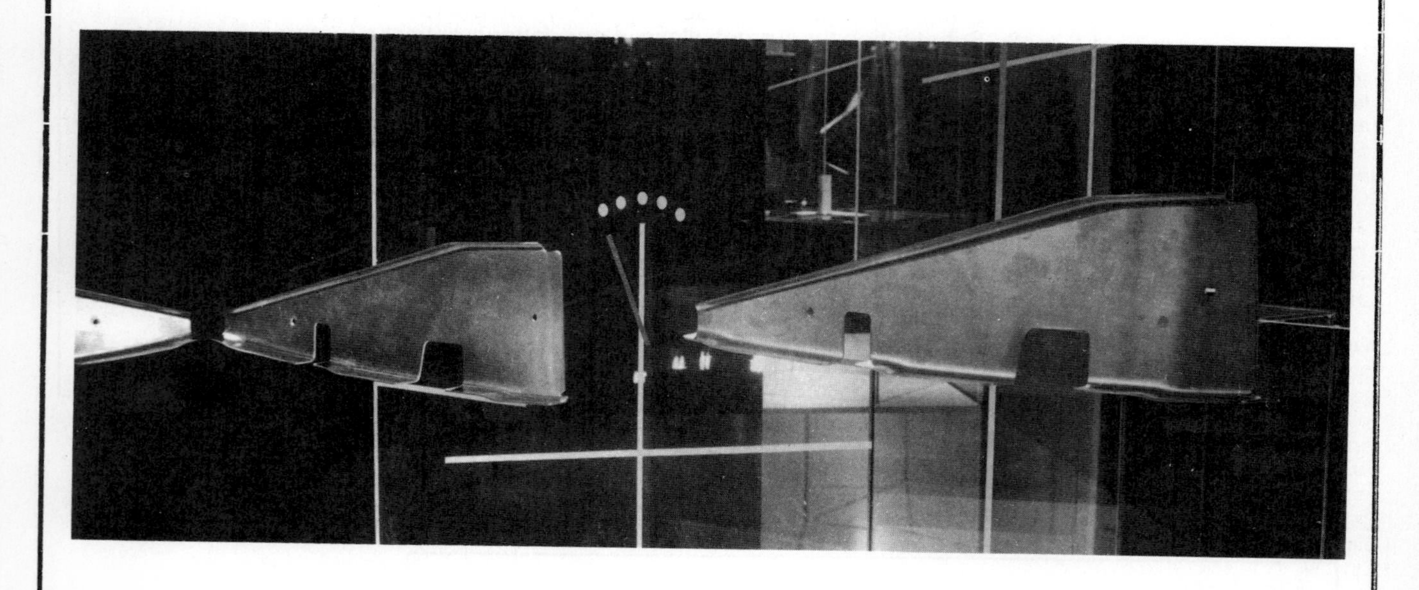

10 A310 test pieces in Al–Li (from MBB Bremen)

servicing vehicles) or components, such as control surfaces or fairings, which are readily replaced.

The joining of composite components to metal structures is another area which can cause problems; I am thinking in particular of attachment lugs and bushings which can create difficulties both in design and production. But we have to be careful also in selection of fasteners not to crush the material, and we must avoid the trap of galvanic corrosion at faying surfaces.

Overall we have a wealth of experience of aluminium alloy structures and when in doubt this is what we should choose. But we are learning fast in the realm of composites and I personally would expect to see even greater use of them in future aircraft.

Within the Airbus Industrie partnership we are actively investigating all technologies that promise weight savings.

For aluminium-lithium we have a four step program (see Fig. 7) with the aim of introducing it progressively to all our aircraft. The work within the partner companies is split up as shown in Fig. 8 with each partner working on material from all suppliers; British Alcan, Alcoa, Cegedur-Pechiney and others.

Investigations have been made into fundamental material composition and microstructure. Study of design properties include tension tests both at room temperature and at elevated temperature, compression tests and notch tensile tests. Fatigue tests are being made both with constant amplitude and random loading and crack propagation testing includes humid, salt laden and cold environments as well as the more normal laboratory atmosphere. Extensive corrosion testing includes accelerated salt spray and stress corrosion bend tests. We are putting a strong emphasis on all forms of corrosion resistance of metals (exfoliation, stress corrosion, filiform, etc).

Manufacturing trials include the full range of machining, bonding, surface treatments; automatic forming of sheet metal and heat treatments. The effects of these manufacturing processes on the design properties will also be fully explored. Full scale testing of aircraft component assemblies using aluminium-lithium alloys are the next step. Extensive use of these alloys for primary structure can only be contemplated when all their characteristics are fully known and understood.

While our principal objective is to have material of consistent quality readily available from our suppliers, we are also very concerned about estimates for the price of these new materials. We regularly hear statements that their cost will be two to three times that of conventional alloys. If we take the mean of these two numbers and apply it to today's alloy prices we arrive at an average increase of approximately $8/kg.

If for an aircraft component the buy:fly ratio is 4:1, the additional cost is then $32/kg of manufactured weight. If the weight saving achieved is 10% it means that we are paying $320/kg of weight saved - a high figure and one which will make us think very carefully about our choice of material.

It has been said that the quoted two to three factor on current material prices is based on recycling of scrap at the rolling mills only and it is not planned to recycle scrap from the aircraft manufacturer. Fig. 9 shows that this should nonetheless be very seriously considered as it could significantly reduce the demand for raw materials and in particular lithium; the most costly element.

Of course the buy:fly ratio varies considerably from one component to another, being lowest for forgings and extrusions and highest for integrally machined thick plate, where it can exceed 10:1.

Clearly there have to be discussions between the aluminium suppliers and the airframe manufacturers to find the most cost-effective ways of using these new alloys, otherwise their price will act as a brake to their widespread use for new design.

We have some other concerns; for example the substitution of conventional alloys on a repair basis in aluminium-lithium structures, because it will take some time before these new materials are readily available all over the world. What I have in mind is the need to replace a broken or corroded stringer in an aluminium-lithium skin and stringer panel, where the only material available is in conventional alloy. Because of the different stiffnesses involved we shall need to exercise care in the instructions we give to the airline personnel responsible for such repairs.

In spite of these concerns we are "bullish" about aluminium-lithium. We believe that it will make a major contribution to improving the efficiency of our current designs and that it will make up a very significant part of the structure of new aircraft.

Western world lithium reserves and resources

R K EVANS

The author is with
Amax Exploration Inc.,
Golden, CO 80401, USA

ABSTRACT

The last comprehensive review of Western World lithium reserves and resources was presented in 1977 when reserves of all classes were estimated to total 10.6 million tonnes of lithium metal equivalent. The total did not allow for beneficiation and chemical processing losses.

Although the massive imbalance between the resource base and the current level of Western World demand at about 6000 tonnes/year Li offers little inducement to further exploration, the evaluation of new sources has continued. In part, this had been encouraged by projections of a possible quantum increase in demand for lithium in batteries and thermo-nuclear power generation and, in part, a quest for lower cost sources than the current ones.

Since 1977 large-scale lithium mineralisation has been discovered in a pegmatite in Western Australia which previously had only been regarded as a source of tin and tantalite. In addition, massive reserves of lithium in brine have been identified at the Salar de Uyuni in Bolivia and geothermal brines in Southern California have been shown to contain large lithium reserves. Finally, encouraged by earlier work by the United States Geological Survey, lithium reserves in clays in the western United States have been shown to be large.

Using criteria comparable to those used in compiling the 1977 estimates, reserves and resources are re-estimated at 36.7 million tonnes.

Few of the sources are fully competitive with the existing chemical operations in North Carolina, Nevada and northern Chile at current costs and prices. The likeliest new high volume source is from lithium recovered as a co-product from a potash and boric acid solar evaporation project in the northern part of the Salar de Atacama in Chile. Annual quantities of lithium in feed to the operation are projected to total 50,000 tonnes and a high percentage is amenable to recovery.

Introduction

Two symposia have been held in recent years at which much emphasis was placed on whether lithium reserves and resources would be adequate to support projected large-scale demands.

At the first symposium held in Golden, Colorado, nearly ten years ago (Anon, 1976), industry representatives were confronted with a series of exponential graphs projecting massive demands for lithium in batteries for both electric vehicles and load levelling, and for fusion energy generation.

At a time when Western World annual demand, as chemicals, was approximately 3,200 tonnes of lithium metal equivalent, the projections were mouth-watering. Annual requirements for lithium in lithium-iron sulphide batteries, assuming 90% recycle, were projected to increase from 2,200 tonnes/year in 1985, to 39,000 tonnes/year in 1995 and 94,000 tonnes/year by Year 2000. Projections for lithium demand in lithium-water-air batteries were not so modest - 425,000 tonnes/year by Year 2000.

Estimates of lithium requirements for fusion were also spectacular with cumulative requirements, through Year 2030, as high as 706,000 tonnes.

In the light of these projections the availability of adequate reserves was seriously questioned, and in an Introduction to the Symposium proceedings reference was made to the "gravity of the impending shortage of lithium forecast by representatives from the United States Geological Survey".

The apparent crisis was manifested in a number of papers where, in one, reference was made to the need to look at other molten salt systems for batteries and another seriously examined lithium recovery from sea water.

The second symposium was held in Corning, New York in October, 1977 (Anon, 1978) with less emphasis on demand and with lithium industry representatives having an adequate opportunity to reply to the reputed supply crisis.

Reserve and Resource Estimate, 1977

Data presented in 1977 were based on a report prepared in 1976 on behalf of the National Academies of Sciences and Engineering for the National Research Council Committee on Nuclear and Alternative Energy Systems. The Lithium Sub-Panel comprised representatives of industry, current and former members of the United States Geological Survey.

Reserve and resource estimates, in tonnes of Li to beneficiation and process (recoveries being regarded as proprietary) are summarized below (Table 1). Various classes of reserves were defined with Class A being substantially proved, Class B being probable, Class C being resources and Class D covering by-product resources.

The reserves and resources included generally had grades in excess of 0.5% Li in the case of pegmatites and in excess of 300 ppm Li in the case of brines other than those brines where lithium could possibly be recovered as a by-product. At the, then, current world demand of about 4,000 tonnes/year Li, as chemicals, the reserve base seemed adequate for millenia.

Not included in the estimates were large tonnages of lithium in oilfield brines in North Dakota, Wyoming, Oklahoma, east Texas and Arkansas, where brines grading up to 700 mg/lt are known to exist. Collins (1978) estimated a possible resource in a tenth of the area underlain by the Smackover Formation at 0.75 million tonnes. Also excluded were the lithium reserves contained in the clay mineral hectorite which were being examined at that time and, also, lithium contained in geothermal brines. Finally, due to lack of quantitative data, all Andean salares, with the exception of the reserves in the Salar de Atacama in northern Chile, were excluded from the tabulation.

Subsequent Developments

The first adjustment to the tonnage estimate occurred when Gyongyossy and Spooner (1978) suggested that the data presented by Evans (1978) underestimated Canadian reserves and resources by 0.4 million tonnes. Whether the CONEAS committee would, though, have accepted the assumptions on which the increase was based, is questionable.

Table 1
Lithium Reserves & Resources
(Evans, 1978)

Pegmatites	Tonnes Li
Lithium Corp. of America, N. Carolina (A+B+C)	334,000
Foote Mineral Co., N. Carolina (A+B+C)	208,000
Undeveloped, N. Carolina (C)	2,608,000
Rest of United States (C)	7,000
Val D'Or, Quebec (A+B+C)	109,000
Bernic Lake, Manitoba (A)	47,000
Rest of Canada (A+B+C)	112,000
Bikita Minerals (Pvt), Zimbabwe (A+B+C)	113,000
Manono, Zaire (A+B+C)	2,340,000
Karibib, Namibia (A+B+C)	11,000
Western Australia (A+C)	41,000
Brazil (A+B+C)	18,000
Argentina (A)	1,000
Total Pegmatites	5,949,000
Brines	**Tonnes Li**
Foote Mineral Co., Silver Peak, Nevada (A+C)	118,000
Searles Lake, California (D)	24,000
Great Salt Lake, Utah (D)	260,000
Salar de Atacama, Chile (A+C)	4,290,000
Total Brines	4,692,000
Stockpile	
U.S. Government	7,000
Total Stockpile	7,000
Total All Classes	10,648,000

Other increases have been more substantial:

a) Western Australia

In the early 1980's (Anon, 1983) Greenbushes Tin Co. announced the discovery of large-scale lithium mineralisation at their property located south of Perth. Previously the pegmatite concerned had been exploited as a source of tin and tantalite with mining activities concentrated in the upper, strongly weathered, portion of the pegmatite.

With the development of a decline to evaluate tantalite reserves at depth, fresh intersections revealed a spodumene rich zone exploitable by open-pit methods containing an estimated 42.7 million tonnes grading 1.85% Li_2O. The reserve block is over an 800 metre strike length in a pegmatite with a total length of about 3 kms. Proved reserves total approximately 380,000 tonnes Li and the potential is, at least, double this.

The company currently produces a spodumene concentrate but the property is being actively evaluated as a producer of lithium chemicals.

b) Salar de Uyuni, Bolivia

A joint French/Bolivian research team has recently concluded an evaluation of the economic potential of the brines of those Andean salares within Bolivia (Ballivian and Risacher, 1981). The largest of these internal drainage basins, covering an area of 9,000 Km^2, is the Salar de Uyuni.

In situ reserves were calculated at 110 million tonnes of potassium grading 8.7 gms/lt, 3.2 million tonnes of boron grading 247 mg/lt, and 5.5 million tonnes of lithium grading 423 mg/lt.

Lithium grade is significantly higher than the current United States brine source but much lower, on average, than that of the Salar de Atacama, to the southwest and at a much lower elevation in Chile. However, an area of about 30 km^2 with lithium values in the 900-4,000 mg/lt range could provide initial feed for any operation that may be initiated.

c) Geothermal Brines in Southern California

Schultze and Bauer (1984) have reported on test work aimed at recovery of lithium from geothermal brines in the Salton Sea/Brawley area of southern California. Their paper is one of several describing various possible recovery techniques. The area of interest is a large one with lithium concentrations in the range of 50 to 400 ppm.

Assuming a feed grade of 150 ppm, each 100 MW electric power generating unit is theoretically capable of co-producing 5,700 tonnes Li annually.

Total field potential is estimated at between 1,600 and 5,200 MW, with a total lithium potential of between 91,000 and 296,000 tonnes/year.

d) Hectorites in the Western United States

Hectorite is a clay mineral, specifically a magnesium lithium smectite, carrying up to 0.5% Li. Current uses are limited to cosmetics and as organophyllic thickeners in paints and drilling muds. The potential of the clay as a lithium source was first evaluated, seriously, by the United States Geological Survey at a time when the lithium reserve situation was regarded as "grave".

Reserves are vast. At the McDermitt Caldera, straddling the Nevada/Oregon border, proved and inferred reserves averaging 0.2% Li in grade are estimated, respectively, at 5.2 million and 3.9 million tonnes. Elsewhere in the western U.S. reserves in both categories total 6.0 million tonnes (Chevron Resources, Personal Communication).

e) Others

With the massive surplus of reserves and resources over conceivable demands in the medium term, the inducement for mining companies to explore for additional lithium should be lacking. Despite this, preliminary evaluations continue of sources which, by virtue of their location or grade, represent an attractive opportunity. Andean salares continue to receive attention, and about a dozen in Argentina and several in Chile contain interesting lithium values.

Areas of lithium pegmatite mineralisation, particularly in Canada, remain highly prospective, and it seems inconceivable that only the western U.S. contains lithium rich geothermal brines and large reserves of hectorite. The fact that they have been discovered there is mainly a function of the exploration effort that has been expended for a variety of reasons.

Reserve & Resource Summary

Whether or not a resource is economic is a function of many parameters and in the tabulation below (Table 2) no attempt is made (as was the case with Table 1) to distinguish between uneconomic or potentially viable sources. In the case of any demand that materialises for fusion power generation, the cost of lithium is projected to be such a small percentage of total costs that strategic, rather than economic, conditions will determine which source(s) will be developed. To a lesser degree lithium costs, at this time, do not appear to be a major factor in battery development. When secondary batteries are developed the situation may change, but current primary batteries, millions of which are produced annually, typically use one tenth of a gram of metal.

Table 2

Updated Reserve & Resource Estimate	Millions Tonnes Li
1977 Estimate (Evans, 1978)	10.6
Canadian addition	0.4
Greenbushes, Australia	1.1
Salar de Uyuni, Bolivia	5.5
S. California (geothermal) (Say 200,000/yr for 20 years)	4.0
Hectorites, Western U.S.A.	15.1
Total	36.7

Commercial Developments

Apart from the consumption of, possibly, 1,000 tonnes/year of lithium metal equivalent as ores and concentrates for direct usage in glasses and ceramics, and the import of erratic, but modest, tonnages of lithium in the form of carbonate and hydroxide from Russia and China, lithium chemical and metal demand in the Western World (about 5,000 tonnes/year) continues to be met by primary production by two producers.

Foote Mineral Company has facilities based on pegmatites in North Carolina, brines in Nevada and, in a joint venture with Corporacion Fomento de la Produccion, from brines in Northern Chile. Lithium Corporation of America's production is based on spodumene pegmatites in North Carolina.

Both operations in North Carolina open-pit pegmatite ore grading approximately 0.7% Li and beneficiate this by flotation to a concentrate grading about 3% Li. The spodumene is then converted from its alpha form to an acid-leachable form by calcining at 1150°C. The decrepitated concentrate is attacked with sulphuric acid, and lithium sulphate, together with other sulphates, is taken into solution for subsequent purification and concentration. Lithium sulphate is reacted with sodium carbonate to yield lithium carbonate, the starting chemical for all other lithium salts and metal, with by-production of sodium sulphate. The process is inherently expensive.

At Silver Peak in Nevada, lithium containing subsurface brines are concentrated by solar evaporation and treated for magnesium and sulphate removal. The resultant lithium chloride brine is reacted with sodium carbonate to yield lithium carbonate. A similar process is utilised at the Salar de Atacama where feed grade is more than seven times higher, but where the magnesium/lithium ratio is also higher. At neither of these brine operations are there any significant by-products.

Other pegmatite sources have been evaluated in recent years as possible bases for lithium chemical production. Most pegmatites, though, have grades which are not conspicuously better than those currently worked by the entrenched producers. Operations based on such sources could expect to incur operating costs comparable with those in North Carolina but amortisation costs would be significantly higher.

There is no evidence at this time that any of the currently known geothermal or oilfield brines are fully competitive sources at current lithium prices. Limited published estimates of costs, using hectorite clays as lithium chemical feedstock, indicate that they offer no cost savings.

The likeliest new lithium source is the northern portion of the Salar de Atacama in Chile. Amax and Molibdenos y Metales Ltda., a Chilean private sector company, in a proposal made to Corporacion de Fomento de la Produccion, intend to co-produce lithium from a large-scale potash project based on brines occurring at shallow depths. The target sized project is one that will produce, annually, 500,000 tonnes of potassium chloride, approximately 200,000 tonnes of potassium sulphate, and 30,000 tonnes of boric acid. The 30.0 million tonnes/year of annual brine feed will contain approximately 50,000 tonnes of lithium - a high percentage of which can be precipitated as a double salt of lithium potassium sulphate. Recovery of lithium in carbonate form will result in additional production of potassium sulphate.

References

Anon (1976) "Lithium Resources & Requirements by the Year 2000", Geological Survey Professional Paper 1005. James D. Vine, Editor.

Anon (1978) "The Lithium Symposium, 1977" Energy Vol. 3, No. 3. S.S. Penner, Editor.

Anon (1983) Greenbushes Tin Ltd. Annual Report for 1982.

Ballivian O., & F. Risacher (1981) "Los Salares del Altiplano Boliviano", O.R.S.T.O.M. Paris.

Collins, A.G.(1976) "Lithium Abundance in Oilfield Waters", U.S.G.S. Professional Paper 1005.

Evans, R.K. (1978) "Lithium Reserves & Resources", Energy Vol. 3, No. 3.

Gyongyossy, Z.D., & E.T.C. Spooner (1978). "Is There a Bright Future for Canadian Lithium Deposits?" Northern Mines, March 2nd, 1978.

Schultze, L.E., & D.J. Bauer (1984) "Recovering Lithium Chloride from a Geothermal Brine", Bureau of Mines Report of Investigation, R1.8883.

Current status of UK lightweight lithium-containing aluminium alloys

C J PEEL, B EVANS and D McDARMAID

The authors are in the
Materials and Structures Department,
Royal Aircraft Establishment,
Farnborough, Hants, UK

SUMMARY

A series of aluminium alloys containing lithium have been developed and are now produced in the United Kingdom by Alcan. The alloys are designed to meet specific targets based upon the concept of substitution of the new alloys for conventional alloys in the 2000 and 7000 series with a density reduction of 10% and a stiffness increase of 10%. The first of these alloys is now registered as 8090 and in the 'A' and 'C' conditions meets the requirements of most 2000 series alloys whilst the higher strength B alloy is to meet those of the 7000 series. A wide range of properties is reviewed for these alloys in sheet, plate, extruded and forged forms including super-plastically formed material. Research within RAE on known problem areas is outlined including a brief description of alloy developments towards improved higher strength and damage tolerant alloys.

INTRODUCTION

The UK development programme for improved aluminium-lithium ingot alloys has been well published.[1-3] Alloys have been designed to meet specific targets based upon the concept of substitution of the new alloys for conventional alloys in the 2000 series and the 7000 series with a density reduction of 10% and a stiffness increase of at least 10%. The concept of substitution has developed to some extent into two philosophies. Firstly, a literal substitution of the new alloys for old on a piece-meal basis with minimal re-sizing of the exchanged parts. Emphasis naturally falls on the achievement of minimum density for this category with excessive increases in elastic modulus being a disadvantage in so far as the distribution of load throughout a structure may be adversely affected with particular respect to the fatigue performance. Limited fatigue experiments have started for a particular application with hybrid joints of 8090 alloy and 2014 and although the initial results look promising more tests are required. Secondly, there is the philosophy of substitution by type with complete or partial re-design using the appropriate aluminium-lithium alloy substitute of the selected type eg damage tolerant, compression critical etc. This latter technique should lead to greater mass savings because the structure is designed to exploit the reduced density and increased stiffness of aluminium-lithium alloys.

Despite the fact that substitution clearly means different things to different people the principle has provided a very attractive set of alloy design targets. Those originally set for the UK development by BAe and RAE were of the straight substitution type with the following requirements:-

Medium Strength A	≡	2014-T6 sheet, plate, extrusion and forgings
High Strength B	≡	7075-T6 sheet plate, extrusion and forgings
Damage Tolerant C	≡	2024-T3 sheet, plate, extrusion

Draft specifications of the form DTD XXXA, YYYA, ZZZA were originally issued by RAE for sheet, plate and extrusions respectively as requirements based on the performance of the patented RAE F92 alloy. Alcan Int. have introduced a registered trade name Lital for their products, Lital A equating to the original F92 type A alloy, Lital B being a further Alcan development for the high strength category. The A type alloy, slightly modified, has been registered with International agreement as 8090.

As the development has progressed more

TABLE 1

Requirements and Properties for Lital A (8090)

	Sheet (T6) XXXA		Plate (T651) YYYA			Extrusion (T651) ZZZA		
	L	T	L	T	ST	L	T	ST
0.2% PS MPa								
Specified minima	380	380	410	390	360	430	-	-
Provisional minima†	350	355	410	380	335	440	395	355
Typical	365	375	450	420	365	480	420	385
TS MPa								
Specified minima	440	440	460	450	420	480	-	-
Provisional minima	430	450	455	455	400	495	470	420
Typical	465	470	495	480	435	550	490	470
Elongation %								
Specified minima	6	6	5	5	2.5	6	-	-
Provisional minima	5	4	4	4	1	3	3	1
Typical	5.5	7.5	6	7	2	5	5	3.5
Fracture Toughness MPa $\sqrt{m}$								
Specified minima	70*	-	30	25	18	30	25	-
Provisional minima			30	26	15	31	16	-
Typical	66	-	37	33	16	38	20	-
Elastic Modulus GPa								
Typical tension	78	78.5	79	79	-	79.5	79	-
Typical compression	-		81.5	82.5	-	-	-	-

* On 400 mm wide CCT panels. K_{app}.

† Based on statistically derived 'A' values wherever possible otherwise worst results.

refined design targets seem to be required for two reasons. Firstly, the metallurgists have developed a better understanding of the capabilities and limitations of these new alloys and, secondly, the designers have begun to identify means to increase the potential weight saving achievable by designing ab initio with aluminium-lithium.

This paper reviews the progress achieved to date in meeting the original A, B and C targets, identifies critical areas in which performance has exceeded the requirements, or failed to meet them, and suggests adjustments to both the requirements and the alloy specifications on the basis of this assessment of performance. This will essentially review the properties of Lital A, B and C alloys but more detailed expositions of selected properties of these alloys will be found in this Conference.

The nominal compositions of the two principal Lital alloys are as follows, all values are in weight per cent:-

	Li	Cu	Mg	Fe	Si	Zr	Specific Gravity
Lital A (8090)	2.5	1.3	0.7	$\ngtr$0.2	$\ngtr$0.1	0.12	2.54
Lital B	2.6	1.9	0.9	$\ngtr$0.2	$\ngtr$0.1	0.12	2.55
2090 (8091)	2.2	2.7				0.12	2.60

The nominal composition of the newly registered 2090 alloy is included for comparison. Lital C is a variant of 8090 presently achieved by a short ageing treatment although this technique is under review. Currently, for stretched plate, Lital A and B would be aged to peak strength at 190°C typically requiring 16 hours. Shorter ageing time (6 hours at 190°C) is used for Lital C plate. Sheet is aged at 185°C for Lital A and B and at 175°C for Lital C. For the data presented in this paper the plate and extrusion samples were consistently stretched for stress relief after quenching whilst sheet samples received no cold work prior to ageing at all. The results quoted in Tables 1 - 3 are for Lital alloys produced by Alcan International Ltd and are drawn from a large population of results for pre-production metal cast on both a 300 Kg and 1000 Kg scale. Results for the development alloys are quoted for RAE casts produced in 20 Kg lots. It can be stated that little effect of ingot size has been noted to date.

The Medium Strength A Requirement

The original UK requirement was for an aluminium-lithium alloy that generally met the performance of 2014-T6 but that was 10% lighter (Fig. 1) and 10% stiffer. Table 1 summarises the performance of Lital A in meeting the required goals. Generally, it can be seen that there is little problem in meeting the tensile strength requirements to the extent that statistically derived minima may eventually exceed all the specified targets. However it was previously reported[2] that the 0.2% proof strength (PS) of T6 material (that is metal that is not stretched or cold worked after quenching) was below requirement and this trend has persisted. Alloy developments designed to improve this feature will be considered further in this paper. This continuing difficulty in producing high levels of 0.2% PS in T6 sheet (Table 1) has resulted in a review of the specified requirements. These were initially chosen as the highest levels specified in the UK for 2014-T6 sheet. In the UK mandatory rules require that for civil aircraft structure the 0.2% PS should be not less than 67% of the tensile strength (TS), ie structures are stressed using either the minimum 0.2% PS or 67% of the minimum TS, whichever is lowest. For military aircraft the ratio should be not less than 75%. These ratios of 0.2% PS to TS are based upon the prescribed relationships between limit loads and ultimate design loads. It can be seen (Fig. 2) that the levels of this ratio chosen initially for A and B type alloys lie approximately at the 85% level, essentially making the tensile strength of the material the critical feature. Since in general the Lital A meets the specified minimum TS in the T6 condition (440 MPa) and this minimum value matches that of 2014 T6, structures designed on tensile properties will essentually be stressed to the same level and the value of the 0.2% PS of either alloy is unimportant unless it falls below 330 MPa for military structure or 295 MPa for civil aircraft structure. In this context the achieved 0.2% PS levels of approximately 370 MPa seem quite adequate and a more critical improvement to achieve further weight saving would be an increase in the allowable level of tensile strength. This critical assessment is most important since high levels of 0.2% PS may well be unuseable and will result in an unnecessarily low level of fracture toughness. For example, the large panel fracture toughness of the medium strength material is acceptable currently being approximately equivalent to 2014-T6 material (Fig. 3). However, this acceptable toughness was achieved at a typical 0.2% PS of 370 MPa. A further increase in 0.2% PS may result in a reduction in toughness.

It can be seen (Table 1) that the short transverse properties of the Lital A plate are somewhat below requirement. The low value of 0.2% proof strength typically achieved has never been considered a problem for reasons essentially the same as those for the sheet material but the scatter in short transverse ductility has produced low levels of tensile strength. However, it has been established that metal quality is affecting short transverse ductility and samples with high ductility (>4%) and matching high tensile strength (>500 MPa) have been produced such that improved values may be anticipated. Variability in short transverse fracture toughness has also been a concern. Typical values of 16 MPa $\sqrt{m}$ achieved for plate 25 mm to 40 mm thick are probably adequate although a value of 20 MPa $\sqrt{m}$ at near peak strength (T651) would seem possible. The short transverse toughness falls with ageing time until a plateau level is reached (Fig. 4). This is a behaviour typical of aluminium alloys and represents an increasing concentration of

TABLE 2

Requirements and Properties for Lital B

	Sheet T6 XXXB*		Plate T651 YYYB			Extrusion T651 ZZZB	
	L	T	L	T	ST	L	T
0.2% PS MPa							
Specified minima	420	420	455	455	420	560	510
Provisional minima	385	385	510	470	415	-	-
Typical	410	400	530	500	430	610	510
TS MPa							
Specified minima	490	490	525	525	490	600	580
Provisional minima	470	460	540	535	465	-	-
Typical	505	490	565	550	490	630	540
Elongation %							
Specified minima	6	6	6	6	-	-	-
Provisional minima	5	8	-	-	-	-	-
Typical	8	9	5	5	1.5	4	4
Fracture Toughness MPa $\sqrt{m}$							
Specified minima			-	-	-		
Provisional minima			†16	14			
Typical			24	21	10		
Elastic Modulus GPa							
Typical tension	78.5	78.5	80	80.5	-		
Typical compression	-	-	82	82.5	-		

* Extended ageing variant.

† Considerably improved toughness is obtained at slightly reduced strength levels.

TABLE 3

Requirements and Properties for Lital C (8090)

	Sheet XXXC		Plate YYYC			Extrusion ZZZC		
	L	T	L	T	ST	L	T	ST
0.2% PS MPa								
Specified minima	290	290	300	300	260	320	-	-
Provisional minima	290	300	370	310	260	-	-	-
Typical	330	335	390	340	290	370	325	-
TS MPa								
Specified minima	420	420	430	430	380	440	-	-
Provisional minima	400	410	430	420	380	-	-	-
Typical	440	450	460	440	415	460	440	-
Elongation %								
Specified minima	6	6	7	7	3.5	6	-	-
Provisional minima	4	7	5	6	2	-	-	-
Typical	6	9	6	9	3.5	5	6.5	-
Fracture Toughness MPa $\sqrt{m}$								
Specified minima	83*	-	35	28	25	-	-	-
Provisional minima			36	32	20	-	-	-
Typical	76	-	>40	>35	25	>45	>45	-

* 400 mm wide CCT panels. K_{app}.

deformation at the grain boundaries with the increasing strength of the matrix. The nearly horizontal nature of the plateau indicates that the growth and coarsening of grain boundary particles is not particularly embrittling. Unlike high strength 7000 series alloys however the lithium alloys do not overage rapidly and excessively large ageing times are required to soften the matrix and so increase the toughness. For this reason short ageing times of an iso-thermal nature have been used to date in an attempt to combine peak longitudinal strength with good short transverse toughness. However operating near the knee in the curve (cf Fig. 4) tends to introduce scatter if the ageing hardening response of different casts is variable. There would seem to be some scope for an ageing practice optimised for longitudinal strength and short transverse fracture toughness. For example reducing the time of ageing at 190°C from the standard 16 hours to 12 hours increases the short transverse toughness by 2 MPa $\sqrt{m}$ with no loss in longitudinal strength.

The High Strength B Requirement

In general terms the high strength Lital B alloy has been aimed at the replacement of the 7000 series high strength aerospace alloys but with a 10% density reduction and a 15% stiffness increase. The composition of the Lital B has been designed to give a similar low density to that of Lital A ie 2.55 g/ml and a similarly improved modulus of approximately 80 GPa. It will be seen that the density of 7075 is slightly greater than that of 2014 (eg 2.81 g/ml compared to 2.80 g/ml) and that its elastic modulus is slightly less (eg 70 GPa compared to 73 GPa) allowing a greater improvement in the comparison of Lital B against 7075 (26% improvement in specific stiffness) than in the comparison of Lital A with 2014 (21% improvement in specific stiffness). The exact choice of strength level for the B requirement has led to some debate. The initial draft MOD requirements were for plate strength levels similar to 7075-T651 and 7010-T7651 and for sheet strength levels equivalent to the slightly weaker 7475-T76. A further requirement has now been introduced (Table 2) potentially to replace the ultra-high strength 7150-T651 particularly in extruded form but possibly also as thin plate in due course.

It can be seen (Table 2) that standard ageing of plate for 16 hours at 190°C produces very high strengths but possibly fracture toughness that is too low. Shorter ageing times, eg as little as 4 hours at 190°C, produces material that will meet the currently specified plate requirements but have toughness of the order of 35 MPa $\sqrt{m}$ in both L-T and T-L test directions and improved short transverse performance. However, if the very high strength levels currently achieved with Lital B were to be exploited, their use would probably have to be limited to thin section material (eg $\ngtr$12 mm) to avoid short transverse problems. Thin covers for upper wing skins are an obvious target. For example the potential compression performance of the new DTD ZZZB requirement against the very advanced high strength 7000 series extrusions, typified by 7150-T6 or 7010-T6, is illustrated (Fig. 5). It can be seen that at low loading indices ZZZB will achieve a 12.5% weight saving reducing to approximately 10% at the higher loading indices.

The actual performance of Lital B against requirement is indicated in Table 2. As with Lital A, the 0.2% PS of unstretched (T6) sheet is below requirement. However, it will be noted that the results presently quoted show a significant increase over the strength of the Lital A sheet. At present this increase is only achieved with extended ageing. This problem is considered in more detail subsequently. Once again it can be seen (Fig. 2) that the specified minimum for 0.2% proof strength is 85% of the specified ultimate and the low levels of 0.2% PS are therefore probably acceptable, with the TS being the critical feature. In general terms the Lital B plate appears to be capable of producing the high strength required of it but heat treatments designed to improve fracture toughness in all three directions are needed. The same comment can be made of the limited results for Lital B extrusions. It is judged that there is still scope for the improvement of heat treatment and fabrication practices for Lital B.

The Damage Tolerant C Requirement

The largest category for potential exploitation of Lital alloys in aerospace structure has always been in the damage tolerant regime, the 'C' requirement. In this category the balance between strength and fracture toughness is more important and the increased stiffness possibly of less benefit. For the present Lital C is a variant of 8090 alloy heat treated to increase toughness at the expense of a reduction in strength. Other options including compositional changes are under review. It is clear that to save sufficient weight to make the use of Lital C attractive a large density reduction is required, possibly combined with an improved balance between strength and fracture toughness.

Table 3 presents the current performance of 8090 aged to Lital C versus the provisional requirements. Most properties seem to be adequate with the exception of thin sheet toughness where the minimum required value of K_e (or K_{app}) of 83 MPa $\sqrt{m}$ at 400 mm panel width (equivalent to 93 MPa $\sqrt{m}$ on 500 mm wide panels) is not met by approximately 10 MPa $\sqrt{m}$ (Fig. 6). However, the metal used for these early large panel test results was approximately 10% stronger than the typical values quoted. Further tests with slightly softer material are in progress.

Other Properties of Lital A, B and C

Detailed papers in this conference will cover topics such as fatigue performance and corrosion behaviour. Fatigue testing to date has not revealed any problems indeed the Lital alloys normally out perform the conventional alloys in this respect.[4] In corrosion tests an area of uncertainty remains in the significance of the exfoliation blistering attack observed on un-

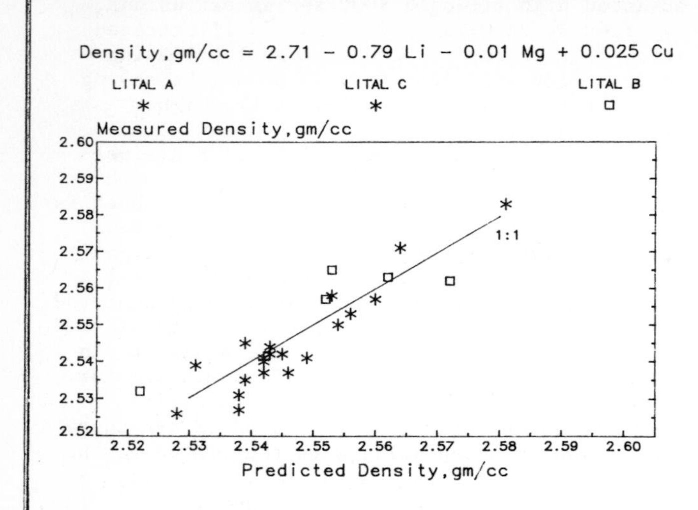

Fig. 1 Measured and predicted densities for Lital alloys.

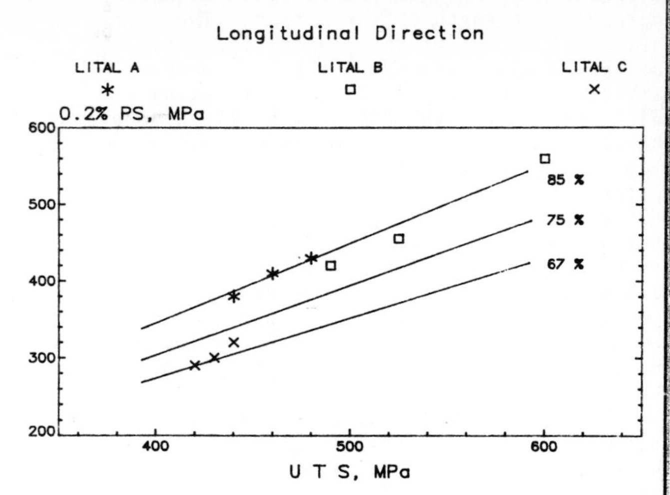

Fig. 2 Specified ratios of minimum 0.2% PS: minimum TS for the A, B and C targets.

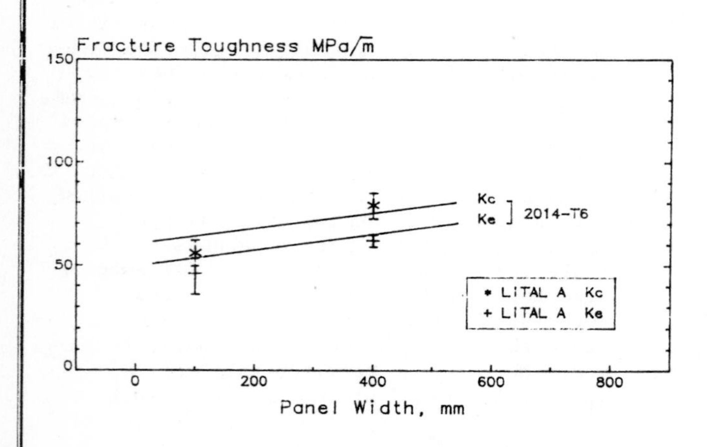

Fig. 3 The effect of CCT panel size on the fracture toughness of Lital A and 2014 T6 sheet. K_e is given by maximum load and initial crack length, K_c is given by maximum load and instantaneous crack length.

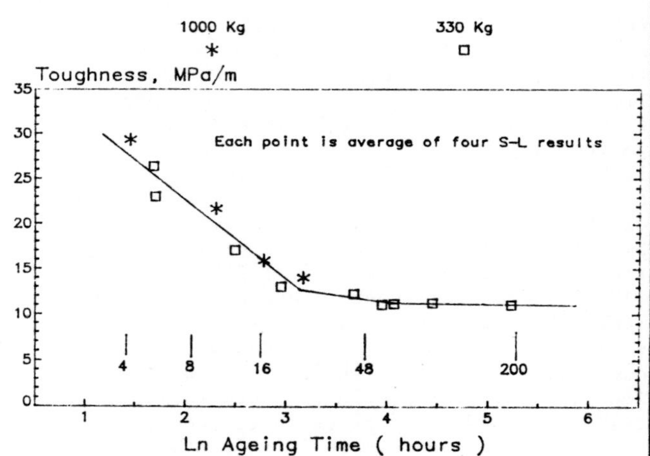

Fig. 4 The plane strain fracture toughness (S-L) on Lital A as a function of ageing time at 190°C.

recrystallised sheet particularly when exposed to acidic environments. This subject will be covered in depth in Paper 52 of this Conference. Stress corrosion testing has been extensive but has revealed a large degree of scatter (Fig. 7). Casts of similar composition have given results that have been barely acceptable eg little better than 2014-T6 on one hand or at least to the required 7000 series T76 standard on the other. There is evidence that the source of this scatter has been identified (cf Papers 6 and 63). In the RAE tests LITAL alloys appear to respond favourably to overageing.

Work has started on the elevated temperature performance of the Lital A 8090 alloy by determining both strength at elevated temperatures up to 250°C and strength at ambient temperatures after selected pre-exposure at elevated temperatures for up to 1000 hours. In general terms 8090 appears to at least match the performance of 2014 alloy in the range tested up to 250°C and to match 2618 alloy at temperatures up to 165°C. However, full creep tests have yet to be conducted.

Fabrication, Forming and Joining

A considerable effort has been expended in the UK assessing the ability of 8090 to be machined, formed and fabricated using a wide range of conventional techniques to produce detailed parts. Few problems have been encountered although certain process adjustments have been made. In particular the use of salt baths has been researched in depth and the results of this work will be reported in Paper 38 of this Conference. The processing of early sheet samples produced anisotropic material and although the sheet was formable the effects of this anisotropy were detected in forming trials. Current material is more isotropic. A systematic study of dimensional changes during heat treatment has been conducted. A significant shrinkage (∿0.4%) was discovered in material that had undergone its first solution treatment from the as-rolled condition. Subsequent solution treatments produced much smaller acceptable dimensional changes.

A significant quantity of demonstration parts for civil and military applications have been identified some of which have been produced and will be flown in critical applications.

The super-plastic forming capabilities of both Lital A and Lital B alloys have been assessed.[5,6] Both alloys prove to be readily formed and a variety of demonstration and test parts have been produced (Fig. 8). Back pressure, or super-imposed hydrostatic pressures of the order of half the flow stress are required to prevent cavitation and obtain good achievement of shape. The properties of the Lital alloys after super-plastic forming have been assessed. It has been found that the forming process leads to changes in grain structure and that the ageing kinetics change to a sufficient extent to require modification of ageing practice. With modified ageing practice formed components yield 95% of the strength of unformed material heat treated to specification. A particular advantage of the 8090 alloy is its low quench sensitivity in that acceptable properties can be achieved by simply ageing after forming that is without a re-solution treatment.

Adhesive bonding trials using conventional UK pre-treatments of pickling followed by either phosphoric acid or chromic acid anodising have not revealed any major problems with joint strengths matching those produced with conventional bare 2014 material. Welding trials (MIG and EB processes) have met with sufficient success to justify extended research. There is clearly some scope for the investigation of optimised filler materials, post welding heat treatment and a detailed examination of post weld properties. Mechanical fastening with a variety of selected fasteners and rivets has yet to produce any significant problems. However, detailed mechanical property evaluations of joints are still required.

Forgings

Work is continuing with forgings in both Lital A and Lital B and significant improvements have been effected in the control of anisotropy. Whilst strengths in the range of 2618 and 2014 alloy forgings can now be achieved in the T6 condition, a requirement for higher strength 7010-T74 equivalent material is developing. Ideally these strengths should be achieved without the use of cold work before ageing.

Further Developments

In stating the current position of Lital alloys A, B and C certain problem areas have been identified in this Paper. Some of these problems such as anisotropy in sheet and forgings are proving to be amenable to improvements in fabrication techniques but others such as the balance between strength and fracture toughness and the improvement of T6 strength require further development of alloy composition and heat treatment. Accepting the need for greater strength in the T6 condition, histograms of tensile properties are shown (Fig. 9) for Lital A and Lital B alloys tuned to yield greater strength. The appropriate compositions are shown. More detailed studies are presented in Paper 64 and similarly the zinc variant is considered in detail in Paper 66 of this Conference. It would seem from preliminary investigations that the zinc containing variant may be capable of producing greater T6 strength, particularly TS (Fig. 9).

It has been explained that the plane stress toughness of Lital sheet is approximately to target with some modification to Lital C required. There is certainly little prospect of increasing strength levels without compromising these fracture toughness values. However, the plane strain toughness of both plate and extrusion has proved very good at least in the L-T and T-L

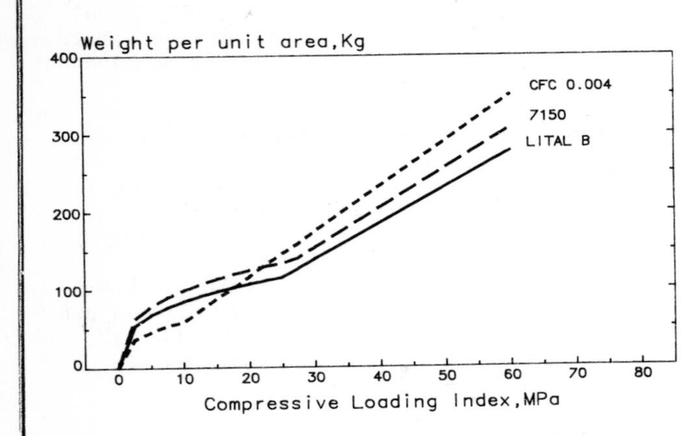

Fig. 5 Predicted weights of rectangular panels that will not buckle, yield or break under compressive loading.

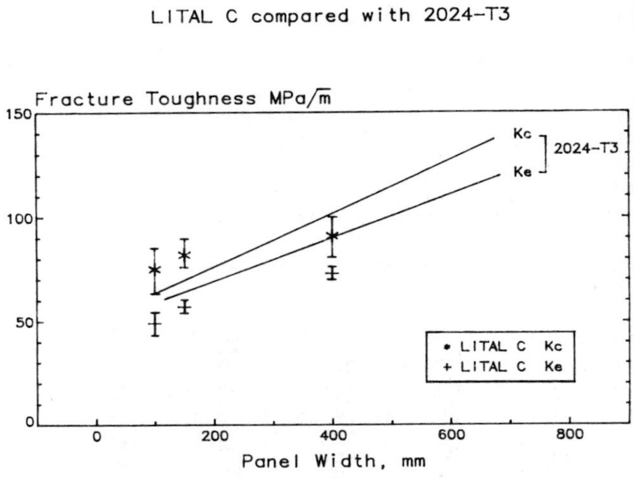

LITAL C compared with 2024-T3

Fracture Toughness MPa$\sqrt{m}$

Kc
Ke
2024-T3
* LITAL C Kc
+ LITAL C Ke
Panel Width, mm

Fig. 6 The effect of CCT panel size on the fracture toughness of Lital C and 2024-T3 sheet.

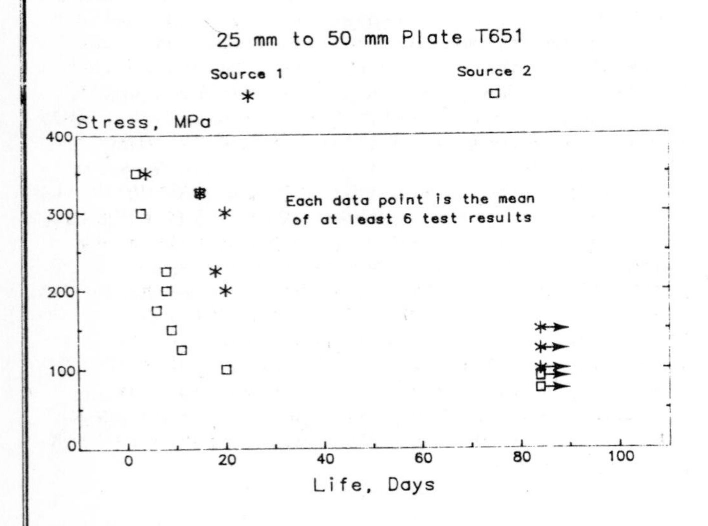

Fig. 7 The short transverse stress corrosion performance of Lital A alloy tested as 'C' rings by alternate immersion in 3.5% NaCℓ and drying in air.

Fig. 8 Superplastically formed demonstration parts produced in Lital A. Courtesy of BAe Warton Division.

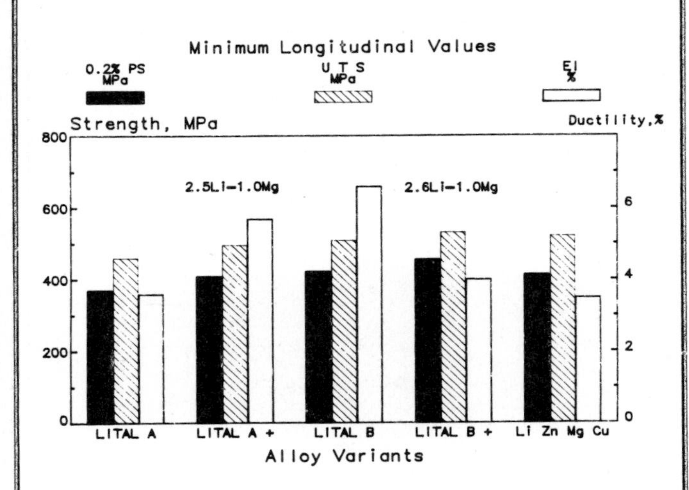

Fig. 9 Typical properties of conventional Lital A and Lital B sheet in the T6 condition compared with those of 'tuned' versions of the Aℓ-Li-Cu-Mg system and the Aℓ-Li-Zn-Mg-Cu system.

First Generation			Second Generation		
2024	T3, T4	LITAL C	2324	T39	LITAL AC
2014	T6, T8	LITAL A	7010	T76	LITAL AB
7075	T6	LITAL B	7150	T6	LITAL D

Fig. 10 Schematic representation of the first and second generation specified requirements.

directions and improved toughness is anticipated in the ST-L direction. It would seem possible to produce a second generation requirement with approximately the fracture toughness of 2024-T351 plate but the strength levels of 2014-T651. The suggested levels are defined in the following draft specification defined as DTD YYYAC.

DTD YYYAC Provisional Requirements

	0.2% PS (MPa)	TS (MPa)	K_{IC} (MPa $\sqrt{m}$)	PS: TS
Min L	385	455	40	85%
Min T	370	440	35	84%
Min ST	340	400	20	85%

These requirements may allow Lital to be used in a damage tolerant context at a higher working stress than 2024-T351 with an associated density reduction. It can be seen that the specified ratios of 0.2% PS to TS are still approximately 85% indicating that the tensile strength levels will be critical. Further toughness improvements could be achieved by relaxing the 0.2% PS requirement. Similarly, there may be a requirement to tune the strength-toughness balance of the Lital B alloy, reducing the strength to increase fracture toughness and to introduce a new very high strength alloy Lital D requirement (Fig. 10).

CONCLUSIONS

The development of the Aℓ-Li-Cu-Mg alloys within the United Kingdom has proceeded following the original concept of substitution alloys to the extent that demonstrator parts have been built and many potential applications identified. This paper has attempted to identify certain new requirements such as the damage tolerant 'C' category and an ultra-high strength extrusion and thin plate category. Second generation targets have also been identified, in particular a new medium strength high toughness category. The properties determined with the ALCAN Lital alloys have been reviewed and critical areas still receiving attention have been identified as increased strength in the T6 category for sheet and forgings, higher fracture toughness for damage tolerant sheet and higher improved short transverse properties particularly for very high strength plate.

ACKNOWLEDGEMENTS

The authors are grateful for the many contributions from colleagues in ALCAN INT and

British Aerospace and from the many associated contractors in Universities and Industry.

REFERENCES

1 Peel, C. J., and Evans, B., March 1983, Spring Residential Conference, Institution of Metallurgists, Loughborough, Publication 1601-83-Y, No 20.

2 Peel, C. J., Evans, B., Baker, C. A., Bennett, D. A., Gregson, P. J. and Flower, H. M., April 1983, Second International Aluminium-Lithium Conference, Monterey, California, published by Met. Soc. AIME.

3 Miller, W. S., Cornish, A. J., Titchener, A. P. and Bennett, D. A., April 1983, Second International Aluminium-Lithium Conference, Monterey, California, published by Met. Soc. AIME.

4 B. Evans, D. S. McDarmaid, C. J. Peel, Fifth International SAMPE Conference, Materials and Processes, Montreux, Switzerland, June 1984.

5 R. Grimes, W. S. Miller, Proc. Second International Aluminium-Lithium Conference, Monterey, California, Met. Soc. of AIME, 1983.

6 D. McDarmaid, A. J. Shakesheff, Proc. International Conference "Superplasticity in Aerospace-Aluminium", Cranfield, July 1985.

Production of aluminium–lithium alloy with high specific properties

P MEYER and B DUBOST

The authors are with
Cegedur Pechiney,
Centre de Recherches et Développement,
BP 27 - 38340 Voreppe, France

SYNOPSIS

Reducing the weight of aircraft structure through the use of lighter advanced materials has become a major goal. PECHINEY's extensive Research and Development program has resulted in a new family of aluminium-lithium alloys. Thanks to the combination of efficient D.C. casting, alloy design coprecipitation strengthening and structural control, these CP 27X alloys overcome the previously identified brittleness of Al-Li alloys and already match most current goals for a wide use on aircraft.

INTRODUCTION

Increasing payload and fuel efficiency of aircraft has become a major issue for the aerospace industry, which has boosted the quest of metallurgists for ever more advanced materials with high specific properties[1]. Among candidate materials, the new generation of low density aluminium-lithium alloys made by ingot metallurgy looks particularly attractive, provided these alloys can match the strength properties, damage tolerance, corrosion resistance and working ability of conventional aluminium alloys. With these goals fulfilled, mass savings of about 10 % are already achievable by simple density reduction and could reach 15 % to 18 % by taking advantage of the substantial stiffness increase[2]. Moreover, no basic changes of manufacturing method or design and technical maintenance would be necessary for the end users.
In order to ensure this high level of material properties, PECHINEY has launched an aggressive research and development program of aluminium-lithium compositions, processing and thermo-mechanical treatments via the ingot metallurgy route[3].

Two casting units were built at the Voreppe Research Centre with ingot weights of 150 kg (1982) and 1500 kg (1984) to meet the growing demand ; a large scale industrial casting unit is already planned.

GOALS FOR NEW ALUMINIUM LITHIUM ALLOYS :

The extensive research program is focused on the following metallurgical goals:
- Maximum weight saving (at least 8 % density reduction)
- Strength levels covering the whole range of strength of conventional alloys (from 2024 T4 to 7150 T651)
- Adequate ductility
- Good damage tolerance and corrosion resistance.

In addition, the selected aluminium-lithium alloys must fulfill the following technological goals :
- Recycling compatibility implying the use of only one family of alloys
- Casting ability into large size ingots
- Good hot-and cold-working ability.

This family of alloys must possess these qualities regardless of the type of semi-products (ie : indifferently plates, sheets, forgings, extrusions, tubes ...)

HIGHLIGHTS OF THE NEW ALUMINIUM-LITHIUM METALLURGY

. Casting procedure

In order to reach these goals a semi-continuous direct chill method was selected for production of round and rectangular ingots. The special equipment used is considered proprietary as much technological know-how is necessary to ensure :
- Sufficient purity without porosity and cracks, with low oxide, gas and inclusion content, and low segregation effects

- Good geometry of ingots (with smooth surface)
- Safe casting conditions.

Concerning the latter point, a series of experiments was carried out to check that the casting process presents no major risk.

. Metallurgical optimization

It was well known that binary aluminium-lithium alloys could not be used because of their brittleness and insufficient strength [4,5]. The need for additional strengthening through coprecipitation has been clearly established [6,7].

Sufficient hardening through matrix precipitation, together with good ductility through grain boundary reinforcement and sufficient slip homogenization appeared however feasible by the ingot metallurgy route.
Taking the previously published results into account [8,9], the alloy design was first based on our knowledge of both equilibrium phase diagrams and precipitation mechanisms within the Al-Li-Cu-Mg system.
As a guideline, the research focused on alloys with favorable coprecipitation hardening without occurrence of coarse constituent particles.
In order to reach high strength levels on semi-products through substructure strengthening, unrecrystallized structures may be obtained by using zirconium as the preferential secondary alloying element.
For such unrecrystallized alloys slight underaging or peakaging provide a good combination of high strength, reasonable ductility, good toughness, excellent crack propagation and fatigue behaviour thanks to copious precipitation of Cu and Mg rich phases in addition to solid solution strengthening and δ'-Al_3Li precipitation.

However for some particular semi-products like sheets a high level of strength may not be essential, whereas ductility, forming ability, in as treated conditions, isotropy of mechanical properties are of major interest.
It has been observed that highly textured unrecrystallized aluminium-lithium alloys might exhibit strongly anisotropic properties. One solution to this problem consists in obtaining a fully recrystallized structure which provides increased ductility and isotropy together with macroscopic slip homogenization through grain misorientation effects. This implies a dedicated thermo-mechanical processing.
Under these conditions, slightly underaged alloys have attractive combinations of strength, ductility, forming ability, crack propagation resistance in addition to increased isotropy and low density.

PROPERTIES MEASURED ON CP27X ALLOYS

Four alloys have been developed at the Voreppe Research Centre covering the whole range of strength of conventional alloys with a density reduction of 8 to 12 %.
All these alloys belong to the Al-Li-Cu-Mg-Zr family which appeared as the most interesting after extensive experiments on small size products.

Alloys CP271, 2091-CP274 and CP276 are now currently cast on the 1500 kg unit and processed in the plants of the PECHINEY group.

Table 1 summarizes the different combination of properties for the three CP27X alloys under industrial production with the available ingot sizes. Although PECHINEY's alloy CP271 and R.A.E.'s alloy DTDXXXA[10] have been developed independently, their close compositions made it possible to register the common European alloy 8090.

Table 1

Typical properties of AlLiCuMg CP27X alloys under production and capacity of casting units located at the Voreppe Research Centre.

PROPERTIES			ALLOYS
* density reduction	* specif. modulus increase	other main characteristics	
10 %	20 %	high strength	8090 CP271
8 %	15 %	high strength high ductility	2091 CP274
8 %	17 %	very high strength	CP276
INGOT SIZES			
Weight (Kg)		Ingot sections (mm)	
1500 kg		800 X 300 Ø 450	
150 kg		350 X 100 Ø 200	

* to be compared to 2XXX 7XXX conventional alloys

Table 2 shows the commercial ranges of composition for these CP27X aluminium-lithium alloys. Patent applications have been filed for 2091-CP274 and CP276 as well as for other alloys not shown here and including CP277, the extra light alloy (12 % density reduction) under limited experimental production.

Table 2

Alloy composition ranges (weight percent)

	8090 - CP271	2091 - CP274	CP276
Cu	1.0 - 1.6	1.8 - 2.5	2.5 - 3.3
Li	2.2 - 2.7	1.7 - 2.3	1.9 - 2.6
Mg	0.6 - 1.3	1.1 - 1.9	0.2 - 0.8
Fe	⩽ 0.30	⩽ 0.30	⩽ 0.30
Si	⩽ 0.20	⩽ 0.20	⩽ 0.20
Cr	⩽ 0.10	⩽ 0.10	⩽ 0.10
Mn	⩽ 0.10	⩽ 0.10	⩽ 0.10
Zn	⩽ 0.25	⩽ 0.25	⩽ 0.25
Ti	⩽ 0,10	⩽ 0.10	⩽ 0.10
Zr	0.04 - 0.16	0.04 - 0.16	0.04 - 0.16

Table 3

8090 - CP271

Typical longitudinal properties measured on extruded flat bars (100 x 13 mm^2)

Alloys	Al-Li alloy 8090 - CP271			conventional (same product)	
				2024	7475
Temper	T6	T61	T651	T351	T7351
YS (MPa)	445	435	490	400	460
UTS (MPa)	555	535	540	530	530
El (%)	7	6	7	13	12
K_{1C} ($MPa\sqrt{m}$)	37	40	37	39	42
d(g/cm^3)	2.52 - 2.54			2.79	2.81
E(GPa)	81.2			75.7	72.8

In the following typical properties resulting from the application of previous metallurgical principles are given with emphasis on structure-properties relationships.

All results given are typical and cannot be considered as guaranteed values. The comparisons with conventional alloys are performed on the same products.

8090 - CP271

This low density alloy (10 % density reduction) may give various combinations of properties, strongly depending on its structure. Unrecrystallized structures, e.g. extrusions, lead to high strength levels thanks to Cu and Mg solid solution strengthening (for underaged tempers) and copious coprecipitation of S'-Al_2CuMg needles (in the vicinity of peak aged conditions) along <100> Al directions, together with δ'-Al_3Li. The favourable effect of dense S' precipitation is strongly enhanced by cold-working between quenching and aging.

Fig. 1 illustrates the tremendous increase of the strength-ductility combination thanks to Cu and Mg alloying.

Even in peak aged conditions ductility loss is avoided by precipitation of semi-coherent S' needles up to δ' precipitate-free zones adjacent to grain boundaries.

Moreover peak-aged conditions (T651 and T6 : with or without controlled stretch between quenching and aging) lead to good levels of fracture toughness, which can even be improved by a slight underaging, e.g. T61 (Table 3).

Fatigue and crack propagation resistance of 8090 - CP271 alloy compare well with those of conventional alloys. Wöhler curves are almost superimposed on those of 2024 T351 for both smooth and notched specimens (Fig. 2) whereas crack propagation behaviour is better over the whole ΔK range under monotonic loading and at low and medium ΔK under spectrum loading (Fig. 3).

Recrystallized structures appeared helpful for 8090 - CP271 to provide a good combination of ductility, isotropy, formability and strength on sheets by associating the aforementioned favorable influence of copper and magnesium alloying and controlled grain misorientation effects. This leads to adequate slip behaviour.

Figure 4a illustrates the macroscopic slip dispersion arising from randomly oriented recrystallized grains at the surface of a tensile specimen just before fracture. Thanks to this additional misorientation effect, it is not necessary to strongly overage the alloys in search of isotropy and ductility (Table 4) through development of extensive S' precipitation[11] ; as a matter of fact, severely overaged aluminium-lithium alloys undergo significant grain boundary embrittlement because of grain boundary precipitate coarsening and the widening of precipitate free zones (Fig. 4b)

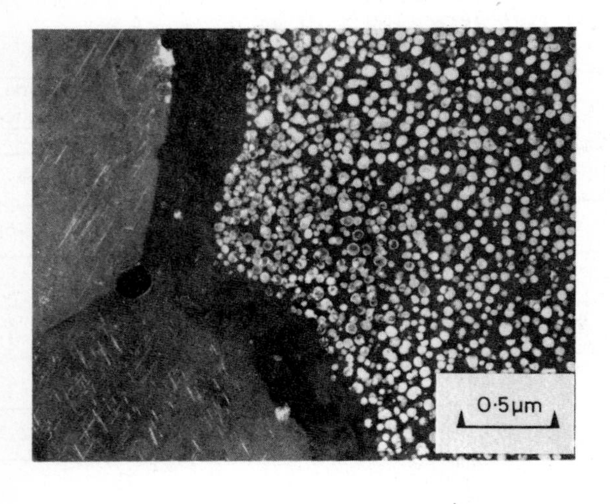

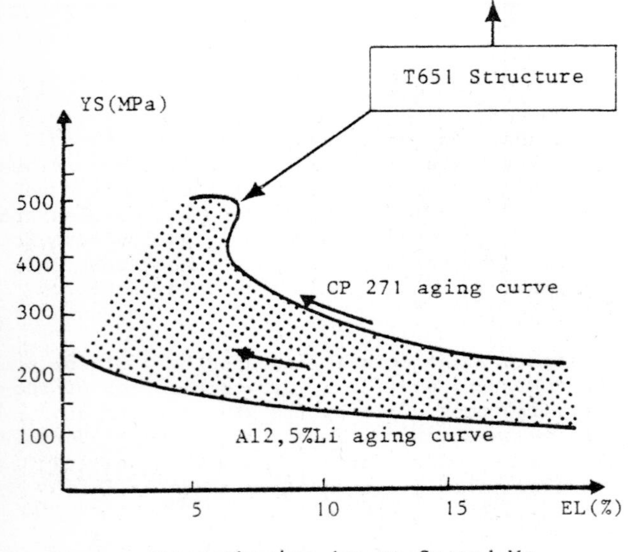

Fig. 1

Improvement of the tensile strength-ductility combination of extrusions through Cu Mg Zr alloying.

Typical T.E.M. dark field micrograph of 8090-CP271 T651. δ' reflexion (right) together with S' reflexion (left).

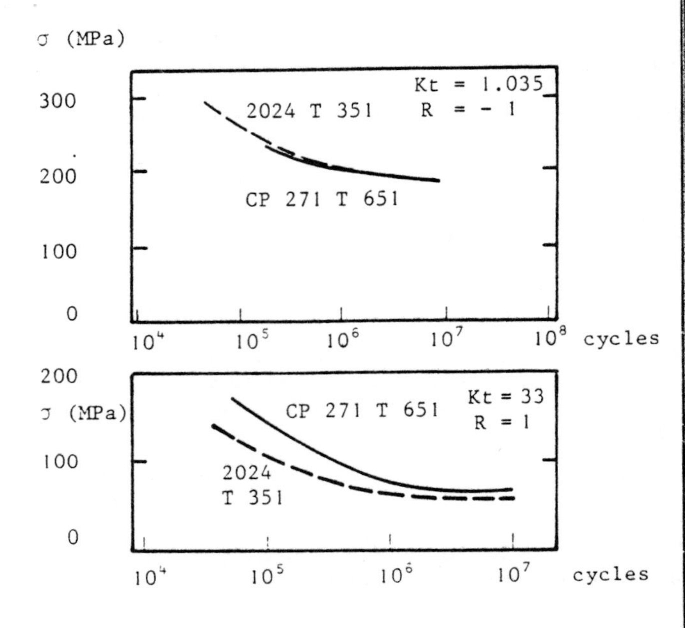

Fig. 2

Wöhler curves of 8090-CP271 T651 and 2024 T351 extrusions. Longitudinal direction; smooth (top) and notched (bottom) specimens.

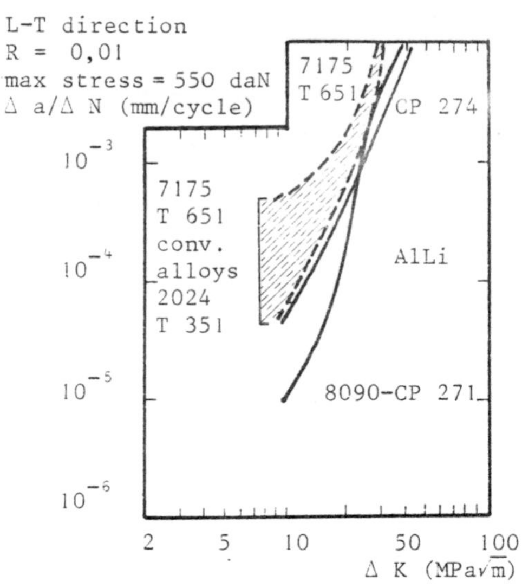

a

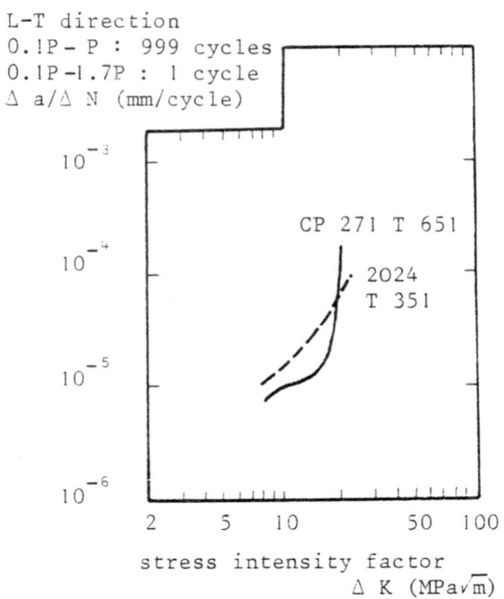

b

Fig. 3

Fatigue crack propagation curves of 8090-CP271, 2091-CP274, 2024 T351 and 7175 T651 extrusions.

L-T direction, monotonic (top) and spectrum loading (bottom).

Fig. 4

In situ S.E.M. micrographs of tensile specimens ; 8090-CP271 recrystallized sheets just before fracture

a) underaged temper : macroscopic slip dispersions through grain misorientation

b) overaged temper : intergranular failure

Table 4

8090 - CP271

Typical longitudinal properties measured on 1.6 mm sheets (small scale production)

Temper	T4 *	T351 *	T6	T651
YS (MPa)	255	200	380	375
UTS (MPa)	395	305	500	480
El (%)	15	15	10.5	13
$d (g/cm^3)$	2.52 - 2.54			
E (GPa)	81.2			

* after one month natural aging

2091-CP274

Alloy 2091-CP274 provides high strength and high ductility in both T6 and T651 tempers together with a 8 % density reduction. These properties are achieved thanks to homogeneous coprecipitation of a relatively large volume fraction of fine $S'-Al_2CuMg$ needle like precipitates and relatively low volume fraction of $\delta'-Al_3Li$. Additional transgranular precipitation of rounded laths of $T1-Al_2Cu[Li,Mg]$ phases also occurs in the T6 temper together with heterogeneous precipitation of S' on dislocations (Fig. 5).

Magnesium is thought to modify the usual plate-like morphology of $T1-Al_2CuLi$ phase in accordance with previous work. [12]

Near net shape forgings have been processed. The elaborate shapes of these products illustrate the overall good hot working ability of 2091-CP274 and more generally of the CP27X family.

In addition to good tensile strength, CP274 T6 may exhibit greater elongations than conventional alloys 7175 T73 and 2214 T6 (Fig. 6).

Tests performed on extrusions confirm the good damage tolerance of 2091-CP274, which provides high fracture toughness and good crack propagation resistance [3]

Recrystallized structures are of interest, as in the case of 8090-CP271 alloy. The excellent isotropy of tensile properties together with high ductility as measured on 1.6 mm sheets is well established (Fig. 7 and Table 5).

Hence 2091-CP274 allows better yield strength - ductility combinations than CP271 with 8 % density reduction over conventional alloys.

CP276

The CP276 alloy has been designed to reach the highest possible strength level within the whole Al-Li-Cu-Mg family without the need for cold working between quenching and aging.

The magnesium content of CP276 was optimized, as this element has a significant influence on microstructure and properties (Fig. 8) :

- inhibition of T'-(AlCuLi) [13] coarsening on <100> plane in underaged and T6 tempers,

Table 5

2091 - CP274

Typical longitudinal properties measured on 1.6 mm sheet

Alloys	Aluminium lithium alloy	conventional (same product)	
	CP274	2024	2214
Temper	T651	T3 - T4	T651
YS (MPa)	430	340	415
UTS (MPa)	480	480	460
El (%)	12	18	11
$d (g/cm^3)$	2.57 - 2.59	2.79	2.79
E (GPa)	78.8	75.7	75.2

Table 6

CP276

Typical longitudinal properties measured on extruded flat bars

Alloys	Aluminium Lithium alloy			conventional
	CP276			7075
Temper	T6	T651	T6511 *	T651
YS (MPa)	490-590	575-625	565	560
UTS (MPa)	590-640	600-655	625	610
El (%)	5	5	5.5	11
$d (g/cm^3)$	2.57 - 2.60			2.81
E (GPa)	80.2			72.8

* K_{IC} L-T : $39 MPa\sqrt{m}$

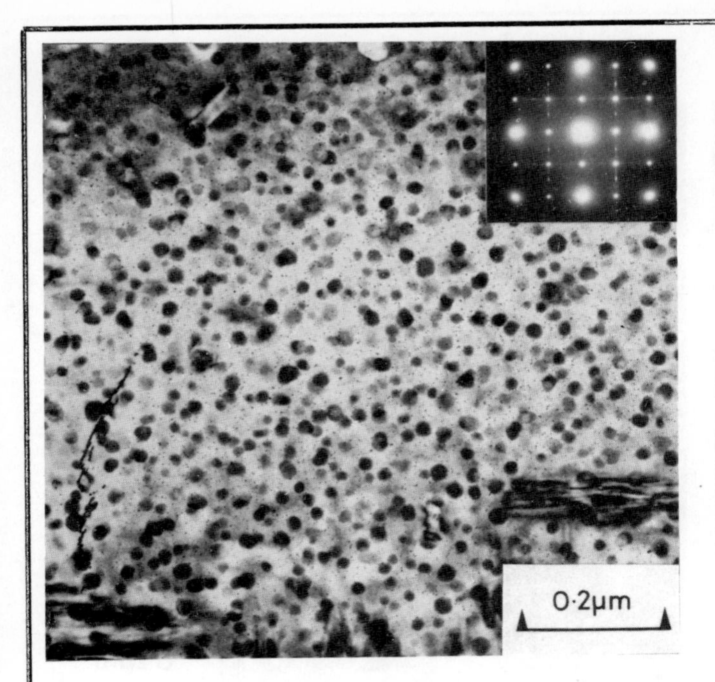

Fig. 5

Typical T.E.M. micrograph and selected-area diffraction pattern of 2091-CP274 T6 : homogeneous δ'-Al_3Li + fine S' precipitation.

Coarser T_1 rods (Top- left) and heterogeneous S' precipitations on dislocations (bottom)

Fig. 8 a-d (OVERLEAF) ▶

Dark field T.E.M. micrographs illustrating effects of Mg on matrix precipitation of CP276 alloy.

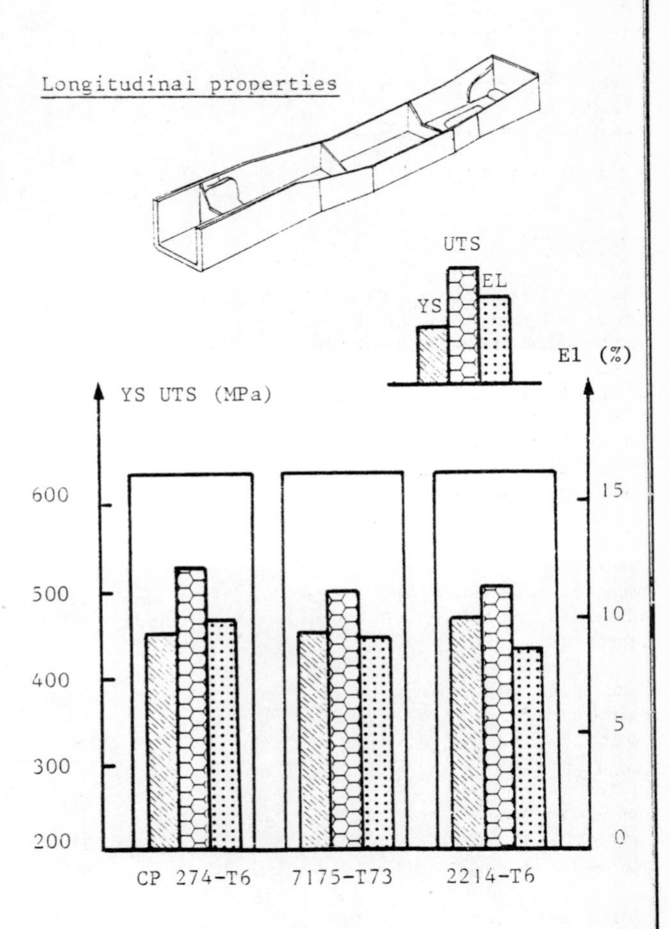

Fig. 6

Compared tensile properties of 2091-CP274 and conventional alloys, as measured on a near net shape forging.

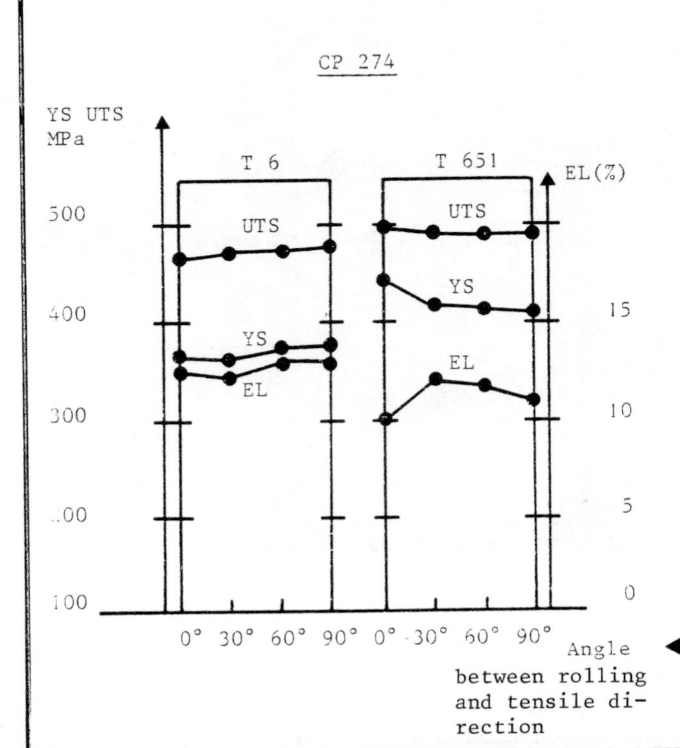

◀ **Fig. 7**

Isotropy of tensile properties as measured on 1.6 mm sheets of 2091-CP274 T6 and T651.

CP 276	Modified CP 276 (Mg free)

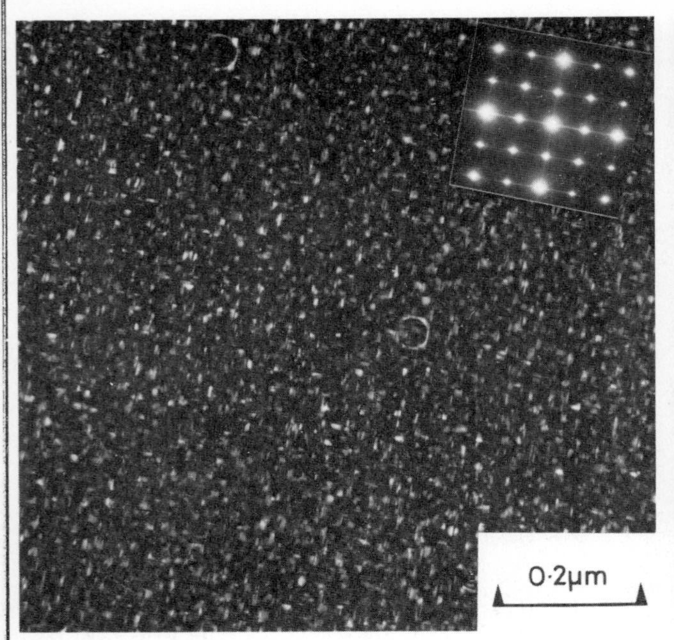

Fig. 8a

Fine δ' and T' precipitates in slightly underaged CP276-T61
$(001)_{Al}$ foil orientation

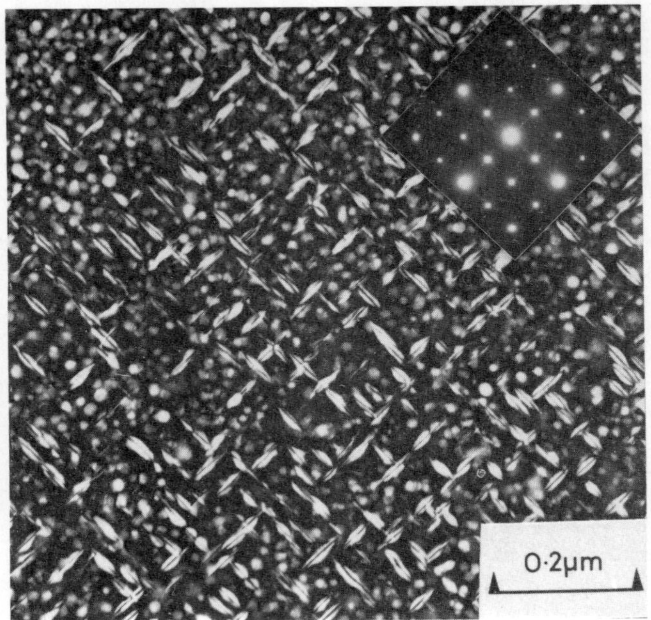

Fig. 8b

δ' precipitation in matrix and along coarser T' plates in Mg-free CP276-T61. Same underaging conditions as Fig. 8a
$(001)_{Al}$ foil orientation

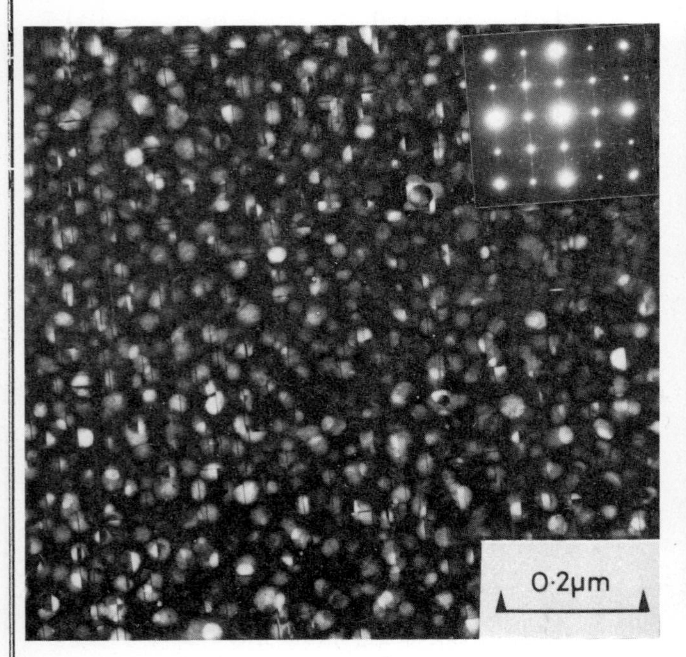

Fig. 8c

Dense δ + T' coprecipitation in CP276-T6.
S' + coarser T1 plates are out of contrast
$(001)_{Al}$ foil orientation

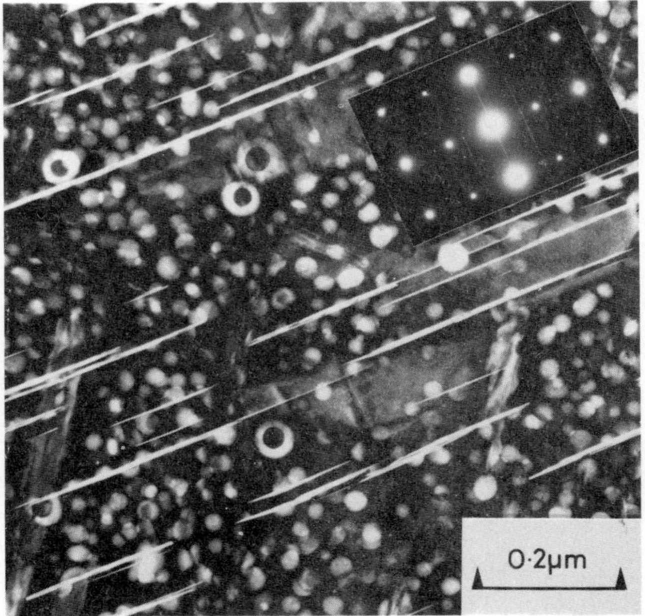

Fig. 8d

δ' + coarser T1 precipitates in Mg-free CP276-T6
Rare T' plates are out of contrast
$(112)_{Al}$ foil orientation

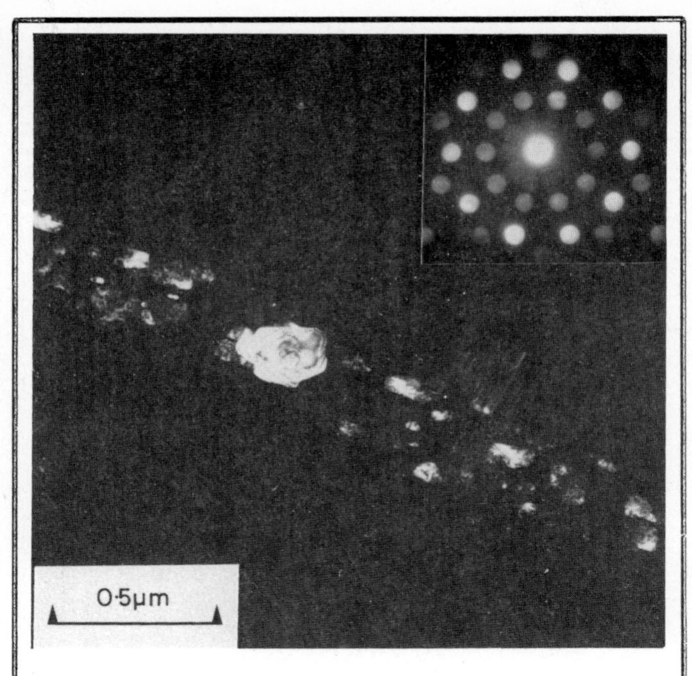

Fig. 9

T.E.M. dark field micrograph showing quasi-crystalline-like grain boundary precipitates in peak-aged 2091-CP274 alloy.

Typical converging beam SAD pattern showing 5-fold symmetry.

- enhancement of maximum hardening in T6 temper through complex coprecipitation of T'-(AlCuLi) $T1-Al_2CuLi$ and $S'-Al_2CuMg$ in addition to $\delta'-Al_3Li$.

This leads to very high tensile properties and attractive combination of strength and toughness in slightly underaged tempers (Table 6).

It is worth noting that T.E.M. examinations performed on CP276 and 2091-CP274 in peak and overaged conditions revealed rounded grain boundary precipitates with quasi-crystalline-like structures and apparent icosahedral symmetry (Fig.9)[14]. This phase might be closed to $T2-Al_6Cu(LiMg)_3$ [15].

CONCLUSION

The extensive research program launched by CEGEDUR PECHINEY in order to take up the challenge of lower aircraft operating costs led to the development of promising new aluminium-lithium alloys with high specific properties.
It is now established that ingot metallurgy is able to provide satisfactory casting conditions for large size products.

Dedicated alloy design and structural control make it possible to achieve, with the CP27X family, attractive combinations of mechanical properties on various kinds of semi-products.

Furthermore, preliminary results of corrosion tests performed in marine and industrial environment appear encouraging, as well.

Thorough evaluation of service properties on a large scale is going on, as a decisive step towards the extensive use on aircraft of these new materials for progress.

ACKNOWLEDGEMENTS

The authors whish to thank all the contributors to this program, in the Voreppe Research Center, in the plants, in the head offices of PECHINEY Branche Aluminium.

The technical support of J. Domeyne and the electron microscopy work of E. Janot and B. Bès for this paper are greatly acknowledged.

REFERENCES

1 - G. Hilaire
Enjeux n° 29, Oct 82, p 31

2 - W.E. Quist, G.M. Narayanan, A.L. Wingert
Aluminium Lithium alloys II
ed. T.H. Sanders, Jr and E.A. Starke, Jr
Publ. Met. Soc. AIME 1984, p 313

3 - J. Moriceau, B. Dubost, G. Le Roy, Ph Meyer
WESTEC'85 conference, March 18, 1985
CEGEDUR PECHINEY

4 - W.R.D. Jones
J. Inst. Met., 1959-60, vol 88, p 435

5 - B. Noble, S.J. Harris, K. Dinsdale
Met. Sc., vol 16, Sept 1982, p 425

6 - T.H. Sanders, Jr
Aluminium Lithium alloys
ed. T.H. Sanders, Jr and E.A. Starke, Jr
Publ. Met. Soc. AIME 1981, p 63

7 - E.A. Starke, Jr
Strength of Metals and Alloys
(ICSMA 6), ed. R.C. Gifkins,
Pergamon Press, 1983, p 1025

[8] - K.K. Sankaran, N.J. Grant
Mat. Sc. Eng. 44, 1980, p 213

[9] - E.A. Starke, Jr, T.H. Sanders, Jr, I.C. Palmer
J. of Metals, Aug 1981, p 24

[10] - C.J. Peel, B. Evans, C.A. Baker, D.A. Bennett, P.J. Gregson, H.M. Flower
Aluminium Lithium Alloys II,
ed. T.H. Sanders, Jr and E.A. Starke, Jr
Publ. Met. Soc. AIME 1984

[11] - P.J. Gregson, H.M. Flower
Acta Metall., vol 33, n° 3, p 527, 1985

[12] - R. Crooks
Ph. D. thesis, Georgia Inst. of Techn., 1982

[13] - P. Sainfort and P. Guyot
Phil. Mag. A, 1985, vol 51, n° 4, p 575

[14] - D. Shechtman, I. Blech, D. Gratias, J.W. Cahn
Phys. Rev. Letters, vol 53, n° 23, 12 Nov 1984, p 1951

[15] - P. Sainfort, B. Dubost and A. Dubus
Comptes rendus Acad. Sc. II
July 1 1985, to be published

□

'Alithalite' alloys: progress, products and properties

P E BRETZ and R R SAWTELL

The authors are in the
Alloy Technology Division of
Alcoa Laboratories,
Alcoa Center,
PA 15069, USA

SYNOPSIS

ALITHALITE® Al-Li alloys are being developed by Alcoa to meet four first-generation objectives: low-density replacements for 2024-T3X, 7075-T6X, and 7075-T73X products, and a moderate-strength, minimum density alloy. One alloy, recently registered as 2090 (Al-2.7wt%Cu-2.2%Li-0.12%Zr), has been introduced as a replacement for 7075-T6X plate, sheet, and extrusion product forms. This paper will concentrate on discussing the metallurgical and engineering property characteristics of this alloy. Progress toward developing alloys to meet the other objectives also will be outlined, with emphasis given to property achievements.

INTRODUCTION

The marriage of lithium with aluminum offers the promise of substantially reducing the weight of aerospace alloys, since each 1 wt.% Li added to Al reduces density by 3%. This promise is particularly attractive to the aerospace industry, since structural weight reduction is a very efficient means of improving aircraft performance. As this conference and its two predecessors [1,2] demonstrate, Al-Li alloys have attracted extraordinary interest within both the aerospace industry and the metallurgical community.

While reduced density is a principal offering of Al-Li alloys, there are other significant advantages as well. The same 1% addition of Li increases elastic modulus by 6%, which will allow greater load-carrying capability in stiffness-critical components. Fatigue crack growth (FCG) resistance in Al-Li alloys generally is very high [3-4]; this is important in damage-tolerant structures such as lower wing surfaces.

In addition to structural design considerations, there are potential economic benefits to utilizing Al-Li alloys. In contrast to new materials systems such as fiber-reinforced composites, low-density Al alloys do not require large capital investments by the aircraft producer in new fabricating facilities. This cost savings can more than offset the greater performance increment which composites may offer, resulting in Al-Li alloys being substantially more cost-effective than composites in some applications [5,6]. In fact, at least one study concluded that low-density Al alloys could provide a performance advantage over composites in an advanced fighter application [7].

In response to this substantial potential, the aluminum industry has embarked on a very dynamic alloy/product development program to bring low-density Al alloys to the marketplace. Alcoa's first generation Al-Li product development goals entail the synthesis of low-density functional replacements for several existing high-performance Al alloys. These functional replacements are designed to have engineering property characteristics which are at least equivalent to those of incumbent aerospace alloys, but at density reductions of 7-10%. This property equivalence will allow rapid implementation of the new alloys, since little or no redesigning of aircraft components would be required. Thus, significant weight savings can be realized

Table 1
Composition Limits for 2090, wt %

	Li	Cu	Mg	Zr	Fe	Si
Min.	1.9	2.4	--	0.08	--	--
Max.	2.6	3.0	0.25	0.15	0.12	0.10

Table 2
Comparison of Achievements Against Minimums
Alloy 2090-T8, Plant-Fabricated Plates
(L Orientation)

	Minimum*	Obtained	
		44 mm	38 mm
UTS, MPa	524	558	569
TYS, MPa	476	510	530
CYS, MPa	455	--	--
Elongation, %	6(LT)	12	7.9
K_{Ic}, MPa$\sqrt{m}$	22	38.5	42.5

*Values taken from MIL-HDBK-5D. Tensile minimums are "A" values; toughness value is minimum from range of test data.

	Target**	Obtained
E, GPa	76	78.6
Density, g/ml	<2.60	2.59
Exfoliation Rating, MASTMAASIS†	EB	EA
SCC Classification, ST Orientation	C	C

**Alcoa first generation goals
†After 21 months exposure at Pt. Judith, Rhode Island (seacost environment), 2090 plate shows no evidence of exfoliation

Table 3
Comparison of Achievements Against Minimums
Alloy 2090-T8, Plant-Fabricated 2.0 mm Sheet
(L Orientation)

	Minimum*	Obtained
UTS, MPa	532	563
TYS, MPa	483	512
Elongation, %	8(LT)	5.2

*"A" values taken from MIL-HDBK-5D.

Table 4
Comparison of Achievements Against Minimums
Alloy 2090-T8, Plant-Fabricated Extruded "Tee" (48 mm thick leg)

	L		LT		ST	
	Minimum*	Obtained	Minimum*	Obtained	Minimum*	Obtained
UTS, MPa	558	546	483	494	--	449
TYS, MPa	496	514	420	453	--	416
Elongation, %	7.0	7.0	--	5.7	--	3.1
K_{Ic}, MPa$\sqrt{m}$	25.3	27.2	19.8	15.6	18.7 (S-L)	15.3

*Values taken from MIL-HDBK-5D. Tensile minimums are "A" values; toughness value is minimum from range of test data.

simply by direct substitution of first generation ALITHALITE alloys.

Specifically, these first generation Al-Li alloys are targeted as replacements for various mill products of the following alloys: 2024-T3X (Goal A), 7075-T6X (Goal B), 7075-T73X (Goal D), and a minimum density, medium-strength alloy (Goal C). Alloys 2024 and 7075 have been the backbone of aerospace structures for many years; low-density replacements for these alloys will find immediate and wide-spread applications and provide substantial weight reductions. The Goal C alloy is aimed at secondary structure which is not crucial to airworthiness but, nevertheless, contributes to the overall weight of the aircraft. The specific mechanical and physical property targets for these four first generation goals have been presented numerous times, most recently in reference [8]. It is sufficient for the purposes of this paper to note, as mentioned previously, that these targets are essentially for properties equivalent to those of incumbent alloys, but with density reductions of 7-10%. Significantly, though, the targets for 2024-T3X and 7075-T6X products aim at improved exfoliation corrosion and stress corrosion cracking (SCC) resistance over that available in the incumbent alloys. Also, all of the low-density alloys would be supplied to controlled fracture toughness levels, a practice common to newer aerospace alloys but not applicable to 2024 and 7075.

Al-Li ALLOY PRODUCTION EXPERIENCE

Commercial-size product development activities began in mid-year 1983 with the scale-up of ingot casting capability at the Alcoa Technical Center to a 4500 kg (10,000 lb.) furnace. This early leap to full-size ingot capability was made in the belief that such a strategy would minimize the time necessary to bring to the marketplace products truly representative of commercial fabrication practices. More than 18 months of valuable experience at several Alcoa fabrication plants has been gained as a result of this early scale-up.

Most of the production experience to date has been on alloy 2090, which is the low-density replacement for 7075-T6X; the properties on several product forms of this alloy will be discussed in detail in the next section. This experience includes fabrication of sheet and plate at Davenport, Iowa; extrusions at Lafayette, Indiana and Vernon, California; and forgings at Cleveland, Ohio and Vernon, California. Sheet gauges from 1.6 mm to 4.6 mm and plate thicknesses of 12.7 mm to 76 mm have been fabricated, along with a total of seven different extrusions (shapes and rectangular bars) and ten forgings (hand, die, and precision).

For Alcoa Goals A, C, and D, product fabrication studies have been conducted at the Alcoa Technical Center, including full-scale ingot casting trials. Properties on Goals A and D products fabricated from full-scale ingot sections are described below. The Goal C alloy has received less emphasis, and only properties from laboratory-size (45 kg) ingot are available. These properties are described in reference [8], and will not be discussed further in this paper.

ALLOY 2090-T8: 7075-T6X REPLACEMENT

In 1984, Alcoa registered an Al-Cu-Li-Zr alloy with the Aluminum Association which was assigned the designation 2090. The purpose of this section of the paper is to briefly describe the microstructure of this alloy and to present engineering property data for various product forms.

The composition of alloy 2090 is listed in Table 1. In addition to Cu and Li, which are the principal strengthening elements, Zr is added as a dispersoid-forming element to control the final grain structure. Fe and Si are controlled to very low levels to minimize the formation of insoluble constituent phases which reduce fracture toughness in all Al alloys.

The grain structure of this alloy generally is completely unrecrystallized as a result of the presence of Al_3Zr dispersoids throughout the microstructure. The precipitation sequences in Al-Cu-Li alloys are not well-defined, particularly for the various metastable phases which may precede the equilibrium binary and ternary precipitates in this system. This subject will be discussed in greater detail by several authors in this conference. It is believed, however, that alloy 2090 in the peak aged temper contains a mixture of δ' (Al_3Li), T_1' (Al_2CuLi), and a Θ'-like precipitate which may be a precursor to the T_2 phase (Al_5CuLi_3). The interested reader is referred to the paper by Rioja and Ludwiczak in this conference [9] for details on this phase characterization.

The remainder of this section is devoted to presenting engineering property information on various mill-fabricated product forms of alloy 2090. Where possible, the properties achieved will be compared against MIL-HDBK-5 minimum values for equivalent product forms of alloy 7075-T6X. The listed tensile property minimums will be "A" values; that is, the 99% confidence limit values. For fracture toughness, the listed minimum is the lowest value in the range of data available for the product form, but is not a statistically-derived minimum. Where MIL-HDBK-5 values are not

Table 5
Goal A, 2024-T3X Replacement

A. Composition, wt%

	Li	Cu	Mg	Zr	Fe	Si
Min.	2.1	1.1	0.8	0.08	--	--
Max.	2.7	1.6	1.4	0.15	0.15	0.10

B. Comparison of Properties Against Minimums
Lab-Produced 38 mm Plate (L Direction)

	Minimum*	Obtained
UTS, MPa	427	476
TYS, MPa	324	400
CYS, MPa	269	--
Elongation, %	6(LT)	9
K_{Ic}, MPa$\sqrt{m}$	29.7	45.6(K_Q)

*Values taken from MIL-HDBK-5D. Tensile minimums are "A" values; toughness value is minimum from range of test data.

	Target**	Obtained
E, GPa	80	78.6
Density, g/ml	<2.58	2.55
Exfoliation Rating, MASTMAASIS	EB	P
SCC Classification, ST Orientation	C	C

**Alcoa first generation goals

Table 6
Goal D, 7075-T73X Replacement
(Plate and Extrusions)

A. Composition, wt%

	Li	Cu	Mg	Zr	Fe	Si
Min.	2.1	0.5	0.9	0.08	--	--
Max.	2.7	0.8	1.4	0.15	0.15	0.10

B. Comparison of Properties Against Minimums
Lab-Produced 38 mm Plate (L Direction)

	Minimum*	Obtained
UTS, MPa	455	488
TYS, MPa	379	406
CYS, MPa	372	--
Elongation, %	6(LT)	7.5
K_{Ic}, MPa$\sqrt{m}$	29.7	45.3(K_Q)

*Values taken from MIL-HDBK-5D. Tensile minimums are "A" values; toughness value is minimum from range of test data.

	Target**	Obtained
E, GPa	80	78.6
Density, g/ml	<2.58	2.55
Exfoliation Rating, MASTMAASIS	EA	P
SCC Classification, ST Orientation	A	C

**Alcoa first generation goals

available, the property achievements are compared to Alcoa's first generation targets. The properties shown will be, except as specifically noted, for tests in the longitudinal orientation (parallel to the direction of grain flow).

Much of the mill fabrication experience on 2090 has been in the production of plate. Table 2 lists properties obtained on two different plate gauges: 44 mm and 38 mm thick. The upper portion of the table shows that strength, elongation, and fracture toughness are well in excess of minimums for 7075-T651 plate. In fact, these combinations of strength and toughness are at least as good as that available in any ingot metallurgy Al alloy. This is shown in Figure 1 as a graph of fracture toughness versus tensile yield strength for various alloys; the values for both 2090 plate gauges are the equal of that obtainable in the current generation of high strength, high toughness aerospace alloys. The lower portion of Table 1 indicates that modulus, density, and exfoliation and SCC resistance targets have been met or exceeded. Despite this success, short-transverse (ST) elongation in plate often has been very low (0-1%); this problem is being actively investigated.

It is important to amplify the significance of the corrosion results in Table 2. With regard to SCC resistance, a rating of "C" indicates the ability to sustain a stress of about 25% of the ST yield strength without encountering SCC failures, and is the rating for 7075 products in the T76 temper. In contrast, 7075-T6 products have only a "D" rating, which shows that any sustained stress in the ST direction may cause SCC. The observed "EA" rating for exfoliation resistance using the MASTMAASIS test indicates that only slight exfoliation occurred during the test, and is better than the target of "EB". Of greater significance, though, is the note regarding preliminary atmospheric exposure results for 2090-T8 plate. To date (21 months), no exfoliation has been observed on mill-fabricated plate samples in test at a seacoast environment site. This encouraging trend is similar to the performance of the first Al-Li alloy, 2020-T651, which was used on the RA-5C Vigilante aircraft flown by the U. S. Navy. In this application, no corrosion degradation was observed, and the alloy also was rated EA in the MASTMAASIS test. In contrast, other exfoliation tests such as EXCO and ASSET rate both 2020-T651 and 2090-T8 as EC (severe attack). Thus, although the initial indication of exfoliation resistance for 2090 is promising, the choice of an appropriate accelerated exfoliation test for Al-Li alloys remains in doubt.

Very recently, 2090-T8 plate at a 12.7 mm gauge was fabricated and supplied to the U. S. Navy as part of an extensive test and evaluation program (this contract effort is outlined below). Figure 2 compares tensile and toughness data on this plate in both L and LT orientations with the appropriate minimums. As was the case for the thicker plate gauges previously discussed, these properties are above minimum values. The longitudinal (L-T orientation) toughness value is actually a K_{R25} value determined in an R-curve test; as such, it is not strictly comparable with plane strain (K_{Ic}) toughness. Nevertheless, the very high R-curve toughness is indicative of a substantial toughness capability in this material.

Preliminary results on plant-fabricated sheet cold rolled to 2.0 mm gauge are listed in Table 3. These strength levels are comfortably above minimum values, and we would expect the exfoliation resistance of this product to be equivalent to that in plate.

Several plant-fabricated extruded shapes have been evaluated. Table 4 lists minimums and properties achieved in the 48 mm thick leg of an aerospace "tee" extrusion. Strength values generally are at or above minimums, while elongation in the L direction is equivalent to minimum. Toughness exceeds the minimum in the L orientation and is somewhat below the minimum observed values for similar 7075-T6X extrusions in the other orientations. Notice that, in contrast to plate, the ST elongation is 3.1%; other values as high as 4.7% were obtained at equivalent strength levels in this same extrusion. The tensile properties for a second extruded "tee" with a 22.6 mm thick leg are plotted in Figure 3, showing the range of values measured from test locations at the front and rear of eight separate extrusions. The lowest strengths measured in both orientations are above appropriate minimums, and maximum values are substantially higher. Some of the measured elongations in the L direction are below the 7075-T6 minimum of 7%, while others exceed it; elongations in the LT orientation span the same range. This latter extrusion also has been supplied to the Navy as a second product for the alloy 2090 test and evaluation contract (see below).

In certain aerospace components, fatigue performance is a primary design criterion; in the case of damage tolerant design philosophies, crack growth behavior is relevant, particularly under spectrum loading conditions. Northrop Aircraft Division has tested a sample from mill-fabricated 44 mm thick plate, using a tension-dominated lower wing load spectrum from the F/A-18 aircraft. This is the same spectrum which has been used to evaluate a number of commercial Al alloys as part of an ongoing Northrop/Alcoa program to investigate fatigue behavior [10,11]. As shown in Figure 4,

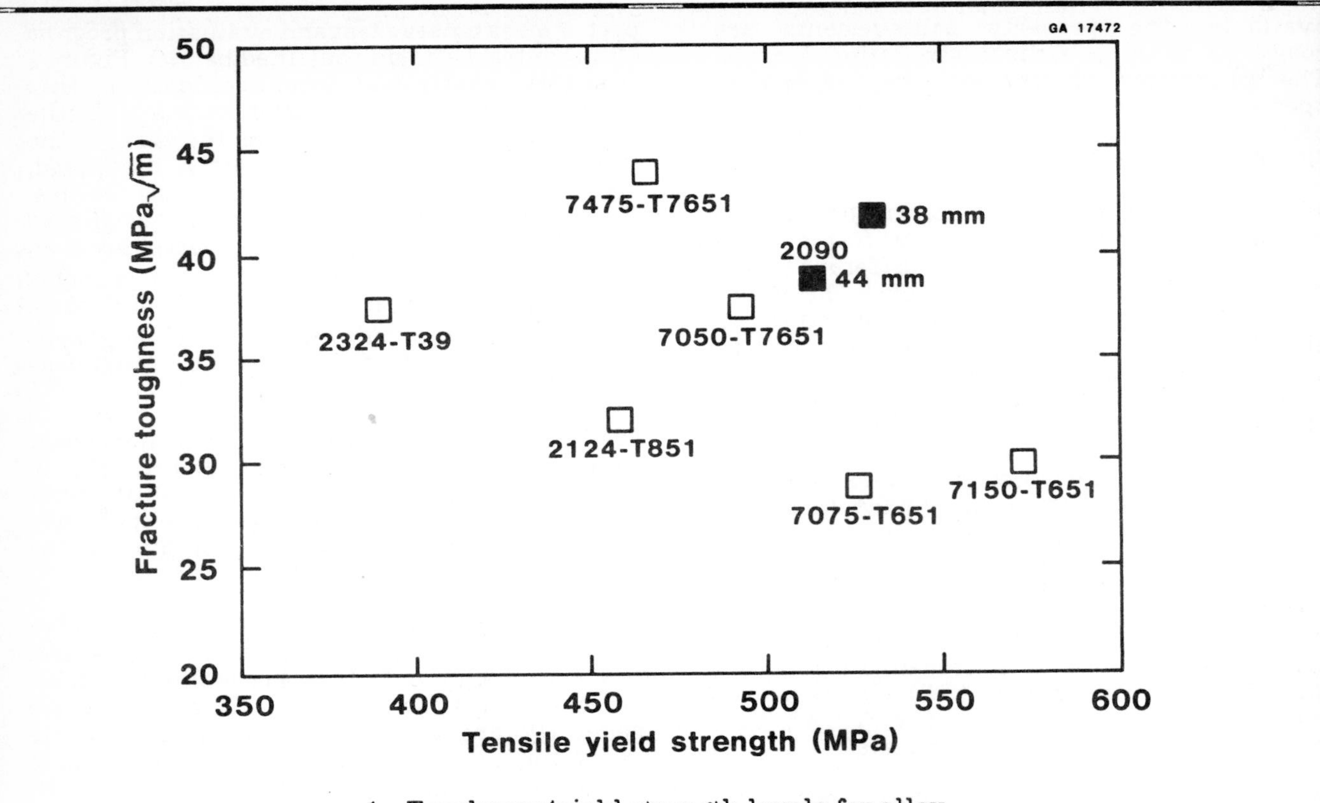

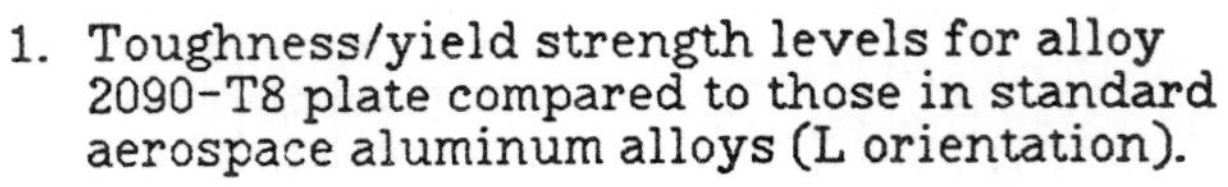

1. Toughness/yield strength levels for alloy 2090-T8 plate compared to those in standard aerospace aluminum alloys (L orientation).

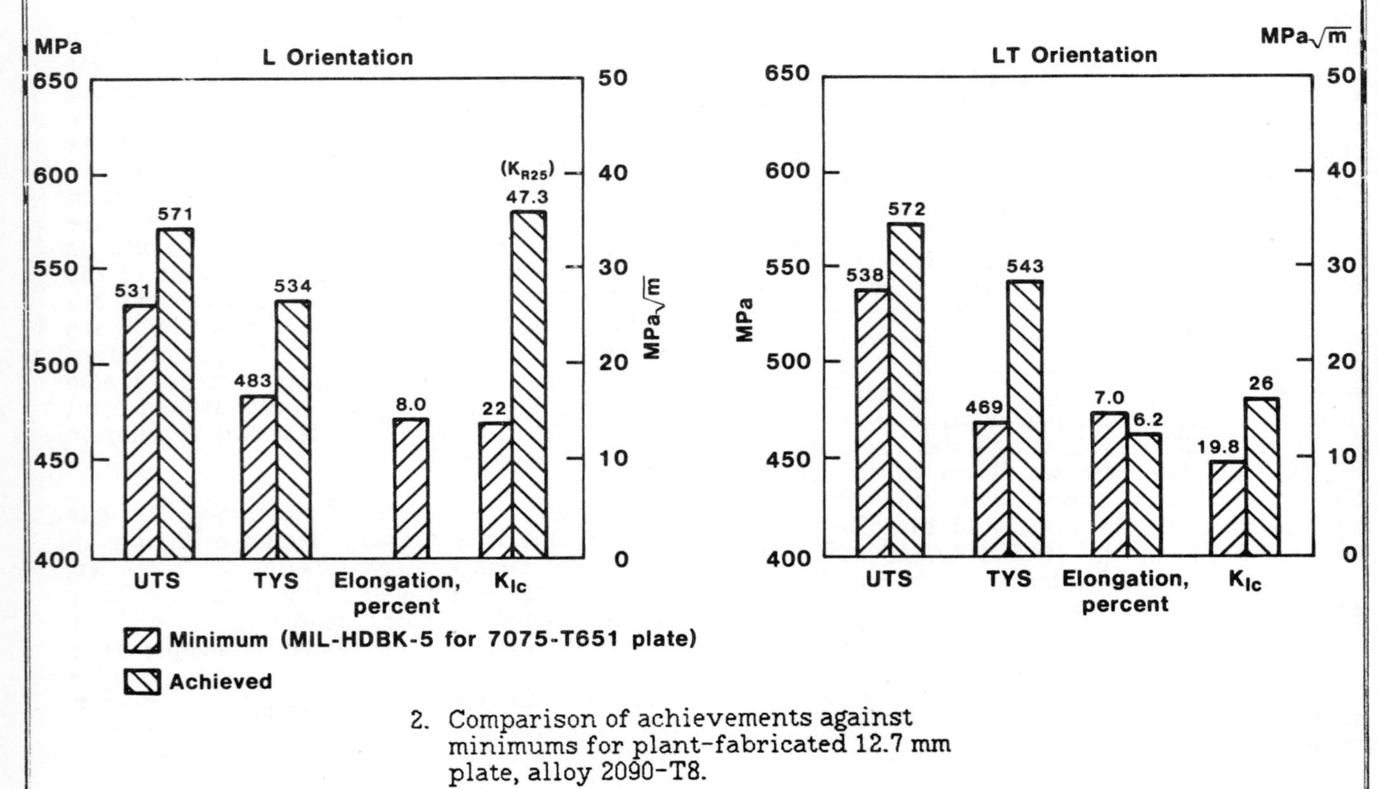

2. Comparison of achievements against minimums for plant-fabricated 12.7 mm plate, alloy 2090-T8.

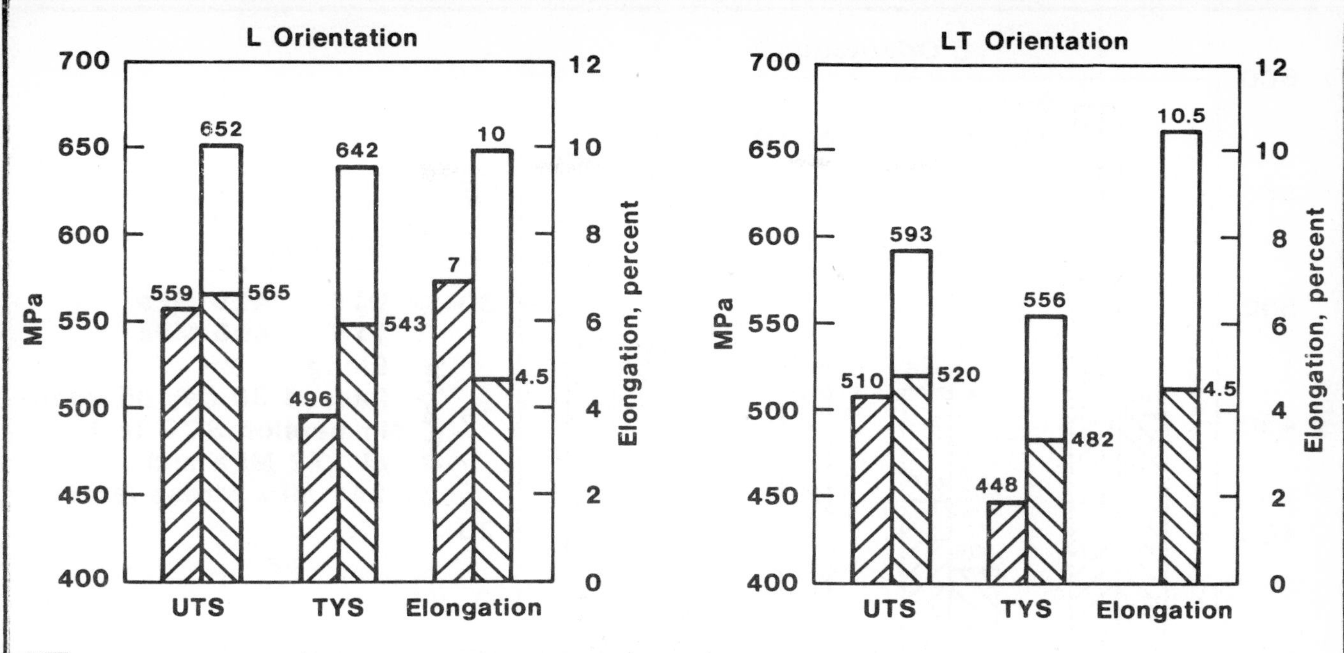

3. Comparison of achievements against minimums for plant-extruded "tee" shape (22.6 mm leg), alloy 2090-T8.

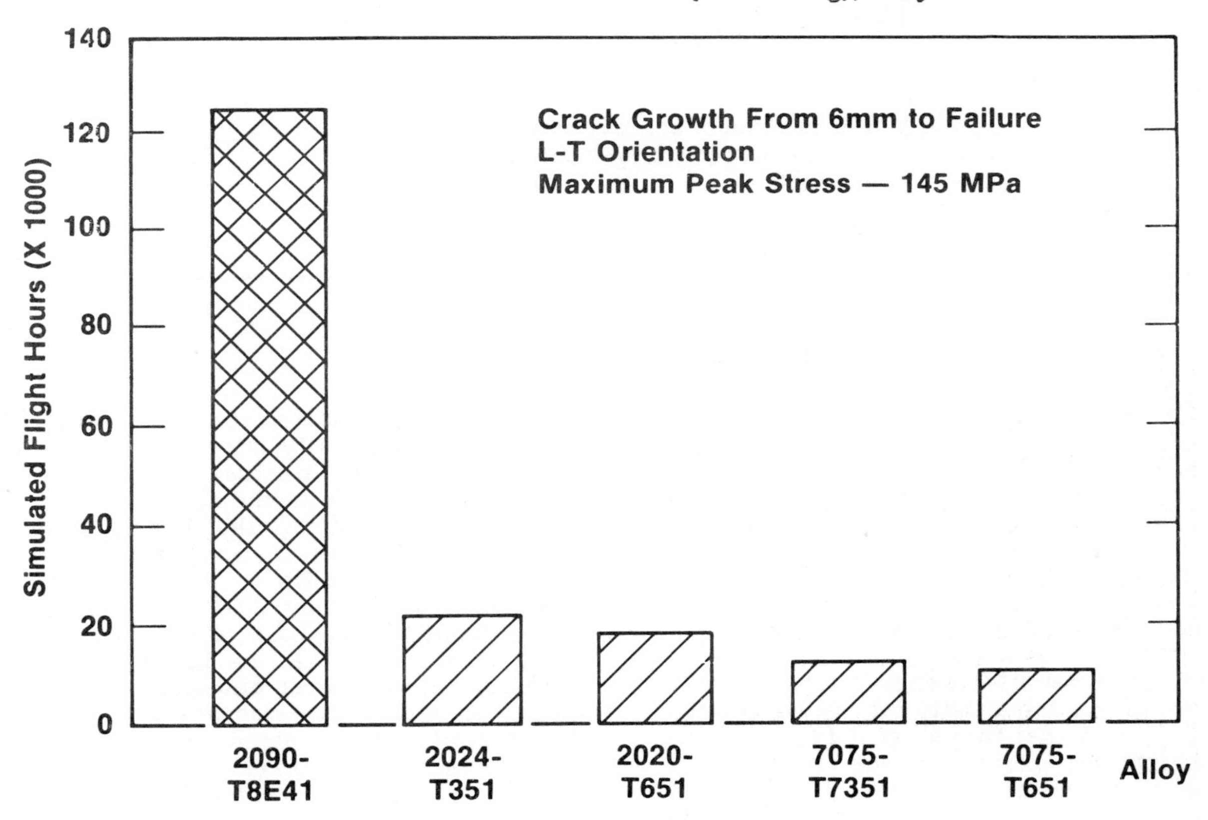

4. Comparison of fatigue crack growth lives for 2090-T8 plate and other Al alloys, using F/A-18 lower wing spectrum.

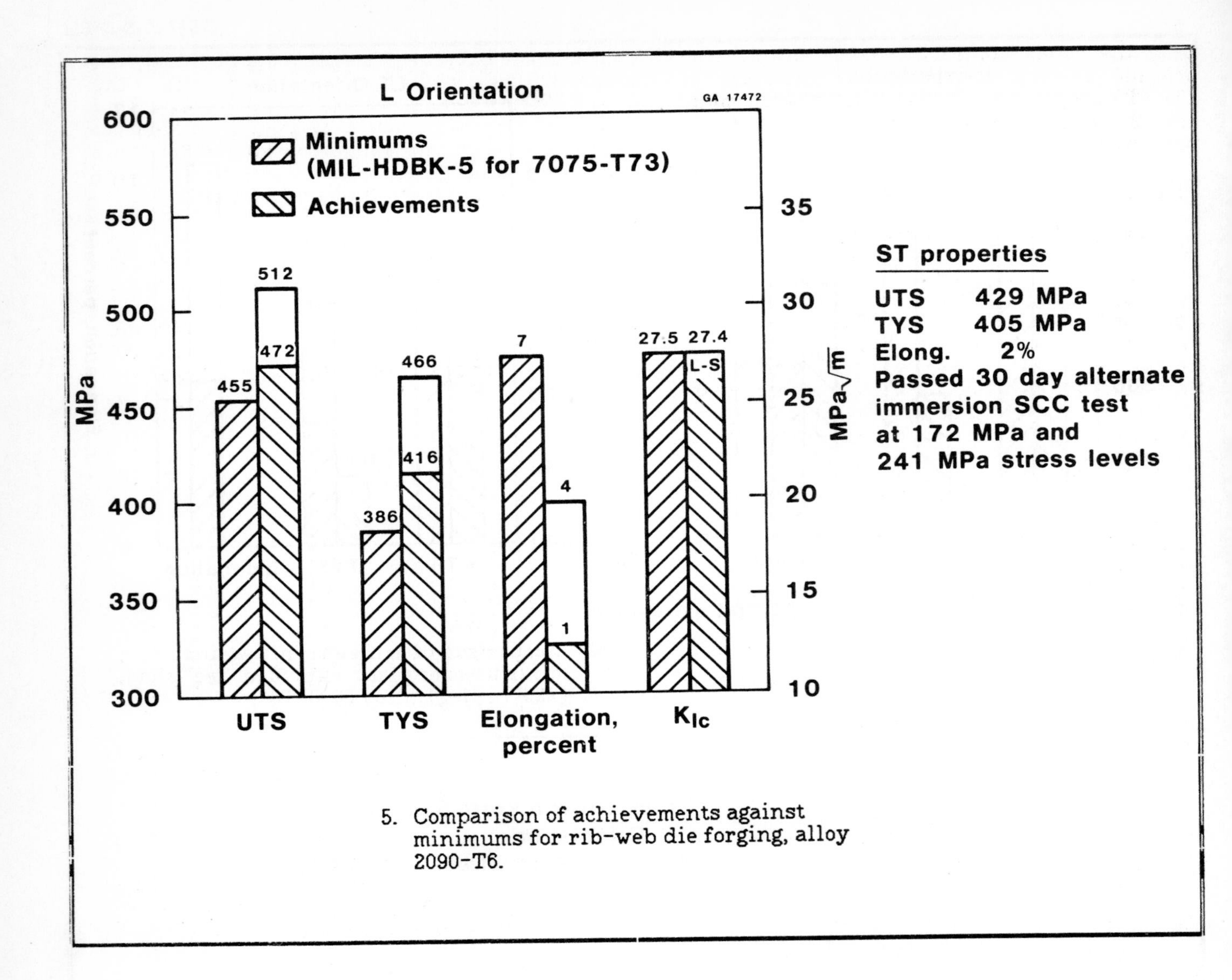

5. Comparison of achievements against minimums for rib-web die forging, alloy 2090-T6.

the spectrum crack growth life of alloy 2090-T8 exceeds by more than a factor of 5 the performance of other Al alloys including 2024-T351, which is widely considered to be the alloy most resistant to fatigue crack growth. Other data on laboratory-fabricated versions of alloy 2090 similarly show excellent crack growth resistance in both constant amplitude and simple overload spectra [4,8].

In summary, the available engineering property data on 2090 in various mill-fabricated product forms indicated that this alloy is fully capable of meeting or exceeding the performance standards necessary to replace 7075-T6X products and provide an immediate 7% density reduction upon substitution.

While the data available on alloy 2090 is sufficient to indicate its potential as a low-density replacement for 7075-T6X products, substantially more information will be required prior to its actual implementation in aircraft. To accelerate the acquisition of this important information, the U. S. Navy is funding a test and evaluation contract through Alcoa [12]. This activity will generate user experience with alloy 2090 and provide an initial property data base. Four products will be supplied for evaluation: 12.7 mm and 38 mm plate, 1.6 mm bare sheet, and an extruded "tee" shape. A total of 31 aerospace companies and U. S. government laboratories are participating. A basic evaluation package will be performed by all participants, including tensile, compression, shear, bearing, modulus, fracture toughness, and fatigue crack growth tests. Some participants also will be evaluating corrosion and SCC resistance. All data will be gathered into the final contract report, which will provide a very comprehensive summary of 2090-T8 properties. To date, the 12.7 mm plate and the extrusion have been supplied to all participants.

GOAL A, 2024-T3X REPLACEMENT

Several products have been fabricated at the Alcoa Technical Center from full-scale ingot sections of an Al-Li-Cu-Mg-Zr alloy. The composition range selected for this Goal A alloy is listed in Table 5, along with properties measured on a 38 mm thick plate. Strength, elongation, and toughness all exceed the minimum values for 2024-T351 plate, while target values for the remaining properties all essentially have been met or surpassed.

GOAL D, 7075-T73X REPLACEMENT

A second Al-Li-Cu-Mg-Zr alloy has been evaluated in the laboratory for this first generation target in both plate and extrusion product forms; the composition range chosen to meet this goal is listed in Table 6. An alloy of this type also was fabricated into 38 mm thick plate, the properties for which are shown in Table 6. Strength, elongation, and toughness minimums are exceeded while meeting modulus, density, and exfoliation targets. The SCC resistance of this sample was below the target for a 7075-T73X replacement, but further alloy modifications have been made since this particular material was fabricated to substantially improve SCC resistance, which is an especially important characteristic of the incumbent alloy.

Forgings are an important product form for 7075 in the SCC-resistant, T73-type temper. Several forgings of various types (hand, die, and precision) have been fabricated for evaluation, using 2090-type Al-Cu-Li alloys. Longitudinal tensile and toughness properties are shown in Figure 5 for a rib-web die forging common in the aerospace industry. The strength ranges measured are above minimums, while elongations are below the minimum value for 7075-T73; nevertheless, toughness in the L-S direction is on par with the minimum encountered in 7075-T73 forgings in the L-T orientation (normally the test direction exhibiting maximum toughness). Furthermore, ST specimens from this same forging passed the 30-day alternate immersion SCC test at sustained stresses of 172 and 241 MPa. Additional forging trials are in progress to optimize the properties of this product.

REMAINING ALITHALITE PRODUCT DEVELOPMENT ACTIVITIES

Casting and fabrication technology development for alloy 2090 is nearly complete. Availability of samples in all product forms (plate, sheet, and extrusions) has been initiated with the supplying of 12.7 mm plate and an extruded shape for a NAVAIR test and evaluation contract [12]; additional products for this activity and other evaluations will become available throughout the remainder of 1985.

Casting and fabrication studies on Al-Li-Cu-Mg-Zr alloys for Goals A, C, and D are proceeding, with initial availability of sample materials expected later in 1985 or early in 1986. We anticipate that government-sponsored test and evaluation programs similar to the 2090 contract will be organized for these alloys.

REFERENCES

[1] Aluminum-Lithium Alloys, T. H. Sanders, Jr. and E. A. Starke, Jr., Eds., TMS-AIME, 1981.

[2] Aluminum-Lithium Alloys II, T. H. Sanders, Jr. and E. A. Starke, Jr., Eds., TMS-AIME, 1984.

[3] P. E. Bretz, L. N. Mueller, and A. K. Vasudévan, "Fatigue Properties of 2020-T651 Aluminum Alloy," Aluminum-Lithium Alloys II, T. H. Sanders, Jr. and E. A. Starke, Jr., Eds., TMS-AIME, 1984.

[4] A.K. Vasudévan, R. C. Malcolm, W. G. Fricke, and R. J. Rioja, "Resistance to Fracture, Fatigue, and Stress-Corrosion of Al-Cu-Li-Zr Alloys," Final Report, Naval Air Systems Command Contract N00019-80-C-0569, 1985 May 30.

[5] A. P. Hays, et al., "Integrated Technology Wing Design Study," NASA Contract Report 3586, 1982 August.

[6] I. F. Sakata, "Systems Study of Transport Aircraft Incorporating Advanced Aluminum Alloys," NASA Contract Report 165820, 1982 January.

[7] D. S. Van Putten and S. W. Averill, "Advanced Aluminum Fighter Structure," AFWAL Contract F33615-83-C-3230.

[8] R. R. Sawtell, P. E. Bretz, J. I. Petit, and A. K. Vasudévan, "Low Density Aluminum Alloy

Development," Proceedings of the 1984 SAE Aerospace Congress and Exposition, Long Beach, CA, 1984 October 15-18.

[9] R. J. Rioja and E. A. Ludwiczak, "Identification of Metastable Phases in an Al-Cu-Li Alloy (2090)," paper 68 in this conference.

[10] G. R. Chanani, I. Telesman, P. E. Bretz, and G. V. Scarich, "Methodology for Evaluation of Fatigue Crack-Growth Resistance of Aluminum Alloys Under Spectrum Loading," Final Report, Naval Air Systems Command Contract N00019-80-C-0427, 1982 April.

[11] G. V. Scarich and P. E. Bretz, "Investigation of Fatigue Crack-Growth Resistance of Aluminum Alloys Under Spectrum Loading," Final Report, Naval Air Systems Command Contract N00019-81-C-0550, 1983 April.

[12] "Cooperative Test Program on Al-Li Alloy 2090-T8EXX Plate, Sheet, and Extruded Products," Naval Air Systems Command Contract N60921-84-C-0078.

□

Processing and properties of Alcan medium and high strength Al–Li–Cu–Mg alloys in various product forms

M A REYNOLDS, A GRAY, E CREED, R M JORDAN and A P TITCHENER

The authors are with
Alcan International Limited,
Southam Road,
Banbury,
Oxon OX16 7SP
UK

SYNOPSIS

The background to the Alcan development of two Al-Li-Cu-Mg alloys and the processing of rolled and extruded material under commercial conditions are outlined. Current typical properties of sheet, plate and extrusions in Lital A, (≡ 2014), Lital B (≡ 7075) and Lital C (≡ 2024) are shown together with their relationship to fabrication and heat treatment parameters. Details of the environmental behaviour of the alloys are included. Reference is made to further scale-up in facilities.

INTRODUCTION

The intention of this paper is to indicate the current situation regarding the development of Al-Li-Cu-Mg alloys within Alcan, to give the latest properties achieved and to indicate the progress being made with further scaling up in facilities. The development of the Al-Li-Cu-Mg alloy has been reported previously[1,2] and initial data has been presented to show the properties which might be obtained. All, or most, of these data has come from small melts with products manufactured by non, or semi-commercial routes. Gradually however the increase in the scale of the effort within Alcan has allowed the casting and fabrication of material which gives a realistic feel for what can be achieved. Material is now being produced on commercial mills and presses and data from this source will be presented.

BACKGROUND

Alcan's aims have been documented previously[2]. In outline they involve the development of alloys or tempers to allow substitution of 2014 T651, 7075 T651 and 2024 T351 with material which is 10% higher and 10% stiffer. In current nomenclature the lithium alloy equivalents are Lital A, Lital B and Lital C respectively. Target properties have been based on the properties obtained in the alloys and tempers they are intended to replace.

The previously published 'A' alloy composition[3] is not strong enough to fulfill the high strength role and a second 'B' alloy composition has been formulated[4]. These two compositions are shown in Table 1. Lital A is internationally registered as 8090 and Lital B has been provisionally allocated the designation 8091. The damage tolerant 2024 T351 temper, it is believed, can be obtained by lightly ageing the 'A' alloy composition.

During the last eighteen months the casting of 1000 kg D.C. ingots has allowed the fabrication of plate under commercial conditions at Alcan Plate Limited: Kitts Green Works and strip rolling of sheet under commercial conditions at British Alcan Sheet Limited: Falkirk. Extrusion data has been obtained from billet up to 300 mm dia. on commercial presses at Alcan High Duty Alloys: Distington. Further scale-up is currently taking place and this will be dealt with later in the paper.

The aim initially has been to develop the 'A' alloy in the medium strength temper, and hence most of the data is in that area. The scale-up to 1000 kg 950 mm x 300 mm ingots has been predominantly with the 'A' alloy. The 'B' alloy has been D.C. cast successfully as 500 kg rolling ingots and some modifications to casting practice are under development to allow scale up to larger ingots. Data presented on the 'B' alloy is therefore from 300 - 500 kg ingots for plate and sheet fabricated under

Table 1 Composition (weight percent) ranges for the Lital alloys.

ELEMENT	LITAL 'A' MIN.	LITAL 'A' MAX.	LITAL 'B' MIN.	LITAL 'B' MAX.
LITHIUM	2·3	2·6	2·4	2·8
COPPER	1·0	1·6	1·6	2·2
MAGNESIUM	0·5	1·0	0·5	1·2
ZIRCONIUM	0·08	0·16	0·08	0·16
IRON	–	0·3	–	0·3
SILICON	–	0·2	–	0·2
ZINC	–	0·25	–	0·25
OTHERS: EACH	–	0·1	–	0·1
TOTAL	–	0·2	–	0·2

Table 2 Typical tensile properties and plane stress fracture toughness (K_c) values with specification minimum values for Lital 'A', Lital 'B' and Lital 'C' sheet in different temper conditions.

Specification	Temper	Test Direction	0·2% P.S. (MPa)	T.S. (MPa)	El (%)	K_c (MPa$\sqrt{m}$)
LITAL 'A'	T6 TYPICAL	L	365	455	4·0	80+
		T	375	465	7·5	60
	T6 Spec. min.	L & T	380	440	6·0	70
	T8 TYPICAL	L	415	485	4·0	60
		T	395	480	7·5	55
	T8 Spec. min.	L & T	380	440	6·0	70
LITAL 'B'	T6 TYPICAL	L	390	480	4·5	55
		T	390	485	6·5	50
	T6 Spec. min.	L & T	450	500	6·0	55
	T8 TYPICAL	L	440	505	4·5	50
		T	410	495	6·5	45
	T8 Spec. min.	L & T	450	500	6·0	55
LITAL 'C'	TYPICAL	L	360	430	5·0	90+
		T	320	430	7·5	–
	Spec. min.	L & T	290	420	6·0	75

+ K_c VALUES OBTAINED WITH 400 mm. WIDE TEST PANELS; ALL OTHER K_c VALUES RELATE TO 100/150 mm. WIDE TEST PANELS

Table 3 Typical tensile and fracture toughness (K_Q) values with specification minimum values for Lital 'A', Lital 'B' and Lital 'C' plate.

Specification	Test Direction		0·2% P.S. (MPa)	T.S. (MPa)	El. (%)	K_Q (MPa$\sqrt{m}$)
LITAL 'A'	L	Typical	450	500	5·5	36
		Spec. min.	410	460	5·0	30
	LT	Typical	425	485	6·5	36
		Spec. min.	390	450	5·0	25
	ST	Typical	350	425	1·5	16
		Spec. min.	360	420	2·5	18
LITAL 'B'	L	Typical	520	560	4·0	28
		Spec. min.	455	525	6·0	
	LT	Typical	465	530	4·0	18
		Spec. min.	455	525	6·0	
	ST	Typical	420	490	1·0	7
		Spec. min.				
LITAL 'C'	L	Typical	400	450	5·0	45
		Spec. min.	300	430	7·0	35
	LT	Typical	340	440	6·5	45
		Spec. min.	300	430	7·0	28
	ST	Typical	270	355	4·0	20
		Spec. min.	260	380	3·5	–

plant conditions at Aluminium Corporation: Dolgarrog. Work on Lital C has been a fairly recent venture but the data has been obtained from material produced from 1000 kg ingots.

Considerable work has gone on, and is continuing, in the liquid metal area. Improvements in casting technique involving degassing, filtration and grain refinement have been made but not yet been fully optimised.

Further improvements will be obtained in the new casting facility at Alcan Plate. Problems with ingot cracking during scale-up have occurred with both the Lital A and B compositions. This was initially overcome with the 'A' alloy by modification to the composition and casting practice. Latest casting trials appear to have overcome this problem but the data on initial 1000 kg ingots is with a slightly below mid-range composition.

PRODUCT FORMS

Sheet

Lital A 1000 kg ingots have been successfully hot and cold strip rolled on commercial mills currently used for airframe material production. No problems, other than those due to the handling of the small size of the slab, were encountered in hot strip rolling down to 4 mm. Cold rolling down to 0.8 mm was achieved.

Virtually all of the data obtained on sheet material has so far come from 300 kg ingots hot rolled to 4 mm plate at A.C.L.: Dolgarrog and cold hand rolled to gauge. The considerable amount of testing has resulted in data which shows how the material produced by this route meets the target specifications. Provided the sheet is stretched after solution heat-treatment the target tensile properties of the medium strength 2014 T851 and the high strength 7075 T651 can be comfortably exceeded. Data from 1000 kg strip rolled sheet is not significantly different. A different situation exists if stretching cannot be carried out and in this event close control of the 'A' alloy composition is required to achieve the minimum 2014 T851 properties but the 'B' alloy will not achieve 7075 T651 properties. Table 2 shows the latest typical properties compared with the specification minima obtained on Lital A, B and C. Work on the effect of delay time between solution heat treatment and stretching has shown that holding at sub-zero temperatures is required to prevent natural ageing which manifests itself in a reduction in ductility. Final aged properties are unaffected by natural ageing time.

Scaling up from 300 kg to 1000 kg ingots has allowed the development of a commercial strip rolling practice. This has removed the possibility of controlling anisotropy by the use of cross-rolling and interannealing.

techniques of compositional control and heat-treatment are being developed which result in the same, or an improved, balance of properties[5]. The new heat-treatment technique involves the use of salt-bath solution heat-treatment. Very extensive work has been carried out, partially funded by the Ministry of Defence, in which it has been shown[6] that it is safe to solution heat-treat the Aluminium-Lithium based alloys considered in this paper in salt baths, although all the normal salt bath safety precautions must of course be observed. The rapid rate of heating obtained in salt bath heat-treatment has been shown to be beneficial in the control of anisotropy as shown in Figure 1.

K_c values obtained on sheet have been somewhat lower than those obtainable on the alloys they are intended to replace. Typical values (at mean proof stress) of 60 MPa $\sqrt{m}$ and 50 MPa $\sqrt{m}$ for Lital A and Lital B respectively have been obtained on 100 mm wide centre cracked panels. However the use of wide panel sizes i.e. 400 mm results in a value of 80 MPa $\sqrt{m}$ for Lital 'A'.

K_c values increase with a reduction in tensile strength such that a damage tolerant version of the Lital A composition would be expected to have a value of 100 MPa $\sqrt{m}$ at a 300 MPa proof stress on a 400 mm wide panel test. This is only marginally lower than those obtainable on 2024 T351.

Plate

Lital A plate in the gauge range 6 to 60 mm has been produced from 1000 Kg ingots at the Kitts Green plant of Alcan Plate. The rolling of larger cast blocks has allowed more cross-rolling to be employed, especially on the thicker plates while still maintaining sufficient length of plate to permit stretching after solution heat-treatment. The distribution of the longitudinal and long-transverse properties obtained from these plates is shown in Figure 2. No difficulty is envisaged in meeting the initial target values in these two directions. Elongation results have been determined autographically which, at these elongation levels, reduces the values obtained by approximately 1% compared with the values obtained from taking measurements from the broken test pieces. Therefore the typical values of 5.2 and 5.9% in the longitudinal and long-transverse directions respectively are considered acceptable. However, it is recognised that the short-transverse tensile properties of plate require some improvement. The average short-transverse elongation has been 1.5% at a proof stress of 350 MPa. Developments in-house however, still undergoing assessment, appear to have considerably improved the situation and values of 4% at higher strength levels have been obtained.

Table 4 Through thickness variation of tensile properties in rolled plate.

50 mm. PLATE

DEPTH (mm.)	0·2%P.S. MPa	U.T.S. MPa	ELONG. %
2·5	388	439	6·5
5·0	381	443	8·3
7·5	377	434	5·4
10·0	380	434	5·5
¼t → 12·5	399	456	6·3
15·0	408	464	5·9
17·5	420	471	4·6
20·0	424	474	4·4
½t → 25·0	424	473	4·4

25 mm PLATE

DEPTH (mm.)	0·2%P.S. MPa	U.T.S. MPa	ELONG. %
2·5	398	459	6·8
5·0	398	453	5·4
≃¼t → 7·5	413	468	4·7
10·0	424	472	3·9
½t → 12·5	448	493	3·7

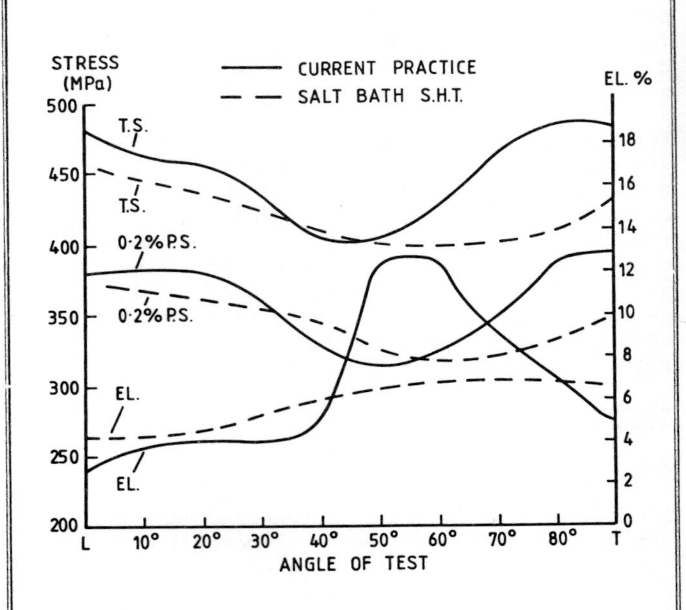

Figure 1 Shows the effect of salt bath solution heat-treatment on the anisotropy of Lital A sheet.

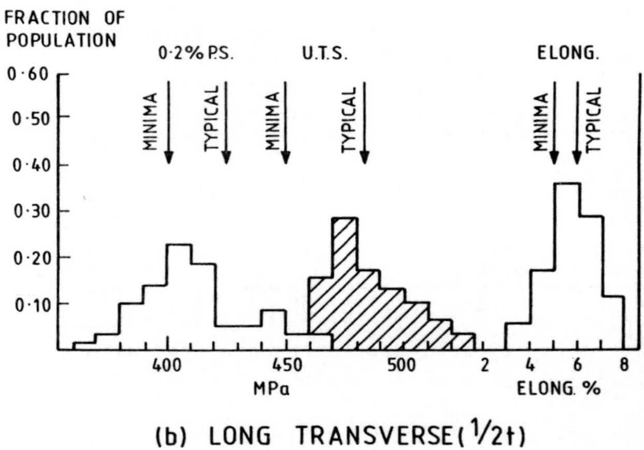

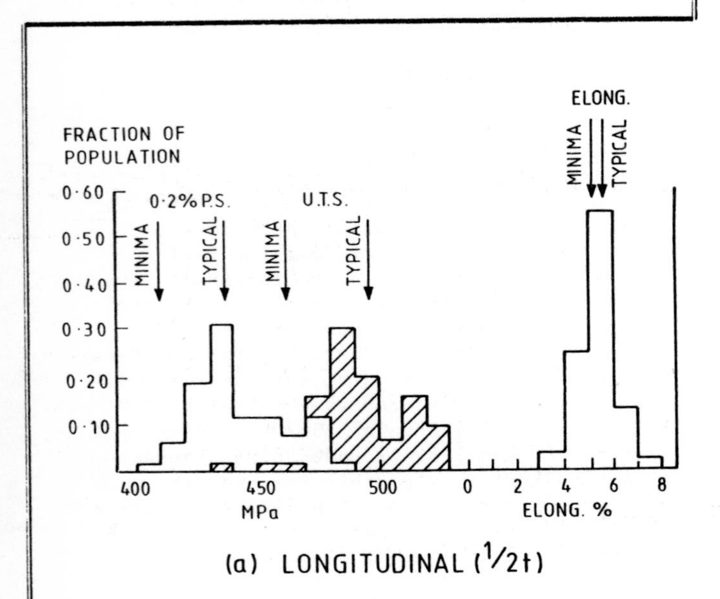

Figure 2 Statistical distribution of longitudinal and long transverse tensile properties for Lital A T651 plate rolled from 1000 kg ingots.

Lital B matches the tensile properties of 7075 T651.

The Lital A composition can produce properties in an underaged temper to match the damage tolerant requirement. However, only limited testing has been done in this temper on plate. Table 3 shows the typical tensile and fracture toughness of the Lital A, B and C plate compared to the respective specification minimum.

The tensile properties of plate show a variation through the plate thickness, as shown in Table 4. This has implications for material specifications, notably whether to test at ¼t or ½t position. Most of the data has been quoted from tests carried out at the ½t position because the plate gauge produced was mostly 25 mm or below. The reason for the larger than normal variation is attributed to the texture development in the plate[7]. It is interesting to note however, that the fracture toughness variation through the plate does not vary in the same way. Fracture toughness has generally been found to be marginally higher at the ½t compared with the ¼t position even though the latter position had a 50 MPa lower proof stress. The solute content at the ingot centre has been found to be lower (by approximately 0.1% in each of the three major alloying elements) but not sufficient to account for the improved toughness/strength relationship.

The fracture toughness of both Lital A and Lital B is affected by iron and silicon content, as for standard aerospace alloys[8,9], and by the degree of ageing. Increasing the iron content from 0.09% to 0.27% reduces the toughness by up to 25%. Underageing, without significantly affecting the proof stress, can increase toughness by 5-10%, but stress-corrosion resistance may be reduced.

Initial compression, shear and bearing data produced on Lital A show the normal trends. Figure 3 shows the data compared with 2014 T651.

The good fatigue characteristics of aluminium-lithium alloys have been well demonstrated[10]. Figure 4 shows a modified Goodman diagram produced from tests on plate fabricated from 1000 kg ingots. The improved performance compared with 2014 T651 is again illustrated. The earlier good fatigue crack growth rate resistance data has been confirmed on 50 mm plate rolled from 1000 kg ingots, as shown in Figure 5.

Extrusion

A large number of extruded sections have been produced, as shown in Figure 6. No difficulties have been experienced in the extrusion of these sections. Data obtained have indicated that both Lital A and Lital B can achieve their respective target strength levels. Typical properties compared with 2014 T651 and 7075 T651/T7651 are shown in Figure 7. The Lital alloys have demonstrated that they can match the strength and fracture toughness of the conventional alloys in the longitudinal direction but have lower ductility. In the transverse direction the fracture toughness of Lital B requires some improvement. However, underageing of these alloys to reduce the proof stress by 5% increases the fracture toughness of Lital A from typically 35 and 19 MPa √m in the L-T and T-L directions to 45 and 32 MPa √m respectively. This approach can be used to produce better T-L toughness in Lital B. Ageing Lital A to a damage tolerant strength level has produced both L-T and T-L toughness values of near to 50 MPa √m.

Extrusions show a large variation in properties with test position and direction but the lowest strengths appear to be associated with the usual test position demanded by test standards. These strength differences, which can be up to 100 MPa between the edge and centre of an extruded bar, appear to be due to grain structure and texture variations. In contrast many conventional alloys show lower properties at the edge and surface due to a peripheral coarse recrystallised grain structure. Extrusion aspect ratio has a considerable effect. The continuing work on extrusions will be aimed at giving a better control of grain structure and thus anisotropy by control of composition and process conditions.

The tensile properties of thin sections are approximately 25 MPa, lower than those of thicker shapes.

An understanding of how process parameters affect the balance of strength and fracture toughness has been developed and this will allow the specific combinations of properties for particular applications to be obtained.

ENVIRONMENTAL BEHAVIOUR

Stress-Corrosion

The stress-corrosion data presented in this paper is limited mainly to crack initiation studies in an aqueous sodium chloride electrolyte (ASTM G44) using constant strain specimens i.e. C-rings or tension bars. The minimum performance curves (M.P.C.) for both the medium strength Lital A and the high strength Lital B alloys in peak aged plate form are shown in Figure 8. The SCC performance of Lital A shows a significant improvement over 2014 T651 which it is designed to replace. The higher strength Lital B alloy exhibits a slightly greater susceptibility than Lital A. However for short transverse stresses below 150 MPa its performance falls between that of the 7075 T651 and T7651 tempers. The crack initiation behaviour of Lital A and B alloy extrusions, in the peak aged temper, generally shows an improvement over the plate alloys in a

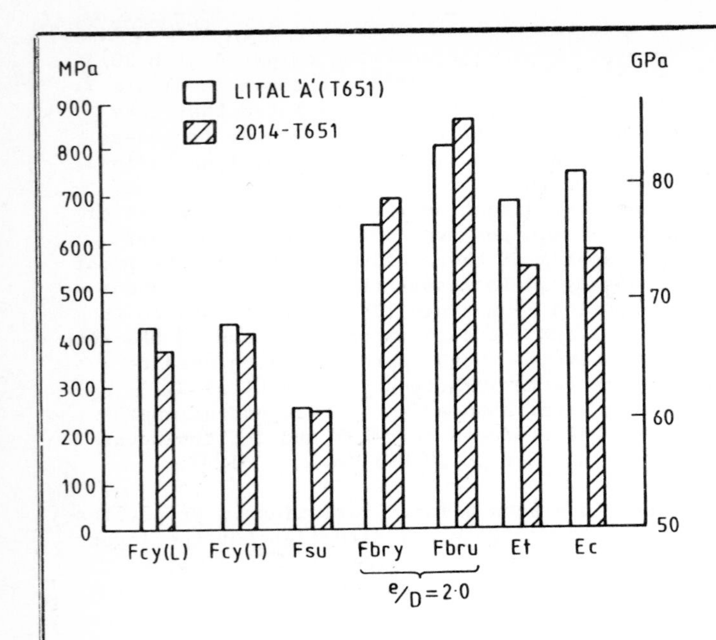

Figure 3 Compression, shear, and bearing properties of Lital A T651 plate compared with 2014 T651.

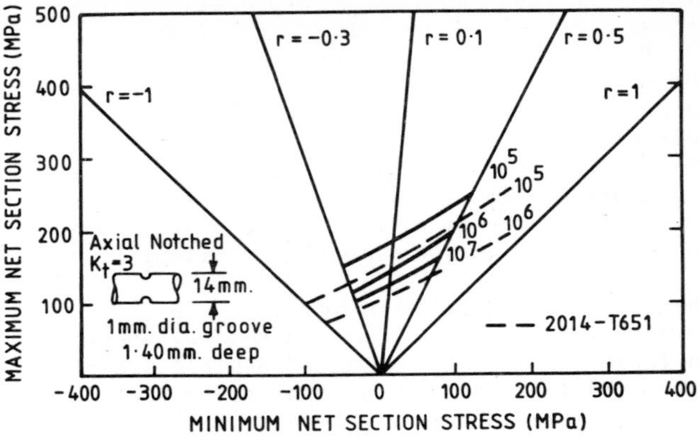

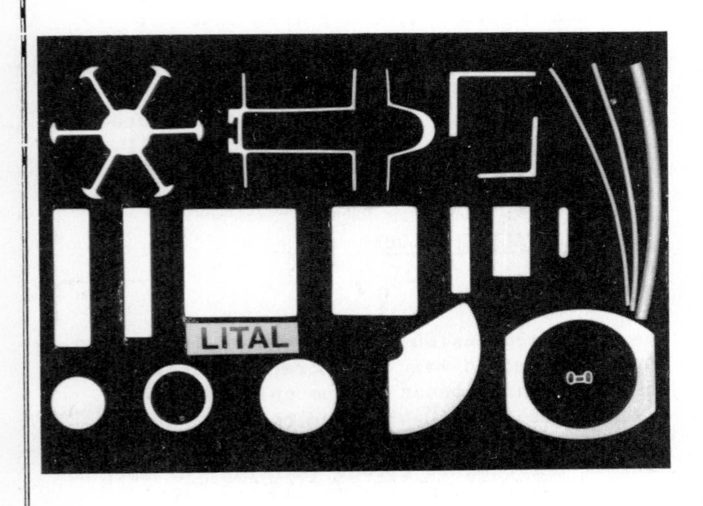

Figure 6 Range of extruded sections produced in Aluminium Lithium alloys.

MAXIMUM NET SECTION STRESS (MPa)
500
400
300
200
100
r = −1
r = −0·3
r = 0·1
r = 0·5
r = 1
10^5
10^6
10^7
Axial Notched
$K_t = 3$
14mm.
1mm. dia. groove
1·40mm. deep
— — 2014-T651
−400 −300 −200 −100 0 100 200 300 400
MINIMUM NET SECTION STRESS (MPa)

Figure 4 Modified Goodman diagram for Lital A plate.

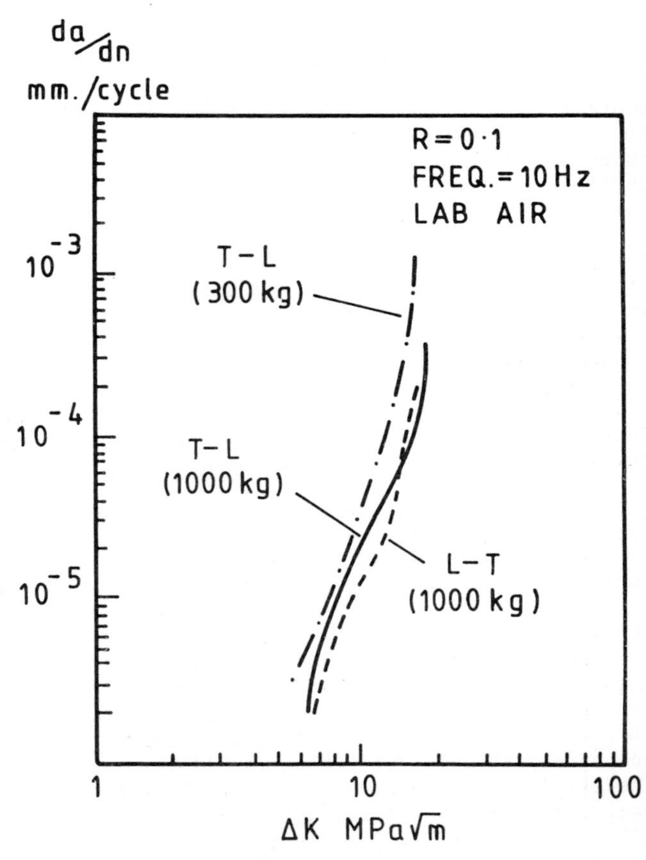

Figure 5 Fatigue crack growth rate curves obtained on 50 mm thick plate produced from 1000 kg ingots compared to earlier data from 300 kg ingot material.

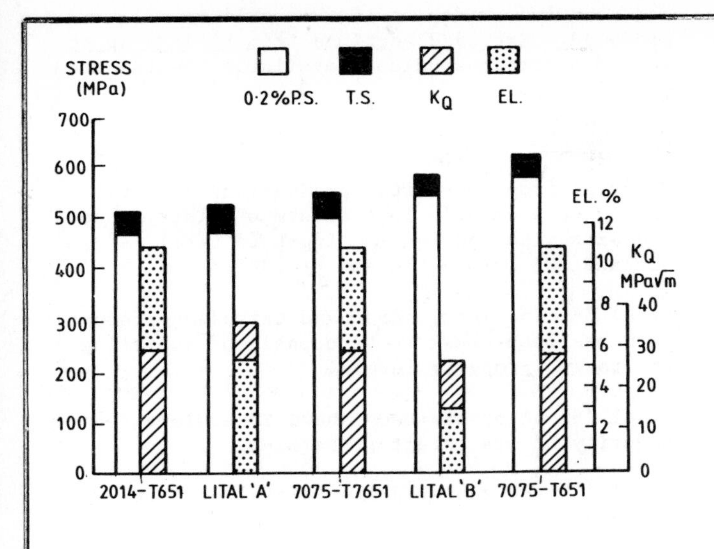

Figure 7 Typical longitudinal tensile and fracture toughness, K_Q, values for T651 Lital 'A' and Lital 'B' extruded 100 mm x 25 mm bar section compared with standard alloys.

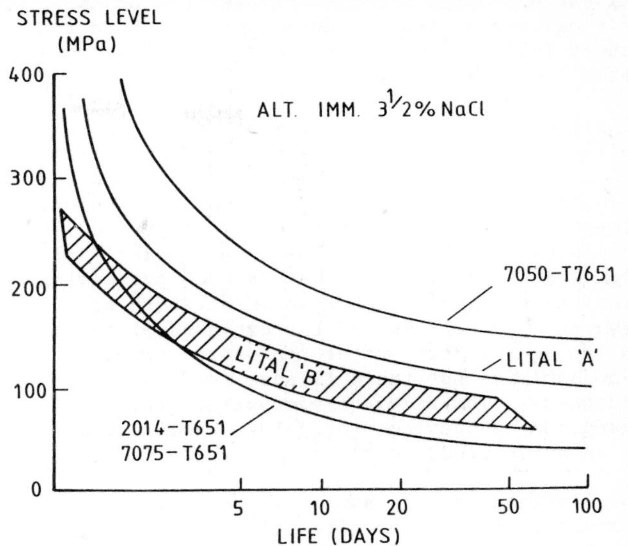

Figure 8 Minimum performance curves for T651 Lital 'A' and Lital 'B' plate compared to some conventional alloys.

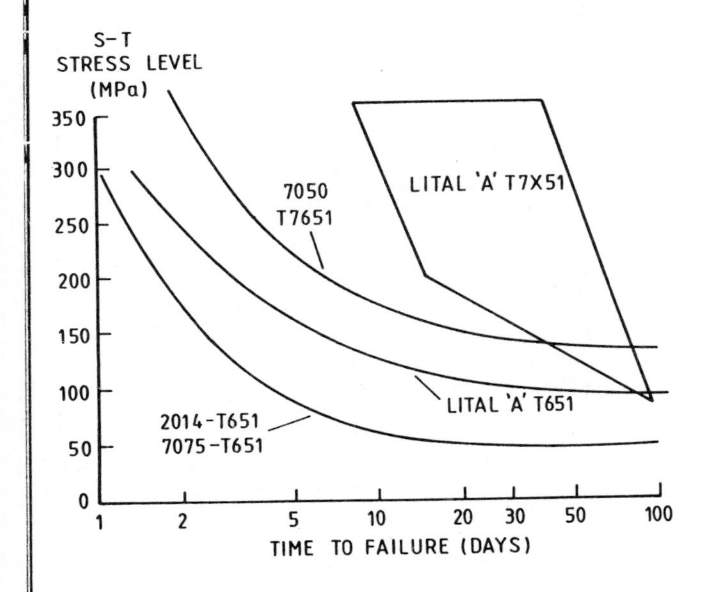

Figure 9 Stress-corrosion data for Lital 'A' plate in an overaged T7x51 temper.

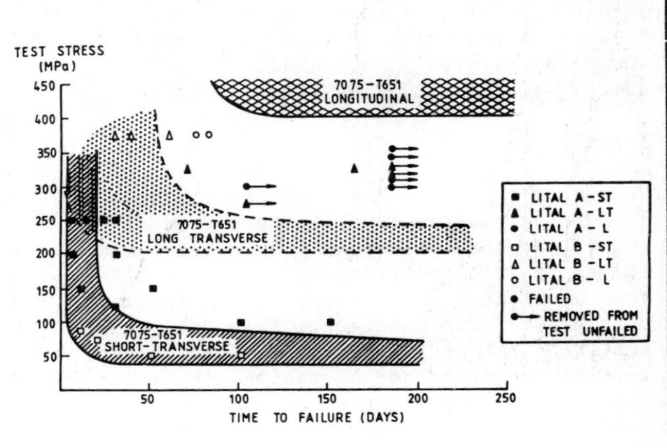

Figure 10 The effect of stress-direction on the stress-corrosion performance of Lital 'A' and Lital 'B' plate.

comparable temper. However, as is common with all extruded high strength alloys, the crack initiation behaviour is dependent upon both the aspect ratio of the profile and the extrusion ratio i.e. grain shape.

Significant increases in the period to crack initiation of the Lital A medium strength alloy can be achieved by overageing. This effect is illustrated in Figure 9. Following overageing, minimum stress-corrosion lives are increased to well above that of 7050 T7651, particularly at high stress levels. The use of such overageing treatments gives a reduction in strength of less than 10% because of the extended plateau in the ageing curve after peak strength has been achieved but there is some trade off in toughness. The relevant microstructural changes accompanying overageing are discussed in an associated paper[11].

Only limited data is available regarding the susceptibility of the Lital alloys to SCC in other than the short-transverse direction. A comparison of the Lital A and B (peak aged) together with 7075 T651 in the short-transverse, long transverse and longitudinal directions is shown in Figure 10.

The Al-Li-Cu-Mg alloys can behave in a significantly different manner from 2XXX and 7XXX type alloys with respect to their response in either constant or alternate immersion environments respectively. This behaviour and its implications is discussed in greater detail elsewhere[11].

Crack propagation rates under conditions of alternate immersion have been measured for Lital 'A' using fatigue pre-cracked double cantilever beam (D.C.B.) specimens. Plateau velocities of the order of 2.0×10^{-9} ms^{-1} were obtained on short-transverse loaded specimens. This compares with a published[12] plateau velocity of 2.0×10^{-8} ms^{-1} for 7075 T651.

Corrosion Behaviour

The overriding trend observed in all the accelerated testing is that the degree of attack is temper dependent. Both Lital A and Lital B exhibit greatest resistance to all forms of attack, irrespective of environment, in the underaged condition. This fact could facilitate the optimum balancing of strength, fracture toughness and corrosion, in particular exfoliation. Irrespective of the test environment the high strength Lital B alloy exhibits greater susceptibility to both intergranular and exfoliation attack than Lital A.

Lital A and B thin gauge sheet is found to exhibit least resistance to exfoliation attack, as measured by the EXCO test (ASTM G34). Ratings of EA-EB are obtained in the peak aged temper but underageing Lital A to the damage tolerant temper improves the rating to EA. In and a maximum width of 1000 mm will be produced. Extruded sections from billets up to 400 mm diameter should be available towards the end of 1985.

CONCLUSIONS

Material from the inital production facility should be available in the form of plate, sheet and extrusion and forging ingot in Lital 'A', 'B' and 'C' before the end of 1985.

Lital 'A' in plate, sheet and extrusion product form has been shown to be capable of achieving its target property levels.

Lital 'B' alloy has been shown to achieve the majority of its target properties.

Lital 'C' has achieved its initial property requirements.

ACKNOWLEDGEMENTS

The authors are grateful to Alcan International Limited for permission to publish this work.

REFERENCES

1. C.J. Peel, B. Evans, C. Baker, D.A. Bennett, P.J. Gregson and H.M. Flower. Aluminium-Lithium Alloys II ed. E.A. Starke Jr., J.H. Sanders Jr. TMS -AIME Warrendale, P.A. 1984, p. 363.

2. W.S. Miller, A.J. Cornish, A.P. Titchener and D.A. Bennett. Aluminium-Lithium Alloys II ed. E.A. Starke Jr., J.H. Sanders Jr., TMS-AIME Warrendale, P.A. 1984, p. 335.

3. C.J. Peel. U.K. Patent, 211-5836-A.

4. W.S. Miller. U.K. Patent Application GB 2137227A.

5. I.G. Palmer, W.S. Miller, D.J. Lloyd and M.J. Bull. "The Effect of Grain Structure and Texture on the Mechanical Properties of Al-Li Base Alloys". This Conference.

6. E.R. Clarke, P. Gillespie and F.M. Page. "The Heat Treatment of Lithium/Aluminium Alloys in Salt Baths". This Conference.

7. D.J. Lloyd and M.J. Bull. "The Textures developed in an Al-Li-Cu-Mg Alloy". This Conference.

8. J.G. Kaufman. Specialists' Meeting on alloy design for fatigue and fracture resistance, AGARD-CP-185, p. 2-1, January 1976.

9. C.J. Peel and P.J.E. Forsyth. R.A.E. TR 70162, September 1970.

10. B. Evans, D.S. McDarmaid and C.J. Peel, Fifth International SAMPE Conference, Materials and Processes, Montreux, Switzeland, June 1984.

11. N.J.H. Holroyd, A. Gray and G.M. Scamans, "Environment-Sensitive Fracture of Al-Li-Cu-Mg Alloys". This Conference.

12. J.G. Rinker, M. Marek and J.H. Sanders Jr. Materials Science and Engineering 64 (1984), 203-221.

□

Processing Al–Li–Cu–(Mg) alloys

R F ASHTON, D S THOMPSON,
E A STARKE Jr and F S LIN

RFA and DST are with the
Metallurgy Laboratory,
Reynolds Metals Company.
EAS is with the School of
Engineering and Applied Science,
University of Virginia.
FSL, formerly with Reynolds
Metals Company, is now with
Boeing Commercial Airplane Company,
Seattle, Washington, USA

SYNOPSIS

The effect of environmental/thermomechanical processing on the structure and properties of Al-Li-Cu-(Mg) alloys is discussed. The processing steps examined include homogenization, rolling, solution heat treatment, stretching, and aging.

INTRODUCTION

Microstructural features of importance for property control of aluminum alloys include: (a) coherency and distribution of strengthening precipitates, (b) degree of recrystallization, (c) grain size and shape, (d) crystallographic texture, (e) size and distribution of intermetallic particles, including the dispersoids (present by design) and insoluble phases (which result from iron and silicon impurities), and (f) surface condition, e.g. oxide characteristics and solute concentration of the near surface region. Most of these features can be controlled by alloy composition and primary processing, from casting to the final aging treatment.

Alloy composition and primary processing are particularly important when attempting to achieve desired properties with the new Al-Li-X alloys. Low ductility and fracture toughness of Al-Li alloys have been associated with the shearable nature of the metastable Al_3Li precipitates that form during aging and the presence of precipitate-free zones (PFZ) along both high and low angle grain boundaries, both of which lead to strain localization.[1] These problems may be overcome by the selection of alloy additions that co-precipitate with the Al_3Li phase, that precipitate up to the grain boundaries, thus eliminating the PFZ, and so promote homogeneous deformation. Al-Li-Cu-(Mg)-Zr alloys show the greatest promise.

Several different strengthening precipitates can be formed in the family of Al-Li-Cu-(Mg) alloys, depending upon alloy composition and thermomechanical processing.[2-5] Alloys containing more than 1.5 weight percent lithium can be expected to form a homogeneous distribution of Al_3Li (δ') and varying amounts of the other phases. These may include all of the phases occurring in binary Al-Cu alloys (i.e. GP zones, θ'', θ', Al_2Cu) and those occurring in the ternary alloys Al-Li-Cu and Al-Cu-Mg (Al_2CuLi [T_1], Al_6CuLi_3 [T_2], Al_2CuMg [S'], and various equilibrium phases). Of the phases occurring in Al-Li-Cu-Mg alloys, T_1 and S' provide major strengthening in addition to that produced by the homogeneous precipitation of δ'. Both of these phases have a large effect on ductility, since they may change the deformation mode. The size and distribution of T_1 and S' can be greatly affected by heat treatment and processing conditions, and precipitate can form in the δ' PFZ.[5]

Processing begins with homogenization to reduce segregation, remove the low-melting nonequilibrium phases, and thus improve workability. This thermal treatment also serves to precipitate the dispersoid-forming elements, such as those containing zirconium, so that they may perform their role of grain control during subsequent processing. Since lithium is a particularly reactive element, the atmosphere used in the homogenization treatment may be very important to minimize oxidation, lithium loss from the surface, and, possibly, hydrogen pickup. The homogenization treatment is followed by hot working for ingot breakdown and shape change to the appropriate product form. The wrought product is then solution heat treated, quenched, possibly worked, and aged to develop the desired microstructure. The temperatures, amount of deformation prior to aging, etc., depend on the

Table 1. Depletion of Li and Mg From Ingot Surface After Homogenization in Dry Argon

Depth Below Ingot Surface (Inch)	Homogenization Time: 12 Hours Li (%)	12 Hours Mg (%)	48 Hours Li (%)	48 Hours Mg (%)
0.00-0.01	2.47	2.0	1.22	2.18
0.01-0.02	2.74	1.98	1.98	1.75
0.02-0.03	2.85	2.0	2.54	1.90
0.03-0.05	2.92	--	2.83	--
0.05-0.07	2.91	--	2.87	--
0.07-0.09	2.91	--	2.89	--

Table 2. Effect of Homogenization Environment on Mechanical Properties of Al-Li Alloy Sheet and Plate

	Homo and SHT in Dry Nitrogen					Homo and SHT in Wet Air				
Aging Condition	TS (ksi)	YS (ksi)	El. (%)	K_Q (ksi$\sqrt{in}$)	Tear/ Yield	TS (ksi)	YS (ksi)	El. (%)	K_Q (ksi$\sqrt{in}$)	Tear/ Yield
0.50" Plate										
375°F/6 hours	74	66	9	31		76	66	10	30	
375°F/24 hours	78	72	7	27		79	74	6	26	
0.10" Sheet										
375°F/6 hours	78	75	2		0.80	80	73	4		0.88
375°F/24 hours	80	75	5		0.61	81	75	4		0.68

Longitudinal test direction. Al-2.5Li-1.3Cu-0.9Mg-0.11Zr

Table 3. Chemical Compositions (Wt. %)

	Li	Cu	Mg	Zr	Fe
2090	2.3	2.5	-	0.13	0.07
8091	2.3	2.1	0.7	0.14	0.08
H3	2.9	1.5	2.0	0.16	0.06
8090	2.6	1.4	1.0	0.14	0.05

alloy composition and the final microstructure and properties that are desired.[7] This paper will describe primarily the effect of the various processing steps on the microstructure and properties of two new commercial Al-Li-Cu-(Mg) alloys, 2090 and 8091.

EXPERIMENTAL PROCEDURES

Melting and casting in a permanent mold were carried out in an inert atmosphere. The molten metal was fluxed before the final addition of pure lithium metal. After homogenizing, the ingots (30-pound) were scalped to 3.25" and hot rolled to 0.50" thick plate. Sharp notch ($\leq$0.001" radius) Charpy impact tests were used as a measure of toughness. This test effectively correlates with toughness over a wide yield strength range. Aging was done using a heating rate of 25°F/hour.

EFFECT OF ATMOSPHERE ON ELEVATED TEMPERATURE PROCESSES

Aluminum and most of its alloys form a protective oxide film on the surface immediately on exposure to air. At low temperatures, this layer of oxide film decreases the amount of oxygen diffused into the metal, so that oxidation is initially rapid and then drops off logarithmically, forming a thin impermeable oxide layer. At high temperatures, as in the case of homogenization (>900°F), oxidation rate is parabolic or a combination of parabolic and logarithmic, resulting in a more porous oxide layer.

Lithium has an adverse effect on the formation of a protective film, since lithium alloys develop a porous oxide film of Li_2O and/or $LiAlO_2$.[8] This increases the oxidation rate for Al-Li alloys homogenized in an air environment. Also, Al-Li alloys were reported to have a higher tendency to pick up hydrogen when the alloys are in contact with water vapor.[9]

HOMOGENIZATION STUDY

An ingot with a composition (weight percent) of Al-2.9Li-1.5Cu-2.0Mg-0.15Zr was used. Coupons cut from the ingot were homogenized at a temperature of 1010°F for a variety of times using a heating rate of 50°F/hour from 600°F to 1010°F. Dry argon, dry nitrogen, wet nitrogen (saturated with water vapor), dry air, laboratory air, and wet air were used as homogenization environments. Thickness of oxidation layers and/or lithium depletion depths after homogenization were determined by chemical analysis and optical microscopy. Corrosion products after homogenization in different environments were identified by x-ray diffraction.

Tensile and toughness properties were determined of another alloy (Al-2.5Li-1.3Cu-0.9Mg-0.11Zr) homogenized and solutionized in dry nitrogen or wet air.

Homogenization in Dry Argon

Since argon is inert, samples homogenized in this environment would be expected to primarily suffer depletion of lithium and possibly magnesium from the surface due to sublimation. The sublimation energies for pure aluminum, copper, lithium, and magnesium at the melting point of the alloy are 75, 79, 38, and 34 Kcal/mole, respectively.[10] Thus magnesium would be sublimated more readily than other elements. However, experimental results showed that only lithium in the alloy was severely depleted (Table 1). This is possibly due to a higher binding force between aluminum and magnesium atoms[11] and also a slower diffusion rate associated with a large atomic size. Lithium content is plotted against distance from the ingot surface in Figure 1. Depletion depth is consistent with that determined by optical microscopy, namely 0.015" and 0.035" for 12 and 48 hours homogenization, respectively.

Homogenization in Dry and Wet Nitrogen

Ingots homogenized in dry nitrogen (Figure 2) exhibited the smallest depletion depth. Initially, it was believed that this is associated with the formation of a thin layer of nitride. Free energy (ΔG^{o}) for the formation of nitrides for aluminum, lithium, and magnesium are -77, -47, and -110 Kcal/mole, respectively[10], indicating that the formation of nitride is possible. A thin layer near the surface could be clearly observed optically; however, it was identified by x-ray photon spectroscopy as lithium carbonate.

The thickest oxidation layer was observed for the samples homogenized in wet nitrogen (Figure 2). These samples were covered with a thick layer of white compound identified by x-ray diffraction as a type of γ-phase oxide, $AlLiO_2$. This is predictable from the thermodynamic viewpoint, since the free energy of formation for $AlLiO_2$ is more negative than that of Al_2O_3 or Li_2O.[10] The $AlLiO_2$ oxide was found to be a very porous product. This explains why the oxidation rate in wet nitrogen is much faster than that in any other environment. $AlLiO_2$ is porous and allows water vapor to easily penetrate and react continuously with the fresh metal.

Homogenization in Dry, Laboratory, and Wet Air

While oxide thickness increased with increasing humidity (Figure 2) in the air environments, even in wet air the layer was thinner than that in wet nitrogen. The primary chemical reaction between Al-Li alloys and dry air is oxidation. Oxidation behavior of Al-Li alloys homogenized in dry air was localized along grain boundaries, which act as diffusion pipes for oxygen atoms to migrate into the alloy, even though the homogenization temperature (1010°F) is higher than 2/3 of the melting point. This means that the activation energy for grain boundary diffusion is still less than that of lattice diffusion at this homogenization temperature.

Oxidation behavior for the alloy homogenized in wet air is different from that in dry air. For the wet air environment, two chemical reactions can occur. One is that oxygen directly forms an oxide with aluminum, lithium, and magnesium; another is that water vapor reacts

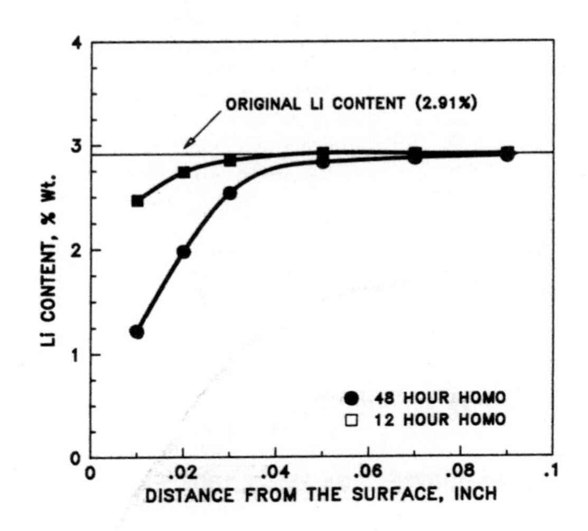

Figure 1. Lithium depletion profiles for Al-Li-Cu-Mg ingot after homogenizing at 1010°F in dry argon.

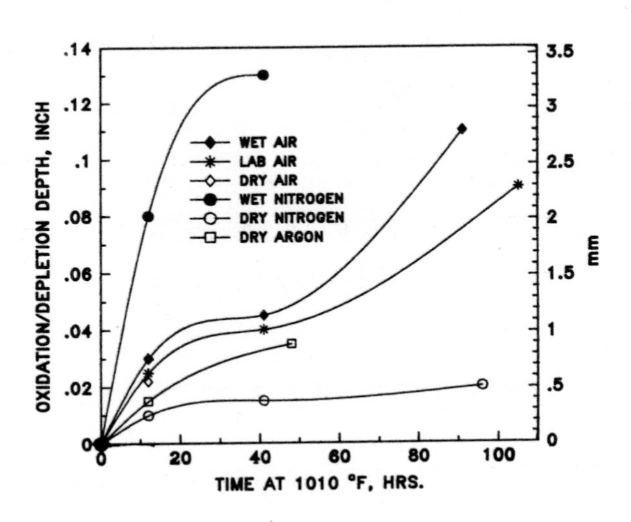

Figure 2. Effect of homogenization atmosphere and time on thickness of the oxidation/depletion layer of an Al-Li-Cu-Mg alloy.

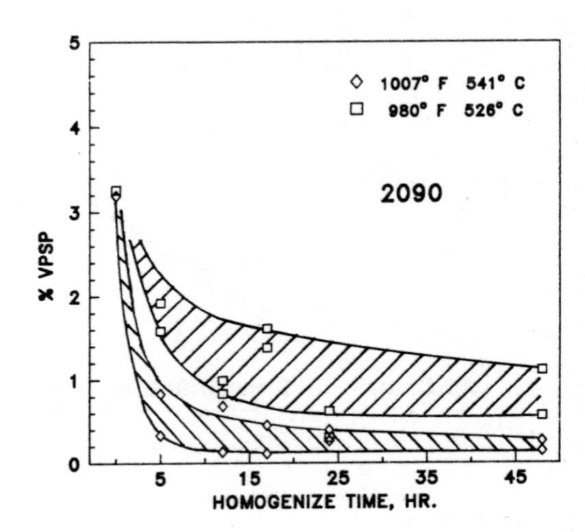

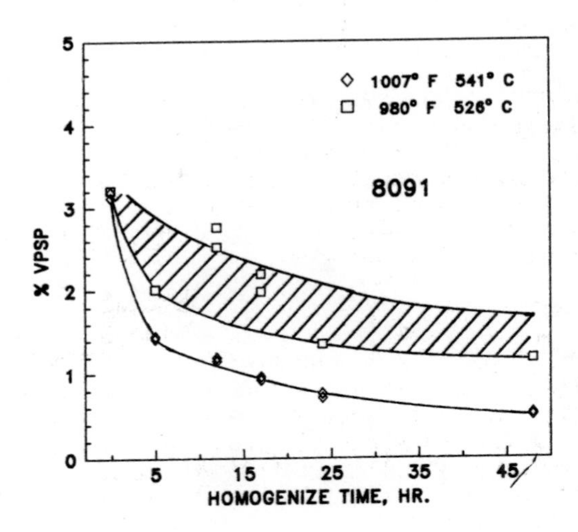

Figure 3. Effect of homogenize time and temperature on volume percent second phase in 2090 and 8091 alloy ingots.

2090

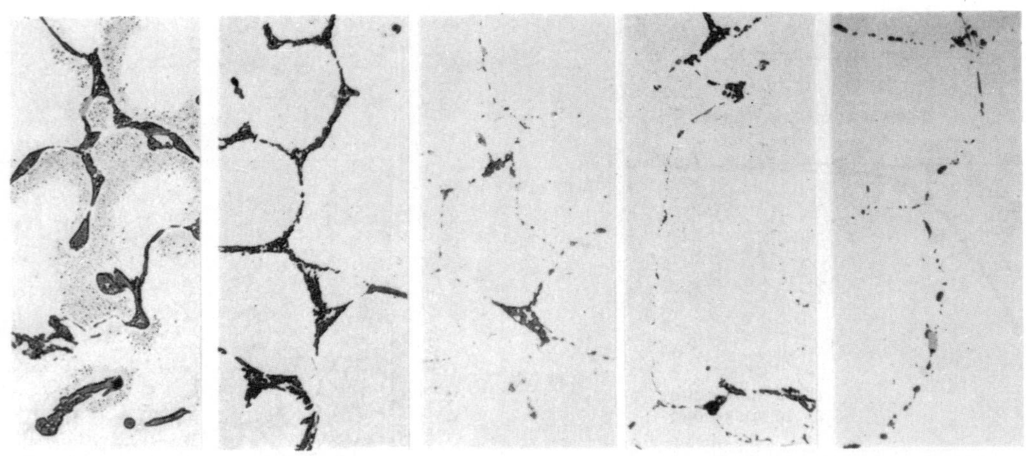

As Cast | **5 Hr @ 980°F** | **24 Hr @ 980°F** | **5 Hr @ 1007°F** | **24 Hr @ 1007°F**

8091

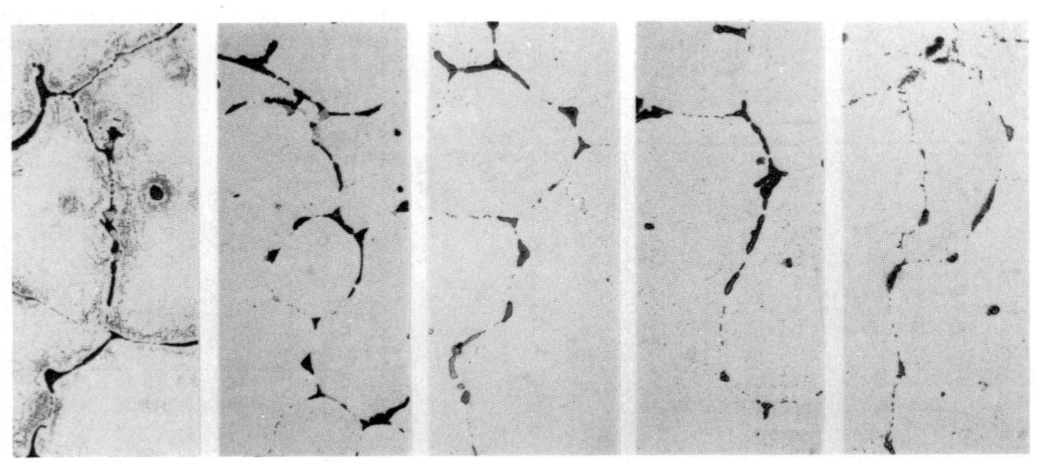

As Cast | **5 Hr @ 980°F** | **24 Hr @ 980°F** | **5 Hr @ 1007°F** | **24 Hr @ 1007°F**

Figure 4. As-polished macrostructures of 2090 and 8091 ingots showing the reduction of soluble phases with increasing homogenize time and temperature.

with these elements to form oxide and evolve hydrogen. The chemical reaction which dominates the oxidation process depends on the reaction energy, since air and water vapor are continually supplied. The reaction energy (ΔH^o) between aluminum, lithium, magnesium, and oxygen is -400, -143, and -149 Kcal/mole, respectively.[10] On the other hand, ΔH^o between aluminum, lithium, magnesium, and water vapor is -342, -85, and -91 Kcal/mole, respectively.[10] Obviously, the reaction of oxygen with metal directly to form oxide prevails, and this protective oxide film can decrease oxidation rate. Apparently, during exposure to wet nitrogen, the protective oxide layer is not formed. Consequently, the reaction layer is thicker than that found in wet air. This can also explain why there is no appreciable difference in the oxidation thickness for the alloy homogenized in dry, laboratory, and wet air.

Effect of Homogenization Environment on Mechanical Properties

The mechanical properties of an Al-2.5Li-1.3Cu-0.9Mg-0.11Zr alloy, which was homogenized and solutionized in dry nitrogen and in wet air, are listed in Table 2 for 0.50" plate and 0.10" sheet, using both an underaging and a peak aging practice. Both the tensile and fracture toughness properties for both plate and sheet products of the alloy homogenized in these two environments were almost identical. The fracture features of the alloy homogenized in dry nitrogen and in wet air were identical.

EFFECT OF TIME/TEMPERATURE PRACTICES

Homogenization Time/Temperature

Differential scanning calorimetry (DSC) indicated that eutectic melting started in as-cast alloys (Table 3) from 1022°F to 1027°F. Several samples of 2090 and 8091 ingots were homogenized at either 980°F or 1007°F for a variety of times. The amount of second phase as a function of time is given in Figure 3. It can be seen that volume fraction of eutectic phase decreased dramatically as homogenization time increased from 0 to 16 hours, and, after that, little change was observed. Alloys containing higher magnesium levels (1.5-2.0%) were found to level out within about 8 hours. A series of micrographs are given in Figure 4, which illustrates the change of microstructure as a function of time and temperature. After 5 hours both the magnesium-free 2090 and the low-magnesium 8091 alloys show advanced dissolution of the eutectic phase at 1007°F but much less at 980°F. At either temperature, little further solutionizing occurs beyond a 24-hour soak. The highest magnesium alloy (H3) showed some undissolved eutectic after homogenization at 990°F, while the intermediate level magnesium alloy (8090) did not. Thus across the whole alloy range, homogenization at about 1010°F is most effective with a soak time of 24 hours. This time is also probably the minimum practical time for large commercial size ingots.

Effect of Homogenization Temperature on Hot Working

Hot workability of each alloy after homogenization at different temperatures can be closely related to microstructures. To assess workability, very low homogenization temperatures were applied to alloys 8090 and 8091 only. After a very low temperature (880°F) homogenization, the alloys cracked severely during hot rolling (within five passes, 1/8" reduction per pass), because of the presence of continuous eutectic phase along boundaries. After a higher temperature (945°F) homogenization, alloy 8090 was successfully rolled to the 0.50" plate, since much of the eutectic phase was dissolved and that remaining in grain boundaries was discontinuous. However, cracking still occurred in alloy 8091 during hot rolling, since the microstructure still contained continuous grain boundary precipitation. After homogenization at 990°F or 1010°F, the eutectic phases were almost completely dissolved, and all the alloys were successfully rolled to 0.50" plate.

Solution Heat Treatment

A combination of differential scanning calorimetry (DSC) plus actual solution heat treatment (SHT) of samples at various temperatures was used to optimize SHT temperature. DSC showed that 1010 ± 10°F should ensure adequate solutionizing without melting for all alloys examined.

In addition, samples of alloys 2090 and 8091 were solutionized at temperatures between 930°F and 1010°F for 30 minutes and then quenched in cold water. After stretching and aging, tensile strength and toughness (Charpy impact energy) were determined. Yield strength increased with increasing SHT temperatures up to the highest practice of 1010°F. Somewhat surprisingly, toughness (Figure 5) levels off at about 950°F for both 2090 and 8091 and even decreases slightly at 1010°F. It is possible that the decrease in impact energy is due to the increasing yield strength. The relationship of Charpy impact energy to yield strength shows that, at lower SHT temperatures, the expected relationship between these parameters is not found. Normally, as yield strength increases, toughness decreases. Yet, at lower SHT temperatures, the opposite relationship occurs. This effect would be consistent with the presence of undissolved soluble phases. Toughness is more sensitive to such a presence than is yield strength.[11] Thus, since the decrease in toughness at the highest SHT temperature is small and yield strength is higher, then 1010°F is a good selection for SHT.

EFFECT OF STRETCH AND AGING PRACTICE

Half-inch gauge plates of alloys 2090 and 8091 were stretched 0, 3, and 7% and aged at 250, 297, and 347°F for 12-60 hours. The combined effect of stretch, age time, and age temperature on yield strength and toughness is shown in Figures 6 and 7. Figure 6A shows that, for the longitudinal direction maximum Charpy impact energy (CIE) at a given yield strength, a low aging

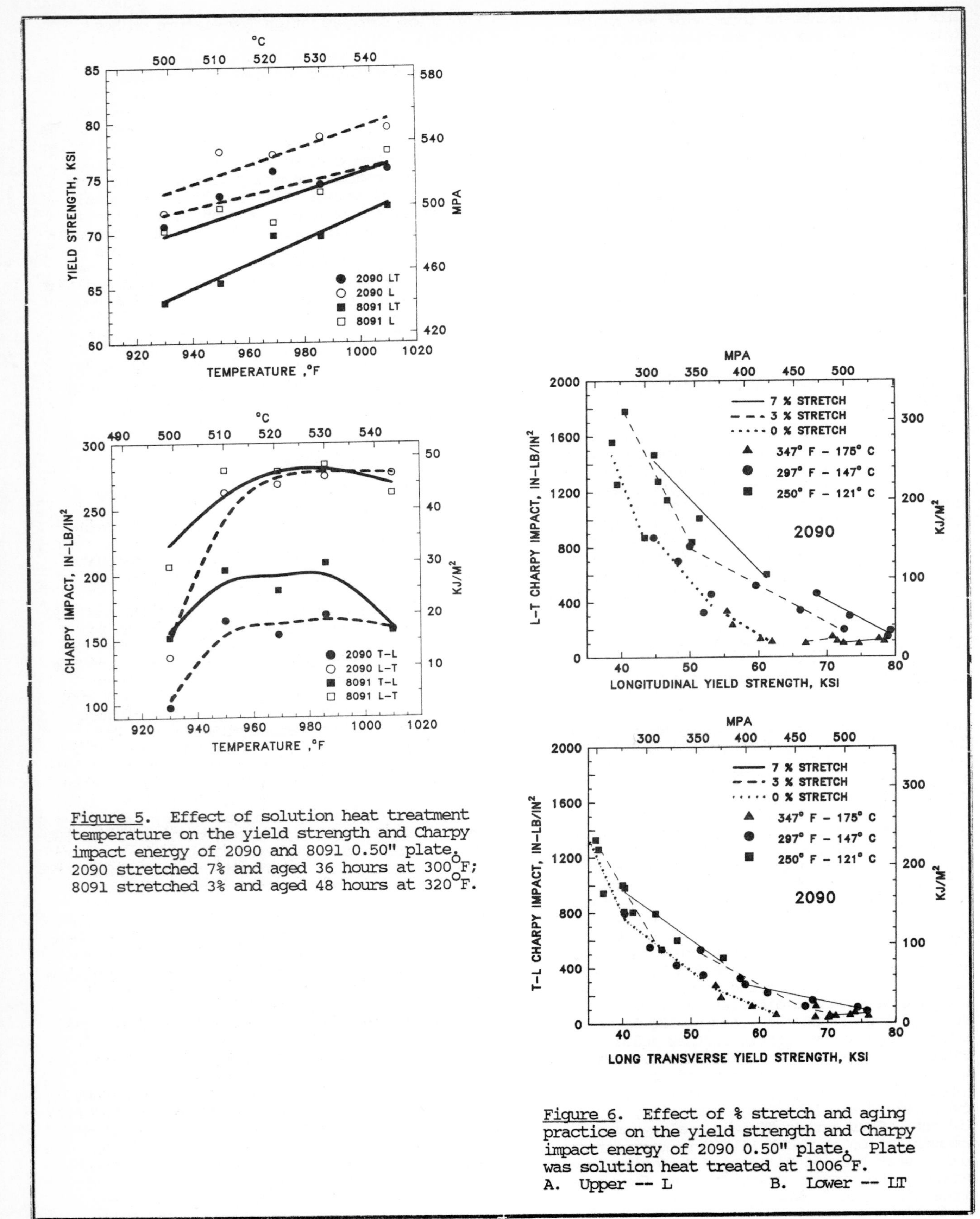

Figure 5. Effect of solution heat treatment temperature on the yield strength and Charpy impact energy of 2090 and 8091 0.50" plate. 2090 stretched 7% and aged 36 hours at 300°F; 8091 stretched 3% and aged 48 hours at 320°F.

Figure 6. Effect of % stretch and aging practice on the yield strength and Charpy impact energy of 2090 0.50" plate. Plate was solution heat treated at 1006°F.
A. Upper -- L B. Lower -- LT

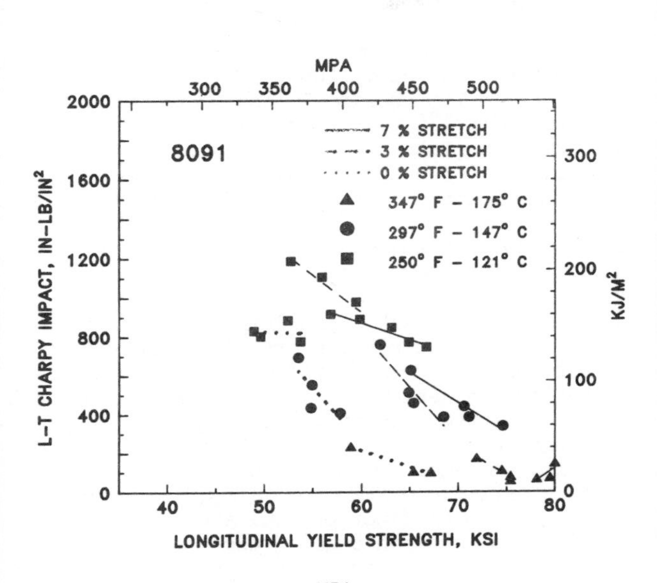

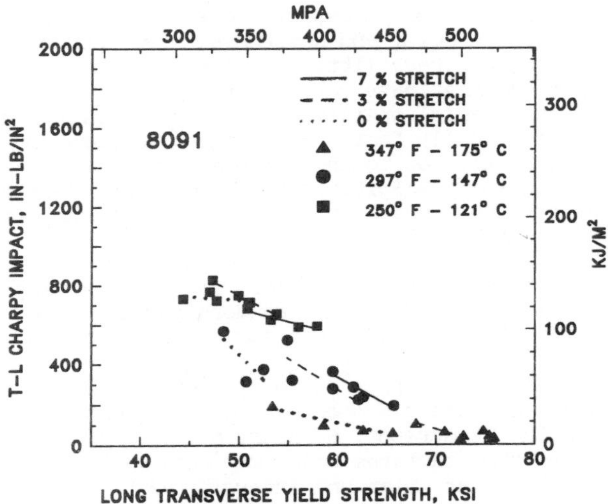

Figure 7. Effect of % stretch and aging practice on the yield strength and Charpy impact energy of 8091 0.50" plate. Plate was solution heat treated at 1006°F.
A. Upper -- L B. Lower -- LT

temperature and a high stretch level are desirable for 2090. For example, an approximate 60 ksi yield strength could be achieved by any of the three following practices:

Stretch (%)	Age: Temp. (°F)	Age: Time (Hrs.)	L-T CIE (in-lb/in²)	Long. Yield Strength (ksi)
0	347	36	135	60.3
3	297	24	520	59.6
7	250	60	600	61.8

Thus the higher stretch/lower aging temperature combination has four times the toughness of the no-stretch/high temperature age at about the same yield strength. There are obvious practical limits on how high the stretch can be and how long an aging practice at low aging temperature can be tolerated. The greatest increase in toughness, on increasing stretch, is achieved by raising the stretch from 0 to 3%. A lesser effect on toughness accompanies increasing stretch from 3 to 7%.

Alternatively, the data of Figure 6A can be seen as a series of contours at each given stretch level; one aging temperature curve tends to blend into the next. Note that this is particularly true for the 0% stretch when the 250°F age data leads to the 297°F age and then the 347°F. This implies that, for a given percent stretch, it does not matter what combination of time and temperature is used to achieve a given yield strength/toughness combination. Though, again, the practicality of extremely long ages provides a natural limit. It is obvious from Figure 6A, where the highest yield strength with 0% stretch and an age of 60 hours at 250°F only resulted in a 43 ksi yield strength, that it is unlikely to be able to achieve a 60 ksi yield strength within a practical time. Significantly, Figure 6A shows that increasing stretch leads to better combinations of strength and toughness at strength levels below about 78 ksi. Above this level the data tend to converge. Also, stretch is most beneficial at aging temperatures of 250 and 297°F. With the 347°F aging temperature, the higher levels of stretch accelerate aging, so that overaging occurs and no increase in toughness is obtained.

The above effect of stretch is limited to the longitudinal direction of 2090. In the long transverse direction, the effect is not as marked (Figure 6B). All age and stretch combinations fall into a narrower band. There is still a toughness benefit in using a 3% stretch but no further improvement in using a 7% stretch. However, it does not appear to matter what aging temperature is used--only practical consideration of length of time is important.

As shown in Figure 7A, alloy 8091 tested in the longitudinal direction shows the same general trends as 2090, i.e., stretching and lower aging temperatures are beneficial. The effect of aging temperature at a given yield strength is more dramatic. A yield strength of 65 ksi can be achieved by any of the four following practices:

Stretch (%)	Age Temp. (°F)	Age Time (Hrs.)	L-T CIE (in-lb/in²)	Long. Yield Strength (ksi)
0	347	36	105	65.4
3	297	24	510	64.9
7	297	12	625	65.1
7	250	36	770	64.9

Thus for 8091, a six- or seven-fold increase in toughness at 65 ksi yield strength is achieved by raising the stretch level from 0 to 7% and reducing the aging temperature. Again, the greatest effect of stretch is seen on going from 0 to 3%.

The long transverse behavior of 8091 (Figure 7B) is somewhat different from 2090. Again, stretch does not have a major effect above 3%, but now aging practice appears to be more important. At 60 ksi yield strength we see that, by extrapolation of the 250°F curve, three practices result in the following toughness values:

Stretch (%)	Age Temp. (°F)	Age Time (Hrs.)	T-L CIE (in-lb/in²)
0	347	~30	100
3	297	36	300
7	250	~100	540

Thus the higher toughness is achieved by using a low aging temperature. Some stretch is needed to achieve a given yield strength within a reasonable aging time.

Structure Examinations

As shown above, the amount of stretch prior to aging and the aging temperature can have a significant effect on the properties of both alloys. Three types of metastable precipitates, which affect the strength, deformation, and fracture behavior, were observed to form in the 2090 alloy. These include Al_3Li (δ'), Al_2CuLi (T_1), and Al_2Cu (θ'). Because of the low interfacial energy and coherency strain associated with the δ' precipitate, it forms homogeneously when the alloy is aged below the metastable-phase solvus temperature. As the aging temperature is decreased, the degree of supersaturation and the driving force for nucleation both increase. This results in a decrease in the critical particle radius needed for stable nucleation and an increase in the frequency of nucleation.[12]

Both T_1 and θ' phases have large coherency strains and nucleate preferentially on dislocations when aged below their metastable-phase solvus temperatures, since the strains associated with the dislocations reduce the overall strain energy. Consequently, the primary role of deformation prior to aging is to increase the dislocation density and, thereby, the number of nucleating sites for precipitation of those phases that contain large coherency strains.[7] Since the solvus lines for the equilibrium phases lie above those for the metastable phases, precipitation of equilibrium phases is possible and competitive with the metastable phases during low temperature aging. Due to the large interfacial energies of the equilibrium phases, they normally nucleate heterogeneously on planar interfaces, e.g. grain and/or subgrain boundaries.[13]

Figure 8 shows the effect of both the stretch and aging temperature on the precipitation behavior of alloy 2090. Figure 8A is a bright field TEM showing T_1 and θ' precipitates distributed relatively homogeneously throughout the grain in plate that was not stretched prior to aging for 60 hours at 347°F. A large equilibrium precipitate is observed in the high angle grain boundary. The grain boundary precipitate has denuded the adjacent region of lithium, resulting in a very large δ' precipitate-free zone (PFZ) adjacent to the grain boundary, as observed in the dark field TEM of δ' of Figure 8B. Some T_1 precipitates are also observed in this figure, since a T_1 reflection was near the δ' reflection used to produce the dark field TEM. The precipitate size and distribution were quite different when the alloy was stretched 7% prior to aging and aged for 60 hours at 250°F. Figure 8C is a dark field TEM of δ' after this treatment and shows a much finer precipitate and no δ' PFZ adjacent to the high angle grain boundary. The observed difference in the δ' precipitate size and distribution is associated with the lower temperature aging and not the stretch, since dislocations have little or no effect on precipitates that have very low coherency strains and interfacial energies. The δ' precipitation is sufficiently fast and the aging temperature sufficiently low to inhibit the formation of the equilibrium precipitates at the grain boundaries. The stretch has a significant effect on the size and distribution of the T_1 precipitate, as observed in the dark field TEM of Figure 8D of one variant of the T_1 phase. The size is very much smaller and the number density very much larger than that observed for the unstretched condition (see Figure 8B).

SEM examination of the fractured Charpy impact specimens supported the TEM observations. With the 7% stretch, 60 hours at 250°F aged plate, the fractures were combinations of flat transgranular shear and intergranular, with the majority of fracture being flat shear (Figure 8E), similar to that reported by Wert and Lumsden[14]. As the aging temperature was increased, the fractures became largely intergranular (Figure 8F), with dimpling present on the intergranular facets, suggesting fracture through PFZs.

Two different strengthening precipitates were observed in the 8091 alloy: the δ', and the Al_2CuMg S' metastable phases. The S' phase has large coherency strains, and it normally nucleates on dislocations similar to the T_1 phase. Figure 9A is a dark field TEM showing S' phase in the 8091 alloy that was not stretched and was aged at the relatively high temperature of 347°F. Although the S' precipitates appear to be homogeneously distributed throughout the grains, a large amount of grain boundary precipitates and a very wide PFZ are observed. Figure 9B is a dark field TEM of δ'. A wide δ' PFZ is also observed

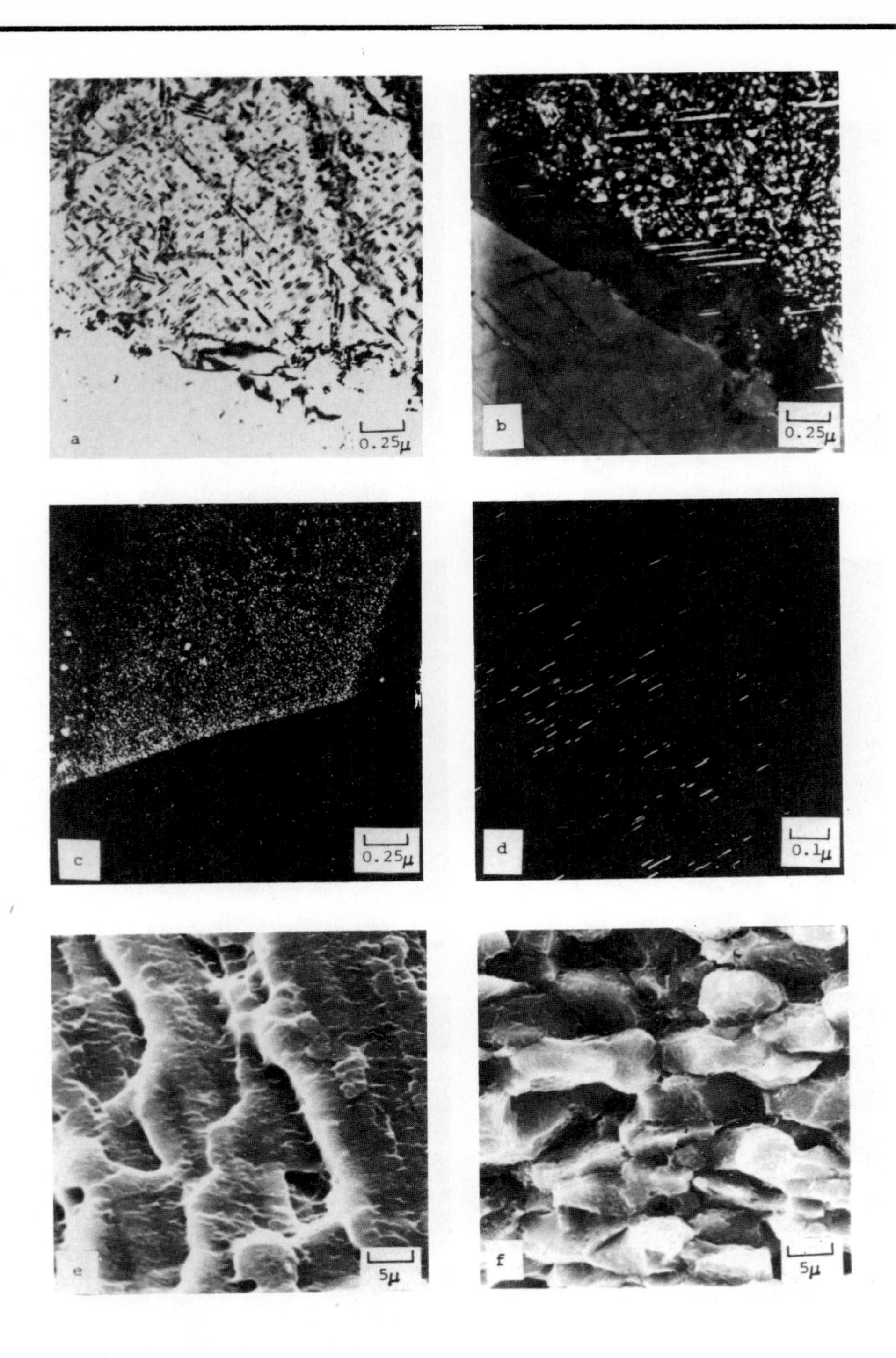

Figure 8. Fine structure and fracture features of 2090 alloy plate.
A. Bright field TEM, T_1 and θ' in 0% stretch, 60 hrs. at 347°F plate.
B. Same as A, only dark field, T_1 and δ'.
C&D. Dark field of 7% stretch, 60 hrs. @ 250°F plate showing δ' & T_1, respectively.
E&F. SEM images of T-L Charpy fractures of 250° and 347°F aged plate, respectively.

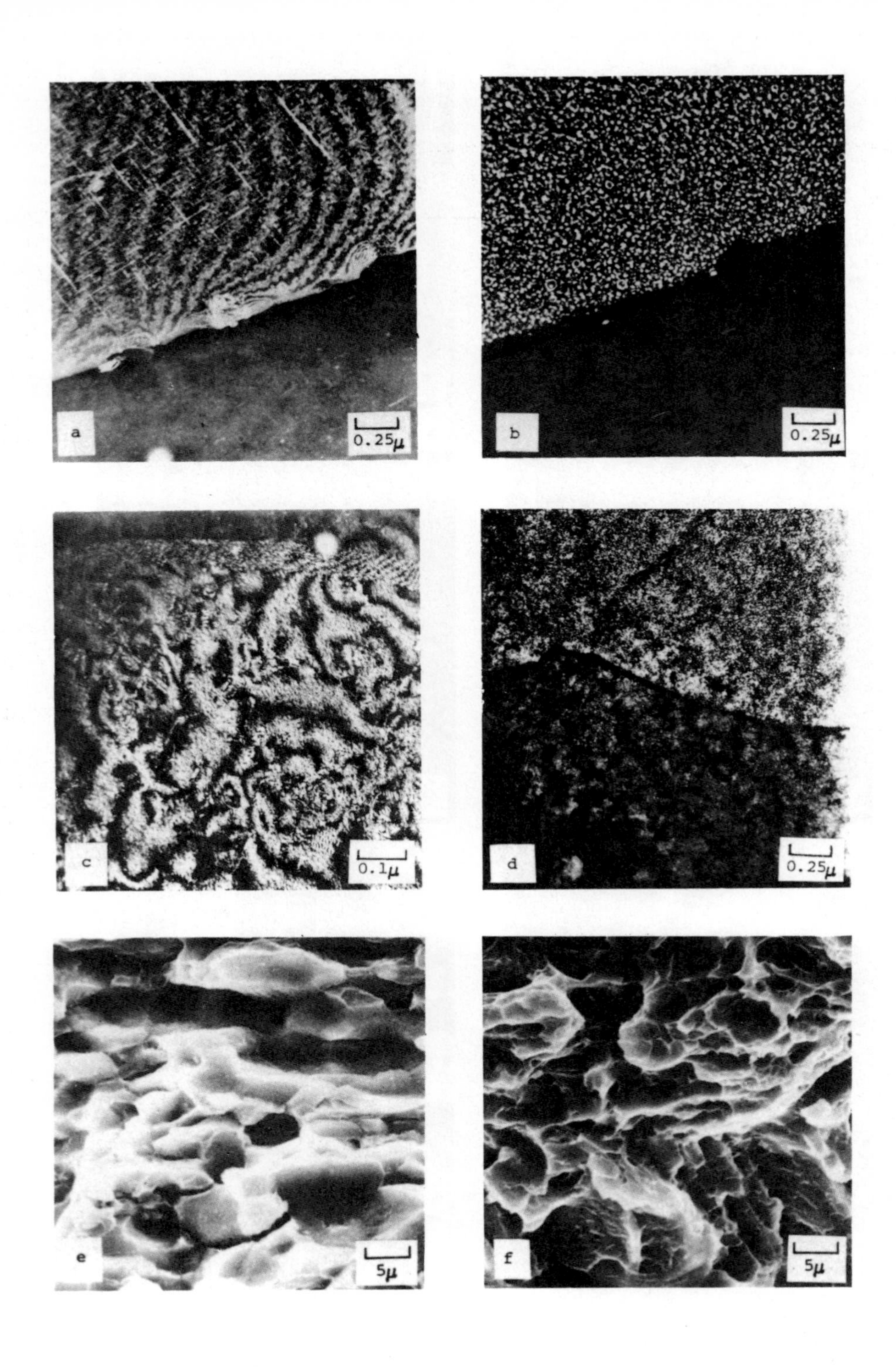

Figure 9. Fine structure and fracture features of 8091 alloy plate.
A&B. Dark field TEMs of 0% stretch, 60 hrs. at 347°F age, S' and δ', respectively.
C&D. Same as A and B only 7% stretch and 36 hrs. at 250°F.
E&F. SEM images of T-L Charpy fractures of 347° and 250°F aged plate, respectively.

in this micrograph. In 8091 stretched 3% prior to aging at the intermediate temperature of 297°F, some grain boundary precipitation and a narrow PFZ were observed. After this treatment the S' was somewhat finer than observed in Figure 9A. Increasing the stretch to 7% and aging at the same temperature, i.e. 297°F, further reduced the amount of grain boundary precipitation and refined both metastable precipitates. As observed for the 2090 alloy, the 7% stretch and 250°F age gave the finest precipitate structure and eliminated both grain boundary precipitates and PFZs (Figures 9C and 9D). All of these samples had been aged to obtain a yield strength of approximately 61 ksi.

The sample that was not stretched and aged at the highest temperature had the lowest fracture toughness. During plastic deformation, strain was localized in the wide PFZ, voids nucleated at the grain boundary precipitates, and low energy intergranular fracture resulted (Figure 9E). As the stretch prior to aging was increased and the aging temperature decreased, the number and size of grain boundary precipitates were reduced, the PFZ width was reduced, and the matrix precipitate structure was refined. All of these factors aided in homogenizing the deformation, leading to higher energy transgranular fracture with dimpling, as observed in Figure 9F. This resulted in the observed increase in fracture toughness.

In summary, it has been shown that the final combination of strength and toughness is a function of aging temperature, aging time, and level of stretch. Broadly, lower aging temperatures yield better combinations of strength and toughness, and, particularly in the case of longitudinal properties, higher stretch levels result in a higher toughness at a given yield strength.

CONCLUSIONS

- Al-Li-Cu-(Mg) alloys can be homogenized in air since the oxidized surface layer can be removed completely during the normal scalping process, and the use of ambient air does not significantly affect toughness or strength.
- Al-Li-Cu-(Mg) ingots homogenized in dry nitrogen displayed the thinnest layer of oxidation/surface depletion; wet nitrogen resulted in the thickest layer.
- High strength Al-Li alloys can be solution heat treated at 1010 ± 10°F.
- The combined effects of stretch and aging time and temperature must be considered in selecting a processing practice. The best combinations of strength and toughness (Charpy impact) are obtained by using the lowest practical aging temperature to achieve a fine hardening precipitate distribution and minimal PFZ widths and grain boundary precipitates.

ACKNOWLEDGMENTS

The help of William A. Cassada in performing much of the TEM work and Barbara Dunbar in assembling the manuscript is gratefully acknowledged. Reynolds Metals Company is thanked for permitting publication.

REFERENCES

(1) T. H. Sanders Jr. and E. A. Starke Jr., Acta Met, Vol. 30, 1982, p. 927.
(2) G. E. Thompson and B. Noble, J. Inst. Metals, Vol. 101, 1973, p. 111.
(3) E. A. Starke Jr. and F. S. Lin, Met. Trans. A, Vol. 13A, 1982, p. 2259.
(4) W. X. Feng, F. S. Lin, and E. A. Starke Jr., Met. Trans. A, Vol. 15A, 1984, p. 1209.
(5) R. E. Crooks and E. A. Starke Jr., Met. Trans. A, Vol. 15A, 1984, p. 1367.
(6) E. Nes, Metal Science J., Vol. 13, 1979, p. 211.
(7) J. C. Williams and E. A. Starke Jr., Deformation Processing and Structure, George Krause, Editor, ASM, Metals Park, OH, 1983, p. 279.
(8) D. J. Field, G. M. Scamans, and E. P. Butler, Aluminum-Lithium Alloys II, E. A. Starke Jr. and T. H. Sanders Jr., Editors, TMS-AIME, Warrendale, PA, 1984, p. 657.
(9) I. G. Palmer, R. E. Lewis, D. D. Crooks, E. A. Starke Jr., and R. E. Crooks, Aluminum-Lithium Alloys II, E. A. Starke Jr. and T. H. Sanders Jr., Editors, TMS-AIME, Warrendale, PA, 1984, p. 91.
(10) O. Kubaschewski, E. Ll. Evans, and C. B. Alcock, Metallurgical Thermochemistry, Pergamon Press, London, 4th Edition, 1967.
(11) A. Kelly and R. B. Nicholson, "Precipitation Hardening," Progress in Materials Science, Vol. 10, Bruce Chalmers, Editor, Pergamon Press, New York, 1963, p. 151.
(12) I. J. Polmear, J. Aust. Inst. of Met., Vol. 11, 1966, p. 246.
(13) E. A. Starke Jr., J. Met., Vol. 22, 1970, p. 54.
(14) J. A. Wert and J. B. Lumsden, Scripta Metallurgica, Vol. 19, 1985, p. 205.

Paper No. 20 "An analysis of powder metallurgy and related techniques for the production of Al-Li alloys" by W E Quist et al. was received at a late stage of compilation and is published in the Appendix (page 625).

Development of low density aluminum–lithium alloys using rapid solidification technology

N J KIM, D J SKINNER, K OKAZAKI and C M ADAM

NJK, DJS and CMA are with Allied Corporation, Morristown, NJ, USA, KO is now at the University of Kentucky, KY, USA

Aluminum-lithium alloys containing 3-3.5 wt% lithium and additions of copper, magnesium, and zirconium have been produced by the melt spinning technique. Composition and heat treatment have been varied to improve the tensile properties of alloys. In particular, a modified heat treatment has been developed to promote the development of composite precipitates. These composite precipitates have been shown to be quite effective in modifying the deformation behavior, with an improvement in ductility.

INTRODUCTION

The addition of lithium to aluminum not only decreases density but also increases elastic modulus, making this alloy system extremely attractive to the aerospace industry. In this regard, the addition of lithium up to the solubility limit (approx. 4.0 wt%) is certainly necessary for achieving maximum possible weight savings. However, current aluminum industry technology utilizing modified DC casting practices have reported only approximately 3.0 wt% Li containing alloys. This is due in part to the formation of coarse brittle intermetallic phases as a result of microsegregation, which will ultimately affect ductility and toughness of the alloy.

The aluminum-lithium alloy program currently under investigation incorporates lithium contents from 3.0 to 4.0 wt% and is primarily concerned with the attainment of the high strength goals (strength levels of 7075-T6/T73). The material has been made using rapid solidification technology via the melt spinning-pulverization-powder processing route. The Al-Li program at Allied is being performed in two phases. Phase I incorporates 3.0 to 3.5 wt% Li with additions of Cu, Mg and Zr in order to gain a basic understanding of the physical metallurgy of the alloy system while still achieving a potential weight saving of 10 to 12%. Phase II will include 3.5 to 4.0 wt% Li (12 to 15% weight saving) and, based on the results of phase I, may include more novel alloy chemistries designed exclusively for rapid solidification. This paper addresses selected results from phase I of the investigation.

MATERIALS AND EXPERIMENTAL PROCEDURE

The compositions of the alloys used in this investigation are listed in Table I. The alloys were rapidly quenched from the melt into continuous ribbons using the jet casting process These ribbons were mechanically comminuted to -60 mesh powder and then consolidated into bulk compacts by vacuum hot pressing and hot extrusion to a final rectangular shape of 63.5 mm by 3.8 mm (18:1 extrusion ratio, 6:1 aspect ratio).

The extrusions were solution treated at 540°C, quenched in cold water, and were given various aging treatments to determine the aging response of the alloys. Based on these studies, several aging conditions were selected to represent different degrees of aging, i.e., under-aging, peak-aging, and over-aging. Longitudinal tensile properties were measured using round specimens with 19.1 mm (0.75 in.) gauge length and 3.2 mm (0.125 in.) gauge diameter at a strain rate of $4X10^{-4}$/sec. Thin foils for transmission electron microscopy (TEM) were made by jet polishing with the electrolyte of 33% HNO_3 and 67% methanol.

Table I. Alloy Compositions (wt%)

Alloy	Li	Cu	Mg	Zr
629	3.10	2.10	1.00	0.45
678	3.22	0.54	3.07	0.83
688	3.20	2.10	1.90	0.51

Table II. Tensile Properties of Alloys Given Single-Aging Treatment

Alloy	Aging Treatment	Y.S.(MPa)	U.T.S.(MPa)	Elong.(%)
629	170°C/4h	541.3	606.1	4.0
	190°C/2h	573.0	629.5	4.5
	190°C/16h	526.8	593.7	3.6
678	160°C/4h	564.0	625.4	3.9
	170°C/4h	581.2	648.8	4.1
	190°C/16h	490.9	577.8	3.9
688	170°C/4h	555.0	621.2	4.0
	190°C/2h	575.7	646.8	4.4
	190°C/16h	521.3	595.0	3.6

Table III. Tensile Properties of Alloys Given Double-Aging Treatment

Alloy	Aging Treatment	Y.S.(MPa)	U.T.S.(MPa)	Elong.(%)
629	170°C/4h + 190°C/16h	530.9	606.8	6.1
678	160°C/4h + 180°C/16h	518.5	601.2	4.7
688	160°C/4h + 180°C/16h	554.4	631.6	5.5

RESULTS

a) Solution Treated

In general, the microstructure of the alloys was unrecrystallized and exhibited a well defined subgrain structure. One of the noticeable features observed in all three alloys was the presence of undissolved constituent within the matrix (Figure 1). Analysis of diffraction patterns showed that this phase was Al_2MgLi. The distribution of this phase was not uniform throughout the microstructure, but an important aspect of this phase might be that it helps disperse dislocation motion due to its non-shearable nature, as shown in Figure 1. There was also the occasional presence of unidentified particles at grain boundaries after solutionizing at high temperature (555°C for 4 hours), as shown in Figure 2. It was difficult to identify this phase because of its reactive nature with the polishing solution; however, the formation of δ' precipitate free zone (PFZ) during aging suggested that it was one of the Li containing intermetallics. Solution treatment at lower temperatures (e.g., 540°C) did not produce this constituent at grain boundaries.

b) Single-Aged

Tensile properties of three alloys with various aging conditions are shown in Table II. Irrespective of alloy chemistry, all three alloys showed the best combination of strength and ductility at peak-strength condition. The variation in the Mg/Cu ratio resulted in different aging characteristics. Alloy 678 with the highest Mg/Cu ratio (~6) showed a more rapid hardening response than the other alloys. It showed a peak-strength after aging for 4 hours at 170°C, while peak-strength conditions were achieved in the other alloys after aging for 2 hours at 190°C. Alloy 678 also showed a more rapid softening at over-aging conditions than the other alloys. The exact cause of this behavior is not clear, but it may be due to the effect of Mg on promoting the development of PFZ, resulting in the deterioration of mechanical properties.

Stress-strain curves of alloy 629 showed some interesting features. Both under- and peak-aged conditions produced serrated stress-strain behavior, which became continuous after aging at 190°C for 16 hours. It has been observed that aging treatments of an Al-Li-Cu-Mg-Zr alloy that led to copious precipitation of S' phase inhibited serrated yielding[1] and widespread precipitation of S' phase in such alloy systems has been shown to strongly influence the deformation behavior, with a significant improvement in ductility.[2,3] In the present study, however, disappearance of serrated yielding was accompanied by deterioration in both strength and ductility. Thus, the microstructure and deformation behavior of alloy 629 was characterized in detail and is described below.

Figure 3 shows the development of δ' precipitate and PFZ in alloy 629 with various aging conditions. As observed in many other alloy systems, in under-aged condition (170°C for 4 hours) the δ' existed as fine precipitate and there was essentially no development of PFZ (Figure 3a). One of the interesting features observed in this aging condition was the relatively homogeneous distribution of composite precipitates, which have been observed to occur in Zr containing Al-Li alloys.[4-7] There was also formation of T_1 phase in this condition. However, the formation of this phase was very sluggish and its volume fraction was very small so that it could be identified only by trace analysis. Aging at 190°C for 2 hours coarsened the δ' precipitates, otherwise the microstructure was essentially the same as that after aging at 170°C for 4 hours. As the aging proceeded to an over-aged condition (190°C for 16 hours), δ' precipitates grew in size and PFZs developed (Figure 3b). The development of composite precipitates still persisted, but could be found mostly around large core particles and their number density was small compared to that in the under-aged condition. The formation of T_1 phase became favorable in the over-aged condition (Figure 4). In contrast to the other investigators' observations, however, precipitation of S' phase was not apparent. Although trace analysis of highly magnified bright field images suggested the presence of S' phase, the volume fraction was so small that any useful dark field and diffraction information could not be obtained. The deformation behavior of the over-aged alloy was classical in that there was a pronounced planar slip and a strain localization at PFZs (Figure 5), indicating that the volume fraction of T_1 and S' phase was not sufficient to induce dispersion of slip during deformation. It was also found that coarsening of the δ' precipitates was not effective in dispersing the dislocation motion. Precipitate size was not large enough to induce Orowan bypassing, and δ' precipitates were sheared by moving dislocations, as shown in Figure 6.

c) Double-Aged

As has been discussed in the previous section, the under-aged microstructure after single-aging showed an interesting feature, i.e., formation of composite precipitates (Figure 3a). It has been suggested that this composite precipitate may be effective in dispersing dislocation movement.[5] However, its small size and small volume fraction in under- and peak-aged alloys may render its effectiveness for slip dispersion as marginal. Changing the aging conditions (i.e., increasing the aging time and/or temperature) did not promote the development of this composite precipitate. Rather, it produced the formation of composite precipitates only around large core precipitates, resulting in small volume fraction and heterogeneous distribution of composite precipitates, as shown in Figure 3b.

In this regard, new aging treatments were given to the alloys to promote the formation of these composite precipitates. This process is basically a double-aging heat treatment, consisting of

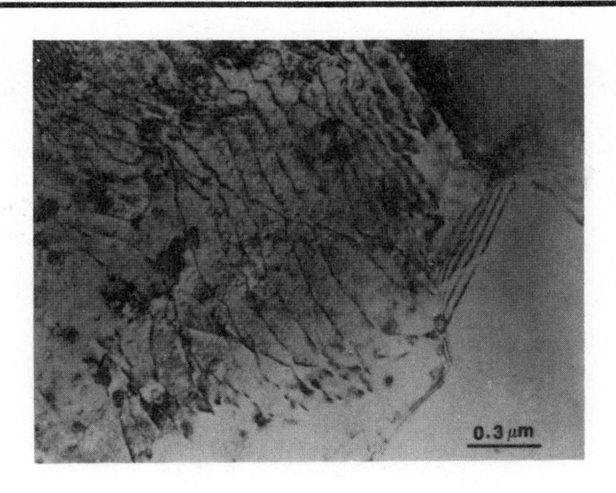

1. Transmission electron micrograph (TEM) showing undissolved Al_2MgLi phase within matrix in the solution-treated condition.

2. TEM showing the presence of unidentified particle at grain boundary formed during high temperature solution treatment.

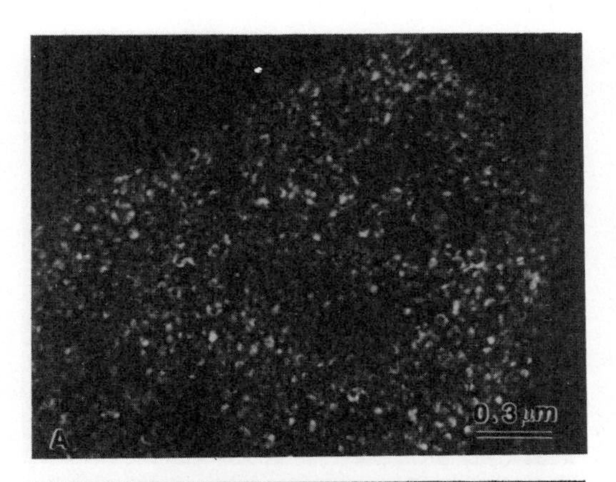

0.3 μm
B

3. Dark field (DF) micrographs showing morphology of δ' precipitate and the development of precipitate free zone (PFZ), (a) aged 170°C/4h, (b) aged 190°C/16h.

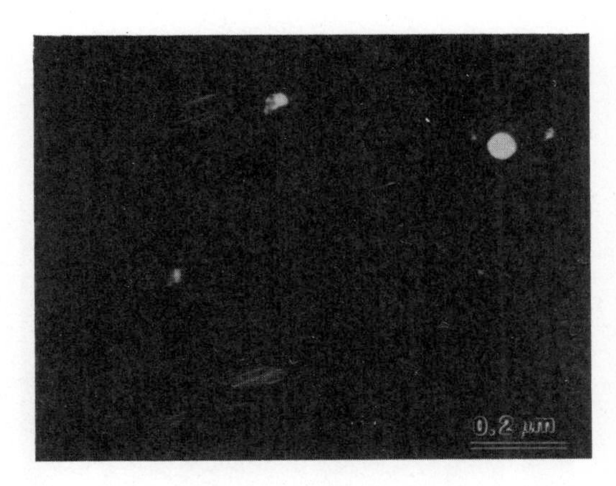

4. DF image of T_1 phase in specimen aged 170°C/4h.

5. Weak beam dark field (WBDF) image of over-aged specimen (190°C/16h) showing planar slip and strain localization at PFZs.

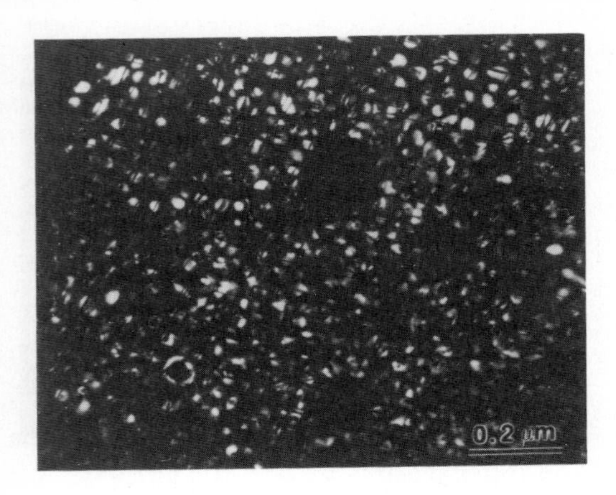

6. DF micrograph showing shearing of δ' in over-aged specimen (190°C/16h).

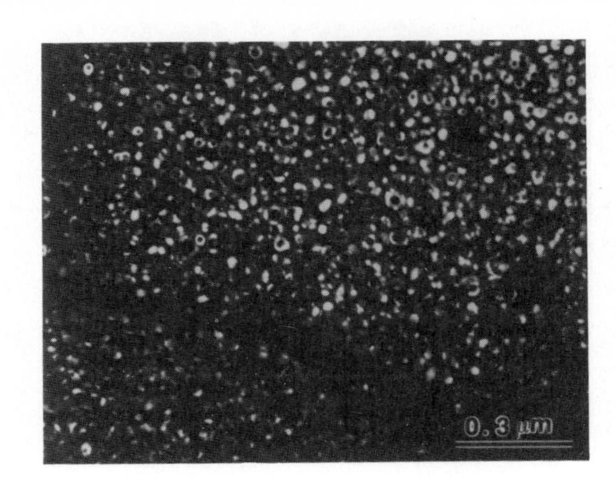

7. DF micrograph of double-aged specimen (170°C/4h + 190°C/16h) showing homogeneous distribution of composite precipitate.

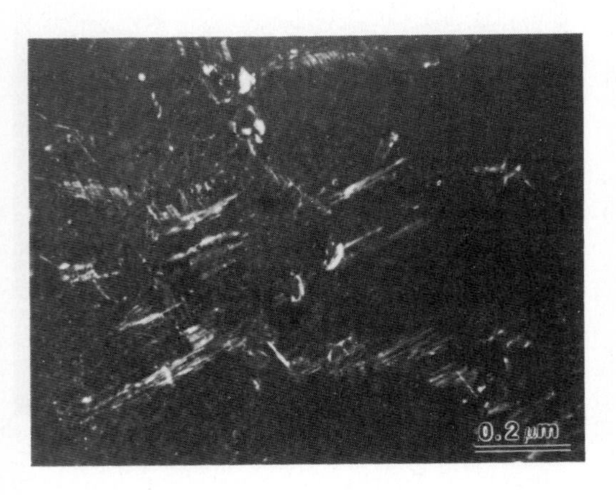

8. DF image of T_1 phase in double-aged specimen.

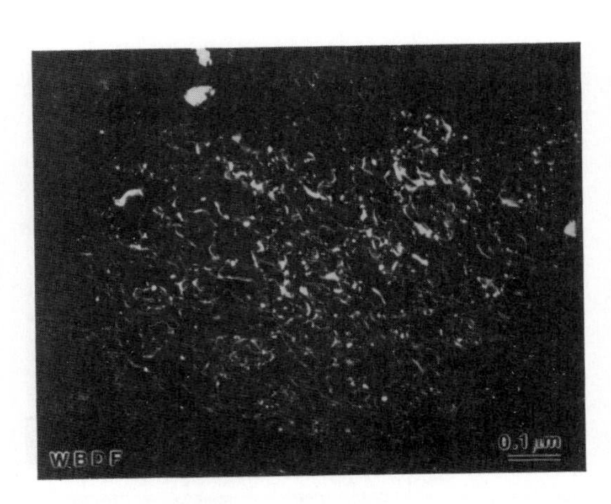

9. WBDF micrograph of double-aged specimen showing homogeneous deformation.

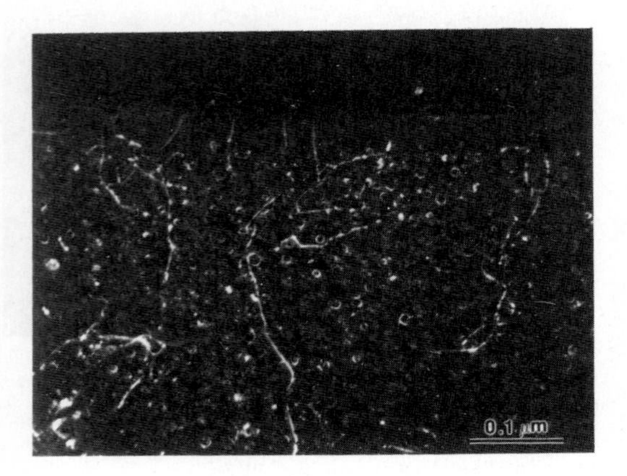

10. WBDF micrograph of ternary Al-3.7Li-0.5Zr alloy showing dispersion of moving dislocations around composite precipitates.

under-aging followed by an over-aging treatment. Figure 7 shows a typical microstructure of a double-aged alloy, consisting of a homogeneous distribution of the composite precipitates. It appears that during the double-aging treatment, nucleation and growth of δ′ precipitates were favored around the core precipitates, while those δ′ precipitates whose nucleation was not aided by Zr became unstable and eventually dissolved.

Other than the predominance of composite precipitates, the microstructure of the double-aged alloy was essentially the same as that of the over-aged alloy. There was precipitation of T_1 phase (Figure 8) and development of PFZs (Figure 7) with a limited amount of S' phase in the over-aged alloy.

As expected, the unique microstructure of the over-aged alloy gave a quite distinctive deformation substructure. There was a dispersion of dislocation arrays and the deformation mode became homogeneous, as shown in Figure 9. There was essentially no localized deformation at PFZs, in contrast to over-aged alloys, which showed extensive strain localization at PFZs (Figure 5). The distinctive deformation behavior of double-aged alloys is well reflected in the mechanical properties as shown in Table III. All of the double-aged alloys had better combinations of strength and ductility than the over-aged alloys.

DISCUSSION

Recently there have been several investigations dealing with the development of composite precipitates in Al-Li-Zr based alloy systems. Although there were some differences in the processing of alloys, these investigations resulted in some conflicting conclusions regarding the nature of core precipitates. Gayle and Vander Sande[5] studied the formation of composite precipitates in a 1 wt% Zr containing alloy, and suggested that weak contrast of core particles in dark field images was due to the interdiffusion of Li into the Al_3Zr dispersoid resulting in the core particle of composition of about $Al_3(Li_{0.6}Zr_{0.4})$.

On the other hand, Makin and Ralph[6] claimed that there was little interaction between Li and Zr in solution and that the structure of the core particle is essentially Al_3Zr. This is supported by the recent observation of Makin et al.[7] showing that some coarse core particles could have contrast comparable with that of δ′ precipitates under certain conditions. Furthermore, they stated that weak contrast of the core was to be expected when the ratio of the total thickness of the shell to that of the core was such that the "average" composition became one which gave a zero structure factor.

Although there exists a discrepancy as to the nature of the core precipitates in the above described observations, both observations are based on the assumption that the formation of composite precipitates is a result of heterogeneous precipitation of δ′ upon pre-existing Al_3Zr dispersoids. However, the present study shows that the presence of Al_3Zr dispersoids in as-solution treated alloys is not the necessary condition for the formation of composite precipitates and different nucleation/growth mechanisms may operate for different aging conditions. It is quite probable that pre-existing Al_3Zr dispersoids are favorable nucleation sites for the heterogeneous precipitation of δ′ for all aging conditions. In such a case, the volume fraction and morphology of composite precipitates should depend on those of Al_3Zr dispersoids formed prior to aging. The present observation of large variations in the volume fraction and morphology of composite precipitates with varying aging conditions, does not suport the conception that formation of composite precipitates is entirely a result of heterogeneous nucleation of δ′ around Al_3Zr dispersoids. Rather, it suggests that diffusion of Li and Zr plays the important role in the formation of composite precipitate at certain aging temperatures, such as at 170°C. At this time the detailed mechanism of composite precipitation during low temperature aging is not clear but may be attributed to the existence of clustered Zr in as-solution treated alloy, which, during aging, reacts with Li atoms to form precursor type $Al_3(Li,Zr)$ precipitates. This subject certainly needs more detailed study and will be discussed in a future paper.

Although double-aging treatments have previously been used to improve the mechanical properties of Al-Li alloys, those treatments were based on the promoted precipitation of T_1 or S' phases.[2,3] The present investigation has shown that the enhanced mechanical properties that were obtained were mainly due to the promoted precipitation of composite precipitate. As has been discussed previously, there was essentially no difference in microstructure between over-aged and double-aged alloys. In the present study, double-aging did not promote the precipitation of S' phase in contrast to previous observations. The cause of the sluggish formation of S' phase in the alloys of this study is not clear, but it may be due to the absence of a stretching operation and also to the relatively high volume fraction of undissolved Al_2MgLi phase which has taken most of the Mg available for the formation of S' precipitate. Nevertheless, to clarify the effect of composite precipitate on the deformation behavior and mechanical properties, a study of a simple ternary Al-3.7Li-0.5Zr alloy has been conducted. Some preliminary results indicate that there is an improvement in both strength and ductility due to the promotion of composite precipitates by double-aging treatments and that composite precipitate is, indeed, effective in inducing the dislocation bypassing, as shown in weak beam dark field image (Figure 10). These observations in the ternary Al-Li-Zr alloy strongly support the idea that the enhancement of mechanical properties in the alloys of the present investigation by double-aging was primarily due to the development of composite precipitates in the microstructure.

CONCLUSION

It has been shown that Allied's rapidly solidified Al-Li alloys have the potential to meet the property goals for high strength application. Good combinations of strength and ductility can be obtained by double-aging heat treatments which produce unique microstructures consisting of composite precipitates in an aluminum matrix. The concept utilized in low lithium alloy development (Phase I) will be similarly applied to the development of high lithium (3.5-4.0 wt%) alloys.

REFERENCES

1. J.A. Wert and P.A. Wycliffe, Scripta Metall., 19, p. 463 (1985).
2. P.J. Gregson and H.M Flower, Acta Metall., 33, 3, p. 527 (1985).
3. B. van der Brandt, P.J. van der Brink, H.F. de Jong, L. Katgerman, and H. Kleinjan, in "Aluminum-Lithium Alloys II", ed. by T.H. Sanders, Jr. and E.A. Starke, Jr., p. 433, TMS-AIME (1984).
4. P.J. Gregson and H.M. Flower, J. Mater. Sci. Letters, 3, p. 829 (1984).
5. F.W. Gayle and J.B. Vander Sande, Scripta Metall., 18, p. 473 (1984).
6. P.L. Makin and B. Ralph, J. Mat. Sci., 19, p. 3835 (1984).
7. P.L. Makin, D.J. Lloyd and W.M. Stobbs, Phil. Mag., 51A, 5, p. L41 (1985).

□

Microstructure and properties of rapid solidification processed (RSP) Al—4Li and Al—5Li alloys

P J MESCHTER, R J LEDERICH and J E O'NEAL

The authors are with McDonnell Douglas Research Laboratories, St. Louis, Missouri, USA

ABSTRACT

Microstructures and mechanical properties of RSP Al-(4-5)Li-(1-2)(Cu,Mg)-0.2Zr extrusions have been determined and correlated. Alloy composition/solution-treatment combinations which yield ⩽ 5 vol% brittle constituent particles are required for the alloys to have significant ductility; in contrast, 42 vol% precipitated δ′ can be tolerated in fine-grained Al-4Li alloys without loss of ductility owing to intense planar slip. Al-4Li-0.2Zr, Al-4Li-1Cu-0.2Zr, and Al-4Li-1Mg-0.2Zr have peak-aged mechanical properties similar to those of the widely used commercial alloy 7050-T76 at 13-14% lower density. Al-4Li-1Mg-0.2Zr is a highly promising alloy for further development as a substitute for commercial 7XXX alloys and near-commercial ingot-metallurgical (I/M) Al-Li alloys because of its favourable combination of strength and corrosion resistance.

INTRODUCTION

Investigations at McDonnell Douglas Research Laboratories (MDRL) (1-5) and elsewhere (6-10) have shown that RSP, powder-source aluminium-3 wt% lithium alloys have attractive properties for aerospace structural applications. For instance, a peak-aged RSP Al-3Li-1Mg-0.2Zr extrusion has a longitudinal yield stress (YS) = 453 PMa (66 ksi), ultimate tensile stress (UTS) = 542 MPa (79 ksi), and elongation = 9% (4). These tensile properties are comparable to those of a commercial 7050-T7651 extrusion (YS = 476 MPa [69 ksi], UTS = 545 MPa [79 ksi], elongation = 7%), but the lithium-containing alloy has an 11% lower density [2.486 g/cm^3 (0.090 lb/in^3)] than 7050-Al.

Although development of RSP Al-3Li alloys has advanced understanding of Al-Li powder production, consolidation processing, and physical metallurgy, these powder-metallurgical alloys have been rendered obsolete for structural applications by lower-cost ingot-metallurgical (I/M) Al-(2.2-2.5)Li alloys. One such alloy, Alcan DTDXXXB (Al-2.5Li-2.0Cu-0.7Mg-0.1Zr) has a density of 2.54 g/cm^3 (9.3% smaller than that of 7050-Al) and a peak-aged YS = 554 MPa (80.4 ksi), UTS = 589 MPa (85.4 ksi), and elongation = 6% in plate form. Such properties make the near-commercial alloy more attractive than the experimental RSP Al-3Li-1Mg-0.2Zr alloy.

For an RSP Al-Li alloy to become competitive with I/M Al-Li alloys, further density reductions are required. These can be achieved either by additions of beryllium (11) or by further Li additions. Because Be poses toxicity problems, and large Be concentrations are required to obtain significant additional density reductions in Al-Li alloys, research at MDRL has concentrated on RSP Al-(4-5)Li alloys, which are 13-17% less dense than commercial 7XXX alloys and which presently can be produced only by RSP. This paper reports determinations of microstructures and properties of RSP Al-(4-5)Li-(1-2)(Cu,Mg)-0.2Zr alloys as functions of composition and heat treatment.

ANALYSIS OF CONSTITUENT AND PRECIPITATE PHASES

The solubility limit of Li (and Cu and Mg, if present) in Al-(4-5)Li alloys is approached or exceeded even at the highest attainable solution-treatment temperatures. In binary Al-Li (or ternary Al-Li-Zr) alloys the phase diagram (Figure 1) (12,13) shows that whereas 3 wt% Li can be taken into solution over temperature range 530-620°C, 4 wt% Li can be dissolved only between 588 and 605°C, and the solubility limit is exceeded for any solution treatment temperature

TABLE 1

Compositions and Densities of RSP Al-(4-5)Li Alloys

Nominal Composition	Analyzed Composition (wt%) Li	Cu	Mg	Zr	Density (g/cm^3)	% Reduction Relative to 7050-Al
Al-4Li-0.2Zr	3.95	--	--	0.20	2.419	13.6
Al-4Li-1Cu-0.2Zr	3.95	1.00	--	0.22	2.429	13.2
Al-4Li-2Cu-0.2Zr	3.90	2.00	--	0.20	2.437	13.0
Al-4Li-1Mg-0.2Zr	3.88	--	1.00	0.20	2.410	13.9
Al-4Li-2Mg-0.2Zr	3.98	--	2.00	0.20	2.401	14.2
Al-5Li-0.2Zr	4.70	--	--	0.20	2.359	15.8
Al-5Li-1Cu-0.2Zr	4.89	0.98	--	0.21	2.347	16.2
Al-5Li-2Cu-0.2Zr	5.00	2.00	--	0.24	2.354	15.9
Al-5Li-1Mg-0.2Zr	4.73	--	0.98	0.22	2.345	16.2
Al-5Li-2Mg-0.2Zr	4.90	--	1.95	0.22	2.332	16.7

at 5 wt% Li. Composition/solution-treatment combinations which do not dissolve all the Li leave brittle δ (AlLi) constituent particles in equilibrium with the matrix. If the volume fraction of such particles is sufficiently large the ductility of the subsequently peak-aged alloy is degraded.

Successive applications of the lever rule to matrix-δ equilibrium at solution treatment temperatures between 394 and 588°C and to matrix-δ' metastable equilibrium at a uniform aging temperature of 160°C enables the δ and δ' volume fractions in peak-aged Al-(3-5)Li alloys to be calculated as functions of initial Li concentration and solution treatment temperature (Figure 2). Minimization of δ by solutionizing at 588°C results in a δ' concentration in Al-4Li of 42 vol%, which is expected to enhance strength but may degrade ductility by promoting intense planar slip. Conversely, lowering the δ' concentration by decreasing the prior solution-treatment temperature lowers strength but may enhance ductility if the δ concentration is not too high.

Additions of Cu or Mg to Al-(4-5)Li alloys increase strength by solution-strengthening, increasing the δ' concentration by decreasing the matrix Li solubility, and (for Cu) co-precipitating T_1 (Al_2LiCu). Since the one-phase field in Al-4Li-0.2Zr has a limited temperature range of stability, further 1-2 wt% Cu or Mg additions result in relatively large constituent particle concentrations even at high solution-treatment temperatures. According to the Al-Li-Cu and Al-Li-Mg phase diagrams (12), the equilibrium constituent phases are δ and T_2 (Al_5Li_3Cu) in Al-Li-Cu alloys and δ and S (Al_2LiMg) in Al-Li-Mg alloys.

ALLOY SELECTION, POWDER PRODUCTION, AND CONSOLIDATION

Nominal compositions of the ten Al-(4-5)Li alloys selected for this investigation are listed in Table 1. The concentrations of Li, Cu, and Mg were systematically varied to determine the specific effects of each elemental addition on microstructure and properties. A small amount of Zr was added to each alloy to prevent recrysallization during consolidation and promote formation of fine, unrecrystallized substructures in the extrusions.

Approximately 10 kg (22 lb) of vacuum-atomized powder of each composition were obtained from Homogeneous Metals, Inc., Clayville, NY. Previous work (5) has shown that these powders have a typical geometric mean particle diameter of 40-45 µm and a typical mean solidification rate of ~ 10^4 K/s. Chemical analyses of the powders performed by Homogeneous Metals, Inc., show good agreement between nominal and analyzed compositions (Table 1). The powders were packaged under inert gas for shipping to the extrusion vendor.

The powders were consolidated to rectangular extrusions at the Kaiser Center for Technology, Pleasanton, CA. Since Homogeneous Metals, Inc., had observed that Al-4Li and (particularly) Al-5Li powders were significantly more reactive with the atmosphere than Al-3Li powders, as-received powders were handled under inert gas cover. The powders were loaded into rubber containers and cold isostatically pressed to approximately 70%-dense green compacts. The compacts were canned and degassed using Kaiser's proprietary depurative degassing process at temperatures up to 510C (950F) and then hot pressed into billets of 100% density at 510°C. The billets were decanned and extruded at 400°C to 1.27 x 5.72 cm (0.5 x 2.25 in) extrusions at an extrusion ratio of 19:1.

RESULTS AND DISCUSSION

Densities

Pycometrically measured densities (accurate to ± 0.005 g/cm^3) of the alloys are listed in Table 1. The 13-14% and 16-17% reductions in density of Al-4Li and Al-5Li alloys, respectively, relative to 7050-Al are consistent with the rule-of-mixtures prediction that Li should reduce the density of Al by ~ 3% for each wt% added. The small further effects of Mg in reducing and of Cu in increasing the density are likewise expected on a rule-of-mixtures basis. The slight anomaly in the densities of Al-5Li-Cu-Zr alloys is the result of variations in Li concentration.

Microstructures

Since the Al-(4-5)Li powders were observed to be highly reactive, oxide particle densities in the extrusions were expected to be high. Optical micrographs of solution-treated, unetched Al-4Li-0.2Zr which had been heat-treated to dissolve all of the Li (Figure 3a) show large concentrations of inclusions, many of which are located in stringers. Scanning electron microscope (SEM) examination (Figure 3b) reveals that typical inclusions are 10-20 µm in diameter and that they charge readily in the electron beam. Energy dispersive X-ray analysis of the inclusion shows only aluminium, but this analytical method cannot detect either Li or O. The topography of the inclusions (Figure 3b) suggests that they had reacted further with the atmosphere after polishing. Al-3Li alloy extrusions from vacuum-atomized powders show much lower inclusion densities (4,5). Based on this evidence, it is concluded that the inclusions are probably aluminium-lithium oxides which have been hydrolyzed upon exposure to ambient water vapour. No evidence was found of foreign-element inclusions (e.g., Fe- or Ni-rich particles from cross-contamination).

The observed inclusions could have originated from reaction in the melt prior to atomization or by scavenging of oxygen by the powders during handling, packaging, and shipping. Microscopic examination of RSP Al-3Li alloy extrusions (4,5) and solution-treated, etched Al-4Li-0.2Zr (Figure 4a) shows a homogeneous distribution of 50-100-nm diameter oxides which are separated in rows rather than being gathered into stringers. These oxides presumably originated by disintegration of oxide layers on the powder particles. The large size and relatively inhomogeneous distribution of the inclusions suggests that they were formed by reaction of the melt with the atmosphere or with the crucible liner, and were atomized into discrete particles which were then incorporated into the extrusions. The inclusions are expected to significantly reduce ductility and fracture toughness.

Optical micrographs of solution-treated Al-4Li-0.2Zr and Al-5Li-0.2Zr extrusions are shown in Figure 4. Besides the fine oxide particles, a few 3-5-µm diameter δ particles are visible in Al-4Li-0.2Zr. Approximately 10 vol% of similar particles appear in Al-5Li-0.2Zr, in qualitative agreement with Figure 2. The geometric mean subgrain diameter is 2.7 ± 1.1 µm, or about 50% larger than the average subgrain diameter in RSP Al-3Li-0.2Zr.

Although numerous high-angle boundaries are visible in Figure 4, (111) pole figures of as-extruded and 588°C/1h solution-treated Al-4Li-0.2Zr (Figure 5) show that the microstructure is always unrecrystallized. Previous research (4,5) has shown that extrusions of Zr-containing RSP Al-Li alloys have a fully-recovered mosaic substructure in which low-angle boundaries predominate. As the extrusion is solution-treated at successively higher temperatures, subboundary motion and rearrangement occur to produce a more equiaxed microstructure in which high-angle boundaries predominate, with only minor changes in grain size and orientation. These changes are reflected in the slight sharpen-

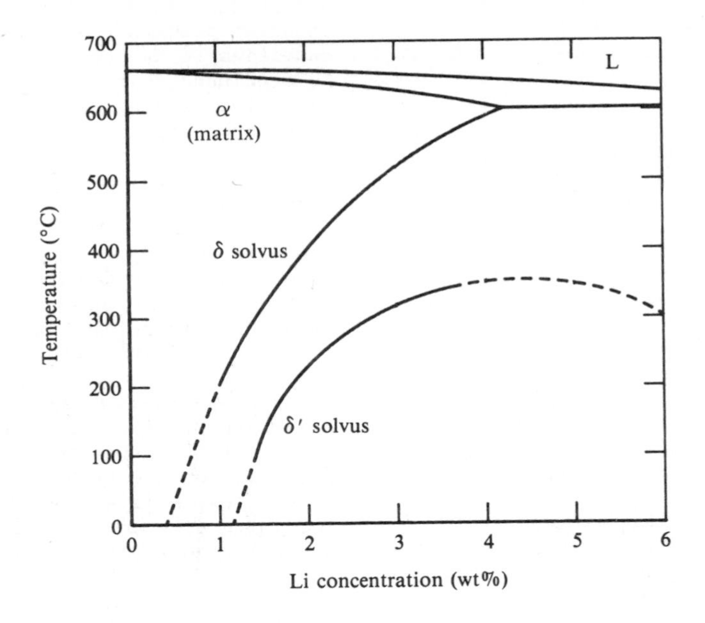

1. Aluminum-lithium phase diagram with metastable δ′ (Al_3Li) solvus.

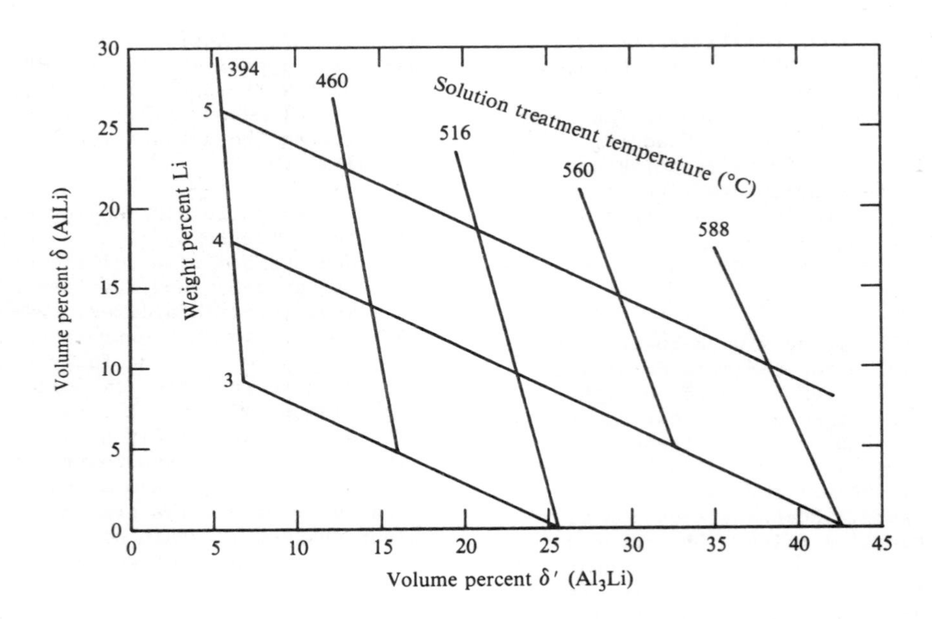

2. Trade-off between δ (AlLi) and δ′ (Al_3Li) concentrations in solution-treated and 160°C-aged Al-(3-5)Li alloys.

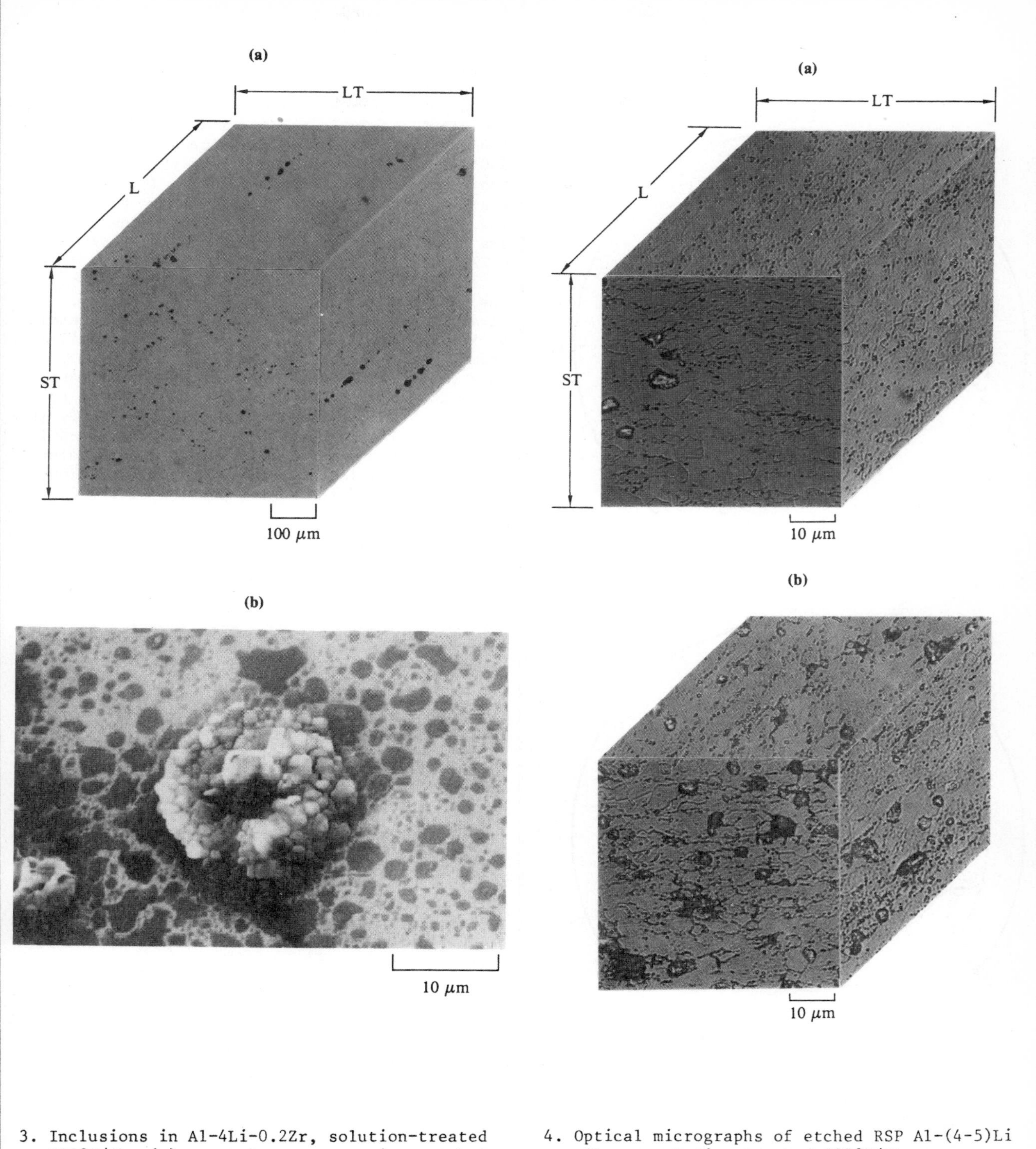

3. Inclusions in Al-4Li-0.2Zr, solution-treated 588°C/1h: (a) general appearance in unetched extgrusion (optical micrograph), and (b) scanning electron micrograph of typical inclusion.

4. Optical micrographs of etched RSP Al-(4-5)Li alloys, solution-treated 588°C/1h: (a) Al-4Li-0.2Zr and (b) Al-5Li-0.2Zr.

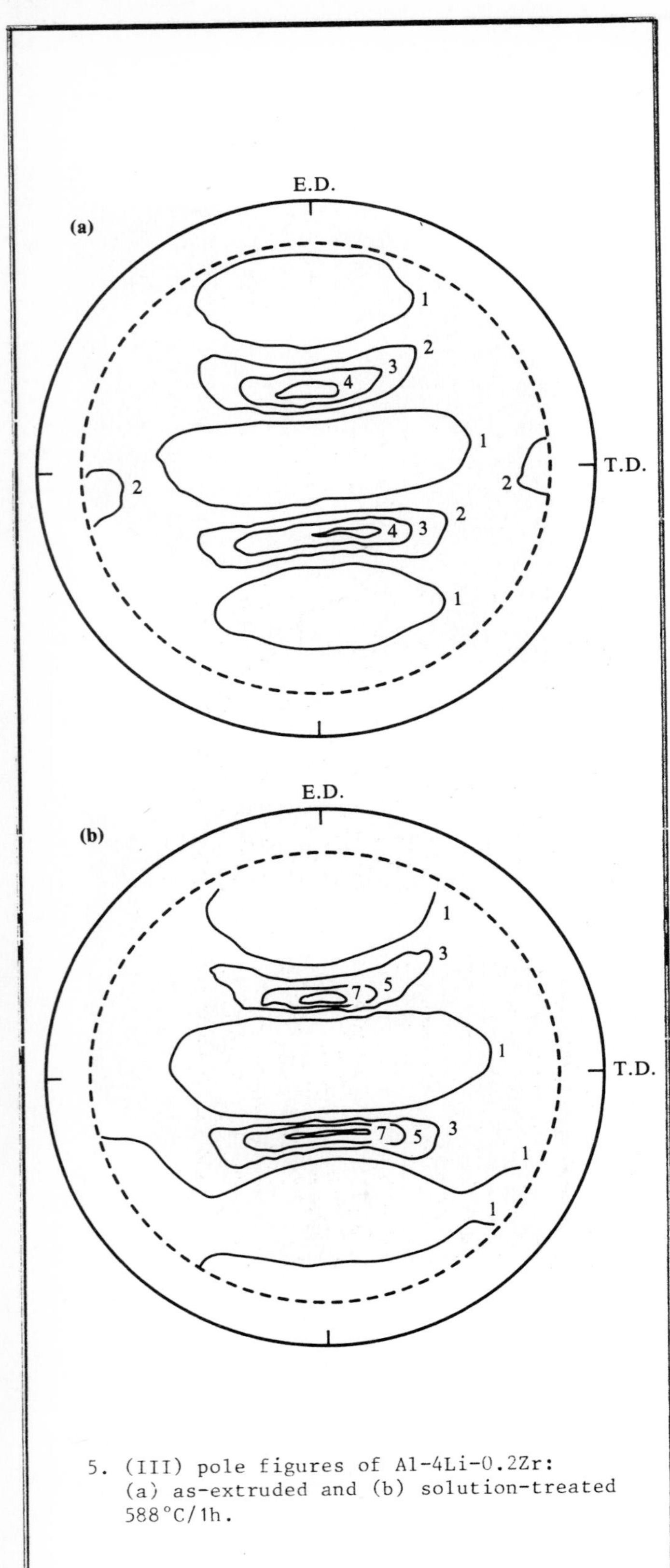

5. (III) pole figures of Al-4Li-0.2Zr: (a) as-extruded and (b) solution-treated 588°C/1h.

ing of texture during solution treatment evident in Figures 5a and 5b. This subtle microstructural transition has little effect on ambient-temperature behaviour, but it is critical for superplastic formability.

The highest safe solution-treatment temperature for all Al-Li-Cu-Zr and Al-Li-Mg-Zr alloys was found to be 560°C. Typical microstructures of 560°C/1h solution-treated alloys (Figure 6) show the combinations of constituent δ, T_2, and S phases predicted by the ternary phase diagrams. The combined concentrations of constituent phases are $\geqslant$ 10 vol% in all alloys except Al-4Li-1Cu-0.2Zr and Al-4Li-1Mg-0.2Zr.

The precipitation behaviour of Al-4Li-0.2Zr, Al-4Li-1Cu-0.2Zr, and Al-4Li-1Mg-0.2Zr, determined by transmission electron microscopy, is familiar from numerous prior investigations (6-10). The only precipitate phase is δ' except for Al-4Li-1Cu-0.2Zr which had been stretched 2% prior to aging, in which a small volume fraction of T_1 is observed. In contrast to Al-3Li alloys, Al-4Li alloys undergo natural aging, and a large volume fraction of δ' is observed in samples solution-treated and aged at room temperature for $\geqslant$ 24 h.

Fracture surfaced of under- and peak-aged Al-4Li-0.2Zr (Figure 7) show features on the order of the subgrain diameter. Small ductile dimples predominate on the fracture surface of the peak-aged alloy (Figure 7b), while the underaged specimen (Figure 7a) shows more direct inter-subgranular cracking. The two specimens had virtually the same ductility, however.

Mechanical Properties

To determine the optimum balance of δ and δ' volume fractions in Al-4Li-0.2Zr and Al-5Li-0.2Zr, mechanical properties of peak-aged specimens were determined after solution-treatments between 394 and 588°C. The yield stress, ultimate tensile stress, and elongation to failure of peak-aged Al-4Li-0.2Zr decrease rapidly with decreasing prior solution-treatment temperature below 560°C (Figure 8). Specimens with δ volume fractions $\geqslant$ 10% exhibit little or no work-hardening, and $\leqslant$ 5 vol% δ is required for an acceptable strength/ductility combination. Since reasonable ductilities are associated with δ' concentrations as large as 42 vol%, 588°C was chosen as the solution-treatment temperature for Al-4Li-0.2Zr. Al-5Li-0.2Zr, which always has $\geqslant$ 10 vol% δ, showed 0.8-4% elongation and no work-hardening in the peak-aged condition even after 588°C/1-h solution treatment. No further measurements were performed on this alloy.

The 160°C aging behaviour of Al-4Li-0.2Zr is shown in Figure 9. The yield stress increases approximately 20% from the naturally-aged to the peak-aged condition owing to an increase in the average δ' diameter from 7 to 18 nm. The strength maximum corresponds to any aging time between 24 and 72 h. As the alloy is overaged, the strength drops slightly and the ductility falls off noticeably, probably as the result

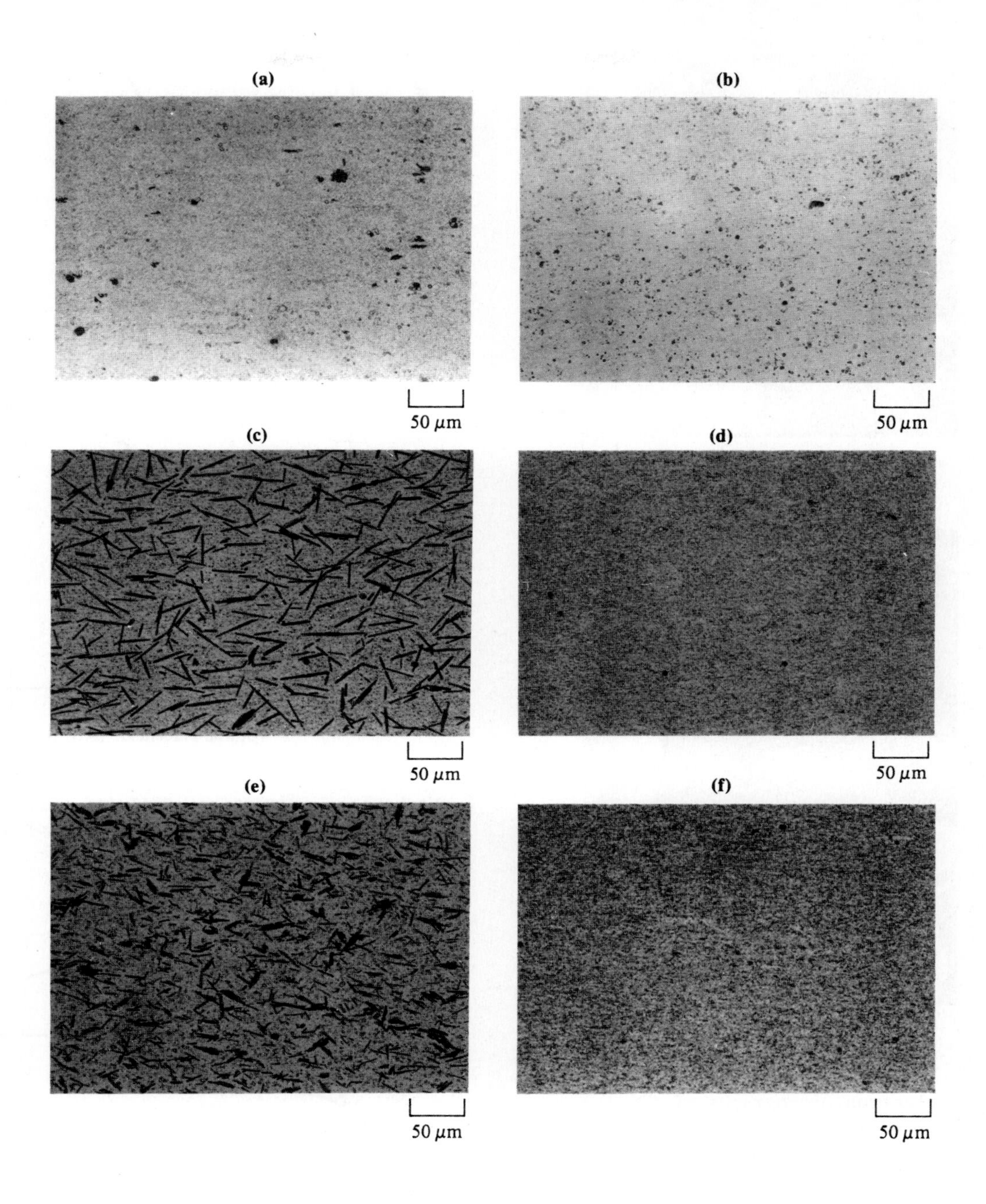

6. Optical micrographs of (a) Al-4Li-1Cu-0.2Zr (constituent particles δ[AlLi] + $T_2[Al_5Li_3Cu]$), (b) Al-4Li-1Mg-0.2Zr (δ), (c) Al-4Li-2Cu-0.2Zr (T_2), (d) Al-4Li-2Mg-0.2Zr (S[Al_2LiMg]), (e) Al-5Li-2Cu-0.2Zr ($\delta + T_2$), and (f) Al-5Li-2Mg-0.2Zr (δ +S); all solution-treated 560°C/1h.

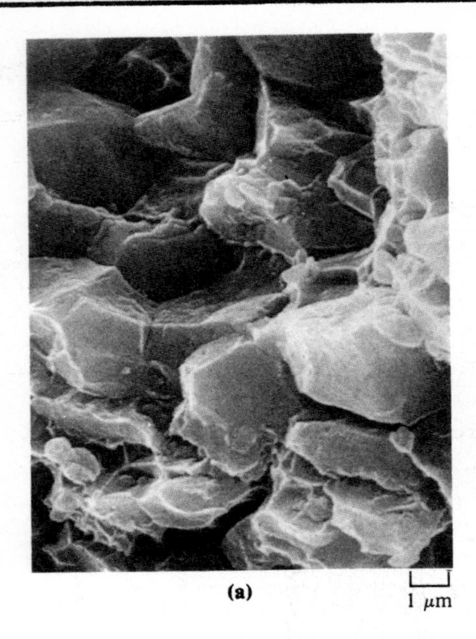

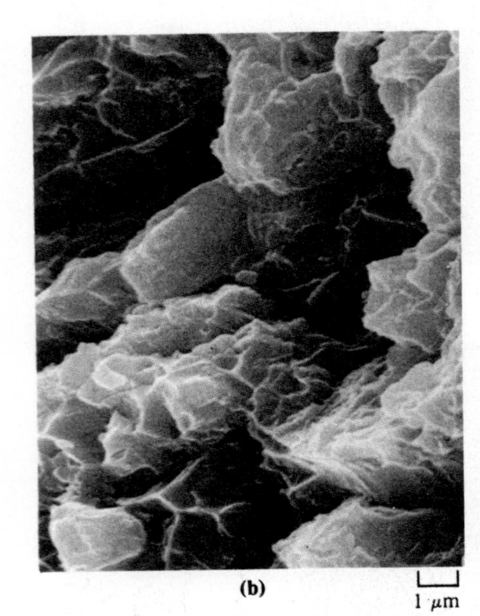

7. Scanning electron micrographs of fracture surfaces of Al-4Li-0.2Zr tensile specimens, solution-treated 588°C/1h and aged at 160°C for (a) 16h and (b) 70h.

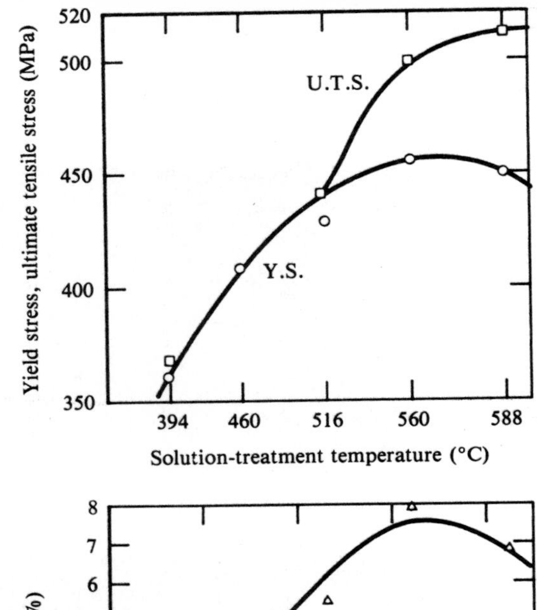

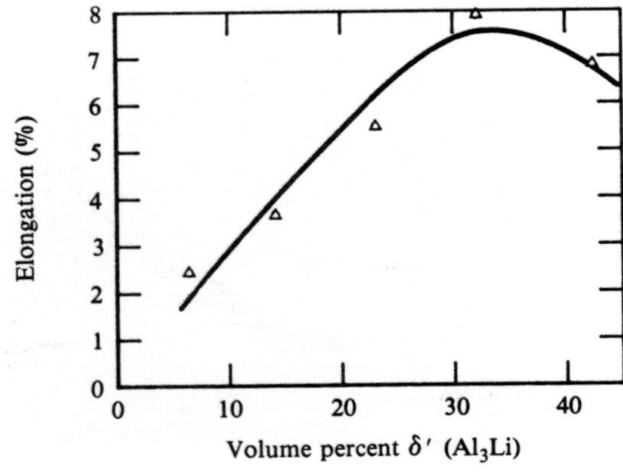

8. Mechanical properties of Al-4Li-0.2Zr peak-aged at 160°C as a function of prior solution-treatment temperature.

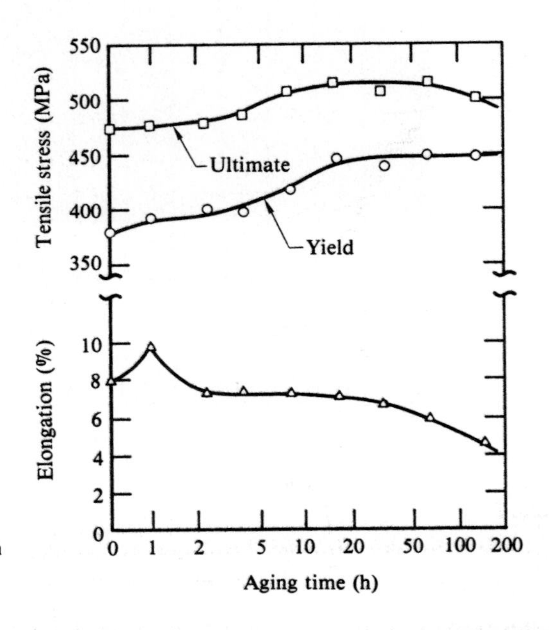

9. Mechanical properties of Al-4Li-0.2Zr, solution-treated 588°C/1h, as a function of 160°C aging time.

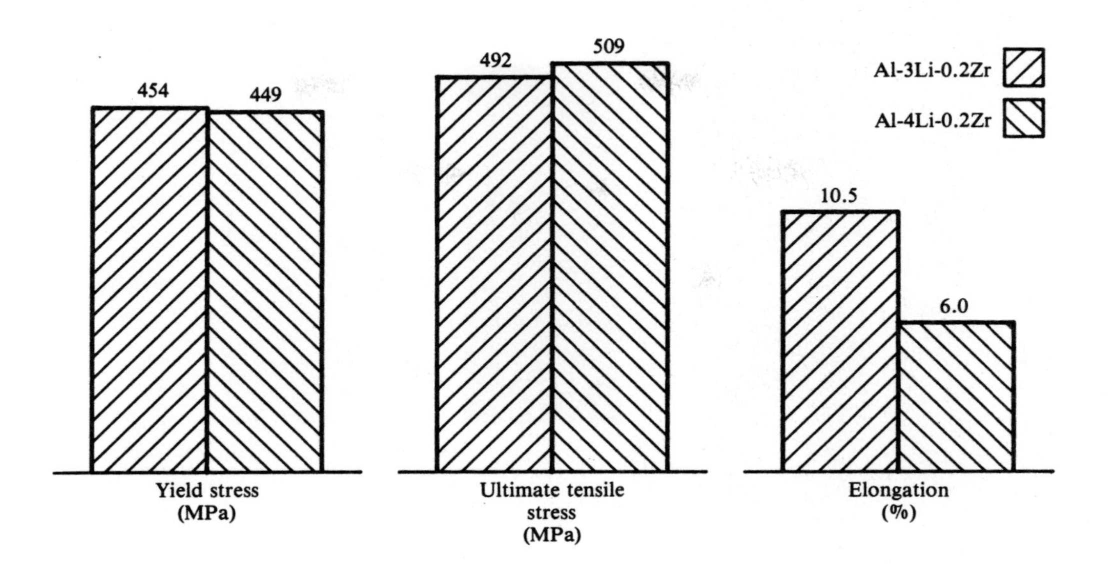

10. Mechanical properties of peak-aged Al-3Li-0.2Zr and Al-4Li-0.2Zr.

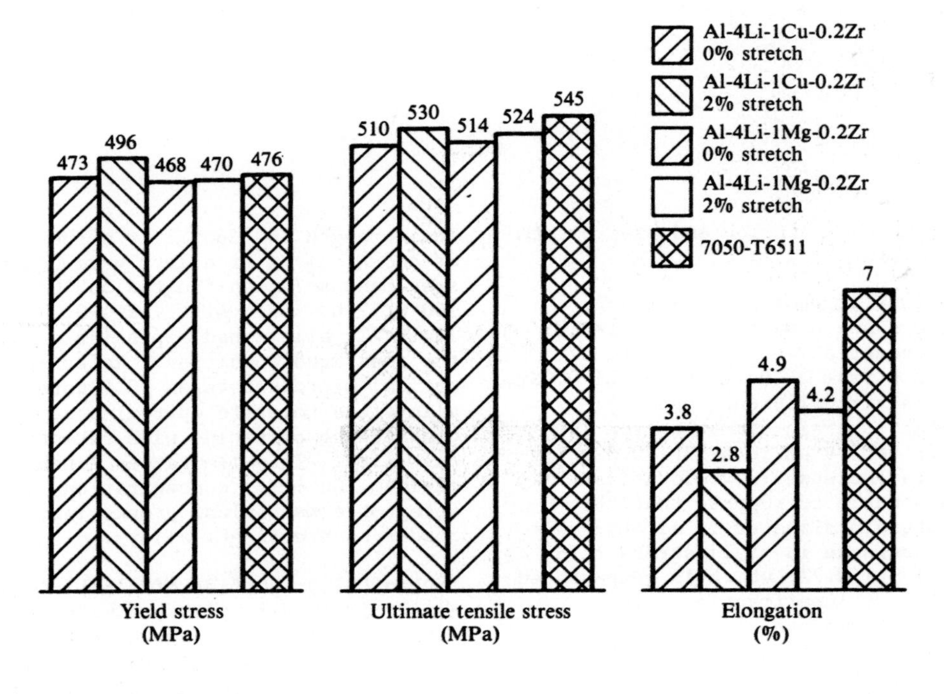

11. Mechanical properties of Al-4Li-(Cu,Mg)-0.2Zr alloys, solution-treated 560°C/1h and aged 160°C/32h.

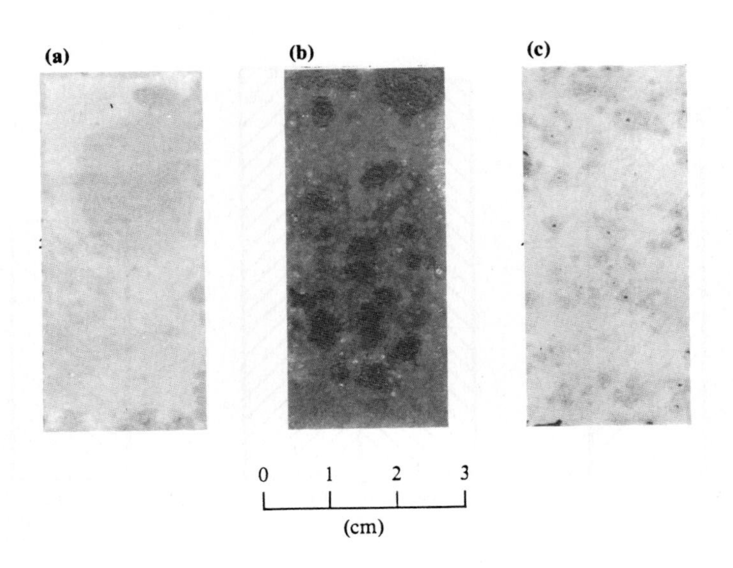

12. EXCO exfoliation test coupons of solution-treated (a) Al-4Li-0.2Zr, (b) Al-4Li-1Cu-0.2Zr, and (c) Al-4Li-1Mg-0.2Zr.

of grain-boundary δ precipitation and precipitate-free zone (PFZ) formation.

Mechanical properties of peak-aged Al-3Li-Zr and Al-4Li-Zr alloys are compared in Figure 10. Although the δ´ concentration increases to 42 vol% in Al-4Li-0.2Zr from 25 vol% in Al-3Li-0.2Zr, the yield stress is virtually unchanged. Decreasing the mean subgrain diameter from 1.9 μm in Al-3Li-0.2Zr to 1.4 μm in Al-3Li-0.5Zr results in a 30 MPa increase in yield stress (4,5). It is reasonable to suppose that the 2.7 μm average subgrain diameter in Al-4Li-0.2Zr corresponds to a decrease in yield stress relative to that of Al-3Li-0.2Zr which is compensated for by the higher δ´ volume fraction. In addition, significant amounts of Li may be diverted to oxides and inclusions in Al-4Li-0.2Zr, lowering the Li concentration available for δ´ formation and decreasing the strength. Increasing the Li concentration from 3 to 4 wt% decreases ductility from 8 to 6%, but the latter value is still sufficient for structural applications.

The strength and ductility of peak-aged Al-Li-Cu-Zr and Al-Li-Mg-Zr alloys were determined on specimens solution-treated 560°C/1h and aged 160°C/32 h. Only Al-4Li-1Cu-0.2Zr and Al-4Li-1Mg-0.2Zr alloys had appreciable ductility in the peak-aged condition. Referring to the optical microstructures in Figure 6, these alloys can tolerate approximately 5 vol% constituent particles with retention of appreciable ductility. This limitation establishes the maximum (Cu + Mg) concentration, consistent with acceptable mechanical properties, as 1 wt% in powder-source Al-4Li alloys.

Mechanical properties of Al-4Li-1Cu-0.2Zr and Al-4Li-1Mg-0.2Zr, with and without an intermediate process stretch, are shown in Figure 11. Typical uncertainties are ± 15 MPa (2.2 ksi) in YS and UTS, and ± 1 percentage point in the ductility. These uncertainties are larger than for Al-3Li alloys (4,5), probably as the result of the larger inclusion density in the present alloys. The strengths of similarly heat-treated Al-4Li-1(Cu,Mg)-0.2Zr and Al-3Li-1(Cu,Mg)-0.2Zr alloys are similar. The YS and UTS of Al-4Li-1Cu-0.2Zr are raised 20-25 MPa by an intermediate stretch with a 1% sacrifice in ductility, while the strength of Al-4Li-1Mg-0.2Zr is unaffected by a process stretch, in agreement with previous results on Al-3Li-Mg alloys (4,5).

The properties of Al-4Li-(Cu,Mg)-0.2Zr alloys, and in particular Al-4Li-1Mg-0.2Zr, are attractive relative to those of current commercial 7XXX alloys. For instance, Al-4Li-1Mg-0.2Zr has a similar yield stress, slightly lower ductility, and 14% lower density in comparison with 7050-T76. Modest improvements in consolidation processing to lower the oxide and inclusion densities are expected to further improve mechanical properties. The 50-100% larger density savings in RSP Al-4Li-1Mg-0.2Zr relative to I/M Al-Li alloys makes it an attractive competitor, particularly in over-strength thin sections where the higher strength of the I/M alloy does not confer design advantages and density is the major criterion for substitution.

Corrosion Resistance

Visual observation of Al-(4-5)Li extrusions shows that alloy/heat-treatment combinations yielding significant δ volume fractions correspond to poor atmospheric corrosion resistance. Al-4Li alloys solution-treated to minimize δ concentrations do not corrode in ambient air. Results of preliminary exfoliation corrosion (EXCO) tests on solution-treated Al-4Li alloys (Figure 12) show that Al-4Li-0.2Zr and Al-4Li-1Mg-0.2Zr are virtually immune except for pitting at inclusions, while Al-4Li-1Cu-0.2Zr is moderately attacked (EA rating). These results, which are in agreement with similar results on Al-3Li alloys (11), render Al-4Li-1Mg-0.2Zr attractive for further development.

SUMMARY

1. RSP Al-(4-5)Li alloy extrusions can be successfully produced from vacuum-atomized powders using conventional powder consolidation methods.
2. Since Al-4Li melts and powders are more reactive than RSP Al-Li alloys previously investigated, improved melting, powder-making, and particulate processing techniques are required to minimize inclusion densities and improve notch-sensitive properties of powder-source mill products.
3. Combined (Li + Cu + Mg) concentrations well in excess of the solubility limit cannot be tolerated in RSP Al-(4-5)Li alloys. Only those alloys which can be solution-treated to retain $\leqslant$ 5 vol% constituent (insoluble) particles have usable ductilities and work-hardening coefficients in the peak-aged condition.
4. Peak-aged Al-4Li alloys containing up to 42 vol% δ′ have acceptable ductilities despite the potential for intense planar slip. The constituent particle concentration is more important than the δ′ concentration in determining ductility
5. RSP Al-4Li-0.2Zr, Al-4Li-1Cu-0.2Zr, and Al-4Li-1Mg-0.2Zr extrusions have mechanical properties competitive with those of 7050-T76 but have 13-14% lower densities. Al-4Li-1Mg-0.2Zr is the most attractive composition for further development based on its favourable mechanical properties, low density, and corrosion resistance.

ACKNOWLEDGEMENT

This research was conducted under the McDonnell Douglas Corporation Independent Research and Development program.

REFERENCES

1. K.K. Sankaran, S.M.L. Sastry, and J.E. O'Neal, "Microstructure and Deformation of Rapidly Solidified Al-3Li Alloys Containing Incoherent Dispersoids", in Aluminum-Lithium Alloys, ed. by T.H. Sanders, Jr., and E.A. Starke, Jr. (The Metallurgical Society of AIME, Warrendale, PA, 1981), pp. 189-203.
2. K.K. Sankaran, J.E. O'Neal, and S.M.L. Sastry, "Effects of Second-Phase Dispersoids on Deformation Behavior of Al-Li Alloys", Met. Trans. 14A, 2174 (1983).
3. P.J. Meschter, J.E. O'Neal, and R.J. Lederich, "Effect of Rapid-Solidification Method on the Microstructure and Properties of Al-3Li-1.5Cu-0.5Co-0.2Zr", in Aluminum-Lithium Alloys II, ed. by T.H. Sanders, Jr., and E.A. Starke, Jr. (The Metallurgical Society of AIME, Warrendale, PA, 1984), pp. 419-432.
4. P.J. Meschter, P.S. Pao, R.J. Lederich, and J.E. O'Neal, "Effect of Compositional Variations on the Microstructures and Properties of Rapidly Solidified Al-3Li Alloys", in Rapidly Solidified Powder Aluminum Alloys, ed. by E.A. Starke, Jr., and M. Fine (ASTM, Philadelphia, in press).
5. P.J. Meschter, R.J. Lederich, and J.E. O'Neal, "Rapid Solidification and P/M Processing of Aluminum-Lithium Alloys", in Modern Developments in Powder Metallurgy, Vols. 15-17 (Metal Powder Industries Federation, American Powder Metallurgy Inst., Princeton, NJ, in press).
6. N.J. Grant, S. Kang, and W. Wang, "Structure and Properties of Rapidly Solidified 2000 Series Al-Li Alloys", in Aluminum-Lithium Alloys, ed. by T.H. Sanders, Jr., and E.A. Starke, Jr. (The Metallurgical Society of AIME, Warrendale, PA, 1981), pp. 171-188.
7. K.K Sankaran and N.J. Grant, "Structure and Properties of Splat Quenched 2024-Aluminum Alloy Containing Lithium Additions", in Aluminum-Lithium Alloys, ed. by T.H. Sanders, Jr., and E.A. Starke, Jr. (The Metallurgical Society of AIME, Warrendale, PA, 1981), pp. 205-227.

8. I.G. Palmer, R.E. Lewis, and D.D. Crooks, "The Design and Mechanical Properties of Rapidly Solidified Al-Li-X Alloys", in Aluminum-Lithium Alloys, ed. by T.H. Sanders, Jr., and E.A. Starke, Jr. (The Metallurgical Society of AIME, Warrendale, PA, 1981), pp. 241-262.
9. I.G. Palmer, R.E. Lewis, D.D. Crooks, E.A. Starke, Jr., and R.E. Crooks, "Effect of Processing Variables on Microstructure and Properties of Two Al-Li-Cu-Mg-Zr Alloys", in Aluminum-Lithium Alloys II, ed. by T.H. Sanders, Jr., and E.A. Starke, Jr. (The Metallurgical Society of AIME, Warrendale, PA, 1984), pp. 91-110.
10. R.J. Kar, J.W. Bohlen, and G.R. Chanani, "Correlation of Microstructures, Aging Treatments, and Properties of Al-Li-Cu-Mg-Zr I/M and P/M Alloys", in Aluminum-Lithium Alloys II, ed. by T.H. Sanders, Jr., and E.A. Starke, Jr. (The Metallurgical Society of AIME, Warrendale, PA, 1984), pp. 255-285.
11. A.E. Vidoz, D.D. Crooks, R.E. Lewis, I.G. Palmer, and J. Wadsworth, "Ultra-Low Density, High Modulus and High Strength RSP Aluminum-Lithium-Beryllium ALloys", in Rapidly Solidified Powder Aluminum Alloys, ed. by E.A. Starke, Jr., and M. Fine (ASTM, Philadelphia, in press).
12. L.F. Mondolfo, Aluminum Alloys: Stucture and Properties (Butterworths, London, 1976).
13. D.B. Williams, "Microstructural Characteristics of Al-Li Alloys", in Aluminum-Lithium Alloys, ed. by T.H. Sanders, Jr., and E.A. Starke, Jr. (The Metallurgical Society of AIME, Warrendale, PA, 1981), pp. 89-100.

□

Production of fine, rapidly solidified aluminium–lithium powder by gas atomisation

R A RICKS, P M BUDD, P J GOODHEW,
V L KOHLER and T W CLYNE

The first three authors are in the Department of Materials Science and Engineering at the University of Surrey, Guildford, UK. VLK and TWC are in the Department of Metallurgy and Materials Science. University of Cambridge, UK

Synopsis

This paper is concerned with the production of aluminium alloy powder (composition Al-2.5%Li-1.2%Cu-0.6%Mg-0.12%Zr) by the process of high-pressure gas atomisation. It has been shown (1) that such a processing route offers unique possibilities for the production of bulk quantities of rapidlysolidified material, provided the particle size is kept below ≈25μm. Results are presented, relating to the operation of a twin-fluid inert gas atomiser at the University of Surrey, which indicate that such aims are entirely feasible. Solidification conditions pertinent to such particles are considered and experimental evidence in support of the expected solute segregation is presented.

Introduction

The advantages to be gained by the choice of a processing route involving the rapid solidification of the required alloy from the molten state are now well documented. From a technological viewpoint, the most important of these advantages are the removal of macrosegregation, reduction in scale of microsegregation and possible enhancement of solute levels by the extension of solid solubility. These advantages are now being exploited in many high strength aluminium-based alloys, especially those hoped to meet the requirements for high strength at elevated temperatures.

Rapid solidification of aluminium-lithium alloys has yet to gain the same momentum as that observed for the high temperature alloys. This is partly associated with the extreme reactivity of the molten alloy, together with the significant levels of improvement already achieved via ingot technology routes. However, it is expected that significant strength improvements should be attainable using rapid solidification processing, since alloys containing higher lithium contents may be employed, and the chemical homogeneity of the microstructure further improved.

The choice of processing route used to achieve rapid solidification conditions for aluminium based alloys is now limited to gas-atomisation for all practical applications. It has now been convincingly demonstrated (e.g. 1-2) that atomisation involving the use of inert gas at high pressure can achieve heat extraction rates comparable with established rapid solidification techniques. However, in order to obtain such conditions the resultant particle size must be very fine (<25μm) and work is currently in progress in many research centres towards this goal. This necessity for ultrafine alloy powders is associated with the combined needs of achieving a high melt undercooling prior to nucleation, together with high heat extraction rates via convective heat flow to the atomising gas. This combination of conditions ensures high crystal growth velocities with a concomitant improvement in the solidification microstructure.

In the present paper the operation and performance of a high pressure inert gas atomiser is described and results are presented relating to the production of ultrafine Al-2.5%Li-1.2%Cu-0.6%Mg-0.12%Zr alloy powder. Chemical segregation of the alloying elements, including lithium, is examined and the results obtained are compared with some quantitative theoretical predictions.

Experimental procedure

The equipment used for the production of aluminium-alloy powder for the present investigation is shown schematically in Figure 1.

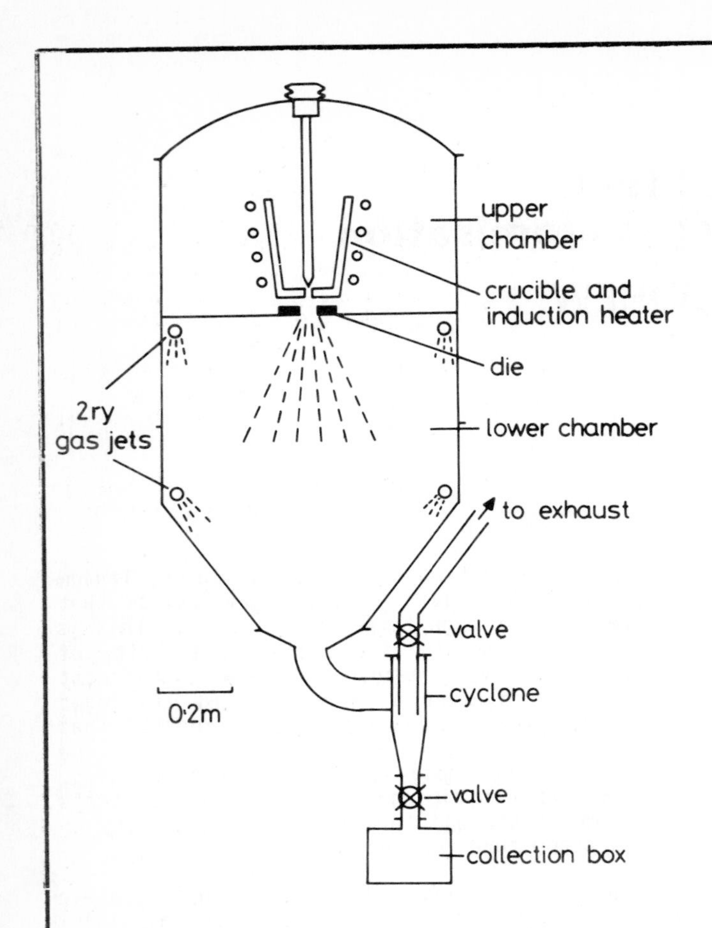

Figure 1
Schematic diagram showing construction of the high pressure inert-gas atomiser used in the present investigation

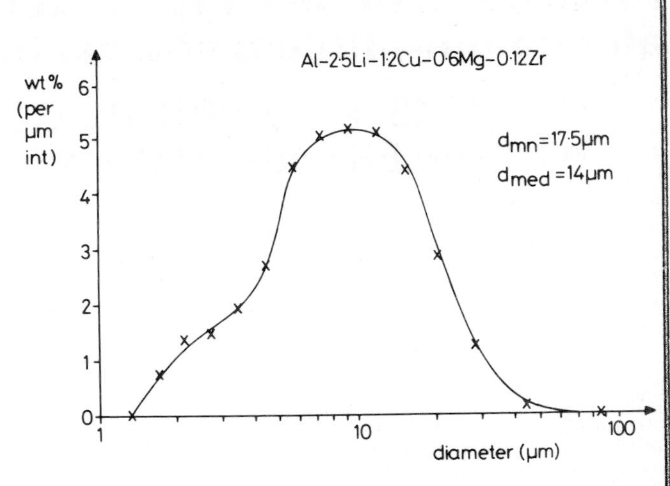

Figure 2
Particle size distribution obtained from powder produced by atomisation at ≈4 MPa inlet gas pressure

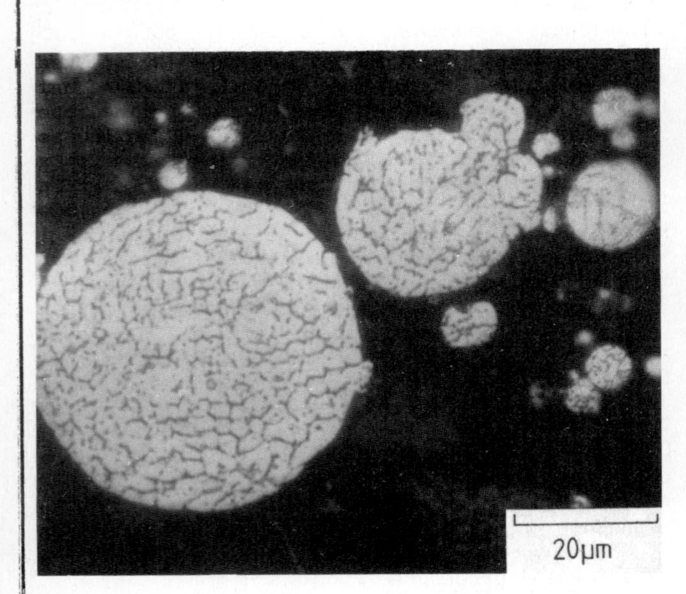

Figure 4
Light optical micrograph showing the cellular solidification structure

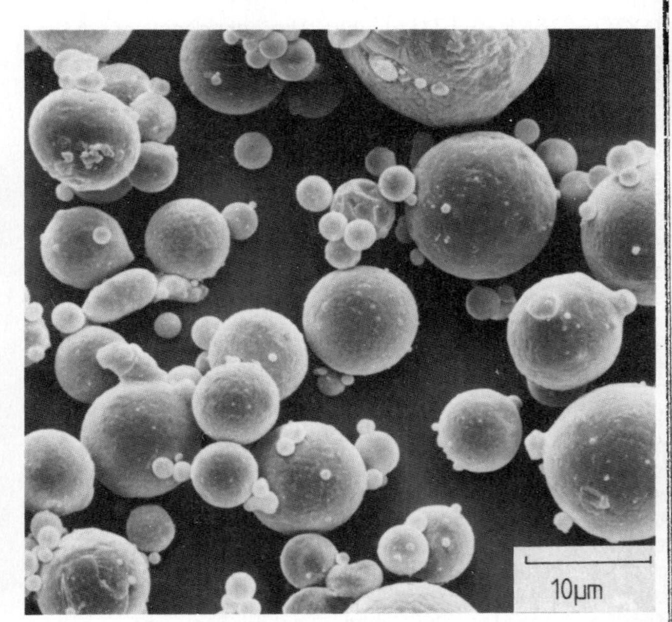

Figure 3
Scanning electron micrograph showing morphology of atomised powder

All melting and atomising operations were carried out in a stainless steel chamber which could be initially evacuated and back-filled with argon. A partition separates the basic chamber into a melting zone and a lower, atomising chamber, thus enabling atomisation to be carried out with a melt overpressure if required (see below). The actual atomising die used in the equipment is not designed to incorporate ultrasonic pressure or shock wave modulations into the atomising gas jets, as is frequently claimed in other high pressure atomisation units (3,4). Instead, the die used for this investigation was deliberately designed to reduce gas/die frictional effects as much as possible, thus ensuring that maximum gas exit velocities were achieved with minimum gas pressures. It should be noted that the gas ducts within the die were not of the converging/diverging design necessary to achieve supersonic gas velocities. Thus operating gas pressures were limited to those which simple mass balance calculations indicate should provide gas exit velocities approaching the transonic regime (i.e. compressible fluid flow). For the present die configuration these operating pressures are in the range 3-4 MPa which are modest compared to some recent investigations involving the use of atomising gas pressures up to 18 MPa (5).

Under most atomising conditions a back pressure could be recorded in the die caused by the direct impingement of the atomising gas jets when no melt stream was present. This effect has been noticed previously (6,7) and may lead to the prevention of the attainment of stable atomisation conditions. To overcome this problem most atomisation was carried out using a slight positive melt overpressure to compensate for the back pressure. This procedure enables atomisation to be performed at any desired gas pressure with a degree of control over the melt flow rate through the nozzle.

The atomised powder is swept from the chamber into the collection cyclone with the atomising gas. During the time spent in the cyclone the powder encounters an atmosphere containing approximately 2% oxygen, fed into the base of the atomiser via a distributor, such that the resultant powder arrives in the collection box in a safely oxidised condition. It should be appreciated that this design allows for the maximisation of heat transfer to the atomising gas during the solidification stage since the droplet is essentially oxide free. The final, controlled oxidation of the solid powder enables the product to be handled safely under normal laboratory conditions. This facility is optional and oxide-free powder may be produced if desired.

Powder characterisation was performed using a Malvern laser granulometer with the sample dispersed in filtered tap water containing a suitable dispersant. Particle morphology was examined by standard scanning electron microscopy techniques whilst thin foil transmission electron microscope specimens of the powder were prepared by the use of an ultramicrotome.

Results

Particle size and morphology

During the initial transient conditions at the start of any atomising run a small amount of flake product is formed which can be easily removed by sieving the sample through a 200 μm mesh sieve. Once this transient has subsided steady state atomisation follows and the bulk of the fine powder observed in the product is formed during this period. A typical size distribution for material produced at higher gas pressures is shown in Figure 2, with the distribution having a median diameter of 14 μm. It should be noted that this would appear to be the finest Al-Li alloy powder yet reported.

The morphology of the powder is illustrated in Figure 3 which shows a scanning electron micrograph of the specimen material. The particles are highly spherical indicating that solidification has taken place during flight. A small but significant satellite population could be detected and although these fine particles are normally undesirable, for the purposes of this present study they allowed the collection of some of the very fine material that would otherwise pass through the collection system.

Particle microstructure

The vast majority of particles exhibited a cellular microstructure easily observable by optical microscopy, as shown in Figure 4. The scale of this cell structure varied with particle size as expected from heat flow calculations (1) and typical cell spacings ranged from ≈5μm for 100μm diameter particles to ≈0.2μm for a 5μm diameter particle. The finer particles were observed more easily using transmission electron microscopy of ultramicrotomed sections. This method of specimen preparation was found to be most suited to the study of <5μm particles, and although severe deformation was inevitably induced into the sections, the method has the advantage of preserving unaltered any chemical segregation present. Figure 5 shows a transmission electron micrograph illustrating the typical cellular structure observed in a 5μm particle. For the very fine particles (<2-3μm) totally segregation free structures were observed (Figure 6). This solidification structure has been associated (8) with the stabilisation of the advancing growth interface by the action of capillarity forces, thus forming a planar interface. Once stabilised in this manner, the rapid crystal growth velocities associated with the high heat extraction rates experienced by such fine particles will almost certainly lead to solute trapping (for this particular alloy).

Chemical segregation and detection

The number of techniques available for the detection of lithium segregation is limited and in the present investigation only two techniques, namely electron energy loss spectroscopy (EELS) and laser-induced ion mass analysis (9) (LIMA) will be considered. The former technique,

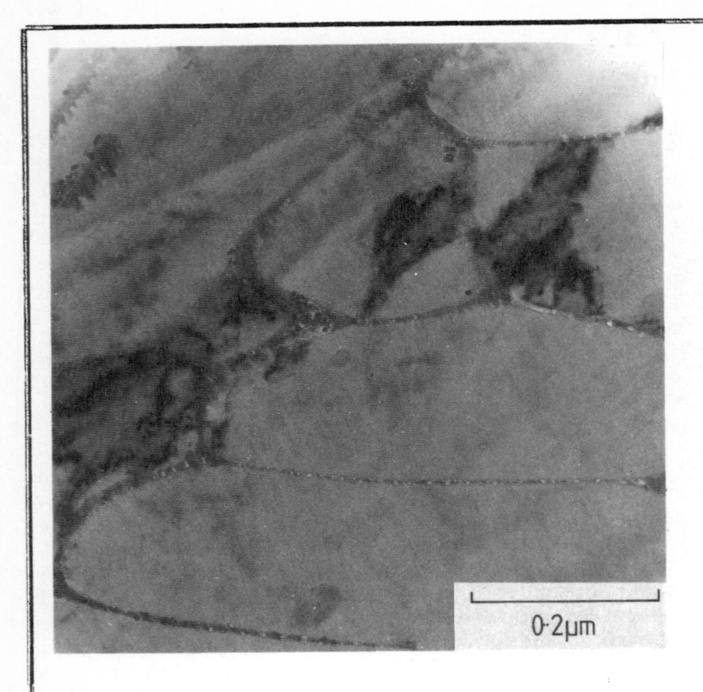

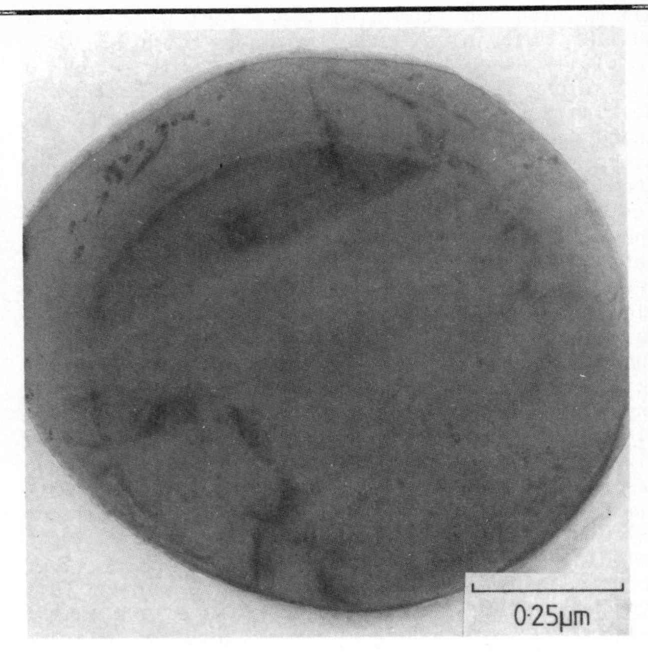

Figure 5
Transmission electron micrograph showing cellular solidification structure in a 5 µm diameter particle

Figure 6
Transmission electron micrograph showing a 1 µm powder section which is completely free from contrast associated with solute segregation

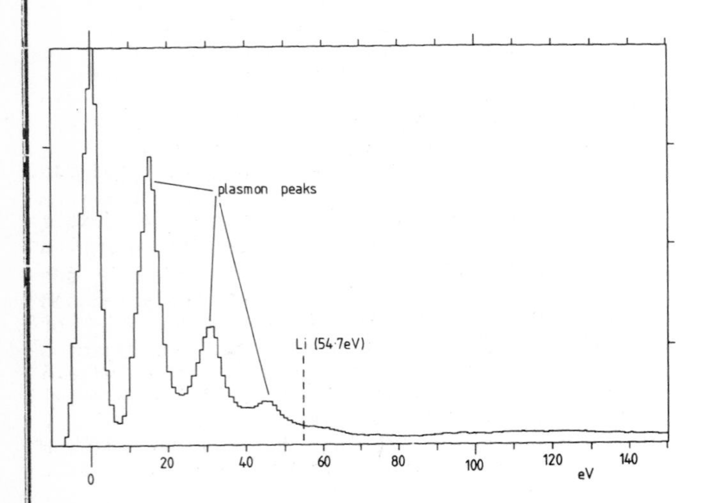

Figure 7
Electron energy loss spectrum illustrating the proximity of the plasmon peaks to the Li edge

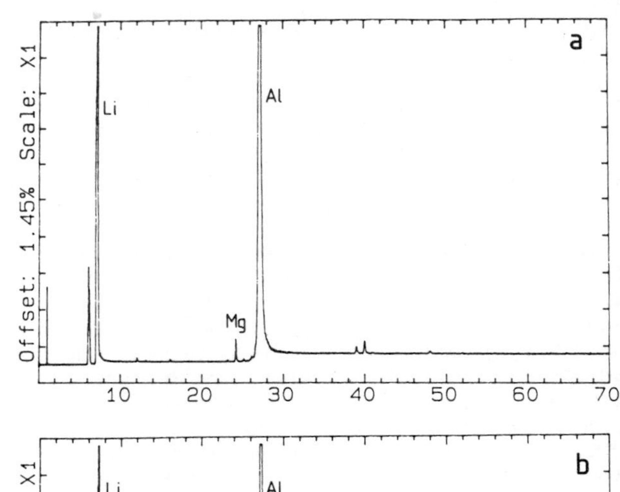

Figure 8
L.I.M.A. spectra obtained from cell centre (a) and intercellular region (b)

although having an adequate spatial resolution for the present purposes, suffers from the major problem of multiple scattering of the incident electrons in all but extremely thin sections (typically <30-50 nm). In addition, the detection of lithium is complicated further by the presence of the plasmon peaks, the third of which usually masks any lithium edge which might be present (Figure 7).

The technique of laser-induced ion mass analysis does not suffer from these specimen preparation problems and, although the spatial resolution was limited (1-2 µm), LIMA proved capable of detecting lithium segregation, especially in the larger particles. LIMA involves the use of a high powdered laser to evaporate a small region of the specimen and form a plasma. The ions from the plasma are then analysed in a time of flight mass spectrometer. Thus the technique has the advantage of simultaneously detecting all elements present in the plasma although to date quantification of an individual spectrum is dubious. Typical LIMA spectra obtained from a 50 µm particle are shown in Figure 8 for plasma formed from a cell centre (8a) and from an intercellular region (8b). Significant enrichments of lithium and magnesium are detected at the intercellular regions whilst copper and iron (impurity) peaks are present only in Figure 8b. Since the two spectra were obtained at equivalent laser power outputs it is possible to obtain some idea of the degree of segregation by comparing the elemental peak areas between the spectra. The results obtained for the elements Li and Mg are as follows:

$$\text{Li} - \frac{\text{peak area cell centre}}{\text{peak area cell wall}} = 0.73$$

$$\text{Mg} - \frac{\text{peak area cell centre}}{\text{peak area cell wall}} = 0.23$$

These results will be discussed in more detail in the following section.

The heavy atomic number elements such as Cu and Fe (presumably an impurity) could also be detected in the intercellular regions using back scattered imaging in the scanning electron microscope. Figure 9 shows a typical image illustrating the enhanced electron yield from the intercellular regions caused by the presence of these segregants.

The relatively poor spatial resolution associated with LIMA precludes the use of this technique for particles less than 50 µm in diamter since the cell spacing is less than the size of the ionised crater formed. It would therefore seem that for the finer particles EELS is the only analytical technique capable of detecting lithium with sufficiently high spatial resolution. Work is currently in hand to improve specimen preparation techniques to enable this examination to take place. Heavy element (Cu and Fe) segregation is readily detected in all particles having a cellular structure using energy dispersive spectrometry (EDS).

Discussion

Solidification condition

Heat transfer during atomisation is a highly complex phenomenon and at best only qualitative predictions may be made. If quasi steady-state conditions are assumed it has been shown that heat transfer is highly dependent on particle size and less dependent on relative velocity between gas and particle and atomising gas (1). However an order of magnitude improvement in heat transfer coefficient may be expected by the use of helium as the primary atomising gas and for this reason helium has been used exclusively during this investigation.

Once a value of the heat transfer coefficient has been established, the solidification time may be estimated as follows. For a superheated melt the total heat to be removed is given by

$$\Delta H = \frac{\pi d^3}{6} (C_L \; \Delta T_S + \Delta H_f)$$

where d = particle diameter
C_L = specific heat of liquid
ΔT_S = superheat
ΔH_f = latent heat of fusion

This heat removal will take place over a time interval t_f given by:

$$t_f \approx \frac{\Delta H}{h_i \; \Delta T_i \; \pi d^2} = \frac{d(C_L \; \Delta T_S + \Delta H_f)}{6 \; h_i \; \Delta T_i}$$

where h_i = interfacial heat transfer coefficient
ΔT_i = temperature difference between droplet (assumed isothermal) and gas.

The value of t_f as a function of droplet diameter is given in Figure 10 for the case of He and for comparison, for Ar as the atomising gas. It is difficult to estimate the values of the relative velocity (u) between the droplet and gas but it should be noted that the value of t_f is weakly dependent on this value and typical solidification times for a 10µm droplet would be 30 µs in helium and 200 µs in argon (assuming $10 \text{ ms}^{-1} < u < 100 \text{ ms}^{-1}$). For a larger 100 µm droplet solidification would take place over a few milliseconds.

Knowledge of the solidification time allows some idea to be obtained of the expected departure (if any) from equilibrium conditions at the solidification interface. Detailed analysis of the solidification history required the combined solution of the various transport equations involved. These include partitioning of solute at the interface, the dependence of growth velocities on both interfacial temperature and composition and the diffusive redistribution of heat and solute. It is clear that any such analysis will be highly complex, particularly since none of the equations have analytical solutions under the boundary conditions concerned. Despite this some simplified analyses are possible. For the limiting case of total equilibrium partitioning at

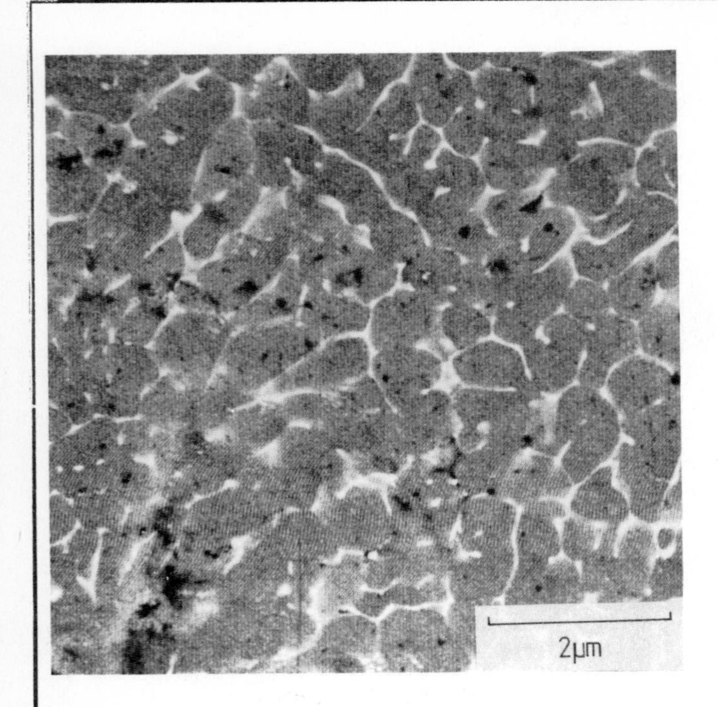

Figure 9
Back scattered electron image showing segregation of heavy atomic number elements to intercellular regions

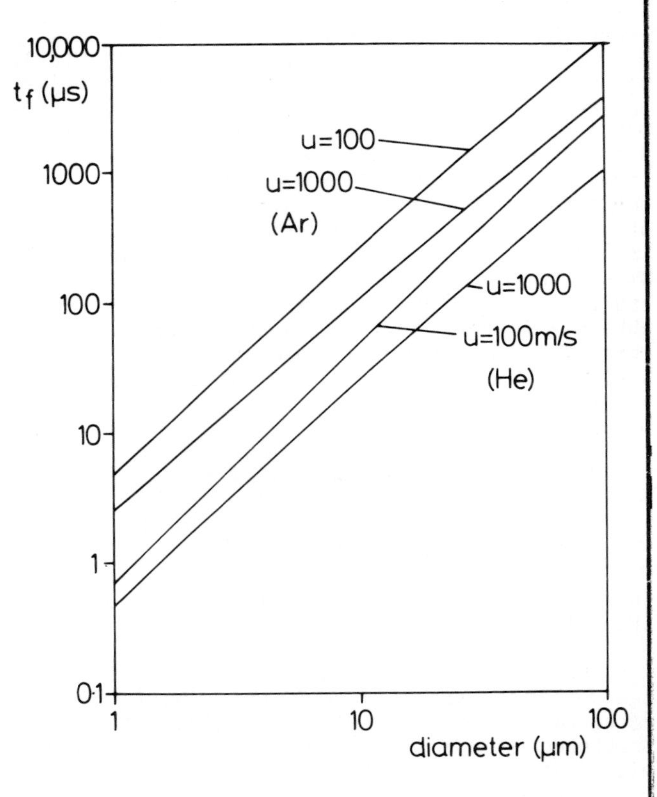

Figure 10
Predicted freezing time as a function of particle diameter

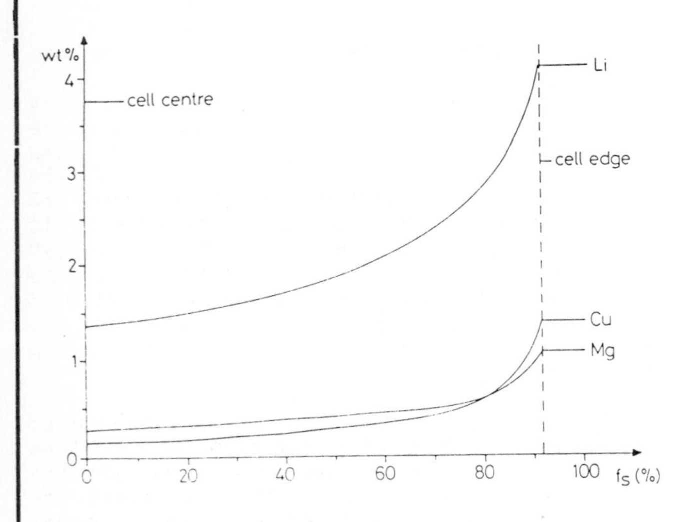

Figure 11
Calculated (Scheil) composition profiles across a cellular solidification structure

the interface but with no back diffusion in the solid (which would remove any segregation effects) the Scheil equation is valid and may be used to predict expected solute profiles arising from solidification. It is pertinent to examine the validity of the Scheil equation for the two growth morphologies observed in the powder sections, i.e. cellular growth and planar growth.

a) cellular growth

An estimate of the lateral growth rate of the cells may be obtained from dividing the cell spacing by the solidification time. Thus for a 100 μm diameter particle

$$u \text{ (growth velocity)} = \approx 10^{-3} \text{ m s}^{-1}$$

At the solidification temperature the extent of the solute enriched layer in the liquid may be estimated by D_L/u where D_L is the diffusivity of the solute in the liquid alloy. Typical values of D_L yield distances ≈ 10 μm and comparison with cell spacings of 4 μm indicate that during lateral growth the liquid is homogeneous - as required for any analysis using the Scheil equation. Back diffusion in the solid may be shown to be negligible in this case and thus the predictions of the Scheil analysis are expected to hold for composition profiles across the cells. The above analysis may be performed with equivalent results for all particle diameters exhibiting cellular microstructures (i.e. down to 5 μm).

b) planar growth

Planar growth induced by absolute stability (i.e. capillarity stabilised interface) was observed in all particles exhibiting a sectional diameter of less than 2-3 μm. Growth velocities associated with this form of solidification will be much higher than that for cellular solidification and calculations indicate that velocities should exceed 1 m s^{-1}. The extent of the solute enriched layer at the interface may again be estimated by the value D_L/u, and for the high velocities pertaining during planar growth this layer is expected to be only ≈10 nm wide. Thus solute diffusion in the liquid may not be fast enough to ensure a totally homogeneous liquid and the Scheil analysis will not apply. One outcome of this solute enriched region, which will travel through the material with the interface, is that the interface is constantly in contact with a solute concentration in excess of that normally present in the liquid. The expected outcome of these phenomena will be the formation of a solid of uniform concentration (probably close to the overall alloy composition) even if equilibrium partitioning takes place at the interface and no solute trapping occurs.

Comparison of experimental and predicted segregation levels from Scheil analysis

The Scheil equation may be written:

$$C_s^i = k\, C_o\, (1-f_s)^{(k-1)}$$

where C_s^i = interfacial solid composition (i.e. instantaneous composition of solid formed)
C_o = average alloy composition
F_s = fraction of solid
k = partition coefficient

and may be used to predict the solid composition profiles across the cellular structure as described above. Figure 11 shows the predicted solute profiles across a cell with the cell boundaries being defined at the (solid) lithium concentration equivalent to a liquid of eutectic composition. The values of the partition coefficient used for these plots are as follows:

Li K = 0.55 (10)

Mg K = 0.49 (11)

Cu K = 0.13 (11)

The curves predict an intercellular region having ≈10% of the width of the cells. For a cell spacing of ≈5 μm (i.e. 50 μm diameter particle) the width of the intercellular region would therefore be ≈0.5 μm which is slightly below the resolution of the LIMA technique. Thus any analysis of the intercellular region would be affected by dilution effects since some of the cell would also be ionised. However, the predicted ratios of solute at the cell centre and edges compare favourably with those predicted, i.e.:

Li predicted ratio = 0.35

measured ratio = 0.73

Mg predicted ratio = 0.25

measured ratio = 0.23

The lack of correlation between the measured and predicted lithium ratios could indicate solute enrichment over and above that expected although in view of the aforementioned diffulties concerning spatial resolution no definite conclusions are possible.

Conclusions

The results presented in this paper confirm the possibility of production of fine aluminium-lithium alloy powder by high pressure inert gas atomisation and using optimised die designs it is hoped that median diameters below 10μm should be possible with this production route. In the present study the atomised powder was deliberately oxidised in the solid state to avoid complications with handling the product in the laboratory. However, in view of the problems often encountered in compacted powder components, caused by oxide films present on the powder feedstock, it is hoped that future atomised powder can be made essentially oxide free and handled, prior to canning or compaction, under inert gas conditions.

In any study concerning the production of advanced alloys using rapidly solidified powder processing routes, it is essential to investigate the structure of the as atomised powder such that microstructural development during consolidation may be monitored. In this paper the microstructure of the powders was shown to consist mainly of a cellular solidification structure for all particles above ≈3 μm. Below this, segregation free structures were observed indicative of capillarity induced interfacial stability together with possible solute trapping or solute enriched liquid at the interface. The cellular structure has been shown to produce equilibrium segregation of the major alloying elements (Li, Mg) in the coarse particles (i.e. 5 μm) although it is possible that enhanced solute compositions in the body of the cells may result from the more rapid freezing of the finer particles. Further work is in progress to determine solute profiles in the fine powder which forms the bulk of the material produced at the higher gas pressures.

References

1. T.W. Clyne, R.A. Ricks and P.J. Goodhew, Int. Journal of R. Solid., 1 (1984) 59

2. T.W. Clyne, R.A. Ricks and P.J. Goodhew, Proceedings of 5th Conf. on Rapidly Solidified Metals (RQ5), Wurzburg (1984) eds. Steeb & Warlimont, p.903

3. V. Anand, A.J. Kaufman and N.J. Grant in "Rapid Solidification Processing, Principles and Technologies", eds. R. Mehrabian, B.H. Kear and M. Cohen, Claitors Baton Rouge, 1980, p.273

4. E.O. Nilsson, S.I. Nilsson and E.G. Hagelin, U.S. Patent No. 2997245 (1961)

5. I.E. Anderson, J.D. Ayers, W.P. Robey and R.G. Hughes, Submitted to Met. Trans.B (1984)

6. M.J. Cooper and R.F. Singer, Proceedings of 5th Conf. on Rapidly Solidified Metals (RQ5) Wurzburg (1984) eds. Steeb and Warlimont

7. R.A. Ricks and T.W. Clyne, Submitted to J. Mat. Sci. Letters

8. W.J. Boettinger, S.R. Coriell and R.F. Sekerka, Mat. Sci. Eng., 65 (1984) 27

9. M.J. Southon, M.C. Witt, A. Harris, E.R. Wallach and J. Myatt, Vacuum, 34 (1984) p.903

10. A.J. McAlister, Bull. Alloy Phase Diagrams, 3 (1982) 177

11. L.F. Mondolfo in "Aluminium Alloys - Structure and Properties", Butterworths, London

Microstructural characterization of rapidly solidified Al–Li–Co powders

F H SAMUEL

The author is Visiting Professor at the Ecole des Mines, Institut National Polytechnique de Lorraine, Parc de Saurupt, Nancy, France

SYNOPSIS

A study of the combined effect of alloying elements and melt superheat has been carried out on the as-solidified structure of rapidly solidified Al-Li-Co powders. Three alloys, viz., Al-3wt% Li, Al-3wt% Li-0.4wt% Co and Al-3wt% Li-0.8wt% Co were chosen and the liquid melt in each alloy atomized from the temperatures 1073 and 1173 K, using the centrifugal atomization technique. The microstructural characterization has been done using light and transmission electron microscopy. The effect of ageing at 473 K for times up to 100 h has also been investigated.

INTRODUCTION

Rapid solidification of powders is an attempt to maximize the physical and chemical properties through control of the distribution of solute elements. The homogeneity of the powders can be explained in terms of extremely small casting of the order of 4 µg. This mass is allowed to cool rapidly from a molten state at a rate of 10^4 to 10^6 $K.s^{-1}$. These rates are many orders of magnitude greater than those employed in conventional casting.

The present investigations were aimed at studying the combined effect of alloying elements and melt superheat on the as-solidified structure. Therefore, three alloys, viz., Al-3wt% Li, Al-3wt% Li-0.4wt% Co and Al-3wt% Li-0.8wt% Co were selected and two initial pour temperatures, viz., 1073 and 1173 K, were employed for each alloy.

EXPERIMENTAL PROCEDURE

The alloys used in the present study were prepared from high purity metals (99.9 %) under an inert atmosphere of argon. Rapidly solidified powders were produced by the conventional centrifugal atomizing process, with helium used as the cooling medium to obtain high speed convective cooling. The two batches of powders for each alloy were carefully screened and separated into different fractions according to their size. The yield of 80-100 µm sized particles was about 50-61 % in high superheat (HS) alloys i.e. atomized from 1173 K, and 35-41 % in low superheat (LS) alloys i.e. atomized from 1073 K. On the other hand, the yield of particles of 50-80 µm size was only ∽10 % in HS alloys and ∽ 24 % in LS alloys.

The microstructures of loose powders were examined by light and transmission electron microscopy using similar techniques mentioned elsewhere (1).

RESULTS & DISCUSSION

The microstructure of Al-3wt% Li powders reveals well defined colonies of dendrites for particles ∽ 200 µm as shown in Fig. 1a. These dendrites are comprised of cylindrical arms probably elongating along the three equivalent ⟨100⟩ directions. In this case the solidification is presumably proceeding as an advancing front that radiates from the initial point of nucleation. Coarse particles > 200 µm exhibit a mixed dendritic (area A)-equiaxed type structure (area B). The equiaxed structure was characterized by the absence of directional growth in a polished section as can be seen in Fig. 1b.

A direct relation between secondary dendrite arm spacing (SDAS) and powder particle diameter is obtained for Al-3wt% Li alloy as shown in Fig. 2. Each reading represents the average of the values for both HS and LS for a given particle diameter. In the atomized condition, Webster et al (2) have reported that binary Al-Li alloys with Li content in the range of 3 to 5 % and particle dimensions ranging from 5 to 200 µm, show a dendritic structure. Their powders were

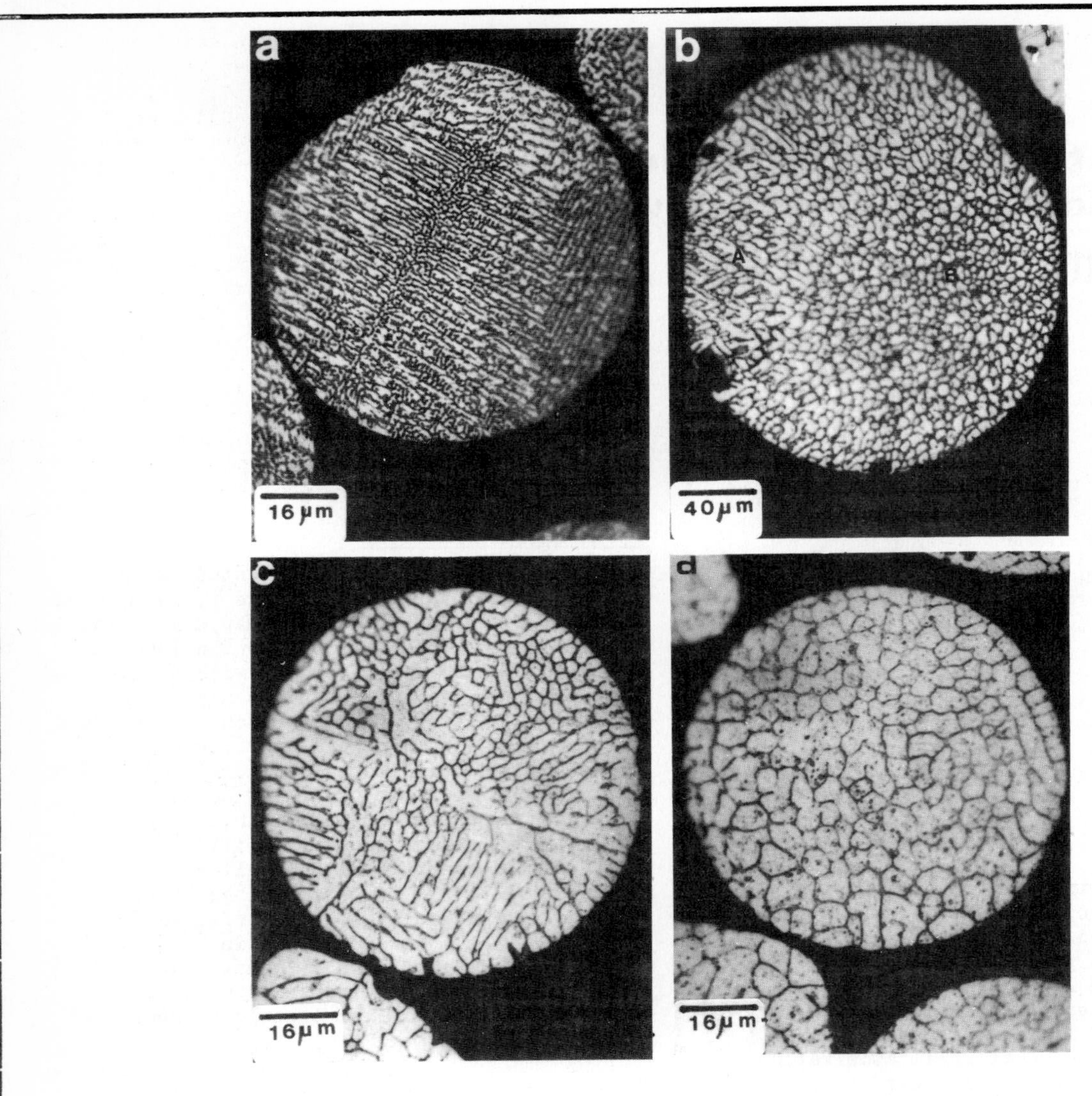

Fig. 1 Light microstructures of as-quenched powders: (a) and (b) Al-3wt% Li, (c) Al-3wt% Li-0.4wt% Co, (d) Al-3wt% Li-0.8wt% Co.

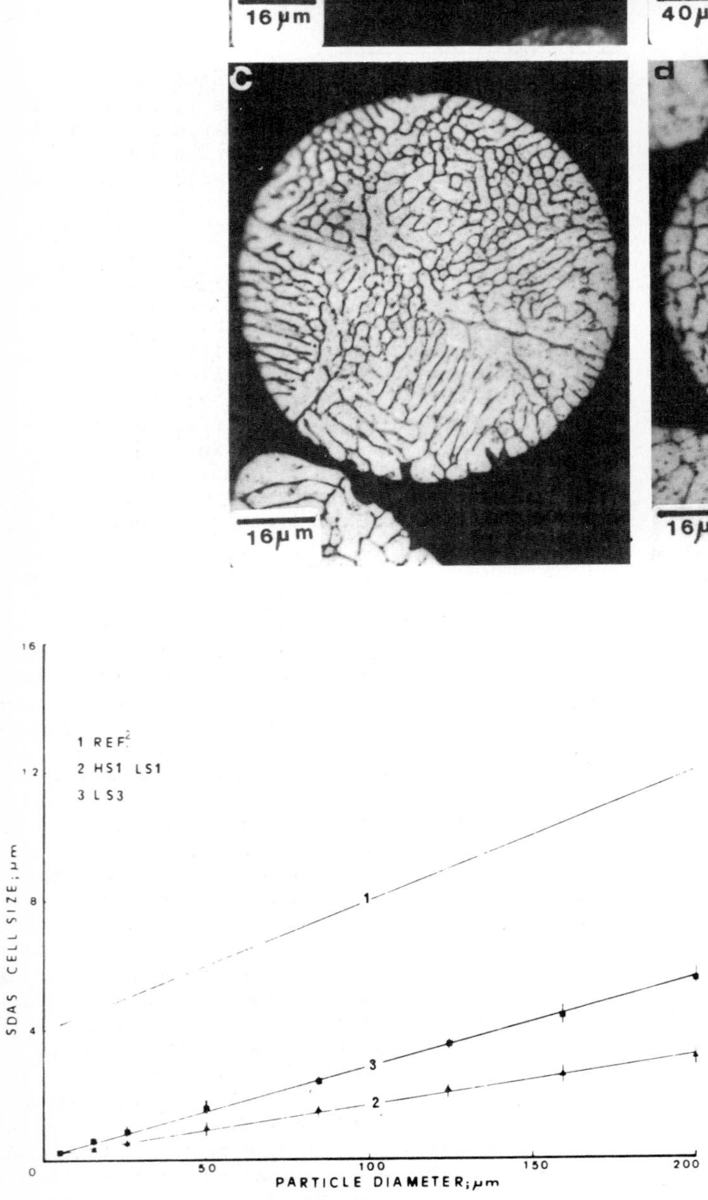

1 REF.[2]
2 HS1 LS1
3 LS3

SDAS CELL SIZE; μm
PARTICLE DIAMETER; μm

Fig. 2

Variation in secondary dendrite arm spacing (SDAS) and cell width as a function of powder particle diameter deduced from light microscopy.
HS1/LS1: Al-3wt% Li, LS3: Al-3wt% Li-0.8wt% Co.

produced by argon gas atomization (GA method) yielding a cooling rate in the range of 10^2 to 10^4 Ks^{-1}. The results of Webster et al on Al-2.7% Li are superimposed in Fig. 2. As can be seen, their values are almost four times the present case. The difference between the present results and theirs is explicable in terms of the gas thermal conductivity (1.7×10^{-2} and 15×10^{-2} W/mK for argon and helium, respectively), gas velocity and gas-particle relative velocity. From the plot of Matyja et al (3), which relates the measured dendrite arm spacing to the average cooling rate in binary Al-alloys, the average cooling rate for the present HS and LS alloys is determined to be about 10^4 -10^6 Ks^{-1}.

In the Al-3wt% Li-0.4wt% Co powder with a particle diameter larger than 25 µm, a large proportion of the microstructure is cellular, Fig. 1c. Also, in the Al-3wt% Li-0.8wt% Co alloy quenched from 1073 K, LS powder, the cellular structure is predominant, irrespective of particle size, Fig. 1d. We are assuming here that all cobalt will be precipitated in the form of Al_9Co_2 which has a density of 3.6×10^3 Kg/m^3 Under this condition the volume fraction of Al_9Co_2 precipitates is about 2.25 %. What needs to be emphasized is that the partitioning of intermetallic Al_9Co_2 particles may reduce the stability of a planar liquid-solid interface, leading to cellular undulations during cooling of a superheated liquid alloy droplet. For this the G_L/R ratio should be high enough to allow transition from dendritic to cellular structure for a given cooling rate.

Fig. 3a is a typical electron micrograph from Al-3wt% Li, HS powder, showing a dendritic-like structure with an arm spacing ∽ 0.3 µm in colonies of ∽ 2-3 µm diameter. Selected area diffraction and dark field microscopy identified the dendritic-like phase as α-Al solid solution with a common orientation throughout each colony. The interdendritic network phase could be indexed on the basis of Al_3Li (δ') phase. Fig. 3b shows a region consisting of images of a primary dendrite with associated secondary dendrite arms. The width of the primary dendrite is approximately 1-1.5 µm. The secondary arms have a spacing of ∽0.5-0.6 µm. The cooling rate estimated from Fig. 3b is about $10^6/10^7$ Ks^{-1} against $10^4/10^5$ Ks^{-1} as determined from Fig. 2.

The microstructure of Al-3wt% Li-0.8wt% Co reveals a cellular structure. Reduction in melt superheat does not appear to exert much influence on either the morphology or the intercellular spacing or on the constitution of the cellular structure for a given particle diameter. Fig. 4a corresponds to the LS powder of this alloy and consists of two high-angle cells where the rows of low-angle cells are only seen in one of them (area A). Since the second phase particles (Al_9Co_2 and Al_3Li) are uniformly distributed throughout the other cell (area B), these oppose the undulating movements of the interfacial boundary between A and B, resulting in a rather smooth boundary. The disappearance of the Al_9Co_2 phase in B, Fig. 4b, is attributed to the higher-angle misorientation existing between A and B.

When the region in front of the advancing cell boundary is free from precipitates, B in Fig. 4c, this contributes to the wavy shape of the cell boundary since each cusp is pinned by the precipitates at the cell wall in A. Fig. 4d is a dark field micrograph produced from B in Fig. 4c, and illustrates the perturbed shape of the A/B boundary and the preferential precipitation of Al_9Co_2 within B. In all cases examined, the particle size of Al_9Co_2 (on the cell wall) is about 30-40 nm whereas that for Al_3Li (throughout the matrix) is not more than 4 nm. A comparison between light and TEM micrographs emphasizes the fact that the cell spacing tends to be an order of magnitude smaller than the cell length e.g. 0.5 µm and 25 µm.

Fig. 5a represents the light micrograph of ∽6 µm particle diameter of Al-3wt% Li-0.8wt% Co, LS powder, showing three nucleation events in the form of a spherical cap occurring at the droplet surface. Each event is comprised of two parts, one being the featureless "white zone" without response to etching (marked H) and the other a cellular zone characterized by radial growth. The microstructural details of the white zone are revealed in Figs. 5b and 5c. The former is a bright field micrograph exhibiting the presence of columnar grains of ∽ 0.2 µm width and ∽1.5 µm length. The latter micrograph indicates the decomposition of these grains giving rise to ultra-fine precipitates (a mixture of Al_3Li and Al_9Co_2 phases) having diameters in the range of 5-10 nm. This structure is developed in two stages: first, the supersaturated solid solution of α-Al is formed from the melt by partitionless solidification, and second, the solid decomposes by precipitation during continuous cooling immediately after solidification (4). The achievable undercooling for this structure is about 360 K (5). Immediately at the end of this micrograph, the interface is seen breaking down, leading to a cellular structure, Fig. 5d.

As the droplet recalesces, the interface velocity and undercooling are reduced. When the interface temperature goes above the temperature required for supersaturation, solute is rejected ahead of the interface. This process breaks the stability of the interface and favours cell formation. Now assuming thermal conditions as in Fig. 6a, the last liquid solidifies at the distance X from the tip given by:

$$X = \Delta T/G_L \quad \text{(where } \Delta T \text{ is the undercooling)}$$

The average cell widths as deduced from Fig. 4 for Al-3wt% Li-0.8wt% Co, LS powders, is about 0.3-0.4 µm. From the dependence of cell size and tip radius (r_t) on the solidification front velocity (R) for rapidly solidified aluminium alloys (6), in the present case, for a cell size of 0.3-0.4 µm, the corresponding value of R is calculated to be about 5×10 μms^{-1}. Taking G x R equal to 5×10^4 Ks^{-1} (average cooling rate), we obtain a value of 100 K/ µm for G_L. The distance X is found to lie in the range 0.5-0.6 µm (Fig. 4). Therefore, the corresponding average tip undercooling is about 55 K.

The solute in the intercellular regions has higher supersaturation. This gives rise to nucleation of Al_9Co_2 in the liquid ahead of the interface. For the trapping of an Al_9Co_2 particle

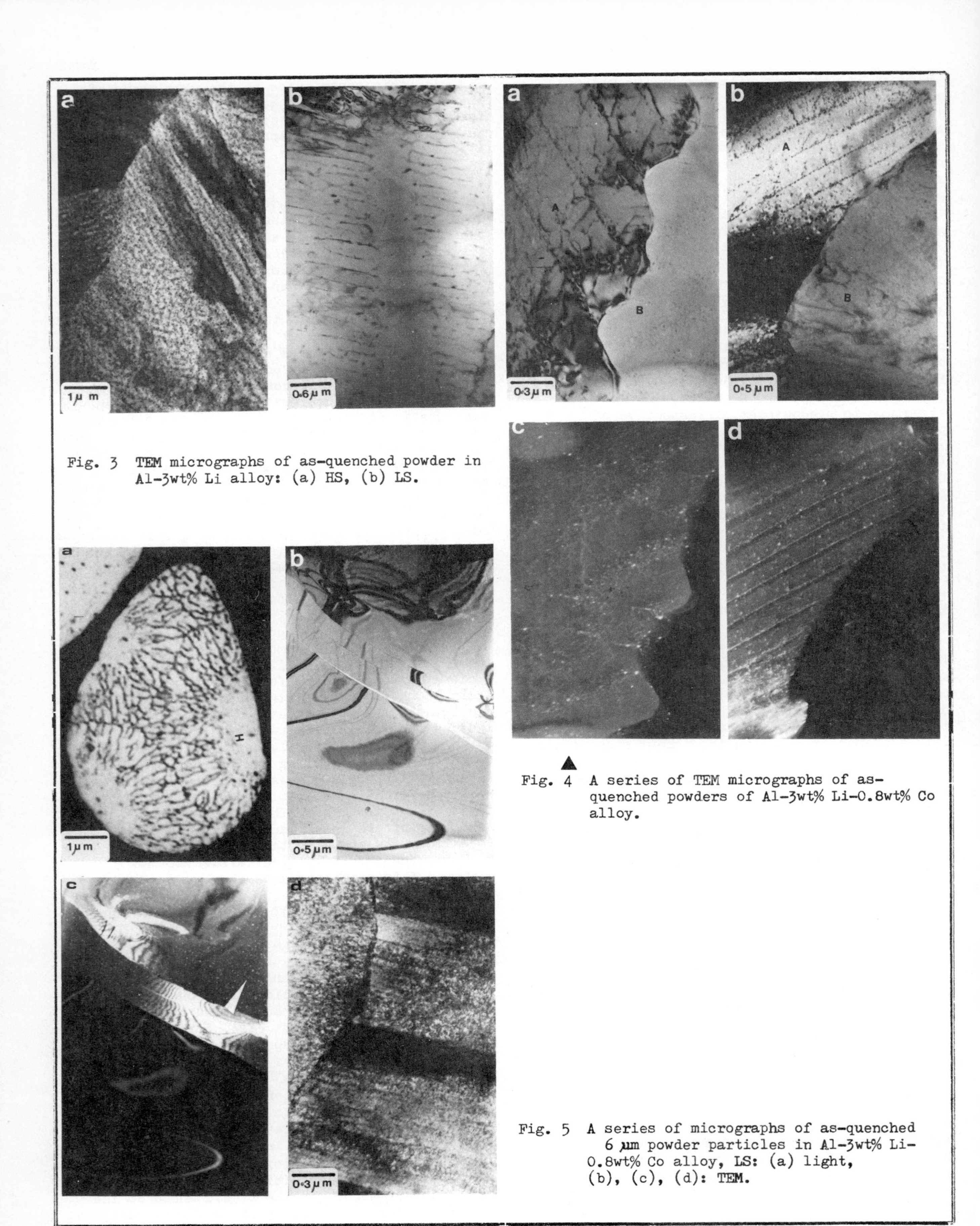

Fig. 3 TEM micrographs of as-quenched powder in Al-3wt% Li alloy: (a) HS, (b) LS.

▲ Fig. 4 A series of TEM micrographs of as-quenched powders of Al-3wt% Li-0.8wt% Co alloy.

Fig. 5 A series of micrographs of as-quenched 6 μm powder particles in Al-3wt% Li-0.8wt% Co alloy, LS: (a) light, (b), (c), (d): TEM.

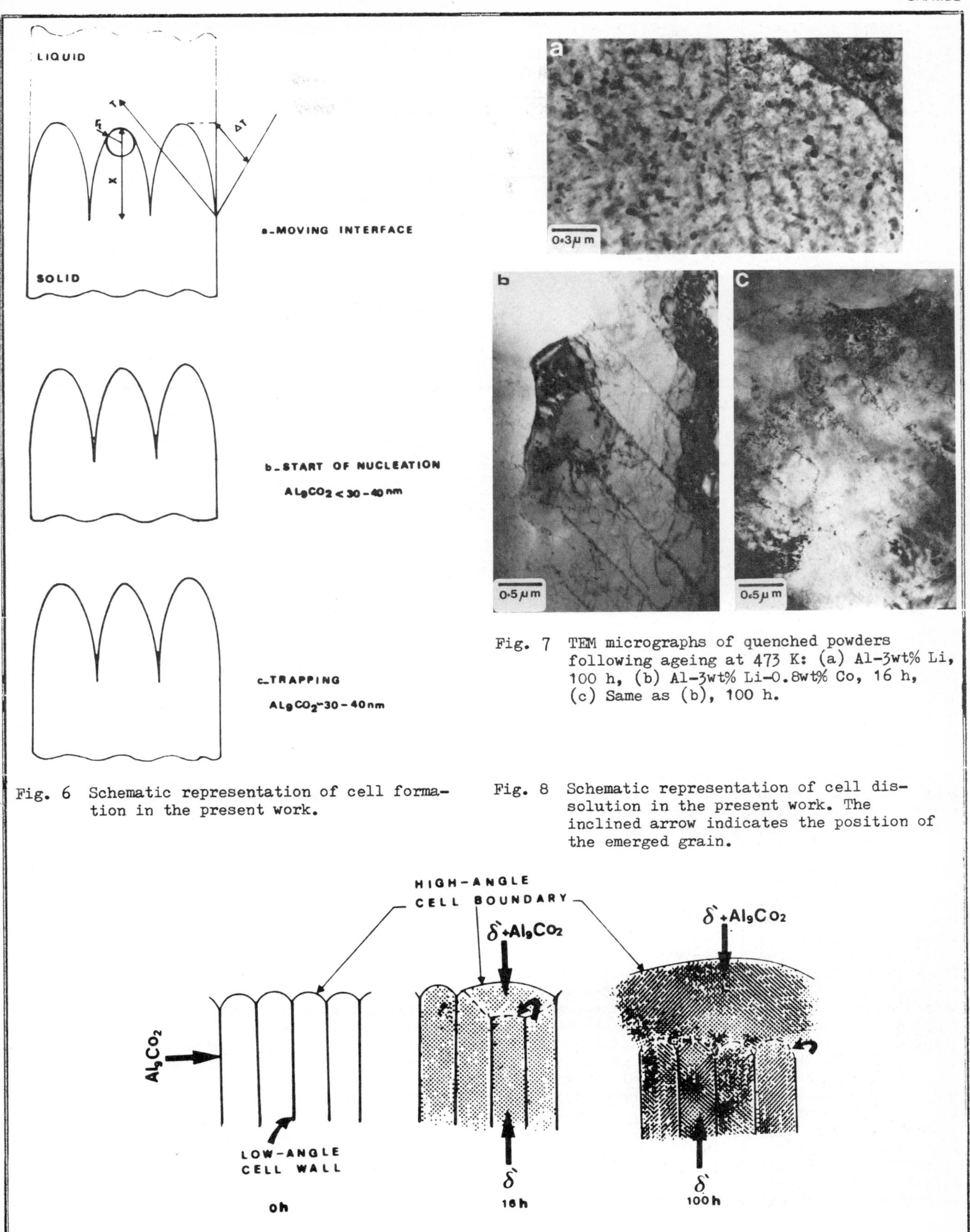

Fig. 7 TEM micrographs of quenched powders following ageing at 473 K: (a) Al-3wt% Li, 100 h, (b) Al-3wt% Li-0.8wt% Co, 16 h, (c) Same as (b), 100 h.

Fig. 6 Schematic representation of cell formation in the present work.

Fig. 8 Schematic representation of cell dissolution in the present work. The inclined arrow indicates the position of the emerged grain.

Table 1 : Variation of 2θ for (hkl)α-Al as a function of ageing treatment

Condition (hkl) -Al	2θ (deg) As-quenched 0 h	Under-aged 1 h	Peak-aged 16 h	Over-aged 100 h	$\Delta\theta^*_{max}$
(111)	45.109	45.034	45.035	45.082	+ 0.043
(200)	52.432	52.450	52.484	52.518	+ 0.086
(022)	77.364	77.389	77.392	77.435	+ 0.071
(113)	94.247	94.275	94.271	94.298	+ 0.051
(222)	99.888	99.920	99.897	99.935	+ 0.047
(004)	124.198	124.216	124.211	124.241	+ 0.043
(133)	148.686	148.770	148.779	148.850	+ 0.164

$\Delta\theta^*_{max} = 2\theta\ (100\ h) - 2\theta\ (0\ h).$

by a moving interface, the nucleated particles must possess a certain critical size for a given interfacial velocity (7). In our case these are, respectively, 30-40 nm and 5 x 10^4 μms^{-1}.

The homogeneous δ' formation during the quenching of Al-3wt% Li, LS and HS powders, is followed by its coarsening throughout the matrix during artificial ageing at 473 K. Fig. 7a shows the persistence of the dendrite structure after an ageing time as long as 100 h. Since the precipitation reaction in Al-3wt% Li-0.4wt% Co and Al-3wt% Li-0.8wt% Co was found to be almost the same, only observations on the latter alloy have been reported here to bring out the comparison between the dendritic and cellular reactions after ageing at 473 K. A heavy precipitation of the δ'-phase with a nearly uniform particle size is occurring on the cell walls as well as within their interiors. After 16 h annealing at 473 K the essentially observed process consists of a continuous dissolution of the Al_9Co_2 phase. This process is accompanied by the emergence of a new grain. The emerging grain grows into the cellular structure by boundary-boundary migration, Fig. 7b. This type of dissolution mechanism is described as cellular dissolution (8). Considering the simultaneous precipitation and coarsening of δ' (Al_3Li) particles, this leads to a slowing down of the movement of the newly formed grain boundary and, in turn, reduces the rate of cellular dissolution as shown in Fig. 7c, following ageing at 473 K for 100 h.

Pawlowski and Truszkowski (8) have reported that the degree of dissolution can be determined from X-ray measurements of the angle Θ_x, according to the formula

$$X = \frac{\Theta_x - \Theta_{min}}{\Theta_{max} - \Theta_{min}}$$

where Θ_{min} denotes the value of Θ in equilibrium at a given ageing temperature, Θ_{max} is the respective value in equilibrium at the dissolution temperature. The change in the angle 2Θ in the course of ageing Al-3wt% Li-0.8wt% Co, LS powder, for times between 1 h and 100 h is shown in Table 1. As can be seen, the maximum deviation in 2Θ for (200)α-Al is obtained for the overaged condition (473 K/100 h as determined from microhardness measurements) and is about 0.86. On the other hand, the increase in the angle 2Θ for (200)α-Al (as an example) is found to be associated with an increase in the relative intensity of (200)α-Al from 44 % to 71 % for the same conditions of treatment.

Based on our metallographic observations, the phase transformation occurring during the ageing of Al-3wt% Li-0.8wt% Co is the outcome of three successive processes: 1) the dissolution of the Al_9Co_2 phase and the formation of grains of solid solution, 2) homogenization of the solid solution and then migration of subgrain boundaries of the newly formed grains, 3) formation of a stable solid and its grain growth. The mechanism of cellular dissolution in the present investigations is depicted schematically in Fig. 8.

CONCLUSIONS

The important microstructural changes as viewed by light microscopy are:

1) The presence of four different structures a) dendritic, b) cellular without secondary arms, c) equiaxed type without directional growth and d) featureless zones with no response to etching.

2) The cooling rate is determined to lie in the range of 10^4 -10^6 Ks^{-1} for a powder particle size ranging between 15-200 µm.

Based on TEM studies, it is found that:

3) The dendritic structure is predominant in Al-3wt% Li regardless of the particle size whereas low-angle cells with perturbed interfaces comprise the main structure observed in Al-3 wt% Li-0.8wt% Co.

4) In ultra-fine powder ∽ 6 µm, zones of supersaturation formed by the partitionless homogeneous mechanism are seen, comprising of elongated columnar grains.

5) From interdendritic and intercellular spacing measurements, the highest achievable cooling rate can reach 10^6 /10^7 Ks^{-1}.

6) Ageing of quenched powders at 473 K for times up to 100 h shows the persistence of the dendritic structure in Al-3wt% Li powders and the dissolution of cellular structure in Al-3wt% Li-0.8wt% Co powders.

The difference in the superheats chosen in the present work is found not sufficient to cause drastic microstructural changes.

ACKNOWLEDGEMENTS

The author wishes to thank Prof. G. Champier, Director, Laboratoire de Physique du Solide, ENSMIM, for providing the necessary research facilities and for his constant encouragement.

REFERENCES

1. F. H. Samuel, J. Hinojosa-Torres and G. Champier: 11th Conference on Applied Crystallography, September 10-14, 1984, Kozubnik, Poland.
2. D. Webster, G. Wald and W. S. Cremens: Metall. Trans. A, 1981, vol. 112, pp. 1495-1502.
3. H. Matyja, B. C. Gissen and N. J. Grant: J. Inst. Met., 1968, vol. 96, pp. 30-57.
4. S. R. Coriell and R. F. Sekerka: Rapid Solidification Processing, Principles and Technologies II, R. Mehrabian, B. M. Kear and M. Cohen, eds., Claitor's, Baton Rouge, LA, 1980, pp. 35-49.
5. G. G. Levi and R. Mehrabian: Metall. Trans. B, 1978, vol. 98, pp. 221-229.
6. W. Kerz and D. J. Fisher: Acta Metall., 1981, vol. 29, pp. 11-20.
7. W. A. Tiller: J. Appl. Phys., 1962, vol. 35, pp. 10-11.
8. A. Pawlowski and W. Truszkowski: Acta Metall., 1972, vol. 30, pp. 37-50.

High temperature tensile properties of mechanically alloyed Al–Mg–Li alloys

P S GILMAN, J W BROOKS
and P J BRIDGES

PSG is with Novamet Aluminum,
Wyckoff, New Jersey, USA
JWB and PJB are with INCO Alloy
Products Ltd, Birmingham, UK

SYNOPSIS

The high temperature mechanical properties of low density mechanically alloyed Al-4Mg-1.5Li strengthened with various combinations of inert dispersoids have been evaluated to explore the potential of extending the application of this alloy system to higher temperatures. Three combinations of inert second phase particles were added to improve high temperature strength: Al_4C_3 and Al_2O_3, and Al_4C_3 and Al_2O_3 with either CeO_2 or SiC. The Al-4Mg-1.5Li alloy matrix controls room temperature tensile strengths of all three alloys at about 550 MPa yield strength and 585 MPa ultimate tensile strength, ductility decreasing with increasing volume fraction of second phase particles. The three alloys display excellent stability after long time exposures at 400°C. There appears to be a critical level of dispersoids needed to affect high temperature properties. Second phase additions are most beneficial for improved very high temperature tensile and creep properties.

INTRODUCTION

Mechanical alloying (MA) has certain benefits for the production of aluminum lithium alloys, particularly the fabrication of solution strengthened Al-Li-Mg alloys that are additionally strengthened by oxide and carbide dispersoids and a fine approximately one micrometer grain size (1,2). The low lithium level (1.5%) is chosen to avoid or minimize embrittling precipitates that may form during age hardening. MA Al-Li-Mg alloys are less susceptible to stress corrosion cracking than equivalent ingot metallurgy alloys and have good resistance to general corrosion.

The commercial alloy IN-905XL (MA Al-4%Mg-1.5%Li-1.2%C-0.4%O) which has a density of 2.58 Mg/m^3 (8.2% lower than 7075) has been fabricated into structural components by both die and precision forging operations(3). IN-905XL has a small grain size, average of 1.5 micrometers, with a fine aluminum carbide and aluminum oxide dispersoid, 5 to 30 nm in diameter, Figure 1. The high dislocation density, both dislocation loops and jogged dislocations, results from the deformation imparted by powder processing, consolidation and extrusion. The IN-905XL forgings exhibit excellent mechanical properties: yield strengths of 468 MPa, 467 MPa, and 420 MPa; tensile strengths of 525 MPa, 497 MPa, and 481 MPa; ductilities of 9%, 6% and 6% elongation to failure, in the longitudinal, long transverse and short transverse orientations, respectively and fracture toughnesses of 31.8 MPa $m^{1/2}$ and 32.4 MPa $m^{1/2}$ in the L-T and S-T directions, respectively.

The main thrust of this work was to assess the high temperature mechanical properties of this promising MA Al-4Mg-1.5Li alloy with various hard particle additions.

MATERIALS

Standard IN-905XL (MA Al-4Mg-1.5Li-1.2C-0.4O) contains approximately 6 volume percent of aluminum oxide and aluminum carbide dispersoids. The Al_2O_3 and Al_4C_3 particles are formed insitu as an inherent part of the mechanical alloying process. The Al-4Mg-1.5Li composition was reinforced with two additional inert second phases: cerium oxide and silicon carbide. The CeO_2 and SiC were added as particulate during processing. Mechanical alloying is one technique that allows various second phases to be added to aluminum. This is especially difficult by other processing techniques used to produce Al-Li alloys. The CeO_2 is a high surface area particle, typically less than 10 micrometers in size and very friable. The angular SiC is approximately 2 micrometers in diameter and is not expected to be comminuted during processing.

The volume percent of second phases contained in the three Al-4Mg-1.5Li alloys are

TABLE I **905XL FORGINGS**

Orientation	Typical Properties	Hot Water Quenched	Annealed
L	Y.S. (MPa)	468	468
	UTS (MPA)	525	518
	% El	9	8
	K_{Ic}(MPa $m^{1/2}$)	31.8	27.5
T	Y.S. (MPa)	469	457
	UTS (MPa)	497	510
	%El	6	7
ST	Y.S. (MPa)	420	414
	UTS (MPa)	481	478
	%EL	6	7
	K_{Ic} (MPa $m^{1/2}$)	32.4	27.5

TABLE II **TENSILE PROPERTIES**

Alloy	Temp °C	YS MPa	UTS MPa	%El
MA Al-4Mg-1.5Li	20	562	584	7
	150	258	388	17
	200	128	233	27
	300	55	82	29
	400	42	44	3
MA Al-4Mg-1.5Li-0.1 v/o CeO_2	20	575	587	6
	150	260	387	14
	200	135	237	22
	300	59	76	20
	400	42	53	6
MA Al-4Mg-1.5Li-5 v/o SiC	20	-	581	0
	150	334	423	1
	200	197	266	3
	300	97	113	8
	400	68	68	0

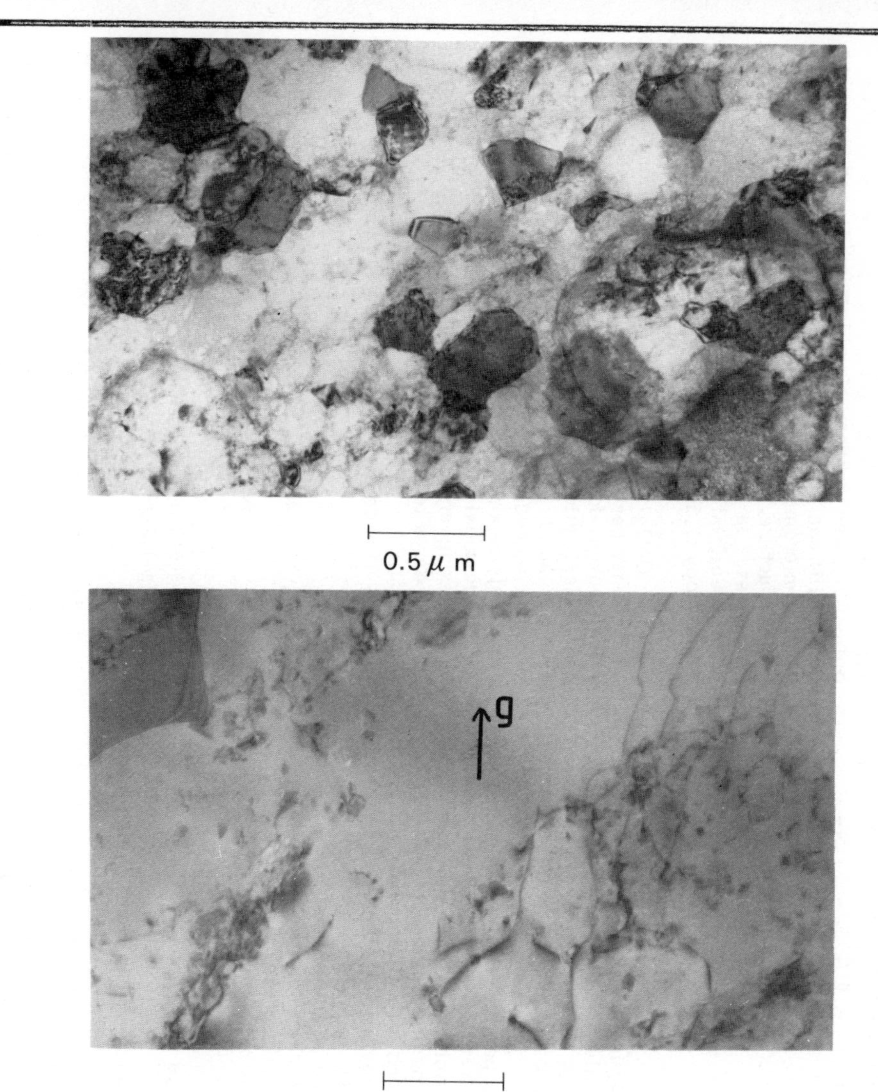

Figure 1. Transmission election micrographs of IN-905XL (MA Al-4Mg-1.5 Li) showing:

TOP - fine grain size, and aluminum carbides and oxides
BOTTOM - High dislocation density

summarized below:

Alloy	Al_2O_3	Al_4C_3	CeO_2	SiC
905XL	0.6	5.2	-	-
905XL-0.1 v/o CeO_2	0.6	5.2	0.1	-
905XL-5 v/o SiC	0.6	5.2	-	5

For the purpose of analysis it was assumed that all insitu formed oxides are aluminum oxide, Al_2O_3. It is possible that other oxides can form in situ such as MgO, Li_2O and/or complex Al,Li and Mg oxides.

The 0.1 volume percent CeO_2 was added to determine if a small amount of fine refractory can stabilize alloy microstructure and possibly enhance direct dispersion strengthening without degrading the room temperature properties of the matrix. The 5 volume percent SiC alloy is a composite material consisting of a low density Al-Mg-Li matrix reinforced by a high modulus second phase. SiC should improve the modulus to density ratio of the material and stabilize the alloy microstructure with greater high temperature strength with some reduction in alloy ductility.

MATERIAL PRODUCTION

All three alloys were made by Novamet Aluminum's established practice of mechanical alloying, powder degassing and vacuum hot pressing 91kg, 0.28m diameter billets (2,4). The three billets were extruded to 0.098m diameter at 260°C, (extrusion ratio of 8:1), and re-extruded from 0.09 m diameter to 0.05m x 0.02m at 371°C, (extrusion ratio of 8.2:1.) All tensile specimens were taken longitudinally parallel to the extrusion axis. Prior to tensile testing all speciments were heat treated at 482°/2 hr/66°C water quench. The alloys tested in this program are extrusions, consequentially it is expected that these alloys will have higher tensile strengths compared to the forging data previously discussed.

EXPERIMENTAL DETAILS

An indication of the stability of these alloys was obtained by measuring room temperature hardness after exposure at 400°C for times up to 36 hours. Room temperature tensile tests were performed on the three alloys exposed for 32 hours at 400°C.

Baseline tensile tests were performed at 20, 150, 200, 300 and 400°C. Change in strain rate tests at 200 and 300°C were used to generate strain rate vs stress curves. To study the effect of strain rate on tensile strength and ductility at 400°C various tests were performed at increasing cross - head displacement rates.

RESULTS

Material

Increases in the amount of second phase particle additions are evident in the alloy microstructures, Figure 2. The microstructure of MA Al-4Mg-1.5Li is very similar to other MA aluminum alloys. The fine 1.5 micrometer average grain size cannot be resolved optically. The 5-30nm aluminum oxide and carbide dispersoids can be seen as a homogeneous distribution of 0.5 micrometer etched artifacts throughout the microstructure. Previous studies have indexed δ' precipitates in this alloy, even though they have not been specifically identified in transmission electron microscopy (1,5).

CeO_2 is seen as a fine distribution of one micrometer particles. The friable CeO_2 clusters have been broken down and homogeneously distributed by mechanical alloying. The 5 volume percent SiC remains as 2 to 12 micrometer particles homogeneously embedded in the Al-Mg-Li matrix. As previously reported for Al-Mg and Al-Cu-Mg matrices (6), mechanical alloying and consolidation by vacuum hot pressing enables complex Al-Li alloys with hard second phases to be fabricated.

The densities of the three extrusions are as follows:

Alloy	Density Mg/m^3	% Decrease Compared To 7075
Al-4Mg-1.5Li	2.57	8.2%
Al-4Mg-1.5Li-0.1 v/o CeO_2	2.59	7.5%
Al-4Mg-1.5Li-5 v/o SiC	2.60	7.1%

High Temperature Stability

All three alloys show good hardness retention after exposures up to 36 hours at 400°C, Figure 3. As expected the 5 v/o SiC alloy is initially the hardest and softens the least as the SiC stabilizes the microstructure. The softening response of IN-905XL and the CeO_2 containing alloy are almost identical. All three alloys show a slight softening followed by hardening at short exposure times, 1 to 4 hours. The high dislocation density remaining after processing and consolidation allows the alloys to recover. The subsequent slight hardness increase may be due to coarsening of any residual δ' or Al_2MgLi precipitates.

The yield strengths and tensile strengths of MA Al-4Mg-1.5Li before and after 400°C, 32hr. exposure are equivalent, Figure 4. There is some reduction in ductility.

The MA Al-4Mg-1.5Li-0.1 v/0 CeO_2 loses little strength after exposure, 575 MPa to 529 MPa and 587 MPa to 557 MPa for yield and tensile strengths, respectively. Ductility is unaffected. The slight loss in strength may result from the alloy recovering, or even possibly overaging or dissolution of the precipitates. The SiC-containing alloy is uneffected by the 400°C 32 hour exposure, confirming the stabilizing contribution of the SiC distribution. Also the densities of the alloys are unchanged by the 400° C 32 hour exposure indicating no change in chemistry, i.e. lithium loss, from the prolonged exposure.

Al-4Mg-1.5Li

50 μ m

Al-4Mg-1.5Li-0.1v/oCeO₂

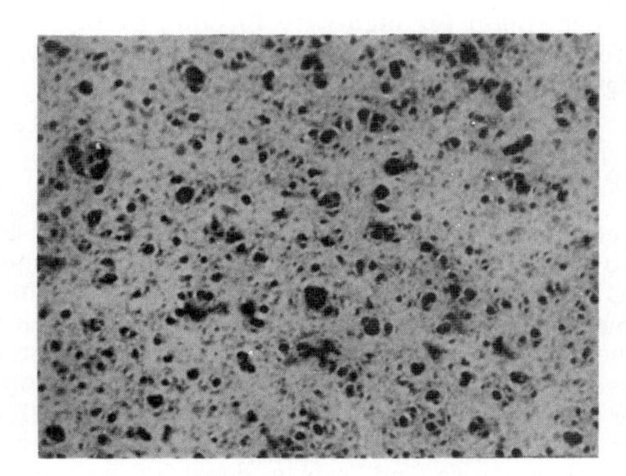

Al-4Mg-1.5Li-5v/oSiC

Figure 2. Optical microstructure of MA Al-4Mg-1.5 Li Alloys showing different levels of second phase particle additions.

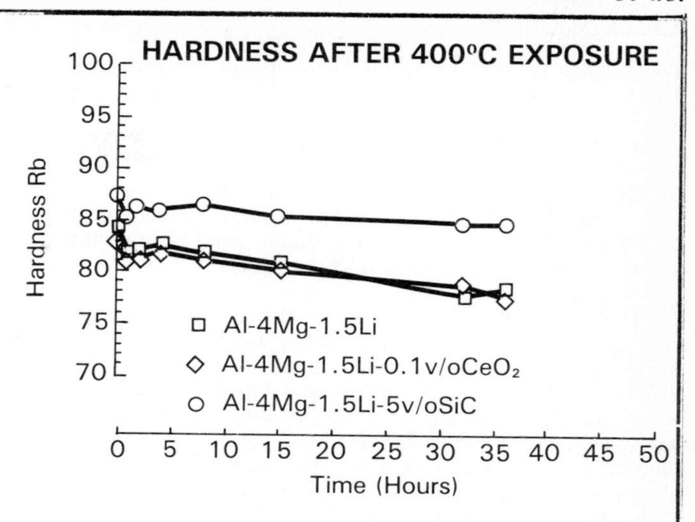

Figure 3. Thermal stability of MA Al-4Mg-1.5 Li alloys at 400°C as measured by room temperature hardness.

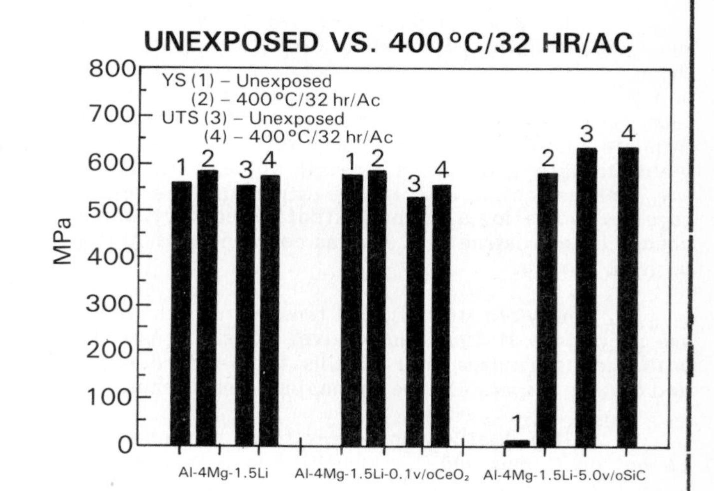

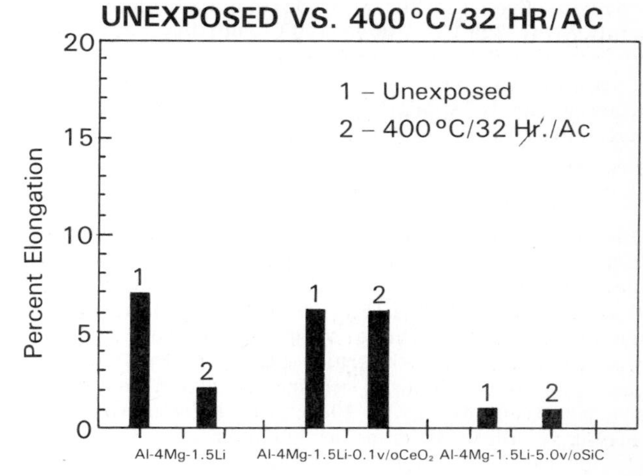

Figure 4. Tensile properties of MA Al-4Mg-1.5Li alloys after 32 hour exposure at 400°C.
TOP - Yield strength and tensile strength
BOTTOM - % Elongation to failure

Tensile Properties

The tensile properties of the three alloys are listed in Table II. The matrix strengths of all three alloys are similiar. The 0.1 v/o CeO_2 has little effect on room temperature tensile properties. Adding 5 v/o SiC does not affect matrix strength, but makes the alloy very brittle. These findings agree with previous studies that investigated adding various volume fractions of particulate SiC to the non-lithium containing MA aluminum alloys IN-9021 and IN-9052 (7).

The yield strengths of the alloys decrease rapidly with increasing temperature up to about 200°C, Figure 5. The rate of decrease is the same for the IN-905XL and CeO_2 containing alloy. The SiC containing alloy has higher yield strengths at the elevated temperatures. At temperatures above 150°C, the rate of decrease in yield strength is the same for all three alloys with the SiC-containing alloy having a 35 MPa yield strength advantage. At temperatures above 200°C the rate of yield strength decrease with increasing temperature diminishes. From 300°C to 400°C the rate of decrease is very slow.

The tensile strength behaves differently from the yield strength at the test temperatures. At 150°C there exists a large difference between yield strength and tensile strength. In this temperature range the rate of work hardening is greater than the rate of dynamic recovery causing a large difference between yield strength and tensile strength values. At 200°C the yield strength and tensile strengths are close together indicating a further extent of recovery, and a change in microstructure, such as coarsening of Al_4C_3 or grain growth.

The yield strength and tensile strength values are very close at 300°C and above. This is consistent with the assumption that at this temperature oxide and carbide dispersoids are the major strengtheners.

Alloy ductility increases to a maximum between 200°C and 300°C. Ductility increases as the alloy softens due to carbide coarsening and dynamic recovery. Even the SiC alloy develops 8% elongation to failure at 300°C. Above 300°C ductility decreases because of the predominance of intergranular failure. No attempt was made to minimize grain boundary area perpendicular to applied stress by thermomechanical treatments for the purpose of improving tensile properties.

Elevated temperature tensile properties of these alloys display a trend similar to that observed in other oxide dispersion strengthened alloys, particularly other MA alloys such as the nickel-chrome alloy INCONEL Alloy MA754 (8). Though the MA Al-Mg-Li type alloys have desirable room temperature tensile properties, due to both grain size strengthening and dispersion strengthening, their intermediate temperature strengths, up to 200°C, are inferior to ingot metallurgy alloys of similiar compositions (9). No attempt has been made to retain the heavily worked structure of processing or to purposely precipitation strengthen the alloy for better intermediate temperature strength. At temperatures greater than 200°C the benefits of MA are more prevalent as the second phase dispersoids stabilize the microstructure and provide direct high temperature strengthening. These alloys do not soften by overaging or dissolution of precipitates as is usually the case for high strength Al-Li alloys.

The small addition of CeO_2 appears to have little or no effect on the elevated temperature tensile properties of Al-4Mg-1.5Li. The 0.1 v/o CeO_2 did not increase the stability of the microstructure as predicted.

The 5 v/o addition of SiC does improve intermediate and elevated temperature strength with a concurrent decrease in ductility. Coupling this good elevated temperature strength and resistance to softening after 400°C make this alloy promising for high temperature creep-resistance.

Comparing the CeO_2 and SiC results suggests that a critical volume fraction of hard second phase particles is needed before a substantial increase in elevated temperature strength is realized. Also no investigation of the efficiency of the particle - matrix bonding of the two types of hard particles and its effect on high temperature properties has been made.

Strain Rate Effects

The three alloys were retested at 400°C with increasing initial strain rates, $8.3x10^{-5}$/sec to $8.3x10^{-2}$/sec, Figure 6. Tensile strength increases by approximately 30 MPa from the slowest to fastest strain rates tested. The improvement in ductility is more dramatic. MA Al-4Mg-1.5Li increases its ductility by about 2000% from 3% to 63% elongation to failure. The ductility of the SiC-containing alloy increases from 0% to 8%. The increase in ductility with increasing strain rate occurs because of the greater fraction of transgranular fracture. At a strain rate of $8.3x10^{-5}$/sec the fracture is predominantly intergranular. As strain rate increases the ratio of transgranular to intergranular fracture increases. This effect may prove beneficial for high rate forming of these alloys and for applications requiring good high temperature strength for short exposure times.

Creep

The creep properties of the three alloys were determined from change in strain rate tests performed at 200°C and 300°C; and the multiple tests performed at various strain rates at 400°C. The data for each alloy at the three test temperatures are represented as stress versus strain rate plots, Figure 7. The slopes of these curves, $d\sigma/d\dot{\varepsilon}$, is the strain rate sensitivity, m, in the equation $\sigma = k\dot{\varepsilon}^m$: where $\dot{\varepsilon}$ is the strain rate, σ is the flow stress and k is a constant incorporating structure and temperature dependancies. The strain rate sensitivities are very low and decrease with increasing temperature. The CeO_2 and SiC containing alloys have negative strain rate sensitivities at 300°C. It is believed that this negative value does not in actuality indicate negative creep, increasing strain rate with decreasing stress; but is a result of the inherently low strain rate sensitivity and insufficient data points for proper representation.

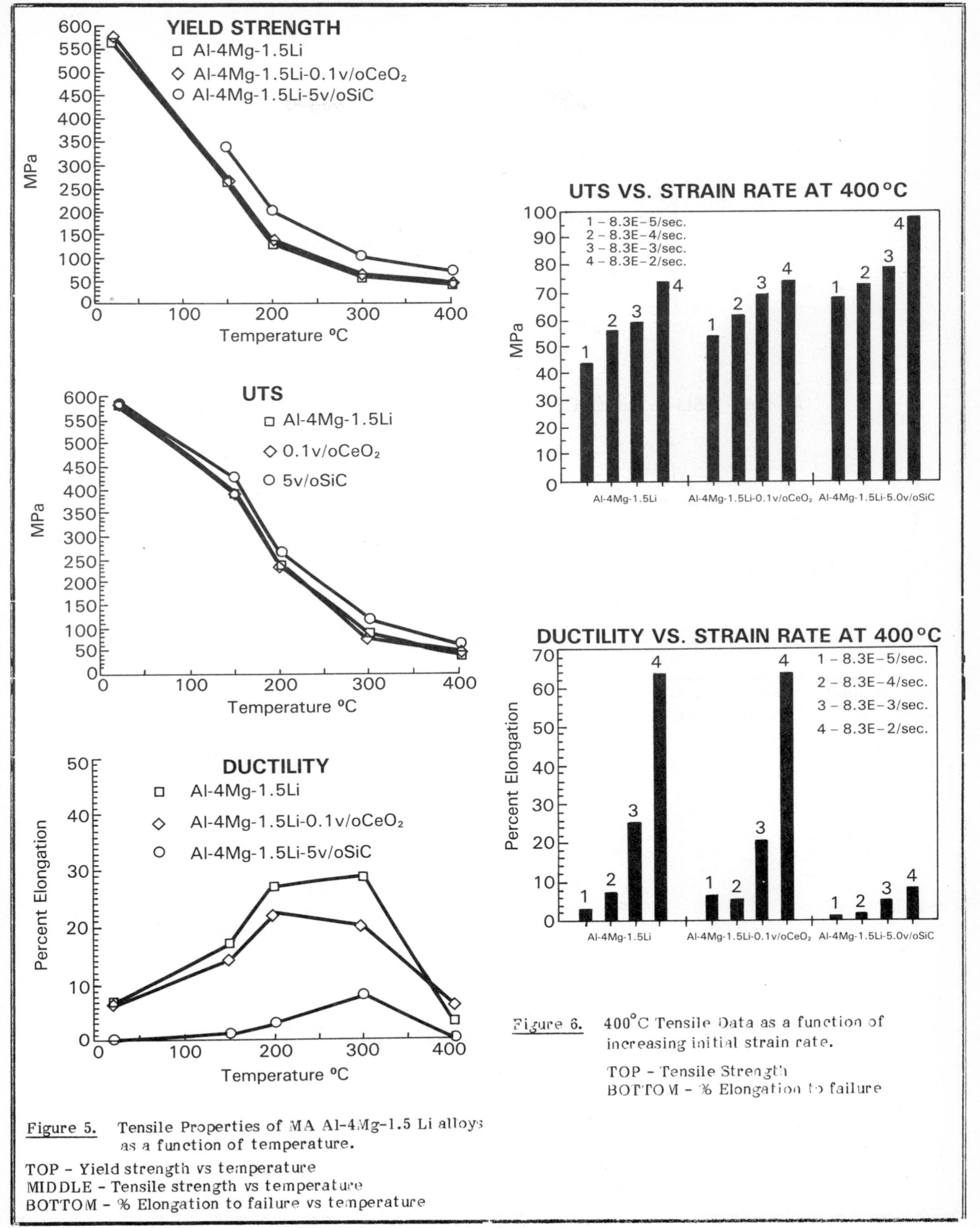

Figure 6. 400°C Tensile Data as a function of increasing initial strain rate.

TOP - Tensile Strength
BOTTOM - % Elongation to failure

Figure 5. Tensile Properties of MA Al-4Mg-1.5 Li alloys as a function of temperature.

TOP - Yield strength vs temperature
MIDDLE - Tensile strength vs temperature
BOTTOM - % Elongation to failure vs temperature

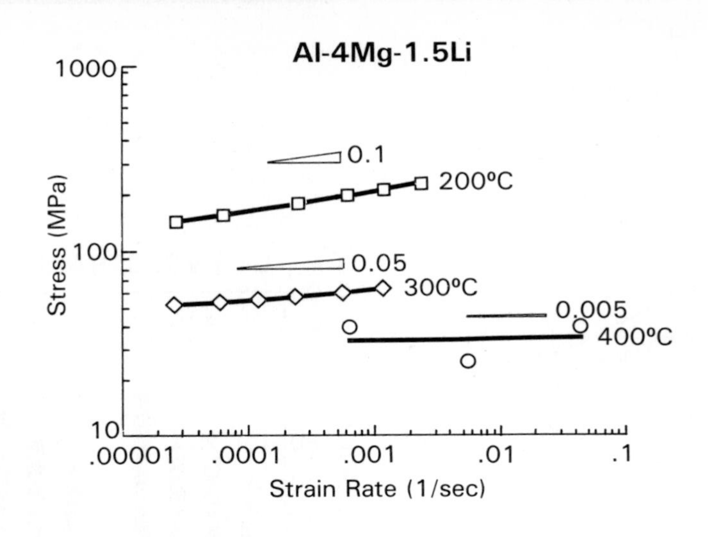

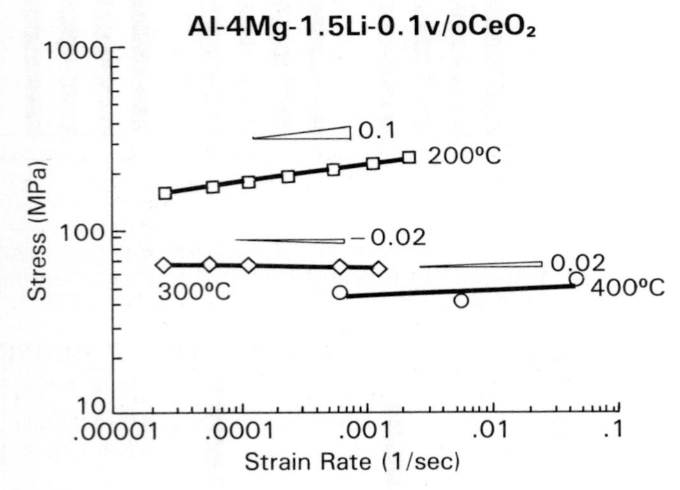

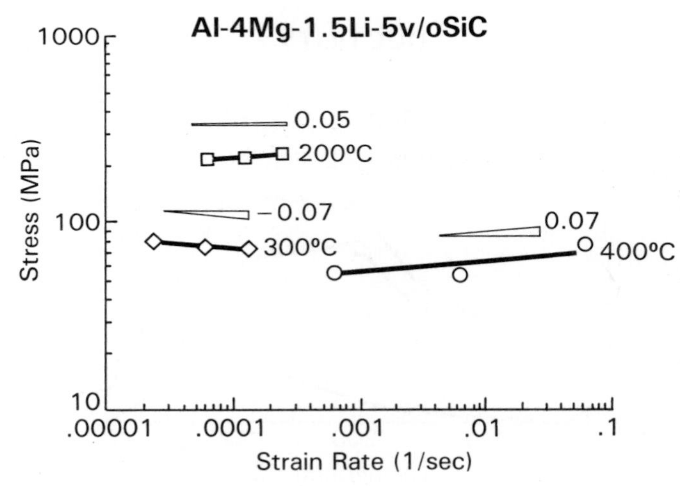

Figure 7. Creep properties of MA Al-4Mg-1.5Li alloys at various temperatures.

TOP - MA Al-4Mg-1.5 Li
MIDDLE - MA Al-4Mg-1.5 Li-0.1 v/o CeO_2
BOTTOM - MA Al-4Mg-1.5 Li-5.0 v/o SiC

The strain rate sensitivities, less then 0.05, are indicative of an apparent threshold stress. The data for the MA Al-4Mg-1.5Li alloys are similiar to that for MA Al-4Mg, i.e. IN-9051 (10). From the creep data for IN-9051 it was postulated that the threshold stress is not a threshold stress in the truest sense where dislocation motion stops below this value, but rather a change in creep mechanism. Above the threshold stress the deformation is climb controlled, below the threshold stress the dislocations actually interact with dispersoids.

The small volume percent addition of CeO_2 and larger amounts of SiC slightly improves the intermediate temperature creep response of the alloy. At 400°C the additional second phase particles cause a substantial improvement in creep response. For example to achieve a strain rate of $6.4x10^{-4}$/sec at 400°C the steady state flow stresses are 39 MPa, 45 MPa and 54 MPa for Al-4Mg-1.5Li, the CeO_2 alloy and the SiC alloy, respectively. At these high temperatures the additional hard second phases directly strengthen the alloy.

DISCUSSION

None of the alloys evaluated were optimized for either high temperature tensile or creep properties. However this evaluation shows that mechanical alloying can be used to fabricate Al-Mg-Li alloy combinations containing various second phase additions that are difficult or impossible to make by other methods.

The 400°C stability of all three alloys tested is excellent. The aluminum carbides, oxides, cerium oxide or silicon carbide stabilize the microstructures and resist softening. The second phase particles stabilize the fine grain size. Also the alloys appear chemically stable, i.e. no lithium loss, after 400°C exposure.

The tensile properties of the three MA Al-4Mg-1.5Li alloys appear similiar to those of non-lithium containing MA Al-4Mg alloys. The stable second phase particles stabilize the small 1.5 micrometer average grain size. The greater than 560 MPa and 580 MPa yield and tensile strengths, respectively, of the three alloys results from direct dispersion strengthening and Hall-Petch type grain boundary strengthening. Also the mechanically alloyed material contains a high dislocation density remaining from the MA processing, consolidation and extrusion even after annealing. The alloys tested in this program are extrusions which account for the increase in room temperature tensile properties over the forging data previously discussed.

At intermediate temperatures, 150°C to 300°C, the tensile strengths drop off while ductility increases. Another interesting result is the large difference between yield strengths and tensile strengths in the 150°C to 200°C range. For example at 200°C MA Al-4Mg-1.5Li has a yield strength of 128 MPa and a tensile strength of 233 MPa. The work hardening rate is greater than the dynamic recovery rate keeping the tensile strength high.

Ductility increases in the intermediate temperature range as the alloys are better able to accommodate cavitation because of the ease of diffusion due to the large grain boundary area.

At 400°C the tensile properties are controlled by a combination of direct dispersion strengthening and grain boundary weakening. Increasing initial strain rate reduces the grain boundary effect and consequently improves strength levels and ductility. The decrease in strength from 300°C to 400°C is much less than at intermediate temperatures. The second phase particles preserve strength. The more the second phase addition the better the strength retention. Further improvements in strength and ductility may be achieved by optimizing the size and volume fraction of second phase particles and creating a more favorable grain structure for high temperature strength, i.e. high aspect ratio grain morphology.

Increasing strain rate at 400°C increases alloy strength and ductility. The increased strain rate reduces the intergranular fracture component of the alloy. Possibly even faster strain rates (greater than $8.3x10^{-2}$/sec) and deforming between 200°C and 300°C would further improve strength and greatly increase ductility.

The strain rate sensitivities of these alloys over the range of temperatures and strain rates studied are very low, less than 0.05, indicating some type of threshold stress or change in creep mechanisms. The creep behavior is similiar to other MA Al-4Mg alloys. Increasing the amount of second phase particles improves creep strength especially at 400°C. Further improvements in creep properties may be attained by microstructural modifications such as stabilization of the initial high dislocation structure and formation of high grain length to width aspect ratios to minimize grain boundary area perpendicular to the applied stress.

High strain rate sensitivities that are indicative of superplasticity, might be expected from these alloys because of their ultra fine grain size and the existance of fine second phase particles, (11). However Al_4C_3, Al_2O_3, CeO_2 and SiC particles are sufficiently harder than the matrix and can cause cavitation, i.e. lower than expected final elongations to failure. As previously shown, at 400°C increasing strain rate increases ductility. Possibly high strain rates, greater than 0.1/sec, in the temperature range from 200°C to 300°C may overcome the premature cavitation problems caused by the strength difference between the Al-Mg-Li matrix and the second phase particles. Similiar results were observed during the high temperature deformation of aluminum reinforced with SiC whiskers(12).

CONCLUSIONS

1. Mechanical alloying is one successful method for producing Al-Mg-Li alloys with prescribed additions of second phase particles, e.g. aluminum carbides, aluminum oxides, cerium oxide and silicon carbide, with appropriate properties.

2. The microstructural stability of these alloys after extended exposure at 400°C is excellent.

3. Alloys combine very good room temperature and high temperature (400°C) tensile properties. Properties quickly fall off at intermediate temperatures, but are stable at high temperatures. A minimum amount of second phase addition must be present to cause beneficial tensile property improvements.

4. Increasing deformation rate improves alloy strength with large improvements in ductility. Potential benefits in fabrication and high temperature short time exposures.

5. Second phase additions improve creep properties with improvements most effective at high temperatures. The more second phase particles the better the creep resistance.

Acknowledgements

The authors would like to thank W.E. Mattson for implementing the testing program and making many helpful suggestions and analyses. This research was performed as part of Novamet Aluminum's IRAD activities.

References

1. P.S. Gilman, Aluminum-Lithium Alloys II, T.H. Sanders, et al ed., AIME, Warrendale, PA 1984, pp. 485-506.

2. S.J. Donachie and P.S. Gilman, Aluminum-Lithium Alloys II, T.H. Sanders, et al ed., AIME, Warrendale, PA 1984 pp. 507-515.

3. S.J. Donachie, "Mechanically Alloyed Al-Mg-Li Alloys", WESTEC, Los Angeles, CA. March 19, 1985.

4. R.D. Schelleng and S.J. Donachie, Metal Powder Report 38, 1983 pp. 575-576.

5. G.H. Narayanan, B.L. Wilson and W.E. Quist, Aluminum-Lithium Alloys II, T.H. Sanders, et al ed., AIME, Warrendale, PA 1984 pp. 517-541.

6. A.D. Jatkar, R.D. Schelleng and S.J. Donachie, "Mechanically Alloyed SiCp -Reinforced Composites," DOD/NASA Advanced Composites Work Group, Cocoa Beach, FLA. January 24-25, 1985.

7. R.D. Schelleng, A.D. Jatkar and S.J. Donachie, "Mechanical Alloying as a Means of Composite Consolidation in the Absence of a Liquid Phase," Proc. Seventh Annual Discontinuous Reinforced Aluminum Material Working Group Meeting, Park City, Utah, DOD MMC/AC April 1985.

8. INCONEL alloy MA 754 DATA Bulletin, IncoMAP, Huntington, W. VA.

9. B. Noble, S.J. Harris, and K. Harlow, Aluminum - Lithium Alloys II, T.H. Sanders et al ed., AIME Warrendale, PA 1984 pp. 65-77.

10. W.C. Oliver and W.D. Nix, Acta Metall, 30, 1982, pp. 1335-1347.

11. J. Wadsworth, I.G. Palmer, D.D. Crooks, and R.E. Lewis, Aluminum - Lithium Alloys II, T.H. Sanders et al ed., AIME, Warrendale, PA 1984 pp. 111-135.

12. T.G. Nieh, C.A. Henshall and J. Wadsworth, Scripta Metall, 18, 1984, pp 1405-1408.

Fatigue crack propagation in mechanically alloyed Al–Li–Mg alloys

W RUCH and E A STARKE Jr

The authors are with, respectively, the Department of Materials Science at the University of Virginia, USA and the School of Engineering and Applied Science, University of Virginia, USA

SYNOPSIS

A total of four Al-Li-Mg alloys manufactured by the mechanical alloying technique have been investigated with respect to their tensile properties and fatigue crack propagation behavior at room temperature in air. The lithium, magnesium and carbon contents were varied from 1.5 to 2.5, 2.0 to 4.0 and 0.7 to 1.1%, respectively. Mechanically alloyed materials derive their strength mainly from a fine dispersion of oxide and carbide particles and an ultrafine grain size, and, therefore, have not been tested in artificially aged conditions. Important microstructural features as well as crystallographic texture differed with composition. Variations in the carbon content which controls the carbide dispersion proved to be very effective in changing strength and ductility. The differences in mechanical behavior among the alloys will be interpreted in terms of composition, microstructure, texture, and deformation mode. Comparison will be made to conventionally produced Al-Li-Mg alloys.

INTRODUCTION

Much current research is being directed toward developing low density Al-Li-X alloys because of their potential in weight savings in aircraft. Of particular concern is the low ductility and fracture toughness that these materials often exhibit. Various modifications in alloy chemistry and fabrication techniques have been used in an attempt to address these problem areas. One approach involves the use of the novel mechanical alloying fabrication technique since this method introduces a large volume fraction of dispersoids and a fine grain size both of which have been shown to improve ductility in Al-Li-X alloys (1,2).

Mechanical alloying is an entirely solid state process consisting of repeated cold welding and fracturing of powder particles. The uniformly distributed small particles, such as Al_2O_3 and Al_4C_3 in the aluminum alloys, together with the fine grain size are responsible for the high strength of the mechanically alloyed products. Oxide dispersions result from both the cold welding process and the fracture of oxides present in the original powder charge. Carbides result from the residual hydrocarbon which is introduced as a process control agent during grinding in the attrition mill (3).

Although there are many beneficial aspects associated with a fine-grained material, e.g. increased homogenization of deformation, increased fatigue crack initiation resistance, increased corrosion resistance, etc., a decrease in fatigue crack growth resistance often accompanies a decrease in grain size (4). This paper will describe a recent study of the fatigue crack growth behavior of a number of Al-Li-X alloys that have been produced by the mechanically alloyed process, and which contain an ultra-fine grain structure.

EXPERIMENTAL PROCEDURES

The investigated alloys were received from the Boeing Commercial Airplane Company in form of extrusions with rectangular cross sections of 50 × 12 mm^2 (alloy 1), and 100 × 25 mm^2 (alloy 2, 3 and 4). More details about the alloys are listed in Table 1.

Previous investigations showed that the tensile fracture is controlled by volume fraction of dispersoids and a reduction of the dispersoid level suggested an increase in ductility. For that reason the dispersoid volume fractions have been varied by changing the carbon content which in turn is controlled by the amount of organic process control agent added during attrition (5). All specimens were examined either in the T3 or T4 condition. Alloy MA1 has been solution heat treated for 2 hrs @ 535° C and the other alloys about 3 hrs @ 520° C. It was attempted to stretch all the extruded plates after solution heat treatment in order to provide stress relief, but only for alloy 2 could this be done properly.

Microstructural analysis was carried out using optical, scanning and transmission electron microscopy. A Siemens texture goniometer set up in back reflection was used to measure preferred orientation among the grains.

Tensile and fatigue crack propagation testing (LT orientation) were carried out with specimens according to ASTM-Standards using a servohydraulic closed loop MTS testing machine. Compact tension specimen thickness was ca. 7 to 12 mm. The initial strain rate in tensile tests was 0.04% per second. Fatigue testing took place in

Table 1. Investigated aluminum alloys

	Mg	Li (w%)	C	0	Vol%Disp.	Grain Size (μm)
MA1	4	1.5	1.1	0.6	5.6	0.6
MA2	4	1.75	0.7	0.6	3.9	1.2
MA3	2	2.5	0.7	0.6	3.9	0.8
MA4	2	2.5	1.0	0.6	5.0	0.7

Table 2. Tensile properties

	YS (MPa)	UTS (MPa)	ε_f (%)	E (GPa)
MA1	440	520	8	80
MA2	430	485	15	80
MA3	465	485	10	79
MA4	495	530	5	82
7075 T651	505	570	11	72

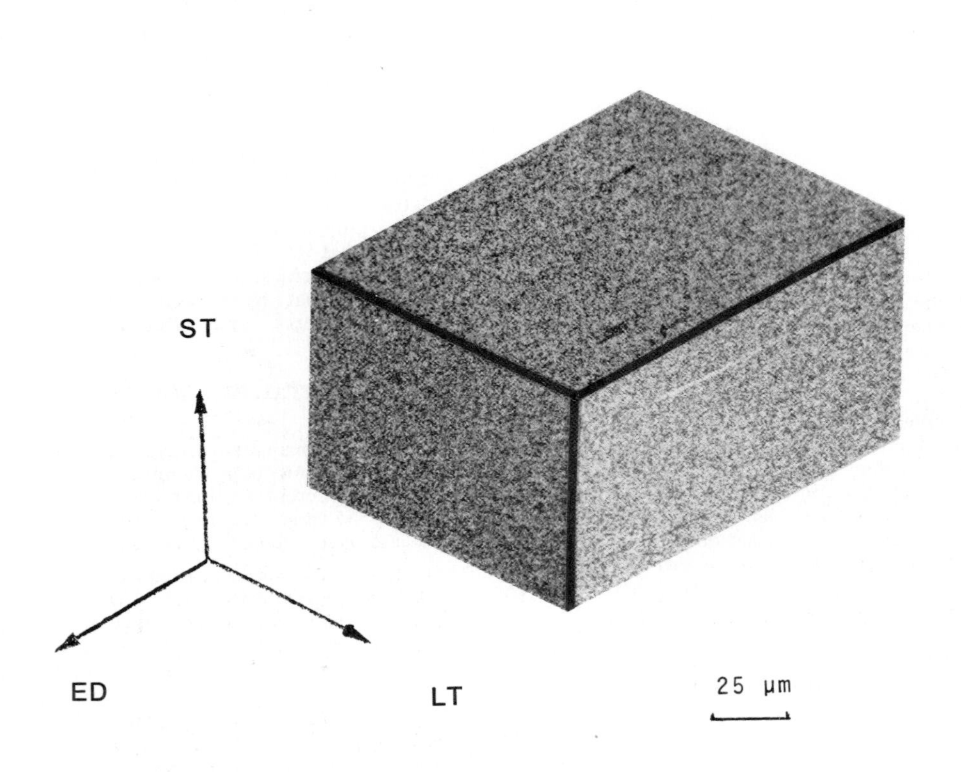

Figure 1. 3-D optical micrograph of grain structure

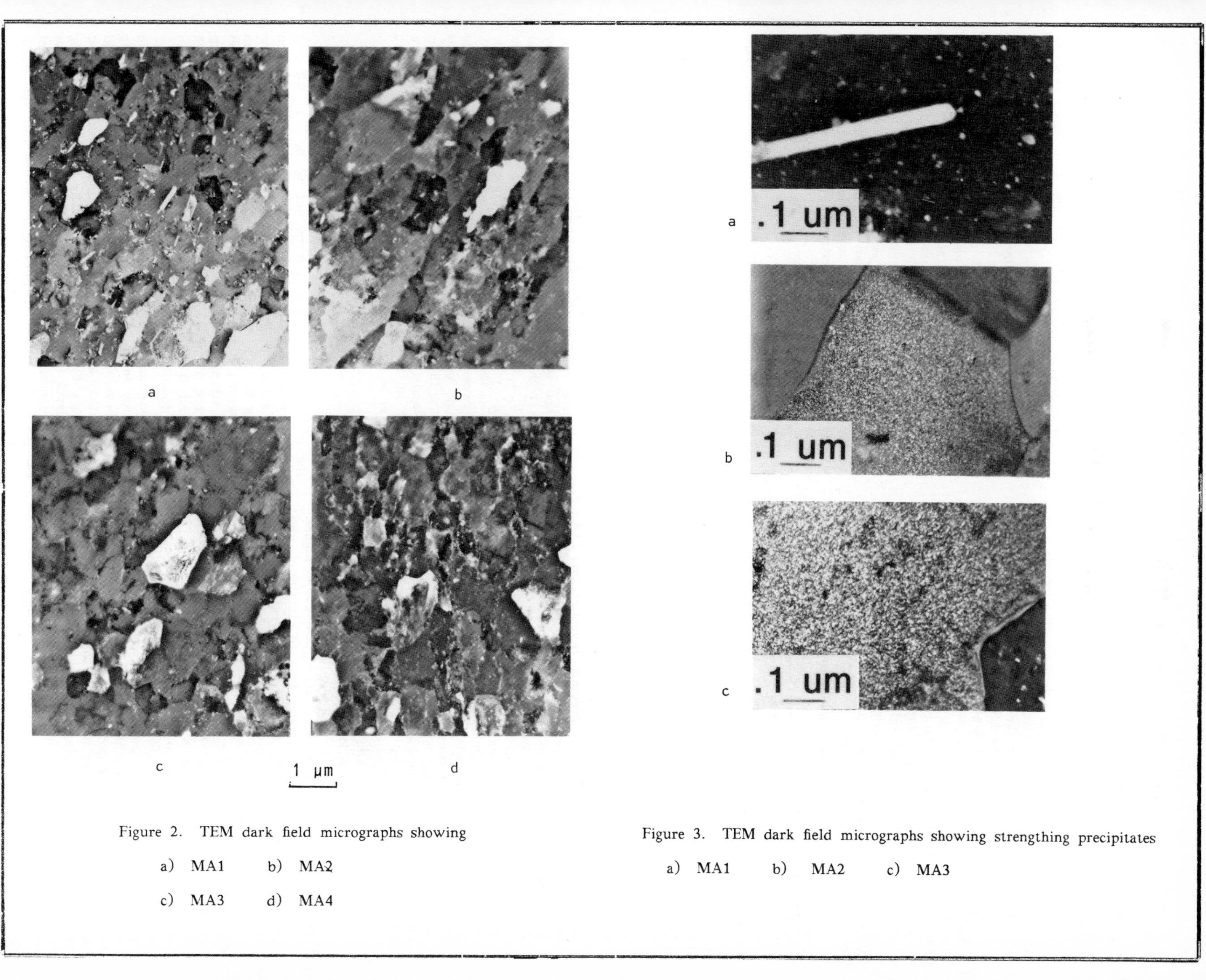

Figure 2. TEM dark field micrographs showing

a) MA1 b) MA2

c) MA3 d) MA4

Figure 3. TEM dark field micrographs showing strengthing precipitates

a) MA1 b) MA2 c) MA3

laboratory air, using an R-ratio of 0.1 and a frequency of 30 Hz (sinusoidal loading). The load shedding technique was used to determine crack propagation rates close to threshold.

RESULTS AND DISCUSSION

Microstructure

Figure 1 shows the optical microstructure of alloy 2 which is representative for all alloys. The grain size is too small to be resolved in the optical microscope. Some white stringers appear on the LS-plane and are believed to be recrystallized, dispersoid free areas which may result from sintering necks formed during vacuum degassing of the powder particulate prior to vacuum hot pressing (5). The difference in microstructure becomes visible in the TEM pictures (figure 2). These dark field micrographs show some grains and small spherical and elongated oxide and carbide particles in bright contrast. Such fine dispersion of nonmetallic particles is typical for mechanically alloyed products.

These nonmetallic dispersoids provide strength to the fact that they have to be looped by dislocations and by keeping the grain size on a micron level. Alloy 1 (figure 2a) shows the smallest grains with an average of 0.6 μm whereas alloy 2 (figure 2b) shows the largest average grain size with about 1.2 μm. Alloy 3 and 4 show a quite similar grain structure. The grains are elongated in the extension direction with an aspect of about 2 in all alloys. The grain sizes are listed in Table 1. Besides rod needle shaped oxides and carbides, varying from 10×200 nm^2 to 40×400 nm^2 in figure 2, there are more equiaxed, finely distributed dispersoids in the 10 nm range present (figure 3a). The distribution of large, rod shaped dispersoids, one of whom is not completely pictured in figure 3a can be seen in figure 2 at lower magnification level. It has been shown by previous investigations (3,6) that such fine microstructures deform very homogeneously even under cyclic loading at small amplitudes. Assuming that all the oxygen is tied up as Al_2O_3 instead of combinations of Al, Li and Mg oxides, and all carbon is tied up as Al_4C_3(5), volume fractions of dispersoids have been calculated and are listed in Table 1. The smaller grain sizes are associated with the higher dispersoid levels. The assumption that Li containing oxides are not present is not correct, because no $Al_3Li(\delta')$, AlLi (δ), or Al_2LiMg precipitates could be detected in alloy 1. Even after aging the 1.5w%Li alloy at 121° C no lithium containing ppt's appeared. It is fair to assume that the solubility limit of Li in Al-Li-Mg alloys is well below 1.5w%, and, therefore, it is quite clear that some lithium will be in the dispersoids, presumably as Li_2O_2 (7), and the actual dispersoid volume fraction may consequently be higher. From X-ray diffractometer scans only α-aluminum and Al_4C_3 could be positively identified. Several diffraction peaks could not be matched unambiguously with oxides of aluminum or mixed oxides or carbides of the elements in the alloys.

In contrast to alloy 1, which has the lowest Li content, alloys 2, 3 and 4 showed the presence of δ' precipitates in the solution heat treated and quenched conditioning. The driving force for δ' is too strong to be suppressed by cold water quenching to room temperature. This is also found in AlLiMg-alloys produced by powder and ingot metallurgy. Figures 3b and c are dark field micrographs imaging δ' in alloys 2 and 3 with a 100 superlattice spot. The volume fraction and size of the precipitates are somewhat larger in alloy 3 (figure 3c) compared to alloy 2 which has a lower lithium content. Figure 3c is also representative for alloy 4 with the same lithium content of 2.5w% as alloy 3. The coherent precipitates are about 5nm in size and due to their small size and their shearable nature their contribution to strength is not expected to be too significant. No Al_2LiMg or AlLi precipitates were detected.

Texture

The texture developments in the four alloys can be seen in figure 4. They all show the same basic type of (110)[001] texture but it is weakest in alloy 1 and strongest in alloy 4, according to the pole figures and the ratio of maximum divided by random intensity which is listed with the pole figures. It is necessary to keep in mind that only alloys 2, 3 and 4 have been processed exactly the same way and are suited for direct comparison. The ideal orientation in the smaller extrusion of alloy 1 yet is the same as in the other alloys. Although the alloy with high lithium, low magnesium and high dispersoid content exhibits the most pronounced texture, its sharpness, specified by the normalized maximum intensity of 4.2, can still be considered medium. Also noticeable is a fibring component in all pole figures which is usually found in extrusions of these aspect ratio ranges.

Tensile Tests

Table 2 lists the tensile properties of the alloys investigated. The highest strength levels are reached by alloy 4, which has a high lithium content, high volume fraction of dispersoids and the sharpest texture. By applying a Hall-Petch relationship with parameters established by (8) for peakaged Al-2Li-2Mg ingot alloys, a yield strength of 430MPa is calculated for MA4. The difference to the measured yield strength of 495MPa can on a first approximation be attributed to the strengthening effect of dispersoids.

The Young's Moduli have been determined with an extensometer during the tensile tests. The accuracy of this method is about $\pm$ 2%, i.e., the low value for alloy 3 with a high lithium content is well within experimental error. It is interesting to note that in MA1 no lithium containing precipitates can contribute to the elastic modulus. The increase in modulus compared to 7075 must be caused by lithium in solid solution and by oxide and carbide particles. The best ductility is displayed by alloy 2 with 15% strain to fracture. A low lithium content and a low volume fraction of carbides are combined in this alloy. If one expresses the strain to fracture e_f as a function of Li, Mg and C in wt%:

$$e_f = a\ Li + b\ Mg + c\ C$$

the coefficients a, b and c turn out to be 5.1, 4.4 and -16.7, respectively. Only alloys 2, 3 and 4 have been used to solve the equation because they have the same processing history. It is obvious that varying the carbon content which is related to the dispersoid volume fraction has the strongest influence on ductility. The way the carbides are dispersed in these alloys suggests that 0.7w% carbon and 0.6w% oxygen are enough for a combination of good ductility and strength.

Magnesium is supposed to act as a solid solution strengthener in these alloys. Alloys 3 and 4 with low Mg levels show smaller differences between UTS and yield strength than alloys 1 and 2 which is a measure for their lower work hardening capability. The effect of decreasing the Mg level from alloy 2 to 3 by about 2.2 at% is more than balanced by the increase in lithium content (~ 2.5 at %) where strength is concerned but the strain to fracture is less in alloy 3. A reduction in strain to

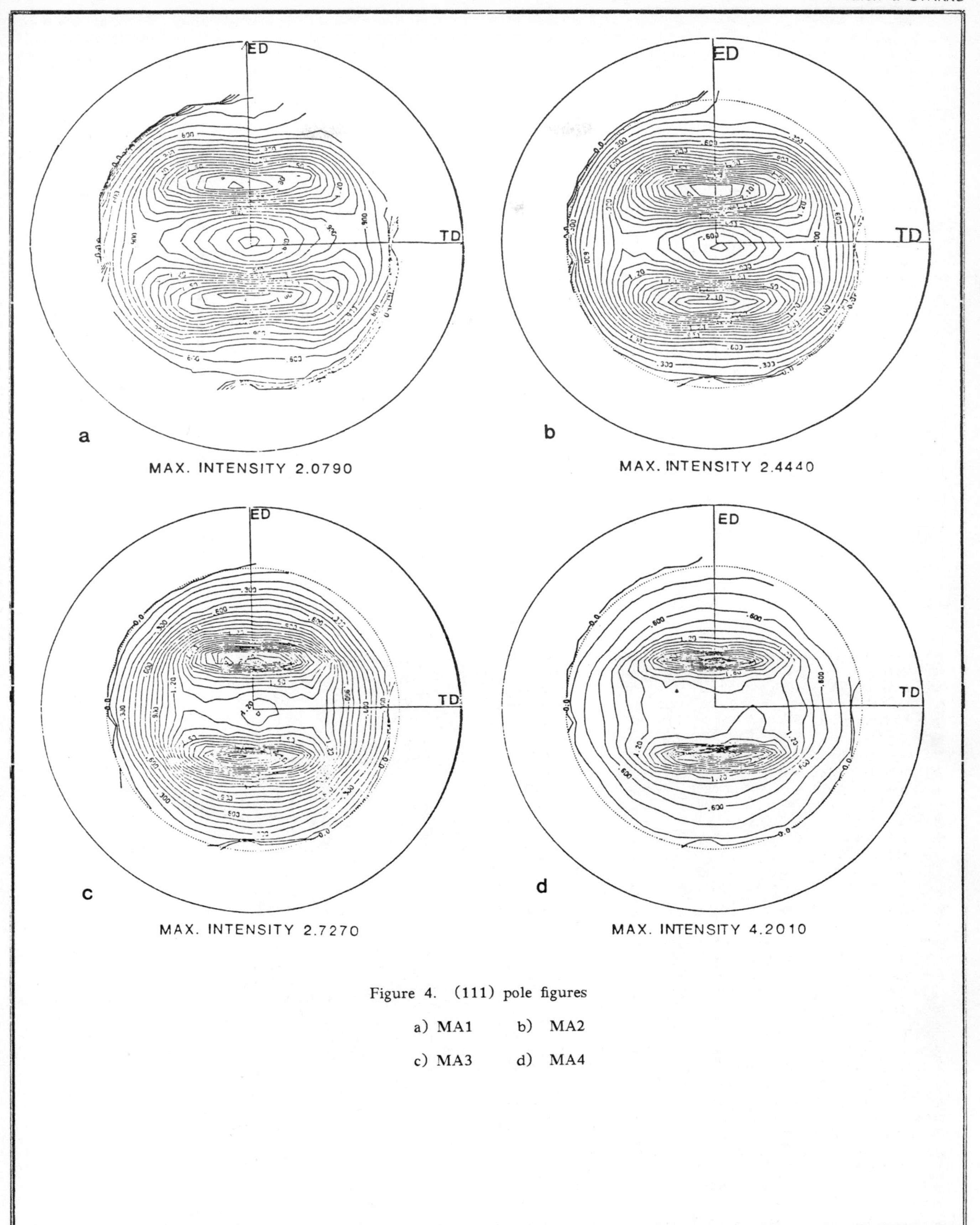

Figure 4. (111) pole figures

a) MA1 b) MA2

c) MA3 d) MA4

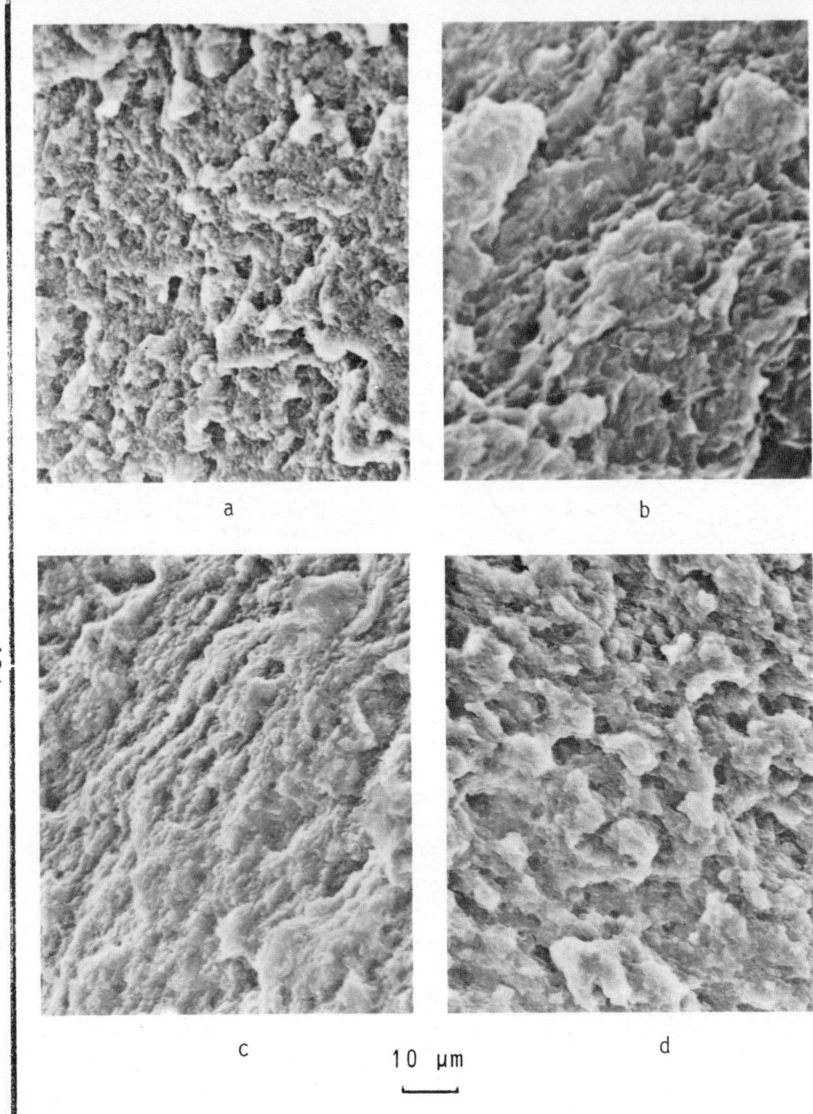

Figure 5. SEM fracture surfaces from tensile tests.

a) MA1 b) MA2

c) MA3 d) MA4

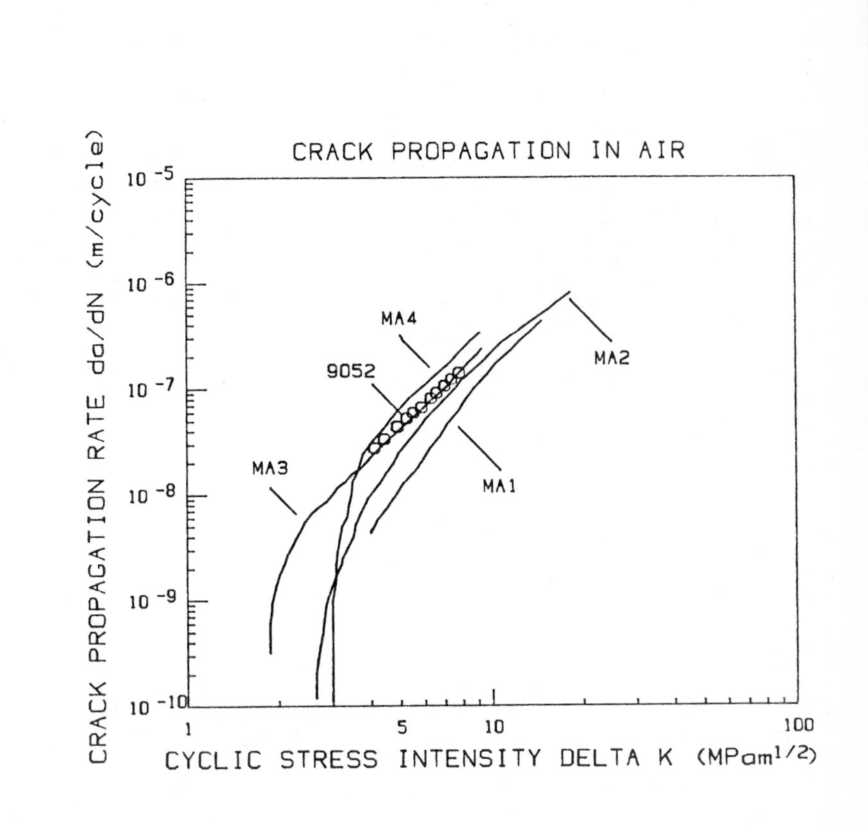

Figure 6. Fatigue crack propagation curves.

fracture has also been found in ingot Al-Li-Mg alloys by Dinsdale et al (8) if one decreases the magnesium content or increases the lithium content. Again, it is not known to what extent magnesium is present in form of dispersoids, i.e., its amount in solid solution remains undetermined.

Tensile fracture surfaces of the investigated alloys display ductile fracture (figure 5). Macroscopically all samples failed with a 45° shear mode. Alloy 2 shows a pronounced shearing on the fracture surface. This is probably promoted by the presence of shearable δ' precipitates and the large grains.

Fatigue Crack Propagation

Figure 6 represents the results of fatigue crack propagation testing in form of $\frac{da}{dN}$ versus ΔK plots. Two specimens have been tested for each material and good coincidence between specimen pairs was observed. For comparison alloy 9052, whose data have been taken from Erich and Donachie (9), is plotted in this diagram. 9052 is a commercially available mechanically alloyed Al-4Mg alloy with a slightly higher dispersoid level than the lithium containing alloys of this investigation.

At ΔK values above 6 $MPa\sqrt{m}$ the largest difference in crack propagation rates is about a factor of three. Alloy 9052 falls in the same range of fatigue crack growth resistance as the Li containing MA alloys.

At ΔK levels below $5MPa\sqrt{m}$ the curves separate and at lower propagation rates the differences become quite pronounced. Alloys 2 and 4 perform best with threshold levels close to $3MPa\sqrt{m}$. Alloy 3 shows the highest crack velocities at low ΔK values and a threshold below $2MPa\sqrt{m}$. Alloy 1 has only been tested at $\frac{da}{dN} = 4 \cdot 10^{-8}$ m/cycle and above.

Crack closure was measured from load-displacement curves which have been taken at slow loading rates using a back face strain gage. The occurrence of crack closure results in a decrease of the effective cyclic stress intensity amplitude and consequently better crack propagation behavior. According to Beevers (10), the applied stress intensity can be divided into an intrinsic and a closure component. The closure component at a certain ΔK may vary, for instance, with environment, deformation and residual stresses depending on what type of closure (oxide, roughness, or plasticity induced) is present. Only alloy 2 in our research did not exhibit any crack closure. This is the only alloy which could be stretched properly (~2%) after solution heat treatment. This seems to indicate that crack closure which was found in the other alloys is caused by residual stresses in the material. The applied load amplitude was reduced due to closure in alloys 3 and 4 by about 32% at threshold levels. Only 75% of the applied stress amplitude were effective in MA1 at $\sim 5 \cdot 10^{-8}$m/cycle, the lowest propagation rates where this alloy has been investigated. With increasing ΔK the closure component decreased rapidly and was close to zero at about $5MPa\sqrt{m}$ in MA3 and MA4.

Figure 7 represents the fatigue crack propagation behavior after correction for closure, i.e., what it is expected from properly stretched material in this case. The curves for the different materials are now closer to each other in the high crack propagation regime. Alloy 9052 is not incorporated in this graph because no information on crack closure was given in the cited reference (9). The curve of alloy 2 remains unshifted because no closure was present in this alloy. The data of alloy 1 also follows this curve and is represented by the dotted line. There is virtually no difference between the two alloys in the investigated regime ($\frac{da}{dN} \geqslant 5 \cdot 10^{-8}$ m/cycle), although a separation between the two curves may occur at lower crack propagation rates. Figure 7 also shows that the threshold stress intensity of alloy 4 with $2MPa\sqrt{m}$ is below the value for alloy 2 with ≃ 2.5 $MPa\sqrt{m}$. Alloy 3 still exhibits the fastest FCP-rates at low ΔK levels and the lowest threshold.

It is generally accepted that the size of the plastic zone ahead of the crack tip and its relation to size and spacing of relevant microstructural features plays an important role in fatigue crack propagation behavior (11). If the size of the plastic zone is much larger than certain microstructural dimensions, then the material will behave more like a continuum and the influence of microstructure on propagation rates will not be very pronounced. The most important microstructural features with considerable influence on plastic deformation are the grain size and the dispersoids in the investigated alloys. If one calculates the cyclic plastic zone size v_p according to the following equation (12),

$$v_p = (\Delta K/\sigma_y)^2/(8\pi)$$

where σ_y is the yield strength, it turns out that at ΔK values below $\sim 2.4MPa\sqrt{m}$ for alloy 2 and below $\sim 2.1MPa\sqrt{m}$ for alloys 3 and 4 the plastic zone is smaller than the grain size. At high crack propagation rates the small differences between the alloys show that microstructural influences are secondary and that the material behaves like a continuum. This is also true for alloy 9052.

Lindigkeit et al (13) demonstrated that an increase in grain size improves the fatigue crack propagation resistance in aluminum alloys by allowing an increase in slip length, resulting in more reversible slip. A higher degree of slip reversibility enables a large portion of dislocations ahead of the crack tip not to contribute to the crack growth process (14). This may explain the fact that alloy 2 with the largest grain size shows the best overall fatigue crack propagation behavior. One might argue with the fact that aluminum alloys containing nonshearable precipitates usually show only a negligible effect of grain size on mechanical properties because slip length is controlled by the precipitates. However, the mechanically alloyed materials show fairly low volume fractions of nonshearable dispersoids and therefore a much larger interparticle spacing compared to an overaged aluminum alloy with nonshearable precipitates.

Alloys 3 and 4 both have a higher Li content and a lower Mg content where the latter results in a lower work hardening capability. A lower work hardening rate in alloys 3 and 4 may indicate easier cross slip, i.e., less localized and, in turn, less reversible slip which yields lower ΔK values for constant crack growth rates. The grain size is almost the same for alloys 3 and 4 and cannot explain the difference in thresholds. Also alloy 3 shows both lower threshold and a slightly larger grain size.

Besides the different volume fractions of dispersoids in alloys 3 and 4, whose role at threshold is not clear, there is a very pronounced difference in preferred orientation. Texture in alloy 4 is over 1.5 times more pronounced than in alloy 3 which can be seen from figure 4. Kuo and Starke (15) have shown that in a high strength P/M aluminum alloy sharp texture increases the effective grain size by making slip transfer across grain boundaries easier because of smaller misorientation angles between neighboring grains. Fatigue crack propagation

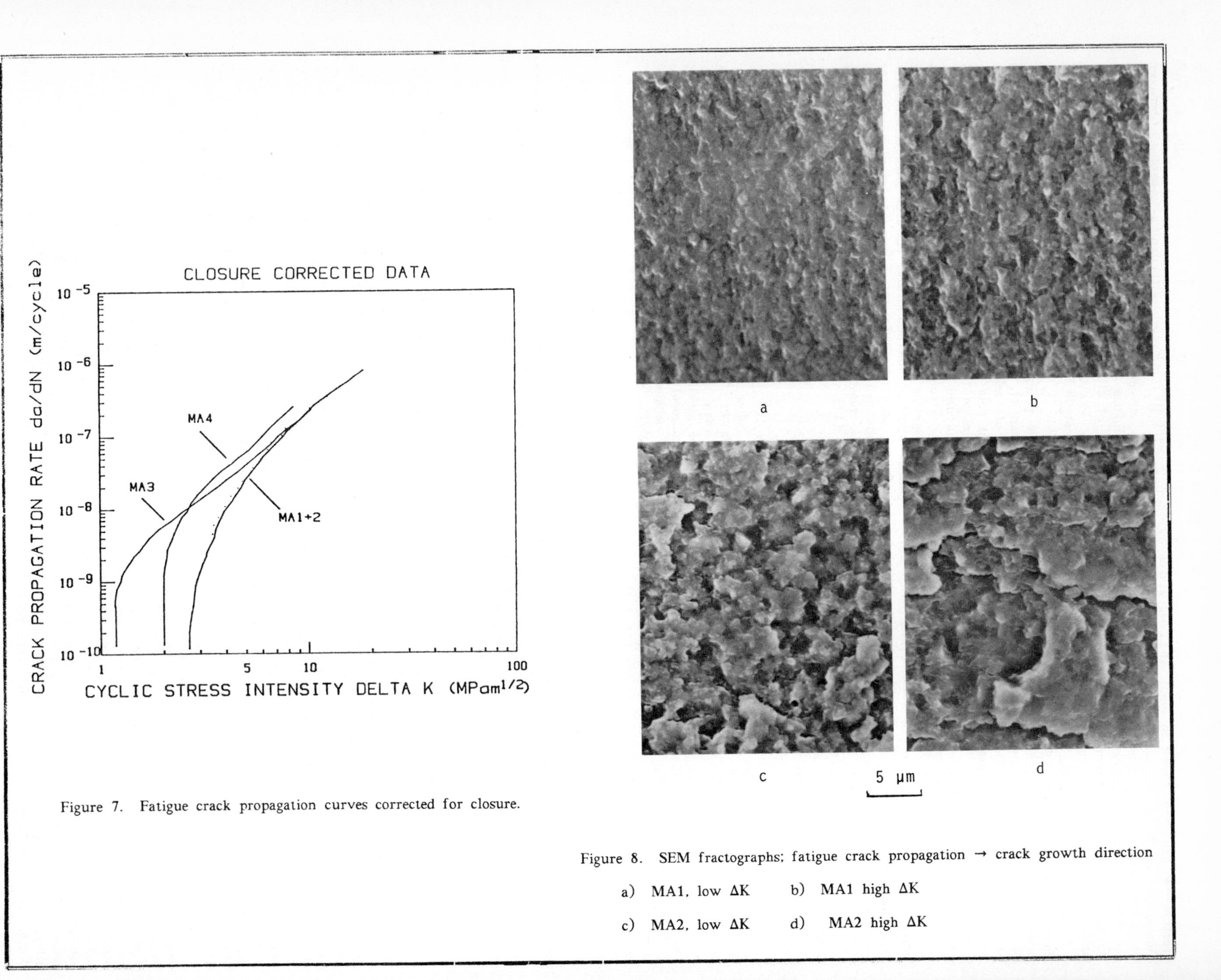

Figure 7. Fatigue crack propagation curves corrected for closure.

Figure 8. SEM fractographs; fatigue crack propagation → crack growth direction

a) MA1, low ΔK b) MA1 high ΔK

c) MA2, low ΔK d) MA2 high ΔK

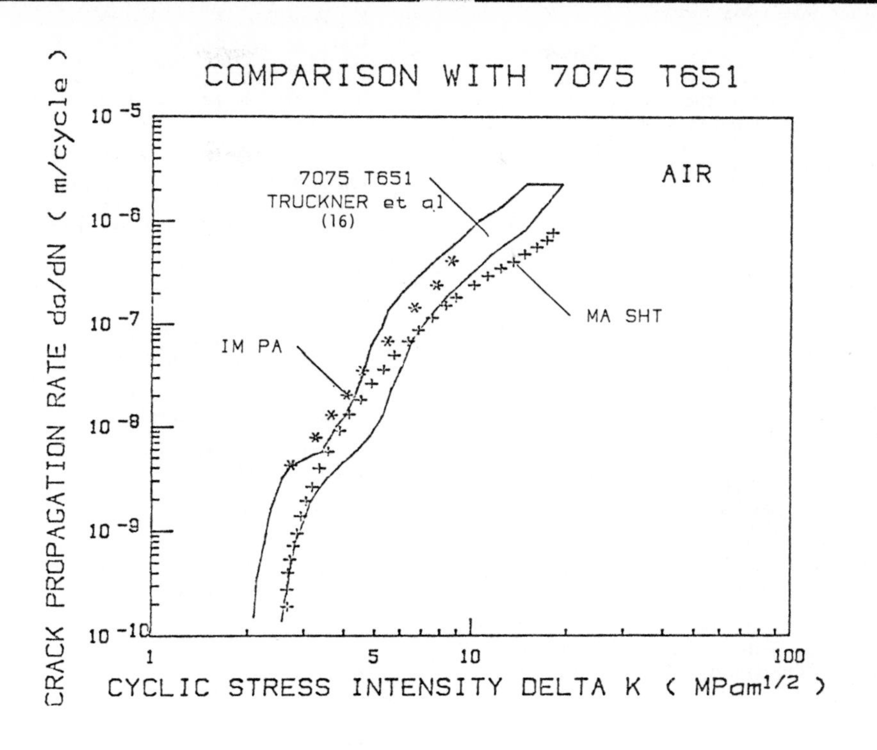

Figure 9. Comparison of FCP behavior with other alloys.

behavior was improved. This effect may explain the differences in propagation rates and threshold values.

All the alloys exhibited very straight cracks running 90° to the rolling direction. No signs of crack branching, crack path tortuousity or the occurrence of wear debris, indicating a Mode II component in crack opening, at the sides of the cracks could be found. The complete absence of microscopic indicators for pronounced slip localization, among them extrusions and extense slip lines which are usually observed close to the crack path on the polished side of the specimen, confirms again the homogeneous deformation behavior of these alloys compared to conventionally processed ones. Localized slip may take place on microscopic levels.

The fatigue crack propagation fracture surfaces of all four alloys show essentially the same type of non-crystallographic crack propagation. The overall fineness of the microstructure makes it hard to identify individual features in the fractographs in figure 8. These fracture surfaces are representative for alloys 3 and 4, too. At low ΔK values the crack seems to propagate transgranularly and lots of dispersoids are displayed as clearly indicated in figure 8c. At higher ΔK there is not much difference in the fracture appearance of alloy 1, but alloy 2 exhibits fairly large, sheared regions. This phenomenon is even more pronounced in alloys 3 and 4 and probably associated with the presence of shearable δ' precipitates.

Figure 9 shows how MA2 with the best overall fatigue crack propagation compares to a peakaged Al-2.2Li-5.4Mg-0.12Zr (IM) alloy, produced by ingot metallurgy, and 7075 in the T651 temper. Considering the fact that the 7075 data are from experiments with a slightly higher R-ratio (0.3) but not closure corrected, the MA alloy falls over a wide range in the 7075 scatterband. The IM alloy has a much lower yield strength (330 MPa) whereas the Young's modulus is about the same as in MA. One can see that the mechanically alloyed material is even slightly better than the IM alloy where crack propagation is concerned. Details of this comparison have been discussed elsewhere (6).

CONCLUSIONS

1. Mechanically Al-Li-Mg alloys show good strength and ductility.
2. The dispersoids control the mechanical behavior directly by acting as slip homogenizers and indirectly by determining the grain size.
3. A dispersoid level of about 4vol% combined with 1.75w%Li and 4w%Mg gives best overall properties.

4. Fatigue crack propagation rates are comparable to 7075 T651 and other Al-Li-Mg alloys.

ACKNOWLEDGEMENTS

This work was sponsored by the Air Force Wright Materials Laboratory Contract F33615-81-C-5053, Dr. T. M. F. Ronald, Program Manager, under subcontract with the Boeing Commercial Airplane Company.

REFERENCES

1. E. A. Starke, Jr. and F. S. Lin, Met. Trans. A, **13A,** 1982, p. 2259.
2. W. A. Cassada, G. J. Shiflet and E. A. Starke, Jr., "The Effect of Germanium on the Precipitation and Deformation Behavior of Al-2Li Alloys," Acta Met, (in press).
3. R. T. Chen and E. A. Starke, Jr., Mater. Sci Engr. **67,** (1984), p. 229.
4. R. D. Carter, E. W. Lee, E. A. Starke, Jr., and C. J. Beevers, Met. Trans. A., **15A,** 1984, p. 555.
5. P. S. Gilman, Aluminum-Lithium Alloys II, ed. T. H. Sanders, Jr. and E. A. Starke, Jr. TMS-AIME, Warrendale, PA, 1984, p. 485.
6. W. Ruch, K. V. Jata, and E. A. Starke, Jr., "Fatigue 84," Proc. of the 2nd International Conference on Fatigue and Fatigue Thresholds, Birmingham, U.K., September 1984, Vol. 1, p. 145.
7. J. R. Pickins, private communication.
8. K. Dinsdale, S. J. Harris, and B. Noble, Aluminum-Lithium, ed. T. H. Sanders, Jr., and E. A. Starke, Jr., Warrendale, PA, 1981, p. 101.
9. D. L. Erich and S. J. Donachie, Met. Progress, **121,** (2), 1982.
10. C. J. Beevers, 1982, in Fatigue Thresholds: Fundamental and Engineering Applications, Vol. I, J. Backhund, A. Blom and C. J. Beevers, eds., Engr. Mat. Advisory Services, Ltd., West Midlands, UK, 257.
11. J. Telesman in NASA Technical Memorandum 83626, April 1984.
12. P. C. Paris, Proc. 10th Sagamore Conference, Syracuse University Press, Syracuse, NY, 1962, p 107.
13. J. Lindigkeit, G. Terlinde, A. Gysler, and G. Lutjering, Acta Metall., **27,** 1979, p. 1717.
14. E. Hornbogen and K. H. Zum Gahr, Metallography, **8,** 1975, p. 181.
15. V. Kuo and E. A. Starke, Jr., Met. Trans. A., **16A,** 1985, p. 1089.
16. W. G. Truckner, J. T. Staley, R. J. Bucci, and A. B. Thakker, AFML-TR-76-169, October 1976.

Microstructures of Al–Li–Si alloys rapidly solidified

G CHAMPIER and F H SAMUEL

The authors are with Ecole des Mines, Institut National Polytechnique de Lorraine, Parc de Saurupt, 54042 Nancy, France. FHS is on leave from the National Research Center, Cairo, Egypt

SYNOPSIS

An effective way of improving the ductility of aluminium-lithium alloys is to introduce dispersoid particles and to refine the particle size using the rapid solidification process. The addition of silicon to Al-Li alloys leads to the precipitation of the AlLiSi compound. The different microstructures produced by the melt-spinning technique are observed as the concentration of silicon changes. When the silicon atomic concentration is lower than that of lithium, all the silicon is included in the AlLiSi precipitates ; the lithium remains in the solid solution and, if the lithium concentration is large enough, in the Al_3Li precipitates. An improvement in the mechanical properties and in the density can be expected from these microstructures.

INTRODUCTION

In the binary aluminium-lithium alloys the strengthening effect is due mainly to the long-range-order of the Al_3Li precipitates. Nevertheless this strengthening is reduced when the particles are sheared by the dislocations and this produces a slip concentration leading to cracking in the extended slip bands and/or across grain boundaries. Furthermore the precipitate free zones which appear along the grain boundaries are sites of preferential deformation leading to high stress concentration at grain boundaries : cracks can nucleate and propagate in an intergranular way within the precipitate free zone. These mechanisms reduce the ductility of the Al-Li binary alloys (Starke and Sanders (11)).

A method for making the deformation less non-homogeneous is to add elements that form incoherent dispersoid during solidification and/ or heat treatments. The plastic deformation becomes therefore more homogeneous and the early crack nucleation due to the extended slip bands is inhibited. Several types of additive have been studied, for example copper, magnesium, cobalt, manganese, zirconium,... (2-10).

In this paper we report preliminary results on the microstructure of Al-Li-Si alloys. From the liquidus surface for the Al-Li-Si system determined by Hanna and Hellawel (11), we expect precipitation of the AlLiSi ternary compound. This has a $F\bar{4}3m$ cubic structure with a lattice parameter a = 0.593 nm and a density ρ = 1.95 ± 0.01 g/cm³ (12).

In cast alloys the size of the AlLiSi precipitates is important; several tens of microns, and the largest ones act as stress concentrators leading to a crack initiation at the interface matrix-precipitate, and so to a reduction in the ductility. Rapid solidification is an effective method of refining the size of the precipitates. We have used the melt-spinning technique in order to prepare the Al-Li-Si alloys.

EXPERIMENTAL PROCEDURE

Four aluminium-lithium-silicium alloys were prepared by induction melting of Al-Li alloy and silicon in a crucible of graphite coated with boron nitride in the open air ; they were chill cast into graphite molds. The atomic concentrations of lithium and of silicon were respectively ALS1 : 8.97 % and 0.98 % ; ALS2 : 8.90 % and 1.79 % ; ALS4 : 8.74 % and 3.78 % ; ALS8 : 8.41 % and 7.28 %.

Melt-spun ribbons were produced by induction melting and ejecting the molten alloys onto a copper-2 % chromium wheel of 28 cm in diameter and having a surface velocity of 33 m/s. The sample, 25 g in weight, was ejected from a silica glass crucible coated with boron nitride into a helium atmosphere under a pressure of 0.4 bar.

Four annealing ribbons were put in a box furnace under an argon atmosphere ; the pressure was 1 bar and the temperature 200 ± 2°C.

X-ray diffraction spectra were obtained with a cobalt target using the $K\alpha$ beam. The optical mi-

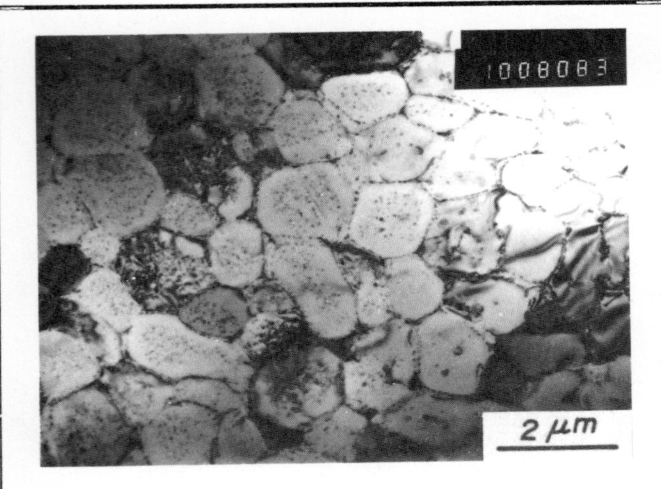

Fig.1 : Bright field electron micrograph of the melt-spun ALS1 alloy. The area observed is in the transition part of the ribbon. The upper left zone is near the wheel side, the lower right zone is near the gas side.

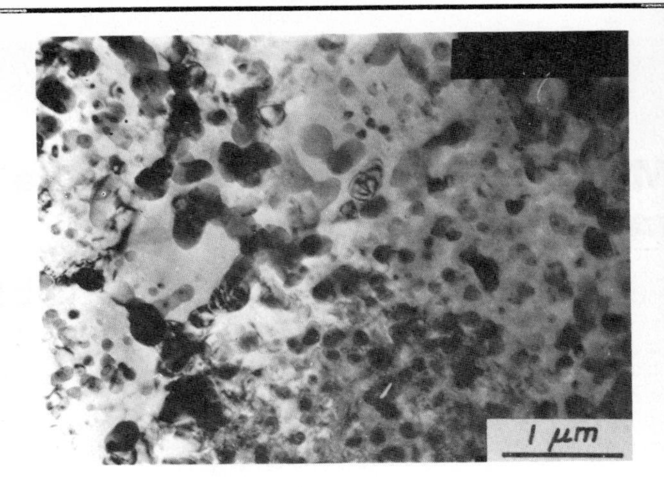

Fig.2 : Bright field electron micrograph of the melt-spun ALS4 alloy. The area observed is in the middle part of the ribbon.

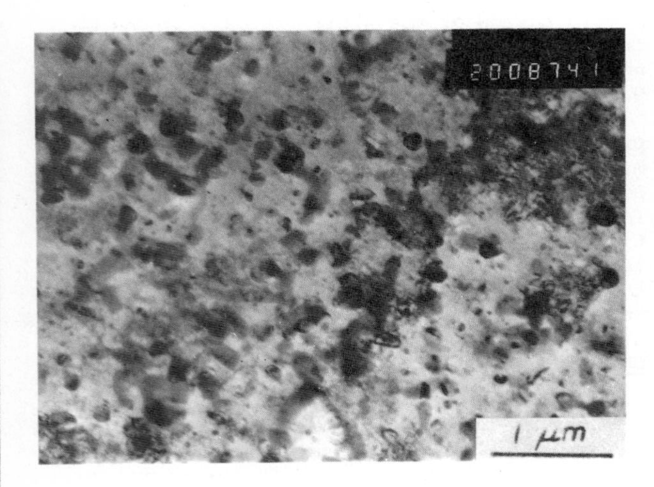

Fig.3 : Bright field electron micrograph of the melt-spun ALS8 alloy. The area observed is near the wheel side.

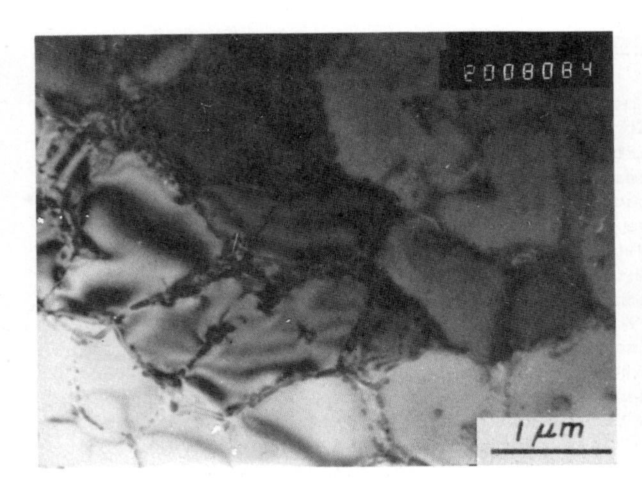

Fig.4 : Bright field electron micrograph of the cell microstructure on the gas side of the melt spun ALS1 alloy ribbon

crography was made after mechanical polishing and chemical etching by Keller's reagent. Thin foil specimens were prepared by electropolishing 3 mm discs extracted from the ribbons. A solution of ethanol (80 %) and perchloric acid (20 %) at 12 V - 1 A produced acceptable thin foils for transmission electron microscopy using a JEOL - 200 CX electron microscope.

RESULTS

The optical micrograph of the cross section of the ALS1 alloy ribbon reveals two zones : on the wheel side a featureless microstructure with a thickness equal to a half or two thirds of the ribbon thickness, and on the gas side a cell microstructure. When the silicon content increases the thickness of the featureless zone near the wheel side decreases, the cell microstructure becomes thicker and thicker and a coarse microstructure appears on the gas side.

By X-ray diffraction two phases are detected in the melt-spun ribbons : aluminium and AlLiSi ternary compound.

Electron micrography shows that the featureless microstructure corresponds to a grain microstructure with precipitates inside the grains and on the grain boundaries (Fig.1). In the ALS1 alloy the grain size is equal to 1-2 µm ; the precipitates inside the grains are 20-40 nm in diameter and a precipitate free zone, 200 nm in width, runs along the boundaries. In the ALS2 alloy the microstructure is about the same, only the precipitate density is larger. In the ALS4 and ALS8 alloys the grain size is larger, 2-4 µm, and the precipitates are respectively 50-70 and 150 nm in size (Fig.2 and 3). From the electron diffraction patterns we can identify the precipitate phase as the cubic AlLiSi compound. Inside a grain all the precipitates have the same orientation and we obtain a point pattern for the α-Al phase and the AlLiSi phase.

Electron micrography shows that the cells of the zone following from the wheel side are about 1 µm in diameter and they are free of precipitates (Fig.4). Some precipitation is visible in the intercellular regions. In the ALS1 and ALS2 alloys the precipitates are about 100 nm in size ; they are 150-250 nm in alloy ALS4. The electron diffraction reveals a point pattern for the α-Al matrix and a ring pattern for the AlLiSi phase.

In the coarse zone on the gas side, the AlLiSi precipitates are larger, from 0.4 to 1 µm for example in the case of the ALS4 alloy.

During the annealing of the ALS1 alloy precipitates of the Al_3Li phase appear inside the grains ; they are 10-30 nm in diameter and they let a precipitate free zone, 400 nm in width, along the grain boundaries (Fig.5). Some precipitate free zones appear also around the initial AlLiSi precipitates. The size of these precipitates remains constant but new very fine particles of AlLiSi phase precipitate inside the grains ; they are 15 nm in diameter and they let a narrow precipitate free zone along the grain boundaries. The annealing produces the same effects in the cell microstructure, only the precipitates are somewhat larger (Fig.6).

In the ALS2 alloy the precipitation of Al_3Li is again visible but the density of these precipitates is lower (Fig.7). The initial AlLiSi precipitates remain practically unchanged and very fine new ones appear with greater density. In the ALS4 and ALS8 alloys no more Al_3Li phase precipitates and we note chiefly a very small enlargement of the previous AlLiSi precipitates (Fig.8).

DISCUSSION

The difference in the microstructure observed on both sides of the ribbon is due to the difference in the solidification rate. On the wheel side the cooling rate is the highest ; at solidification, the extension of the solubility (Jones (13)) leads to α-Al grains supersaturated with lithium and with silicon ; during cooling in the solid state a part of the lithium and silicon precipitate in AlLiSi particles ; these have the same orientation inside a grain and they give an electron diffraction point pattern. On the gas side the cooling rate is the lowest ; at the solidification a part of the lithium and silicon is rejected from the cells into the intercellular regions ; in these regions the concentrations become high and AlSiLi particles precipitate with random orientations ; during cooling in the solid state the supersaturation remains low inside the cells and we observe only α-Al cells at the room temperature. The randomly oriented AlLiSi particles in the intercellular regions give an electron diffraction ring pattern. The coarse microstructure observed in the ALS4 and ALS8 alloys is due to the low cooling rate and to the high silicon concentration.

After the melt-spinning the ALS1 and ALS2 alloys are supersaturated with lithium and silicon. During annealing at 200°C a part of lithium remains in solid solution, another part reacts with aluminium giving Al_3Li precipitates and the last part reacts with aluminium and silicon giving very fine AlLiSi particles. After the melt-spinning the ALS4 and ALS8 alloys are only slightly supersaturated with lithium and silicon, the volume fraction of the AlLiSi precipitates is higher than for the previous alloys. During annealing at 200°C the lithium remains in solid solution and we do not observe any new precipitates.

We deduce from these observations that all the silicon is involved in the formation of the AlLiSi precipitates. If the atomic concentration of lithium is higher than the atomic concentration of silicon, no silicon precipitates are formed during the melt-spinning and the annealing processes. The binding forces between lithium, silicon and aluminium atoms in the AlLiSi compound are probably very strong and this explains the formation of this compound at the solidification of the alloys.

From the previous results we can calculate the expected density of an Al-Li-Si alloy. For this we assume that the atomic concentration of silicon is lower than that of lithium, the silicon is wholly included in the AlLiSi compound and the remaining lithium is in the solid solution or in the compound Al_3Li, the lattice parameters of both are practically equal. The calculation shows therefore that the density of the

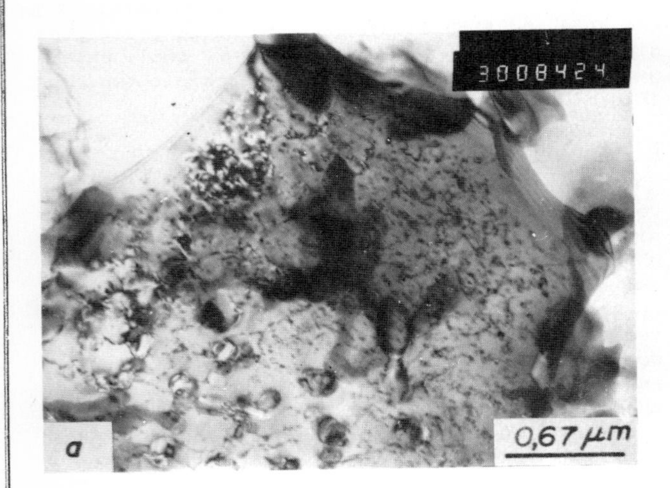

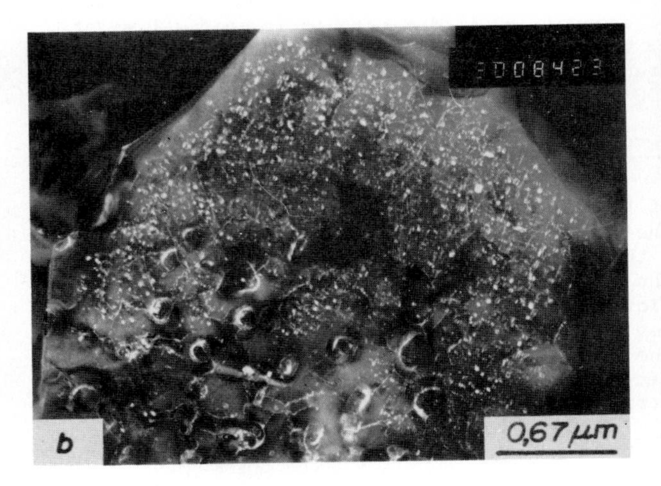

Fig.5 : Electron micrographs of the grain micro-structure on the wheel side of the ALS1 alloy annealed at 200°C for 100 hours (a) bright field, (b) dark field image using an AlLiSi reflection.

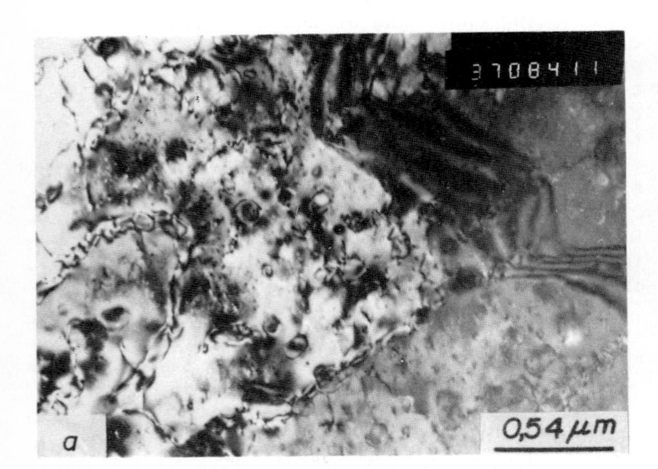

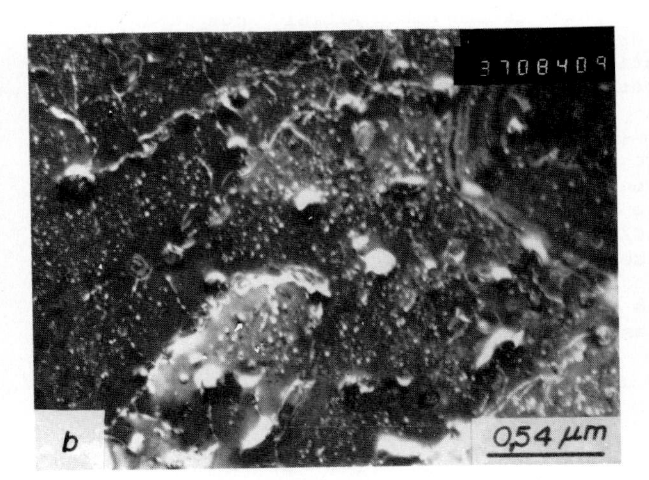

Fig.6 : Electron micrographs of the cell micro-structure on the gas side of the ALS1 alloy annealed at 200°C for 100 hours (a) bright field, (b) dark field image using an AlLiSi reflection.

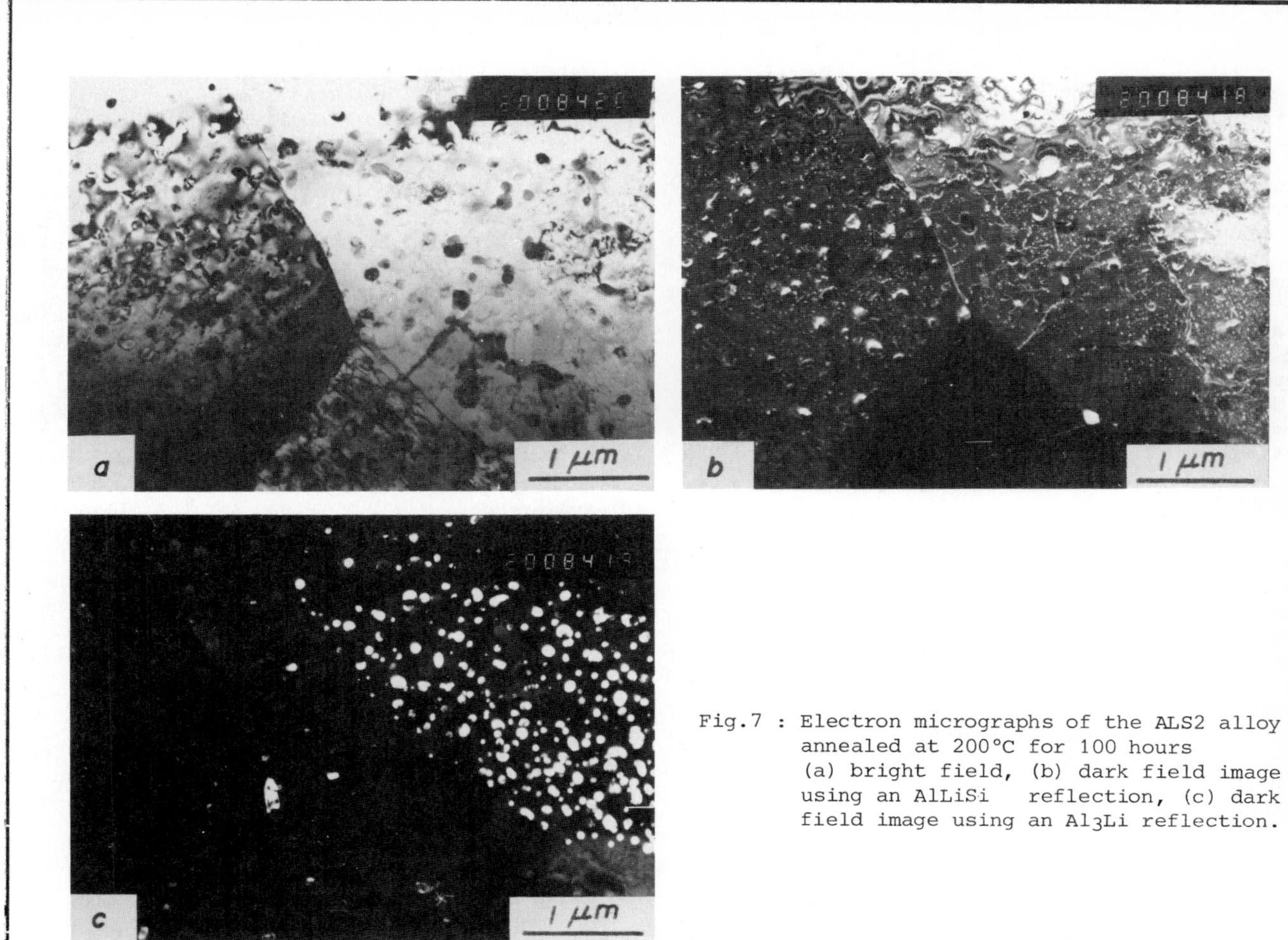

Fig.7 : Electron micrographs of the ALS2 alloy annealed at 200°C for 100 hours (a) bright field, (b) dark field image using an AlLiSi reflection, (c) dark field image using an Al_3Li reflection.

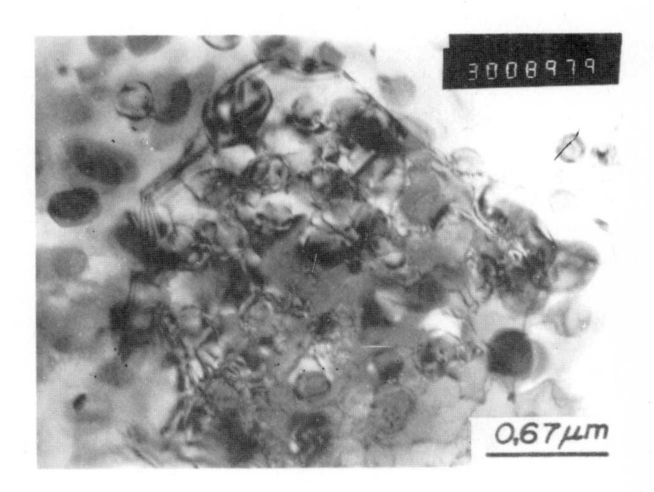

Fig.8 : Bright field electron micrograph of the ALS4 alloy annealed at 200°C for 100 hours. The area observed is in the middle part of the ribbon.

alloy can be expressed by the relationship :

$$\rho(g/cm^3) = 2.70 - 0.24\, b\, C_{Si} - 2.008\, C_{Li}$$

C_{Si} and C_{Li} are respectively the atomic concentrations of silicon and of lithium. This relation ship is to be compared with that proposed by Peel and Evans (14) for more complex alloys. In our case the addition of silicon produces a slight decrease in density instead of an increase.
The measured value of the density for the ALS1 alloy, 2.526 g/cm³, is close to the value calculated by the previous expression, 2.517 g/cm³.

In our study it would be interesting to increase the concentration of lithium in the ALS4 and ALS8 alloys in order to precipitate the Al_3Li compound with an appreciable volume fraction. So we can expect from this an improvement in the density of the alloy and in its mechanical properties.

CONCLUSIONS

The addition of silicon to the Al-Li alloys rapidly solidified leads to different microstructures. If the silicon atomic concentration is lower than that of lithium, all the silicon is included in the AlLiSi particles. If the lithium concentration is too low, the lithium remains in the solid solution and the Al_3Li compound does not precipitate. For a given value of the volume fraction of the Al_3Li precipitates, the addition of silicon permits an increase in the concentration of lithium in the alloy and this produces a decrease in the density.

REFERENCES

(1) Starke Jr E.A. and T.H. Sanders Jr. J. of Metals 33 (Aug.1981) 24.

(2) Sanders Jr T.H. and E.A. Starke Jr. Aluminium Lithium Alloys II. The Metallurgical Society/AIME (1984) 1.

(3) Kulwicki J.H. and T.H. Sanders Jr. Aluminium Lithium Alloys II. The Metallurgical Society/AIME (1984) 31.

(4) Parson N.C. and T. Sheppard. Aluminium-Lithium Alloys II. The Metallurgical Society/AIME (1984) 53.

(5) Noble B., S.J. Harris and K. Harlow. Aluminium Lithium Alloys II. The Metallurgical Society/AIME (1984) 65.

(6) Sastry S.M.L. and J.E. O'Neal. Aluminium-Lithium Alloys II. The Metallurgical Society/AIME (1984) 79.

(7) Gilman P.S. Aluminium-Lithium Alloys II. The Metallurgical Society/AIME (1984) 485.

(8) Avalos-Borja M., P.P. Pizzo and L.A. Larson. Aluminium-Lithium II. The Metallurgical Society/AIME (1984) 287.

(9) Feng W.X., F.S. Lin and E.A. Starke Jr. Aluminium-Lithium II. The Metallurgical Society/AIME (1984) 235. Metall. Trans. 15A (1984) 1209.

(10) Gayle F.W. and J.B. Vander Sande Scr. Metall. 18 (1984) 473

(11) Hanna M.D. and A. Hellawell. Metall. Trans. 15A (1984) 595.

(12) Schuster H.U., H.W. Winterkeuser, W. Schaffer and G. Will. Z. Naturforsch. 31 (1976) 1540.

(13) Jones H., Rapid Solidification of Metals and Alloys. The Institution of Metallurgist. London (1982).

(14) Peel C.J. and B. Evans. The Metallurgy of Light Alloys. The Institution of Metallurgists London (1983) 32.

Weldability of Al–5Mg–2Li–0.1Zr alloy 01420

J R PICKENS, T J LANGAN and E BARTA

JRP is in the Materials Science Department of Martin Marietta Laboratories, Baltimore, MD, USA. TJL is in the same Department of Martin Marietta Laboratories. EB is with Martin Marietta Baltimore Aerospace

SYNOPSIS

The weldability of Soviet-developed Al-5%Mg-2%Li-0.1%Zr (wt%) alloy 01420 was studied using Tungsten-Inert Gas (TIG) welding with either an Al-Mg or 01420 filler. Alloy 01420 is indeed weldable and tensile strength joint efficiencies as high as 85% were obtained using 01420 filler, in conjunction with post-welding heat treatment. Although the alloy is susceptible to weld-zone porosity, milling the alloy surface prior to welding, either chemically in a NaOH solution or mechanically, reduced porosity to acceptable levels.

Surface science studies revealed Li depletion in the weld bead when using 01420 filler which, along with microstructural coarsening, account for the softening in the weld bead with respect to the base-material. Nevertheless, the attractive joint efficiencies obtained suggest that the alloy should receive greater attention in the Western world for applications requiring a low density, weldable alloy.

INTRODUCTION

Aluminum-lithium alloys are currently receiving much attention because of their reduced density and increased elastic modulus.(1) In general, the alloys being developed have been for aircraft applications, where the weight savings effected by using these low-density alloys greatly reduce vehicle fuel costs and also increase performance.(2) Because most aircraft parts are mechanically fastened, the weldability of Al-Li alloys has received relatively limited attention. If weldable Al-Li alloy variants were available commercially, this alloy system could potentially be used for many non-aircraft applications, such as marine hardware, lightweight pressure vessels, and lightweight armor.

A recent literature survey(3) of weldable Li-containing aluminum alloys revealed that Soviet-developed Al-5%Mg-2%Li-X* alloy 01420(4-8) is the most widely used such alloy in the world. Its density, 2.47 g/cm^3,(4) is lower than that of any commercial aluminum alloy. Alloy 01420 was developed by Fridlyander(5) and several alloy variants bear the 01420 designation. These variants all contain the nominal 5%Mg and 2%Li, but can contain either 0.05-0.3%Zr, 0.2-1.0%Mn, 0.05-0.3%Cr, 0.05-0.15%Ti, or 0.05-0.3%Cr + ≤0.1%Ti for grain refinement.

The fusion weldability of 01420 has been demonstrated in numerous Soviet investigations. For example, Mironenko et al.,(9-12) have extensively studied the weldability of 01420 and claimed that the alloy has "good weldability".(9) They also found that Al--5-6.5%Mg--0.0021-0.0037%Be filler materials containing various amounts of Mn, Ti, Zr, or Cr produce the strongest joints, and parent material filler produced joints about 15% lower in strength.

Several investigators have reported attractive joint efficiencies (i.e., strength of weld divided by strength of parent material) for 01420 weldments.(11,13,14) For example, Fridlyander et al.,(13) obtained joint efficiencies of 70% without post-weld heat treatment and as high as 99.6% after re-solutionizing, air cooling, and artificial aging. In addition, Ischenko and Chayun(14) observed joint efficiences of 80% on Ar-TIG (tungsten-inert-gas) welds that received no post-weld heat treatment using an Al-6.3Mg-0.5Mn-0.22Zr filler alloy.

Numerous Soviet investigators have claimed

---

* Unless stated otherwise, all compositions in this paper are in weight%.

Table I

COMPOSITION OF ALLOY 01420 (wt%)

(by spark spectrographic technique, unless stated otherwise)

Mg	Li	Zr	Si	Fe	Cu	Ti	Pb	Mn	Cr	Ni	Zn
5.09	2.05	0.075	0.06	0.06	0.07	0.011	0.01	0.00	0.00	0.00	0.00

Na[†]	Ca[†]	K[†]	H[*]
0.0007	0.0015	0.0004	0.00016 (sample 1) 0.00013 (sample 2)

† By Atomic Absorption
* Performed by Leco Corp.

Table II

Surface Preparation Techniques used that Produced Weldments that Passed Radiographic Inspection

Weldment No.	Preparation Technique	Amount removed from surface
3	Powered wire brushing	0.013 mm (0.0005 in.)
5	Machining	0.13 mm (0.005 in)
6	Chemical milling (30% NaOH at 50-60°C followed by rinsing in 30% HNO_3)	0.13 mm (0.005in)
7	Chemical milling (30% NaOH at 50-60°C followed by rinsing in 30% HNO_3)	0.25 mm (0.010 in)

TABLE III

Tensile Properties of 01420 Base-Material

Orientation	UTS MPa(Ksi)	YS MPa(Ksi)	El %
L	425 (61.7)	291 (42.2)	16.6
LT	425 (61.7)	263 (38.2)	16.9

a) Each datum is the mean of 3 tensile test results.

that alloy 01420 is particularly susceptible to weld-zone porosity.[9,10,15-20] In a systematic study of surface preparation techniques, Fedoseev et al.,[15] found that mechanically milling 0.5mm from the 01420 surface, or chemically milling 0.3mm from the surface in a 200 g/ℓ NaOH aqueous solution prior to welding would reduce weld-zone porosity to acceptable levels. Likewise, Mironenko et al[9] found that chemically milling the surface in an alkaline solution [20-30% NaOH (6.5-9.7M) aqueous solution 40-60°C, followed by cleansing in a 30% HNO_3 (4.5M) desmutting solution] would greatly reduce weld-zone porosity.

Thus, Soviet researchers claim that alloy 01420 is weldable and that good joint efficiencies can be obtained, but great care must be exercised during surface preparation. The goal of the present work is to verify these claims, and to establish preliminary TIG welding parameters for alloy 01420.

MATERIALS

The 01420 variant that uses only Zr for grain refinement was selected for this study for the following reasons: 1) the alloy would be as simple a commercial alloy as practical, i.e., a nominal quarternary alloy, 2) studies of other aluminum alloys suggest that Zr improves weldability,[21,22] 3) Zr improves the corrosion resistance of 01420;[8] and 4) the current trend in Al-Li alloy development favors Zr over other grain refiners.[23]

The alloy was direct-chill cast into 15.2 cm (6.0 in.) diameter billets at Martin Marietta Aluminum, Torrance Division, Torrance, CA; its composition is in Table I. Although the impurity content is typical of commercial alloys, the hydrogen and sodium contents are low for a Li-containing aluminum alloy. The billet was extruded without lubrication (i.e., hydrocarbons not present) to 7.6 x 0.63 cm (3.0 x 0.25 in.) plate, solutionized at 450°C, water-quenched (WQ), stretch-straightened with 2.5% plastic strain, and artificially aged at 120°C for 12 h. This heat treatment was claimed to produce "high strength" in the Soviet literature.[24,25] Material in this condition will be referred to as "base-material."

EXPERIMENTAL PROCEDURES

Aging studies measuring Rockwell B hardness were performed at 120°C and 170°C to determine whether the strength of the alloy could be increased over that attained by the aforementioned heat treatment. Tensile tests were performed on the base-material in the longitudinal (L) and long transverse (LT) orientations.

The following surface preparation techniques were performed on various specimens prior to welding:

1. Remove 0.013 mm (0.0005 in.) by powered wire brushing.
2. Chemically mill in a 30% NaOH (6.5 M) aqueous solution at 50-60°C removing either:
   a. 0.13 mm (0.005 in) or
   b. 0.25 mm (0.010 in),
   then rinse in 30% HNO_3 aqueous solution (4.8 M).
3. Machine 0.13 mm (0.005 in.) from the surface in a milling machine.

The preparation techniques that were used for the weldments that subsequently passed the radiographic inspection are in Table II.

TIG welding was performed using a Miller Model #BWC-500 MAPA welding machine at Martin Marietta Baltimore Aerospace, Baltimore, MD. The 01420 extruded plate was welded along its L axis using a bevelled butt design (see Fig. 1). Base-material filler material was used for most welding and was fabricated by machining the 01420 extruded plate to 3.2mm square wires. None of the Soviet fillers claimed to be superior to 01420 filler by Mironenko et al.[9] could be obtained, so a somewhat similar filler, 5356, was used (composition: 5%Mg-0.4%Fe-0.25%Si-0.2%Mn-0.2%Ti-0.2%Cr-0.1%Zn, balance Al). Approximately 0.25 mm (0.010 in.) was removed from the filler wires by chemical milling in a 20-30% NaOH (6.5-9.7M) aqueous solution at 20°C, and cleansing in the aforementioned HNO_3 desmutting solution prior to welding. The filler wires were wiped with reagent grade methyl-ethyl-ketone immediately prior to welding, and handled only with clean gloves.

All welding was performed using 75% He + 25% Ar cover gas at a flow rate of 425 ℓ/h. The 01420 was unrestrained during welding and alternating current was used at a potential of 15 volts. Each TIG weld was made manually using three passes. Radiographic inspection was performed on each weldment and those containing severe porosity were discarded. Welding parameters found to produce weldments worthy of further study are detailed in the Appendix (p. 142), whereas those found to be totally unsatisfactory are not reported.

Optical microscopic examinations were performed on cross sections of several weldments in the as-polished condition, and after etching in a 0.5% HF aqueous solution (0.25 M) for 1 min. Vickers hardness (Hv) measurements were made across selected transverse weld sections using a 2.5-kg load.

Triplicate tensile tests of the weldments were performed (LT orientation) in both the as-welded condition, and after re-heat treating the weldment [solutionize at 450°C(2 h) - WQ - age at 120°C (12 h)].

Because alloy 01420 is claimed to be extremely susceptible to weld-zone porosity, surface science studies were performed on the as-received surface oxide film, and on material with this film removed. It was hoped that these studies could determine the amount of material that needs

TABLE IV

Weld-Zone Porosity

Weldment No.	Pore Size (μm)	# Pores/area (Pores/cm²)	Comments
3	20-50	20	Pores only in weld bead
5	10-15	20	Pores only in weld bead
6	25 75	30 10	Duplex pore distribution Worst porosity of weldments that used 01420 filler
7	<10	<15	Lowest porosity observed

TABLE V

Long Transverse Tensile Properties of Weldments and Joint Efficiencies With No Post-Welding Heat Treatment[a]

Weldment No.	Tensile Specimen No.	UTS MPa(Ksi)	Apparent YS MPa(Ksi)	El (%)	UTS Joint Eff. (%)	Apparent YS Joint Eff.(%)
3	-1	239 (34.6)	86 (12.5)	18.7	56.2	32.7
	-2	245 (35.6)	112 (16.3)	17.1	57.6	42.6
	-3	235 (34.1)	114 (16.6)	14.0	55.2	43.3
5	-1	209 (30.3)	113 (16.4)	14.0	49.2	43.0
	-2	261 (37.9)	113 (16.4)	14.0	61.4	43.0
	-3	268 (38.9)	107 (15.5)	n/a	63.1	40.7
6	-1	262 (38.0)	98 (14.2)	17.1	61.6	37.3
	-2	239 (34.7)	105 (15.3)	10.9	56.2	40.0
	-3	261 (37.8)	114 (16.5)	12.5	61.4	43.3
7	-1	268 (38.9)	112 (16.3)	14.0	63.1	42.6
	-2	251 (36.4)	105 (15.3)	15.6	59.1	40.0
	-3	272 (39.4)	119 (17.3)	14.0	64.0	45.2

a) The weldments were air-cooled after welding, and tensile testing was performed about 1 week later.

TABLE VI

Long Transverse Tensile Properties of Weldments and Joint Efficiencies After Re-Heat Treating the Weldments

Weldment No.	Tensile Specimen No.	UTS MPa(Ksi)	Apparent YS MPa(Ksi)	El %	UTS Joint Eff.(%)	YS Apparent Joint Eff.(%)
5	-1	363 (52.7)	179 (25.9)	9.3	85.4	68.8
	-2	341 (49.4)	132 (19.1)	9.3	80.0	50.0
	-3	341 (49.4)	114 (16.6)	9.3	80.0	43.4
6	-1	352 (51.0)	156 (22.6)	9.3	82.7	59.2
	-2	343 (49.8)	128 (18.6)	12.5	80.7	48.7
	-3	301 (43.7)	89.6 (13.0)	9.3	70.8	34.0
7	-1	337 (48.9)	---- ----	14.0	79.3	----
	-2	291 (42.2)	130 (18.8)	10.9	68.4	49.2
	-3	300 (43.5)	128 (18.5)	9.3	70.5	48.4

to be removed during welding pre-treatment, and possibly identify the contaminants that cause this porosity. Secondary Ion Mass Spectroscopy (SIMS) profiling and Auger depth profiling were performed on 1) a polished cross section of the base-material, and 2) the as-received surface oxide film. In addition, SIMS was used to assess lithium depletion in the weld-zone and heat-affected-zone (HAZ) of weldment No. 7, which was made using base-material filler. The SIMS measurements were performed at various locations along a polished cross section of the weld, from the parent material far from the weld, through the HAZ, and into the weld bead of weldment No. 7 itself. Similar measurements were made from the top to the bottom of the weld bead along the weld centerline. Before each SIMS measurement was taken, material was sputtered at ~50 Å/min for 10 min. to ensure that measurements were made on uncontaminated material. Lithium, hydrogen, and aluminum concentrations were measured.

RESULTS

The tensile strength of the 01420 base material was the same — 425 MPa (61.7 Ksi) — in both the L and LT orientations. However, the yield strength was about 10% lower in the LT orientation -- 263 MPa (38.2 Ksi) vs. 291 MPa (42.2 Ksi) for the base-material, see Table III. The aging study revealed that the strength of the 01420 could be increased by lengthening the aging time at 120°C, or by aging at 170°C for the 12 h period (see Fig. 2).

Alloy 01420 is indeed susceptible to weld-zone porosity. Several weldments displayed severe porosity, as observed by radiographic inspection and consequently received no further investigation. Weldment No's. 5, 6, and 7, which were made using base-material filler, each contained about four isolated pores, ranging from 0.13 mm (0.005 in.) to 0.25 mm (0.010 in). However, sufficient material in each of these three weldments was free from this macroporosity so they, along with weldment No. 3, passed the radiographic inspection. These four weldments were sectioned metallographically to look for microporosity and to observe the weld-zone microstructures.

Weldment No. 3 was made using material that was power wire brushed to remove 0.013 mm (0.0005 in) prior to welding with 5356 filler. Metallographic examination revealed pores in the weld bead that measured 20-50 μm, at about 20 pores/cm^2, (Table IV). A similar density of finer pores was observed in weldment No. 5, which used base-material filler and had been prepared by machining 0.13 mm (0.005 in.) from the surface.

Chemically milling the 01420 surface before welding with base-material filler produced a low micro-porosity weld-zone when 0.25 mm (0.010 in.) was removed (weldment No. 7) but not when 0.13 mm (0.005 in.) was removed (weldment No. 6), (Table IV). We believe that the difference in porosity resulted from variations in welding parameters or inadvertent contamination and not from the difference in milling depth. This conclusion is based on our SIMS results which will be presented later.

The joint efficiency values were scattered, which we attribute to non-steady-state welding conditions present during the manual welding. The joint efficiencies of the weldments that did not receive post-weld heat treatment were low for apparent yield strength--typically 40%, but at times were above 60% for tensile strength joint efficiency (see Table V). However, specimens that received the entire heat treatment (re-solutionize at 450°C, WQ, age at 120°C for 12 h) were significantly better. Apparent yield strength joint efficiencies were higher; the highest being 67.8% (see Table VI). The tensile strength joint efficiencies were often over 80%--the highest being 85.4%.

Hardness profiles across the weldments are consistent with the joint efficiency values, showing a decrease in Vickers hardness in the weld bead (see Fig. 3). The HAZ's showed hardness variations only over a narrow range. The microstructures of all the weld-zones had grain boundaries decorated with second-phase particles. In all cases, the particles were selectively attacked by the HF etchant. The grains were generally coarser in the weld bead than in the base-material.

The Auger depth profiling revealed a significant concentration of Mg in the surface oxide film, and relatively little aluminum (see Fig. 4a). This information, coupled with the Auger line shapes, (not shown) revealed that the surface oxide film is composed largely of magnesium containing oxides, as opposed to alumina. In addition, significant carbon concentration is present in the surface oxide film, which extends to a depth of about 6000 Å (0.6 μm) This carbon is part of the oxide film (see Fig. 4a) and is not related to the thin contaminate layer that is removed after 15-20 min. (750-1000Å) of sputtering for both the as-received oxide film (see Fig. 4a) and the polished cross section (see Fig. 4b). Lithium was detected by Auger electron spectroscopy, but its level was obscured by a Mg peak and noise.

SIMS profiling revealed both extensive hydrogen (Fig. 5a) and lithium (Fig. 5b) concentrations in the surface oxide film. The lithium and hydrogen profiles follow the same general trend. The hydrogen concentration is slightly lower on the as-received surface film than in the film after 15 min. of sputtering, and is slightly higher on the surface of the polished cross-section than after sputtering (see Fig. 5a). We suggest that this anomaly results from adsorption of contaminants (i.e., water and carbon dioxide. The contaminant layer is thin and is removed by sputtering for about 15-20 minutes (750-1000 Å) from both the as-received oxide film and the polished cross-section (Fig. 5a). After removal of this layer, high concentrations of hydrogen are still present on the as-received oxide surface, which shows that hydrogen is present in the oxide

APPENDIX

WELDING PARAMETERS

(all used 75%He-25%Ar cover gas at 425 ℓ/h, interpass temperature of 107°C, alternating current 60 cycles/s, voltage=15V, 3.97mm (0.56 in) diameter electrode

Weldment No.	Filler	Pass No.	Current (A)	Speed (mm/s)
3	5356 (2.4 mm round)	1	125-130	1.02
		2	150-155	1.81
		3	135-145	1.66
5	Base-material (3.2 mm square)	1	95-105	NA[a]
		2	90- 95	NA
		3	90- 95	1.69
6	Base-material (3.2 mm square)	1	90-105	1.27
		2	90- 95	1.69
		3	90- 95	1.27
7	Base-material (3.2 mm square)	1	100-105	1.27
		2	90- 95	1.69
		3	90- 95	1.69

a) NA = Not available

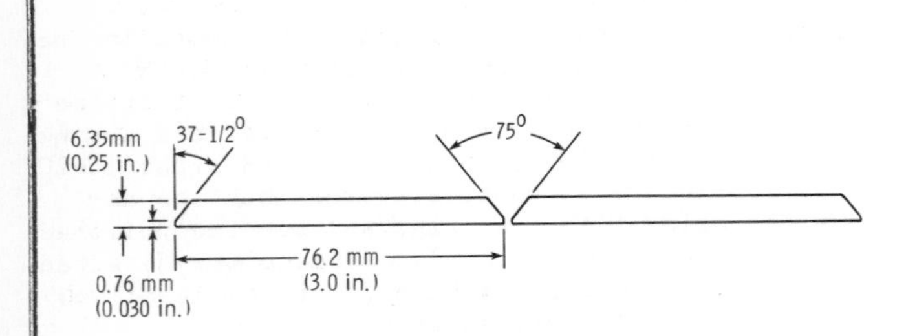

1. Schematic of bevelled butt joint used for TIG welding alloy 01420.

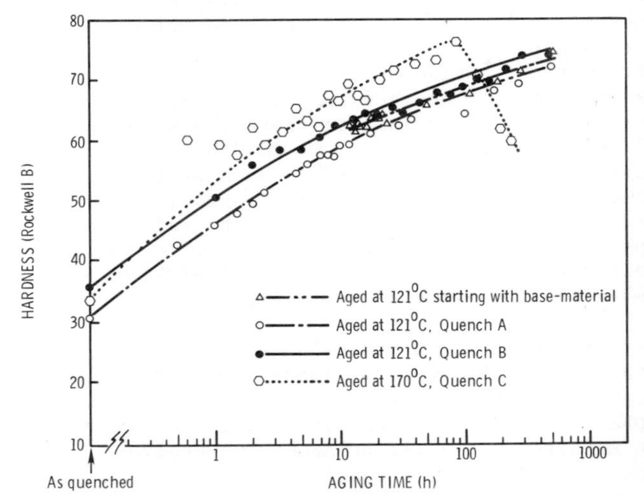

2. Aging curves for alloy 01420 at 121°C and 170°C.

film. After chemical milling in the alkaline solution, the hydrogen content dropped to undetectable levels.

Lithium depletion occurred in the weld-zone of weldment No. 7, which was made using 01420 filler (see Fig. 6). In addition, the lithium concentration in the weld bead increases from the root (bottom) to the crown (top) (see Fig. 7). It is interesting to note that in both cases, the hydrogen concentration follows the same trend as the lithium concentration.

DISCUSSION

Our aging study revealed that the heat treatment used for the alloy 01420 does not produce the highest possible strength. In fact, the 01420 base material probably could be heat treated to about 15% higher strength, while still using practical aging times. Although peak hardness was not reached in the aging study at 120°C, it is clear that greater strength can be obtained at this temperature than at the higher temperature, 170°C. This behavior is typical for precipitation-strengthened aluminum alloys, because at lower aging temperatures, a greater number of nuclei have sufficient time to grow large enough to provide strengthening than at higher aging temperatures, where the more rapidly growing larger particles consume the smaller nuclei.

The susceptibility to weld-zone porosity can be greatly reduced by either mechanical or chemical milling prior to welding. The chemical milling technique offers greater practicality in that irregular sections could be more readily milled chemically than mechanically.

The joint efficiencies in the as-welded condition were measured shortly after welding; ~1 week. They could perhaps be increased by natural aging for a longer period. Nevertheless, the tensile strength joint efficiencies obtained after re-heat treating the weldments were at times over 80%, which is consistent with values reported for 01420 using base material filler.[9] Note, however, that joint efficiency values can sometimes be inflated by testing the weldments with the weld reinforcement intact, which increases the cross-sectional area of the tensile specimen. In the present work, all weld beads were machined off to produce tensile specimens of constant cross-sectional area. Consequently, the joint efficiencies reported were measured conservatively.

It is curious that Mironenko et al.,[9] claimed that joint efficiencies of weldments made using the aforementioned Soviet fillers produced joints about 15% stronger than those made using base-material filler. Such an improvement would increase the highest tensile strength joint efficiency we measured to about 100% of base-material properties.

The fact that the as-received surface oxide film contains a significant amount of magnesium oxide is consistent with surface science studies on Al-Zn-Mg alloys.[26-29] During solution-heat-treatment of Al-Zn-Mg alloys, Mg atoms diffuse to the free surface where they are either oxidized by ambient air, or reduce aluminum oxide, thereby replacing aluminum atoms in the film. It appears that Mg atoms behave in a similar fashion in 01420. In addition, both lithium and carbon were detected in the as-received surface film by SIMS, and Varakina et al.[30] have claimed that primarily Li_2CO_3 and Li_2O, and to a lesser extent, $LiCHO_2$ and LiCOOH, exist in the surface oxide film of 01420 after heat treatment. Varakina et al.[30] also observed MgO in the film. Field et al.[31,32] and Field[33] have observed both Li_2O and Li_2CO_3 on the surface of the aluminum lithium alloys, and claim that the Li_2O adsorbs CO_2 from ambient air during heat treatment to form the lithium carbonate. The Li_2O and, to a lesser extent, the Li_2CO_3 hydrate by adsorbing water from ambient air. Although this water-of-hydration presumably causes the weld-zone porosity, it might be possible that the Li_2CO_3 decomposes, forming CO_2 gas, which also contributes to pore formation.

The SIMS profile for hydrogen content was terminated after 145 minutes at a sputtering rate of 50Å/min and did not fall to the level in the base-material. However, by extrapolating the curve, it apears that the hydrogen content of the as-received surface oxide film would decrease to base-material levels after sputtering away about 13,000Å (see Fig. 5). Although depths estimated by sputtering time are at best only semi-quantitative, it is likely that the depth at which the hydrogen falls to the base-material value is greater than the depth for the carbon to fall to zero (Fig. 4). At present, we can not place too much significance on this observation because the bulk values of hydrogen (Fig. 5b) concentration were measured from a polished cross-section that was ~1 cm below the surface oxide film, and it is possible that hydrogen content varies throughout the cross section of the extrusion.

In any event, it is likely that to reduce weld-zone macro-porosity, significantly less material need be removed from the surface than the 0.13 - 0.25 mm chemically milled or machined off in the present work, and certainly less than the 0.3-0.5mm removed by Feedosev et al.[15] However, even after removing 0.25 mm, some weld-zone microporosity was observed. It is possible that the alloy surface may adsorb contaminants after the milling operation. It would be interesting to perform surface science studies on chemically or mechanically milled surfaces after exposure periods to ambient air to determine the extent and nature (i.e. H, C) of contamination that occurs during the period between pre-treatment and the actual welding operation.

The lithium depletion in the weld-zone (Fig. 6) apparently results from the volatility of lithium in the molten aluminum alloy, which is a well known problem in casting lithium-containing aluminum alloys. It is interesting that the root of the weld, which experiences heating during each

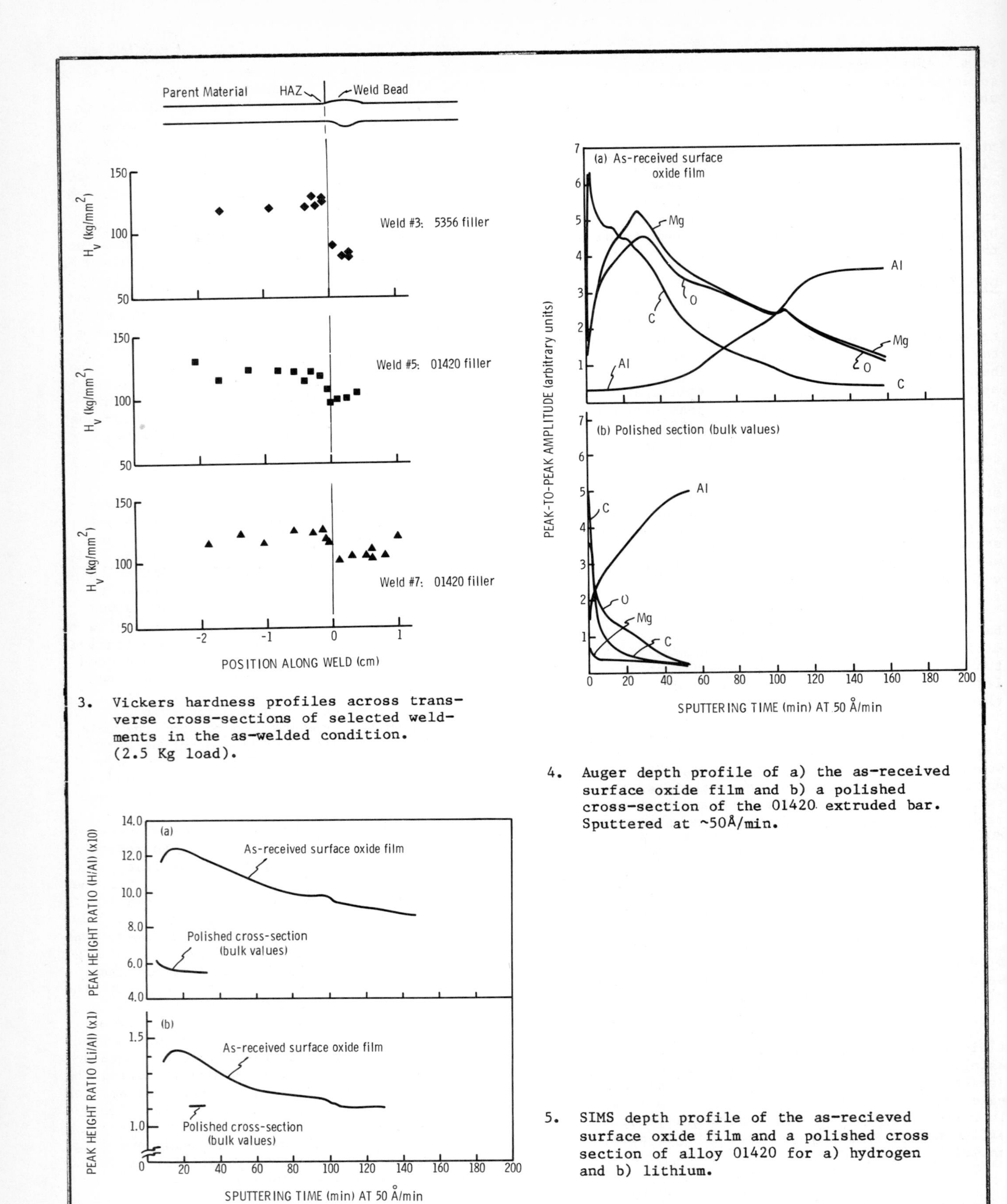

3. Vickers hardness profiles across transverse cross-sections of selected weldments in the as-welded condition. (2.5 Kg load).

4. Auger depth profile of a) the as-received surface oxide film and b) a polished cross-section of the 01420 extruded bar. Sputtered at ~50Å/min.

5. SIMS depth profile of the as-recieved surface oxide film and a polished cross section of alloy 01420 for a) hydrogen and b) lithium.

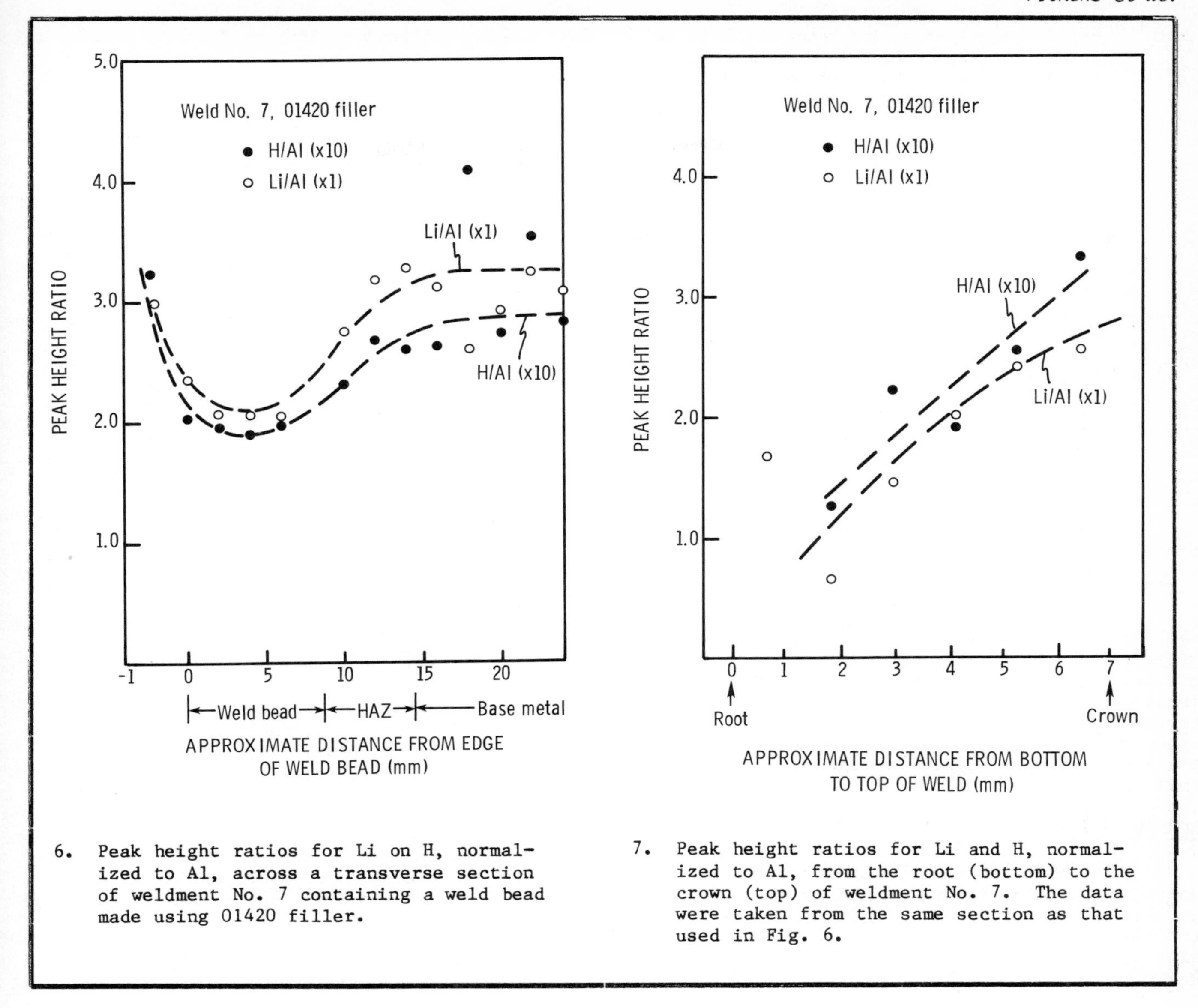

6. Peak height ratios for Li on H, normalized to Al, across a transverse section of weldment No. 7 containing a weld bead made using 01420 filler.

7. Peak height ratios for Li and H, normalized to Al, from the root (bottom) to the crown (top) of weldment No. 7. The data were taken from the same section as that used in Fig. 6.

of the three passes, suffers more lithium depletion than the crown of the weld, which experiences only one welding pass--and consequently one heating cycle. The fact that the joint efficiencies are less than unity are likely in part a result of this lithium depletion. These results suggest that when using base-material filler for welding 01420, rapidly cooling the weldment is likely to be beneficial for reducing lithium depletion in the weld-zone, and the resulting loss in mechanical properties.

It is interesting that the lithium and hydrogen composition gradients follow the same trends (Fig. 5-7). One explanation for this behavior is that both elements are mobile and independently diffuse towards free surfaces. However, lithium is an extremely electropositive, reactive element, and perhaps lithium and hydrogen interact by forming an Li-H complex in lithium-containing aluminum alloys.

CONCLUSIONS

- Alloy 01420 is indeed weldable and tensile strength joint efficiencies of as high as 64% were obtained using base-material filler, without post welding heat treatment. By repeating the entire heat treatment of the alloy, the tensile strength joint efficiency could be increased to as high as 85%.

- Alloy 01420 is particularly susceptible to

weld-zone porosity, but removing 0.13 mm (0.005 in) from the alloy's surface by either chemical milling in a 30% NaOH solution (6.5 M) followed by rinsing in a 30% HNO_3 solution (4.8 M) at 50-60°C, or by machining, reduces weld-zone porosity to acceptably low levels. SIMS data indicate that significantly less material could be milled off to reduce the surface contamination, i.e., hydrogen and carbon, to base-material levels.

- The surface oxide film of 01420, as supplied in the solutionized, quenched, and artificially aged condition, contains high concentrations of magnesium oxide, lithium, carbon, and hydrogen.

- Lithium depletion in the weld bead occurs when welding 01420 with base-material filler and, along with microstructural coarsening, results in a softening of the weld bead with respect to the base-material.

Recommended Work:

Welding studies should be performed using the Soviet filler compositions found to produce the highest joint efficiencies for alloy 01420. In addition, the surface composition should be measured for 01420 as a function of exposure time to ambient air of constant temperature and humidity after milling the surface. This would determine the rate, and possibly the nature of contamination that occurs after welding pretreatment.

ACKNOWLEDGEMENTS

This research was sponsored under a Martin Marietta Baltimore Aerospace IRAD program. The authors are grateful to J. Dougherty for performing the TIG welding, K. Anderson for performing the SIMS analyses, and Drs. J.D. Venables and W.C. Moshier for their helpful comments.

REFERENCES

1. G.G. Wald, NASA Contractor Report 16576, May 1981.

2. W.E. Quist, G.H. Narayanan, and A.L. Wingert, "Aluminum-Lithium Alloys for Aircraft Structure - An Overview," Aluminum-Lithium Alloys II, T.H. Sanders and E.A. Starke, Jr., eds., TMS-AIME Conference Proceedings, NY, Dec. 1983, pp. 313-334.

3. J.R. Pickens, "A Review of the Weldability of Lithium-Containing Aluminum Alloys," Accepted for publication in J. Mater. Sci.

4. I.N. Fridlyander, S.M. Ambartsumyan, N.V. Shiryayeva, and R.M. Gabidullin, Metaloved. Term. Obrab. Met., No. 3, 1968, pp. 211-212.

5. I.N. Fridlyander, British Patent # 1,172,736, Feb. 27, 1967.

6. I.N. Fridlyander, Metaloved. Term. Obrab. Met., No. 4, 1970, pp. 44-51.

7. I.N. Fridlyander, S.M. Ambertsumyan, N.V. Shiryaeva, and R.M. Gabidullin, "New Light Alloys of Aluminum Lithium and Magnesium," Met. Sci. Heat Treat. Met., No. 3, 1968, pp. 211-212, Translated from Metalloved. Term. Obrab. Met., No. 3, pp. 50-52, March 1968.

8. R.W. Rogers, Jr., "Technical Analysis of Soviet Technology, Research and Development for Aluminum Alloys," Alcoa working paper report on Contract F33657-81-C-2070 for Batelle Columbia Laboratories, Jan. 7, 1983.

9. V.N. Mironenko, V.S. Evstifeev, and S.A. Korshunkova, "The Effect of Filler Material on the Weldability of Alloy 01420," Weld. Prod., Vol. 24, No. 12, Dec. 1977, pp. 44-46.

10. V.N. Mironenko, V.S. Evstifeev, G.I. Lubenets, S.A. Karshukova, V.V. Zakherov, and A.I. Litvintsev, "The Effect of Vacuum Heat Treatment on the Weldability of Aluminum Alloy 01420," Weld. Prod., (Engl. Transl.) Vol. 26, No. 1, Jan. 1979, pp. 30-31.

11. V.N. Mironenko, V.S. Evstifeev, and S.A. Korshunkova, "The Weldability of Aluminum Alloy 01420 by Argon-Arc and Electron Beam Welding," Weld. Prod., Vol. 24, No. 10, Oct. 1977, pp. 30-32.

12. V.N. Mironenko and I.F. Kolgarova, "Conditions for Vacuum Heat Treatment of the 01420 Alloy Before Welding," Automatic Welding, (Avt. Svarka), No. 8, 1979, pp. 61-64. Translation pp. 44-48.

13. I.N. Fridlyander, N.V. Shiryaeva, T.I. Malinkina, I.F. Anokhin, and T.A. Gorokhova, "Properties of Welded Joints in Alloy 01420," Met. Sci. Heat Treat. Met., No. 3, 1975, pp. 240-241, Translated from Metalloved. Term. Obrab. Met. No. 3, pp. 53-54, March, 1975.

14. A. Ya. Ishchenko and A.G. Chayun, "Mechanical Properties of Welded Joints Made with a Tungsten Electrode in the 01420 Aluminum Alloy," Autom. Weld. (Avtom. Svarka.), No. 6, No. 6, pp. 56-58. Translation pp. 43-46.

15. V.A. Fedoseev, V.I. Ryazansev, N.V. Shiryaesa, and Yu. P. Arbuzov, Svar. Proizvod., No. 6, 1978, pp. 15-17, Translated pp. 19-22 in Weld. Prod., Vol. 12, 1979.

16. G.E. Kainova and T.I. Malinkina, "Weldability of Aluminum Alloy 01420," Met. Sci. Heat Treat. Met., No. 2, 1969, pp. 104-105, Translated from Metalloved. Term. Obrab. Met., No. 2, pp. 22-23, Feb. 1969.

17. C.A. Bokshtein et al. "Improving the Quality of Welded Joints in Aluminum Alloy 01420,"Automatic Welding, (Avt. Svarka) No. 6, 1978, pp. 34-36 and p. 39. Translation pp. 2426.

18. A. Ya. Ishchenko, "The Weldability of Modern High-Strength Aluminum Alloys," (Review of Published Works), Automatic Welding, Vol. 32, No. 2, Feb. 1979, pp. 18-22.

19. A.G. Chayun, V.V. Syrovatka, and V.I. Matyash, "Arc Welding of 01420 Aluminum Alloy with Electromagnetic Agitation," Autom. Weld., (Avtom. Svarka), 1981, No. 6, pp. 19-21. Translation pp. 15-17.

20. M.A. Abralov, R.U. Abdurekhmanov, and A.T. Iuldashev, "The Effects of Electromagnetic Action on the Properties and Structure of Welded Joints in the 01420 Alloy," Autom. Weld. (Avtom. Svarka), No. 5, pp. 21-24 and p. 29. Translation pp. 15-18, 1977.

21 K.G. Kent, "Weldable Al:Zn:Mg Alloys," Review 147, Metall. Rev., pp. 135-146, 1970.

22. G.M. Young, "Development and Commercial Use of an Al-Zn-Mg Alloy" Metals Mater., 5 (2) pp. 71-75, 1971.

23. Ref 2 op cit., pp. 235-251, 335-362, 407-418.

24. V.G. Kovsizhnykh, V.A. Tikhomirov, N.I. Zaitseva, and M.V. Ermanok, "Mechanical Properties of Extruded Panels of Alloy 01420," Met. Sci. Heat Treat. Met., (Engl. Transl.) No. 2, 1969, pp. 103-ff; Translated from Metalloved. Term. Obrab. Met., No. 2, pp. 20-21, Feb. 1969.

25. A. Ya. Chernyak, E.A. Kosakovski, and A.A. Mezhlum'yan, "Effect of Heat Treatment on the Low Cycle Fatigue Strength of Aluminum Alloy 01420," Met. Sci. Heat Treat. Met., Vol. 20, No. 1, 1978, pp. 63-65, Translated from Metalloved. Term. Obrab. Met., No. 1, pp. 59-61, Jan. 1968.

26. T.S. Sun, J.M. Chen, R.K. Viswanadham, and J.A.S. Green "Surface Activities of Mg in Al Alloys," J. Vac. Sci. Technol. Vol. 16, No. 2, Mar/Apr, 1979, pp. 668-671.

27. R.K. Viswanadham, T.S. Sun, and J.A.S. Green, "Influence of Moisture Exposure on the Composition of Oxides on Al-Zn-Mg Alloy: An Auger Electron Spectroscopy Study," Corrosion, Vol. 36, No. 6, 1980, p. 275.

28. J.R. Pickens, J.R. Gordon, and L. Christodoulou: "Stress-Corrosion Cracking and Hydrogen Embrittlement in P/M X7091 and I/M 7075," in High Performance Aluminum Powder Metallurgy, M.J. Koczak and G.J. Hildeman, eds. (TMS-AIME), Nov. 1982, pp. 177-192.

29. J.R. Pickens, D. Venables, and J.A.S. Green: "Improved SCC Resistance of Al-Zn-Mg Alloys by Control of Mg Content in the Bulk Metal and in the Oxide Film," in Hydrogen Effects in Metals, I.M. Bernstein and A.W. Thompson, eds. (AIME, 1981), pp. 513-523.

30. L.P. Varakina, V.N. Mironenko, and V.M. Polyanski, "The Effect of Aging 01420 Alloy at Room Temperature on the Charge in the Phase Constitution of the Surface Layer," Tekhnol. LegK. Splavcv, (Engl. Transl.) Vol. 6, 1979, pp. 10-13.

31. D.J. Field, E.P. Butler, and G.M. Scamans, "High Temperature Oxidation Studies of Al-3wt%Li, Al-4.2wt%Mg and Al-3wt%Li2wt%Mg Alloys", Aluminum Lithium Alloys, T.H. Sanders, Jr. and E.A. Starke, Jr., eds, AIME-TMS Conference Proceedings, NY, 1981, pp. 325-346.

32. D.J. Field, G.M. Scamans, and E.P. Butler, "The High Temperature Oxidation of Al-Li Alloys", Ref 2 op cit., pp. 657-66.

33. D.J. Field, Alcan Banbury Research Laboratories, Barbury, Oxon, United Kingdom, Private Communication with J.R. Pickens, September 1984.

Adhesive bonding of aluminium–lithium alloys

D J ARROWSMITH, A W CLIFFORD,
D A MOTH and R J DAVIES

DJA and RJD are with the Department of Mechanical and Production Engineering at the University of Aston in Birmingham, UK.
AWC and DAM are at the Admiralty Research Establishment, Portland, Dorset, UK

SYNOPSIS

Al-2·35%Li-1·13%Cu alloy was adhesively bonded using established chromic acid anodizing and phosphoric acid anodizing processes. A new process consisting of sulphuric acid anodizing followed by a phosphoric acid dip was also used. The initial bond strengths and durabilities of lap shear joints of aluminium-lithium and aluminium-magnesium alloys were found to be similar. The anodizing characteristics of aluminium-lithium and aluminium-magnesium alloys were also found to be similar.

INTRODUCTION

Aluminium-lithium alloys are likely to have a considerable impact on the future design, manufacture and operating economics of aircraft[1]. The structure and properties of commercial aluminium-lithium alloys currently available have been given recently by Grimes et al[1]. The addition of lithium to aluminium simultaneously reduces the density and increases the elastic modulus of the resultant alloy. A simultaneous 10% density reduction and 10% stiffness improvement can be achieved with mechanical properties comparable with those of the high strength aluminium alloys not containing lithium. Solution heat treatment and ageing of the alloys are required to give high strength, and, as with all heat treatable alloys, joining by welding can present difficulties, but these have been overcome in the aircraft industry through the design and use of adhesively bonded joints.

Structural adhesives are now extensively used in airframe construction. Metal-to-metal adhesively bonded joints were first used in aircraft in 1944[2]. Progressive development and proven performance of adhesive bonding of aircraft in service led to the design of the current BAe 146 jet aircraft which is based on the use of metal-to-metal adhesively bonded structures throughout and incorporates the largest bonded wing panels in production[3]. Adhesively bonded joints have a number of advantages. The load is more evenly distributed and stress concentrations are minimised. As a result, longer fatigue lives can be obtained with adhesively bonded joints than can be obtained with spot welds or mechanical fasteners especially as adhesively bonded joints tend to dampen vibrations. Dissimilar materials, metals of different thickness, and large areas may be easily and inexpensively joined with weight reductions achieved and streamline flow improved.

Essential pretreatment of the surface of aluminium prior to adhesive bonding has been developed to give not only an initial high bond strength but also durability throughout the long service life of an aircraft[4]. A chromic acid / sulphuric acid etch followed by chromic acid anodizing and the use of vinyl phenolic adhesives has been favoured in Europe. In the USA, the FPL etch based on sodium dichromate / sulphuric acid has been optimised and is used either alone, or is followed by phosphoric acid anodizing, or is replaced with phosphoric acid anodizing. Such replacement is followed by the use of corrosion inhibiting adhesive primers and moisture-resistant epoxy adhesives to give the required durability in humid service environments[5]. Recently[6] a new pretreatment has been developed and preliminary results have been published[7]. Essentially the new process consists of hard anodizing in sulphuric acid followed by a phosphoric acid dip. The hard anodizing, which may be applied over the entire surface (and not solely the joint area), produces an alumina coating with the high resistance to corrosion needed for long term durability of structural adhesively bonded joints. Also essential for adhesive bond durability is resistance to hydration of the alumina coating. This is achieved by the phosphoric acid dip which alters the surface chemistry of the hard anodized coating with a surface layer of adsorbed phosphate to inhibit hydration of the alumina. The phosphoric acid has a second function; it etches the alumina coating to increase the surface area wetted by the adhesive and to distribute loads in service over a thicker layer within the joint.

The aim of the work detailed in this paper was firstly to find out if aluminium-lithium alloys could be joined by adhesive bonding using the established processes. The new pretreatment process was then assessed and compared with the established processes. Finally, the bonding process behaviour of the aluminium-lithium alloy was compared with that of non-lithium-containing alloys.

EXPERIMENTAL PROCEDURE

Two alloys of aluminium were used, Al-Li alloy to DTD XXXA and for the purpose of comparison Al-Mg alloy 5251. The Al-Li alloy was supplied by British Alcan Aluminium in the form of 1·6 mm thick sheet in the solution treated condition and had the following composition:-

	Li	Cu	Mg	Zr	Fe	Si	Ti	Al
%	2·35	1·13	0·60	0·12	0·18	0·04	0·025	balance

The sheet had been solution heat treated for 15 minutes at 530°C, quenched in cold water and roller levelled. The surface had then been cleaned by dipping in dilute nitric acid followed by a light etch in dilute sodium hydroxide and then desmutted in dilute nitric acid.
The 5251 alloy was commercial quality sheet, 2 mm thick, and of nominal composition, 2% Mg, 0·3% Mn, balance Al.
3" x 1" samples were cut from the aluminium alloy sheet degreased in acetone and etched for 20 minutes at 68°C in FPL etch made up with 1·44 l sulphuric acid, 8 l distilled water, 264 g sodium dichromate and optimised with 12 g aluminium and 8 g copper predissolved as sulphates in distilled water. After the FPL etch, the specimens were rinsed thoroughly in tap water
Adhesive joints were made using the following three processes:-
(1) The chromic acid anodizing process (CAA)
The CAA process was carried out in accordance with MOD specification DEF-151.
After the FPL etch, specimens were anodized in chromic acid made up with 35 g/l chromium trioxide in distilled water and held at 40°C. The voltage was raised in 2V steps to 40V in the first 10 minutes, held at 40V for 20 minutes, gradually raised to 50V in the next 5 minutes and then held at 50V for 5 minutes.
(2) The phosphoric acid anodizing process (PAA)
The PAA process was carried out in accordance with the Boeing specification BAC 5555.
After the FPL etch, specimens were anodized in 10%wt phosphoric acid at 20°C. The voltage was raised to 10V in the first minute and held at 10V for 20 minutes.
(3) The new process (SAA + PAD)
Prior to anodizing, the specimens were etched in sodium hydroxide instead of the FPL etch in order to simplify the process, reduce costs and avoid the use of chromic acid. The specimens were etched in 1·5 M sodium hydroxide at 65°C for 5 minutes, rinsed, dipped in 10% nitric acid at room temperature for 1 minute, rinsed and dried with an air blower. The specimens were then hard anodized to MOD specification DEF-151 using 10%(vol) sulphuric acid at -5°C at $2A/dm^2$ after raising the voltage slowly in the first two minutes.
Anodizing was followed by a dip in dilute phosphoric acid to etch the anodic oxide and render it both chemically receptive to adhesive and resistant to hydration.
After anodizing, the specimens were thoroughly rinsed in tap water and dried in air at room temperature with an air blower.
Rubber toughened epoxy adhesive Permabond ESP 110 and rubber toughened acrylic adhesive Permabond F241 were used to bond the treated aluminium alloy specimens.
Single lap shear joints were made from 3" x 1" strips with ½" joint overlap and fractured in tension using an Instron 1195 machine. Environmental performance of the adhesively bonded joints was assessed using the Boeing wedge-test immersing the joints in distilled water at 50°C and measuring crack growth with a travelling microscope. Long term durability was assessed on lap shear joints after immersion in distilled water at 50°C for 1500 hours.
At each stage of pretreatment and bonding, the surface of the Al-Li alloy was examined by scanning electron microscopy. Cross-sections of the anodic coating were examined after bending specimens sharply through 180° to expose a fracture surface.

RESULTS AND DISCUSSION

From a practical anodizing point of view, it was found that the anodizing of the aluminium-lithium alloy was similar to the anodizing of aluminium alloys not containing lithium. Care was necessary at the start of anodizing to raise the voltage slowly during the initial formation of the barrier layer. If the voltage was raised too quickly, rapid dissolution occurred at localised spots on the surface, especially at the edges, giving the condition known to anodizers as 'burning'. This was probably due to the presence of copper in the Al-Li alloy DTD XXXA, as it is known[8] that alloys containing copper are more prone than others to 'burning' on anodizing. Indeed, the anodizing behaviour of DTD XXXA was no different in this respect from that of 2024 and other high strength aluminium alloys containing copper.

At each stage, the surface of the aluminium-lithium alloy was examined by scanning electron microscopy. The cross-section of the anodic oxide formed under hard anodizing conditions in sulphuric acid is shown in Fig. 1 and is very similar to that obtained previously[9] when anodizing pure aluminium. Fracture round the cell boundaries is clearly seen, and measurements from micrographs gave a cell diameter of 60 nm which is very close to the cell diameter of 55 nm obtained[9] on hard

TABLE 1 Adhesive Bond Strength and Durability

Alloy:- Al-Mg 5251

Adhesive:- F241 Rubber toughened acrylic

Pretreatment	Mean Bond Strengths Initial MPa	After 1500 h water immersion MPa	% of initial
FPL + PAA	22.48	22.76	101
FPL + CAA	24.29	23.07	95
NaOH + SAA + PAD	23.97	24.90	104

TABLE 2 Adhesive Bond Strength and Durability

Alloy:- Al-Li DTD XXXA

Adhesive:- F241 Rubber toughened acrylic

Pretreatment	Mean Bond Strengths Initial MPa	After 1500 h water immersion MPa	% of initial
FPL + PAA	23.02	25.60	111
FPL + CAA	25.10	23.49	94
NaOH + SAA + PAD	25.49	23.54	92

TABLE 3 Adhesive Bond Strength and Durability

Alloy:- Al-Mg 5251

Adhesive:- ESP 110 Rubber toughened epoxy

Pretreatment	Mean Bond Strengths Initial MPa	After 1500 h water immersion MPa	% of initial
FPL + PAA	22.80	17.96	79
FPL + CAA	23.85	not tested	
NaOH + SAA + PAD	25.82	19.01	74

TABLE 4 Adhesive Bond Strength and Durability

Alloy:- Al-Li DTD XXXA

Adhesive:- ESP 110 Rubber toughened epoxy

Pretreatment	Mean Bond Strengths Initial MPa	After 1500 h water immersion MPa	% of initial
FPL + PAA	26.20	18.73	72
FPL + CAA	28.19	19.04	68
NaOH + SAA + PAD	27.33	18.08	66

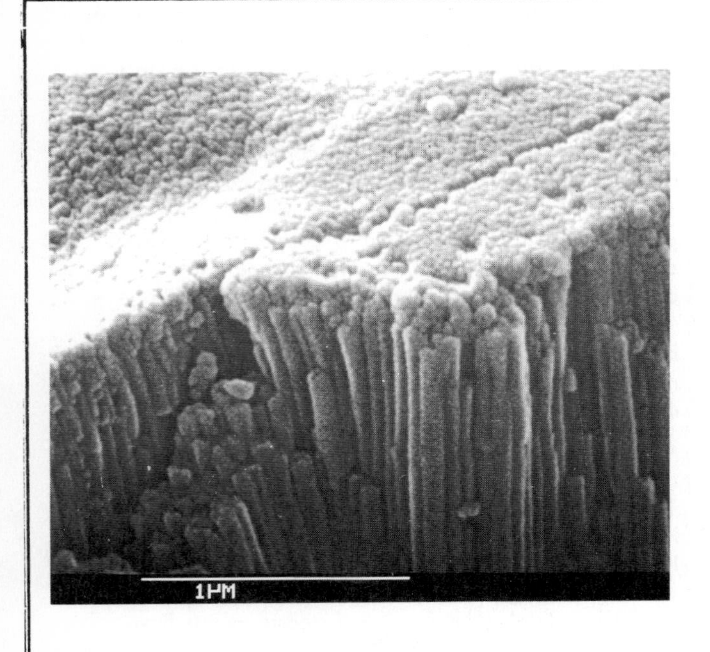

Fig. 1 Scanning electron micrograph of the anodic oxide on Al-2·35% Li-1·13% Cu showing surface topography (top half of figure) and cross-section (lower half). Hard anodized in sulphuric acid at -5°C.

anodizing commercially pure aluminium in 10% (vol) sulphuric acid at 24V for 10 minutes at -5°C and the cell diameter of 58·5 nm measured by Wood et al[10] from transmission electron micrographs of carbon replicas of anodic films formed on pure aluminium in 1·5 M sulphuric acid at 23·5V and 20°C.

The upper half of Fig. 1 shows the topography of the outer surface of the anodic oxide formed on Al-Li under hard anodizing conditions at -5°C where dissolution of the outer surface is minimised. The spherical protrusions formed in the initial stages of anodizing are clearly seen. They have been observed previously[11,12] in phosphoric acid anodizing, and recently[12] in the FPL etch and NaOH etch similar to those seen on conversion coatings on aluminium[13,14].

Adhesive bond strengths are given in Tables 1-4. Each figure is the mean of four or more tests. The locus of failure of all the initial bond strength tests was cohesive (within the adhesive). After immersion for 1500 hours in water, the locus of failure was either cohesive or mixed cohesive and failure near the interface between oxide and adhesive.

The results show that the adhesive bonding of Al-Li is similar to that of Al-Mg (Compare Tables 2 & 4 with Tables 1 & 3 respectively). The new pretreatment gives similar initial bond strengths and durabilities to the established PAA and CAA pretreatments. There is a difference, however, in the performance of the two adhesives. The joint durability after 1500 hours water immersion was excellent when using the rubber toughened acrylic adhesive with final bond strengths in several cases greater than the initial bond strength. This is thought to be due to the extended curing and strengthening of the adhesive during the 1500 hours immersion in water at 50°C. The loss of strength when using the rubber toughened epoxy was considered to be due to incomplete wetting and filling of the rough surface of the anodic oxide. This view is supported by preliminary results using BR 127 primer prior to ESP 110. Use of the primer also reduces the scatter of bond strength values obtained while giving a similar mean value.

Minford[15] has discussed the use of the Boeing wedge test and shown its value for assessing the relative durability potential of different pretreatments. The results obtained for aluminium-lithium were very similar for all three pretreatments (FPL + PAA, FPL + CAA, and NaOH + SAA + PAD). The crack growth in each case grew to only 1 mm in the first hour and remained unchanged for 24 hours. These results are better than those obtained previously[7] for the adhesive bonding of 5251 alloy.

CONCLUSIONS

Al-2·35% Li-1·13% Cu in the solution treated condition can be anodized without difficulty. Thick anodic oxide coatings can be formed by hard anodizing in sulphuric acid. The presence of lithium makes little or no difference to the structure of the anodic oxide formed. The presence of dissolved lithium in the anodizing solution appears to make no difference to subsequent anodizing of pure aluminium or of alloys of aluminium not containing lithium.

The established chromic acid anodizing and phosphoric acid anodizing processes for the adhesive bonding of aluminium alloys as specified for aerospace applications can be used for Al-2·35% Li-1·13% Cu in the solution treated condition and give high bond strengths and bond durabilities similar to alloys not containing lithium. A new pretreatment process consisting of hard anodizing in sulphuric acid followed by a phosphoric acid dip can also be used.

As with all adhesive bonding of aluminium alloys, correct joint design, selection of adhesive and pretreatment procedure are essential.

ACKNOWLEDGEMENTS

The authors are grateful to Dr.R.Grimes and British Alcan Aluminium for supplying the aluminium-lithium alloy sheet.

REFERENCES

1. R.Grimes, A.J.Cornish, W.S.Miller and M.A.Reynolds, Metals and Materials 1985, 1, 357.
2. N.A.de Bruyne, Symp on Adhesives for Structural Applications, ed:M.J.Bodnar, Interscience 1962, 1.
3. P.K.Nelson and W.D.Sanders, Proc SAMPE Conf 1982, 967-977.
4. P.J.Thompson and H.B.Heaton, Trans Inst Met Fin 1980, 58, 81.
5. D.B.Arnold and E.E.Peterson, Proc SAMPE Conf Oct 1981, 162-176.
6. A.W.Clifford and D.J.Arrowsmith, UK Pat 2 141 139 (1984).
7. D.J.Arrowsmith and A.W.Clifford, Int J Adhesion & Adhesives 1985, 5, 40.
8. D.A.Thompson, Trans Inst Met Fin 1976, 54, 97.
9. D.J.Arrowsmith and A.W.Clifford, Int J Adhesion & Adhesives 1983, 3, 193.
10. G.C.Wood and J.P.O'Sullivan, Electrochim Acta 1970, 15, 1865.
11. G.E.Thompson, R.C.Furneaux, G.C.Wood, J.A.Richardson and J.S.Goode, Nature 1978, 272, 433.
12. D.J.Arrowsmith, A.W.Clifford and D.A.Moth, Trans Inst Met Fin 1985, 63, 45.
13. J.A.Treverton and M.P.Amor, Trans Inst Met Fin 1982, 60, 92.
14. D.J.Arrowsmith, J.K.Dennis and P.R.Sliwinski, Trans Inst Met Fin 1984, 62, 117.
15. J.D.Minford in "Durability of Structural Adhesives" edited by A.J.Kinloch (Applied Science, London and New York 1983) pp 158-159.

Grain refining of aluminium–lithium based alloys with titanium boron aluminium

M E J BIRCH

The author is with
London & Scandinavian
Metallurgical Co. Ltd,
Rotherham, UK

Summary

The grain refining of LITAL 'A' and aluminium alloys containing lithium, copper, magnesium and zirconium has been studied. It has been shown that zirconium reduces the efficiency of the grain refiner, and that this effect is accentuated by the presence of lithium. Areas for further work are outlined, and a grain refining practice for lithium-containing aluminium alloys is proposed.

Introduction

Grain size control of lithium-containing aluminium alloys is of great importance in the development of their properties. At one time the powder metallurgical route appeared to be the most promising way to produce aluminium-lithium alloys with the required properties, since the product would have been inherently fine-grained. However, the ingot route is now seen as the one most likely to be followed in initial commercial application of the alloys (1). The Royal Aircraft Establishment (RAE) played a major part in establishing that satisfactory alloys could be developed for the ingot route. To produce the required properties careful control of the microstructure is essential. This is achieved by mechanical working and the use of zirconium to act as a recrystallisation inhibitor (2). Lithium-containing aluminium alloys were developed by the RAE in close cooperation with British Alcan. These are now marketed as the 'LITAL' series of alloys (3,4).

A number of beneficial effects of small grain size in lithium-containing aluminium alloys have been identified, among them improved strength, stiffness, ductility, fracture toughness, fatigue life, and resistance to crack growth. Reported work indicates that zirconium additions are currently the best way to regulate the microstructure of the final product. Attempts to control the as-cast grain size of aluminium-lithium alloys with the widely used commercial range of titanium boron aluminium (TiBAl) grain refining master alloys have not previously been reported.

Refinement of the as-cast grain structure, however effectively carried out, would not eliminate the need for working and heat treatment. It would however still be advantageous if it reduced the amount of work needed, or allowed higher tolerance in the treatment processes without loss of properties. As long as thermo-mechanical treatment is needed, it is unlikely that the need for a small addition of zirconium to the alloy can be avoided. We would expect that even where working is required to develop properties, a small as-cast grain size would make a small uniform grain size easier to develop in the final product, and would lead to reduced reject levels.

Grain refining would also be expected to reduce some of the difficulties encountered during the casting of larger ingots.

The work reported here examines the use of TiBAl master alloy additions to aluminium-lithium alloys at the levels generally accepted in the aluminium industry for grain size control of wrought alloys.

Experimental Programme

The alloys investigated are shown in Table 1. LITAL 'A' was chosen as representative of a new

TABLE 1 — COMPOSITION OF ALLOYS EVALUATED

ALLOYS	Li	Cu	Mg	Zr	Al
	COMPOSITION wt%				
Commercial Alloys					
Aluminium					99.7
Aluminium					99.85
LITAL 'A'	2.3	1.25	0.89	0.13	Bal
Binary Alloys (max. analysis shown)					
LiAl	2.3				Bal
CuAl		3.0			Bal
MgAl			5.0		Bal
ZrAl				0.3	Bal
Ternary Alloys					
LiCuAl	2.3	1.25			Bal
LiMgAl	2.3		0.89		Bal
LiZrAl	2.3			0.13	Bal
Quaternary Alloy					
LiCuMgAl	2.3	1.25	0.89		Bal

TABLE 2 — LITHIUM LOSS DURING GRAIN REFINING TESTS

ALLOY	TYPICAL Li ANALYSIS (wt%) INITIAL	60 MINUTES
LITAL 'A'	2.3	1.8
2.3 LiAl	2.2	1.7
2.3 Li 1.25 CuAl	2.1	1.7
2.3 Li 0.9 MgAl	2.3	2.1
2.3 Li 0.13 ZrAl	2.4	2.0

TABLE 3 — APPROXIMATE GRAIN SIZES OF UNREFINED ALLOYS

ALLOY	UNREFINED GRAIN SIZE (Microns) TYPICAL
99.7 aluminium	1200
99.85 aluminium	1400
LITAL 'A'	1000
2.3 LiAl	850
1.25 CuAl	300
1.0 MgAl	1000
0.13 ZrAl	2000

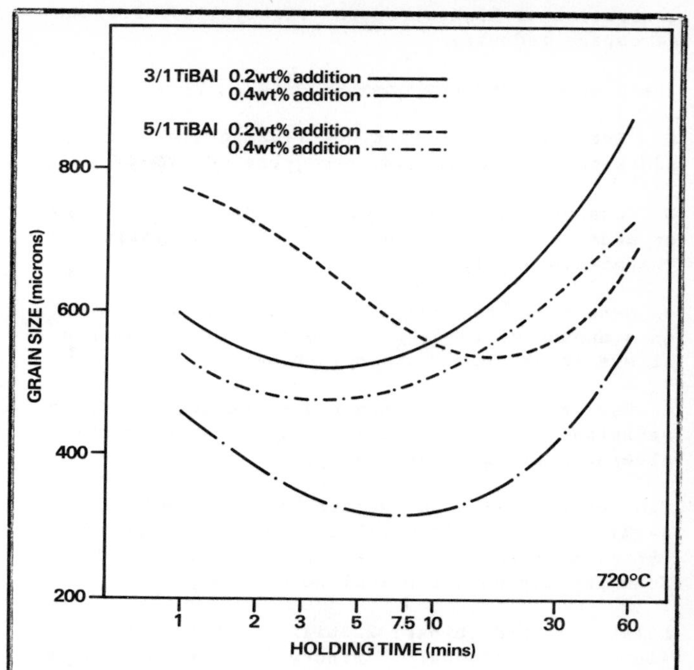

Figure 1.
Effect of addition rate and titanium content of TiBAl alloy on grain refinement of LITAL 'A'.

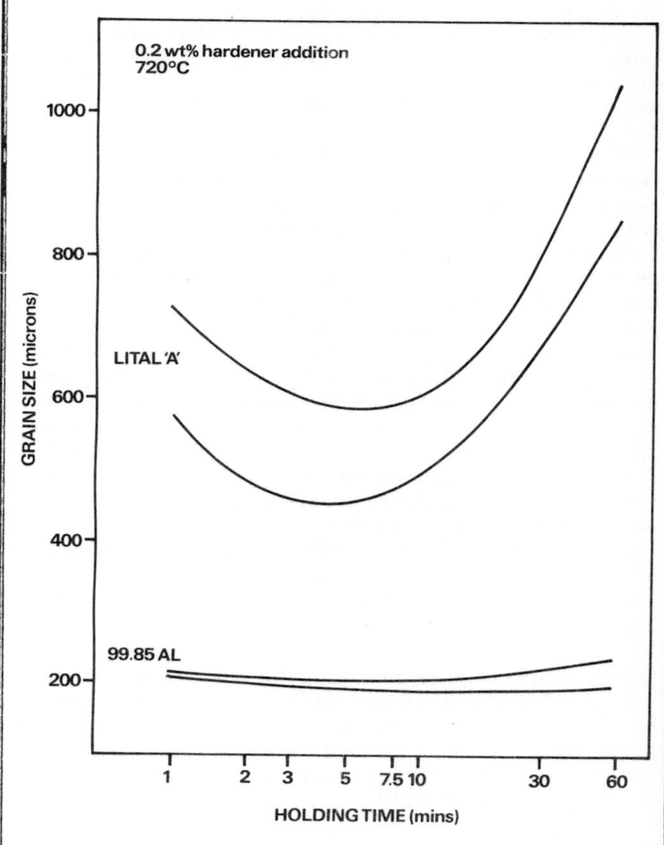

Figure 2.
Effect of 3/1 TiBAl on LITAL 'A' and 99.85% aluminium.

generation of high strength light alloys for the aerospace industry.

The aims of the programme were to:

a. Establish the effect of TiBAl grain refining additions on the as-cast structure of LITAL 'A'.

b. Determine the effect of TiBAl on binary alloys of aluminium containing the individual alloying components of LITAL 'A'.

c. Determine the grain refining effect of TiBAl on a number of ternary and more complex aluminium alloys which also contain lithium.

d. Use the results obtained to suggest a grain refining practice for commercial aluminium-lithium based alloys.

Two grain refining master alloys were used, 3wt%Ti1wt%BAl (3/1 TiBAl) and 5wt%Ti1wt%BAl (5/1 TiBAl). These were produced by London & Scandinavian Metallurgical Co Limited.

LITAL 'A' and binary 2.3LiAl were supplied by Alcan International. Binary aluminium-lithium alloys of other compositions were made either by dilution of the alloy supplied with 99.85% aluminium, or by direct addition of lithium metal to 99.85% aluminium. Other alloying additions were made with 34wt%CuAl, 15wt%ZrAl, (both based on 99.7%Al) and with magnesium metal. The grain sizes obtained with unalloyed 99.85% aluminium under standard test conditions were about twenty microns greater than with 99.7% aluminium used in previous work in this laboratory (5,6,7).

The Alcan grain refining test was used to evaluate the effect of TiBAl additions on the grain size of the alloys. This test, which has been shown to correlate well with results found in casthouse practice, has already been described in some detail (8). Tests were conducted at a temperature of 720C. To overcome problems of rapid oxidation and attack on the refractory washed cast iron pot, changes were made to the test procedure for use with lithium-containing alloys. The modified test used a LiCl/KCl flux cover to the melt which was held in a silicon carbide crucible (9), but was in other respects similar to the published test. Even so, grain refining tests were limited to one hour because of lithium loss by oxidation (Table 2). The one hour test time was sufficient for measurement of contact time (time to reach maximum efficiency) and ultimate grain size (smallest grain size achieved).

Test specimens were prepared by linishing down to 800 grit and finishing with a commercial metal polish. The samples were etched with cupric chloride solution, cleaned with nitric acid, and washed in hot water. Grain sizes were measured by an intercept method.

Results

Figure 1 shows the effect of grain refining LITAL 'A' alloy with 0.2wt% and 0.4wt% 3/1 and 5/1 TiBAl, equivalent to two and four kilogrammes per tonne. The holding time after grain refining addition is shown on a logarithmic scale.

The effect of grain refining LITAL 'A' with 0.2wt% 3/1 TiBAl is shown in more detail in Figure 2. The upper and lower curves for this alloy and for 99.85% aluminium indicate the range of values found in a series of nominally similar tests. The susceptibility of commercial alloys to uncontrolled random effects is clearly seen.

The effect of grain refining additions to binary alloys of aluminium with lithium, copper, magnesium and zirconium at the levels found in LITAL 'A' alloy are shown in Figure 3. Figures 4 to 7 summarise the effects of different alloying addition rates at two and sixty minute holding times. Only the two minute curve is shown for the lithium alloy since after about thirty minutes erratic lithium losses appear to affect the grain refining results.

Figure 8 illustrates the effect of grain refining the ternary and quaternary alloys shown in Table 1.

Discussion

It has been shown that under most circumstances 3/1 TiBAl is a more cost effective grain refining agent than 5/1 TiBAl (6). However, there are theoretical reasons for believing that 5/1 TiBAl, with its greater ratio of titanium to boron, may often be more effective than 3/1 TiBAl in refining zirconium containing alloys. A comparison was therefore made between the two alloys at two addition rates (Figure 1). The tests showed that 3/1 TiBAl was as efficient as 5/1 TiBAl for the refinement of LITAL 'A'. The remainder of the work was done with 3/1 TiBAl alloy.

As shown in Table 3, under the test conditions used the grain size of the unrefined commercial alloy was about 1000 microns (though this varied considerably from test to test). Unrefined 99.85% aluminium had a grain size of approximately 1400 microns. Figure 2 indicates that an addition of 0.2wt% 3/1 TiBAl produced a grain refining effect which lasted for at least fifteen minutes. It can be seen however, that the commercial alloy was more difficult to grain refine than 99.85% aluminium. Minimum grain size was achieved during the first ten minutes after the grain refining addition.

Figure 3 shows the effect of grain refining binary alloys of aluminium with the individual components of LITAL 'A' using 3/1 TiBAl. The presence of copper and magnesium enhanced grain

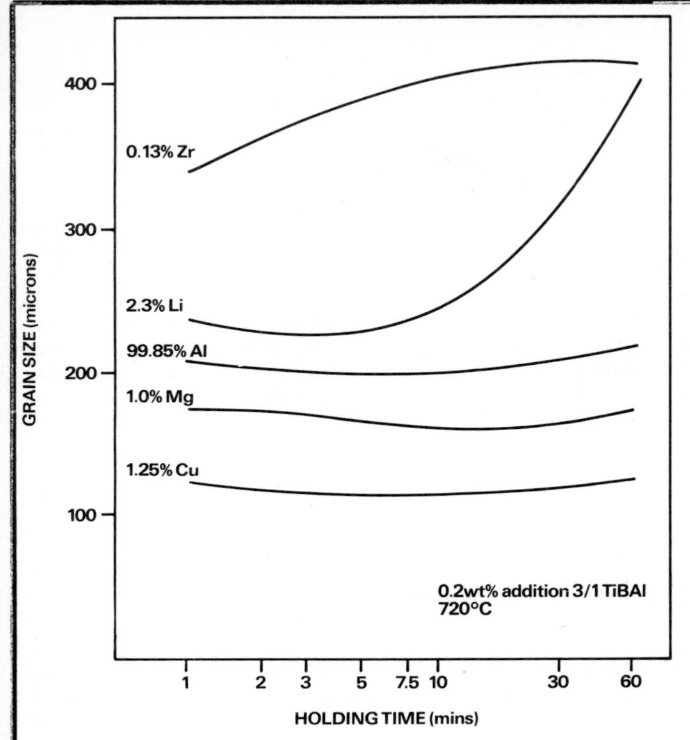

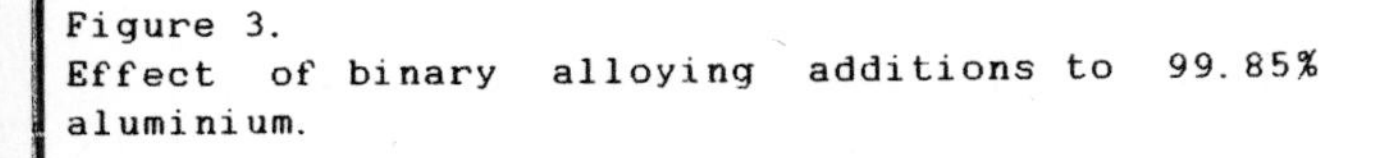
Figure 3.
Effect of binary alloying additions to 99.85% aluminium.

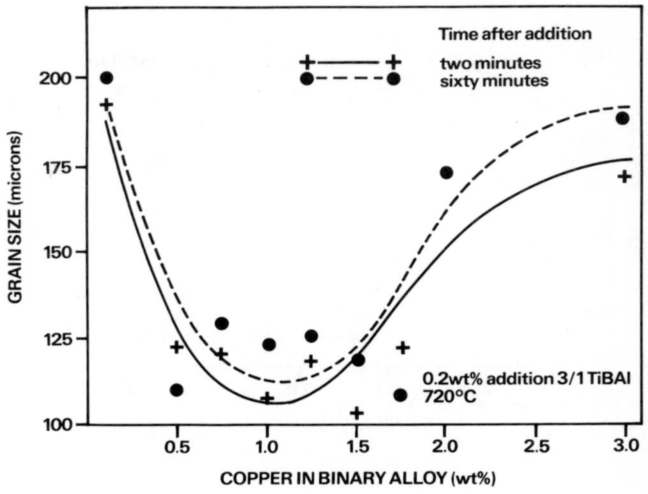

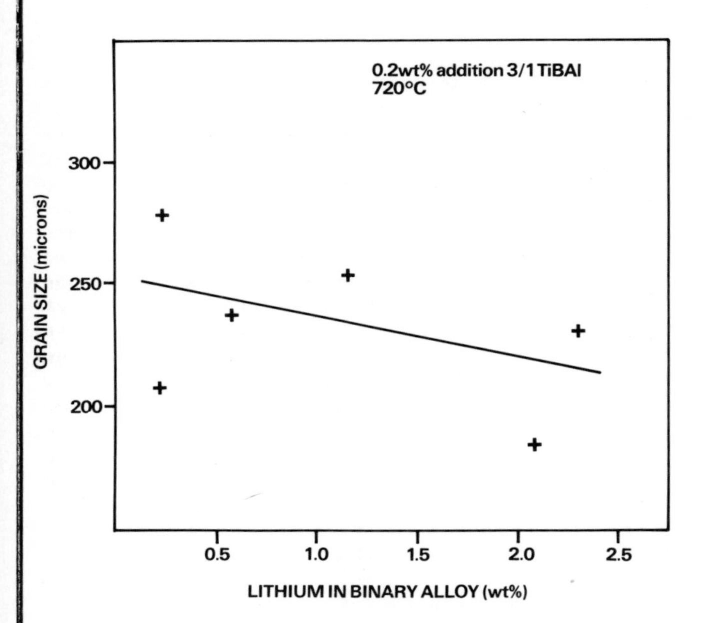

Figure 4.
Effect of lithium additions on grain size of 99.85% aluminium after two minutes holding time.

Time after addition
two minutes
sixty minutes
0.2wt% addition 3/1 TiBAl
720°C
GRAIN SIZE (microns)
COPPER IN BINARY ALLOY (wt%)

Figure 5.
Effect of copper additions on grain size of 99.85% aluminium.

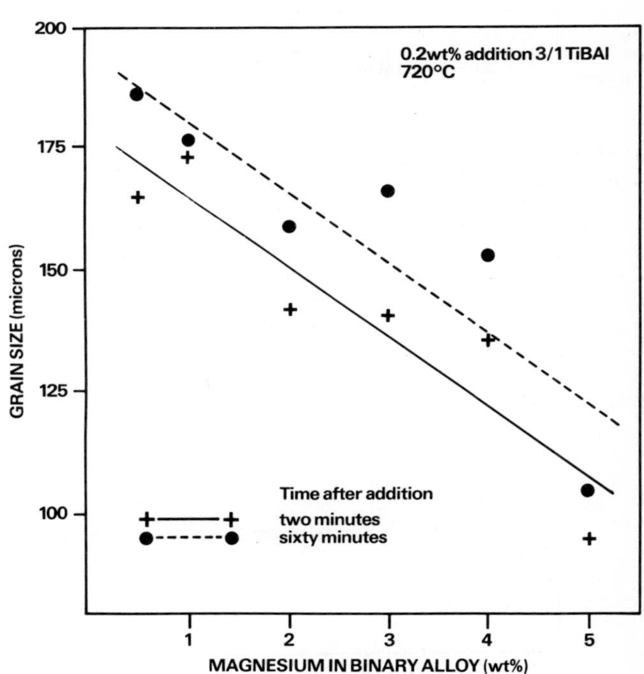

Figure 6.
Effect of magnesium additions on grain size of 99.85% aluminium.

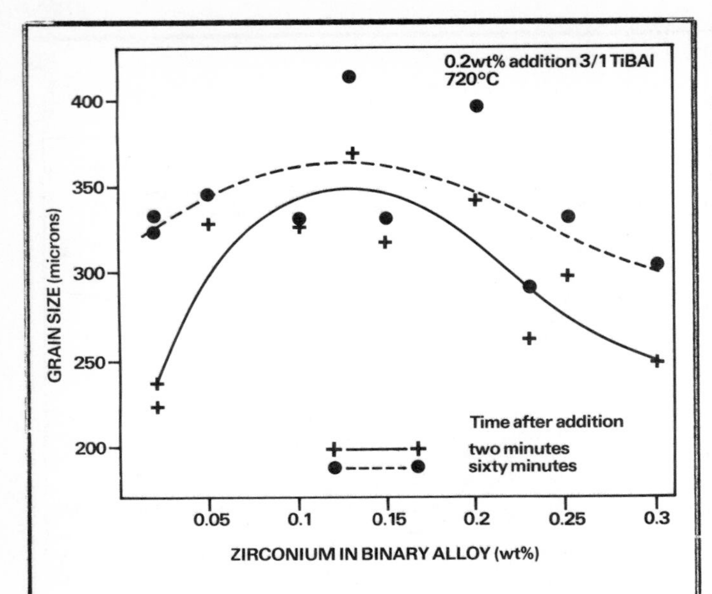

Figure 7.
Effect of zirconium additions on grain size of 99.85% aluminium.

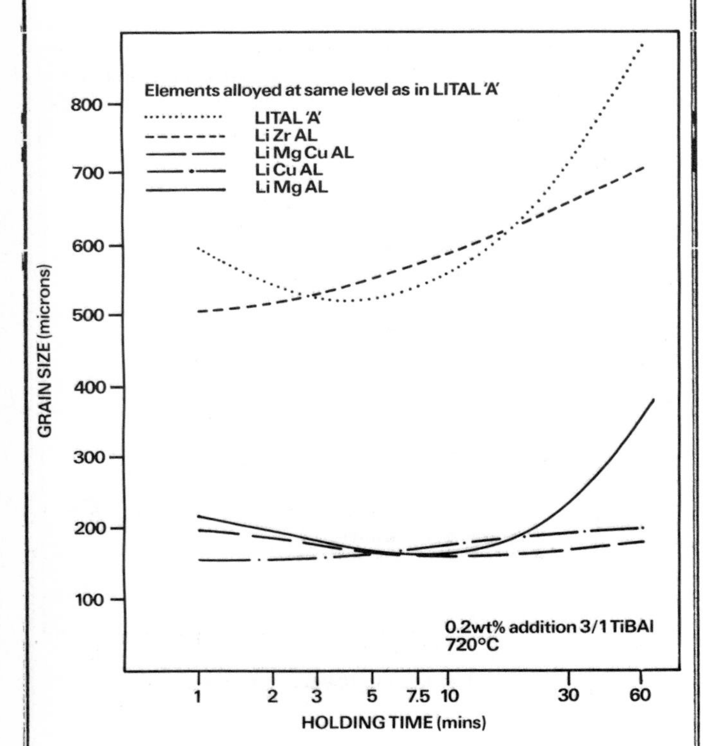

Figure 8.
Effect of other alloying elements on grain refining of aluminium-lithium based alloys.

refinement. The effect of copper was substantial.

Lithium and zirconium increased the grain size of aluminium treated with TiBAl. The effect of zirconium was considerable but remained fairly constant after the first few minutes. The effect of the lithium on the other hand was at first only slight, but led to a rapid loss of grain refining properties after about ten minutes. Due to loss of lithium during the tests, this result should be treated with caution.

Levels up to 2.3wt% lithium appear to have little effect on the grain size of a binary alloy two minutes after the grain refining addition (Figure 4). The figures showed a slight reduction as the lithium level was increased but are within the normal range of experimental error. Further work is needed to establish the effect of lithium at longer holding times using an experimental technique which is not prone to loss of lithium.

Increasing copper content in the binary alloy up to about 0.5wt%Cu gave a reduction in grain size (Figure 5). The grain refining effect remained fairly constant until about 1.75wt%Cu was reached. Above this figure the grain size began to increase. Although there is some indication that the grain size after an hour was slightly coarser than at two minutes, the fade was not marked.

Figure 6 shows that increasing additions of magnesium to aluminium up to 5wt%Mg resulted in decreasing grain sizes at times up to one hour after TiBAl addition. The grain size after an hour was marginally coarser than after two minutes.

The curves in Figure 7 show that zirconium poisoned the grain refining efficiency of TiBAl. Grain refinement reduced as zirconium levels increased to about 0.13wt%Zr, above this level there was some recovery of grain refining properties, which may have been due to the peritectic reaction which starts in pure zirconium-aluminium at 0.11wt%Zr. The poisoning effect of the zirconium was more marked after one hour than after two minutes, particularly at lower (0.02-0.1wt%) addition rates. Although we are not aware of published work showing the recovery of grain refinement above the peritectic this confirms our experience on the effect of zirconium additions to 99.7% aluminium. The poisoning effect below the peritectic has been explained by Jones and Pearson (8), who showed that when the ratio of zirconium to titanium is sufficiently high a layer of zirconium boride can form on the surface of the TiB_2 nuclei.

The grain refining curves in Figure 8 show that when aluminium was alloyed with 2.3wt% lithium and either 1.25wt% copper or 0.9wt% magnesium, the grain sizes were of the same order as those found with 99.85% aluminium (Figure 2). However, when 0.13wt% zirconium was present the grain size was similar to that found in LITAL 'A'. It appears therefore that difficulty in refining the commercial alloy with TiBAl was due to the presence of zirconium. The concentration of zirconium in the alloy is exactly that found to be most deleterious to grain refining properties in the binary alloy.

A grain refining curve has been published for the zirconium-containing alloy 7050 (7). In that case the ultimate grain size was slightly over 200 microns, considerably finer than with equivalent grain refining additions to the commercial alloy used in this work. Of the two, alloy 7050 has a slightly higher zirconium content (0.16wt%Zr), but hardly enough to explain the difference in grain size. The grain sizes recorded for the binary ZrAl alloy (Figure 7), although larger than found with 7050, were smaller than those found with LiZrAl, and it would therefore appear that the presence of lithium increased the poisoning effect of zirconium.

The grain refining of aluminium-lithium based alloys in the as-cast state is clearly not a simple matter. The results presented here show that a 0.4wt% addition of 3/1 TiBAl produces a degree of grain refinement, though not of the order expected with the more conventional aluminium alloys. Since the zirconium is present as a grain growth inhibitor but appears to be responsible for the poor as-cast grain size of the commercial alloy, further work may be justified to show whether a TiBAl grain refining treatment, possibly at higher addition rates, together with a lower addition rate of zirconium, would be a cost-effective way of developing the properties required in the final product.

Since minimum holding times for lithium-containing alloys are desirable, and fade of the grain refining properties appears to set in rapidly, a rod grain refining practice is essential. With this practice grain refiner is fed continuously to the casting launder a few minutes before solidification. Rod additions to aluminium-lithium based alloys may cause practical difficulty in the casthouse because of the need to protect the alloy from oxidation and hydrogen pick-up.

To understand fully the grain refining of aluminium-lithium based alloys further work is clearly needed. At this time we recommend the following practice:

a. maximum grain refining addition rate without exceeding the impurity specification.

b. minimum zirconium within the alloy specification.

c. addition of TiBAl grain refiner as rod to the launder during casting.

Conclusions

Although this work used LITAL 'A' as representative of the new generation of high strength light alloys based on aluminium-lithium, published specifications for all similar alloys also contain zirconium. It is believed therefore that these conclusions have a general application.

In the commercial alloy system studied;

a. The unrefined grain size showed considerable cast-to-cast variation even under nominally identical casting conditions.

b. Refinement of the grain size was achieved by the addition of 0.4wt% 3/1 TiBAl.

c. A rod practice was shown to be essential for effective grain refinement.

d. An alloy without zirconium but otherwise to the commercial specification proved easy to refine with 3/1 TiBAl. Difficulty in refining the commercial alloy with TiBAl is attributed to the presence of zirconium.

Binary aluminium alloys of each of the major components of the commercial alloy when grain refined with 3/1 TiBAl showed;

a. increasing lithium content had little effect on grain size.

b. copper additions up to 1.5wt% reduced grain size. In the 1.5 to 3.0wt% range grain size increased again.

c. magnesium additions up to 5.0wt% reduced grain size.

d) at zirconium levels up to about 0.13wt% the grain size increased. Above this level the grain size gradually decreased.

References

1. Grimes R. and Miller W.S., 'Current status of the development of aluminium-lithium alloys for aerospace application.', ASM Metals Congress, Philadelphia, 1983.

2. Evans B., McDarmaid D.S. and Peel C.J., 'An evaluation of the properties of improved aluminium-lithium alloys for aerospace applications.', SAMPE Conference, Montreux, June 1984.

3. Miller W.S., Cornish A.J., Titchener A.P. and Bennett D.A., 'Development of lithium-containing aluminium alloys for the ingot metallurgy production route.', Proceedings of the Second International Conference on Aluminium-Lithium Alloys, Monterey, Calif., April 1983, 335-362.

4. Peel C.J. et al., 'The development and application of improved aluminium-lithium alloys.', Proceedings of the Second International Conference on Aluminium-Lithium Alloys, Monterey, Calif., April 1983, 363-392.

5. Pearson J. and Birch M.E.J., 'Effect of the titanium:boron ratio on the efficiency of aluminium grain-refining alloys.', Journal of Metals, 1979, Vol 31, 27-31.

6. Pearson J. and Birch M.E.J., 'Improved grain refining with TiBAl alloys containing 3% titanium.' Light Metals 1984, 113rd Annual Meeting AIME Proceedings, Feb 27th - March 1st, 1984, 1217-1229.

7. Pearson J., Birch M.E.J., and Hadlet D., 'Recent advances in aluminium grain refinement.', Solidification Technology in the Foundry and Casthouse Conference, University of Warwick, UK, Sept 1980.

8. Jones G.P. and Pearson J., 'Factors affecting the grain refinement of aluminium using titanium and boron additives.', Met. Trans. B., 7B 1976, 223-234.

9. J. Worth, private communication.

□

Heat treatment of Li/Al alloys in salt baths

E R CLARK, P GILLESPIE
and F M PAGE

The authors are at the
University of Aston in Birmingham,
UK

The behaviour of various alloys of aluminium with either lithium and magnesium have been examined when heated in molten sodium nitrate, both on the laboratory scale and in field tests. It is concluded that lithium/aluminium alloys may be safely heat-treated in such baths at temperatures up to 600°C.

There is a long-standing tradition (1) that aluminium/magnesium alloys containing more than 3% Mg will explode violently when heated in molten sodium nitrate, and this is attributed (2) to the reaction:-

$$5Mg + 2NaNO_3 = Na_2O + 5MgO + N_2$$

While there is no doubt that such reaction, if it occurs, will be strongly exothermic, it has not proved possible to find an instance of the reaction occurring in industrial practice, or to discover any incident on which the tradition could be based. After exhaustive enquiries in the United Kingdom, only two explosions of nitrate salt baths have been located in the past 45 years, and in neither case was there any material being treated in the bath at the time, both explosions occurring overnight, when the baths were on stand-by.

The first experiments were directed to establishing the behaviour of sodium nitrate when heated alone, or in the presence of sodium nitrite, using the technique of thermogravimetric analysis (TGA) in which the weight of a sample is followed as it passed through a heating cycle. It was found that the sample lost weight continually at temperatures above 575°C, due to the slow evaporation of the salt, but that at higher temperatures there was a tendency for the loss of weight to increase, as the salt started to decompose to the less volatile oxide. A similar behaviour was observed with the nitrite, formed from the nitrate as the first step in decomposition, but here it was possible to detect a slow increase in weight at lower temperatures, as the nitrite was oxidised by the air to reform the nitrate.

The so-called stabilising effect of nitrite on nitrate baths is no more than the superposition of these two effects.

Having established the basic behaviour of the salt systems to be used, the next step was to attempt to draw an occurrance diagram, delineating the temperatures and compositions where explosions were to be expected. It was decided to carry out this preliminary study with magnesium alloys, since they are rather more easy to prepare and handle, and since the behaviour of the lithium alloys had been deduced from them. It was with some surprise that we found that the supposedly dangerous 5%Mg/95%Al alloy could readily be melted under sodium nitrate without any noticeable effect. A range of alloys were therefore prepared, between 5%Mg, and 90%Mg/10%Al, and small samples were heated in the TGA apparatus in 10 gm. of sodium nitrate held in a nickel crucible. The results are shown in Figure 1.

The disposition of the points is strongly reminiscent of a phase diagram, and when the points are superimposed on the Al-Mg diagram, the fit is impressive. We therefore adopt, as a working hypothesis, the idea that the metal must melt before it can react. This offers a criterion for safety, but it is unsatisfactory in so far as melting is a necessary, but not a sufficient condition for reaction. Before proceeding further, it is desirable to substantiate the general obser-

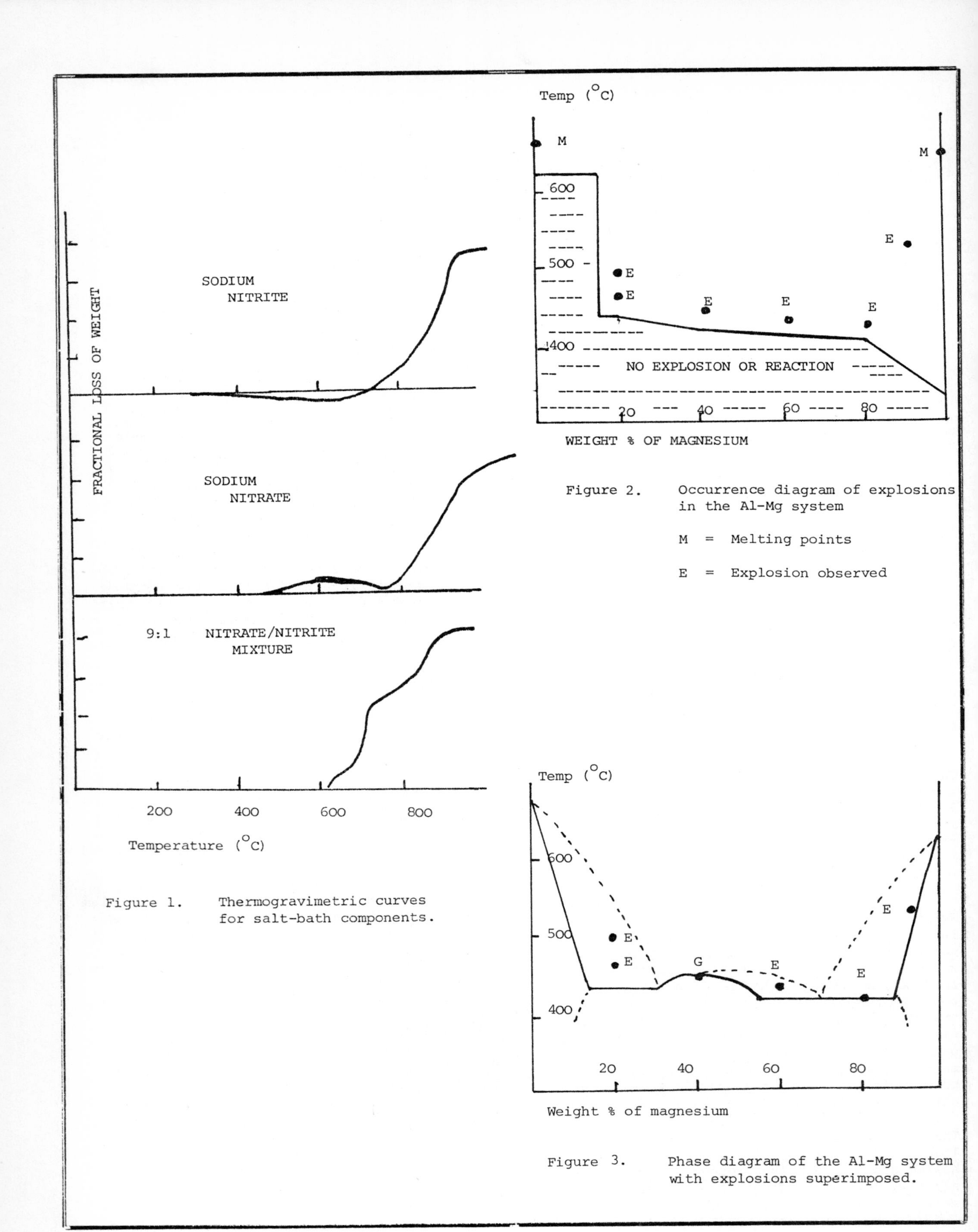

Figure 1. Thermogravimetric curves for salt-bath components.

Figure 2. Occurrence diagram of explosions in the Al-Mg system

M = Melting points

E = Explosion observed

Figure 3. Phase diagram of the Al-Mg system with explosions superimposed.

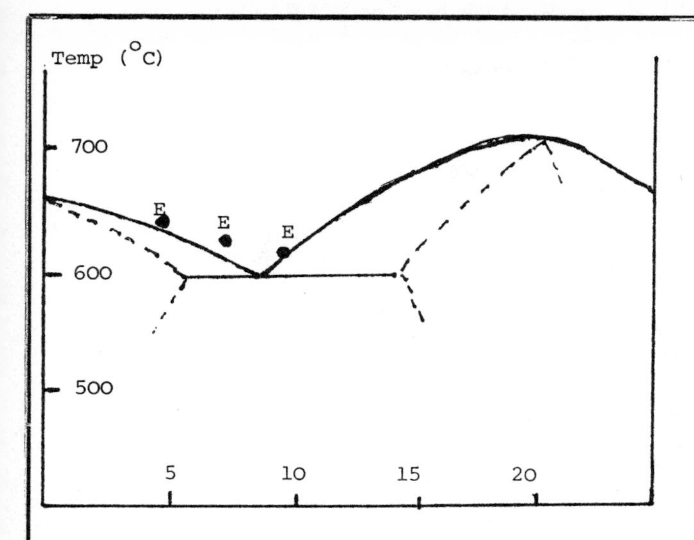

Weight % Lithium

Figure 4. Occurrence diagram for explosions in the Al-Li system.

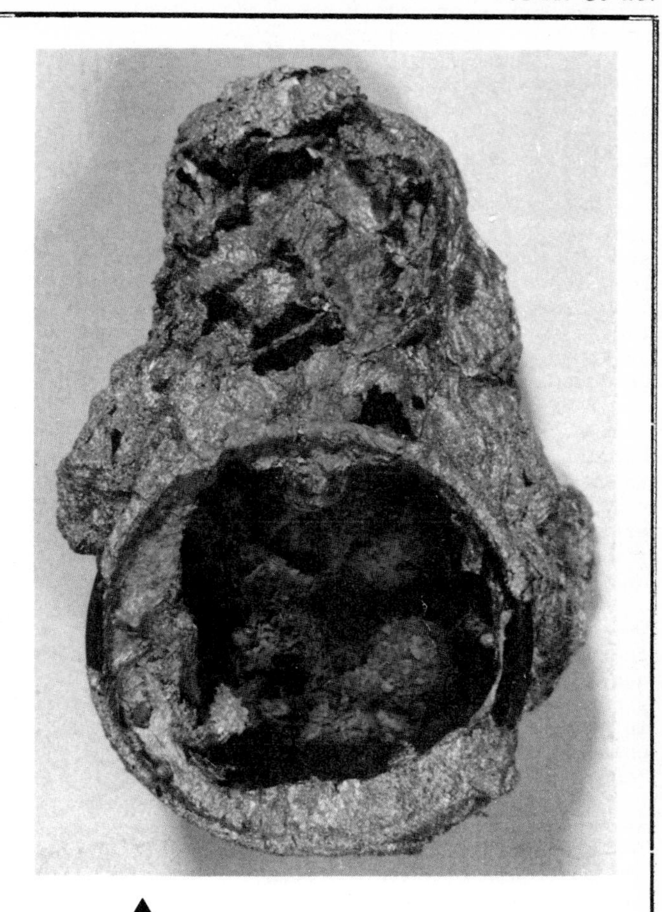

▲ Figure 5. Remains of a nickel crucible in which 0.2g of 9.5% Li/Al had reacted with sodium nitrate at 630 °C.

Figure 6. Salt-bath containing 100g 9.5% Li/Al in eruption. Reaction started at 616 °C.

vations and to establish that, for example, there is no effect of scale.

The tests were repeated on the same, and on other alloys, and only on the wings of the diagram were differences ever observed. The melting point is so dependent on composition in the region of 20%Mg, that local variations in composition between the tiny samples used could well explain the variation in the temperatures at which reaction occurred. A forthfold increase in the mass of the sample, from 7 mg. to 260 mg. showed no systematic change in the onset of reaction, which always occurred at 410+5°C. Increasing the charge to 530 mg. did produce a significant drop in the temperature of onset of reaction, but it was noted that this sample protruded above the salt, and was simply burning in air. So far from being dangerous, the molten salt was actually exerting a protective effect ! It was noticed that the temperature of ignition in air was always less than the temperature to which the material could be heated in the salt bath without major effect.

The word "major" is used advisedly, because there was often a slow evolution of bubbles of gas, or loss in weight, detectable but not significant, which did not presage a more violent interaction. This evolution of gas was the two-phase reaction between liquid sodium nitrate and solid metal. The theory of two-phase reactions is well known, and the rate depends markedly on the relative motion (stirring) of the two phases. The basic process is the diffusion of liquid reactant through a static boundary layer, whose thickness is determined by the shear due to stirring. Rates of reaction may therefore be compared on the basis of the viscosities of the liquid phase. It so happens that the viscosity of molten sodium nitrate at 500°C is close to that of water at 25°C, and this slow attack in the salt bath may therefore be modelled by the dissolution of zinc in dilute acid ! In point of fact, the reaction in the salt bath is much slower, because of the protective film of aluminium oxide, which offers a second barrier to diffusion, and one which may persist even after melting.

When these observations are transferred to the Lithium/Aluminium system, where the phase diagram is somewhat different, difficulties arise. In the first place, the 20%Mg w/w alloy contains 18.2 moles% Mg, which would correspond to 5.5% by weight of a lithium alloy, which is far beyond the normal commercial range of alloys. Secondly, the lithium/aluminium system shows a simple eutectic at about 9.5% lithium, and no plateau as does the magnesium system. Thirdly, this eutectic is at 608°C, some 200°C above the magnesium plateau. It would therefore not be expected that any normal alloy would show reaction below 600°C, and then only if the lithium concentration were over 5%.

These predictions have been fully borne out in practice. Numerous specimens of commercial alloys have been melted under sodium nitrate with no sign of any reaction. At 4½% Li, a steady loss of weight was observed, more rapid than usual, which indicated the onset of reaction at a temperature of over 630°C. At 5% Li, the reaction was rapid enough to throw a small amount of material out of the crucible, while a near-eutectic sample (9.7% Li) reacted violently at 620°C. This was repeated in a sample of sodium nitrate held in a nickel crucible in an electrically heated furnace. The temperature was not known with precision, but was of the same order as before, and the reaction which occurred was violent enough, and the temperatures produced high enough, to melt the nickel crucible, which was razed to the level of the slag.

As a final check on the validity of the conclusions drawn from the laboratory studies, a pilot scale salt bath was set up, capable of handling 1 kg. quantities of sheet. A number of such samples, usually in the form of four 1 foot square sheets, 1.6 mm thick spaced 3 mm apart, were successfully heat treated at temperatures up to 575°C, without the slightest indication of any reaction occurring. Two further samples were deliberately heat-treated under conditions predicted to be dangerous. Firstly a slab of 30% Mg alloy, weighing 100 g. was heated at 2 deg/min. At a temperature of 442°C, measured by a type K thermocouple with one junction in ice, there was a pyrotechnic display, which melted a hole in the stainless steel bath. A video record from a camera overhead showed that the reaction started as bubbles of gas, but that the slab ignited at the edges, followed by a complete eruption. Secondly, a similar slab of 9.7% Li/Al was heated in a similar manner to 610°C. There was considerable gas evolution, some of it internal since the slab swelled up into a nearly circular form. The experiment was terminated at 610°C through a misunderstanding between the operators, and was therefore repeated. The observed behaviour up to 610°C was the same, though the form of the sample could not be seen. On further heating, the sample erupted in a pyrotechnic display similar to that seen with magnesium at a temperature of 616°C.

These temperatures 442°C and 616°C lie precisely on the liquidus of the phase diagrams, and confirm the validity of the laboratory experiments.

CONCLUSION

The laboratory tests, supplemented by the pilot scale observations, indicate that there is no foundation to the fear of salt baths when used with aluminium alloys. Magnesium alloys containing less than 15% magnesium may safely be heat treated at temperatures up to 575°C, though this is not true for alloys with over 30% magnesium. More importantly, since no lithium/aluminium alloy melts below 600°C, all such alloys may be heat treated in nitrate salt baths up to this temperature with safety.

REFERENCES:

(1) "Guidelines for Safety in Heat Treatment, pt. 1 - The use of Salt Baths".

Wolfson Heat Treatment Centre, University of Aston, Birmingham. 1984.

(2) "Magnesium and its Uses", American Chemical Society Monograph 57.

□

Microstructure of cast aluminium–lithium–copper alloy

A F SMITH

The author is in the Materials Laboratory of Westland Helicopters Ltd, Yeovil, Somerset, UK

SYNOPSIS

Examination of an Al-Li-Cu casting has shown an interdendritic network of a previously undocumented Al-Cu-Li-Mg phase indexed as bcc with a_o = 12.11 Å. Additionally present were particles of (Fe, Cu) Al_6 and an undocumented Al-Mg-Si-(Li) phase. Effective solution treatment was not possible due to incipient melting of the cell/grain boundary precipitate. (Fe, Cu) Al_6 transformed to Cu_2 Fe Al_7 at solution treatment temperatures of ~520°C.

For comparison, a casting of modified composition was examined in which the interdendritic phase was indexed as bcc with a_o = 10.73 Å. Solution treatment was achieved without incipient melting.

Lithium/magnesium surface loss was not problematic with the as-cast component although noticeable depletion occurred on solution treatment in air.

INTRODUCTION

Although considerable progress has been made with the development of wrought aluminium-lithium alloys, development of casting alloys has received relatively little attention. The development of successful aluminium-lithium casting alloys would enable the full potential of these low density materials to be realized because of the high material utilization of castings and the inherent weight penalty associated with meeting design casting factors.

The unacceptably low fracture toughness of binary Al-Li alloys prevents their use as structural materials and hence much attention has been paid to the effect of ternary and higher order additions (1,2). Wrought alloys of the Al-Li-Cu-Mg system have shown particular promise (3) and a number of papers have been published concerned with the nature of submicroscopic hardening precipitates (4,5,6). In contrast, little work has been reported on Al-Li based castings where the presence of coarse equilibrium phase particles assumes a greater importance as dissolution must rely solely upon heat treatment without the aid of thermomechanical processing.

A relatively small number of ternary equilibrium phases have so far been established in Al-Li based alloys, notably T_B (Al_7Cu_4Li), T_1(Al_2CuLi) and T_2 (Al_6CuLi_3), (7), S ($MgLiAl_2$), (8) and $Li_3Si_2Al_2$ (9). No quaternary phase containing lithium has been reported.

The original purpose of the present work was to carry out a detailed examination of an Al-Li-Cu-Mg casting with particular emphasis on the nature of the constituent phases and the effectiveness of solution treatment on their dissolution. During the progress of the investigation, a solution treated casting of modified composition became available and was incorporated in the work for comparison purposes.

EXPERIMENTAL PROCEDURE

The chemical composition of the two alloys examined are given in Table 1 (identified as Y and Z)

Characterisation of the microstructure was carried out using both optical and scanning microscopy, the latter employing an energy dispersive x-ray analysis facility (EDAX) for detection of elements greater than atomic no. 7. Lithium (atomic no. 3) is therefore undetectable using this equipment and the composition at specific points so determined exclude this element, although a 100% read-out is nevertheless given. Elemental weight ratios are consequently

Table 1. Chemical Composition of alloys Y and Z

Element	Weight %	
	Alloy Y	Alloy Z
Li	1.53	2.5
Cu	2.9	1.61
Mg	2.1	1.06
Si	0.07	0.14
Fe	0.10	1.1
Ti	-	0.1
Zr	0.17	0.02
Al	Bal.	Bal.

Table 2 Etching Characteristics of phases in alloys Y and Z

Ppte Type	Particle Appearance		
	As Polished	20s, 0.5% HF (optical)	20s, 0.5% HF (SEM)
A	Light grey	Light brown	White, rough
B	Light grey	Dark brown	White, rough
C	Medium grey	Medium brown	White, smooth
D,G	Bluish-grey, easily tarnishes black	Black	Black
E	Light grey	Light grey	White
F	Light grey	Light, watery brown. Outlined black	White, rough

Table 3 Composition of intermetallic particles in alloys Y and Z

Alloy condition	Phase Type	Composition (w/o)					Elemental weight ratio				
		Al	Cu	Mg	Fe	Si	Al:Fe	Al:Si	Al:Cu	Cu:Mg	Cu:Fe
Alloy Y as cast	A	51	32	14	0.3	3	-	-	1.6	2.3	-
	B	44	38	13	0.6	3	-	-	1.2	2.9	-
	C	56	19	0.6	15	3	3.9	-	2.4	-	1.3
	D	29	9	22	0.2	40	-	0.73	-	-	-
Alloy Y 2h, 510°C	A	46	33	17	0.4	2	-	-	1.4	1.9	-
	B	48	31	16	1.2	3.2	-	-	1.6	1.9	-
	C	55	19	0.6	16	3.2	3.4	-	2.8	-	-
	D	30	8	20	0.5	42	-	0.71	-	-	-
Alloy Y 2h, 530°C	A	49	33	14	0.8	3.3	-	-	1.5	2.4	-
	B	50	34	13	0.3	2.2	-	-	1.5	2.6	-
	E	51	31	0.3	10	1.7	5.1	-	1.7	-	3.1
	G	24	34	22	0.4	16	-	1.5	0.71	1.6	-
Alloy Z 2h, 520°C	F	65	24	8	0.3	2	-	-	2.7	-	-

Table 4. Elemental weight ratios of established phases

Phase	Theoretical Elemental Weight Ratio							
	Al:Cu	Cu:Fe	Cu:Mg	Al:Fe	Al:Mg	Al:Si	Fe:Si	Mg:Si
Cu_2FeAl_7	1.49	2.28	-	3.3	-	-	-	-
Al_2CuLi (T_1)	0.85	-	-	-	-	-	-	-
Al_6CuLi_3 (T_2)	2.55	-	-	-	-	-	-	-
Al_7Cu_4Li (T_B)	0.74	-	-	-	-	-	-	-
$Cu_2Mg_8Si_6Al_5$	1.06	-	0.65	-	0.69	0.80	-	1.15
Al_2CuMg (S)	0.85	-	2.61	-	2.22	-	-	-
$Al_7Cu_3Mg_6$ (Q)	0.99	-	1.31	-	1.30	-	-	-
$CuMg_4Al_6$	2.55	-	0.65	-	1.66	-	-	-
$FeSiAl_5$	-	-	-	2.44	-	4.76	1.99	-
$Al_{12}Fe_3Si_2$	-	-	-	1.93	-	5.76	2.98	-
Fe_2SiAl_8	-	-	-	1.93	-	7.68	3.98	-
Fe_3SiAl_{12}	-	-	-	1.93	-	11.5	5.97	-
(Fe,Cu)Al_6	2.55	1.14	-	2.9	-	-	-	-
$Li_3Si_2Al_2$	-	-	-	-	-	0.96	-	-
$LiMgAl_2$(S)	-	-	-	-	2.22	-	-	-
Mg_2Si	-	-	-	-	-	-	-	1.73

Table 5. X-ray diffraction data from the predominant phase in alloys Y and Z

Alloy Y			Alloy Z		
d Å	I/I_1	hkl	d Å	I/I_1	hkl
5.664	8	200	7.596	100	110
4.890	8	211			
3.832	29	310	4.409	44	211
3.493	10	222			
2.714	10	420	3.790	88	220
2.380	100	510, 431			
2.085	86	530, 433	2.490	71	411, 330
1.961	83	611, 532			
1.922	18	620	2.101	62	510, 431
1.887	16	541			
1.761	8	444	1.898	47	440
1.544	16	732, 651			
1.387	10	662			
1.321	18	842			
1.277	11	754, 851			
1.188	14	862			

quoted in order to partially circumvent this problem. Limited attempts to locate lithium within specific particles were made using an Auger microprobe.

Crystallographic data from the predominant cell/grain boundary phases was obtained by x-ray diffraction analysis of samples prepared by an electrochemical extraction technique. For this, round samples of 20 mm diameter and 2 mm thick were polished to 1 μm diamond finish and electrolytically etched in a solution of 2% KI in methanol using a pure aluminium cathode. Application of a voltage of 3V (giving 0.2A) for 24h resulted in preferential dissolution of the aluminium matrix. The intermetallic network remaining on the surface was removed as a powder, mounted on a diffractometer stage and examined using Ni-filtered CuKα radiation.

Samples for solution treatment were heated in a circulating air furnace for times of 2-42h at temperatures of 490°-530°C followed by still air cooling. The response to heat treatment was in each case monitored by microhardness measurements and metallographically.

RESULTS

As cast microstructure

Figure 1 shows the microstructure typical of as-cast alloy Y in which the cell and grain boundaries are delineated by a network of coarse second phase particles. Figures 2a-b are higher magnification micrographs before and after etching in 0.5% HF and clearly indicate the presence of four distinct intermetallic phases. For convenience these are designated A-D, type A being predominant but duplex with smaller amounts of B; their etching characteristics are detailed in Table 2. X-ray maps corresponding to the areas in Figures 2a-b provided a semi-quantitative guide to the predominant distribution of elements relative to the intermetallics, Figures 2c-f. Whilst generating no quantitative information, the Auger microprobe nevertheless confirmed the presence of lithium within particle types A and B.

To facilitate identification of the intermetallics, a number of X-ray spot analyses were taken and averaged results are shown in Table 3 together with mean values for predominant elemental weight ratios. Consideration of these for particles A and B show no essential difference, notwithstanding the differing etching response of each phase. No correlation is apparent with the elemental ratios of phases previously encountered in Al-Li-Cu or Al-Cu-Mg alloys (Table 4). Similar analysis for particle type C is indicative of the (Fe, Cu) Al_6 phase. Relatively small amounts of particle type D were noted which contained appreciable amounts of aluminium, magnesium and silicon.

Reference to Figure 1 additionally shows darkened regions adjacent to the cell/grain boundary network which is indicative of coring. Although these appear to be composed of discrete second phase particles in Figure 2b, the SEM electron image in Figure 2c suggests that they are fine surface depressions which may or may not be due to preferential dissolution during etching of fine intermetallics.

Solution Treated Material

The 'coring' exhibited by as-cast alloy Y was eliminated after solution treatment for 2h at 510°C although dissolution of the cell and grain boundary intermetallics did not occur, Figure 3a. Metallography and EDAX spot analysis (Tables 2 and 3) indicated the predominant phases to be those designated A and B of the as-cast material with smaller quantities of phase type C present. Needle-like particles of a phase designated type E and corresponding to Cu_2FeAl_7 were also observed, as were occasional particles of the as-cast phase type D, essentially unchanged by the heat treatment.

The 'coring' was similarly eliminated upon solution treatment at 530°C for 2h as well as the intercellular second phase, whilst retaining a grain boundary network, Figure 3b. However, the appearance of grain boundary cracks together with rosette formations within the grains indicated that incipient melting had occurred at this temperature, Figure 3c. Again, phase types A and B constituted the predominant intermetallic particles, both at grain boundaries and in rosettes. Particles of phase type C were not detected and appeared to have been completely replaced by phase type E (Cu_2FeAl_7). Occasional particles were additionally observed, type G, which in appearance and etching response were essentially identical to phase type D although EDAX spot analysis showed that considerable compositional modifications had occurred, with significant apparent enrichment in copper, accompanied by silicon depletion.

Solution treatment of 2h at 520°C was carried out which resulted in incipient melting to give microstructures essentially identical to those after 2h at 530°C.

Extended solution treatment times of up to 42h at 510°C appeared to little affect the amount of cell/grain boundary intermetallics. This was not the case when solution treated at 530°C for similar times where, in addition to almost complete elimination of the intercellular phase network after 2h, increasing the soak time was found to progressively reduce the amount of precipitate remaining at grain boundaries, accompanied by progressive hardening of the aluminium matrix, Figure 4. However, incipient melting again occurred in all material solution treated at this temperature.

Duplex solution treatments were attempted in order to circumvent the problem of incipient melting. Samples heated for 40h at 490°C exhibited microstructures essentially identical to those of 2h at 510°C, while additional heating

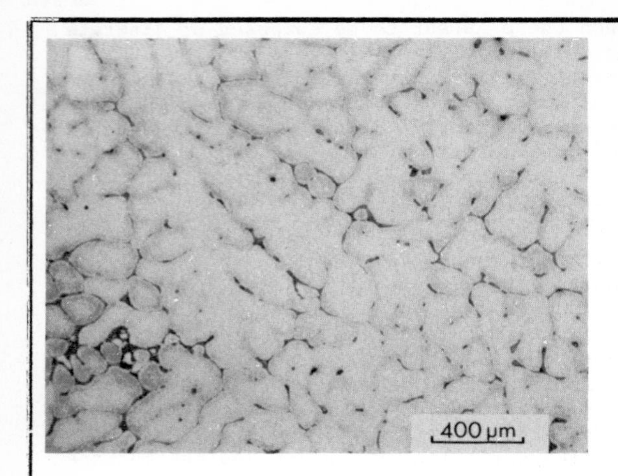

Figure 1. Typical as-cast microstructure of alloy Y.

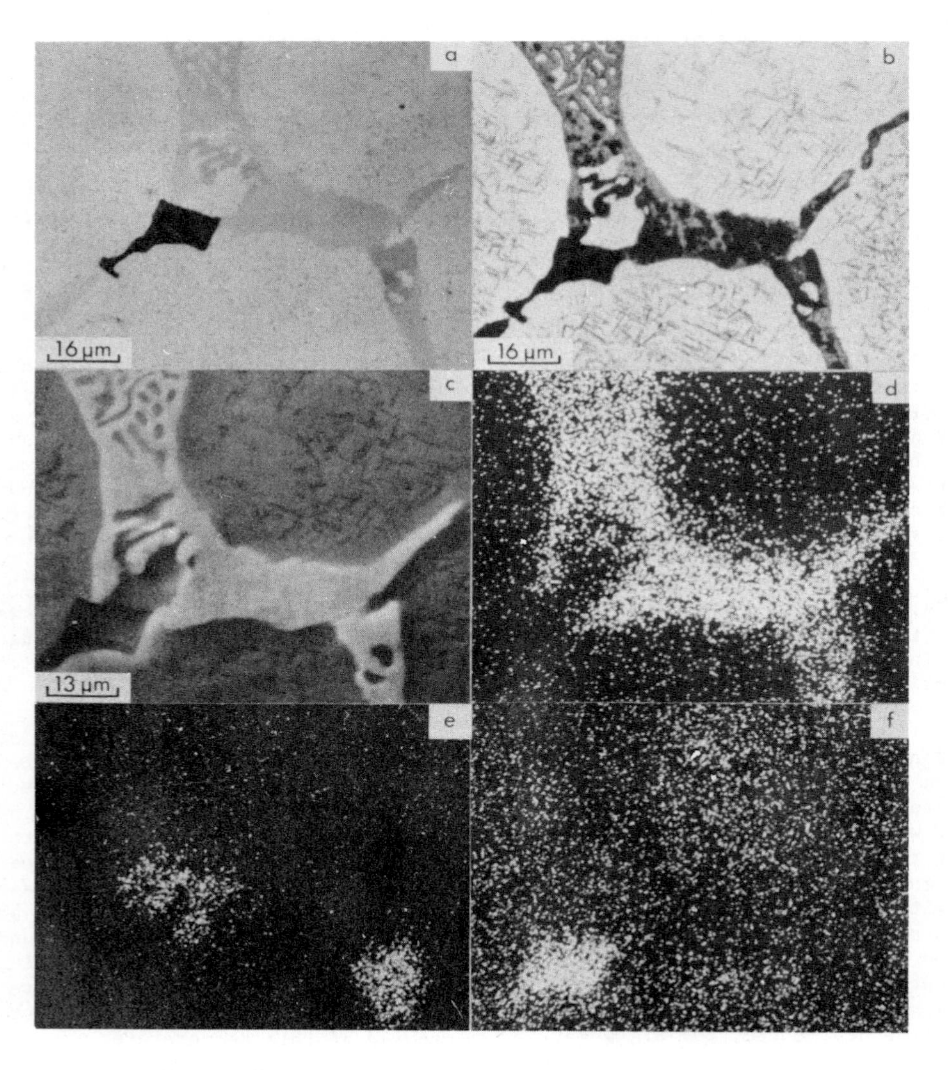

Figure 2. Grain boundary precipitates in as-cast alloy Y. (a) As polished (b) Etched in HF (c) Electron image (d) Cu X-ray map (e) Fe X-ray map (f) Si X-ray map.

at 530°C for 2h resulted in incipient melting, giving structures similar to those after a single stage treatment of 2h at 530°C.

Samples of alloy Z were examined having been solution treated for (a) 3h at 500°C, (b) 3h at 500°C + 16h at 510°C and (c) 3h at 500°C + 16h at 520°C. Metallographic examination showed that extensive cell and grain boundary precipitate dissolution had occurred with no evidence of incipient melting. EDAX spot analysis indicated the predominant undissolved intermetallics type F to be compositionally independent of solution treatment temperature and time, and to differ significantly from corresponding precipitates in alloy Y, in that lower copper and magnesium levels were apparent, Table 3. Auger microprobe investigations indicated the presence of lithium.

X-ray diffraction analysis of extracted predominant cell/grain boundary precipitates from as-cast alloy Y and solution treated alloy Z indicated two distinctly different phases, that from the former metal being indexed as bcc with a_o = 12.11 Å while that of the latter was also bcc but with a_o = 10.73 Å. Measured values of d spacings are shown in Table 5; no correlation was found with established phases.

Lithium/Magnesium Loss

Figure 5 indicates that although there is no significant loss from the surface of as-cast alloy Y as monitored by microhardness measurements, noticeable and progressive depletion occurs with increasing time at 510°C.

DISCUSSION

The coarse-grained microstructure of as-cast alloy Y was typical of non-grain refined sand cast aluminium alloys. In the initial stages of solidification relatively pure aluminium dendrites formed, the cored structure after etching being indicative of progressive enrichment of alloying elements at cell/grain boundaries. Optical microscopy revealed clusters of needle-like features within these regions which resembled fine precipitates although the SEM revealed fine surface depressions. It was not clear if these were due to the preferential dissolution of specific particles during etching or to a form of etching effect, possibly caused by localised solute concentrations. Similar features are observed in micrographs of as-cast 2014 alloy (11) although the authors did not comment upon them. However, this does indicate that such structures are not confined to the Al-Li-Cu-Mg system.

A number of intermetallic compounds were detected in both alloy compositions studied. In alloy Y the predominant precipitate at cell/grain boundaries appeared to be a quaternary Al-Li-Cu-Mg phase, (designated type A) the identity of which is uncertain and, as far as the author is aware, undocumented. However, an isomorphous relationship is known to exist

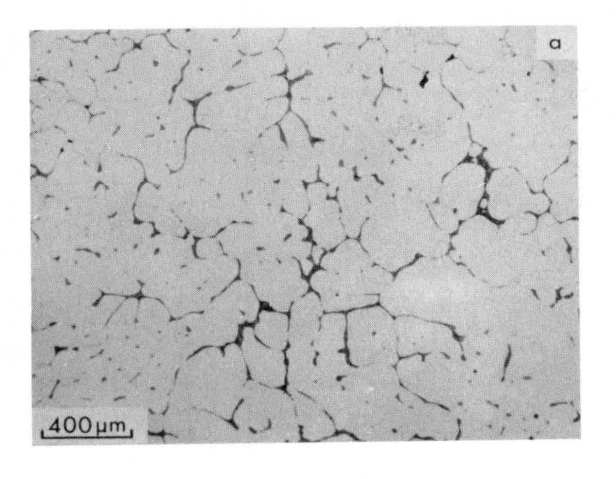

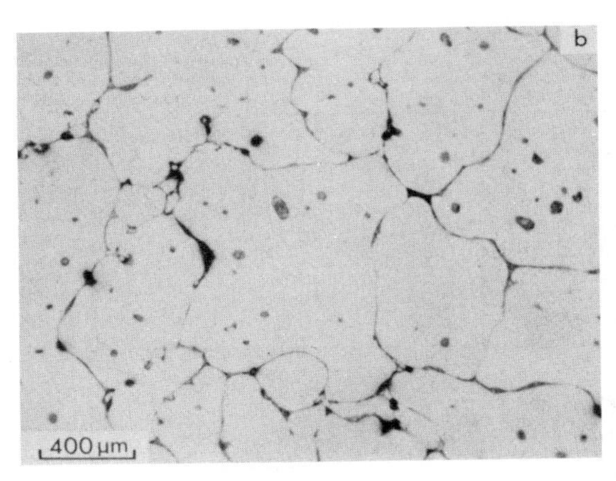

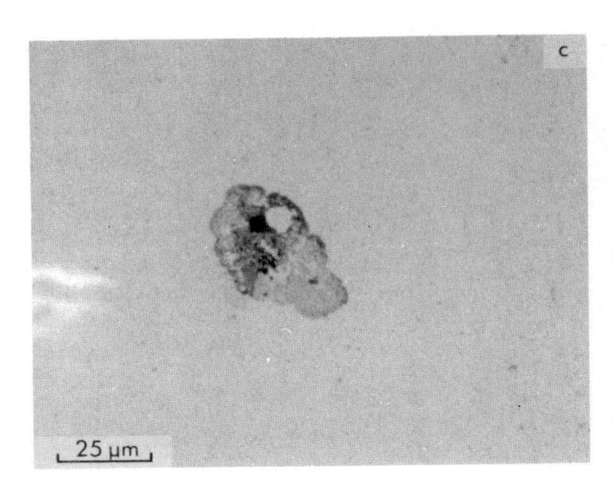

Figure 3. Solution treated alloy Y (a) 2h at 510°C (b), (c) 2h at 530°C.

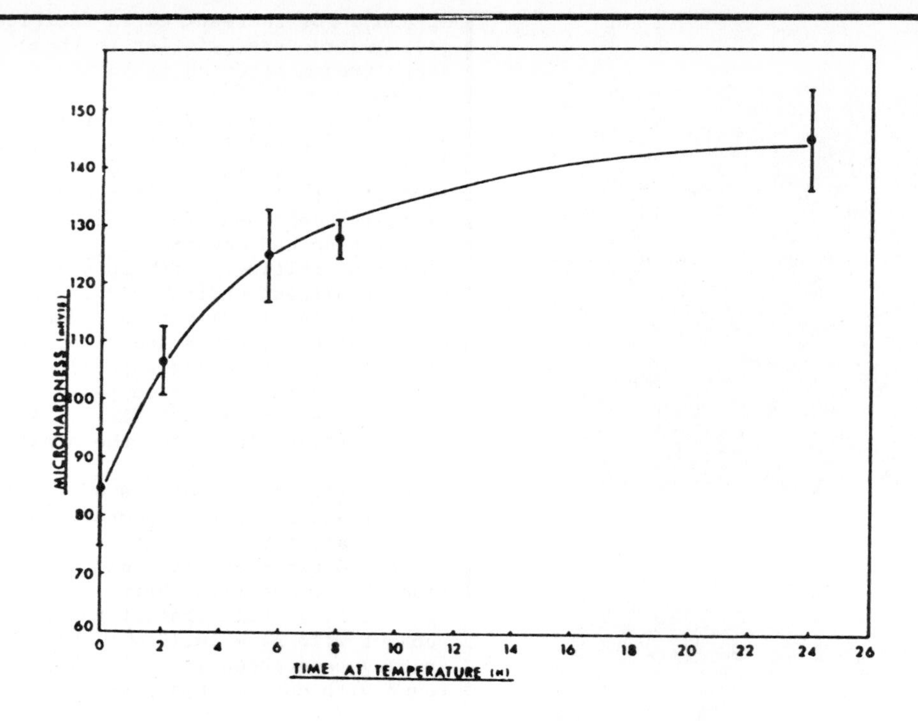

Figure 4. Microhardness as a function of solution treatment time at 530°C.

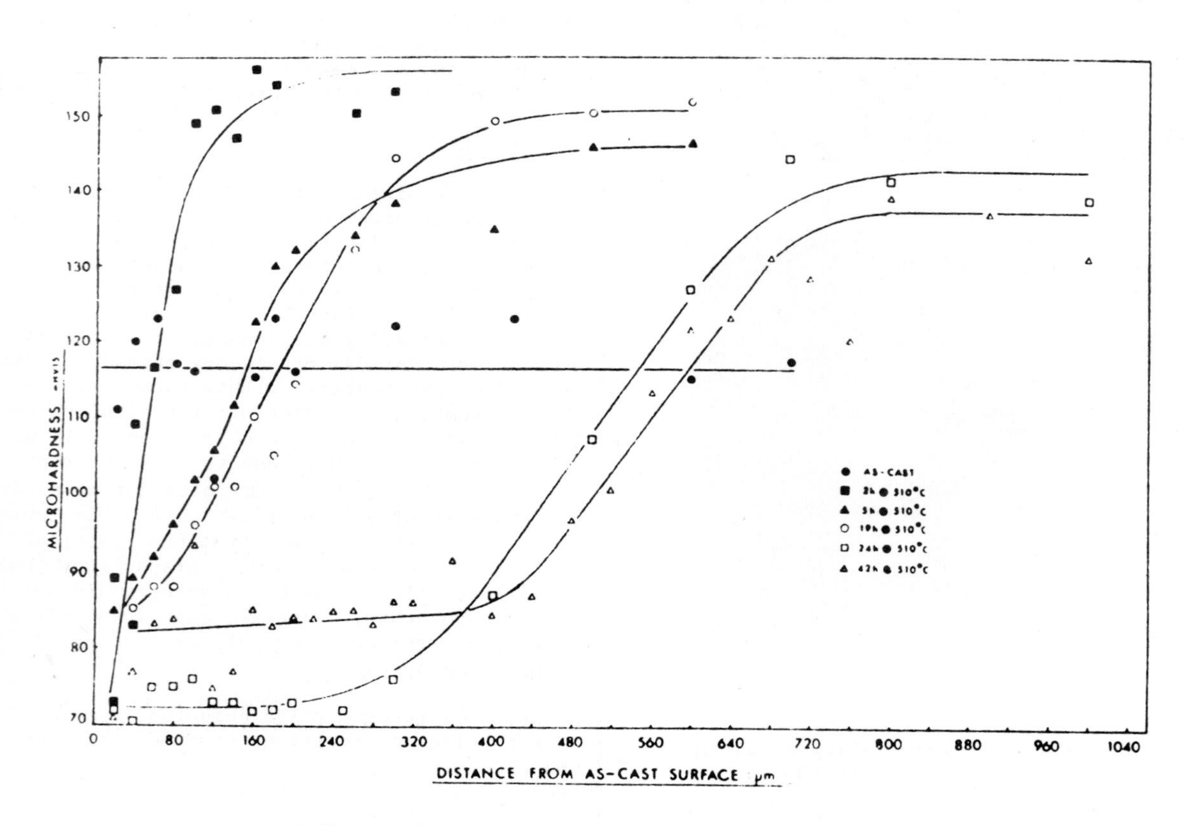

Figure 5. Lithium and magnesium loss from casting surface as monitored by microhardness.

between the ternary phases Al_6CuLi_3 (T_2) and Mg_{32} $(Cu,Al)_{49}$, both bcc with a_o = 13.914 Å and 14.35 Å respectively. The current phase may be indexed as bcc with a_o = 12.11 Å and although not conclusive, it is possible that it may constitute part of a series of intermetallic compounds bounded by the above isomorphous phases with possible interchangeability between Li and Mg atoms. A variable Li content may explain the difference between phases A and B which exhibited very similar Al, Cu and Mg levels, although etching differently.

The predominant cell/grain boundary phase in solution treated alloys Y and Z differed significantly. It was noticeable that the lower Cu and Mg contents of alloy Z (1.61 w/o and 1.06 w/o respectively) compared to those of alloy Y (2.9 w/o and 2.1 w/o respectively) have modified the precipitate composition from ~35 w/o Cu,~17 w/o Mg, ~48 w/o Al (alloy Y) to ~21 w/o Cu, ~8 w/o Mg, ~64 w/o Al (alloy Z). Although the latter Cu and Al levels broadly correspond with those of T_2 (Al_6 $CuLi_3$), the X-ray diffraction data, although limited at present, clearly show 'd' spacings which are not consistent with this or the phase in alloy Y. However, they may be indexed as a bcc structure with a_o = 10,73 Å which is a phase also believed to be undocumented.

Occasional small particles of a phase, (type D), metallographically resembling Mg_2Si and which was compositionally unaffected by solution treatment at 510°C were detected in alloy Y. However, the presence of aluminium together with inappropriate Mg:Si ratios indicated a different phase and since no ternary Al-Mg-Si compound is known, it is postulated that Li is also present. Solution treatment at 530°, however, resulted in apparent enrichment of Cu and depletion of Si and it is unclear whether significant elemental substitution of a phase of variable composition has occurred or whether transformation to a distinctly different precipitate has taken place.

As-cast alloy Y also contained particles, (type C), tentatively identified as a phase designated (Fe,Cu) Al_6 [12]. This describes approximately a ternary phase in the Al-Cu-Fe system which is isomorphous with $MnAl_6$ and probably exists over a range of compositions. This phase was present after solution treatment at 510°C together with particles of Cu_2FeAl_7 while the former appeared to be eliminated after heating at 520°C. This accords with the findings of Sperry (13) who reported the transformation of $(Fe,Cu)Al_6$ to Cu_2FeAl_7 at high temperatures. Iron containing precipitates were also present in solution treated alloy Z but in this case they appeared to be $FeAl_6$.

Studies on Li/Mg surface loss indicated that this did not occur during cooling of the casting but that solution treatment in air caused noticeable depletion, the extent of which was a function of time at temperature. Differences in actual microhardness values obtained for different times are likely to be attributable to compositional variations along the casting surface. A notable feature of the curves shown in Figure 5 is the formation of a band of relatively low but approximately constant hardness at the surface after ~20h heating, rather than a continuing decrease in slope of the curve found at shorter times. This implies that use of solution treatment times often employed for Al alloys (ie. up to ~48h) would result in depleted surface layers up to 0.6 mm deep. The use of inert gas or reducing atmospheres may be beneficial in this respect.

CONCLUSIONS

1. As cast alloy Y exhibited a cell/grain boundary precipitate network which appeared to be of a previously undocumented Al-Cu-Li-Mg quaternary phase indexed as bcc with a_o = 12.11 Å. This could not be dissolved by solution treatment without incipient melting.

2. As cast alloy Y exhibited small amounts of a phase believed to be an undocumented Al-Li-Mg-Si quaternary. This appeared to be unaffected by solution treatment at 510°C but significant apparent copper enrichment and silicon depletion occurred at 530°C.

3. As cast alloy Y contained particles of (Fe,Cu) Al_6 which transformed to Cu_2FeAl_7 after solution treatment at 520°C.

4. The predominant cell/grain boundary phase in alloy Z was indexed as bcc with a_o = 10.73 Å and is believed to be undocumented. Extensive dissolution of this phase was achieved by solution treatment with no evidence of incipient melting.

5. Li/Mg loss was not apparent from the as-cast surface but significant depletion occurred during solution treatment in air.

ACKNOWLEDGEMENTS

The author wishes to thank Messrs B C Gittos, R L Cudd, and G Swinyard for their encouragement and helpful discussions which have contributed to the preparation of this paper.

REFERENCES

1. T.H.Sanders Jnr, "Aluminium-Lithium Alloys". Proceedings of the First International Aluminium-Lithium Conference, Stone Mountain, Georgia, May 1980. The Metallurgical Society of AIME, Warrendale, Pa, 1981.

2. F.W.Gayle, ibid

3. C.J.Peel, B.Evans, C.A.Baker, D.A.Bennett, P.J.Gregson and H.M.Flower, "Aluminium-Lithium Alloys II". Proceedings of the Second International Aluminium-Lithium

Conference, Monterey, California, April 1983. The Metallurgical Society of AIME, New York, 1983.

4. F.S.Lin, S.B.Chakrabortty and E.A.Starke Jnr, Met.Trans.A. 13A, 1982, 401-410

5. J.M.Silcock, J.I.M. 1959-60, 88, 357-364.

6. B.Noble and G E Thomson, Met. Sci. Journal, 1971, 5, 114-120.

7. H.K.Hardy and J.M.Silcock, J.I.M. 1955-56, 84, 423-428

8. F.I.Shamrai, Bulletin of the Academy of Sciences of the U.S.S.R, Section of Chemical Science, Part II, 1948, p83.

9. D.W.Levinson and D.J.McPherson, Transactions of the Metallurgical Society of A.I.M.E. 224, 1962, 970-974.

10. M.E.Drits, E.S.Kadaner and N.I.Turkina. Russian Metallurgy (Metally), Part 4, 1971, 163-166.

11. A.Munitz, A.Zangril and M.Metzyer, Met.Trans.A, 11A, 1980, 1863-1868.

12. G.Phragmen, J.I.M. 1950, 77, 489-552.

13. P.R.Sperry, Trans. ASM, 1956, 48 904-918.

□

Effect of cold deformation on mechanical properties and microstructure of Alcan XXXA

D T MARKEY, R R BIEDERMAN and A J McCARTHY

DTM is with the
Wyman-Gordon Company,
North Grafton, MA 01536, USA.
RRB and AJMcC are at the
Worcester Polytechnic Institute,
Worcester, MA 01609, USA

ABSTRACT

The effect of prior cold work on the aging response of Alcan XXXA (2.5%Li-1.2%Cu-0.7%Mg-0.12%Zr bal. Al) is presented. This investigation was an attempt to achieve strength levels comparable to 7075-T73 forgings. A cold reduction range of 3% to 6% was necessary to achieve adequate strength and ductility. Optical Metallography (OM) and Transmission Electron Microscopy (TEM) were utilized to characterize the microstructural changes as a function of aging time and cold reduction. Dislocations introduced during cold coining prior to aging served as sites for nucleation of the S' phase,(Al_2CuMg). This results in a substantially finer heterogeneous precipitation of this phase on aging. Closed die forgings were produced from this alloy, with and without cold reduction, and the results are compared to 7075-T73 forgings.

INTRODUCTION

The development of aluminum - lithium alloys dates back to the Scleron alloys (Al-Zn-Cu-Li) developed in Germany in the 1920's (1). With the exception of the brief commercialization of alloy 2020 (Al-Cu-Li-Cd) by Alcoa (2) in the 1960's, the aluminum - lithium alloys have been confined to research and development. However, due to the energy crisis in the 1970's as well as the accelerated development of advanced polymer composites, interest in aluminum - lithium alloy development has been rekindled. Extensive research (3,4) in both alloying and processing of aluminum - lithium alloys has been undertaken over the last five years. The benefits associated with these new alloys generally include: 1. a density reduction on the order of 10% ; 2. an increase in stiffness of approximately 10%; and 3. compatibility with existing metal working facilities when compared with currently used aluminum alloys.

Much of the high strength development work has centered around the Al-Li-Cu-Mg system, with more emphasis placed on sheet, plate and extrusion products which comprise the bulk of aluminum usage in aerospace applications (5). Other product forms such as forgings, which often have complex shapes and varying section sizes, present a greater challenge. Differences in section size can result in property variations as well as distortion problems associated with heat treatment. To accommodate unique problems encountered with forged products, forging alloys are typically developed with deep hardening capability or utilize special thermomechanical processing treatments to minimize property variation with section size. This has yet to occur in the development of aluminum - lithium alloys.

This study was conducted to determine the potential use of a cold coining operation to increase the strengthening response of a currently available aluminum - lithium alloy, Alcan XXXA. This alloy was not developed as a forging alloy. However, since it was available in sufficient quantity and appropriate size it was selected for this study to get some forging experience with a high strength aluminum - lithium alloy. Two major objectives of this work were: 1) the determination of mechanical properties for a wide range of cold reduction prior to aging and 2) comparison of XXXA closed die forgings, both with and without cold coining, to a currently produced 7075-T73 forging. Although variation of section size in a forging makes control of mechanical properties more difficult if cold coining is required, this procedure has been successfully used in the past with large aluminum forgings to relieve machining problems due to residual stresses and to increase the strength in 2000 series alloys (6).

Table I Summary of Chemistry from Various Regions in the Ingot*

Ingot Slice	Location	Li	Cu	Mg	Zr	Fe	Si	Ti	Al
Top RJS-2A	center	2.52	1.00	.56	.12	.093	.030	.022	bal.
	mid-rad.	2.56	1.08	.59	.11	.109	.033	.019	bal.
	surface	2.58	1.10	.59	.11	.103	.032	.018	bal.
Bottom RJS-2D	center	2.56	1.01	.57	.11	.093	.030	.027	bal.
	mid-rad.	2.62	1.08	.60	.11	.105	.032	.024	bal.
	surface	2.68	1.10	.59	.11	.101	.032	.023	bal.

* weight % determined by atomic absorption.

Table II: Summary of Tensile Results from Forgings

	S/N 7-2 7075-T73			S/N 2A-1 XXXA w/o Coin			S/N 2D-1 XXXA w/Coin		
S/N LOCATION & DIRECTION	YLD. Ksi (MPa)	ULT. Ksi (MPa)	ELON. %	YLD. Ksi (MPa)	ULT. Ksi (MPa)	ELON. %	YLD. Ksi (MPa)	ULT. Ksi (MPa)	ELON. %
1-RAIL SHORT TRANS.	59.2 (408)	70.2 (484)	9.0	50.4 (348)	63.4 (437)	4.0	60.0 (414)	68.0 (469)	2.5
2-RAIL SHORT TRANS.	58.4 (403)	70.2 (484)	8.0	50.6 (349)	61.4 (423)	3.0	61.6 (425)	67.8 (467)	2.5
3-RAIL LONG.	64.4 (444)	74.6 (514)	13.0	59.0 (407)	73.0 (503)	8.0	68.0 (469)	77.1 (532)	10.0
4-RAIL LONG.	62.8 (433)	73.4 (506)	15.0	59.0 (407)	73.0 (503)	10.0	64.2 (443)	75.2 (519)	8.0
5-RAIL LONG.	64.4 (444)	74.0 (510)	13.5	57.0 (393)	72.8 (502)	9.0	56.8 (392)	72.6 (501)	10.0
6-RAIL LONG.	62.6 (432)	72.2 (498)	14.0	53.4 (368)	70.0 (483)	10.0	64.2 (443)	75.4 (520)	10.0
7-WEB LONG.	64.0 (441)	75.0 (517)	14.5	58.8 (405)	72.6 (501)	9.0	65.0 (448)	77.0 (531)	8.5
8-WEB TRANS.	62.0 (427)	72.2 (498)	13.5	56.0 (386)	74.0 (510)	8.5	62.0 (427)	75.3 (519)	7.0

Table III: Summary of Closed Die Forging Fracture Toughness Results.

S/N	MATERIAL	SPECIMEN THICKNESS in.(cm.)	KQ Ksi(in.)$^{1/2}$ (MPa(m.)$^{1/2}$)	KIC Ksi(in.)$^{1/2}$ (MPa(m.)$^{1/2}$)	COMMENTS
21	7075-T73	.5(1.27)	25.2 (27.5)	---	crack plane
22	7075-T73	.5(1.27)	23.8 (25.9)	---	loading
23	7075-T73	.5(1.27)	29.1 (31.7)	---	size
21	XXXA w/o	.5(1.27)	13.6 (14.8)	---	precrack
22	XXXA w/o	.5(1.27)	17.0 (18.5)	---	precrack
23	XXXA w/o	.5(1.27)	22.4 (24.4)	---	precrack/plane
21	XXXA w/	.5(1.27)	failed in precracking		
22	XXXA w/	.5(1.27)	13.5 (14.7)	---	precrack
23	XXXA w/	.5(1.27)	15.4 (16.8)	---	precrack/plane
24	7075-T73	.75(1.91)	---	28.0 (30.5)	valid
25	7075-T73	.75(1.91)	---	26.9 (29.3)	valid
24	XXXA w/o	.75(1.91)	18.8 (20.5)	---	precrack
25	XXXA w/o	.75(1.91)	22.4 (24.4)	---	precrack/plane
24	XXXA w/	.75(1.91)	16.6 (18.1)	---	precrack
25	XXXA w/	.75(1.91)	18.1 (19.7)	---	precrack/plane

NOTE: w/ - coined
w/o - without coining

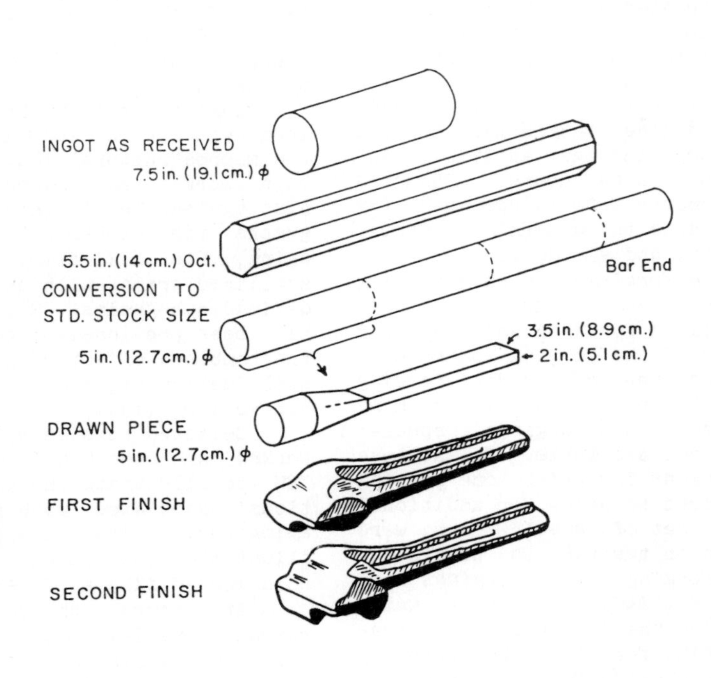

Figure 1. Schematic of the Closed Die Forging Operations.

EXPERIMENTAL PROCEDURE

An 8.0 in. (20.3cm.) diameter homogenized ingot was scalped to 7.5in. (19.1cm.) and cut into four 24in.(61cm.) lengths. For this study the top and bottom sections of the ingot were used. Macrostructures prepared from both sections showed a fine uniform structure with a slight coarsening in the center of the bottom section. Chemistry from the center, mid-radius and surface of the slice are summarized in Table I. Chemical analysis shows good uniformity from center to edge as well as from top to bottom of the ingot. Density samples were taken from the same locations in each macroslice and tested in accordance with ASTM E12 (7). Again the uniformity of the as-received material was demonstrated with the density at all locations determined to be .091 lbs/cu.in.(2.521g/c.c.). Remnants from the macroslices were sectioned into cubes approximately 0.5in. (1.27cm.) on a side and exposed at various times at 820F(438C) to evaluate the extent of lithium loss to the environment during processing.

Two types of forgings were made. Simple open die pancake forgings and more complex closed die forgings. For the pancake forgings the as-received ingot was cogged at 820F (438C) on open dies in a series of passes which reduced the stock from 7.5in. diameter to 5in. diameter (19.1cm. to 12.7cm.) billet. For convenience, the billet was cut into three 5in. diameter x 4.5in. high (12.7cm x 11.4cm.) mults and upset forged at 820F (438C) to 8.6in. diameter x 1.5in. thick (21.8cm. x 3.8cm.) pancakes. The pancakes were sectioned in half and solution treated at 975F (524C) for 2 hours then water quenched. The half sections were then coined by cold forging on flat dies utilizing stop blocks to produce a range in cold reduction from 0% to 10% in 2% intervals. The half pancakes were sectioned to yield nine tangential tensile blanks each. These were then aged at 375F (191C) for times ranging from 2 to 64 hours. Standard 0.252in.(0.640 cm.) diameter tensile specimens (8) were machined and tested to determine tensile strength, yield strength, and elongation.

For the closed die forgings,the same forging dies, deformation sequence and forge temperature, 820F (438C), were utilized for XXXA and the 7075 material. A schematic of the forging sequence is shown in Figure 1. Since the closed die forging chosen does not currently utilize a cold coining step in the production of 7075 forgings, special cold coining dies were not available. To achieve the desired cold reductions for XXXA, some parting line material was removed to allow for additional metal movement and a set of steel shims were fabricated and placed on the part to provide a cold reduction to the forging. XXXA forgings were solution treated at 975F (524C) for 2 hours, water quenched and, following the cold coining operation, aged at 375F (191C) for 16 hours, then air cooled. The XXXA and 7075-T73 forgings were sectioned to provide fracture toughness (9), stress corrosion (10) and smooth and notched tensile specimens. Sectioning with specimen locations are given in Figure 2.

Optical and electron metallography studies were performed on a large variety of XXXA samples to correlate the observed behavior with microstructure. Standard metallographic procedures were used. Electron diffraction techniques were used to identify all phases present.

RESULTS AND DISCUSSION

Surface Effects

The loss of lithium from surfaces during elevated temperature exposure is of concern if these alloys are to be forged (11,12). This loss results in decreased strength near the surface as evidenced by a microhardness gradient. A typical example is shown in Figure 3 along with the time dependence of lithium loss at 820F(438C) for XXXA. With conventional closed die aluminum forgings this surface reaction effect would be removed. However, it presents concern for parts with as-forged surfaces, particularly precision forgings.

Pancake Evaluation

A summary of the yield strength and ductility values generated as a function of the percent cold work is presented with the aging curves shown in Figure 4. Yield strengths increase with increasing cold deformation. Although 2% cold work resulted in an improvement over no reduction, it did not result in the desired level of strength. Cold reductions in the 4% to 10% range resulted in adequate strengths. However, further strength improvement was minimal with additional cold work. The introduction of cold deformation prior to aging also appears to increase the kinetics of the age hardening response of XXXA. This is shown by a shift in the aging curves, presumably due to a lowering of the activation energy associated with precipitation.

Tensile specimen threaded sections were examined via optical and TEM in order to characterize the microstructural features associated with aging both worked and unworked XXXA. There was no distinguishable difference in the optical micrographs for either condition as a function of aging. The XXXA alloy exhibited large unrecrystallized grains with an extensive network of well defined sub-grains. See Figure 5. This condition has been previously reported and is attributed to the combined effect of zirconium inhibiting recrystallization as well as sub-grain formation during hot working (13).

Selected tensile specimens representing both worked and unworked conditions were examined via TEM to illustrate the microstructural transformations associated with various points along the aging curve. The sequence of pictures in Figure 6 illustrates the microstructures of both 0% and 8% cold worked XXXA at 2, 16, and 64 hours.

At 2 hours, the cold worked specimen already revealed precipitation of the lathe shaped S'phase (Al_2CuMg) while the unworked specimen shows only isolated areas of dislocation tangles. Increased aging time for the unworked XXXA alloy resulted in heterogeneous nucleation and growth of the S' phase at grain boundaries as well as at high dislocation density regions. In comparison, the worked material results in a more uniform nucleation and growth of the S' phase throughout the matrix. This uniform precipitation is attributed

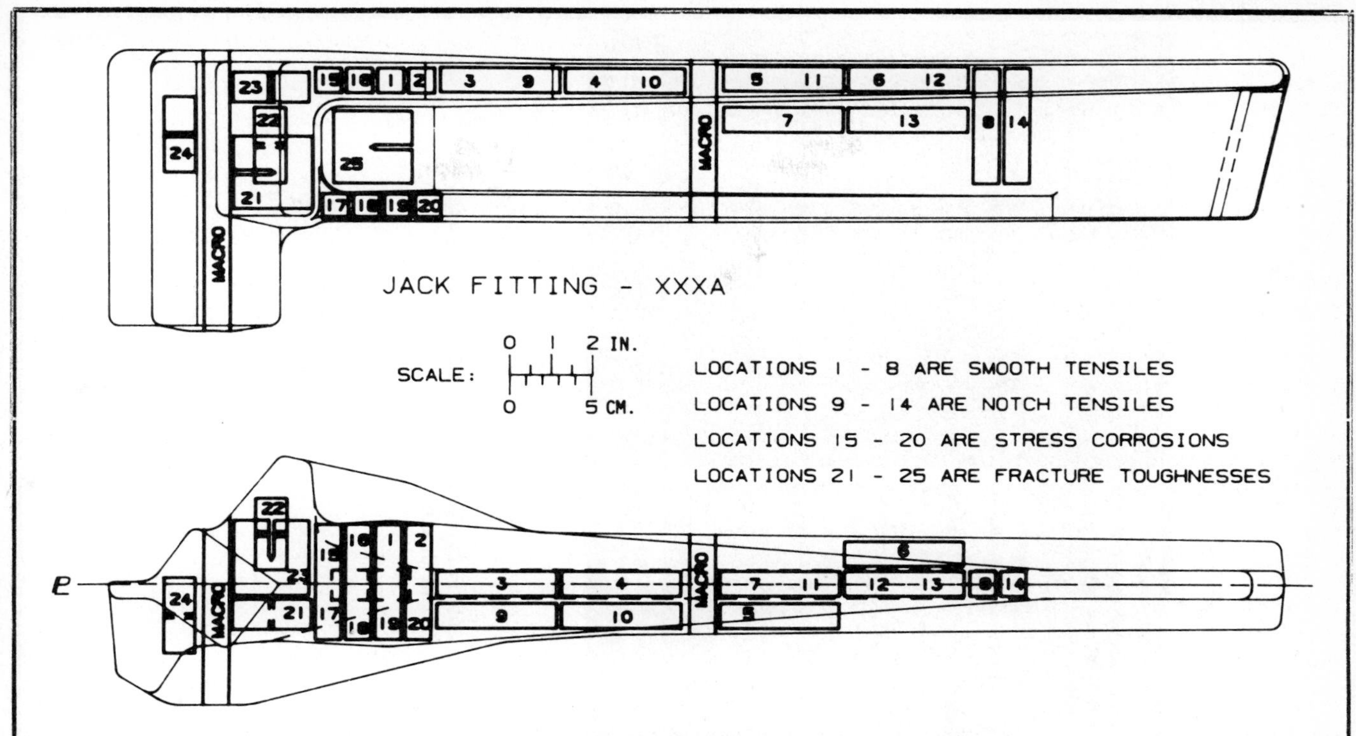

Figure 2. Sectioning Diagram for Specimen Location and Identification.

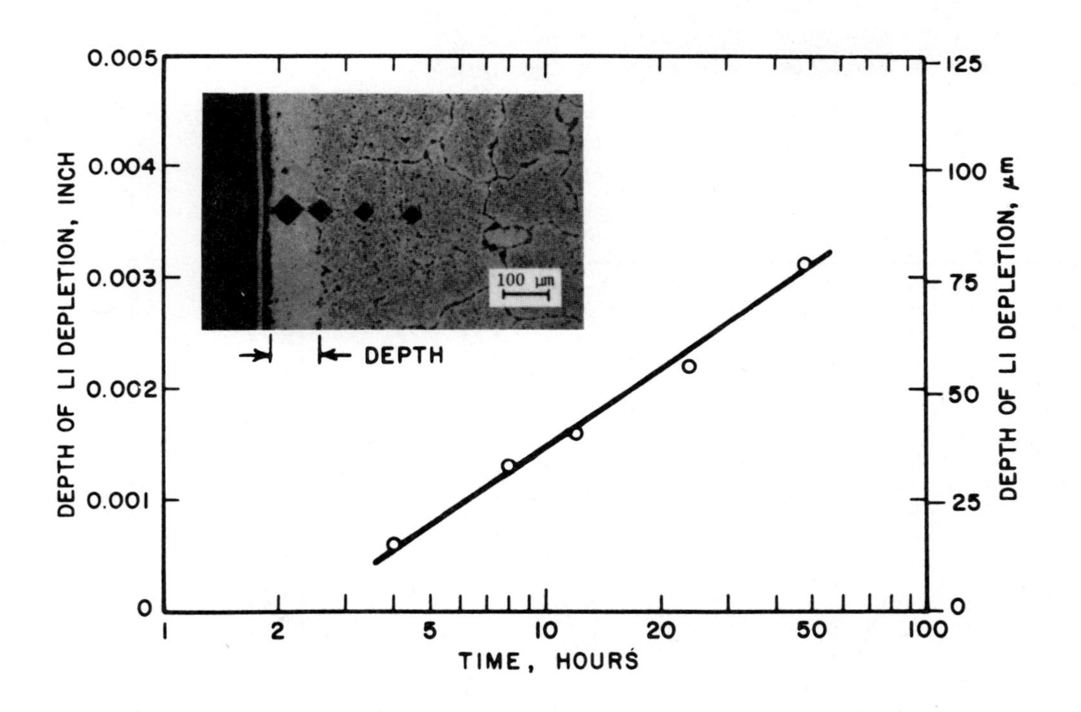

Figure 3. The Time Dependence of Lithium Loss in Air for XXXA at 820F (438C).

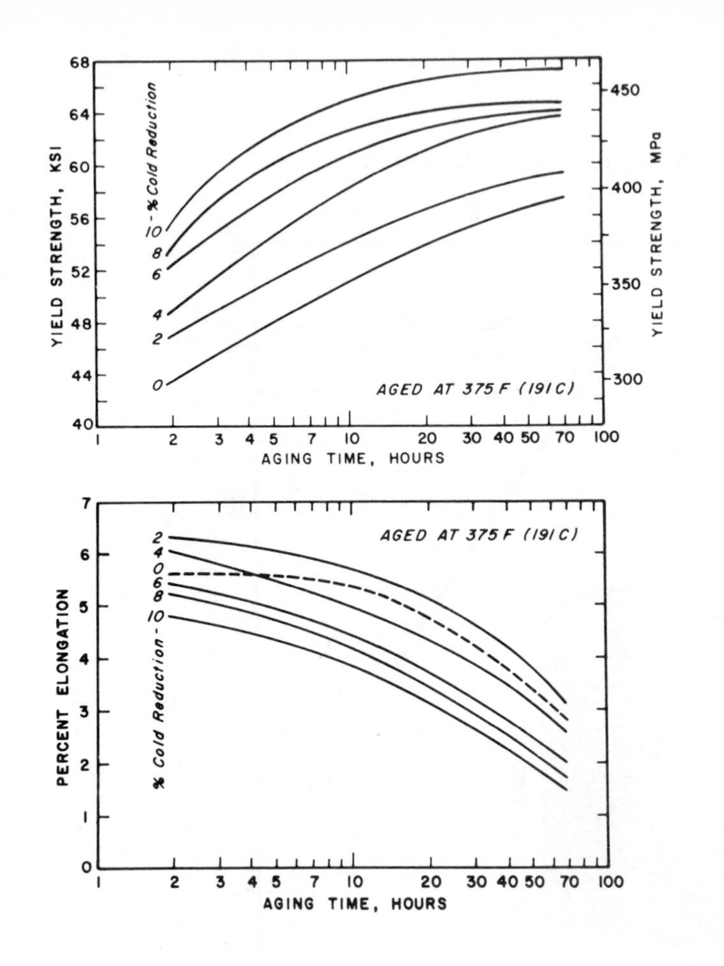

Figure 4. The Effect of Cold Deformation on the Aging Response of XXXA.

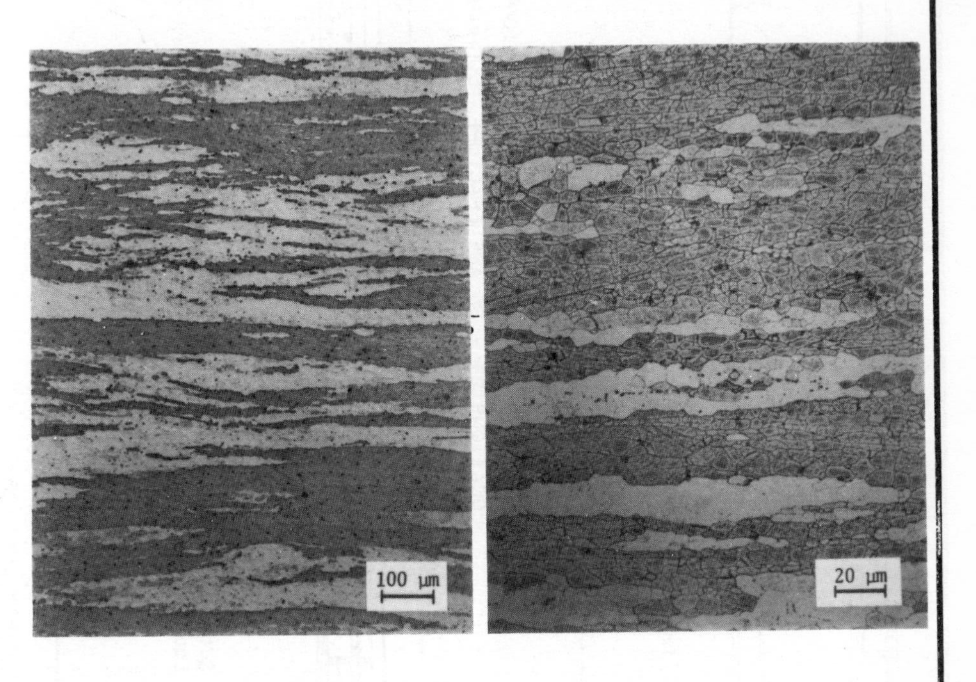

Figure 5. Subgrain Formation Within Unrecrystallized Grains.

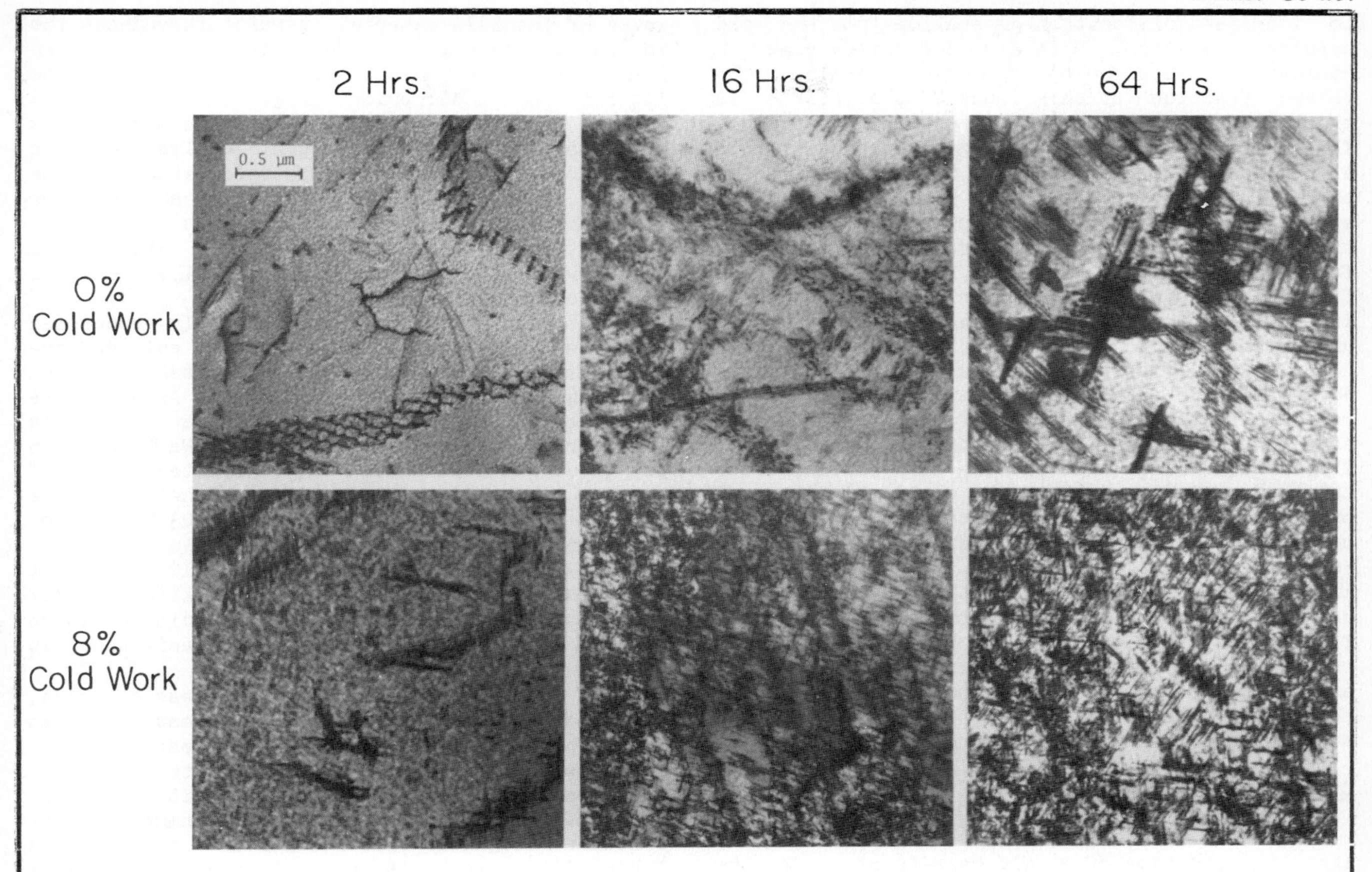

Figure 6. Comparison of Worked vs. Unworked XXXA Microstructures as a Function of Aging Time at 375F (191C).

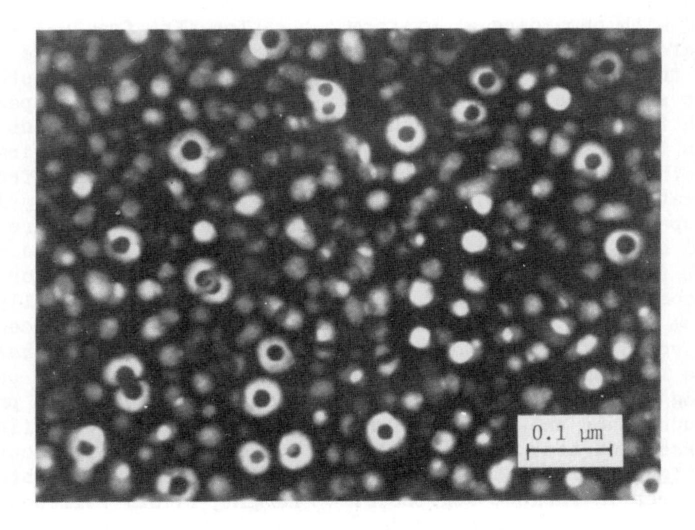

Figure 7. Duplex Delta Prime Phase Precipitation.

to increased dislocation generation from the cold coining operation, with dislocations serving as nucleation sites for S' formation (14-16). The volume fraction of this finer S' precipitate is greatly increased over that in unworked material. This is reflected in the strength response of this alloy during aging. Also, a duplex delta prime phase, (Al_3Li), precipitation was observed in all specimens. The delta prime phase coarsens with time, with the larger delta prime heterogeneously forming around $ZrAl_3$ particles,(17-19). See dark field micrograph, Figure 7. Unlike the S' precipitation, delta prime precipitation appears to be unaffected by cold deformation.

The difficulty in accurately controlling deformation in forgings with varying section sizes requires that a wide range of reductions be available to produce a similar strengthening response. The results from this pancake study demonstrate that adequate strength values could be obtained over a wide range of cold reductions. Although a continuous strength increase was associated with increasing cold deformation, the incremental increases in strength beyond 4% were small. The ductility decreased with additional cold working. Therefore, a cold reduction range of 3-6% was chosen for the closed die forging evaluation. Construction of shims for the closed die portion of this study based on this 3%-6% cold reduction actually imparted 2-5% cold work.

Closed Die Forging Evaluation

Review of the tensile data in Table II shows that the addition of a cold coining step in the processing of XXXA results in strengths which were equivalent to those for 7075-T73, although the ductilities were reduced (particularly in the short transverse direction). As previously seen with the pancake study, the yield strength of XXXA could be increased 10-20% with equivalent ductilities via a cold reduction operation. The potential problem associated with providing a uniform distribution of cold deformation in a forged product was illustrated by the lack of strength response in specimen 5. It should be noted that this specimen was located in the bottom die portion of the forging rail. Due to the use of top die shims, this region was trapped thereby preventing a sufficient amount of deformation.

The XXXA tensile specimens were subsequently examined in the threaded section via TEM to correlate the microstructural features associated with the properties. It was observed that for any location, the specimens from the cold reduced forging demonstrated a higher volume fraction of the S' phase precipitate. The introduction of dislocations in the cold coining step provided sites for the nucleation of a much finer and more uniform dispersion of the S' phase than that found in the unworked forging. See Figure 8 for a typical comparison. The microstructures in various regions of the cold coined forging were very similar except for specimen S/N 5. A much lower volume fraction of S'precipitation was observed in this region of the forging, see Figure 8 , although preferential nucleation on the (100) slip planes was still observed.

The large unrecrystallized grains, characteristic of the XXXA material, persist throughout the entire processing. A recovered grain structure with low angle grain boundaries occurs in the interior of these prior large unrecrystallized grains which allows slip to propagate easily from grain to grain. Microstructural analysis in this region confirms a lack of cold deformation in the forging, in agreement with previously shown lower strength. The deformation variations observed in this study utilizing only a simple steel shim demonstrate the need for a specifically designed set of cold coining dies.

Although this study was primarily concerned with the strength response of XXXA to cold deformtion, the alloy's toughness and stress corrosion cracking behavior was also evaluated. An alternate immersion stress corrosion test was performed with short transverse specimens in a 3.5% NaCl solution for 30 days. Even though only a limited amount of testing was conducted, the SCC threshold for XXXA appears to be approximately 15Ksi (103.5MPa) regardless of cold deformation, while that for 7075-T73 is approximately 45Ksi (310.5MPa). It should be noted, however, that the XXXA material was heat treated to a peak-age condition which historically is undesirable for SCC resistance in aluminum alloys.

To evaluate the relative toughness of XXXA, both compact tension fracture toughness specimens and notched tensile specimens were tested. Forging section size dimensions limited the fracture toughness specimens thickness to either 0.5in. (1.27cm.) or 0.75in. (1.91cm.). A summary of the fracture toughness test results are provided in Table III, along with comments regarding reasons for invalid KIC tests. All XXXA specimens were considered KQ values because the fatigue precrack length exceeded specified limits. Also, some samples exhibited a non-parallel crack plane. See Figure 9. Although only KQ values were obtained, the data indicate that cold deformation in XXXA will result in decreased toughness.

The XXXA fracture toughness specimens appeared to be very sensitive to nonplanar crack growth. Test specimens from both locations 24 (planar) and 25 (nonplanar) were specially prepared to show the microstructure adjacent to the fracture surface to determine the underlying microstructural features associated with this fracture behavior. The crack path is preferentially intergranular with a strong dependence on the unrecrystallized grain orientation of the alloy. See Figure 10. This undesirable crack growth behavior can be minimized in XXXA sheet by using specially developed thermomechanical processing sequences (20). Additional forging studies of XXXA are needed to examine the effects of thermomechanical processing sequences (e.g. cross working) on the properties of forging.

The notched tensile/yield strength ratio was averaged from six locations within each forging. The highest value obtained was for the 7075-T73 forging (1.62) followed by the XXXA forging without cold coining (1.53) and then the XXXA forging with cold coining (1.37). This relative measure of toughness correlated with the trend previously demonstrated with the fracture toughness values of XXXA.

In summary, the utilization of a specifically designed set of cold coining dies should be con-

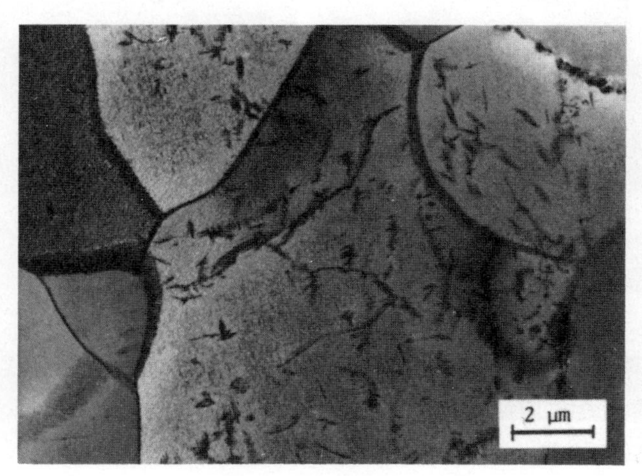

NO COINING

40/8

LIMITED COINING   EFFECTIVE COINING

Figure 8. The Effect of Cold Reduction on the Nucleation and Growth of S' in Closed Die Forged XXXA.

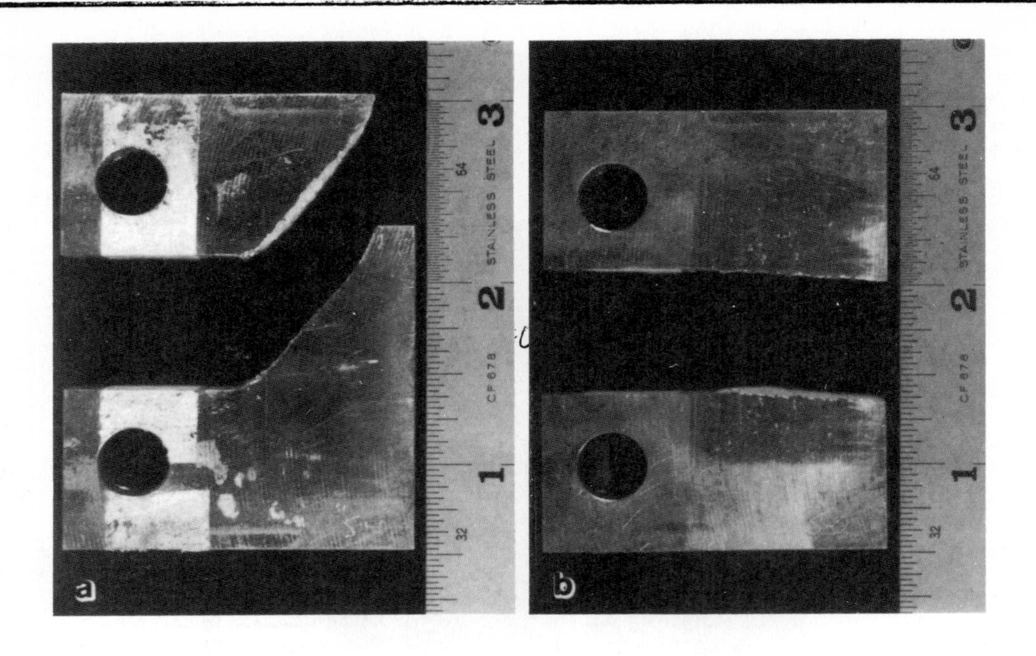

Figure 9. Irregular Crack Propagation in XXXA Following Unrecrystallized Grain Boundaries.

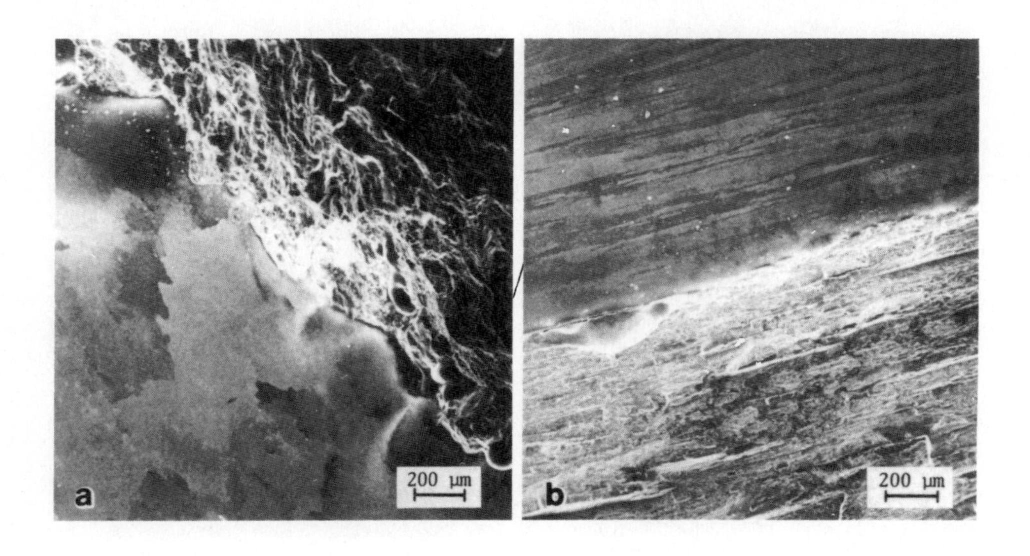

Figure 10. Correlation of Fracture Path with Microstructural Features.

sidered as a viable alternative in the fabrication of closed die aluminum-lithium forgings. Although variations in cold deformation will exist in complex closed die forgings, the demonstration of adequate strength levels over a wide range of reductions should minimize the property variations. Increasing the copper content in aluminum -lithium alloys should result in equivalent strengths to those demonstrated with the cold reduction approach. However, the density penalty associated with copper additions severely reduces the most important benefit for the use of aluminum - lithium alloys in aerospace applications.

CONCLUSIONS

1. The as-received XXXA ingot demonstrated good microstructural and chemical uniformity.
2. The loss of lithium at the surface during high temperature exposure presents some problems for parts to be used with as-forged surfaces, particularly precision forgings.
3. The addition of 3% - 6% cold deformation resulted in a 10-20% increase in yield strength for XXXA over unworked XXXA, with esentially equivalent ductility. This wide range of useful cold reductions makes XXXA amenable to a cold coining operation in closed die forgings.
4. Cold deformation prior to aging increases the age hardening kinetics for XXXA. This results from the introduction of dislocations during the cold coining which serve as sites for precipitation of S' phase.
5. The increased volume fraction as well as the more uniform nature of the S' precipitation with cold worked XXXA is responsible for the strength increase over unworked XXXA.
6. The variability of cold deformation imparted by the steel shim demonstrates the necessity for specifically designed cold coining dies.
7. No attempt was made to optimize stress corrosion cracking resistance. However, the short transverse SCC threshold for XXXA appears to be approximately 15Ksi (103.5MPa) regardless of cold deformation.
8. Fracture toughness testing of XXXA was shown to be very dependent on grain texture, with a slight reduction in the KQ values resulting from the use of cold deformation.

ACKNOWLEDGMENTS

The authors express their appreciation for financial and technical support from the Wyman - Gordon Company, particularly from John Mc Keogh, Steve Reichman and Robert Sparks. Also, technical discussions with William Miller and Roger Grimes of Alcan Ltd. were helpful in understanding XXXA. A special thanks are extended to George Schmidt, Jr. for help in preparation of the manuscript and figures.

REFERENCES

1. Scheuer, E.,"Scleron and Aeron," Z. Metallkunde, Vol. 19, 1927, pp. 16-19.
2. Spuhler, E. H., Knoll, A.H.and Kaufman, J. G., "Lithium in Aluminum-X2020," Met. Prog., Vol. 79, 1960, pp. 80-82.
3. Saunders, T. H., and Starke, E. A., Editors, "Aluminum - Lithium Alloys," 1981 AIME Conference Proceedings of the 1st International Aluminum - Lithium Conference Stone Mountain, Georgia, May,1980.
4. Saunders, T. H., and Starke, E. A., Editors, "Aluminum - Lithium Alloys II,"1984 AIME Conference Proceedings of the 2nd International Aluminum -Lithium Conference, Monterey, California, April, 1983.
5. Quist, W.E., Narayanan, G. H., and Wingert, A. L.,"Aluminum - Lithium Alloys for Aircraft Structure - An Overview," Aluminum - Lithium Alloys II, AIME Conference Proceedings, 1984.
6. Hull, J.H., and Erwin, S. J., "The Effect of Mechanical Deformation on the Tensile Properties and Residual Stresses in Aluminum Forgings," ASM Report System Paper No. W72-53.1, WESTEC 1972.
7. ASTM Standards E12-70, "Density and Specific Gravity of Solids, Liquids and Gases," ASTM 1984, Vol. 4.02,pg. 806.
8. ASTM Standards E8-83, "Tension Testing of Metallic Materials," ASTM, 1984, Vol. 3.01, Pg. 130.
9. ASTM Standards E399-83,"Plane Strain Fracture Toughness of Metallic Materials," ASTM, 1984, Vol. 3.01, Pg.519.
10. ASTM Standards G47-79, "Determining Susceptibility to Stress - Corrosion Cracking of High-Strength Aluminum Alloy Products," ASTM, 1984, Vol. 3.02, Pg. 274.
11. Peel, C. J., Evans, B., Baker, C. A., Bennett, D. A., Gregson, P. J., and Flower, H. M.,"The Development and Application of Improved Aluminum - Lithium Alloys," Aluminum - Lithium Alloys II, AIME Conference Proceedings, 1984, Pg. 363.
12. Adb El-Salam, F., Eatah, A.I., and Tawfik, A., "Effect of Li Loss on Some Physical Properties of Al-Li Alloy," Phys. Stat. Sol. Vol.75, 1983, Pg.379.
13. Sanders, T. H.,Jr., and Starke, E. A, Jr., "Overview of the Physical Metallurgy in the Al-Li-X Systems, "Aluminum - Lithium Alloy II, AIME Conference Proceedings, 1984, Pg. 1.
14. Ibid. Ref.13.
15. Ibid. Ref.11.
16. Chen, R. T., and Starke, E. A., Jr., "Microstructure and Mechanical Properties of Mechanically Alloyed, Ingot Metallurgy and Powder Metallurgy Al-Li-Cu-Mg Alloys," Materials Science and Engineering, Vol. 67, 1984, Pg. 229.
17. Makin, P. L., and Ralph, B., "On the Ageing of an Aluminum-Lithium-Zirconium Alloy," Journal of Materials Science, Vol. 19, 1984, Pg. 3835.
18. Gayle, F. W., and Vander Sande, J. B., ""Composite" Precipitates in an Al-Li-Zr Alloy," Scripta Metallurgica, Vol.18, 1984, Pg. 473.
19. Gregson, P. J., and Flower, H. M., "Delta Prime Precipitation in Al-Li-Mg-Cu-Zr Alloys, "Journal of Material Science Letters 3, 1984, Pg. 829.
20. Ibid 11.

Role of hydrostatic pressure on cavitation during superplastic flow of Al–Li alloy

J PILLING and N RIDLEY

The authors are in the Department of Metallurgy and Materials Science at the University of Manchester/UMIST, Grosvenor St., Manchester M1 7HS, UK

ABSTRACT

The influence of hydrostatic pressure on the cavitation behaviour of an Al-Li-Cu alloy during superplastic flow in uni-axial and equi-biaxial tension has been examined. Precision density measurements have been used to measure the volume fraction of cavities, while the number of cavities and their size distributions have been determined using quantitative metallographic techniques. Cavities were observed to be associated with regions of the microstructure having a greater than average grain size. The application of an increasing level of hydrostatic pressure during superplastic deformation led to a progressive fall in the level of cavitation at a given strain, to a reduction in the rate of increase of the volume fraction of cavities with superplastic strain and to an increase in the superplastic strain to failure.

An analysis of cavitation during superplastic flow based on the experimental observation that cavity growth is controlled primarily by plastic flow predicts that cavitation should be greatly reduced by pressures between 1/3 and 2/3 of the effective flow stress of the material. This was consistent with experimental observation and because of the low flow stress associated with the high superplastic deformation temperature (520°C), relatively low hydrostatic pressures would be required to substantially reduce cavitation levels. However, since cavity nucleation was relatively insensitive to pressure within the range examined, pressures in excess of the flow stress would be required to eliminate cavitation completely.

INTRODUCTION

High strength aluminium alloys which have been thermomechanically processed to develop a fine grain size, can form internal voids when they are deformed superplastically. The presence of voids, whilst limiting the ultimate strain obtainable at failure, can have a detrimental effect on the post forming mechanical properties of a component (1). The average strain in most superplastically formed components is low relative to the elongation to failure, and thus the cavitation damage is also low. However, certain parts of the component can undergo much higher strains and as a consequence cavitation damage is locally more severe. Attempts to reduce cavitation have involved the design of forming processes to limit the strain attained and to distribute that strain more evenly throughout the component (2); hot isostatic pressing following forming to sinter up any voids (3-5), or the superposition of a confining hydrostatic pressure during the forming process (6,7). The first process redistributes the cavitation damage and prevents poor properties on a local scale, but does not eliminate the cavitation. Similarly, although hipping has the potential to remove the cavities completely it is possible that in aluminium alloys the voids are not actually removed but instead just the surfaces of the void are brought into contact. There is the possibility that they could re-appear during subsequent heat treatment. On the other hand, the superposition of a hydrostatic pressure during the forming process would appear to have the greatest potential for keeping the levels of cavitation low, by inhibiting nucleation and growth processes.

In the present paper the effects of hydrostatic pressure on the volume fraction of cavities, on cavity nucleation and growth and on the strain to failure during superplastic flow of an Al-Li-Cu alloy are examined. Although material which had been processed to give a uniform fine grain size was available to the authors, the experimental work described was carried out on a conventionally rolled alloy which contained a large proportion of fine grains interspersed with regions of coarser grains. In this way it was

Table 1. Comparison of cavitation levels at different pressures at typical SPF strains.

Strain	Pressure (MPa) 0	0.7	1.4	2.1	3.05
1.09 (200%)	1.6%	0.25%	0.2%	0.08%	none
1.39 (300%)	5.0%	0.5%	0.35%	0.1%	none
1.61 (400%)	failed	1.1%	0.55%	0.11%	0.02%
1.79 (500%)	failed	failed	failed	0.13%	0.02%
1.95 (600%)	failed	failed	failed	failed	0.12%

Table 2. Summary of experimentally determined cavity growth rate factors.

Geometry	P, MPa	P/σ_e	η_{obs}	η/η_o	m	η eqn. 2
uni-axial	0	0	3.61	1	0.48	1.94
	0.70	0.11	2.56	0.71		1.28
	1.40	0.22	1.74	0.49		0.64
	2.10	0.32	0.46	0.13		0.08
	3.05	0.47	no detectable cavitation			-.78
bi-axial	0	0	5.31	1	0.39	5.42
	1.40	0.19	3.13	0.59		3.67

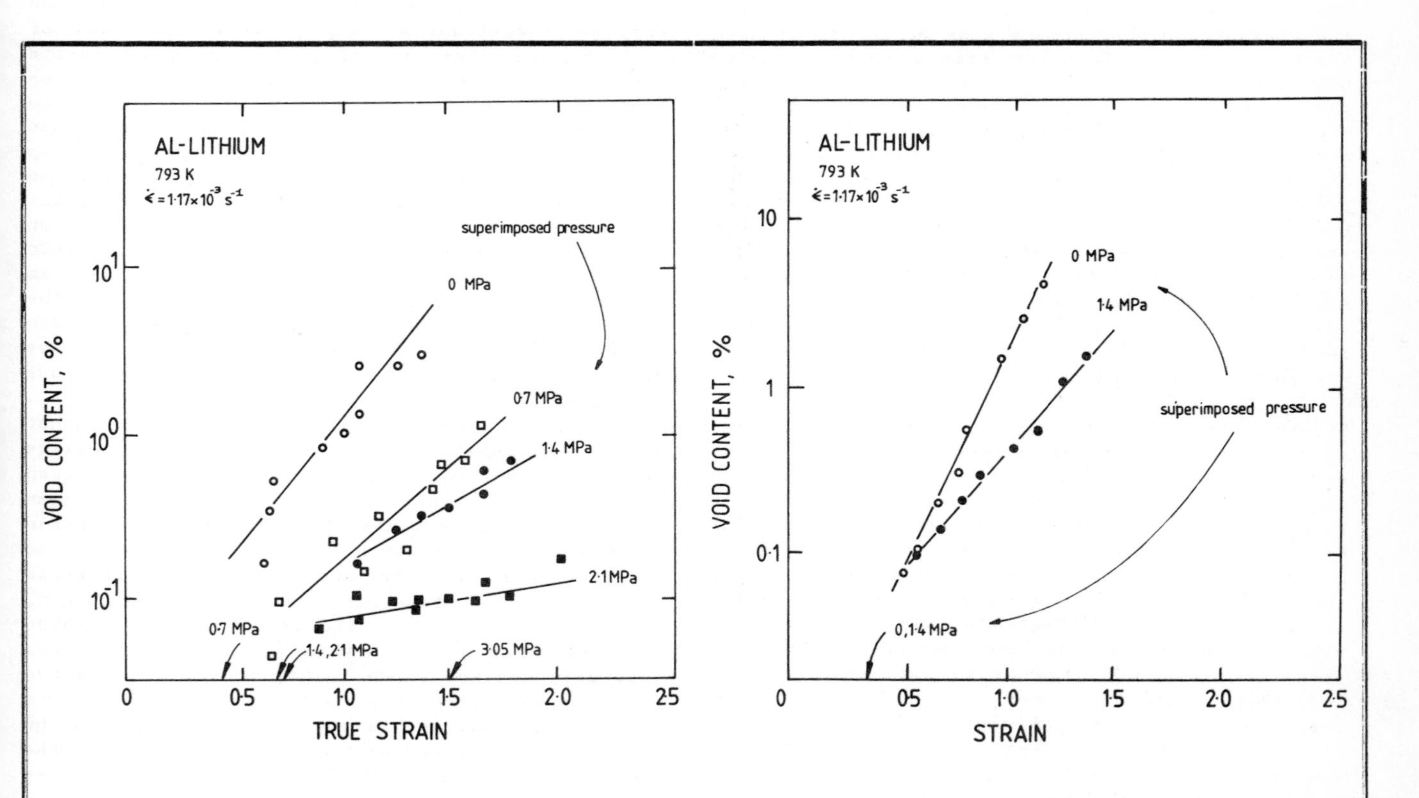

Figure 1. The variation of the volume fraction of voids with effective strain at different superimposed pressures (uni-axial).

Figure 2. The variation of the volume fraction of voids with effective strain at different superimposed pressures (bi-axial).

possible to investigate the ability of hydrostatic pressure to influence cavitation behaviour in material which did not have a fully ideal superplastic microstructure.

EXPERIMENTAL

Uni-axial tests: Tensile specimens with gauge lengths of 10 mm were prepared from 2 mm thick sheets of Al-2.5%Li- 1.2%Cu-0.6%Mg-0.12%Zr alloy. The tests were carried out at a constant strain rate of 1.17 x 10^{-3} s^{-1} at 520°C in a capsule which could be pressurised to 6.7 MPa (1000 psi) using argon gas. The capsule was mounted within a 3-zone furnace supported on the crosshead of an Instron universal testing machine. The volume fraction of cavities in the gauge lengths of samples deformed to a pre-determined strain, under a constant superimposed pressure, was measured by precision densitometry (9). The densities were determined by the displacement method using iodoethane as the high density liquid. An undeformed gauge head of each sample was used as a reference specimen.

Bi-axial tests: Discs 65 mm in diameter were prepared from the same sheet of material as the tensile specimens, and then blow formed into 'top hat' shapes at 520°C. The strain rate in the pole of each forming dome was monitored using a displacement potentiometer. The information was then used within a feed back circuit to control the pressure on the forming side of the sheet such that a constant effective strain rate equal to that used in the uni-axial tests was maintained in the pole of the dome. Once contact with the strain limiting insert within the forming die was attained, the pressure was kept constant until the 'top hat' had formed. By varying the back pressure against which the forward forming pressure acted to form the 'top hat', the superimposed hydrostatic pressure could be varied. The volume fraction of cavities in each sample was measured from a 20 mm diameter disc which was spark eroded from the flat section of the 'top hat' using densitometry. The undeformed rim of the top hat was used as a reference sample.

Strain rate sensitivity: The strain rate sensitivity of the alloy was determined using specimens with tensile axes parallel to, at 45° and at 90° to the rolling direction. The strain rate sensitivity at each orientation was evaluated by deforming samples to a strain of 0.33 at a constant velocity of 0.5 mm/min and then reducing the crosshead speed to 0.05 mm/min. After allowing the load to equilibrate, the crosshead velocity was repeatedly incremented in 11 small steps to 10 mm/min, the steady state or maximum load being measured at each step. The stress and instantaneous strain rate were then calculated and plotted in log-log form. The strain rate sensitivity was evaluated at 1.17 x 10^{-3} s^{-1} from the derivative of a polynomial fit to the log stress-log strain rate data.

Quantitative metallography: The distribution of void sizes in each sample was measured using a Quantimet 720 image analysing computer operating with a spatial resolution of 0.35 μm and a sizing interval of 1.4 μm, for specimens which had been metallographically prepared but were unetched. The measured distributions of void sizes were corrected to remove the weighting effect of sectioning using the Schwarz-Saltykov method (10). An average of 120 fields of view, yielding approximately 10,000 cavities, were examined for each specimen. The accuracy of the Quantimet data was then checked by comparing the volume fraction of voids calculated from the corrected size distributions with those determined by densitometry.

RESULTS AND DISCUSSION

The initial microstructure of the alloy could not be discerned metallographically, there being few high angle boundaries in the heavily worked structure. Electron diffraction from thin foil specimens showed that the grains were orientated with ⟨111⟩ in the plane of the sheet. However after a small amount of SP deformation at elevated temperature, the structure appeared to recrystallise and consisted of alternating bands of fine (<5 μm) and coarse (>30 μm) grains running parallel to the original rolling direction.

The variation of the volume fraction of voids with strain at different superimposed pressures is shown in Figure 1 for samples deformed uni-axially and in Figure 2 for samples deformed bi-axially. It can be seen from these figures that for a given strain the application of hydrostatic pressure can lead to a substantial reduction in the level of cavitation. For example, for a superplastic strain of 1.39 (300% elongation) the volume fraction of cavities is reduced from 5% at zero superimposed pressure to 0.1% at a superimposed pressure of 2.1 MPa (300 psi). Therefore it is evident that substantial benefits can be achieved by the application of relatively low pressures. The effects of various superimposed pressures on the cavitation damage at typical superplastic strains are summarised in Table 1. It is also evident from Figures 1 and 2 that as the superimposed pressure increases, the strain attained before the cavities become detectable by densitometry increases (as indicated by the arrows intersecting the horizontal axis).

It can be seen from both Figures 1 and 2 that an approximately linear relationship exists between the void volume fraction (plotted logarithmically) and true strain. The linear relationship is indicative of plasticity or strain controlled void growth. Furthermore, it can also be seen that as the superimposed pressure is increased, the slope of the linear region, η, decreases, data points for the uni-axial samples deformed under a superimposed pressure of 3.05 MPa (450 psi) are not shown in Figure 1. In only one specimen were any voids detected (ϵ = 1.5; v = 0.02%) the remainder giving approximately zero or marginally negative cavitation. The slope of the data at a pressure of 3.05 MPa is therefore presumed to be zero.

If it is assumed that void growth occurs as a result of plastic deformation within the sample then (11,12):

$$v = v_o EXP(\eta \epsilon) \qquad (1)$$

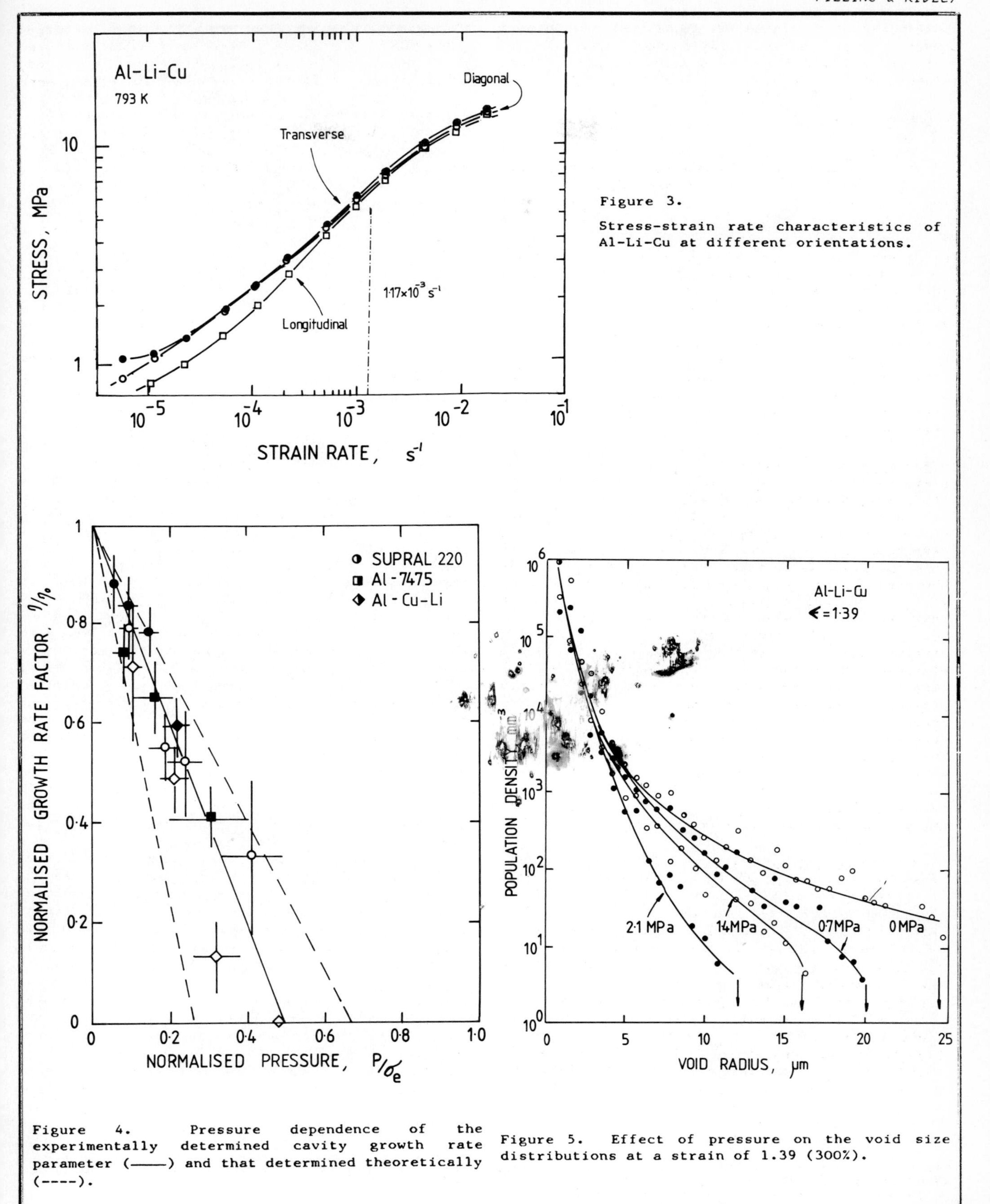

Figure 3. Stress-strain rate characteristics of Al-Li-Cu at different orientations.

Figure 4. Pressure dependence of the experimentally determined cavity growth rate parameter (——) and that determined theoretically (----).

Figure 5. Effect of pressure on the void size distributions at a strain of 1.39 (300%).

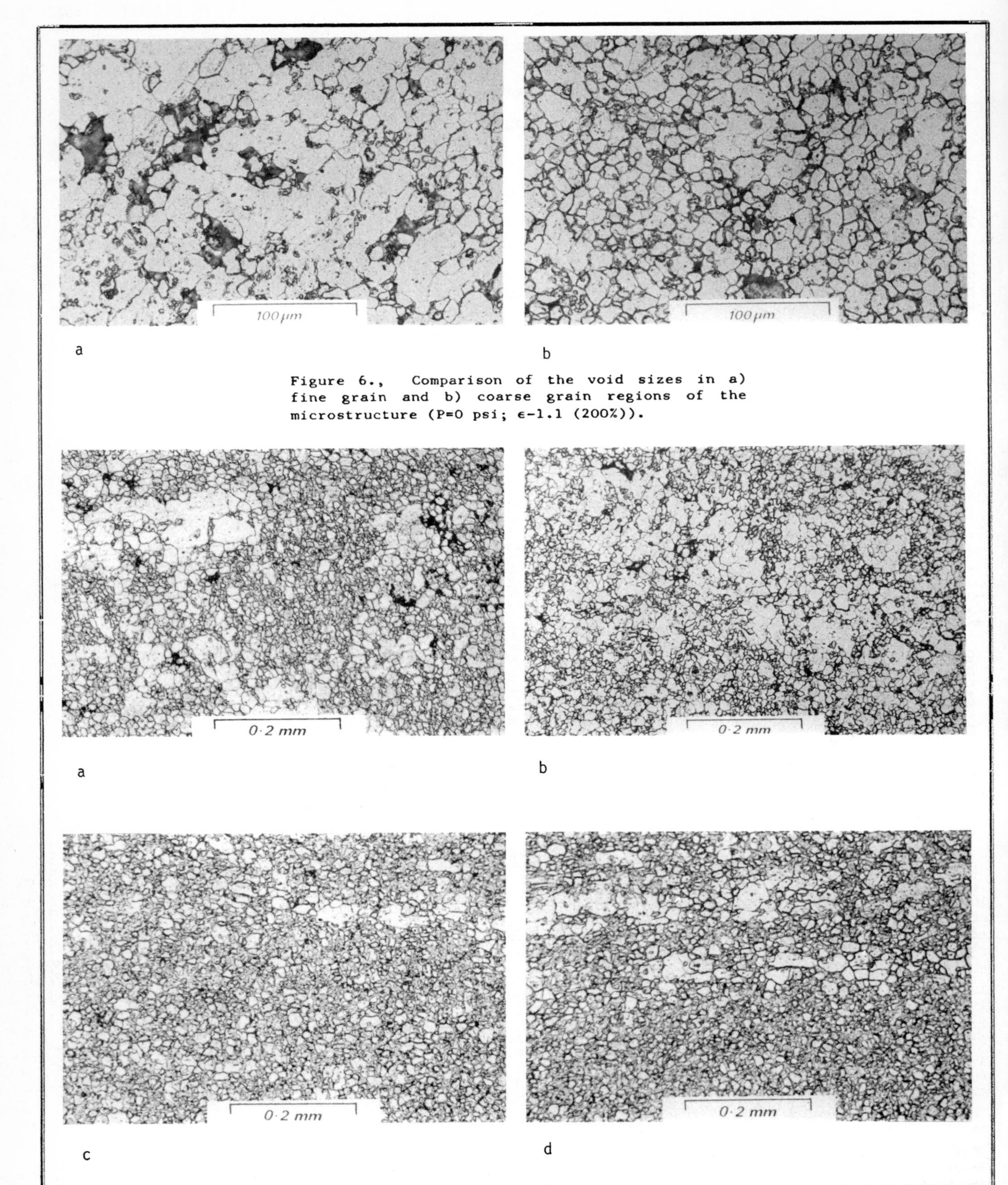

Figure 6., Comparison of the void sizes in a) fine grain and b) coarse grain regions of the microstructure (P=0 psi; ϵ-1.1 (200%)).

Figure 7. Cavity distributions in samples deformed uni-axially to a strain of 1.39 (300%) at a) 0 MPa; b) 0.7 MPa; c) 2.1 MPa; d) 3.05 Mpa.

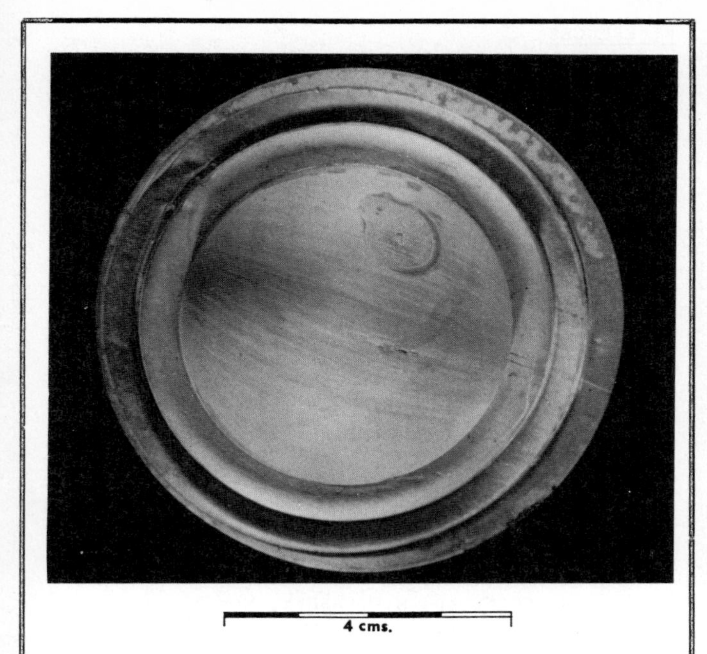

Figure 8. Surface of blow formed dome.

where η is the cavity growth rate parameter given by

$$\eta = \frac{3}{2}\left\{\frac{m+1}{m}\right\}\sinh\left\{2\left[\frac{2-m}{2+m}\right]\left[\frac{k}{3}-\frac{P}{\sigma_e}\right]\right\} \quad (2)$$

in which m is the strain rate sensitivity of the material, P the superimposed pressure, σ_e the equivalent uniaxial flow stress and k a geometric factor. The stress-strain rate characteristics of the Al-Cu-Li alloy at each orientation are shown in Figure 3. From this figure the uniaxial flow stresses and strain rate sensitivities at a strain rate of 1.17×10^{-3} s^{-1} were found to be as follows:- σ_e = 6.5 MPa, m = 0.48 parallel to the rolling direction, σ_e = 7.2 MPa, m = 0.39 at 45° to the rolling direction; σ_e = 7.9 MPa, m = 0.35 perpendicular to the rolling direction.

In Figure 4, the variation of η (normalised against its value at zero superimposed pressure) with pressure (normalised against flow stress) predicted by equation (2), is plotted and compared to that determined from experiment. The experimental values of η were obtained using a least squares fit to the data points and are summarised in Table 2. It can be seen from Figure 4 that the experimental points, whilst showing a pressure dependence similar to that predicted by equation (2), cannot be differentiated in terms of test type; and are in broad agreement with those obtained on other single phase superplastic aluminium alloys (8).

Measurement of the size distributions of cavities at different strains in samples deformed with different superimposed pressures showed that the reduction in the volume fraction of voids with increasing pressure was achieved by reducing the growth rate of the voids rather than through a reduction in the total number of voids nucleated. It can be seen from Figure 5 that as the superimposed pressure was increased, the population density of the larger voids ($r > 10$ μm) decreased, whilst the number of small voids ($r < 2$ μm) remained approximately constant.

After etching the uni-axially deformed specimens in Keller's reagent, it was found that the voids were concentrated within the coarse grain regions, and that the voids in those areas, were much larger than those in the fine grain regions of the microstructure, Figure 6 & 7. The inhomogeneous distribution of the voids would not be unexpected in view of the established inverse dependency of superplastic deformability on grain size. The increased grain size results in a locally higher flow stress and a reduced capacity to accommodate grain boundary sliding. The inhomogeneity of deformation in the aluminium-lithium alloys examined is clearly demonstrated by the presence of corrugations parallel to the original rolling direction in the bi-axially deformed specimens, shown in Figure 8.

In the uni-axial specimens failure occurred as a result of cavity interlinkage within the coarse grained regions of the microstructure, such areas being compelled to deform at the same rate as the fine grained regions. In the bi-axial specimens deformation parallel to the original rolling direction occurred in the same manner as that in the uni-axial specimens. However, perpendicular to the original rolling direction, the strain was concentrated in the fine grain regions and failure resulted from separation at the interface between the fine and coarse grained areas.

In the absence of any superimposed pressure failure, and hence the maximum usefully attainable strain in a conventionally processed Al-Li alloy is governed by coalescence of spatially inhomogeneously distributed voids. However, whilst the superposition of a hydrostatic pressure during forming has little effect on the nucleation and hence the spatial distribution of voids within the material, the substantial reduction in the kinetics of growth of those voids can increase significantly the maximum attainable superplastic strain in uni-axial deformation (310% at 0 MPa; 672% at 3.05 MPa). The improvement in the bi-axial samples was less because of the inhomogeneous nature of the deformation.

CONCLUSIONS

Conventionally processed aluminium-lithium alloys can be deformed superplastically, though the strain at failure is limited by the inhomogeneously distributed cavitation damage, the cavities developing primarily in the regions of large grain size. The superposition of a hydrostatic pressure during superplastic deformation has been found to have several effects.

1) As the pressure is increased the volume fraction of cavities at a given strain decreases.
2) As the pressure is increased the void growth rate is decreased.
3) As the pressure is increased the strain at failure increases toward a limiting value.
4) Increasing the pressure has little effect on void nucleation within the range of pressures examined.

Despite the non-ideal superplastic microstructure of the alloy, the relatively low flow stress at the deformation temperature allows low pressures (150-300 psi) to be superimposed during deformation enabling high strains to be attained with only minimal cavitation damage.

ACKNOWLEDGMENTS

The work was carried out with the financial support of British Aerospace PLC, Manufacturing Group, Warton.

REFERENCES

1. M.J. STOWELL, in 'Superplastic forming of structural alloys', Ed. N.E. PATON and C.H. HAMILTON, The Metallurgical Society AIME, Warrendale, USA (1982) 321-336.
2. D.B. LAYCOCK, ibid., 257-336.
3. P.J. MESCHETER, P.S. PAO and R.J. LEDERICH, Scripta Met. 18 (1984) 833-836.
4. R.A. STEVENS and P.E.J. FLEWITT, Acta Met. 27 (1979) 67-77.
5. W. BEERE and G.W. GREENWOOD, Metal Sci. 5 (1971) 107-113.
6. C.C. BAMPTON and R. RAJ, Acta Met. 30 (1982) 2043-2053.
7. A.K. GHOSH, see ref. 1 85-103.
8. J. PILLING and N. RIDLEY, submitted to Acta Met. (1985).
9. D.W. LIVESEY and N. RIDLEY, Met.Trans. 13A (1979) 1619-1626.
10. E.E. UNDERWOOD, in 'Quantitative Microscopy' Ed. R.T. DEHOFF and F.N. RHINES, McGraw-Hill, New York (1968) 151-181.
11. A.C.K. COCKS and M.F. ASHBY, Metal Sci. 16 (1982) 465-474.
12. M.J. STOWELL, D.W. LIVESEY and N. RIDLEY, Acta Met. 32 (1984) 35-42.

Thermomechanical processing of two-phase Al–Cu–Li–Zr alloy

J GLAZER and J W MORRIS Jr

The authors are with
Materials and Molecular Research Division,
Lawrence Berkeley Laboratory, and
Department of Materials Science,
University of California,
Berkeley, CA 94720, USA

ABSTRACT

The effect of large second-phase particles on the workability, mechanical properties, and response to thermomechanical processing of an Al-2.49Cu-2.37Li-0.13Zr alloy was investigated. The solution-treated alloy was compared with a dispersoid-containing alloy fabricated by aging the base alloy at an intermediate temperature to bring out second-phase particles. Extensive deformation and short heat treatments that cause some of the solute to be incorporated into dispersoid particles were both found to be beneficial to the mechanical properties of the alloy. In specimens which failed intergranularly, the presence of dispersoid particles did not significantly affect the fracture properties.

I. INTRODUCTION

Thermomechanical treatments of aluminum alloys are usually designed to avoid precipitation of large incoherent second-phase particles. These particles, which often appear in heavily alloyed materials, are generally thought to be harmful to the mechanical properties of the alloy [1,2]. However, the effect of second phase particles is alloy dependent. The properties of these second-phases and their effects on the microstructure, workability, and response to thermomechanical processing have not yet been characterized for aluminum-lithium alloys. In this study we have investigated the effect of various thermomechanical treatments on the microstructure and mechanical properties of an Al-Cu-Li-Zr alloy.

A number of incoherent phases exist in the aluminum-copper-lithium phase diagram. This study focussed on the T_2 phase (Al_5CuLi_3, cubic structure, isomorphous to $Mg_{32}(ZnAl)_{49}$ [3]). To facilitate comparison between alloys with and without the T_2 phase, an Al-2.49Cu-2.37Li-0.13Zr alloy was selected which could either be solutionized or processed to bring out the T_2 phase as large incoherent particles.

II. EXPERIMENTAL PROCEDURE

The specimens used in this study were fabricated from an ingot cast at the Alcoa Technical Center. Prior to receipt, the ingot was homogenized, hot rolled to one half inch plate and then solution treated and stretched 2%. The composition of the alloy was Al-2.49Cu-2.37Li-0.13Zr-0.06Fe-0.03Si-0.02Ni-0.0008Na-0.0007Ca in weight percent. The general appearance of the as-received material is shown in Figure 1. The grains are elongated as a result of hot-rolling and unrecrystallized due to the presence of Zr in the alloy.

Six thermomechanical processing sequences were considered. The processing schedules are summarized in Table 1.

The dispersoid-containing alloy (450 in Table 1) was prepared by heat treating for four hours at 450°C to precipitate the T_2 phase. This heat treatment temperature was chosen on the basis of a linear extrapolation from isothermal sections of the Al-Cu-Li phase diagrams given by Mondolfo [3] and Hardy and Silcock [4]. However, X-ray diffraction results indicate that the T_1 phase and trace amounts of δ phase were also present.

To study the effect of these incoherent phases on the response of the material to

Table 1. Final thermomechanical treatments. The as-received condition is solution treated and stretched 2%.

Specimen	450°C	Form Roll/Anneal	Stretch	Peak-age
Base	--	--	--	190°C/8hrs
Base-D	--	1000% + 520°C/6hrs	2%	190°C/15hrs
Base-A	--	520°C/6hrs	2%	190°C/24hrs
450	4 hrs	--	2%	190°C/20hrs
450-D	4 hrs	1000% + 450°C/6hrs	2%	190°C/16hrs
450-A	4 hrs	450°C/6hrs	2%	190°C/20hrs

Table 2. Mechanical properties of processing sequences described in Table 1. All strengths are given in MPa.

Specimen	σ_{YS}	σ_{UTS}	σ_{NTS}	σ_{NTS}/σ_{YS}	Elong. %
Base	430	475	571	1.3	3.2
Base-D	453	486	646	1.4	3.0
Base-A	418	448	590	1.4	2.5
450	469	520	662	1.4	3.8
450-D	341	381	558	1.6	4.0
450-A	378	412	521	1.4	2.1

1. Optical micrograph of grain structure of as-received material.

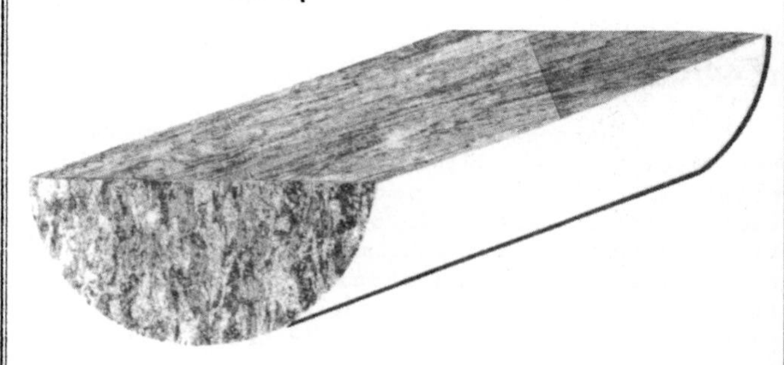

2. Optical micrograph of grain structure of alloy after 1000% deformation to rod.

3. Optical micrograph of particles present in the alloy after 450°C/4hrs.

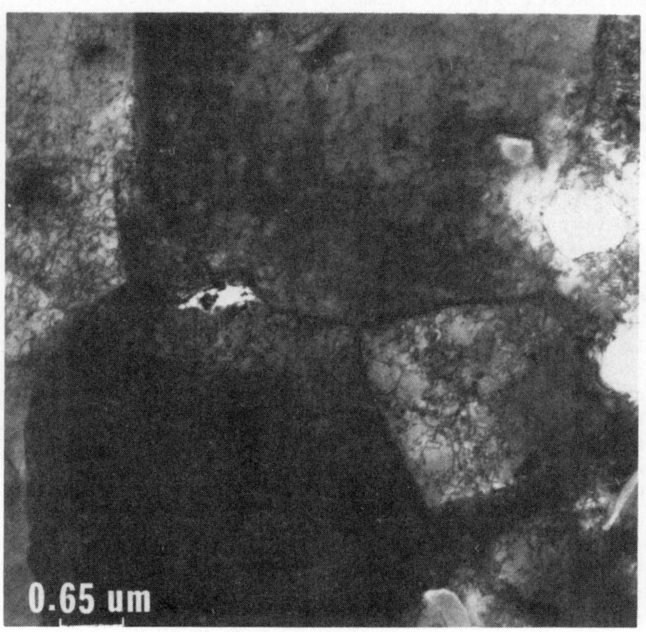

5. Transmission electron micrographs of the subgrain structure after 450°C/4hrs.

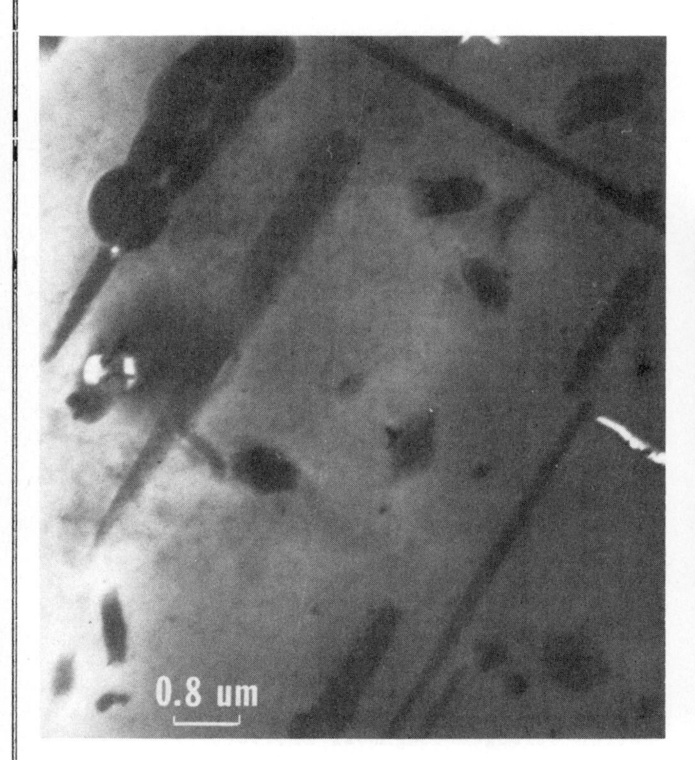

4. Transmission electron micrograph of particles present in the alloy after 450°C/4hrs.

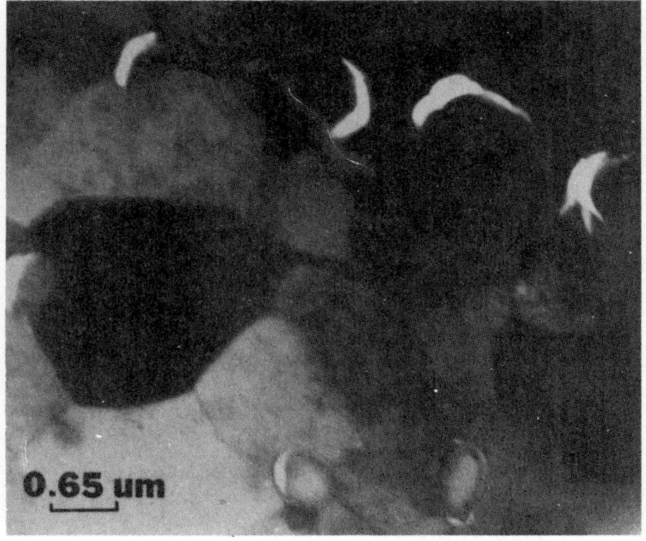

6. Transmission electron micrographs of the subgrain structure after 450°C/4hrs + deformation.

thermomechanical processing, both alloys were extensively cold-worked. Extensive cold-forming without annealing (>150% reduction) caused the alloy to fail in shear, probably due to the fact that the Zr in the alloy prevented easy dynamic recrystallization. This problem was particularly pronounced in the dispersoid alloy. However, it was possible to design a processing sequence which allowed extensive deformation. A form rolling process in which single large deformation steps were alternated with 1 hour anneals at the "solutionizing temperature" (520°C and 450°C, respectively) was used to deform the specimens a total of nearly 1000% in six steps. In the last step the material was swaged into rod. The rod axis corresponds to the rolling direction of the original plate. The grain structure of the deformed material is illustrated in Figure 2. After the final anneal, both alloys (Base-D and 450-D) were stretched 2% before final aging.

For comparison, additional specimens (Base-A and 450-A) were prepared that received the same heat treatment as the deformed specimens above, but were not deformed.

Rockwell B hardness was used to determine the peak-aged condition for each processing sequence. All specimens were peak-aged at 190°C before mechanical testing, with the exception of the deformed base alloy which was tested in a slightly overaged condition.

The mechanical properties of the specimens were compared in the peak-aged condition using subsize tensile and notched tensile bars machined in the longitudinal direction. The specimens were similar in design to those used by Rack and Edstrom [5], with a 3.2mm gauge diameter and a 25.4mm gauge length. The notches were 60° with a root radius of less than 50μm. All tests were conducted using at a strain rate of 1.7×10^{-4} sec^{-1}.

Microstructural examination was carried out using a combination of techniques. Polished sections were examined by optical and scanning electron microscopy after etching by Keller's etch (for grain size) and 1% HF (for second-phase particles) [4]. Samples were prepared for transmission electron microscopy by jet polishing at 12V, -30°C in a 1:3 mixture by volume of 70% nitric in methanol. The foils were then examined in a Philips 301 at an operating voltage of 100kV.

III. RESULTS

1. Microstructure.

a. Base Alloy

The base alloy provided a benchmark for comparison with other alloys. Before aging, optical microscopy showed the material to be relatively free of large inclusions and second-phase particles. At peak strength the alloy was hardened by δ', T_1' and T_2' phases.

The main difference between the deformed base alloy (Base-D) and the undeformed alloy was the smaller and more elongated grains in the deformed alloy. The subgrain size in the deformed alloy was also substantially smaller than in the undeformed alloy.

b. Dispersoid Alloy

An optical micrograph of the alloy after aging at 450°C for 4 hours is shown in Figure 3. A number of particles are visible. Transmission electron microscopy (Figure 4) indicates that several different types of particles exist, including T_1, T_2 and Al_3Zr. The particles are relatively uniformly dispersed and do not tend to be concentrated along grain or subgrain boundaries. The subgrain structure, similar to that observed in the base alloy, is shown in Figure 5.

The particles in the deformed dispersoid alloy (450-D) are almost the same size as those in the undeformed alloy despite the additional six hours at 450°C during the deformation process. It is not clear whether the particles deformed or fractured during deformation, or if their growth was inhibited. Preliminary transmission electron microscopy results indicate that the particles tend to be aligned so that they present a large cross section in planes parallel to the rod axis and a small cross section perpendicular to it. Figure 6 illustrates that as a result of the extensive deformation the subgrain size in the deformed alloy was again much smaller than in the undeformed alloy.

2. Mechanical Properties

The mechanical properties of all six samples are shown in Table 2. All processings result in very low elongations. The ratio of notched tensile strength to yield strength (NYR) is used as a measure of toughness in these small samples. The notch yield ratio is plotted against yield strength in Figure 7. The strength-toughness characteristic of some commercial 2XXX and 7XXX alloys is shown for comparison [6]. As noted by Feng et al. [7] for alloys similar to 2020, the tensile elongation and fracture toughness do not correlate well.

As shown in Figure 7a, the dispersoid and base alloys have similar mechanical properties in spite of the fact that their microstructures are quite different. Perhaps surprisingly, the dispersoid alloy seems to ride a better strength-toughness characteristic. However, after an additional six hours at 450°C, its properties drop off significantly. The deformed dispersoid alloy (450-D), which received the same heat treatment as the 450-A specimen, shows significantly better properties.

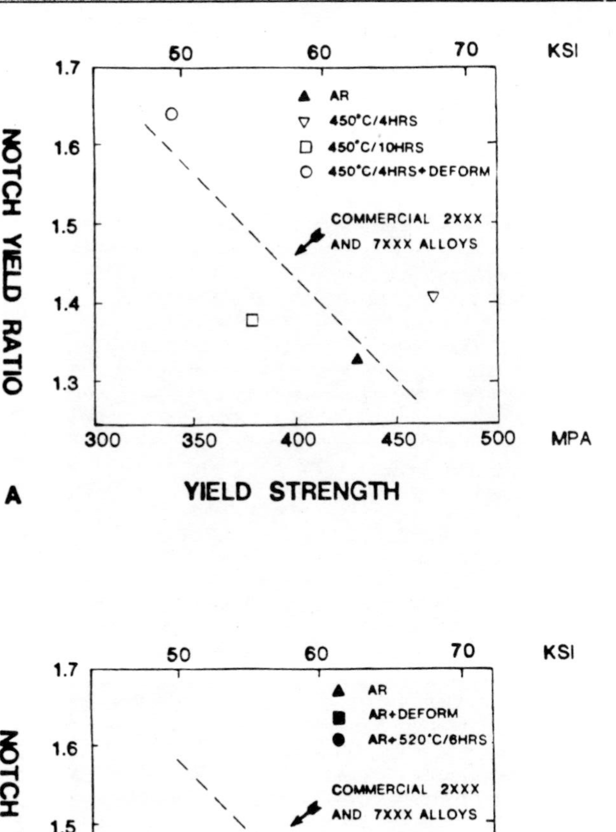

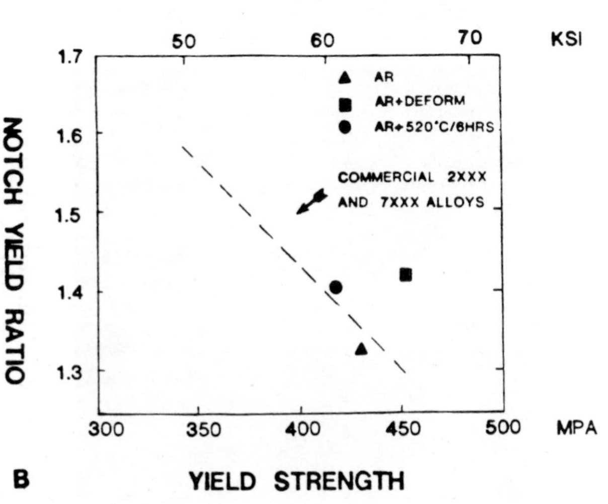

7. Notch yield ratio versus yield strength for (a) dispersoid-containing samples and (b) base alloy processings.

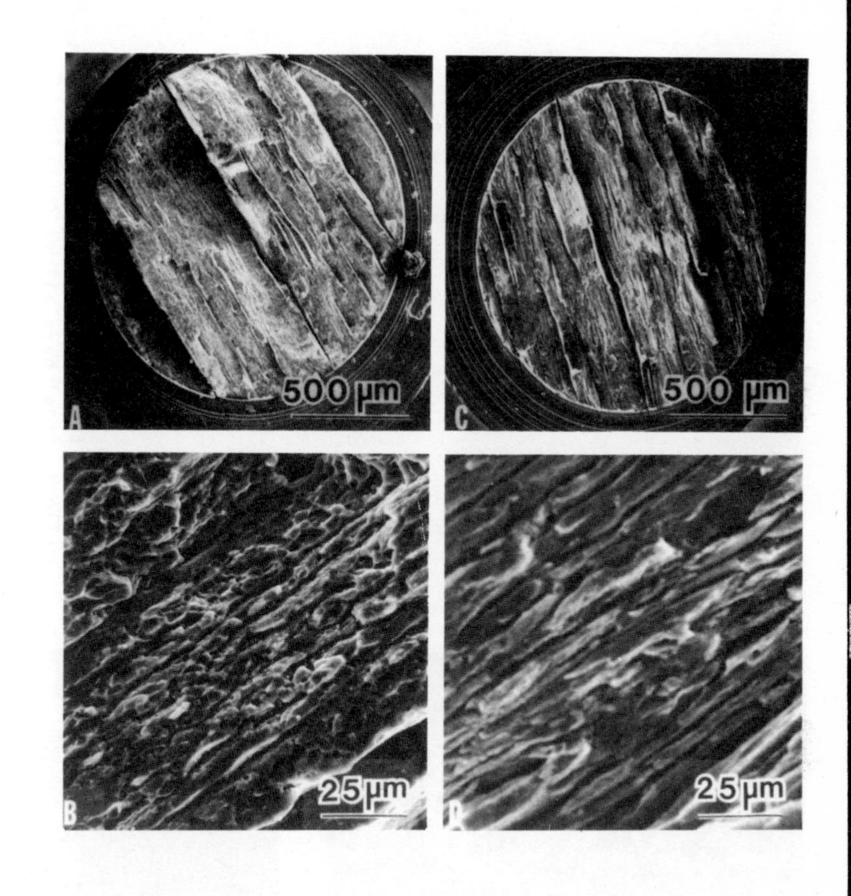

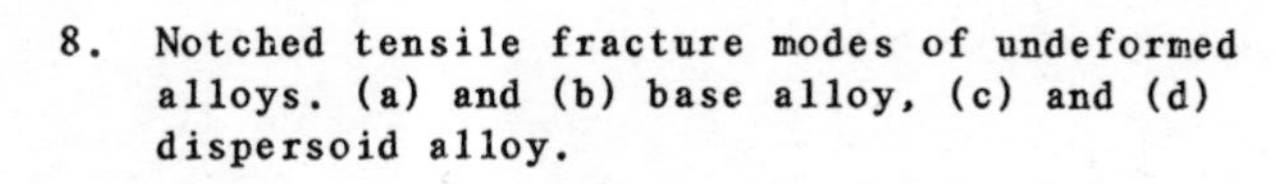
8. Notched tensile fracture modes of undeformed alloys. (a) and (b) base alloy, (c) and (d) dispersoid alloy.

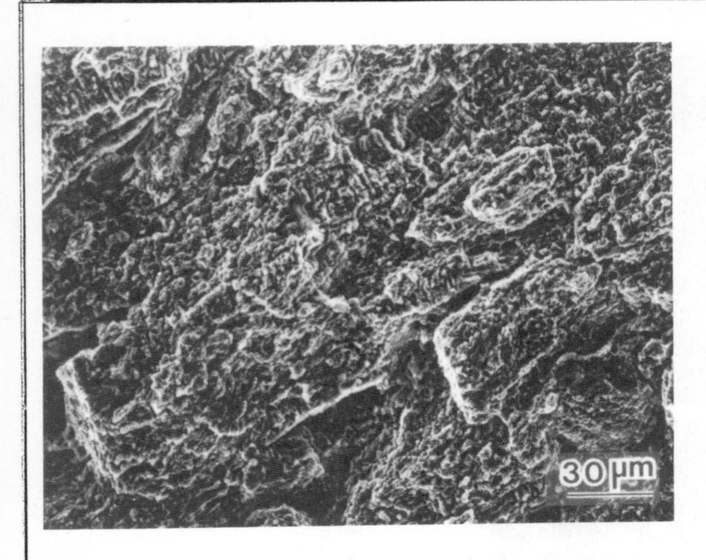

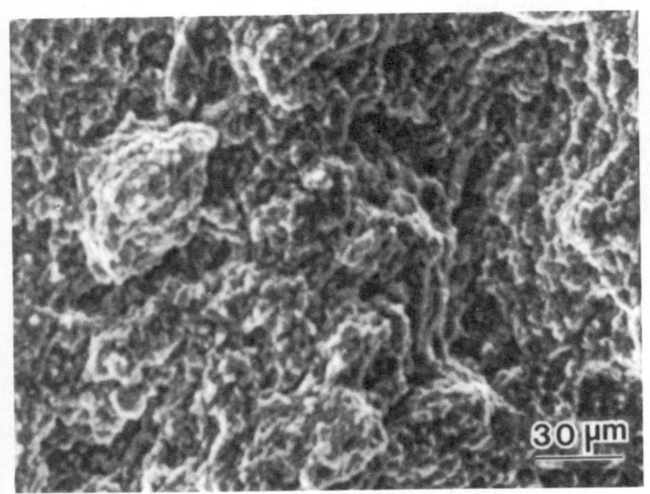

9. Notched tensile fracture modes of deformed alloys. (top) base alloy, (bottom) dispersoid alloy.

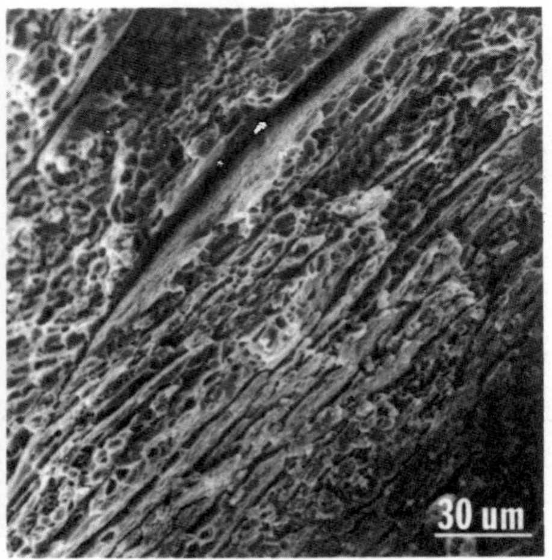

10. Notched tensile fracture mode of 450-A alloy.

As expected, the properties of the solutionized base alloy, plotted in Figure 7b, were much less sensitive to thermomechanical processing of this kind. The deformation processing improves the strength-toughness relation slightly for the base alloy. The Base-A treatment results in both lower strength and higher toughness.

The notch yield ratio has been found to correlate well with plane stress fracture toughness in sheet specimens of alloys of similar composition and mechanical properties to those tested here [8]. Although the stress-state varies among these samples, the trends indicated by the NYR are probably valid over relatively small ranges of strength. The apparent toughness of the low strength samples is almost certainly spuriously high with respect to the other specimens since these specimens are nearer plane stress. Nevertheless, the deformed dispersoid specimen (450-D) is clearly considerably tougher than the other samples.

3. Fracture Behavior

a. Undeformed alloys.

The notched tensile fracture modes of the undeformed alloys are shown in Figure 8. The macroscopic failure mode was nearly identical for both undeformed alloys. Both exhibited significant longitudinal cracking. The longitudinal cracks are intergranular, as observed by Wert and Lumsden [9] for an Al-Li-Cu-Mg-Zr alloy. The microscopic failure modes were relatively similar although not identical. The base alloy exhibited considerable ductility. The fracture surface of the dispersoid alloy contains some ductile regions and some regions of transgranular shear. The smooth tensile specimens failed in shear at low elongations.

b. Deformed Alloys.

The notched tensile fracture of the deformed dispersoid alloy was completely ductile as shown in Figure 9. No indication of the grain or subgrain boundaries is present in the fracture surface. The deformed base alloy (Base-D) appears to be at a transition point with respect to fracture mode. Although the two specimens tested had almost identical yield strength and notch toughness, they exhibited two different fracture modes. One specimen (pictured in Figure 9) fractured in a ductile manner although there were some regions of intergranular fracture. The other specimen showed significant longitudinal intergranular cracking like that seen in the undeformed alloys in Figure 8 and did not show significant ductility. The smooth tensile specimens all failed in shear.

c. Annealed Alloys.

These specimens (Base-A and 450-A) both exhibited a macroscopic fracture mode similar to the undeformed specimens (Base and 450). The fracture surface of the 450-A specimen is shown in Figure 10. Considerable longitudinal intergranular cracking was observed. Consistent with the lower strength of the annealed samples, these specimens appeared to be more ductile than the Base and 450 samples.

IV. DISCUSSION

The similarity of the mechanical properties and fracture appearance of the undeformed dispersoid and base alloys indicates that the particles present in the microstructure of the dispersoid alloy do not significantly affect the fracture properties of the alloy because the 'weak link' which leads to failure remains unchanged. As long as intergranular weakness controls the fracture mode, the material is relatively immune to inclusions in the matrix.

Deformation of the base alloy raises both the yield strength and the notched tensile to yield strength ratio slightly. However, deformation of the dispersoid alloy results in an entirely ductile fracture mode and high toughness, albeit at much lower strength. The low yield strength in this condition is probably a consequence of the large quantity of solute incorporated into the equilibrium phases precipitated at 450°C. The large volume fraction of second-phase particles leaves little solute to be precipitated as coherent strengthening phases. Once the alloy is weak enough to eliminate the laminar intergranular fracture mode seen in the undeformed alloys, the dispersion of precipitates in the matrix becomes important. An additional factor which may contribute to the more ductile behavior exhibited by the deformed specimens is the reduced grain size. The control specimen 450-A, which received the same heat treatment but was not deformed shows relatively poor strength and toughness. Its poor toughness confirms that the deformation process, as well as the precipitation of solute, plays a role in developing the good properties of the deformed dispersoid alloy.

V. CONCLUSIONS

The influence of the dispersoid phases present in this alloy may be summarized as follows. Large equilibrium phases are tolerable in relatively brittle alloys, and in small amounts may even have a beneficial effect on the mechanical properties of the alloy. Although these phases may make deformation more difficult, this problem can be overcome, at least for cold-forming. For alloys which fail intergranularly, precipitates in the bulk do not significantly influence the fracture mode or the mechanical properties. If other microstructural changes (e.g. change of grain size, lack of hardening precipitates) promote a ductile fracture mode, then the large inclusions will control the fracture mode. In this case the inclusions lower the strength-toughness relation. Finally, large equilibrium phases can incorporate a large fraction of the total solute in the alloy and thereby significantly reduce the hardenability of the alloy.

ACKNOWLEDGEMENTS

The authors are grateful to S.L. Verzasconi, Materials and Molecular Research Division, Lawrence Berkeley Laboratory for experimental assistance, to the Alcoa Technical Center for supplying the material used in these experiments and to W.H. Hunt, Jr., Alcoa Technical Center, for helpful advice. This work was supported by the Director, Office of Basic Energy Sciences, Materials Science Division of the U. S. Department of Energy under Contract No. DE-AC03-76SF00098. One of the authors, JG, was supported by an ATT Bell Laboratories Scholarship.

REFERENCES

1. J.T. Staley, Met. Eng. Q., **16**, 1976, 52.

2. J.G. Kaufman, Agard Conf. Proc. No. 185, 1975.

3. L.E. Mondolfo, **Aluminum Alloys, Structure and Properties**, Butterworths, 1979, 495.

4. H.K. Hardy and J.M. Silcock, J. Inst. Metals, **84**, 1955-56, 423.

5. H.J. Rack and C. Edstrom, in **Thermomechanical Processing of Aluminum Alloys**, J.G. Morris, ed., TMS-AIME, Warrendale, PA, 1979, 86.

6. T.H. Sanders, Jr., NAVAIR Contract No. N62269-74-C-0438, Final Report, June 1976.

7. W.X. Feng, F.S. Lin, and E.A. Starke, Jr., Met. Trans. A, **15A**, 1984, 1209.

8. P.J. Gregson and H.M. Flower, Acta Metall., **33**, 1985, 527.

9. J.A. Wert and J.B. Lumsden, Scr. Met., **19**, 1985, 205.

Superplastic aluminum–lithium alloys

J WADSWORTH, C A HENSHALL
and T G NIEH

The authors are with the Lockheed Palo Alto Research Laboratory, Metallurgy Department, 0/93-10, B/204, 3251 Hanover Street, Palo Alto, CA 94304, USA

SYNOPSIS

Superplasticity is now an accepted manufacturing technique for Ti-, Ni-, and Al- based alloys. Recent work at Lockheed Palo Alto Research Laboratory has led to the development of superplasticity in a number of novel aluminum-lithium based alloys manufactured by ingot metallurgy and powder (rapid solidification) metallurgy techniques. These alloys are of considerable current interest in the aerospace and aircraft industries because lithium improves the modulus and decreases the density of aluminum. Considerations of design of Al-Li alloys for superplastic properties are discussed and the interactions of compositional requirements with thermomechanical processing are described. The superplastic properties of a range of Al-Li alloys are presented and a detailed comparison with Al-7475 alloy and commercial Al-Cu-Zr based alloys is made. The ultrafine microstructures of the Al-Li alloys are illustrated and the interrelationship of these microstructures with the high strain rates at which superplasticity is found in the alloys is highlighted. Phenomenological equations for superplastic flow have been developed for several of the alloys. Finally, room temperature properties of the alloys are presented and a comparison is made with other superplastic Al-based alloys.

INTRODUCTION

Superplastic forming of alloys is now a commercial reality in three major alloy groups. These are the Ti-, Ni-, and Al- based systems[1]. Although considerable effort has been expended in studying superplasticity in aluminum alloys, the only commercially successful systems at the present time are the SUPRAL alloys. These alloys are primarily based on the Al-Cu-Zr system although, in very recent years, superplastic alloys based on Al-Mg-Zr have also become available. Because the superplastic forming capability already exists for these alloys, and because there is an evident advantage in cost savings using superplastic technology, it is only natural to attempt to develop superplasticity in developmental Al alloys such as those based on Al-Li.

A research and development program to study the possibility of developing superplastic Al-Li based alloys was initiated at the Lockheed Palo Alto Research Laboratory in 1981. This program, coupled with the extensive development work at Lockheed on Al-Li alloys[2], has led to the demonstration of superplasticity in a number of Al-Li based alloys[3-11]. In this paper some of the key features of this work are presented.

EXPERIMENTAL

A number of Al-Li based alloys has been produced and studied over the last five years. The compositions of these alloys are shown in Table I and the details of their manufacture are shown in Table II.

As may be seen, a range of alloys have been manufactured, processed, and evaluated. These range from laboratory castings (Al-3Li-0.5Zr, Al-4Cu-3Li-0.5Zr), to developmental alloys made by ingot metallurgy (IM Al-3Cu-2Li-1Mg-0.15Zr) as

TABLE I COMPOSITIONS OF Al-Li BASED ALLOYS

NUMBER	NOMINAL COMPOSITION	ACTUAL COMPOSITION (weight %, bal. Al) Cu	Li	Mg	Zr	Other
1	Al-3Li-0.5Zr	--	2.92	--	0.57	--
2	Al-4Cu-3Li-0.5Zr	3.92	2.56	--	0.58	--
3	Al-3Cu-2Li-1Mg-0.15Zr	2.9	1.9	1.0	0.15	--
4	Al-3Cu-2Li-1Mg-0.2Zr	2.8	1.9	0.9	0.20	--
5	Al-2.5Li-1.5Cu-1Mg-0.1Zr	1.4	2.3	1.01	0.07	--
6	Al-2.5Cu-2.2Li-0.1Zr	2.50	2.16	--	0.10	--
7	Al-2.4Li-1.1Cu-0.8Mg-0.1Zr	1.12	2.42	0.76	0.12	--
8	Al-2.5Li-1.1Cu-0.7Mg-0.14Zr	1.06	2.45	0.72	0.14	--
9	Al-2.2Cu-2.1Li-0.8Mg-0.16Zr	2.19	2.08	0.83	0.16	--
10	Al-4Cu-2Li-0.15Zr-0.04Cd	3.56	1.70	--	0.15	0.04Cd
11	Al-2.5Li-1.2Cu-0.5Mg-0.1Zr	1.2	2.5	0.5	0.10	--
12	Al-2.6Cu-2.4Li-0.2Zr	2.62	2.42	--	0.20	--

well as rapid solidification powder metallurgy processing (PM Al-3Cu-2Li-1Mg-0.2Zr, PM Al-2.5Li-1.5Cu-1Mg-0.1Zr). Also included are ingot metallurgy alloys produced by Reynolds Aluminum Company (Al-2.5Cu-2.2Li-0.1Zr, Al-2.4Li-1.1Cu-0.8Mg-0.1Zr, Al-2.5Li-1.1Cu-0.7Mg-0.14Zr, Al-4Cu-2Li-0.15Zr-0.04Cd, Al-2.2Cu-2.1Li-0.8Mg-0.16Zr, Al-2.6Cu-2.4Li-0.2Zr) and the British alloy XXXA (nominal composition Al-2.5Li-1.2Cu-0.5Mg-0.1Zr).

In Table III, the optimum superplastic behavior, that is, the strain rate at which the greatest elongation is observed, is presented for each alloy.

RESULTS AND DISCUSSION

A wide range of investigations have been carried out and published on the alloys described in the previous section. These results, as well as new findings, can be described under the following general sections.

- o Alloy Design and Processing for Superplasticity
- o Superplastic Properties
- o Microstructure
- o Phenomenological Description of Superplasticity
- o Room Temperature Properties

It is not the purpose of this paper to review in detail all of these sections but, rather, to present the important observations together with new data.

1. Alloy Design and Processing for Superplasticity

Recent work[6,7,9] has described in detail the development of superplasticity in Al alloys. The original work on superplastic aluminum alloys, which started in 1966, was based on the Al-Cu eutectic system. This was because the prevailing philosophy of the time was that a large amount of second phase was necessary to maintain the very fine grain size required for superplasticity. Indeed, this philosophy continues to prevail in many alloy systems other than those based on aluminum[12]. In aluminum alloys, it was soon recognized that the fine grain sizes necessary for superplasticity could be maintained by small but deliberate additions (less than 0.5 wt. %) of elements such as Zr. In the major work in this area, that of the SUPRAL alloys, an amount of Zr from about 0.4 to 0.5 wt. % was considered to be necessary. This is in excess of that usually found in large scale ingot alloys (less than 0.2 wt. %) and therefore these alloys need to be chill cast in production. It is now clear that smaller amounts than 0.4 wt. % Zr are adequate, at least in some aluminum alloys. However, the processing steps required have to be modified and the precise physical mechanism leading to fine grain sizes may be different; i.e., in some cases discontinuous

TABLE II

ALLOY	PRIMARY MANUFACTURE	SECONDARY PROCESSING FOR SUPERPLASTICITY
Al-3Li-0.5Zr	high purity Al, Li + Al-6Zr chill cast from 843°C into Cu molds 1" x 1/2" x 4"	i. 300°C isothermal rolling ii. age 370°C/4hr + 300°C isothermal rolling iii. solutionize 520°C/20 min, CWQ, age 370°C/4 hrs, AC, 300°C isothermal rolling
Al-4Cu-3Li-0.5Zr	high purity Al, Cu, Li + Al-6Zr chill cast from 843°C into Cu molds 1" x 1/2" x 4"	i. 370°C isothermal rolling ii. age 370°C/4 hrs AC + 370°C isothermal rolling iii. solutionize 520°C/20 min, CWQ + age 370°C/4 hrs, AC + 370°C isothermal rolling
Al-3Cu-2Li-1Mg-0.15Zr	direct chill cast, homogenized 460°C/16 hrs + 500°C/16 hrs in air, extruded 370°C, ratio 25:1	i. solutionize 538°C/0.5 hrs, CWQ, age 400°C/0.5-136 hrs, AC, isothermal rolling at 300°C to 90% reduction, recrystallize at test temperature. ii. as extruded + isothermal rolling at 300°C to 90% reduction iii. as extruded + 500°C, 1 hr, CWQ + isothermal rolling at 300°C to 90% reduction iv. as extruded + 475°C, 1 hr, CWQ + isothermal rolling at 300°C to 90% reduction
Al-3Cu-2Li-1Mg-0.2Zr	chill cast ingots centrifugally atomized to -100 mesh in helium, vacuum degassed 538°C canned in 6061 Al preheat 382°C/5 hr, upset in blind die, extruded at 382°C ratio 22:1	i. solutionize 538°C/0.5 hrs, CWQ, age 400°C/0.5-136 hrs, AC, isothermal rolling at 300°C to 90% reduction, recrystallize at test temperature. ii. as extruded + isothermal rolling at 300°C to 90% reduction iii. as extruded + 500°C, 1 hr, CWQ + isothermal rolling at 300°C to 90% reduction iv. as extruded + 475°C, 1 hr, CWQ + isothermal rolling at 300°C to 90% reduction
Al-2.5Li-1.5Cu-1Mg-0.1Zr	same as Al-3Cu-2Li-1.0Mg-0.2Zr	i. solutionize 538°C, 0.5 hrs, CWQ + age 400°C, 16 hrs, AC, isothermal rolling at 300°C to 90% reduction, recrystallize at test temperature ii. as extruded + isothermal rolling at 300°C to 90% reduction

TABLE CONTINUES OVERLEAF ▷

TABLE II (continued)

ALLOY	PRIMARY MANUFACTURE	SECONDARY PROCESSING FOR SUPERPLASTICITY
Al-2.5Cu-2.2Li-0.1Zr Al-2.4Li-1.1Cu-1Mg-0.1Zr Al-2.5Li-1Cu-0.7Mg-0.15Zr Al-2.2Cu-2.1Li-0.8Mg-0.16Zr	induction melted and cast into permanent water-cooled mold 4" x 8" x 10", homogenized 538°C/24 hrs in N_2, AC, hot rolled to 1.25" thick at 400°C	i. solutionize 538°C/1 hr, cold water quench, aged 400°C/4 hrs, cooled 25°C/hr to 260°C; isothermal rolling to 92% reduction ii. Reynolds proprietary practice iii. two-step rolling at 466°C to 0.25 in. thick + 260°C down to 0.10" thick (92% reduction)
Al-4Cu-2Li-0.15Zr--0.04Cd	induction melted and cast into Al book mold, homogenized in 3 steps: a) 400°C for 48 hrs b) 490°C for 18 hrs c) 515°C for 0.5 hrs, fan air cooled, isothermally rolled at 455°C to 1.25"	i. Reynolds proprietary practice ii. solutionize 538°C, 0.5 hr, cold water quench, age 400°C/32 hrs, air cool, isothermal rolling at 300°C to 89% red. iii. isothermal rolling at 300°C to 82% red. iv. isothermal rolling at 250°C to 89% red.
Al-2.5Li-1.2Cu-0.5Mg-0.1Zr	Proprietary Practice	Proprietary Practice
Al-2.6Cu-2.2Li-0.2Zr	cast into 30 lb. ingots 4"x8"x10", homogenized 24 hrs 538-543°C, scalped to 3-1/4" thick, hot rolled to 12" wide x 1" thick	i. solutionize 1 hr 538°C CWQ, overage 16 hr 400°C, cool to 288°C, isothermal rolling at 288°C to 90% reduction

recrystallization is observed, in others continuous recrystallization occurs[6].

Initial attempts at Lockheed[3,4] to make Al-Li alloys superplastic used model systems containing 0.5 wt. % Zr and processing steps based on those used in SUPRAL alloys (Table II). These model systems, Al-3Li-0.5Zr and Al-4Cu-3Li-0.5Zr, were shown to be superplastic and to recrystallize by a continuous method. Subsequent studies[4,5,6] were based on Al-Cu-Li-Mg-Zr alloys developed by Palmer and Lewis[2] for room temperature properties. These alloys were manufactured by both ingot metallurgy and powder metallurgy (via a rapid solidification route) and were restricted to having levels of 0.15 and 0.2 wt. % Zr, respectively. The alloys were processed by an overage practice (which was a small modification of that developed for superplastic Al-7475 alloys). These Al-Li alloys were also shown to be superplastic. In these cases, however, the alloys recrystallized by a discontinuous method.

The most recent work[8] has been on ingot metallurgy alloys produced by Reynolds Aluminum Company that contain the typical levels of Zr (0.12 to 0.18%) found in commercial ingot alloys. These alloys were also processed by an overage technique and were observed to recrystallize discontinuously.

The history of the development of commercial superplastic alloys[13] has shown clearly that the likelihood of success is strongly dependent on the prior establishment and availability of the alloy in its own right. For example, the superplastic Ti-6Al-4V alloy and the IN 100 Ni-based alloys pre-existed their development as superplastic alloys. The exception to this general case is that of the SUPRAL alloys which were designed based on the requirements for superplastic properties as well as room temperature properties. In the case of Al-Li alloys it is apparent from Table III that superplastic properties can be developed in some compositions that already have

TABLE III OPTIMUM SUPERPLASTIC BEHAVIOR

NUMBER	ALLOY	TEMPERATURE (°C)	STRAIN RATE (%/min)	ELONGATION (%)
1	Al-3Li-0.5Zr	450	20	1035
2	Al-4Cu-3Li-0.5Zr	450	40	825
3	Al-3Cu-2Li-1Mg-0.15Zr	500	8	878
4	Al-3Cu-2Li-1Mg-0.2Zr	500	20	654
5	Al-2.5Li-1.5Cu-1Mg-0.1Zr	450	1	210
6	Al-2.5Cu-2.2Li-0.1Zr	500	5	760
7	Al-2.4Li-1.1Cu-0.8Mg-0.1Zr	500	10	222
8	Al-2.5Li-1.1Cu-0.7Mg-0.14Zr	500	20	339
9	Al-2.2Cu-2.1Li-0.8Mg-0.16Zr	500	10	948
10	Al-4Cu-2Li-0.15Zr-0.04Cd	500	20	1182
11	Al-2.5Li-1.2Cu-0.5Mg-0.1Zr	500	10	476
12	Al-2.6Cu-2.4Li-0.2Zr	500	20	911

TABLE IV SUPERPLASTIC ALUMINUM ALLOYS ROOM TEMPERATURE PROPERTIES

	YS, MPa	UTS, MPa	% EL	E, GPa	ρ, Mgm^{-3}
MEDIUM STRENGTH					
S 100	300	420	8	73.8	2.84
Al-3Li-0.5Zr	334	416	7	78.7	2.51
HIGH STRENGTH					
Al-4Cu-3Li-0.5Zr	409	516	7	78.2	2.59
S 210	420	480	8	73.8	2.84
S 220	450	510	6	73.8	2.84
7475	460	525	12	70.0	2.80
IM Al-3Cu-2Li-1Mg-0.15Zr	445	553	9.5	78.1	2.65
PM Al-3Cu-2Li-1Mg-0.2Zr	448	531	9.5	77.4	2.60
After Superplastic Forming:					
IM Al-3Cu-2Li-1Mg-0.15Zr	448	503	6.5	77.2	2.65
PM Al-3Cu-2Li-1Mg-0.2Zr	455	537	7.2	77.9	2.60

been created for room temperature applications. The greatest likelihood of superplastic forming being used in Al-Li alloys would be on these alloys that already exist. The Zr level is typically less than 0.2 wt. % in these compositions.

Superplastic behavior can be developed in Al-Li alloys containing amounts of Zr in this range but there are other compositional considerations also. The use of an overage practice to develop coarse second phase particles appears to necessitate additions of copper over a certain minimum level. For example, in Figure 1, a plot of elongation-to-failure as a function of copper content is shown for the alloys listed in Tables I and III. As may be seen, those compositions containing less than 2% Cu are difficult to process into a highly-superplastic condition, whereas, those containing 2% Cu or greater are clearly superplastic (all of these alloys, with the exception of alloy #5, contain over 0.1% Zr). This dependence on copper content is related to the distribution of second phase particles which act as sources of inhomogeneous plastic flow during the deformation step of the overage practice. These zones of high defect density are nucleation sites for recrystallizing grains and are therefore a key element in the thermomechanical processing treatment.

The thermomechanical treatments such as the overage practice are complex (as shown in Table II). It would be advantageous to simplify the additional processing so that conventional microstructures could be converted readily to ones that are fine-grained and superplastic. A limited study was carried out on IM and PM Al-3Cu-2Li-1Mg-0.15/0.2Zr alloys. This minimum processing consisted of rolling the alloys from the as-extruded (as-received) condition. As shown in Figure 2, a wide range of superplastic properties can be developed in IM and PM Al-3Cu-2Li-1Mg-0.15/0.2Zr alloys depending upon the precise thermomechanical processing route. In the typical as-received or as-extruded condition, as shown in the bottom of Figure 2, the alloys show good elongations (100-200%) but these values are only marginal from the viewpoint of superplasticity. After the overage practice (top of Figure 2) excellent elongations (650-900%) are observed. The minimum processing (center of Figure 2) shows intermediate ductilities (400-500%) but this is probably sufficient for many forming operations.

One of the important observations from the Lockheed study on Al-Li alloys is that the RSR PM Al-Li alloys are not superplastic in the as-received condition but require additional processing[5]. This is summarized in Figure 3 in which the properties of the PM alloy are shown before and after such processing. After processing the alloy shows high ductility, high strain rate sensitivity (m) and low strength by comparison with the as-received alloy. Furthermore, a change in activation energy from lattice diffusion (Q_L) to grain boundary diffusion (Q_{gb}) is also observed. The microstructural changes accompanying these mechanical property changes are discussed in Section 3.

2. Superplastic Properties

In general the optimum superplastic properties of the Al-Li alloys in the present study meet or exceed those developed in other aluminum alloy systems tested under identical conditions at the Lockheed Laboratory. For example, in Figure 4 the superplastic properties of several superplastic Al-Li alloys are compared with those of SUPRAL 100 and SUPRAL 210. As may be seen, essentially equivalent results are observed amongst these groups of alloys.

As shown in Figure 5, a key difference between the optimum formability of the Al-Li alloys and the Al-7475 alloy is the relatively high strain rate at which the optimum formability of the Al-Li alloys is observed (20-50%/min) by comparison with the Al-7475 alloy (1-2%/min). This improvement is a direct result of the ultrafine microstructure developed in the Al-Li alloys (1-2 μm) by comparison with Al-7475 alloys (10-15 μm). In Figure 5 the optimum superplastic formability of the alloys listed in Tables I and III are shown with those of Al-7475 and the SUPRAL 100 and 210 alloys. All the materials were tested in the same apparatus and under the same conditions.

For the case of the Al-Li alloys, the influence of temperature on superplastic elongation-to-failure was determined. Tests were carried out over the temperature range 400-525°C on each alloy. In the temperature range 400-500°C superplasticity was observed in each alloy and the elongations-to-failure for each alloy showed an improvement with increase in temperature. The IM alloy showed slightly better properties than the PM alloy, but both alloys were superplastic over the quoted temperature range. At temperatures greater than 500°C, the alloys showed a marked decrease in elongation-to-failure. Interestingly, the PM alloy had quite good ductility (400%) at 520°C whereas the IM alloy showed a pronounced drop in ductility (to less than 200%) at 525°C. This may reflect the greater stability of microstructure in the PM alloy compared to the IM alloy at this very high temperature, possibly as a result either of its higher Zr content (0.2 wt.% compared to 0.15 wt.%) or of the presence of oxide particles.

3. Microstructure

The microstructure of the Al-Li alloys in the present study has been investigated by Pelton and this work has been described in several publications[5,8,9]. In general the alloys have been observed to recrystallize by a discontinuous mechanism after an overage thermomechanical processing treatment. Examples of the ultrafine microstructures developed in IM and PM Al-3Cu-2Li-1Mg-0.15/0.20Zr alloys by these techniques are shown in Figures 6 and 7, respectively. The grain sizes of these recrystallized alloys have been

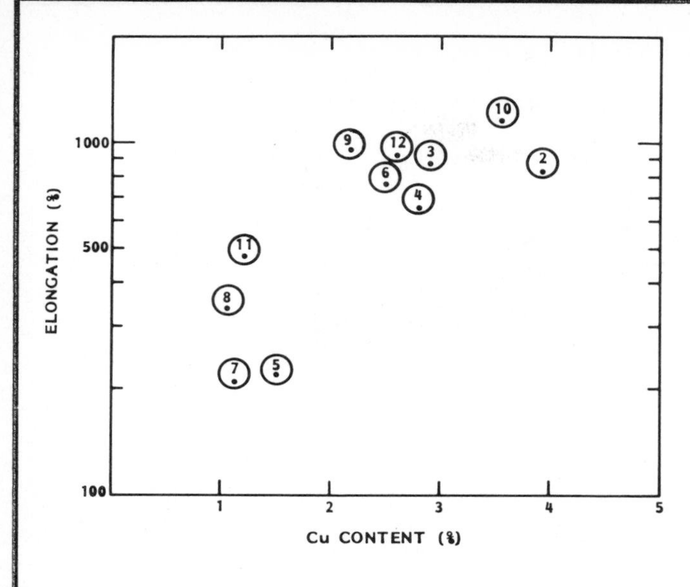

1. Relationship between copper content and elongation-to-failure for superplastic aluminum-lithium alloys. The numbers adjacent to the data points refer to the alloys listed in Table I.

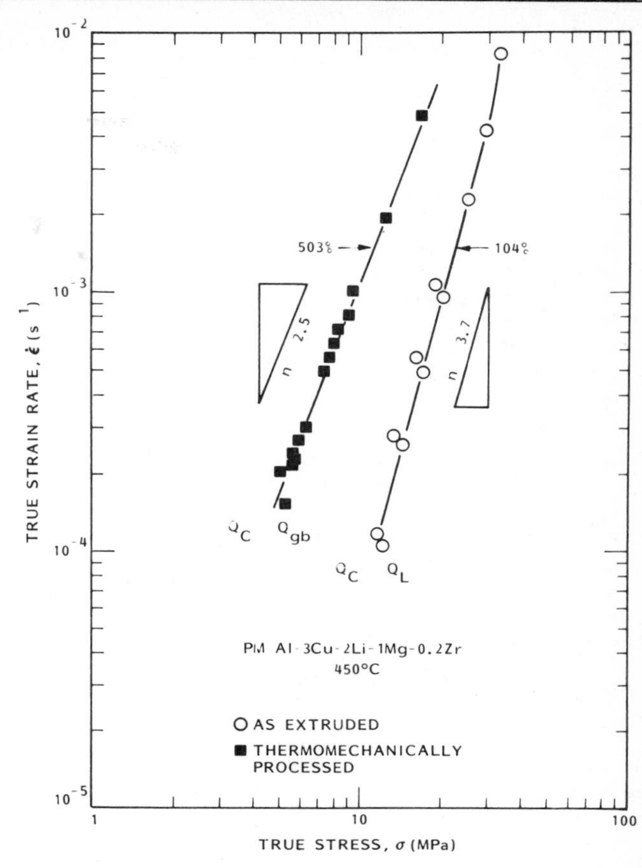

3. Log strain rate versus log stress for PM Al-3Cu-2Li-1Mg-0.2Zr alloy for both the as-extruded and overage-processed conditions. The values of elongation-to-failure refer to individual tests carried out at the approximate strain rate indicated by the arrows.

2. Elongation-to-failure as a function of strain rate and thermomechanical processing for Al-3Cu-2Li-1Mg-0.15/0.2Zr alloys. As is shown, minimum processing techniques significantly improve the ductility of these alloys over the as-extruded microstructure; however, an overage practice is required to develop optimum superplastic behavior.

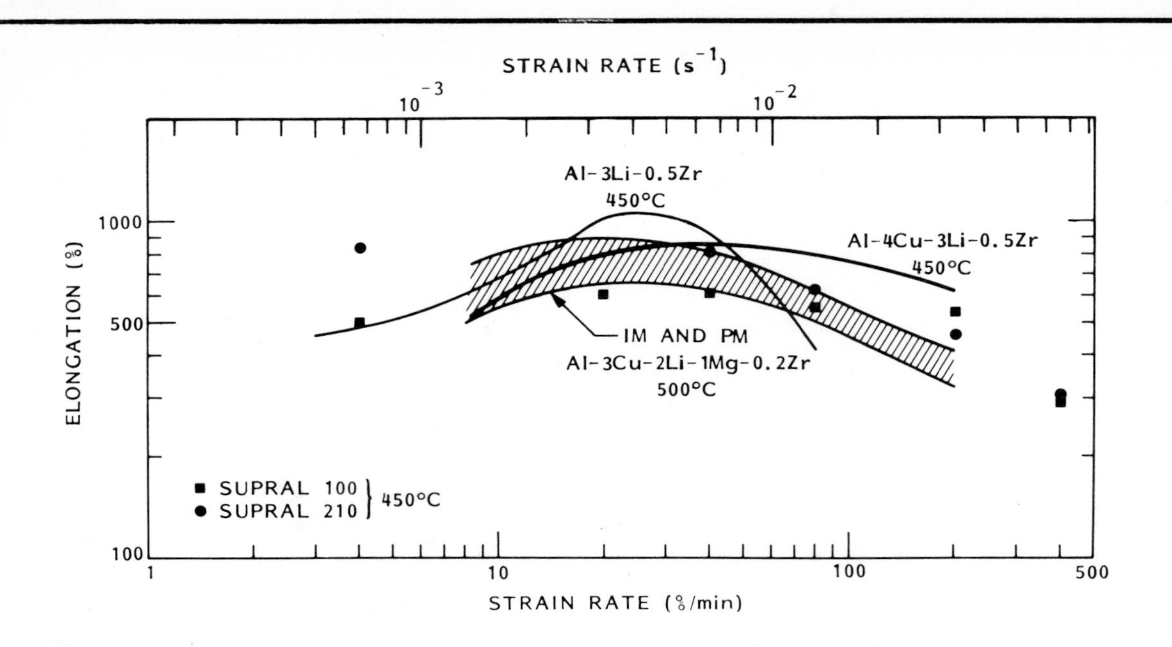

4. Comparison of superplastic behavior between SUPRAL alloys and several Al-Li alloys. The Al-Li alloys show maximum ductility in essentially the same strain-rate range as do the SUPRAL alloys.

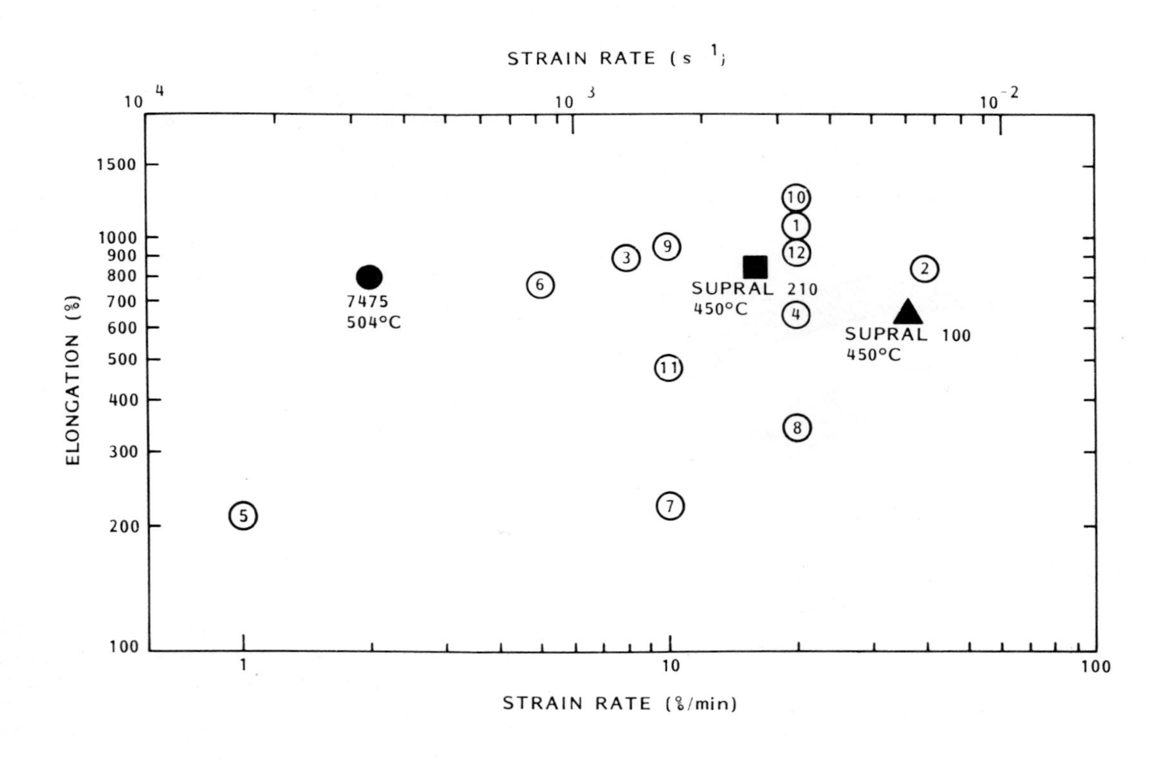

5. Comparison of superplastic behavior in Al-Li alloys, SUPRAL alloys and 7475. The 7475 alloy showed maximum ductility at 2%/min, whereas the SUPRAL alloys and the majority of the Al-Li alloys exhibited maximum ductility at strain rates of 10%/min or greater.

6. Transmission electron photomicrograph of the microstructure of the IM Al-3Cu-2Li-1Mg-0.15Zr alloy in the overage-processed condition. In this recrystallized condition, ultrafine grains (1-2 μm) with high angle boundaries are observed.

measured to be between 1 and 2 μm prior to deformation. After deformation, a small degree of grain growth is observed to occur but this is limited to several microns[9,14].

Very fine microstructures are developed in both continuously and discontinuously recrystallized Al-Li based alloys[3,5,6]. The similar microstructures of these two groups of alloys result in rather similar superplastic properties. This is illustrated in Figure 8 in which elongation-to-failure is plotted as a function of strain rate for discontinuously and continuously recrystallized Al-Li alloys. High elongations-to-failure at high strain rates are observed in both groups.

The transition from non-superplastic to superplastic behavior in the PM alloy as a result of thermomechanical processing as described in Section 2 (Figure 3) has been explained by a transmission electron microscopy study[5]. In this study, it was shown that in the as-received condition the fine microstructure consists of medium-angle subgrains and not true (high-angle) grain boundaries. These medium-angle boundaries are not able to undergo grain boundary sliding, a principal mechanism in superplastic flow[1]. After appropriate processing, involving recrystallization, a fine-grained structure consisting of high-angle grain boundaries is developed and superplastic behavior is observed.

4. Phenomenological Description of Superplasticity

Phenomenological equations describing the superplastic flow of alloys requires (1) measurement of the strain rate sensitivity, m, in the equation $\sigma = k\dot{\epsilon}^m$ (where σ is the flow stress and $\dot{\epsilon}$ is the true strain rate) and (2) measurement of the temperature dependence.

The strain rate sensitivity of thermomechanically processed PM and IM Al-3Cu-2Li-1Mg-0.2Zr alloys has been measured. For the thermomechanically processed materials, strain rate change tests were performed at 400, 425, 450 and 500°C for the PM alloy and for the IM alloy. A value of strain rate sensitivity of m = 0.4 was found at each of the temperatures for both alloys. A value of the activation energy for plastic flow was also determined. The activation energy for the PM alloy is about 93 kJ/mole, close to that for grain boundary diffusion. Similar results were found for the IM alloy.

Although grain boundary sliding is usually considered to be the principal deformation mechanism in superplastic flow other mechanisms have been found to contribute to the deformation process. Activation energies corresponding to lattice diffusion and grain boundary diffusion as well as values in between these two extremes have been reported for superplastic alloys in general and for superplastic aluminum alloys in particular[15]. Aluminum alloys that have a very fine grain size (<5 μm) tend to show activation energies corresponding to grain boundary diffusion whereas aluminum alloys that have a relatively large grain size (>5 μm) show activation energies corresponding to lattice diffusion. Examples of alloys that have a fine grain size and exhibit activation energies for grain boundary diffusion include the commercial SUPRAL Al-Cu-Zr alloys, Al-Mg-Zr alloys, and Al-Li-Zr alloys[3,4]. An example of an alloy that has a relatively coarse grain size and exhibits an activation energy for lattice diffusion is 7475[6].

The measured value of activation energy in the present study can be used to write a phenomenological equation for superplastic flow in the PM and IM alloys. Specifically, the data can be represented by:

$$\dot{\epsilon} = A\,(\sigma/E)^n D_{gb}$$

where A is 1.5×10^{17} for the PM alloy and 7.8×10^{16} for the IM alloy, n = 2.5, and $D_{gb} = D_o \exp(-Q_{gb}/RT)$. D_o is assumed to be equal to $10^{-4} m^2 s^{-1}$ and Q_{gb} is the measured value of 93 kJ/mole. In Figure 9 the data are plotted as $\ln \dot{\epsilon}/D_{gb}$ vs. $\ln \sigma/E$ where the lines through the data represent the above equations. As shown, the equations fit the data quite well. As discussed above, the IM alloy shows consistently higher strength than the PM alloy and this is attributed to a minor difference in grain size between the PM and IM alloys.

Despite the fact that the grain size is finer in the PM alloy than in the IM alloy, overall values of elongation-to-failure are generally observed to be greater in the IM alloy than the PM alloy. It is tempting to attribute this reduction in elongation-to-failure in the PM alloy to an inherent likelihood of a greater propensity for cavitation nucleation in PM alloys due to their relatively high oxide content. On-going studies of cavitation behavior of the two types of alloys do not show a marked difference; however, this area of study will be reported at a later date[11].

5. Room Temperature Properties

Tensile, elastic modulus (E), and density (ρ) data were measured for some of the Al-Li based alloys and for the Al-3Li-0.5Zr and IM and PM Al-3Cu-2Li-1Mg-0.15/0.20Zr alloys in particular. The measurements were made on thermomechanically treated material that was solution treated, water quenched and aged to the T6 condition. Although the T8 treatment (i.e., solution treatment plus 2% prestrain) results in even higher strength in these alloys[2], it was not used in the present work, since such treatment is probably not feasible for superplastically formed material. Detailed room temperature results for the IM and PM Al-3Cu-2Li-1Mg-0.15/0.20Zr alloys are shown in Figure 10. In this figure, ultimate tensile strength, 0.2% yield strength, hardness and tensile elongation are shown as a function of aging time at 171°C, after solution treatment at 538°C. This results in recrystallization in these particular alloys that have been subjected to the overage thermomechanical processing practice.

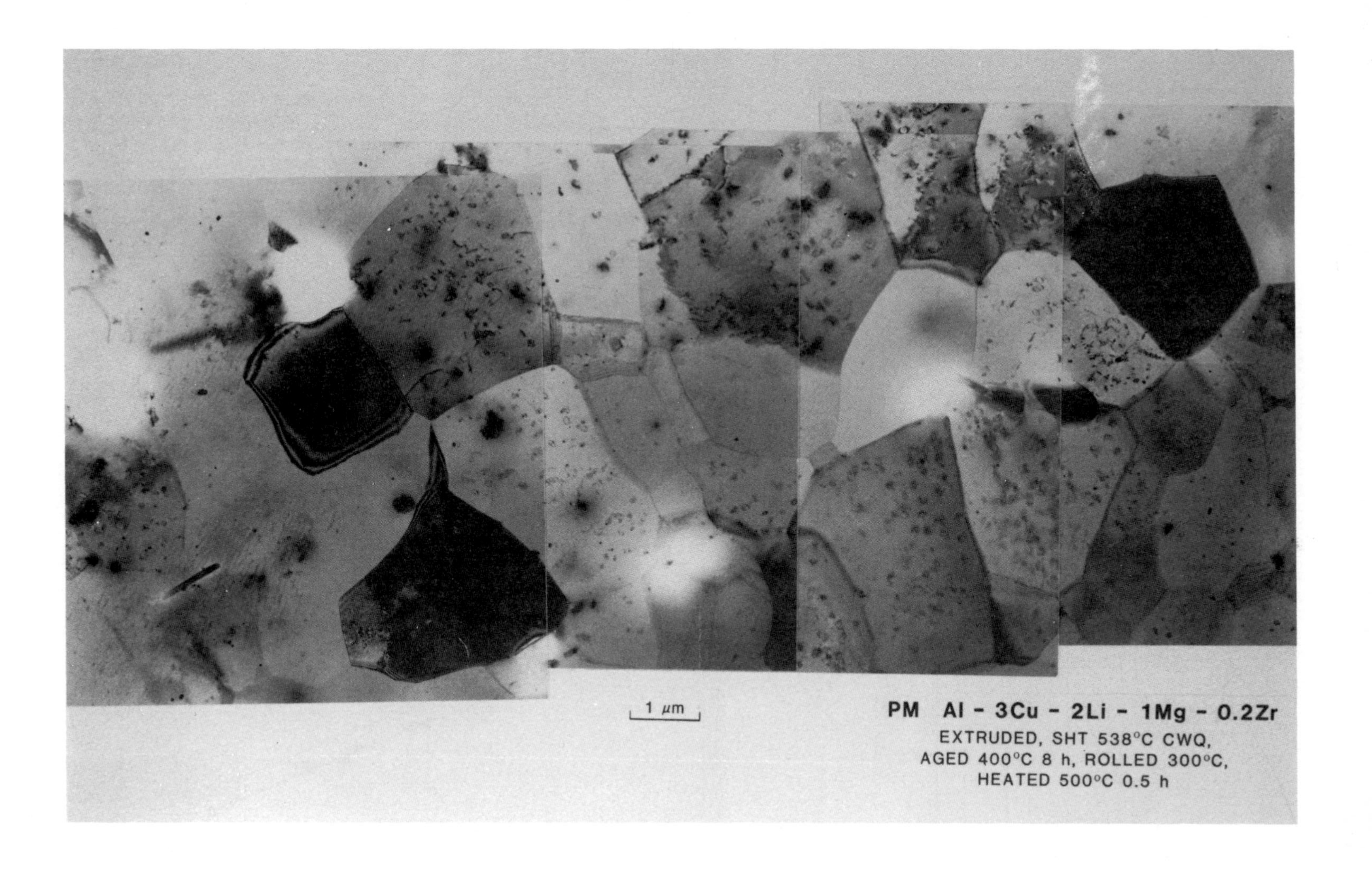

7. Transmission electron photomicrographs of the PM Al-3Cu-2Li-1Mg-0.2Zr alloy microstructure. After overage processing, this alloy was also in the recrystallized condition, and exhibited ultrafine grains with high angle boundaries.

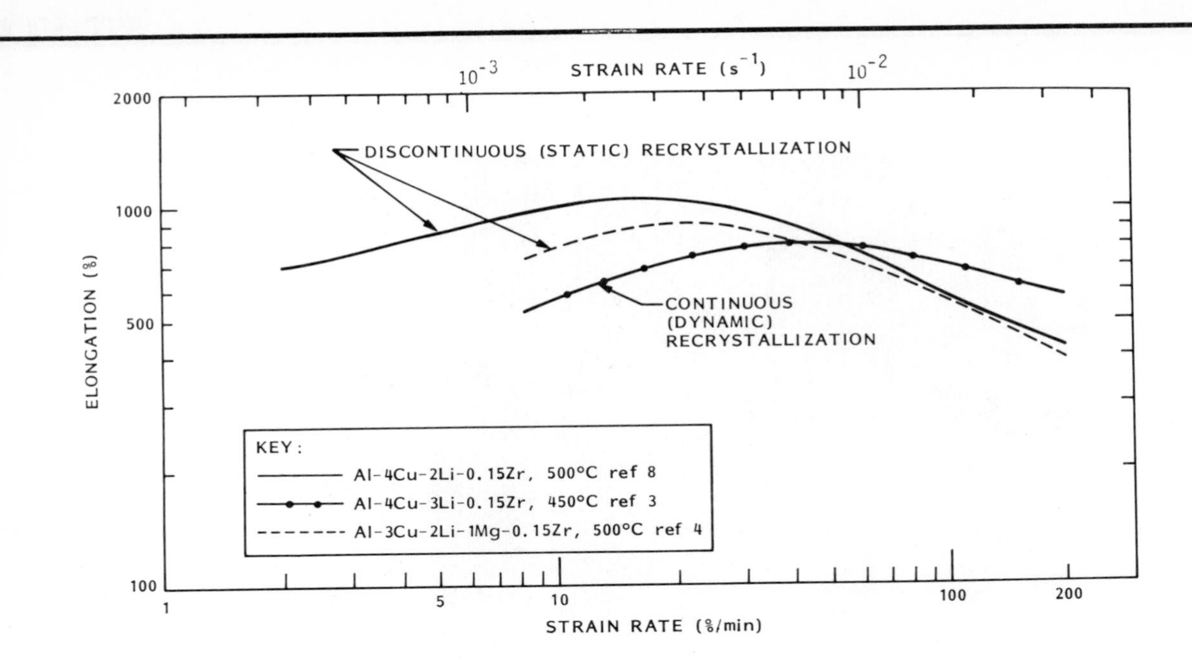

8. Elongation-to-failure plotted as a function of strain rate for discontinuously and continuously recrystallized Al-Li alloys. Similar superplastic behavior was observed for both types of alloys.

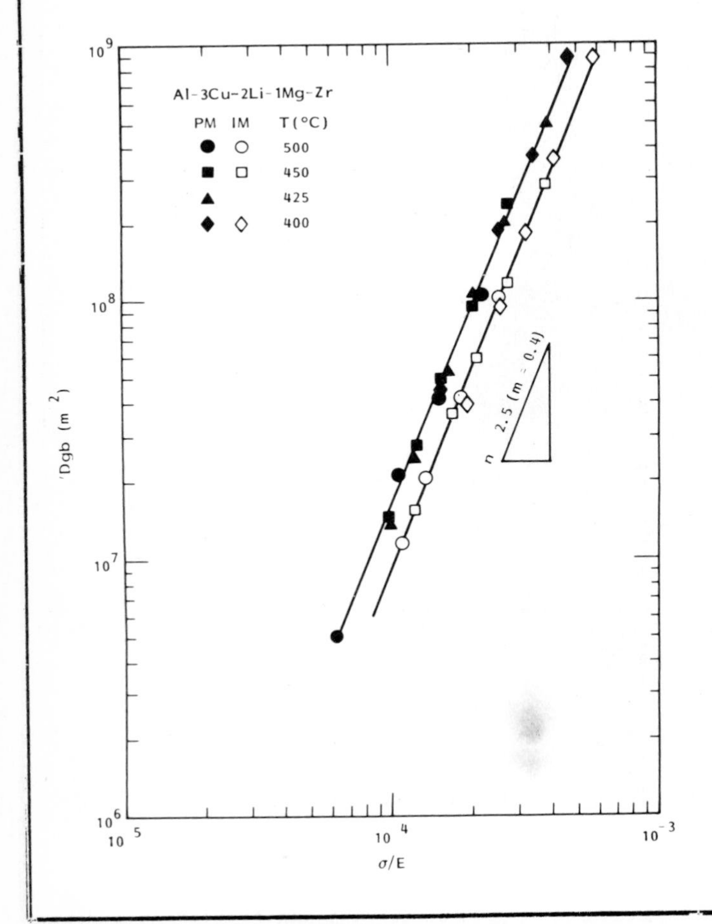

9. Diffusion-compensated strain rate versus modulus-compensated stress for Al-3Cu-2Li-1Mg-0.15/0.2Zr alloys.

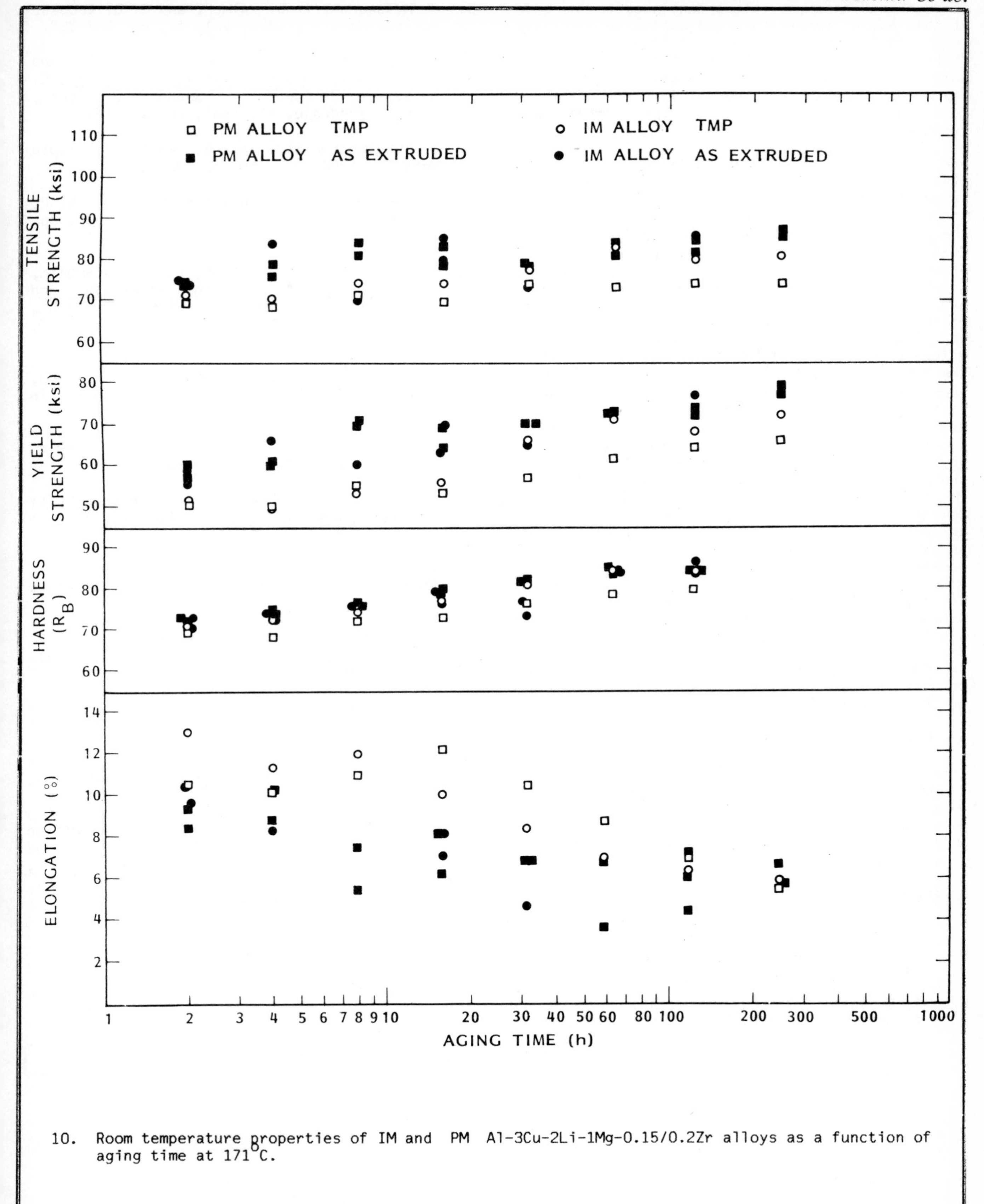

10. Room temperature properties of IM and PM Al-3Cu-2Li-1Mg-0.15/0.2Zr alloys as a function of aging time at 171°C.

Strength and hardness values continued to increase at this aging temperature but ductility decreased suggesting an optimum aging condition of about 4-20 hrs at 171°C. Comparison can be made with the same alloys that are in the as-extruded condition prior to heat treatment (i.e., in which recrystallization does not occur during heat treatment) by reference to Palmer et al (2).

The results are compared in Table IV with data for other superplastic aluminum alloys of commercial interest and with the alloys listed in order of increasing yield strength. All the data are for the peak aged (T6) condition except for the non-heat-treatable Al-5Ca-5Zn alloy which was tested after superplastic forming. The high-Zr alloys, Al-3Li-0.5Zr and Al-4Cu-3Li-0.5Zr, show strength and ductility values which are comparable to those of the SUPRAL 100 and 210 alloys, respectively. The low-Zr alloys show strength and ductility values comparable to SUPRAL 220 and 7475. All of the lithium-containing alloys show significantly lower density and higher elastic modulus than any of the other alloys. This is very important for weight saving in aerospace structural applications. The actual weight saving achievable for any given structure or component will depend on the particular failure modes involved. Both the PM and IM alloys were evaluated for room-temperature properties after superplastic forming. In this case, samples of each alloy were deformed to a level of 100% at 500°C and then solution treated and aged. The results of this evaluation are also presented in Table IV, both before and after superplastic forming. As may be seen, the strength properties of the PM and IM alloys are unaffected by prior superplastic deformation, and the elongation-to-failure is reduced from about 9% to 6%. This result is similar to published results on the influence of such prior deformation on the properties of 7000-series Al alloys.

ACKNOWLEDGEMENTS

This work was supported by the Lockheed Independent Research and Development Program. Valuable discussions and contributions from Dr. A. R. Pelton, Dr. M. C. Pandy and Professor A. K. Mukherjee are gratefully acknowledged.

REFERENCES

1. Superplastic Forming of Structural Alloys, N. E. Paton and C. H. Hamilton, Eds., AIME, Warrendale, PA, 1982.

2. I. G. Palmer, R. E. Lewis, D. D. Crooks, E. A. Starke, Jr., and R. E. Crooks: Aluminum-Lithium Alloys II, E. A. Starke, Jr., and T. H. Sanders, Jr., Eds., AIME, Warrendale, PA, 1984, p. 91-110.

3. J. Wadsworth, I. G. Palmer, and D. D. Crooks, Scripta Metall., 1983, Vol. 17, pp. 347-352.

4. J. Wadsworth, I. G. Palmer, D. D. Crooks, and R. E. Lewis, Aluminum-Lithium Alloys II, E. A. Starke, Jr., and T. H. Sanders, Jr., Eds., AIME, Warrendale, PA, 1984, pp. 111-135.

5. J. Wadsworth and A. R. Pelton, Scripta Metall., 1984, Vol. 18, pp. 387-392.

6. J. Wadsworth, Superplastic Forming, S. P. Agrawal, Ed., ASM, Metals Park, OH, 1984, pp. 43-57.

7. C. A. Henshall, T. G. Nieh, and J. Wadsworth, Advancing Technology in Materials and Processes, 30th National SAMPE Symposium, Anaheim, CA, SAMPE, Covina, CA, Vol. 30, 1985, pp. 994-1004.

8. J. Wadsworth, C. A. Henshall, A. R. Pelton, and B. Ward, Jour. Mater. Sci. Lett., In Press, 1985.

9. J. Wadsworth, A. R. Pelton, and R. E. Lewis, Metall. Trans., Ser. A, In Press, 1985.

10. T. G. Nieh and J. Wadsworth, Superplasticity in Aerospace-Aluminum, R. Pearce, Ed., In Press, 1985.

11. M. C. Pandey, J. Wadsworth, and A. K. Mukherjee, Submitted to Mater. Sci. Eng., 1985.

12. J. W. Edington, K. N. Melton, and C. P. Cutler, Progr. Mater. Sci., 1976, Vol. 21, pp. 61-170.

13. O. D. Sherby and J. Wadsworth, Materials Science and Technology, In Press, 1985.

14. M. C. Pandey, J. Wadsworth, and A. K. Mukherjee, submitted to Scripta Metall., 1985.

15. O. D. Sherby and J. Wadsworth, "Deformation Processing and Structure in Metals and Alloys", G. Krauss, Ed., ASM, Metals Park, OH, 1984, pp. 355-389.

Hot deformation behavior in Al–Li–Cu–Mg–Zr alloys

M NIIKURA, K TAKAHASHI
and C OUCHI

The authors are in the Technical Research Center of Nippon Kokan KK in Japan

Synopsis

As a fundamental study of thermo-mechanical processing in Aℓ-Li alloys, hot deformation behavior and the microstructural changes accompanied by dynamic and static restoration processes were systematically investigated, focusing on the effect of solute Li and δ precipitates on this behavior. Li-free alloy exhibited a typical stress-strain curve for dynamic recovery type of materials, while Li-containing alloys showed a peak stress behavior. This was suggested to be due to drag effect by solute Li against dislocation motion. Dynamically restored microstructure in Li-containing alloys was revealed as equiaxed subgrain structure. Dynamic and static restoration was retarded in hot deformation under a presence of undissolved or dynamically precipitated δ, resulting in elongated pan-cake structure. Possibilities for new thermo-mechanical processing specific Aℓ-Li alloys were also discussed.

Introduction

Aℓ-Li alloys are well known as potential candidate materials for future aircraft structure, since the alloys significantly contribute to weight savings in aircraft due to lower density and higher stiffness. However, relatively poor qualities in ductility, toughness, and hot-workability seem to decelerate the progress of commercialization in these alloys; several causes, such as slip localization behavior relevant to δ' precipitates and intergranular fracture due to impurity segregation, have been pointed out for those poor qualities.[1] Then slip behavior alteration through alloy design and impurity control through new techniques in ingot making have been proposed to overcome such problems.[2][3]

Microstructural control through thermo-mechanical processing is also expected as a measure improving those qualities as well as acquiring superplasticity.[4] Several types of Intermediate Thermo-mechanical Treatment were successfully developed in 2000 and 7000 series alloys, but only a few works[5] seem to have been done on Aℓ-Li alloys. In this view, as the fundamentals for thermo-mechanical processing, hot deformation characteristics and microstructural changes specific in Aℓ-Li alloys should be made clear. In this study, with the use of computerized hot deformation equipment, hot deformation behavior and microstructural changes accompanied by dynamic and static restoration processes were systematically investigated in Aℓ-Li-Cu-Mg-Zr alloys, focusing on the effect of solute Li and δ precipitates on their behavior.

Experimental Procedures

Chemical composition of the materials used in this study is shown in Table 1. The basic composition was Aℓ-Li-1%Cu-1%Mg-0.1%Zr, and Li content was varied up to 4 wt%. Most studies were conducted using Li free and 2%Li alloys. 800g ingots were prepared in a vacuum induction melting furnace, using high purity aluminum, magnesium, lithium, and aluminum-copper master alloy, aluminum-zirconium master alloy as raw materials.

Ingots were homogenized at 500°C for 120 hrs, and then hot-rolled to 11 mm^t thickness plates. The plates were solution-treated at 550°C for 4 hrs as a standard condition. In addition, some other plates were solution-treated at 500°C or at 400°C. δ precipitates remained partially undissolved for 500°C and largely

Table 1 Chemical Composition of Used Materials

wt%

	Li	Cu	Mg	Zr	Fe	Si	Na	K
Li-free	tr.	1.18	0.92	0.14	0.03	<0.01	<0.005	0.019
1Li	1.24	1.20	1.25	0.14	0.03	<0.01	<0.005	0.024
2Li	2.16	1.28	1.24	0.14	0.03	<0.01	<0.005	0.027
3Li	3.13	1.30	1.32	0.13	0.03	<0.01	<0.005	0.022
4Li	3.92	1.24	1.18	0.14	0.03	<0.01	<0.005	0.038

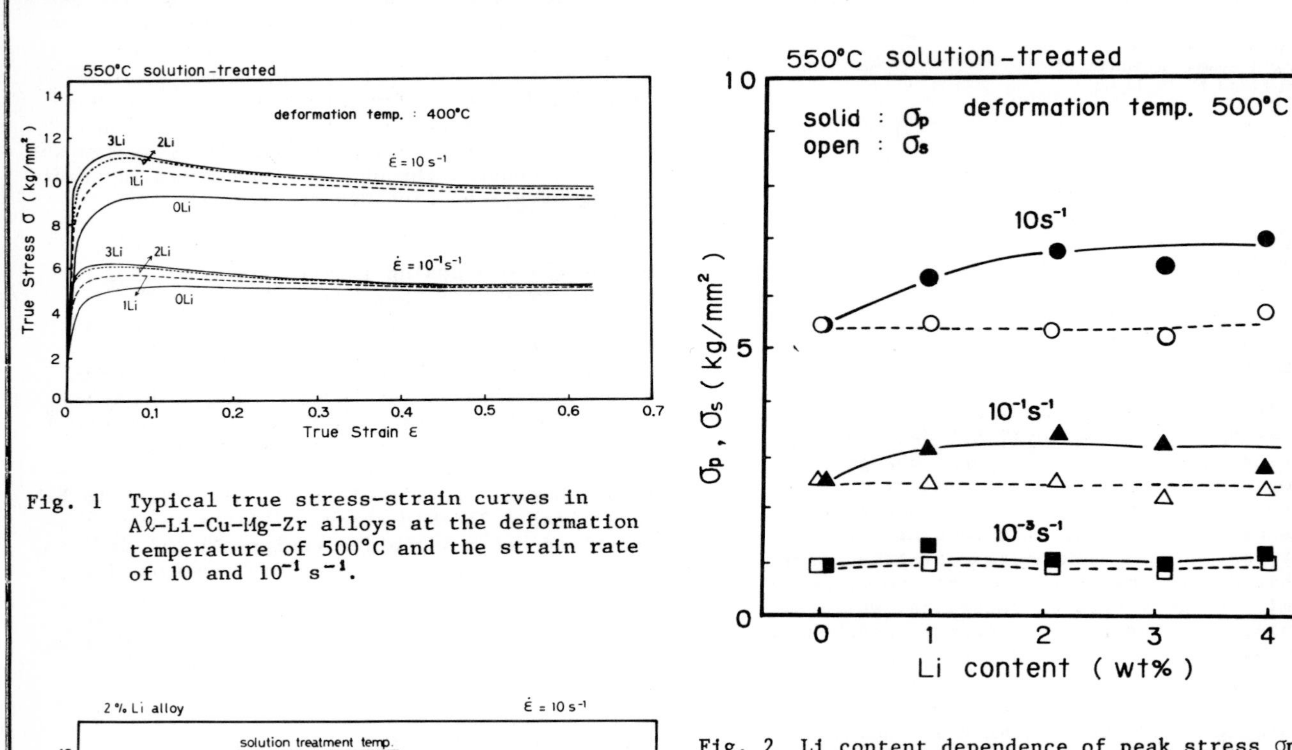

Fig. 1 Typical true stress-strain curves in Aℓ-Li-Cu-Mg-Zr alloys at the deformation temperature of 500°C and the strain rate of 10 and 10^{-1} s^{-1}.

Fig. 2 Li content dependence of peak stress σp and steady state flow stress σs at the deformation temperature of 500°C.

Fig. 3 Effect of lowering the solution-treatment temperature on the true stress-strain curves at the strain rate of $10s^{-1}$.

undissolved for 400°C treatment respectively. The effect of δ precipitates was investigated in these materials. For hot deformation tests cylindrical specimens of 6 mm^{ϕ} x 10 mm^{h} were machined from these plates.

A fully computerized hot deformation equipment (THERMECMASTOR-Z) developed by the authors was used in this study. This equipment enables one to vary strain rate in a wide range from 10^{-4} to $50s^{-1}$ under a maximum load of 5 ton. The specimen is heated by a high frequency induction coil in vacuum. Various conditions of heating, hot deformation sequence and subsequent cooling are completely programmed. An He gas quenching unit installed in site of the induction coil provides on-site rapid quenching capability on deformed specimens in any time interval over 10ms. This quenching unit enables delay time to be minimized and is critically important to separate strictly microstructural changes between dynamic and static restoration processes. Load and displacement data are recorded and processed by an on-line microcomputer to provide a true stress-strain curve on a plotter.

The specimens which were prepared from the plates solution-treated at 550°C were reheated at 550°C and cooled down to various deformation temperatures in the range from 175°C to 500°C, and subsequently compressed to a strain of 0.7. Strain rate was varied from 10^{-4} to $10s^{-1}$. True stress-strain curves were examined in detail under a wide variety of deformation conditions. The specimens prepared from the plates with the lowered temperatures for solution-treatment were reheated at the same temperature as the deformation temperature and deformed in the same conditions as described above.

Dynamic microstructural change taking place during deformation was observed in specimens quenched during a delay time of less than 10ms after deformation, while static microstructural change after the deformation was examined by holding the deformed specimens at 500°C in various holding time from 10ms to 2000s, followed by quenching.

Results

(1) Hot deformation behavior

Typical true stress-strain curves for the alloys are shown in Fig.1. Li free alloy exhibited a typical stress-strain curve for dynamic recovery type of materials, which was characterized by work hardening in the early stage of straining, followed by steady state flow in the strain above 0.1. On the other hand, Li containing alloys showed a peak stress at a strain of around 0.1, and then flow stress decreased, reaching to a steady state flow at a strain of around 0.5. This peak stress behavior in Li containing alloys was observed for a wide range of deformation temperatures from 250 to 500°C and all the strain rates investigated here. Fig.2 shows the effect of Li content on the peak stress σp and the steady state flow stress σs at the deformation temperature of 500°C. σp increased with Li content up to 2%, but stayed unchanged above 2%. σs was not influenced by Li content. In as-solution-treated condition at 550°C, the alloys with Li content up to 2% exhibited microstructures containing no undissolved precipitates, while 3 to 4% Li alloys showed evidence of undissolved precipitates which appeared to be δ phase. These results indicate that the increase of σp in the range of Li content below 2% is brought about by the increase of solute Li, and that the undissolved δ phase formed in the alloys with higher Li content does not influence σp. Temperature increase in the specimen during the adiabatic deformation could be the cause for the peak stress behavior. However, this is not the case here, because Li free alloy did not show any peak stress behavior, even though it exhibited almost the same flow stress as Li containing alloys.

The effect of lowering the temperature of solution-treatment was examined in 2%Li alloy as shown in Fig.3. Although the materials solution-treated at 500°C exhibited almost same peak stress behavior as the materials solution-treated at 550°C, the materials solution-treated at 400°C did not show peak stress behavior any more. The solution-treatment at 400°C may have made a large fraction of Li added in this alloy precipitate as coarse δ phase and the amount of Li as solute decrease. This could be the cause for the disappearance of peak stress behavior in the materials solution-treated at 400°C.

Deformation temperature and strain rate dependence of the steady state flow stress σs is summarized in Fig.4 for the 2%Li alloy solution-treated at 550°C. Strain rate dependence appeared markedly at the temperature above 300°C, but not below 200°C. Dynamic restoration progress during the deformation should be small in the lower temperature range. Hot deformation above 300°C is controlled by a thermally activated process, and the relation between the flow stress and the deformation parameters is expressed by the following equation.[6)]

$$\dot{\varepsilon} = A\sigma^{m} \exp(-Q/RT) \quad \ldots\ldots\ldots\ldots\ (1)$$

here, A and m are constants, Q is activation energy, R is gas constant. Q is derived from eq.(1) as follows,

$$Q = -R(\partial \ln\dot{\varepsilon}/\partial 1/T)_{\sigma} \quad \ldots\ldots\ldots\ldots\ (2)$$

Q=37500 cal/mol was obtained for 2%Li alloy. This value is a little higher than the activation energy Q_{SD}=34200 cal/mol for self-diffusion in pure Aℓ. With this activation energy, m in eq.(1) was obtained as m=4.8.

(2) Microstructural change

Microstructural changes during the dynamic restoration process in 2%Li alloy solution-treated at 550°C was investigated by interruption of straining and the subsequent rapid quenching with the minimized delay time. Photo 1 shows the case for the deformation temperature of 500°C and strain rate of $10^{-1}s^{-1}$. Subgrain formation in the initial coarse microstructure began to develop from the early stages of straining, and

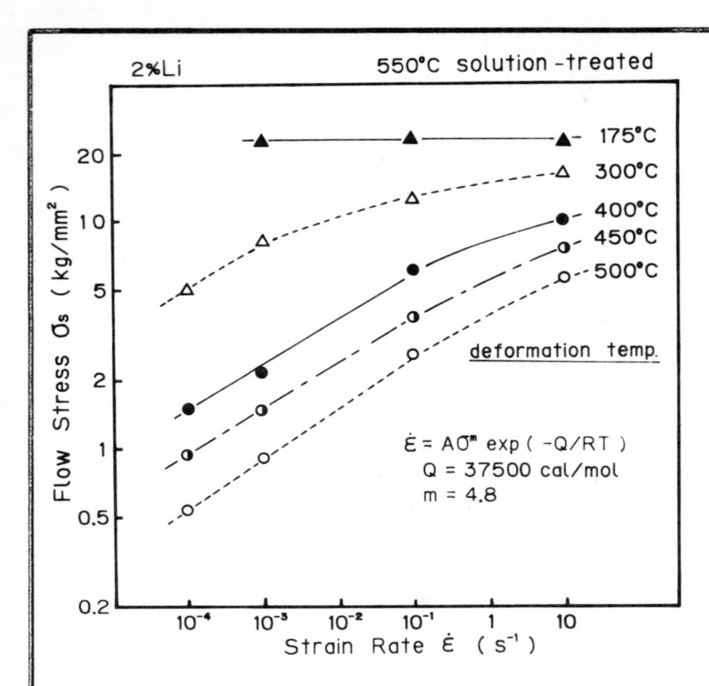

Fig. 4 Effect of the deformation temperature and the strain rate on the steady state flow stress σ_s in 2%Li alloy solution-treated at 550°C.

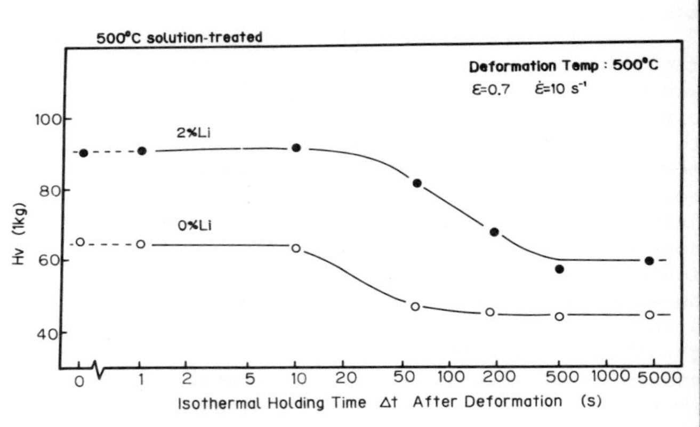

Fig. 5 Hardness change with isothermal holding time after hot deformation at 500°C in the alloys solution-treated at 500°C.

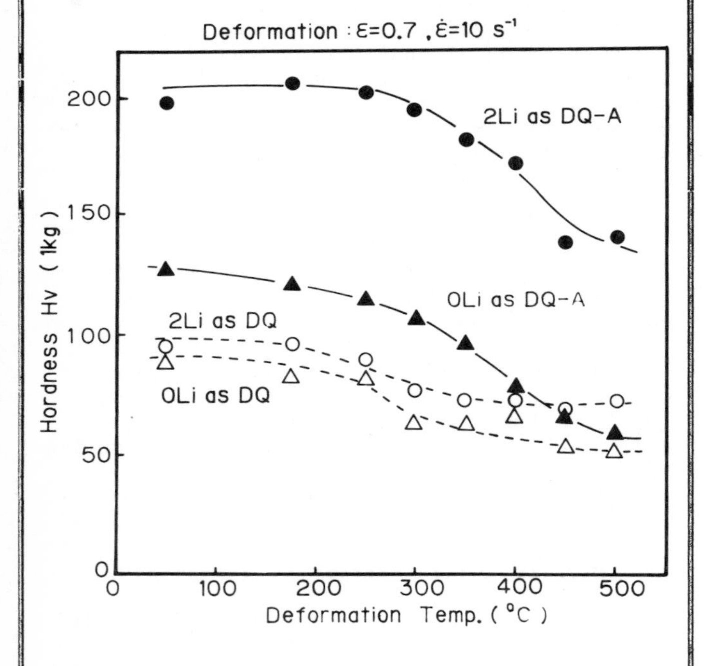

Fig. 6 Hardness change in as-direct-quenched materials and direct-quenched and aged materials with deformation temperature.

equiaxed subgrain structure was obtained as the steady state microstructure in the higher strain range above 0.5. In the steady state range, subgrain size was around 20μm and the initial grain boundaries disappeared.

Photo 2 shows the microstructures of the 2%Li alloy deformed in several deformation conditions with a fixed strain of 0.7. The equiaxed subgrain structure became finer with the deformation temperature decrease and the strain rate increase. On the other hand, for low temperature deformation below 400°C, an elongated pan-cake structure was developed instead of the equiaxed subgrain structure. A few finer equiaxed subgrains were formed in the vicinity of such elongated grain boundaries. Precipitates of submicron size were also observed in the subboundaries of deformed matrix and grain boundaries. These precipitates appeared to be δ precipitates. As the deformation conditions of slower strain rate and lower deformation temperature enhanced this precipitation, these precipitates may be due to dynamic precipitation.

The equiaxed subgrain structures formed in the dynamic recovery process shown in Photo 1 and Photo 2 are relatively coarse, the subgrain size ranging from 15 to 30μm, because of the higher deformation temperature. With the lowering of deformation temperature below 400°C, such coarse equiaxed subgrain formation was suppressed, resulting in elongated pan-cake structure. Dynamic precipitation observed there could be partly responsible for the formation of the

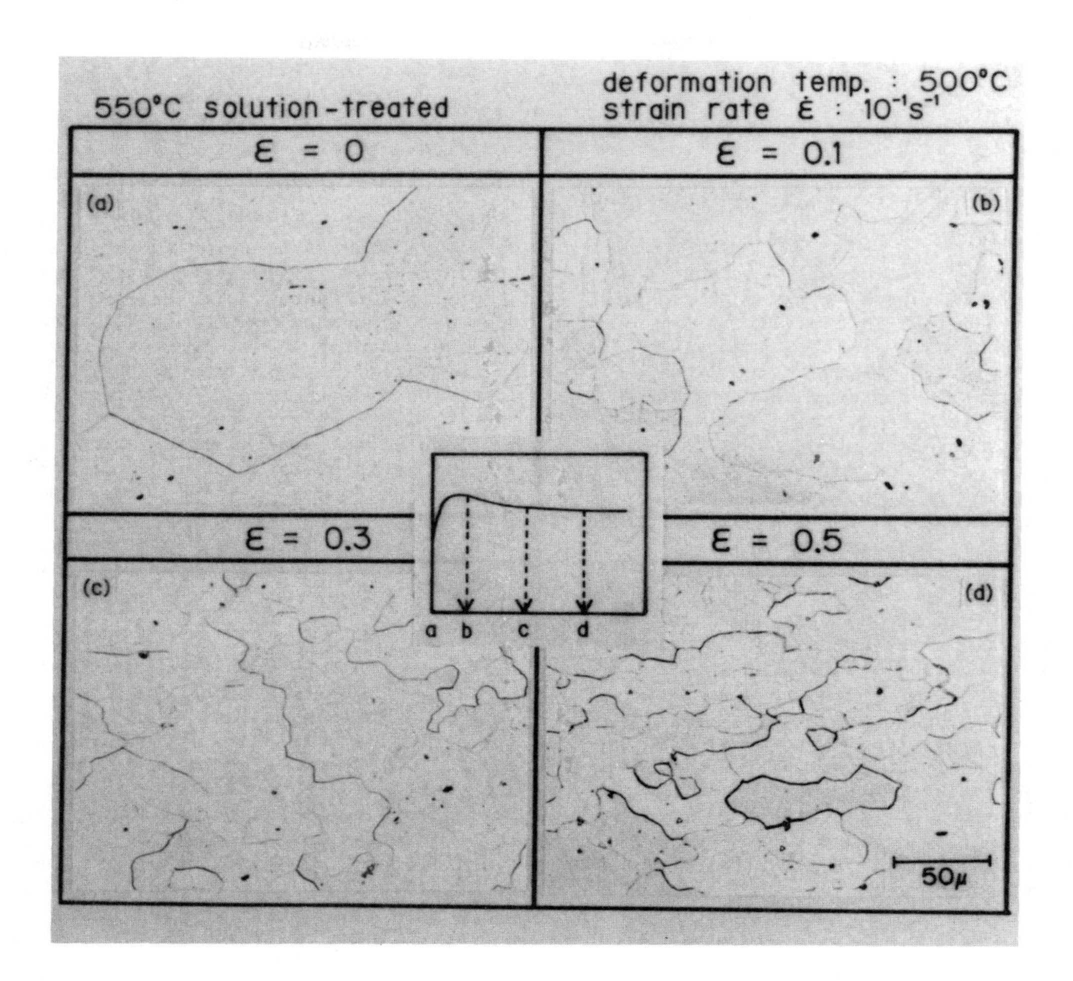

Photo 1 Dynamic microstructural change during hot deformation at 500°C for 2%Li alloy solution-treated at 550°C (a) as reheated at 550°C (b) ε=0.1 (c) ε=0.3 (d) ε=0.5.

pan-cake structure through suppressing dynamic restoration.

Equiaxed subgrain size developed by dynamic recovery process was related to Zener-Hollomon parameter in the following equation.

$$Ds^{-1} = G + F \log Z = G + F \log [\dot{\varepsilon} \exp(Q/RT)] \quad .... (3)$$

In the present case, constants in eq.(3) were obtained as $G=-0.0135\mu m^{-1}$, $F=0.0065\mu m^{-1}$.

(3) Effects of δ precipitates on the restoration process

In the previous section, it was shown that dynamic recovery and coarse equiaxed subgrain formation proceeded very intensively in the hot deformation of 2%Li alloy in the temperature range above 450°C. Effects of undissolved δ phase on the dynamic and static restoration processes were investigated, using 2%Li alloy which was solution-treated at 500°C. The micro-structures of the deformed materials are shown in Photo 3 for the deformation temperature of 500°C and the strain rate of 10, 10^{-1}, and $10^{-3}s^{-1}$ All of those basically exhibited elongated pan-cake structure and contrasted with the ones of the materials which were solution-treated at 550°C and deformed at 500°C as shown in Photo 2. Since the grain size in the short transverse direction was around 10μm, subgrain size here would be less than 10μm, which was much finer than the equiaxed subgrain structure. Partially undissolved δ precipitates in the materials solution-treated at 500°C might be responsible for the pan-cake structure formation through suppressing the dynamic restoration. Although undissolved δ precipitates had no influence on the flow stress as shown in Fig.2, they markedly affected the microstructure formed by dynamic recovery.

Static restoration kinetics was studied in the elongated pan-cake structure containing partially undissolved δ precipitates, which was formed by the deformation at 500°C in 2%Li alloy solution-treated at 500°C. Hardness change during the isothermal holding at 500°C after deformation is given in Fig.5 in comparison with Li free alloy. Hardness continuously decreased in the holding time between 60 to 500s due to the progress of recrystallization in 2%Li alloy, while it decreased in the holding time between 10 to 60s in Li-free alloy. This means that static recrystallization was retarded by the presence of partially undissolved δ phase. Photo 4 shows the microstructural change during isothermal holding. Recrystallized grains in 2%Li alloy appeared in a holding time longer than 180s, but the shape of these remained somewhat elongated with the short transverse grain size of around 15μm. On the other hand, Li-free alloy showed equiaxed recrystallized grain structure in a holding time longer than 60s with a grain size of more than 100μm.

(4) Effect of the hot-deformed microstructure on age hardenability

It was made clear that the microstructure in the hot deformed Aℓ-Li alloys remarkably changed depending not only on deformation conditions such as deformation temperature or strain rate, but also on the existing form of Li, whether in solution or in precipitates. The decrease of deformation temperature changed the microstructure from equiaxed subgrain structure to elongated pan-cake structure as shown in Photo 2. Since this change in the microstructure is accompanied by the change in dislocation density or subgrain size, it may influence the aging characteristics of δ' precipitation. Fig.6 shows the effect of the deformation temperature on hardness of materials quenched immediately after hot deformation and subsequently aged. Deformation took place in the materials solution-treated at 550°C with a condition of strain of 0.7 and strain rate of $10s^{-1}$. Although hardness increased with the decrease of deformation temperature, the increment of hardness is much higher in aged materials than as-quenched materials. This indicates that suppression of progress in dynamic restoration due to the lowering of the deformation temperature enhances the precipitation hardening due to δ' phase.

Discussion

Li-containing alloys showed peak stress in the strain of 0.1 in the true stress-strain curve. This peak stress behavior was observed only in the presence of solute Li and disappeared in conditions where a large fraction of Li content was lost from solution due to δ phase precipitation. Yoshinaga et al[7)8)] observed high temperature yielding phenomena in hot deformation of Aℓ-Mg alloy between 300 and 500°C. They suggested that the drag effect of solute Mg was responsible for the work softening through the decrease of effective stress due to the increase of dislocation density. The peak stress behavior in this study might be due to the same mechanism by solute Li as in Aℓ-Mg alloy. According to Yoshinaga et al, high temperature yielding phenomena are generally remarkable in alloy systems such as Aℓ-Mg alloy or Aℓ-Cu alloy where intensive solution hardening is operative, and not so great in Aℓ-Li alloy. However, distinct phenomena could appear even in Aℓ-Li alloy in case of high strain rate deformation as in this study. Multiple alloying of Cu and Mg in addition to Li also seems to help the tendency towards high temperature yielding phenomena.

Hot deformation under conditions where Li remained in solution without precipitating dynamically produced an equiaxed subgrain structure with a subgrain size of 15 to 30μm. On the other hand, hot deformation under conditions where δ phase remained undissolved or where δ phase precipitated dynamically gave elongated pan-cake structure due to the retardation of dynamic and static restora-

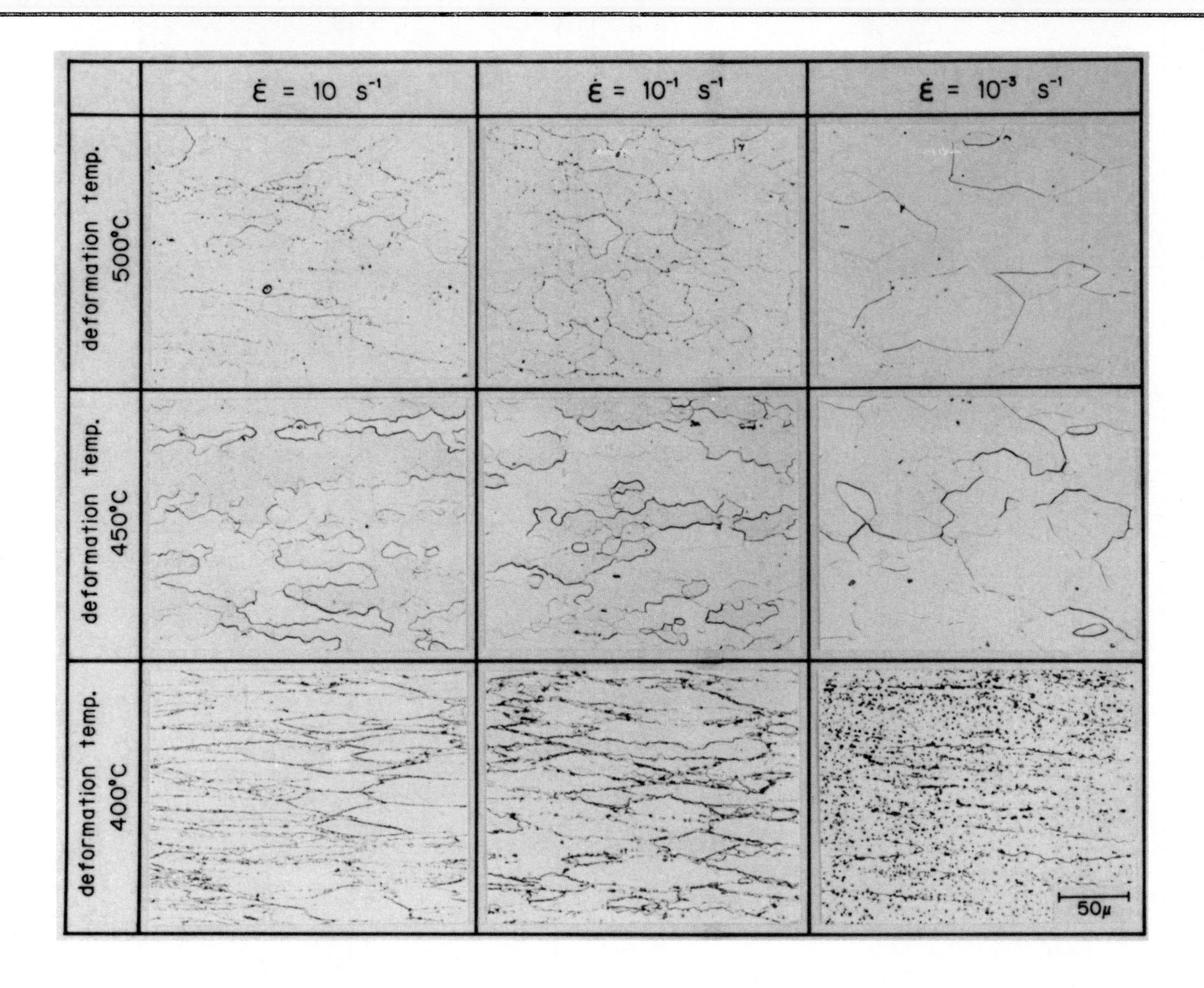

Photo 2 Effect of deformation temperature and strain rate on the microstructure in 2%Li alloy solution-treated at 550°C.

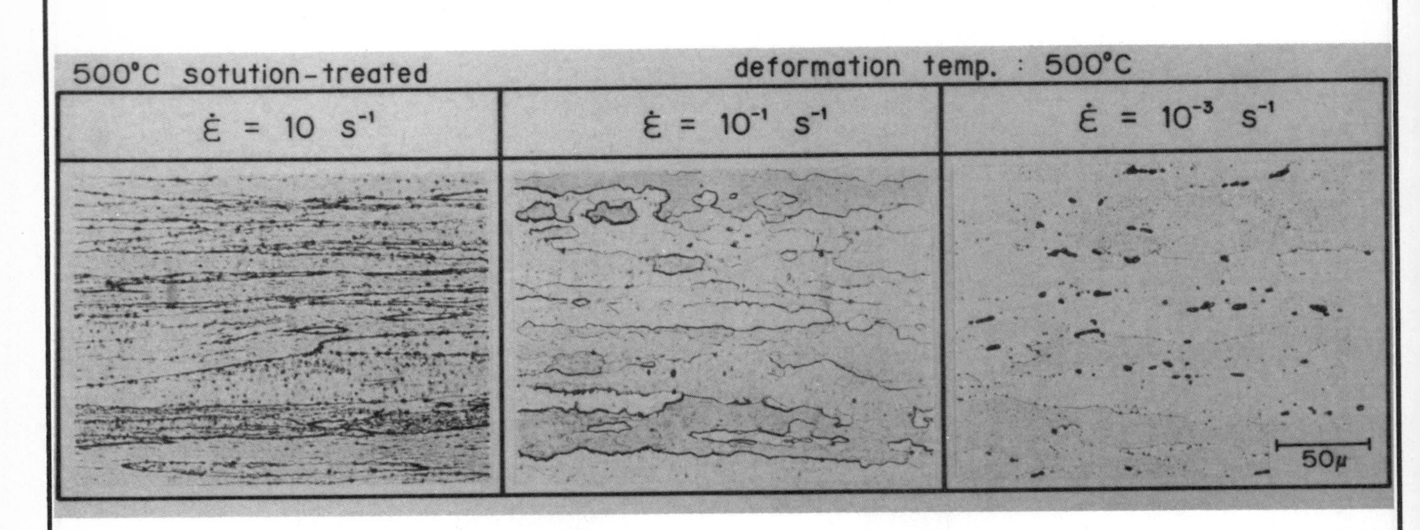

Photo 3 Microstructure of deformed 2%Li alloy which was partially solution-treated at 500°C.

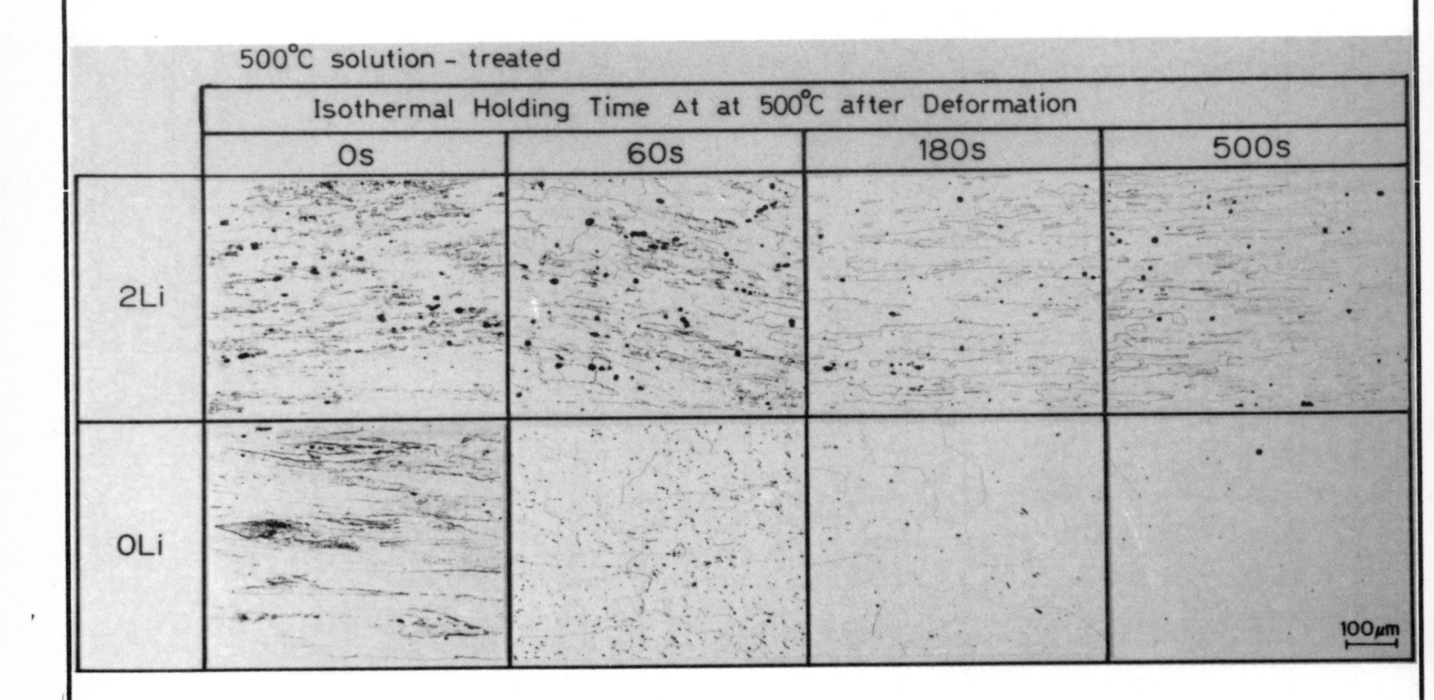

Photo 4 Microstructural change during isothermal holding after hot deformation at 500°C in the alloys partially solution-treated at 500°C.

tion. Subgrain size in this elongated structure was presumably less than 10μm. With such a structure, direct quenched and aged materials had an advantage in attaining high strength, because suppression of the progress in both dynamic recovery and static restoration enhanced the precipitation hardening due to δ' phase.

Several types of Intermediate Thermomechanical Treatment were successfully developed in 2000 and 7000 series alloys.[9)10)11)] For obtaining fine recrystallized grains, warm deformation in overaged materials are effectively utilized yielding uniform dispersion of the localized and severely deformed zone for nucleation site of recrystallization. The microstructure obtained in Aℓ-Li alloy for deformation under the presence of undissolved δ precipitates or dynamical precipitation has metallurgical aspects similar to those in Intermediate Thermomechanical Treatment, since it has rather a fine subgrain structure yielding homogeneous dispersion of locally strained subgrain boundaries. Thus potential capability for static recrystallization to finer grain size is expected in such pan-cake structure, while only subgrain growth to coarser grain size may occur in an equiaxed substructure obtained in deformation under conditions where Li remains in solution. A static recrystallized structure starting from such a pan-cake structure was actually found to be rather fine with a short transverse grain size of around 15μm in 2%Li alloy as shown in Photo 4. It might be possible that this kind of hot deformation utilizing δ precipitates provides an idea for a new thermo-mechanical processing suitable specifically for Aℓ-Li alloys with the advantage of allowing a rather higher deformation temperature compared with the existing Intermediate Thermomechanical Treatment.

Conclusions

(1) Li-free alloy exhibited a typical stress-strain curve for dynamic recovery type of materials, while Li-containing alloys showed a peak stress behavior. It was suggested that this was due to the drag effect by solute Li against dislocation motion.

(2) Steady state flow stress σs was not influenced by Li addition, but showed a prominent dependence on both deformation temperature and strain rate above 300°C. σs is governed by the following equation, $\dot{\varepsilon}=A\sigma^{m}\exp(Q/RT)$, Q=37500 cal/mol, m=4.8.

(3) Dynamically restored microstructure in Li-containing alloys was revealed as equiaxed subgrain structure under conditions where Li remained in solution without precipitating dynamically. Subgrain size varied from 15 to 30μm depending on Zener-Hollomon parameter.

(4) Dynamic and static restoration was retarded in hot deformation under conditions where δ phase remained undissolved or that δ phase precipitated dynamically. The microstructure in this case was shown as elongated pan-cake structure with subgrain size presumably less than 10μm, yielding potential capability for static recrystallization to finer grain size. It was suggested that such hot deformation utilizing δ precipitates might possibly provide an idea for a new thermo-mechanical processing suitable specifically for Aℓ-Li alloys.

(5) Hardness in direct-quenched and aged alloys remarkably increased with the deformation temperature decrease. It was suggested that suppression of the progress in both dynamic and static restoration enhanced the precipitation hardening due to δ' phase.

References

1) T.H.Sanders, Jr.; Proceedings of the 1st International Aluminum-Lithium Conference, ed. by T.H.Sanders,Jr. and E.A.Starke,Jr. (1980), P.63 [The Metallurgical Society of AIME]

2) S.F.Banmann and D.B.Williams; Proceedings of the 2nd International Aluminum-Lithium Conference, ed. by T.H.Sanders,Jr. and E.A. Starke,Jr. (1983), P.17 [The Metallurgical Society of AIME]

3) A.P.Divecha and S.D.Karmarkar; Proceedings of the 1st International Aluminum-Lithium Conference, ed. by T.H.Sanders,Jr. and E.A. Starke,Jr. (1983), P.17 [The Metallurgical Society of AIME]

4) H.J.McQueen; Proceedings of a Symposium on Thermomechanical Processing of Aluminum Alloys, ed. by J.G.Morris (1978), P.1 [The Metallurgical Society of AIME]

5) J.Wadsworth, I.G.Palmer, D.D.Crooks and R.E.Lewis; Proceedings of the 2nd International Aluminum-Lithium Conference, ed. by T.H.Sanders,Jr. and E.A.Starke,Jr. (1983), P.111 [The Metallurgical Society of AIME]

6) C.M.Sellars and W.J.McG.Tegart; Mem. Sci. Rev. Met., 63 (1966), 731.

7) H.Asada, R.Horiuchi, H.Yoshinaga and S.Nakamoto; Trans. J.I.M., 8 (1967) 159.

8) H.Yoshinaga and S.Morozumi; Phil. Mag., 23 (1971), 1351.

9) E.Di Russo, M.Conserva, M.Buratti and F.Gatto; Mat. Sci. Eug., 14 (1974), 23.

10) J.Waldman, H.Sulinski and H.Markus; Met. Trans., 5 (1974), 573.

11) J.A.Wert, N.E.Paton, C.H.Hamilton and M.W. Mahoney; Met. Trans., 12A (1981), 1267.

Extrusion processing of Al–Mg–Li alloys

N C PARSON and T SHEPPARD

NCP is with Alcan International,
Banbury, Oxon, UK.
TS is in the Department of Metallurgy
and Materials Science, Imperial
College, South Kensington,
London SW7, UK

Summary

This communication reports work covering the variation in structure and properties with extrusion processing of an Al-Mg-Li alloy and an Al-Mg-Cu-Li alloy. Extrusion was performed over a wide range of temperature compensated strain rates and it was established that the pressure required could be related by a single expression in terms of the constitutive equation constants. Limit diagrams were constructed for the alloys which indicated that the working range was reduced when the Cu addition was made. Both alloys were observed to be extremely strain-rate sensitive. The surface region of the extrudates was fully recrystallised over the entire process range investigated whilst the core exhibited recrystallisation at all but the highest extrusion temperatures. The extruded structure was thus influenced by the Al_3Zr dispersoids in particular and by second phase particles in general. Both alloys were difficult to recrystallise further after extrusion and the extruded structure remained essentially unchanged through the solution soak sequence. Consequently in the fully heat treated condition the strength and ductility showed a direct relationship with the extrusion variables. In both alloys the properties were related to the extrusion ratio as well as the temperature compensated strain rate. The addition of Cu to the ternary alloy did not result in any considerable solid solution strengthening but altered the ageing response such that the proof stress was raised but at the expense of ductility. Both alloys were capable of 7075 T6 strength levels under appropriate deformation conditions, but, particularly in the quarternary alloy, with reduced ductility. The modulus of the two alloys were similar and both had a density equivalent to 90% of that of AA 2014 alloy. Both alloys were insensitive to an intermediate cold stretching operation and both could be directly aged after press quenching to yield properties similar to those after solution soaking. In the peak aged conditon the Cu bearing alloy exhibited superior strength but with poor ductility and lower fracture toughness than the ternary alloy. Both alloys showed superior toughness properties in the under-aged condition.

Introduction

The effects of lithium additions to aluminium alloys were first studied in the 1920's when considerable effort was expended in an attempt to produce heat treatable alloys similar to duralumin. Throughout the 1940's and 1950's the development of the Al-Cu-Mn and Al-Zn-Mg-Cu systems dominated alloy research but in 1957 the commercial Al-Li-Cu alloy X2020 was introduced in the U.S.A; the only production application however being in the wing skins of a naval aircraft. Due to low ductility and fracture toughness it was withdrawn from service. Work in the U.S.S.R. during the 1960's led to the development of the commercial Al-Mg-Li alloy 01420 but the extent to which this was used in practical applications is unclear. Recent research has studied both Al-Mg-Li and Al-Li-Cu systems in some detail utilising both ingot and powder metallurgy routes and the latest results in the United Kingdom have indicated that the Al-Mg-Li-Cu system might offer the best combination of properties.

Dinsdale [1] studied the effects of Mg additions to the binary Al-2%Li alloy and reported a hardening rate of 50 MPa/wt%Mg in aged material indicating that Mg was producing more than a solid solution strengthening effect. Thompson and Noble [2] determined that the precipitation sequence in this alloy was

$$SSS \rightarrow \delta' \rightarrow Al_2MgLi$$

TABLE 1. Composition of Alloys, wt%

Li	Mg	Cu	Zr	Fe	Si	Al
2.17	3.82	-	0.17	0.12	01.0	bal
2.24	3.64	-	0.14	0.10	0.04	bal
2.3	3.53	-	0.17	0.11	0.05	bal
2.40	3.46	0.86	0.15	0.12	0.02	bal
2.46	3.45	0.84	0.15	0.12	0.02	bal
2.43	3.47	0.82	0.13	0.12	0.02	bal

TABLE 2. Hot Working Constants

$$Z = \dot{\varepsilon} \exp \frac{\Delta H}{RT} = A\,(\sinh[\alpha\sigma])^{n}$$

	Al-Li-Mg	Al-Li-Mg-Cu	Units
ΔH	168	153	kj/mole
α	0.033	0.025	m^2/MN
n	2.15	2.28	-
lnA	24.49	24.01	s^{-1}

TABLE 3. Variation in the T6 Tensile Properties with the Extrusion Conditions

Alloy	Proof Stress (MPa)	UTS (MPa)	% El	Conditions
Al-Li-Mg	490	540	5	R = 10, T = 325°C
	340	475	14	R = 60, T = 325°C
Al-Li-Mg-Cu	565	620	4	R = 10, T = 450°C
	490	580	6	R = 30, T = 350°C

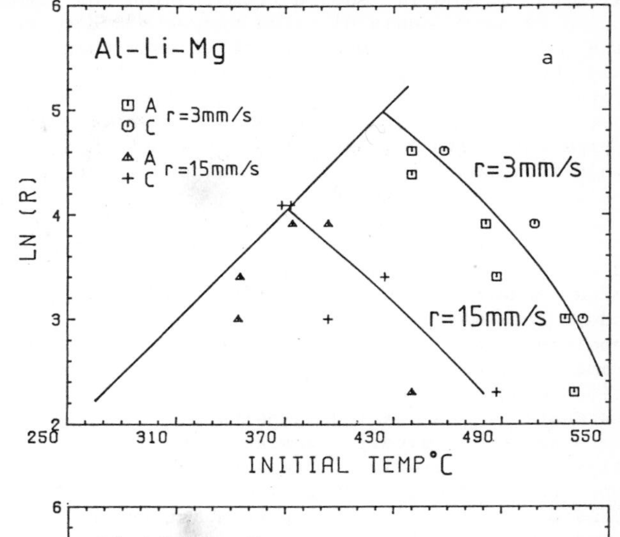

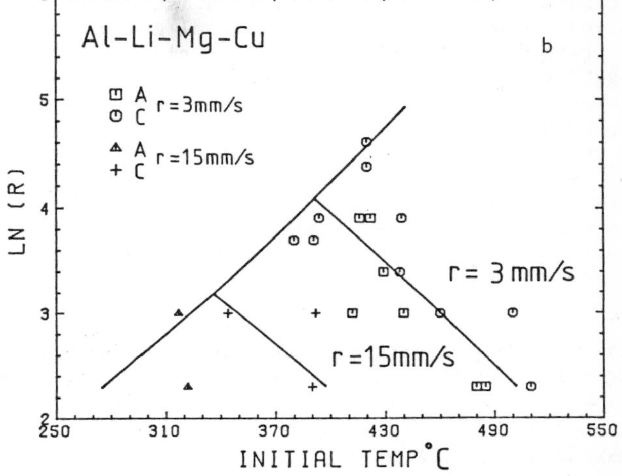

1. a. b. Experimental extrusion limit diagrams.

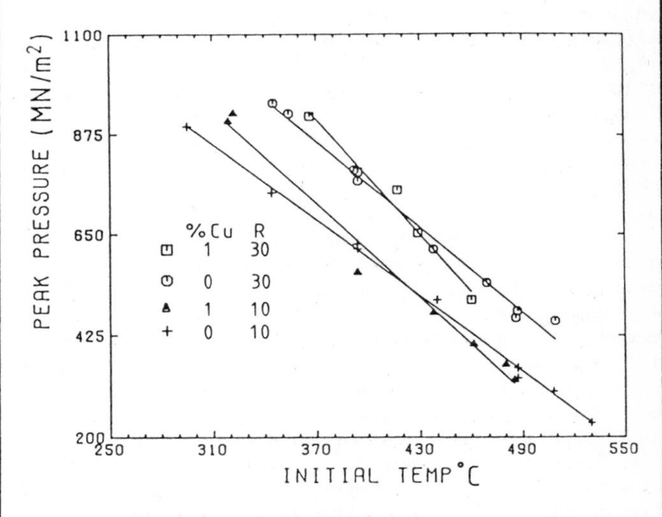

2. Variation in breakthrough pressure with initial billet temperature.

Work on the Al-Mg-Cu-Li quaternary alloy [3-7] has been extensive and a combination of two lithium ternary phase diagrams suggests that the alloy utilised in the present work lies within the $S(Al_2CuMg)$ and $T(Al_6CuMg_4)$ phase region. The reported precipitation sequences are predominantly those producing δ', S' and T, where S' is the intermediate form of the S phase. Strengthening is mainly from the Al_3Li phase with S' and T acting as slip dispersing agents.

Direct extrusion is simple in operation and complex shapes can easily be produced. In a typical modern transport aircraft the use of this product form may be as high as 28% [8]. Material properties are dependent upon the process parameters (temperature, ram speed and extrusion ratio) even after solution soak and ageing treatments. The various aspects of the process have been extensively reviewed together with analyses to predict breakthrough pressure, temperature rise and strain rate [9,10]. The extrusion of Al-alloys produces a structure with marked directionality [11]. Original grains are elongated parallel to the extrusion direction together with any inclusion particles present. Consequently material properties are strongly dependent on the test direction. The distribution of deformation is not uniform across the section and surface layers undergo considerably higher strains than elsewhere. This feature can give rise to recrystallised annuli. The fibre texture generated in the extrusion direction can add a marked contribution to strength.

The work reported in this communication was initiated to determine the effect of extrusion variables on the properties of Al-Mg-Li-Zr and Al-Mg-Li-Cu-Zr alloys.

Experimental Procedure.

Material was supplied by Alcan International in the form of logs machined from DC cast blocks. The compositions of the alloys are given in Table 1. Billets were homogenised in an air circulating furnace; the treatment for the Al-Mg-Li having been established in earlier work [12] to be 500°C for 24 hours. The Al-Li-Mg-Cu alloy contained a large volume fraction of coarse S-type particles and the optimum heat treatment was found to be 36 hours at 510°C. This treatment did not completely solutionise the S phase.

Extrusion was performed on a direct 5MN press with a 75mm diameter container. Ram speeds utilised were in the range 3-15mm s^{-1} and the extrusion ratio from 10:1 to 100:1 giving exit speed varying between 30 and 1500mm s^{-1}. Except in the cases where mechanical testing required square or strip sections all the extrusions were of round bar. Heating for extrusion was by induction giving a heat-up rate of 125°C/min and the electrically heated press container was maintained at 50°C below the initial billet temperature. All extrusions were press quenched.

A standard solution treatment of 1 hr at 510°C was applied to both alloys and ageing was performed at 170°C. Heat treatments were established from hardness/time data to give three tempers, underaged (UA), peak aged (PA) and overaged (OA). The UA and OA tempers correspond to 80% of the hardness increment produced by the peak strength temper.

Tensile testing was performed at a strain rate of 5×10^{-4} s^{-1}. Toughness measurements were made in the L-T direction using single edge notch bend specimens produced from 12mm extruded square sections. The data produced did not therefore correspond to valid K_{1C} results.

Results and Discussion

Extrusion

The range of process conditions under which an alloy can be extruded is determined by the available pressure and the onset of undesirable surface features such as incipient melting, surface tearing or unacceptable recrystallisation. Hence the effect of processing upon structure and properties must be determined after these conditions have been established. Figs.1a and 1b are experimental extrusion limit diagrams for the ternary and quaternary alloy respectively, constructed for a maximum press capacity of 1000 MPa and conveniently combining the basic process parameters (ram speed, temperature and extrusion ratio). The temperature limits have been constructed using the onset of surface cracking as the criterion and loci are shown for the extremes of ram speed utilised. The variation in ram speed permissible does not affect the strain rate sufficiently for it to be discernable in the pressure loci. Surface finish was simply classified as either cracked or acceptable. The most common mechanism of cracking is hot shortness when excessive localised heating of the surface causes liquation of non-equilibrium low temperature phases but cracking occurs before such localised temperatures are reached because friction in the die land area results in material exhibiting low flow stress characteristics being unable to resist the shear stresses. At low ram speeds heat dissipation, both in the shear area of the dead metal zone and in the die load area is possible and hence the material is particularly sensitive to ram speed as shown in Fig.1. The working limits of both alloys is considerably reduced when ram speed is raised from 3 mm s^{-1} to 15 mm s^{-1}.

A comparison of Figures 1 indicates that the working limits of the quaternary alloy are considerably more restricted than those for the ternary. The Al-Mg-Li-Cu alloy cannot be extruded at ratios greater than 50:1 even at the lowest ram speed. This can be attributed to the rapid softening of the alloy at high temperatures due to the large fraction of S phase remaining after homogenisation. The pressure limits, determined by the flow stress of the alloys at

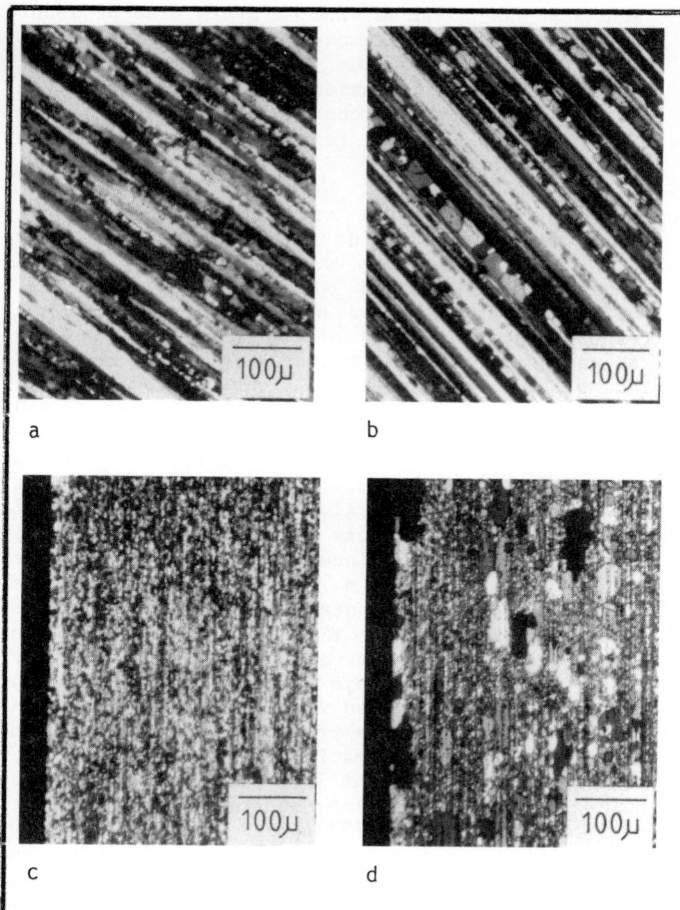

3. As-extruded grain structure, Al-Li-Mg, R = 20:1.

a. 500°C, 5mm/s, extrudate core
b. 325°C, 15mm/s, extrudate core
c. 500°C, 5mm/s, extrudate surface
d. 325°C, 15mm/s, extrudate surface.

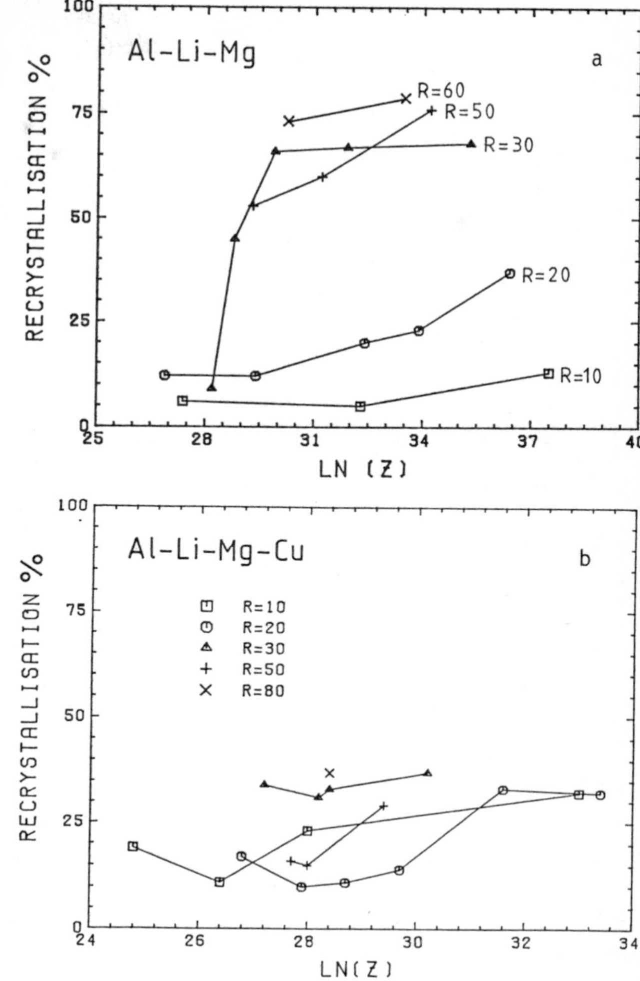

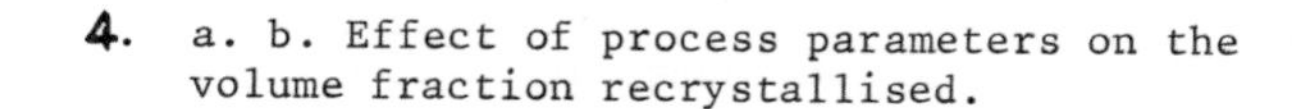

4. a. b. Effect of process parameters on the volume fraction recrystallised.

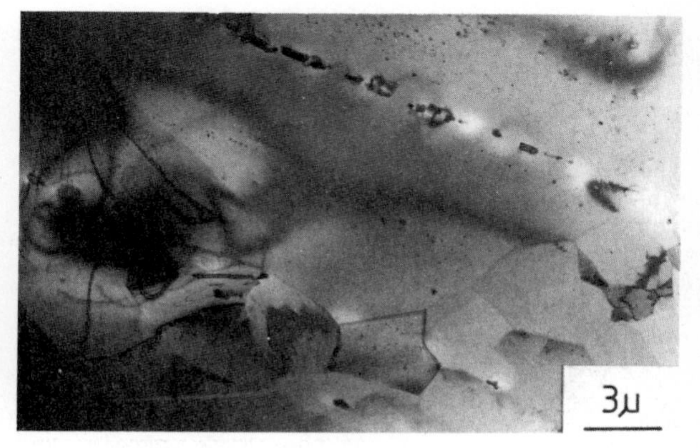

a

b

5. a. b. TEM micrographs showing recrystallised grains.

a. Al-Li-Mg T6, 450°C, 5mm/s, 80:1.
b. Al-Li-Mg as-extruded, 5mm/s, 30:1.

lower temperatures were similar for both alloys; breakthrough pressure requirements being about the same in each case. Fig.2 defines the variation in extrusion pressure required at varying initial billet temperature for the two alloys: at extrusion ratios of 10:1 and 30:1. The data may be fitted linearly. For any given process conditions the breakthrough pressure for each alloy is similar but it can bé observed that the Cu addition has the effect of raising the required pressure at lower temperatures and lowering this requirement at high temperatures.

The flow stress variation with strain rate and temperature during steady state may be described by the constitutive equation which was established by hot torsion testing using a method previously described [13]. The constants in the constitutive equation are given in Table 2 for each alloy. The desired constants should be reviewed as an average for the temperature-strain rate range investigated (300°C - 500°C and 0.03 - 29 s^{-1}) and the similarity of the constants for the two alloys suggests that during hot working the addition of 1% Cu to the ternary has little effect on the frequency with which obstacles are surmounted, the obstacles encountered or the relative ease of dislocation movement during high temperature deformation. Consequently the activation energies for mechanical deformation are also similar. These activation energies may now be utilised to calculate the temperature compensated strain rate, Z, thus enabling the process variables to be incorporated in a single parameter and hence generalise data from differing temperature/ram speed experiments.

The Effect of Varying Process Conditions on the Extruded Structure

Fig.3 indicates the variations in the extrudate core and surface structures produced in the Al-Li-Mg alloy extruded at a ratio of 20:1. Similar variations were observed in the Cu-containing alloy.

The core structures were partially recrystallised to varying extents for all process conditions. At the highest extrusion temperatures (Fig.3a) very little recrystallisation occurred and when observed was in the form of small equiaxed grains of about 5μm generally located near original grain boundaries. Reducing the deformation temperature (Fig.3b) increased the volume fraction recrystallised and grain sizes were as large as 40μm. The recrystallisation appeared as a banded structure corresponding to the extrusion direction.

The periphery of the extrudate recrystallised under all imposed process conditions; a result of the high strain and strain rate experienced by the surface material as it traverses the deformation zone. The morphology of the layers varied with the extrusion processing conditions At extrusion temperatures above 450°C(Fig.3c) a very fine equiaxed structure was produced with a grain size of about 5μm whilst at lower temperatures this coarsened, as shown in Fig.3d, due to discontinuous grain growth resulting in considerable variations in the grain size. The Cu-containing alloy exhibited similar trends but in general the surface grain sizes were larger.

The extrusion ratio had an independent and strong influence on the extruded structure which could not be related directly to the temperature compensated strain rate as is normal for Al-alloys. Figs. 4a and 4b illustrate volume fraction recrystallised of the core material and the variation of this factor with extrusion ratio and with processing conditions. The figures represent the analysis of the as-extruded structure but, since recrystallisation did not proceed further during heat treatment nor was any grain growth observed, the results after heat treatment were identical. This indicates that recrystallisation occurs very rapidly after the extrusion process since products were press quenched or, that a dynamic mechanism is responsible for recrystallisation. It should also be noted that similar structures were produced when the extrusion was allowed to air cool. In both alloys the volume fraction recrystallised increased with increasing extrusion ratio and decreasing extrusion temperature. Both effects were more dramatic in the ternary alloy. At an extrusion ratio of 30:1 extrudates processed below 500°C contained very substantial fractions of recrystallisation (75%). This was significantly greater than when extruding at 20:1. Raising the reduction ratio above 30:1 produced more softening suggesting that the event may be strain induced. The most likely cause of the recrystallisation would appear to be a dynamic mechanism but no evidence of this was observed and hence further study is clearly required. If the event is static then the most important factor determining the volume fraction recrystallised is the stored energy due to deformation since contrarily the factor increases with decrease in temperature. Comparison of the structures produced at high and low temperatures suggests that nucleation in each case occurs near original grain boundaries when there is a substantial driving force in the form of a higher dislocation density. Thermal activation at high temperature does not appear to promote high angle grain boundary migration and this must be attributed to the pinning action of the Al_3Zr precipitates.

The unrecrystallised regions of the extrudate contained well recovered subgrains which were observed to decrease with increasing Z and to follow the usual form of relationship;

$$\text{for Al-Li-Mg} \quad d^{-1} = 0.037 \ln Z - 0.67 \ (\mu^{-1})$$
$$\text{for Al-Li-Mg-Cu} \quad d^{-1} = 0.041 \ln Z - 0.77 \ (\mu^{-1})$$

An interesting feature of the recrystallisation was its banded nature which can be observed in Fig.3b. Transverse electron micrographs of this feature are shown in Figs.5a and 5b for differing process conditions. The newly formed grains occur in laths between what

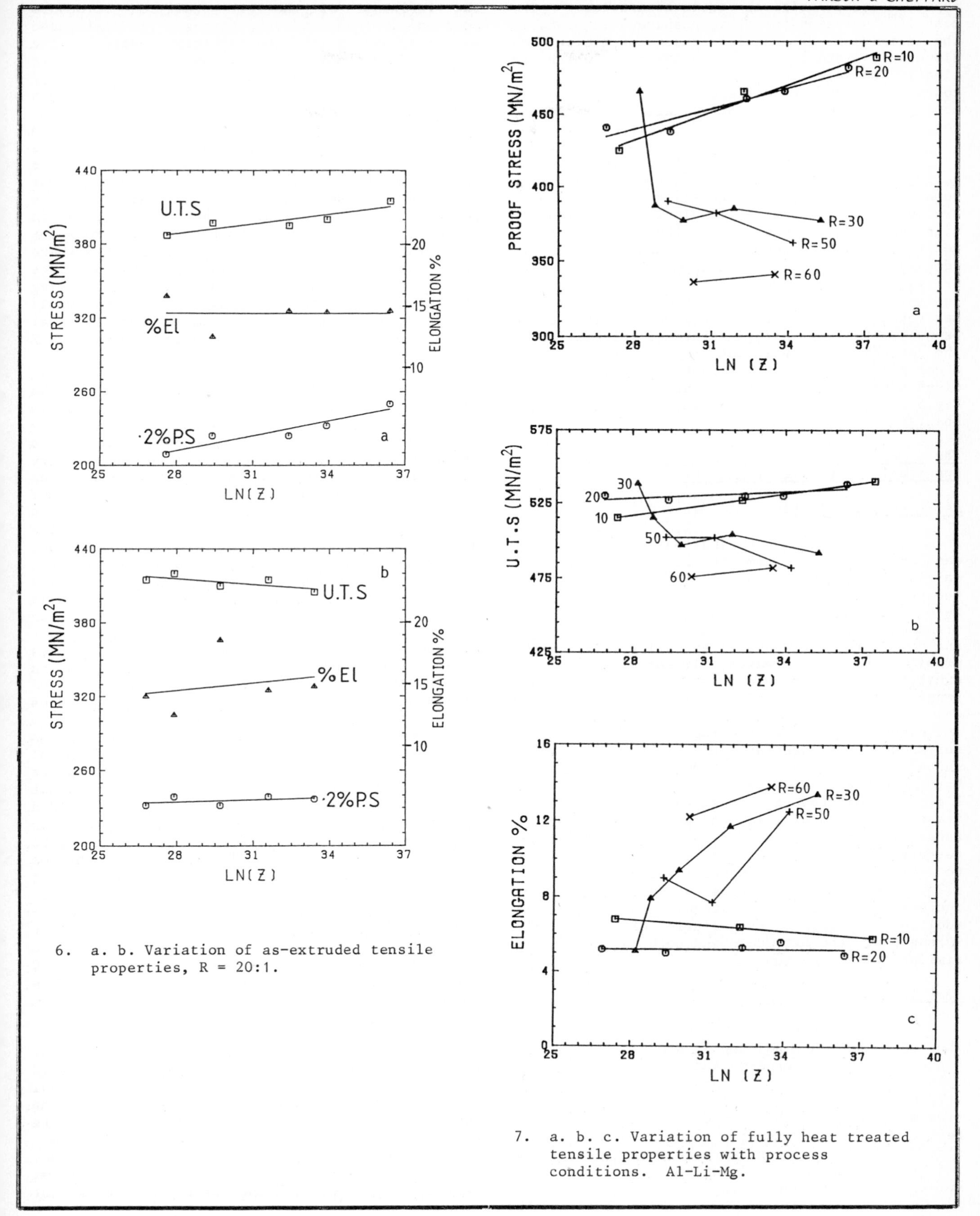

6. a. b. Variation of as-extruded tensile properties, R = 20:1.

7. a. b. c. Variation of fully heat treated tensile properties with process conditions. Al-Li-Mg.

appears to be original grain boundaries. The most likely reason for this phenomena is probably the alignment of second phase particles and dispersoids into the extrusion direction which can clearly be seen in the figures. There is of course a texture difference across a grain boundary which can also influence recrystallisation.

The extruded structures were stable during solution treatment, even the subgrain size remaining unchanged. The Al_3Zr precipitates are hence very effective in pinning both low and high angle grain boundaries. The question of how recrystallisation occurs in these alloys is certainly not clear. It appears that localised driving forces promote an initial surge of recrystallisation and when these areas have been destroyed the structure stabilises. This could be due to the non-uniform nature of deformation in the extrusion process but is not observed in other alloy systems. Alternatively a non-uniform dispersion of Al_3Zr precipitates as shown in Fig.5 could allow the localised recrystallisation observed.

The Effect of Process Parameters on Longitudinal Tensile Properties

As-Extruded Properties

Mechanical testing was performed for each alloy extruded from the 20:1 extrusion ratio matrix. The contribution to final properties of the extruded structure could thus be established before being partially masked by the heat treatment. The results are presented in Figs.6a and 6b. For the copper-free alloy (Fig.6a) the proof stress increased linearly with lnZ exhibiting a variation of 40 MPa over the working range. At this extrusion ratio recrystallisation was relatively low and any softening has been outweighed by strengthening due to subgrain refinement. In contrast the quaternary alloy showed only a small variation in strength with varying process conditions although reference to Fig.4b indicates a consistently low volume fraction recrystallised throughout the working range. The proof stress was 15 MPa higher than the ternary alloy at low Z conditions, presumably due to solid solution strengthening but the ternary alloy was capable of this strength level with increased substructural strengthening. Thus the addition of copper has little effect on the strength of the extrudate and both alloys showed similar 14% elongation.

Heat Treated Properties

The variation in tensile properties in the heat treated condition is shown in Fig.7 and Fig.8 for the ternary and quaternary alloy respectively. As indicated above the extruded structure is substantially retained after heat treatment and for the copper-free alloy this is reflected in the tensile results. The proof stress shows a linear increase with lnZ at low reduction ratios indicating the retained substructural strengthening. At higher ratios the influence of the recrystallised structure becomes dominant and the strength is diminished with increasing lnZ and extrusion ratio R. The amount of softening at high extrusion ratios is quite considerable; at a lnZ value of 34 the proof stress falls from 450 MPa at R = 10:1 to 350 MPa at R = 60:1. This difference is not entirely due to substructural effects because recrystallisation also destroys the contribution from fibre strengthening. The UTS varied in a similar manner but the range of values encountered was narrower because greater work hardening was possible in the softer structures. Elongation showed an opposite trend to the proof stress but it is worth noting that at low values of R where recrystallisation is limited strength increases are possible without loss in ductility. Processing of this alloy for optimum properties would therefore appear to be more complex than is usual in Al-alloys; low extrusion ratios and a high Z producing the best results.

The variation in proof stress of the Al-Mg-Li-Cu alloy is quite different (Fig.8a). In the ternary alloy the proof stress appeared to be the sum of a fixed age hardening term and a variable substructure term but in the quaternary both terms seem to vary with processing conditions. For all extrusion ratios the proof stress increases as the temperature is raised (lower Z) although in the extruded condition the strength was invariant with lnZ. The effect must therefore be attributed to a feature of the hot worked structure not connected with subgrain size or volume fraction recrystallised. It is probable that the process heating solutionises a proportion of the cast S phase particles remaining after homogenisation. Higher extrusion temperatures would allow more Mg and Cu to go into solution and thus produce a greater ageing response. The diffusion may even be aided by the high stresses encountered in the extrusion process. Fig.8a also indicates a very rapid drop in strength at high temperatures and low extrusion ratios which might also suggest that stress is an important factor since under these conditions we would expect solutionising to be enhanced. However TEM did not reveal any significant change in structure for these conditions. The UTS values shown in Fig.8b showed a similar trend. The range of properties obtained by varying the process conditions whilst still maintaining a sound product are given in Table 3 for each of the alloys. The copper addition raised the proof stress by 50 MPa over the condition in which the ternary alloy exhibited peak strength. It was possible to achieve a 100 MPa increment but this was associated with a severe ductility loss.

The additional strengthening achieved in the quaternary alloy must be caused by greater precipitation hardening since it was not observed in the as extruded condition. Additional precipitation sequences may be induced by the copper addition or it is possible that the existing Al_3Li kinetics may be affected. The ductility of the system is however not improved. Fig.9a and 9b are ageing curves for the alloys at varying temperatures. The curves are similar but

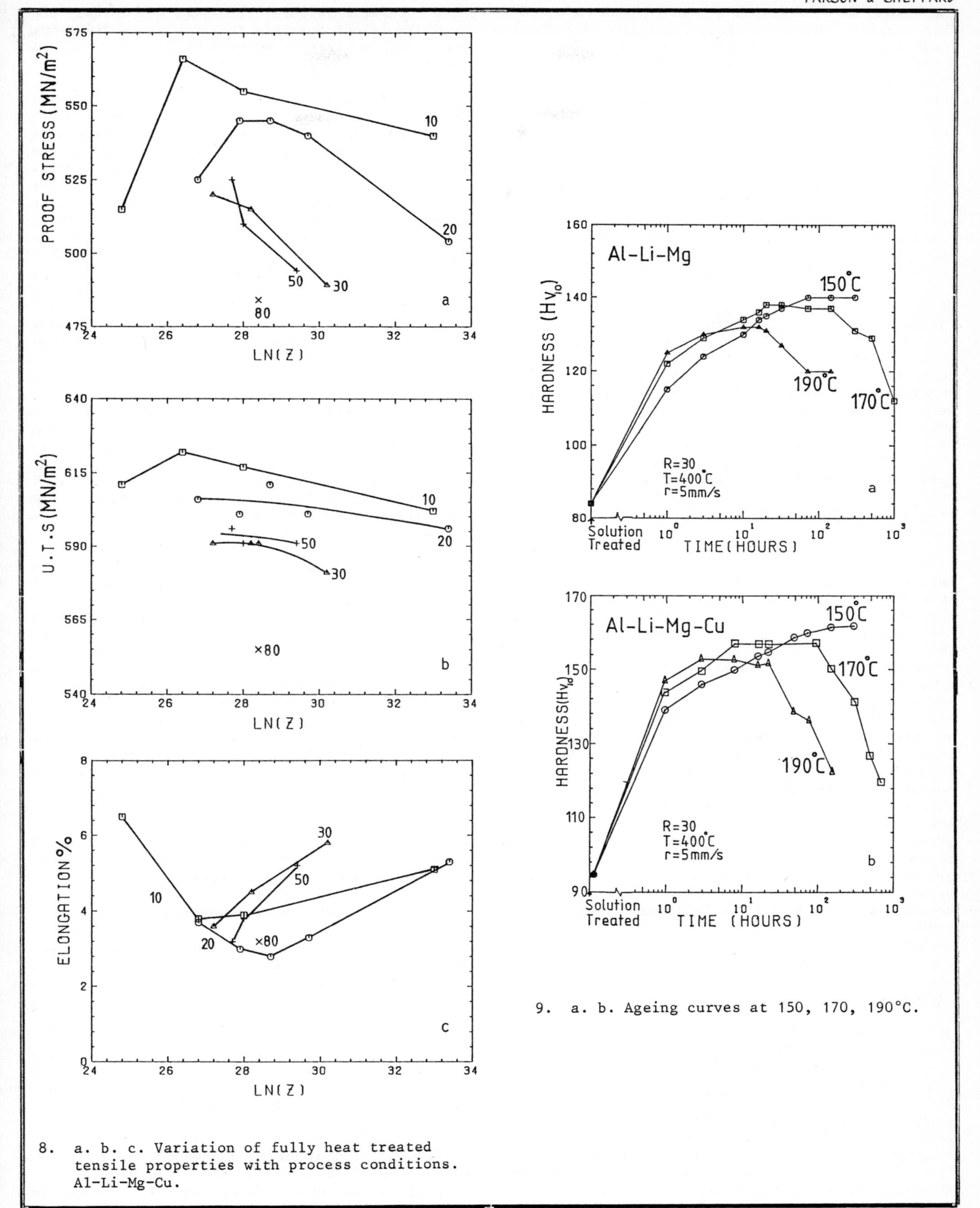

8. a. b. c. Variation of fully heat treated tensile properties with process conditions. Al-Li-Mg-Cu.

9. a. b. Ageing curves at 150, 170, 190°C.

the quaternary exhibits a higher peak hardness. This suggests that the ageing process is similar in both alloys and the TEM micrographs of the two alloys shown in Figs.10a and 10b support this thesis showing almost identical microstructures. Extensive δ' precipation in the matrix and coarse Al_2MgLi particles preferentially formed at high angle boundaries can readily be observed. There does not appear to be additional precipitation in the Al-Mg-Li-Cu alloy and the strength increase must therefore be attributed either to a change in the δ' distribution or to the precipitate structure. Further work is clearly required to clarify this point but this study suggests that the precipitation sequence in the Al-Mg-Li alloy is also dominant in the quaternary alloy.

Effect of Process and Heat Treatment Parameters on the L-T Toughness

The results of the toughness test performed on the two alloys in the UA, PA and OA conditions are shown in Figs. 11a and 11b. Tests were performed on samples which covered the possible limits of extrusion. The extrusion of square bar considerably narrowed the working range of the quaternary alloy thus limiting the quantity of data which could be generated. For comparative purposes a 2014 alloy extrusion was processed and tested together with the Al-Li specimens.

Fig.11a reveals that overaging the ternary alloy results in serious embrittlement even though the proof stress is lowered and the matrix plasticity improved. This is caused by the high density of Al_2MgLi precipitates found in the matrix and at grain boundaries in the overaged specimen. Harris et al [14] were able to reach the same conclusion when working on this alloy system. The results for this alloy system are somewhat confusing because two sets of values for the UA condition are shown. In this temper a large load drop or 'pop-in' was observed on the load-displacement trace whereas the other tempers showed a gradual increase to the maximum load. This was associated with an audible click during the test. Fractography revealed that the phenomenon was being caused by the crack front being diverted 90° into the extrusion direction as shown in Fig.12a. The extruded grains have separated in a completely brittle manner suggesting an inherent grain boundary weakness in these alloys. Presumably this 'splitting' mechanism was only observed in the UA condition because the toughness in the transverse direction was sufficiently high in this temper to allow the directional change to be activated. The values of K_Q calculated for this condition were significantly lower than those for the PA temper even when calculated using the maximum loads which are also shown in the figure. This is mainly because the rapid crack growth altered the specimen geometry. Without this peculiar mode of fracture there is no doubt that toughness values would have exceeded those in the PA condition because in Al-alloys the UA temper is generally toughest due to the yield stress effect. However the redirection of the crack front could lead to quite dramatic failure in large components and hence it is not possible to draw any conclusions about the relative toughness of PA and UA tempers.

In all three tempers the material exhibited reduced toughness at the lowest extrusion temperatures which initially seems incongruous since this is also the temperature which corresponds to the lowest proof stress values. However both effects are due to the increased volume fraction recrystallised experienced under these processing conditions. At extrusion temperatures of 500^oC, 450^oC and 400^oC the recrystallisation was observed to be 5, 25 and 55% respectively. Fig.12b and 12c depict the fracture surfaces of high and low temperature extrudates in the PA temper. Fig.12b (high temperature) shows clear evidence of ductile tearing whilst in Fig.12c smooth intergranular facets due to the brittle separation of favourably orientated recrystallised boundaries can be identified. This also suggests the thesis that these alloys contain a fundamental grain boundary weakness which clearly is not confined to original grain boundaries. The toughness of the ternary alloy can approach that of 2014 T6 but we should note that this is in the L-T direction and short transverse properties could be expected to be significantly poorer due to the observed grain boundary weakness.

The Al-Mg-Li-Cu alloy (Fig.11b) showed a clear trend of decreasing toughness with increasing ageing time. In this alloy also the crack front was redirected into the extrusion direction in the UA temper but complete specimen failure was so rapid in all tests that this feature did not distort the results. The toughness is much lower than that encountered in the ternary alloy with a difference of 18 MPa in the PA temper. This is partly caused by the higher inclusion content of the alloy but the increased yield stress must be the major factor.

Conclusions

1. The extrusion limits of the two alloys have been established. The addition of copper to the Al-Li-Mg ternary reduced the working range primarily due to the presence of the S phase. Conversely the addition of copper had little effect on the high temperature flow stress.

2. All extrusions were partially recrystallised. The volume fraction recrystallised was affected by the extrusion parameters; increasing with decreasing temperature and increasing extrusion ratio.

3. The extruded structures were stable during heat treatment presumably due to the presence of Al_3Zr. Thus the heat treated tensile properties could also be related to the extrusion conditions.

4. The addition of copper raised the proof stress whilst reducing ductility. The ageing sequence in both alloys corresponded to that established for

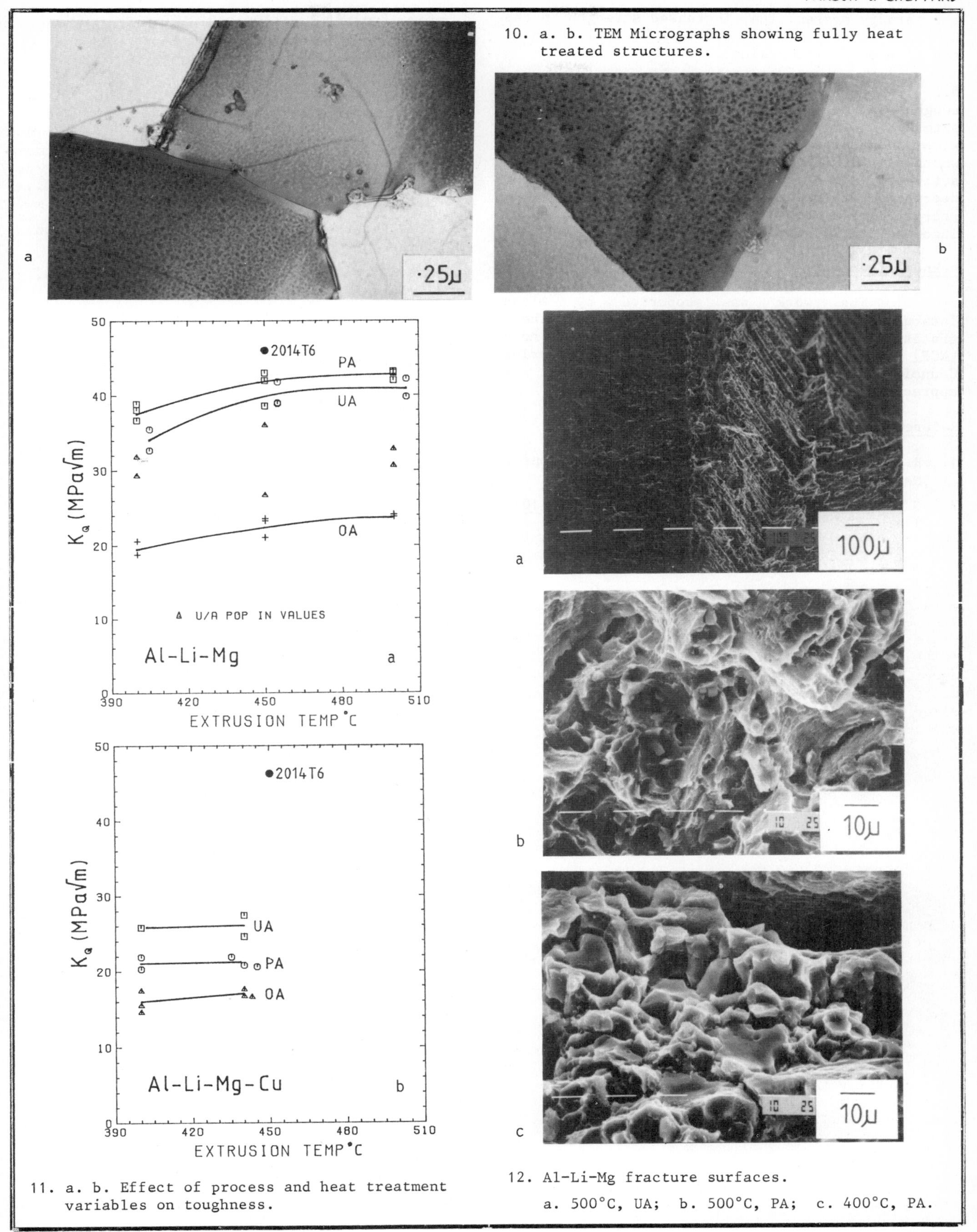

10. a. b. TEM Micrographs showing fully heat treated structures.

11. a. b. Effect of process and heat treatment variables on toughness.

12. Al-Li-Mg fracture surfaces.
a. 500°C, UA; b. 500°C, PA; c. 400°C, PA.

the ternary system. The increased strength in the Cu alloy appears to be due to a modification of the δ' precipitation kinetics.

5. The quaternary alloy exhibited poorer toughness, mainly due to the increased yield stress. In both alloys toughness decreased with increasing ageing time. Grain boundary weakness was detected in both alloys and was activated in the UA condition. Toughness decreased with decreasing extrusion temperature as recrystallised grain boundaries were introduced into the crack direction.

Acknowledgements

The work was supported by Alcan International Ltd. and the authors express their gratitude for this support. One of the authors (NCP) was in receipt of a Science and Engineering Council research grant and this support is appreciatively recognised.

References

1. K. Dinsdale et al, Proc. 1st Int. Al-Li Conf., Met.Soc., AIME, 1982, p.219.
2. G.E. Thompson and B. Noble, J.I.M., 1973, 101, 111.
3. C.S. Peel et al, Proc. 2nd Int.Al-Li Conf., Met. Soc., AIME, 1984, p.363.
4. W.S. Miller et al, ibid, p.335.
5. I.G. Palmer et al, ibid, p.91.
6. R.J. Kav et al, ibid, p.255.
7. I.J. Polmear, Light Alloys, publ. Arnold, (1981), p.127.
8. W.E. Quist et al, Proc. 2nd Int. Al-Li Conf., Met. Soc, AIME, 1984, p.363.
9. S.J. Paterson, Ph.D. Thesis, Univ. of London, (Imperial College), 1981.
10. M.G. Tutcher, Ph.D. Thesis, Univ. of London, (Imperial College), 1979.
11. P.J.E. Forsyth, C.A. Stubbington, Met.Tech., 1975, 2, 158.
12. N.C. Parson and T. Sheppard, Proc. 2nd Int. Al-Li Conf., Met.Soc., AIME, 1984, p.53.
13. T. Sheppard and D.S. Wright, Met. Tech., 1979, 6, 215.
14. S.J. Harris et al, Proc. 2nd Int. Al-Li Conf., Met.Soc, AIME, 1984, p.219.

Cyclic deformation of binary Al–Li alloys

J DHERS, J DRIVER and A FOURDEUX

The authors are with
Ecole Nationale des Mines de
Saint-Etienne, 158 Cours Fauriel,
42023 Saint-Etienne Cédex 2, France

SYNOPSIS

The mechanical and microstructural cyclic behaviour of two Al-Li alloys has been investigated:
(i) a solid solution, Al-0, 7 wt % Li
(ii) a precipitation hardened Al-2.5 wt % Li.

The latter was examined in two conditions after a solution treatment :
(a) room temperature aged to precipitate very fine (2-5 nm diameter) δ' ;
(b) aged 1 hour at 200°C, to form both homogeneous δ' ($\simeq$ 20 nm diameter) and δ precipitates along the grain boundaries.

The solid solution exhibits different hardening-softening behaviours according to the plastic strain amplitude. The overall results on Al-0.7 Li are quite similar to those obtained on pure Al. It is concluded that Li additions in solid solution have only a small influence on the fatigue properties.

The cyclic response of the precipitation hardened alloys is very sensitive to the ageing treatment. The alloy with fine δ' precipitates exhibits permanent hardening and planar slip. The aged alloy containing both δ' and δ precipitates exhibits intergranular brittleness and substantially reduced fatigue lifetimes.

INTRODUCTION

There has been a rapidly increasing interest, over the last few years, in adding lithium to high strength aluminium alloys. As is well known, Li additions have significant beneficial effects (lower density, higher elastic modulus) but often lead to low ductility. In order to understand the complex deformation mechanisms of Al-Li-X alloys fundamental studies of the deformation behaviour of simple binary alloys are required. For this reason we have carried out a basic study of the cyclic plastic behaviour of binary Al-Li alloys.

EXPERIMENTAL TECHNIQUE

Two alloy compositions were used in this study :
- Al 0.7 wt % Li (2.7 at % Li) : a solid solution and
- Al 2.5 wt % li (9.1 at % Li) : containing coherent and shearable δ' precipitates.

The Al-Li alloys were extruded to 20 mm diameter bars at 430°C. The bars were solution heat treated for 2h at 533°C in a molten salt bath, and then quenched in cold water. Some of the Al-2.5 Li specimens were then aged at room temperature (underaged) and others aged for 1h at 200°C, to increase the size of δ' precipitates (aged specimens).

The tensile characteristics of these alloys are given in Table 1.

The grain size was quite large in all cases: 0.5 mm for the low lithium and 0.8-1 mm for the high lithium alloy . Cylindrical fatigue specimens, 6 mm in diameter, with 15 mm gauge length, were machined from the bars. Flats were machined on a few specimens to provide a surface for in-situ optical microscopy observations. All specimens were mechanically polished and electropolished before testing. Symmetrical push-pull fatigue tests were carried out under constant plastic strain control, using a servohydraulic MTS machine. Axial strain was measured with LVDT transducers attached to the specimen shoulders. Fatigue tests were performed in the plastic strain range : $5.10^{-5} < \frac{\Delta\varepsilon_p}{2} < 10^{-2}$, and at frequencies : 0.3 < f < 1 Hz, strain rate having only a small influence on the cyclic stresses and lifetimes.

An optical microscope was also set up on the fatigue machine in order to observe surface evolution during cycling. Complementary surface observations have also been carried out by SEM. The microstructure was characterized by TEM, on foils prepared from discs cut perpendicular to the stress axis, using a JEOL 100C microscope.

TABLE 1

	E (MPa)	$\sigma_{0,2\%}$ (MPa)	σ_{max} (MPa)	Elongation (%)
Al-0.7 Li	71040	45	65	26
Al-2.5 Li underaged	74930	67	157	33
Al-2.5 Li 1h at 200°C		185	220	2.6

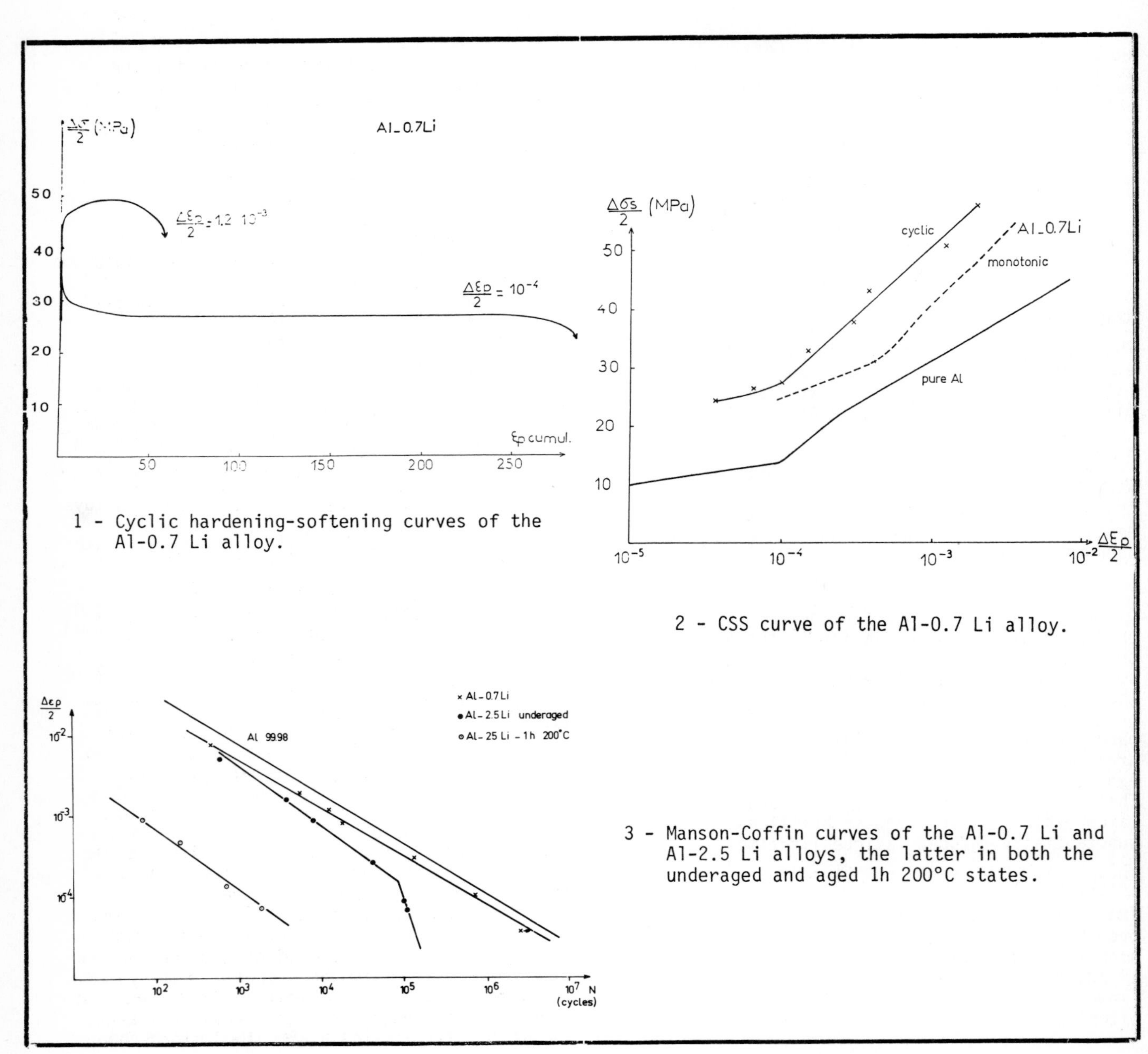

1 - Cyclic hardening-softening curves of the Al-0.7 Li alloy.

2 - CSS curve of the Al-0.7 Li alloy.

3 - Manson-Coffin curves of the Al-0.7 Li and Al-2.5 Li alloys, the latter in both the underaged and aged 1h 200°C states.

Al-0.7 Li

Mechanical behaviour

Two typical cyclic hardening and softening curves of the Al-0.7 Li alloy are given in Figure 1.

For $\frac{\Delta\varepsilon_p}{2} < 10^{-3}$ rapid initial hardening is followed by slight softening and then saturation.

For $\frac{\Delta\varepsilon_p}{2} > 10^{-3}$ hardening is permanent, and no saturation occurs before softening due to crack initiation.

A cyclic-stress-strain curve has been established plotting the saturation stress, or, if no saturation occurs, the maximum stress, versus the plastic strain. As shown in Figure 2 the CSS curve appears quite similar to that of pure aluminium polycristals[1]. There is a distinct decrease in the slope at plastic strains below $\simeq 10^{-4}$, but no plateau behaviour characteristic of Cu and Al single crystals[5,10].

The plastic strain amplitude ($\frac{\Delta\varepsilon_p}{2}$) versus the cycles to failure (Nf) (Manson-Coffin curve) is shown in Figure 3. The relation deduced from this curve is $\frac{\Delta\varepsilon_p}{2} = 0.31\ (Nf)^{-0.6}$. The plot is parallel and close to that of pure Al[1].

Fatigue crack nucleation sites in Al-0.7 Li are similar to those of ductile fcc polycristals such as Copper[2] and Al[1] :

For $\frac{\Delta\varepsilon_p}{2} < 10^{-3}$ cracks nucleate in the intrusions of slip band. For $\frac{\Delta\varepsilon_p}{2} > 10^{-3}$. Cracks nucleate along grain boundaries. Small cracks are also observed in the slip bands which impinge on the grain boundaries.

Microstructure

The absence of detectable diffraction spots (other than those of Al) by TEM, confirmed that Al-0.7 Li is essentially a solid solution.

Observations on specimens fatigued to failure over the entire range of plastic strain amplitudes revealed a dislocation cell structure, with somewhat coarser walls, i.e. less well polygonized, than in pure Al (Figure 4). At $\frac{\Delta\varepsilon_p}{2} = 10^{-3}$ the cells tend to be elongated, but neither veins nor ladder-like PSBs are observed.

Dislocations cells have previously been observed in fatigue Al polycrystals[3,4], and more recently elongated cells have been observed in fatigued Al single crystals[5]. It would appear that Al-0.7 Li exhibits similar microstructural behaviour to that of pure Al, which is typical of high stacking fault energy fcc metals. However the rather coarse walls found in Al-0.7 Li and the higher cyclic flow stresses indicate a slight influence of Li atoms in solid solution which may be due to a short range order effect or to a reduced stacking fault energy.

Underaged Al-2.5 Li

Mechanical behaviour

All specimens of the underaged Al-2.5 Li alloy exhibit rapid initial hardening followed by slow but persistent hardening, up to failure (Figure 5). The CSS ($\frac{\Delta\sigma}{2}$ max versus $\frac{\Delta\varepsilon_p}{2}$) is given in Figure 6.

The cyclic softening behaviour previously observed on other precipitation hardened Al alloys, due to strain localization within the bands of sheared precipitates[6] has not been observed.

The Manson-Coffin curve of the underaged Al-2.5 Li alloy is given in Figure 3. A change in slope seems to occur at plastic strain amplitudes, $\frac{\Delta\varepsilon_p}{2} < 2.10^{-4}$. The strain life relation deduced from the curve (for $\frac{\Delta\varepsilon_p}{2} > 2.10^{-4}$) is :

$$\frac{\Delta\varepsilon_p}{2} = 0.39\ (Nf)^{-0.68}.$$

The alloy shows a loss of fatigue resistance compared to the solid solution, particularly at low plastic strains.

For $\frac{\Delta\varepsilon_p}{2} > 2.10^{-4}$ cracks nucleate essentially along grain boundaries, although a few cracks have also been observed within the slip bands.

For $\frac{\Delta\varepsilon_p}{2} < 2.10^{-4}$ the alloy is very sensitive to small surface defects on which cracks initiate. This may explain the slope change in the Manson-Coffin curve for $\frac{\Delta\varepsilon_p}{2} < 2.10^{-4}$, where relatively early failure occurs.

Observations carried out by in situ microscopy and SEM reveal that very few cracks appear during cycling, particularly at low plastic strains. The low density of secondary cracks tends to confirm the high fatigue notch sensitivity of this alloy ; the first microcrack that appears often propagates rapidly to final rupture, before the initiation of secondary cracks.

MICROSTRUCTURE

TEM of the room temperature aged alloy before cycling indicates the presence of a high density of homogeneously distributed very fine (2-5nm diameter) δ' precipitates. The diffraction pattern exhibits superlattice reflections, at positions corresponding to those of the δ' ordered structure, and from which precipitates could be just resolved in dark field.

During fatigue intense slip bands are developed parallel to the (111) planes (Figure 7). Slip is essentially planar. Many paired superdislocations are observed together with a few loops. The deformation mechanism is almost certainly δ' shearing by dislocations, but this is very difficult to confirm directly. Dark field images from δ' superlattice spots do not reveal precipitate denuded slip bands[7] ; δ' precipitates may be sheared without dissolution, or alternatively, if dissolution does occur δ' precipitates may reform at low temperatures.

DISCUSSION

The underaged Al-2.5 Li alloy exhibits a lower fatigue resistance than that of the solid solution, although qualitatively, the strain seems less localized in the former alloy : PSBs are not observed, and no softening occurs during cycling deformation. The quite homogeneous deformation may be related to the large volume fraction of fine δ' precipitate (15 %) and eventually to the competition between δ' shearing and δ' reprecipitation. The loss of fati-

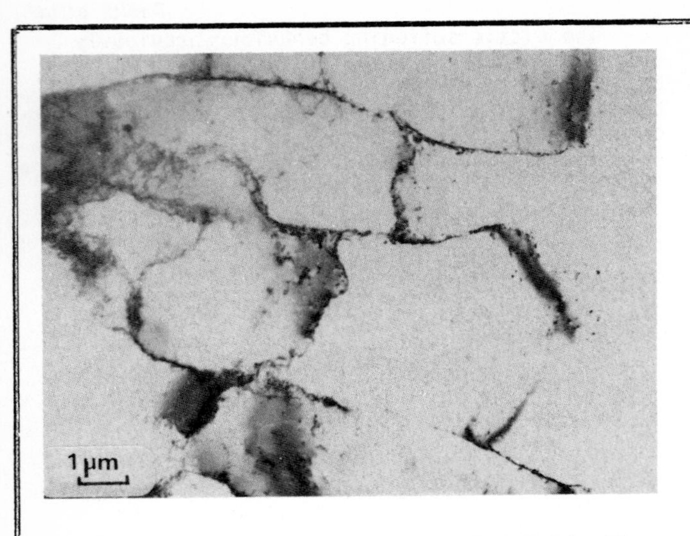

4 - Cells observed by TEM in the Al-0.7 Li alloy, cycled to failure at $\frac{\Delta\varepsilon_p}{2} = 10^{-3}$.

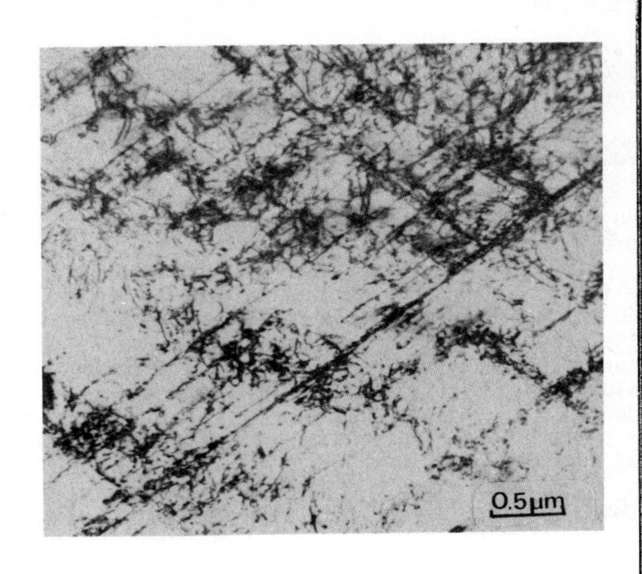

7 - Bands observed by TEM on the underaged Al-2.5 Li cycled to failure at $\frac{\Delta\varepsilon_p}{2} = 8.5\ 10^{-5}$.

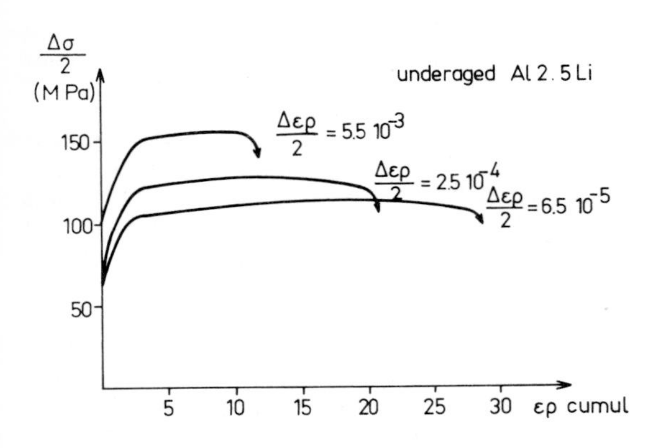

5 - Cyclic hardening curves of the underaged Al-2.5 Li alloy.

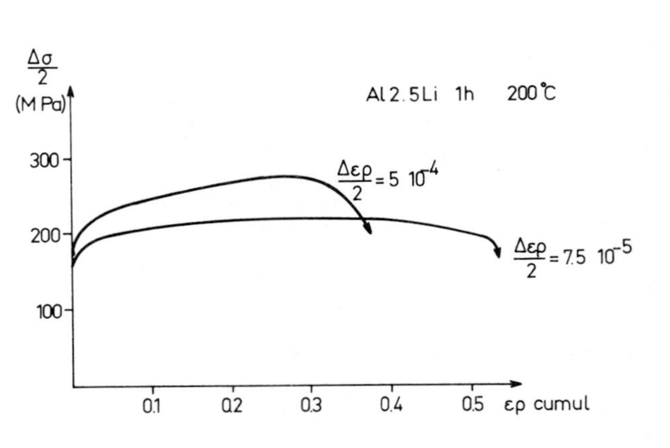

8 - Cyclic hardening curves of Al-2.5 Li aged 1h at 200°C.

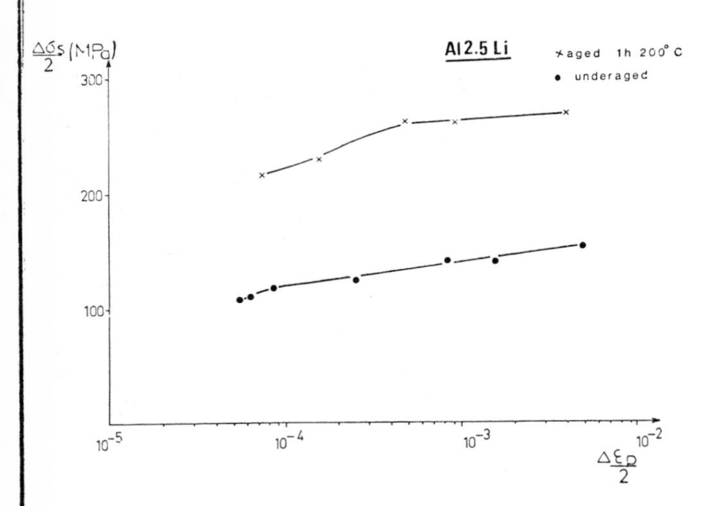

6 - CSS curves of Al-2.5 Li ; underaged and aged 1h at 200°C.

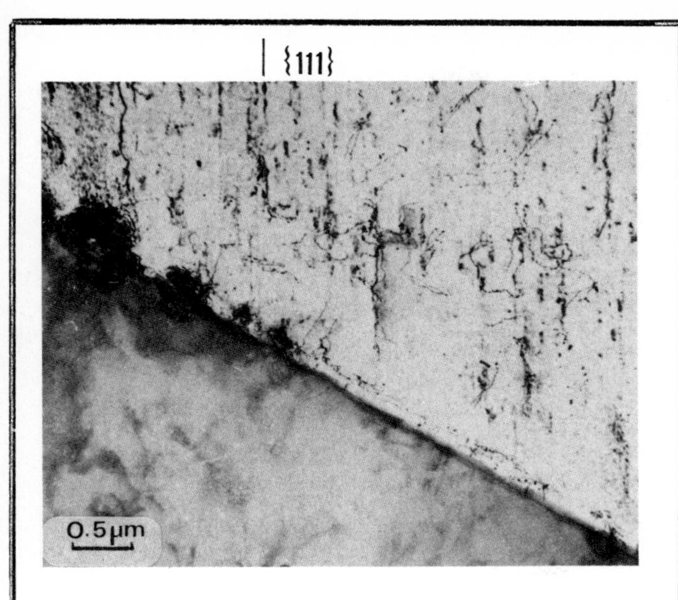

9 - Slip bands and apparent stress concentration around δ precipitates observed in the Al-2.5 Li alloy aged 1h at 200°C, cycled to failure at $\frac{\Delta\varepsilon_p}{2} = 10^{-4}$.

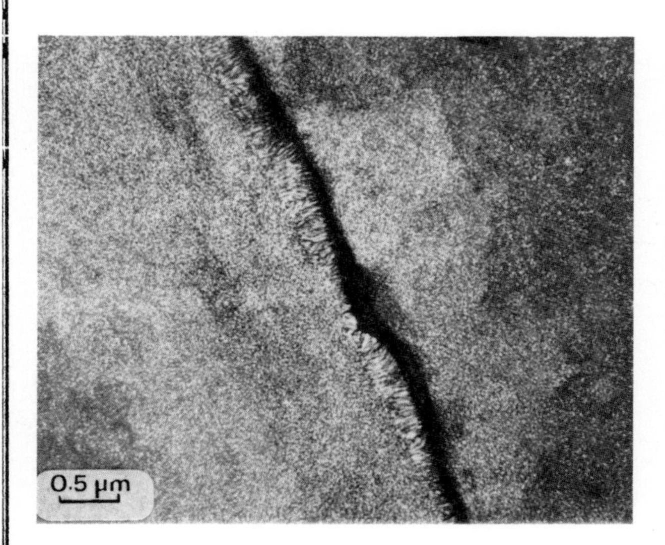

10 - Same specimen as in Fig. 9, PFZs and δ' discontinuous precipitation observed in dark field.

gue resistance, compared with the low Li alloy, is probably due to the onset of grain boundary failure ; this crack initiation mode is favoured by the relatively large grain size which allows the development of extensive dislocation pile-ups and local stress concentrations at the boundaries.

Aged Al-2.5 Li

Mechanical behaviour

At all plastic strain amplitudes the aged 2.5%Li alloy exhibits rapid initial hardening followed by slower hardening as in the underaged state ; there is no saturation and no softening (see Figure 8). The CSS curve of the aged material is given in Figure 6 ; comparison with the cyclic flow stresses of the underaged alloy indicates a stress difference of about 100 MPa.

The Manson-Coffin curve of the aged Al-2.5Li alloy, given in Figure 3, shows that the lives are substantially lower than those of the underaged Al-2.5 Li alloy. The corresponding Manson-Coffin equation is : $\frac{\Delta\varepsilon_p}{2} = 0.023\ Nf^{-0.76}$. The fatigue lifetimes also appear to be lower, by a factor of 10 than those of an equivalent (8.9 at % Li) alloy tested by Sanders and Starke[8] (after extrapolation from the high $\Delta\varepsilon_p$ strain range used by these authors). This drastic decrease in fatigue resistance is due to grain boundary embrittlement. Cracks nucleate at grain boundaries for all plastic strains. There are in fact very few secondary cracks - as soon as a grain boundary fails the primary crack propagates rapidly along the adjacent boundaries.

MICROSTRUCTURE

Before fatigue testing the aged Al-2.5 Li contains a homogeneous distribution of δ' precipitates (15-20 nm diameter) in the grains. The stable δ precipitates are observed along grain boundaries, and are associated with relatively narrow PFZs of width less than 0.1µm.

After cycling (Figure 9) TEM reveals a slip band structure , parallel to the (111) planes . Slip is planar, and δ' precipitates are sheared by superdislocations. However, on dark field images from δ' spots no precipitate denuded zones are detected within the grains (as in the underaged state). The slip bands are less intense than those of the underaged Al-2.5 Li ;this could be due to either the relatively small cumulative plastic strain of the aged alloy, or to a size effect of δ' precipitates.

A particularly interesting observation is shown in Figure 10 : along a grain boundary one can see, not only the usual PFZs, but also discontinuous δ' precipitation of larger width than the PFZs. δ' discontinuous precipitation has also been reported by Williams and Edington[9] in a 12.9 at % Li alloy, aged 10 min at 200°C.

DISCUSSION

The fatigue resistance of the Al-2.5 Li alloy is clearly very sensitive to the microstructure created by heat treatment after ageing 1h at 200°C;

fatigue lives are two orders of magnitude lower than those of the underaged alloy.

The ageing treatment develops a microstructure significantly different from the underaged state and which enhances grain boundary brittleness. PFZs and δ' cellular precipitation are regions softer than the matrix ; in addition the non-shearable δ precipitates in the grain boundaries act as stress concentration sites (see Figure 9). The following crack initiation mechanism seems to be consistent with these observations : when the stress field level, around the grain boundary δ precipitate, becomes too high, a crack nucleates around δ, and then propagates quickly along the soft PFZ, or alternatively along the boundary regions of discontinuous precipitation.

CONCLUSIONS

The present study of the cyclic deformation of binary Al-0.7 Li and Al-2.5 Li alloys leads to the following results :

- Al-0.7 Li exhibits similar cyclic behaviour to that of pure Al ; Li additions in solid solution appear to have only a small influence on the fatigue properties ;

- the fatigue properties of Al-2.5 Li alloys are very sensitive to the heat treatment ; the aged alloy exhibits a substantial loss of fatigue resistance due to intergranular brittleness.

REFERENCES

[1] DHERS J., DEA report, ENSMSE (1983).
[2] FIGUEROA J.C. et al, Acta Met., 29, 1667 (1981).
[3] KAYALI E.S., PLUMTREE A., Met. Trans., 13A, 1033, (1982).
[4] KONIG G., BLUM W., Acta Met., 28, 519, (1980).
[5] DHERS J., DRIVER J., to be published.
[6] CALABRESE C., LAIRD C., Mat. Sci. Eng., 13, 141, (1974).
[7] WILHEM M., Mat. Sci. Eng., 48, 91, (1981).
[8] SANDERS T.H., STARKE E.A., Acta Met. 30, 927, (1982).
[9] WILLIAMS D.B., EDINGTON J.W., Acta Met., 24, 324, (1976).
[10] MUGHRABI H., Mat. Sci. Eng., 33, 207, (1978).

Fatigue behavior of Al–Li–Cu–Mg alloy

M. PETERS, K WELPMANN,
W ZINK and T H SANDERS Jr

MP and KW are with the
Institut für Werkstoff-Forschung,
DFVLR, 5000 Köln 90, FRG.
WZ is with MBB-Bremen,
2800 Bremen 1, FRG.
THS is with the
School of Materials Engineering,
Purdue University,
West Lafayette, IN 47907, USA

Introduction

Aluminium-Lithium alloys have recently gathered strong interest especially in the aircraft industry [1-4]. Compared to conventional high strength aluminium alloys of the 2000- or 7000-series it is anticipated that these alloys offer a 10 % increase in stiffness along with a 10 % decrease in density, thus making them rather competitive to new upcoming non-metallic materials like carbon fiber reinforced composites.

The major drawbacks of these alloys, however, are their low ductility and toughness, which had caused the first commercial lithium-containing aluminium alloy - the Alcoa 2020 - to be withdrawn in the early seventies. However, renewed interest in low-density, high-modulus aluminium alloys, stimulated by fuel price increases and the introduction of composites, led to the development of a second generation of aluminium-lithium alloys in the early eighties. The first of these alloys was the Alcan International DTD XXXA, originally developed at the Royal Aircraft Establishment [5,6]. This semi-commercial Al-Li-Cu-Mg alloy was the basis of this investigation of static and cyclic mechanical properties. Results are compared to conventional high strength aluminium alloys.

Experimental Procedure

The investigation was performed on 1.6 mm sheet material. The Al-Li alloy DTD XXXA supplied by Alcan International Ltd. (GB) had the chemical composition of 2.49 wt.%Li, 1.21 wt.% Cu, 0.69 wt.% Mg and 0.12 wt.% Zr and a density of 2.54 g/cm^3. It is important to note that this alloy was produced on laboratory scale. Combinations of various levels of stretching (0 %, 2 % and 4 %) and artificial aging treatments led to four different strength levels designated as LOW (T6), LOW, MEDIUM and HIGH (Table I). Comparative tests were performed on 1.6 mm Alclad 2024 sheet material in T3 and T8 tempers.

Table I: Alloy conditions investigated

ALLOY	STRETCH	AGING
DTD XXXA LOW (T6)	0%*	8h 170°C**
DTD XXXA LOW	2%	1h 170°C***
DTD XXXA MEDIUM	2%	5h 187°C***
DTD XXXA HIGH	4%	24h 187°C***
2024-T3 (Alclad)	2%	-
2024-T8 (Alclad)	2%	9h 190°C**

* straightened
** heat treated at DFVLR
*** heat treated by Alcan International

Tensile tests were performed on 1.6 mm thick flat specimens with a gage length of 25 mm and a gage width of 8 mm. A crosshead speed of 1 mm/min was chosen and strain gages were employed to determine the elastic modulus. Tests were done with the loading axis parallel (L), perpendicular (T) and under 45° to the rolling direction (45°). At least two tests were performed for each condition.

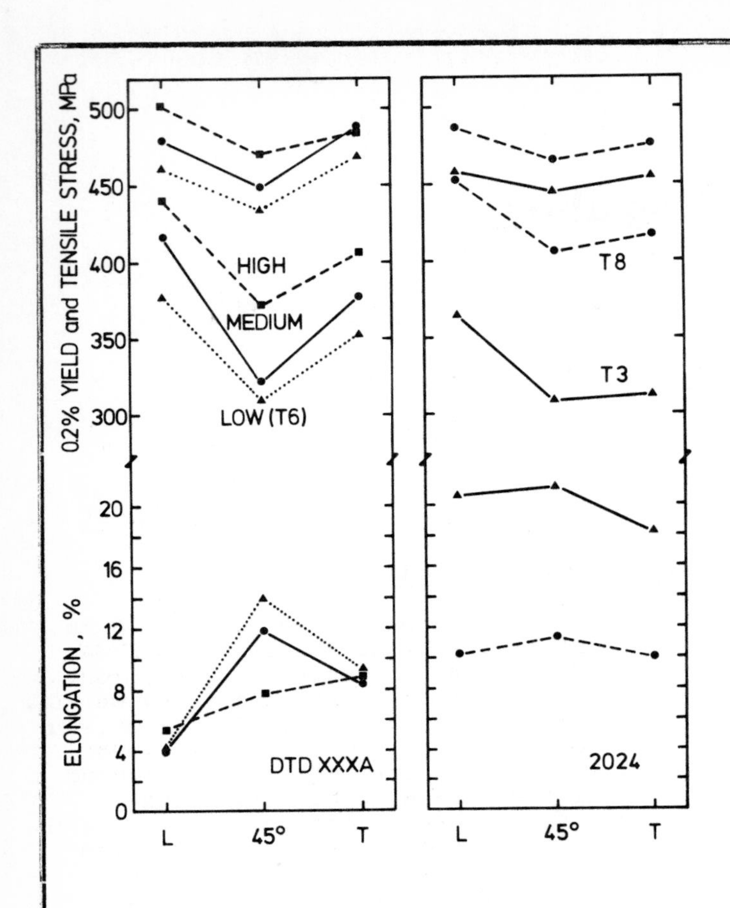

Fig. 1: Results of tensile tests of the DTD XXXA conditions LOW (T6), MEDIUM and HIGH and 2024-T3 and T8 parallel (L), perpendicular (T) and under 45° to the rolling direction (45°).

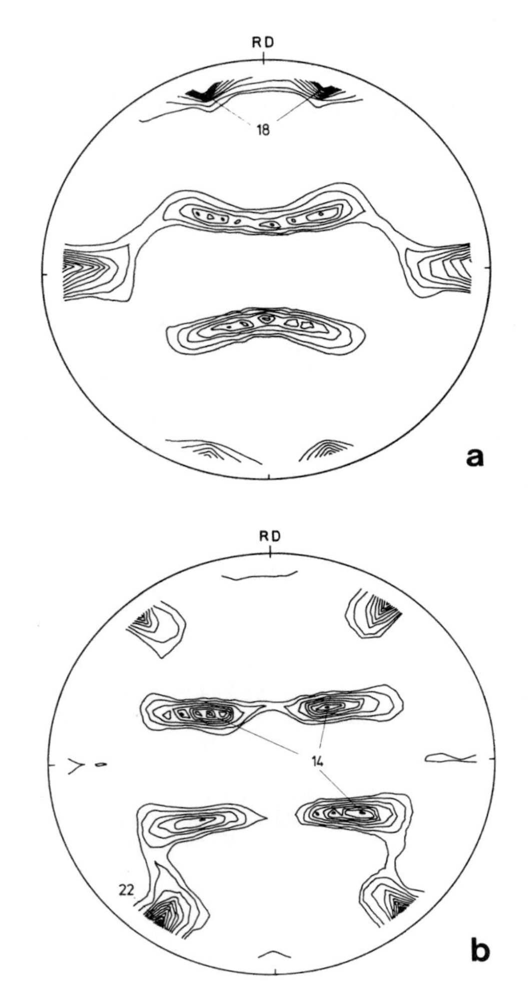

Fig. 2: {111} (a) and {200} pole figures (b) of DTD XXXA MEDIUM.

Pole figures were made from the center of the sheet by employing the Schulz-reflection method. Intensities were normalized to a random pure aluminium standard.

Fatigue crack growth investigations were performed on a hydraulic testing system with a frequency of 30 Hz using two types of specimens:
1.6 mm thick single-edge-notched (SEN) specimens (W = 20 mm) were tested with R = 0.2 in vacuum, laboratory air and a 3.5 % NaCl solution, whereas same thickness center-cracked-tension (CCT) specimens (160 mm width) were used in laboratory air to compare longitudinal and transverse fatigue crack growth behavior at R = 0.38.

High cycle fatigue tests in ambient environment were performed on 1.6 mm thick alternate bending specimens with a frequency of 20 Hz and an R-ratio of 0.1. Smooth specimens ($\alpha_K = 1$) had a width of 16 mm while notched specimens ($\alpha_K = 2.5$) had a width of 40 mm with a center-drilled 8 mm diameter hole.

Fractography on fatigue crack growth specimens was done by scanning electron microscopy.

Results

The results of the tensile tests of specimens tested in the rolling direction (L) are shown in Table II.

Table II: Tensile properties (L-direction)

ALLOY	E GPa	0.2%YS MPa	UTS MPa	EL. %
DTD XXXA LOW (T6)	79.5*	377	461	4.1
DTD XXXA LOW	79.5*	388	446	5.4
DTD XXXA MEDIUM	79.5*	416	481	4.2
DTD XXXA HIGH	79.5*	440	502	5.3
2024-T3 (Alclad)	72.4	365	458	20.5
2024-T8 (Alclad)	72.4	452	487	10.3

*average value out of more than 80 measurements on differently aged specimens

For the various conditions of the DTD XXXA an average value for the elastic modulus of 79.5 GPa was calculated while the yield strength ranged from 377 MPa for the LOW (T6) condition to 440 MPa for the high strength version. For the tensile strength a similar ranking was observed, except, here, the LOW-condition showed the lowest value of 446 MPa. The HIGH-version revealed an ultimate tensile strength of 502 MPa.

The ductility of the four DTD XXXA conditions did not show a correlation to the strength values. Elongation values ranged between 4.1 and 5.4 %.

2024 had an elastic modulus of 72.4 GPa for both, the T3 and T8 temper. Strength levels of the T3 condition were similar to the DTD XXXA LOW (T6) and LOW versions while the T8 condition was comparable to DTD XXXA HIGH. Elongation was found to be 20.5 for 2024-T3 and 10.3 for 2024-T8.

Results of tensile tests of the DTD XXXA and 2024 alloy conditions performed in L-, T-, and 45°-direction are plotted in Fig.1. For the various conditions of the Al-Li alloy normally strength levels were highest in L-direction and lowest tested at 45° to the rolling axis. Intermediate values were found in transverse direction. This ranking was opposite for the ductility (except condition HIGH). Generally, for the 2024 alloy a similar behavior was observed. Effects were, however, less pronounced. For both alloys the yield strength values showed a stronger anisotropy than the tensile strength values.

{111} and {200} pole figures of the DTD XXXA MEDIUM and 2024-T3 are shown in Figs. 2 and 3. The highest intensities measured for the Al-Li sheet material were 18 and 22 for the {111} and {200} pole figure, respectively, while these values were 4 for 2024.

Figure 4 shows the fatigue crack growth behavior of the three DTD XXXA conditions LOW (T6), MEDIUM and HIGH, and 2024-T3. Results are plotted as fatigue crack growth rate, da/dN, versus cyclic stress intensity factor, ΔK, for tests performed in vacuum (Fig. 4a), laboratory air (Fig. 4b) and 3.5 % NaCl solution (Fig. 4c). Generally, 2024-T3 turns out to be slightly superior over the various tempers in the Al-Li alloy investigated. For the DTD XXXA material, the higher strength versions showed a slightly faster propagation rate than the low strength condition. "Threshold" values ranged between 5 and 6 $MPa\cdot m^{1/2}$ in vacuum and 3 and 4 $MPa\cdot m^{1/2}$ in air and sodium chloride solution.

Figure 5 shows the influence of testing direction on fatigue crack growth in air. For 2024-T3 propagation rates in LT and TL orientation were similar, whereas for the DTD XXXA LOW version a much faster crack growth rate was measured in LT compared to the TL orientation.

Figure 6 shows fracture surfaces on LT oriented specimens of DTD XXXA MEDIUM and 2024-T3 tested in air at low crack growth rates ("threshold"-regime). The Al-Li alloy shows a strong "fiberlike" crack path with 5 to 30 µm wide layers oriented

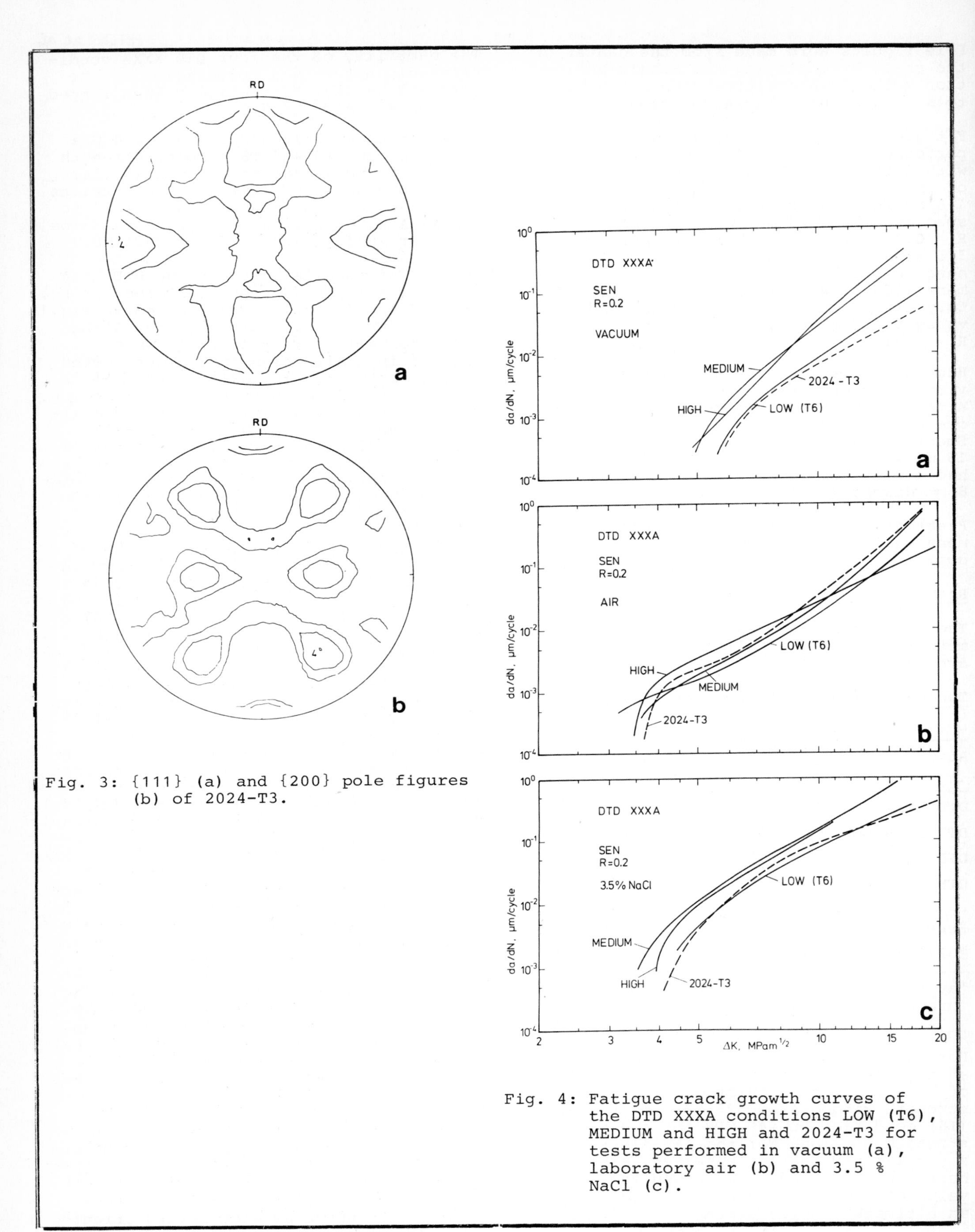

Fig. 3: {111} (a) and {200} pole figures (b) of 2024-T3.

Fig. 4: Fatigue crack growth curves of the DTD XXXA conditions LOW (T6), MEDIUM and HIGH and 2024-T3 for tests performed in vacuum (a), laboratory air (b) and 3.5 % NaCl (c).

perpendicular to the ligament of the specimen in the growth direction (Fig.6a). The crack morphology of 2024 appears more ductile with cleavage-like features in-between and without any preferred orientation similar to the one observed for the DTD XXXA (Fig. 6b).

The high cycle fatigue behavior of the DTD XXXA conditions LOW, MEDIUM and HIGH is shown in Fig. 7 for smooth and notched specimens. The dashed SN-curves represent corresponding results of 2024-T3. For both specimen types the Al-Li alloy does not reveal a superiority of either one of the three differently strengthened conditions for the number of specimens investigated. 2024-T3 seems to show slightly superior fatigue behavior for smooth specimens (Fig. 7a), whereas this superiority nearly diminishes for the notched specimens (Fig. 7b).

Discussion

a) Tensile Properties

The DTD XXXA alloy is Alcan International's "all purpose", medium strength Al-Li alloy. Thus, it is intended to replace conventional Al-alloys like 2014-T6, 2024-T8 or 7075-T73 [7-9]. The anticipated goal for this new alloy is to reduce density and increase stiffness, both, by about 10 % compared to the conventional alloys to be substituted and by keeping other properties constant. As far as this investigation on 1.6 mm sheet material laboratory production DTD XXXA alloy has revealed, both, density and stiffness goals were essentially met: Compared to the 2000- and 7000-alloys mentioned above, the measured density of 2.54 gcm^{-3} for the DTD XXXA represents an 8.3 to 9.3 % decrease while the elastic modulus of 79.5 GPa of the Al-Li alloy is about 9.8 to 12 % higher than the corresponding values for the conventional alloys (Reference: Metals Handbook [10]).

Looking closer at the tensile results of the different aging conditions of DTD XXXA presented in Table II, it is interesting to note that low strength is not necessarily associated with high ductility. For instance, the lowest ductility was found for condition LOW (T6) which also revealed the lowest yield strength, whereas the material aged to high strength showed relatively high ductility. This behavior is due to the aging characteristics of the DTD XXXA alloy [11]. During aging, ductility first decreases, since precipitation and growth of Al_3Li-particles promote increased planarity of slip. Then ductility increases before maximum strength is reached. This could be attributed to accelerated precipitation of the S'-phase, which - besides further strengthening - tends to homogenize slip and therefore increases both, strength and ductility [11]. As a consequence, the maximum in yield strength does not coincide with the minimum in ductility, which can explain the somewhat untypical observations of strength/ductility combinations in Table II.

As far as ductility values are concerned, it is obvious that they are low for the DTD XXXA conditions when compared to equivalent strength levels of 2024 (Fig. 1 and Table II). This is, of course, known and primarily related to the presence of ordered, coherent, Al_3Li precipitates. They cause a strong planarity of slip which results in high local stress concentrations at grain boundaries, thus decreasing ductility [5,12,13]. Also embrittlement of grain boundaries due to segregation of alkali impurities like sodium, potassium or calcium is known to lead to reduced ductility [14,15]. The present study has, however, revealed that the strong crystallographic texture has to be considered as an additional source for the low ductility. Since orientations in neighbouring grains are similar, the impact of grain boundaries to act as obstacles for dislocation pile-ups is reduced. This tends to increase the "effective" slip length and reduces ductility.

Similar observations were recently made by Gregson and Flower on a variety of Al-Li base alloys. They claimed that the poor toughness of some of these alloys was a result not of planar slip per se, but of its occurrence in combination with a strong crystallographic texture [16].

Besides its influence on ductility, the strong texture has also to be considered as a major cause for the anisotropy of the yield strength of the DTD XXXA [5] which becomes evident by comparing Figs. 1 to 3. The onset of plastic deformation - e.g. yielding - is controlled by the critical resolved shear stress of a slip plane and by the relative orientation of its slip system to the loading axis. The two strongest poles of the {111} pole figure of the DTD XXXA MEDIUM condition (Fig. 2a) indicate that {111} planes are nearly perpendicular to the rolling direction. Therefore, the associated slip systems are unfavourably oriented which accounts for the highest yield strength to be found in L-direction (Fig. 1). The lower intensity of the {111} pole in transverse direction explains the slightly lower yield strength measured in T-direction (Fig. 1). All {111} poles discussed above are, however, nearly ideally oriented for early slip under 45° to RD (Fig. 2a) which is in accordance with the minimum of yield strength observed in that direction (Fig. 1). The pole figures of the

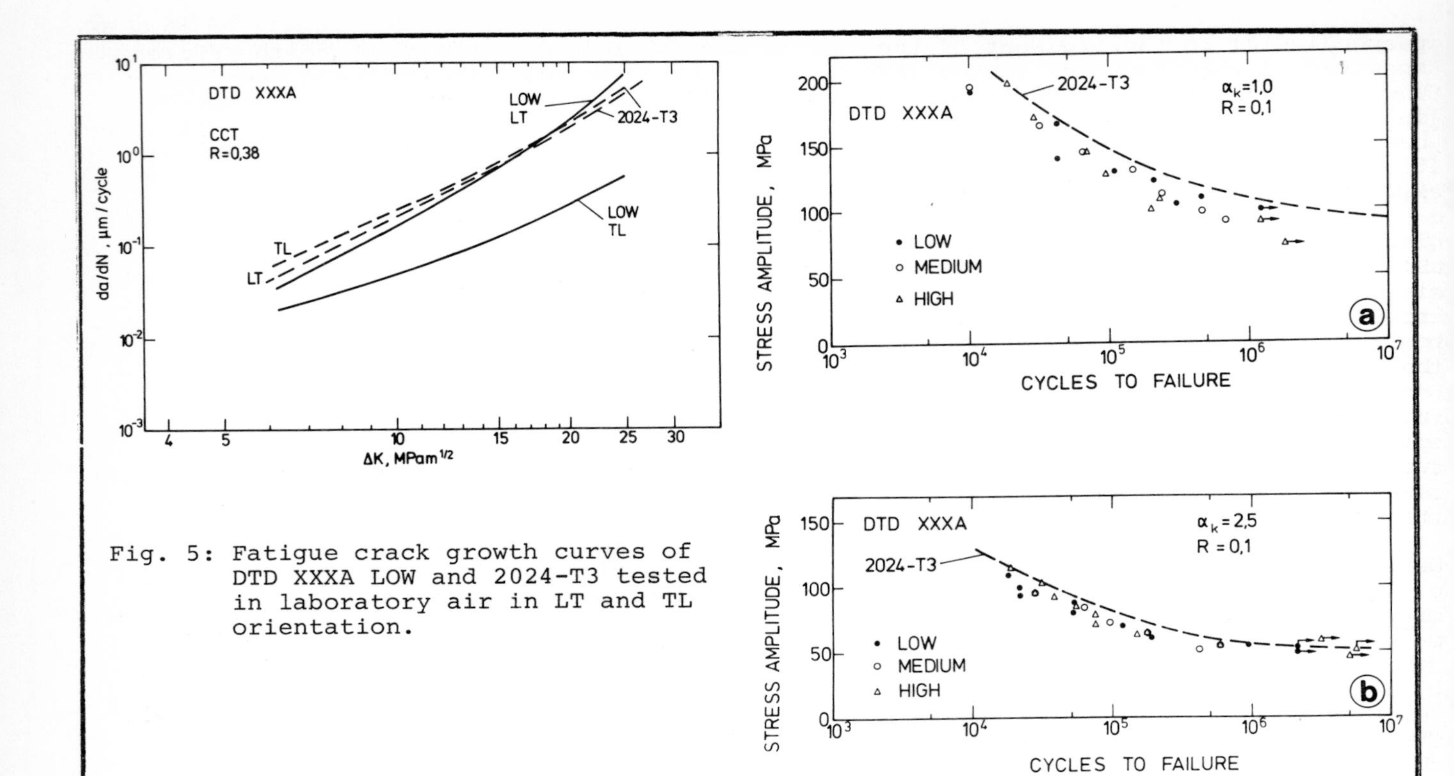

Fig. 5: Fatigue crack growth curves of DTD XXXA LOW and 2024-T3 tested in laboratory air in LT and TL orientation.

Fig. 7: High cycle fatigue tests of the DTD XXXA conditions LOW, MEDIUM and HIGH for unnotched (a) and notched specimens (b).

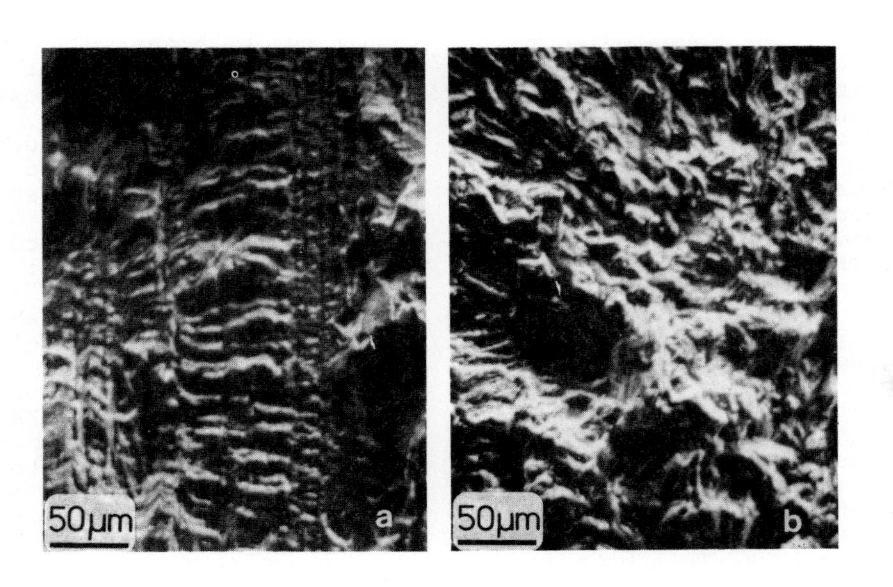

Fig. 6: Fatigue crack growth fracture surfaces of DTD XXXA MEDIUM (a) and 2024-T3 (b) specimens at low crack growth rates (1-5x10^{-3} μm/cycle) tested in laboratory air. Crack growth direction from bottom to top.

other DTD XXXA conditions were similar to the ones shown in Fig. 2, which explains the similar yielding behavior of the LOW (T6) and HIGH version (Fig. 1).

It should also be mentioned that these explanations are relatively simple, because they do not take into account the orientation of the slip direction as well as other poles in the pole figure. However, a similar yielding behavior of textured Al-Li-Cu-Mg sheet metal was observed by other investigators [5,16]. They found the minimum of yield strength to be about 60° off the rolling direction. The slight tilt off RD of the two high intensity split {111} poles in Fig. 2a indeed support that the minimum in yield strength should be found at angles slightly higher than 45° away from the rolling direction.

Ways to reduce the anisotropy might include changes in forming procedure, like cross rolling, and/or special aging treatments [16].

b) Fatigue Properties

The fatigue crack growth behavior of Li-containing Al-alloys has been found to be good or excellent when compared to conventional Al-alloys [5,7, 16-19]. The high resistance to crack propagation for low ΔK's was attributed to the higher elastic modulus associated with Al_3Li precipitates [17]. Also extensive crack deflection processes induced by the δ' precipitates improved the crack growth behavior [19]. In a number of investigations a rough fracture surface gave rise to a component of crack closure which has the added effect of reducing crack growth rates.

In the present investigation it was apparent that Al-Li alloys are not always superior over conventional alloys such as 2024 (Fig. 4). However, it should be mentioned that the comparison to 2024-T3 is not entirely appropriate, since this is the damage tolerant alloy which probably exhibits the highest fatigue propagation resistance of all conventional high strength Al-alloys. In view of this, DTD XXXA clearly has more than acceptable crack propagation resistance in both, inert and corrosive environments (Fig. 4).

Beside the influence of the higher elastic modulus, the combination of planar slip and pronounced texture is considered to be responsible for the good fatigue propagation behavior of the DTD XXXA in this investigation. The "fiber-like" fracture morphology (Fig. 6a) reveals a strong crystallographic crack advance within the fine subgrain structure due to similar orientations of subgrains. But also high angle grain boundaries, which create the "fiber-like" morphology, do not represent strong obstacles for crystallographic crack advance, as a result of the strong texture. This can be seen in the center part of Fig. 6a where the crack front crosses several grains. In essence, the strong crystallographic crack advance due to δ' and texture leads to improved crack growth behavior. This can be attributed to closure or deflection effects [19,20] or simply to the fact that - on a micro-scale basis - crack growth rates are underestimated while stress intensities are overestimated [21].

The beneficial effect texture has on fatigue crack growth, has to be paid for by a strong anisotropy of crack growth behavior which becomes evident by comparing DTD XXXA and 2024-T3 tests in LT and TL orientation (Fig. 5).

Looking at environmental effects, it is interesting to note that the "threshold" values in air (Fig. 4b) are slightly lower than in the 3.5 % sodium chloride solution (Fig. 4c). This can probably be attributed to corrosion induced closure effects, as recent investigations on 25 mm thick DTD XXXA plate material have revealed [22].

Often factors which improve resistance to fatigue crack propagation tend to have a detrimental influence on fatigue crack initiation [23]. For the Al-Li alloy investigated the combined appearance of planar slip and texture can cause locally high stress concentrations at grain boundaries. This leads to early crack initiation which is known to decrease the high cycle fatigue strength. In view of this, the fatigue results in Fig. 7 have to be evaluated. Generally, the fatigue strength values of the DTD XXXA conditions are not too much different from those of 2024. Further investigations on texture and the response to a corrosive environment should give more insight into the fatigue behavior of DTD XXXA.

Summary and Conclusions

The investigation on 1.6 mm thick DTD XXXA laboratory production sheet material has shown that tensile and fatigue properties are to a large extent dominated by the combined occurrence of planar slip caused by the coherent Al_3Li particles and a strong crystallographic texture. Compared to conventional high strength Al-alloys this leads to a more pronounced anisotropy of strength and is thought to play a key role for the low ductility.

Besides the higher elastic modulus texture and Al_3Li precipitates are responsible for the good to excellent fatigue crack growth behavior of the Al-Li alloy. The crack path is crystallographically oriented which leads to a rough fracture surface appearance. Crack closure effects or crack deflection considerations are, therefore, possible explanations for the high crack propagation resistance.

First results on high cycle fatigue tests indicate that planar slip combined with texture might lead to earlier crack nucleation which slightly reduces fatigue strength.

Acknowledgements

The experimental assistance provided by H. Frauenrath and J. Eschweiler (DFVLR) and H. Krüger (MBB) is gratefully acknowledged.

References

1. T.H. Sanders, Jr. and E.A. Starke, Jr. (editors), "Aluminum-Lithium Alloys", Proc. 1st International Aluminum-Lithium Conference, Stone Mountain,CA, May 1980, TMS-AIME, Warrendale, PA, 1981.

2. T.H. Sanders, Jr. and E.A. Starke, Jr. (editors), "Aluminum-Lithium Alloys II", Proc. 2nd International Aluminum-Lithium Conference, Monterey, CA, April 1983, TMS-AIME, Warrendale, PA, 1984.

3. W.E. Quist, G.H. Narayanan and A.L. Wingert in: "Aluminum-Lithium Alloys II", eds.: T.H. Sanders, Jr. and E.A. Starke, Jr., p. 313, TMS-AIME, Warrendale, PA, 1984.

4. K. Welpmann, M. Peters and T.H.Sanders, Jr., Aluminium English 60 (1984), p. E641 and E709.

5. C.J. Peel, B. Evans, C.A. Baker, D.A. Bennett, P.J. Gregson and H.M. Flower in: "Aluminum-Lithium Alloys II", eds.: T.H. Sanders, Jr. and E.A. Starke, Jr., p. 363, TMS-AIME, Warrendale, PA, 1984.

6. W.S. Miller, A.J. Cornish, A.P. Titchener and D.A. Bennett in: "Aluminum-Lithium Alloys II", eds.: T.H. Sanders, Jr. and E.A. Starke, Jr., p. 335, TMS-AIME, Warrendale, PA, 1984.

7. N.N. "Aluminium Lithium Alloys Development", British Alcan Aluminium Ltd., 1983.

8. J. Fielding, "The Present Status and Future Potential of Aluminium Alloys in Aerospace", Alcan International Ltd., Chalfont Park, England, 1984.

9. J.F. Hawkins, Proc. Airmec 85, Düsseldorf, FRG, 1985.

10. N.N., Metals Handbook, Ninth Ed., Vol. 2, ASM, Metals Park, OH, 1979.

11. K. Welpmann, M. Peters and T.H. Sanders, Jr., Proc. 3rd Intern. Aluminium-Lithium Conference, Oxford, England, July 9-11, 1985.

12. T.H. Sanders, Jr. and E.A. Starke, Jr., Acta Metall. 30 (1982) p. 927.

13. E.A. Starke, Jr., T.H. Sanders, Jr. and I.G. Palmer, Journal of Metals 33 (8/1981) p. 24.

14. A.K. Vasudêvan, A.C. Miller and M.M. Kersker in: "Aluminum-Lithium Alloys II", eds. T.H.Sanders, Jr. and E.A. Starke, Jr., p. 181, TMS-AIME, Warrendale, PA. 1984.

15. J.A. Wert and J.B. Lumsden, Scripta Metall. 19 (1985) p. 205.

16. P.J. Gregson and H.M. Flower, Acta Metall. 33 (1985) p. 527.

17. E.J. Coyne, Jr.,T.H. Sanders, Jr. and E.A. Starke, Jr. in: "Aluminum-Lithium Alloys", eds. T.H. Sanders, Jr. and E.A. Starke, Jr., p. 293, TMS-AIME, Warrendale, PA, 1981.

18. S.J. Harris, B. Noble and K. Dinsdale in: "Aluminum-Lithium Alloys II", eds. T.H. Sanders, Jr. and E.A. Starke, Jr., p. 219, TMS-AIME, Warrendale, PA, 1984.

19. A.K. Vasudêvan and S. Suresh, Metall. Transactions 16A (1985) p. 475.

20. E. Zaiken and R.O. Ritchie, Mat.Sci. Eng. 70 (1985) p. 151.

21. M. Peters, K. Welpmann and H. Döker in: "Titanium Science and Technology" eds. G. Lütjering, U. Zwicker and W. Bunk, Vol. 4, p. 2267, DGM, Oberursel, FRG, 1985.

22. M. Peters, J. Eschweiler, V. Bachmann and K. Welpmann, DFVLR-Report, to be published.

23. E.A. Starke, Jr. and G. Lütjering in: "Fatigue and Microstructure" ed. M. Meshii, p. 205, ASM, Metals Park, OH, 1978.

□

Fatigue crack growth and fracture toughness behavior of Al–Li–Cu alloy

K V JATA and E A STARKE Jr

The authors are, respectively, with the Department of Material Science at the University of Virginia, USA, and the School of Engineering and Applied Science, University of Virginia, USA

A high purity Al-Li-Cu alloy with Zr as a dispersoid forming element has been studied as a function of aging time. The slip behaviour and fatigue crack growth and fracture toughness results in both air and vacuum indicate that irregularities in the crack profile and the fracture surfaces do not totally account for the improved fatigue performance for the underaged conditions, and suggest that slip reversibility is a major contributing factor. The fracture toughness variation for the aging conditions studied can be related to the slip planarity and the accompanying changes in the strain hardening exponent. Although the current alloy exhibits planar slip for all aging conditions examined, the crack initiation fracture toughness, K_{1c}, compares favorably with those of 2XXX and 7XXX aluminum alloys.

I. INTRODUCTION

Various modifications in alloy chemistry and fabricating techniques have been used in an attempt to improve the ductility of Al-Li-X alloys, while maintaining a high strength. Copper, Mg and Zr solute additions have been shown to have beneficial effects (1). Magnesium and Cu improve the strength of Al-Li alloys through solid solution and precipitate strengthening, and can minimize the formation of precipitate free zones (PFZ) near grain boundaries. Zirconium, which forms the cubic Al_3Zr coherent dispersoid, stabilizes the subgrain structure and suppresses recrystallization.

The goals of most Al-Li-X alloy development programs include improvements in density and modulus with equivalent or improved damage tolerance and corrosion properties compared with currently used materials, e.g. 7075 and 2024 (2). Although there have been numerous reports on the relationship between composition, microstructure and monotonic properties of Al-Li-X alloys (3,4), there have been few studies on the cyclic properties and fracture toughness of these materials. This paper describes the fatigue crack propagation and fracture toughness of a new alloy based on the Al-Li-Cu system which is somewhat related to the Al-Li-Cu alloy 2020 that was commercially available in the 1960's.

II. EXPERIMENTAL PROCEDURES

The material was obtained from Reynolds Metals Company in the form of 27.7 mm thick plates. The original cast ingots were homogenized in an argon atmosphere as follows: (i). Heated at 523 K/hour to 673 K, held 48 hours, (ii). heated at 298 K/hour to 763 K, held 18 hours and (iii). heated at 298 K/hour to 788 K held for 30 minutes and fan cooled. The ingots were scalped on the surface to 69.8 mm thick and then cleaned.The hot rolling was performed in three steps, preheated to 733 K, held for one hour, hot rolled from 69.8 mm to 57.1 mm, reheated to 733 K, hot rolled to 44.5 mm, reheated to 671 K and hot rolled to a final thickness of 27.7 mm. The composition of the alloy is shown in Table 1.

The alloy was solutionized in a salt bath at 788 K for 30 minutes, quenched in ice water, stretched 2 percent and aged in an oil bath at 433 K for different periods of time. Fracture toughness was measured on 11 mm thick compact tension samples in the L-T orientation and a conditional K_{1c} value for fracture toughness was obtained from the load-crack opening displacement plots by choosing the 5 percent secant offset line. Of all the samples tested only the two underaged samples did not meet the ASTM thickness criterion, $b \geqslant 2.5\ (K_{1C}/\sigma_{ys})^2$. A thickness of 54 mm would have been required for the underaged samples for a valid K_{1C} test.

Crack propagation tests were conducted on compact tension specimens in laboratory air (R. H. ~45percent) and a vacuum of 10^{-5} torr at 295 K using a R ratio of 0.1 and a frequency of 30 Hz. The crack propagated in the long transverse direction on a plane normal to the rolling direction. The microstructures were examined by optical, scanning and transmission electron microscopy.

III. RESULTS AND DISCUSSION

A. Microstructure and Tensile Properties

Optical metallography revealed a predominantly unrecrystallized structure in the as-rolled condition with large elongated unrecrystallized grains. However, some recrystallization occurred during solution treatment. TEM studies revealed subgrains with an average diameter of 5

TABLE 1

Chemical Compositions in weight percent

Cu	Li	Cd	Zr	Fe	Si	Al
3.6	1.68	0.16	0.16	0.01	0.02	bal.

TABLE 2

Aging time	σ_{ys}	σ_{uts}	percent elongation	n	K_{1C}
8	300	431	20	0.13	44
12	350	452	18	0.098	40
17	520	534	8	0.055	32.2
19	500	526	8.6	0.062	30
22	484	500	6.6	0.056	27
26	472	496	4.0	0.062	30

TABLE 3

Fatigue Threshold Stress Intensities

Condition	Environment	ΔK_{th} Mpa $^{1/2}$	ΔK^{i}_{th} Mpa $m^{1/2}$	ΔK^{Cl}_{th} Mpa $m^{1/2}$
UA	Air	3.7	3.1	0.6
PA	Air	2.8	2.4	0.4
OA	Air	2.3	2.3	negligible
UA	Vacuum	5.5	5.3	0.2
PA	Vacuum	4.3	4.0	0.3

$$\Delta K_{th} = \Delta K^{observed}_{th} \text{ or } \Delta K^{apparent}_{th}$$

$$\Delta K^{i}_{th} = \Delta K_{th} - \Delta K^{Cl}_{th}$$

$$\Delta K^{Cl}_{th} = K^{Cl}_{th} - K^{min}_{th}$$

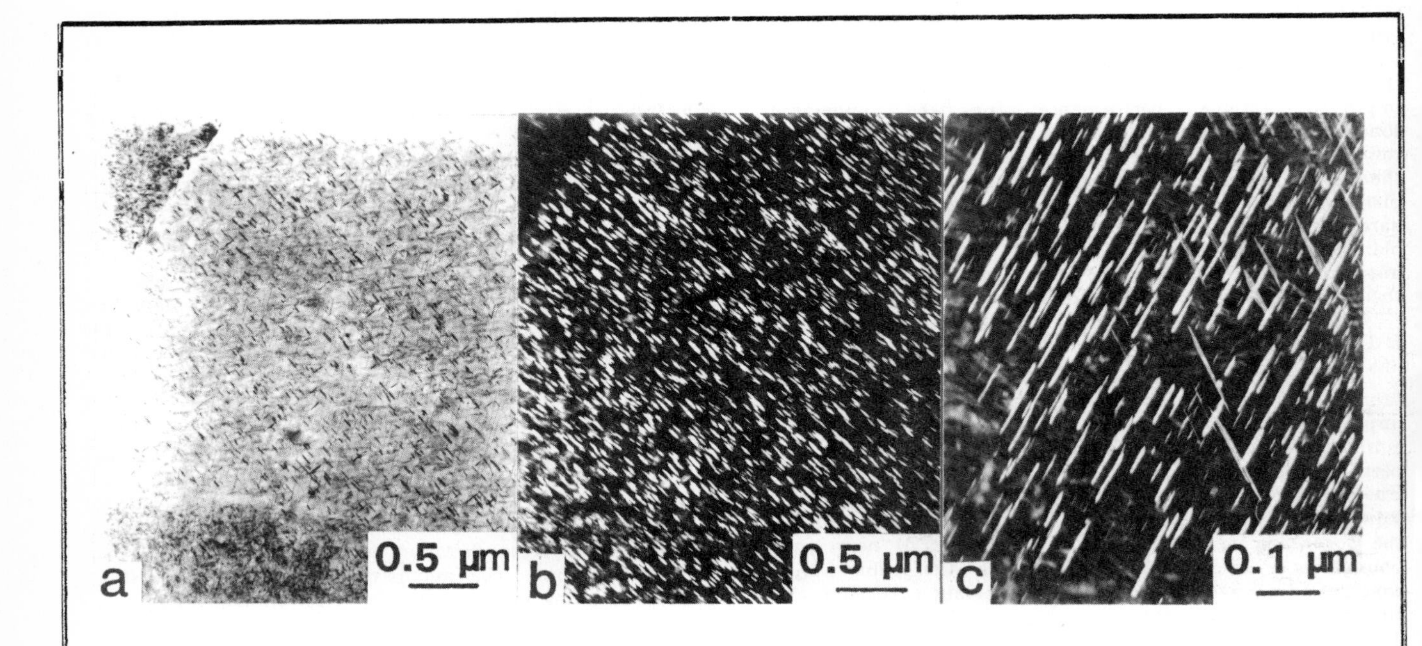

Fig. 1: (a) Bright field TEM of θ' and T_1 precipitates in the {112} foil plane, (b) dark field TEM using θ' reflection, and (c) dark field TEM using T_1 reflection.

microns. The major strengthening precipitates θ' (Al_2Cu), and T_1 (Al_2CuLi) are shown by the bright and dark field TEM's in Figure 1, for the peakaged condition. The tensile properties obtained at various aging times are listed in Table 2. The slip behaviour was studied by TEM analysis of thin foils obtained from regions adjacent to the fracture surface of the tensile samples. Representative micrographs of the slip character obtained with a [100] Zone axis show planar slip behavior observed for all the aging conditions employed, Figure 2(a-d). In the as-quenched condition only a few planar slip bands were observed at near fracture strain and only in certain grains as most of the deformation was homogeneously distributed. These slip bands may be associated with some $Al_3Li(\delta')$ being present in the as-quenched condition. It is observed that with progressive aging the fine relatively homogeneous slip changes to well defined intense slip bands in the peakaged condition. These bands become more prominent, the separation between them increases, and their width decreases as aging progresses. The slip bands still persist in the slightly overaged condition, Figure 3d, due to the presence of coherent and particially coherent precipitates. However, they are more diffused when compared to the peakaged condition suggesting a slight reversal to relatively homogeneous slip. The slip bands are most often contiguous across the subgrains. The changes in the monotonic properties and slip behaviour with aging show that the increase in the yield strength is accompanied by increasing amounts of strain localization in the slip bands and a concomitant decrease in the strain hardening exponent and ductility level.

The model developed by Gleiter and Hornbogen (5, 6) to describe dislocation shearing of misfit-free ordered particles may be applied to the present results. This model essentially states that the increase is in the critical resolved shear stress, $\Delta\tau_0$, associated with the strengthening precipitates is proportional to the product of their volume fraction, f, and their radius, r_0, given by

$$\Delta\tau_0 = 0 \cdot 28\ G^{-1/2} b^{-2} \gamma^{3/2} f^{1/2} r_0^{1/2} \qquad [1]$$

where γ is the antiphase boundary energy, G is the shear modulus and b is the Burgers vector. The increase in the strength of this alloy with aging qualitatively conforms to this model since both the volume fraction and radius of the precipitates would increase with aging. Furthermore, the present experimental results of slip behaviour and previous observations on Al-Li alloys indicate that the degree of work softening increases with $f^{1/2}r_0^{1/2}$. Due to the complexity of the alloy no quantitative measurements of the volume fractions, diameters, and coherency of the strengthening precipitates have been made.

The lower degree of strain localization associated with the underaged conditions results in a larger capacity for strain accumulation. This is reflected both by the high strain hardening exponents and strains to fracture. With aging the strain hardening exponents decrease due to the easier movement of the dislocations in the slip bands, and since the majority of the deformation is localized, the net effect is a lower ductility level. Beyond the peakaged condition, when the particle radius increases and coherency of θ' and T_1 is lost, there is a slight tendency for the strain hardening exponent to rise as the dislocations by-pass the particles and more homogenous slip occurs. Figure 2 (d) shows that, although planar slip bands were prevalent in this aging condition slip homogenization is substantial in between the slip bands.

The effect of slip localization on the tensile fracture can be observed in the fractographs, Figure 3, for an underaged and overaged alloy. In both cases transgranular shear fracture is observed along with some voids associated with the precipitate particles or slip band intersections. There was little evidence of intersubgranular fracture and voids associated with large equilibrium precipitates along the subgrain boundaries. TEM examination of a sample aged for 22 hours revealed extremely long slip bands due to only a slight misorientation of the subgrains. Thus, the transfer of plasticity to adjacent subgrains is easily attained and this minimizes intersubgranular cracking. However, as discussed later, the intersubgranular ductile fracture process was more prevalent in the notched fracture toughness samples. The ductile intergranular cracking associated with the equilibrium precipitates was often observed in the overaged condition as shown in Figure 3(b). The intense secondary cracking observed may be associated with the small recrystallized grains, but is most likely due to the presence of coarse equilibrium precipitate at the high angle grain boundaries, Figure 3b. The decrease in ductility with aging seems to be simply due to the extensive shear localization and accompanying slip band softening up to the peakaged condition. Beyond the peakaged condition larger equilibrium precipitates on the grain boundaries give rise to intergranular cracking which further reduces the ductility.

B. FATIGUE CRACK PROPAGATION

The results obtained from the fatigue crack propagation tests in air (R. H. 45 percent) and in vacuum ($<10^{-5}$ torr) are shown in Figures 4 (a) and (b) as plots of crack propagation rates da/dN versus the stress intensity factor range ΔK. The near threshold fatigue crack growth rates increased and the threshold stress intensities (Table 3) decreased with aging in both air and vacuum. Furthermore, there is an enhancement of the fatigue crack growth resistance in vacuum for similar aging conditions. The crack closure data obtained from the compliance versus load curves has been expressed (7) in the form of the closure stress intensity factor $\Delta K_{th}^{Cl} = K_{th}^{Cl} - K_{th}^{min}$ and the intrinsic (or effective) stress intensity factor $\Delta K_{th}^{i} = \Delta K_{th}^{obs} - \Delta K_{th}^{Cl}$, as shown in Table 3. These values indicate that the intrinsic fatigue crack growth thresholds follow the same trend as the apparent or observed values, for the aging conditions tested here.

The crack deflections for the alloy tested in the air in the under and overaged conditions appear to be due to the crack propagating along the slip bands. It was not surprising to observe extensive crack deflections in the overaged alloy since slip planarity is maintained and intense slip bands are available for crack propagation paths.

The fracture surface features shown in Figure 5, reveal that the mode of crack growth is highly crystallographic in the low crack growth rate region. Extensive faceting was observed in vacuum and the fracture surfaces were considerably rougher than those observed on the samples tested in air. The lower crack path tortuosity in air as compared to that in vacuum could be due to the aggressive air environment reducing the amount of irreversible plasticity needed for the crack to propagate (8). There is the possibility for oxide formation which might be expected to enhance closure and thus give an apparent higher threshold. However, Vasudevan et. al. (9) showed by SIMS that the thickness of the oxide for a similar 2020 alloy is considerably smaller than the crack opening displacement, suggesting that the oxide contribution, if any, would be negligible.

Figure 6 compares the results obtained for the Al-Cu-Li alloy with those of a thermomechanically processed 7475 (10). The fatigue crack growth rates and thresholds of the Al-Cu-Li alloy are better than those of the 80 μm

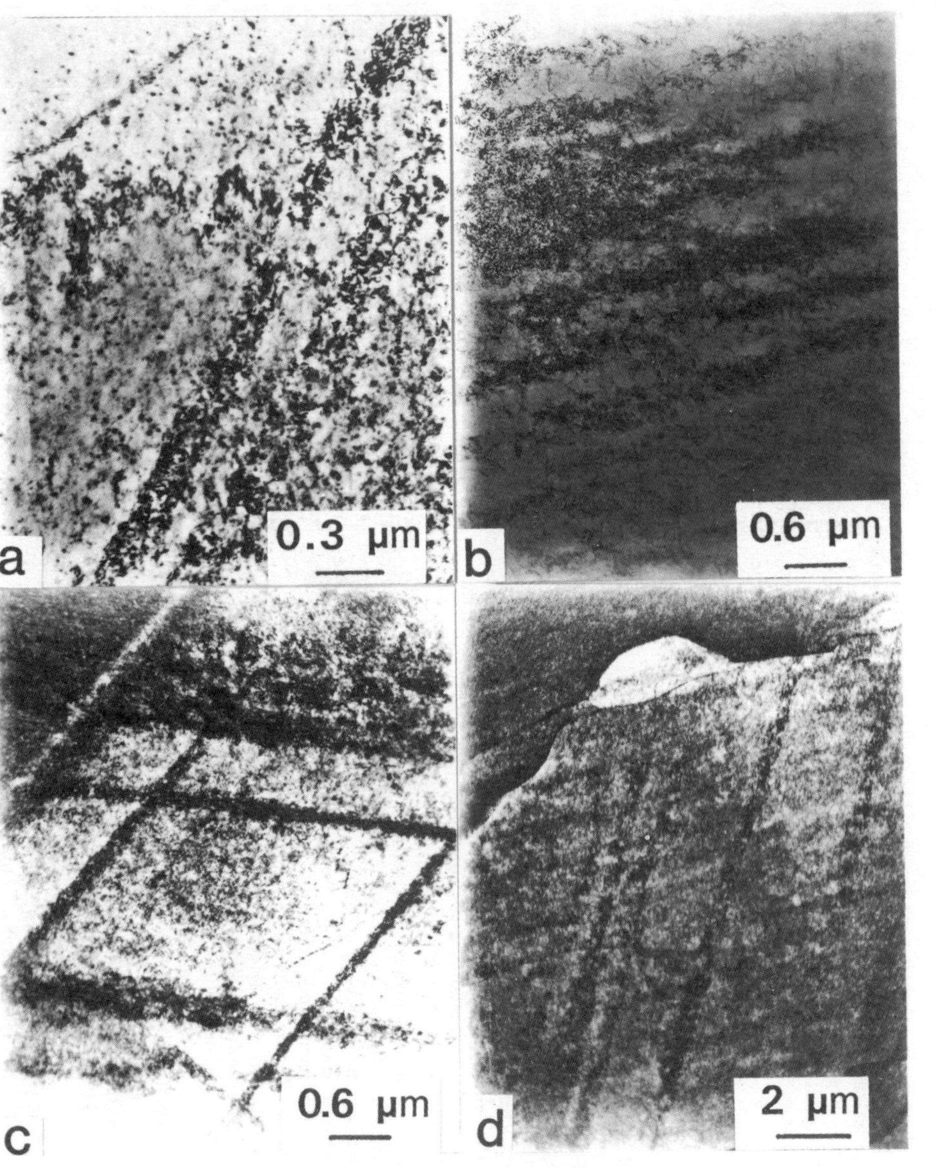

Fig. 2: Bright field TEMs with <100> zone axis of planar slip bands in fractured tensile specimens aged at 434K for (a) 0, (b) 10, (c) 17, and (d) 22 hrs.

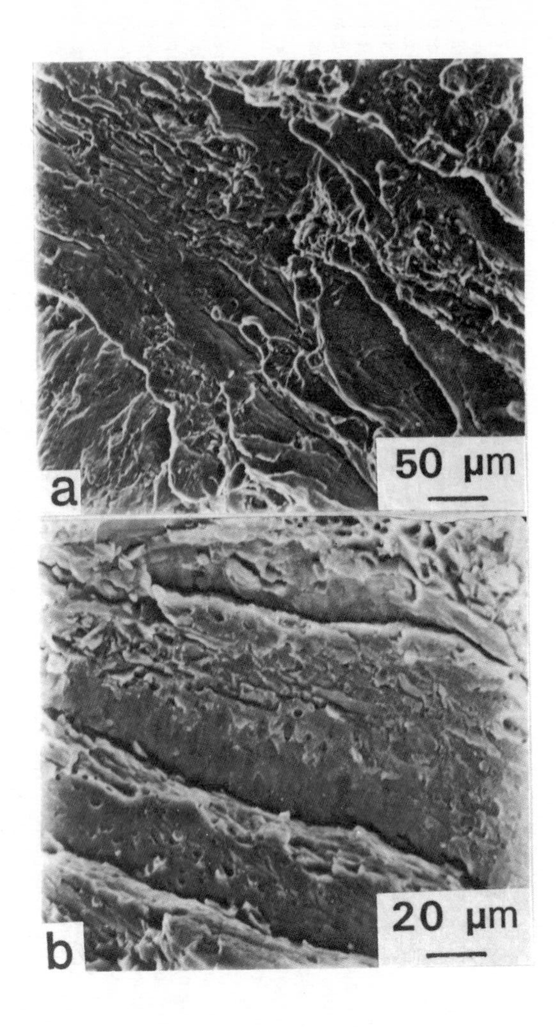

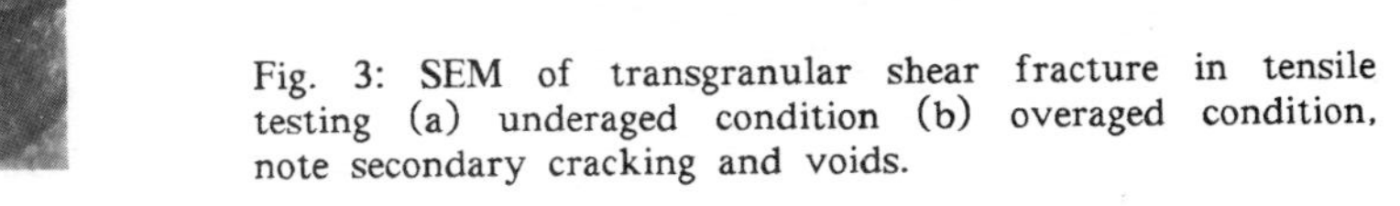

Fig. 3: SEM of transgranular shear fracture in tensile testing (a) underaged condition (b) overaged condition, note secondary cracking and voids.

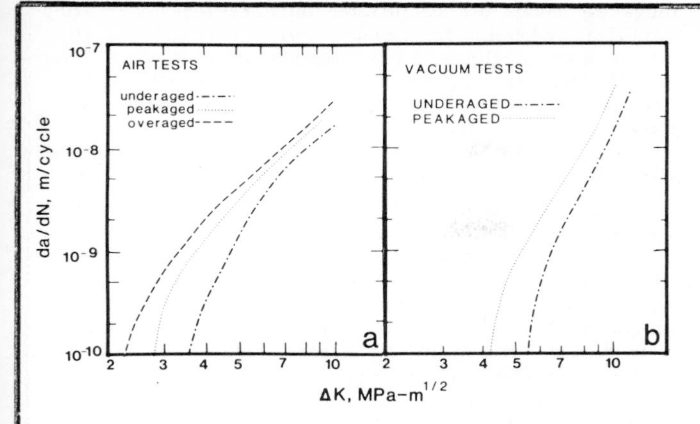

Fig. 4: Fatigue crack growth rates of the microstructural variants tested in (a) laboratory air (a R.H of 45 percent) and (b) vacuum (10^{-5}torr).

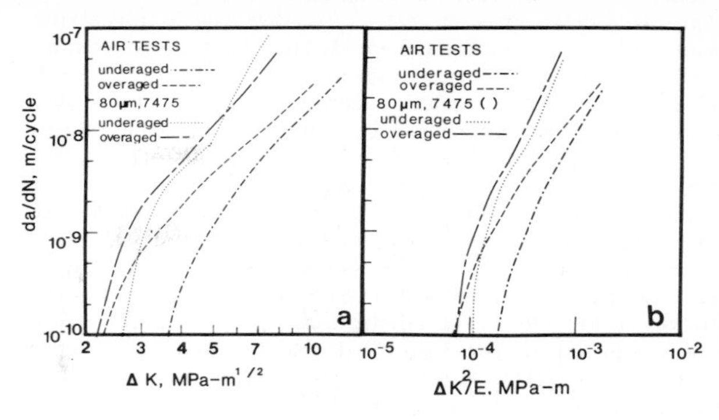

Fig. 6: A comparison of the fatigue crack growth rates to 7475 alloy as a function of ΔK (a), and as a function of normalized ΔK^2 with respect to Young's modulus E, (b).

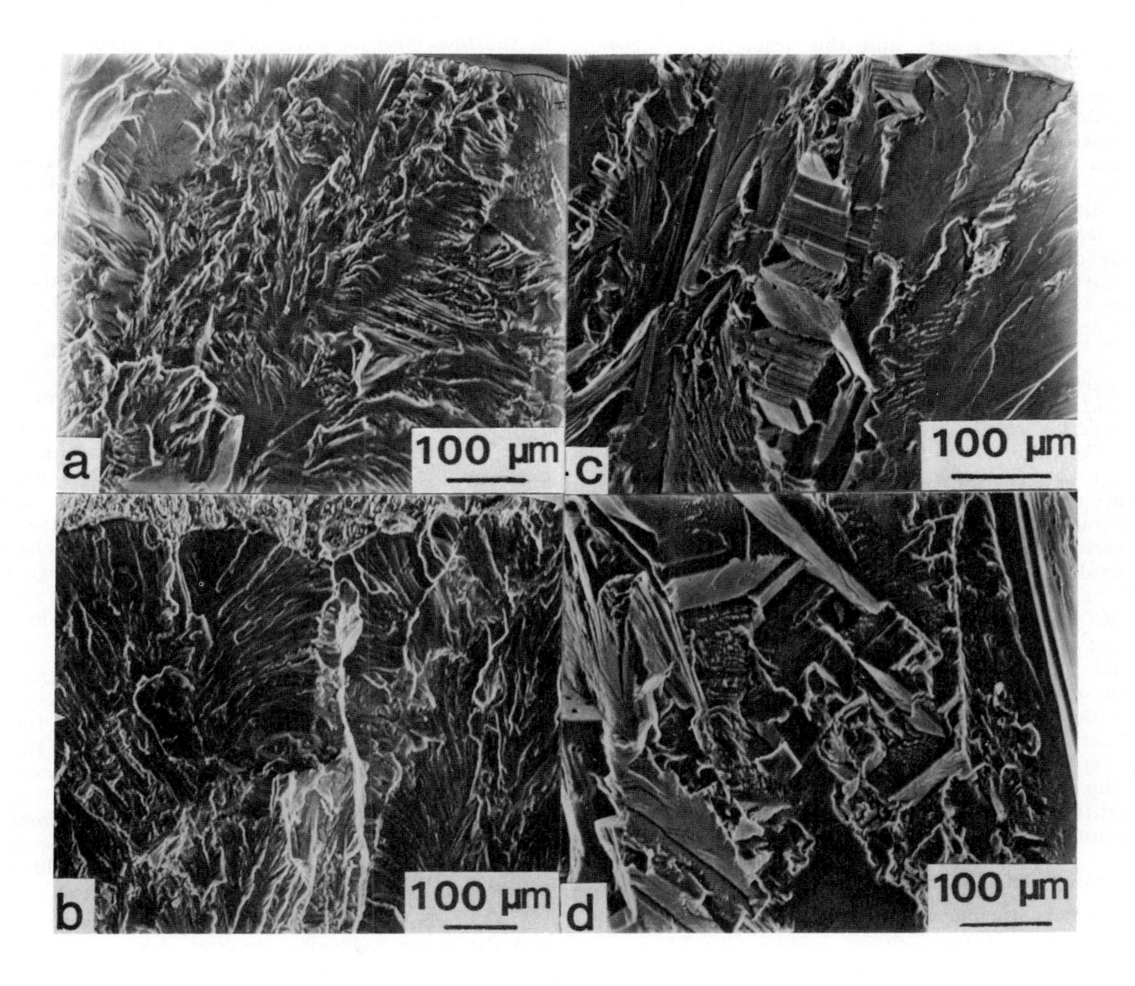

Fig. 5: Fractography in fatigue at near threshold (a) and (b), underaged and peakaged, tested in air, (c) and (d) underaged and peakaged, tested in vacuum.

grain size 7475. Several investigators have proposed that the fatigue crack growth rates decrease with increasing modulus (11), thus indirectly predicting that Li-containing Al alloys would have improved FCP resistance. Therefore, a more realistic comparison of the two materials may be obtained by normalizing the stress intensity range with Youngs' modulus since the yield stresses are approximately the same. As shown in the normalized curves of Figure 6 (b), the fatigue growth rate curves remain above the 7475 alloy suggesting that the improvement is not due to the modulus effect alone. In a similar comparison, Vasudevan et. al. (9) came to the conclusion that the crack deflections in the 2020 alloy were a major cause for improved crack growth resistance over that of a 7075 alloy. A comparison of the crack deflections observed in 7475 (10) with those observed here also show that this mechanism may account for the improved crack growth resistance observed, although differences in slip reversibility may also be a contributing factor.

The slip reversibility would be higher in the underaged alloy and the plastic strain accumulation would be lower for a given number of cycles. For the overaged alloys, the slip reversibility is reduced because of the precipitate bypass mechanism and this leads to a higher accumulation of plastic strain for the same number of cycles. This implies that for a given ΔK an overaged material would exhibit faster crack growth rates compared with an underaged material. The lowering of the crack growth rates in the underaged alloy can be directly attributed to the above slip processes since there is no substantial difference in the crack deflections. For the 7475 alloy tested in vacuum, Carter et. al. (10), showed that in spite of a major difference in the crack deflections between the under and overaged conditions the contribution from this microgeometrical effect (12) could not account for the overall differences in the crack growth resistance.

Our current results also suggest that the ΔK_{th}^{Cl} decreases and ΔK_{th}^{i} increases in vacuum. Firstly, the reduction in ΔK_{th}^{Cl} in vacuum may be due to a decreased propensity for load bearing oxide asperities. Secondly, although a rougher fracture surface or extensive faceting is observed in vacuum, the absence of oxide induced asperities could substantially decrease the mismatch between crack faces. In laboratory air, a possibility for crack tip oxidation is more likely and if the thickness of the oxide layer is comparable to the crack opening displacement, oxide induced crack closure may occur. Carter et. al. (11) proposed that slip reversibility could also increase in vacuum due to less oxide formation. Thus, although the fractography indicates extensive faceting it does not result in an increase of ΔK_{th}^{Cl} in vacuum. The effects of Mode II displacements behind the crack tip appear to be reduced in vacuum due to improved slip reversibility and the absence of load bearing oxide asperities. Thus intrinsic crack growth resistance, ΔK_{th}^{i} in vacuum is improved.

C. FRACTURE TOUGHNESS

The fracture toughness data collected over a wide range of aging times from compact tension specimens, are summarized in table 2. The results show decreasing fracture toughness with aging, typical of many other aluminum alloys such as 2020 (13). The accompanying change of yield stress with aging, Figure 7, shows that the K_{IC} decreases as the yield stress increases until the peakaged condition is attained beyond which the strength toughness combination decreases. A comparison of the K_{IC} values of this alloy to the commercial 2020 alloy containing Cu, Li, Mn, and Cd (Figure 7) shows that the toughness levels are comparable only at the lower aging condition, whereas for longer aging the present alloy retains high toughness. For example, at a strength level of 475 MPa, the K_{IC} values are 35 and 20 MPa—m$^{1/2}$ for the present alloy and the commercial 2020 alloy, respectively. In the same Figure, K_{IC} values recently obtained for an Al-2Li-4Mg alloy (13) and a modified 2020 alloy (14) are also shown for comparison purposes.

Extensive fractography of the fracture toughness samples revealed five different fracture processes, Figure 8. These are (i) transgranular shear band failure, Figure 8a (ii) secondary cracking along the recrystallized grains, Figure 8a,(iii) voids associated with necking of the subgrains, Figure 8b, (iv) some very rare occurrences of transgranular microvoid coalescence Figure 8c, and (v) ductile intergranular failure, Figure 8d. The major mechanism of failure in the under to peakaged conditions is transgranular shear band failure along with secondary cracking associated with recrystallized grains. In many instances failure was found to occur over several microns along the shear bands, Figure 8a. These observations suggest that the slip bands emanating from the crack tip have often extended unimpeded transgranularly across many subgrains and possibly several grains. This is due to the low misorientation between subgrains and the sharp deformation texture of the plate. Some transgranular microvoid coalescence associated with iron and silicon constituents is shown in the micrograph 8c, however, this occurrence was rare and did not appear to control the overall fracture. There was an entire change in the mechanism of fracture for the overaged condition, Figure 8d. This micrograph shows the presence of small microvoids along grain boundaries suggesting a ductile intergranular type of failure, associated with a weakening of the grain boundaries in the overaged condition. Starke, Sanders and Palmer (1) suggest that for Al-Li alloys the slip bands impinging on the grain boundaries transfer the strain to the soft PFZ's along the grain boundaries. These regions deform with ease and hence strain is localized here, giving rise to a more easily attainable fracture route along the grain boundaries.

The dependence of K_{IC} on the strain hardening exponent (15, 16, 17) for the Al-Cu-Li alloy under study is suggested by the micrographs of Figure 9, taken after unloading the compact tension samples at the onset of crack extension. These micrographs clearly illustrate that, in the underaged conditon the plastic zone wings are much wider than in the peakaged condition and furthermore roughly proportional to the square of the strain hardening exponent.

Due to the dispersion of the plasticity, the crack tip in the underaged condition experiences a lower strain concentration than in the peakaged condition for a given load level. Thus the load bearing capacity in the presence of a notch is higher in the underaged conditions. If the crack extension took place after a critical strain has accumulated (critical strain criterion) in front of the crack tip (or when the strain hardening capacity is exhausted) then the underaged samples would exhibit a higher fracture toughness.

The above relationship between the fracture toughness and the strain hardening exponent does not, however, offer an explanation at a microscopic level, since the plastic zones in front of the crack tip include the plasticity in intense slip bands as well as the plasticity between the bands. In order to incorporate microstructural features and/or the process zone it might be necessary to consider the extent of strain localization which is controlled by the character of the strengthening precipitates. Going from the under to the peakaged condition, the fracture toughness is limited by the slip band decohesion process

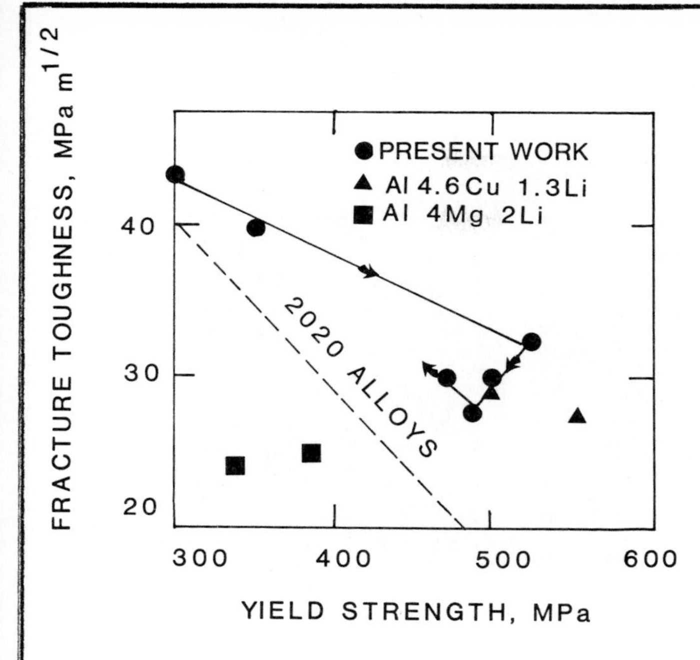

Fig. 7: Variation of fracture toughness, K_{IC}, with yield strength for the present alloy and a comparison to the data of other Al-Li-X alloys.

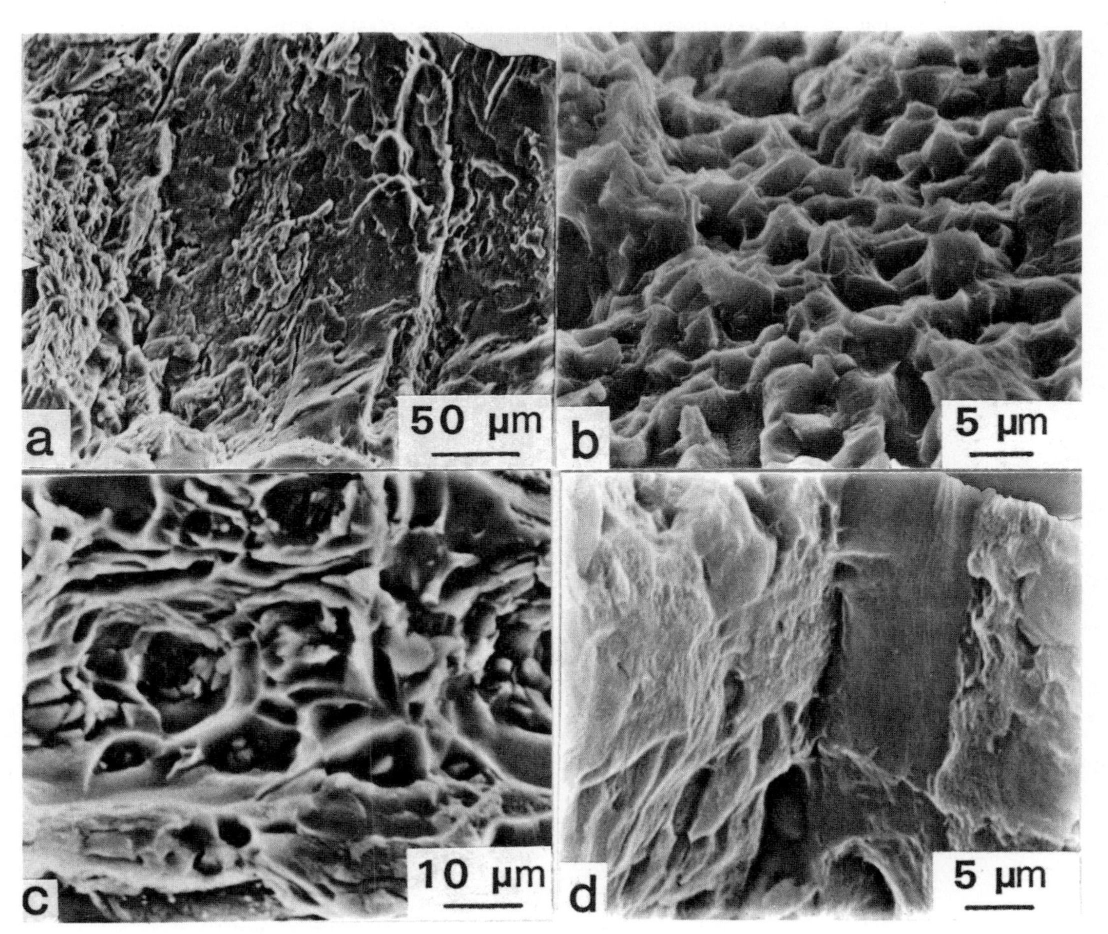

Fig. 8: SEM of K_{Ic} samples (a) shear band failure and secondary cracking, (b) necking of subgrains, (c) voids and (d) ductile intergranular fracture.

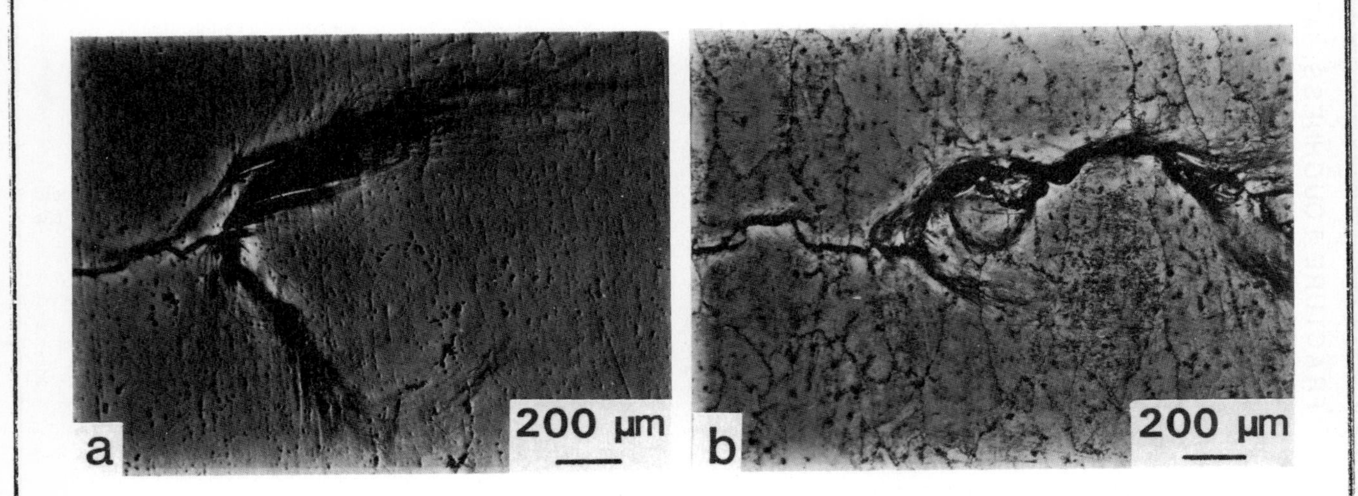

Fig. 9: Optical micrographs comparing the plastic zones, taken after unloading at the onset of initial crack extension in (a) underaged, and (b) peakaged conditions.

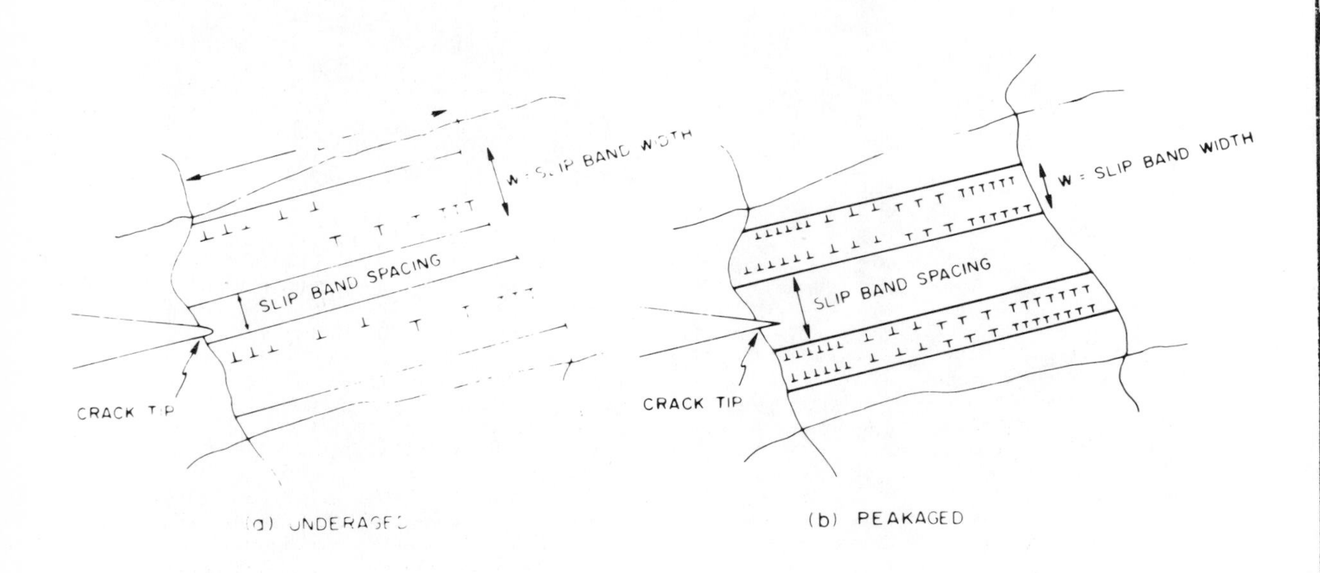

Fig. 10: Schematic illustrating that the slip band developed in front of the crack tip is wider in the (a) underaged condition as compared to the (b) peakaged condition.

without the involvement of any void growth. Thus with aging, as the strength increases and strain hardening capacity decreases the fracture process could occur due to an increase in the ease of strain localization. The extent of the slip localization inherently takes into account the effects of the strain hardening exponent, and the attainment of a critical strain in the slip band before decohesion occurs.

The critical strain for crack extension by transgranular slip band fracture is produced in the plastic zone in front of the crack tip when the crack is opened. The crack opening displacement, δ, may be equilibrated to the total number of dislocations, n, that are emitted from the crack tip during crack opening, times the Burgers vector (18), i.e.

$$\delta = nb \quad [2]$$

Some of these dislocations may be homogeneously distributed throughout the plastic zone and some may be concentrated in intense slip bands that lie within the plastic zone. Consequently, δ may be written as the sum of these contributions,

$$\delta = (n)_u b + (n)_{SB} b \quad [3]$$

where $(n)_u$ b represents the uniform displacement in the plastic zone and $(n)_{SB}b$ represents the displacement in the slip bands. The relative magnitude of each contribution would be dependent on the aging conditions since the tendency for strain localization increases as aging progresses up to the peakaged condition. Only those dislocations in the intense slip bands should be considered to contribute to the critical strain that is necessary for transgranular slip-band fracture. The density of these dislocations may be defined as

$$\rho_{SB} = (n)_{SB}/LW \quad [4]$$

where W is the slip band width and L the slip band length. Following Orowan, the shear strain in the slip band may be written as

$$\gamma_{SB} = \rho_{SB} bL$$

$$= \frac{(n)_{SB}}{LW} \cdot bL = (n)_{SB} b/W \quad [5]$$

A critical value of γ_{SB} would be necessary for fracture to occur. As aging proceeds a smaller δ is required to reach the critical strain since, for a given δ, $(n)_{SB}$ increases and W decreases with aging. We may express δ in terms of the stress intensity factor, K, the yield stress, σ_{ys}, and Young modulus, E, by the relationship (19)

$$\delta = \frac{K^2}{E\sigma_{ys}} \quad [6]$$

Substituting for δ and ignoring the contributions of those dislocations that are homogeneously distributed, we obtain

$$\frac{K^2}{E\sigma_{ys}} = \gamma_{SB} W \quad [7]$$

Initial crack extension occurs when γ_{SB} reaches a critical value, γ_{SB}^c, and the fracture toughness may be expressed as

$$K_{1c} = (E\sigma_{ys}\gamma_{SB}^c W)^{1/2} \quad [8]$$

Although the yield strength increases with aging, thus suggesting an increase in fracture toughness, the yield strength is accompanied by an increase in strain localization and a decrease in W, which, as equation [8] predicts, decreases the fracture toughness.

It is interesting to note here, that the above equation is similar to the equation 13(c) in reference (20), originally proposed by Ritchie et. al. (21) and recently reexamined by Ritchie and Thompson (20). In that equation, the authors proposed that crack initiation occurs when a critical fracture strain ϵ_f^*, is attained over a characteristic fracture distance l_0^* equal to half the distance between void initiating particles.

In the present equation fracture initiation takes place when a critical fracture strain is achieved over a slip band width W characteristic of the aging condition. The equation proposed here suggests that the precipitates that increase the strength also cause the critical fracture strain to be achieved over a decreasing slip band width resulting in a lower fracture toughness, until the peakaged condition is reached. This is schematically depicted in Figure 10.

CONCLUSIONS

The Al-Li-Cu alloy studied exhibits comparable strength and ductility levels to commercial high strength Al alloys. The yield strength in the peakaged condition is 520 Mpa and is accompanied by an 8 percent elongation. The strength increase is, however, accompanied by strain localization which curtails ductility. The general trend of increasing fatigue crack growth resistance with increasing slip localization or decreasing slip homogeneity remains after considering both crack closure and crack deflections. The fracture toughness in the peakaged condition is 32 Mpa-m$^{1/2}$. This value together with yield strength compares well with the currently available data on 2XXX and 7XXX alloys. It has been demonstrated that increasing slip localization accompanied by decreasing strain hardening exponents decrease the fracture toughness through a lower strain accumulation at the crack tip.

ACKNOWLEDGEMENTS

This research was sponsored by the Air Force Office of Scientific Research, United States Air Force Systems Command, under Grant AFOSR-83-0061, Dr. Ivan L. Caplan, program manager.

REFERENCES

1. E. A. Starke, Jr., T. H. Sanders, Jr., and I. G. Palmer: J. of Metals, 1981 vol. 33, No. 8, p. 24.

2. W. S. Miller, A. J. Cornish, A. P. Titchener, and D. A. Bennet: in Aluminum-Lithium Alloys II, E. A. Starke, Jr., andT. H. Sanders, Jr., eds., TMS - AIME, Warrendale, PA, p. 355, 1984.

3. T. H. Sanders, Jr., and E. A. Starke, Jr., eds., Aluminum-Lithium Alloys, TMS - AIME, 1981.

4. E. A. Starke, Jr., and T. H. Sanders, Jr., eds., Aluminum-Lithium Alloys II, TMS - AIME, Warrendale, PA, 1984.

5. H. Gleiter and E. Hornbogen: Phys. Status. Solidi. 1964, 12, p. 235.

6. H. Gleiter and E. Hornbogen: Mater. Sci. Eng. 1967/68, 2, p. 285.
7. C. J. Beevers: in Fatigue Thresholds, Fundamentals and Engineering Applications, J. Backlund, A. F. Blom, and C. J. Beevers, eds., Engineering Materials Advisory Services, Ltd., West Midlands, UK, 1981, Vol. 1, p. 257.
8. J. Petit, B. Bouchet, C. Goss and J. de Fouguet: Fracture, proceedings 4th Int. Conf. on Fracture (ICF4), D. M. R. Taplin, ed., University of Waterloo Press, Waterloo, Canada, 1977, vol. 2, p. 687.
9. A. K. Vasudevan, P. E. Bretz, A. C. Miller, and S. Suresh: Mater. Sci. and Eng., 1984, 64, p. 113.
10. R. D. Carter, E. W. Lee, E. A. Starke, Jr., and C. J. Beevers: Metall. Trans. A, 1984, vol. 15A, p. 555.
11. R. L. Donahue, H. M. Clarke, P. Alonanio, R. Kumble, and A. J. McEvily: Inter. J. of Fract. Mech., 1972, 8, p. 209.
12. S. Suresh: Metall. Trans., 1983, vol. 14A, p. 2375.
13. S. J. Haris, K. Dinsdale, B. Noble: in Al-Li Alloys II, E. A. Starke, Jr., and T. H. Sanders, Jr., eds., TMS - AIME, Warrendale, PA.
14. W. X. Feng, F. S. Lin, and E. A. Starke, Jr.: Metall. Trans. A, 1984, 15, p. 1209.
15. G. G. Garret, and J. F. Knott: Metall. Trans. A., 1978, vol. 9A, p. 1187.
16. G. T. Hahn and A. R. Rosenfield: Metall. Trans. A, 1975, 6A, p. 653.
17. J. G. Rinker, M. Marek and T. H. Sanders, Jr.: in Aluminum-Lithium Alloys, E. A. Starke, Jr., and T. H. Sanders, Jr., eds. TMS - AIME, Warrendale, PA, p. 597, 1984.
18. Johaness Weertman: in Fatigue and Microstructure, M. Meshii, ed. American Society for Metals, Metals Park, Ohio, 1979, p. 279.
19. J. F. Knott: in Fundamentals of Fracture Mechanics, Butterworths & Co. (Publishers) Ltd. London, UK, 1973.
20. R. O. Ritchie and A. W. Thompson: Metall. Trans. A, 1985, 16A, p. 233.
21. R. O. Ritchie, W. L. Server and R. A. Wullaert: Metall. Trans. A, 1979, 10A, p. 1557.

Constant amplitude and post-overload fatigue crack growth in Al–Li alloys

J PETIT, S SURESH,
A K VASUDÉVAN and R C MALCOLM

JP and SS are with the Division of Engineering, Brown University, Providence, Rhode Island 02912, USA.
AKV and RCM are with the Alloy Technology Division, Alcoa Laboratories, Alcoa Center, Pennsylvania 15069, USA.
JP was formerly with ENSMA, UA CNRS 863, 86034 Poitiers Cedex, France

SYNOPSIS

The influence of lithium and copper compositions (1.1 Li/4.5 Cu, 2.1 Li/3 Cu and 2.9 Li/1 Cu in wt pct) on constant amplitude fatigue crack growth rates and post overload crack growth retardation is examined in alloys belonging to the Aℓ-Li-Cu-Zr system tested *in vacuo* and moist air environments. Substantial improvements in fatigue crack growth resistance, threshold level and retardation are observed with increasing Li content. Several intrinsic mechanisms associated with intrinsic microstructural and environmental effects contribute to the differences in fatigue behavior among the alloys studied. An important aspect of such beneficial crack growth resistance stems from the crystallographic crack growth mechanism promoted by the presence of δ' precipitates in alloys of high lithium content.

INTRODUCTION

Previous work on the high cycle fatigue behavior of aluminum alloy 2020-T651 (Aℓ, 4.5 Cu, 1.1 Li, 0.52 Mn, 0.2 Cd) [1-4] showed that this alloy is characterized by a superior crack growth resistance as compared to widely used 7075-T651. Such beneficial behavior has been shown to be even more pronounced in the near threshold regime [3]. Initial studies [1-4] have attributed this to the highly inhomogeneous slip behavior induced by ordered δ' ($Aℓ_3Li$) precipitates. A recent study carried out by Vasudevan and Suresh [5] on high purity Aℓ-Li-Cu-Zr-X alloys has shown that increasing the Li/Cu ratio in aluminum alloys leads to an improved resistance to cyclic crack growth in room temperature moist air environment. The intent of this paper is to document some new results obtained on the same high purity alloys on near threshold crack growth *in vacuo* and on retardation following single periodic overloads in a moist air environment at room temperature.

EXPERIMENTAL PROCEDURE

The nominal compositions (wt pct) of the three lithium-containing high-purity aluminum alloys investigated in this study are listed in Table I. The 12.74 mm thick plates of alloys A, B and C had been hot rolled at an initial temperature of 520°C (after preheating at 520°C for 8 hrs) and were solution-treated at 528°C (Alloy A) or 551°C (Alloys B and C) for one hour, cold-water quenched and stretched by 2 percent. Corresponding room temperature mechanical properties are reported in Table II. The microstructures of the three materials were predominantly unrecrystallized with subgrain size of 2 to 3 μm with some partially recrystallized (< 10 percent) grains. The spacing between high angle grain boundaries was about 50 to 100 μm in the direction of crack advance. Detailed descriptions of microstructures are provided in reference [5]. Constant load amplitude fatigue crack growth experiments were performed on 6.35 mm thick compact (CT) specimens (W = 50 mm) machined in L-T orientation, in room temperature (~23°C) 30% RH moist air and vacuum (< 10^{-4}Pa) environments, at frequency of 25 Hz and load ratio of 0.05 and 0.5. Single periodic overload tests were performed on 6.35 mm thick CCT specimens machined in the L-T orientation, in room temperature (~95 percent relative humidity) moist air environment, at a frequency of 25 Hz, a load ratio of 0.33, and an overload ratio (overload peak load/baseline peak load) of 1.8. The crack length was monitored using an optical method for vacuum tests and by compliance method for moist air tests.

RESULTS AND DISCUSSIONS

Near Threshold Crack Growth

Figure 1 shows the variation of fatigue crack propagation rates da/dN with stress intensity factor

TABLE I

Composition (WT %) of Alloys Studied

Material	Li	Cu	Zr	Aℓ	(Li/Cu) Ratio*
A	1.1	4.6	0.17	Bal	2.2
B	2.1	2.9	0.12	Bal	6.5
C	2.9	1.1	0.11	Bal	25.2

Fe (0.06%), Si (0.04%), Ti (0.01%), for all materials.

*The (Li/Cu) ratio is expressed as a ratio of the atomic fractions.

TABLE II

Room Temperature Mechanical Properties

Material	Peak Aged Treatment	σ_y (MPa)	σ_{UTS} (MPa)	El. Mod. (GPa)	Elong. (Pct)
A	64h at 160°C	531	593	76.5	12
B	18h at 191°C	490	544	79.2	10
C	32h at 191°C	434	524	82.7	6

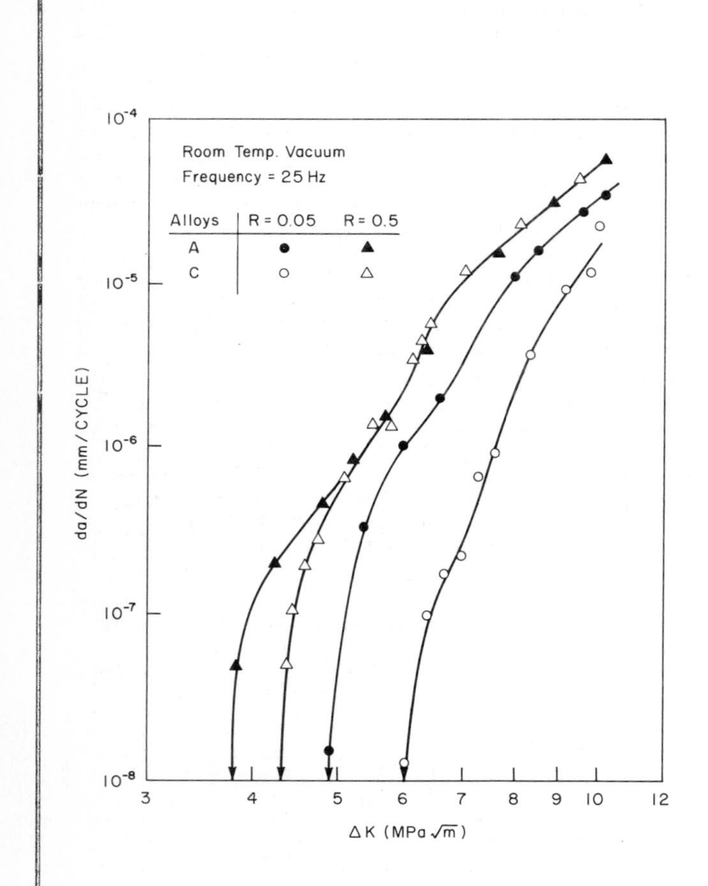

Figure 1. Constant amplitude fatigue crack growth data in vacuum for the A and C alloys at R=0.05 and R=0.5.

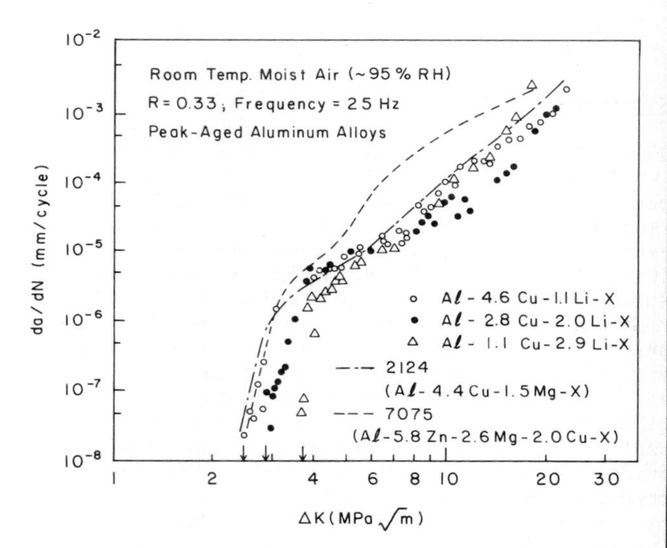

Figure 2. Constant amplitude fatigue crack growth data in 95% RH moist air at R=0.33 for the A, B and C alloys. The dashed lines indicate typical upper bound fatigue crack growth behavior of peak-aged alloys 2124 T851 and 7075 T651.

range ΔK for the A and C alloys tested in vacuum. It is seen that the resistance to constant amplitude fatigue crack growth for alloy C is superior to that of alloy A. Figure 2 reports results previously obtained in 95 percent RH moist air, at R=0.33, compared to typical crack growth data for two widely known alloys, i.e., 7075-T651 and 2124-T851 [5]. The variation of threshold stress intensity factor range ΔK_o is plotted in Figure 3 as a function of Li/Cu ratio expressed in atomic fractions. This figure reveals that ΔK_o increases with increasing values of Li/Cu ratio in both vacuum and moist air.

Prior experimental work has shown that while lithium additions may enhance fracture surface oxidation during near-threshold fatigue crack growth [3], the presence of Cu tends to inhibit oxide formation [3,6,7]. Consistent with this notion, our Secondary Ion Mass Spectroscopy (SIMS) analyses of fatigue fracture surfaces for the present alloys reveal a greater propensity for the formation of oxide layers within near-threshold cracks in alloy C than in alloy A. Such differences in oxidation behavior and the resulting oxide-induced crack closure processes [15], however, can account for only a fraction of the observed differences in fatigue threshold ΔK_o values for the alloys studied. This inference is based on the comparison of measured excess oxide thickness values with the computed values of maximum crack opening displacements corresponding to threshold ΔK_o. Furthermore, the large apparent difference in the crack growth data for alloys A and C (at R=0.05) cannot be rationalized based on the variations in elastic modulus (see Table II). There is a substantial reduction in the difference between the ΔK_o values for alloys A and C at R=0.5 from those at R=0.1. Our closure measurements (described in detail elsewhere [16]) suggest that this reduction is accompanied by a significant drop in closure level at R=0.5. It is of interest to note in Figures 1 and 3 that there is a precipitous drop in ΔK_o with increasing R ratio. This trend is different from the one generally observed *in vacuo* on high strength (non-lithium-containing) aluminum alloys where ΔK_o is almost independent of R [7-10].

Optical micrographic observations of the crack growth profiles for the A and C alloys [5] are shown in Figures 4 and 5. Material A (Li/Cu ratio of 2.2) shows a very linear crack growth path (Fig. 4) while material C (Li/Cu ratio of 25.2) exhibits a highly nonlinear path where deflections in crack profile are clearly associated to high-angle grain boundaries. Qualitatively, these results are consistent with the assumption of an enhancement in crystallographic crack growth mechanism with increasing Li content, i.e., with increasing amounts of coherent and shearable δ' precipitates in the matrix.

In Figure 6 typical crack growth results in vacuum at R=0.1 and 35 Hz for 7075-T351, 7075-T651 and 2024-T351 alloys [11,12,13] are plotted for comparison purposes with the A and C data at R=0.1 in vacuum. Near threshold data for the C alloy is very similar to the one of the 7075-T351 (an underaged microstructure) which presents a crystallographic transgranular mode of failure in a matrix containing shearable hardening GP zones of very small size (about 10 Å diameter) [12] and where Cu and Mg in solid solution lower SFE and restrict cross-slip [14]. In such cases a predominantly planar slip mode is observed leading to very low crack growth. A similar mechanism induced by δ' precipitates can be expected in the C alloy. The similarity in the behavior of alloy A. 7075-T651 and 2024-T851 is consistent with a smoother mode of failure in the presence of noncoherent particles of larger size than GP zones resulting in a more wavy slip mode.

In the mid-rate range (da/dN > 10^{-6} mm) the analysis of the crack growth data for the three reference alloys appears to be more complex. Detailed discussions [16] which include closure measurements and crack surface observations would give a more precise insight into the fatigue behavior of these lithium-containing alloys.

The present results, however, exhibit an increase in fatigue resistance with increasing Li/Cu ratio which is qualitatively consistent with an increase in crack growth path tortuosity. The beneficial effects of lowered crack tip stress intensity due to crack path deviations persist even at high R values although the associated closure levels diminish. This is evident from both the crack path observations as well as the measured ΔK_o and closure values. While surface observations need not necessarily reflect crack kinking behavior all along the specimen thickness, models for fatigue crack deflection [17,18] appear to provide a theoretical justification for the role of crack path during fatigue.

It should be noted that near-threshold crack growth characteristics of Al-Li alloys are influenced by several concurrent mechanisms involving intrinsic microstructural and environmental effects and extrinsic deflection and closure phenomena. For the present materials, it appears that microstructurally-influenced crack path changes provide one of the important contributions to the differences in fatigue behavior among the alloys studied. A detailed discussion of the individual contributions from all the mechanisms is beyond the scope of this short communication and will be considered in [16].

Single Periodic Overload Fatigue Crack Growth Tests

Figure 7 shows the fatigue crack length plotted against the number of post-overload cycles for the three lithium-containing alloys and for the 7075-T651 alloy. Tensile overloads (ratio = 1.8) were applied during constant amplitude cycles at ΔK = 7 MPa$\sqrt{m}$ in room temperature (95% RH) moist air. Increasing Li/Cu ratio leads to a very large increase in retardation. Based on micromechanical modelling of the experimental evidence available in the literature [19-22], it has been proposed [23,24] that post-overload retardation characteristics in aluminum alloys are influenced by the micromechanisms of near-threshold (constant amplitude) fatigue crack

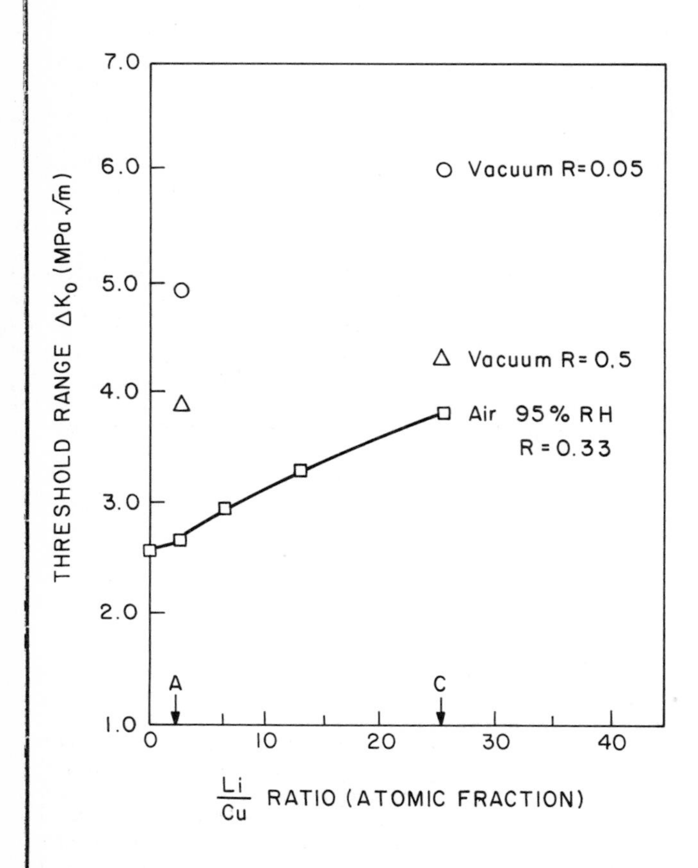

Figure 3. Variation of threshold stress intensity range ΔK_o with Li/Cu ratio expressed in terms of atomic fractions.

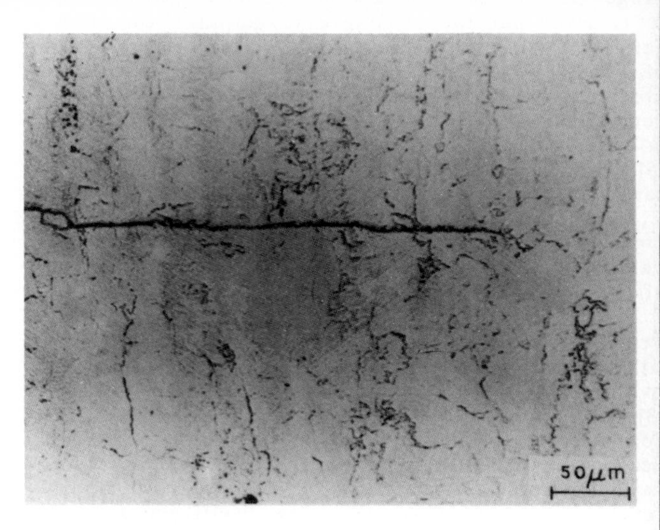

Figure 4. Fatigue crack profile for Alloy A; typical ΔK range $\approx$ 2.6 to 6.0 MPa$\sqrt{m}$.

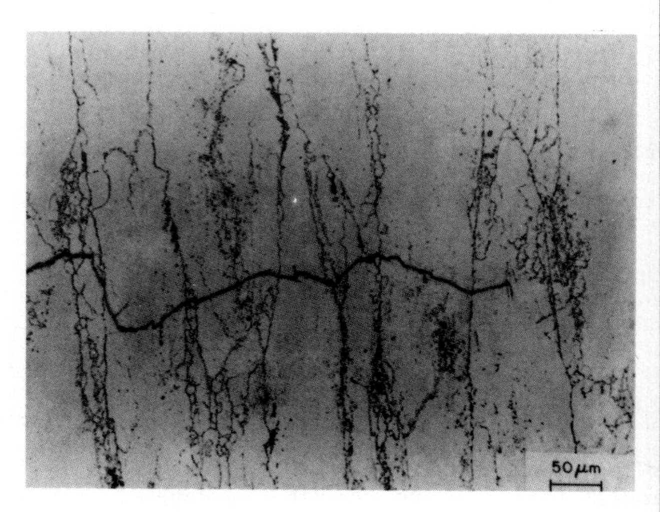

Figure 5. Fatigue crack profile for Alloy C; typical ΔK range $\approx$ 3.8 to 8.0 MPa$\sqrt{m}$.

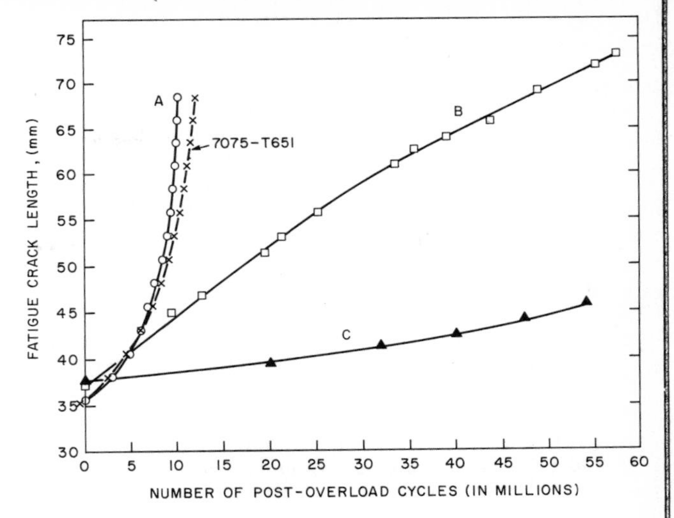

Figure 7. Variation of fatigue crack length with the number of post-overload cycles for the Aℓ-Li alloys and 7075-T651 tested in moist air.

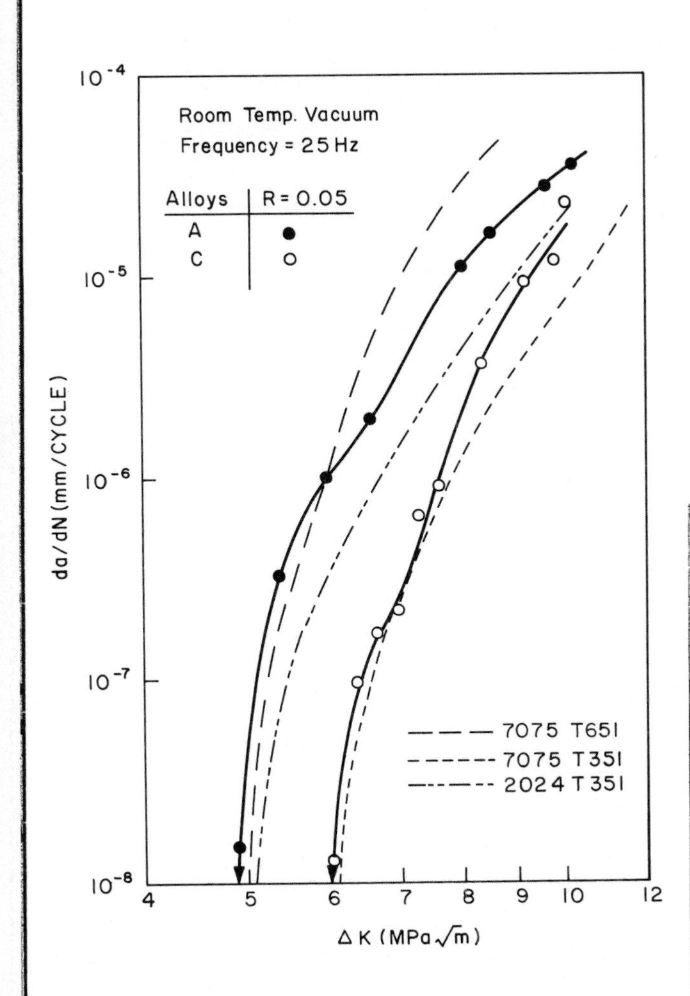

Figure 6. Constant amplitude fatigue crack growth data in vacuum at R=0.05 for the A and C alloys (reported from Fig. 1) as compared to fatigue crack growth behavior in vacuum at a low R ratio (0.1) of 2024 T351, 7075 T651 and 7075 T351 alloys tested in nearly similar conditions from [11] and [13].

growth (when post-overload growth rates are in the near-threshold regime). Indeed, a careful examination of the microstructural effects on post-overload growth in wrought aluminum 7075 corroborates this hypothesis [25]. The present results on alloys A, B and C are in accordance with these results; the most marked retardation occurs in alloy C where a strong propensity for crystallographic crack advance (and for crack deflection) is observed in both the near-threshold and post-overload regimes.

CONCLUSIONS

Constant amplitude fatigue crack growth and post-overload retardation in Aℓ-Li alloys are strongly dictated by several concurrent mechanisms involving intrinsic microstructural effects as well as extrinsic (mechanical) effects due to crack deflection and closure. Increasing Li/Cu ratio results essentially in an enhancement in crystallographic crack growth mechanism for near-threshold fatigue growth cycling. The occurrence of such a mechanism induced by the coherent and shearable δ' precipitates existing in the higher lithium content alloys results in:

(a) A substantial increase of the threshold stress intensity range and of near-threshold crack growth resistance with increasing Li/Cu ratio, both *in vacuo* and moist air for all R values studied.

(b) A very large increase in retardation (induced by single periodic overload) with increasing Li/Cu ratio.

REFERENCES

[1] T. H. Sanders, Jr., Final Report NAVAIR Contract No. 62269-76-C-0217, Alcoa, Pittsburgh, PA (1979).

[2] E. J. Coyne, Jr., T. H. Sanders, Jr. and E. A. Starke, Jr., *Aluminum-Lithium Alloys*, TMS AIME, Warrendale, PA (1984) p. 293.

[3] A. K. Vasudevan, P. E. Bretz, A. C. Miller and S. Suresh *Mat. Sci. Eng.* (1984) 62, p. 113.

[4] S. J. Harris, B. Noble and K. Dinsdale, *Aluminum-Lithium Alloys*, TMS AIME, Warrendale, PA (1984) p. 219.

[5] A. K. Vasudevan and S. Suresh, *Met. Trans.* (1985) 16A, p. 475.

[6] F. S. Lin and E. A. Starke, Jr., *Mat. Sci. Eng.* (1980) 43, p. 65.

[7] S. Suresh, A. K. Vasudevan and P. E. Bretz, *Met. Trans.* (1984) 15A, p. 1.

[8] B. R. Kirby and C. J. Beevers, *Fat. Eng. Mat. and Struct.* (1979) 1, p. 203.

[9] J. Petit and J. L. Maillard, *Scripta Met.* (1980) 14, p. 163.

[10] M. C. Lafarie-Frenot and C. Gasc, *Fat. Eng. Mat. and Struct.* (1984) 6, p. 329.

[11] J. Petit, P. Renaud and P. Violan, "Effect of Microstructure on Crack Growth in a High Aluminum Alloy", K. L. Maurer and F. E. Matzer, eds., EMAS Pub., Warley (UK) (1983) 2, p. 426.

[12] J. Petit, "Fatigue Crack Growth Threshold Concepts", D. Davidson and S. Suresh, eds., TMS AIME Pub. (1984) p. 3.

[13] J. Petit, B. Bouchet, C. Gasc and J. de Fouquet, "Fracture", Proc. ICF4, University of Waterloo Press, Canada (1977) 867.

[14] R. J. Selines, "The Fatigue Behavior of High Strength Aluminum Alloys", Ph.D. Thesis, MIT (1971).

[15] S. Suresh, G. F. Zamiski and R. O. Ritchie, *Met. Trans.* (1981) 12A, p. 1435.

[16] S. Suresh, J. Petit and A. K. Vasudevan, manuscript in preparation.

[17] S. Suresh, *Met. Trans.* (1985) 14A, p. 2375.

[18] S. Suresh, *Met. Trans.* (1985) 16A, p. 249.

[19] S. W. Hopkins, C. A. Rau, G. R. Leverant and A. Yuen, ASTM STP 595 (1976) p. 125.

[20] P. J. Bernard, T. C. Lindley and C. Richards, *Metal Sci.* (1977) 12, p. 390.

[21] N. Ranganathan, J. Petit and A. Nadeau, Proc. ECF2, Darmstadt, VDI-Z (1979) 18, p. 337.

[22] N. Ranganathan, J. Petit and B. Bouchet, *Eng. Fract. Mech.* (1979) p. 775.

[23] S. Suresh, *Scripta Met.* (1982) 16, p. 995.

[24] S. Suresh, *Eng. Fract. Mech.* (1983) 18, p. 557.

[25] S. Suresh and A. K. Vasudevan, "Fatigue Crack Growth Threshold Concept", D. Davidson and S. Suresh, eds., TMS AIME Pub. (1984), p. 361.

Formation of solute-depleted surfaces in Al–Li–Cu–Mg–Zr alloys and their influence on mechanical properties

S FOX, H M FLOWER and D S McDARMAID

SF and HMF are with Imperial College, London.
DSMcD is with the Royal Aircraft Establishment, Farnborough, UK

SYNOPSIS

The formation of Li and Mg depleted surfaces, in an Al-Li-Cu-Mg-Zr alloy, after solution heat-treatment in air have been investigated. The depletion layer has been characterised by microhardness measurement and electron probe microanalysis together with light and electron microscopy. Li and Mg depletion profiles were found to be similar in extent, as predicted by a simple model for depletion assuming solid state diffusion controlled migration of Li and Mg to the free surface which behaves, via oxidation, as an infinite solute sink.

Solute depletion has been found to be responsible for a slight decrease in S-N fatigue properties. Polishing after heat-treatment causes an improvement in fatigue strength in the high life regime. Conventional pickling and anodising processes applied to material after heat-treatment degrade fatigue resistance as a consequence of the formation of corrosion pits formed preferentially at grain boundaries.

INTRODUCTION

Aluminium based alloys containing lithium and magnesium may be expected to suffer from considerable solute depletion at free surfaces, when exposed to air in the temperature range encountered during solution treatment, as a result of selective oxidation. The free energy changes [1,2] for possible primary oxidation reactions at the typical solution treatment temperature of 530°C are shown in table 1. The thermodynamic driving force for the oxidation of lithium and magnesium can be seen to be much greater than that for copper and slightly greater than that for oxidation of the parent aluminium. This together with the similar diffusivities of lithium and magnesium [3], as shown in table 2, indicates that preferential oxidation of these solutes will occur during solution treatment, leading to a lithium/magnesium depleted surface layer. Certainly previous studies of the oxidation reactions [2,4,5] confirm that oxides of magnesium and lithium combined with aluminium form preferentially on exposure to air in the range 480 to 575°C.

The depletion of lithium must affect the mechanical properties of the product sheet or plate. The removal of lithium, and hence the capacity to age harden by δ' formation, has been shown to greatly reduce the surface strength of the material as determined by microhardness measurements [6,7]. Additionally the presence of a soft surface layer is known to have an adverse effect on fatigue properties. Lithium loss would thus be expected to have the same effect on fatigue crack initiation as the cladding of aluminium alloys [8,9,10], resulting in a reduction in fatigue strength.

It is the purpose of this work to investigate the formation and compositional profiles of the depletion layer and to relate its presence to the fatigue resistance of pre-production quality sheet material prepared to the DTDXXXA specification [6]. Also the effect of modifying the surface by processes widely used in aerospace applications - pickling and anodising are determined.

EXPERIMENTAL

The sheet material used was of DTDXXXA specification (see table 3 for detailed composition) and supplied in the as rolled condition. Measurements of solute depletion profiles were carried out on 3 and 5mm thick sheet but fatigue and tensile testing were confined to 3mm sheet where the effects of surface depletion should be greatest.

The recommended heat-treatment practice for this material is solution treatment for 20 min. at 530°C quenching in a poly-alkalene glycol/water mixture at 30°C followed by artificial ageing for 5 hours at 187°C. In this work solution treatment times ranging from 20 min. to 4 hours and temperatures of between 480°C and 575°C were

TABLE 1

Elements	Oxide	Free Energy Per Mole of Oxygen at 530°C (kJ)
Cu	Cu_2O	-223
Al	Al_2O_3	-950
Li + Al	$LiAl_5O_8$	-962
Li	Li_2O	-988
Li + Al	$LiAlO_2$	-1018
Mg	MgO	-1029

TABLE 2

Element	Do(cm^2/s)	Q(KJ/mol)
Li	4.5	139.2
Mg	4.4	140.2
Al	1.7	142.3

TABLE 3 Alloy Composition (wt%)

SHEET	Li	Cu	Mg	Zr	Si	Fe	Ti
3mm	2.57	1.24	0.68	0.13	0.06	0.06	0.015
5mm	2.62	1.22	0.70	0.09	0.03	0.05	0.013

Table 4 Surface Treatments Applied to Fatigue Testpieces

Surface Treatment	Testing Condition
Prior depletion removed before solution treatment (S.T.) in dry argon	Ground Surface, as heat-treated (minimal depletion).
Standard S.T. (20 min) Prolonged S.T. (4 hrs.)	As received surface as heat-treated.
S.T. then polished retaining depletion layer.	Polished surface
S.T. depletion removed then polished	Polished surface
S.T. then pickled.	Pickled surface
S.T. then pickled and anodised.	Pickled and anodised surface.

Table 5 Width of precipitate depleted surface layer after S.T. for 1 hr and overageing for 17 hours at 300°C.

Temperature (°C)	Width of Precipitate depleted layer (μm)	Width of microhardness depleted layer (μm)
480	16	60
520	24	120
575	35	550

Table 6 Effect of solution treatment time on tensile properties.

Time at 530°C	0.2% Proof Stress (MPa)	Tensile Strength (MPa)	Young's Modulus (GPa)	Elongation on 36mm (%)
20 min.	387	472	76.9	3.6
4 hrs.	362	456	75.6	3.6

employed to investigate the formation of solute depleted surfaces and their influence on mechanical properties. Light microscopy of standard and overaged material was carried out to determine the total depletion depth. Microhardness traverses (50g and 100g loads) were used to measure loss in surface hardness. Magnesium depletion was measured directly by electron probe microanalysis. Thin foils were prepared for transmission electron microscopy at selected depths through the depletion layer and in the undepleted interior to examine precipitate distributions.

In order to determine any loss in strength due to prolonged solution heat-treatment (S.H.T.), 50mm tensile specimens were cut from material heat-treated for 20 min. and 4 hours and tested in the standard aged condition.

Fatigue testing was carried out in tension on Haigh machines operating at 40Hz. The overall testpiece dimensions were 152 x 38 x 3mm. The range of maximum stress applied to the specimens was between 175 MPa and 350 MPa with a stress ratio of R=0.1. In all cases the stress axis was in the rolling direction. After testing, the fracture surfaces were examined by scanning electron microscopy. In addition to the 20 min. and 4 hour heat-treatments, which were carried out in air, some testpieces were solution heat-treated in dry argon to minimise lithium/magnesium depletion.

A range of surface treatments were applied to the specimens both before and after heat-treatment, these were grinding, polishing, chromic/sulphuric acid pickling and chromic acid anodising. The pickling and anodising treatments were the same as those which are applied to conventional aluminium alloys. The details of all these surface treatments are shown in table 4.

RESULTS

In order to determine the effect of increasing solution heat-treatment temperature on the through thickness hardness of 5mm sheet the initial "as received" solute depleted surface layer (from prior thermomechanical processing) was removed before exposure for 1 hour at 480, 520 and 575°C respectively. This was followed by water quenching and ageing for 16 hours at 190°C. Microhardness measurements were made and normalised to the centreline, or bulk, hardness of the material heat-treated at 520 and 575°C. The sheet heat-treated at 480°C was not fully solutionised before ageing and therefore has a lower overall hardness. The results are presented in Figure 1 which shows in each case a soft surface layer with hardness rising to a constant "plateau" value in the interior. The plateau region of the three profiles appears further into the sheet with increasing temperature, indicating more extensive lithium depletion. Heat-treatment for 1 hour at 530°C (the recommended solution treatment temperature for this alloy) gave a profile essentially similar to that at 520°C as shown in Figure 1. The existence of a solute depleted surface layer was also confirmed by light microscopy of transverse sections of the solution treated material that had undergone an ageing treatment of 17 hours at 300°C. Light microscopy of etched material (Figure 2) reveals a narrow band depleted in coarse precipitates at the surface. The presence of magnesium and copper in the alloy both serve to reduce the solid solubility of lithium [11,12]. Examination of the Al-Li-Cu ternary diagram [12] shows that the copper level of 1.2% will reduce the solubility of lithium to below 1 wt% at 300°C, therefore the precipitate depleted band is expected to be a region where the lithium level is below 1 wt%. In the example shown in Figure 2 it can be seen that the combined effect of solute depletion and high temperature has caused recrystallisation at the surface. A smaller recrystallised zone was observed after treatment at 520°C. The extent of the precipitate depleted region was measured for the three solution treatment temperatures and the results are shown in table 5. The width of this zone is much less than that indicated by microhardness profiles.

Lithium loss could not be directly measured to provide correlation with the microhardness and metallographic observations. However, magnesium loss after 1 hour at 530°C was measured by wavelength dispersive X-ray microanalysis across a transverse section, the resulting profiles were similar to those obtained from microhardness testing. The normalised magnesium concentration profile is shown in Figure 3(a).

If the free surface is assumed to behave as an infinite solute sink for lithium and magnesium, concentration profiles can be modelled using the following relationship:

$$C_{x,t} = C_i \operatorname{erf}\left[\frac{x}{2\sqrt{Dt}}\right]$$

where
- x = distance from surface
- t = time
- D = diffusivity
- C_i = initial (bulk) concentration of solute
- $C_{x,t}$ = concentration of solute at x and time t

It is assumed that D is not dependent on concentration. Using values of D calculated from the standard diffusion data noted earlier, both lithium and magnesium theoretical concentration profiles were calculated. They were normalised for comparison with measured magnesium and microhardness profiles. Figure 3(a) shows that agreement between theoretical and actual magnesium profiles is excellent, indicating that the simplifying assumptions used in the calculation are appropriate and that depletion is controlled by solid state diffusion through the metal. The calculated lithium profile is shown in Figure 3(b) and is very similar to that for magnesium as expected given the similar thermodynamic driving forces for oxidation and similar diffusivities of the two species. Additionally Figure 3(b) shows that the lithium concentration profile correlates well with the microhardness profile. Hardness declines slightly faster than the lithium concentration, presumably reflecting the concomitant loss of magnesium.

Light micrographs of transverse sections of 3mm sheet, retaining the "as received" surface, after standard (20 min) and prolonged (4 hour) solution treatments at 530°C are shown in Figure 4. The directionality of the unrecrystallised, hot worked

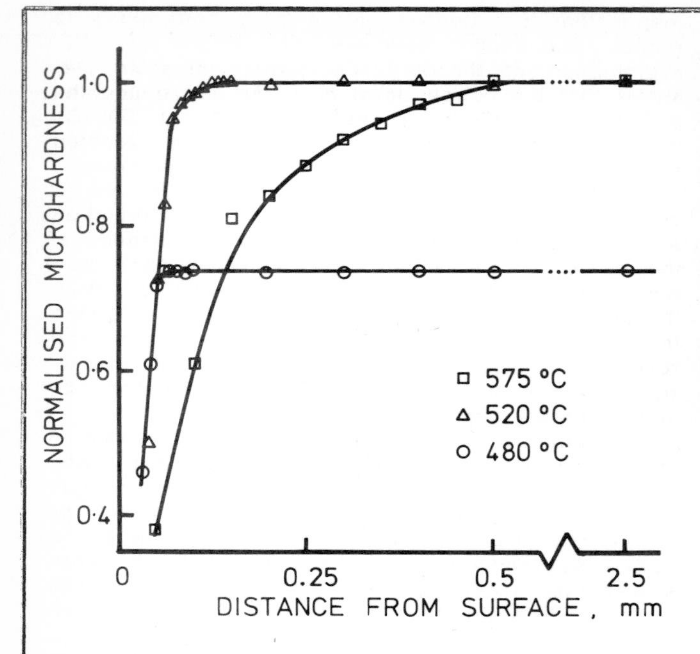

Figure 1. Effect of temperature on depletion, as measured by microhardness, after 1 hour at temperature.

Figure 2. Transverse section through the surface of sheet solution treated at 575°C for 1 hour and aged at 300°C for 17 hours.

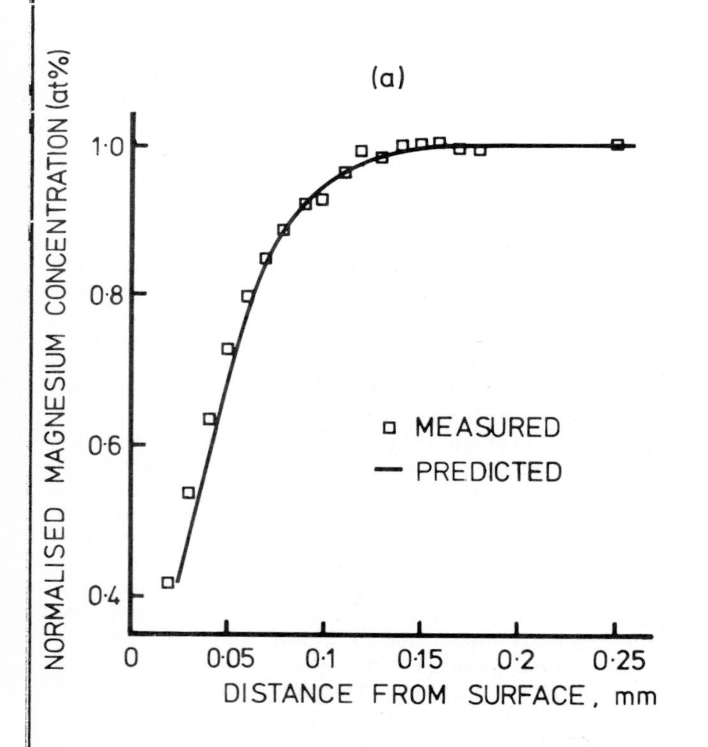

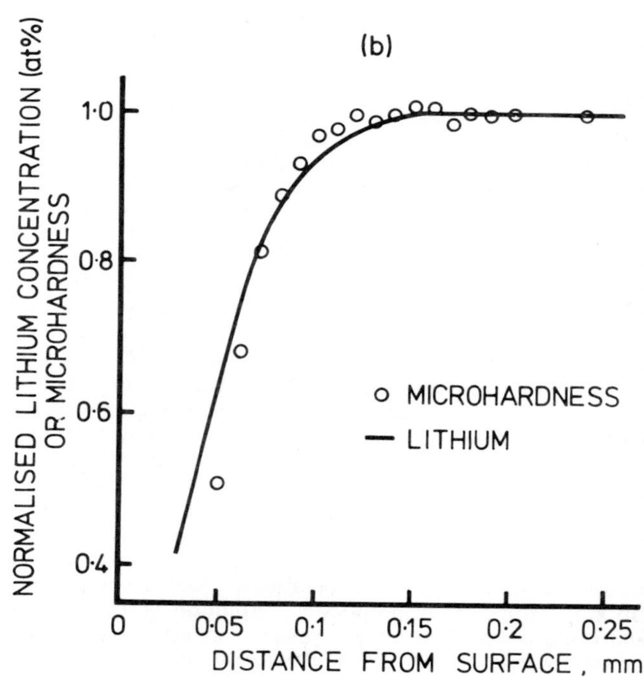

Figure 3. Comparison of normalised Mg/Li profiles with theoretical predictions (S.T. I hr. at 530°C, aged 5 hrs. at 187°C).

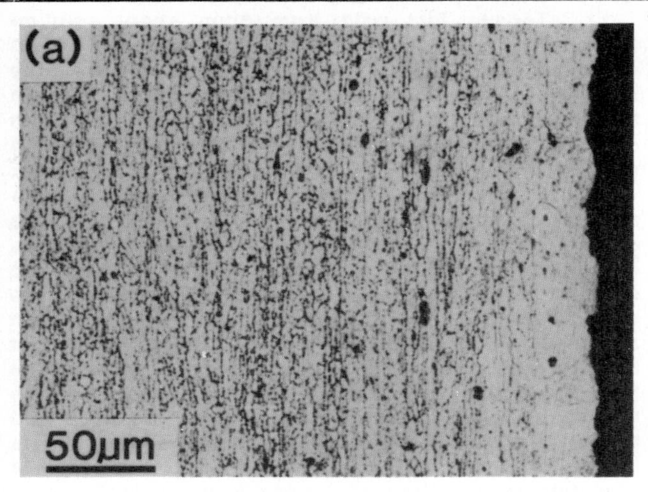

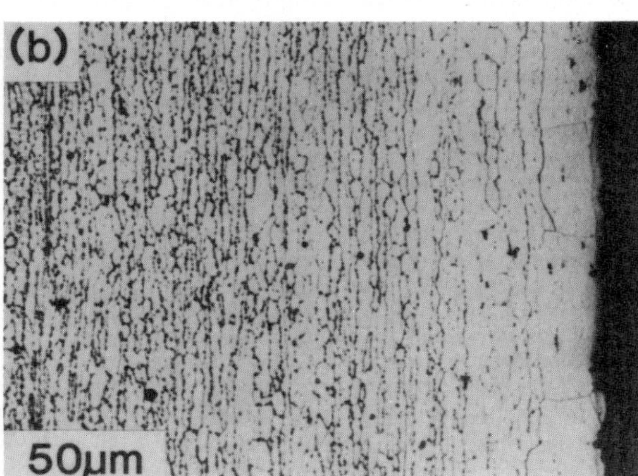

Figure 4. Microstructure of material solution treated at 530°C for (a) 20 min. and (b) 4 hrs., both aged at 187°C for 5 hrs.

NORMALISED MICROHARDNESS
1·0
0·8
0·6
0·4
20 min
4 hrs
● AS RECEIVED SURFACE REMOVED PRIOR TO HEAT TREATMENT
○ AS RECEIVED SURFACE
0 0·05 0·1 0·15 0·2 0·25
DISTANCE FROM SURFACE, mm

Figure 5. Microhardness traverses of samples with and without the 'as received' depleted surface layer removed.

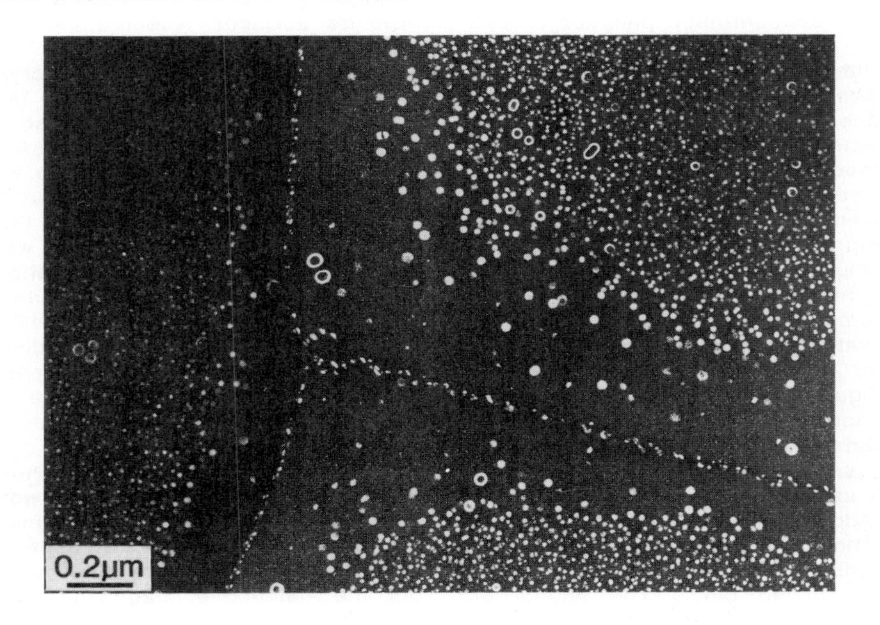

Figure 6. H.R.D.F. showing δ' depleted zones at a depth of 75µm in sheet S.T. for 4 hrs. at 530°C, aged for 5 hrs. at 187°C.

structure is apparent in the interior of the sheet. At the surface, however, recrystallisation has occurred and has increased in the sheet subjected to the very long solution treatment, confirming that recrystallisation is a function of both temperature and solute depletion.

Microhardness traverses (Figure 5) on both of these samples and also on surfaces from which the initial "as received" depleted layer was removed prior to heat-treatment show a considerable difference between 20 minutes and 4 hours at temperature, as expected for a depletion process controlled by solid state diffusion through the parent metal. The results also indicate that processing from billet to sheet produces depletion comparable to that occurring in 20 minutes at 530°C but that the influence of the prior processing diminishes at longer solution treatment times.

The microhardness profiles obtained from the 3mm sheet which underwent the prolonged solution treatment were used to select depths at which to prepare thin foils for transmission electron microscopy. At a depth of 15μm from the surface no δ' precipitation or associated superlattice reflections were evident. At 50μm a very fine precipitate, with a particle size of 2-4nm, was present throughout the material. This was probably formed during quenching from 530°C. The normal precipitate of between 15 to 30nm diameter was present in the interior of subgrains at 75μm but a depleted zone existed around subgrain boundaries (although the fine "as quenched" precipitate was present throughout (Figure 6). At this depth fine scale precipitation of a second phase, tentatively identified as θ' was observed. The extent of the depleted zones at subgrain boundaries decreased progressively from 75 to 125μm until at 150 to 175μm the microstructure was no diferent from that observed in material taken from the centre of the sheet. This depth is slightly less than the edge of the lithium/magnesium concentration plateaux (215 μm).

To minimise lithium and magnesium depletion during solution treatment, but retain an "as heat-treated" surface for fatigue testing, some testpieces were heat-treated in dry argon. The resulting microhardness profile, together with those for which the "as received" depleted surface had been either removed by machining followed by mechanical polishing or simply mechanically polished without removal, are shown in Figure 7. The argon atmosphere greatly reduces lithium depletion when compared with the profiles for material solution treated in air and lightly polished ((3) in Figure 7) or material solution treated in air and unpolished (Figure 5). The remaining profile ((1) in Figure 7) indicates that sufficient material has been removed from the testpiece after heat-treatment, to entirely eliminate the depleted surface layer – the specimen does however have a polished surface. Additionally it is common practice to pickle and anodise aluminium alloys for aerospace applications and fatigue testing in these conditions was also carried out. The effects of such conventional processes on the surfaces of this material have been examined by light microscopy of transverse sections (Figure 8). The pickling treatment has clearly attacked exposed grain boundaries and subsequent anodising has caused further erosion from the original pickling pits.

Tensile test data from 3mm sheet solution treated and aged for 20min and 4 hours at 530°C and aged for 5 hours at 187°C are shown in table 6. These results indicate that prolonged solution treatment leads to a reduction in the overall strength of sheet product attributable to the increasing thickness of the soft solute depleted surface layers.

The results from fatigue life testing of the 3mm sheet solution treated for 20min. and 4 hours are shown in Figure 9a. The results obtained indicate that prolonged solution heat-treatment has no deleterious effect. However the testpieces that were heat-treated in dry argon (with minimal depletion after 20min.) show increased fatigue lives, at 350MPa maximum stress, compared to material heat-treated in air. Also shown in Figure 9a are the results for polished testpieces – as expected these lives are generally improved at all stress levels. The polished specimens retaining a depletion layer displayed longer lives at each stress level than did the testpieces from which the depletion layer was removed by machining after heat-treatment.

Fatigue life data for the pickled and pickled and anodised testpieces are shown in Figure 9b. The scatter band of results for material solution treated (20 min. or 4 hours) in air is included for comparison. The S-N curves for the three conditions show pickling to reduce the fatigue life of the sheet so that it falls approximately into the lower half of the scatter band of the plain testpieces. Anodising after pickling further reduces fatigue life, especially at higher levels of maximum stress ($\geq$ 225MPa).

Scanning electron micrographs of two typical fatigue fractures are shown in figure 10. The testpieces had been solution treated in air for 20 min. and 4 hours respectively and were subjected to a maximum stress of 225MPa. Examination of the fracture surfaces revealed the initiation sites at or near a corner of the testpiece. Both initiation sites contain a region of smooth, crystallographic crack growth suggesting stage I type fatigue. The extent of the smooth area in each case corresponds to the recrystallised surface layer seen in Figure 4 (a & b). This type of initiation site was common in the testpieces heat-treated in air but was only seen to a very limited extent in one of the three testpieces heat-treated in dry argon. It was more difficult to establish a definite initiation site in the testpieces with polished surfaces and limited fractography has been carried out so far. Initiation sites in the pickled and pickled and anodised condition were obviously associated with the pits produced by the chemical attack shown in Figure 8.

DISCUSSION

Microhardness and magnesium concentration profiles measured in this work are in good agreement with lithium and magnesium concentration profiles predicted for oxidation induced depletion controlled by solid state diffusion through the metal. As the calculated magnesium profile (Figure 3a) agrees well with experimental results, the calculated lithium profile (Figure 3b) is also expected to correctly predict the lithium concentration profile at the sheet surface. This shows lithium to fall to 1wt% at a depth of 30μm after 1 hour at 530°C.

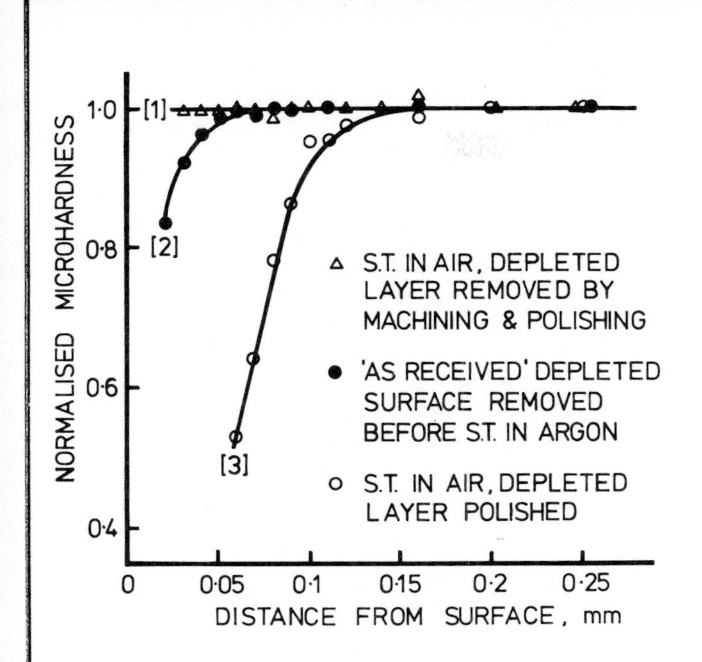

Figure 7. Microhardness of samples with different surface conditions and solution treatment atmospheres.

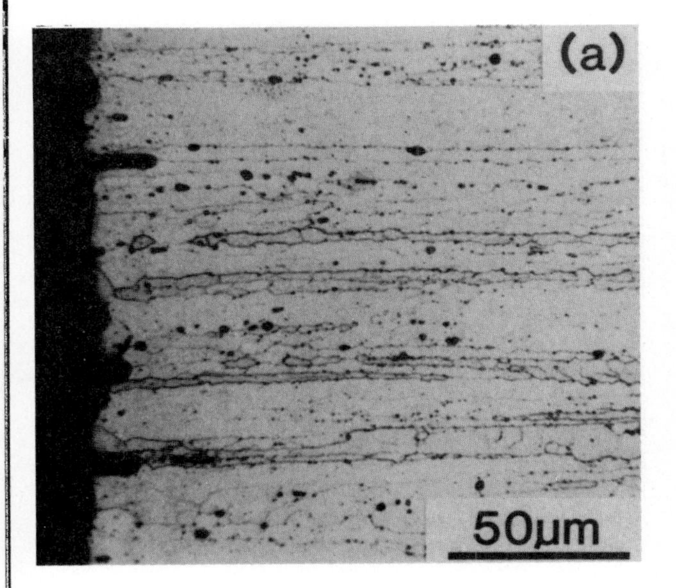

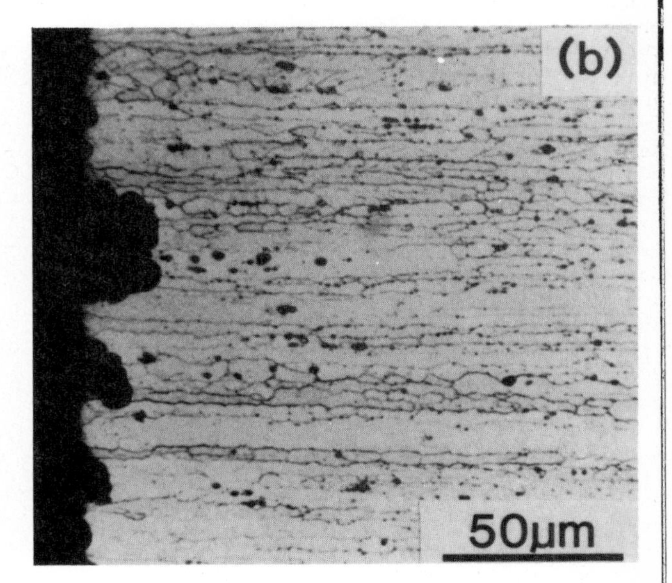

Figure 8. Microstructure of pickled and pickled and anodised material, sections transverse to the rolling direction (Keller's Etch).

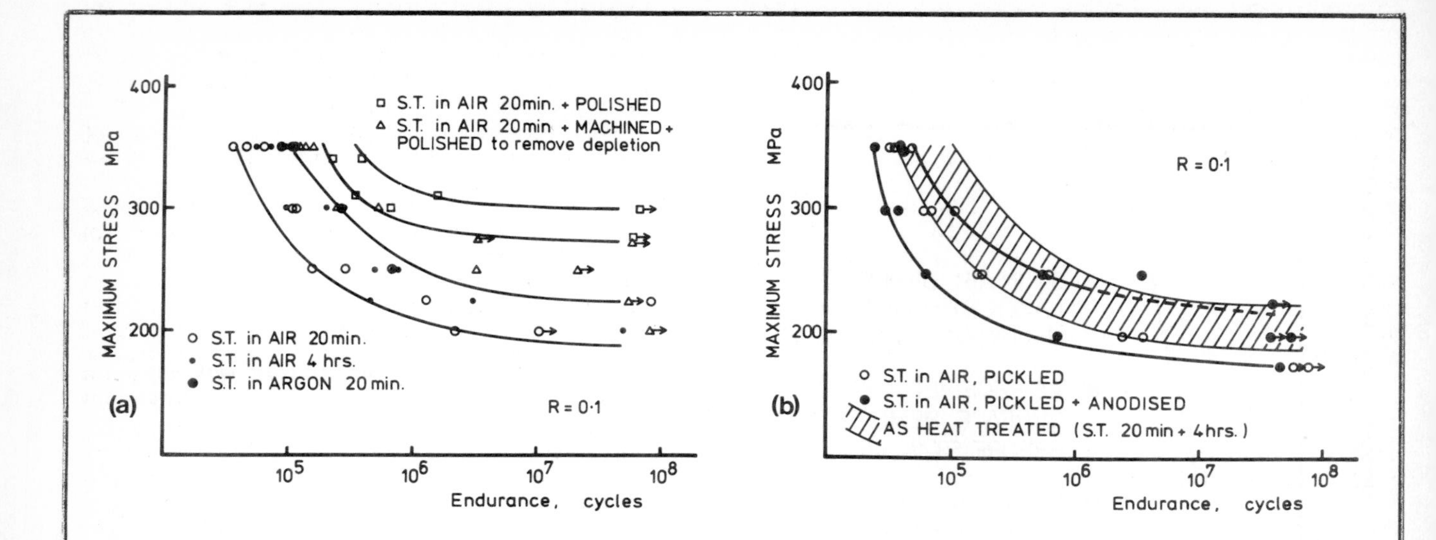

Figure 9. S–N Fatigue curves for 3mm sheet.
(a) Effect of surface depletion and mechanical polishing.
(b) Effect of pickling and pickling and anodising.

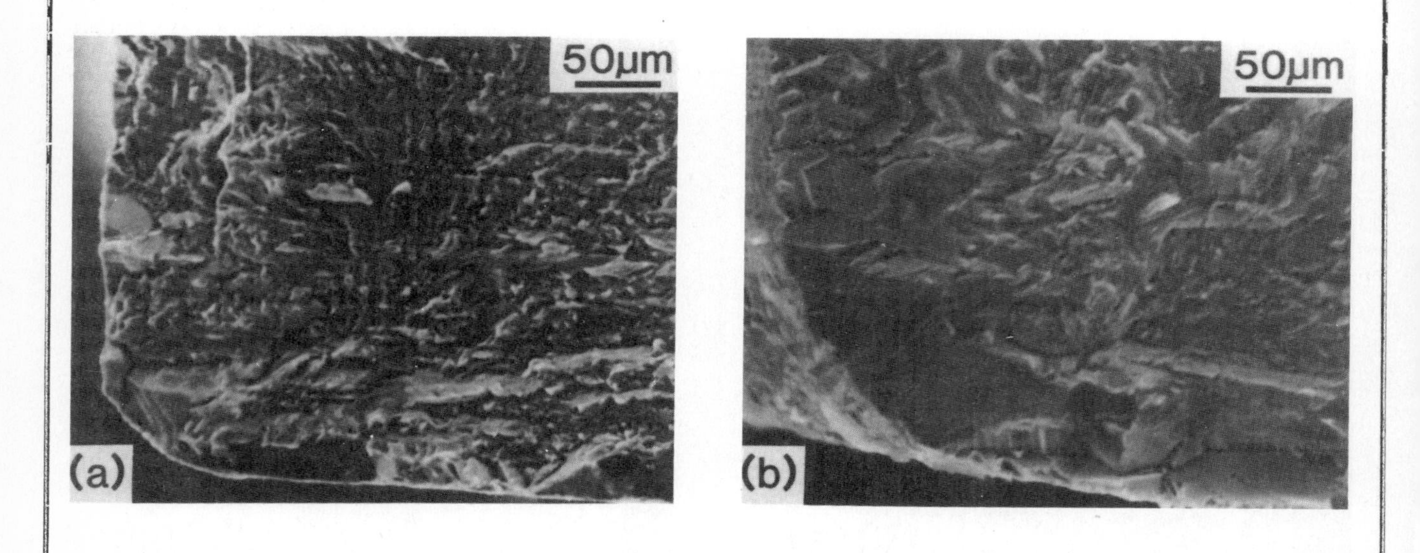

Figure 10. S.E.M. of fractures; (a) 20 min. (b) 4 hrs., at 530°C, aged 5 hrs. at 187°C. Maximum stress = 225 MPa (R= 0·1).

This correlates well with the coarse precipitate-free zone depth of 24μm, seen in the material solution treated for 1 hour at 520°C and aged at 300°C.

Detailed transmission electron microscopy studies through the depleted surface layer after 4 hours at 530°C have shown substantial changes in the volume fraction and distribution of δ' precipitation, which corresponds well with microhardness profiles as the δ' phase is largely responsible for strengthening in this alloy.

Clearly the absence of any δ' at the surface indicates that the lithium content is reduced below the metastable solvus at 187°C or even lower since no δ' forms during cooling. The very fine δ' observed at a depth of 50μm (lithium concentration ≈ 0.9 wt%) is undoubtedly produced on cooling or on natural ageing and is not a product of ageing at 187°C. As lithium concentration increases with depth δ' does form at 187°C but precipitate free zones are observed at subgrain boundaries. The heterogeneous nucleation of δ' on $ZrAl_3$ within the precipitate free zone and the increased size of homogeneously nucleated δ' on the edge of the P.F.Z. (Figure 6) clearly indicates that the P.F.Z. is not lithium depleted compared with subgrain interiors. Recent work by Baumann and Williams [14] has shown that at very low lithium supersaturations even the small surface energy of δ' can result in inhibition of δ' formation in vacancy depleted regions close to dislocations. This is attributed to the fine δ', present at room temperature, being too small to resist reversion on heating to the ageing temperature. Where the vacancy concentration is undiminished the as quenched δ' is significantly larger: it can coarsen rather than revert at the ageing temperature. In that work also, heterogeneous nucleation within the P.F.Z. was observed together with enlargement of δ' on its periphery as observed here. It is therefore believed that the P.F.Z.'s observed here have the same origin as discussed by Baumann and Williams.

At a depth of 150μm the lithium supersaturation is sufficiently increased (lithium concentration 2.1 wt%) to reduce the critical radius size at 187°C to a value below that of the as quenched δ' throughout the microstructure so a uniform homogeneous dispersion of δ' is produced and no P.F.Z.'s are observed.

The appearance of a second precipitate phase at a depth of 75μm corresponds to a region where the Mg/Cu ratio is reduced to such a level that additional precipitation from the ternary Al-Cu-Mg system is possible. The isothermal section at 190°C [13] indicates that, with a constant copper level of 1.2 wt%, magnesium levels of between 0.25 and 0.5 wt% (cf 0.33 wt% measured at 75μm) permit the possibility of θ/θ' precipitation in addition to δ' phase.

Increasing the solution treatment time at 530°C from 20 minutes to 4 hours resulted in depletion layers of 125 and 215μm respectively. Calculation shows the resulting decrease in cross-sectional area of undepleted material (9%) to be largely responsible for the observed drop in tensile properties (6.5% drop in yield stress).

Comparison of the S-N data from testpieces heat treated for 20 minutes at 530°C in air and argon (Figure 9a) shows the depleted layer to have a small adverse effect on fatigue life. However, the total overlap of S-N data from testpieces heat-treated for 20 minutes and 4 hours at 530°C implies that increasing the depth of depletion from 125 to 215μm does not further adversely affect fatigue life. Increasing the depletion does however increase the width of the apparent stage I initiation observed (Figure 10).

The effects of depletion can be totally masked by polishing the sample surfaces, mechanical polishing will remove surface blemishes, introduce residual stresses and work harden the surface. Therefore direct comparison between polished and plain testpieces is not possible. The improvement in fatigue life due to minimised lithium depletion (argon heat-treatment) was less than that obtained by polishing testpieces whilst retaining the depletion layer. The polished samples for which the depletion layer had been removed showed lives intermediate between the as heat-treated and heat-treated and polished testpieces. It is interesting to note that polished samples containing a depletion layer actually exhibited superior S-N characteristics to those polished after removal of this layer. At present no complete explanation can be given for this observation although it should be noted that depleted samples contain a recrystalised surface microstructure and contain an additional precipitate, tentatively identified as θ'.

Texture measurements are being carried out on both the surface and interior of fully heat-treated sheet, these will then be related to the depletion profiles and more detailed fractography.

As with conventional alloys standard pickling and anodising processes have been found to reduce S-N fatigue properties although the effect in this case is not large. Light and scanning electron microscopy have shown large pits at exposed grain boundaries which act as initiation sites for fatigue cracks. Reduced attack may be achieved by slight modifications to the standard pickling and anodising treatments.

CONCLUSIONS

1. Thermomechanical processing of Al-Li-Cu-Mg-Zr alloys in air results in selective oxidation of Li and Mg. This leads to surface layers which are equally depleted in these elements.

2. Lithium depletion results in a reduced δ' volume fraction and inhomogeneous precipitate dispersions.

3. Magnesium depletion lowers the Mg/Cu ratio and results in the formation of additional precipitate phases from the Al-Cu-Mg system.

4. Prolonged solution treatment (4 hours at 530°C) causes a slight reduction in tensile properties of the 3mm sheet as a result of increased lithium depletion.

5. The depleted surface layer is associated with stage I type fatigue.

6. Reducing lithium depletion by heat-treatment in an argon atmosphere produces a small improvement in S-N fatigue properties though this is outweighed by the effects of mechanical polishing.

7. Conventional pickling and anodising treatments result in a degradation of S-N fatigue properties compared to as heat-treated material.
This effect is common to all aluminium alloys and is relatively small in this case.

ACKNOWLEDGEMENTS

This work was supported by the S.E.R.C. as part of a C.A.S.E. project in collaboration with the Procurement Executive, Ministry of Defence.

REFERENCES

[1] E.T. TURKDOGAN, "Physical Chemistry of High Temperature Technology" (Academic Press, 1980).

[2] D.J. FIELD, G.M. SCAMANS and E.P. BUTLER, Aluminium Lithium Alloys 11, T.H. SANDERS and E.A. STARKE, eds, (Met Soc. AIME, 1984.

[3] "Smithell's Metals Reference Book", ed. E.A. BRANDES, (Butterworths, 1983).

[4] D.J. FIELD, E.P. BUTLER, G.M. SCAMANS, Aluminium Lithium Alloys 1, T.H. SANDERS and E.A. STARKE, eds, (Met. Soc. AIME, 1981).

[5] P.J. GREGSON, PhD thesis, University of London (1983).

[6] C.J. PEEL, B. EVANS, C.A. BAKER, D.A. BENNETT, P.J. GREGSON, H.M. FLOWER, Aluminium Lithium Alloys 11, T.H. SANDERS and E.A. STARKE, eds, (Met. Soc. AIME, 1984).

[7] J.A. WERT, A.B. WARD, Scripta Metall, 19 (1985) 367.

[8] R.M. BRICK, A. PHILIPS, Trans. A.S.M. 29 (1941) 435.

[9] "Strength of Metal Aircraft Elements", MIL HDBK 5 (Armed Forces Supply Support Center, Washington DC, 1961).

[10] G.E. NORDMARK, J. Spacecraft and Rockets, 1(1964) 125.

[11] D.W. LEVINSON and D.J. McPHERSON, Trans. A.S.M., 48 (1956) 689.

[12] H.K. HARDY and J.M. SILCOCK, J. Inst. Metals, 84 (1955-56) 423.

[13] "Light Alloys", I.J. POLMEAR (Edward Arnold, 1981).

[14] S.F. BAUMANN and D.B. WILLIAMS, Acta Metall. 33 (1985) 1069.

□

Comparison of corrosion behaviour of lithium-containing aluminium alloys and conventional aerospace alloys

P L LANE, J A GRAY and C J E SMITH

The authors are in the Materials and Structures Department, Royal Aircraft Establishment, Farnborough, UK

SYNOPSIS

Marine exposure and accelerated laboratory corrosion tests have been used to compare the corrosion behaviour of some Al-Li-Mg-Cu alloys with conventional aerospace alloys. The pitting resistance of Al-Li-Mg-Cu alloys in neutral salt fog and under total immersion conditions is superior to that of conventional alloys. In a marine environment, Al-Li-Mg-Cu alloys exhibit exfoliation corrosion; the sheet material is particularly susceptible and resistance to exfoliation is improved by overaging. Stress corrosion cracking properties are similar to conventional alloys. Al-Li-Mg-Cu alloys are galvanically compatible with other aluminium alloys.

1 INTRODUCTION

Considerable progress has been made recently in the development of aluminium-lithium alloys for aerospace applications. The production of Al-Li-Mg-Cu alloys has made possible a 10% density reduction; a 10% increase in elastic modulus and strength properties which match established aircraft alloys. To date, there has been only limited evidence of corrosion properties of aluminium-lithium alloys. Previous experience with an Al-Cu-Li alloy (2020) developed in the USA was reported to show good corrosion resistance[1] whereas an Al-Mg-Li alloy (Russian alloy designation 01420) was susceptible to exfoliation[2].

The work presented here concerns the corrosion behaviour of Al-Li-Mg-Cu alloys. Conventional, high strength aluminium alloys are not immune to corrosion but are susceptible to various forms of attack, namely pitting, intergranular corrosion, exfoliation and stress corrosion cracking. In service corrosion problems are limited only by correct design, alloy selection and application of appropriate protection schemes. The aims of this work are to generate basic corrosion data for the Al-Li-Mg-Cu alloys and compare the corrosion performance with that of conventional aerospace alloys. The tests undertaken were designed to evaluate the susceptibility of Al-Li-Mg-Cu alloys to corrosion and stress corrosion cracking.

2 EXPERIMENTAL PROCEDURE

2.1 Materials

Aluminium alloys containing lithium are currently being developed to give high and medium strengths. The medium strength alloy was available in the form of 1.6 mm sheet (DTD XXXA), 11 and 25 mm plate (DTD YYYA) and 25 mm extrusion (DTD ZZZA). The high strength alloy was only available in the form of 25 mm plate. Typical compositions are given in Table 1. Control alloys used included 2014-T6 in the form of sheet to BSL150 and plate to BSL93; 7010-T76 plate to DTD 5120; 7075-T6 in the form of sheet to BSL88 and plate to BSL95; 2024-T3 plate to BSL97 and clad sheet alloy 7475-T761.

2.2 Corrosion Tests

Coupons of the sheet aluminium-lithium alloys and sheet control alloys were prepared by either pickling in 10% sodium hydroxide solution at 60°C for 30 secs, rinsing in water and desmutting in nitric acid or by lightly abrading. Coupons cut from the plate and extruded alloys were either lightly abraded or machined to expose the half thickness surface. Before corrosion testing coupons

TABLE 1

Typical chemical compositions (wt %) of Al-Li-Mg-Cu sheet, plate and extrusion.

	Li	Mg	Cu	Zr	Fe	Si
XXXA SHEET	2.48	0.60	1.28	0.12	0.09	0.04
YYYA PLATE	2.41	0.61	1.16	0.11	0.14	0.10
YYYB PLATE	2.36	0.53	2.16	0.11	0.09	0.06
ZZZA EXTRUSION	2.49	0.52	1.25	0.10	0.08	0.08

TABLE 2

Weight loss (mg/cm^2) and mean depth of pitting after 42 days exposure to a neutral salt fog.

	SHEET 1.6mm	PLATE 11mm	PLATE 25mm	EXTRUSION 25mm
Al-Li 'A' (16h 190 C)	0.27* (0.021)	0.40	0.45 (0.030)	0.65 (0.037)
Al-Li 'B' (16h 190 C)	-	-	0.71 (0.025)	-
2014-T6	3.40 (0.100)	-	-	-
7075-T6	3.29 (0.050)	-	2.52 (0.022)	-
7010-T76	-	-	2.81 (0.161)	-

* aged 8h. at 170 C

Figures in brackets show mean depth of pitting (mm)

TABLE 3

Weight loss (mg/cm^2) after 21 days total immersion in sodium chloride solutions of various concentrations for DTD XXXA sheet alloy (aged 8h at 170°C) and for a range of conventional aluminium alloys.

mM/l NaCl	Al-Li XXXA	7075-T6	2014-T6	2014-T3	7475-T761 CLAD
0	0.82	0.12	1.92	1.36	0.24
1	0.83	0.96	1.82	1.16	0.26
10	1.10	1.40	1.28	1.00	0.38
100	0.94	0.94	1.22	1.00	0.18
600	0.60	0.64	1.05	0.64	0.20

TABLE 4

Visual assessment of exfoliation corrosion in DTD XXXA sheet alloy exposed to a marine environment and to various accelerated corrosion tests.

AGED CONDITION	280 DAY MARINE EXPOSURE	ALTERNATE IMMERSION (30 DAYS)	INT. ACID. SALT FOG (14 DAYS)	EXCO 48 HOURS	EXCO 96 HOURS
NAT. AGE	N	P	P	N	N
HOURS AT 170 C 1.5	E	E	EA	EA	EA
HOURS AT 170 C 4	EA	EB	EA	EA	EB
HOURS AT 170 C 8	EB	EA	EC	EA	EC
HOURS AT 170 C 16	EA	EA	EC	EA	ED
HOURS AT 170 C 64	E	E/P	EA	EB	ED
HOURS AT 185 C 1.5	EB	EB	EC	EA	EB
HOURS AT 185 C 4	EA	EB	EC	EA	EB
HOURS AT 185 C 8	EA	EA	EB	EA	EC
HOURS AT 185 C 16	E	EA	EA	EA	ED
HOURS AT 185 C 64	E	E/P	EA	EA	ED

TABLE 5

Total charge passed in 48 hours during galvanic corrosion of various aluminium alloy couples in 3½% sodium chloride solution.

	Al-Li 'A' PLATE	Al-Li 'B' PLATE	7075-T6	2024-T3
2014-T6	+13.0 C	+11.1 C	+10.0 C	+7.5 C
2024-T3	+8.5 C	+9.3 C	+7.2 C	-
7075-T6	+8.6 C (-0.1 C)	+4.2 C (-0.1 C)	-	-

were weighed. After testing coupons were examined microscopically and photographed before cleaning in a chromic phosphoric acid solution and then reweighed. Coupons were encapsulated in epoxy resin and after sectioning and polishing, the mean depth and width of the corrosion attack was determined.

2.2.1 Marine Exposure Trials

Coupons of the sheet, plate and extruded aluminium lithium alloys together with coupons of the control alloys were tested at the Admiralty Research Establishment, Exposure Trials Station at Eastney. The coupons were mounted on wooden racks which faced south and were inclined at an angle of 45°. Inspections of the coupons were made at approximately 2-monthly intervals, and samples returned to RAE Farnborough for photographing and more detailed studies.

2.2.2 Accelerated Corrosion Tests

A number of accelerated corrosion tests were performed to evaluate the resistance of aluminium-alloys containing lithium to pitting and exfoliation corrosion. The tests employed were exposure to neutral salt fog (BS 5466 part 1), alternate immersion in 3.5% neutral salt solution; exposure to intermittent acidified salt fog[3]; total immersion in EXCO solution (ASTM G34) and total immersion in distilled water and 1 mM to 600 mM salt solutions. The neutral salt fog, alternate immersion and total immersion tests were primarily intended to evaluate the resistance to pitting corrosion whilst the EXCO and intermittent acidified salt fog tests were included to examine the resistance to exfoliation corrosion.

2.3 Stress Corrosion Cracking

The resistance of the 25 mm plate and extruded material to stress corrosion cracking was evaluated using constant strain tests. 19.05 mm diameter C-ring test pieces were machined so that the stress was applied in the short transverse direction. After stressing the C-rings were either tested by alternate immersion in 3½% neutral salt solution in accordance with ASTM G44 for up to 84 days or were exposed to a marine environment at Eastney.

2.4 Open Circuit Corrosion Potential Measurements

The variation in open circuit corrosion potential with respect to a saturated calomel electrode was monitored using a data logger for a period of 21 days. Details of the test procedure have been published elsewhere[4]. Tests were made on coupons of sheet aluminium lithium alloy DTD XXXA immersed in 1 mM, 100 mM and 600 mM sodium chloride solution. Measurements were also made on other aluminium-lithium alloys and a range of conventional alloys immersed in 600 mM/L salt solution. In this case readings were taken for up to 2 hours.

2.5 Galvanic Corrosion Measurements

The galvanic corrosion currents flowing between various aluminium-lithium and conventional alloy couples were measured using a zero resistance ammeter. Galvanic couples were composed of electrodes 2 mm x 2 mm in area and polished to a 1/4 μm finish. The electrodes were separated by a distance of 1 cm in an aerated 600 mM/L sodium chloride solution.

3 RESULTS AND DISCUSSION

3.1 Pitting Corrosion

In the neutral salt fog and total immersion tests in salt solution, the principal form of corrosion attack developed on the aluminium-lithium alloys was pitting. The weight loss data presented in Table 2 give an indication of the extent of the attack in comparison with the control alloys after exposure to neutral salt fog for 42 days. Under these conditions, the aluminium-lithium alloys compare very favourably with the clad 7475-T761 sheet material and appear to be more resistant to corrosion attack than the conventional 2XXX and 7XXX alloys. More important perhaps, the mean depth of attack is appreciably lower than that experienced with materials such as 2014-T6 and 7075-T6.

Under total immersion conditions, the weightloss after 3 weeks appears to be dependent on the chloride concentrations of the test solution. Table 3 indicates that as the chloride concentration increases, the extent of the attack is decreased. Similar behaviour has been found with the control alloys 2014-T6 and 2014-T3.

A further insight into the corrosion behaviour of aluminium-lithium alloys under total immersion conditions can be obtained by considering the results presented in Fig 1 which show the variation of weightloss and open circuit corrosion potential with time for DTD XXXA sheet immersed in solutions containing 10,100 and 600 mM/L sodium chloride. In the 10 mM/L salt solution, the weightloss data indicates that there is an initial very high rate of corrosion followed by a period where there is little further corrosion. This behaviour is reflected in the form of the potential-time curve which exhibits an initial slow fall in corrosion potential followed by a large drop and eventual recovery to approximately the initial value. The first stage has been interpreted as a period where pits are initiated[5] whilst the second stage is thought to be associated with pit growth. During pit growth there will be a high rate of corrosion causing an increase in the ratio of the relative anode to cathode areas. This will shift the potential to a more negative value. In the third

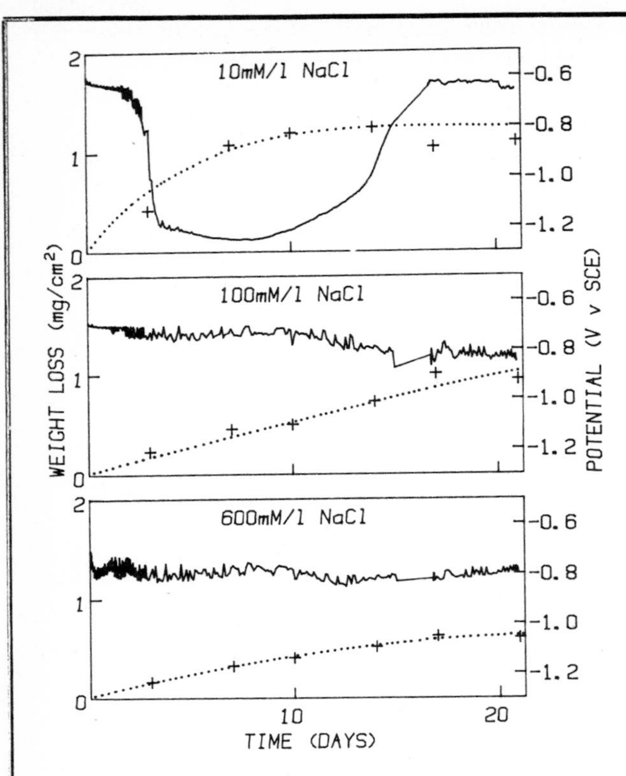

FIGURE 1

The variation of weight loss and open-circuit potential with time for DTD XXXA sheet (aged 8h at 170°C) immersed in various sodium chloride solutions.

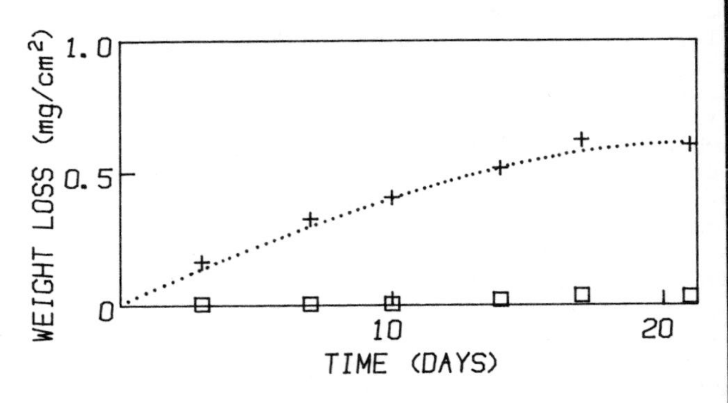

FIGURE 2

Weight loss data for DTD XXXA sheet in 10 mM/L sodium chloride solution showing the inhibiting effect of an addition of 2 mM/L sodium chromate.

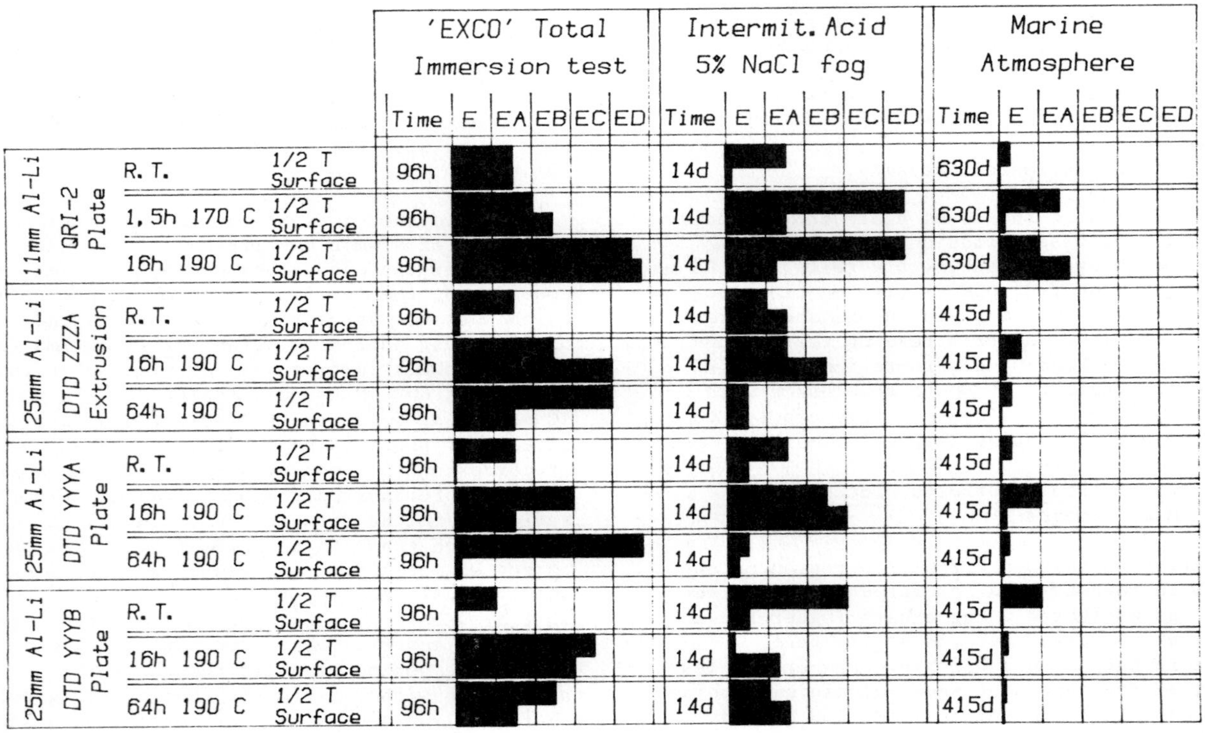

			'EXCO' Total Immersion test						Intermit. Acid 5% NaCl fog						Marine Atmosphere					
			Time	E	EA	EB	EC	ED	Time	E	EA	EB	EC	ED	Time	E	EA	EB	EC	ED
11mm Al-Li QRI-2 Plate	R. T.	1/2 T Surface	96h						14d						630d					
	1.5h 170 C	1/2 T Surface	96h						14d						630d					
	16h 190 C	1/2 T Surface	96h						14d						630d					
25mm Al-Li DTD ZZZA Extrusion	R. T.	1/2 T Surface	96h						14d						415d					
	16h 190 C	1/2 T Surface	96h						14d						415d					
	64h 190 C	1/2 T Surface	96h						14d						415d					
25mm Al-Li DTD YYYA Plate	R. T.	1/2 T Surface	96h						14d						415d					
	16h 190 C	1/2 T Surface	96h						14d						415d					
	64h 190 C	1/2 T Surface	96h						14d						415d					
25mm Al-Li DTD YYYB Plate	R. T.	1/2 T Surface	96h						14d						415d					
	16h 190 C	1/2 T Surface	96h						14d						415d					
	64h 190 C	1/2 T Surface	96h						14d						415d					

FIGURE 3 Visual assessment of exfoliation in Al-Li-Mg-Cu alloy plate and extrusion.

stage repassivation occurs, further pit growth ceases and the corrosion rate will fall.

The behaviour in 10 mM/L salt solution may be compared with that in more concentrated solutions. The weightloss data indicates that the corrosion rate is relatively constant throughout the test period and is lower than that experienced initially in the dilute salt solution. The fluctuations in the potential-time curves are thought to indicate that pit initiation and growth continues throughout the exposure period.

The results obtained here parallel closely work published earlier on total immersion testing of 2014-T6 sheet to BSL150[4]. In this case the change in behaviour occurred at a lower concentration of salt.

The pitting corrosion of 2XXX and 7XXX series alloys used on aircraft is largely controlled by the use of epoxy based primers inhibited with leachable chromates. The weight-loss data in Fig 2 was obtained from total immersion tests made on sheet aluminium-lithium alloy to DTD XXXA. The addition of 2 mM/L of sodium chromate to the test solution effectively inhibits the corrosion process. The results suggest that the current paint schemes used should also be effective on the aluminium-lithium alloys.

3.2 Exfoliation Corrosion

Exfoliation corrosion is a well known phenomenon in high strength aluminium alloys. It is a lamellar form of attack which runs parallel to the metal surface and can be attributed to the combined effects of an elongated grain structure and susceptibility to intergranular attack. The insoluble corrosion products which form occupy a greater volume than the parent metal giving rise to wedging stresses which cause the surface grains to lift. Previous experience with 2XXX alloys indicates that material is most susceptible to exfoliation when rolled into thin plate 4-20 mm in section and naturally aged. Sheet material is generally insensitive to exfoliation attack. Mahoon[6] demonstrated the relationship between susceptibility to exfoliation of 2014-T6 and 2024-T3 and grain structure. Lightly worked, 16-40 mm thick plate was unrecrystallised the grains being elongated, and was found to be relatively susceptible to exfoliation. In heavily worked material, 2-6 mm thick, the structure was recrystallised consisting of small equiaxed grains and showed better resistance to exfoliation corrosion. The behaviour of the aluminium-zinc-magnesium (7XXX) series alloys is similar except that they are very susceptible when peak aged (T6) but may be rendered immune to exfoliation by overaging to the T73 condition.

Al-Li-Mg-Cu alloys exhibit a markedly elongated grain structure[7] that is preserved in thin sheet material due to the grain refining additions of zirconium which inhibits recrystallisation. Consideration should, therefore, be given to the possibility of exfoliation corrosion in these alloys. Fig 3 shows the results of visual assessment of exfoliation corrosion in 11 mm plate, 25 mm plate (DTD YYYA and DTD YYYB) and 25 mm extrusion (DTD ZZZA). The severity of exfoliation attack was rated according to the exfoliation ratings of ASTM standard G34[8]. Coupons were tested with either the surface or the machined half thickness exposed to the corrosive environment. Exposure of the 25 mm plate and extrusion to a marine atmosphere for 415 days generated only slight exfoliation attack. The 11 mm plate exhibited slightly more severe exfoliation after exposure for 630 days, which may be attributed to the longer exposure period or may be associated with a more elongated grain structure in the thinner section. The EXCO test and intermittent acidified salt fog test generated severe exfoliation corrosion, more so than in the marine exposure trials. The 11 mm plate was particularly susceptible to attack in the artificially aged conditions when the ½T was exposed to the intermittent acidified salt fog test. Artificially aged test coupons of the thicker 25 mm plate were less suceptible in this test. The wider scatter in the exfoliation ratings from the accelerated tests makes it difficult to draw any firm conclusions as to the roles of ageing and exposed surface on exfoliation corrosion of Al-Li-Mg-Cu plate and extrusion.

Conventional aluminium-sheet alloys are fully recrystallised exhibiting a fine equiaxed grain structure and are not susceptible to exfoliation corrosion but undergo pitting and intergranular attack. In marked contrast, Al-Li-Mg-Cu sheet alloys do not recrystallise and exhibit an elongated grain structure. The grain aspect ratio increases with decreasing sections and is much higher than that observed in Al-Li-Mg-Cu plate material. Test coupons of 1.6 mm aluminium-lithium sheet alloy (to DTD XXXA) were exposed to a range of corrosive environments to establish the susceptibility of sheet material to exfoliation attack. Table 4 summarises the results from a variety of corrosion tests and for a range of aged conditions. The marine exposure trials show that the aluminium-lithium sheet is susceptible to exfoliation corrosion which took the form of discrete blisters on the surface of the test coupons. Comparing the susceptibility of the sheet alloy to that of the plate (Fig 3) the sheet alloy was rated E-EB, depending on ageing treatment, after exposure for just 280 days. In contrast the 25 mm DTD YYYA plate was rated E-EA after exposure for 415 days. The greater susceptibility of the sheet alloy to exfoliation can be attributed to a much higher grain aspect ratio than that in the plate.

The exfoliation susceptibility of the aluminium-lithium sheet alloy in a marine

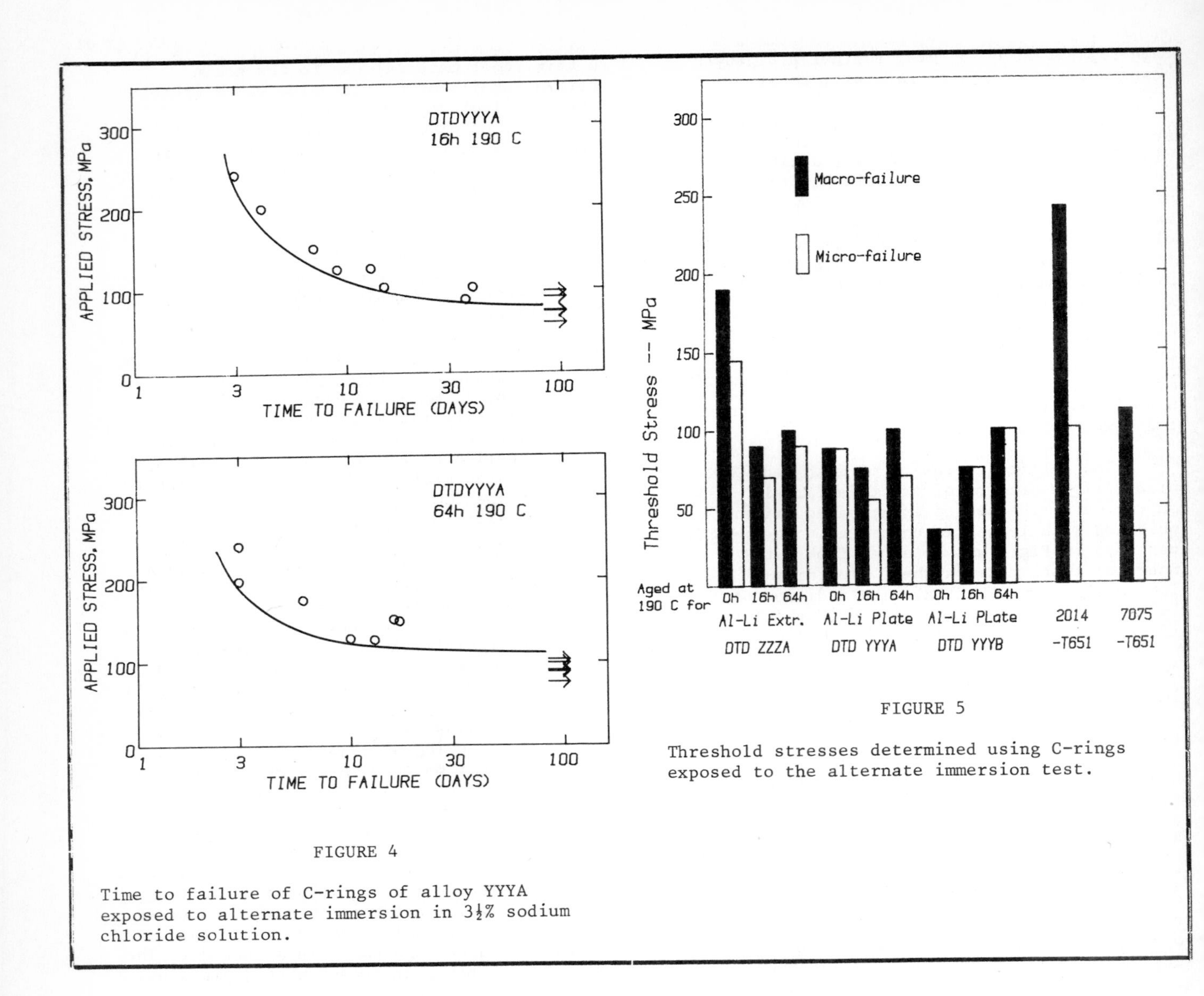

FIGURE 4

Time to failure of C-rings of alloy YYYA exposed to alternate immersion in 3½% sodium chloride solution.

FIGURE 5

Threshold stresses determined using C-rings exposed to the alternate immersion test.

environment was dependent on the ageing treatment. The highest rating was given to material aged at 170°C for 8 hours or at 185°C for 1.5 hours. Ageing for longer times at these temperatures considerably improved resistance to exfoliation and can be attributed to a reduction in the susceptibility to intergranular corrosion. The sheet alloy studied also exhibited exfoliation corrosion in the alternate immersion test, the intermittent acidified salt fog test and in the EXCO test. A reliable accelerated corrosion test for these alloys should reflect the dependence of exfoliation susceptibility on ageing treatment, as observed in the marine exposure trials. Table 4 indicates that the exfoliation ratings for the 48 hour EXCO test were similar for all artificially aged conditions, unlike the trends observed in a marine environment. The 96 hour EXCO test showed a clear dependence of exfoliation rating on ageing treatment, the severity of attack increasing with ageing time. This was not the case in marine exposure trials where overaging imparted some resistance to attack. Exfoliation generated in the EXCO test generally took the form of flaking across all the surface. In contrast, marine exposure produced isolated, discrete blisters where much of the surface remained unattacked.

The alternate immersion and intermittent acidified salt fog tests reproduce the general dependence of exfoliation susceptibility on ageing treatment observed in marine tests. The peak aged conditions are susceptible to exfoliation whereas the underaged and overaged treatments impart some resistance to attack. However the alternate immersion test produces some pitting tendency in the overaged condition. Both the alternate immersion test and the intermittent acidified salt fog tests are relatively long term and require specialised equipment. In contrast, the EXCO test is a simple, inexpensive short term test favoured by the aerospace industry for conventional alloys. Its poor correlation with real time tests here makes it unsuitable for use with aluminium-lithium alloys.

3.3 Stress Corrosion Cracking

The susceptibility to stress corrosion crack initiation may be assessed by considering either the time to failure at a given stress level or the SCC threshold stress defined as the stress at and below which failure does not occur. Whilst the first aim of the work was to compare the SCC threshold stress for Al-Li-Cu-Mg alloys with conventional aluminium plate alloys, wherever possible times to failure were recorded. Fig 4 shows plots of data obtained for the DTD YYYA alloy tested by alternate immersion in 3½% salt solution. Material which had been overaged at 190°C for 64h exhibited a higher threshold stress than the peak aged material. However the time to failure at a stress level of 150 MPa is the same, approximately 7 days and falls short of the 30 days which has been suggested for this material[9]. Similar times to failure at this stress level were found for the DTD YYYB material in both aged conditions. The extruded material DTD ZZZA showed rather longer times to failure in the overaged condition (64h at 190°C), approximately 30 days at 150 MPa.

The SCC threshold stresses determined from the alternate immersion tests and marine exposure trials are presented in Figs 5 and 6 respectively. For tests made in the alternate immersion two threshold stress levels have been determined. The macro threshold stress refers to the stress below which a loaded specimen did not fail after 84 days of testing. However sectioning and microscopic examination of a number of C-rings which had apparently not failed revealed the presence of intergranular cracks running perpendicular to the direction of application of stress. Thus although the threshold values given in Fig 5 give an order of merit for the alloys they cannot be regarded as threshold values in the sense that at stresses below the threshold no stress corrosion cracks will develop. Previous work on L97 for example revealed that deep cracks could develop in samples tested at only 6% of the 0.2% proof stress[10]. A second threshold value has therefore been included in Fig 5 for each alloy which corresponds to the stress below which no evidence of microcracking is found. If the aluminium-lithium alloys are compared on this basis with the two control alloys it is clear that they approach 2014-T6 in performance and are rather better than the 7075-T6. Tests in a marine environment are still in progress and data for the micro-threshold levels are not yet available. The data in Fig 6 represents the position after one year's exposure. Indications are that the macro-threshold stresses are likely to be less than 2024-T3 material.

3.4 Galvanic Corrosion

Galvanic corrosion occurs when two dissimilar metals are electrically coupled and exposed to a corrosive environment but is not considered a problem for couples between different conventional aluminium alloys. Fig 7 shows a galvanic series constructed for a range of lithium-bearing and conventional alloys ranked according to the open-circuit potential in 3½% NaCl solution. In a galvanic couple, the more negative member will suffer enhanced corrosion whilst partial or complete protection of the more positive member will result. The open-circuit potentials of the Al-Li-Mg-Cu alloys fall within a narrow potential range, increasing with the degree of artificial ageing, and are similar to the potentials of 2014-T6 and 7075-T76. The Al-Li-Mg (Cu free) alloy was the most electronegative of the lithium-bearing alloys tested. On the basis of the galvanic series, the lithium-bearing alloys should be compatible with conventional aerospace aluminium alloys since their potentials fall well within the wide range of potentials exhibited by the commercial alloys. To confirm this assumption, the galvanic current flowing between lithium-bearing and conventional alloys was measured. Al-Li-Mg-Cu alloys (to DTD YYYA and DTD YYYB) aged 16 hours at 190°C were coupled with 2014-T6, 2024-T3 and 7075-T6. Couples between the three conventional alloys were included for comparison purposes. Table 5 shows the total charge passed during each test, which is a quantitative measure of the overall extent of galvanic corrosion. In all cases, the lithium-bearing alloys corrode preferentially to the conventional alloys. This is not predicted from the galvanic series since alloy 7075-T6 is electronegative relative to the lithium-bearing alloys. However, during the test there is a short, initial period when preferential corrosion of 7075-T6 does occur but, after about one hour the direction of the corrosion current reverses, indicating a change in polarity of the couple. Fig 8 shows the galvanic current and coupled potential measured when Al-Li DTD YYYB was coupled to 2014-T6 and 7075-T6 shows the current reversal. The galvanic currents generated between lithium-bearing alloys and conventional alloys are not significantly greater than those generated between coupled conventional alloys.
These measurements confirm the assumption that aluminium-lithium alloys are galvanically compatible with conventional aluminium alloys.

4 CONCLUSIONS

1 The pitting resistance of Al-Li-Mg-Cu alloys in neutral salt fog and total immersion conditions is generally superior to conventional alloys. The dependence of open circuit potential on chloride concentration is broadly similar to that reported for conventional

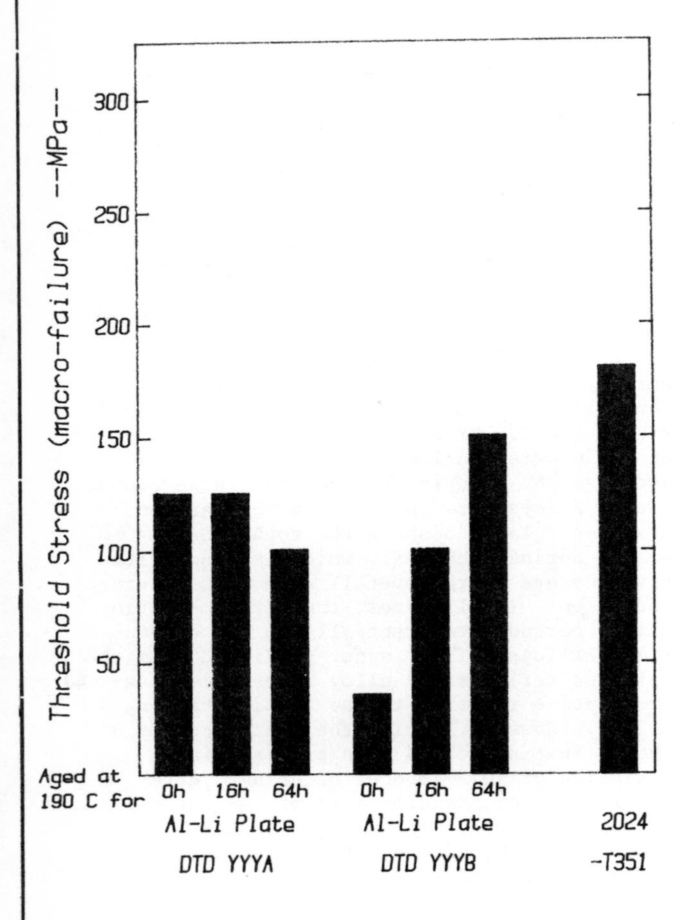

FIGURE 6

Threshold stresses determined using C-rings exposed to a marine environment.

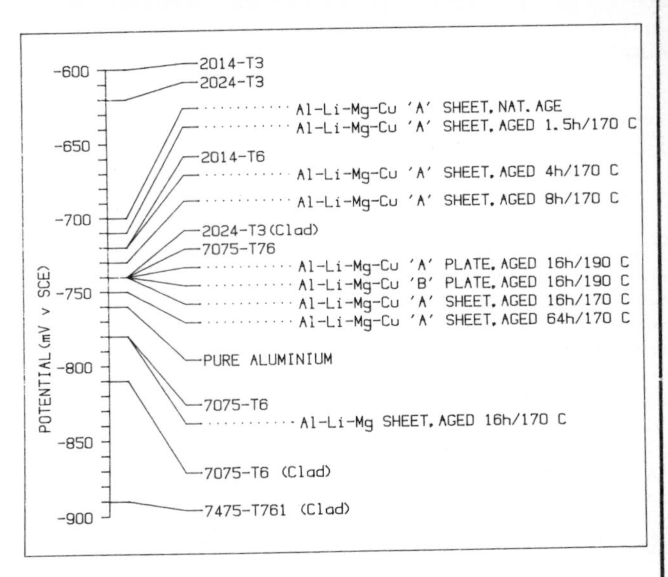

FIGURE 7

Galvanic series for a range of lithium-bearing and conventional aluminium alloys in 3½% NaCl solution.

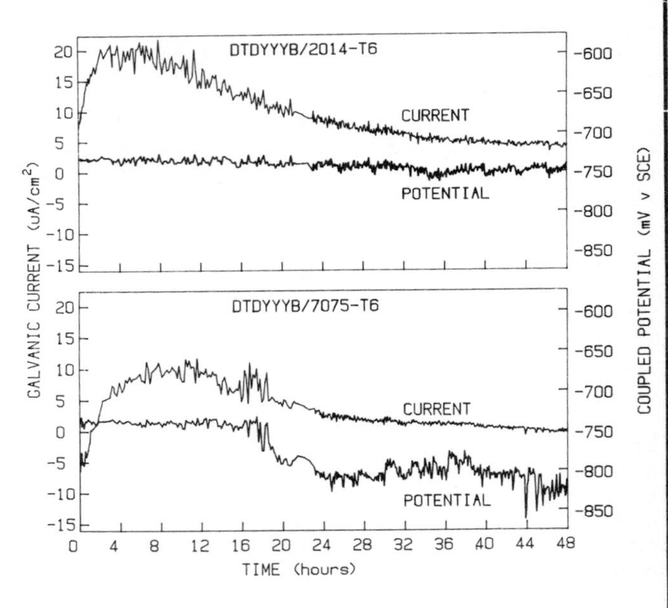

FIGURE 8

Galvanic corrosion current and coupled potential measured for DTD YYYB alloy coupled to 2014-T6 and 7075-T6.

alloys and can be attributed to pit initiation and growth.

2 Al-Li-Mg-Cu alloys are susceptible to exfoliation corrosion in a marine environment. Sheet materials are generally more susceptible than plate. Overageing improves the resistance to exfoliation attack. The intermittent acidified salt fog and alternate immersion tests are considered to give better correlation with marine exposure trials in generating exfoliation corrosion than the EXCO test.

3 The stress corrosion threshold stresses for Al-Li-Mg-Cu alloys studied are higher than that for 7075-T6 and approach that for 2014-T6 and 2024-T3.

4 Al-Li-Mg-Cu alloys are galvanically compatible with conventional aluminium alloys.

REFERENCES

1. T.H. Sanders, E.S. Balmuth. Metal Progress, March, 32, 1978

2. V.P. Batrakov, S.A. Karimova, V.S. Komissarova. Zashchita Metallov 17, 6, 1981

3. B.W. Lifka, D.O. Sprowls. Corrosion 22, 7, 1966

4. C.J.E. Smith, V.C.R. McLoughlin. Proceedings of Conference on 'Control and Exploitation of the Corrosion of Aluminium Alloys', 1983. To be published.

5. K. Nisancioglu, H. Holtan. Werkstoffe and Korrosion 30, 105-113, 1979

6. A. Mahoon. Unpublished Data

7. C.J. Peel, B. Evans, R. Grimes, W.S. Miller. Proc of 9th European Rotocraft Forum, Stresa, Italy, 1983

8. Standard Test Method for Exfoliation Corrosion Susceptibility of 2XXX and 7XXX Series Aluminium Alloys (EXCO Test). ASTM G34-79, Philadelphia, Pa, 19103, American Society for Testing and Materials (1979)

9. Draft Specification for Aluminium-Lithium Alloy DTD YYYA, January 1985

10. J.A. Gray. RAE Technical Report TR 79117, 1979

Effect of heat treatment on corrosion resistance of Al–Mg–Li alloy

B A BAUMERT and R E RICKER

The authors are in the
Department of Metallurgical Engineering
and Materials Science,
University of Notre Dame,
Indiana 46556, USA

ABSTRACT

The precipitation of the anodic phases during heat treatment of aluminum alloys containing lithium may have a deleterious influence on the corrosion resistance of these alloys. As a result, an investigation was conducted into the corrosion resistance of an Al-Mg-Li alloy as a function of heat treatment. The resistance of the alloy to pitting and general attack was investigated through electrochemical polarization measurements. The pitting potentials and the corrosion potentials did not change significantly with aging time while the corrosion rates estimated from the polarization curves and from polarization resistance measurements increased with aging time.

INTRODUCTION

Lithium-containing aluminum alloys are attractive because of their lower densities coupled with increased strength and elastic modulus. However, these alloys may become susceptible to corrosive attack if lithium-rich anodic phases precipitate and grow during heat treatment. Niskanen et al. (1,2) reported that overaging Al-Li, Al-Li-Mn and Al-Li-Zr alloys resulted in the precipitation of δ phase (AlLi) at the grain boundaries which results in increased corrosion rates. However, for Al-Li-Mg alloys, they found only slight changes in the corrosion behavior with overaging. Since Noble and Thompson (3,4) reported that overaging Al-Mg-Li alloys resulted in the precipitation of Al_2MgLi inhibiting the formation of δ, Niskanen et al. concluded that the improved corrosion resistance of these alloys was the result of the formation of the Al_2MgLi phase hindering the formation of the anodic δ phase. However, Tsao and Pizzo (5) conducted an investigation into the electrochemical behavior of Al-Li P/M alloys with varying Cu and Mg contents and found that neither heat treatment nor Mg content influenced the electrochemical behavior. As a result, this study of the electrochemical polarization behavior of an Al-Mg-Li P/M alloy was undertaken to determine if aging influences the corrosion behavior of this type of alloy and to identify the causes of the observed behavior.

EXPERIMENTAL

An Al-Mg-Li P/M alloy was selected for this study; its composition is given in table 1. This same alloy, in the peak aged condition, was used in a previous study of the effect of environmental variables on the corrosion behavior (6). Samples of this alloy were solution treated at 465°C for 1.5 hours, quenched in cold water and aged in silicone oil for varying times to determine the aging response of this alloy. Figure 1 is a plot of the hardness of this alloy as a function of aging time at 190°C and figure 2 is a micrograph of a peak aged sample showing the typical microstructure of this alloy.

Cylindrical samples were machined from the alloy for electrochemical polarization testing and solution treated (1.5 hours at 465°C), quenched in cold water and aged at 190°C for 1, 2, 5.75, 21.5 and 45 hours as indicated in figure 1. A sample was solution treated, quenched and tested in the as-quenched or naturally aged condition. After heat treatment, the polarization samples were ground with successively finer grades of silicon carbide paper to 600 grit. The samples were cleaned and degreased in an ultrasonic cleaner with acetone and distilled water.

Polarization studies were conducted in a standard glass polarization cell (6). The samples were mounted on a stainless steel rod encased in glass with a teflon seal between the glass and the sample. Two amorphous carbon counter electrodes were used to provide the polarizing current and a saturated calomel reference electrode with a Luggin capillary was used to monitor the potential. All potentials are given versus the saturated calomel electrode. A deaerated 0.5 molar sodium chloride solution was selected for the electrolyte. The solution was mixed from standard laboratory chemicals and low conductivity distilled, deionized water. The solutions were thoroughly deaerated by bubbling nitrogen gas (<0.5 ppm oxygen) through the solution before and after insertion of the sample.

Two types of polarization tests were performed. First, potentiodynamic polarization curves were obtained by sweeping the electrochemical potential of the sample from -1.3 V(sce) toward more noble or positive potentials, while recording the current required to maintain the potential with a log converter. Once pitting was initiated, the direction of potential scanning was reversed and the potential was scanned in the active direction until

Table 1. Chemical Composition of the Al-Mg-Li Alloy after Heat Treatment

Element	Wt. %	Element	Wt. %
Al	Bal.	Ni	0.006
Mg	4.24	Zn	0.004
Li	2.13	Mn	0.002
Si	0.10	Cr	< 0.002
Cu	0.030	Ti	< 0.001
S	0.012	Na	< 0.001
Fe	0.01	K	< 0.001

Table 2. Average Corrosion, Pitting and Repassivation Potentials

Aging (hrs.)	E_P (mV)	E_{RP} (mV)	E_{icorr} (mV)	i_p^* (mA/m^2)
0	-774	-803	-982	3.17
1	-773	-803	-1121	8.1
2	-778	-800	-1094	6.81
5.75	-784	-817	-1077	4.40
21.5	-796	-814	-1122	12.4
45	-788	-808	-1082	12.8

*Passive current density measured at $E=0.5\ (E_P+E_{corr})$

Table 3. Corrosion Rate Estimates From Polarization Resistance Measurements

Aging (hrs.)	b_a (mV)	b_c (mV)	B	R_p	$i_{corr.}$ (mA/m^2)
0	305	196	51.8	43.4	1.19
1	441	163	51.7	29.2	1.77
2	405	163	50.5	21.3	2.37
5.75	319	168	47.8	18.2	2.63
21.5	328	154	45.5	13.7	3.32
45	328	121	38.4	9.99	3.84

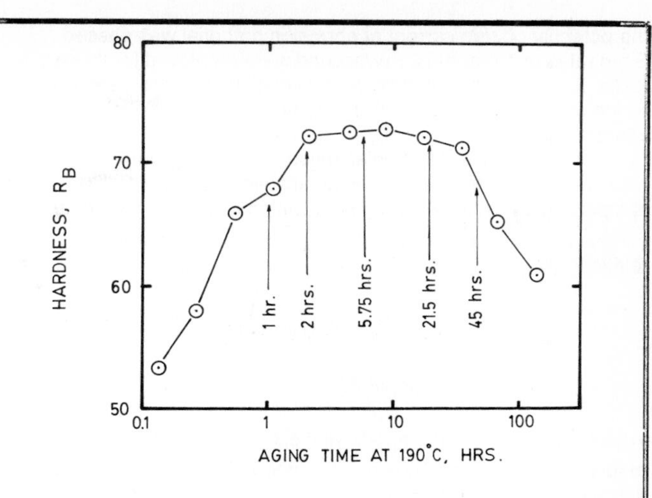

FIGURE 1 Hardness of the Al-Mg-Li alloy (Rockwell B Scale) versus Aging Time at 190°C.

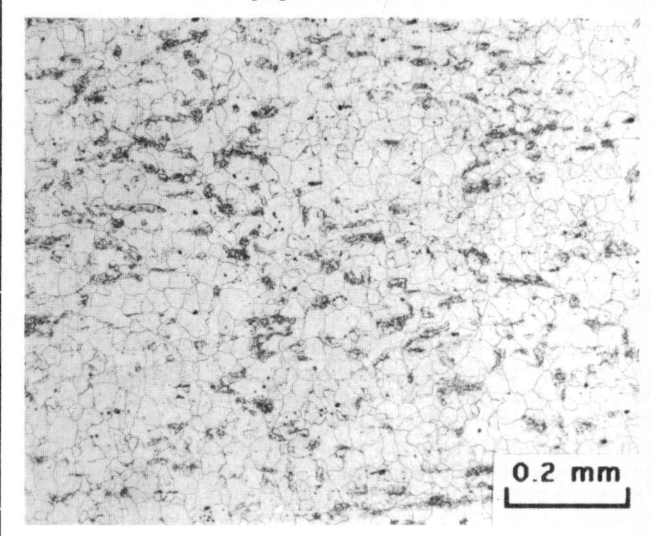

FIGURE 2 Microstructure of the Al-Mg-Li alloy aged 5.75 hours at 190°C (Keller's etch).

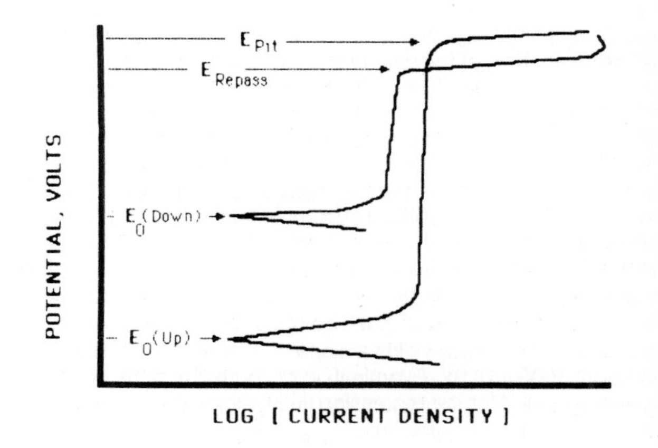

FIGURE 3 Schematic of a polarization diagram identifying the corrosion, pitting and repassivation potentials.

the potential of zero current or corrosion potential was passed. Scan rates of 1.0 and 0.2 mV/second were employed for these tests. Figure 3 is a schematic representation of a polarization curve which would be obtained in this manner with the potentials of interest labeled.

The second type of polarization experiments were linear polarization tests from which the polarization resistance of the sample can be estimated. The polarization resistance (R_p) of a sample is inversely related to the corrosion current density by the relationship:

$$R_p = B/i_{corr}$$

$$B = (b_a b_c)/[\, 2.303\ (\, b_a + b_c)\,]$$

where b_a and b_c are the anodic and cathodic Tafel slopes respectively. The polarization resistance was estimated by cycling the potential (at 1 and 0.2 mV/second) to 100 mV on either side of the corrosion potential to obtain a hysteresis loop of the form illustrated in figure 4. The polarization resistance was then estimated from the slope of the linear portion of the loop as the end of the forward or reverse sweep is approached (7).

X-ray diffraction experiments were performed on six flat samples which received the same heat treatments and preparation as the polarization samples. These experiments were performed with a Diano diffractometer at a scan rate of 1 degree/minute from 20 to 80 degrees and then repeated at greater detector sensitivities over the 2θ range of 20 to 35 degrees. The minimum detectable volume fraction of the phases by this technique is unknown.

RESULTS

Figure 5 shows the average pitting, repassivation and corrosion potentials determined for each aging time. The pitting and repassivation potentials are essentially constant over the range of aging times studied as shown in table 2. Except for the as quenched-naturally aged sample, the variations in the measured corrosion potentials with aging time were insignificant as compared to the range of the measurements as indicated in figure 5. For the as quenched-naturally aged sample, the corrosion potentials measured were, on the average, higher than those determined for the other aging conditions.

The corrosion currents were estimated from the potentiodynamic polarization curves. The mean and the probable error were determined for each aging condition and are graphed in figure 6 versus the logarithm of the aging time (8). A linear regression analysis of the data in this figure (for aging times greater than zero) yields a slope of 0.099 and a correlation coefficient 0.28 (8,9). The positive slope of this line and examination of figure 6 indicates that the rate of corrosion is increasing with aging time.

The corrosion currents were calculated from the polarization resistance measurements and the measured Tafel slopes. Table 3 gives the average polarization resistance, tafel slope and corrosion current determined for each aging time. The mean and the probable error of the corrosion rates calculated for each aging time are graphed versus the logarithm of the aging time in figure 7 (8). A linear regression analysis of the points on this graph (for aging times greater than zero) yields a slope of 0.147 and correlation coefficient of 0.57 (8,9). Examination of figure 7 and the positive slope determined for the regression line indicate that the corrosion rate is increasing with aging time.

The X-ray diffraction examinations identified Al_3Li in the samples aged at 190°C; however, the diffraction peaks indicative of this phase were not found in the as quenched-naturally aged (one week) sample. No diffraction peaks indicative of δ (AlLi) or Al_2MgLi were found in any of the samples.

DISCUSSION

Electrochemical polarization measurements are not a reliable technique for predicting the actual corrosion rates that will be observed in service. While this is due in part to interpretational and reproducibility problems, it is also the result of the difficulty of accurately simulating the service environment and the influence that the cyclic perturbations in this environment have on the corrosion performance of the alloy. Since the objective of this study was not to predict actual corrosion rates but to observe what effect heat treatment may have on the corrosion behavior of the alloy, a single environment was selected which contained a reproducible level of chloride ions and dissolved oxygen (deaerated with nitrogen). Increasing the dissolved oxygen content by aeration will cause the free corrosion potential to increase toward the pitting potential and may result in pitting.

The pitting resistance, as measured by the magnitude of the pitting potential, and the ability of the alloy to repassivate a growing pit, as measured by the repassivation potential, did not change over the range of heat treatments studied. This indicates that the ability of the passivating layer to resist pitting is not influenced by the aging treatments. Also, since the free corrosion potentials did not vary significantly with aging time, the probability of the free corrosion potential exceeding the pitting potential and causing pitting is unchanged over the range of heat treatments studied.

The corrosion rates estimated by two different electrochemical techniques increased with aging time as shown in figures 6 and 7. Because the scatter in the measurements was relatively large as compared to the indicated trend, the statistical significance of the trend was evaluated using Student's t-distribution to test the significance of the positive slope determined for the regression lines (9). This analysis assumes that all of the deviations from the regression line are due to statistical fluctuations and measurements errors and then estimates the probability of measuring the positive slope when the parameters are actually unrelated (a slope of zero). This analysis determined that the probability of observing the positive slope of figure 6 if the corrosion rate actually does not increase with aging is less than 5 percent ($t=1.836$ with $\nu=40$) while for the line in figure 7 the probability becomes vanishingly small ($t=4.135$ with $\nu=36$). If the actual relationship between the parameters is not linear, this technique will over-estimate the variance of the measurement errors. As a result, this is a conservative estimate of the probability of error and we are confident (to the 99.5 % confidence level based on the polarization resistance measurements) that the corrosion rate increases with aging time (9).

The aging sequence for Al-Mg-Li alloys was determined by Thompson and Noble (4) to be:

$$\text{S.S.} \rightarrow \delta'\ (Al_3Li) \rightarrow Al_2MgLi$$

Reportedly, the Al_2MgLi forms as a coarsely dispersed precipitate which consumes the δ' precipitates as it grows inhibiting the formation of AlLi (4). The X-ray diffraction examinations, of this investigation, did not detect the presence of either δ (AlLi) or Al_2MgLi in any of the samples. This does not mean that these phases are not present but that if they are present it is in quantities which are still below the minimum detectable volume fraction in even the longest aging treatment (10). Since these phases are

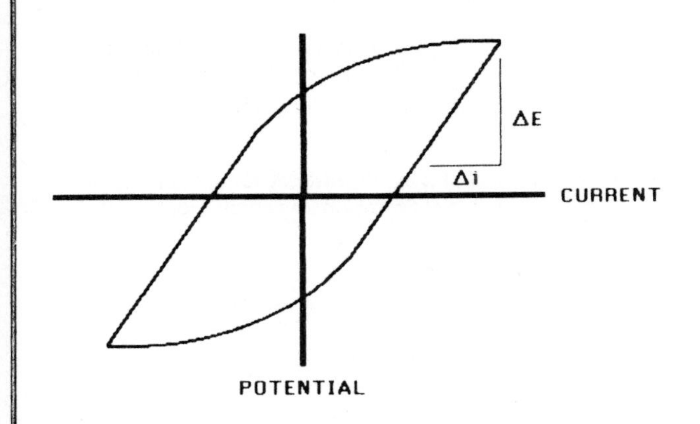

FIGURE 4 Schematic of a polarization resistance scan. The slope of the hysteresis loop yields the polarization resistance.

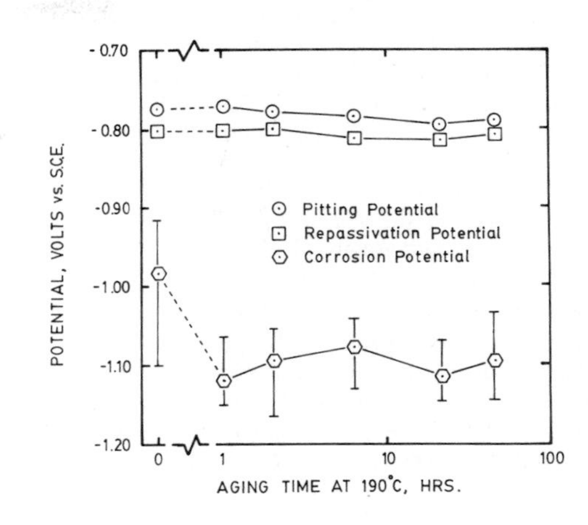

FIGURE 5 Corrosion, pitting and repassivation potentials measured for the Al-Mg-Li alloy in deaerated 0.5 molar sodium chloride versus the aging time. The average and the range of values measured are indicated.

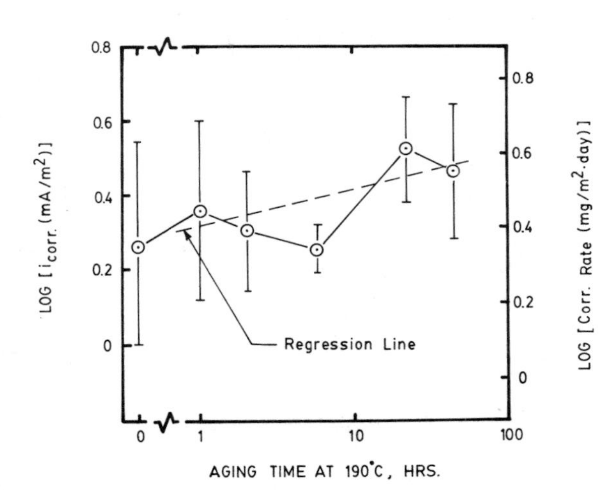

FIGURE 6 Corrosion current densities estimated from potentiodynamic polarization curves in deaerated 0.5 molar sodium chloride versus aging time.

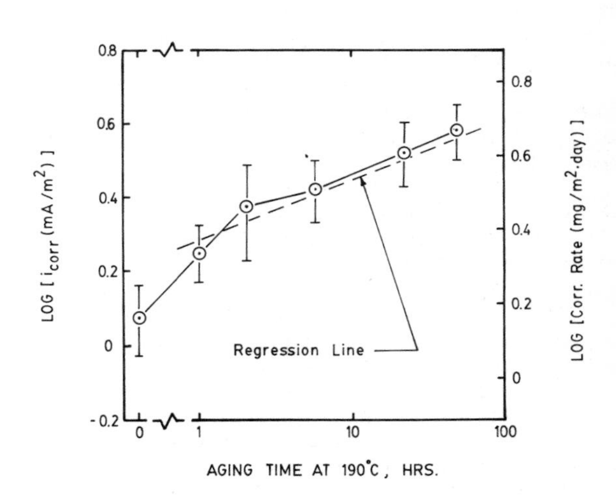

FIGURE 7 Corrosion current densities calculated from the polarization resistance measurements in deaerated 0.5 molar sodium chloride versus aging time.

not present in sufficient quantities to be detected by X-ray diffraction, it seems unlikely that they would significantly influence the corrosion behavior of the samples. As a result, we must conclude that the growth of δ' precipitates is the only metallurgical change occurring during aging that can influence the electrochemical or corrosion behavior of the alloy.

The only electrochemical change observed with aging is the increase in the estimated corrosion rates. The x-ray diffraction results indicated that the only metallurgical change occurring during the aging process is the coarsening of the δ' precipitates. As a result, we must conclude that the increase in the rate of corrosion is the result of the coarsening process. After nucleation is complete, the δ' precipitates grow by Ostwald ripening (10,11). During Ostwald ripening, the volume fraction of precipitate remains constant while the mean diameter of the precipitates increases and the number of precipitates decreases. Since the volume fraction of δ' is constant, then the surface area fraction of δ' exposed to the environment will also be constant (12). However, the size distribution of the exposed regions of δ' will vary with aging time in accordance with the precipitate size distribution. As a result, the corrosion rate must be increasing with either the mean precipitate diameter or with the number of precipitates above some critical diameter.

CONCLUSIONS

1. The pitting, repassivation and corrosion potentials of this alloy did not vary significantly over the aging times studied.

2. The corrosion current densities estimated from the potentiodynamic polarization curves increased with the aging times over the range of times studied.

3. The corrosion rates calculated from the polarization resistance measurments also increased with aging time over the range of times studied.

4. No evidence of either Al_2MgLi or δ (AlLi) was detected by X-ray diffraction. These phases may be present at quantities below the volume fraction that can be detected.

5. The increase in corrosion rates with aging time is attributed to the increasing size distribution of δ' precipitates since the volume fraction of this phase is constant over the aging times studied.

ACKNOWLEDGEMENT

The authors are grateful to Mr. J. E. Coyne and the Wyman-Gordon Foundation for their support.

REFERENCES

1. P. Niskanen, T. H. Sanders, J. G. Rinker, M. Marek. Corro Sci, 22, 4, p. 238 (1982).

2. P. Niskanen, T. H. Sanders, M. Marek and J. G. Rinker. Al-Li Alloys, edited by T. H. Sanders and E. A. Starke, AIME, New York (1981), p. 347.

3. B. Noble and G. E. Thompson. Met. Sci., 5, p. 114 (1971).

4. G. E. Thompson and B. Noble. J. Inst. of Metals, 101, p. 111 (1973).

5. C. T. Tsao and P. P. Pizzo. CORROSION/85, Paper No. 69, National Association of Corrosion Engineers, Houston, Texas (l985).

6. R. E. Ricker and D. J. Duquette. Al-Li Alloys II, T. H. Sanders and E. A. Starke edts., AIME, New York (1984), p. 581.

7. D. D. MacDonald. J. Electrochem. Soc., 125, 9, p. 1443 (1978).

8. P. R. Bevington. Data Reduction and Error Analysis for the Physical Sciences, McGraw-Hill, New York (1969).

9. R. E. Walpole and R. H. Myers, Probability and Statistics for Engineers, Macmillan, New York (1972).

10. D. B. Williams and J. W. Edington. Met. Sci., 9, p. 529, (1975).

11. J. H. Kulwicki and T. H. Sanders, Al-Li Alloys II, E. A. Starke and T. H. Sanders edts, AIME, New York (1984), p31.

12. E. E. Underwood, Quantitative Stereology, Addison-Wesley, Reading MA (1970).

Elevated temperature oxidation of Al–Li alloys

M BURKE and J M PAPAZIAN

MB is at the
Massachusetts Institute of Technology,
Cambridge, Massachusetts, USA;
JMP is with the
Corporate Research Center, Grumman Corporation,
Bethpage, NY 11714, USA
where the work was performed

SYNOPSIS

The elevated temperature oxidation behavior of several high purity Al-Li, Al-Li-Zr and Al-Li-Cu-Zr alloys and the commercial British Alcan DTD XXXA and XXXB alloys was studied in wet and dry air. The rate of oxidation was found to be primarily dependent on the lithium content of the alloy and the moisture content of the atmosphere. The rate of oxidation increased linearly with lithium content. Magnesium and copper additions caused an increase in the rate of oxidation, with magnesium being more potent than copper. Low levels of zirconium (approximately 0.15%) were found to have little or no effect. Mechanical testing and metallographic analyis of samples that were solution treated in various atmospheres suggest the presence of a lithium depleted surface layer that modified the mechanical properties of the samples. Such a layer may have important implications in the industrial use of these alloys.

INTRODUCTION

An attractive feature in the industrial use of aluminum-lithium alloys is their ability to be substituted for current aluminum alloys with minimal changes in processing and working procedures. In many aspects of the aircraft production route this appears to be true, but some properties of aluminum-lithium may require substantial modifications to existing processing routines. An important example of such a property is the tendency of aluminum-lithium alloys to oxidize rapidly at high temperatures. This property may make these alloys more vulnerable in situations where elevated temperatures are encountered, such as in fires or in solution heat treatments. Solution heat treatments are an integral part of the processing cycle of all age hardening alloys, and all current aluminum-lithium alloys are age hardenable. Rapid oxidation during solution treatments may lead to degradation of some properties, requiring either the introduction of special post heat treatment and cleaning procedures or special protective atmospheres in solution treatment facilities. As a first step toward evaluation of these possibilities, this study of high temperature oxidation of aluminum-lithium alloys was undertaken.

Our objective in this work was to develop a basic understanding of the rate of oxidation of aluminum-lithium alloys and of the effects of several common alloying elements on this rate. The elements we are concerned with are lithium, copper, magnesium and zirconium, all of which are common additions to commercial Al-Li alloys. Previous work has shown that Al-Li alloys undergo rapid oxidation at solution treatment temperatures and that the rate of oxidation is primarily a function of the water vapor content of the furnace atmosphere regardless of the carrier gas (Ref. 1, 2), but has not established the effects of Li, Cu, Mg or Zr content on the oxidation rate. Previous workers also report that trace element additions of Be, Ca and Bi, known to inhibit oxidation in certain aluminum alloys, do not reduce oxidation in Al-Li alloys (Ref. 1, 2).

An additional objective of this work was to make an initial evaluation of a possible loss of strength or formability during solution treatment of thin sections of Al-Li alloys. Previous work has shown that substantial lithium loss may occur during solution treatment (Ref. 3) and that this loss may result in a degradation of mechanical properties near the

Table 1. Alloy Compositions

ALLOY	Li	Cu	Mg	Zr
L-1	0	0	0	0.12
L-2	1.0	0	0	0.10
L-4	4.0	0	0	0.16
L-8	2.5	3.6	0	0.16
L-11	3.5	2.6	0	0.10
L-12	2.9	0	0	0
L-13	3.1	0	0	0.15
L-30 (XXXA)	2.6	1.1	0.6	0.08

Table 2. Tensile Test Results

ALLOY	ATMOSPHERE	UTS, KSI	ELONGATION, %
XXXA	ARGON	59.2 ± .7	6.3 ± 1.3
XXXA	DRY AIR	57.5 ± .8	8.3 ± 1.0
XXXA	WET AIR	55.8 ± .3	8.6 ± 1.5
XXXB	ARGON	62.1 ± .2	6.3 ± 1.3
XXXB	DRY AIR	62.0 ± 1.0	7.6 ± 0.6
XXXB	WET AIR	57.1 ± .5	8.6 ± 0.4

Table 3. Bend Test Results

ALLOY	ATMOSPHERE	RADIUS	BEND TO 90°	SURFACE CRACKING
XXXA	ARGON	1t	YES	NO
XXXA	ARGON	2t	YES	NO
XXXA	DRY AIR	1t	YES	NO
XXXA	DRY AIR	2t	YES	NO
XXXA	WET AIR	1t	YES	LARGE CRACKS
XXXA	WET AIR	2t	YES	
XXXB	ARGON	1t	NO	LARGE CRACKS
XXXB	ARGON	2t	YES	LARGE CRACKS
XXXB	DRY AIR	1t	NO	—
XXXB	DRY AIR	2t	YES	NO
XXXB	WET AIR	1t	NO	—
XXXB	WET AIR	2t	NO	—

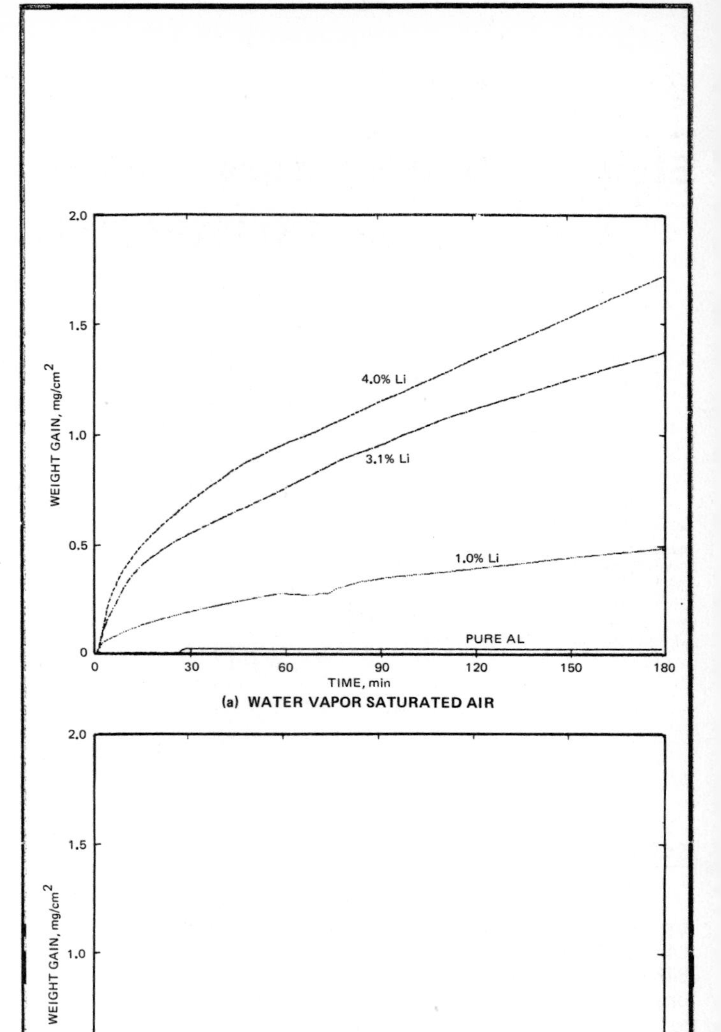

Fig. 1. Weight Gain of Al-Li-Zr alloys at 500°C in High Purity Air

surface (Ref. 4). Other work showed that this loss could also affect bulk properties (Ref. 5).

EXPERIMENTAL PROCEDURE

The experimental approach to this work centered on thermogravimetric analysis, TGA, by means of which the weight of a sample can be continuously monitored as a function of time, temperature and atmosphere. Additional experiments involved tensile and bend properties and metallographic analysis.

The starting materials were 3 in. diameter x 5 in. long ingots of high purity alloys that had been cast in an inert atmosphere by the Naval Surface Weapons Center. The ingots were chemically analyzed by atomic absorption spectroscopy; their compositions are listed in Table 1. After casting, the ingots were wrapped in several layers of aluminum foil, homogenized for 24 h at 510°C and furnace cooled. They were then scalped and extruded to 0.25 in. x 2 in. bars. The extrusion dies were preheated to 315°C, the liner to 370°C and the billets to 400°C; ram speed was approximately 6 in./min. Pieces of the extruded bar were subsequently hot rolled to approximately 0.015 in. thick. TGA samples, 0.625 in. diameter, were punched from this sheet and a 0.08 in. hole was drilled in each disc to allow positioning on the balance arm. The discs were then electropolished for 4 min at 7 V in a 1:4 perchloric acid: ethanol electrolyte at -10°C. These discs were then placed in a desiccator for 4 weeks to establish a stable room temperature oxide.

Thermogravimetry was performed using a DuPont Series 99 thermal analyser with a 951 thermogravimetric module. The furnace was preheated to 500°C prior to introduction of the sample, which reached temperature within 2.5 min. Two atmospheres were used, dry or wet air, at a flow rate of 200 ml/min. Dry conditions consisted of dry scientific grade air (99.999% nitrogen plus 99.999% oxygen) passed through a molecular sieve dryer; wet conditions were obtained by bubbling the same feed gas through water at room temperature. A typical TGA run lasted 3 h at 500°C during which the sample weight was continuously monitored.

Samples were also made from British Alcan Aluminum DTD-XXXA and XXXB materials. These alloys were received as 0.278 in. plate in the T31 condition. They were hot rolled and prepared in the same manner described above. Tensile and bend test samples were also prepared from these alloys, but rolling was stopped at 0.05 in. thickness. Dogbone tensile specimens with a 0.25 in. x 1.5 in. gage length were cut from this sheet, electropolished under the same conditions as the TGA samples, and stored in a desiccator for 3 weeks. These specimens were solution treated for 1 h in a 500°C drop furnace in three atmospheres: dry argon, dry air, and wet air, and subsequently aged 4 h at 190°C in an oil bath. In order to simulate an unoxidized condition, samples treated in argon were electropolished again prior to tensile testing.

Bend test samples (2 in. x 0.5 in. x 0.05 in.) were cut and heat treated (dry argon, dry air, wet air, 1 h at 500°C) from the same XXXA and XXXB rolled sheet as the tensile specimens. Bend tests were performed in accordance with ASTM standard E290: "Semi-Guided Bend Test for Ductility of Metallic Materials." Samples were held in a vise and bent through a 90° arc using a rawhide mallet over 1/4 in. and 1/8 in. diameter bending dies. This corresponded to an approximate 1t and 2t radius bend.

Metallographic and microhardness measurements were made using standard techniques. The microhardness tests employed a Knoop diamond indenter with a 50 g load. Kellers etch was used to reveal grain boundaries.

RESULTS

A. THERMOGRAVIMETRIC ANALYSIS

1. Effect of Lithium Concentration

The results of TGA runs on four alloys of various lithium contents oxidized at 500°C in wet air are shown in Fig. 1a. The curve for pure aluminum has a zero slope and would not be distinguishable from the abscissa except for an instrumental transient which caused a vertical displacement of the curve after 30 min. As can be seen in Fig. 1a, increasing lithium contents results in increasing oxidation rates, and the extent of oxidation is very substantial. All of the weight gain curves appear to be parabolic. Data for the same alloys oxidized in dry air are shown in Fig. 1b. The weight gains in this case are substantially reduced, as had been expected from previous work (Ref. 2), but a monotonic increase in oxidation rate with lithium concentration is still observed.

2. Effect of Copper Concentration

Many of the aluminum-lithium alloy systems that are being considered for commercial application contain substantial amounts of copper; thus the oxidation behavior of several ternary aluminum-lithium-copper alloys was investigated. TGA results from oxidation of these alloys in wet and dry air are plotted in Fig. 2a and 2b, respectively. In Fig. 2a, the oxidation rate of the 2.6% Cu, 3.5% Li alloy in wet air was somewhat greater than that of the 3.1% Li binary, while the 3.6% Cu, 2.5% Li alloy oxidized at a slower rate. This implies that copper enhances oxidation, because if copper had no effect, the higher and lower lithium content would have been expected to fall almost symmetrically above and below the binary data. The data in Fig. 2b for dry air also support this line of reasoning, because the 2.5% Li, 3.6% copper curve is again slightly below the binary data while the 3.5 Li, 2.6% Cu curve is

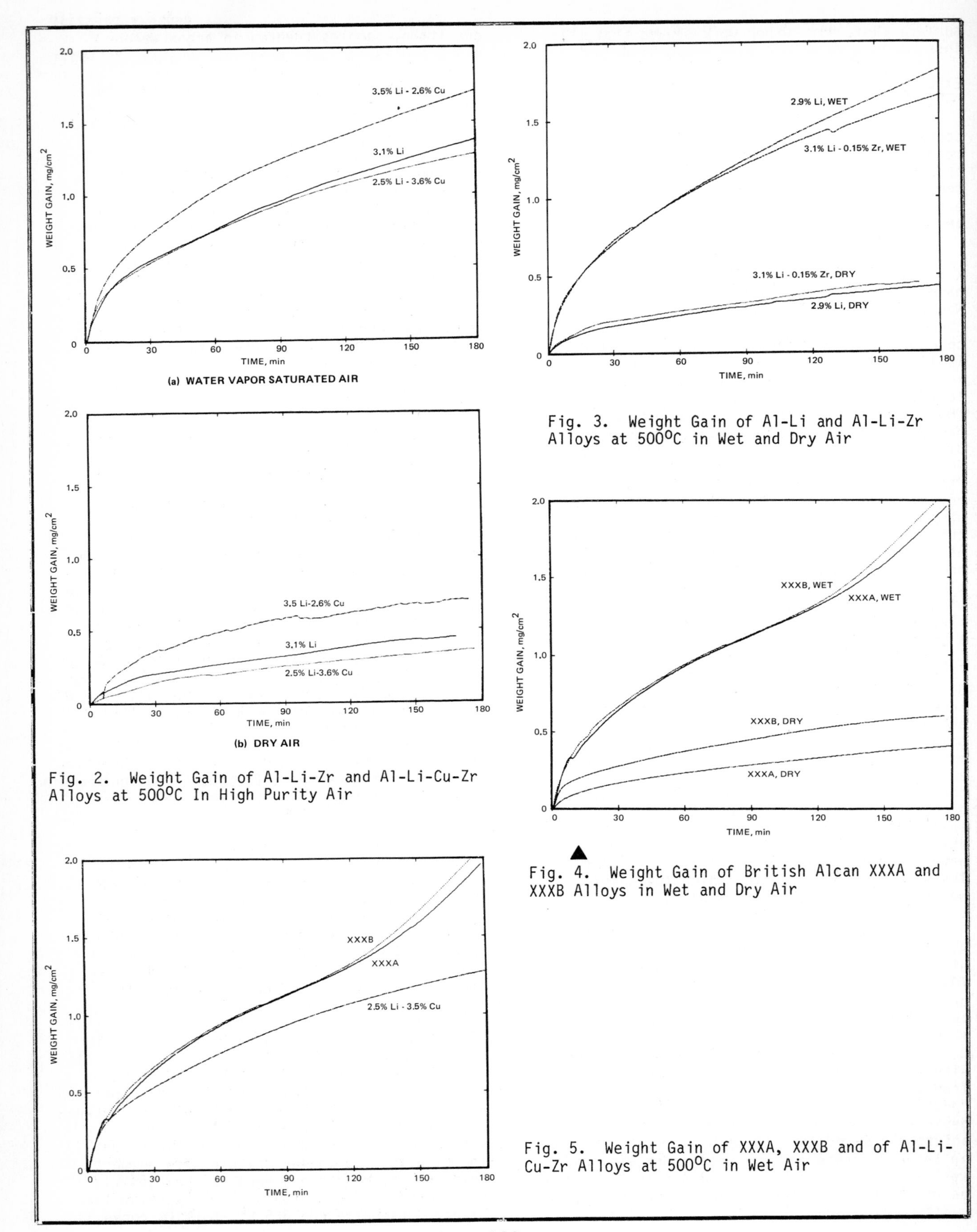

Fig. 2. Weight Gain of Al-Li-Zr and Al-Li-Cu-Zr Alloys at 500°C In High Purity Air

Fig. 3. Weight Gain of Al-Li and Al-Li-Zr Alloys at 500°C in Wet and Dry Air

Fig. 4. Weight Gain of British Alcan XXXA and XXXB Alloys in Wet and Dry Air

Fig. 5. Weight Gain of XXXA, XXXB and of Al-Li-Cu-Zr Alloys at 500°C in Wet Air

significantly above. Thus, it appears that copper enhances the high temperature weight gain of aluminum-lithium alloys to a moderate extent.

3. Effect of Zirconium Concentration

The data from a 2.9% Li alloy and a 3.1% Li-0.15% Zr alloy are plotted in Fig. 3. There does not appear to be any effect of this small amount of zirconium on the oxidation rate.

4. Commercial Alloys

TGA data from British Alcan Aluminum alloys DTD XXXA and XXXB are shown in Fig. 4 for oxidation in wet and dry air. In dry air, the B alloy seems more oxidation prone, while in wet air they show similar characteristics. In wet air, both alloys showed a departure from parabolic oxidation kinetics after 2 h and suffered accelerated oxidation. The total weight gained by these two alloys in a 3 h run was greater than that of any of the other alloys investigated in this program.

The same data for wet oxidation of XXXA and XXXB are replotted in Fig. 5 along with data from a 2.5% Li - 3.6% Cu binary alloy. The lithium contents of all three alloys are similar, while the copper contents of the pure ternary is higher than that of XXXA or XXXB. In addition, both XXXA and XXXB contain 0.6% Mg. The XXXA and XXXB alloys oxidize more rapidly than the ternary, presumably because of the presence of Mg. Previous work (Ref. 2) has shown that Mg promotes the high temperature oxidation of aluminum alloys; thus, it is likely that the addition of Mg is responsible for the accelerated oxidation of XXXA and XXXB when compared to similar binary Al-Li alloys or ternary Al-Li-Cu alloys.

B. MECHANICAL TESTING

In order to assess the potential effects of oxidation during solution treatment of commercial Al-Li alloys on their subsequent utility, several sets of tensile and bend samples were subjected to a 1 h solution treatment at 500°C in wet air, dry air and argon, quenched, aged 4 h at 190°C and then tested. The argon atmosphere employed was not pure or dry enough to suppress oxidation, so these samples were given a slight electropolish after aging in order to provide a suitable control condition.

1. Tensile Tests

Results from tensile tests of the three sets of samples are listed in Table 2. For both alloys, the ultimate tensile strength decreased and the elongation increased as the solution treatment atmosphere became more aggressive. These results are the average of triplicate tests. The 6 to 8% decrease in tensile strength represents a significant loss in properties and might be expected as a logical consequence of surface oxidation and possible lithium loss from the sample, but the increase in elongation to failure is a puzzling result. Perhaps the same lithium loss from the sample surface also resulted in a more ductile surface layer and promoted deformation.

2. Bend Tests

Results from the bend tests are listed in Table 3. The XXXA alloy survived the test in all cases with no cracking evident except for the 1t, wet air specimen, which showed surface cracking. The XXXB alloy could not be bent around the 1t mandrel in any condition, but survived the 2t test after the argon atmosphere and dry air solution treatments although large cracks appeared in the argon sample. These data might be interpreted as showing a trend toward poorer formability after solution treatment in oxidizing conditions.

C. METALLOGRAPHY

Polished cross sections of the fracture region of the tensile samples were examined. A macroscopic zig zag shape of the fracture was typical of both the XXXA and XXXB alloys after wet air solution treatment. The shape of the fracture suggested the presence of a more ductile surface layer. This shape was less evident in the dry air samples and disappeared completely in the argon samples, which showed a straight 45° failure.

Microhardness measurements were made on several samples in order to investigate the possible loss of lithium from the surface layer that was suggested by the tensile results. The microhardness results from the dry air and argon atmosphere samples are shown in Fig. 6 and 7. It is apparent from these figures that the microhardness of the near surface region is lower than the interior for the dry air samples, while the microhardness is uniform for the argon atmosphere plus electropolish sample. Data from the wet air sample were very similar to those of the dry air. For both the wet and dry air samples, the hardness rose to the bulk valve within the first 50 to 60 µm from the surface. Simple estimates based on the diffusivity of lithium in aluminum (Ref. 6) show that 1 h at 500°C will result in a mean square diffusion distance of approximately 30 µm. Thus, the apparent loss of lithium from the bulk to the surface and simultaneous redistribution of the internal lithium could easily have occurred in the time allowed.

After etching, an apparently recrystallized surface layer was observed in some areas. This recrystallization is not thought to be due to lithium loss, however, because similar layers were also seen further into the sample. Surface pitting was also visible. Pits were observed in all three types of samples, but were less severe in the sample treated in argon. The largest pit was observed in the wet air sample and measured approximately 200 µm diameter by 30 µm deep. These pits were not observed in as-

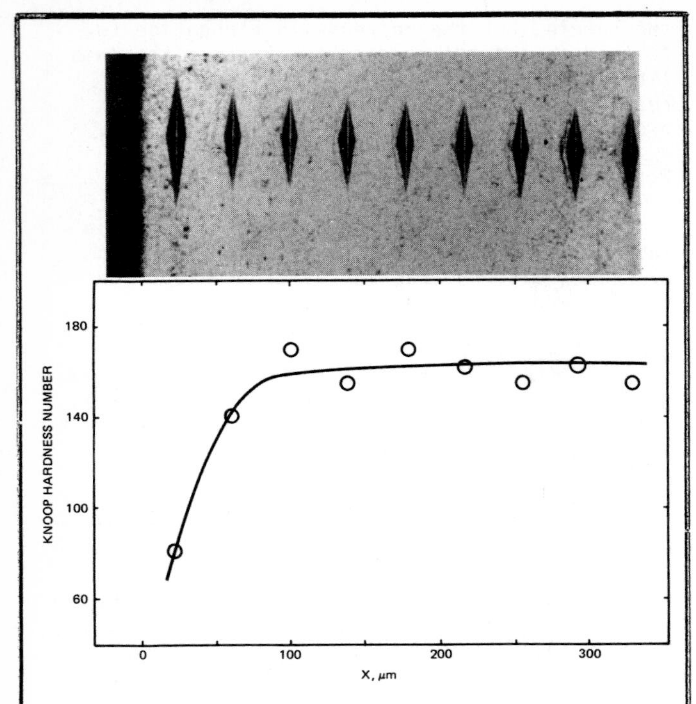

Fig. 6. Optical Micrograph and Knopp Hardness as a Function of Distance from the Surface of an XXXB Tensile Sample after Solution Treatment in a Dry Air Atmosphere

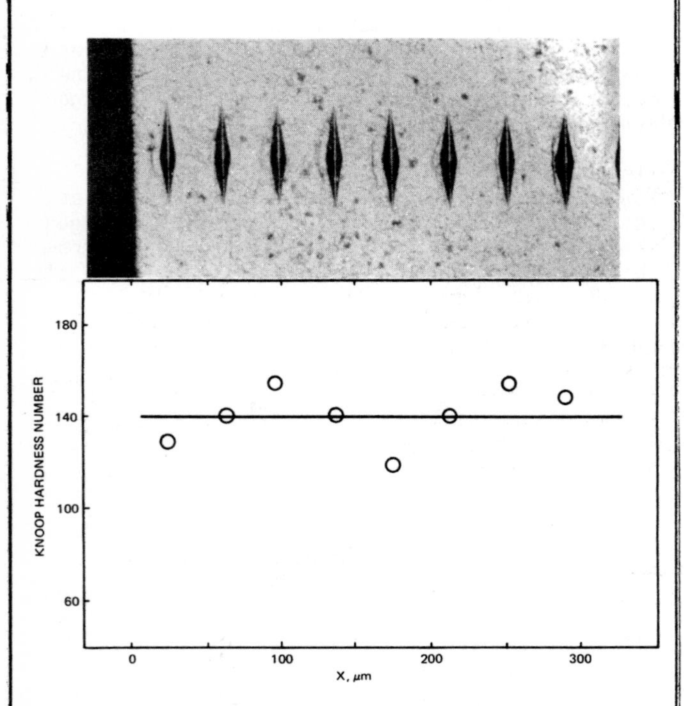

Fig. 7. Optical Micrograph and Knopp Hardness as a Function of Distance from the Surface of an XXXB Tensile Sample after Solution Treatment in an Argon Atmosphere and Subsequent Light Electropolish

electropolished samples. In the oxidized samples, there appeared to be pronounced intergranular attack, and in some cases whole grains were lost due to intergranular undercutting.

DISCUSSION

These results show that Al-Li alloys oxidize rapidly at high temperatures and that the rate of oxidation is directly proportional to lithium content. The presence of lithium causes the protective film normally found on aluminum to break down and results in continuous oxidation, possibly until all of the lithium in solution has been removed by surface oxidation. This behavior is severely aggravated by the presence of water vapor in the furnace atmosphere. Magnesium appears to cause an additional enhancement of the oxidation rate of Al-Li alloys, as does copper. The effect of copper appears to be the smallest of the three. Small amounts of zirconium do not appear to have much of an effect.

Oxidation results in the formation of a powdery gray surface deposit that possesses a characteristic odor and that can be irritating to the lungs. The oxidation also results in an apparent loss of lithium from the near-surface region. For a 1 h solution treatment, this layer did not extend deeper than 60 μm (0.0025 in.) in the British Alcan alloys and would be completely removed if any subsequent machining operations were performed. If this surface layer were not removed, however, the strength of a thin section would be degraded while its ductility might be increased. Our efforts to evaluate the effects of oxidation on post heat treatment formability were not conclusive.

Subsequent experiments (Ref. 7) have confirmed the loss of lithium from surface regions during elevated temperature oxidation. A surprising result of those experiments was the observation that the depth of lithium loss was unaffected by moisture in the atmosphere, despite the pronounced difference in weight gains shown above. This highlights a possible trap in interpretation of weight gain results. The weight gain data merely show the additional sample weight due to the presence of the oxide. Without detailed analysis of the composition of the oxide, the amount of solute lost from the sample and the concomitant loss of mechanical properties cannot be estimated. Thus, the relationship between weight gain data and the severity of a particular atmosphere is tenuous.

SUMMARY

As a result of this work, we have concluded the following:

- The weight gain during oxidation of high purity and commercial Al-Li alloys is primarily dependent on the lithium content of the alloy and the water vapor content of the atmosphere. Increase in either results in increased weight gained

- Magnesium and copper additions increase the weight gain of Al-Li alloys, with Mg being more potent than Cu
- Zirconium additions of up to 0.15% do not have an effect on the oxidation rate
- The ultimate tensile strength decreases and the elongation to failure increases after thin samples of British ALCAN XXXA and XXXB are solution treated for 1 h at 500°C in wet or dry air
- Lithium is lost during solution treatment, resulting in a weaker, more ductile surface layer.

REFERENCES

1. Field, D.J., Scamans, G.M. and Butler, E.P., "The High Temperature Oxidation of Al-Li Alloys," in Aluminum-Lithium Alloys II, Starke, Jr., E.A. and Sanders Jr., T.H., TMS-AIME, Warrendale, PA, 1984, pp 657-67.

2. Field, D.J., Butler, E.P. and Scamans, G.M., "High Temperature Oxidation Studies of Al-3% Li, Al-4.2% Mg and Al-3% Li-2% Mg Alloys," Aluminum-Lithium Alloys, by Starke, Jr., E.A. and Sanders Jr., T.H. TMS-AIME, Warrendale PA 1981, pp 325-47.

3. Willams, D.B., and Edington, J.W., "The Precipitation of δ' (AL_3Li) in Dilute Aluminum-Lithium Alloys," Metal Science, Vol 9, pp 529-32.

4. Peel, C.J., Evans, B., Baker, C.A., Bennett, D.A., Gregson, P.J., and Flower, H.M., "The Development and Application of Improved Aluminium-Lithium Alloys," in Aluminum Lithium Alloys II, ibid, pp. 363-392.

5. EL-Salam F.A., Eatah, A.I., and Tawfik, A., "Effect of Li Loss on Some Physical Properties of Al-Li Alloys", Phys. Stat. Sol (a) Vol 75, pp 379-384, 1983.

6. Williams, D.B., and Edington, J.W., "Microanalysis of Al-Li Alloys Containing Fine δ' (Al_3Li) Precipitates", Phil Mag., 30, No. 5, pp. 1147-53, 1974.

7. R.L. Schulte, J.M. Papazian and P.N. Adler, "Application of Nuclear Reaction Analysis to the Investigation of Lithium Loss During Oxidation of Al-Li Alloys", poster session, this conference. Also published in Procedings of International Conference on Ion Beam Analysis, 7-12 July 1985.

Hard anodising and marine corrosion characteristics of 8090 Al–Li–Cu–Mg–Zr alloy

K R STOKES, D A MOTH and P J SHERWOOD

The authors are with the MOD(PE), Admiralty Research Establishment, Portland, Dorset DT4 9RE, UK

SYNOPSIS

The sulphuric acid hard anodising process has been successfully applied to 8090-T651 alloy in the form of 5 mm thick sheet. Dense adherent coatings up to 60 μm thickness were obtained on test coupons. The corrosion characteristics of the alloy in both the unprotected and hard anodised condition in natural seawater and synthetic salt spray are described and compared with 7010 Al-Zn-Mg-Cu-Zr alloy.

INTRODUCTION

The combination of higher elastic modulus and lower density of the latest aluminium-lithium alloys which has made them so attractive for weight-saving in aerospace structures when compared to current 7000 series alloys, also makes them attractive for deep-diving hydrospace structures where compressive buckling is the dominant mode of failure. However, little is yet known of the corrosion resistance of these alloys in seawater, or of the effect of thick, hard anodising on corrosion rates under bimetallic or crevice corrosion conditions. Results reported so far for Al-Li alloys under accelerated salt-spray conditions(1) and real-time marine atmosphere conditions(2) indicate that overall corrosion rates are very dependent on the amount of δ' (Al_3Li) precipitated, but are not significantly higher than for conventional aerospace alloys, although there is an increased tendency for exfoliation corrosion to occur with the heat treatments developed so far.

In this paper, preliminary corrosion data is reported for the 8090 alloy in a peak aged temper, in both the unprotected and hard anodised condition, and the data is compared with that for 7010 alloy in an overaged, T736 temper.

EXPERIMENTAL

The 8090 Al/Li alloy was obtained from RAE Farnborough in the form of 5 mm thick sheets, approximately 1220 mm long x 150 mm wide identified REG. The chemical composition is given in Table 1 and falls within the RAE/Alcan DTDXXXA specification. The sheet had been solution treated at 530°C for 20 minutes, water quenched and controlled stretched 2½% prior to ageing at 190°C for 16 hours to give a peak strength T651 temper. Tensile properties, hardness and electrical conductivity of the sheet were measured and the data is given in Table 2 together with other properties supplied by RAE. Test coupons 100 mm x 50 mm were cut from the sheet and the rolled surfaces were milled to remove the usual lithium-depleted layer on heat treated material (see Fig 1). The typical grain structure and a random inclusion, probably of oxide or flux, is shown in Fig 2. Two examples of the 7010 alloy were corrosion tested for comparison with the 8090 alloy; their compositions and properties are summarised in Tables 1 and 2. The first example was a back extruded tube of ~50 mm wall thickness manufactured by Pechiney Ugine Kuhlmann S.A., France, and the second example was 50 mm thick rolled plate manufactured by Alcan Plate Ltd, Kitts Green, UK. Both materials had been overaged to a T7365 temper. Test coupons 150 mm x 50 mm x 6 mm were cut from the forged tube so that the grain flow was parallel to the 6 mm thickness. In the case of the plate, coupons 100 mm x 50 mm x 5 mm were cut with the 100 mm and 50 mm dimensions parallel to the longitudinal and long transverse directions respectively.

Table 1. Chemical Composition of the Alloys (Weight Per Cent)

	Li	Zn	Cu	Mg	Zr	Si	Fe	Al
8090-T6 (5 mm Sheet)	2.41	-	1.18	0.70	0.10	0.04	0.10	Balance
7010-T73652 (PUK Forging)	-	5.99	1.52	2.41	0.11	0.04	0.90	"
7010-T73651 (ALCAN 50 mm Plate)	-	6.11	1.64	2.26	0.12	0.06	0.11	"

Table 2. Typical Properties

	8090-T6 Sheet	7010-T73652 Forging	7010-T73651 Plate
Tensile Properties			
0.2% Proof Stress (MPa)	411-421 (L) 411-414 (LT)	474 (L) 499 (LT) 491 (ST)	450 (L) 445 (LT) 435 (ST)
Ult. Tensile Strength (MPa)	471-479 (L) 497-503 (LT)	556 (L) 545 (LT) 545 (ST)	519 (L) 514 (LT) 506 (ST)
Elongation (%)	10 (L) on 25 mm 10-12 (LT) "	9 (L & LT) on 25 mm 7 (ST)	11 (on SD) (L) 12 " (LT) 7 " (ST)
Hardness	148-157 HV10	171 HB	-
Young's Modulus (GPa)	78	71	71
Density (Mg/m^3)	2.53	2.83	2.83
Electrical Conductivity (% IACS)	20.3	38.8-39.5	40.0
Modulus/Density Ratio	30.8	25.1	25.1
Proof Stress/Density Ratio (L Direction)	165	167	159

Table 3. General Corrosion Results of Unprotected 7010 and 8090 Alloys

Alloy	Test Condition	Av. Weight Loss (mg)	Specimen Area (cm^2)	Av. Uniform Corrosion Rate		Localised Corrosion	
				m.d.d.	μm/y	Max. depth (μm)	Corrosion Rate (μm/y)
8090-T651 Sheet	Total Immersion (120 days)	52	100	0.43	6	190	578
7010-T73651 Plate	Total Immersion (120 days)	59	80	0.66	9	280	852
7010-T73652 Forging	Total Immersion (120 days)		30	0.42	5.5		
8090-T651 Sheet	Salt Spray (28 days)	Negligible (blistering)	100	-	-	90 (isolated pit)	1173
7010-T73652 Forging	Salt Spray (28 days)	64	160	1.32	17	22	287

Table 4. Perspex Crevice Corrosion Results of Unprotected 7010 and 8090 Alloys

Alloy	Test Condition	Bold:Crevice Ratio	Av. Weight Loss (mg)	Specimen Area (cm^2)	Av. Uniform Corrosion Rate		Max. Depth of Crevice Corrosion	
					m.d.d.	μm/y	μm	μm/y
8090-T651 Sheet	Total Immersion (120 days)	1.5:1	113	100	1.0	14.5	40	122
7010-T73651 Plate	Total Immersion (120 days)	2:1	91	75	1.07	14.0	80	243
7010-T73652 Forging	Total Immersion (267 days)	2:1	685	150	1.67	22.0	310	424
7010-T73652 Forging	Salt Spray (28 days)	2:1	-	150	1.09	14.0	15	196

Table 5. INCO 'Castellated Nut' Crevice Corrosion Results of Unprotected 8090 Alloy after Total Immersion in Natural Seawater

Sample Number	Bold:Crevice Ratio	Av. Weight Loss (mg)	Percentage Probability of Crevice Corrosion	Av. Uniform Corrosion Rate		Localised Corrosion	
				m.d.d.	μm/y	Max. depth (μm)	Rate (μm/y)
1	12.5:1	85	27.5%	0.91	13	Pitting: 170 Crevice: 140	630 560
2	12.5:1	83	45.0%	0.91	13	Pitting: 200 Crevice: 130	760 490
3	12.5:1	94	77.5%	1.01	15	Pitting: 160 Crevice: 90	610 340

Table 6. Galvanic Corrosion Results for Anodised 8090 and 7010 Alloys

Alloy	Test Condition	Anode:Cathode Ratio	Av. Weight Loss (mg)	Av. Uniform Corrosion Rate		Localised Corrosion	
				m.d.d.	μm/y	Max. depth (μm)	Rate (μm/y)
8090-T651 Sheet	Total Immersion	9:1	38	0.33	5	420 (edge site)	1280
7010-T73652 Forging	Total Immersion	11:1	81	0.56	7	1150 (edge site)	3500
8090-T651 Sheet	Salt Spray	9:1	41 (Worst Case)	1.37	20	70	850

Anodising Conditions

Prior to sulphuric acid hard anodising the coupons were degreased in acetone, rinsed in tap water, etched for 5 minutes (minimum) in 1.5M sodium hydroxide at 65° ± 3°C, rinsed in tap water, dipped in 10 vol % nitric acid for 1 minute (minimum) at room temperature to de-smut the etched surface and rinsed thoroughly in tap water. Hard anodising was carried out at a constant current density of $2A/dm^2$ in 10 volume per cent sulphuric acid at -5°C for 40 minutes to give a coating thickness of 85 ± 20 µm on 7010 alloy and 58 ± 7 µm on 8090 alloy.

The surface of the oxide on 8090 alloy is shown in Fig 3 where a fractured edge reveals the cell structure of the anodic film. The cell size has been measured to be approximately 65 nm which is similar to that produced on NS4 and other aluminium alloys. However, the pore size appears to be considerably smaller than that obtained with other alloys. Measurements made after ion-beam etching to remove the self-sealing surface layer gave a pore diameter of approximately 10 nm as shown in Fig 4. For comparison the pore and cell structure of the sulphuric acid anodic film on NS4 alloy is shown in Fig 5. No reason can yet be advanced for this marked difference in pore size, and the effect of lithium content on morphology of the anodic film obviously warrants further investigation.

The microhardness values for the hard anodic films on 8090 and 7010 alloys were found to be very similar, 467 HV and 452 HV respectively.

Corrosion Testing

Hard anodised and unprotected coupons of the 8090 and 7010 alloys were degreased in 1:1:1 trichloroethane prior to weighing, after which, depending on material availability, the crevice and galvanic, as well as general corrosion performance of the alloys were evaluated under total immersion and salt spray conditions.

The former evaluation was undertaken over a 120 day period in natural seawater (pH 8.1) at laboratory temperature, whereas the salt spray test, carried out to Defence Standard 07-55 (ie synthetic salt sprayed for 2 hours, followed by storage at 90-95% relative humidity at 30°C for 7 days) was repeated for four complete cycles.

Perspex and nylon artificial crevice formers were used to evaluate the susceptibility of the two alloys to this type of corrosion, and in addition the effect of a metal crevice, in the form of type 316 stainless steel discs coupled to the anodised samples was also appraised. The discs were attached to the coupons using nylon nuts and bolts to avoid galvanic coupling.

To further simulate possible future in-service use of the 8090 alloy the galvanic compatibility of the anodised material with stainless steel was assessed. The test carried out was a modified version of ARE Table Drawing 1318 in which the integrity of the anodising is appraised after a total immersion period of 90 days. Similar tests were carried out on the 7010 alloy.

On completion of the tests, the coupons were photographed, chemically cleaned in 10% nitric acid, degreased and re-weighed. Weight differences were then determined, from which uniform corrosion rates in $milligrams/dm^2/day$ (mdd) and equivalent penetration rates in µm/year were calculated. As the corrosion attack however was invariably localised, where applicable, the depth was directly measured using a Talysurf instrument.

RESULTS

a) General Corrosion

Under total immersion conditions both the unprotected 7010 and 8090 material exhibited pitting corrosion, although in the former alloy it was more random and often deeper in nature. Optical microscopical examination of sectioned, polished and etched pits indicated an intergranular mode of attack (figures 6 and 7) in both alloys.

After 28 days exposure to salt spray conditions the unprotected 8090 sheet coupons were generally in poor condition with much blistering and exfoliation corrosion evident, as well as incipient pitting. In contrast however the unprotected forged 7010 alloy exhibited a uniform salt-water etched appearance with only limited pitting.

The hard anodised coupons of both alloys showed some removal of the anodic film at sharp edges, although subsequent attack of the exposed metal had not occurred. Under total immersion conditions the 7010 samples had a 'crazed' appearance while the base 8090 alloy exhibited some underlying blistering.

There did not appear to be any significant weight change on the anodised samples but the corrosion rates for the unprotected material are given in Table 3.

b) Crevice Corrosion

Both the 7010 and 8090 alloys showed susceptibility to crevice corrosion under total immersion conditions. The 7010 material was also clearly prone to severe intergranular attack particularly under very narrow crevice gap conditions (figure 8). Results are given in Table 4.

Crevice tests (which were only carried out on the unprotected 8090 alloy) using nylon castellated nuts, revealed the random occurrence of corrosion attack. The results, which are presented in Table 5, also suggested that the depth of corrosion under the nylon was greater than that under Perspex. Examination of the

1. Transverse section through as-received 8090 sheet showing solute-depleted layer at rolled surface.

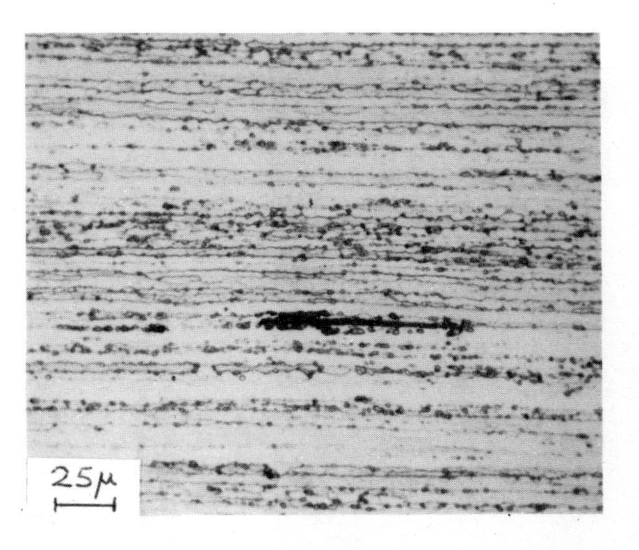

2. Longitudinal section of 8090 sheet showing elongated grain structure and typical inclusion.

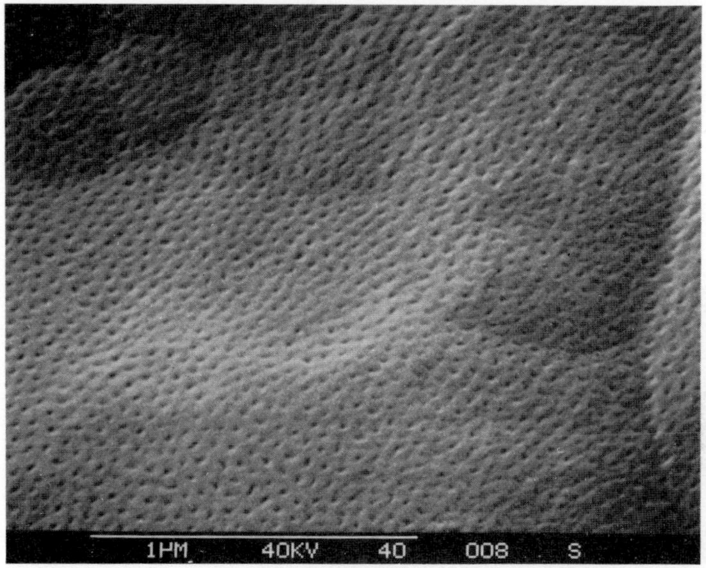

3. Morphology of 5 micron anodic film on 8090 sheet showing cell size (65 nm) and self-sealing effect on surface pores.

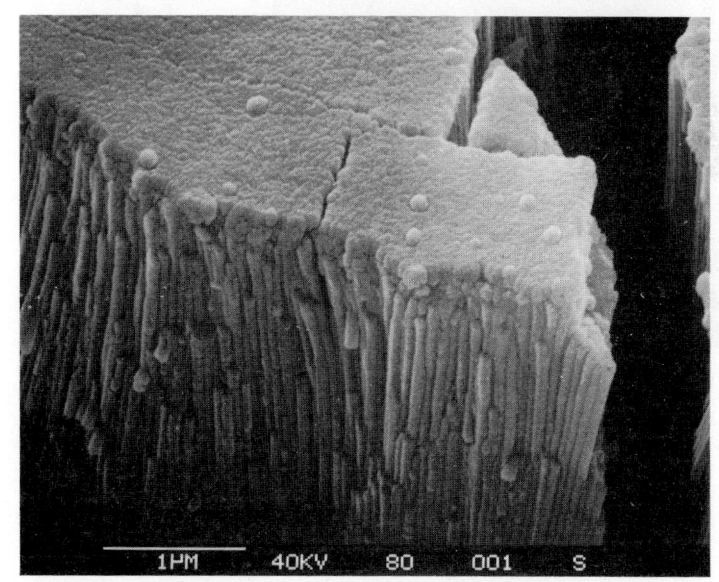

4. Surface of anodic film on 8090 alloy after ion beam etching to a depth of 0.5 micron, revealing a pore diameter of approximately 10 nm.

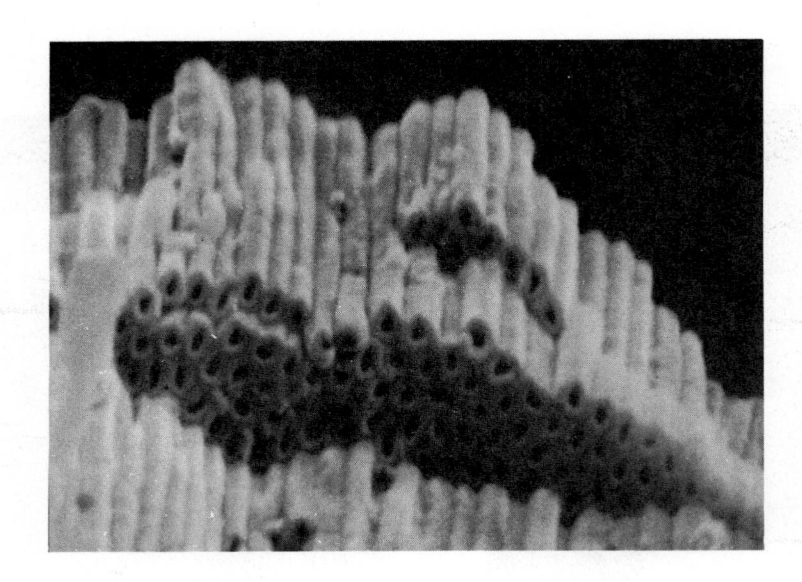

5. Fracture surface of hard anodic film on NS4 aluminium alloy showing a cell size of 75 nm and pore diameter of 20-25 nm.

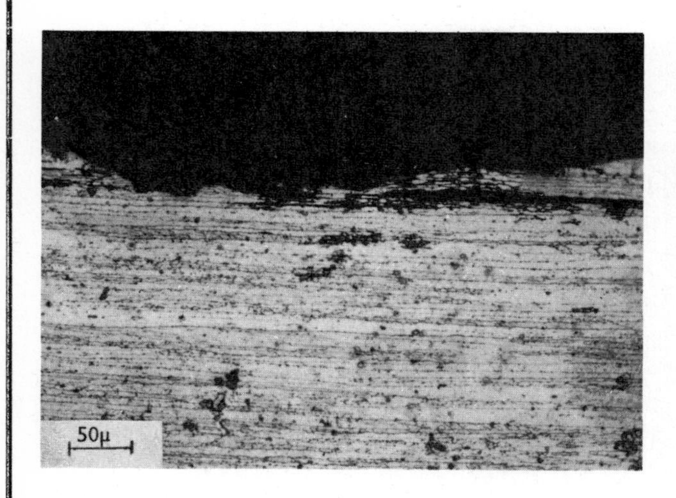

6. Pitting corrosion of 8090 sheet showing intergranular mode of attack.

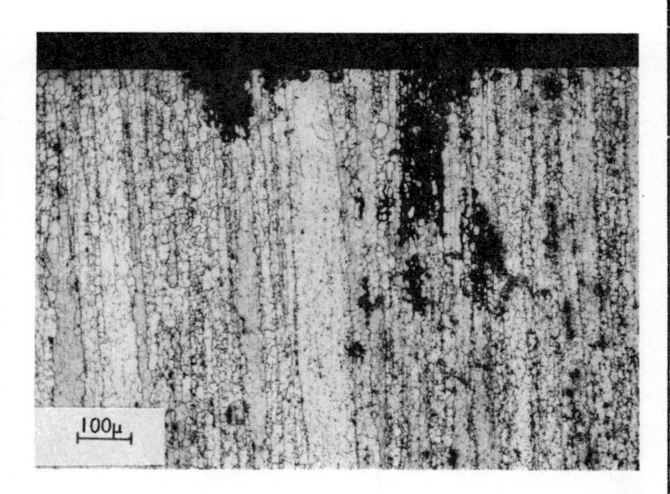

7. Pitting corrosion of 7010 forging showing intergranular mode of attack.

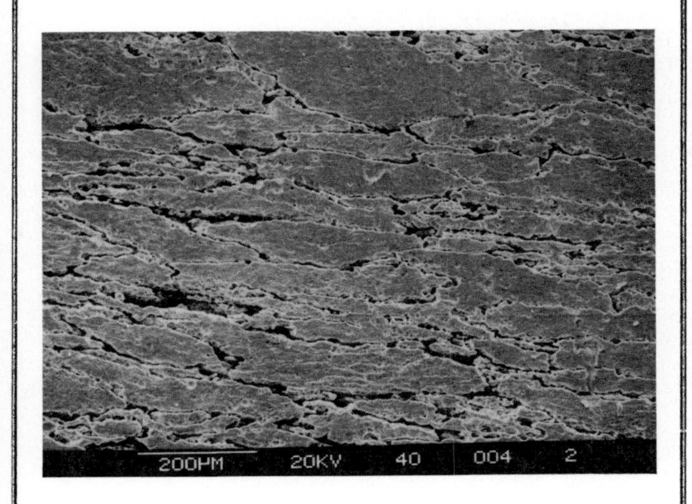

8. Intergranular corrosion of forged 7010 alloy under tight crevice conditions after 21 days total seawater immersion.

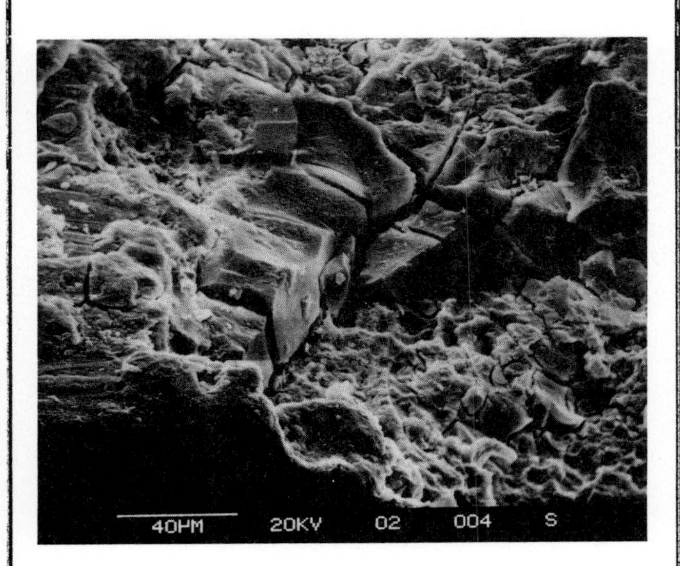

9. SEM of sectioned corrosion pit on 8090 sheet showing intergranular mode of attack.

crevice corroded 8090 alloy on a scanning electron microscope again revealed an intergranular mode of attack (figure 9).

The integrity of the anodised film under the stainless steel crevices appeared to be unimpaired in both alloys, although on one 7010 coupon subsequent attack of exposed metal at an edge site yielded a recorded weight loss of 29 mg from which a penetration rate of 3 μm/y was calculated.

Salt spray Perspex crevice tests were only carried out on the anodised and unprotected forged 7010 material. Results revealed that the hard anodising is capable of preventing attack of the aluminium alloy under the crevices, although, in common with other tests, removal of the anodic film at sharp edges was noted. The unprotected alloy exhibited crevice corrosion.

c) Galvanic Corrosion

Examination of the anodised 8090 sheet under the galvanically coupled 316 stainless steel revealed a large amount of blistering and some removal of the anodic film at several sites. In addition, all of the 8090 coupons showed some degradation of the anodising at sharp edges. On close inspection possible incipient pitting of the substrate (without subsequent removal of the anodic film) was also evident in those samples which had undergone total immersion, whilst blistering and exfoliation was apparent on those 8090 samples subjected to salt spray conditions.

The 7010 samples were adequately protected from galvanic corrosion by hard anodising although, particularly on the total immersion samples, quite severe attack was evident where the anodic film had broken down at sharp edges. On one of the samples in particular this had led to a greater weight loss and localised depth of corrosion than recorded on equivalent 8090 samples.

Table 6 gives the corrosion resúlts for the 316/aluminium alloy galvanic samples.

As well as testing the integrity of the anodised samples under galvanic conditions, directly coupled 1:1 ratio combinations of unprotected 7010 and 8090 were assessed under total immersion in natural seawater. Preliminary results indicate that the 7010 is anodic to the 8090 material and that its corrosion rate is accelerated by a factor of four when compared to uncoupled corrosion rates.

CONCLUSIONS

1. No problems were encountered in applying the conventional thick, hard anodising process to the 8090 alloy in a peak-aged condition.

2. An investigation of the morphology of the anodic film on 8090 alloy by high resolution scanning electron microscopy has revealed a much

finer pore structure (approximately 10 nm diameter) than on conventional alloys (20-25 nm diameter). This should confer improved resistance to corrosion and weathering, and warrants further investigation into the effect of lithium content on pore structure. The anodic film on 8090 alloy also develops a self-sealing layer.

3. Both 7010 and 8090 alloys suffer from pitting and crevice corrosion in the unprotected condition, and in addition the 8090 alloy in the peak-aged temper suffers from blistering and exfoliation corrosion.

4. Under conditions of galvanic coupling the 8090 alloy is cathodic to the 7010 alloy.

5. Hard anodising protects 7010 alloy against corrosion under all conditions tested, although breakdown was noted at sharp edges. To prevent breakdown of the anodising on 8090 alloy under galvanic conditions, further work is required to improve the surface pre-treatment or sealing process.

ACKNOWLEDGEMENTS

The authors are grateful to Mr B Evans, RAE Farborough for supplying the 8090 sheet material, and to Mr M S Atherton and Mr F A Chick, ARE Portland for assistance with mechanical testing and metallography.

REFERENCES

1. P Niskanen, T H Sanders, M Marek and J G Rinker. Proc. 1st International Conference on Aluminium-Lithium Alloys, Stone Mountain, Georgia, USA. May 1980, page 347. Edited by T H Sanders and E A Starke, published Met. Soc. AIME 1981.

2. C J Smith, P Lane and J A Gray, Royal Aircraft Establishment, Farnborough. Workshop on Al-Li Alloys, 23-24 April 1985.

Stress corrosion resistance of Al–Cu–Li–Zr alloys

A K VASUDÉVAN, P R ZIMAN,
S C JHA and T H SANDERS Jr

AKV and PRZ are with Alcoa Laboratories,
Alcoa Center, PA 15069, USA.
SCJ and THS are with the School of Materials
Engineering, Purdue University,
West Lafayette, IN 47907, USA

SYNOPSIS

Stress corrosion crack (SCC) growth resistance of several high purity Al-Cu-Li-Zr alloys were studied. SCC experiments were conducted in two parts to determine the effects of aging in Al-2.9Cu-2.1Li alloy and the influence of the (Li/Cu) ratio in T651 temper on SCC resistance. The results indicated that both the apparent K_{ISCC} and the plateau velocity were not affected markedly by aging the alloy Al-2.9Cu-2.1Li. On the other hand, the apparent K_{ISCC} significantly decreased with an increase in the (Li/Cu) ratio. These results qualitatively indicated that the resistance to SCC growth appears to increase with an increase in the amount of grain boundary δ in the alloy, which occurred when the Li-content was increased or as in the case of Al-2.9Cu-2.1Li, the aging time was increased.

INTRODUCTION

In general, SCC resistance of high strength aluminum alloys is strongly affected by both composition and aging treatments [1]. In 7XXX type alloys, underage to T651 tempers are susceptible to SCC, while the T7 type aging treatments are resistant to SCC. In the peak strength temper in 7XXX alloy, increasing Cu-content improves the SCC resistance. In 2XXX alloys, T3 type tempers are more susceptible to SCC than T851 tempers [1]. Such variations in SCC resistance has been interpreted either in terms of hydrogen assisted cracking or as a dissolution mechanism.

Early work by Rinker et al. [2] on alloy 2020 indicated that significant reductions in plateau velocities occur when aging the alloy from the underage to the peak-age condition. The change in SCC behavior was attributed to a reduction in the electrochemical difference between the matrix and the grain boundaries during aging. Christodolou et al. [3] studied the SCC susceptibility in the binary Al-2.8Li alloys and observed that the degree of SCC susceptibility is dependent on the aging condition, the peak-aged temper being the most susceptible. They have also suggested that hydrogen embrittlement may play a role in the SCC mechanism of the binary Al-2.8Li alloy.

The present study was undertaken to systematically characterize the SCC behavior of high purity Al-Cu-Li-Zr alloys in terms of composition and aging treatments.

EXPERIMENTAL PROCEDURE

The compositions of the alloys studied are given in Table 1.

Table 1: The compositions of various alloys investigated (wt%).

Cu	Li	Zr	Fe	Si	Ti	Al	(Li/Cu) Ratio Atomic Fraction
4.6	1.1	0.17	0.06	0.04	0.01	Balance	2.2
2.9	2.1	0.12	0.06	0.04	0.01	Balance	6.5
1.1	2.9	0.11	0.06	0.04	0.01	Balance	25.2

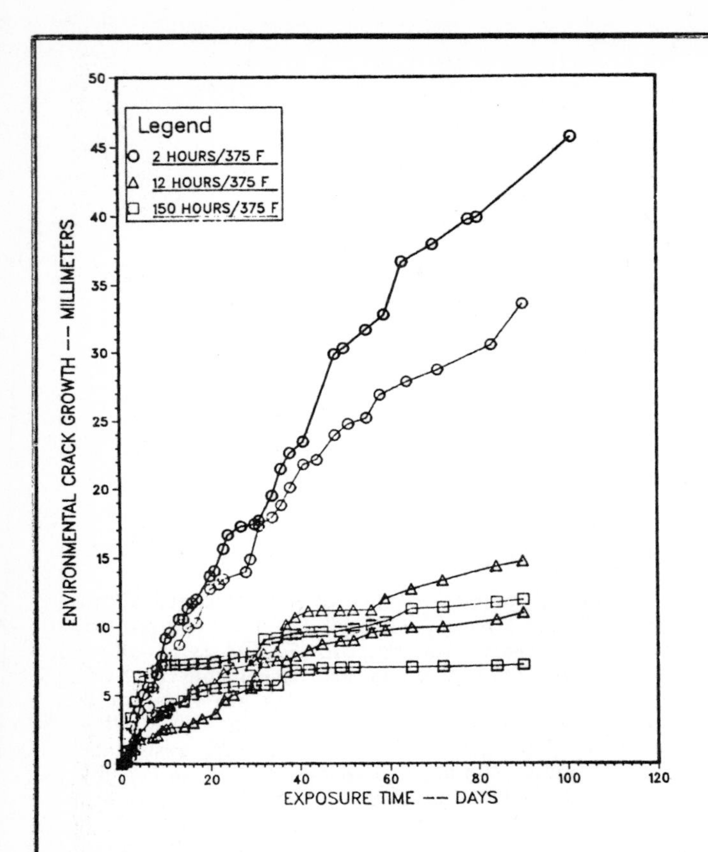

Figure 1: Variation of crack growth with exposure time (t), for UA, PA and OA treatments.

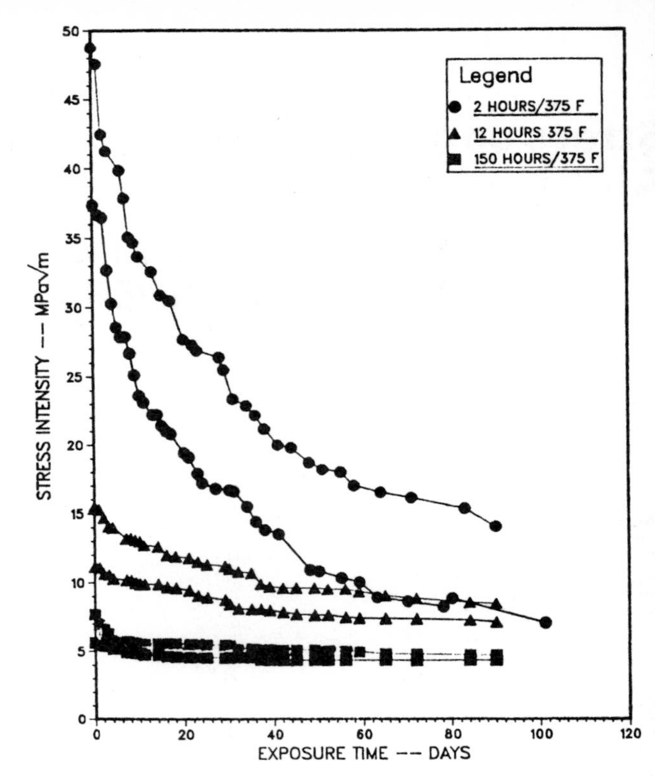

Figure 2: Variation of K1SCC with aging time.

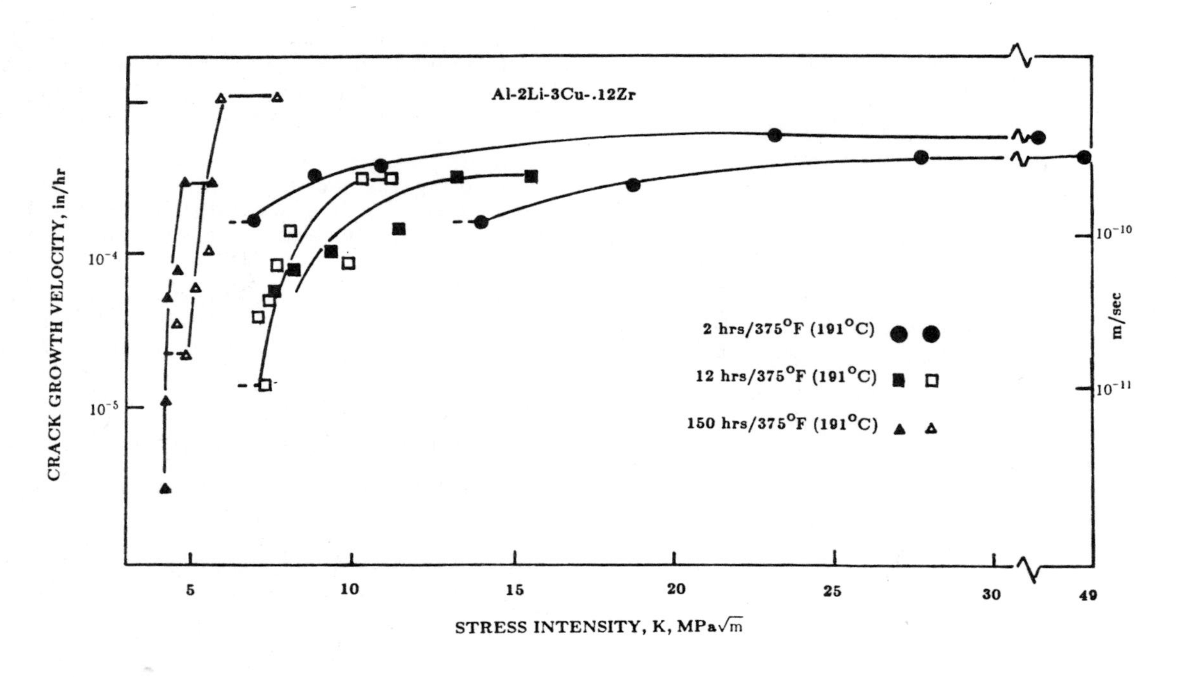

Figure 3: Variation of stress corrosion crack velocity with the stress intensity factor K.

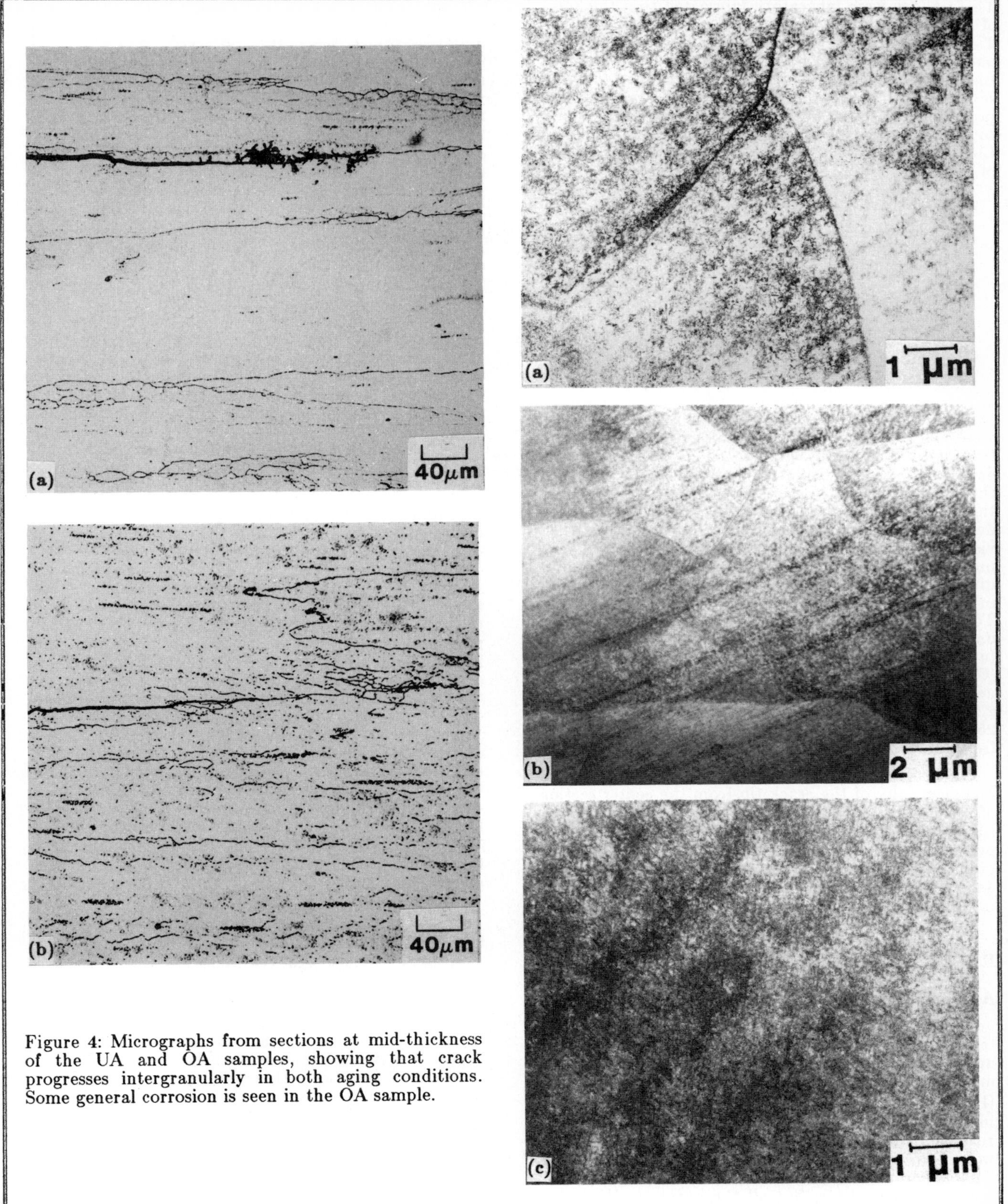

Figure 4: Micrographs from sections at mid-thickness of the UA and OA samples, showing that crack progresses intergranularly in both aging conditions. Some general corrosion is seen in the OA sample.

Figure 5a,b,c: Transmission micrographs from areas near the crack tip in (a) UA, (b) PA and (c) OA samples.

These alloys were laboratory fabricated to a thickness of ≃35 mm, solution heat treated, quenched, stretched 2% and aged to a T651 temper. One of the alloys, namely Al-2.9Cu-2.1Li alloy, was aged for different time intervals from severely underaged (2 hours) to severely overaged (150 hours). All 3 alloys were predominately unrecrystallized, however, a tendency toward recrystallization was observed in the higher Li-containing alloy(s). The S-T mechanical properties of the alloys are listed in Table 2.

Table 2: The S-T mechanical property data for varius alloys (T651 Temper).

Alloy Cu	Alloy Li	Aging Treatment	YS (MPa)	TS (MPa)	% Elongation
4.6	1.1	160/62 hrs	451	465	2
2.9	2.1	191/16 hrs	360	--	0
1.1	2.9	191/32 hrs	329	349	0
7075-T651		121/24 hrs	439	487	4

Duplicate DCB samples 2.5 cm x 2.5 cm x 12.7 cm were machined from the plates along the S-L orientation and mechanically precracked in tension. A few drops of 3.5% NaCl solution were added to the crack three times a day. Crack growth was monitored at the center of the sample with an ultrasonic detection device developed at Alcoa Laboratories. In addition, visual measurements of the crack extension were made on the surfaces for comparison with the interior ultrasonic measurements. The accuracy of the measurements were generally correct to within ≃ 1 mm. Total length of exposure time varied from 30 to 100 days. Stress intensity (K) values were calculated as a function of specimen crack opening displacement (COD) and crack length (a) using an equation developed by Hyatt [4]. DCB SCC tests on the Al-2.9Cu-2.1Li alloys were done on samples taken from 25 mm plate, while the (Li/Cu) experiments were from the 35 mm plates.

In order to characterize the crack profile optical metallography of the crack tip from the center section of selected alloys were obtained after mechanical polishing and etching. In addition, TEM studies were performed on an underage, peak age and overage samples of the Al-2.9Cu-2.1Li alloy to characterize the deformation/corrosion features. Details of experimental procedures are listed in reference [5].

RESULTS & DISCUSSION

Aging Effect in Al-2.9Cu-2.1Li

Figure 1 shows the variation in crack growth (in terms of crack length, a) with exposure time (t), for under (UA), peak (PA) and over (OA) aging treatments. The UA and PA had comparable hardnesses ($R_B \simeq 72$). The data clearly indicate that with increasing aging time, crack growth is markedly reduced; the UA alloy microstructure showing the maximum crack growth compared to the OA alloy. The apparent K_{ISCC} tends to decrease slightly with increased aging time, but all the curves tend to converge at long times of exposure, as shown in the K versus t plot (Figure 2). The results in Figure 2, also appear to suggest that the magnitude of the initial K can have some effect on the changes in the magnitude of K_{ISCC}. Such effects on crack growth curves due to initial K-values were also observed in 7XXX and 2XXX alloys by Sprowls et al. [6]. The initial slope of the curves in Figure 1 is indicative of the plateau velocity, which seems to be less affected by the aging treatments. This can be seen in Figure 3, on a SCC velocity (V) versus K plot, where the plateau velocity appears to remain within a scatter of 2×10^{-6} to 10^{-5} mm/sec, with no systematic trend with respect to increase in aging time. This observation in Al-2.9Cu-2.1Li alloy differs from that observed by Rinker et al. [2] in a lower Li-containing alloy 2020 (Al-4.5Cu-1.2Li) where the plateau velocity decreased by ≃ 3 orders of magnitude as the aging treatment was increased from severely UA to PA (T651) condition. This difference between the two alloys seems to indicate the dominant effect of Cu in the low Li-containing 2020 alloy. In the present case, the small changes in the plateau velocity and K_{ISCC} with various aging treatments appears to indicate that there may be one mechanism operating such as the dissolution of δ phase on the high angle grain boundaries. TEM microstructural observation on a similar composition alloy indicates that both low and high angle grain boundaries are decorated with δ phase, at later stages of aging [7].

Figure 4 shows photomicrographs from sections at mid-thickness of the UA and OA samples, indicating that the crack is progressing intergranularly in both cases, with some general corrosion occurring in the UA sample. The microstructural features near the crack-tip can be seen in the TEM micrographs taken from the UA, PA and OA specimens (Figure 5). The UA microstructure contains fine planar deformation bands. Increasing the aging time to PA results in course, more widely spaced slip bands. However, in the OA condition a more uniform distribution of dislocations was observed in the vicinity of the crack tip. The observed deformation structure could be in part due to the corrosion wedging effect during the SCC test, which is commonly observed in very UA alloys in the form of jumps in crack length at various points on the a versus t curves (Figure 1). Such dislocation features were absent in the undeformed samples (Figure 6). TEM near-crack-tip micrographs were taken after the samples were exposed to the environment for ≃ 80 days.

(Li/Cu) Ratio Effect (T651 Temper)

Figure 7 shows how the crack-growth varied with time as the (Li/Cu) ratio was increased. Even though the YS of these alloys were different (Table 2), the crack growth rate decreased markedly with increasing (Li/Cu) ratio. The plateau velocities estimated from Figure 7 show that the variation lies between 3×10^{-6} to 6×10^{-6} mm/sec. However, the apparent K_{ISCC} decreased significantly with increasing (Li/Cu) ratio, Figure 8. For comparison, K_{ISCC} values of 2024-T851 (PA) and 2024-T351 (UA) alloy data are included. This loss of K_{ISCC} with increasing (Li/Cu) ratio can be partly accounted for by the poor toughness (K_Q) of these alloys (Figure 8). Macroscopic observations of the crack-tip profiles on the low Li-containing alloy show cracks that are transgranular and branching, whereas in the high Li-containing alloy, crack-paths

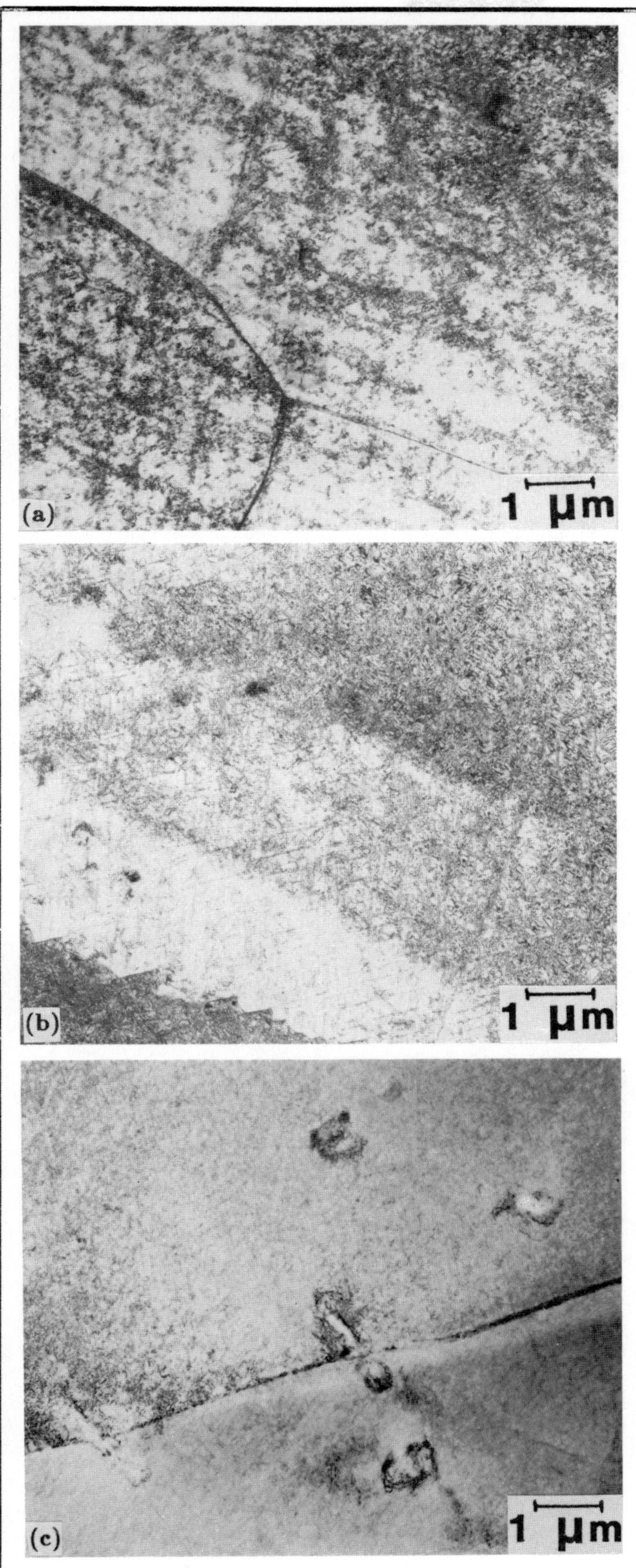

Figure 6a,b,c,: Transmission micrographs from undeformed regions, away from the crack in (a) UA, (b) PA and (c) OA samples.

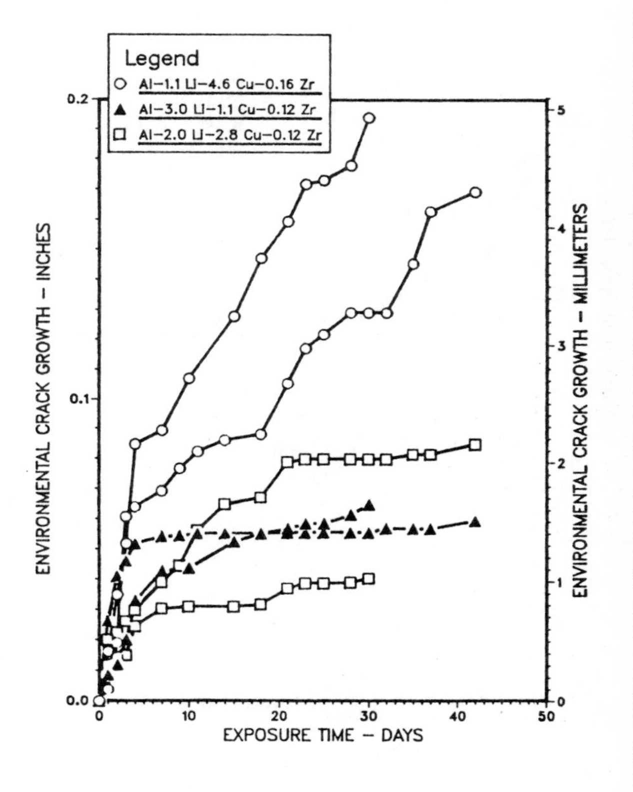

Figure 7: Variation of crack growth with time at increasing (Li/Cu) ratio.

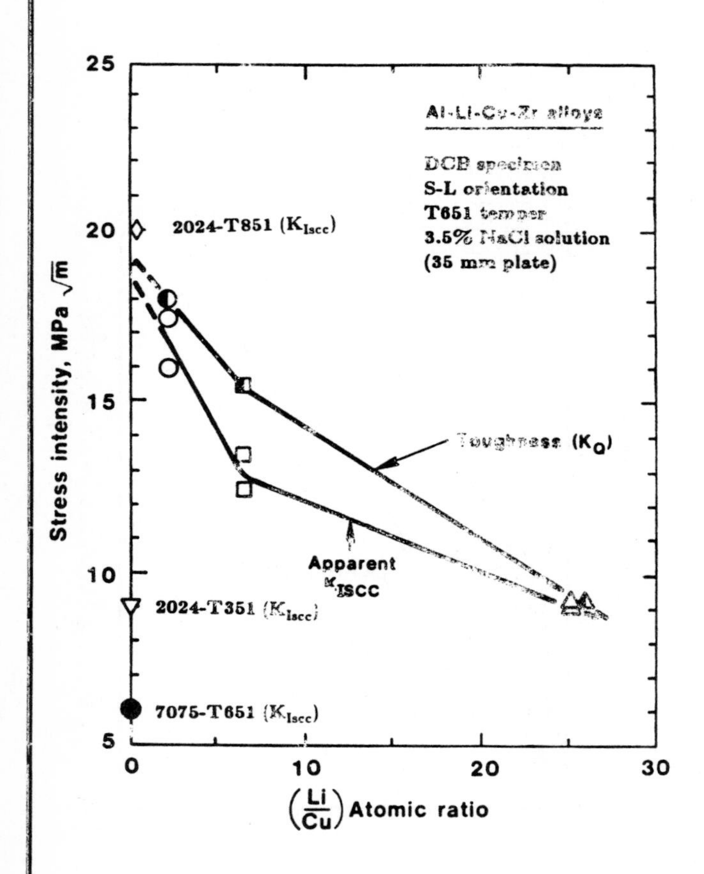

Figure 8: Variation of K_{1scc} with (Li/Cu) ratio.

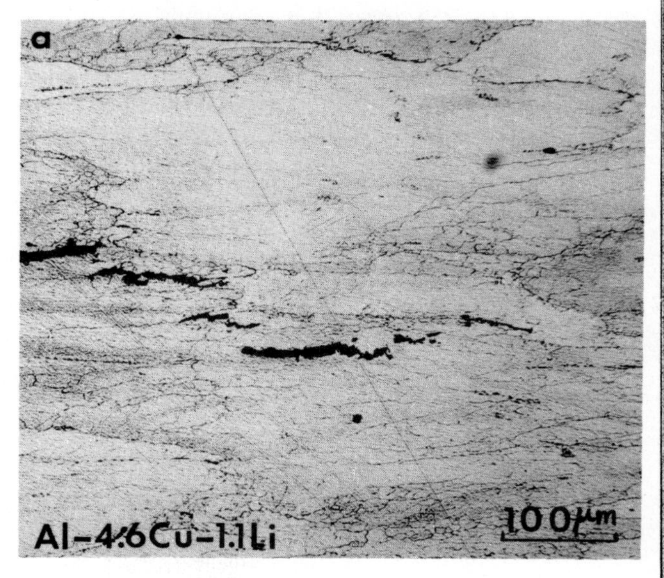

Figure 9a: Microstructure of low Li containing alloy showing transgranular, branching cracks.

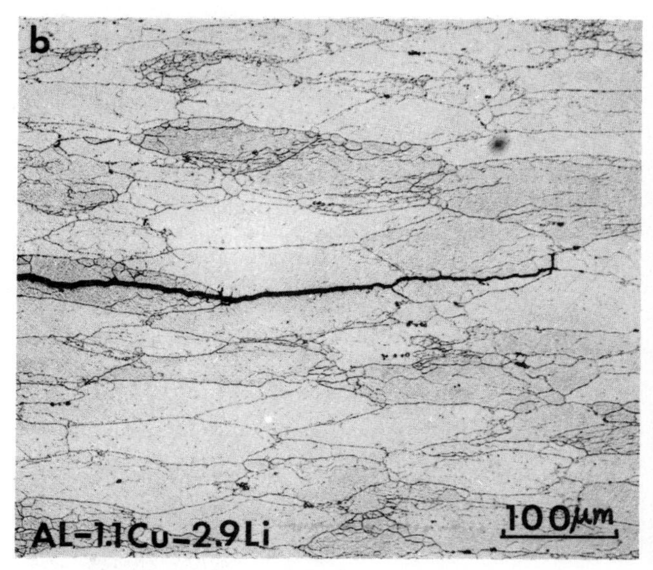

Figure 9b: Microstructure of high Li containing alloy showing crack-path along the grain boundaries.

are observed along the grain boundaries, Figure 9a and 9b, respectively. Since the total crack extension in these alloys is low (Figure 7), one cannot discount the effects of precracking on the crack profiles.

COMPARISON TO 7075 ALLOY

In general, Al-Cu-Li-Zr alloys exhibit improved resistance to SCC growth behavior in the T651 condition compared to 7075-T651, Figure 9. In addition, the plateau velocities of the Al-Cu-Li-Zr alloys are slightly lower than 7075-T651, even though the overall crack extension at long exposure times are significantly lower than 7075-T651 alloy. The crack growth rates of these alloys, particularly higher Li-containing alloys, tends to be comparable to 7075-T351, in spite of the poor S-L fracture toughness (K_Q) of these alloys. Similar comparisons can be made to 2024-T351 and 2024-T851.

SUMMARY

1. Aging Al-2.9Cu-2.1Li alloy had little effect on the plateau velocities. Apparent K_{ISCC} decreased slightly with aging time, the trend being similar to the loss in S-L toughness (K_Q).
2. Increasing (Li/Cu) ratio in the T651 temper showed a decrease in K_{ISCC}, which is in part due to loss in S-L toughness. In addition, small changes in Plateau velocities was observed as the (Li/Cu) ratio was increased.
3. It is observed that the magnitude of the variation in crack growth behavior is affected by the background fracture toughness of the alloys, which could mask the mechanistic understanding of the SCC behavior. As a result, one can only speculate that mechanism such as the dissolution of grain boundary precipitates (possibly δ) plays an important role in the higher Li-containing alloys leading to crack blunting.
4. In general, Al-Cu-Li alloys exhibit better SC growth resistance than 7075-T651, or 2024-T351.

ACKNOWLEDGEMENT

The authors wish to acknowledge thanks to Mr. P. R. Chvala Mr. A. J. Becker, and E. A. Ludwiczak for technical help during the course of this work. Helpful comments on the manuscript by Mr. S. C. Byrne is appreciated. This work was supported by the NAVAIR Contract N00019-80-C-0569 and part of the work was done at Purdue University.

REFERENCES

1. M. O. Speidel, *Met. Trans., Vol. 6a,* p. 631, 1975.
2. J. G. Rinker, M. Merek, and T. H. Sanders, Jr., *Aluminum-Lithium Alloys II,* eds. T. H. Sanders, Jr. and E. A. Starke, Jr., TMS-AIME, Warrendale, PA, pp. 597-622, 1983.
3. L. Christodolou, L. Struble, and J. R. Pickens, *Aluminum-Lithium Alloys II,* eds. T. H. Sanders, Jr. and E. A. Starke, Jr., TMS-AIME, Warrendale, PA, pp. 561-579, 1983.
4. M. V. Hyatt, *Corrosion-NACE, Vol. 26, #1,* p. 487, Nov. 1970.
5. A. K. Vasudevan, R. C. Malcolm, W. G. Fricke, and R. J. Rioja, "Resistance to Fracture, Fatigue and SCC of Al-Cu-Li-Zr Alloys," NAVAIR Contract N00019-80-C-0569, Final Report, June 1983.
6. D. O. Sprowls, M. B. Shumaker, J. D. Walsh and J. W. Coursen, "Evaluation of SCC Susceptibility Using Fracture Mechanics Techniques," Contract NASA 8-21487, Final Report, May 1973.
7. M. Tosten, A. K. Vasudevan and P. R. Howell, these proceedings.

□

Environment-sensitive fracture of Al–Li–Cu–Mg alloys

N J H HOLROYD, A GRAY, G M SCAMANS and R HERMANN

NJHH, AG and GMS are with
Alcan International Ltd,
Southam Road, Banbury,
Oxfordshire, UK.
RH is with The Open University,
Milton Keynes, UK

INTRODUCTION

To date little is known about the susceptibility of alloys of the aluminium-lithium system to environment-sensitive fracture. Stress corrosion cracking (SCC) data available in the literature (1-5) has employed standard test methods developed for conventional aluminium alloys and has involved little mechanistic interpretation. The aim of the reported work programme has been to compare and contrast the SCC behaviour of aluminium-lithium alloys with conventional alloys and to begin to develop an understanding of cracking mechanisms.

Studies have principally involved an Al-Li-Cu-Mg alloy 8090 and its high purity Al-Li and Al-Li-Zr alloy analogues.

EXPERIMENTAL METHOD

ALLOYS

Alloys were chosen for the programme to cover most of the basic aluminium alloy systems at appropriate alloying levels of lithium, copper and magnesium to represent the majority of the potential commercial systems. Purity base and zirconium levels were also set at levels typical of those likely for commercial alloys, (see Table 1).

Alloys of both the commercially established and experimental compositions were produced by an ingot metallurgy route (ingot weights up to 1200 Kg) and fabricated on commercial rolling mills to 25 or 50 mm gauge plate. Isothermal artificial ageing was carried out at 150 or 190°C depending upon the alloy system involved (see Table 2).

STRESS CORROSION TESTING

Standard specimen designs were used both for crack initiation C-ring (ASTM G38), tension-bar (ASTM G49) and crack propagation testing (double cantilever-beam (6)). All specimens were machined so that SCC was assessed in the most susceptible orientation, i.e. short transverse. Strict specimen machining practices were employed to minimise any induced surface stresses and heating effects. Specimens were then degreased prior to testing.

Total immersion tests were conducted at the free-corrosion potential in standardised environments e.g. 3.5% NaCl (ASTM G44), 2% NaCl /0.5% $Na_2Cr_2O_4$ pH=3 using HCl - (LN65666) and additionally in 3.5% NaCl with the pH adjusted between 3 and 9 using HCl and NaOH respectively.

Alternate immersion testing (A.I) was conducted according to specifications in ASTM G44 (10 minute wet/50 minute dry cycle, air temperature 27 ± 1°C and relative humidity R.H. 45 ± 6%).

Time to failure (TTF) of C-ring specimens was defined as the time of detection of the first observable crack using a low-power microscope at a magnification of X 20. Using this criterion a good agreement was obtained between the C-ring data and tension-bar data at high stresses where initiated cracks rapidly caused specimen failure. This involved frequent inspection during the initial period of testing and although many of these cracks subsequently blunted to broad intergranular fissures and became non-propagating cracks the specimens were diagnosed as having failed.

TABLE 1

CHEMICAL ANALYSIS OF EXPERIMENTAL ALUMINIUM ALLOYS IN WT%

Alloy	Li	Mg	Cu	Fe	Si	Ti	Zr	Mn
Al-Li	2.8	< 0.01	-	0.07	< 0.03	0.005	-	-
Al-Li-Zr	2.8	< 0.01	-	0.09	0.03	0.005	0.16	-
Al-Li-Cu-Mg (8090)	2.4	0.61	1.15	0.09	0.05	0.005	0.12	-
Al-Li-Cu-Mg	2.6	0.59	2.03	0.06	0.04	0.014	0.14	-
Al-Mg-Li	2.4	3.8	-	0.06	0.08	0.017	0.10	-
Al-Mg-Li-Mn	2.10	3.7	-	0.06	0.08	0.023	0.12	0.25
Al-Cu-Li	2.20	0.1	2.7			0.005	0.14	-
Al-Cu-Li-Mg	2.20	0.4	2.5			0.005	0.14	-

TABLE 2

ISOTHERMAL AGEING PRACTICE

ALLOY SYSTEM	TEMP °C	TIME IN HOURS		
		UNDER	PEAK	OVER
Al-Li	190	2	16	100
Al-Li-Zr	190	2	16	> 60
Al-Li-Cu-Mg	190	2	16	> 60
Al-Mg-Li	190	-	16	-
Al-Mg-Li-Mn	190	-	16	-
Al-Cu-Li	150	-	36	-
Al-Cu-Li-Mg	150	-	36	-

TABLE 3

TENSION-BAR TEST DATA 8090-T651 AT 240 MNm^{-2}

TEST ENVIRONMENT	SCC LIFE (DAYS)
3.5% NaCl A.I.	2.5
3.5% NaCl T.I.	> 40
2% NaCl/0.5% Na_2CrO_4 (pH=3)	15
3.5% NaCl, pH=3	14, > 30, > 30
3.5% NaCl, pH=9	2.5, > 30, > 30
Lab air	> 150

Double cantilever-beam both testing employed a fully automated multi-specimen facility where crack length was monitored by a potential drop technique (6-8) and single specimen testing using an optical technique to monitor crack growth. Prior to testing specimens of the same design as that used by Smith and Piper (7) were given a 2 mm air fatigue pre-crack (70 Hz, R=0.6 and Kmax = $8MNm^{-3/2}$) and then loaded to a stress intensity just below the predetermined K_{ic} value. Specimens were then exposed to the test environment which was either 95% R.H. at 40°C or in a few cases total immersion in 3.5 wt % NaCl.

PRE-EXPOSURE EMBRITTLEMENT TESTING

Prior to testing in laboratory air (27 ± 1°C, R.H. 45 ± 6% ASTM G44) or vacuum (10^{-1} Torr) 8090 tension-bars both stressed (240 MNm^{-2}) and un-stressed were totally immersed in 3.5% NaCl for various times up to 50 days. After pre-exposure specimens were cleaned using practices ranging from a rinse and ultrasonic wash in de-mineralised water followed by warm air drying to mechanically polishing to remove surface attack and films developed during pre-exposure.

SURFACE ATTACK STUDIES

Mechanically polished sections of peak and over aged 8090 (short transverse plane) were exposed to 3.5% NaCl for 30 minutes under conditions of total or alternate immersion and the resultant attack morphologies were examined using both optical and scanning electron metallography.

SOLUTION CHEMISTRY

Determination of the solution chemistry within restricted geometries involved several specimen designs, namely: 1) carefully machined alloy shavings, 2) flat-bottomed cylindrical holes (dia 1 mm and depths 5,10 and 20 mm) drilled dry into plate samples to provide artificial crevices and 3) modified DCB specimens with glass slides held over the notch/crack-tip region which allowed periodic sampling of the crack-tip environment (9,10).

In a typical shavings experiment 18 g of shavings were compressed into a beaker and then immediately covered by 100 cc of 3.5% NaCl. Shavings were either ultrasonically degreased or degreased and caustic etched in 0.1 M NaOH for 30 seconds at 25°C prior to exposure.

Crevice and crack-tip experiments were conducted both with and without the presence of a bulk electrolyte reservoir. For the crevice experiments a bulk volume of 250 cc was used and when no bulk was involved test volumes were of 9,18 and 35 µl for the 5,10 and 20 mm holes respectively.

In each case pH and dissolved lithium and aluminium concentrations were monitored as a function of time. Solution pH measurements were obtained via calibrated glass pH-electrodes for the shaving experiments and micro-pH papers (±0.1 pH unit) for the crevice and crack-tip experiments. Dissolved lithium and aluminium concentrations were assessed using atomic absorption spectroscopy (ASS) and X-ray fluorescence (XRF) respectively on 5 µl samples extracted from the bulk electrolyte, the crevice or the crack-tip region (9).

RESULTS

CRACK INITIATION

Crack initiation data taken from both C-ring and tension-bar tests is given in Figure 1 for the range of aluminium alloys containing around 2.5% lithium (see Table 1) when subjected to alternate immersion stress corrosion testing. Immediately apparent from this data is the importance of the alloy copper concentration, as is more clearly shown in Figure 2. Crack initiation in peak-aged alloys is rapid for all alloys containing above about 1% copper and occurs within a few days for stress levels above about 200 MPa. However, cracking susceptibility may be significantly reduced by overageing when copper levels are below about 2.0 wt %. Crack initiation in Al-Li-Zr, Al-Mg-Li and Al-Mg-Li-Mn alloys was not apparent after 50 days testing which indicates that lithium additions either singly or in combination with magnesium do not inherently promote susceptibility to stress corrosion crack initiation. Al-Mg-Li-Cu alloys exhibiting a high resistance to SCC failure have been developed within Alcan International (11).

The Al-Li-Cu-Mg (8090) and 2014 initiation data shown in Figure 1 are miniumum performance curves based upon extensive test data for 8090 (12) and published data from a variety of sources for 2014 (13). These results suggest that peak-aged 8090 should have a greater inherent resistance to stress corrosion crack initiation than 2014-T651 at the same mechanical property level.

Tests on peak-aged 8090 under constant rather than alternate immersion in 3.5% NaCl, resulted in no initiation in tension-bars after 40 days testing which contrasts dramatically with the alternate immersion results (see Table 3)

This difference in behaviour is more marked with Al-Li-Cu-Mg alloys than with conventional aluminium alloys although as expected alternate immersion is the more severe test (14).

The influence of bulk solution pH upon SCC initiation for 8090-T651 in 3.5% NaCl has not been fully established. Occasionally tension-bar test failures occurred in solutions with pH's adjusted to 3 or 9, (see Table 3), but the majority of specimens survived test periods exceeding 30 days. Buffered pH 3 solutions (2%

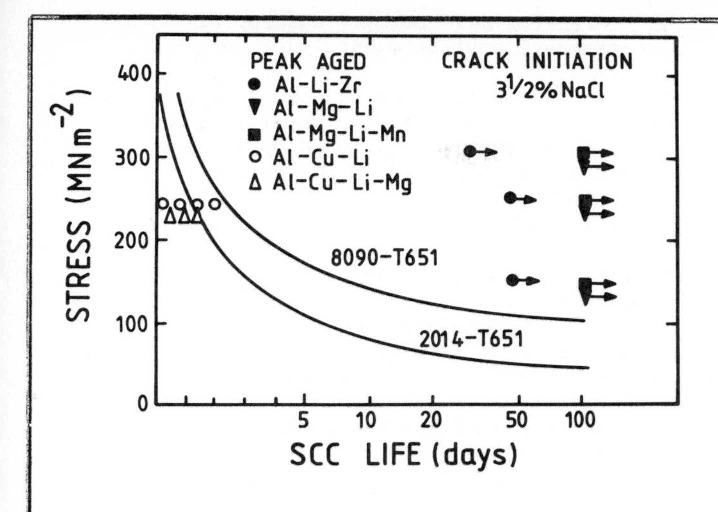

Figure 1. Crack initiation data from C-ring and tension-bar testing for various lithium containing aluminium alloys tested in 3½% NaCl under alternate immersion conditions.

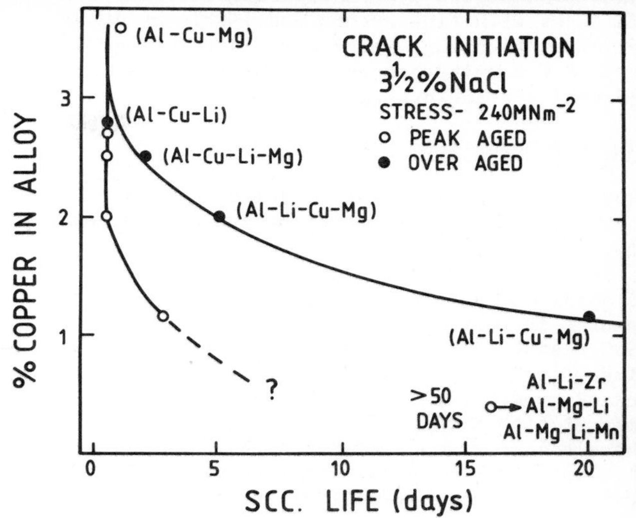

Figure 2. Influence of alloy copper content upon the SCC initiation data given in Figure 1.

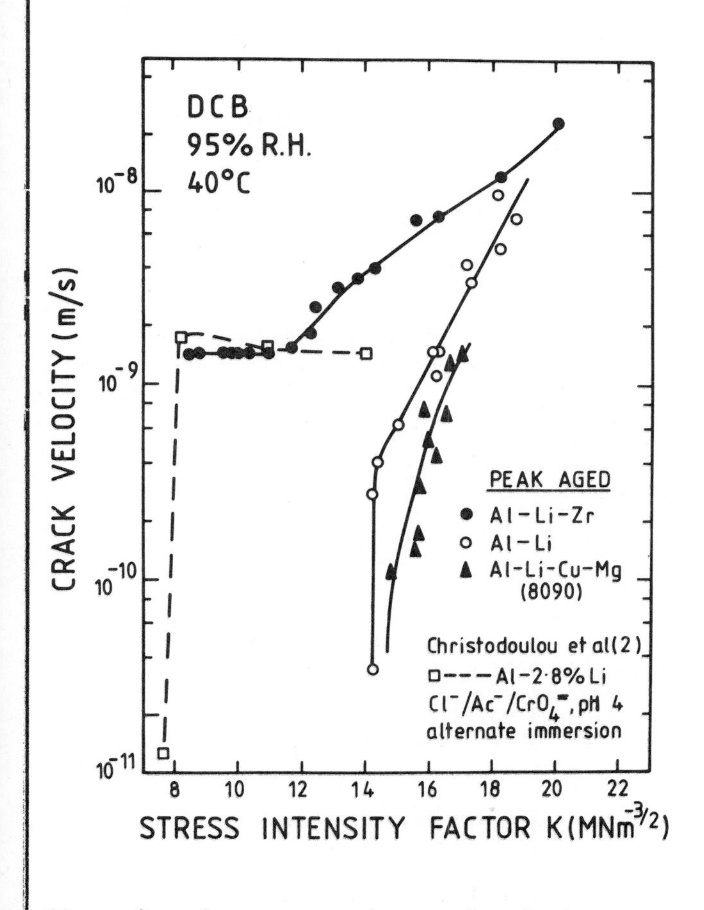

Figure 3. Stress corrosion crack velocity as a function of stress intensity factor for peak aged Al-Li, Al-Li-Zr and 8090 alloys tested in 95% R.H. at 40°C.

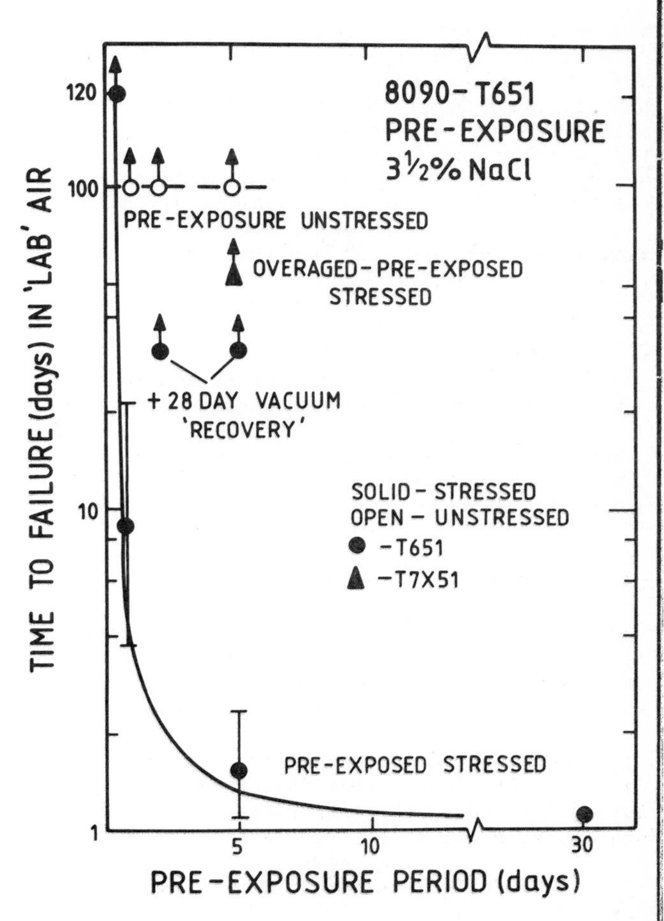

Figure 4. Time to failure for 8090 pre-exposed to 3½% NaCl for various times prior to testing in laboratory air or vacuum.

$NaCl/0.5\% Na_2CrO_4$) promoted reproducible failures with fine cracks and little evidence of crack blunting. Non-buffered solutions with pH's 3 and 9 induced both blunt fissures and general surface attack.

CRACK PROPAGATION

8090-T651 DCB specimens loaded to stress intensity factors approaching short transverse K_{Ic} values (15 $MNm^{-3/2}$) showed no evidence of crack propagation in 3.5% NaCl under total immersion conditions for test times up to 3 months in agreement with the crack initiation SCC tests. Although not tested here alternate immersion and drip-feed conditions promote crack growth (2,15) at rates around 2 x 10^{-9} m/s at high K values. Similar growth rates have been reproduced here for 8090-T651 in 95% R.H. at 40°C, however, observed crack velocities decrease rapidly (to 10^{-11} m/s) as K values fall below 14 $MNm^{-3/2}$ as shown in Figure 3.

Intergranular stress corrosion cracks were also developed in the peak aged Al-Li-Zr ternary alloy and the peak aged Al-Li binary alloy in the 95% R.H. environment at 40°C at rates higher than those observed for the quaternary 8090 alloy. Neither the binary or ternary alloy initiated cracks in the C-ring or tension bar alternate immersion test. Stress intensity values for the large grained recrystallised binary alloy are not meaningful due to the occurrence of significant crack branching. Cracking of a similar Al 2.8% Li binary alloy has also been observed under alternate immersion conditions (2) and these results are reproduced in Figure 3.

PRE-EXPOSURE EMBRITTLEMENT

The occurrence of reversible pre-exposure embrittlement can be used as a means of determining if hydrogen embrittlement can cause cracking in a particular alloy system as has been shown for 7000, 5000 and 2000 series alloys (16-21).

Pre-exposure tests on the 8090 alloy showed that:-

1) Stressed pre-exposure to 3.5% NaCl followed by stressed exposure to laboratory air can lead to rapid failure that would not have occurred in the pre-exposure environment (Figure 4).

2) Pre-exposure un-stressed for periods up to 5 days did not promote test failures on subsequent exposure to laboratory air. Longer pre-exposure times are being examined.

3) Exposure to vacuum for 28 days after stressed pre-exposure to 3.5% NaCl eliminated any subsequent cracking on exposure to laboratory air, demonstrating that pre-exposure embrittlement of 8090 is reversible.

4) Overageing 8090 removed susceptibility to pre-exposure embrittlement.

The fracture morphology induced by stressed pre-exposure embrittlement was intergranular and similar to that developed by stress corrosion crack growth in DCB specimens.

SURFACE ATTACK MORPHOLOGY

Surface attack morphologies on 8090 alloy samples after 30 minutes immersion time in 3.5% NaCl under total immersion or alternate immersion conditions (3 one hour cycles - 10 min wet/50 min dry) were examined. Overaged 8090 suffered general attack whereas the attack on peak aged material was more intergranular in nature and more pronounced for the alternate immersion conditions in agreement with both the increased susceptibility of the peak aged state and the increased severity of alternate immersion testing. Figure 5 shows the attack morphology developed under alternate immersion conditions for the peak aged and overaged microstructures. Under conditions of constant immersion more general attack occurred in both ageing conditions.

SOLUTION CHEMISTRY

Shavings experiments (Figure 6(a)) show that the solution pH developed for 8090 is independent of temper.

Prevention of air contact (Figure 6(b)) using a layer of liquid paraffin (10), had little effect during the early stages of solution chemistry development but did have an effect after 2 to 3 hours. The sudden increase in pH from 9 to 11, which was accompanied by copious evolution of hydrogen gas, occurred earlier and did not decay with time as was observed if air contact was allowed. Caustic etching after degreasing had a minimal effect. The presence of increased solution above the shavings delayed the pH rise to longer exposure times and eventually prevented the pH increase altogether. Similar results were obtained for peak aged Al-Li and Al-Li-Zr alloys.

Dissolved lithium concentration as a function of time and measured pH is shown in Figure 7. When air contact was prevented the lithium concentration behaviour was similar save that the minimum moved to shorter times and after the rapid pH increase lithium levels were maintained.

These results are very different to those found for un-alloyed aluminium, Al-Zn-Mg (-Cu) and Al-Cu-Mg alloys which all generate acidic conditions, pH's 4, 3.5 and 2 respectively when given a caustic wash pre-treatment prior to testing.

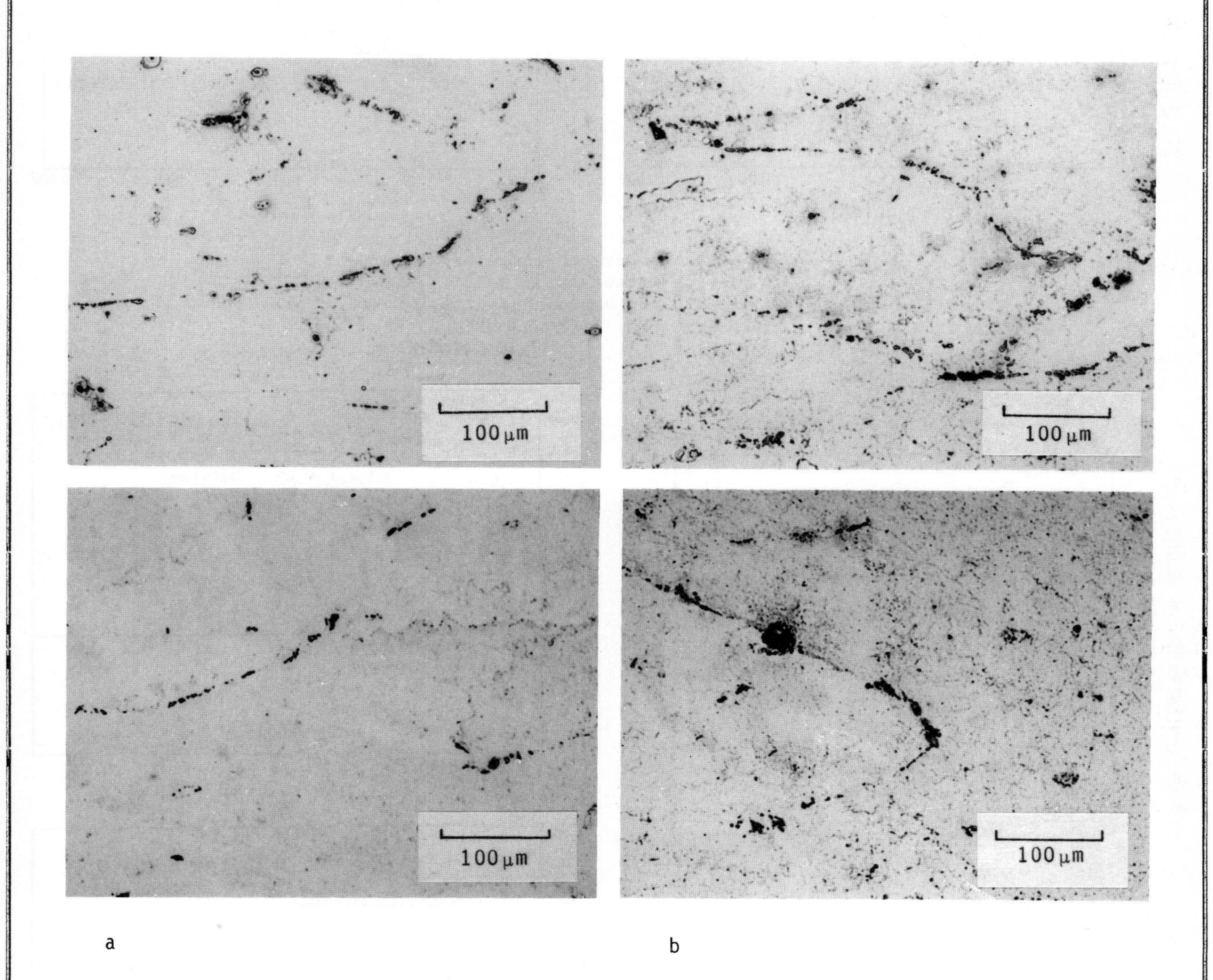

Figure 5. Surface attack on peak and overaged 8090 polished sections after immersion in 3½% NaCl for 30 minutes using a) alternate immersion and b) total immersion conditions.

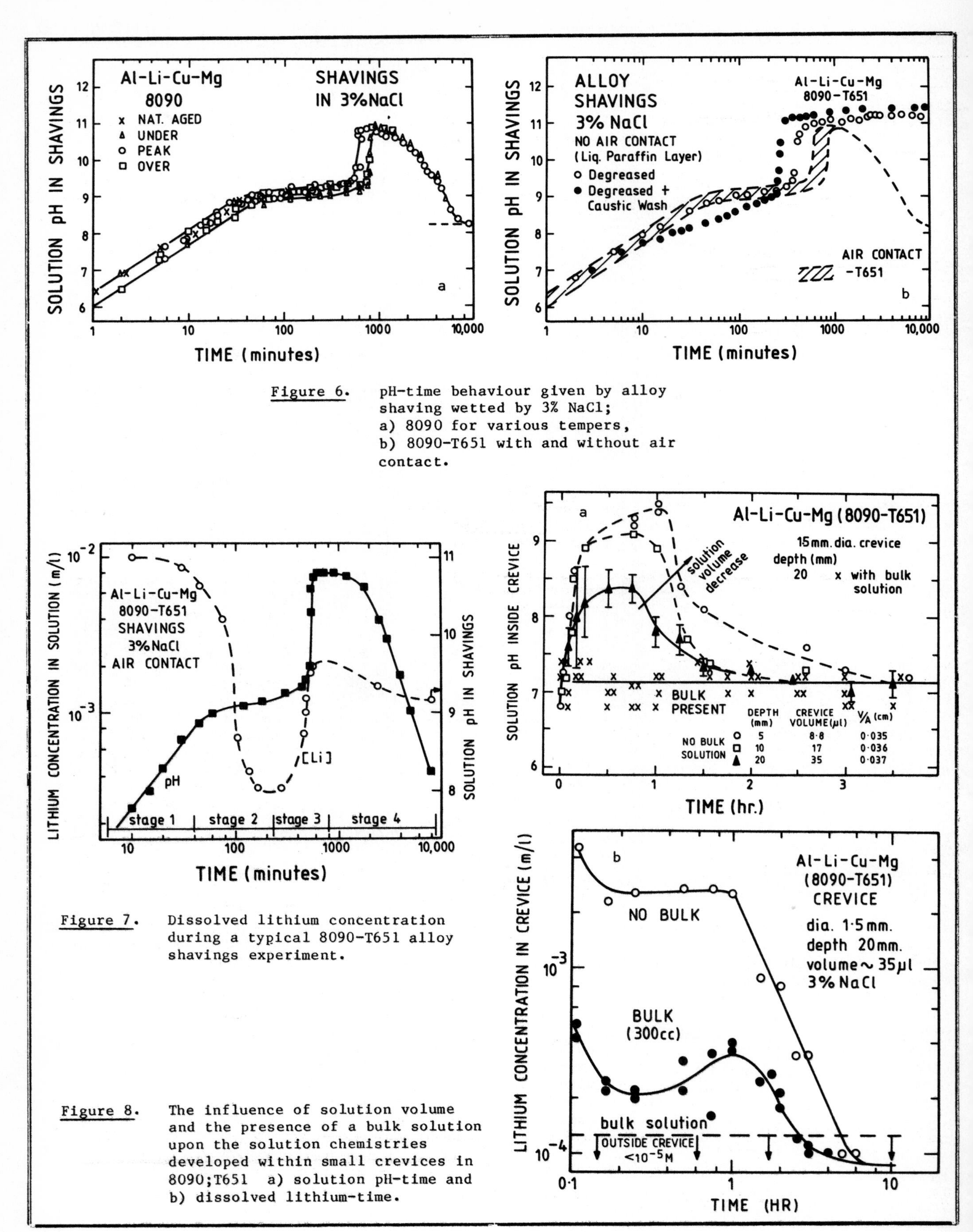

Figure 6. pH-time behaviour given by alloy shaving wetted by 3% NaCl; a) 8090 for various tempers, b) 8090-T651 with and without air contact.

Figure 7. Dissolved lithium concentration during a typical 8090-T651 alloy shavings experiment.

Figure 8. The influence of solution volume and the presence of a bulk solution upon the solution chemistries developed within small crevices in 8090;T651 a) solution pH-time and b) dissolved lithium-time.

CREVICES AND STRESS CORROSION CRACK-TIPS

pH-time behaviour in a given crevice geometry for 8090-T651 (solution volume to surface area ratio (V/A) of 0.036 cm) is shown in Figure 8(a). This V/A ratio for Al-Zn-Mg (-Cu) alloys is known to generate similar solution conditions as found in propagating stress corrosion cracks (10). As with the 8090 shavings experiments (V/A = 0.1 cm) solution pH within crevices initially became alkaline. A volume effect occurs which leads to more alkaline conditions being developed the smaller the test volume. Again as with the shavings experiment the presence of a bulk solution can prevent alkaline conditons developing. Measured lithium concentrations were found to be an order of magnitude higher in the absence of bulk electrolyte (Figure 8(b)).

SOLUTION pH'S OF AQUEOUS LITHIUM SALT SOLUTIONS

The measured solution pH of lithium chloride, carbonate and hydroxide for concentrations in the range 10^{-5} to 10^{-1} M are independent of whether the solvent employed is 3% NaCl or distilled water (Figure 9). Solution pH of lithium chloride solutions is independent of concentration, presumably because the hydrolysis involves a strong acid and a strong base, HCl and LiOH respectively. Hydrolysis of lithium carbonate and hydroxide solutions involves a strong base and a weak acid and hence solution pH's increase with increasing concentration, (Figure 9). If these solutions are exposed to the atmosphere their pH's decrease with time such that for concentrations up to 10^{-2} M they decay to stable values within 40 hours which, for a given lithium concentration are independent of whether the initial solution was the hydroxide or the carbonate. Prevention of air contact stabilised solution pH at its initial value which indicates that the observed decay must be associated with either CO_2 or O_2 pick-up.

DISCUSSION

The important results from the crack initiation studies were that susceptibility is highly dependent on copper content for alloys containing 2 to 2.5 wt % lithium with magnesium either very low or in the 0.4 to 0.6 wt % range. In addition this susceptibility can be moderated by overaging especially for copper contents below about 2.5 wt %. Crack initiation did not occur for the alloys tested without copper either under conditions of alternate or constant immersion in saline electrolytes. This suggests that the presence of one of the copper containing precipitate phases e.g. T_1 (Al_2CuLi), Θ (Al_2Cu) or S (Al_2CuMg) may be responsible for promoting crack initiation rather than the δ (AlLi) or the (Al_2MgLi) of the Al-Li or Al-Mg-Li systems. Increased magnesium in Al-Li-Cu-Mg alloys improves resistance to crack initiation and also favour S phase precipitation suggesting that Θ and/or T_1 are the phases present when crack initiation is readily promoted. The electrochemical characteristics of each of these phases in Al-Li based alloys have not, at present, been well established and hence speculation as to how initiation is promoted is premature.

Overaged microstructures tend to have δ phase precipitated both on and adjacent to sub-boundaries, Figure 10. This leads to a more uniform general corrosion in preference to specific grain boundary attack as confirmed by the surface attack morphology studies.

Initiation of cracking was significantly easier under conditions of alternate immersion than with constant immersion. In fact cracks were only initiated under constant immersion when surface film stability was sufficiently reduced by a pH shift and inhibitors prevented crack blunting, (e.g. LN65666 with chromate).

The propagation results supported the initiation tests in that cracks did not develop under conditions of total immersion but rather under conditions where bulk electrolyte effects were reduced e.g. in this case in water vapour (95% R.H.) at 40°C, or as reported in the literature, under alternate immersion conditions. A significant difference between the initiation and propagation dominated tests under conditions of limited electrolyte was that cracks in Al-Li and Al-Li-Zr alloys initiated and propagated from notches during DCB testing but not from smooth specimen surfaces during C-ring or tension bar testing. It may be that the electrolyte chemistry required for cracking of these alloy systems requires both restricted geometries and limited electrolyte volumes. Alternate immersion testing may hence promote cracking during the "dry" cycle and inhibit and/or blunt out cracking during the "wet" part of the cycle. A further testing programme of varied immersion cycling and water vapour testing is clearly indicated for both smooth and pre-cracked specimens.

The solution chemistry results provide some insight as to why restricted geometries and limited electrolyte volumes promote cracking in that clearly under such conditions lithium ion concentrations are built up and alkaline pH is developed. This process is prevented in the presence of a bulk electrolyte and is assisted by de-aeration of the electrolyte. This second point may also help to explain why cracking from restricted geometries is favoured and can occur in alloy systems which will not crack from smooth surfaces. Particular anodic phases may hence have the ability to localise pitting sufficiently to promote the required lithium solution chemistry for cracking. This suggests that further detailed dissolution morphology studies of the various alloy families are required.

13
12
11
10
9
8
7
6
SOLUTION pH
SOLID - 3%NaCl
OPEN - DISTILLED WATER
LiOH
Li_2CO_3
90 hr. 'LAB AIR' EXPOSURE
LiCl
distilled
3%NaCl
10^{-5} 10^{-4} 10^{-3} 10^{-2} 10^{-1} 1
LITHIUM CONCENTRATION (m/l)

Figure 9. Solution pH of lithium chloride, carbonate and hydroxide as a function of lithium concentration.

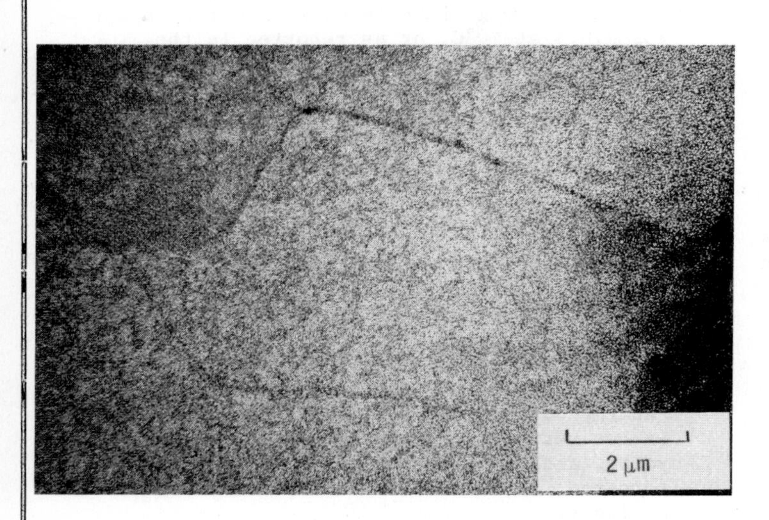

a

Figure 10

Dark field TEM micrographs for 8090 in two tempers, (a) peak aged (T651) and (b) overaged (T7X51).

b

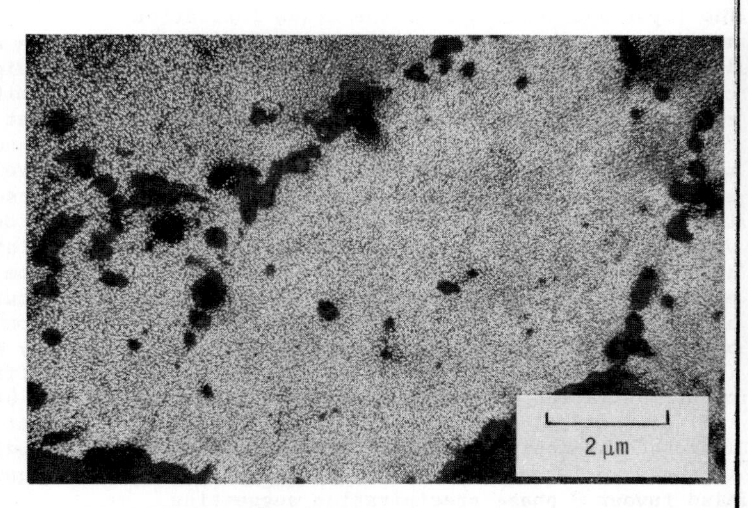

Once lithium ions in local regions have built up sufficiently, passive film stability is reduced and dissolution processes are enhanced which leads both to further pH increase and copious hydrogen evolution in cathodic sites. The solution pH may be maintained by the equilibrium reaction:-

$$Li^+ + H_2O \rightarrow LiOH + H^+$$

This gives a pH of 11.05 for a dissolved lithium concentration of 2×10^{-3} M.

Air contact can reduce pH both for bulk electrolytes and restricted electrolyte volumes by absorption of carbon dioxide and promotion of lithium carbonate and then bicarbonate formation.

The results support the hypothesis that lithium dissolution controls the critical solution chemistry of the crack tip, crevice or pit enclave and over-rides effects of other dissolved species e.g. Al^{3+}, Mg^{2+}, Cu^{2+} etc. In particular dissolved aluminium levels were below 10^{-4} M (measured) and could not be much higher as the pH would shift from 11.05 to 3.5 at 10^{-2} M dissolved aluminium.

Cracking of the aluminium-lithium alloy family is hence promoted by a different electrolyte condition than that of conventional aluminium alloys which are acidic (9). However both dissolution dominated and/or hydrogen dominated cracking mechanisms can still occur. A case for hydrogen embrittlement could clearly be developed from the stress pre-exposure experiments where reversible embrittlement was observed. In addition hydrogen can permeate aluminium-lithium alloy (again 8090-T651) membranes albeit at rates an order of magnitude lower than those reported for 7000 series alloys (6).

To summarise, aluminium-lithium based alloys are certainly not immune to stress corrosion cracking but in terms of alternate immersion crack initiation testing they are less susceptible than extensively used aerospace alloys (e.g. 2014) which do not cause significant cracking problems in service. Survival criteria can readily be achieved particularly if copper concentrations are kept below 2.0% and overageing practices are employed. It should be added that strength penalties are insignificant due to the flat ageing plateau. Future testing and ranking of alloys and tempers should concentrate on the "worst case" condition i.e. thin film electrolytes, aggressive anions, and restricted specimen geometries to determine precise stress intensity crack velocity relationships.

CONCLUSIONS

Crack Initiation: Controlled by copper level and moderated by heat treatment (overaging). Copper-free alloys did not inititate cracks. Alternate immersion testing was considerably more severe than constant immersion testing in 3.5 wt % NaCl where cracks did not develop unless film stability was disrupted by pH shift.

Crack Propagation: Again the presence of a bulk electrolyte prevented cracking although the combination of both restricted geometry and thin film electrolyte conditions promoted cracking even in copper-free alloys.

Pre-exposure Embrittlement: Reversible embrittlement was demonstrated after stressed pre-exposure in a non-cracking total immersion electrolyte and subsequent exposure to thin film electrolyte conditions. This suggests that hydrogen embrittlement may be a possible SCC fracture mechanism.

Surface Attack Morphology: Peak aged microstructures and alternate immersion promote intergranular attack.

Solution Chemistry within restricted Geometries: pH is alkaline and controlled by lithium ions in solution. This is favoured by de-aeration, restricted electrolyte volumes, the absence of bulk electrolytes and is independent of alloy composition and/or temper for the alloys examined.

REFERENCES

1. P.P. Pizzo, R.P. Galvin and H.G. Nelson, "Aluminium-Lithium II", Eds. E.A. Starke, Jr and T.H. Sanders, AIME, Warrendale, 1984, p. 627.

2. L. Christodoulou, L. Struble and J.R. Pickens, As Reference 1, p. 561.

3. J.G. Rinker, M. Marek and T.H. Sanders, Materials Science and Engineering, 64, 203, 1984.

4. C.J. Peel, B. Evans, R. Grimes and W.S. Miller. Ninth European Rotocraft Forum, September 1984, Stressa, Italy.

5. W.S. Miller, A.J. Cornish, A.P. Titchener D.A. Bennett, As Reference 1, p. 335.

6. N.J.H. Holroyd and D. Hardie, Corrosion Science, 23, 527, 1983.

7. H.R. Smith and D.E. Piper, Boeing Report, D6-24872, 1970.

8. N.J.H. Holroyd, G.M. Scamans and J. Hunter, to be submitted to Corrosion (1985).

9. N.J.H. Holroyd, G.M. Scamans and R. Hermann. "Embrittlement by the Localised Crack Environment", Ed. R.P. Gangloff, AIME, Warrendale, 1984, p. 327.

10. N.J.H. Holroyd, G.M. Scamans and R. Hermann, "Chemistry within Pits, Cracks and Crevices", Ed. A. Turnbull, National Physical Laboratory, Teddington, HMSO - in press, 1985.

11. U.K. Patent Application: G.B. 216 + 7915A.

12. M. Reynolds, A. Gray and E. Creed, paper No 6, this conference.

13. B. Evans, D.S. McDarmid and C.J. Peel, Fifth International Sample Conference, Materials and Process, Montreux, Switzerland, June 1984.

14. ASTM STP 610. Sprowls et al on Proposed Standard Method.

15. C.J. Peel, Private Communication, 1984.

16. W. Gruhl, Z. Metallkd., 54, 86, 1963.

17. L. Montgrain and P.R. Swan, "Hydrogen Effects in Metals", ASM, 1974, p. 575.

18. G.M. Scamans, R. Alani and P.R. Swan, Corrosion Science, 16, 443, 1976.

19. D. Hardie, N.J.H. Holroyd and R.N. Parkins, Metal Science, 13, 603, 1979.

20. M. Meuller, I.M. Berstein and A.W. Thompson Scripta Met., 17, 1039, 1983.

21. R.E. Ricker, Ph.D. Thesis, Renneslaer Poly, U.S.A., 1983.

□

CORRECTIONS

To accelerate publication of this volume some of the traditional stages of sub-editing and proof-reading (which considerably protract the completion of a book) were renounced. Hence this 12-page "stop-press" section.

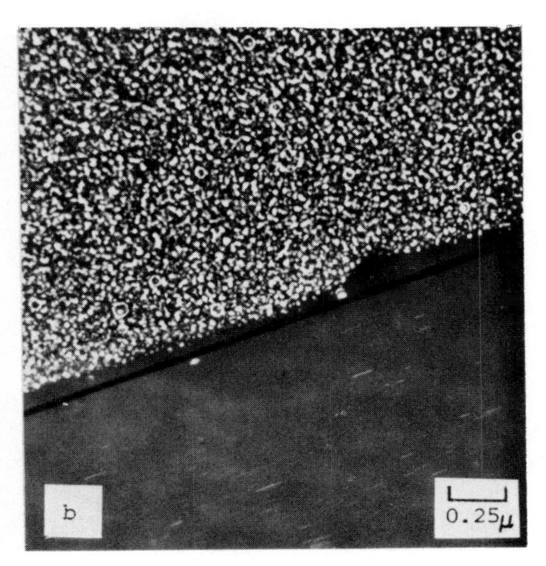

page 76, fig. 9b

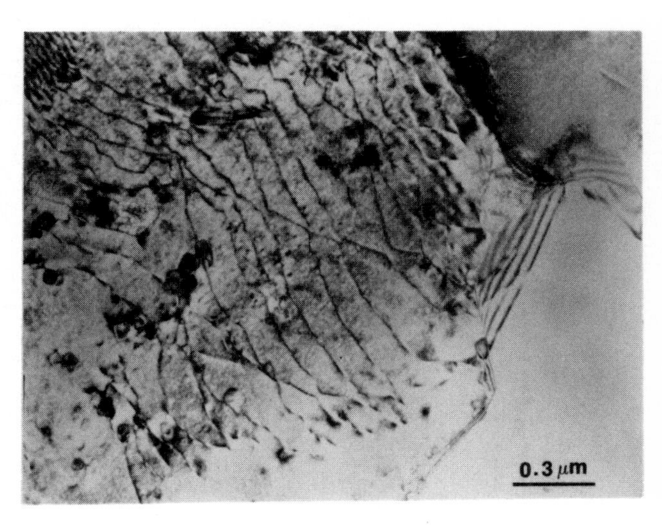

page 81, fig. 1

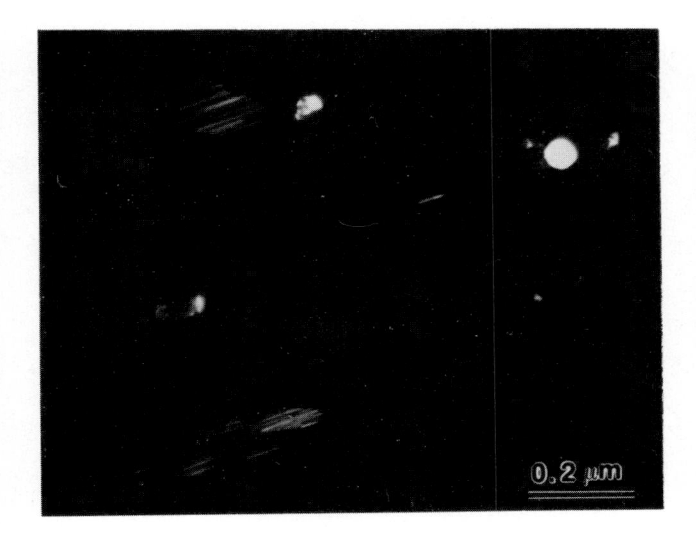

page 81, fig. 4

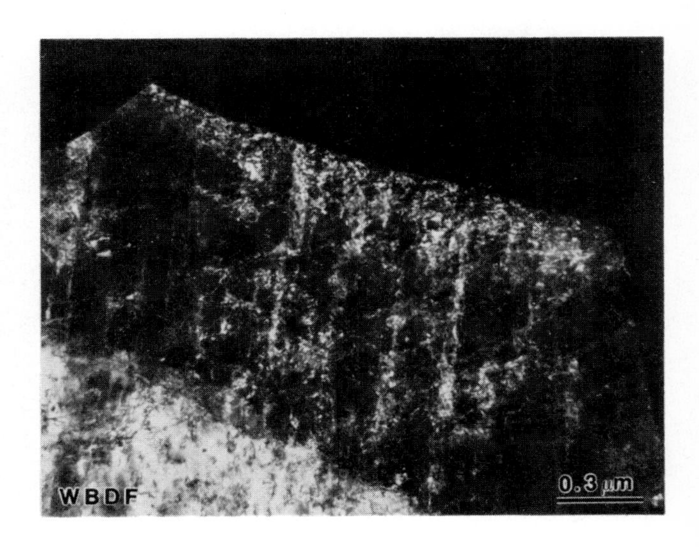

page 81, fig. 5

LIMITED COINING

EFFECTIVE COINING

page 181, fig. 8

page 192, fig. 1

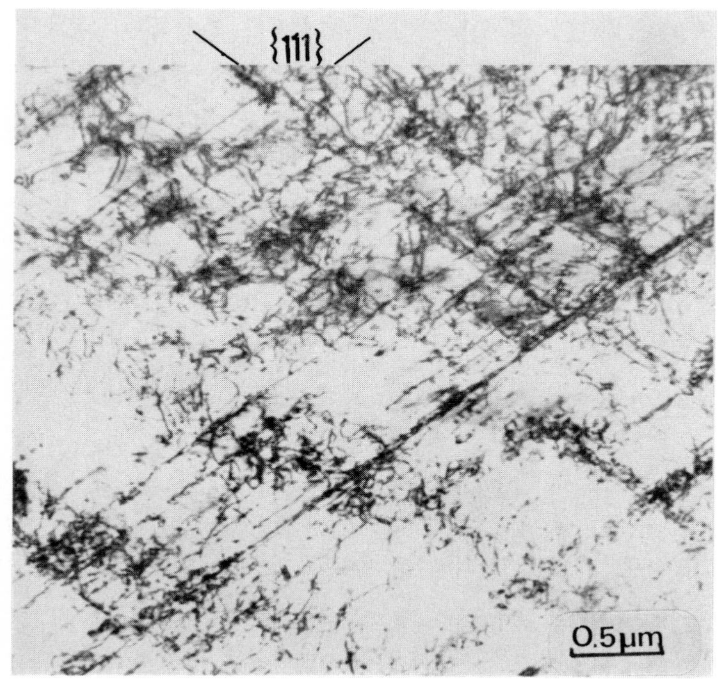

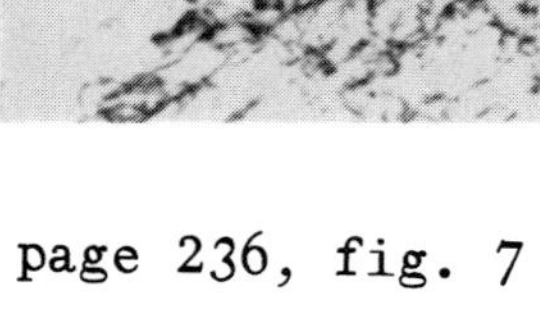

page 236, fig. 7

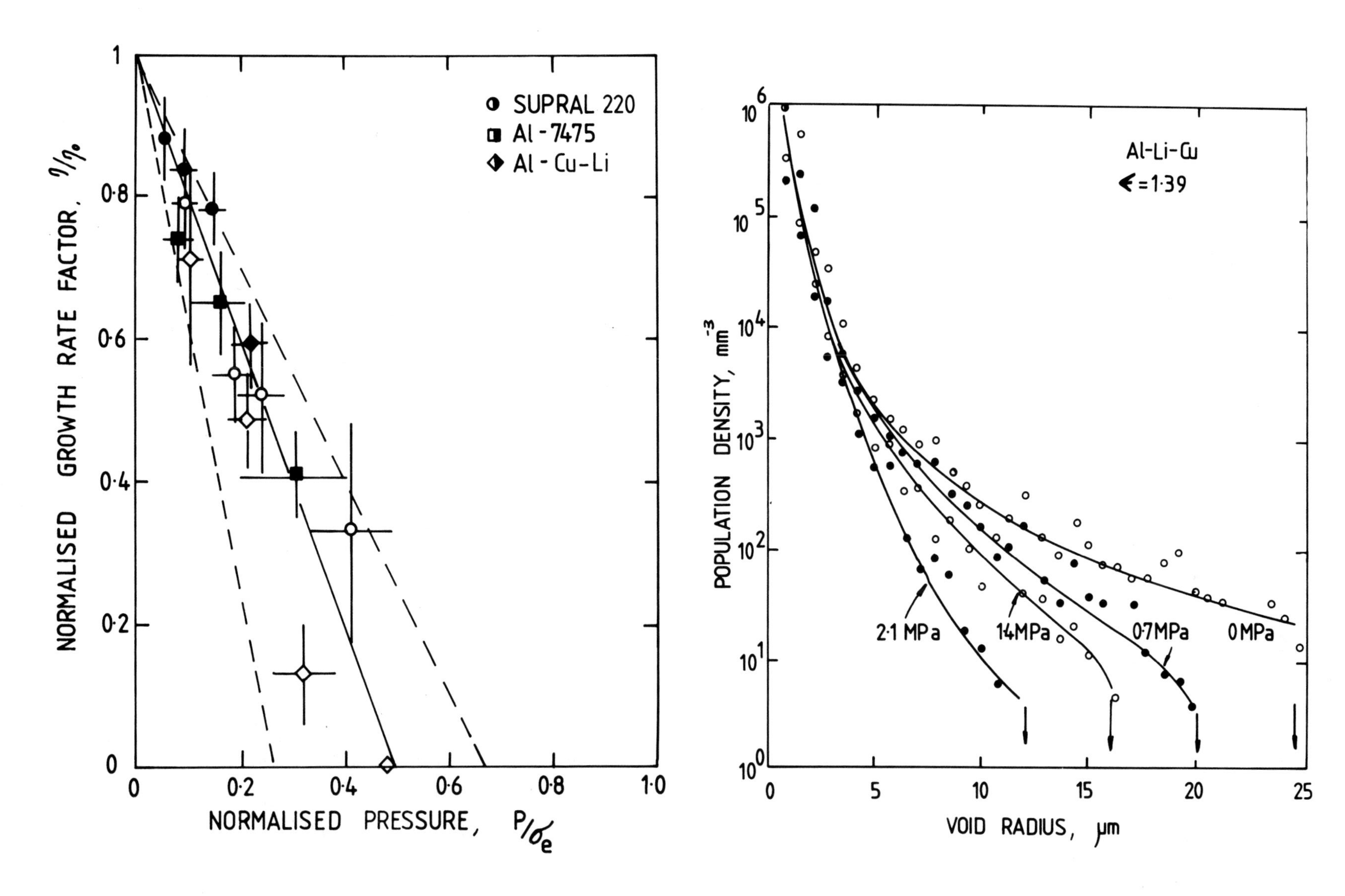

page 187, figs. 4 and 5

page 254, fig. 10

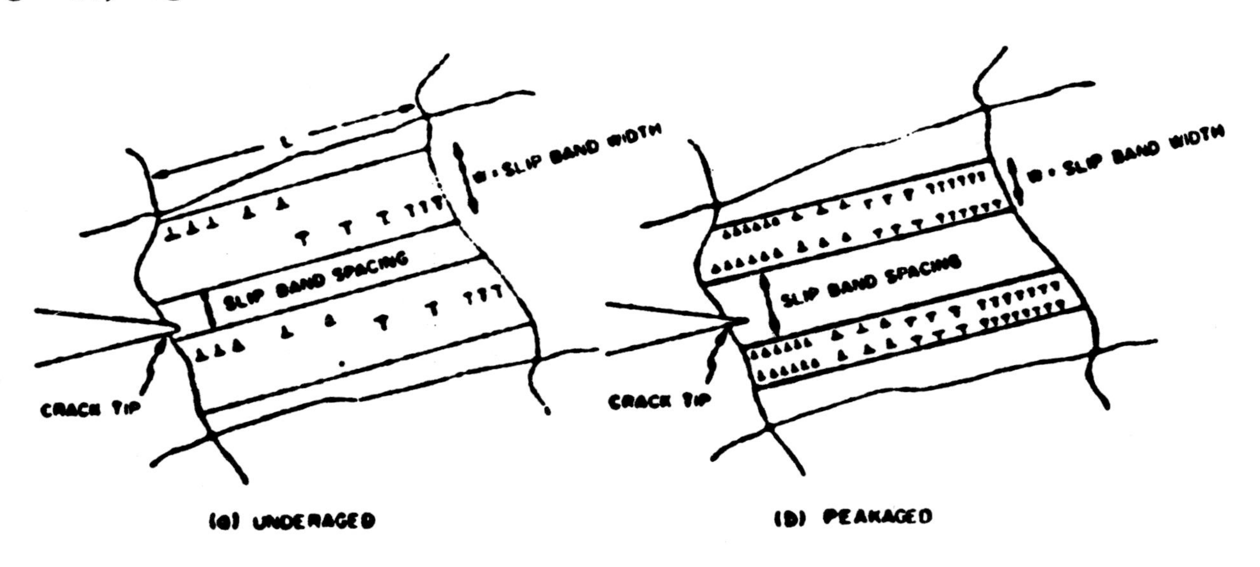

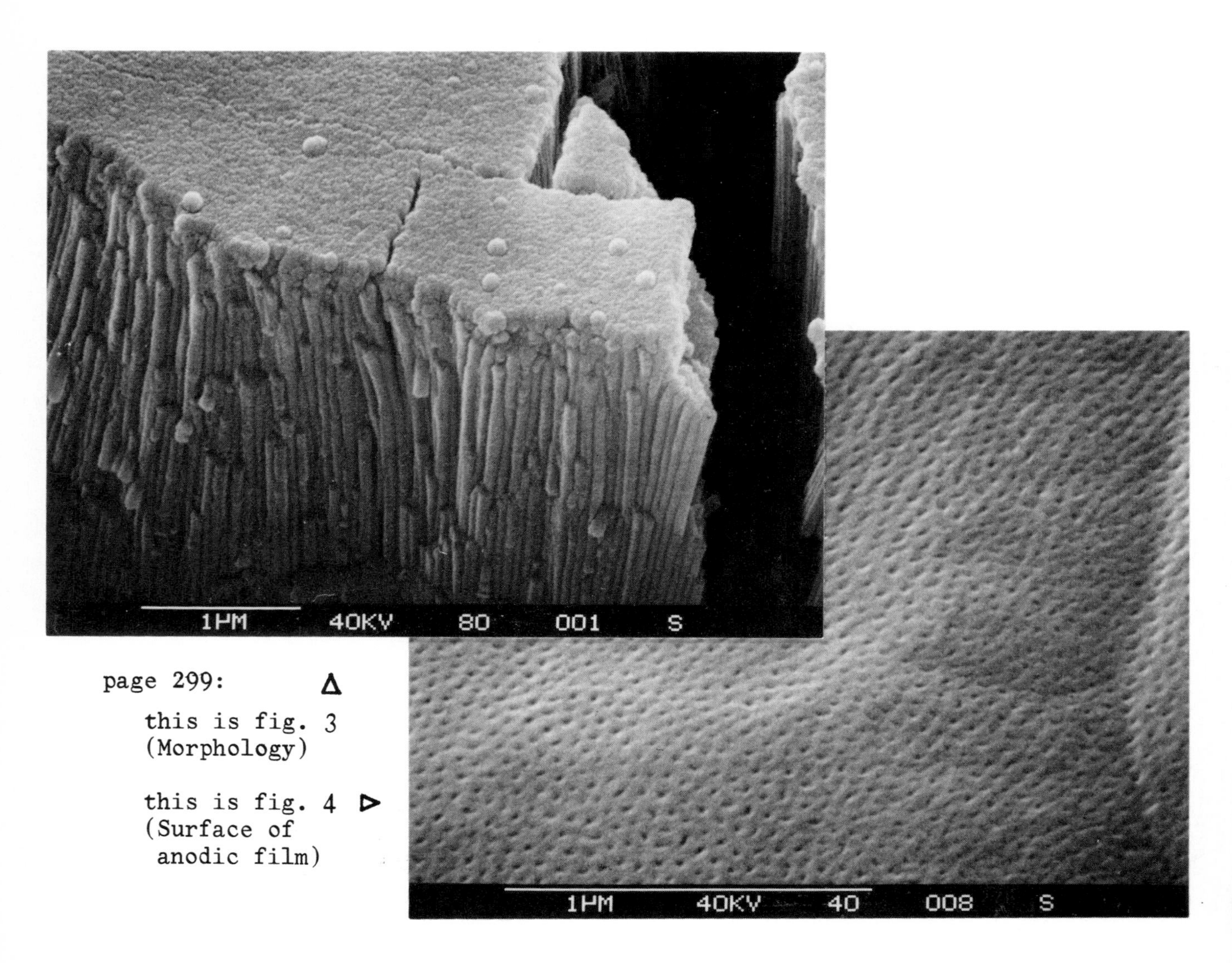

page 299: Δ

this is fig. 3
(Morphology)

this is fig. 4 ▷
(Surface of
anodic film)

page 300, fig. 5
including scale bar etc.

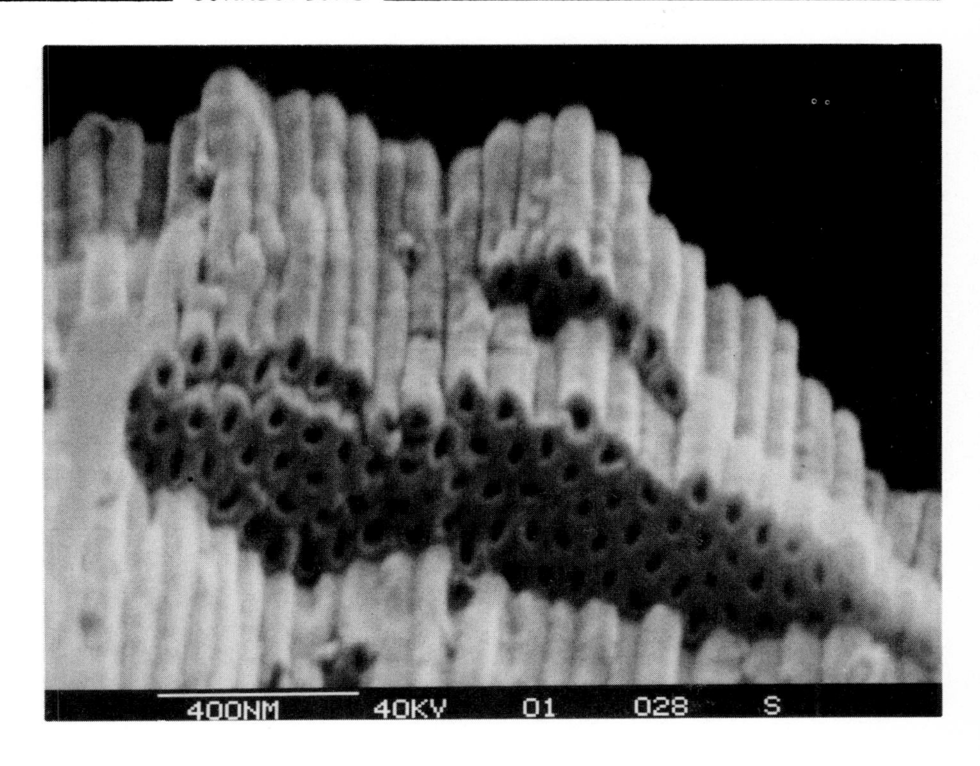

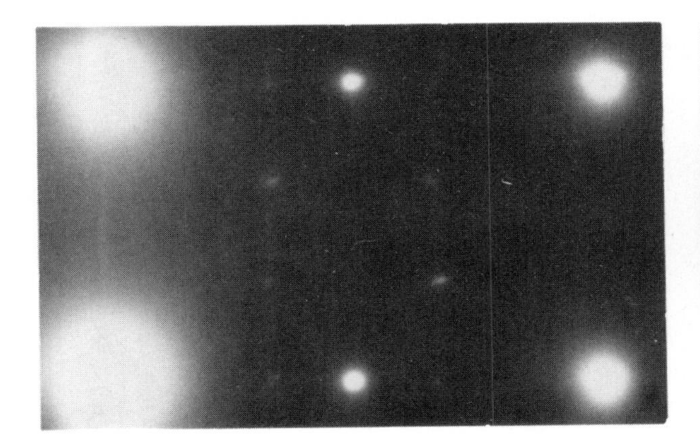

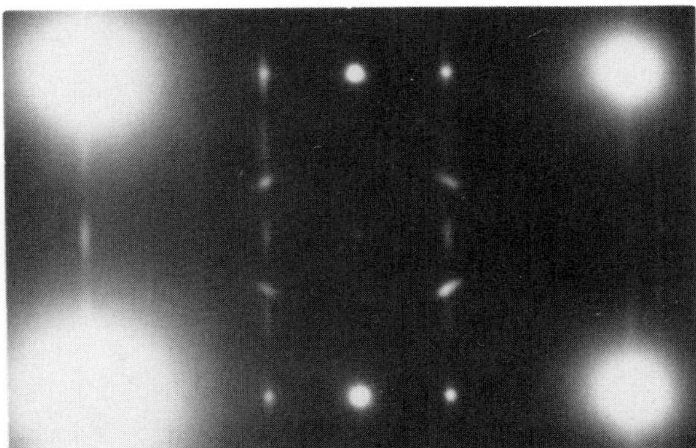

page 462, fig. 6 (d) and (e)

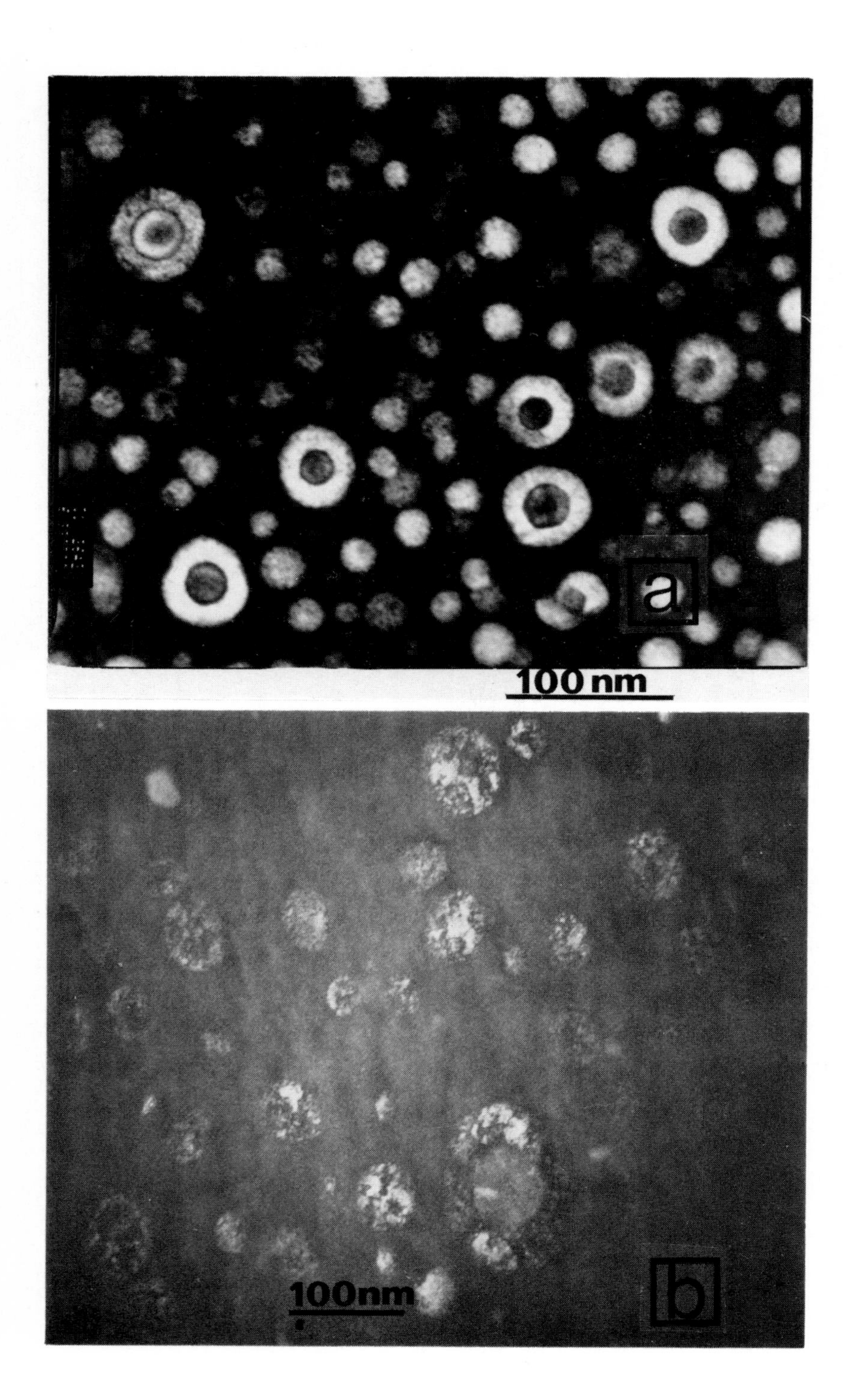

page 350, fig. 4 (a) and (b)

page 432, photo 4

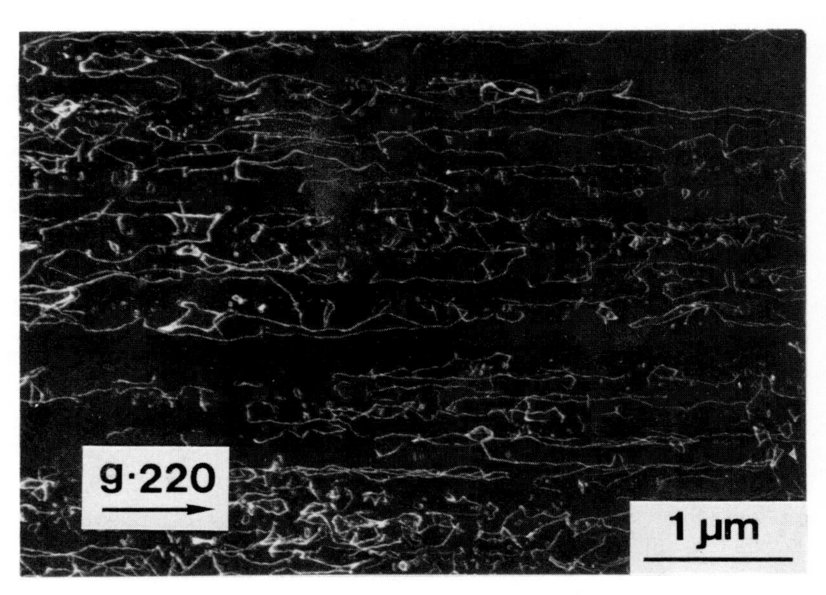

page 432, photo 5

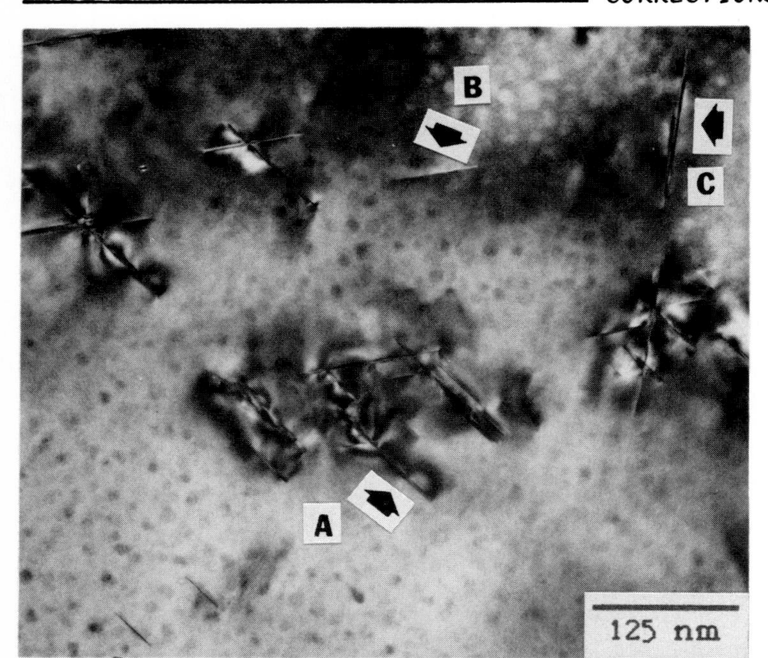

page 484, fig. 1a

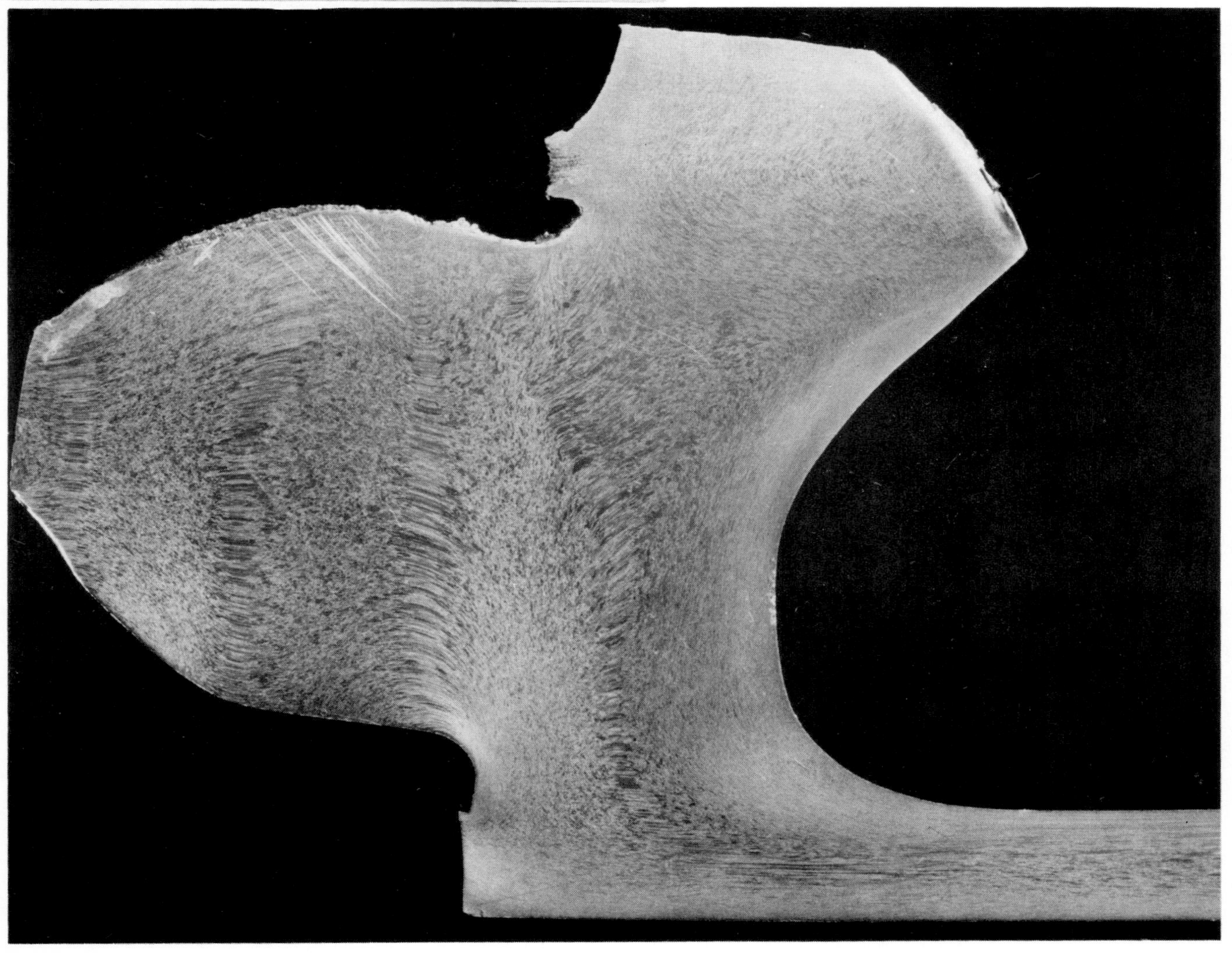

page 506, fig. 17

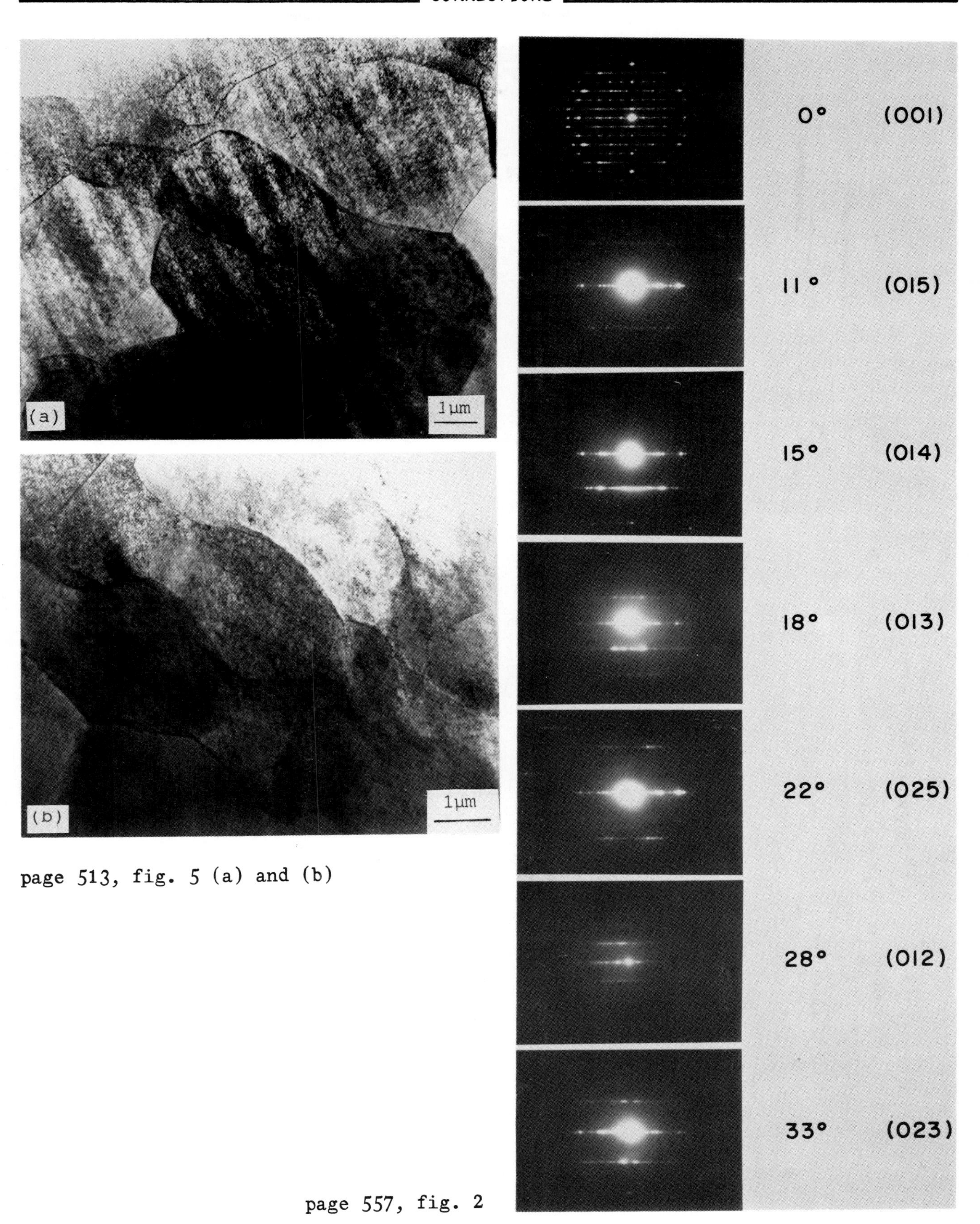

page 513, fig. 5 (a) and (b)

page 557, fig. 2

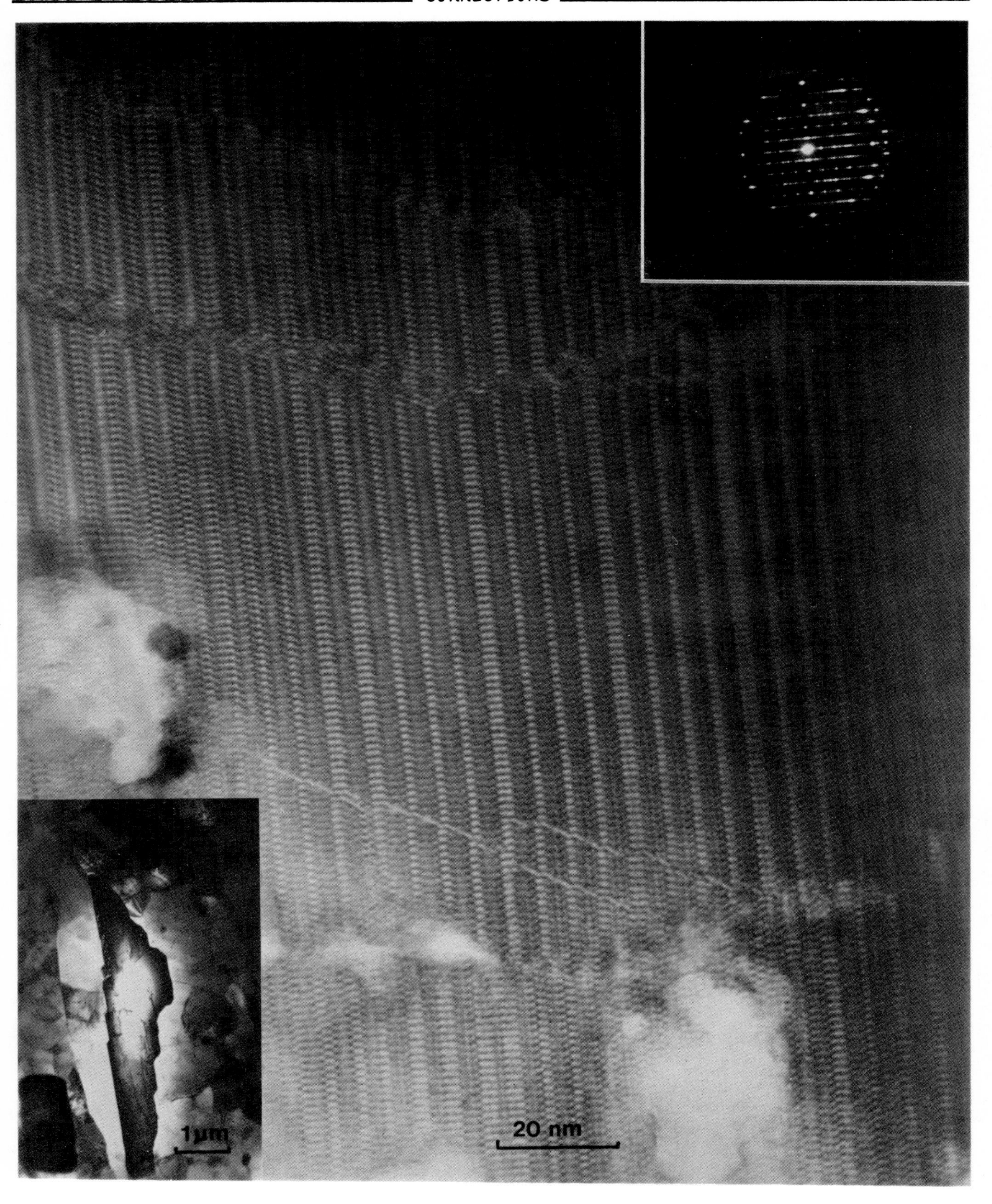

page 556, fig. 1

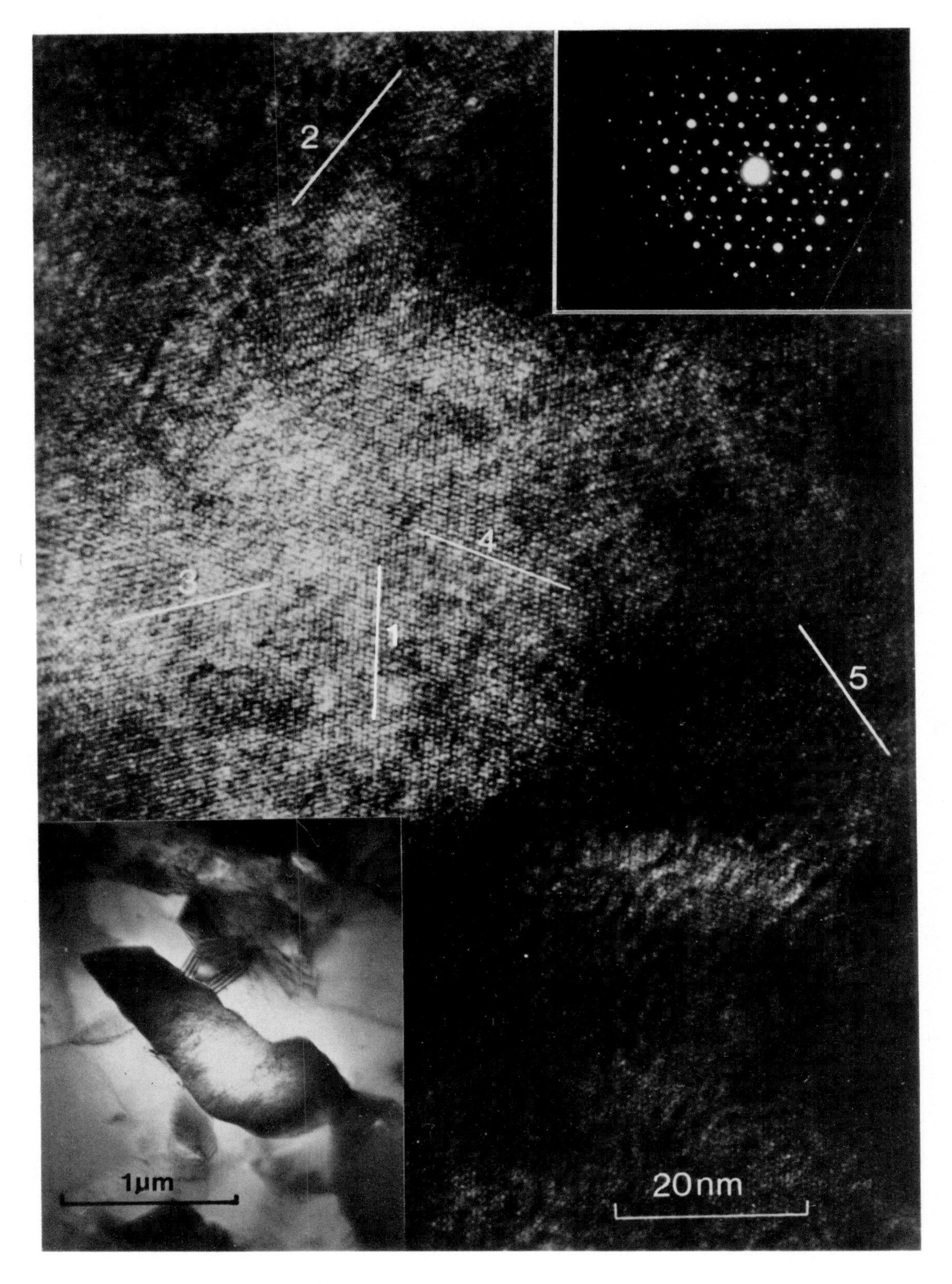

page 562, fig. 6

CORRECTIONS

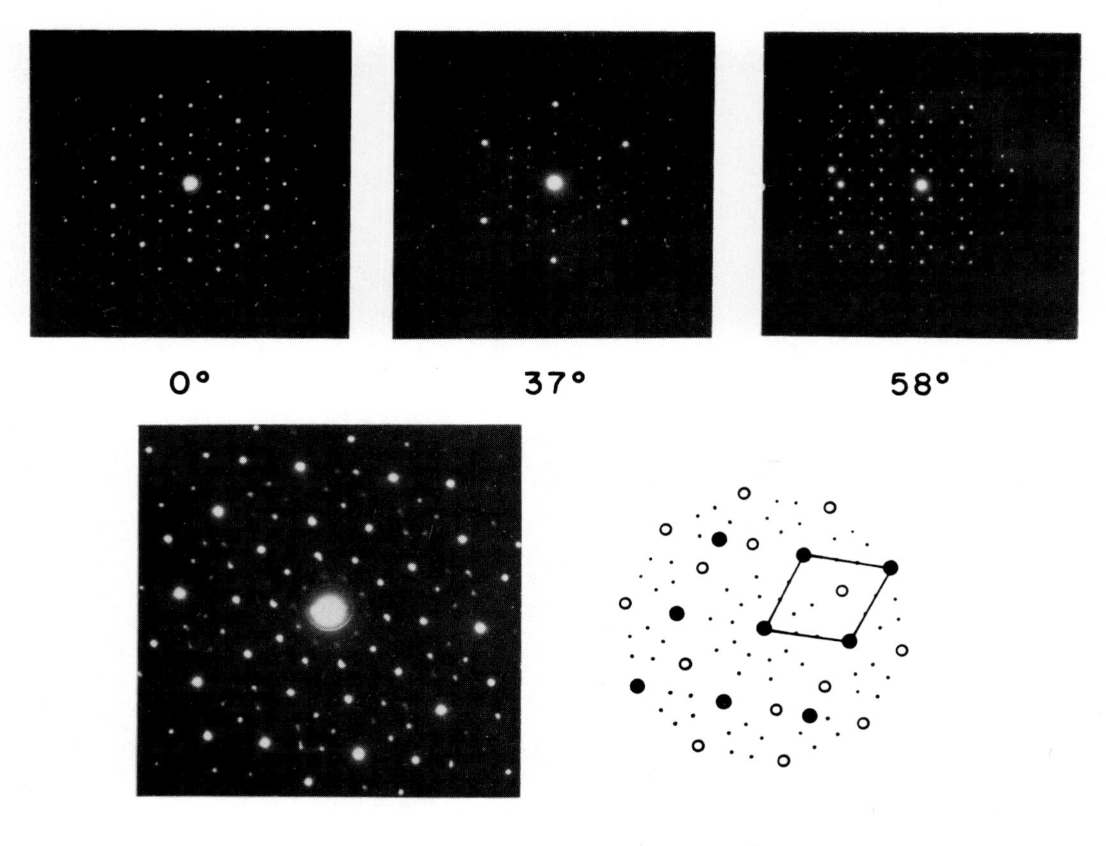

Key

- ● Basic Diffraction Pattern
- o Twin Diffraction Spots
- · Double Diffraction Spots

page 563, fig. 7

page 500, fig. 5

page 501, figs. 6 & 7

page 502, fig. 9

page 503, figs. 13 & 14

page 506 figs. 18 & 19

BOOK II

CONTENTS OF BOOK II

FOREWORD to Book II: Physical metallurgy of Al–Li alloys

P J GREGSON and S J HARRIS

MANY of the developments currently being made within the aluminium–lithium field rely upon an understanding of the relationships between the physical and mechanical properties of the alloys and their microstructures. It is fitting that the large number of papers presented at the conference in this area of research should be collected together in this Book. The subject matter ranges from reports of fundamental and theoretical studies in which Al–Li alloys provide a model for a specific technique, through to property evaluation of alloys produced on the commercial scale.

Detailed studies of the fine particles of metastable δ' (Al,Li) have continued to stimulate fundamental research. Questions are posed concerning the earliest stages of δ' precipitation and its association with spinodal decomposition, or the more conventional nucleation models. X-ray, electron beam, and thermal analysis techniques have been used to shed more light on this matter (page 329). Microanalytical techniques such as Electron Energy Loss Spectroscopy have been exploited in attempts to determine the lithium distribution and characterise the changes in microchemistry during age-hardening (pages 337, 347, 355). The growth of the metastable particles has been studied from both practical and theoretical standpoints (pages 360, 369). Modifications to the δ' precipitation which occur when zirconium is present in the alloys have been subject to further study (pages 360, 376, 386); the combined effect of lithium and zirconium in controlling recrystallisation (page 392); and the texture which results has been the subject of a detailed analysis (page 402).

The mechanisms by which δ' imparts the major contribution of strength to the heat-treated alloys were subjected to scrutiny in a number of presentations (pages 411–447); open discussion, however, stressed the importance of taking account of all possible strengthening mechanisms in this sort of study. The influence of lithium in solution, and the presence of δ' and other precipitate phases upon the elastic properties of the alloy has also been the subject of detailed research employing a variety of techniques (pages 435, 442, 448).

For commercial exploitation, the properties of Al–Li-based alloys must adequately match those of the aluminium alloys currently used in airframe construction. Much research is therefore concerned with developing an understanding of more complex lithium-containing aluminium alloys in which other solute additions enhance mechanical properties via solution strengthening or through further precipitation-hardening reactions. In recent years research has become concentrated upon two alloy systems: the ternary Al–Li–Cu alloys (2090 type) within which δ' precipitation may be supplemented by the formation of θ' ($CuAl_2$) and/or T_1 (Al_2CuLi); and the quaternary Al–Li–Cu–Mg alloys (8090 type) in which the precipitation of S (Al_2CuMg) is feasible.

At this conference the presentations concerning the Al–Li–Cu alloys were limited to the microstructural evolution of matrix and grain boundary precipitation during age-hardening (pages 455–495).

A large number of papers were concerned with developing an understanding of the microstructure/property relationships within the Al–Li–Cu–Mg alloys. The role of matrix precipitation in developing the promising combination of properties within these alloys was covered in some detail (pages 496–538). Elevated temperature properties (pages 539–564) were also evaluated and interpreted. The influence of grain structure, grain boundary particles, intermetallic inclusions, and microchemical segregation upon mechanical properties were described (pages 565–609); in the course of open discussion some of the problems in interpreting microanalytical data were highlighted. An interesting variation on 8090 is provided by the further addition of zinc: encouraging properties were displayed and interpreted in terms of additional strengthening mechanisms (page 610).

The Book concludes with the transcript from the closing address of Dr. C J Peel in which he summarises recent advances and suggests topics for further research.

P J GREGSON
Engineering Materials,
Southampton University

S J HARRIS
Department of Metallurgy and Materials Science,
University of Nottingham

Combined small angle X-ray scattering and transmission electron microscopy studies of Al–Li alloys

S SPOONER, D B WILLIAMS
and C M SUNG

SS is with the National Center for Small Angle Scattering Research, Oak Ridge National Laboratory, Oak Ridge, TN, USA.
DBW and CMS are in the Department of Metallurgy and Materials Engineering, Lehigh University, Bethlehem, PA, USA

SYNOPSIS

Combined small angle X-ray scattering (SAXS) and transmission electron microscopy (TEM) studies have been performed on two binary Al-Li alloys subjected to different pre-aging treatments. Both the SAXS and TEM data indicate that δ' (Al_3Li) in samples of the two alloys directly quenched to the aging temperature is larger in size than in samples quenched to below room temperature prior to aging. SAXS data indicate an initial rise in scattering center size in Al-9.4 at% Li, which is not observed in the more dilute Al-6.8 at% Li, or in the TEM studies of δ' in both alloys. It is considered that these observations are evidence that the initial rise in the SAXS data in Al-9.4 at% Li alloys is due to a phenomenon other than δ' precipitation. The possibility of matrix spinodal decomposition in this alloy is one explanation.

INTRODUCTION

Small angle X-ray scattering (SAXS) is a technique for studying non-periodic structures, such as second phase precipitates in a matrix, voids, pore structures, gels, etc. It is sensitive to these structures (scattering centers) on a scale from ∿0.5 nm to 200 nm in diameter. The SAXS spectrum is usually obtained from a region of specimen ∿3 mm^2 x ∿100 μm thick, and thus gives information characteristic of bulk material (>∿10^9 μm^3), which may contain from (∿10^{11}→10^{18}) scattering centers. While the scattering statistics are excellent, the effects of small sub-micron variations such as particle/void shape changes and the presence of other small scattering centers, cannot currently be accounted for and therefore limit the quantification of the SAXS data. In other words the SAXS spectrum does not give a unique structure identification. A specific model (e.g., spherical precipitates) is essential in order to interpret the spectrum. Deviations from the model (e.g., other precipitates) in the real specimen limit the interpretation. Only when the range and type of scattering centers is known can the separate contributions to the SAXS spectrum be separated. However, these separate scattering contributions can often be imaged using conventional transmission electron microscopy (TEM). Conversely TEM imaging is severely limited in its statistical sampling since a typical volume of specimen examined is ∿<1 μm^3. Therefore it is difficult to obtain quantitative information from the TEM concerning such phenomena as the kinetics of precipitation, the onset of growth or coarsening, the particle size distribution and the volume fraction of second phase. These parameters are more routinely accessible using SAXS. Therefore it seems sensible to combine the strengths of both techniques and this paper reports the initial results of a combined SAXS-TEM study of δ' (Al_3Li) precipitation in two binary Al-Li alloys containing 6.8 at% and 9.4 at% Li.

The effect of pre-aging treatment on the development of aging microstructure was shown in earlier TEM studies (1). The particle size developing after a direct quench to the aging temperature is larger than the particle size which follows an (indirect) quench-and-age treatment for a given aging time. Because pre-aging treatment must influence the nucleation and growth stages of decomposition, investigation of alloys given short aging times at low temperature was undertaken. At the time of designing the experiments no emphasis was placed on investigating the possibility of effects due to a metastable miscibility gap. However, results of the study suggest that miscibility gap decomposition may have an influence on the aging kinetics studied in this work. This possible interpretation is significant given the recent calculations of Sigli and Sanchez (2) which predict a metastable miscibility gap in the low temperature α (Al-Li solid solution) matrix.

From the perspective of selecting a scattering

Table I. Comparison of Guinier Radius (R_g) of Scattering Centers and Radius of δ' from TEM Images of Al-9.4 at% Li Aged at 100°C

Aging Time (hrs.)	Direct Quench		Indirect Quench	
	SAXS (R_g)	TEM (δ' radius)	SAXS (R_g)	TEM (δ' Radius)
2	1.24 ± 0.04 nm	3.0 ± 0.5 nm	1.47 ± 0.04 nm	2.0 ± 0.3 nm
4	1.76 ± 0.03 nm	2.5 ± 0.7 nm	1.71 ± 0.04 nm	3.3 ± 0.5 nm
8	1.60 ± 0.04 nm	3.2 ± 0.5 nm	1.70 ± 0.02 nm	2.1 ± 0.3 nm
16	1.65 ± 0.02 nm	3.0 ± 0.8 nm	1.95 ± 0.03 nm	3.2 ± 0.3 nm
32	2.02 ± 0.02 nm	4.0 ± 0.9 nm	1.61 ± 0.02 nm	3.0 ± 0.3 nm

Table II. Comparison of Guinier Radius (R_g) of Scattering Centers and Radius of δ' from TEM Images of Al-6.8 at% Li Aged at 120°C

Aging Time (hrs.)	Direct Quench		Indirect Quench	
	SAXS (R_g)	TEM (δ' radius)	SAXS (R_g)	TEM (δ' Radius)
1	0.63 ± 0.12 nm	--------- nm	1.34 ± 0.04 nm	2.6 ± 0.5 nm
2	0.92 ± 0.10 nm	--------- nm	1.14 ± 0.04 nm	1.5 ± 0.3 nm
4	0.98 ± 0.07 nm	--------- nm	1.58 ± 0.02 nm	1.9 ± 0.4 nm
6	2.21 ± 0.04 nm	2.7 ± 0.5 nm	1.99 ± 0.01 nm	2.2 ± 0.5 nm
9	2.72 ± 0.02 nm	3.2 ± 0.4 nm	2.25 ± 0.02 nm	2.9 ± 0.4 nm
13	2.98 ± 0.02 nm	3.3 ± 0.4 nm	2.37 ± 0.02 nm	2.5 ± 0.3 nm

roblem, the Al-Li system has the important dvantage of generating spherical precipitates for hich the scattering theory is well known. Were it ot for the influence of high particle density, it hould have been quite simple to extract size istributions using well-established techniques of cattering analysis. However, this particular bjective is hard to achieve, has been treated lsewhere (3) and will be ignored in this study. e will focus on those features of scattering which an be readily interpreted and which shed some ight on processes of δ' precipitation in Al-Li lloys. In addition we have concentrated on aging equences that give rise to δ' distributions that re not routinely accessible to TEM studies (SAXS xtends the size range for structure investigation elow the range where conventional TEM methods are onvenient, i.e., <∿ 5 nm).

XPERIMENTAL

hick sheets of each alloy were cold rolled to pproximately 0.8 mm thickness, and then given an levated temperature anneal at 520°C in a salt bath or 4 hours for the purpose of enlarging the grain ize. The material was reduced in thickness to .25 mm by mechanical polishing to remove the ithium-impoverished layer. The material was then olution treated for 20 minutes at 500°C for the .4 at% alloy and 405°C for the 6.8 at% alloy then ither quenched into water or quenched directly nto a medium at aging temperature. Oil at 100°C as used with the 9.4 at% alloy and fine sand at 26°C was used with the 6.8 at% alloy. Aging times aried from 1 hour to 32 hours.

After aging, 3 mm disc specimens were punched, round and then electrochemically polished to 120 m thickness for X-ray scattering studies. SAXS as carried out on the 10 m SAXS machine (4) at the ational Center for Small-Angle Scattering Research t Oak Ridge National Laboratory. CuK_α radiation rom a rotating anode is scattered from a graphite onochromator to remove K_β radiation. Apertures of mm x 1 mm at the X-ray source and sample separated y 1.62 meters define a highly collimated beam which s scattered into a 20 cm x 20 cm position-sensitive roportional counter. Data collected on a 64 x 64 rray are reduced to a list of radially-averaged ntensities for scattering analysis. In addition o empty sample and dark current (no X-ray beam) ackgrounds, a detector sensitivity experiment ith a ^{55}Fe radioactive source (MnK_α) is done for inal correction. Several sample-to-detector istance were employed in order to accommodate the ifferent ranges of scattering distribution arising rom various heat treatments. The sample trans-mission used in data correction was determined in situ thereby reducing its uncertainty due to ariation in sample thickness arising from sample lacement.

A unique feature in this SAXS investigation is the use of a point-geometry with an area position-sensitive detector. (Linear detectors are now in more common use with Kratky cameras which employ the so-called slit geometry and even with point geometry. By point geometry we mean that the radiation source is a "point"--i.e., 1 mm x 1 mm--and the sample has a similarly small area.) Although not required in this study, the area detector can directly measure variations in scat-tering symmetry (as might be found in the scattering from oriented rod- or plate-like precipitates). Rather, the area detector accumulates data with relatively small statistical error bars at large angles where scattering intensity is low. This is important in examining the limiting intensity in the so-called Porod region where intensity falls rapidly with decreasing scattering angle. (A small additional advantage in the use of a small sample area is that TEM methods of sample preparation can be used.)

TEM was performed on a Philips EM400T at Lehigh University. The same three millimeter disc SAXS samples were electropolished in a 10% perchloric acid 90% ethanol solution at -40°C and 15V applied potential, using a Struers Tenupol system. Centered dark field (CDF) images of δ' were obtained using superlattice reflections. The image magnification was calibrated using a diffraction grating replica standard (5). On average ten to twenty δ' parti-cles were measured, ensuring that the image was of an isolated particle. The largest such particles were selected to minimize the possibility of sec-tioning by the top or bottom of the foil. Measure-ments were made from enlarged prints (∿50,000X - 750,000X) using a 15X magnifier containing a graticule, calibrated to 0.1 mm.

RESULTS AND DISCUSSION OF SAXS EXPERIMENTS

The general form of all the scattering curves is illustrated in Figure 1, which is termed a Guinier plot. Figure 1 plots log. of the scattering intensity versus the scattering vector, k, squared. The magnitude of the scattering vector k is equal to $4\pi \sin\theta/\lambda$ where 2θ is the scattering angle and λ is the X-ray wavelength. Interparticle interference gives an intensity peak removed from the forward direction ($k^2 \cong 7x\ 10^{-4}$) and this is a predominant feature of all the scattering curves in these experiments. (This feature is consistent with a uniform, high density distribution of the precipitate and is in common with spinodally-decomposed alloys, binary liquids, glasses, micelle solutions and other two-phase systems.) The scattering peak in the forward direction ($k^2 \to 0$)) is attributed to scattering from large particles which will not concern us directly in this study. The slope of the linear portion of the curve beyond the maximum is equal to the average precipitate size squared divided by 3. From the slope the particle radius R_g (termed the Guinier radius) is calculated with a least-squares curve fitting.

The trend in particle size evolution is shown in Figures 2a and 2b where the cube of R_g is plotted against time (t). This form of presentation is adopted to emphasize the presence of coarsening which results in a straight line after longer

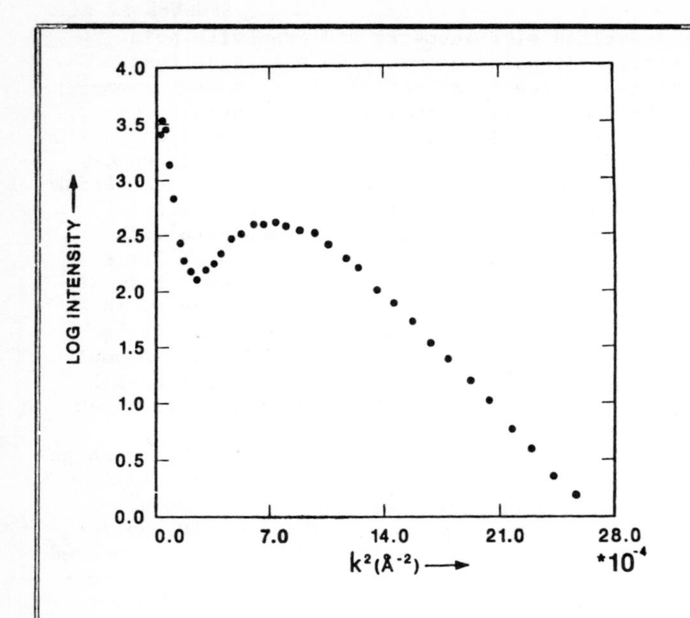

1. Guinier plot of X-ray scattering intensity vs the square of the scattering vector. The form of this curve is representative of all the scattering curves from both Al-9.4 at% Li and 6.8 at% Li.

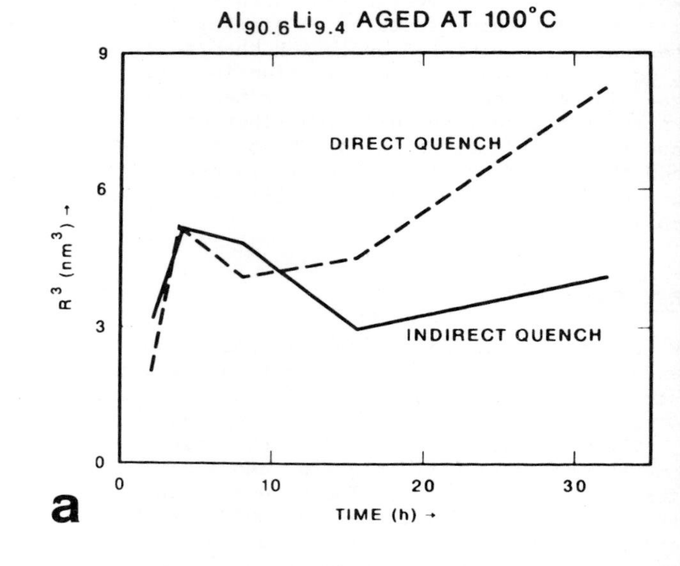

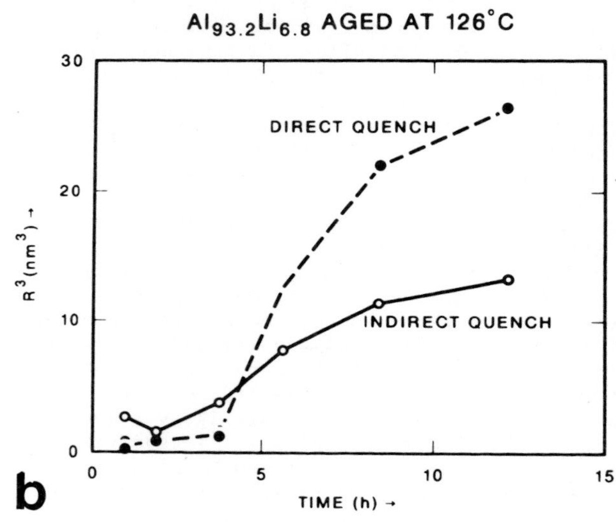

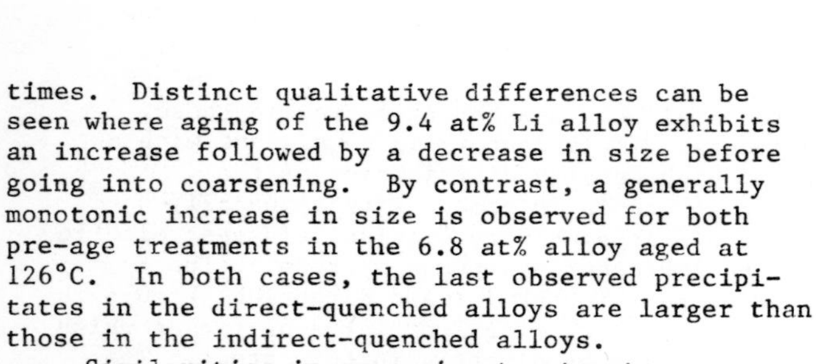

times. Distinct qualitative differences can be seen where aging of the 9.4 at% Li alloy exhibits an increase followed by a decrease in size before going into coarsening. By contrast, a generally monotonic increase in size is observed for both pre-age treatments in the 6.8 at% alloy aged at 126°C. In both cases, the last observed precipitates in the direct-quenched alloys are larger than those in the indirect-quenched alloys.

Similarities in pre-aging treatment response for the two alloys are seen in the cross over from smaller to larger precipitate size with time. In so far as the indirect-quench treatment causes the alloy to experience temporarily higher nucleation rates, the microstructure gets an earlier start with a small scale (high density) structure. The spatial constraint of a high density precipitate structure may then reduce the individual precipitate size. The structure evolving from a low nucleation rate stage might then be expected to have a larger ultimate precipitate size. A question which remains is what differences in coarsening kinetics are to be found? A postulate of coarsening theory is that a steady-state particle size distribution

2. a) Variation in the scattering center size (R) as a function of aging time and pre-aging treatment in Al-9.4 at% Li aged at 100°C.
b) Variation in the scattering center size (R) as a function of aging time and pre-aging treatment in Al-6.8 at% Li aged at 126°C.

develops upon coarsening. It seems contrary to this postulate that such a steady-state distribution should depend on the starting precipitate structure, and that the kinetic constant should depend on starting structure.

The volume fraction of precipitate and hence the solute removed from the matrix is proportional to the integrated small angle scattering. The scattering invariant, Q_o, is calculated from

$$Q_o = \int k^2 I(k)\, dK \qquad (1)$$

and therefore a plot of $k^2I(k)$ vs. k gives a means of estimating Q_o and the average particle size from the peak position. A set of such curves for each aging sequence is shown in Figures 3a and 3b. In the 100°C aging treatment of the 9.4 at% Li alloy the set of curves indicates that all the solute is largely removed from the matrix at early times and only the peak position changes with a coarser particle emerging (shown by the peak moving to smaller k values). Note that areas under the curve are essentially the same for direct and indirect pre-aging treatment. By contrast, the same curves for the 6.8 at% Li alloy aged at 126°C show a different pattern of emergence. For both pre-aging treatments there is an increase in area with indirect quenching giving a larger area in all cases.

Differences in the response of the two alloys to aging may depend largely on differences in nucleation rate--the 9.8 at% alloy being more strongly supersaturated. However, it is possible that the 9.8 at% alloy is being aged within the α matrix miscibility gap proposed by Sigli and Sanchez (2). This might account for the initial rapid decomposition (larger area under the $k^2I(k)$ vs. k) and the increase and decrease of scattering center size prior to coarsening. A kinetically rapid means of removing solute would account for an initially large Q_o. A spinodal decomposition in the matrix with fast coarsening followed by the nucleation of more stable δ' at smaller size may then be consistent with the observed R_g^3 vs t behavior in this alloy.

The effective Guinier radii for spherical precipitates were analyzed in terms of the Lifshitz-Slyozov theory of coarsening:

$$\bar{r}^3(t) - \bar{r}^3(0) = Kt \qquad (2)$$

where $\bar{r}$ is the average radius after time t and at zero time in the coarsening regime, K is the temperature-dependent kinetics constant in cm^3/sec and t is time. The constant is written:

$$K = 8\gamma C_e D v_m^2/9RT \qquad (3)$$

where γ is the precipitate/matrix interfacial energy in Joules/cm^2, Ce is the equilibrium matrix composition in moles/cc, D is diffusivity in cm^2/sec, v_m is molar volume in cm^3, R is the gas constant in J/mol-deg and T is temperature in K. From the indirect quench data at 100°C, a value of γ of .011 J m^{-2} was obtained, after longer aging times (3).

The calculated value of γ is comparable to the value obtained by Baumann and Williams (6) using critical nucleus size observations in TEM equal to 0.014 Jm^{-2}. This is an excellent corroboration of the very low value for the particle/matrix interfacial energy.

RESULTS AND DISCUSSION OF TEM EXPERIMENTS

All specimens exhibited a two phase α-δ' microstructure except at the earliest stages of aging the direct quench samples. The equilibrium δ phase was not visible on the grain boundaries, and no large dislocation densities were observed. Therefore the primary scattering centers are most likely δ'. Typical micrographs are shown in Figure 4 from samples of Al-9.4 at% Li a) directly quenched then aged and b) indirectly quenched and aged. In Figure 5 similar images from the Al-6.8 at% Li sample are shown. It is clear that in the 9.4 at% Li alloy, the δ' volume fraction is such that unambiguous determination of isolated δ' sizes is difficult. The more dilute alloy exhibited somewhat lower densities of δ' making particle size measurement easier.

Table I summarizes the measurements of δ' radii obtained from the Al-9.4 at% Li alloy aged at 100°C and Table II lists the radii obtained from the Al-6.8 at% Li alloy aged at 126°C. The large errors are due to the small sampling and the problems with overlap and sectioning.

Despite the large errors it is clear in the 9.4 at% Li alloy that direct quench samples have a larger radius than the indirect quench samples. Similar conclusions were drawn from the SAXS data. Furthermore this observation agrees with the TEM studies of Baumann and Williams (1) who report average δ' radii of direct quench samples aged in the range 200°C-250°C to be from two to three times the radius of indirect quench samples. At these higher aging temperatures δ' was coarse enough to permit accurate size measurement in the TEM, in contrast to the current investigation.

In the 6.4 at% Li alloy, δ' sizes are too small and the errors too large to permit any unambiguous differences to be determined. The absence of clear δ' images for aging times up to 4 hrs. is probably due to the fact that particles <1 nm in radius, as predicted by the SAXS data (Table II) are probably below the resolution limit of the instrument. In normal thin foils chromatic aberration can easily limit the image resolution to ∿2.5 nm (5). High resolution microscopy is therefore required to observe the smallest δ'. It may thus be concluded that the early δ' aging kinetics in Al-Li alloys containing <∿ 9 at% Li are not susceptible to study by routine TEM--except to observe gross changes in δ' size and distribution. The earliest δ' coarsening kinetics reported in Al-Li alloys are those of Williams and Edington (7) at 140°C. In those studies, coarsening was detected after ∿10 mins. aging, with a δ' particle radius of ∿3 nm. It is perhaps not surprising therefore that the TEM data at 100°C do not

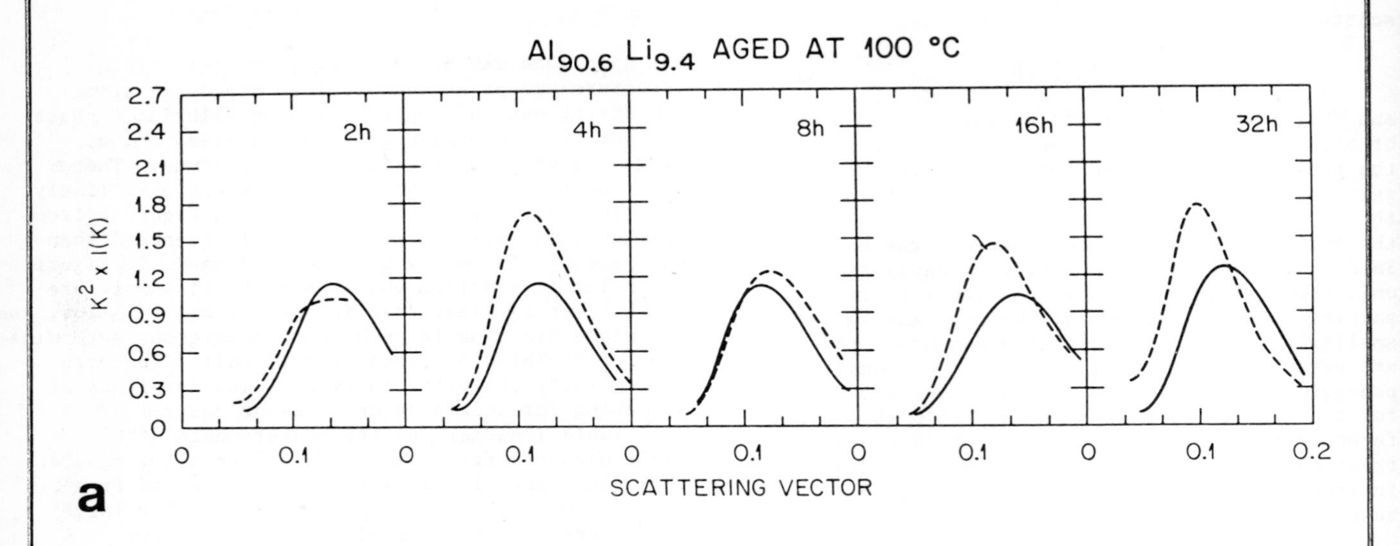

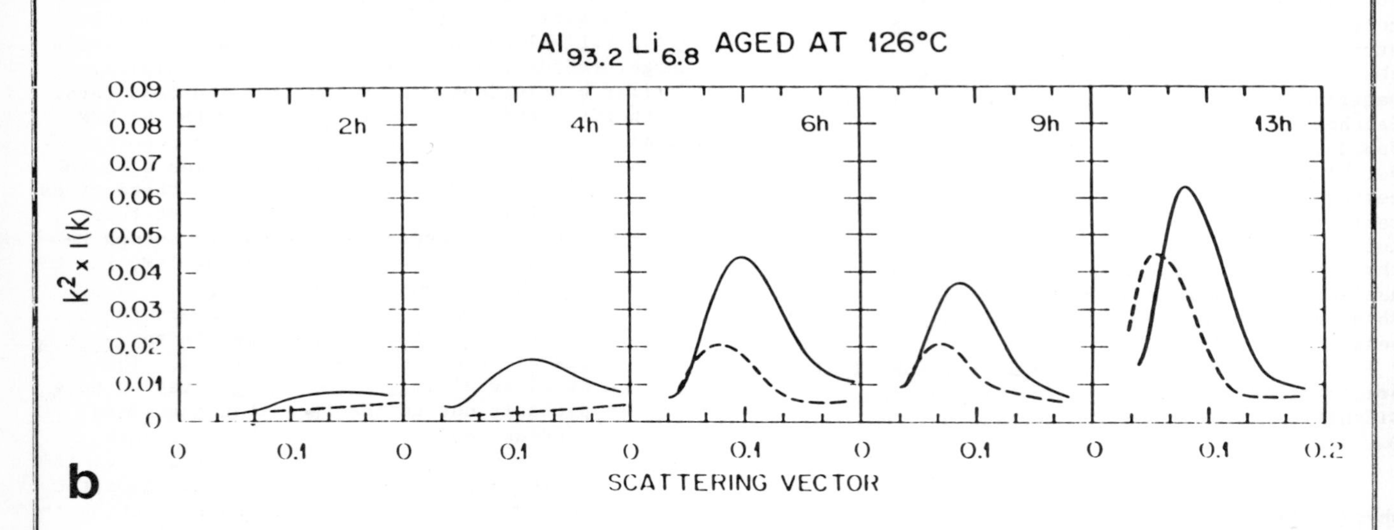

3. a) $k^2I(k)$ vs k curves for Al-9.4 at% Li aged at 100°C showing the relatively constant value of the volume fraction of precipitate (area under the lines) as a function of time. This indicates that the nucleation and growth process is completed relatively early in the aging sequence. The direct quench data (dotted) indicate a larger scattering center size in these specimens than the indirect quench samples.

b) $k^2I(k)$ vs k curves for Al-6.8 at% Li aged at 120°C showing an increase in scattering center volume with time, indicating that the nucleation and growing process is continuing throughout this aging sequence.

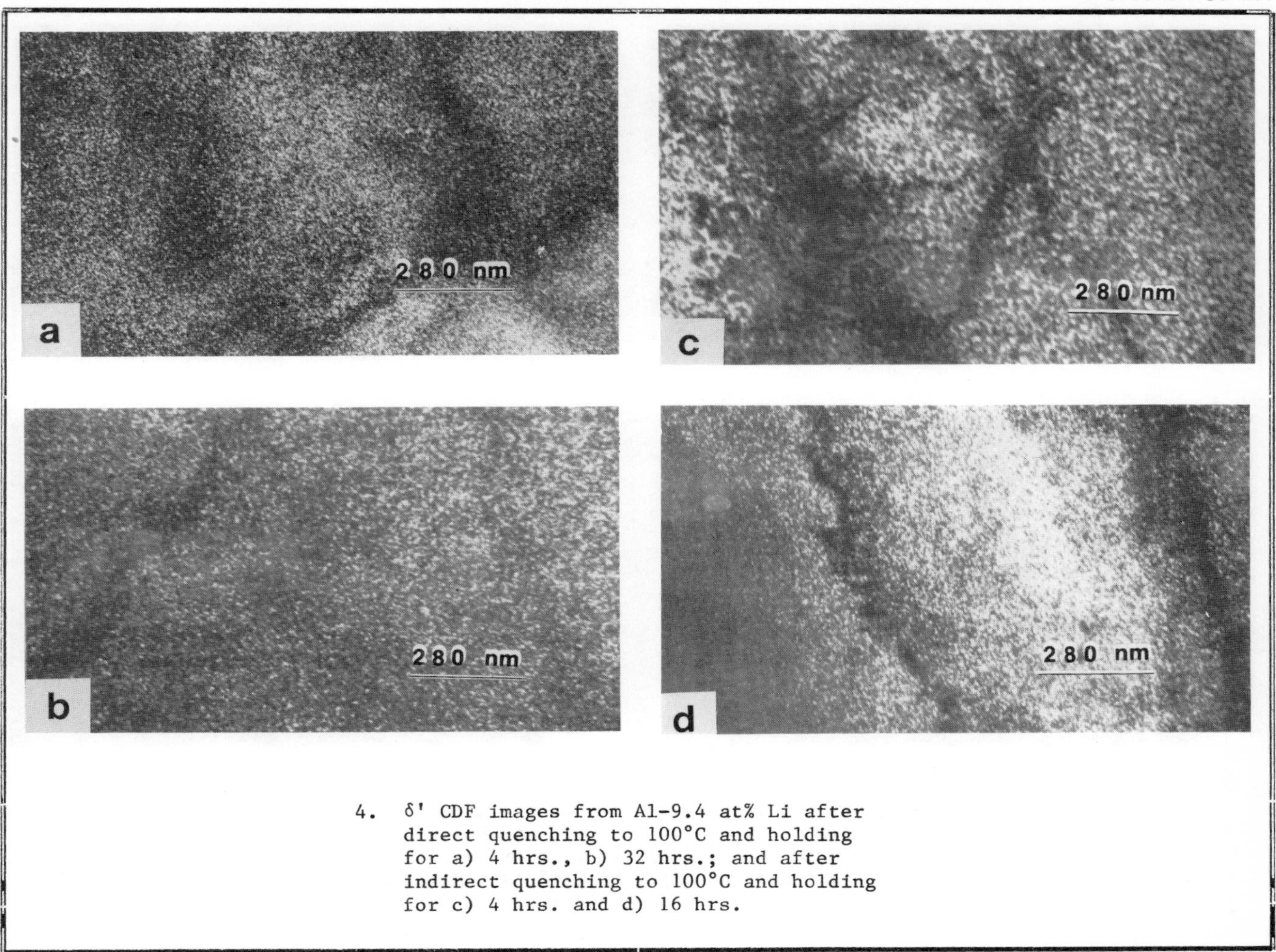

4. δ' CDF images from Al-9.4 at% Li after direct quenching to 100°C and holding for a) 4 hrs., b) 32 hrs.; and after indirect quenching to 100°C and holding for c) 4 hrs. and d) 16 hrs.

register any significant change in δ' size, except for a slight increase in specimens aged for 32 hours. However, it is possible to draw some further conclusions if comparison is made between the SAXS and TEM data. To facilitate this comparison, the SAXS data for R_g are given along side the δ' radii in Tables I and II.

COMPARISON OF SAXS AND TEM DATA

From Tables I and II and Figures 4 and 5 it is clear that both SAXS and TEM give similar order of magnitude results concerning the Guinier radius of the scattering centers and the δ' radius from TEM images. In both cases the radii are <5 nm. This agreement is not unusual and in fact much better agreement has been reported (Windsor et al. (8)) for γ' particles in Ni-base alloys--albeit for particles of a much larger average radius (5-25 nm). The TEM data from the 9.4 at% alloy show no significant apparent change with aging time except after 32 hours aging for the direct quench specimen. This is due to the fact that at 100°C δ' coarsening has barely commenced, and the large errors (∿±50%) in the δ' radii will mask virtually all the slight changes registered in the SAXS spectrum. In contrast the SAXS data show a clear commencement of coarsening, and extrapolations of these data match well with the previously published δ' coarsening data using SAXS (3).

The obvious initial rise and fall in R_g in the 9.4 at% Li sample is not reflected in the TEM δ' data. This leads to the possible conclusion that this rise is not due to δ' scattering only. An alternative (and additive) source of scattering would be some other aging phenomenon such as a matrix spinodal decomposition. This alloy was aged in the miscibility gap region calculated by Sigli and Sanchez (2). The effects of this spinodal decomposition would not be visible in a δ' CDF image. The absence of any rise and fall in the SAXS and TEM data from the 6.8 at% Li alloy further supports this speculation because this alloy was aged outside the calculated miscibility gap of Sigli and Sanchez.

To confirm this possible conclusion, more low temperature measurements are required with better TEM counting statistics in order to lower the large error bars in Tables I and II. It is unfortunate in some respects that the SAXS data need TEM images to identify other possible sources of scattering while in practice the TEM data obtained at those early stages are of very poor quality and cannot

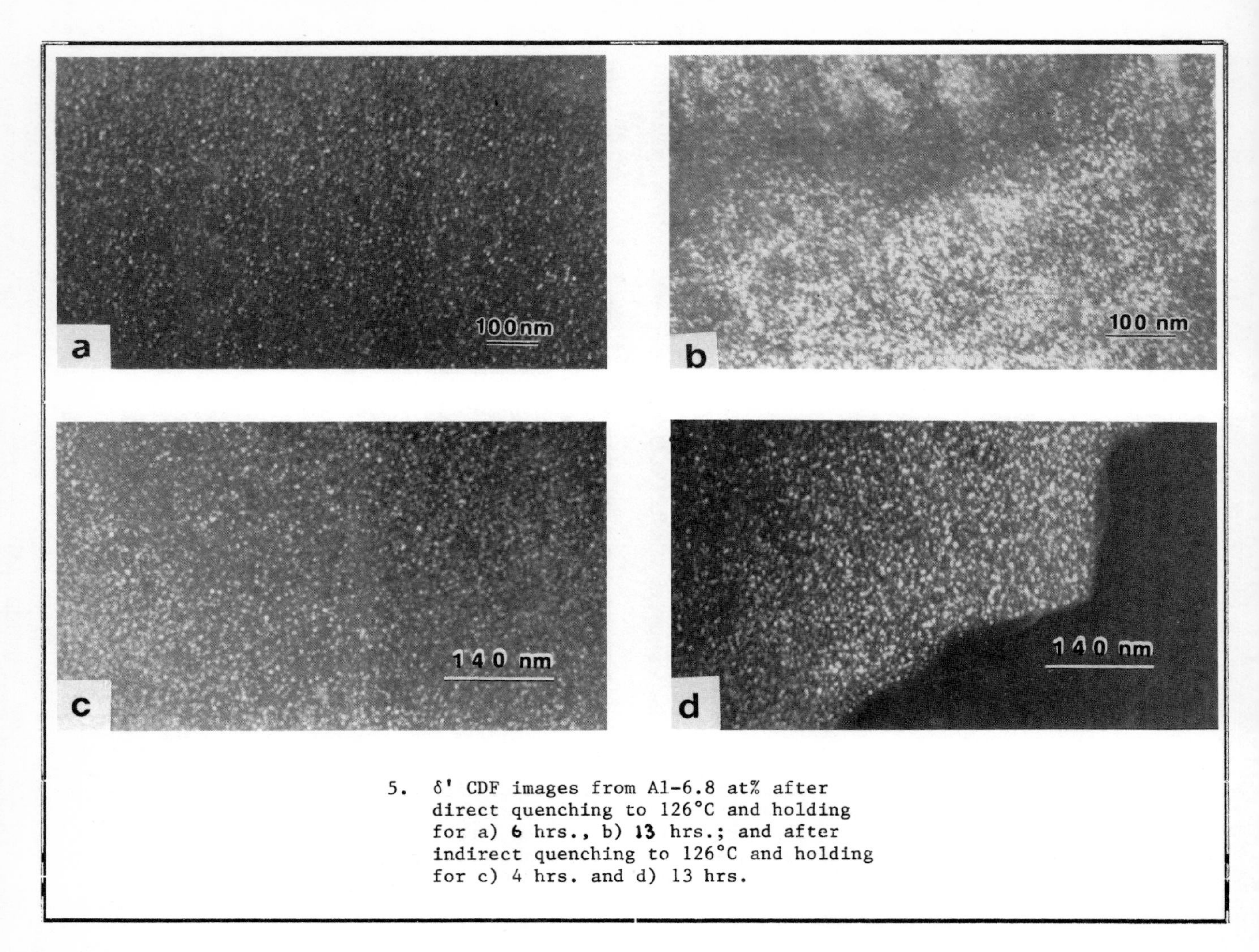

5. δ' CDF images from Al-6.8 at% after direct quenching to 126°C and holding for a) 6 hrs., b) 13 hrs.; and after indirect quenching to 126°C and holding for c) 4 hrs. and d) 13 hrs.

show the most likely alternative source of scatter, i.e., a spinodal decomposition.

CONCLUSIONS

1) SAXS data indicate that indirect quench samples have, in general, a smaller δ' size than direct quench samples. TEM images support this conclusion.
2) SAXS data indicate that in Al-9.4 at% Li there is an initial rise then fall in R_g, which is not reflected in TEM images of δ'.
3) TEM measurements of δ' radii at the early stages of aging are very difficult to perform without incurring large errors, and only serve, qualitatively to augment the inferences from the SAXS data.
4) Extrapolation of the SAXS data into the coarsening regime at 100°C leads to an estimate of the α/δ' interfacial energy γ of 0.011 J m^{-2}.

ACKNOWLEDGEMENTS

S.S. was supported by N.S.F. Grant No. DMR-7724459 through Interagency Agreement No. 40-636-77 with the U.S. D.O.E. under contract DE-AC05-840R21400 with Martin Marietta Energy Systems, Inc. D.B.W. and C.M.S. were supported through a grant-in-aid from ALCOA and the ALCOA Foundation, respectively.

REFERENCES

(1) Baumann, S. F. and Williams, D. B., Metall. Trans., 16A (1985), p. 1203.
(2) Sigli, C. and Sanchez, J. M., Acta Metall. (1985) submitted for publication.
(3) Spooner, S., Proc. Mat. Res. Soc. Symp. 41 "Advanced Photon and Particle Techniques for the Characterization of Defects in Solids," J. B. Roberts, R. W. Carpenter and N. C. Whittels (eds.). (1985), p. 89.
(4) Hendricks, R. W., J. Appl. Phys., 11 (1978), p. 15.
(5) Edington, J. W., 'Practical Electron Microscopy in Materials Science,' (1976) Van Nostrand-Reinhold, New York.
(6) Baumann, S. F. and Williams, D. B., Scripta. Metall. 18 (1985), p. 611.
(7) Williams, D. B. and Edington, J. W., Metal Science, 9 (1975), p. 529.
(8) Windsor, C. G., Rainey, V. S., Rose, P. K. and Callen, V. M., J. Phys. F.-Met. Phys. 14 (1984), p. 1771.

Quantitative microanalysis of Li in binary Al–Li alloys

C M SUNG, H M CHAN and D B WILLIAMS

The authors are in the Department of Metallurgy and Materials Engineering, Lehigh University, Bethlehem, PA, USA

SYNOPSIS

Direct and indirect methods of quantitative microanalysis for Li in Al-Li alloys are discussed. Electron energy loss spectrometry using ionization losses has the desired spatial resolution, but lacks the analytical sensitivity and is subject to large errors. Plasmon loss spectrometry can detect Li indirectly, and has both the necessary resolution and sensitivity. However data generation and reduction problems limit its general applicability. Convergent beam diffraction techniques use higher order Laue zone line shifts to show the effect of Li on the lattice parameter of Al. Other possible contributions to the line shifts must be removed before direct interpretation is possible.

INTRODUCTION

It can be argued that quantitative microanalysis for Li in Al-Li and Al-Li-X alloys is unnecessary. Given a basic understanding of the appropriate phase diagrams and microstructures it can usually be inferred where the Li resides. For example it could be assumed that δ' is Al_3Li, δ is AlLi and precipitate free zones (PFZs) indicate Li depletion. However, in response to these arguments, it is worth noting that:

a) the phase diagram is not well known;--the most recent experimental version proposed by Cocco et al. (1) implying an α+δ' miscibility gap does not indicate the δ' phase field;

b) the δ' stoichiometry is unknown;--the assumption of the composition Al_3Li is based solely on diffraction evidence for the $L1_2$ ordered structure;

c) δ in fact shows a wide stoichiometry range (45-55 at% Li) and

d) it has been shown (2) that vacancy depletion effects play a strong role in δ' precipitation morphology and PFZ formation. Therefore indirect inference of the Li distribution solely from microstructural observation is dangerous.

As well as simply detecting the Li, quantitative analysis is desirable. Quantitative analysis would permit the following studies:

a) Direct determination of the δ' stoichiometry and Bragg-Williams long range ordering parameter would be achievable. This would have a critical influence on determining the amount of Li available for solid solution (modulus) strengthening. Furthermore, from knowledge of the true δ' composition it would be possible to perform accurate calculations of the amount of order (antiphase boundary) strengthening at any alloy composition, and aging temperature.

b) Direct measurement of Li composition profiles at solute depleted PFZs could be performed. This would result in the generation of low temperature diffusion data and pre-empt the current requirement for tedious gathering of kinetic data such as δ' coarsening rates (3,4) and PFZ width measurements (5).

c) Direct determination of the Li content of ternary and higher order intermetallic phases could be carried out, e.g., this would resolve the current controversy concerning the presence or absence of Li in "composite" Al_3Zr/Al_3Li precipitates (6,7). Also, this would permit an understanding of the amount of Li available in these ternary, etc. alloys for solid solution and δ' precipitation strengthening.

d) The binary and ternary phase diagrams could be determined unambiguously. The advantages of this achievement are obvious.

In order to fulfill the above needs, any technique for the quantitative microanalysis of Li should be capable of detecting <∿3 wt% (∿10 at%)Li, with a spatial resolution of <50 nm. The technique should also be able to relate Li content and distribution directly to the microstructure of the alloy. These conditions arise because ∿3% Li is

TABLE I

Aging Temperatures and Times	Experimental δ' Lattice Parameter at −187°C (nm)	Extrapolated δ' Lattice Parameter Value at 20°C* (nm)	Calculated Lithium Content from Lattice Parameter at 20°C** (at% Li)	Calculated Misfit at −187°C (%)
270°C (40 mins.)	0.40235	0.40418	10.5	−0.11
240°C (2.5 hrs.)	0.40150	0.40333	22.8	−0.32
210°C (17 hrs.)	0.40051	0.40233	37.4	−0.57

*Lattice parameters at 20°C were obtained by extrapolation using the thermal expansion coefficient of pure Al.

**Lithium content calculated by extrapolation of Kellington et al's. (22) X-ray data in Al-Li solid solutions.

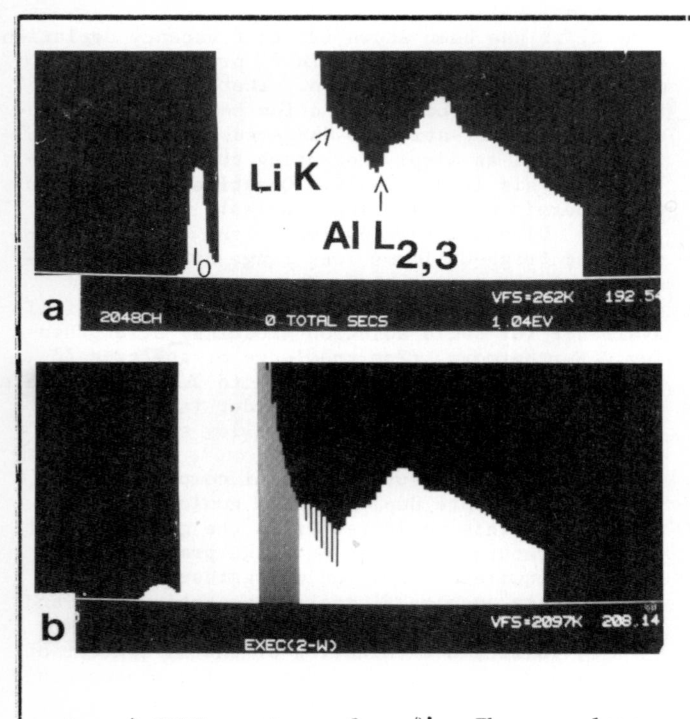

1. a) EELS spectrum, from δ'. The zero loss peak (I_o) is followed by a display gain change of ∿1000X to magnify the low intensity background. At ∿54 eV a change in the background slope (arrowed) signifies the presence of the Li K ionization edge. The Al $L_{2,3}$ edge is also indicated.
b) The method of background subtraction to obtain the integrated ionization edge intensity for quantification. The backgound is fitted over the shaded region and extrapolated under the Li K edge for Δ ∿20 eV (vertical hatched region).

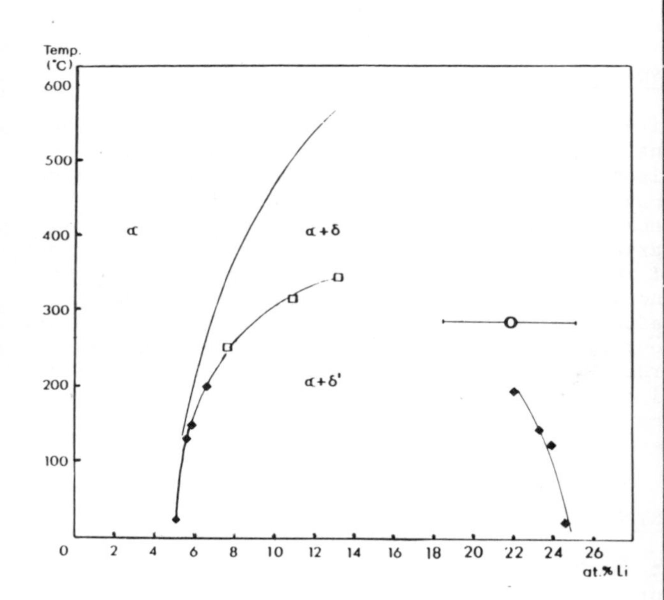

2. Al-Li phase diagram up to ∿25 at% Li showing the α/δ and α/δ' solvus (3) as well as the δ' phase field up to ∿200°C (1). The EELS result for δ' is shown by the single datum point with error bar of ∿±20% relative.

the maximum amount that will be added to commercial alloys in the near future and ∿50 nm represents an upper limit on the dimensions of microstructural features that control many of the alloy properties (particularly δ' and other precipitates). The need to relate microanalysis to microstructure is self-evident.

Consideration of the above criteria leads to the conclusion that a transmission electron microscope (TEM) based technique will fulfill many of the needs. X-ray microanalysis is the most popular and straightforward technique for elemental detection. However the Li K_α X-ray is so weak (∿52 eV) that it would generate only 14 electron-hole pairs if it could get into a solid state energy-dispersive spectrometer. Low fluorescence yields ($<10^{-5}$) and absorption within the specimen and the detector itself make the possibility of Li K_α X-ray detection remote in the foreseeable future. Likewise, even if a crystal spectrometer were interfaced to a TEM a crystal with a d-spacing of ∿23 nm would have to be created. Currently this is impossible. Therefore at Lehigh we have been pursuing two other TEM-based techniques: direct Li measurement using ionization electron energy loss spectrometry (EELS) and indirect Li detection using convergent beam electron diffraction (CBED) patterns. This paper will report primarily on the capabilities of these two techniques and the current technological limitations. In addition prospects for the revitalization of the plasmon EELS technique (8) will be discussed.

EELS EXPERIMENTAL TECHNIQUES

EELS is the measurement of the energy distribution of electrons emerging through a thin foil TEM sample. The energy spectrum contains information about inelastic electron-specimen interactions. Specifically at energy losses >∿50 eV detection of individual ionization events is possible, giving rise to identification of the presence of individual atoms in the sample. In the regime <50 eV, detection of changes in the electron density in the specimen is possible through plasmon peak shifts. The electron density can sometimes be related empirically to the composition of the sample. Both these techniques have the necessary spatial resolution (<50 nm) but both have other drawbacks, as discussed below.

EELS RESULTS AND DISCUSSION

a) Ionization EELS

The Li K-shell electrons are ionized by the input of ≧54 eV. At this point in the energy loss spectrum a discrete intensity change (termed an ionization edge) should be observed. Recent work by Chan and Williams (9) and Sainfort and Guyot (10) has demonstrated that the Li K edge is detectable when isolated, through-thickness δ' particles are examined. The Li K edge is not sharp and well defined and it is not very visible because there is a high background intensity in this low energy region of the spectrum. The background is due to multiple scattering of electrons, i.e., the combination of several different inelastic processes.

A typical spectrum from δ' is shown in Figure 1a. In order to quantify the spectrum it is necessary to extrapolate the rapidly decreasing background intensity underneath the Li K edge over an integration window Δ∿20 eV as shown in Figure 1b. Then the intensity in the K edge ($I_{Li(\beta,\Delta)}$) can be related to the number of Li atoms per sq.cm. of the specimen (N_{Li}) through the relationship (11):

$$N_{Li} = \frac{I_{Li(\beta,\Delta)}}{I_{O(\beta,\Delta)}} \cdot \frac{1}{\sigma_{K_{Li}}(\beta,\Delta)} \qquad (1)$$

where β is the energy loss spectrometer entrance semi-angle, I_O is the intensity in the zero loss region of the spectrum and σ_K is the partial ionization cross section (cm^{-2}) for the Li K shell electrons.

As shown in Figure 1 there is also an ionization edge due to the presence of Al atoms in the δ'. The Al $L_{2,3}$ edge occurs at ∿72 eV and integration over 20 eV from this values gives $I_{Al(\beta,\Delta)}$. It is possible then to ratio the number of Al atoms (N_{Al}) to Li atoms in the δ':

$$\frac{N_{Al}}{N_{Li}} = \frac{I_{Al(\beta,\Delta)}}{I_{Li(\beta,\Delta)}} \cdot \frac{\sigma_{K_{Li(\beta,\Delta)}}}{\sigma_{L_{Al(\beta,\Delta)}}} \qquad (2)$$

The ratio obtained was 3.6/1 (i.e., $Al_{3.6}Li$) which corresponds to ∿22 at% Li in δ' produced by aging at 300°C. This point is reproduced on the experimental phase diagram in Figure 2. The error bars of ± 20% relative represent the combination of errors in the partial ionization cross section, the background intensity extrapolation and the counting statistics. The result indicates a decrease in δ' stoichiometry with increasing temperature in agreement with the small angle scattering data of Cocco et al. (1), but in disagreement with the recent phase diagram calculations of Sigli and Sanchez (12).

It can be shown (9) that given the low peak to background ratio of the Li K edge, a minimum Li detectability of ∿3 wt% is reasonable, using EELS with a magnetic prism spectrometer and LaB_6 electron source. From this investigation it can be concluded that a) Li is detectable in δ' using ionization EELS; b) the accuracy of quantification is low, c) there is evidence for a decrease in δ' stoichiometry as temperature increases and d) the minimum detectability of Li in Al is ∿3 wt% (∿10 at%).

The last conclusion means that using the current capabilities of ionization EELS, quantitative analysis of Li in solution in Al, or segregated into precipitates is not possible if <3 wt% Li is present. Therefore the analytical sensitivity of the technique must be improved before it becomes practically useful, since most commercial Al-Li base alloys have <3 wt% Li at present. Current research is emphasizing the development of deconvolution routines in order to remove the multiple scattering contribution to the energy loss spectrum and thus extract the so-called 'single

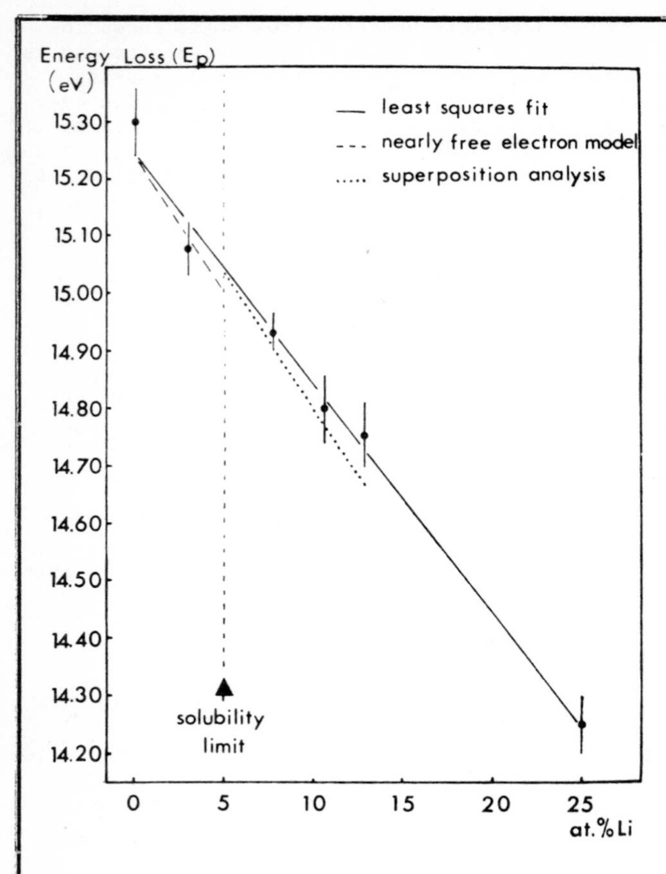

3. Calibration data for the change in Al plasmon energy loss as a function of Li content (13).

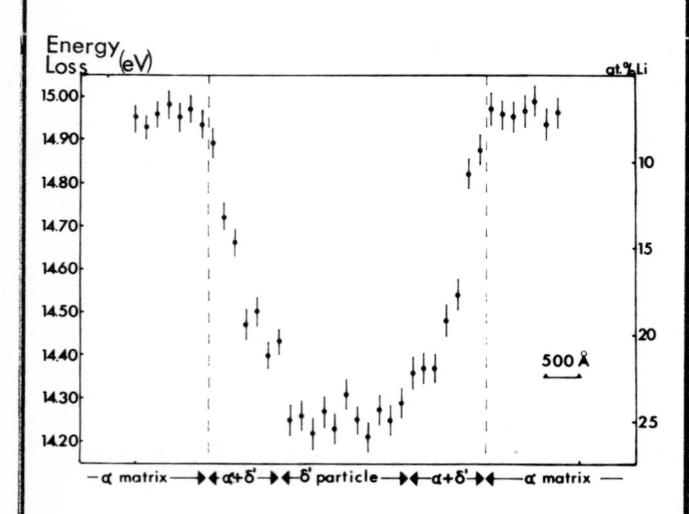

4. Change in plasmon energy loss and inferred change in at% Li across a large δ' particle in Al-8 at% Li aged 100 hrs. at 240°C.

scattering' Li K ionization edge and Al $L_{2,3}$ ionization edge.

b) Plasmon EELS

An alternative method of detecting the presence of Li in Al using EELS involves the indirect microanalysis technique of plasmon energy loss analysis. This technique flourished in the early 1970's (3) and demonstrated a clear ability to detect small changes in Li composition in Al on a scale of <10 nm. Plasmon loss analysis is most easily performed in specimens with a high free electron density (n) such as Al, which gives rise to several well defined Gaussian peaks in the low energy (<50 eV) region of the spectrum. The position of these peaks in terms of the energy loss ΔE is given by:

$$\Delta E = \hbar\left(\frac{ne^2}{\varepsilon\ m}\right)^{\frac{1}{2}} \qquad (3)$$

where $2\pi\hbar$ = Planck's constant, e and m are the electron charge and mass respectively and ε is the permittivity of free space. Thus, if the free electron density (n) changes, e.g., by alloying, then ΔE changes. Using calibration specimens of known Al-Li composition it is possible to generate the calibration curve shown in Figure 3 (13). From this calibration curve shifts in ΔE can be related to the percent Li and an example of the Li composition variation across a large isolated δ' particle is shown in Figure 4. This technique therefore can clearly register small (<1 wt%) changes in Li content in Al on a sub-10 nm scale. The technique has been used to demonstrate Li depletion around grain boundary δ' PFZs (13) and discrete composition changes at the advancing discontinuous reaction front (14).

Plasmon loss spectroscopy has not seen wider use because a) it requires a high resolution, high-dispersion spectrometer usually of the retarding field type (15) which is not available commercially; b) interpretation of plasmon peak shifts is an empirical process requiring the creation of binary standards. There are problems in extending interpretation of peak shifts to results from ternary or higher order alloys; and c) data generation and analysis are extremely time consuming.

However current developments in analytical electron microscopy may result in a resurgence of this technique. For example, recent results (10) using a conventional low dispersion magnetic prism spectrometer, but with a high resolution field emission source have reproduced the early plasmon peak shifts in the δ' phase. Detection of <3 wt% shifts were not shown however, Other techniques such as high resolution X-ray analysis in the AEM can detect all the other ternary additions (e.g., Mg, Cu, Zr) to commercial Al-Li alloys thus permitting any effects of these elements on the plasmon peak position to be accounted for. Finally advances in computer control and on-line spectral manipulation may permit more rapid handling of the plasmon spectrum. Plasmon energy loss spectroscopy, in conjunction with X-ray microanalysis and advanced computer control is thus considered a feasible technique for

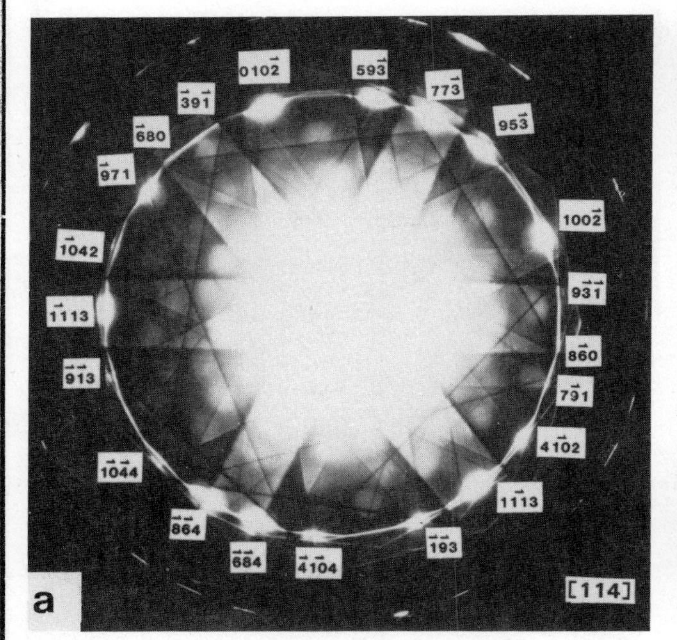

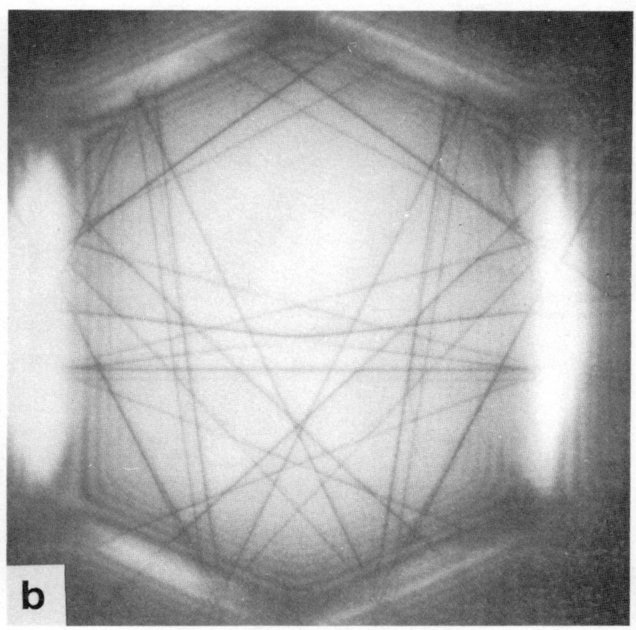

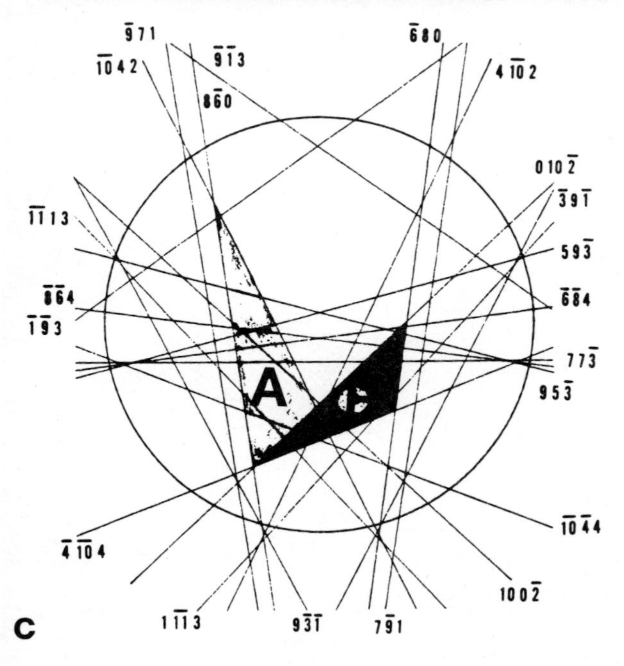

5. a) Low camera length CBDP from pure Al showing HOLZ ring of intensity, and Kikuchi bands radiating from the overexposed central region.
   b) Correctly exposed central region of Figure 5a showing defect HOLZ lines.
   c) Computer simulation of the HOLZ line pattern in Figure 5b.

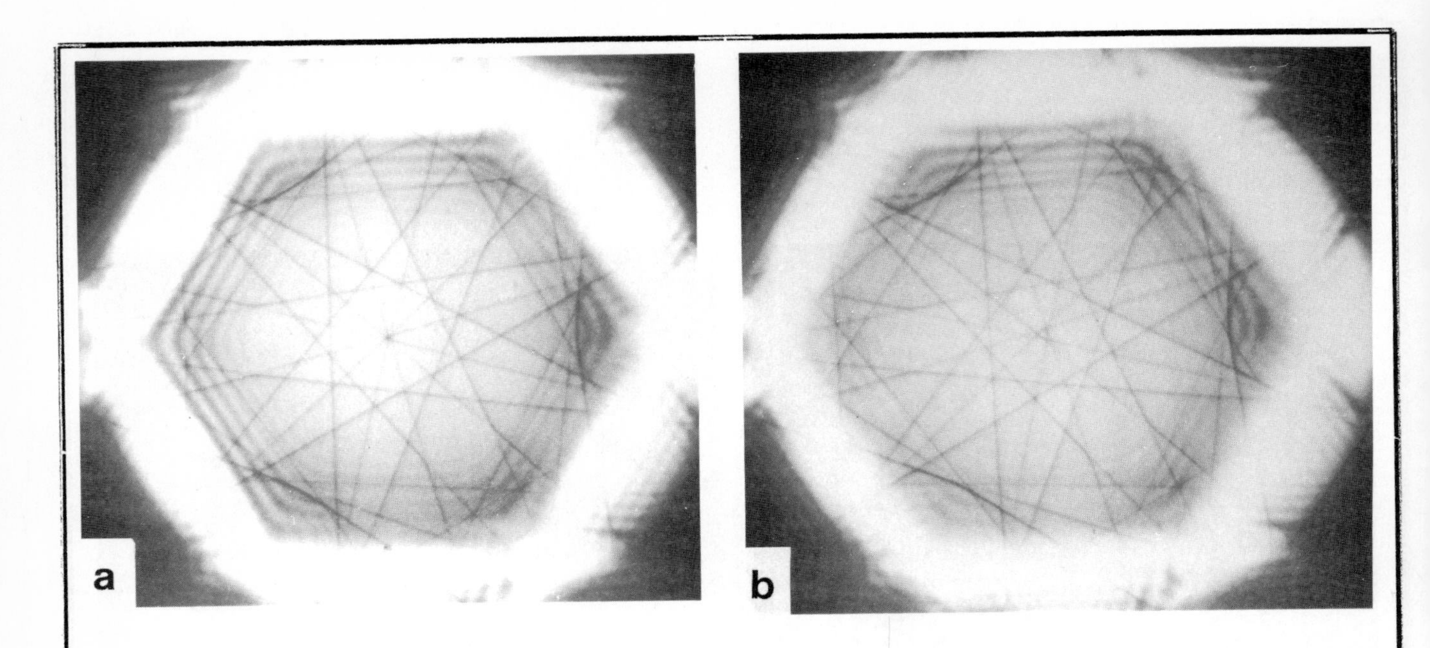

6. The change in HOLZ line positions as a function of accelerating voltage - a) 120 kV, b) 119.2 kV.

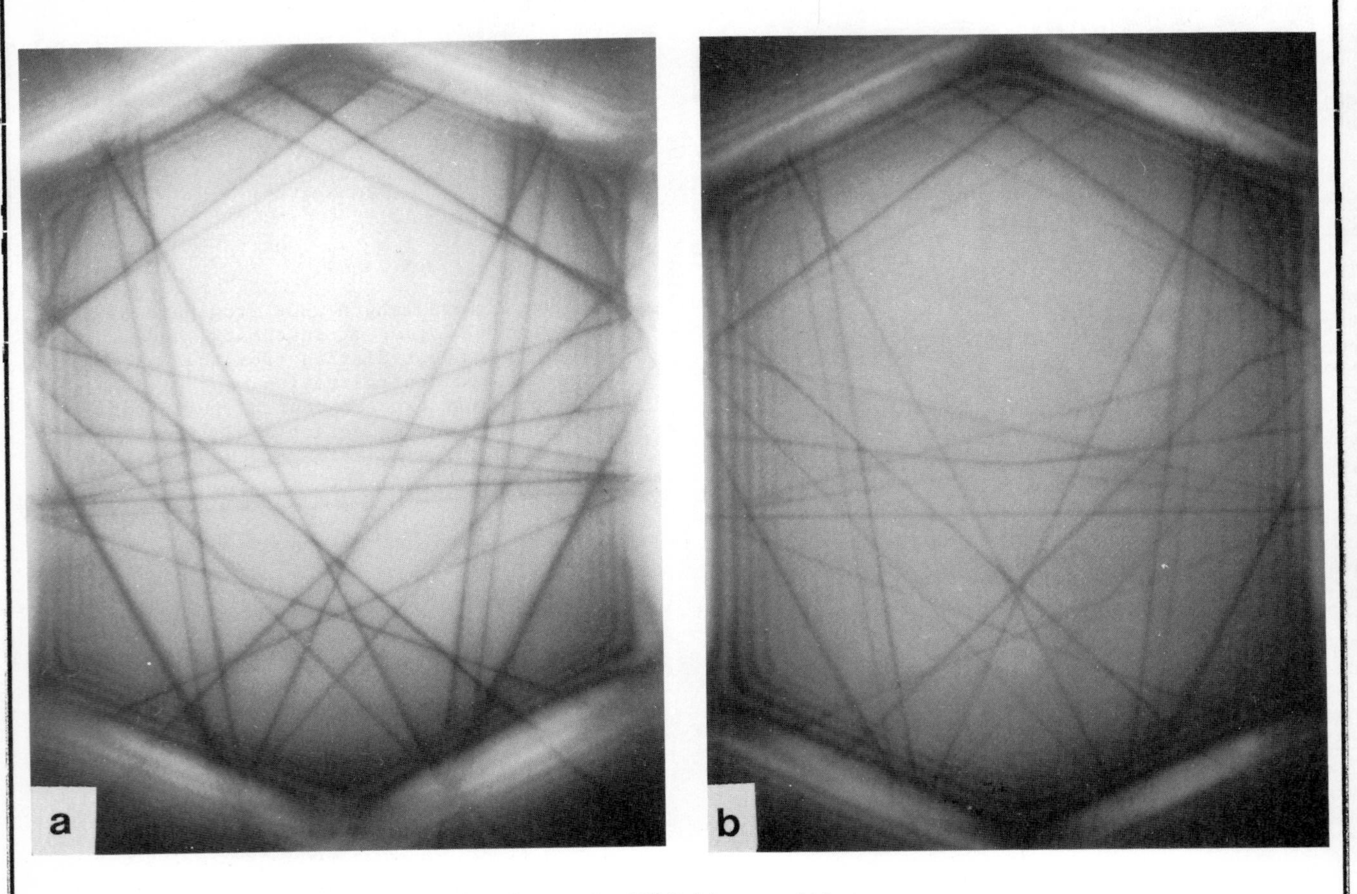

7. The change in HOLZ line positions as a function of temperature - a) -187°C, b) -49°C.

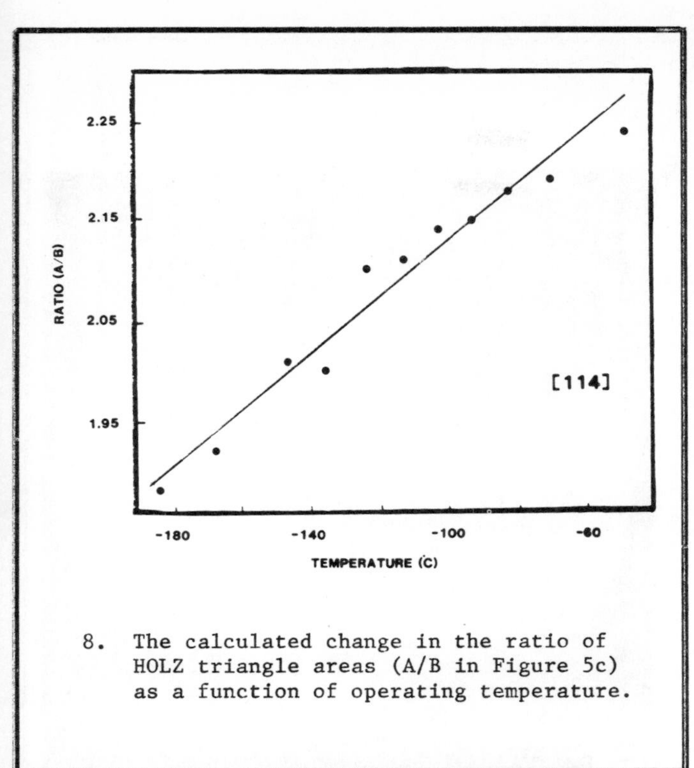

8. The calculated change in the ratio of HOLZ triangle areas (A/B in Figure 5c) as a function of operating temperature.

quantitative determination of the Li distribution at levels <3 wt% and on a scale of <100 nm.

CONVERGENT BEAM DIFFRACTION PATTERNS (CBDPs) EXPERIMENTAL TECHNIQUE

The second, indirect, method of detecting the presence of Li in Al involves the study of CBDPs from thin foil specimens. CBDPs are formed when a convergent beam of electrons is diffracted by a thin single crystal region of the sample. Figure 5a shows a low camera length (400 mm) CBDP of pure aluminum indexed as zone axis [UVW] = [114] at 120 kV. The central over-exposed region consists of the conventional zero order Laue zone (ZOLZ) spot pattern from {hkl} planes where hu + kV + lW = 0, corresponding to the region of reciprocal space usually observed in conventional selected area patterns in the TEM. Each spot is expanded to a disc because the electron probe is convergent. The pairs of lines radiating from the central region are Kikuchi bands. The bright ring of intensity is due to electrons diffracted from higher order Laue zone (HOLZ) planes in which hu + kV + lW = n (n > 0).
Through each HOLZ diffraction maximum is a bright line termed a HOLZ line which arises from elastic scatter of the part of the convergent beam of electrons which is at the exact Bragg angle. (These lines are the elastic analog of Kikuchi lines.) Paired with each bright (excess) line in the HOLZ ring is a parallel dark (defect) line which is visible in a correctly exposed image of the 000 disc in the ZOLZ and this is shown in Figure 5b. Figure 5c is an indexed example of the [114] HOLZ pattern as simulated by a computer (16).

The acquisition of CBDPs from aluminum alloys is difficult due to thermal diffuse scattering which destroys the HOLZ line contrast. However, if the aluminum specimens are cooled, the HOLZ line detail begins to be visible in the transmitted 000 disc. All CBDPs in this study were obtained with the specimen at -187°C using a double tilt cooling holder.

The positions of the HOLZ lines are a sensitive function of lattice parameter and temperature (17) and therefore, using suitable standards and/or computer simulations it is possible to obtain composition information, because of the change in lattice parameter with alloying. This technique has been used to measure lattice parameters (18) and precipitate/matrix misfit parameters (19) to an accuracy of 2 parts in 10^4 (20). In order to use the HOLZ line technique for absolute lattice parameter determination, it is necessary to calibrate the technique with a specimen of a known lattice parameter, a. In this study we used pure aluminum (a = 0.4049 nm at 20°C).

The electron energy has a significant effect on the position of the HOLZ lines. Therefore the value of the accelerating voltage has to be calibrated and this can be carried out using a continuous high voltage control unit. Annealed pure aluminum specimens give easily recognizable HOLZ line patterns from the [111] zone axis orientation at 120 kV. Figure 6 shows the sensitive changes of these HOLZ lines to changes in accelerating voltage. Thus it is possible to pre-set the continuous high voltage control to exactly 120 kV using the pure Al [111] standard pattern, and maintain this value during studies of Al-Li samples.

Using computer simulations of HOLZ line positions for given alloy compositions at exactly 120 kV and measuring the relative changes in the area of certain HOLZ line triangles (e.g., A and B in Figure 5c), we can predict the lattice parameter changes in aluminum as a function of lithium content. The computer predictions can then be compared with experimentally obtained patterns from Al-Li samples in order to determine the lattice parameter of the sample.

It was necessary to determine the exact relationship of the lattice parameter with temperature. Therefore, the HOLZ line technique was used to measure lattice parameters of annealed pure aluminum with changing temperature in the cooling holder. Figures 7a and b show the HOLZ line patterns of pure aluminum at -187°C and -49°C respectively with corresponding computer-simulated patterns showing that the best match occurs for lattice parameters of 0.4031 nm and 0.4042 nm. The relation between the HOLZ line triangle areas A/B and operating temperature in pure aluminum is shown in Figure 8. It reflects the linear decrease in the lattice parameter of aluminum with decreasing temperature. There is reasonable agreement between the literature values and the present data. The HOLZ pattern was obtained at -187°C which through thermal contraction results in a lattice parameter (a_o) of 0.4031 nm according to the coefficient of linear thermal expansion:

$$\alpha = (a - a_o)/(a_o \cdot \Delta T) \quad (4)$$

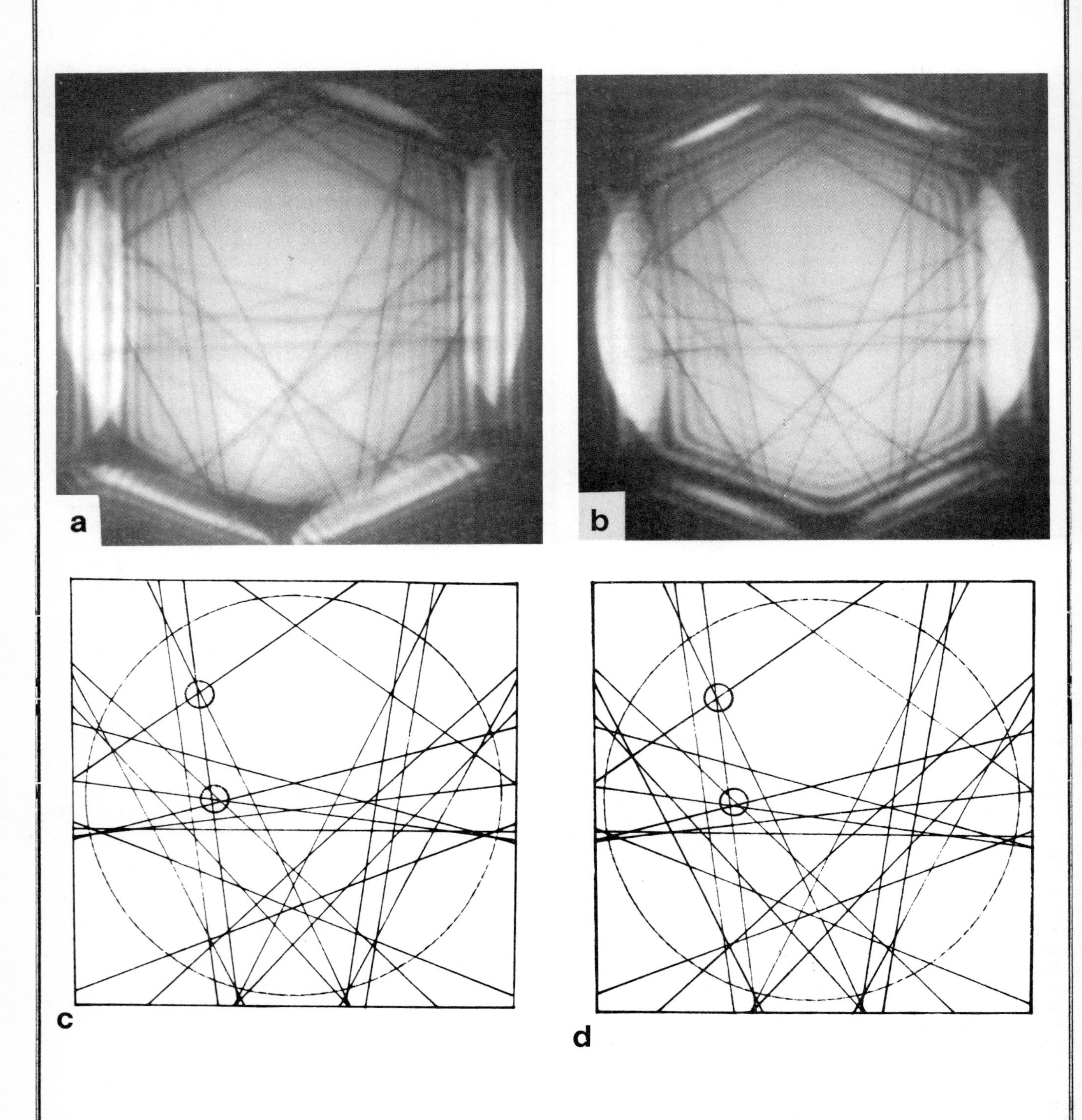

9. HOLZ line patterns from δ' produced by aging at a) 270°C, b) 240°C. The corresponding computer-simulated patterns are shown in c) and d).

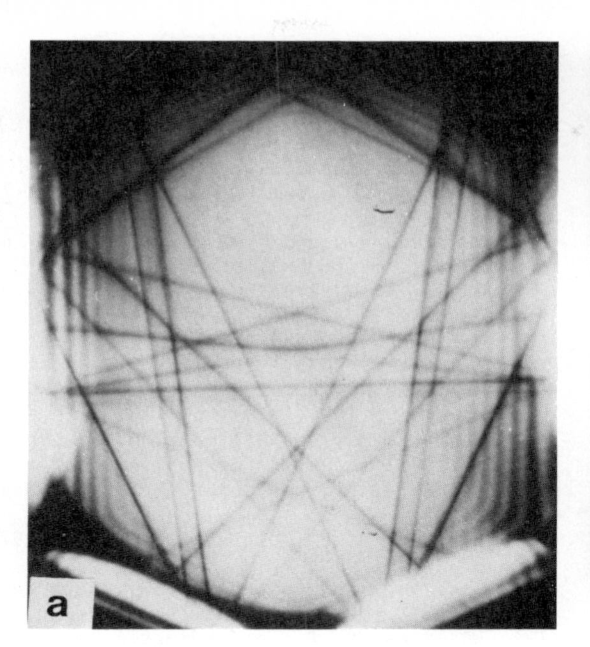

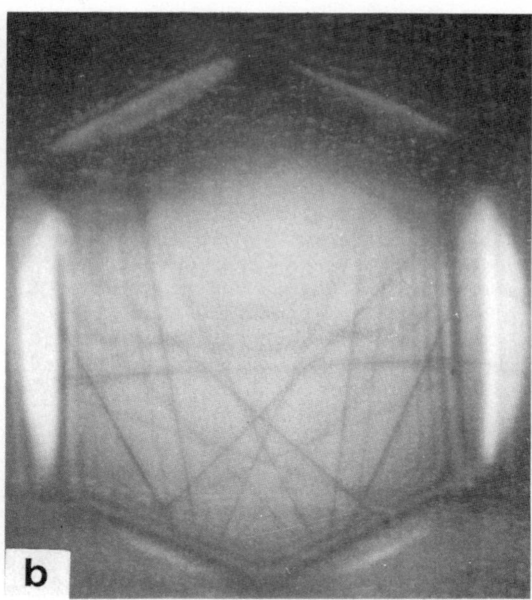

10. a) HOLZ line pattern from α matrix aged at 210°C, b) HOLZ line pattern from δ' aged at 210°C.

where $\alpha = 22 \times 10^{-6}$ cm/°C for pure aluminum, ΔT = 207°C and a = room temperature lattice parameter of pure Al (0.4049 nm). This value of lattice parameter was used in simulations of the absolute lattice constant of Al-Li alloys, as a function of temperature.

CBDP - RESULTS AND DISCUSSION

The sensitivity of the HOLZ line position to the lattice parameter of the specimen makes composition measurement possible. As an example of the changes in the HOLZ line positions with lithium contents we have examined CBDPs from δ' as a function of aging temperature. Specimens of Al-10 at% Li were aged at 270°C, 240°C and 210°C for times sufficient to generate coarse through-thickness δ'. Foil thickness measurement was performed by measuring the spacing (21) of the Kossel-Möllenstedt fringes in a CBDP under two-beam conditions (21). This was necessary to ensure a smaller thickness value than the diameter of the δ' precipitates. Then, a nominal 40 nm probe was positioned on the middle of the δ' precipitates (whose diameters were usually larger than 100 nm). It was easy to position the probe on the middle of the δ' because severely distorted HOLZ lines were observed around the edge of precipitates, due to lattice strain in the surrounding matrix.

According to the Al-Li phase diagram of Cocco et al. (1), Al_3Li should show an increase in lithium content with decreasing aging temperature and therefore the lattice parameter should change. Figures 9a and 9b show [114] HOLZ patterns taken from δ' in specimens aged at 270°C and 240°C, along with the corresponding computer-simulated patterns. Examination of the HOLZ patterns shows that the variation in HOLZ line position is consistent with a decrease in the lattice parameter of δ' precipitates with decreasing aging temperature. The results are summarized in Table I which lists the δ' lattice parameter, measured at -187°C, for δ' formed at 270°C, 240°C and 210°C. The room temperature lattice parameter in Table I was calculated assuming that the δ' obeys the same thermal expansion coefficient as pure Al. However, if the δ' lattice parameter is converted into a value of % Li by extrapolating the Al-Li solid solution lattice parameter data of Kellington et al. (22) then the results are clearly in error, since an enormous range of Li content in δ' is predicted. Therefore, either δ' does not obey the same thermal expansion behavior as pure Al, and/or it is not valid to extrapolate Al-Li solid solution lattice parameters (<6 at% Li) to the Al_3Li composition (25 at% Li). Both these problems are currently under investigation. However the results can still be considered to indicate a change in δ' lattice parameter as a function of aging temperature and this therefore implies a varying δ' stoichiometry as a function of temperature.

The lattice misfit ε between the α matrix and the δ' precipitates is possible to determine according to the following equation:

$$\varepsilon = (a_{\delta'} - a_{\alpha})/a_{\alpha} \quad (5)$$

Thus from CBDP measurements of lattice parameter it is possible to obtain an estimate of the α/δ' misfit. Because the δ' studied by this technique has to be through-thickness (strain distorts and diminishes the HOLZ line contrast), all measurements are of unconstrained δ'.

Figure 10 shows CBDPs of the α matrix and δ' precipitates (produced at 210°C). From corresponding computer-simulated patterns the unconstrained lattice parameters of the matrix and Al_3Li were determined to be 0.40285 nm and 0.40051 nm respectively at -187°C. The unconstrained lattice misfits are shown in Table I. These values should be compared with a value of -0.3% reported by Tamura et al. (23) using X-ray diffraction; and the constrained misfit of ~-0.09% (3). Before the CBDP value can be unambiguously interpreted however it must be demonstrated that no factors other than Li content are contributing to the HOLZ line shifts. The calculated misfits are also subjected to the same problems as described above for the calculated Li compositions of δ'.

CONCLUSIONS

1) Ionization EELS can detect Li directly and quantitatively on a sub-50 nm scale, but its detectability limit must be improved before it can be routinely applied to commercial Al-Li alloys.

2) Plasmon EELS can detect Li indirectly and quantitatively with the necessary spatial resolution and sensitivity to be useful in studies of commercial alloys. The technique should therefore be developed to make it more routine and reproducible.

3) HOLZ line positions in CBDPs can be used to infer changes in Li composition on a sub-50 nm scale. There is no information on the minimum Li detectability as yet and the problem of other contributions to HOLZ line shifts must be clearly resolved before this technique can be applied with confidence.

ACKNOWLEDGEMENTS

DBW and HMC were supported through NSF Grant DMR 84-00427. CMS was supported through an ALCOA Foundation Grant. We are happy to acknowledge the generosity of Dr. Roger Ecob (C.E.G.B. Berkeley) who provided the computer program used to simulate the HOLZ lines.

REFERENCES

(1) Cocco, G., Fagherazzi, S., and Schiffini, L., J. Appl. Cryst., 10 (1977) p. 325.
(2) Baumann, S. F., and Williams, D. B., Acta Metall., 33 (1985) p. 1069.
(3) Williams, D. B., and Edington, J. W., Metal Science, 9 (1975), p. 529.
(4) Gu, B. P., Liedl, G. L., Kulwicki, J. H., and Sanders, T. H., Mat. Sci. and Engrg. 70 (1985) p. 217.
(5) Jensrud, O., and Ryum, N., Mat. Sci. and Engrg., 64 (1984), p. 229.
(6) Gayle, F. W. and Vander Sande, J. B., Scripta Metall. 18 (1984), p. 473.
(7) Makin, P. L., and Ralph, B., J. Mat. Sci. 19 (1984), p. 3835.
(8) Williams, D. B., and Edington, J. W., J. Microsc., 108 (1976), p. 113.
(9) Chan, H. M., and Williams, D. B., Phil. Mag. A (1985) accepted for publication.
(10) Sainfort, P., and Guyot, P., Phil. Mag. 51A (1985), p. 575.
(11) Egerton, R. F., Ultramicroscopy 3 (1978), p. 243.
(12) Sigli, C. and Sanchez, J. M., Acta Metall. (1985), p. 243.
(13) Williams, D. B., and Edington, J. W., Phil. Mag. 30 (1974), p. 1147.
(14) Williams, D. B., and Edington, J. W., Acta Metall., 24 (1976), p. 323.
(15) Metherell, A. J. F., Recent Adv. Optical, Electron Microsc., 4 (1971), p. 263.
(16) Ecob, R. C., Shaw, M. P., Porter, A. J., and Ralph, B., Phil. Mag., 44A (1981), p. 1117.
(17) Steeds, J. W. "Introduction to Analytical Electron Microscopy" (1979) Plenum Press, NY, p. 387.
(18) Shaw, M. P., Ecob, R. C., Porter, A. J., and Ralph, B., Quantitative Microanalysis with High Spatial Resolution (1981), The Metals Society, London, p. 229.
(19) Ecob, R. C., Ricks, R. A., and Porter, A. J., Scripta Metallurgica, 16 (1982), p. 1085.
(20) Steeds, J. W., Quantitative Microanalysis with High Spatial Resolution (1981), The Metals Society, London, p. 210.
(21) Kelly, P. M., Jostsons, A., Blake, R. G., and Napier, J. G., Phys. Stat. Sol., A31 (1975), p. 771.
(22) Kellington, S. H., Loveridge, D., and Titman, J. M., Brit. J. Appl. Phys. 2 (1969), p. 1162.
(23) Tamura, M., Mori, T., and Nakamura, T., J. Jap. Inst. Met., 34 (1970), p. 919.

Characterization of lithium distribution in aluminum alloys

T MALIS

The author is in the
Physical Metallurgy Research Laboratory,
Department of Energy, Mines and Resources,
568 Booth Street, Ottawa,
Ontario K1A OG1, Canada

SYNOPSIS

The detection of Li with high spatial resolution in Al alloys by electron energy loss spectroscopy (EELS) and transmission electron microscopy requires careful consideration of non-standard specimen preparation, such as the use of ultramicrotomed sections, and the method of EELS data processing. It is then possible to obtain quantitative data on Al-Li compounds or solutions and at least semi-quantitative data from more complex Al-Li-X phases.

INTRODUCTION

When a new series of alloys enters the marketplace, it is inevitable that chemical characterization of the microstructure will be useful at some point. It is not always easy to achieve this goal given the fine scale of the microstructure and the multi-element precipitates common to many commercial alloys. The use of an analytical electron microscope (AEM) equipped with an energy dispersive X-ray detector (EDX) has been used extensively for this task in recent years. In cases such as that of ceramics, metal carbonitride precipitates in microalloyed steels, or Li in Al alloys, even the new "windowless" (light element) X-ray detectors may be of limited use. Thus, as pointed out recently by Sung et al. (1), an AEM with a good EELS capability is the only viable technique, certainly for the detection of Li in Al alloys.

This detection is by no means straightforward, since the necessity for specimen thicknesses of 40 nm or less (1) permits examination of only a small fraction of the electron transparent area in a normal electropolished thin foil. Moreover, even after careful background fitting, the achievement of elemental count values from EELS spectra of Al-Li alloys is complicated by a large overlap between the Li-K ionization edge and the Al-L edge (Fig. 1). Figure 1 also points out the need to eliminate the thickness-dependent Al plasmon peaks and the added overlaps if Zr, Fe or Cu are present.

One means of controlling specimen thickness and producing large areas suitable for EELS is that of slicing thin sections with an ultramicrotome equipped with a diamond knife (2). The purpose of this paper is to elaborate on the advantages of ultramicrotomed (UM) sections for Li detection, to check on the ability to produce quantitative data from EELS spectra containing Li and Al, and to ascertain to what extent quantification can be applied to commercial alloy microstructures.

EXPERIMENTAL PROCEDURE

Both UM sections and electropolished thin foils of an Al-3 wt% Li binary, an Al-3Li-0.12Zr ternary and the 8090 alloy (composition 0.7 Mg, 2.5 Li, 1.2 Cu, 0.12Zr) were obtained from Alcan International Ltd. Sections of varying thickness were sliced by a Reichart-Jung Ultracut E microtome into a water bath, then collected on Cu or Ni grids, a few of which had Formvar films covering them for greater section support.

Sections and foils were examined at 120 KeV in a Philips EM400T equipped with an EDAX

Table 1. Al-Li Binary Quantification
- small window Δ= 17eV, large window Δ= 80eV for common bkg

(A) UM section, t = 55 nm, analysis diam.=0.5μm large window (11 analyses) - 11.9 ± 3.8 at %

(B) UM section, t = 40 nm, analysis diam.=0.5μm
large window (5 analyses) - 11.6 ± 3.4%
small window (same pts) - 13.3 ± 6.2%

(C) UM section, t = 30 nm, analysis diam.=4μm
large window (5 analyses) - 11.9 ± 3.3%
small window (same pts) - 11.65 ± 2.2%

(D) UM section (δ′-ppts), t = 30 nm
large window (5 ppts) - 28.4 ± 4.6%
small window (same ppts) - 21.8 ± 1.4%

Table 2. Al-Li-X Precipitates (Semi-Quantitative Estimates)
- large window (Δ= 80eV) only
- all amounts / ratios in atomic %

(A) "Donut" Configurations - core analysis
- #1-4, of peak-aged 8090, foil, t = 40 nm
- #5&6, Al-Li=Zr, section, t = 40 nm
- rings consistently showed 13-17 at % Li

	Zr/Al(EDS)	Li/Al(EELS)	Zr/Li
#1	0.036	0.47	0.076
2	0.089	0.47	0.19
3	0.105	0.2	0.525
4	0.033	0.21	0.16
5	--	0.15	--
6	--	0.18	--

(B) T-Phase (overaged 8090)
- #1 - thin foil, #2to7 - UM sections

	t(nm)	est.Cu (EDS)	est. Li (EELS)	Total Cu+Li
#1	20	13	41	54
2	65	9	21	30
3	65	10	22	32
4	50	12	35	47
5	40	14	20	34
6	40	13	29	42
7	40	13	37	50

(C) Constituent Particles
- EDS gives Mg/Al=0.015, Cu/Al=0.25, Fe/Al=0.35
- EELS indicates Li present

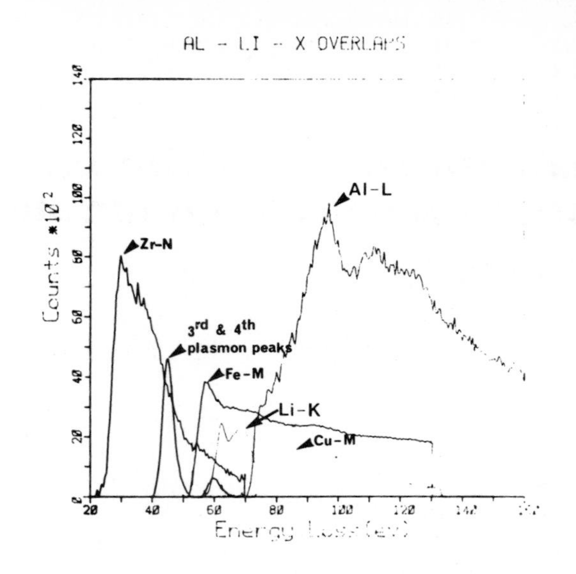

Figure 1. Composite of the elemental ionization edges in a Li-bearing Al alloy which may be found in EELS spectra.

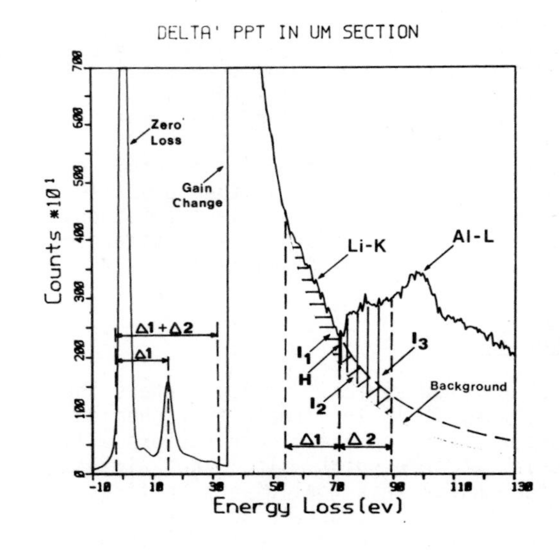

Figure 2. Spectrum and background from $LiAl_3$ in a 30 nm thick UM section showing nomenclature for processing.

X-ray detector, a Gatan 607 energy loss spectrometer and the necessary data processing software. It is worth noting that the latter included the latest, relativistically-corrected elemental partial cross-sections for EELS quantification (3). Serious errors can arise for some edges if non-relativistic cross-sections are used.

Relative specimen thickness as a fraction of the mean inelastic scattering length (λ) was obtained with the EELS, where any quantification should be limited to values of $t/\lambda < 1$ (4). Absolute values of specimen thickness were obtainable using values of λ (Al) obtained experimentally (5).

The most common EELS collection condition was focussed-beam TEM mode with the collection semi-angle (β) = 5.6 mr (a 30 μm objective aperture). Thus the incident beam semi-angle was of the order of 2 mr or less. For larger areas, the beam was defocussed ($\alpha \sim 0$) and spectra collected in Diffraction mode, but with the objective aperture still defining the collection angle.

For high spatial resolution with high signal, a focussed nanoprobe beam of 20 nm (α= 10 mr) was used, with β = 19 mr in TEM mode to avoid any complications due to the convergence factors needed for partial cross-sections if $\alpha > \beta$ (1).

The Al-3wt% Li binary had a bulk ICP analysis of 11.2 at % Li and was examined after aging 12 h at 325°C to produce ($LiAl_3$) precipitates of 20-50 nm in diameter. The ternary alloy was given the same treatment. The 8090 alloy was examined in both the peak-aged and overaged condition (24 h at 190°C and 12 h at 325°C, respectively).

Spectra normally were collected at 0.5eV/ch to view both the Li-K and Al-L edges (at 54 and 71 eV, respectively) along with the C-K edge at 284 eV, should contamination have occurred. Initial calibration was performed on the zero loss peak and the O-K edge at 532eV from spectra collected at 1eV/ch (the O-K was present due to the inevitable oxide layer on all specimens). Spectra at 1eV/ch were sufficient for detection of Fe-L at 708eV, but 2eV/ch was needed to detect Cu-L at 940eV. In thicker specimens where the Al plasmon at 45eV began to limit the background-fitting region, spectra collected at 0.2eV/ch resulted in an expansion of the region so as to meet the software requirements for background-fitting. As determined by Sung et al. (1), an Ae^{-R} background fit was used in most cases, but spectral fluctuations often meant that a segmented log-polynomial fit was preferable.

Since the extrapolation of the background to higher energies could not be observed at 0.5 or 0.2eV/ch (normally a good check on background validity), two further points were applied for the choice of best background fit. They were that the Li-K edge reach a maximum at 60-65eV, then decline to perhaps one-half to two-thirds of the peak value before disappearing under the Al-L edge at 71eV. This is characteristic of K-shell edge behaviour, as can be seen in spectral libraries (6,7) or the Li-K edge from LiF used for Fig. 1.

EDS X-ray analyses were performed in a low background holder at a detector take off angle of 40°, with experimental K-factors used in the quantification. For Cu-bearing precipitates in UM sections on Cu grids, a "hole count" from the adjacent Cu-free Al matrix was subtracted. Some sections were examined on Ni grids, which necessitated the deconvolution of Cu-Kα from Ni-Kβ via the EDS software.

RESULTS

The uniformity of UM sections enabled EELS analysis to be conducted on reasonably large areas of the Al-Li binary so that the bulk composition could be used as a standard. The purpose of this was to examine two methods of handling the Li/Al overlap. In what might be termed the "small window integral" method used by Sung et al. (1), the Li contribution (I_2) to the Al-L integral (I_3) (see Fig. 2) can be estimated from integrals of width $\Delta 1$ and ($\Delta 1 + \Delta 2$) taken at the zero loss peak (I_0) and the calculated values of Li partial cross-section (σ) for those widths. That is, for a given collection angle,

$$I_2 = I_1 (I_0(\Delta 1+\Delta 2)/I_0(\Delta 1)) \times \sigma\text{-Li}(\Delta 1+\Delta 2)/\sigma\text{-Li}(\Delta 1) \quad (1)$$

where $\Delta 1 = \Delta 2 = 17$eV was used in this study. Then the concentration ratio of Al/Li is given by

$$C\text{-Al/Li} = (I_3 - I_2)/I_1 \times \sigma\text{-Li}(\Delta 1)/\sigma\text{-Al}(\Delta 2) \quad (2)$$

where the cross-sections are determined from Egerton's hydrogenic approximation (8, 9).

An alternative method to utilize 80eV windows ("large window integral" method) was devised as follows. Many background-stripped K-shell edges (from analyses where t/λ was less than one) were examined to determine by what factor the channels located just after the large near-edge fluctuations could be multiplied to account for the slow inverse exponential decay of the hidden portions of the edge. Most K-shell edges showed factors of 0.7 to 0.9 for 80eV windows and the Li-K edge of Fig. 1 gave a factor of 0.8. This was used to estimate a 63eV hidden portion of the Li edge from the last visible Li channels as,

$$\text{Li}(80\text{eV}) = I1 + H \times 0.8 \times \text{channels for 63eV} \quad (3)$$

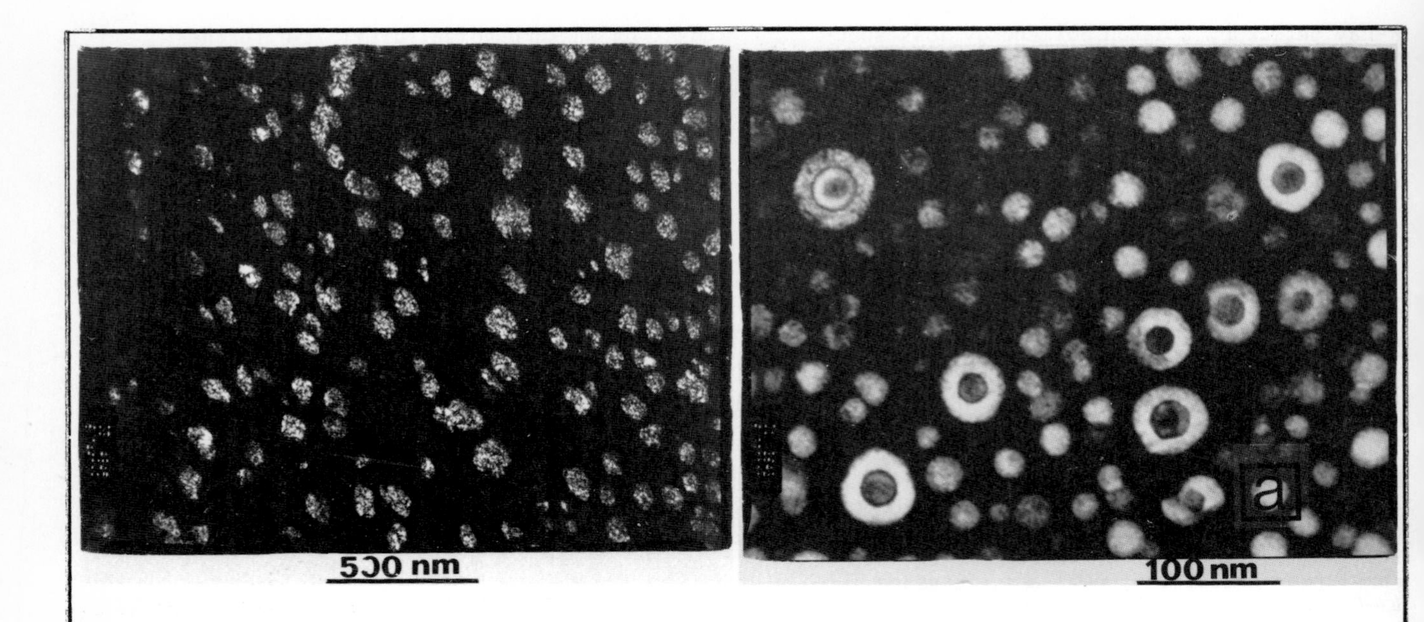

Figure 3. Centered dark field of $LiAl_3$ precipitates in a 30 nm UM section of Al-11.2 at % binary.

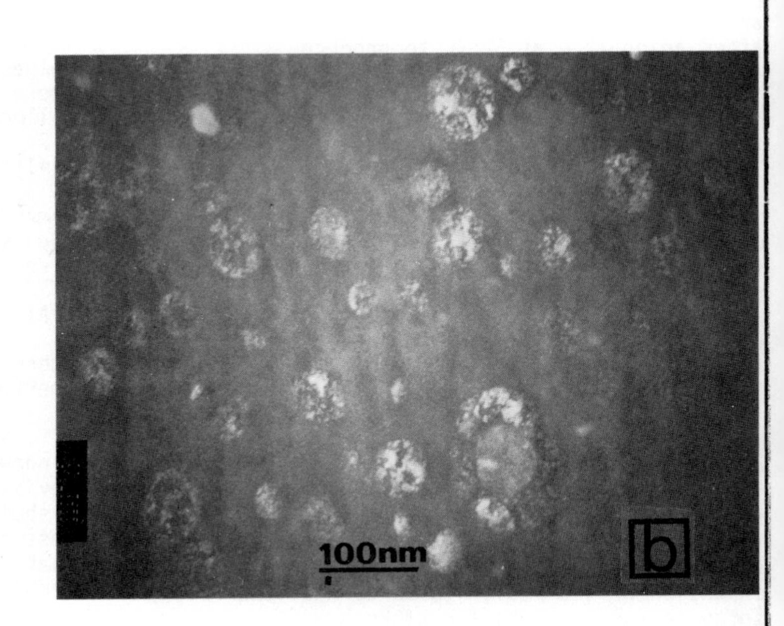

Figure 4. Dark fields of $LiAl_3$ and $ZrAl_3$ precipitates in a) a thin foil of peak-aged 8090, and; b) a UM section of Al-Li-Zr ternary.

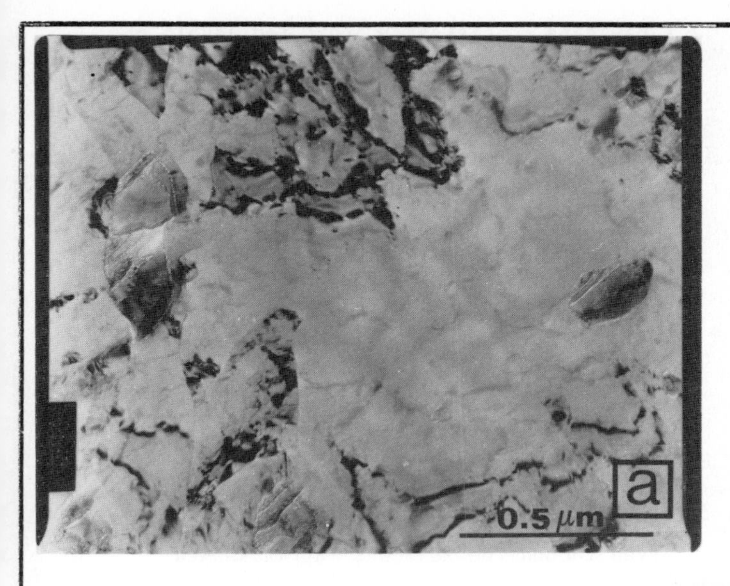

Figure 5. T-phase precipitates in a 40 nm UM section of overaged 8090; a) bright field; b) dark field from conical scan of diffraction pattern.

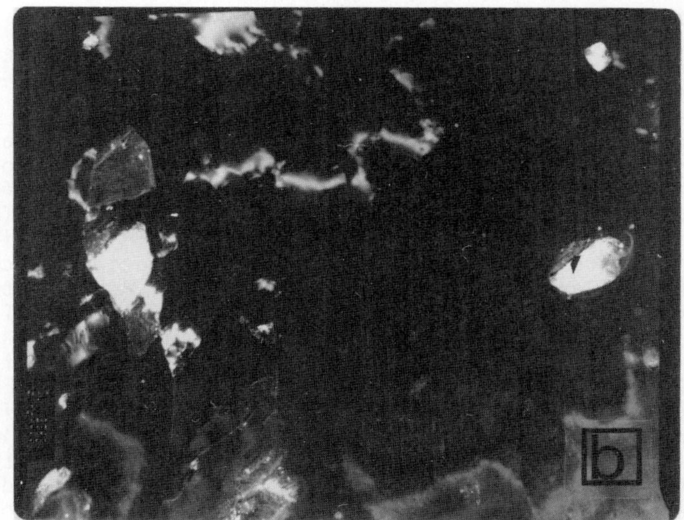

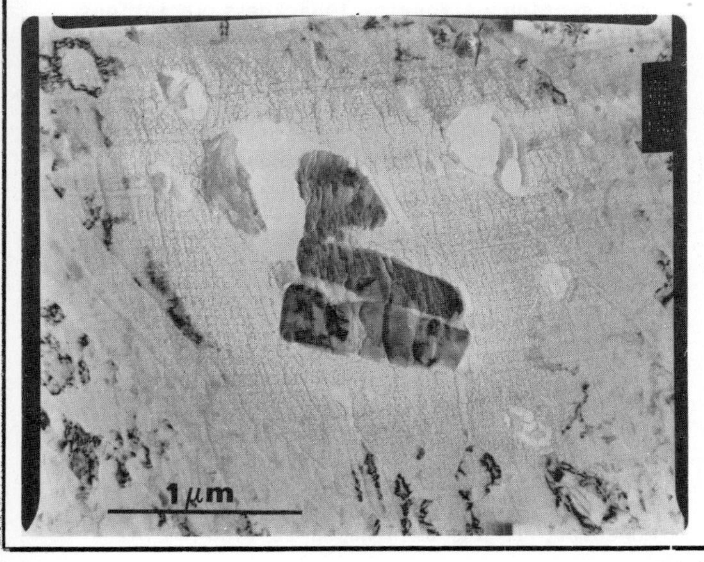

Figure 6. Constituent particle in a 65 nm UM section of overaged 8090. Note oxidized region around particle.

where H = avg. height of last few visible Li channels (Fig. 2). Similarly,

Al(80eV)=Al+Li(80eV)-Hx0.8xchannels for 80eV (4)

The concentration ratios were calculated as per equation (2), except, of course, that partial cross-sections for 80eV integrals were used, cross-sections that should be more accurate than those for 17eV (4,10).

The quantification data from several UM sections of the binary alloy are summarized in Table 1. Included are some analyses of individual $LiAl_3$ precipitates in a 30 nm section which were large enough to likely pass through the entire section (Fig. 3). Note the altered shapes of some precipitates, perhaps from the heavy deformation of the microtoming, and the non-uniformity of the $LiAl_3$ distribution at these low thicknesses.

The 8090 peak-aged alloy showed many of the so-called "donut" configurations (Fig. 4) which have been postulated as being rings of $LiAl_3$ around Li-rich cores of $ZrAl_3$ (11). A few of these co-precipitates near the edge of a peak-aged 8090 thin foil were examined and EDS analyses showed varying Zr in the cores and none in the rings (Table 2). EELS analyses showed varying Li in the cores and approximately 15 at % in the rings. One Al-Li-Zr ternary UM section of similar thickness was examined (Fig. 4b) with similar results. The Zr could not be included in the EELS analysis because none of the Zr cross-sections are known.

Precipitate distribution in the overaged 8090 alloy was monitored quite easily in the UM sections for both the Cu-rich T-phase (Fig. 5) and Fe-rich constituent particles (Fig. 6). Only a few of the former thin enough for EELS analysis were found in thin foils and none of the latter were visible at all in foils. The constituent particles often were surrounded by thinner "oxidized" regions (as in Fig. 6).

Spectra from the T-phase clearly showed Li (Fig. 7) but the Cu-M/Al-L overlap (Fig. 1) presented an additional complication. The overlap is total, hence only an intensity ratio of Li/(Cu+Al) can be obtained. Moreover, M-edge cross-sections are not available from the hydrogenic approximation. However, a plot of hydrogenic cross-sections for L- and K-edges versus edge energy (Fig. 8) shows convergence at the low energy end. Thus, the hydrogenic Li-K partial cross-section for 5.6 mr and an 80eV window is $1.02 \times 10^{-19} cm^2$, while that for Al-L is $1.21 \times 10^{-19} cm^2$. With this in mind, the assumption was made that the Cu-M and Al-L cross-sections were approximately equal. This allowed the Li/(Cu+Al) intensity ratio to be treated as per equation (2) and the Li content of a number of precipitates was estimated as between 20 and 40 at %. (Table 2). At the same time, the EDS X-ray data for Cu/Al indicated Cu contents of 10-15 at %. Mg/Al X-ray data consistently indicated levels of only a few atomic % in all precipitates. This was fortunate in that the amounts were not sufficient to show up on the EELS spectrum, where the Mg-L edge beginning at 50eV would have caused even more problems.

The Fe-bearing constituent particles presented an additional total overlap of Fe-M on Li-K, thus only the intensity ratio of (Fe+Li/Al+Cu) could be obtained from the EELS spectra. If one makes the above assumption that σ(Fe-M) is approximately equal to σ(Li-K) and the assumption that no Li is present, the concentration ratio of Fe/(Cu+Al) from EELS is quite different from that from EDS data. Thus the inference can be made that the constituent particles contain substantial amounts of Li. The apparent oxidized regions around some particles (Fig. 6) indeed showed high O contents, along with Al and substantial C (even though the adjacent matrix showed negligible C-contamination).

DISCUSSION

The use of ultramicrotomed sections appears to be a viable technique of specimen preparation, not only for EELS analysis, but for large-scale TEM examination in general. This was true in particular for precipitates in the overaged 8090 alloy. The Fe-rich constituent particles (Fig. 6) are generally too large to be retained in thin foils. The T-phase precipitates, while smaller (Fig. 5), are nonetheless large enough that very few are found at the foil edge, the only place where EELS analysis of Li is feasible. It is further encouraging that centered dark field imaging of rather weak superlattice reflections can be accomplished (Figs. 3, 4). The details of ultramicrotoming of metals and other inorganic materials are discussed elsewhere (2, 12). However, it can be summarized here that the major advantages of thickness control (down to the order of 10 nm), section uniformity (thickness variations of less than 10% as measured by EELS) and large areas for examination more than offset the principal disadvantage of heavy plastic deformation, which obscures the microstructure and severely limits the amount of diffraction information which may be obtained from a UM section.

One potential area of improvement for Al-Li alloys lies in the liquid bath used to collect the sections as they are sliced. The surface tension, viscosity and chemical reactivity of water make it a fairly universal medium for section collection. However, the reactivity of Li with water becomes a handicap as the Li content increases. Very thin UM sections showed substantial oxide formation on some precipitates immediately after sectioning (2), and there was the apparent oxidation "patches" around the constituent particles. Moreover, it had been hoped to find δ-LiAl

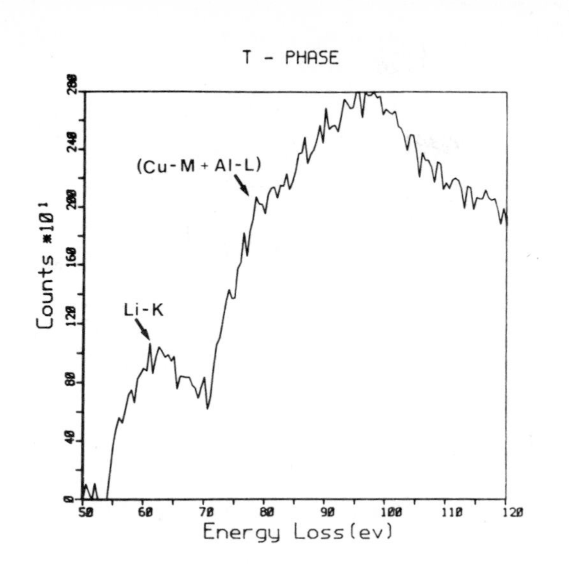

Figure 7. EELS spectrum, after background removal, from a T-phase precipitate in a 40 nm thick UM section.

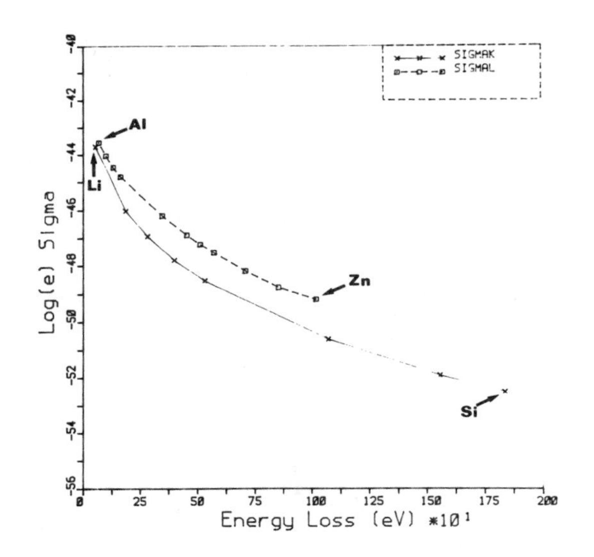

Figure 8. Calculated hydrogenic partial cross-sections for K-shell edges and L-shell edges as a function of energy loss.

precipitates in the overaged binary alloy, but only holes were found in UM sections where such particles may have been (thin foils fared no better due to preferential electropolishing of LiAl).

The quantification data from the Al-Li binary (Table 1) indicates that an accuracy of perhaps ±20% for Li in Al may be attainable on a regular basis by either the small or large window integral methods. It is important that both methods give close to the same results, since there are possible errors in each. Regarding the small window method, the uncertainty of calculated cross-sections for small energy windows has been mentioned, and a recent study comparing experimental to theoretical cross-sections ratios (13) indicates that this can be a valid concern. Also, as the window size decreases, window placement becomes more critical due to the large variations of near-edge fine structure in K- and L-edges. For the large window method, the estimation of H (Fig. 2) and the factor (0.8) to account for the decay of the hidden Li edge are rather operator-intensive, to say the least.

These potential errors, combined with the delicacy of background fitting, make it unlikely that much further improvement in data processing for intensity ratios will occur, though spectral deconvolution may aid in examining somewhat thicker foil regions.

It is, of course, possible that processing errors in both of the above methods occurred but were countered by errors in the hydrogenic cross-section values for Li-K and Al-L. The above-mentioned study (13) indicates, however, that the actual cross-section values are within about 10-15% of the hydrogenic values for these two edges.

Both the EDS and EELS analyses of the "donut" configurations (Fig. 4, Table 2) show sufficient variation that it was unlikely that many (or any) of them were confined only to the core or ring. Thus all that can be said at the present is that both Zr and Li are present in the cores and that the rings are likely $LiAl_3$.

It is clear that substantial Li is present in the T-phase of the overaged 8090 alloy (Fig. 7). However, the total composition is dependent on the assumptions made concerning the value of Cu-M cross-section. An experimental check of this value could be made from Cu-M/L intensity ratios determined from pure Cu and multiplying this value by the Cu-L hydrogenic cross-section. However, there appears to be some major discrepancies among the transition metal L-shell cross-sections when comparing calculated to experimental cross-section ratios (3, 13, 14).

Until these discrepancies are resolved

and/or independent values of M-edges from experimental standards are obtained, it is perhaps best to consider the analyses of Table 2 as semi-quantitative in nature. Trends in the chemical data may show correlations with information obtained by other means and in this sense at least, EELS analysis of the precipitates may prove quite useful.

As mentioned earlier, full quantitative analysis of the Fe-bearing constituent particles (or other Al-Fe-Li-Cu phases) is even more difficult. The relative scarcity of such particles may render such quantification largely unnecessary. Nonetheless, it would be useful to ascertain whether or not the oxidized regions around constituent particles represent a Li solute "cloud" formed during solidification and whether or not such a Li concentration could lead to enhanced corrosion at constituent stringers or clusters.

CONCLUSIONS

(1) Ultramicrotomed sections permit simultaneous visual and chemical characterization on a large scale, yet with high spatial resolution, for commercial Al alloys.

(2) For regions or precipitates involving only Li and Al, quantitative EELS analysis with reasonable accuracy is possible.

(3) Only semi-quantitative EELS analysis is possible at this time for Al-Li-Cu/Fe/Zr precipitates.

ACKNOWLEDGEMENTS

The author is grateful to D. Steele for specimen preparation and wishes to thank Drs. M. Ball and D.J. Lloyd and Prof. R. Egerton for many useful discussions.

LIST OF REFERENCES

(1) C.M. Sung, H.M. Chan and D.B. Williams, these Proceedings.

(2) M.D. Ball, T.F. Malis and D. Steele, Analytical Electron Microscopy - 1984, D.B. Williams and D.C. Joy, Eds., San Francisco Press (1984), p. 189.

(3) R.F. Egerton, Scanning Electron Microscopy/1984/II, SEM Inc., Chicago (1984), p. 505.

(4) R.F. Egerton, Microbeam Analysis - 1982 K.F.J. Heinrich, Ed., San Francisco Press (1982), p. 43.

(5) T. Malis, to be submitted to J. of Electron Microscopy Techniques.

(6) C.C. Ahn and O.L. Krivanek, EELS Atlas, HREM Facility, Center for Solid State Science, Arizona State University, Tempe, Arizona (1983).

(7) N.J. Zaluzec, Argonne National Laboratory, 9700 South Cass Ave., Argonne, Illinois 60439.

(8) R.F. Egerton, J. of Microscopy, Vol. 123, (1981), p. 333.

(9) R.F. Egerton, 39th Proceedings - EMSA, G.W. Bailey, Ed. Claitors Press, Baton Rouge, LA, (1981), p. 168.

(10) D.M. Maher, Introduction to Analytical Electron Micrscopy, J.J. Hren, J.I. Goldstein and D.C. Joy, Eds., Plenum Press, New York, (1979), p. 259.

(11) F.W. Gayle and J.B. Vander Sande, Scripta Met., Vol. 18, (1984) p. 473.

(12) D. Steele, T. Malis, R.S. Timsit and M.D. Ball, to be submitted to J. Mat. Sci.

(13) T. Malis and J. Titchmarsh, Electron Microscopy and Analysis - EMAG/85, Inst. of Physics, to be published.

(14) C. Allison, W.S. Williams and M.P. Hoffman, Ultramicroscopy, Vol. 13, (1984), p. 253.

Metallography and microanalysis of aluminium–lithium alloys by secondary ion mass spectrometry (SIMS)

B DUBOST, J M LANG and F DEGREVE

The authors are with Cegedur Pechiney,
Centre de Recherches et Développement,
BP 27 38340 Voreppe, France

SYNOPSIS :

The severe limitations of conventional investigation techniques (e.g. optical microscopy, electron probe microanalysis) for Al-Li alloys can be overcome by using Secondary Ion Mass Spectrometry (SIMS), which is the sole technique allowing chemical analysis with low detection limits and the direct imaging capability of all elements on the surface of bulk samples.
Using a CAMECA IMS-3F ion microscope with high mass resolution, the following examples of SIMS metallography and microanalysis applications are illustrated on experimental Al-Li-(Cu, Mg, Si) alloys :
- equilibrium-phase recognition in as-cast or as-quenched ingots, showing the existence of quaternary intermetallic compounds with extended composition domains and the occurrence of complex phase reactions.
- quantitative chemical analysis of lithium in the Al matrix of quenched samples.

INTRODUCTION :

The development of new aluminium-lithium alloys with high specific properties requires a prior knowledge of the nature and chemical composition of the equilibrium phases of as-cast or heat treated semi-products.
Conventional metallographic techniques (e.g. Optical Microscopy, Electron Probe Microanalysis) are not completely suitable for low density Al-Li- Cu- Mg alloys, because of the low optical contrast of several intermetallic compounds and the high surface reactivity to humidity of alloys containing intermetallic phases and their strong tendency to oxidize (e.g. δ-AlLi), as well as the inability of EPMA to perform quantitative analysis of lithium. These severe limitations for Al-Li alloys can be overcome by using Secondary Ion Mass Spectrometry (SIMS), which is the sole technique allowing chemical analysis with low detection limits in addition to the capability of direct imaging of all elements ($1 \leqslant Z \leqslant 92$) on the surface of bulk samples[1].
Besides its wide application to surface analysis of oxide layers[2], this technique was also combined with Scanning Auger Microscopy to characterize the oxide stringer particles in powder-metallurgy processed Al-Li-Cu alloys[3].
The purpose of this work was to use SIMS as a major experimental technique for evaluation of the constitution of alloys within the Al-Cu-Li-Mg system, including equilibrium-phase recognition (in combination with conventional techniques) and quantitative analysis of lithium in the matrix.

EXPERIMENTAL PROCEDURE :

Cylindrical ingots were cast at 720°C into Ø18mm x 60 mm molds with solidification times of about 1 minute, using high purity metals (Al, Mg, Li) or master alloys (Al-Cu). The chemical analysis of investigated samples was performed by atomic absorption in the as-cast condition or in as-solution heat treated and quenched temper (for calibration of quantitative analysis of lithium in the matrix). The Fe and Si content of the alloys was below 0,02%, unless Si was deliberately added.
SIMS analyses were carried out in a CAMECA IMS 3F Ion Microanalyser. The polished samples were sputtered with a mass filtered primary ion beam of $^{40}Ar^+$ (8KeV). The primary beam was focused in a 50μm spot which was rastered over the surface in order to produce a uniform current density, i.e. a flat bottomed crater. Different modes of analysis were used (fig. 1) :
- for the recording of elemental distribution (ion micrographs), ionic lenses were adjusted to select a 150 μm wide area displaying the

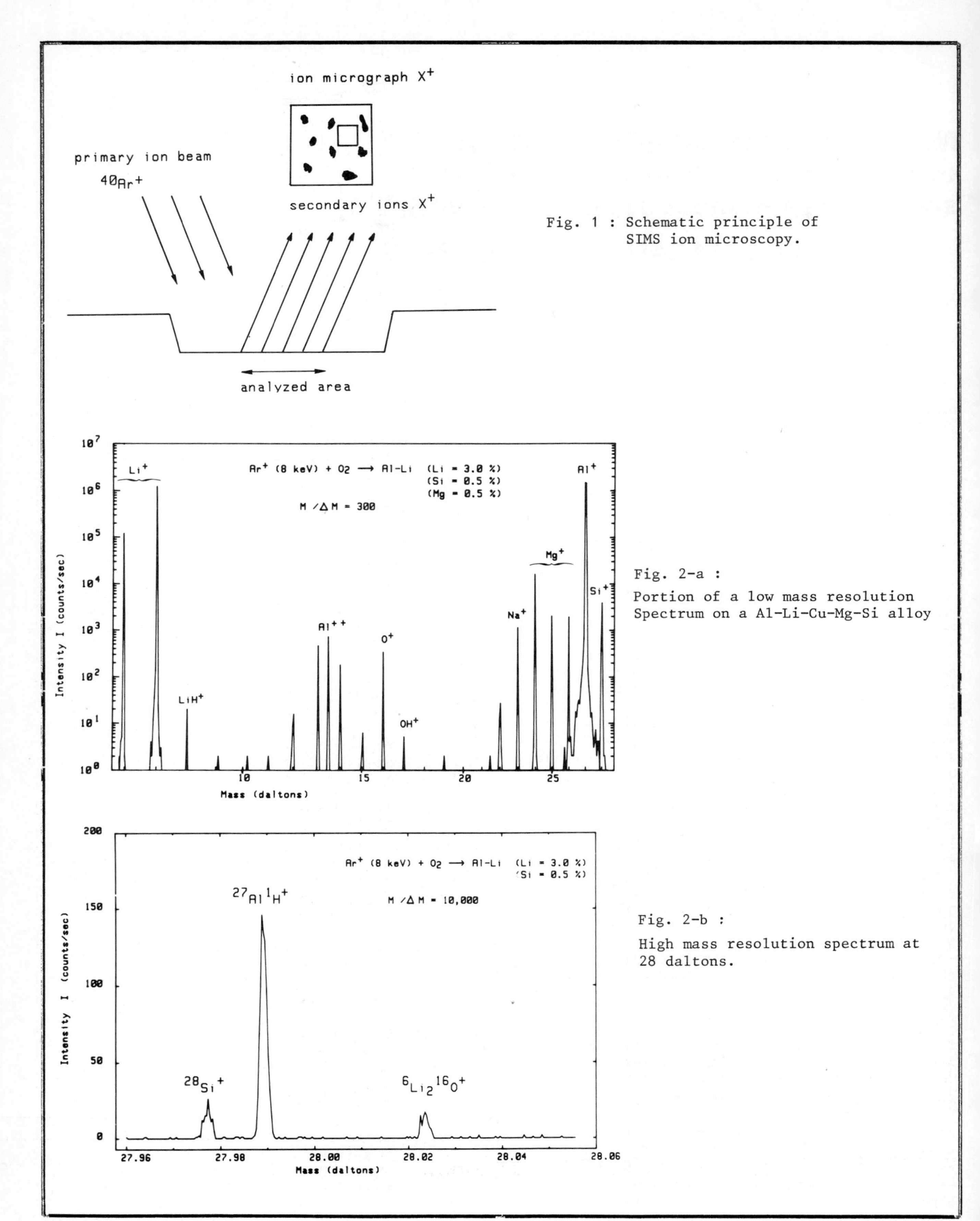

Fig. 1 : Schematic principle of SIMS ion microscopy.

Fig. 2-a : Portion of a low mass resolution Spectrum on a Al-Li-Cu-Mg-Si alloy

Fig. 2-b : High mass resolution spectrum at 28 daltons.

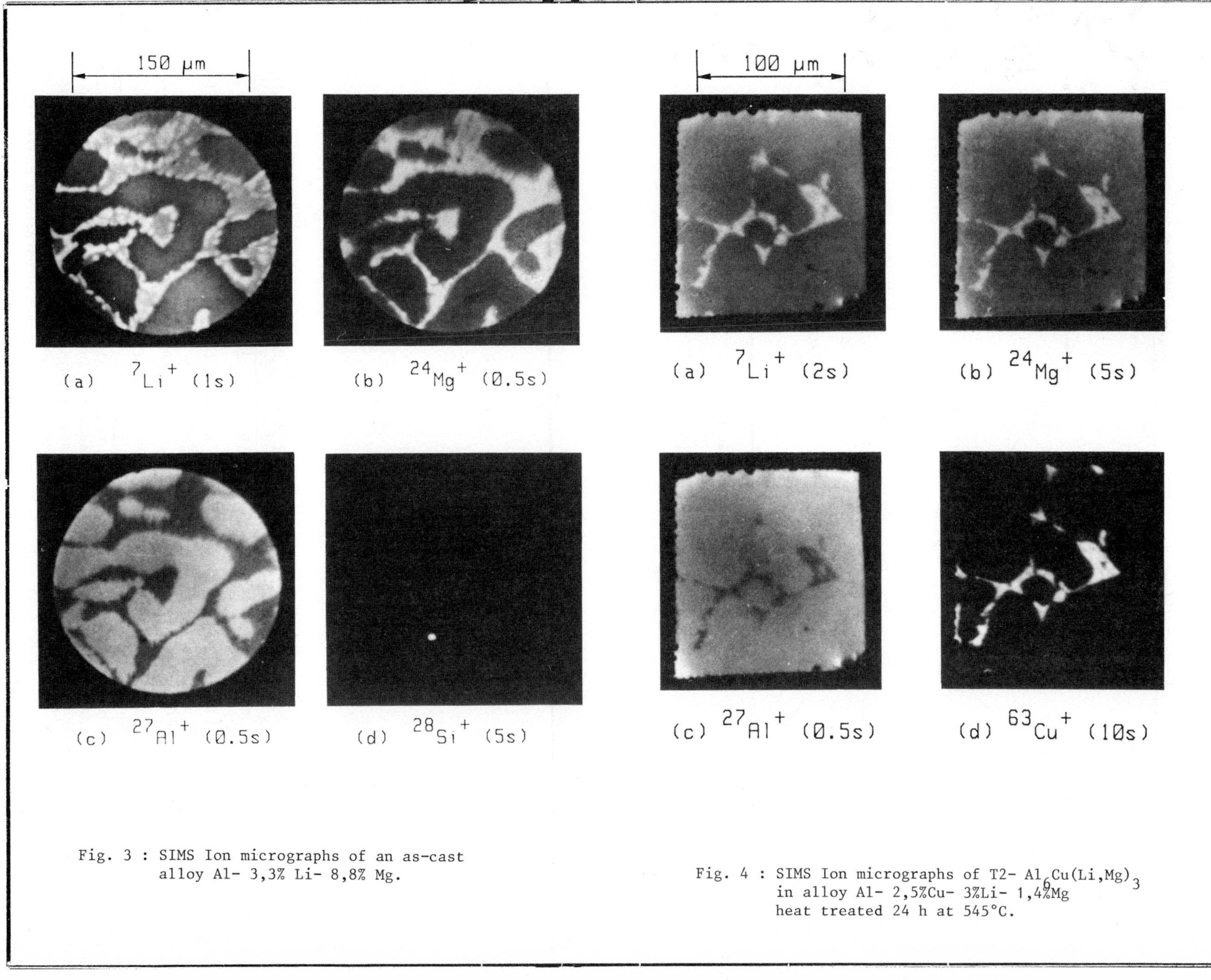

Fig. 3 : SIMS Ion micrographs of an as-cast alloy Al- 3,3% Li- 8,8% Mg.

Fig. 4 : SIMS Ion micrographs of T2- $Al_6Cu(Li,Mg)_3$ in alloy Al- 2,5%Cu- 3%Li- 1,4%Mg heat treated 24 h at 545°C.

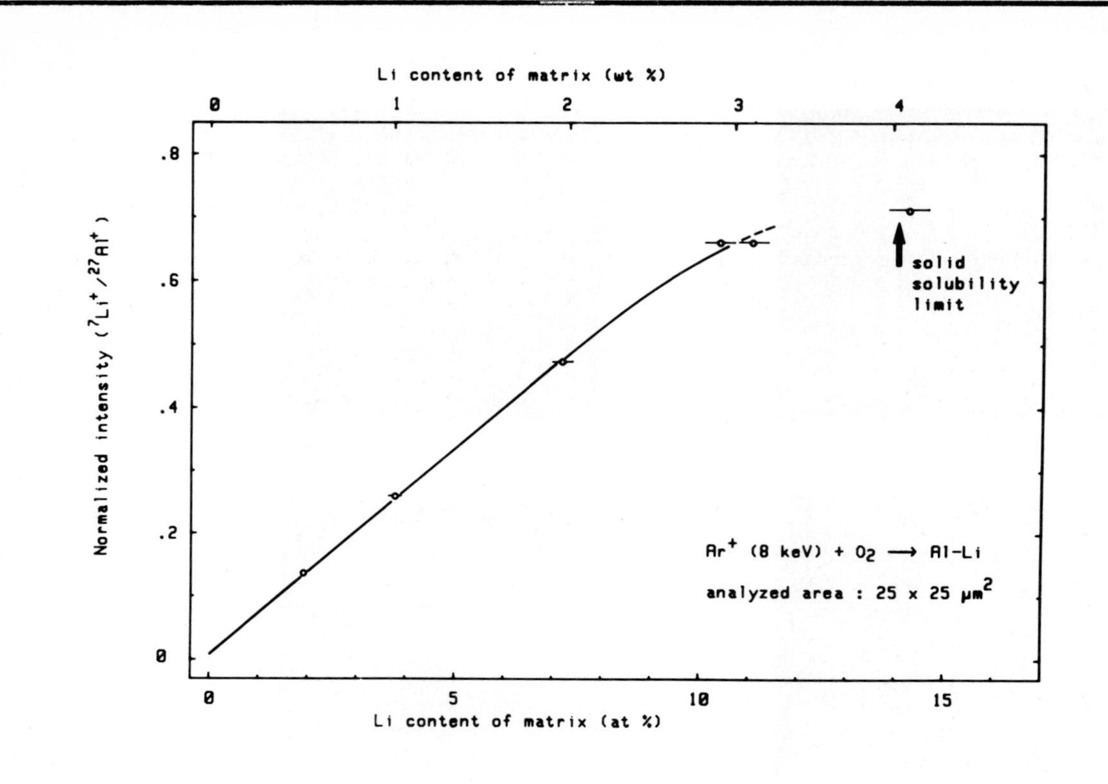

Fig. 5 : Plot of normalized ionic intensity $^7Li^+/^{27}Al^+$ in the matrix versus Li content of as-quenched binary Al-Li alloys.

matrix and the intermetallic phases with a 1 μm lateral resolution.
- for the analyses, areas were selected with a field diaphragm on phases or matrix.
Qualitative analyses were carried out with a mass spectrum at low ($M/\Delta M$ = 300: Fig. 2 a) and high ($M/\Delta M \cong$ 10,000:Fig. 2 b) mass resolution.
Quantitative analyses were performed by recording accurately the intensity of $^7Li^+$ and $^{27}Al^+$ on a Faraday cup.
In all cases, the surface was saturated with an oxygen jet in order to maintain a constant stoechiometry and to both maximize and stabilize the emission of secondary ions.
When necessary, SIMS examinations were combined with X-ray diffraction (using a Seeman-Bohlin transmission camera with Cu-kα radiation) and EPMA (Cambridge MK 9).

RESULTS :

Examples of phase equilibria and composition studies by ion imaging of elemental distribution in Al-Li-(Cu-Mg) alloys :

1. Study of ternary four-phase reactions in the Al-Li-Mg system (Fig. 3) :

As-cast ingots of experimental alloy Al-3,3%Li-8,8%Mg (wt%) exhibit the initial crystallization of primary aluminium-rich dendrites surrounded by a semi-continuous layer of a lithium-rich ternary phase (Al_2LiMg) enclosing an interdendritic magnesium-rich ternary phase which is the last to solidify. This Li-containing phase appears to be an extension of the binary $Mg_{17}Al_{12}$ phase

with lithium atoms mainly substituting for magnesium.
It can be designated as $(Mg, Li)_{17}Al_{12}$, in good agreement with prior results[4,5]. Silicon combines preferentially with magnesium to form Mg_2Si particles. These microstructural features are consistent with the following peritectic reaction in the Al-Li-Mg system :

$$Liq. + Al_2LiMg \longrightarrow \alpha Al + (Mg, Li)_{17}Al_{12}$$

with non-equilibrium casting conditions.

2. Existence of quaternary equilibrium compounds in the Al-Cu-Li-Mg system (Fig. 4) :

Ingots of experimental alloy Al- 2,5%Cu- 3%Li- 1,4%Mg exhibit an interdendritic quaternary T_2 phase[6] in equilibrium with the Al-rich matrix after heat treating 24 h at 545°C and quenching. The copper and magnesium content of this phase (respectively 25 to 29,5% for Cu and 6 to 7,5% for Mg, as given by EPMA) show that it can be considered as an extension of the ternary T_2-$Al_6\,Cu\,Li_3$ phase (Cu = 28,8% - Li = 9,6%) with magnesium atoms substituting for lithium up to the compound formula $T_2 - Al_6\,Cu\,Li_{3-x}\,Mg_x$ with $x \leqslant 1$.

Quantitative analysis of Lithium in the matrix of as-quenched binary Al-Li alloys :

Ingots of binary Al-Li alloys with lithium contents of 0.5 to 4% Li were homogenized at 595°C from 2 to 24 hours and quenched in cold water. All alloys except Al- 4% Li were fully homogeneous with uniform composition at the centre of the ingots, at the resolution level of SIMS (1 µm). Alloy Al- 4% Li was supersaturated and contained undissolved constituent particles of δ - AlLi equilibrium phase after heat treatment of 24 hours at 595°C.
Fig. 5 shows the plot of normalized ionic intensity $^{7}Li^{+}/^{27}Al^{+}$ (as analyzed on a 25 µm x 25 µm area) versus Li content of the same sample (as analyzed by atomic absorption). This plot shows a slight deviation from linearity with increasing Li content (up to the solid solubility limit) and yields a calibration curve for further analysis of lithium in an aluminium-rich matrix, in the same operating conditions. However it does not give a true value of the Li content of the solid solution in alloys containing small metastable δ' - Al_3Li precipitates, the sizes of which are below the lateral resolution of SIMS.

CONCLUSIONS :

SIMS appears to be a major investigation technique for basic metallurgical studies of Aluminium-Lithium alloys with high specific properties. The use of ion microscopy to visualize all elements at the surface of bulk metallographic samples and the ability of SIMS to perform quantitative analysis of Lithium in the matrix contributed to the understanding of phase transformations and the constitution of complex alloys, in combination with conventional techniques.
These preliminary results clearly showed the existence of ternary or quaternary equilibrium intermetallic compounds with extended composition domains within the Al-Cu-Li-Mg system, suggesting a strong affinity between lithium and magnesium.
Although elemental analysis of intermetallic compounds remains purely qualitative and subject to matrix effects, recent improvements of SIMS lateral resolution might make it possible to visualize the distribution of light elements (Li, Mg) in smaller precipitates (e.g. at grain boundaries), and thus allow further understanding of structure-property relationships in industrial Al-Li alloys.

ACKNOWLEDGMENTS :

The authors wish to thank the Directions des Recherches et Etudes Techniques for its financial contribution to this study and are grateful to Mme M. DUSSOUILLEZ, MM. G. DUVERNEUIL, C. BERNARD and N. THORNE for technical support.

REFERENCES :

(1) G.A. MORRISON and G. SLODZIAN, Anal. Chem. 47, 942 A (1975).
(2) F. DEGREVE and J.M. LANG, Spring Meeting of MRS, San Francisco, April 1985.
(3) L.A. LARSON, M. AVALOS-BORJA. P.P. PIZZO, Aluminium-Lithium II. Proc. 2nd int. Aluminium-Lithium Conference, MONTEREY, ed by E.A. STARKE Jr and T.H. SANDERS, Jr, Met. Soc. AIME 1984.
(4) D.W. LEVINSON, D.J. Mc PHERSON, Trans ASM 48, 1956 p. 609.
(5) L.F. MONDOLFO, Aluminum Alloys - Structure and Properties, ed. Butterworths, LONDON, 1976.
(6) H.K. HARDY, J.M. SILCOCK, J. Inst. Met., 1955-56, vol 85, p. 423.

The δ′ (Al_3Li) particle size distributions in a variety of Al–Li alloys

B P GU, K MAHALINGAM, G L LIEDL and T H SANDERS Jr

The authors are in the School of Materials Engineering, Purdue University, West Lafayette, Indiana, USA

SYNOPSIS

This paper reviews the results of a series of quantitative transmission electron microscopy (TEM) investigations of the coarsening behavior of δ′ in a variety of Al-Li alloys. The alloys investigated include: Al-2.8Li-0.3Mn; Al-2.8Li-0.14Zr; and a series of binary alloys whose compositions range from Al-2.4Li to Al-4.5Li.

For all the alloy systems investigated, the rate of growth of δ′ was found to obey Ostwald ripening kinetics. Increasing the lithium content increased the aging kinetics. For alloys which contained zirconium with the same lithium content, the addition of zirconium accelerated the aging kinetics. The Al_3Zr dispersoids acted as nucleation sites for the δ′ particles. The growth of δ′ on the pre-existing Al_3Zr particles affected the initial or ``critical size'' of the composite particles.

The particle size distributions (PSD's) in the binary alloys varied with lithium content. The negatively skewed PSD, predicted by many of the ripening theories, was only observed in more dilute alloys with small δ′ volume fractions (0.12). However, progressively increasing the δ′ volume fraction to approximately 0.55 resulted in a positively skewed PSD. The shape of the PSD was volume fraction dependent changing from negatively to positively skewed. The PSD's were modeled using the Weibull density function.

The δ′ PSD for the zirconium containing alloy is actually a combined or bimodal distribution. One distribution constructed from δ′ particles containing Al_3Zr particles, referred to as composite particles, had PSD'S shifted to larger particle sizes compared to the distribution constructed from particles which did not contain Al_3Zr, the zirconium-free particles. The shift in the combined PSD was determined to be responsible for the accelerated aging kinetics of zirconium containing Al-Li alloys. The accelerated aging kinetics could be seen in both growth rate curves and in the yield strength-aging behavior.

INTRODUCTION

Coarsening Theory

The ideas of precipitate coarsening were first formulated by Ostwald [1] in 1900. The coarsening process or Ostwald ripening as it is often called, describes the change of the average particle size with increasing time while the number of the dispersed phase particles decreases to maintain a constant volume fraction. The driving force for the coarsening process is the reduction in the free energy associated with the decrease in total surface area of the particles during the coarsening process.

Zener [2] and Greenwood's [3] pioneering work on this subject pointed out that growth and coarsening are diffusion-controlled processes, and they developed relationships which showed that the change of particle size is proportional to $(Dt)^{1/2}$ for growth and $t^{1/3}$ for coarsening, respectively. In 1961, Lifshitz and Slyozov [4] and Wagner [5] (LSW theory) established a unified theoretical analysis incorporating the diffusion, continuity, and mass balance equations together, giving a result of $\overline{R} \propto t^{1/3}$. These authors also developed a theoretical, asymmetrical PSD and described the effect of volume fraction, elastic strain, and anisotropy on the coarsening process. Unfortunately, there is disparity between either the theoretical PSD or the effect of volume fraction on the experimentally determined PSD's. Therefore, modifications [6-8] to the LSW theory have been attempted to produce a better fit of theoretical PSD's to actual experimental PSD's.

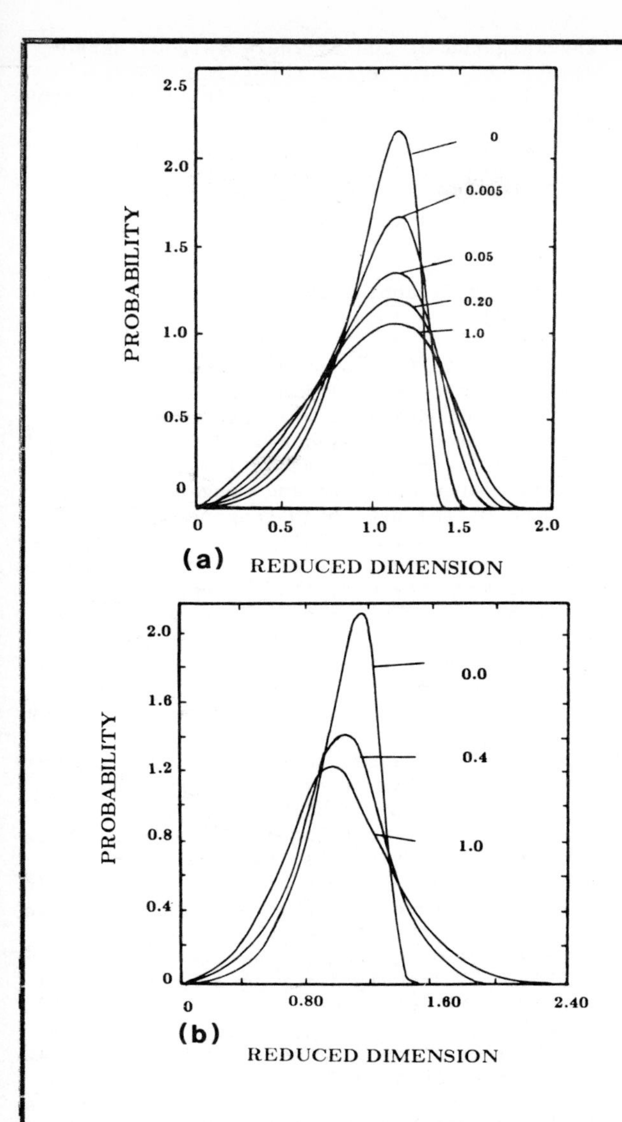

Figure 1. (a) Theoretical PSD based on the MLSW theory with volume fraction of precipitate, from [6]. (b) Theoretical PSD based on the LSEM theory with varying volume fraction of precipitate, from [8].

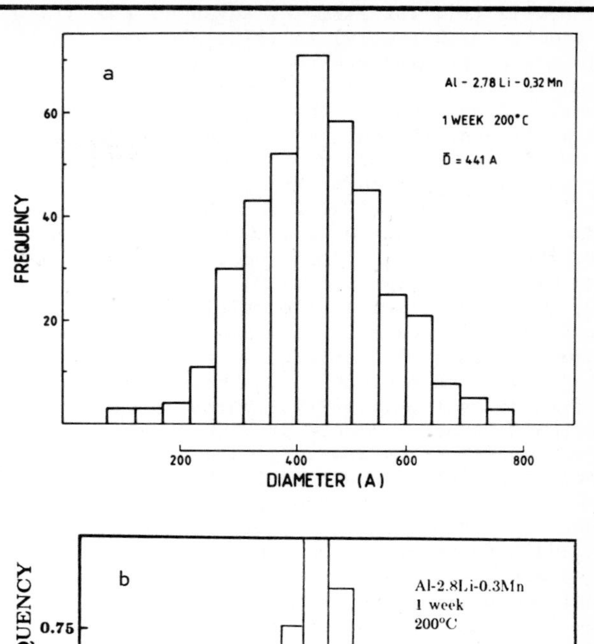

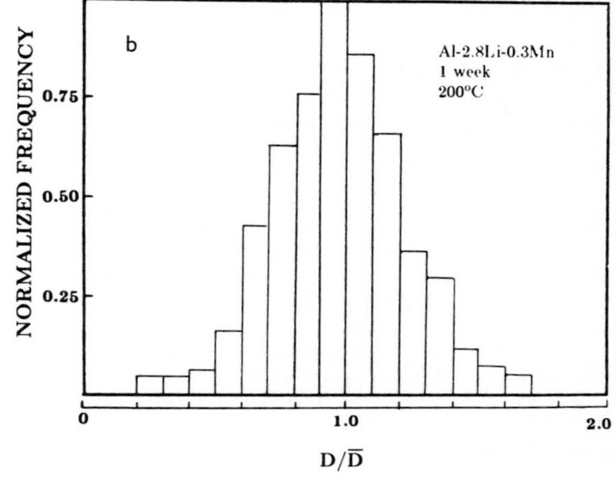

Figure 2. (a) Histogram of δ′ particles aged 1 week at 200°C. (b) Normalized histogram of δ′ particles aged 1 week at 200°C.

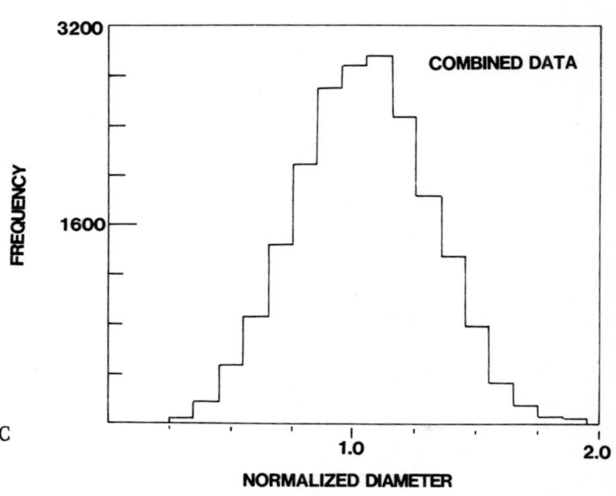

Figure 3. Combined PSD from the 168,200 and 225°C aged microstructures.

Ardell [6], for example, advanced a modified LSW theory (MSLW theory) in 1972. Introducing the influence of precipitate volume fraction into the diffusion equation by using a modified diffusion geometry, Ardell showed that although the change of $\bar{R}$ is still proportional to $t^{1/3}$ the growth rate of the particles increases as the volume fraction increases. Comparison of Ardell's theory with experimental data indicates that his model overestimates the influence of the volume fraction on coarsening. Since 1972 new models and modifications have been presented [7,8]. For example, in 1980, the possibilities of particle encounters (particle coalescence) in systems containing large volume fractions of precipitates was presented by Davies, Nash and Stevens [8]. Their model is referred to as the Lifshitz-Slyozov encounter modified (LSEM) theory. Their predicted PSD's and average growth rates are in good agreement with limited experimental data in the Ni-Ni_3Al system. These PSD's tend to be symmetric about a mean value. The shapes of the theoretical PSD's for the MLSW and LSEM theories are presented in Figure 1. In these figures the abscissa is the ratio of the calculated precipitate size, to the average precipitate size and the ordinate is the frequency of occurrence of a particular ratio. Both theories reduce to the LSW theory in the limit zero volume fraction.

Because of the importance and the number of recently published theoretical analyses on the coarsening process a series of experimental programs investigating the coarsening phenomenon of the metastable δ'(Al_3Li) in the Al-Li system were undertaken [9-11]. This paper will review the results of those investigations.

Coarsening Results of δ' in Al-Li-Mn Alloys [9]

From the particle size measurements at a variety of aging conditions the average particle size, $\bar{R}$, and the standard deviation of the distribution, s, were determined. The rate of coarsening for each temperature was shown to follow the relationship:

$$\bar{R}^3 \propto t$$

which implies that δ' coarsening in the Al-2.8Li-.3Mn follows Ostwald ripening kinetics.

The PSD's were constructed from a minimum of 600 data points for each aging condition. An example of a PSD for an Al-Li-Mn alloy is shown in Figure 2a. An alternative method of plotting the PSD's is to plot the frequency, or normalized frequency, versus the normalized diameter ($D/\bar{D}$). This type of plot is shown in Figure 2b for the same data plotted in Figure 2a. Histograms for aging temperatures ranging from 168° to 225°C and aging times ranging from 12 hours to 8 weeks showed that the resulting normalized histograms were virtually identical in form. A composite histogram was constructed by summing all of the normalized class intervals for the different aging conditions investigated, Figure 3. Figure 4 compares the composite distribution with the minimum and maximum aging times investigated at 200°C. There is good agreement between the form of the composite distribution and the individual distributions. It is clear from these figures that the PSD for this particular alloy is not the negatively skewed distribution predicted by many of the Ostwald ripening theories but a more symmetrical distribution with a cut-off beyond the predicted 1.5.

The Influence of Zirconium on the Coarsening Phenomenon [10]

In addition to the major solute elements all commercial high strength aluminum alloys contain ancillary element additions which control grain size and the degree of recrystallization in wrought products. Traditionally these ancillary element additions have been chromium and manganese although recently some of the new, high toughness 7XXX alloys contain zirconium. Unlike the other ancilliary element additions, the Al_3Zr phase is spherical, coherent and oriented, and is approximately 200(A) in diameter. The crystal structure of the metastable Al_3Zr is $L1_2$ in binary alloys. The Al_3Zr phase is, thus, similar to δ'. It should also be noted that most of the successful Al-Li alloys reported in [12] contain zirconium.

Zirconium, when added to Al-Li alloys, is very effective in inhibiting recrystallization, and the yield strength tends to be higher than a corresponding alloy containing no zirconium. There are two other interesting observations which have been made concerning Al-Li-Zr alloys. The first is that these alloys tend to age more rapidly and have higher yield strengths than similar alloys which contain manganese, for example, as the ancillary element [13,14]. The yield strength for two Al-Li alloys versus artificial aging time at 200°C are plotted in Figure 5. It is clear from the figure that the Al-Li-Zr alloy ages more quickly and to a higher peak strength than does the Al-Li-Mn alloy containing the same lithium content. The second is that Al_3Li precipitates appear to nucleate on the existing Al_3Zr precipitates [15] as well as in the matrix.

The more rapid aging response of the Al-Li-Zr alloy compared to the Al-Li-Mn alloy can be explained by comparing the PSD's for the two alloys at similar aging conditions. For all the aging conditions investigated the PSD's were asymmetrical having a positively skewed distribution. Figure 6 shows a representative PSD for the Al-Li-Zr system. The clear asymmetric nature of the distribution of the PSD's in the Al-Li-Zr alloy contrasts with the more symmetrical distribution found in the Al-Li-Mn alloy, Figure 2. This discrepancy in the shape of the PSD in the Al-Li-Zr alloy compared with the shape of the PSD in the Al-Li-Mn alloy can be interpreted in light of the interaction of the coherent Al_3Zr with the δ' phase.

Centered dark field (CDF) images of an aged Al-Li-Zr alloy showed the presence of small dark circles lying inside of some of the δ' precipitates. As documented in a recently published paper [15] these small interior particles are thought to be Al_3Zr. Since an asymmetry was present in the

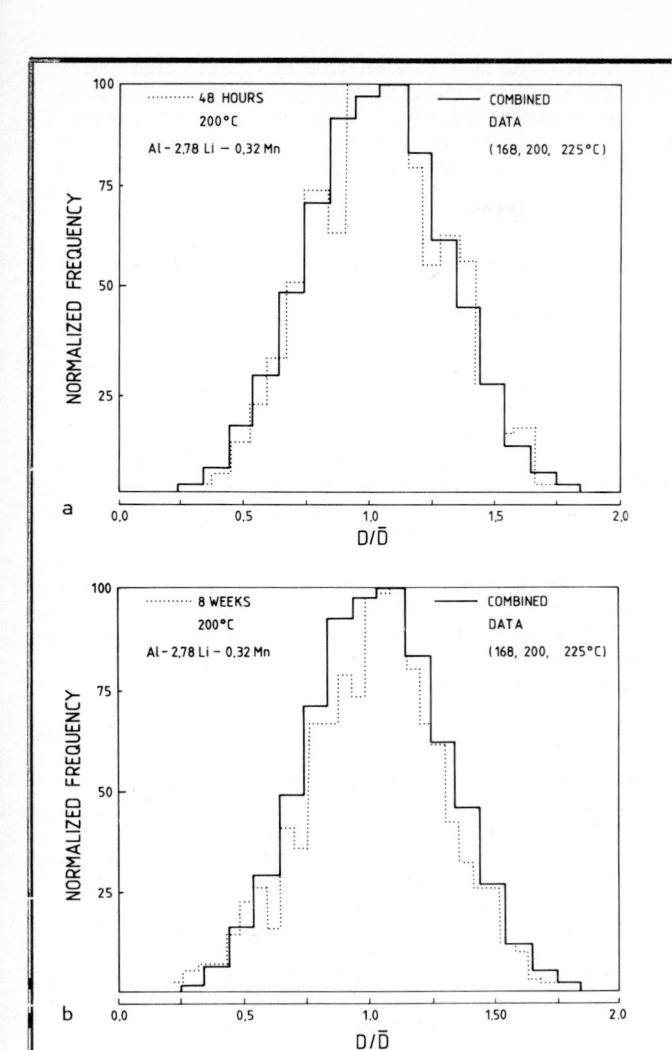

Figure 4. Comparison of the combined PSD with the PSD from the Al-Li-Mn aged (a) 48 hours at 200°C; (b) 8 weeks at 200°C.

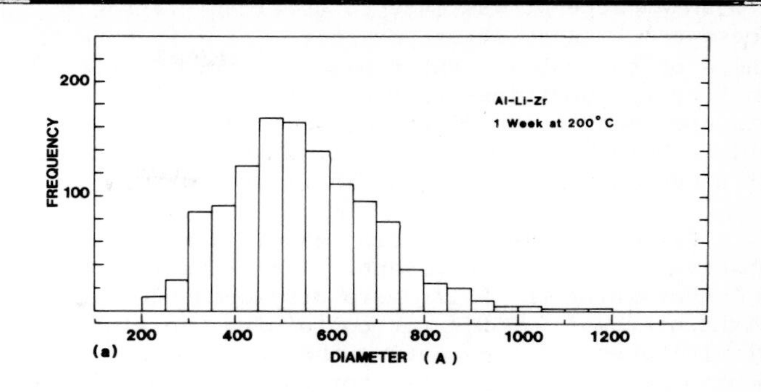

Figure 6. Histogram of δ´ precipitates aged 1 week at 200°C in the Al-Li-Zr.

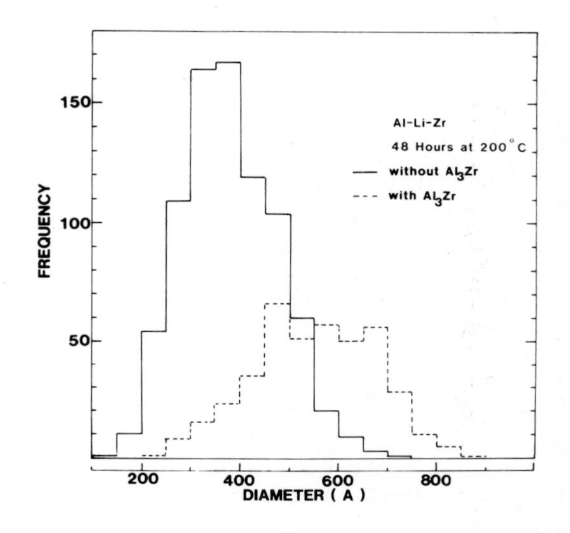

Figure 7. Separated histogram of δ´ precipitates aged 48 hours at 200°C.

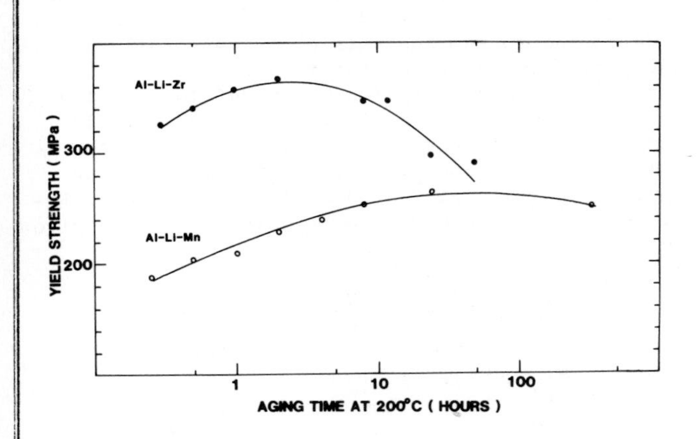

Figure 5. Yield strength versus aging time for Al-Li-Mn and Al-Li-Zr alloys aged at 200°C.

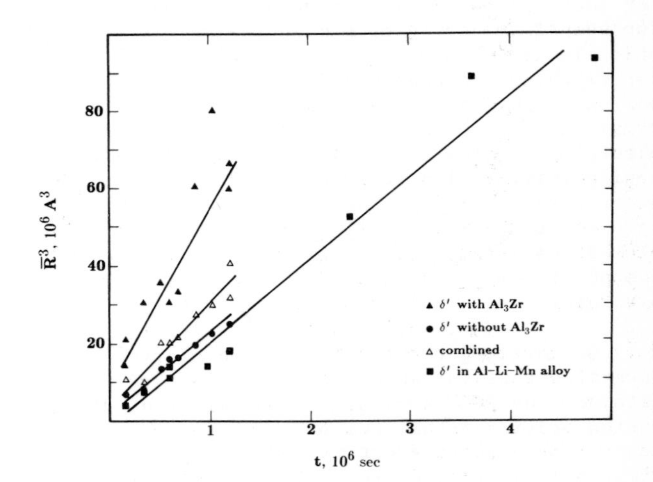

Figure 8. Linear relation between $\overline{R}^3$ and time for Al-Li-Mn and Al-Li-Zr alloys.

PSD´s, an attempt was made to determine if the presence of these Al_3Zr particles was affecting the shape of the PSD´s. PSD´s were constructed for each aging condition using the Al_3Li particle which did not contain the Al_3Zr precipitates and those which contained the Al_3Zr precipitates. An example of a separated PSD is shown in Figure 7.

At every aging condition examined several observations were consistent. The PSD´s for the δ´ containing the Al_3Zr were displaced to the right of the δ´ PSD´s not containing the Al_3Zr precipitates. Consequently, the mean and the median were always greater for those δ´ particles which contained Al_3Zr compared to those δ´ particles which did not contain Al_3Zr particles. Also, the coefficients of variation of the PSD´s containing Al_3Zr were always smaller than those PSD´s constructed from δ´ not containing Al_3Zr precipitates. The growth rate of δ´ in the Al-Li-Zr alloy is plotted in Figure 8 using the relationship $\overline{R}^3 \propto t$. the results of the coarsening of δ´ in the Al-Li-Mn alloy aged at 200°C are included for comparison.

The absolute value of the yield strength is dependent upon the PSD, the degree of recrystallization and the subgrain structure. During aging the rate of increase in yield strength, however, is due to the change in the PSD. Increasing the mean particle size will increase the yield strength on the underaging side of the yield strength-aging time response curve, and increasing the mean particle size will decrease the yield strength on the overaging side of the yield strength - aging response curve. Figure 5 shows that the peak yield strength for the Al-Li-Zr alloy is reached when the alloy is aged at 200°C for 2 hours, for the Al-Li-Mn alloy, after about 50 hours. Using the linear expression of $\overline{R}^3 - R_{CO}^3 = kt$, Figure 8, a calculation of average particle size at peak yield strength for both alloys in their corresponding peak aged conditions shows $\overline{D} \cong 300A$ would be the critical particle diameter at peak yield strength for both alloys. The initial average particle size for the Al_3Zr particles was approximately $\overline{D} = 200A$ which is very close to the critical size required for reaching the peak yield strength. Therefore, only a short amount (< 5 hours at 200°C) of isothermal aging is necessary for a sufficient number of particles to grow such that the mean size of the entire distribution is in the range of 300A. In other words, the more rapid response to artificial aging of the Al-Li-Zr alloy peak as compared to the slower response of the Al-Li-Mn alloy is related to the precipitation of δ´ on the pre-existing Al_3Zr particles.

Summarizing the results of two coarsening investigations for an alloy containing 2.8 wt. % lithium: The PSD is close to a symmetrical distribution providing that the δ´ nucleates homogeneously throughout the matrix. However, if a distribution of particles exists whose crystal structure is the same as δ´, the δ´ can grow on these pre-existing particles. As one would expect, the bimodal distribution affects the shape of the PSD and the kinetics of the δ´ coarsening.

Since the δ´ PSD in the Al-2.8Li-.3Mn was more symmetrical than the PSD´s predicted by many of the Ostwald ripening theories, a quantitative TEM investigation of the influence of composition on the shape of the distribution and the kinetics of coarsening were investigated.

Coarsening of δ´ in Binary Al-Li Alloys [11]

Aluminum-lithium alloys ranging from 2.4 to 4.5 wt % lithium were aged for different periods of time ranging from 12 hours to 10 days at 200 and 225°C. As before, PSD´s were determined.

The growth rate of the average particle as a function of composition and temperature are given in Figure 9. As shown in the Al-Li-Mn and Al-Li-Zr alloys the δ´ obeyed Ostwald ripening kinetics, and representative normalized PSD´s for the 2.4% and 4.5% alloys are shown in Figure 10. The shape of the PSD is clearly a function of alloy content. The more dilute alloy tends to have a long tail on the left side of the distribution, whereas the more concentrated alloy has a long tail on the right side of the distribution.

The coefficient of kurtosis was for most cases < 0, indicating that the distribution was flatter than a normal distribution. The coefficient of skewness (C.S.) tended to be < 0 for the dilute alloys, tail to the left of the maximum, and > 0 for the concentrated alloys. However, a comparison between the C.S. of the experimental values with the value predicted by the LSW theory (C.S. = -0.9200) showed that all the experimental values of C.S. were greater than the theoretical value, indicating that the PSD predicted by the LSW theory is not in agreement with the experimental data. Thus, even for the more dilute alloy (2.4Li), the distribution predicted by the LSW theory is not an appropriate function to model the PSD´s in this system.

Since the form of the PSD´s varied systematically with lithium content, the Weibull statistical distribution function may be used to describe the δ´ PSD´s. The Weibull distribution has been widely used in the analysis of fatigue failure data, and, recently, to describe the steady-state particle size distribution which develops during liquid phase sintering [16]. The power of this function lies in the fact that the shape of the distribution can be systematically varied using only two parameters. A general form of the probability density of the Weibull distribution is given by

$$p(x) = \begin{cases} abx^{b-1}\exp(-ax^b) & \text{for } x>0,\ a>0,\ b>0, \\ 0 & \text{elsewhere} \end{cases} \tag{1}$$

and

$$\int_0^{+\infty} p(x)\,dx = 1 \tag{2}$$

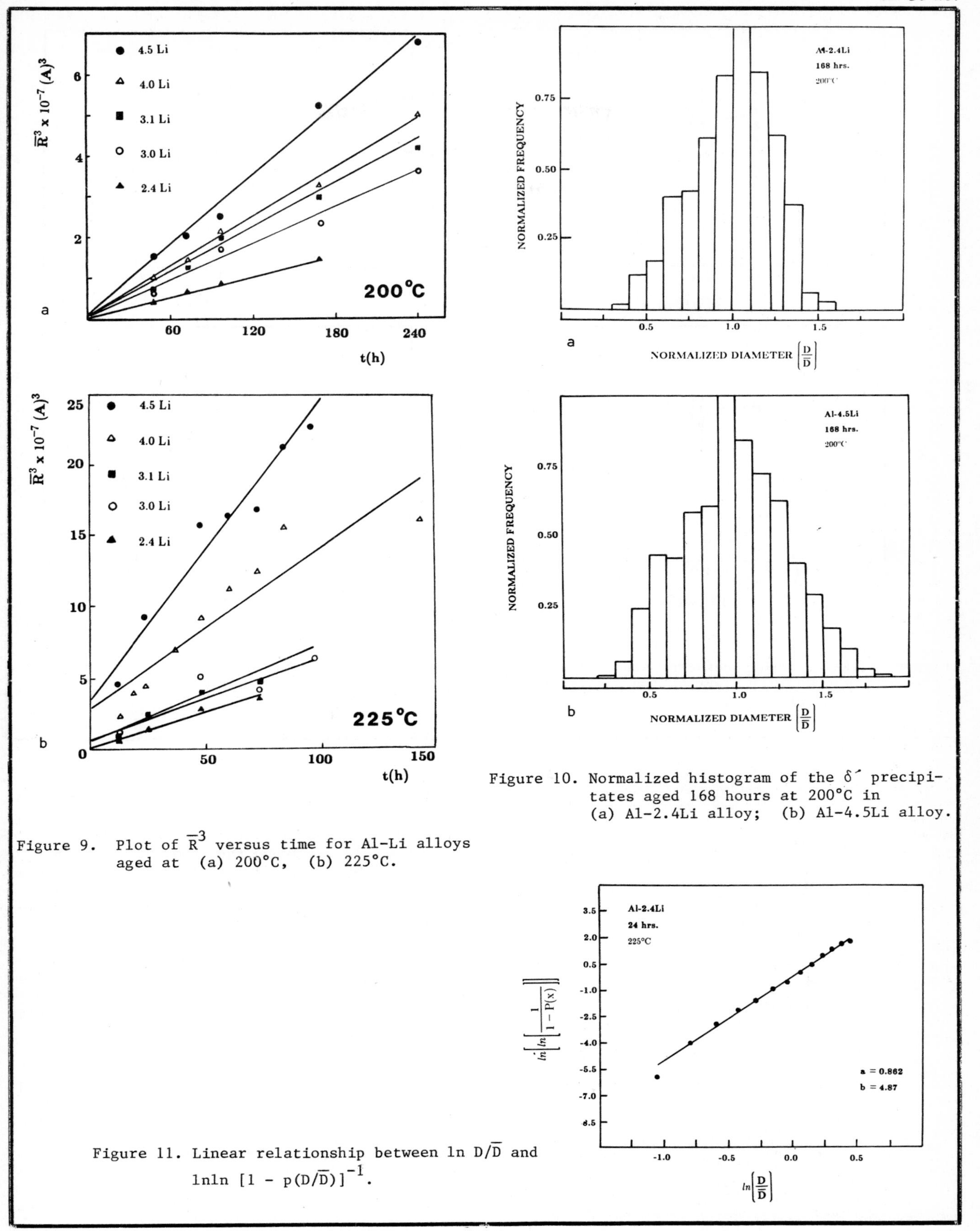

Figure 9. Plot of $\overline{R}^3$ versus time for Al-Li alloys aged at (a) 200°C, (b) 225°C.

Figure 10. Normalized histogram of the δ′ precipitates aged 168 hours at 200°C in (a) Al-2.4Li alloy; (b) Al-4.5Li alloy.

Figure 11. Linear relationship between ln $D/\overline{D}$ and lnln $[1 - p(D/\overline{D})]^{-1}$.

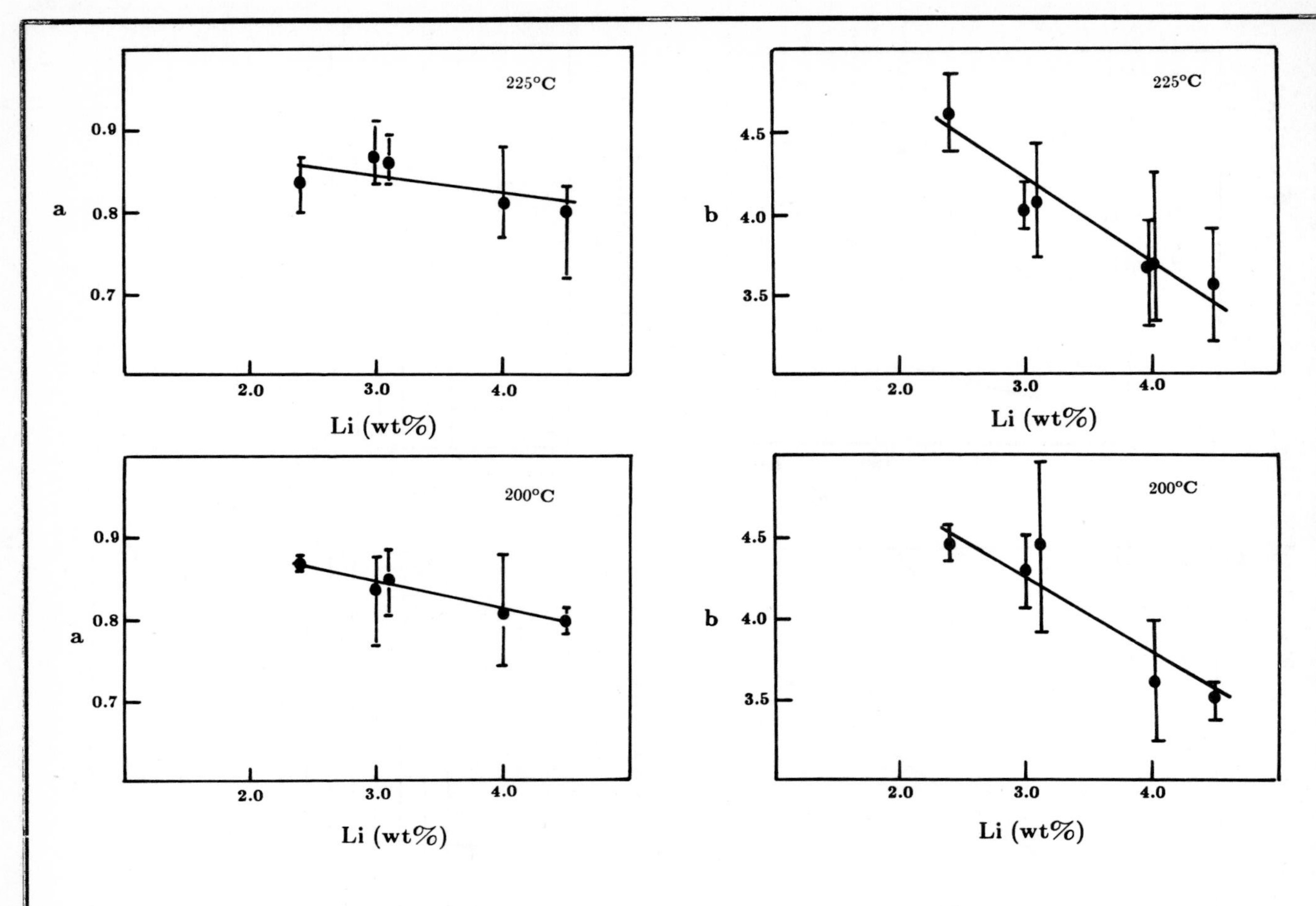

Figure 12. Relationship between parameter a of the Weibull distribution and lithium content.

Figure 13. Relationship between parameter b of the Weibull distribution and lithium content.

By means of a double-logarithmic transformation, the Weibull distribution can be expressed as

$$\ln \ln \frac{1}{[1 - p(x)]} = \ln a + b \ln x \quad (3)$$

The values of a and b can, thus, be determined by the linear regression analysis. Since the shape of the Weibull distribution curve can be altered from a negatively skewed to a positively skewed distribution by adjusting the variables a and b, the Weibull distribution function is suitably tailored to approximate the δ′ PSD′s in binary Al-Li alloys.

In our application of the Weibull distribution, the normalized particle size $D/\overline{D}$ is the variable x, and the cumulative normalized frequency, $p(D/\overline{D})$, is the cumulative probability of the Weibull distribution. If the variables $\ln D/\overline{D}$ and $\ln \ln[1 - p(D/\overline{D})]^{-1}$ are linearly related, it can be assumed that the distribution is of the Weibull type. This method was used for the binary Al-Li alloys, and it was found that the data could be represented by a Weibull distribution. An example from our data is shown in Figure 11. The parameters a and b of this distribution can then be calculated by applying the linear regression method.

The parameters a and b were determined for each alloy and aging condition. The relationships between the parameters and Li content are illustrated in Figures 12 and 13. The results show that for the Al-Li binary system, the values of the parameters a and b of the Weibull distribution decrease with increasing Li content. Using the linear regression method we obtained an expression relating the value of a and b to lithium content and found that the regressed equations were unaffected by aging temperature and time over the ranges investigated. The resulting expressions relating a and b to the lithium content were:

$$a = 0.94 - 0.03\text{Li (wt\%)} \quad (4)$$

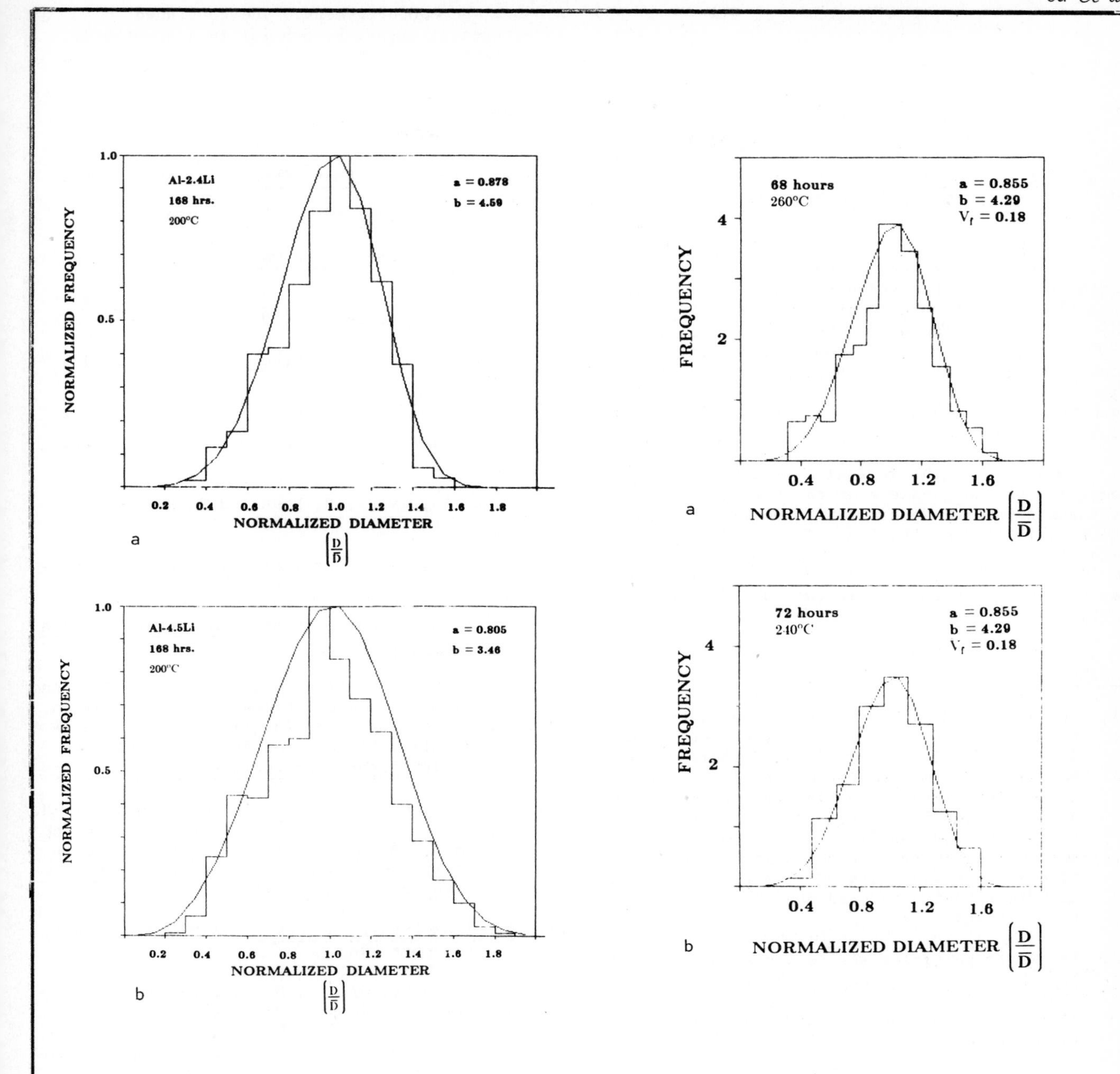

Figure 14. A comparison of the normalized histogram and generated Weibull distribution curve with calculated parameters a and b for (a) Al-2.4Li alloy and (b) Al-4.5Li alloy aged at 200°C for 168 hours.

Figure 15. A comparison of generated Weibull distribution curve with calculated parameters a and b and normalized histograms presented by Jensrud and Ryum [17] for Al-3.07Li alloy aged (a) 68 hours at 260°C and (b) 72 hours at 240°C.

$$b = 5.80 - 0.52Li \text{ (wt\%)} \quad (5)$$

or based upon volume fraction, f:

$$a = 0.88 - 0.12f \quad (6)$$

$$b = 4.68 - 2.17f \quad (7)$$

Two examples of calculated δ´ PSD´s using the regressed values of a and b for the experimental data plotted in Figure 10 are plotted in Figure 14. The comparison between the observed and calculated PSD´s is quite good. Consequently, knowing the Li content of a particular alloy, the above expressions can be used to obtain the parameters a and b, such that the shape of the normalized δ´ PSD can be calculated.

As a test of the general applicability of our model and of the goodness of the fit, Figure 15 contains data from Jensrud and Ryum [17] and our fitted Weibull distribution. The a and b parameters were calculated using expressions (6) and (7). It is clear from these figures that our expressions can be applied to the results of other investigators.

CONCLUSIONS

1. The Weibull distribution can be used to describe the δ´ PSD´s in binary Al-Li alloys. The parameters a and b of the Weibull distribution over the experimental ranges investigated are a linear function of lithium content or δ´ volume fraction. Dilute binary alloys tend to have a coefficient of skewness < 0, and the coefficient of skewness becomes positive in more concentrated alloys. A symmetrical distribution is predicted to occur at Li $\cong$ 3.8 wt%.

2. The Al_3Zr particles act as preferential sites for the nucleation and growth of δ´ in the Al-Li-Zr system and as a consequence shift the distribution of the composite particles to larger sizes.

3. The coarsening behavior of δ´ in binary alloys follows Ostwald ripening kinetics. Increasing the lithium content increases the rate of particle coarsening.

4. The change in the PSD due to the presence of the Al_3Zr phase accelerates the aging kinetics of the alloy. Consequently, the zirconium alloy ages more rapidly than a similar alloy not containing zirconium.

ACKNOWLEDGEMENTS

The research was conducted under NSWC Contract No. #N0009-82-C-0485. We would like to thank A. P. Divecha of the Naval Surface Weapons Center for his support and encouragement. The continued encouragement of Mr. R. Schmidt is also greatly appreciated.

REFERENCES

1. W. Ostwald, Z. Phys. Chem., 34, 495 (1900).
2. C. Zener, J. Appl. Phys., 20, 950 (1949).
3. G. W. Greenwood, Acta Met., 4, 243 (1956).
4. I. M. Lifshitz and V. V. Slyozov, J. Phys. Chem. Solids, 19, 35 (1961).
5. C. Wagner, Z. Electrochem., 65, 581 (1961).
6. A. J. Ardell, Acta Met., 20, 61 (1972).
7. A. D. Brailsford and P. Wynblatt, Acta Met., 27, 489 (1979).
8. C. K. L. Davies, P. Nash and R. N. Stevens, Acta Met., 28, 179 (1980).
9. B. P. Gu, J. H. Kulwicki, G. L. Liedl, and T. H. Sanders, Jr., ``Coarsening of δ´(Al_3Li) Precipitates in an Al-2.8Li-0.3Mn Alloy´´, accepted for publication in Materials Science and Engineering.
10. B. P. Gu, G. L. Liedl, T. H. Sanders, Jr., and K. Welpmann, ``The Influence of Zirconium on the Coarsening of δ´(Al_3Li) in an Al-2.8Li-0.14Zr Alloy,´´ accepted for publication in Materials Science and Engineering.
11. B. P. Gu, G. L. Liedl, K. Mahalingam and T. H. Sanders, Jr., ``The Effect of Lithium Content on the δ´(Al_3Li) Particle Size Distribution in Binary Al-Li Alloys,´´ submitted to Materials Science and Engineering.
12. Aliminum-Lithium Alloys II, AIME, Warrendale, PA, December 1983.
13. E. A. Starke, Jr., T. H. Sanders, Jr., and I. G. Palmer, J. Metals, 33 (1981), pp. 24-33.
14. T. H. Sanders, Jr., ``Factors Influencing Fracture Toughness and Other Properties of Aluminum-Lithium Alloys,´´ Naval Air Development Center, Contract No. N62269-76-C-0271, Final Report (June 1979).
15. F. W. Gayle and J. B. Vander Sande, Scripta Metallurgica, 18 (1984), pp. 473-478.
16. S. Takajo, W. A. Kaysser and G. Petzow, Acta. Met., 32, (1984) p. 107.
17. O. Jensrud and N. Ryum, Materials Science and Engineering, 64, (1984) pp. 229-236.

Theoretical analysis of aging response of Al–Li alloys strengthened by Al_3Li precipitates

J GLAZER, T S EDGECUMBE and J W MORRIS Jr

The authors are with the
Materials and Molecular Research Division,
Lawrence Berkeley Laboratory, and the
Department of Materials Science,
University of California,
Berkeley, CA 94720, USA

ABSTRACT

The possibility of increasing the strength of aluminum-lithium alloys hardened principally by the δ' precipitates was investigated. The elastic theory of phase transformations was used to determine that the δ' precipitate is not expected to deviate from spherical morphology even at very large sizes. The theory of the critical resolved shear stress for dislocation glide through a random array of obstacles was used to predict a moderate strengthening increment if the precipitate size distribution could be made more uniform. Experimental and theoretical hardening curves for an experimentally determined precipitate size distribution were also compared.

I. INTRODUCTION

Although the aluminum-lithium alloys nearing commercialization are strengthened by copper-rich phases, the ideal aluminum-lithium alloy should be primarily strengthened by lithium-rich phases such as Al_3Li, which do not increase the density of the alloy. We have therefore conducted a theoretical investigation of the possibility of achieving significantly greater strengths than realized to date in near-binary alloys strengthened by δ'.

In the nickel-based superalloys it has been possible to obtain considerable strength increments through modification of the composition of the $L1_2$ precipitate γ' either to increase its misfit or to induce precipitation of the tetragonal phase γ''. Both of these avenues have been investigated for aluminum-lithium alloys, but have been unsuccessful thus far. It is possible that the search for elemental additions with these effects will eventually prove fruitful. In the meantime, near-binary aluminum-lithium alloys will be strengthened primarily by δ' precipitates with relatively low misfit strains.

Given that the crystallography of the δ' precipitate is fixed, we have investigated two other microstructural factors which could have a significant effect on the strength at constant precipitate volume fraction: (1) the precipitate morphology, and (2) the precipitate size distribution. These problems may be examined using the elastic theory of structural transformations and the random array theory of dislocation glide, respectively.

II. MORPHOLOGY

In other alloys hardened by $L1_2$ precipitates, three precipitate morphologies are observed: spherical, cuboidal, and plate-like [1]. Since platelet precipitates provide the greatest strengthening for a given amount of precipitated solute, it would be desirable to induce this morphology. We have focussed initially on identifying the conditions under which each of the observed shapes is preferred.

The linear elastic model of phase transformations is based on the recognition that elastic strains often influence the thermodynamics and kinetics of transformation. The theory may be used to predict the preferred shapes and habits of coherent inclusions [2,3]. The important assumptions are as follows:

(1) A coherent precipitate will have the shape and habit which minimize its contribution to the Helmholtz free energy of the material.

(2) The portion of the energy change which is a function of morphology is the sum of the surface energy and the elastic energy introduced by the precipitate. The surface energy is assumed to be isotropic.

(3) Since in most cases the unconstrained elastic constants of the precipitate are unknown, they are assumed to be the same as those of the matrix. This is a good assumption if the anisotropies are similar.

(4) The elastic energy of a misfitting coherent precipitate has been derived by Khatchaturyan using a version of the Eshelby cycle. The solution is the integral over the precipitate volume

$$\Delta F_{el} = (1/2)\int B(\mathbf{e})\ |\theta(\mathbf{k})|^2\ d^3k/(2\pi)^3 \qquad (1)$$

where $B(\mathbf{e})$ is an energy factor which depends on the habit of the precipitate, the elastic constants of the precipitate and the matrix, and the unconstrained transformation strain, and θ (k) is an amplitude which depends only on the shape of the inclusion. The determination of the preferred shape and habit of the precipitate is accomplished through a minimization of this integral.

(5) The above expression for the elastic energy is easily derived if the transformation is modelled using the Eshelby cycle [2,3,4]. In the first step, a body the size and shape of the inclusion is cut out of the material and allowed to transform while unconstrained by the matrix. The strain with respect to its original dimensions is designated ε^0, the stress-free transformation strain. Tractions are then imposed on the inclusion to return it to its original size and shape. The energy associated with this step is given by the product of the strain ε^0 with the stress σ^0 required to reverse the transformation strain integrated over the volume of the precipitate

$$\Delta F_1 = \int 1/2\sigma^o_{ij}\varepsilon^o_{ij}\ dV \qquad (2)$$

If the inclusion is now allowed to relax into another shape, its energy will decrease by an amount given by the integral over the precipitate volume

$$\int_V(\int_\varepsilon \sigma_{ij} d\varepsilon_{ij})dV \qquad (3)$$

where the ε_{ij} are the elements of the relaxation strain tensor. If the stress in (3) is decomposed into σ^0 and a relaxation stress, equations (2) and (3) may be combined into an expression for the elastic energy

$$\Delta F = \int(1/2\sigma^o_{ij}\varepsilon^o_{ij}\theta(r) - \sigma^o_{ij}\varepsilon_{ij}\theta(r) + 1/2\lambda_{ijkl}\varepsilon^o_{ij}\varepsilon^o_{kl})dv \qquad (4)$$

where the integral is now over the total volume of the system , $\theta(\mathbf{r})$ is the particle shape function, which is 1 inside the inclusion and zero outside it, and the λ_{ijkl} are the elastic constants of the precipitate and matrix.

The only unknown parameter in equation (4) is the relaxation strain field. Since the inclusion is a macroscopic defect, this strain field may be determined in the long-wavelength approximation by requiring that it obey static equilibrium. The Fourier transform of the strain must then be

$$\varepsilon_{ij}(\mathbf{k}) = \Omega_{im}(\mathbf{e})\sigma^o_{ml}e_j e_l\theta(\mathbf{k}) \qquad (5)$$

where $\Omega(\mathbf{e})=(k)^2A^{-1}_{ij}(\mathbf{k})$, A_{ij} is the dynamical matrix of the elastic medium and $\mathbf{e}$ is a unit vector in the direction of $\mathbf{k}$. Substituting into (4) gives the equation (1) for ΔF where $B(\mathbf{e})$

$$B(\mathbf{e}) = \lambda_{ijkl}\varepsilon^o_{ij}\varepsilon^o_{kl} - e_i\sigma^o_{ij}\Omega_{jk}(\mathbf{e})\sigma^o_{kl}e_l. \qquad (6)$$

The function $B(\mathbf{e})$ is a scalar energy factor which depends on particle shape only through the direction of the $\mathbf{k}$-vectors. $B(\mathbf{e})$ is positive semi-definite so that the lattice is mechanically stable.

The shape factor $\theta(\mathbf{k})$ is the Fourier transform of the real space shape function $\theta(\mathbf{r})$. Its amplitude depends on the inclusion shape through the allowed values of $\mathbf{k}$.

It can be shown easily that an infinitely thin plate is always the elastically preferred morphology of the precipitate. If the function $B(\mathbf{e})$ depends on $\mathbf{e}$, then there is some value of $\mathbf{e}$ for which $B(\mathbf{e})$ has its minimum value $B(\mathbf{n})$. The lower bound for the elastic energy is then given by

$$\Delta F \geq 1/2B(\mathbf{n})\int|\theta(\mathbf{k})|^2 d^3k/(2\pi)^3 = (V/2)B(\mathbf{n}) \qquad (8)$$

where V is the volume of the precipitate. The energy is minimized when the shape factor $\theta(\mathbf{k})$ approaches a δ-function with $\mathbf{k}$ equal to $\mathbf{n}$, that is, when the precipitate is an infinitely thin plate with normal vector $\mathbf{n}$.

The minimum energy shapes for a cubic inclusion in an anisotropic cubic matrix can be predicted easily in the limits of zero and infinite volume. At small volumes, the precipitate will be spherical because the surface energy term will dominate. At large sizes, the surface-to-volume ratio is low, and the precipitate will be a thin plate with a well-defined habit which minimizes the volumetric elastic energy term.

The regimes in which these morphologies are stable may be determined by comparing their energies. The sphere to plate transition may be modelled by considering the energy of an ellipsoid as a function of aspect ratio. Figures 1-3 illustrate the variation with aspect ratio of the surface, elastic, and total energies as a function of the precipitate volume. (The aspect ratio, K, is defined so that a sphere has an aspect ratio of one and plates and rods have aspect ratios which are greater than, and less than one respectively. The parameter g is a

Table I. Numerical results for the minimum characteristic diameter (in μm) of the cube and plate. Experimental data is given where available.

PPT	cube		plate	
	theory	expt	theory	expt
δ'-Al	1.4	>0.3	>60m	-
γ'-Ni	0.13	0.1-0.2	200μm	-

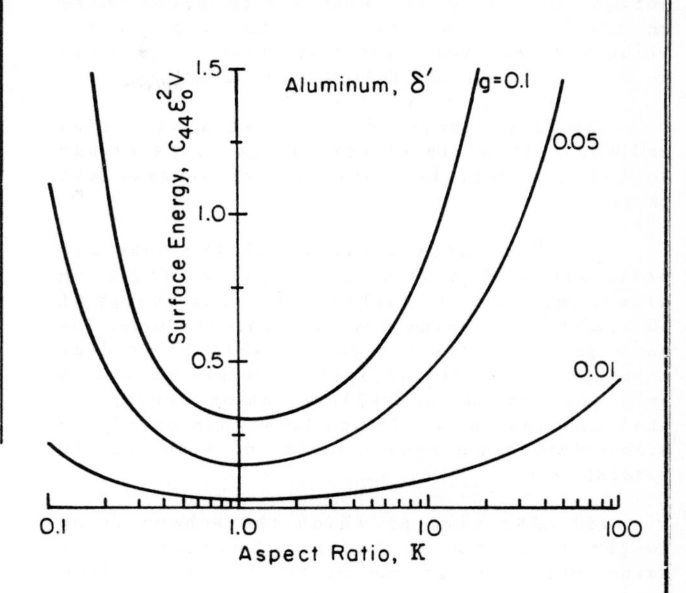

1. Surface energy of δ' precipitate as a function of aspect ratio. The parameter g is given by $(\gamma/c_{44}(\varepsilon^o)^2)(4\pi/3V)^{1/3}$.

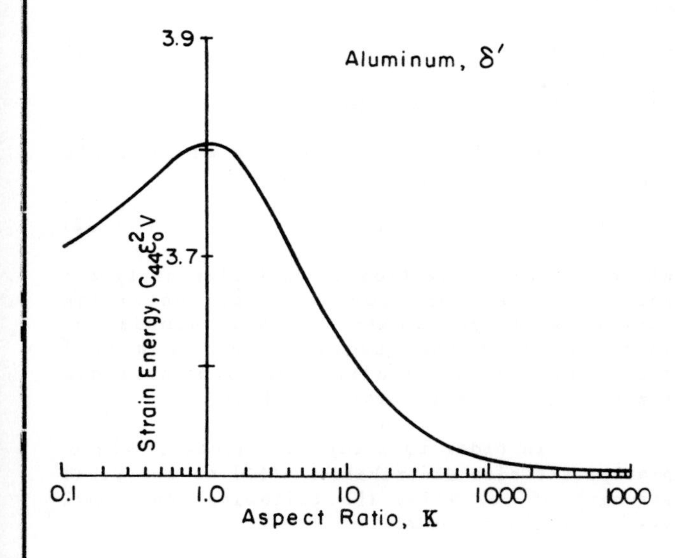

2. Elastic strain energy of the δ' precipitate as a function of aspect ratio and the parameter g.

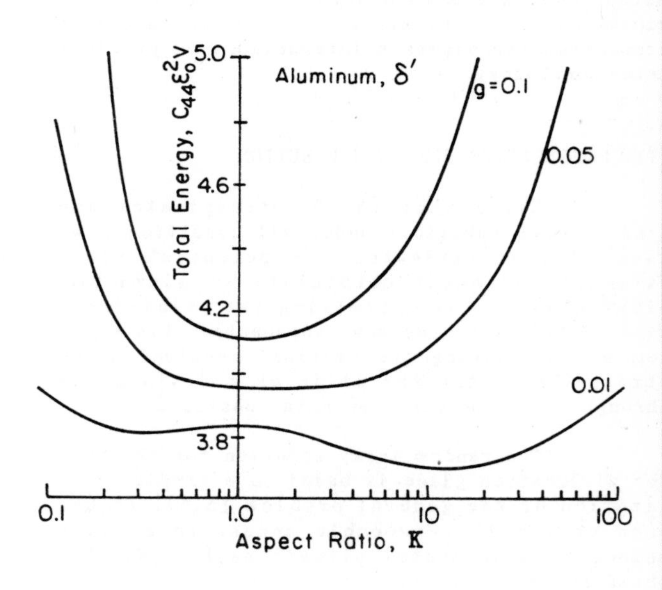

3. Total energy of the δ' precipitate as a function of aspect ratio and the parameter g.

dimensionless variable given by $(\gamma/c_{44}(\varepsilon^o)^2)(4\pi/3V)^{1/3}$ where γ is the surface energy per unit area). Figure 4 compares these energies with the energy of the cube shape, which was computed separately. From these plots we have obtained the following results:

(1) The sphere is stable at small sizes and the thin plate at very large sizes as expected. The cube is stable in an intermediate range.

(2) The sizes above which the cube and plate are preferred were calculated for δ' in aluminum, and γ' in nickel. A surface energy of 10 erg/cm^2 was assumed in all calculations. The numerical results (given in Table 1) predict that spherical and cuboidal γ' should be stable in nickel-based superalloys, as observed, but that cuboidal δ' would not be stable until the precipitate had a mean diameter of more than 60 meters!

(3) The size at which the sphere is no longer preferred is a strong function of the anisotropy ratio of the matrix. Since aluminum is nearly isotropic the spherical shape remains stable to very large sizes; the elastic energy is not significantly lowered by the shape change.

(4) Other results indicate that a sphere is never the stable shape of a tetragonal precipitate in a cubic matrix.

This model can be used to examine a number of other relevant problems such as the effect of external stress on the precipitate morphology, the morphology of a tetragonal precipitate in a cubic matrix, and the influence of precipitate-precipitate interactions on precipitate morphology.

III. PRECIPITATE SIZE DISTRIBUTION

Given that the δ' precipitates are likely to be spherical under all conditions, we have also investigated the potential yield strength increment obtainable in aluminum-lithium alloys by optimizing the precipitate size distribution. We have approached this problem by calculating the critical resolved shear stress (CRSS) for the glide of a dislocation through a random array of point obstacles.

The random array solution for the CRSS for dislocation glide is based on a simple idealization of the general problem [5,6]. It has been used with reasonable sucess to model a number of experimental situations [7,8,9]. The chief assumptions are as follows:

(1) The dislocation is a flexible line of constant line tension T for a given mean square obstacle spacing l_s.

(2) The obstacles to dislocation glide are modelled as a random array of immobile point barriers. The properties of the point obstacle are adjusted so that the interaction of the dislocation with the obstacle is mathematically equivalent to its interaction in the glide plane with the physical obstacle.

(3) The configuration of the dislocation is described by a unique set of pinning points. The CRSS of the array is reached when the dislocation bypasses the weakest point in the strongest configuration.

(4) For a random array of identical obstacles the problem may be solved analytically using standard statistical techniques. It is convenient to define a dimensionless CRSS

$$\tau^* = \tau l_s b/2T \tag{9}$$

where b is the Burgers' vector in the glide plane. The analytic solution is then given by

$$\tau^* = 0.8871(\beta)^{3/2} \tag{10}$$

where β is a dimensionless obstacle strength given by

$$\beta = F/2T \tag{11}$$

and F is the force to bypass the obstacle.

(5) The CRSS for a mixture of obstacle types is a quadratic sum

$$\tau^2 = \Sigma x_\alpha(\tau_\alpha)^2 \tag{12}$$

where x_α is the fraction of obstacles of type α and τ_α is the CRSS for an array containing obstacles of type α only. Computer simulation indicates that the quadratic sum is a good approximation if the distribution of obstacle sizes is relatively continuous [10].

In order to study the effect of the δ' precipitate size distribution (PSD) on the yield strength of the alloy the following additional assumptions were made:

(1) The precipitates coarsen according to the LSW rate law and at constant volume fraction (in the interior of a grain) [11].

(2) The shape of the PSD as a function of $r/\bar{r}$ does not evolve during coarsening [11].

(3) Each spherical precipitate may be reduced to a set of point obstacles whose strengths correspond to the effective radii of the precipitate in the glide planes it intersects. Therefore, even if all precipitates are the same size, there will be a distribution of obstacle sizes in the glide plane.

(4) The strength of the obstacles is assumed to be a function of the radius of the sheared ordered precipitate only. This is equivalent to neglecting the misfit of the precipitate.

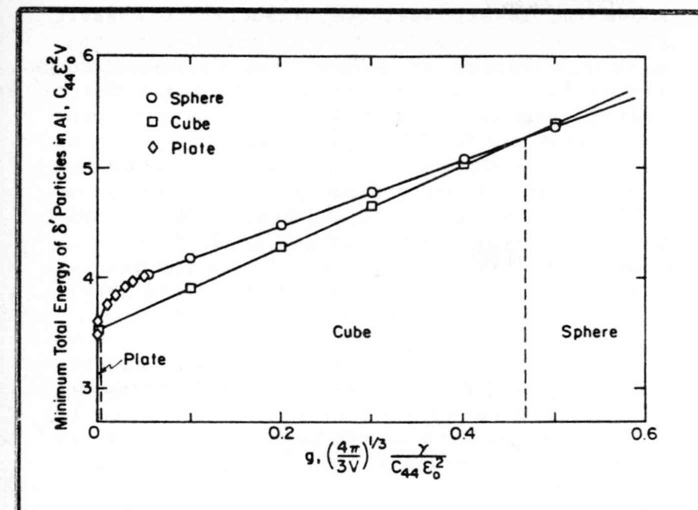

4. **Relative energies of the sphere, cube, and plate morphologies as a function of g.**

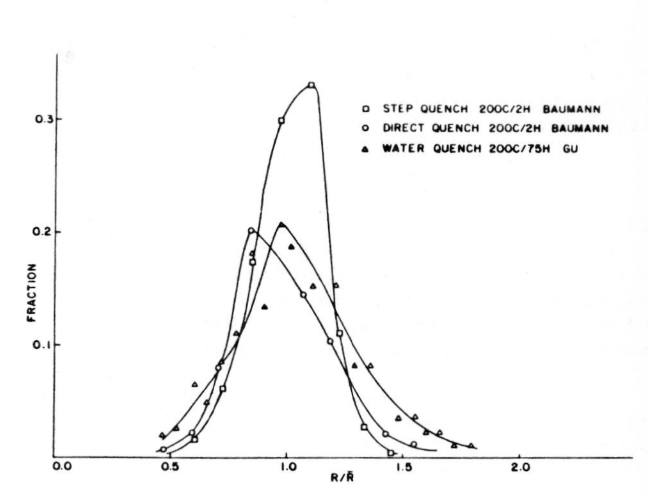

5. **Experimentally determined precipitate size distributions for some aluminum-lithium alloys.**

6. **Comparison of the variation of the critical resolved shear stress with aging time for the PSD's in Figure 5.**

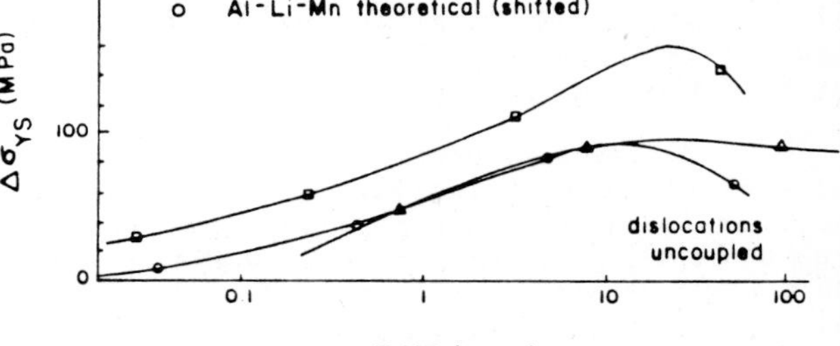

7. **Comparison of the theoretical and experimental aging curves for Al-2.78Li-0.3Mn alloy. Time scale refers to experimental curve.**

(5) The strengths of the obstacle may be related to the maximum obstacle strength which corresponds to the looping radius at which Orowan looping is first preferred over shearing of the precipitate. The Orowan condition puts an upper limit on the strength of the obstacle whatever its physical size. Based on available TEM micrographs, an upper limit on the looping radius in aluminum-lithium is approximately 300Å for a mean square particle spacing of about 1.1μm [12]. The antiphase boundary energy is not required to compute the obstacle strengths.

(6) There is considerable confusion in the literature over the appropriate form of the line tension. We have adopted the approach of Bacon, Kocks, and Scattergood [13] and taken the line tension to be

$$T=(Gb^2/4\pi(1-\upsilon))\ln(l_s/b). \quad (13)$$

This is the correct form of the line tension for screw dislocations. It is not necessary to consider edge dislocations since macroscopic yielding requires the motion of screw, edge, and mixed dislocations and it is harder to move screw dislocations through the array.

We have also adopted the upper limit on β of 0.7 suggested by Bacon, Kocks, and Scattergood.

(7) The calculated CRSS due to the obstacle distribution represents the increment in the total strength of the alloy due to precipitate hardening only and should properly be denoted $\Delta\tau$.

These criteria allow the CRSS for hardening by δ' to be calculated for various PSD's. We have calculated the CRSS for experimentally obtained PSD's measured by Gu et al. [11] and Baumann [14], respectively, and for the case of uniform precipitate size. The PSD's and the corresponding 'aging curves' are shown in Figures 5 and 6 respectively. As can be seen from the figures, the more sharply peaked the distribution, the more effectively it strengthens. This result is in qualitative agreement with the results of a study by Munjal and Ardell [15] of the effect of PSD's on the CRSS of Ni-Al alloys. The theoretical increment in $\Delta\tau_c$ in Figure 6 is approximately 30%, so this result may have alloy design implications.

A direct comparison of the theoretical and experimental strengthening increment from the coarsening of δ' is possible for the PSD from Gu et al. [11] for Al-2.78Li-0.3Mn aged at 200°C which is shown in Figure 5. The experimental and theoretical aging curves are shown in Figure 7. The theoretical coarsening rate has been fixed using the experimentally determined LSW rate constant. Since the quenched alloy undoubtedly contains some atom clusters, the theoretical aging curve has been shifted to slightly shorter times to obtain the best fit. The strength increment for the both yield strength curves has been taken as the strength above the lowest measured strength. To convert the theoretical CRSS values to yield strengths, a Taylor factor of 3 has been assumed. The figure shows that the theoretical and experimental aging curves are in excellent agreement up to peak strength. Beyond peak strength, the model is no longer valid since it does not account for the uncoupling of the paired dislocations after Orowan looping begins. This uncoupling would cause the strength to drop off more gradually after peak strength. The theoretical aging curve for a uniform distribution of precipitates is shown for comparison.

In the future, we hope to extend this model to consider the effect of other strengthening precipitates such as thin plates and to other problems such as the prediction of texture and the superposition of several hardening mechanisms.

ACKNOWLEDGEMENTS

The authors are grateful to Drs. T.H. Sanders, Jr. and S.F. Baumann for helpful discussions, experimental data, and access to unpublished work. This work was supported by the Director, Office of Basic Energy Sciences, Materials Science Division of the U.S. Department of Energy under Contract No. DE-AC03-76SF00098. One of the authors, JG, was supported by an ATT Bell Laboratories Fellowship.

REFERENCES

1. H. Gleiter and E. Hornbogen, Mat. Sci. Eng., **2**, 1967, 285.

2. A.G. Khachaturyan, **Theory of Structural Transformations in Solids**, Wiley and Sons, New York, 1983.

3. J.W. Morris, Jr., A.G. Khachaturyan and S.H. Wen, in **Solid-Solid Phase Transformations**, H.I. Aaronson, D.E. Laughlin, R.F. Sekerka and C.M. Wayman, eds., The Metallurgical Society-AIME, Warrendale, PA, 1983, 101.

4. S.H. Wen, E. Kostlan, M. Hong, A.G. Khachaturyan, and J.W. Morris, Jr., Acta Met., **29**, 1981, 1247.

5. K. Hanson and J.W. Morris, Jr., J. Appl. Phys., **46**, 1975, 983 and 2378.

6. J.W. Morris, Jr. and D.H. Klahn, J. Appl. Phys., **45**, 1974, 2027.

7. A. Melander, Scand. J. Metall., **7**, 1978, 109.

8. A. Melander and P.Å. Persson, Acta Met., **26**, 1978, 267.

9. A. Melander and P.Å. Persson, Metal Science, **12**, 1978, 391.

10. S. Altintas, PhD. thesis, University of California, Berkeley, 1978.

11. B.P. Gu, G.L. Liedl, T.H. Sanders, Jr., and K. Welpmann, Mat. Sci. and Eng., in press.

12. J. Th. M. DeHosson, A. Huis in't Veld, H. Tamler and O. Kanert, Acta Metall., **32**, 1984, 1205.

13. D.J. Bacon, U.F. Kocks, and R.O. Scattergood, Phil. Mag., **28**, 1973, 1241.

14. S.F. Baumann, unpublished work.

15. V. Munjal and A.J. Ardell, Acta metall., **24**, 1976, 827.

□

$Al_3(Li,Zr)$ or α' phase in Al–Li–Zr system

F W GAYLE and J B VANDERSANDE

The authors are in the Department of Materials Science and Engineering at the Massachusetts Institute of Technology, Cambridge, MA 02139, USA

SYNOPSIS

The Ll_2-ordered $Al_3(Li,Zr)$ phase, or α', has been shown to exist in the Al-Li-Zr system with a wide variation of Li:Zr ratios. The α' and δ' (Al_3Li) distributions after various heat treatments are presented. Physical and thermodynamic properties of the α' phase as a function of composition are calculated.

1. INTRODUCTION

Zirconium is used in many aluminum alloys to inhibit recrystallization during alloy processing,[1] afford superplastic properties,[2,3] and provide strengthening.[4] In aluminum alloys, zirconium promotes the formation of a very fine distribution of metastable, coherent, Ll_2-ordered Al_3Zr precipitates. Zirconium is used in Al-Li alloys, as well, to preserve unrecrystallized microstructures[5] and to provide superplastic properties.[6,7] Zirconium additions have also been proposed to promote formation of a strengthening $Al_3(Li,Zr)$ precipitate.[8,9] The Al_3Li phase, or δ', is the strengthening precipitate in Al-Li alloys, and is also a metastable, coherent, Ll_2-ordered phase.

There has been relatively little investigation of phase transformations in the Al-Li-Zr system. In recent studies of the system, workers have assumed that the precipitate phase (meta-)stable at normal solution-heat-treat temperatures is Al_3Zr.[10,11] Other work suggests that the high-temperature phase is a complex $Al_3(Li,Zr)$ phase, based on diffraction contrast[8] and electron energy loss studies.[9] More recent work utilizing image calculations indicates that the phase which precipitates at temperatures above $\sim 450°C$ is an Ll_2-ordered $Al_3(Li,Zr)$ phase whose Li:Zr ratio ranges from approximately 2:1 to 1:4 depending upon heat treatment and precipitation mechanism (i.e., discontinuous versus continuous precipitation).[12] For the purposes of this paper such an $Al_3(Li,Zr)$ phase will be denoted α', whereas the name δ' will be reserved for the Al_3Li phase.

The influence of the $Al_3(Li,Zr)$ phase on properties of Al-Li alloys depends upon the distribution and properties of the phase. These two characteristics are in turn influenced by composition of the phase (i.e., Li:Zr ratio) and processing conditions. In the present paper the α' and δ' distributions in a high-Zr alloy resulting from various heat treatments are presented, along with the calculated thermodynamic and physical properties of the α' phase as a function of Li and Zr content of the phase. An understanding of the properties of the α' phase is necessary both in assessing the effect of the phase on microstructure and alloy properties and in designing ternary or higher order Al-Li-Zr alloys for specific applications. For example, thermodynamic properties of the phase determine high temperature stability, phase composition, and antiphase boundary energy (including ability to disperse slip). The phase composition, in turn, determines the volume fraction of α' which results from a given alloy Zr content. This last factor is important since a large Zr supersaturation is relatively difficult to produce and effectively limits the volume fraction of α'.

2. EXPERIMENTAL PROCEDURE

The alloy used in this study was Al-2.34Li-1.07Zr (wt.%) produced using rapid solidification techniques to retain the Zr in a supersaturated solid solution. The alloy was vacuum atomized at 960°C, vacuum hot compacted at 340°C and extruded at 200-300°C at a 29:1 ratio into a 13 mm rod. Solution heat treatments (SHT) were carried out at 450 to 580°C for various times with samples

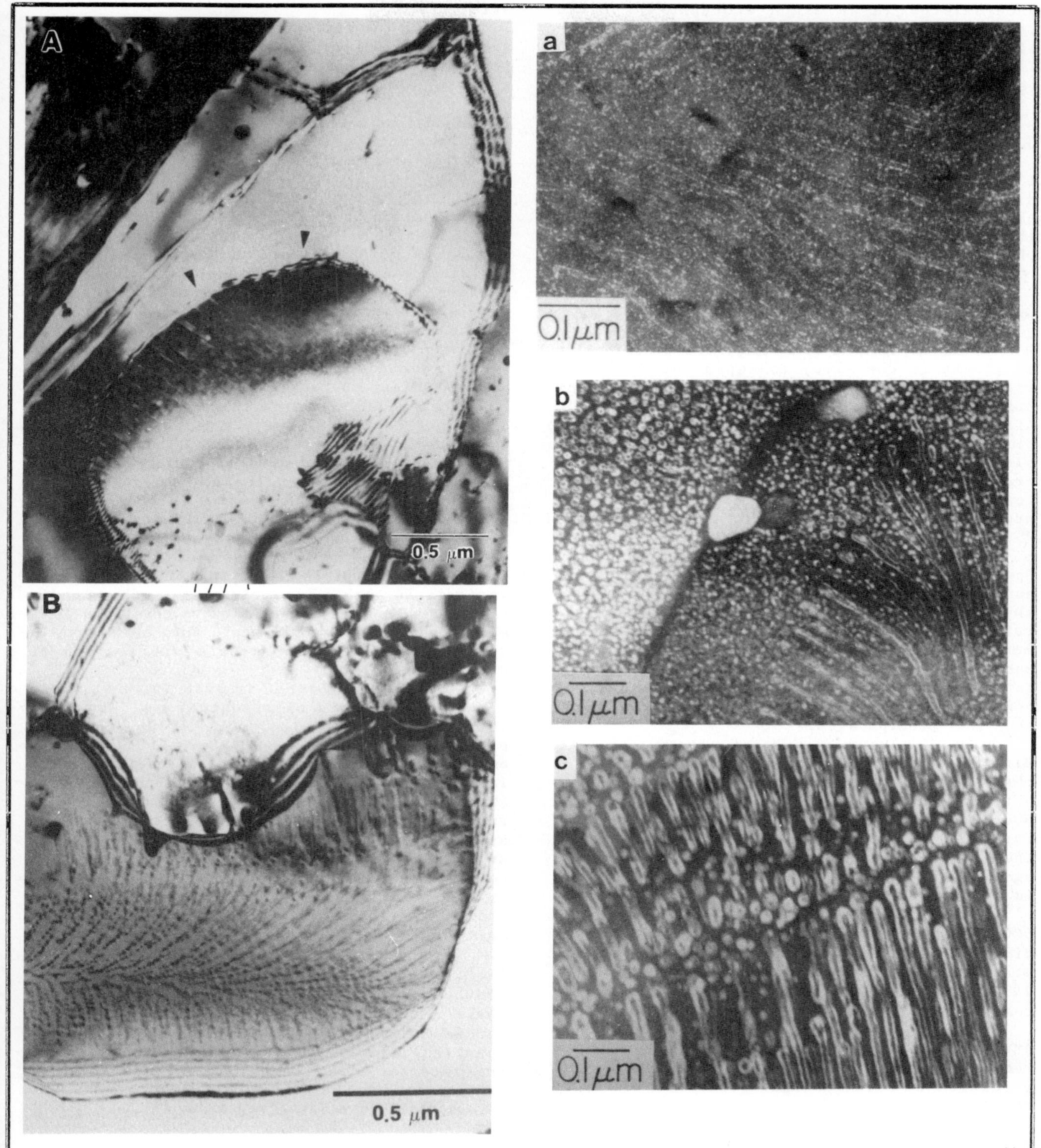

Fig. 1. Discontinuous precipitation of α' following extrusion (a) and solution heat treatment 500°C/2 hours (b).

Fig. 2. Superlattice darkfield image showing δ' nucleating and growing at α'/matrix interface. SHT 500°C/2 h + aged at 190°C for: a) 0 h, b) 0.25 h, and c) 4 h.

encapsulated in an argon atmosphere. Aging treatments were conducted at 190°C. Samples for electron microscopy were electropolished in a 30 % nitric acid in methanol solution using standard techniques.

3. RESULTS

A microstructural study of the precipitation and distribution of α' and δ' phases within the Al-Li-Zr alloy is presented first. Subsequently, thermodynamic and physical properties of the α' phase will be calculated for various compositions of the phase.

3.1. Distribution of α' and δ'.

General Precipitation Behavior. The distribution of α' within the alloy is dependent upon heat treatment. Following extrusion some α' has precipitated discontinuously as shown in Fig. 1a. After a standard 500°C/2 h SHT, further discontinuous precipitation of α' is evident (Fig. 1b), along with normal nucleation of spherical α' in areas not depleted of Zr by the discontinuous reaction. The α'/matrix interface serves as a preferred nucleation site for δ' (Fig. 2). The δ' in Fig. 2a images brightly, and decorates the α'/matrix interface after a SHT and water quench. Upon aging at 190°C, δ' forms complete envelopes surrounding the α'. These envelopes increase in thickness with aging time (Fig. 2b,c) (See also references 8-11 for more on the aging phenomenon.) The δ' envelopes nucleate and grow on both the filamentary and spherical α' particles. In regions devoid of α', δ' undergoes its typical homogeneous nucleation behavior and appears as spherical precipitates.

Distribution of α' After Heat Treatment. The α' phase remains in a very fine distribution even after extended heat treatment at 450°C. After 500 h at 450°C the α' filaments and spheres are typically 12-25 nm in diameter (Fig. 3). (Since the α' phase often images with very low contrast, aged specimens are used here to provide a clearly imaging outline of δ').

The filamentary α' present after 2 hours at 500°C (Fig. 1 and 2) is typically 6 to 10 nm in diameter, whereas the spherical α' appears somewhat smaller, at approximately 5 to 6 nm. After 24 hours at 500°C, both morphologies of α' have coarsened to diameters on the order of 30 nm.

The α' precipitates present after the 580°C/15 h SHT have coarsened and reshaped to reduce interfacial area (Fig. 4). Although some precipitates have retained aspect ratios up to ∿8, in general the filamentary-type precipitates have coarsened to low-aspect ratio shapes. These precipitates are typically 30 to 50 nm in diameter.

Epitaxial Precipitation of δ' on α'. It is of interest to determine the exact spatial relationship of the minor sublattice (i.e., Li-containing) of the α' phase with that of the nucleating δ'. The continuous nature of the sublattice across the α'/δ' interface is demonstrated here.

Within the fcc lattice, lattice sites can be arranged into four distinct sublattices. The preferential occupation of any one of the sublattices by solute atoms results in the $L1_2$-type order, reducing the symmetry to simple cubic. For Al_3Li four different variants correspond to the minor sublattice being situated on the four different atomic sites within the unit cell (see Fig. 5).

There are two possible scenarios to consider for nucleation of the embryonic δ' decorating the α' filaments in Fig. 2a. One is that perfect epitaxy is maintained across the interface, i.e., the minor sublattice for δ' nuclei is of the same variant as that for the α' substrate. In such a case all δ' nuclei on a given filament would be of the same variant and thus form a continuous envelope of δ', free of antiphase boundaries (APB), surrounding the α'. This corresponds to actual micrograph images for aged specimens.

The second scenario to be considered, and shown not to be the case, is as follows. Here, α' serves as a nucleation site for δ', but either no particular variant of δ' is preferred (leaving 4 variants of δ' to nucleate), or the variant matching that of the substrate is energetically unfavorable, as is the case if Li and Zr prefer to be nearest neighbors across the interface (leaving 3 possible variants of δ'). Figure 6 schematically depicts the growth of two different variants in such a situation. If such a condition existed, then the continuous envelopes of δ' which eventually develop would have APB's running through them between sites of nucleation of differing variants.

Actual APB's formed in such a manner would occur in a variety of orientations. Those of low-energy orientation, e.g., {100}, could be expected to exist as APB's within the δ' envelope. As such they would be readily visible in an appropriate superlattice dark field image (such that $g \cdot R \neq 0$, where g is the imaging reflection and R is the displacement vector associated with the APB). APB's of high energy orientation, e.g., lying close to {111}, would heavily groove. Examples of the type of grooving to be expected at the 190°C aging temperature are shown for deformation APB's lying on {111} planes in Figure 7.

In actual micrographs, neither grooving nor APB's are apparent for the aged specimens which had been heat treated at 500-580°C. Thus for these conditions it appears that the epitaxy across the α'/δ' interface is perfect, i.e., the identical minor sublattice is maintained across the interface. The 450°C/500 h SHT samples exhibit somewhat different behavior which will be discussed in a later paper.

3.2. Calculated Properties

For convenience, some characteristics of the Al-Li and Al-Zr systems will be examined before calculation of properties of the α' phase are presented.

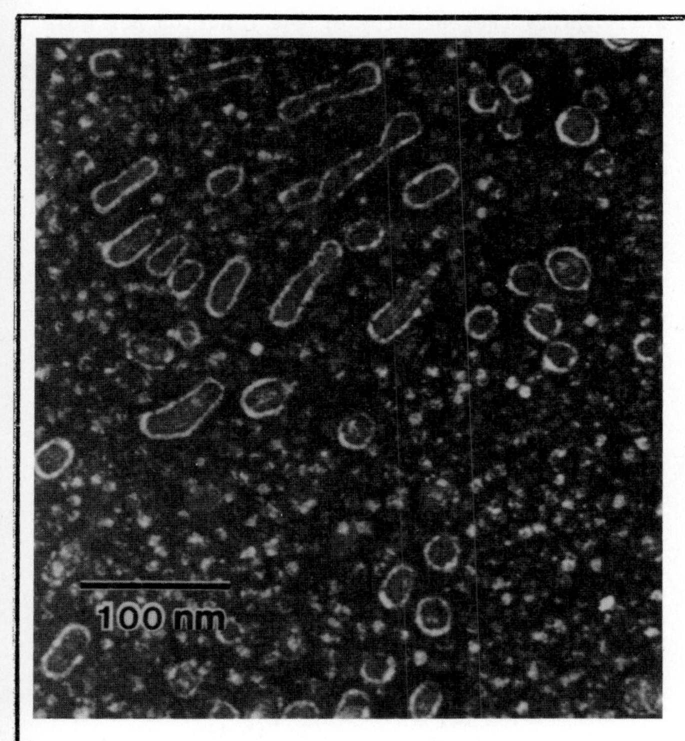

Fig. 3. SHT 450°C/500 h + aged 190°C/15 min. The α' phase, in a fine distribution, is enveloped in a thin layer of δ'.

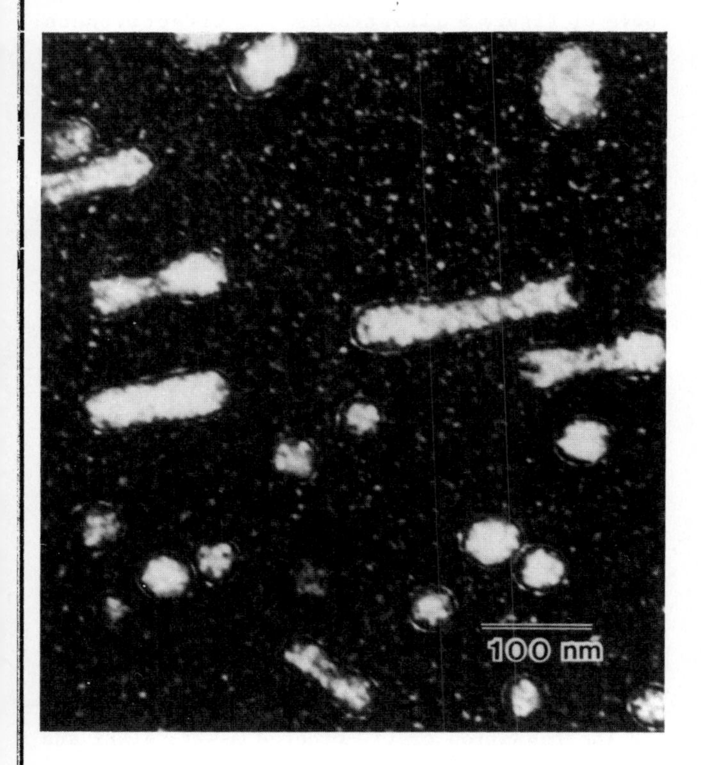

Fig. 4. SHT 580°C/15 h + aged 190°C/15 min. The α' phase images brightly in this condition, and is surrounded by envelopes of δ'.

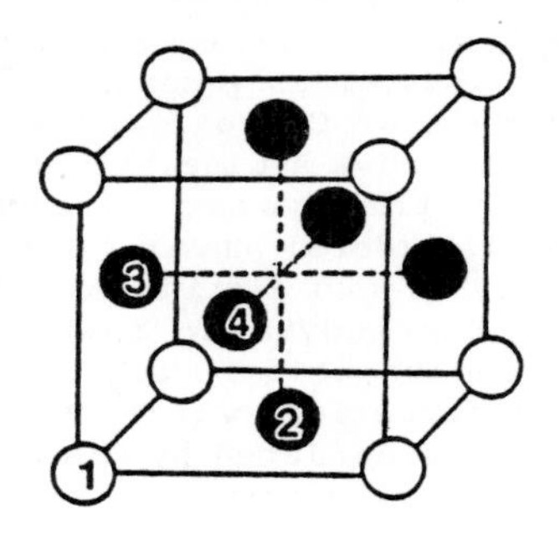

Fig. 5. The $L1_2$ unit cell. In Al_3X, four $L1_2$ variants are obtained when "X" occupies site 1 (as shown) or sites 2, 3, or 4.

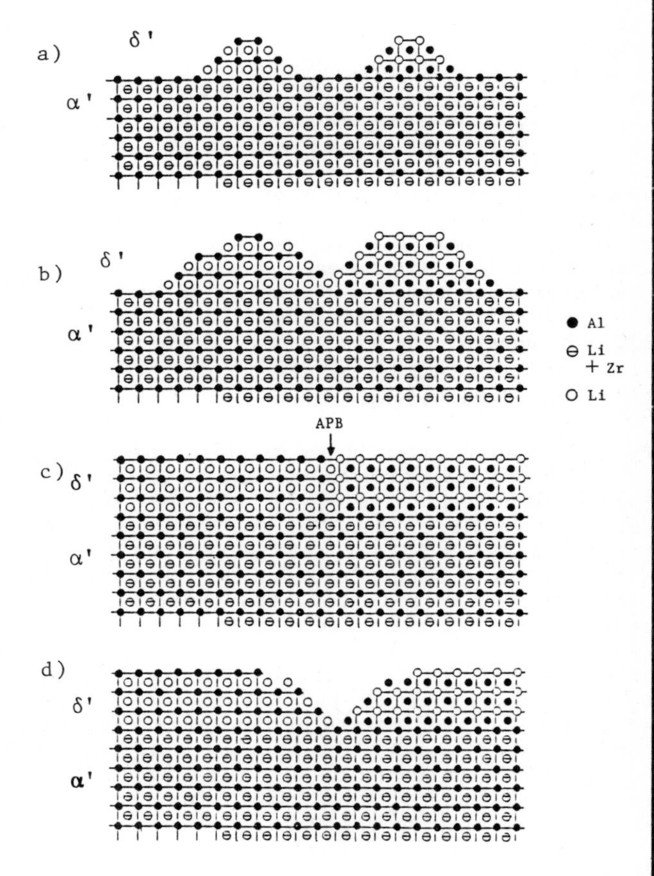

Fig. 6. Schematic showing nucleation of different variants of δ' on α'. a) nucleation of two variants, b) growth, c) growth of continuous δ' layer, requiring APB in δ', and c) grooving of δ' layer due to high APB energy. Such nucleation of differing variants is not observed.

3.2.1. The Al-Li System

The α - δ' System. The phase diagram for the metastable α - δ' system as generally used in the literature is that of Cocco[13] and Williams.[14] Their construction shows a miscibility gap in the α-field whereby the system separates into disordered and $L1_2$-ordered phases. However, such a miscibility gap of two distinct crystal structures violates thermodynamic restrictions on phase diagram construction: $L1_2$ ordering is a first-order transformation, which requires two-phase regions to be bordered by the two single-phase fields in binary diagrams.

Data points for the metastable α - δ' system have been given in the literature.[14,15] Recently a more reliable, higher temperature δ' solvus at 13 at.% Li has been determined.[16] Our interpretation of the existing data points is given in Fig. 8. Also included in this construction is a critical ordering temperature for δ' which has been calculated based on data in the literature for anti-phase boundary energy. The details of this calculation are presented below.

Thermodynamic Properties. For an $L1_2$-ordered phase, antiphase boundary (APB) energy, and critical ordering temperature (T_c) are directly related[17]:

$$\gamma_{APB} = \frac{4\,k\,T_c}{a^2\,\sqrt{3}\,0.82} \qquad (1)$$

where k is the Boltzmann constant, a is the lattice parameter, and 0.82 is the value for a parameter which relates to ordering energy calculations for $L1_2$ structures. Averaging values in the literature for δ' APB energy of 160[18] and 195 $ergs/cm^2$,[2,19] and using Eqn. 1, T_c = 665°K = 392°C is determined.

3.2.2. The Al-Zr System

The $L1_2$-ordered Al_3Zr phase has been observed to be stable at temperatures as high as 600°C.[20,21] The metastable $L1_2$ and equilibrium DO_{23} structures are very similar. The two have identical first nearest neighbor configurations, which accounts for a large fraction of the ordering energy.[22] Thus it is expected that the metastable phase should have a critical ordering temperature (T_c) close to that of the equilibrium DO_{23} phase. T_c is the critical ordering temperature with respect to the disordered α-solid solution, and from thermodynamic considerations would be greater than the melting temperature. For the equilibrium phase, T_m = 1582°C = 1855°K. Thus an estimate of T_c = 1855°K for the metastable phase appears reasonable.

Another approach to estimating T_c is to use the relationship for $L1_2$ structures relating T_c to the enthalpy of disordering (dissolution) derived by Yang and Li[17,23]:

$$\frac{\Delta H}{R\,T_c} = 1.072 \qquad (2)$$

where ΔH is the enthalpy of the reaction per mole of Al_3X, and R is the gas constant. Using a value of ΔH for dissolution of Al_3Zr (implied DO_{23} structure) of 22,970 joule/mole,[24] a T_c = 2577°K = 2304°C is calculated. Thus, for the $L1_2$ structure, a T_c of 1582°C = 1855°K appears to be a reasonable estimate, and will be used for the purposes of this discussion. Using this value for T_c in Eqn.1 gives a {111} APB energy of 447 $ergs/cm^2$.

The diffusivity of Zr in aluminum is very low, and is given as D = 728exp(-240/RT) cm^2/sec, with Q given in kJ/mole.[33] In comparison, D_o = 4.5 cm^2/sec and Q = 138 kJ/mole for the diffusion of Li in aluminum.[33]

3.2.3. Properties of the Ternary $Al_3(Li,Zr)$ Phase

Volume Fraction α'. The substitution of Li for Zr in Al_3Zr provides a means for increasing the volume fraction of hard, Zr-rich second phase without the problems associated with producing large Zr supersaturations. For a given mole fraction of Zr available to precipitate as $Al_3(Li_x,Zr_{1-x})$, the volume fraction second phase increases with x as follows:

$$f_{\alpha'} = \frac{1}{(1-x)}\,f_{Al_3Zr}$$

where f_{Al_3Zr} is the volume fraction Al_3Zr available from the given Zr level of the alloy. The curve of Figure 9 illustrates the effectiveness of Li substitution in increasing the fraction of precipitate.

Lattice Misfit and Interfacial Energy. The lattice misfit for δ' is given as -0.08 percent,[25] whereas that of Al_3Zr is $\sim$+0.8.[26,27] The lattice misfit for α' can be expected to vary with Li:Zr ratio by the rule of mixtures as given in Figure 10. The low misfit associated with the Li-rich compositions will thus lead to a low associated strain energy.

The δ'/matrix interfacial energy is approximately 14 $ergs/cm^2$.[28] The $L1_2$-ordered Al_3Zr/matrix interfacial energy has been given as 66 $ergs/cm^2$.[29] It is expected that the α' phase will have an interfacial energy of intermediate value. Indeed, assuming the random distribution of Zr and Li on the sublattice, and counting bonds across the interface, a simple weighted average for interfacial energy is determined:

$$\gamma_{\alpha'} = 66 - 42x \quad ergs/cm^2 \qquad (3)$$

Thermodynamic Properties. Values of entropy and enthalpy for the formation of α' can be estimated. Thermodynamic values are calculated here for the formation of $L1_2$-ordered Al_3Zr, Al_3Li and α' from Al_3Zr and Al_3Li in the disordered state.

Enthalpy. Several models of first-order type ordering reactions have been developed, and in each case the enthalpy change associated with ordering is directly related to the critical

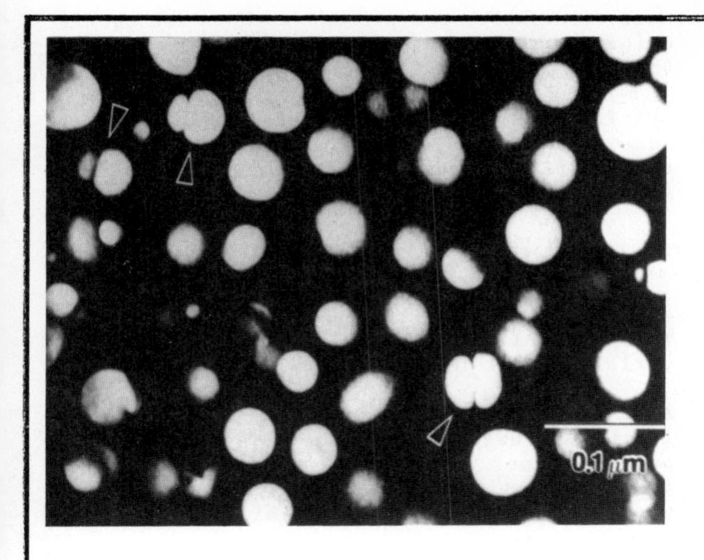

Fig. 7. Low Zr alloy, tensile gauge sample subsequently re-aged 190°C/4 h. Area of high-energy deformation APB is reduced when grooves form in precipitate surface.

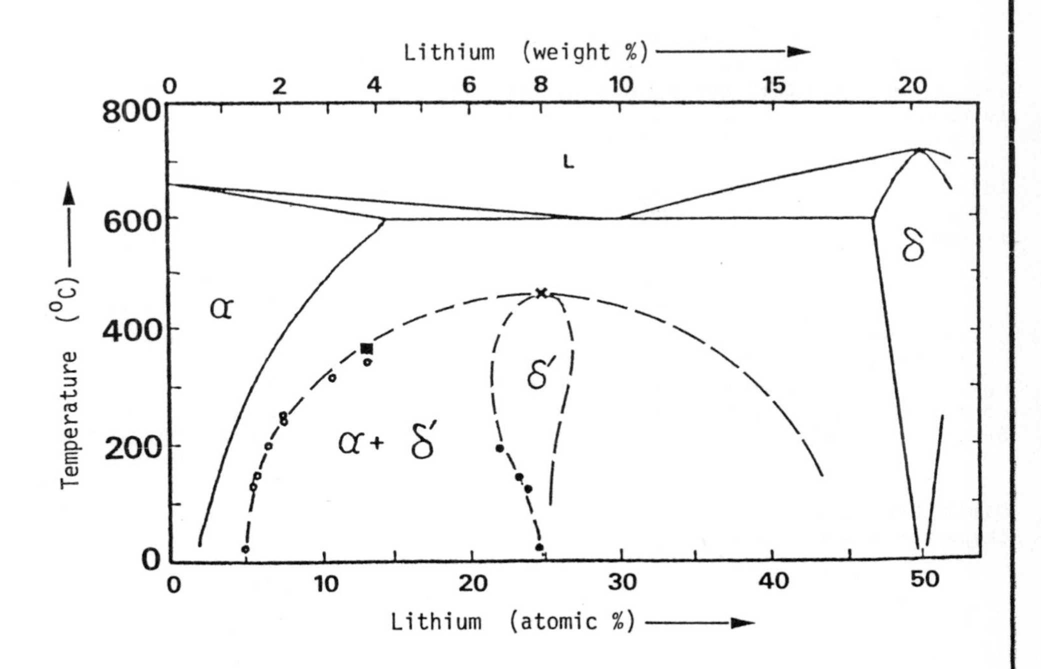

Fig. 8. The α-δ' system as proposed in the present study. T_c (X) for δ' has been calculated (see text). Data points O,● from ref. 14, 15, ■ from ref. 16.

ordering temperature (T_c) of a phase.[17] The best thermodynamic analysis of $L1_2$ ordering appears to be the quasi-chemical, tetrahedral model of Yang and Li,[17,23] which yields the relationship of Eqn. 2. Using T_c = 738°K for δ' and T_c = 1855°K for Al_3Zr, as derived above, the enthalpy of ordering is given by Eqn. 2 as $\Delta H(\delta')$ = 6,580 joules/mole and $\Delta H(Al_3Zr)$ = 16,540 joules/mole.

In the perfectly-ordered $Al_3(Li,Zr)$ phase, there are no Li-Zr nearest neighbor pairs. Thus, considering a nearest-neighbor pair-wise potential model, the ΔH associated with the ordering of α' is simply a weighted average of the ΔH values for the binary phases. The enthalpy associated with the formation of $Al_3(Li_x,Zr_{1-x})$ can then be given, in joules/(mole α'), as:

$$\Delta H_{\alpha'} = -16,540 + 9960x. \qquad (4)$$

Entropy. The critical ordering temperature, T_c, for an $L1_2$-ordered phase is that temperature for which the reaction

$$A_3B_{ord} = (3\underline{A} + \underline{B})_{fcc-ss}$$

has an associated free energy change, ΔG, of zero. Thus

$$\Delta G = \Delta H - T\Delta S = 0. \qquad (5)$$

Combining Eqns. 2 and 5 and setting T = T_c, we arrive at an entropy of ordering, ΔS_o, of 8.92 joules/(mole°K) for the ordering reaction.

The additional configurational entropy term for α', due to the presence of two species, Li and Zr, on the minor sublattice, is positive and is readily calculated. There are an Avogadro's number of Li-Zr sublattice sites per mole of α'. The configurational term is thus equal to the ideal ΔS of mixing for Al_3Li and Al_3Zr. The configurational entropy term for $Al_3(Li_x,Zr_{1-x})$ is then,

$$\Delta S'_{conf} = -R(x\ln x + (1-x)\ln(1-x)) \qquad (6)$$

after Gaskell.[30] The prime is used to indicate that the term is due to the presence of two species on the sublattice rather than to the configurational entropy associated with $L1_2$ ordering.

Gibbs Free Energy. The free energy of the α' phase can now be described as:

$$\Delta G = -16,540 + 9960x$$
$$-T[-8.92 - R(x\ln x + (1-x)\ln(1-x))] \qquad (7)$$

in joules/mole, where x describes the composition of the phase as $Al_3(Li_x,Zr_{1-x})$. The free energy as a function of composition at various temperatures is depicted in Fig. 11. Minima in free energy are associated with increasing Li content as the temperature increases, moving from x ≃ 0.125 at T = 300°C to x ≃ 0.325 at T = 1500°C. Most relevant to the present work is the equilibrium composition at the SHT temperatures of 450°C (x ≃ 0.15) and 500 - 580°C (x ≃ 0.2).

Critical Ordering Temperature. Combination of Eqns. 2 and 3 yields the following relation for the critical ordering temperature for the $Al_3(Li_x,Zr_{1-x})$ phase:

$$T_c = 1855 - 1117x \quad °K$$
$$1582 - 1117x \quad °C \qquad (8)$$

Thus, T_c is linearly dependent on the composition variable "x".

Anti-phase Boundary Energy. The {111} APB energy for $L1_2$ structures is directly related to the ordering energy, and thus to the enthalpy of formation and the critical ordering temperature. Combining Eqns. 1 and 8, we obtain

$$\gamma_{APB} = 440 - 265x \quad ergs/cm^2 \qquad (9)$$

The antiphase boundary plays a strong role with regard to deformation behavior. The critical stress to shear a particle varies with APB energy to the 3/2 power.[32] The shear resistance of δ' has been determined to be due largely to APB energy.[18] Thus Al_3Zr and Zr-rich α' should exhibit significantly higher strength than δ', with Al_3Zr having approximately four times the strength of δ'.

Nucleation Behavior. The free energy change associated with the solid state homogeneous nucleation of a coherent precipitate can be described as

$$\Delta G = (4/3)\pi r^3 \Delta G_v + 4\pi r^2 \gamma \qquad (10)$$

where r is the embryo radius, ΔG_v is a volume free energy term and γ is the interfacial energy. The volume free energy is composed of chemical and strain energy components. The addition of Li to the Al_3Zr phase results in a reduction in each of these terms as shown in Figs. 10 and 11. In addition, the α'/matrix interfacial energy is lower than that of Al_3Zr, the decrease being proportional to the Li content (Eqn. 3). These modifications lead to a lower activation energy for nucleation of the α' phase as compared to Al_3Zr. More rapid nucleation is then to be expected as is a smaller critical radius for nucleation.

4. DISCUSSION

The calculations presented above suggest the strong influence which Li substitution may have on the Al_3Zr phase and on alloys containing α'. Al_3Zr and α' should have high strength relative to δ' based upon APB energy calculations (Eqn. 9). These phases should thus prove useful in dispersing slip and in increasing the strength of Al-Li alloys. Improved mechanical properties have been demonstrated with large Zr additions to Al-Li binary alloys,[9] and microstructural studies indicate the α' phase is indeed more resistant to

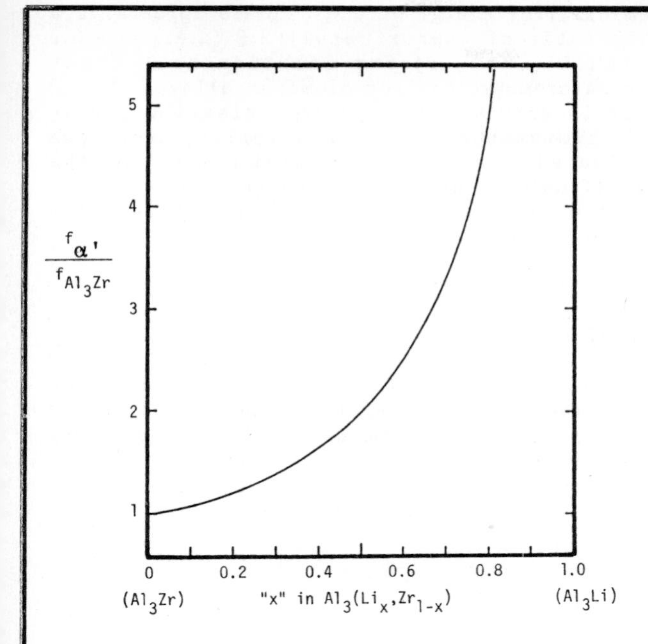

Fig. 9. Volume fraction α' versus α' composition, assuming alloy Zr content controls α'level. Normalized to volume fraction Al_3Zr.

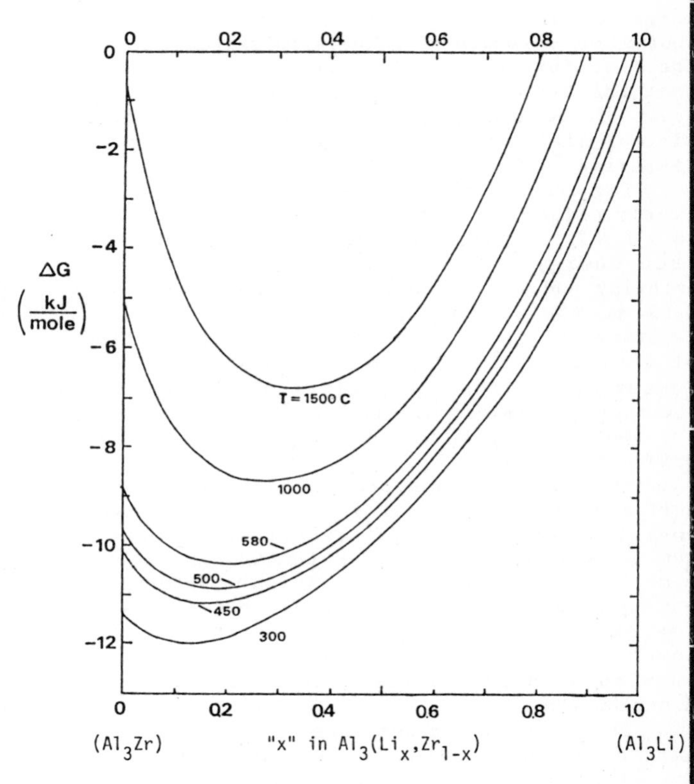

Fig. 11. Free energy of the α' phase as a function of temperature and composition. Standard states are disordered Al-25 at.%Li and Al-25 at.%Zr.

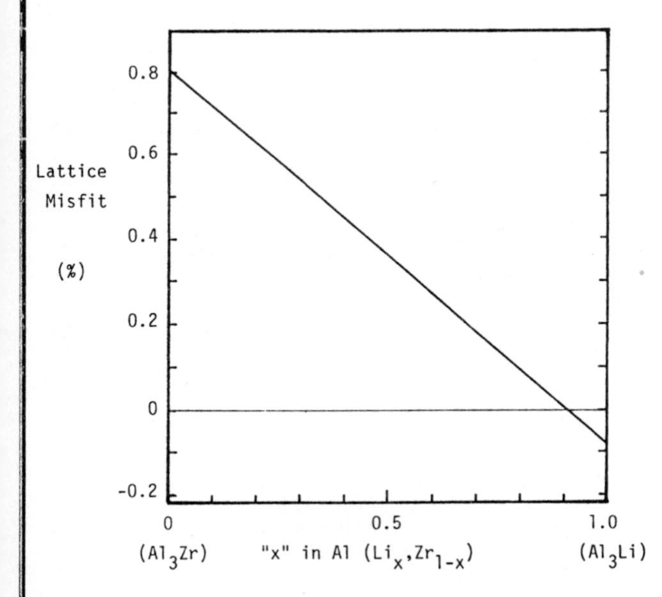

Fig. 10. Lattice misfit of α' with matrix as a function of α' composition.

shear than is δ',[12] in agreement with the implications of the present calculations.

A second basic influence on mechanical properties will be through the volume fraction α'. The beneficial effects of hard second phases in strengthening an alloy and in dispersing slip are related to the volume fraction of the phase. Strengthening effects have been calculated to be proportional to the square root of the volume fraction for both cutting and looping behavior.[31] Thus, there is interest in increasing the volume fraction of hard, Zr-rich precipitate.

Increasing the Li:Zr ratio in α' is an effective means for increasing the volume fraction of α', as seen in Fig. 10. This increase in volume fraction of second phase is provided without the problems associated with producing very large Zr supersaturations, yet may provide equivalent strengthening and slip dispersion. If the distribution and composition of α' are such that the APB energy is high enough to preclude shear of the phase, then the α' will be as effective as Al_3Zr in strengthening an alloy and in dispersing slip.

The free energy of the α' phase is significantly reduced as the Li content goes from zero to $x \simeq 0.2$, as shown in Fig. 11. This minimum in free energy results from a competition of enthalpy and entropy factors, and occurs at $x \simeq 0.15$ at 450°C, and at $x \simeq 0.2$ at 500-580°C. Experimentally, we have measured α' compositions of $x \simeq 0.2$ for 500°C/24 h and 580°C/15 h SHT.[12] Comparison of these experimental findings with the thermodynamical model suggests that the α' in the mentioned conditions was close to its equilibrium composition.

The calculations presented indicate that surface energy, chemical free energy and strain energy effects all should contribute to ease of homogeneous nucleation of α' containing at least some Li, as compared with Al_3Zr. The interfacial energy of such a phase should lie between values for Al_3Li and Al_3Zr as given in Eqn. 3. The lower energy as compared with Al_3Zr would contribute to high-temperature stability of any α' distribution.

The α' phase has been seen to be stable in a very fine distribution for extended periods at the SHT temperatures studied. The distribution is especially stable at 450°C, where even after 500 hours the precipitates have coarsened to diameters of only 12 to 25 nm. This compares favorably with coarsening behavior typical within high temperature aluminum alloys.[29] The high temperature stability of α' can be attributed to a low interfacial energy as discussed earlier and the low solubility and diffusivity of Zr in aluminum. The observed stability of the α' distribution and the calculated low interfacial energy of the phase support the study of the Al-Li-Zr system in high temperature applications.

5. SUMMARY

The presence of Li in α' has been calculated to have profound effects on various properties of the phase when compared to Al_3Zr, including chemical free energy, APB energy, and strain (lattice misfit) energy effects. The minimum chemical free energy of the α' phase appears at a Li:Zr ratio of approximately 2:8 (i.e., $x \simeq 0.2$ in $Al_3(Li_x,Zr_{1-x})$) for solution heat treat temperatures typical for aluminum alloys. The α' phase is very stable in a fine distribution at high temperatures. This stability has been attributed to a low interfacial energy and the low diffusivity of Zr in aluminum. The δ' phase has been shown to nucleate on the α' phase with perfect epitaxy.

ACKNOWLEDGEMENTS

The support of the National Science Foundation-MRL is gratefully acknowledged, as is the support of the Reynolds Metals Co. for alloy production.

REFERENCES

1. Aluminum-Properties and Physical Metallurgy, J.E. Hatch, ed. (ASM, Metals Park, Ohio, 1984).
2. K.N. Melton, J. Mat. Sci. 10 (1975), 1651.
3. R.H. Bricknell and J.W. Edington, Met. Trans. A, 10A (1979), 1257.
4. T. Ohashi and R. Ichikawa, Metall. Trans. A, 12A (1981), 546.
5. F.W. Gayle, in Aluminum-Lithium Alloys (ed. by T.H. Sanders and E.A. Starke, TMS-AIME, New York, 1981), 119.
6. R.J. Lederich and S.M.L. Sastry, in Aluminum-Lithium Alloys II, (ed. by E.A. Starke and T.H. Sanders, TMS-AIME, New York, 1984), 137.
7. J. Wadsworth and A.R. Pelton, Scr. Metall. 18 (1984), 387.
8. F.W. Gayle and J.B. VanderSande, Scr. Metall. 18 (1984), 473.
9. F.W. Gayle and J.B. VanderSande, Proceedings, ASTM Conference on Rapidly Solidified Powder Metallurgy Aluminum Alloys, Philadelphia, April, 1984.
10. P.L. Makin and B. Ralph, J. Mat. Sci. 19 (1984), 3835.
11. P.J. Gregson and H.M. Flower, J. Mat. Sci. Letters 3 (1984), 829.
12. F.W. Gayle, Sc.D. Thesis, Massachusetts Institute of Technology, 1985, 185 pp.
13. G. Cocco, G. Fagherazzi and L. Shiffini, J. Appl. Cryst. 10 (1977), 325.
14. D.B. Williams, in Aluminum-Lithium Alloys (ed. by T.H. Sanders and E.A. Starke, TMS-AIME, New York, 1981), 89.
15. A.J. McAlister, Bull. Alloy Phase Diagrams 3 (1982), 177.
16. S.F. Baumann and D.B. Williams, in Aluminum-Lithium Alloys II, (ed. by E.A. Starke and T.H. Sanders, TMS-AIME, New York, 1984), 17.
17. L. Guttman, Solid State Physics 3 (1956), 145.

18. B. Noble, S.J. Harris and K. Dinsdale, Metal Science 16 (1982), 425.
19. M. Tamura, T. Mori and T. Nakamura, J. JIM 34 (1970), 919.
20. E. Sahin and H. Jones, in Rapidly Quenched Metals III (B. Cantor, ed., The Metals Society, London, 1978), 138.
21. S. Rystad and N. Ryum, Aluminium 53 (3), (1977), 193.
22. P.A. Flinn, Trans. TMS-AIME 218 (1960), 145.
23. C.N. Yang, J. Chemical Physics 13, n.2 (1945), 66.
24. G.M. Kuznetsov, A.D. Barsukov and M.I. Abas, Soviet Non-Ferrous Metals Res. 11, n.1 (1983), 47.
25. D.B. Williams and J.W. Edington, Metal Science 9 (1975), 529.
26. N. Ryum, Acta Metall. 17 (1970), 269.
27. E. Nes, Acta Metall. 20 (1972), 499.
28. S.F. Baumann and D.B. Williams, Scr. Metall. 18 (1984), 611.
29. M.E. Fine and J.R. Weertman, "Synthesis and Properties of Elevated Temperature P/M Aluminum Alloys," Air Force Office of Scientific Research, Contract No. AFOSR-82-0005, Annual Report, Nov. 30, 1984.
30. D.R. Gaskell, Introduction to Metallurgical Thermodynamics, (2nd Ed., Hemisphere, New York, 1981).
31. R.F. Decker, Met. Trans. 4 (1973), 2495.
32. R.K. Ham, Trans. JIM 9, Supplement (1968), 52.
33. D. Bergner, Neue Hutte 29, n.6 (1984), 207.

□

Precipitation and lithium segregation studies in Al–2wt%Li–0.1wt%Zr

W STIMSON, M H TOSTEN,
P R HOWELL and D B WILLIAMS

WS is in the Metal Physics Department at Carpenter Technology Corporation, Reading, PA, USA;
MHT and PRH are in the Department of Materials Science and Engineering, Pennsylvania State University, PA, USA;
and DBW is in the Metallurgy and Materials Engineering Department at Lehigh University, Bethlehem, PA, USA

SYNOPSIS

The co-precipitation of Al_3Zr and Al_3Li (δ') has been studied as a function of pre-aging temperature, in order to discern the effect of the Al_3Zr coherency state on the co-precipitation. It was found that, depending on the temperature of the pre-age heat treatment, Al_3Zr was either semi-coherent or coherent. The coherency state of the Al_3Zr affected the subsequent Al_3Li co-precipitation. At pre-aging temperatures above 500°C, Al_3Zr formed semi-coherently, was invariably associated with dislocations, and showed no matrix strain contrast. Subsequent low temperature aging to produce Al_3Li precipitation resulted in isolated Al_3Li precipitation at the semi-coherent Al_3Zr interface and an Al_3Li precipitate-free-zone (PFZ) around the Al_3Zr. Electron energy loss spectrometry showed no detectable Li segregation to the Al_3Zr, suggesting that the PFZ formed as a result of local vacancy depletion. At pre-aging temperatures below 500°C, the Al_3Zr formed coherently and subsequent aging below the Al_3Li solvus resulted in the formation of an Al_3Li shell around the Al_3Zr precipitate. Factors affecting the formation of the Al_3Li shell are discussed, as well as the possibility of interdiffusion between the Al_3Li and Al_3Zr.

INTRODUCTION

Zirconium additions to aluminum alloys have been shown to be effective in inhibiting recrystallization (1,2,3) and have also been associated with improved toughness and stress corrosion cracking (4,5). These improvements in properties are due to metastable cubic Al_3Zr precipitates, which can effectively pin grain and subgrain boundaries. Cubic Al_3Zr has an LI_2 structure with a = 0.408 nm. In Al-Zr sub-peritectic alloys (Zr<0.18 wt.%), Al_3Zr is normally found to be spherical, nucleating heterogeneously on dislocations and boundaries (2).

When Zr is added to Al-Li alloys, anomalies in Al_3Li (δ') precipitation have been observed (6-9). Commonly in these ternary alloys "duplex" or "composite" precipitates are observed (e.g., Gregson and Flower (9), Makin and Ralph (7)) and these precipitates are assumed to consist of a spherical inner core of Al_3Zr and an annulus or shell of δ'. The spherical morphology of the inner core is a characteristic of alloys containing $\sim$ 0.1 wt.% Zr. However, in both binary Al-Zr and ternary Al-Li-Zr alloys containing $\sim$ 1.0 wt.% Zr, the Al_3Zr and δ' can adopt both spherical and rod-like morphologies; the latter being the result of discontinuous precipitation (Nes and Billdall (10) and Gayle and Vander Sande (8)).

Both Al_3Zr and δ' have the LI_2 ordered structure and separate imaging of each precipitate has therefore proven difficult. There has been some disagreement in recent research concerning the co-precipitation of Al_3Zr and δ'. Gayle and Vander Sande (6) report that there is inter-solubility of Al_3Zr and δ' based on electron diffraction and electron energy loss spectrometry results. Makin and Ralph (7), however, indicate that there is negligible chemical interaction between Li and Zr in either the precipitates or in the matrix.

This paper presents results concerning the co-precipitation of Al_3Zr and δ'. In particular we have investigated the effect of pre-aging temperature above the δ' solvus in order to discern the effect of Al_3Zr coherency state on δ' precipitation. Possible factors contributing to the formation of δ' shells are discussed and further evidence for the absence of chemical interaction between δ' and Al_3Zr is presented.

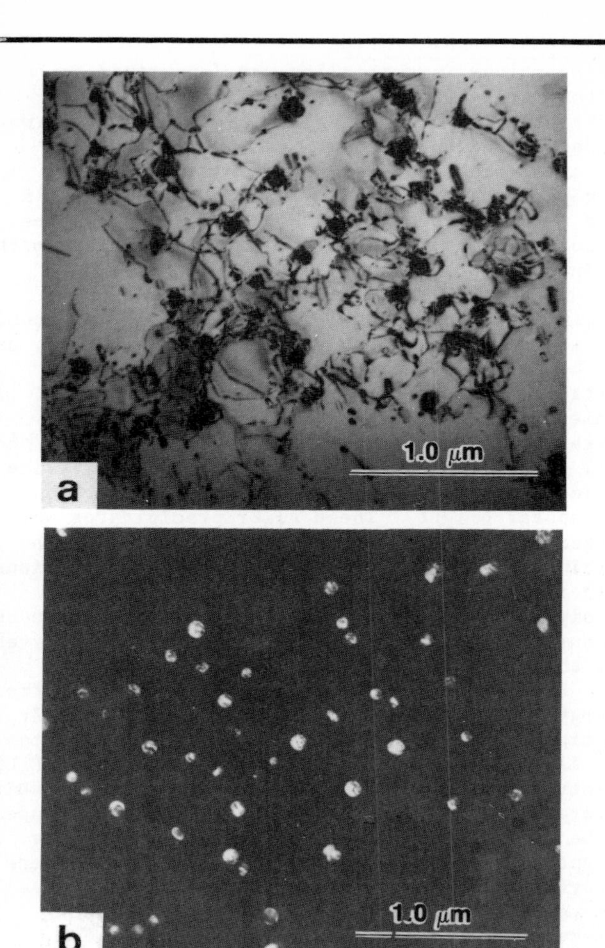

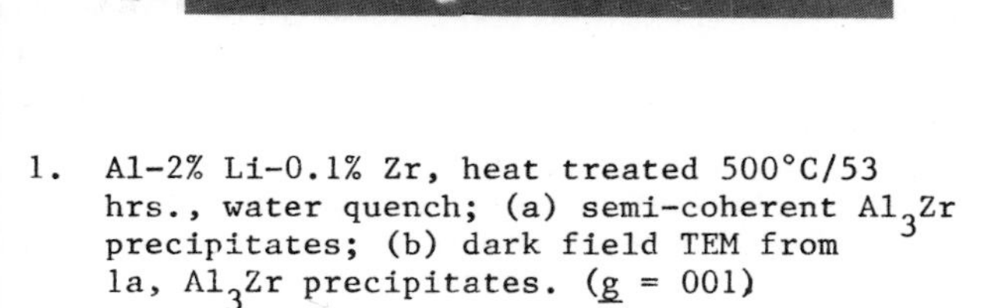

1. Al-2% Li-0.1% Zr, heat treated 500°C/53 hrs., water quench; (a) semi-coherent Al_3Zr precipitates; (b) dark field TEM from 1a, Al_3Zr precipitates. ($\underline{g}$ = 001)

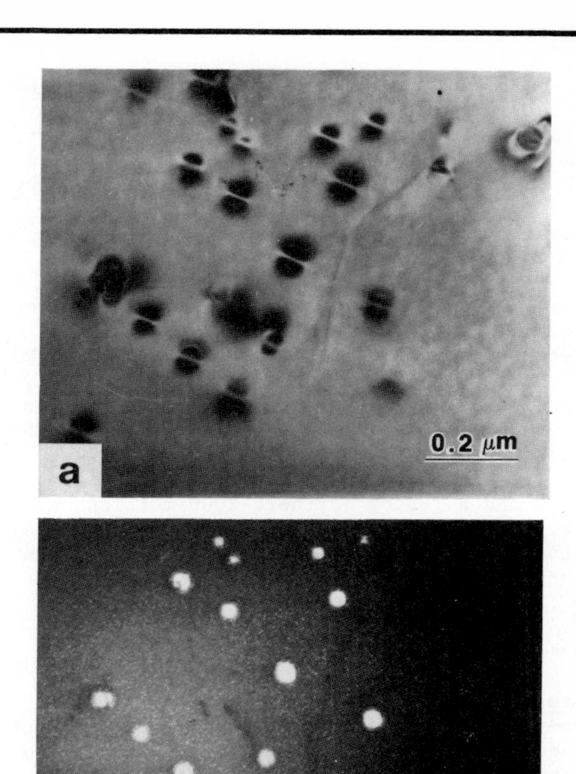

2. Al-2% Li-0.1% Zr, heat treated 450°C/49 hrs., water quench; (a) coherent Al_3Zr precipitates; (b) dark field TEM of Al_3Zr and Al_3Li precipitation. ($\underline{g}$ = 001)

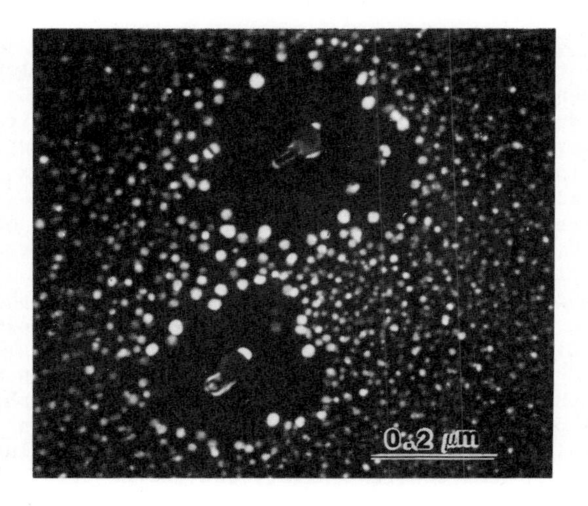

3. Al-2% Li-0.1% Zr, heat treated 500°C/53 hrs., water quench + 140°C/50 hrs., water quench, dark field TEM showing Al_3Li PFZ around Al_3Zr precipitate. ($\underline{g}$ = 001)

EXPERIMENTAL PROCEDURE

A ternary alloy of nominal composition Al-2 wt%. Li 0.1 wt% Zr was received from Lockheed in bar form. This alloy was homogenized for 6 hours in argon at 580°C and then repeatedly cold rolled and annealed at 580°C for 10 minutes and quenched in water. The solutionized specimens were then heat treated in evacuated capsules at 400°C, 450°C, and 500°C for 24-72 hours and water quenched. This constituted the pre-aging treatment which completely precipitated the Al_3Zr phase. Specimens were then aged at 140°C for 50 hours to precipitate δ'.

Specimens were prepared for transmission (TEM) and analytical (AEM) electron microscopy observation by conventional twin-jet electropolishing using an electrolyte of 10% perchloric acid in ethanol. The thin foils were examined in a Philips EM400T AEM equipped with an EDAX energy dispersive spectrometer (EDS) and a Tracor Northern-2000 multichannel analyzer. Electron energy loss spectrometry (EELS) was performed on this same AEM fitted with a Gatan 607 spectrometer.

RESULTS

Al_3Zr Precipitation

Solution-treated specimens exhibited a small amount of semi-coherent Al_3Zr inhomogeneously distributed throughout the specimen. Pre-aging at 500°C resulted in further precipitation of spherical Al_3Zr. These precipitates occurred in random patches throughout the entire matrix containing ∿ 5-40 precipitates. The Al_3Zr particles were invariably associated with matrix and sub-grain boundary dislocations, as shown in Figure 1a. Note that in addition to the association with dislocations there is no matrix strain contrast around the Al_3Zr precipitates, indicating that the precipitate/matrix interface is not coherent. EDS analysis showed these precipitates to be Zr-rich compared to the matrix. Also, these precipitates were imaged in dark field using an 001 superlattice reflection (see Figure 1b). Based on the EDS and electron diffraction results, and the results of other studies on Al-Zr binaries (1,2), it is assumed that these precipitates are semi-coherent cubic Al_3Zr.

Heat treatment at 450°C also resulted in the precipitation of spherical Al_3Zr precipitates (see Figure 2a). Heat treatment at 400°C produced very similar results, and will not be discussed in detail. The precipitation at 450°C also occurred in random patches, but these Al_3Zr precipitates were invariably coherent with the matrix, as shown by the strong Ashby-Brown (11) lobe-type strain contrast in Figure 2a. From this contrast it was possible to deduce the misfit between Al_3Zr and the matrix. The average misfit for several particles ranged from + 0.3% for 10 nm size particles to +0.6% for 25 nm size. EDS analysis and electron diffraction data also confirm that these coherent precipitates are Al_3Zr (see Figure 2b). Both coherent and semi-coherent phases showed identical EDS and electron diffraction characteristics. Very small Al_3Li precipitates can also be seen in Figure 2b, and have precipitated during the quench from 450°C.

Al_3Li (δ') Precipitation

The δ' solvus in this alloy was determined to be approximately 150°C. Due to this low solvus temperature, very long aging times (∿ 50 hrs.) were needed for adequate growth of the δ'. For all pre-aging treatments (500°C, 450°C, 400°C) δ' precipitated homogeneously throughout the matrix. In a few areas near grain boundaries, δ' was seen to coarsen preferentially compared to δ' within the grain interior.

In the 500°C pre-aged specimens, Al_3Li precipitate-free-zones (PFZs) were observed around the semi-coherent Al_3Zr precipitates. However, as can be seen in Figure 3, small individual δ' particles are clearly present at the interface. In addition, it can also be seen in Figure 3 that there was preferential coarsening of δ' around the PFZ. The δ' was imaged using an 001 superlattice reflection. However, the Al_3Zr precipitates did not appear bright. These Al_3Zr precipitates were imaged in dark field, however, after tilting the specimen to the appropriate diffraction conditions. This implies that these semi-coherent Al_3Zr precipitates may not retain the coherent cube-cube orientation relationship with the matrix, observed for the coherent Al_3Zr

Because of the extent of the δ' PFZ, electron energy loss spectrometry was used on these Al_3Zr precipitates to check if Li was segregating into the Al_3Zr precipitates. Figures 4a and b are EELS spectra from Al_3Li in a binary alloy and from semi-coherent Al_3Zr (such as in Figure 3). There appears to be no detectable Li in the Al_3Zr precipitate. In specimens pre-aged at 450°C, shells of δ' were observed around coherent Al_3Zr precipitates (see Figures 5a and b) in agreement with previous observations already referenced (6,7,8). EDS analysis showed the interior of the precipitate to be Zr-rich. However, the outer rings proved to be too thin, even after long aging times, to analyze accurately for Zr content due to beam spreading effects. After long aging times, very slight δ' PFZs were formed around the 'composite' precipitates, probably due to slight solute depletion. These small PFZs are visible in Figure 5b.

DISCUSSION

The Coherency of Al_3Zr

The coherency state of cubic Al_3Zr in this alloy depended on aging temperature. Aging at 500°C produced semi-coherent spherical Al_3Zr precipitates invariably associated with subgrain and matrix dislocations and showing no evidence of matrix strain contrast, as shown in Figure 1. This correlates with the results of Nes' study (3) on an Al-0.18% Zr binary alloy, in which cubic Al_3Zr precipitated on dislocations and subgrain boundaries when aged at 460°C. Makin and Ralph (7) also report similar results in a study of an Al-3% Li-0.2% Zr alloy.

However, in contrast to Nes (3) and Makin and Ralph (7), aging at 450°C produced coherent spherical precipitates, (as evidenced by Ashby-Brown strain contrast in Figure 2). There were isolated cases of semi-coherent Al_3Zr, but the majority of Al_3Zr precipitates were clearly coherent. Coherent spherical Al_3Zr precipitation has been observed

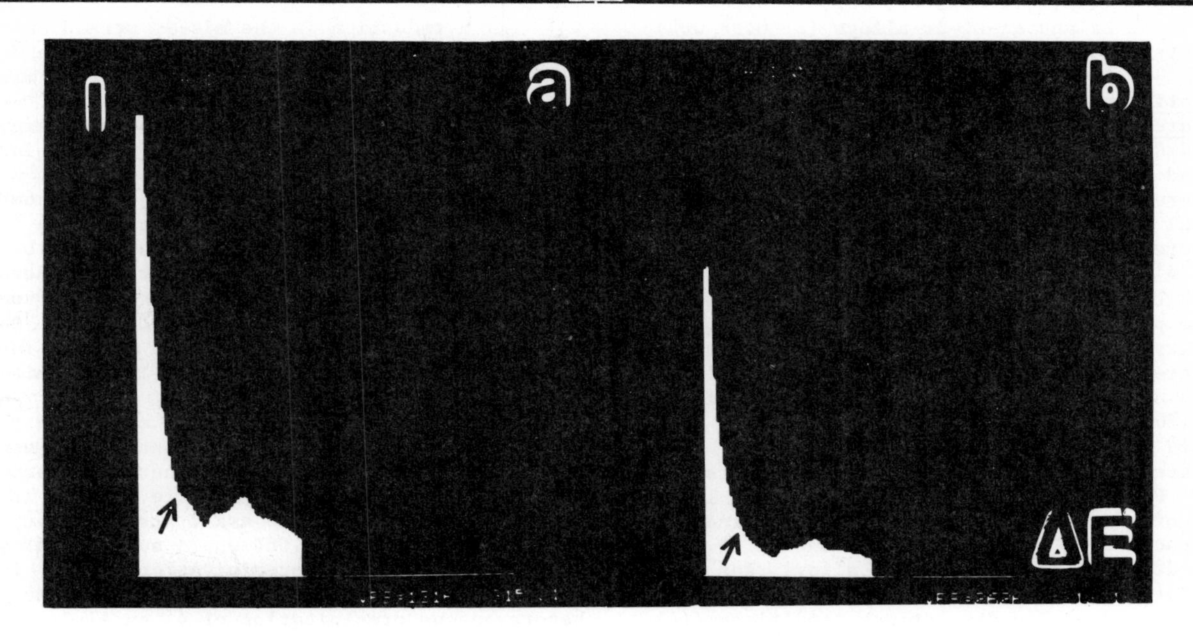

4. Electron energy loss spectra showing plots of intensity (I) vs energy loss (ΔE): (a) spectrum from Al_3Li precipitate in an Al-3% Li binary; The Li K edge is arrowed. (b) spectrum from Al_3Zr precipitate. Notice no Li K-edge at 54 eV (arrow).

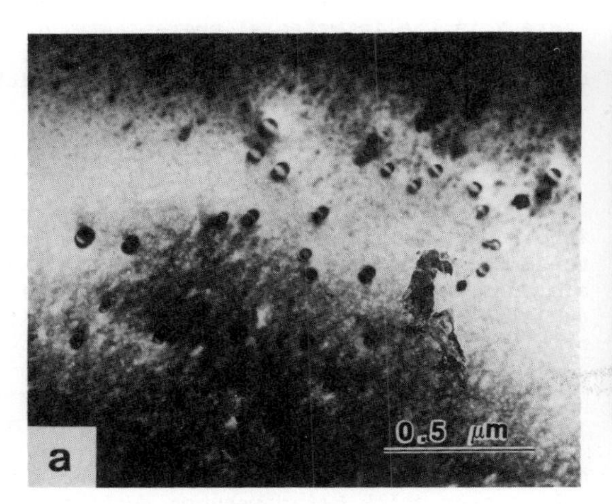

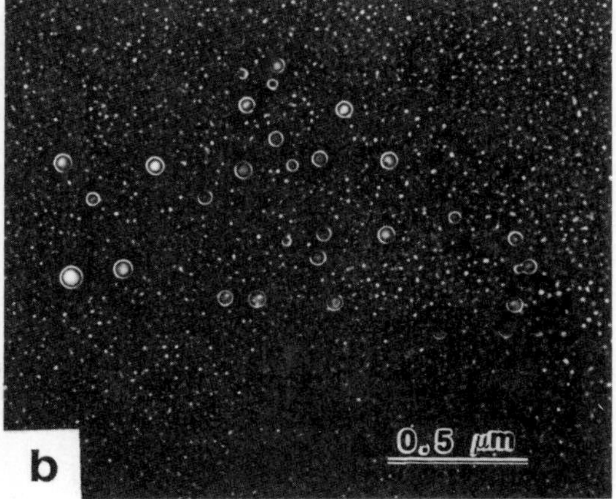

5. Al-2% Li-0.1% Zr, heat treated at 450°C for 49 hrs., water quench + 140°C/50 hrs. followed by a water quench; (a) coherent Al_3Zr precipitates; (b) dark field TEM showing Al_3Zr precipitates with Al_3Li shells, and Al_3Li matrix precipitation. ($\underline{g}$ = 001)

before (e.g., 1,12,13), however, only in hyper-peritectic Al-Zr and Al-Li-Zr alloys (Zr>0.28 wt%) at aging temperatures below 500°C.

Precipitation of Al_3Li in Conjunction with Semi-coherent Al_3Zr

Makin and Ralph (7) have observed δ' PFZs associated with the growth of δ' shells around semi-coherent Al_3Zr. In the present study, individual δ' precipitates were observed at the semi-coherent Al_3Zr interface (Figure 3). However complete δ' shells did not form, possibly due to the lower Li content of our alloys. Extensive δ' PFZs were present (Figure 3) suggesting the possibility of Li diffusing into the Al_3Zr precipitate. The semi-coherent Al_3Zr/matrix interface could enhance diffusion of Zr and allow Li to diffuse into the Al_3Zr, but this is considered unlikely. EDS analysis shows the precipitate to be Zr rich, but Zr was not detected in the δ' PFZ. In addition, electron energy loss analysis of the Al_3Zr precipitate does not indicate the presence of Li in the semi-coherent Al_3Zr. Li has been detected in Al-Li binary alloys, and the minimum detectable limit has been shown to be approximately 3 wt.% under optimum conditions (14). This result implies that Li is not diffusing into the semi-coherent Al_3Zr to any substantial degree. It is considered therefore that the Al_3Li PFZ arises due to local vacancy depletion. In order for δ' to nucleate, there must be a critical vacancy concentration in the matrix. However, the semi-coherent Al_3Zr/matrix interface may act as a vacancy sink, thereby depleting the immediate area of vacancies. This will inhibit the nucleation of δ'. Further evidence for vacancy depletion is the presence of a ring of coarser δ' around the edge of the PFZ in Figure 5. This phenomenon has been seen in Al-Li binary alloys (15) and explained in terms of vacancy profiles and the associated Li super-saturation.

Precipitation of Al_3Li in Conjunction with Coherent Al_3Zr

When the Al_3Zr/matrix interface is coherent a shell of δ' was observed to form around the Al_3Zr (see Figure 3) in agreement with other observations (6-9). This shell of δ' is larger than the surrounding matrix δ' and therefore a solute depleted PFZ might be expected to form. Such PFZs are evident after long aging time (see Figure 3).

The possible causes of the δ' shell formation will now be discussed in some detail. Due to the very low solubility of Zr in the α matrix, Al_3Zr is always present after solution treatment. In the present study the pre-age ensured that all the Al_3Zr had precipitated. Therefore, the duplex δ'/Al_3Zr precipitate can be formed during quenching to or aging at lower temperatures by either:

(i) preferential nucleation of δ' at the Al_3Zr/Al matrix interface, and/or
(ii) impingement of previously existing δ' and the growing Al_3Zr precipitates followed by preferential coarsening of δ' in contact with the Al_3Zr, leading to encapsulation of the Al_3Zr by δ'.

Gregson and Flower (9) suggested that the δ' nucleates heterogeneously on the α/Al_3Zr interface due to:

(i) a reduction in the misfit strain energy (δ' has a negative misfit with the α matrix while Al_3Zr has a positive misfit), and
(ii) a reduction in the interfacial energy due to the elimination of the α/Al_3Zr interface.

Further, the authors proposed that, since complete wetting of the Al_3Zr by δ' is observed, the free energy of the δ'/Al_3Zr interface is negligible.

Gayle and Vander Sande (8) also considered that the reduction in strain energy would promote heterogeneous precipitation but noted that "factors working against such precipitation include increased interfacial energy due to a large lattice mismatch at the δ'/Al_3Zr interface."

Finally, Makin and Ralph (7) concluded that nucleation of δ' on the Al_3Zr/α interface resulted from a decrease in interfacial energy. Considering the strain energy argument in more detail, it appears unlikely that the small strain associated with a δ' misfit of -0.08% would substantially promote heterogeneous precipitation of δ'. In addition, if the strain energy term did result in heterogeneous precipitation of δ' on Al_3Zr, then nucleation on lattice dislocations would also be expected. This was not observed after aging at 450°C and 400°C, as evidenced by Figure 5.

Consider now the interfacial energy argument. If we assume that the shape of the Al_3Zr is unaffected either by a δ' nucleation event, or an impingement event, then the equilibrium shape of the δ' will be given by:

$$\gamma_{\alpha/Al_3Zr} = \gamma_{\delta'/Al_3Zr} + \gamma_{\delta'/\alpha} \cos \theta \qquad (1)$$

where γ is the interfacial energy between the various phases and θ is the contact angle. The condition for complete wetting is θ≤0° which yields:

$$\gamma_{\alpha/Al_3Zr} \geq \gamma_{\delta'/Al_3Zr} + \gamma_{\delta'/\alpha}. \qquad (2)$$

It should be noted that this inequality does not impose any restrictions on the absolute value of the interfacial energy between δ' and Al_3Zr. In addition, complete encapsulation of the Al_3Zr by δ' can also occur as long as the contact angle θ≤180° or:

$$\gamma_{Al_3Zr/\delta'} \leq \gamma_{Al_3Zr/\alpha} + \gamma_{\delta'/\alpha} \qquad (3)$$

Hence, there exists a range of interfacial energy values over which heterogeneous nucleation and/or impingement events will lead to the development of duplex δ'/Al_3Zr precipitates. Examination of a large number of micrographs (see for example, Figure 3 in this paper and Figures 1b and c of Makin and Ralph (7)) indicates that 0<θ<180. Since the preceding argument is valid for both nucleation and impingement, a distinction between the two is difficult. However, reference to Figure 3 shows that nucleation of δ' will occur on the α/Al_3Zr since this interface would provide the only effective δ' nucleation site within the vacancy-depleted PFZ.

Gayle and Vander Sande (6) suggest the possibility of interdiffusion of Li into the Al_3Zr precipitate to accommodate the increase in interfacial energy at the δ'/Al_3Zr interface. However it is unlikely that Li would replace Zr during the Al_3Li aging treatment due to the very slow diffusion of Zr in Al at 140°C. In common with other investigations (6-9), we also observed the "dark-imaging" centers of a number of the duplex precipitates as discussed by e.g., Gregson and Flower (9), and Makin and Ralph (7). In common with previous findings, EDS revealed the presence of Zr in the dark imaging core of the precipitate and little or no Zr in the δ' annulus. The above would tend to suggest that the core is indeed Al_3Zr, which is consistent with the fact the δ' envelopes the inner core; a phenomenon which can only be rationalized if the δ' and the core are isostructural. However, it is evident from Figure 5 that many of the Al_3Zr cores are not illuminated in the δ' dark field image. Gayle and Vander Sande (8) were the first authors to address this problem. They suggested that either:

(i) the Al_3Zr was disordered which would yield a zero intensity in, for example, an 001 Al_3Li CDF, or

(ii) Li was incorporated onto the Zr sublattice in the Al_3Zr phase. This incorporation would reduce the structure factor for e.g., the 001 superlattice reflection and for a composition $Al_3(Li_{0.6}Zr_{0.4})$, the structure factor would be zero (i.e., no intensity in the 001 superlattice reflection).

While our EELS evidence from the semi-coherent Al_3Li argues against Li incorporation, it was not possible to use this technique to discern the presence of Li in Al_3Zr surrounded by a δ' shell. The Al_3Li shell above and below the Al_3Zr will invariably contribute to a Li edge in the EELS spectrum. Using other TEM and diffraction techniques, we find strong evidence that the 'dark imaging' centers are merely Al_3Zr particles of slightly different (<1°) orientation to the δ' shell. These results will be published elsewhere (16).

CONCLUSIONS

1. The coherency state of the Al_3Zr/matrix interface depends on aging temperature.
2. Al_3Li can form continuous shells around coherent Al_3Zr precipitates. The Al_3Li shells coarsen at the expense of matrix Al_3Li.
3. Al_3Li precipitate-free-zones can form around incoherent Al_3Zr precipitates without an Al_3Li shell. This is due to local vacancy depletion that is caused by the structure of the Al_3Zr/matrix interface.
4. There is no evidence for diffusion of Li into Al_3Zr.

ACKNOWLEDGEMENTS

W.S. and D.B.W. wish to acknowledge the generous financial support of the National Science Foundation through Grant DMR 84-00427. M.H.T. and P.R.H. were supported through an ALCOA Grant in Aid and this is gratefully acknowledged. The alloy used in this study was supplied by Lockheed Missiles and Space Company, Palo Alto, California, courtesy of Mr. R. E. Lewis.

REFERENCES

1. Ryum, N., Acta Metall., 17 (1969), p. 269.
2. Nes, E. and Ryum, N., Scripta Metall., 5 (1971), p. 987.
3. Nes, E., Acta Metall., 20 (1972), p. 499.
4. DiRusso, E., Alluminio-Nuova Metall., 37 (1967), p. 349.
5. DiRusso, E., Alluminio-Nuova Metall., 33 (1964), p. 505.
6. Gayle, F. W. and Vander Sande, J. B., Proc. 1984 ASTM Conference on Rapidly Solidified Powder Aluminum Alloys, Philadelphia, PA.
7. Makin, P. L. and Ralph, B., J. Mat. Sci. 19 (1984), p. 3835.
8. Gayle, F. W. and Vander Sande, J. B., Scripta Metall., 18 (1984), p. 473.
9. Gregson, P. J. and Flower, H. M., J. Mat. Sci. Letters, 3 (1984), p. 829.
10. Nes, E. and Billdall, H., Acta Metall. 25 (1977), p. 1039.
11. Ashby, M. F. and Brown, L. M., Phil. Mag. 91, (1963), p. 1083.
12. Thundal, B., Scan. J. of Met. 2 (1973), p. 207.
13. Rystad, S. and Ryum, N., Aluminum, 53 (1977), p. 391.
14. Chan, H. M. and Williams, D. B., Phil. Mag.A, accepted for publication.
15. Baumann, S. F. and Williams, D. B., Acta Metall. 33 (1985), p. 1069.
16. Stimson, W. Tosten, M. H., Howell, P. R. and Williams, D. B., Scripta Metall. (1985) submitted for publication.

□

Comparison of recrystallisation behaviour of an Al–Li–Zr alloy with related binary systems

P L MAKIN and W M STOBBS

The authors are in the Department of Metallurgy and Materials Science at the University of Cambridge, UK

SYNOPSIS

A comparison of the recrystallisation behaviour of an Al-Li-Zr alloy is made with that of related binary systems, the degree of recrystallisation as a function of annealing temperature and time being followed by hardness and texture measurements as well as by microstructural observations. Reasons for the inhibition of recrystallisation in the ternary system are discussed in relation to currently accepted models for the nucleation of recrystallisation.

1. INTRODUCTION

Many of the applications of the light industrial alloys based on the Al-Li-Zr system, centre on their uses as rolled sheet and thus require the optimisation of the properties of the alloy in this form. In principle this can be achieved through an improved understanding of the alloy's recrystallisation behaviour, as a function of prior deformation, this giving scope for the control of grain size and shape and hence of the material's strength and isotropy. Accordingly, while the primary aim of the work described here was simply to provide information on the recrystallisation of the Al-Li-Zr system, the results are also interpreted, as far as this is possible, through a comparison of the behaviour of the ternary alloy with that of related Al-Li and Al-Zr alloys.

2. EXPERIMENTAL

The alloys examined were provided by Alcan International Ltd. (Banbury) and the ternary alloy described here had a composition of Al-3 w/o Li-0.12 w/o Zr, the behaviour of this alloy being compared with that of Al-2.5 w/o Li and Al-3 w/o Li as well as Al-0.13 w/o Zr. The alloys, as hot-rolled to 6mm plate after casting, were further deformed by cold-rolling after a solution treatment, though it should be noted both that Al_3Zr particles are retained after such a treatment and that fine δ' Al_3Li particles precipitate on quenching to room temperature. While most of the recrystallisation characteristics were followed after conventional furnace annealing treatments, salt pot anneals were also used to study the effect of faster heat-up rates given the rapid ageing kinetics of the δ phase. Texture measurements were made of all the alloys, using X-ray techniques, both prior to and after deformation as well as after recrystallisation while the kinetics of the process were followed both by hardness measurements and by microstructural observation. The hardness data were obtained using a Vicker's diamond indenter with a 10 kg load, all specimens being ground to 1200 grit and the microstructure was examined using polarised light techniques on anodised surfaces as well as by both TEM and SEM methods. TEM foils were electropolished by standard methods using a 30% HNO_3/methanol solution and SEM specimens were aged in the δ phase field at 623K after annealing, the resultant grain boundary precipitation facilitating the characterisation of the grain and sub-grain morphology as described by Makin and Ralph.[1]

The hardness and texture data are first presented, these demonstrating the problems associated with the recrystallisation of the ternary alloy, and the possible reasons for the recrystallisation behaviour of Al-Li-Zr are considered after a comparison of its microstructural development with that of the binary alloys.

TABLE I

Alloy (wt. %)	Solution Treatment	Deformation	Method of heating	Annealing Temperatures
Al-2.5Li	853K 15 minutes	20% straight rolling	Furnace	698K, 723K 748K, 773K, 798K
Al-2.5Li	853K 15 minutes	20% straight rolling	Salt Pot	773K, 798K
Al-3.0Li	853K 15 minutes	20% straight rolling	Furnace	773K, 798K
Al-0.13Zr	903K 24 hours	50% straight rolling	Furnace	698K, 723K 748K
Al-3Li-0.12Zr	853K 16 hours	20% straight rolling	Furnace	773K, 798K, 823K
Al-3Li-0.12Zr	853K 16 hours	20% straight rolling	Salt Pot	773K, 798K

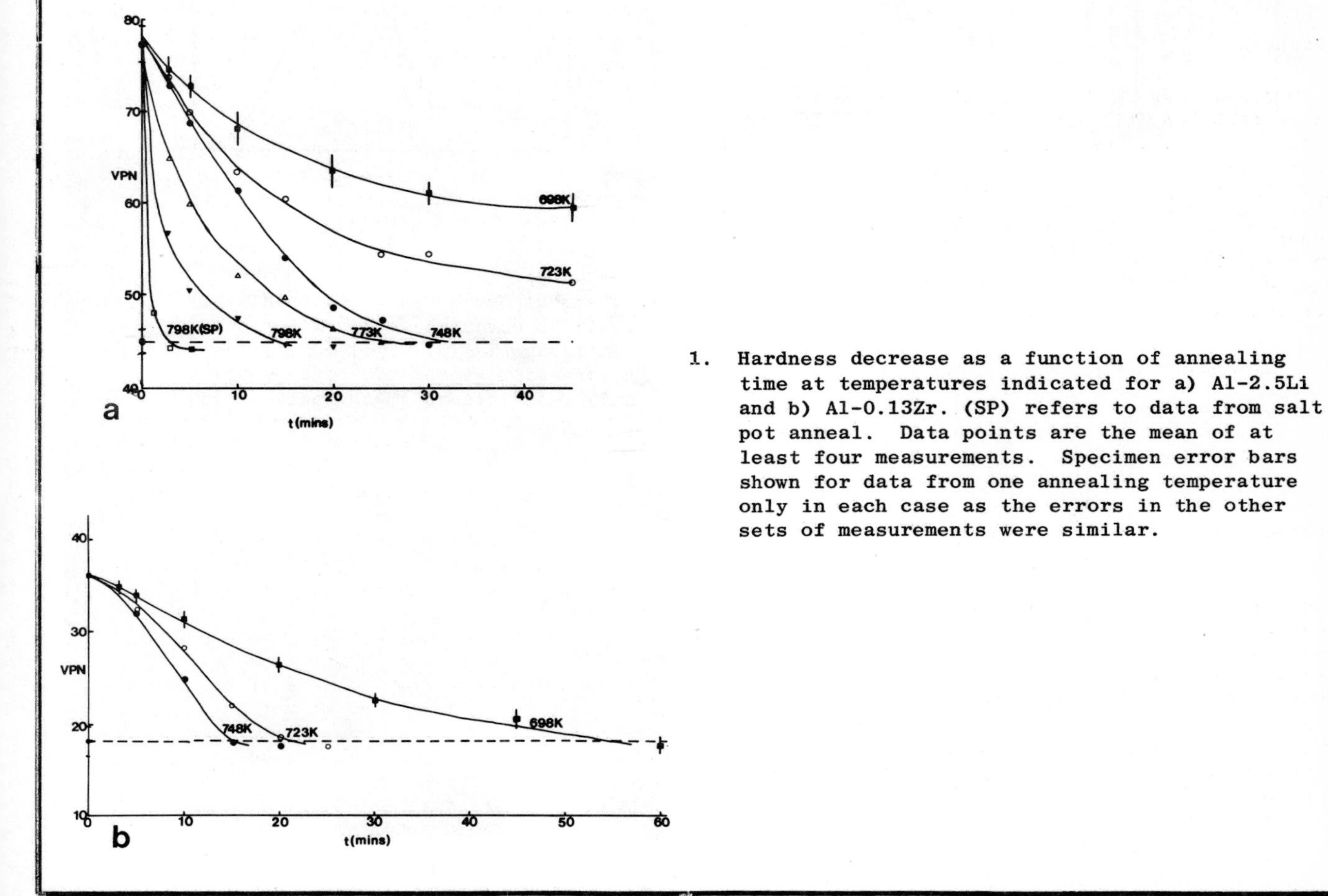

1. Hardness decrease as a function of annealing time at temperatures indicated for a) Al-2.5Li and b) Al-0.13Zr. (SP) refers to data from salt pot anneal. Data points are the mean of at least four measurements. Specimen error bars shown for data from one annealing temperature only in each case as the errors in the other sets of measurements were similar.

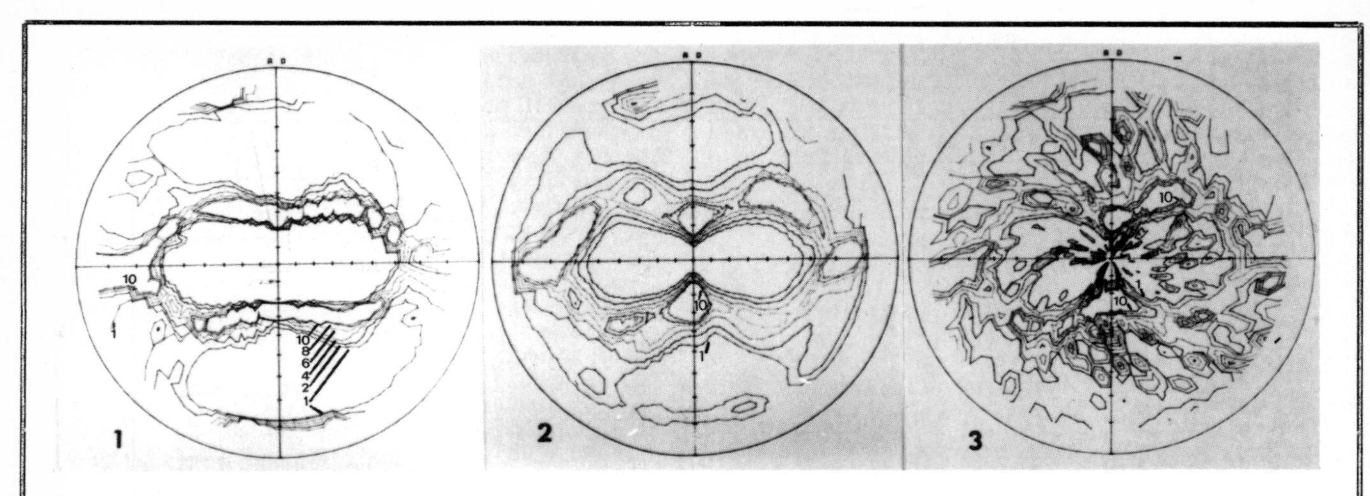

2. (111) pole figures obtained for Al-Li-Zr alloy after
   1) Solution treating at 853K for 16 hours;
   2) Subsequent deformation by 50% cold rolling;
   3) Annealing at 853K for 20 minutes.
   Relative intensities of contours are shown in 1) and are similar for the other pole figures.

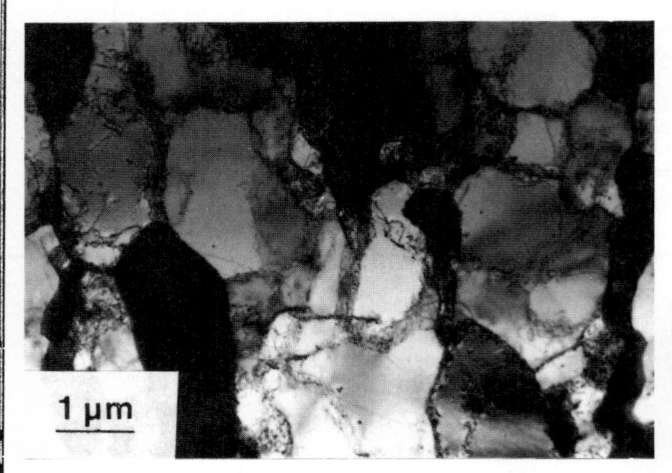

3. TEM micrograph of cell structure formed in solution treated and 50% deformed Al-0.13Zr alloy. The low dislocation density within the cells is apparent.

4. Optical micrographs demonstrating inhomogeneity of recrystallisation observed in Al-0.13Zr alloy after annealing at a) 698K 10 mins to produce partially recrystallised microstructure, and b) 698K 1 hour after which recrystallisation is complete.

3. Hardness and Texture Changes during annealing

The alloys were deformed and solution treated as shown in Table I which also specifies the annealing treatments employed. The ternary alloy did not exhibit a fully recrystallised structure after the solution treatment used but was demonstrated to have reached a "limit of softening" while further, more extended, annealing led both to oxidation and lithium loss problems. More deformation was applied to the Al-Zr alloy than either to the other binary or the ternary alloy with the aim of producing comparable hardening increments for each system this being considered to be a qualitatively reasonable way of giving each a similar driving force for recrystallisation. Both the ternary and the binary lithium containing alloys work hardened more rapidly at low deformation levels than Al-Zr though for reductions of more than about 20%, the hardening rates became similar.

The hardness data for the Al-2.5Li alloy are shown in figure 1a as a function of annealing time for all the temperatures investigated. Within experimental error the softening rates were indistinguishable for the two Al-Li alloys. Some of the problems associated with following the recrystallisation behaviour of an alloy of this type by this technique can be appreciated by the clear way in which the limiting decrease in hardness for the lower temperature curves (for which data were obtained for substantially longer times than shown) is clearly less than that for higher annealing temperatures. It was for this reason that the annealing kinetics were also examined using salt pot techniques with faster initial heating of the specimens. Data obtained in this way for 798K are also shown in figure 1 and comparison of the curves for 798K demonstrates a substantial acceleration of recrystallisation when both recovery at lower temperatures is suppressed and the degree of δ precipitation is reduced. The higher limiting hardnesses at the lower annealing temperature must be primarily due to a reduced degree of recrystallisation as a result of δ precipitation on subgrain boundaries. Similar annealing curves were obtained for the zirconium binary alloy (figure 1b), although now the hardness approached a similar limiting value for both high and low temperature anneals. However, the ternary alloy did not exhibit similarly temperature related softening on annealing. Instead, the hardness decreased from the cold-worked value of $\sim$ 95 VPN to the solution-treated value of 60 $\pm$ 5 VPN within a few minutes at all the temperatures studied, implying a low activation energy for the recovery processes involved. Indeed, no general recrystallisation behaviour was observed, what little local recrystallisation that did occur being too inhomogeneously distributed to be detectable through hardness measurements.

The strikingly different recrystallisation behaviour of the ternary alloy from that of the related binary systems was also evident from the texture data obtained. While the ternary alloy exhibited a very similar texture as solution treated, cold worked and annealed, both binaries developed the characteristic Al rolling texture, i.e. with strong elements of $\{112\}\langle 11\bar{1}\rangle$ which reverted to the cube texture, similarly present in the solution-treated state, on annealing. The only difference in behaviour of the binary systems was a greater rate as a function of strain at which the deformation texture was developed for the Al-Li system reflecting the increased tendency for the formation of shear bands in this alloy. Pole figures obtained for the ternary alloy in the three states examined are shown in figure 2. The slight differences in intensity of the poles are associated with an increased spread in orientation after cold work as subsequently reduced on annealing. Recovery processes involving sub-grain formation, growth and coalescence would account for this, the absence of new orientations on annealing thus suggesting a low density of new recrystallised grains with a significant orientation difference from that of the recovered substructure. Microstructural evidence confirming this inference is presented in §4.3.

4. The Development of the Recrystallised Microstructures

The differences in recrystallisation behaviour of the two binary systems and the ternary alloy were followed using scanning and transmission electron microscopy as well as by the optical examination of anodised specimens. The results for each of the different systems are described in turn.

4.1 The Aluminium-Zirconium system

Specimens of this alloy cold-worked by 50% after solution treatment exhibited a well-defined cell structure of $\sim$ 1 µm cell size in thin foils (figure 3). While the misorientations between the cells are evident in figure 3, at lower reductions little cell misorientation was apparent though the association of the Al_3Zr particles with the partially condensed cell walls could then be observed more readily. Clearly a degree of recovery of the deformation microstructure occurs at room temperature prior to observation and this might partially explain the lack of any local misorientations associated with the microstructure at the hard second-phase particles. Optical examination of the 50% cold-worked specimen showed evidence for similar deformation processes from grain to grain, the Al_3Zr particles being responsible for the refinement of the active slip plane spacing, though transition bands[2] were also evident at the high deformation level these showing no continuity from grain to grain consistently with previous observations of similarly deformed aluminium alloys.[3]

Annealing the deformed zirconium containing alloy resulted ultimately in a condensation of the cell wall structure described above with the reduction of the dislocation density within the cells and an increase in the intercell misorientations. Further annealing caused an inhomogeneous increase in the size of the sub-grains formed as described above; the regions showing a less marked increase in sub-grain size being those with a greater Al_3Zr particle density (this being characteristically inhomogeneous in solution-treated alloys of this type). The

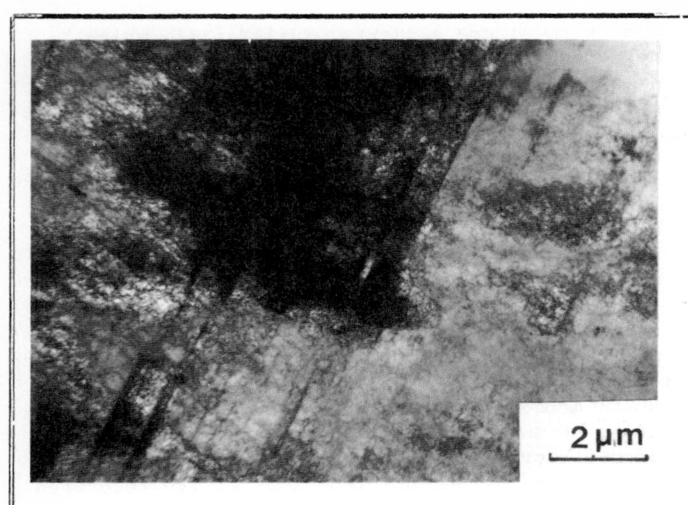

5. TEM micrograph of slip bands formed within the grains of Al-2.5Li alloy deformed 20% by cold-rolling after solution heat-treating.

6. Optical micrograph of Al-2.5Li alloy annealed at 733K for 8 hours after 20% cold rolling. Polarised light contrast shows recrystallised grains growing into deformed grains whilst original boundary positions are delineated by δ particles, e.g. triple point at A.

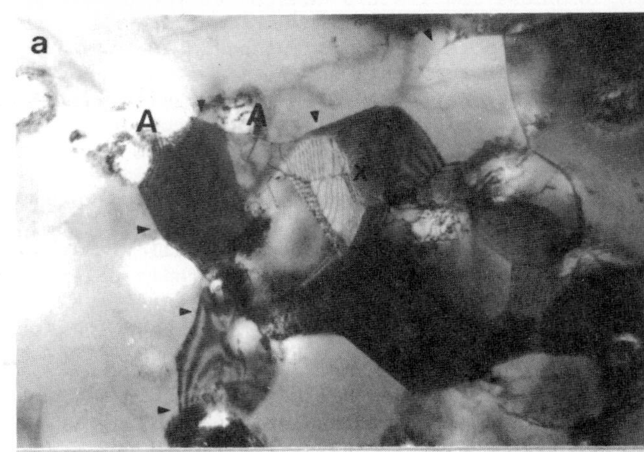

7. TEM micrograph of two segments a) and b) of the same migrating recrystallisation interface in Al-3Li alloy, annealed at 723K for 15 minutes after 75% reduction by cold-rolling. Pinning by δ particle A in a has inhibited migration of the interface (arrowed) into the sub-grain structure. Increase in dislocation spacing at X is indicative of a sub-grain coalescence process occurring to form a new nucleus in front of the advancing interface. Pinning by δ particle B in b has caused formation of the large sub-grain C, which is now acting as a nucleus.

larger grains were often associated with high angle boundaries as may be seen from the optical micrograph of a partially recrystallised specimen as shown in figure 4a; the majority of the nuclei bulging from these boundaries though a few can be seen to be associated with transition band structures. The nucleation inhomogeneity, the effect of which is retained in the fully recrystallised state (figure 4b), is associated with the well documented pinning effect of the Al_3Zr particles whose dispersion was equally inhomogeneous from grain to grain.

4.2 The Aluminium-Lithium System

The deformation structures observed in the aluminium lithium binary alloys differed from those seen in the zirconium binary alloys principally in that no discernible cell structure was formed at any level of deformation. At 20% reduction, well-defined slip bands became evident within the grains, as may be seen from figure 5. Optical examination of the same specimen showed the deformation to be inhomogeneous from grain to grain, with some grains showing a large number of transition bands whilst others showed uniform contrast similar to that seen in the solution-treated material, though all the grains underwent the same shape change as the specimen. A higher level of deformation resulted in a very complex structure of fragmented and relatively misoriented regions, while optical examination of a specimen given a reduction of 75% revealed bands of misorientation that extended across grain boundaries in a similar manner to the shear bands observed by Malin and Hatherly[4] in heavily cold-rolled copper.

The initial stages of annealing of these alloys were associated with the formation of sub-grains, and the concomitant precipitation of δ. Sub-grain formation was observed to occur to a greater extent in regions where δ particles were formed, although it is not clear whether the δ precipitated once the boundaries had formed or whether the presence of δ, and hence a lower concentration of lithium in the matrix, aided the formation of sub-grains. Further annealing resulted in an increase in size of the sub-grains by a combination of sub-grain growth and coalescence with the δ particles interacting with the dislocations from adjacent low angle boundaries which often showed irregular boundary dislocation spacings. An increase in dislocation spacing of a fragmenting low angle boundary adjacent to a high angle boundary was first observed by Jones et al.[5] and is now accepted as evidence for the occurrence of a coalescence process (e.g. 6).

Optical examination of a partially recrystallised specimen after 20% deformation again showed that high angle boundaries were important sites for the nucleation of recrystallisation, the few nuclei observed in transition bands remaining relatively smaller. The δ precipitation on high angle boundaries during the early stage of annealing can be used to delineate their original position in the deformed microstructure. The grain structure formed during annealing is shown in figure 6 by polarised light contrast, the original boundary position being delineated by the δ particles thus allowing the growth into the adjacent grains to be seen clearly. The faceted nature of the new grain boundaries was found to be associated with the pinning of the recrystallisation interface by the δ phase as precipitated on the sub-grain boundaries. Figure 7 shows an example of this type of pinning effect for the 3 $^w/o$ Li alloy given a 75% reduction. The same migrating high angle recrystallisation interface is shown in figures 7a and 7b. The pinning effect at A in figure 7a has allowed renucleation to occur ahead of the original interface by a sub-grain coalescence process as evidenced by the increasing dislocation spacing at the triple point. A more advanced stage of this process is seen in figure 7b where the original interface is shown pinned at B, and the new nucleus formed in front of it is now continuing to grow into the sub-grain structure.

The resulting fully recrystallised microstructures were homogeneous for both the alloys studied with a grain size smaller than that for the solution treated state as is consistent with the multiple nucleation events described.

4.3 The aluminium-lithium-zirconium system

As previously noted, heat treating the as-received ternary alloys at a high temperature for a long time ('solution' treating) does not result in a recrystallised microstructure. The nature of the cold-worked state after 20% deformation was seen to be similar to that of the Li binary alloy after a similar level of deformation. The slip was again, for example, localised into bands, but these exhibited elongated cells of misorientation and were seen to extend across sub-grain boundaries (figure 8). It was difficult to discern any specific dislocation tangling around the Al_3Zr particles after this level of deformation because of the generally high dislocation density surrounding the particles, although observations at a lower level of strain suggested that there was some particle-dislocation interaction.

The initial stages of annealing resulted in the general formation of sub-grains from the "cells" described above with Al_3Zr particles both being incorporated into the low angle boundaries as well as remaining within the sub-grains (figure 9). This figure also shows a recrystallisation interface which was one boundary of a 20 μm wide band of recrystallised material that ran through the specimen in the direction of rolling. It was, however, found to contain large (∿ 4 μm) particles of an impurity intermetallic and was thus the result of particle enhanced nucleation. The interface was pinned along its length by Al_3Zr particles, yet the Al_3Zr particles within the recrystallised region of figure 9 did not have the same orientation as that of the recrystallised grain indicating the way the boundary must have been able to by-pass these particles.

The inhomogeneity of recrystallisation in these alloys was also evident from optical and SEM examination of partially recrystallised microstructures. The SEM micrographs in figure

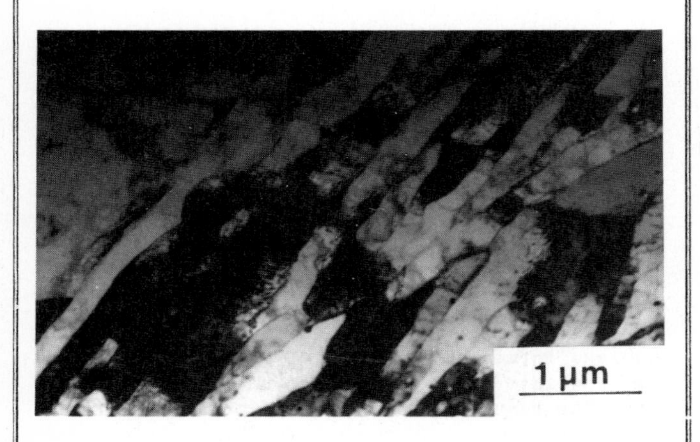

8. TEM micrograph of Al-3Li-0.12Zr alloy deformed 20% by cold work after solution treatment.

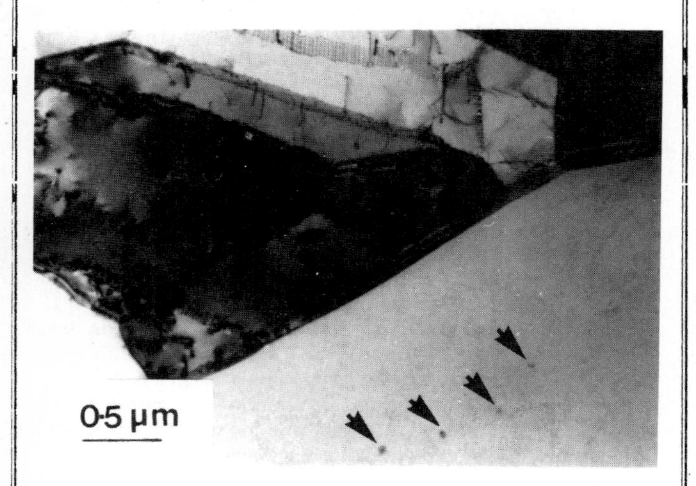

9. TEM micrograph of recrystallisation interface in Al-3Li-0.12Zr alloy that had been furnace-annealed for 5 mins at 823K after 20% cold work. The Al_3Zr particles arrowed were not oriented with the recrystallised grain in which they are situated.

10 are the alloy as deformed 24% by straight rolling before and after annealing just below 873K for 45h. The ageing treatment of 1 hour at 623K, as used to make the grain boundaries apparent through the boundary precipitation of δ, was sufficient to have caused the deformed structure in the specimen unannealed at a higher temperature to recover into sub-grains. The treatment was, however, shown not to have affected the grain structure of the annealed material, so that the comparison to be made is essentially of a recovered microstructure with one that has undergone some recrystallisation. Figure 10a and b both show the grain morphology exhibited in the plane perpendicular to a rolling plane and containing the rolling direction while figures 10c and d show the grain morphology in the rolling plane. Heavy precipitation of the δ phase has occurred on the high angle boundaries present in all the micrographs with apparently broadened regions of δ precipitation seen where the grain boundaries intersect the specimen surface at a low angle, as in figure 10c. The partially recrystallised microstructures of figure 10b and d show an inhomogeneous distribution of sub-grain sizes. The enlarged sub-grains can generally be seen to be associated with high angle grain boundaries, subsequent growth occurring into one or both of the adjacent grains. The light coloured regions are impurity intermetallic particles and are often associated with large recrystallised areas particularly where such intermetallics were themselves formed at a high angle boundary. The recrystallised grains can be seen to be generally elongated in the rolling direction, and if the density of δ precipitation is used as an indicator of dislocation density then there is evidence, as apparent from figure 10b, that a sub-grain coalescence mechanism may be involved in the formation of these grains: the distribution of δ precipitates in the positions arrowed is consistent with a locally reduced density of sub-grain boundaries. However, some grain growth in a direction perpendicular to the rolling direction was also observed as at the bulges in grain A in figure 10b. The curved nature of this and other similar boundaries indicates strong boundary pinning and enlarged sub-grain formation can often be seen ahead of such recrystallisation interfaces, as in figure 10d. The difference in density of δ precipitate imaged in figure 10c and d on the pre-existing high angle boundaries and the recrystallisation interface was a general feature providing evidence that the recrystallised grains are at increased inclinations to the rolling plane surface and are thus exhibiting limited growth perpendicular to the rolling direction.

Optical examination of the specimens used for hardness measurements demonstrated the formation of a few inhomogeneously distributed pancake-shaped recrystallised grains in the furnace-annealed specimens. However, a much greater degree of recrystallisation was observed for specimens that had been heated in a salt pot. The grain structures then observed were very complex as may be judged from figure 11. Bulging of one deformed grain into another was common as

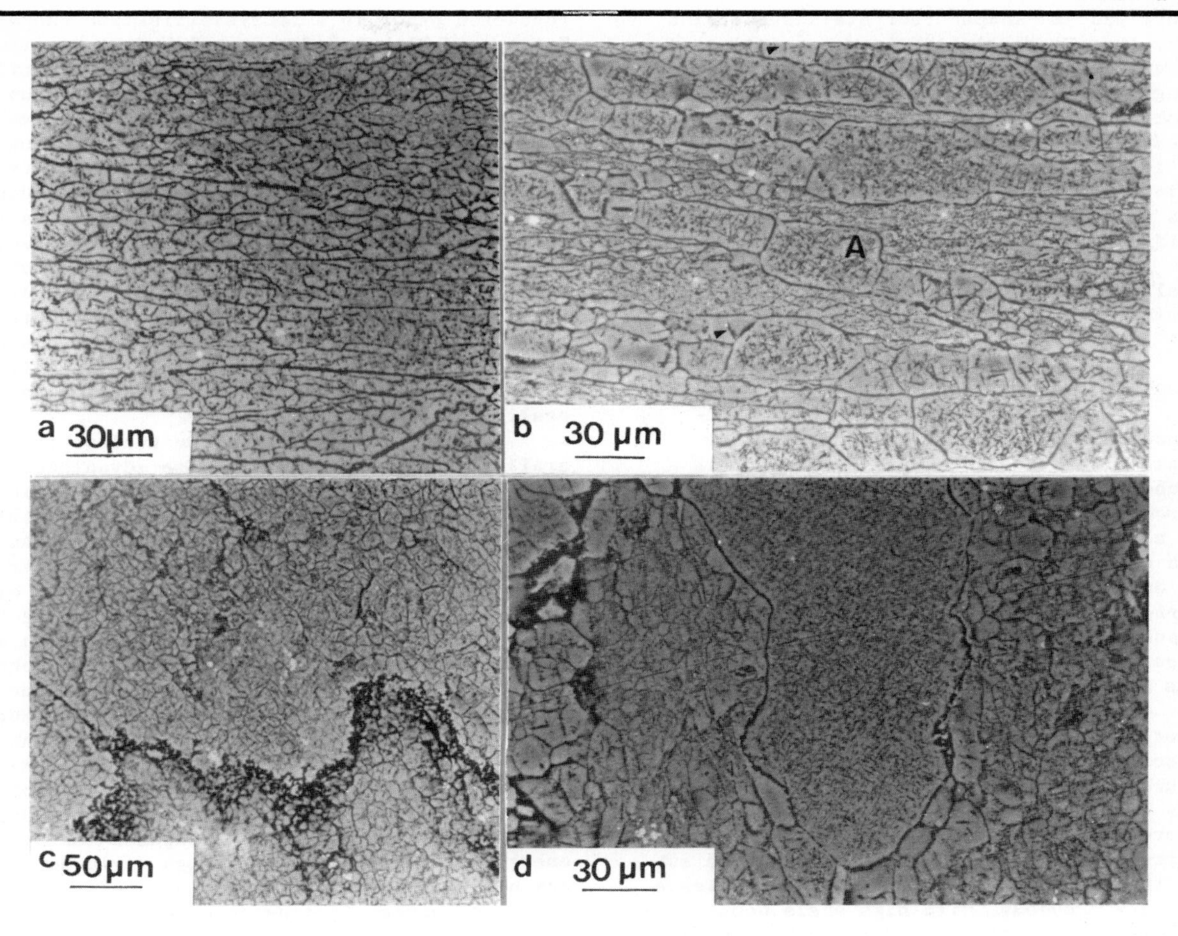

10. SEM micrographs of Al-3Li-0.12Zr alloy that had been deformed 24% by rolling before ageing in δ-phase field at 623K for 1h to delineate boundary structure. a and c are microstructures on the transverse plane and rolling plane respectively, that result from directly ageing the deformed structure, whilst b and d show the effects on these structures of a pre-ageing high temperature anneal of 45 hours at ∿ 820K that results in partial recrystallisation.

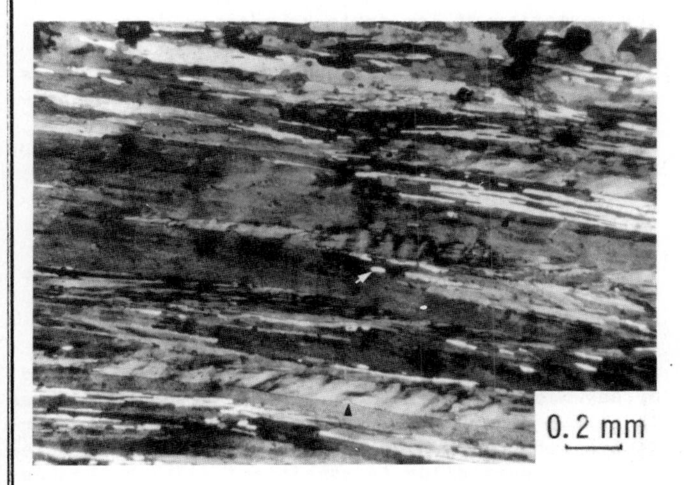

11. Optical micrograph of Al-3Li-0.12Zr that had been annealed for 5 minutes at 798K in a salt pot after 20% deformation. Examples of bulging of one grain into another, complex networks of recrystallised grains and small 'island' grains are shown arrowed.

were long pancake-shaped recrystallised grains. However, there were also regions consisting of a large number of very small grains situated along high angle boundaries, as well as regions where the linking of the long recrystallised grains resulted in the formation of networks. A few small 'island' grains were also seen to have grown in some of the deformed areas and the regions near to the edge of the specimen, and to a specific depth, consisted of large equiaxed grains. Such behaviour was also observed in the Al-Li alloy and is known to be associated with the depth to which lithium is lost from the surface.[7]

5. Summary of results

It is useful at this point to summarise the information that has been obtained from this study about the recrystallisation behaviour of the three alloy systems. The aluminium-zirconium binary alloy forms a well-pronounced cell structure on deformation which develops into subgrains during annealing. Prior high angle grain boundaries are important nucleation sites, and the resulting grain structure is very inhomogeneous, this being related to the inhomogeneous distribution of Al_3Zr particles in the alloy. The recrystallisation reaction can be followed by a change in VPN and a cubic texture is associated with a fully recrystallised microstructure. The aluminium lithium binary alloy deforms to form slip bands at the early stages of deformation and a complex locally misoriented microstructure at higher deformations. The subgrains that form on annealing grow and coalesce as recovery proceeds, with high angle boundaries again being important sites for the nucleation of recrystallisation. δ-phase precipitation has also been demonstrated to have an important rôle in boundary pinning and regions of lithium depletion show enhanced recovery and recrystallisation. The reaction can again be followed by changes in VPN, and a cubic texture has been seen to be associated with a fully recrystallised microstructure.

The ternary alloy deforms to form bands of elongated and misoriented cells which form subgrains during the early stages of annealing. Some growth and coalescence occurs, predominantly in regions adjacent to high angle grain boundaries while nuclei which do not then grow are also formed in these regions. A very inhomogeneous microstructure results, both in the distribution and size of the recrystallised regions, and the reaction cannot be followed either by changes in VPN or by alterations in texture. Regions depleted in lithium content show an equiaxed recrystallised microstructure whereas a large number of the bulk recrystallised grains are elongated in the rolling direction. The ternary alloy recrystallisation is enhanced by increased initial heating rates as well as by impurity particle stimulated nucleation.

6. DISCUSSION

The behaviour of the ternary alloy during annealing is similar to that of an alloy that has been recovered prior to a recrystallisation anneal. Kaspar and Pluhar,[8] for example, demonstrated that an initial high temperature recovery treatment delays the onset of nucleation for aluminium, similarly slowing the boundary migration velocity and thus the recrystallisation kinetics. The ternary alloy considered here has a very slow rate of nucleation, this facilitates the reduction of the driving force for recrystallisation, through gross dislocation microstructure recovery. The increased amount of recrystallisation observed either when recovery is retarded by using a fast initial heating rate to the annealing temperature, or when nucleation is stimulated by the presence of large particles correlates well with this suggestion for the reason why nucleation is generally inhibited.

A viable recrystallisation nucleus is generally considered to need a size advantage and a highly mobile interface. Mechanisms by which these prerequisites might arise in the vicinity of high angle boundaries in an alloy such as Al-Li-Zr, have recently been reviewed by Jones[6] who also described the way nuclei might evolve by subgrain coalescence processes adjacent to the boundary. The strain-induced boundary migration model for the nucleation of recrystallisation[9] requires that neighbouring grains in a deformed structure have substantially different stored energy densities so that a recrystallisation nucleus can then form by the bowing of a section of the high angle boundary into the grain with the higher stored energy density. This model has been developed e.g. Doherty and Cahn[10] who proposed that the energy differences across a high angle boundary may build up during annealing as a result of subgrain coalescence processes adjacent to the boundary.

If the sub-grain structure produced on annealing from the deformed state is resistant to coalescence processes, then nucleation by either of these mechanisms will be difficult.

Considering now the recrystallisation behaviour observed in the two binary alloys studied here we can gain further insight into the reason why recrystallisation is inhibited in the ternary alloy. For example, we have seen how the Al_3Zr dispersion can interact with the dislocation cell structure produced during the recovery stage with the formation of low angle boundaries decorated by the Al_3Zr in a manner very similar to that observed by Jones and Hansen[11] for Al_2O_3 particles in Al. Just as in this case the movement of dislocations on the low angle boundaries, as is necessary for sub-grain coalescence[6], is inhibited by the presence of the particles. Turning to the behaviour of the Al-Li binaries we have also seen how, when the matrix is depleted in lithium (as when the δ-phase is formed), the associated decrease of the solute drag effect of lithium on the boundaries present facilitates enhanced recovery and recrystallisation. It is possible that there is lithium segregation generally to grain boundaries though this has not as yet been demonstrated by any microanalytic technique for grain boundaries in bulk. However, the high density of δ-phase particles formed on both high and low angle boundaries in the ternary alloy is consistent with the boundaries having a high affinity for lithium. That this effect is less pronounced in the binary alloy is of itself interesting but probably relates to the nucleation of δ being enhanced in the ternary by the higher local dislocation densities retained, rather than to any sympathetic effect of lithium segregation as a function of the presence of zirconium. However, whatever the reasons for this effect, any dislocation rearrangements necessary for the production of a recrystallisation nucleus in the ternary alloy would require a higher driving force than if the level of solute segregation had not been so high.

Once the recrystallisation nucleus is formed, the factors that become important in its growth are those relating to the mobility of high angle interfaces and the driving force available in the system. The solute drag effect already considered and the pinning behaviour observed for Al_3Zr particles will both tend to retard grain boundary motion in the ternary alloy, while the extensive recovery which occurs prior to nucleation will reduce the driving force. It would appear that the additive effect in the ternary alloy is sufficient to inhibit general recrystallisation. The experimental observations of high angle boundary sites being favoured for nucleation and the effect of the inhomogeneous distribution of Al_3Zr particles leading to a related inhomogeneity in the annealed microstructures in the zirconium containing alloys, as well as the enhanced recrystallisation due to lithium loss, or when recovery is inhibited, are all consistent with the proposed reasoning. Thus, in the binary alloys, the nucleation is apparently not retarded sufficiently for general recovery to decrease the driving force for recrystallisation sufficiently to inhibit this process, and while recrystallisation can proceed to completeness in both binary alloys, the effects of the lithium and zirconium taken together are sufficient additively to inhibit general recrystallisation

7. ACKNOWLEDGEMENTS

The authors are grateful to Alcan International Ltd., and Professor D. Hull for the provision of laboratory facilities. Financial support for one of the authors (PLM) from SERC(UK) and Alcan International Ltd. is acknowledged. Valuable discussions with Dr. D.J. Lloyd are also gratefully acknowledged.

8. REFERENCES

1. P.L. Makin and B. Ralph, Proceedings of the 42nd Annual Meeting of the Electron Microscopy Society of America, ed. G.W. Bailey, San Francisco Press, San Francisco (1984),500.

2. H. Hu, Recovery + Recrystallisation of Metals, ed. L. Himmel, Gordon and Breach, New York (1963), 311.

3. S.P. Bellier and R.D. Doherty, Acta Met. 25 (1977), 521.

4. A.S. Malin and M. Hatherly, Met. Sci. 13 (1979), 463.

5. A.R. Jones, B. Ralph and N. Hansen, Proc. Roy. Soc., London, A368 (1979), 345.

6. A.R. Jones, Grain boundary structure and kinetics, ASM, Metals Park, Ohio (1980), 379.

7. D.J. Lloyd, private communication (1984).

8. R. Kaspar and J. Pluhar, Met. Sci. 9 (1975), 104.

9. P.A. Beck and P.R. Sperry, J. Appl. Phys. 21 (1950), 150.

10. R.D. Doherty and R.W. Cahn, J. Less Common metals, 28 (1972), 279.

11. A.R. Jones and N. Hansen, Acta Met. 29 (1980), 589.

Textures developed in Al–Li–Cu–Mg alloy

M J BULL and D J LLOYD

The authors are with
Alcan International Limited,
PO Box 8400, Kingston,
Ontario, Canada

ABSTRACT

The crystallographic texture development of alloy 8090 as a function of rolling (hot and cold), cross-rolling (hot and cold) and after various thermal treatments is discussed. Where applicable, comparison will be made with other high strength aluminium alloys subjected to similar thermo-mechanical treatments.

INTRODUCTION

Various Al-Li alloys are presently being developed as the "next-generation" aluminum alloy for aircraft/aerospace applications. As such, the uniformity of properties, in particular the planar anisotropy of mechanical properties, plays an important role in the general applicability of these new materials. The observed planar anisotropy exhibited by these alloys may be generally subdivided into the relative contributions from crystallographic considerations (preferred orientations), and those attributed to morphological features such as precipitates and grain boundaries.

This paper details the development of preferred orientations in alloy 8090 as a precursor to an improved understanding of the relative contributions (preferred orientation/ morphological features) in the planar anisotropy of alloy 8090. The influence of the observed texture components on the mechanical properties are discussed elsewhere[1].

EXPERIMENTAL PROCEDURE

Initial experiments consisting of both laboratory simulations as well as commercial trials have been conducted on alloy 8090 with the following composition:

Li	Cu	Mg	Zr	Fe	Si	Ti
2.58	1.24	0.70	0.12	0.06	0.06	0.02

Samples after specific processing steps have been examined for their preferred crystallographic orientation by means of pole figures. Pole figures have been obtained using the Schulz reflection technique[2]. Pole figures obtained using this technique have been corrected for "defocussing" and absorption using a commercially pure Al compressed powder sample.

Sample preparation for pole figure analysis consists of sectioning the sample into 20×20×6 mm pieces (circular samples have been used when possible) and removing the surface layer by mechanical grinding, and chemical etching in Poulton's etch for several seconds. In the case of transverse pole figures, samples have been prepared from a single thickness of material when possible, otherwise multiple pieces have been glued together and then polished.

The x-ray source is a Cu X-ray tube (Kα) operating at 40 kV and 20 mA. All pole figures are for the (111) pole and have been obtained at ¼ depth (¼ of surface material removed) unless noted otherwise.

Additionally, several samples have been subjected to orientation distribution function (O.D.F.) analysis, employing the series technique developed by Roe[3,4]. A technique employing a composite[5] sample has been used to measure the

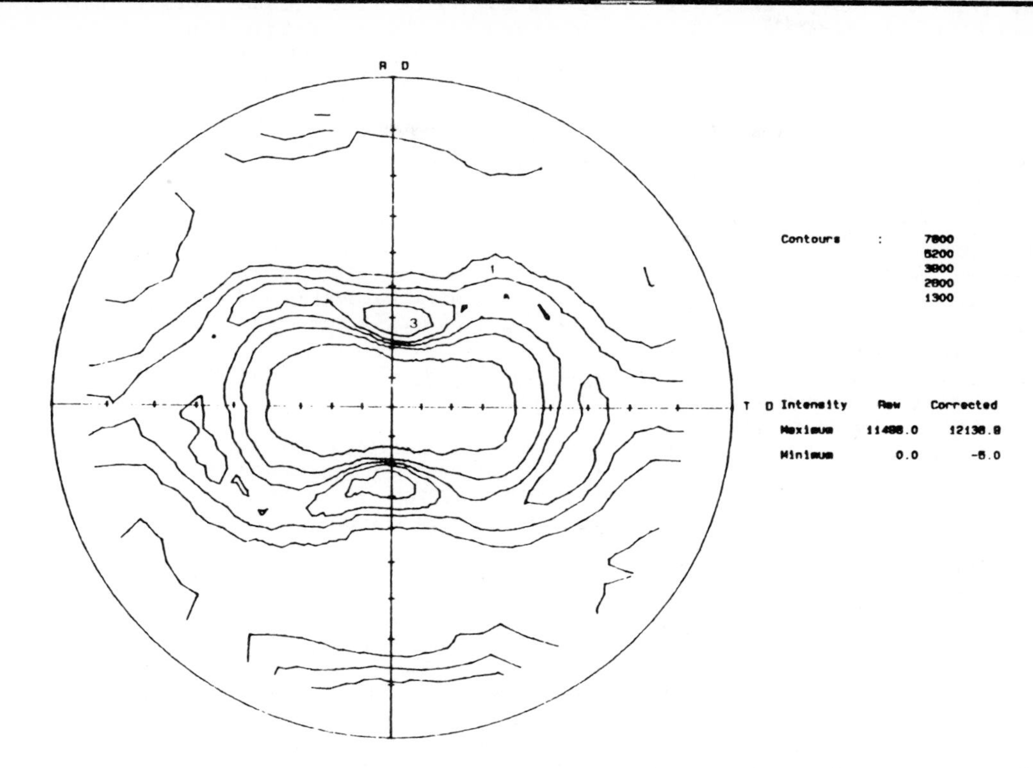

1. Alloy 8090 Hot Rolled ¼ Depth (111) RND = 2600

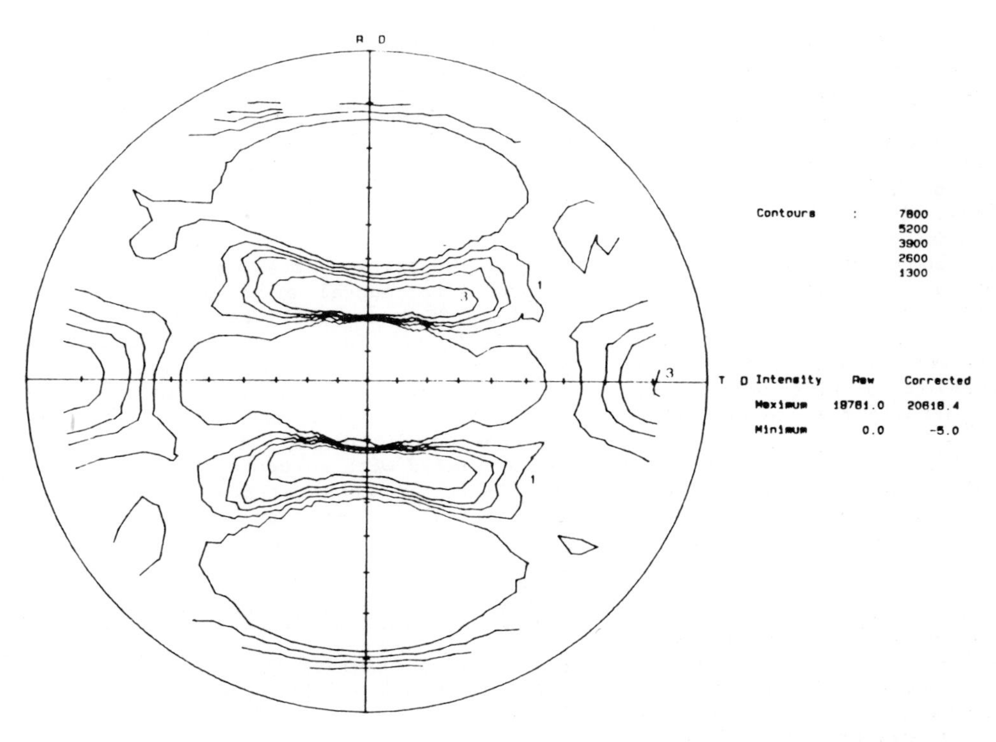

2. Alloy 8090 Hot Rolled ½ Depth (111) RND = 2600

complete pole figures [(111), (200), (220)] as input data into the O.D.F. analysis.

RESULTS

Homogenization

The texture for the as-homogenized alloy 8090 alloy (24 h @ 535°C) is random.

Pole figures for alloy 8090 hot straight rolled to 24 mm are presented in Figures 1-2. A "copper-type" rolling texture which is common to many aluminium alloys after cold rolling (room temperature) and several non-recrystallizing alloys during hot rolling is observed. There is a gradient of observed texture components in the through-thickness direction. Compare for example the texture displayed at ¼ depth with that at mid-plane (½ depth).

It is not possible to precisely "index" the texture components associated with the rolling texture from pole figures, due to the multiplicity of components which lie in close vicinity to the observed poles. However it appears that there is a strong propensity towards a $\{135\}\langle 21\bar{1}\rangle$ and $\{110\}\langle\bar{1}12\rangle$ orientation, and away from a $\{112\}\langle 11\bar{1}\rangle$ orientation.

In order to resolve this ambiguity resulting from the pole figures this hot rolled plate as well as some hot rolled AA-7075 have been subjected to O.D.F. analysis.

The rolling texture of many aluminium alloys is generally best described as "tube" (in Euler space) or as a specific continuum of orientations. This "tube" extends from $\{110\}\langle\bar{1}12\langle$ to $\{4\ 4\ 11\}$ $\langle 11\ 11\ \bar{8}\rangle$ (close to $\{112\}\langle\bar{1}11\rangle$). Hence, one method of describing the rolling texture is to describe the intensity (probability × random) of the components associated with this continuum of orientations. This continuum has been termed a "skeletal line". While the skeletal line does not give the volume fractions directly, it is a convenient method of comparing the relative "strengths" of the preferred orientations.

Skeletal lines for AA-7075 hot rolled plate and alloy 8090 hot rolled plate calculated using a least squares technique to combine the (111), (220) and (200) input pole figures are presented in Figure 3.

Both alloys exhibit a strong $\{110\}$ $\langle\bar{1}12\langle$ orientation "tapering-off" towards the $\{4\ 4\ 11\}$ $\langle 11\ 11\ \bar{8}\rangle$ orientation.

STRAIGHT ROLLED SHEET

Pole figures for alloy 8090 cold rolled from 5 to 1.6 mm without inter-annealing display rolling textures which are intermediate between that texture displayed for the hot rolled plate at ¼ depth (Figure 1) and the texture obtained at ½ depth (Figure 2). It should be noted that a variety of cold rolling textures which display varying amounts of "spreading" of the (111) poles at 20-35° from the plane normal have been observed. This variation is believed to be attributed to a through-thickness variation in texture, rather than any specific influence from a process variation. This through-thickness variation is examined in more detail in the section covering the final gauge solution heat treatments.

The O.D.F. skeletal line for some commercially rolled material is presented in Figure 4. This skeletal line is similar to that obtained with other high work hardening alloys such as AA-5083. A strong $\{135\}\langle 21\bar{1}\rangle$ component is observed.

THERMAL TREATMENTS INTRODUCED DURING COLD ROLLING

This section examines the influence of the solution heat treatment and over-age introduced prior to cold rolling as well as the influence of inter-annealing during cold rolling.

Pole figures for alloy 8090 solution heat treated 30 min @ 530°C following hot rolling, closely resemble the pole figure of the hot rolled material presented in Figure 2. However, in addition to those components associated with the "pure-metal" rolling texture, an $\{100\}\langle 011\rangle$ component has been observed in the transverse pole figures which is not readily apparent in the plan pole figures. Strangely, although this component is generally associated with high shearing conditions, such as found at the surface interface during hot rolling, this component has not been observed in the hot rolled material.

The introduction of an inter-anneal (1 h @ 350°C) at 3 mm (cold rolled 5 mm - 3 mm) has a negligible influence on the cold rolling texture. The texture components after annealing are essentially the same as those for the rolling texture. The $\{100\}\langle 001\rangle$ component which was evident after the solution heat treatment at 5 mm is not observed.

Similar results are obtained for the introduction of the inter-anneal at 2 mm, namely that very little difference could be detected between the as-rolled or inter-annealed material's pole figures, except to note that the maximum diffracted intensity tends to increase following an anneal compared to the rolling texture. This increase in diffracted intensity is believed to be associated with a reduction in internal stress and hence a less strained matrix (improved diffraction) rather than an increase in volume fraction of material with the observed texture components.

THERMAL TREATMENTS INTRODUCED AT FINAL GAUGE

This section examines the influence of the final gauge (1.6 mm) solution heat treatment (15 min @ 535°C and water quench), or anneal (1 h @ 350°C on the materials texture development, for materials which have been straight rolled with or without inter-anneals. Figures 5 and 6 detail the resulting texture for 8090 solution heat

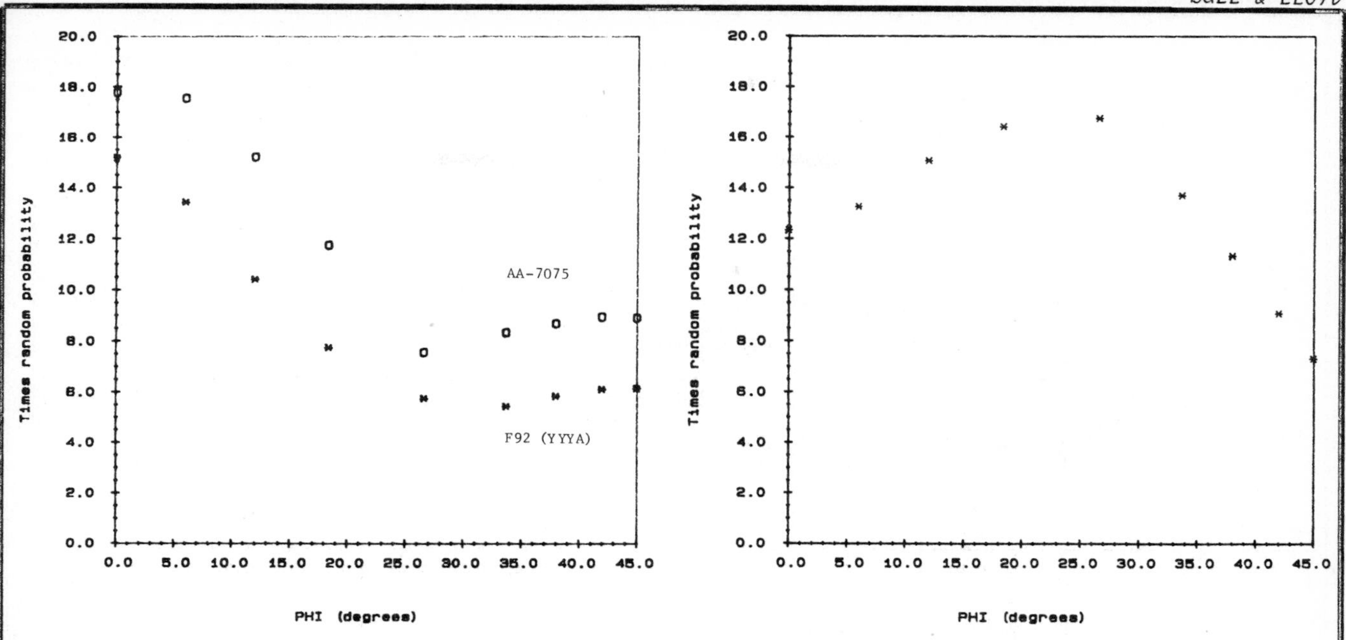

3. O.D.F. Skeletal Line For Hot Rolled 8090 and AA-7075 [{110} <$\bar{1}$12> to {4 4 11} <11 11 $\bar{8}$>]

4. Alloy 8090 Cold Rolled 68% With Inter-Anneals O.D.F. Skeletal Line [{110} <$\bar{1}$12> to {4 4 11} <11 11 $\bar{8}$>]

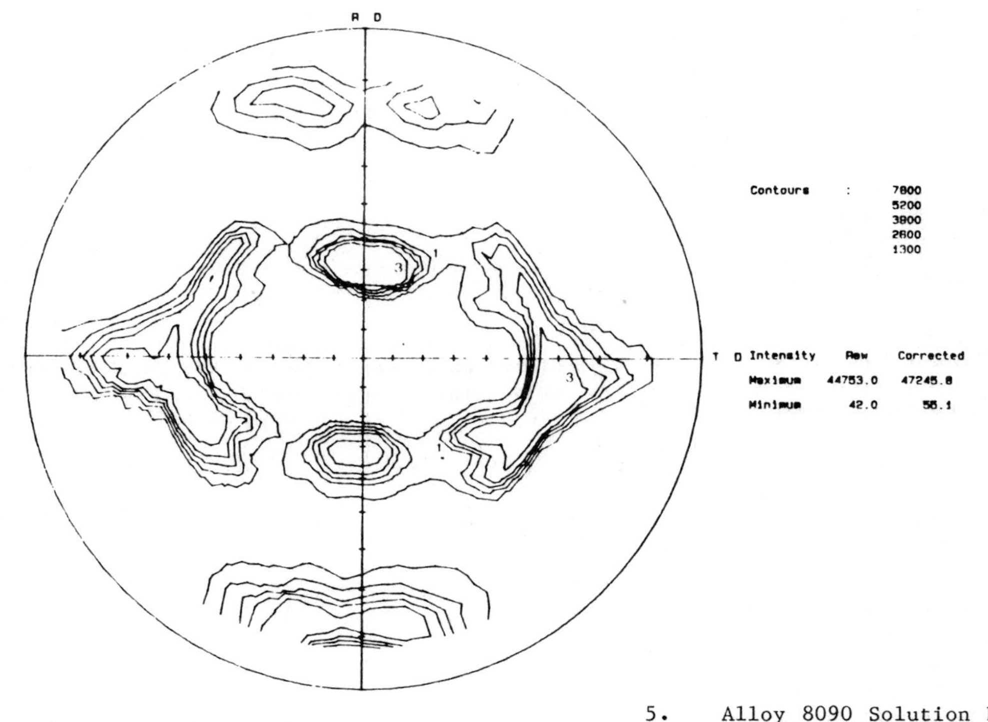

5. Alloy 8090 Solution Heat Treated (15 min @ 525°C) 0.005" Removed From Surface (111) RND = 2600

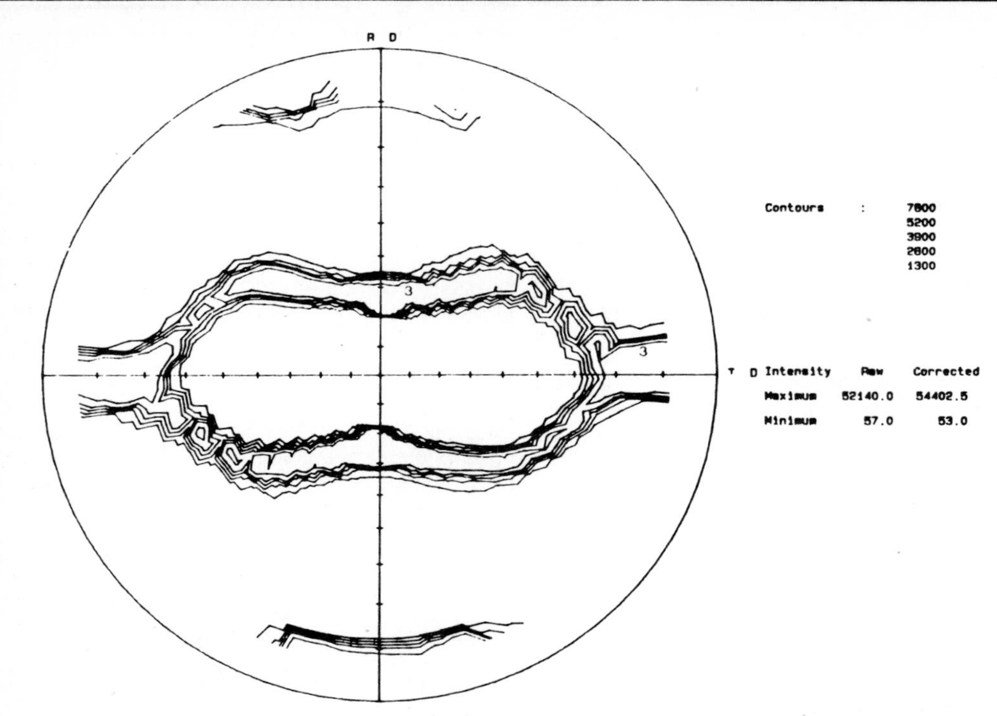

6. Alloy 8090 Solution Heat Treated (15 min @ 535°C) 0.020" Removed From Surface (111) RND = 2600

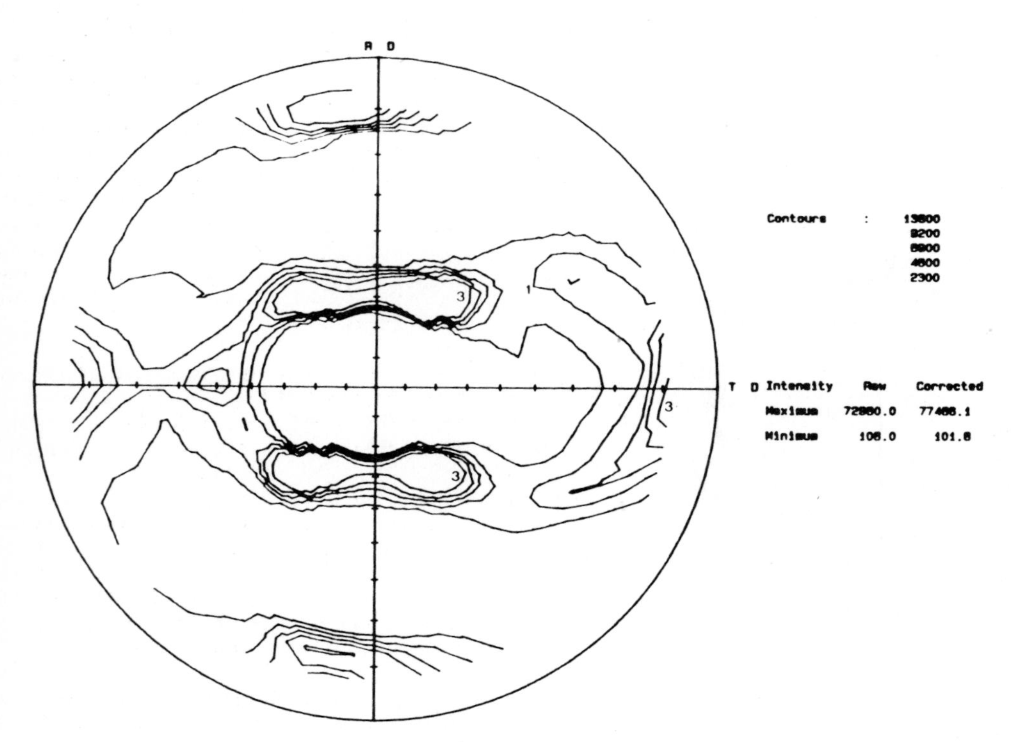

7. Alloy 8090 Hot Cross Rolled 30% (111) RND = 2600

treated (15 min @ 535°C and W.Q.) at various through-thickness depths. Figure 5, which is a pole figure near the outer rolled surface (0.005" removed) has a texture which may be indexed as approximately as $\{113\}\langle 54\bar{3}\rangle$ orientation. As more material is progressively removed from the surface, the texture evolves to the more widely spread rolling texture (0.020" removed from surface) as previously seen in both the hot and cold rolled material.

The solution heat treated materials display similar textures regardless of the introduction of the inter-anneals or not, displaying a range of textures which are encompassed by the pole figures represented in Figures 5 and 6, all of which must be considered as retained rolling textures.

The material annealed at 1.6 mm displays many of the same characteristics as the solution heat treated material, however more closely resembling its prior rolling texture.

CROSS ROLLING (HOT AND COLD)

As shown in the preceding sections, thermal treatments such as a solution heat treatment, or the introduction of an anneal generally result in textures which closely resemble the prior rolling process. As such, the introduction of cross-rolling, and the subsequent influence on the rolling texture have been examined.

A group of experiments was designed to investigate the transition from a well developed rolling texture to that of the cross-rolling texture as a function of strain.

Figures 7, 8 and 9, detail the evolution of the hot (520°C) cross-rolling texture at rolling reductions of 30, 50 and 70%, for a material which was initially hot rolled 90% parallel to the casting direction. The initial rolling texture spreads in a cross-rolling direction (30% reduction), then further evolves into a $\{110\}\langle\bar{1}12\rangle$ texture at 50% reduction. At 70% reduction a rolling texture which is perpendicular to the original rolling texture beings to form.

A similar trend is observed when the original hot rolling texture was formed by hot rolling perpendicular to the casting direction. As before, the rolling texture starts to spread in the cross-rolling direction at 30% reduction. A considerable amount of segmentation towards the $(110)\langle\bar{1}12\rangle$ texture is observed at 50% reduction. However at 70% cross-rolling reduction the "new" rolling texture is not nearly as well developed as in the former example.

The introduction of a 20% cold cross-rolling reduction following one of the inter-anneals at either 3 or 2 mm behaves similarly to that of hot cross-rolling at low reductions (< 30%). Spreading of the predominant poles in the direction of cross-rolling is observed. At higher reductions, the segmentation towards the $\{110\}\langle\bar{1}12\rangle$ orientation is not as pronounced as with the hot rolling experiments, instead developing an orientation which is approximately centro-symmetric about the plan normal.

An alloy which has been alternately straight rolled and cross-rolled exhibited relatively weak rolling texture with a high degree of rotational symmetry about the normal direction. There is however some propensity towards poles developing on the "corners" ($\{110\}\langle\bar{1}12\rangle$). The pole figure for this solution heat treated material is presented in Figure 10. A strong $\{110\}\langle\bar{1}12\rangle$ texture is observed.

DISCUSSION

In terms of their texture development, the Al-Li alloys which have been investigated in this study are unique compared to many other well characterized aluminium alloys, (for example AA-1100, AA-5083).

The high level of Zr and Li[6] generally ensures that no significant amount of recrystallization occurs, even after the final gauge solution heat treatment. Hence, those textures which are produced by rolling or cross-rolling prior to an inter-anneal or SHT are stable during these heat treatments. The one exception is the solution heat treatment and overageing treatment which preceeds the cold rolling. This thermal treatment produces a $\{001\}\langle 110\rangle$ texture component in addition to the remnants of the rolling texture. This texture component is generally characterized as a shear texture in f.c.c. materials and is often present where the roll friction is high, such as in hot rolling. It is not clear why this component is more easily observed after the solution heat treatment and age as compared to the hot rolled material. Two possibilities exist: the component was produced during hot rolling as a result of the rolling conditions, but is masked by the stronger intensity rolling texture components, or that this component is produced during the solution heat treatment and ageing. It should be noted that none of the final gauge solution heat treatments produced an observable $\{001\}\langle 110\rangle$ component.

An additional consequence of this low recovery, non-recrystallizing behaviour due to the high Zr levels may be found by observing the material's response to inter-annealing during cold rolling.

Pole figures for those materials which have been annealed indicate that the texture remains largely unchanged from the previous rolling reduction, either straight rolling or cross-rolling. This has been confirmed with O.D.F. skeletal lines comparing the straight rolled material versus the annealed material.

The texture development during hot rolling for 8090 is in general agreement with that produced with AA-7075 and several other high strength, recrystallization resistant aluminum alloys. O.D.F. analysis indicates that a strong $\{110\}\langle\bar{1}12\rangle$ orientation is produced tapering-off towards the $\{4\ 4\ 11\}\langle 11\ 11\ \bar{8}\rangle$ orientation.

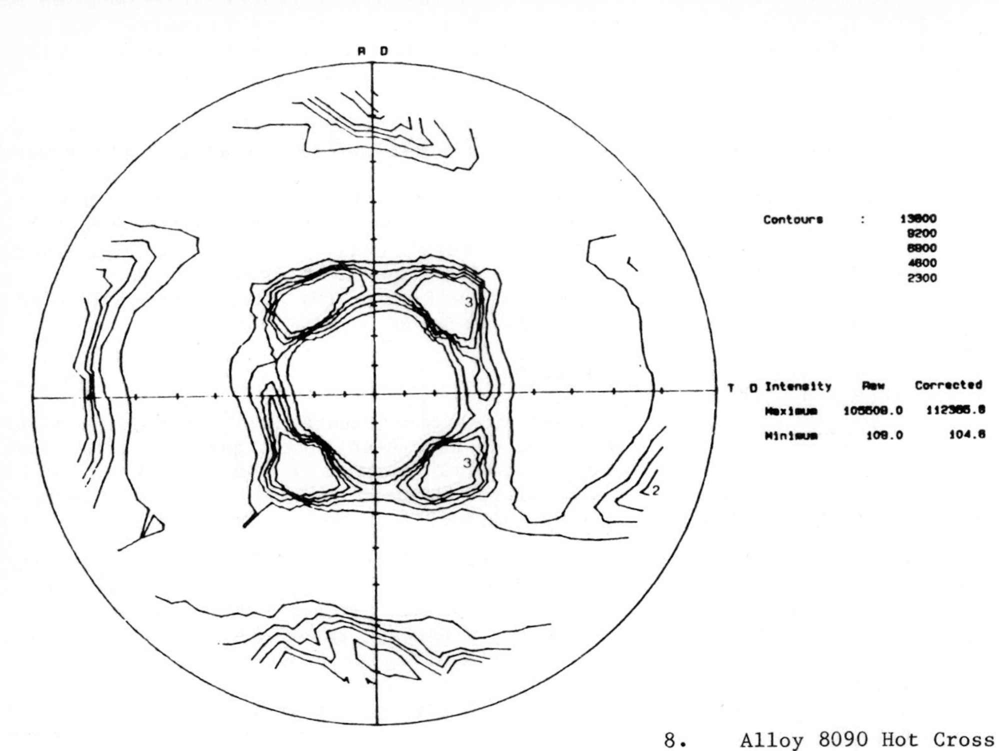

8. Alloy 8090 Hot Cross Rolled 50% (111) RND = 2600

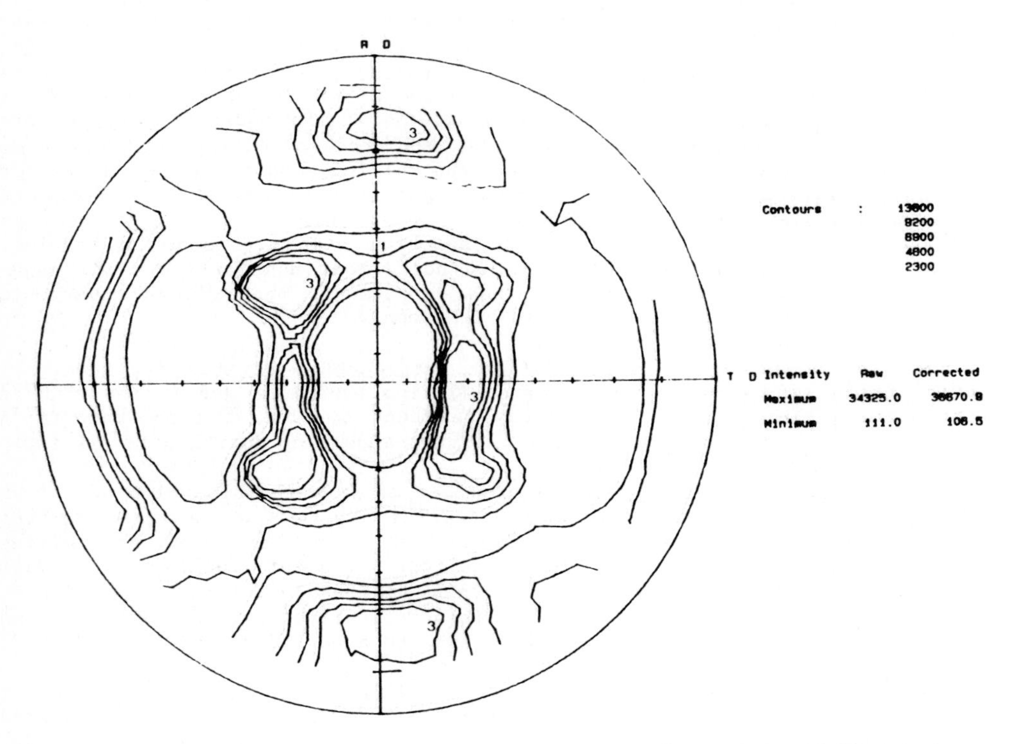

9. Alloy 8090 Hot Cross Rolled 70% (111) RND = 2600

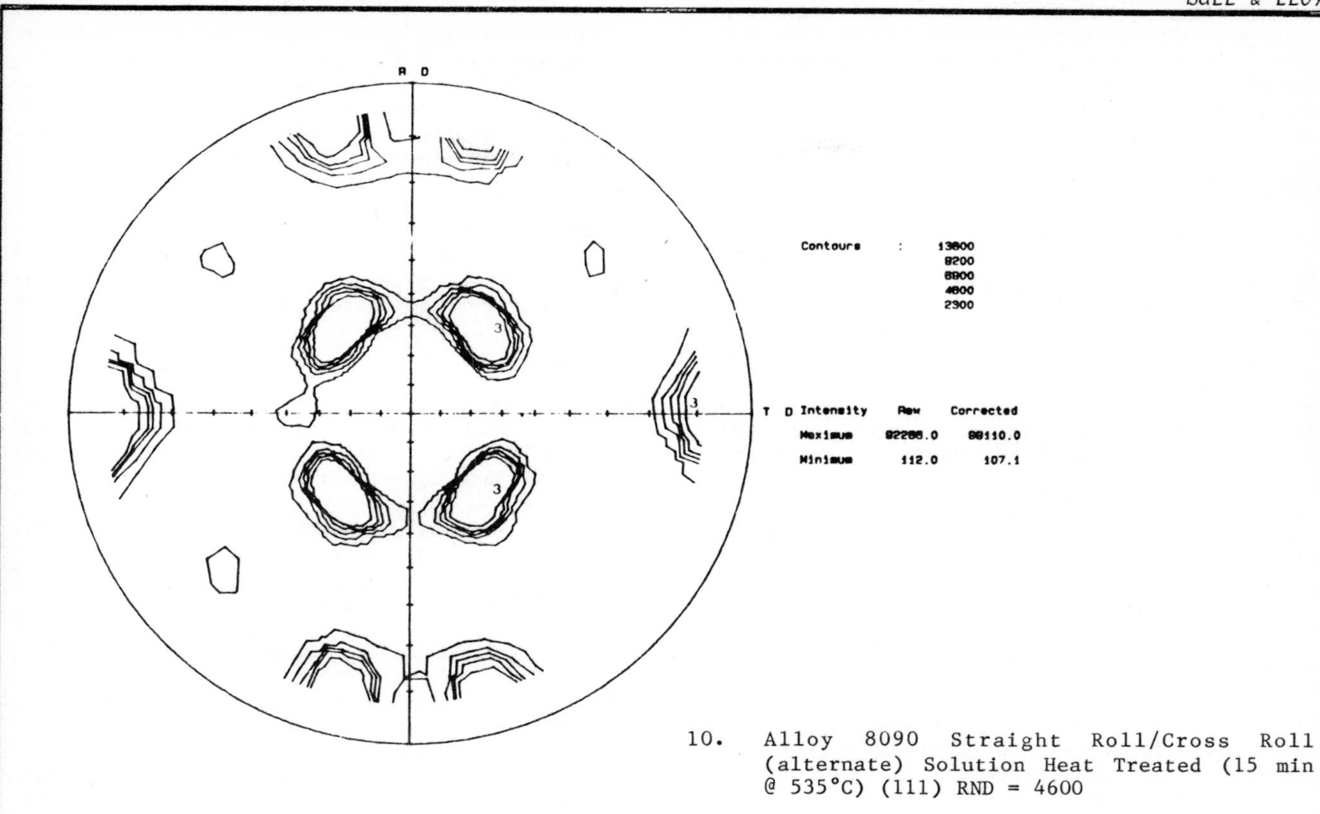

10. Alloy 8090 Straight Roll/Cross Roll (alternate) Solution Heat Treated (15 min @ 535°C) (111) RND = 4600

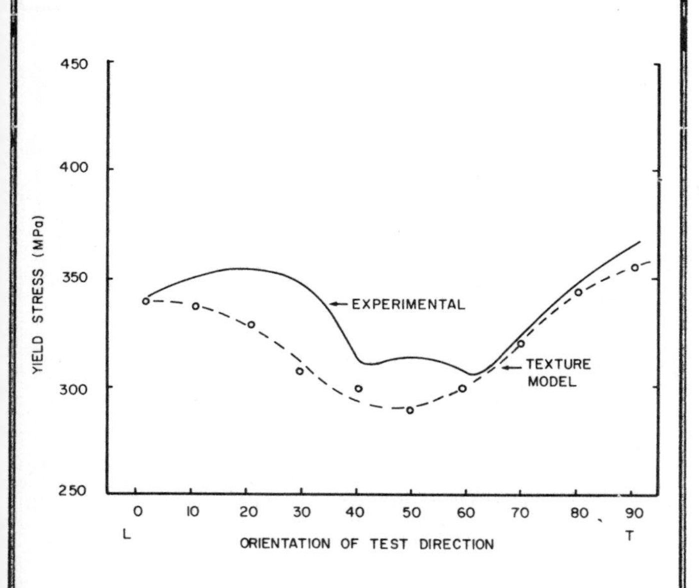

11. Alloy 8090 Observed Planar Yield Stress vs. Texture Model

The texture development during cold rolling is in general agreement with other high work hardening alloys such as AA-5083. These alloys produce a rolling texture with a strong $\{135\}<21\bar{1}>$ orientation.

In an attempt to reduce anisotropy of cold work, and hence anisotropy of fracture strain, moderate amounts of cross-rolling can be introduced in the process route. As shown this produces an initial spreading in the cross-rolling direction, then further evolving into an $\{110\}<\bar{1}12>$ orientation. This $\{110\}<\bar{1}12>$ orientation is quite strong for hot cross-rolled material and less intense for cold cross-rolled material. However, material which had been alternatively cold straight rolled cross-rolled produced a rolling texture which was almost centro-symmetric about the plan normal, however displaying a strong $\{110\}<\bar{1}12>$ orientation after solution heat-treatment.

Although only a limited amount of modelling of the observed textures to the mechanical properties has been performed, so far the correlations have been good. An example of the observed planar yield stress of the straight rolled material compared to the predicted values based upon the observed texture components is presented in Figure 11. Both the straight rolled and cross-rolled material have exhibited planar yield stress values which are in reasonable agreement with the observed textures. Further it should be noted that the planar yield stress for both the straight rolled and cross-rolled material will exhibit minimum yield strengths at 45° to the R.D.

CONCLUSIONS

(1) The homogenized material is random.

(2) Strong, well developed rolling textures are observed after hot rolling (520°C). A strong $\{110\}<\bar{1}12>$ component is observed. Similar results have been obtained for hot rolled AA-7075.

(3) The cold rolling texture is similar to that of other high work hardening aluminium alloys such as AA-5083, with a strong $\{135\}<21\bar{1}>$ component.

(4) Thermal treatment comprising of either inter-annealing during cold rolling or solution heat treatments do not substantially alter the rolling components. The components remain associated with the rolling texture.

(5) Initial (20% reduction) development of the cross-rolling texture consists of spreading the predominant poles (≈ 25° to N.D.) in the direction of cross-rolling. The intermediate (50% reduction) hot cross-rolling texture is $\{110\}<\bar{1}12>$ (with respect to principal rolling direction). Further reductions (70%) develop the straight rolling texture aligned with the cross-rolling direction. The texture development for cold cross-rolling, is similar however more spreading about the predominant poles is observed.

ACKNOWLEDGEMENTS

The authors are grateful to Alcan International Limited for permission to publish this work.

REFERENCES

(1) J.G. Palmer, W.S. Miller, D.J. Lloyd and M.J. Bull, "The Effect of Grain Structure and Texture on the Mechanical Properties of Al-Li Base Alloys", this conference.

(2) L.G. Schulz, J. Appl. Phys. 20, 1949, p. 1039.

(3) R.J. Roe, J. Appl. Phys. 36, 1965, p. 2069.

(4) R.J. Roe, J. Appl. Phys. 37, 1966, p. 2069.

(5) S.L. Lopata and E.B. Kula, Trans. A.I.M.E. 224, 1962, p. 865.

(6) P.L. Makin and W.M. Stobbs, "Recrystallization in Aluminium-Lithium Alloys", this conference.

Hardening mechanisms and ductility of an Al + 3.0 wt% Li alloy

O JENSRUD

The author is in the
Department of Metallurgy,
Sintef,
7034 Trondheim-NTH,
Norway

SYNOPSIS.

The aim of this work is to increase the understanding of the hardening mechanisms and the ductility problem in Al+Li base alloys. An Al+3.0wt%Li alloy was investigated and three particle hardening mechanisms are discussed. The grain size effect on yield stress is negligible, but small grain sizes improve the ductility of the alloy. The inhomogeneity of plastic deformation causes nucleation of intergranular microcracks. The inhomogeneity of plastic strain and stress concentration are due to superdislocation cutting the Al_3Li particles thus causing work-softened slip planes.

1. INTRODUCTION.

Ageing of the solution heat-treated alloy, Al+3Li produces a dispersion of metastable δ′(Al_3Li) precipitates which result in increased strength. The precipitates nucleate during quenching from the solution temperature |1| and coarsen with increasing ageing time. Isothermal ageing results in growth of the metastable precipitates. Together with the growth of the matrix precipitates a preferential coarsening of AlLi precipitates on the grain boundaries has also been found to occur |1|. The only hardening phase is δ′. The δ′ particles are spherical, and form an ordered phase which is coherent with the Al matrix. The misfit has been found to be about 0.1% |12,13|. The particles δ′ are also known to have greater elastic modulus then the Al matrix.

2. EXPERIMENTAL

An Al-3wt%Li alloy was prepared from 99.995%Al metal and 99.5%Li granules by melting together the two metals in a high frequency furnace in a high purity graphite mould and cast in a cylindrical iron mould. Both melting and casting were done in purified argon atmosphere. The chemical analysis gave 3.07±0.003%Li, the main impurities were iron and silicon (less than 0.01wt% for each). The casting was then solution heat treated for 6 hours at 580°C, scalped and extruded at approximately 400°C to a 8 mm rod. The solution heat treatment was done in a saltbath furnace, while the ageing heat treatment was done in oil baths. The temperatures of these baths were controlled to an accuracy of ∓1°C.

The extruded material was thermomechanically treated to give different grain sizes. The solution heat-treatments were done at different temperatures in the range 540°C to 580°C. The thermomechanical treatment was done in the two phase region before the last solution-heat-treatment to give small grains. The material was then machined into tensile test specimens.

The symbols used in this paper are R_o, the yield stress at the end of the elastic region, $R_{p0.2}$ the 0.2% proof stress, and R_m, the ultimate tensile stress. The e_m^P is the plastic deformation of the tested specimen at peak load, and A_m is the elongaton at peak load.

3. RESULTS.

3.1. Precipitation and deformation hardening.

Figure 1 shows the ageing curves of the alloy at 200°C. The curves are labelled with numbers which correspond to the deformation ratio before ageing. The alloy was solution heat treated before deformation by rolling. Peak hardness for undeformed material was reached after 24 hours of ageing.When the material was deformed 70% by rolling, the peak hardness was reached after 8 hours. On fig. 2 the tensile properties are shown as a function of ageing time at 200°C. The undeformed and the 67% deformed alloy are plotted against ageing time on the same figure. The yield stress and the plastic deformation at peak load are shown. The curves indicate that overaged material is more ductile than underaged. Figure 3 shows the increase in yield stress due to deformation hardening of the alloy after solution heat treatment. On the same figure the contribution to the strength from the precipitates as a fuction of the roll deformation before ageing are shown. The gain from precipitation hardening decreases a little as the deformation increases but the total strength of the alloy increases with increasing deformation.

The scanning electron micrographs in figure 8 show the fracture modes for three different ageing conditions. In the as quenched condition the fracture surface is both trans- and inter-granular. The underaged condition is characterized by steps and is mainly intergranular fracture. The overaged condition is typically intergranular fracture with small dimples on the grain facets.

3.2 The grain size effect.

The effect of ageing condition and grain size on tensile properties are shown in the figures 4,5 and 6. Fig. 4 shows the tensile properties in the as quenched condition as a function of inverse square root of the grain size. Yield stress and ductility increases with decreasing grain size. The underaged and the overaged condition show the same effects on the tensile properties, fig. 5 and 6. The hardness in the two ageing conditions are the same and is about 110 kp/mm^2.

The optical micrographs, fig. 7, show the planar slip in the alloy at two different grain sizes, and at two different ageing conditions. The small- grained material having a grain size about 6o μm shows a homogeneous slip pattern. The coarse grained material, grain size 400 μm, has a few and intense planar slips in the underaged condition, but the slip line pattern in the peak aged condition is relatively homogeneous.

Figure 9 shows TEM micrographs from tensile tested specimens. The small grained material shows a homogeneous distribution of dislocations within the grains. The coarse grained material shows only a few paired dislocations inside the grain. The two specimens were deformed to fracture before the TEM investigation. Table 1. summarizes the metallographic investigation.

TABLE 1

Metallographic investigation of tensile tested specimens.

Structure \ Ageing condition		Underaged 2 minutes at 240°C	Peak hardness 2 hours at 240°C
Microstructure in undeformed material		High density of δ'. No PFZ and δ	High density of δ'. PFZ is about 0.15μm, δ as GBP
Fracture mode		Intergranular, Sliplines and Steps	Ductile intergranular fracture,small dimple at grain surfaces.
Slip line distribution		Some slip lines are strongly defined.	High density of slip lines.
Micro structure in deformed material	Small grained	Paired dislocation in matrix grain, some pile ups.	Homogeneous dislocation structure.
	Coarse grained		Paired dislocation in matrix grains, pile ups.

δ = stable phase AlLi
δ' = metastable precipitate Al_3Li
PFZ = precipitate free zone
GBP = grain boundary precipitate

4. DISCUSSION

4.1. Strengthening by precipitation

It is still difficult to understand which particle hardening mechanism dominates in the Al + 3Li alloy. To compare theoretical and experimental values it is necessary to find the single-crystal yield stress corres-ponding to the yield stress for the specimen. In figures 2 and 4 the yield stress (0.2% poof stress) at peak hardness is 225 MPa, but the value in the as quenched condition is 62 MPa. Increased yield stress from the as-quenched condition to the peak hardness is 1/3 (225-62)MPa = 54 MPa.

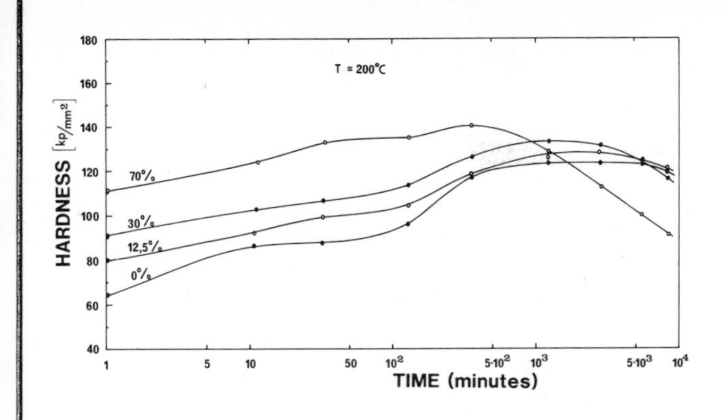

Figure 1.

Ageing curves of an Al + 3Li alloy deformed by rolling and aged at 200°C.

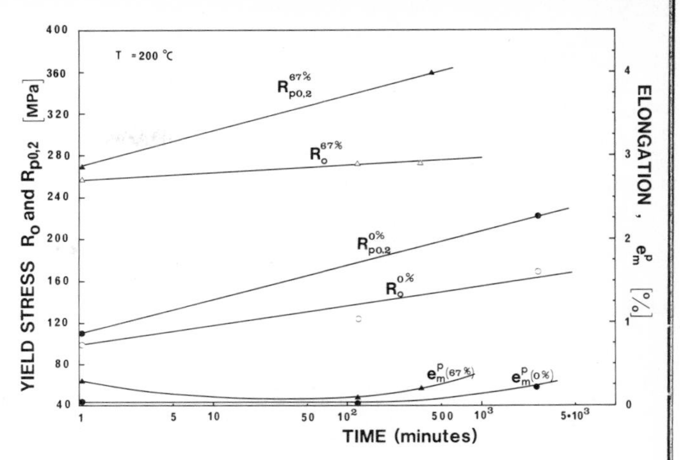

Figure 2.

Tensile properties of an Al + 3Li alloy as a function of ageing time at 200°C, deformed 67% before ageing and not deformed before ageing.

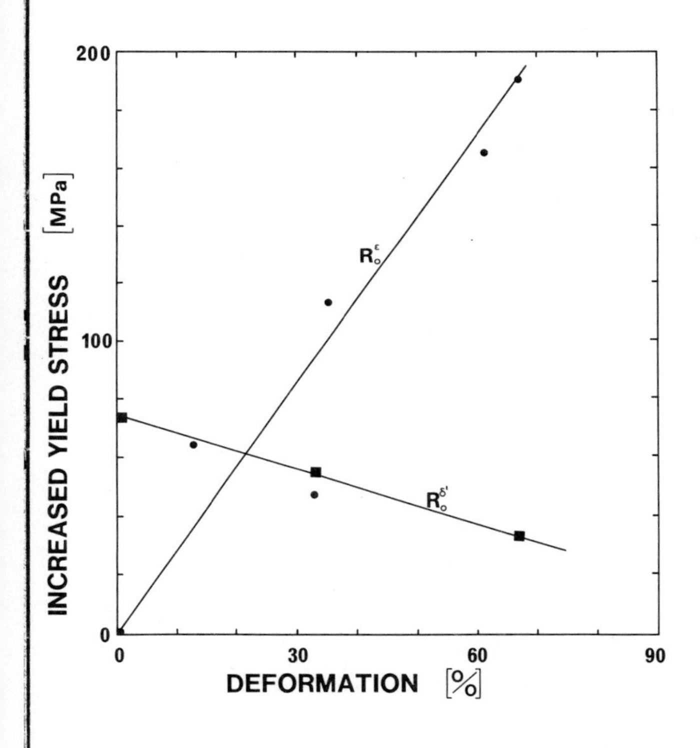

Figure 3.

The contribution to the yield stress from deformation hardening and precipitation hardening as a function of cold-roll deformation. The grain size is about 150 μm. $R_o^{\delta'}$ is the increased yield stress to peak hardness after deformation before ageing.

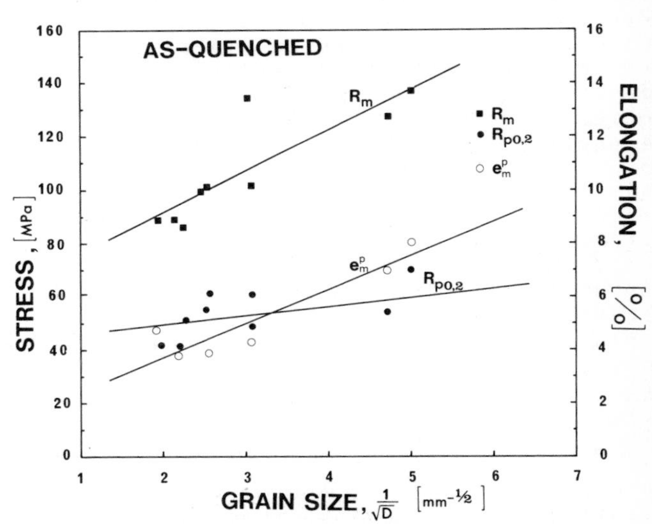

Figure 4.

Tensile properties as a function of grain size, as-quenched condition.

The factor 1/3 converts yield stress for polycrystal to yield stress in single-crystal. The grain size is about 150 μm. In the alloy Al+3,0Li the δ′ phase is present in the as quenched condition |1|. The yield strength due to solid solution hardening must be less than 62/3 MPa. The contribution to the hardening from the particles are proportional to $\sqrt{r}\sqrt{f}$, the square root product of the particle size and the volume fraction. Linear regression on the data points for $R_{po.2}$ versus $\sqrt{r}\sqrt{f}$ gives $R_{po.2}$ = 22 MPa when the particles are not present. The increase in yield stress due to the particles must thus be 67 MPa at peak hardness.

4.1.1 Coherency strengthening

The increase in yield stress due to the presence of coherency strains around the precipitates can be calculated from the relation |6|:

$$\tau_\varepsilon = 11.8\ G_m(\varepsilon)^{3/2}\left(\frac{r}{b}\right)^{1/2} f^{5/6} \qquad (4.1)$$

G_m is the shear modulus of the matrix, b the magnitude of the Burgers vector, ε the misfit strain caused by the precipitate, and f the volume fraction of precipitates. The references |10, 11| indicate the misfit between δ′ and the matrix to be around 0,1%. The shear modulus is 27.6 GPa |8|, and we found about 11% δ′ particles with mean radius 0.02 μm after 48 hours at 200°C. The increase in yield stress due to coherency strengthening was found to be about 8 MPa in the Al + 3Li. Noble et. al. |2| found that coherency strengthening was less then 5 MPa.

4.1.2. Modulus strengthening

The difference between shear moduli matrix and precipitates produces a strengthening increment given by equation 4.2 |4|.

$$\tau = \frac{\Delta G}{4\pi^2}\left[\frac{3\Delta G}{G_m b}\right]^{1/2}\left[0.8 - \frac{\ln(r/b)}{7}\right]^{3/2}\sqrt{rf} \qquad (4.2)$$

In equation 4.2 ΔG is the difference in moduli between matrix and precipitates, f is the volume fraction of precipitates and r is the mean radius. Young modulus has been found to be 81.1 GPa |8| in peak hardness condition, and the modulus of the δ′ precipitates is estimated to be 30.6 GPa. Calculations shows that the contribution from modulus hardening is 31 MPa when ΔG = 3 GPa. The modulus hardening can explain about half of the observed strengthening increment.

4.1.3. Order hardening

An estimate of the yield stress due to order hardening involves several unknown factors. First of all the γ_{APB}, the antiphase boundary energy, is not well known.
Brown and Ham |5| have shown that for the situation where dislocations move in pairs the increased strength is due to an antiphase boundary between the two dislocations and is given by:

$$\tau = \frac{\gamma_{APB}}{2b}\left[\sqrt{\frac{4f}{\pi}} - f\right] \quad \text{if } r > \frac{T}{\gamma_{APB}} \qquad (4.3)$$

Equation 4.3 applies to large particles that are cut by dislocations. Large particles means particles in the peak hardness condition. Chaturvedi |12| gave a theoretical calculation of the antiphase boundary energy for a $L1_2$ structure.

$$\gamma_{APB} = \frac{2U_{L1_2}}{\sqrt{2}a} \quad \text{where} \quad U_{L1_2} = \frac{KT_c}{0.82} \qquad (4.4)$$

In the equations a is the lattic constant, T_c the critical temperature for ordering and K the Boltzmanns constant. T_c is 329°C according to the metastable phase diagram |1| and a is 0.4 nm |13|. The calculated γ_{APB} is then 0.18 J/m^2.

The next factor to be calculated is the line tension T. This is given by:

$$T = \frac{3Gb^2}{8\pi} \qquad (4.5)$$

The shear modulus is 28.6 GPa, and the Burgers vector is $2.8 \cdot 10^{-10}$ m. The line tension is $2.7 \cdot 10^{-9}$ N for the given values. If $r > T/\gamma_{APB}$, equation 4.3 gives the shear stress |5|. $T/\gamma_{APB} = 1.5 \cdot 10^{-9}$ m which is less than r=0.02 μm for max hardness (aged at 200°C). The values f = 0.12, r = 0.02μm and γ = 0.18 J/m^2 give (4.3) τ = 87 MPa, and this is close to the experimental value for the yield stress of the alloy.

In table 4.1 the contribution to particle hardening from the three mechanisms discussed are shown. It can be seen that order hardening correlates best with the observed yield stress in the peak hardness condition. Order hardening is characterized by coupled dislocations which are observed in the alloy, fig.9.

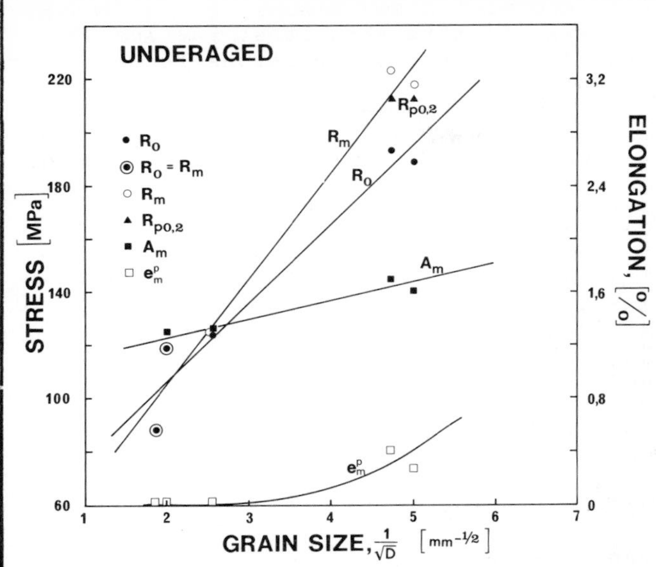

Figure 5.

Tensile properties as a function of grain size, underaged condition (3 hours at 220°C).

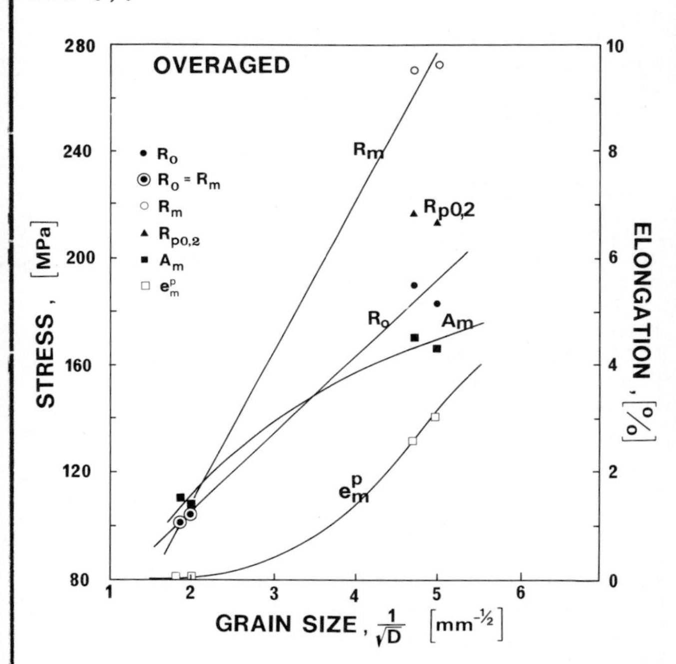

Figure 6.

Tensile properties as a function of grain size, overaged condition. (72 hours at 220°C.)

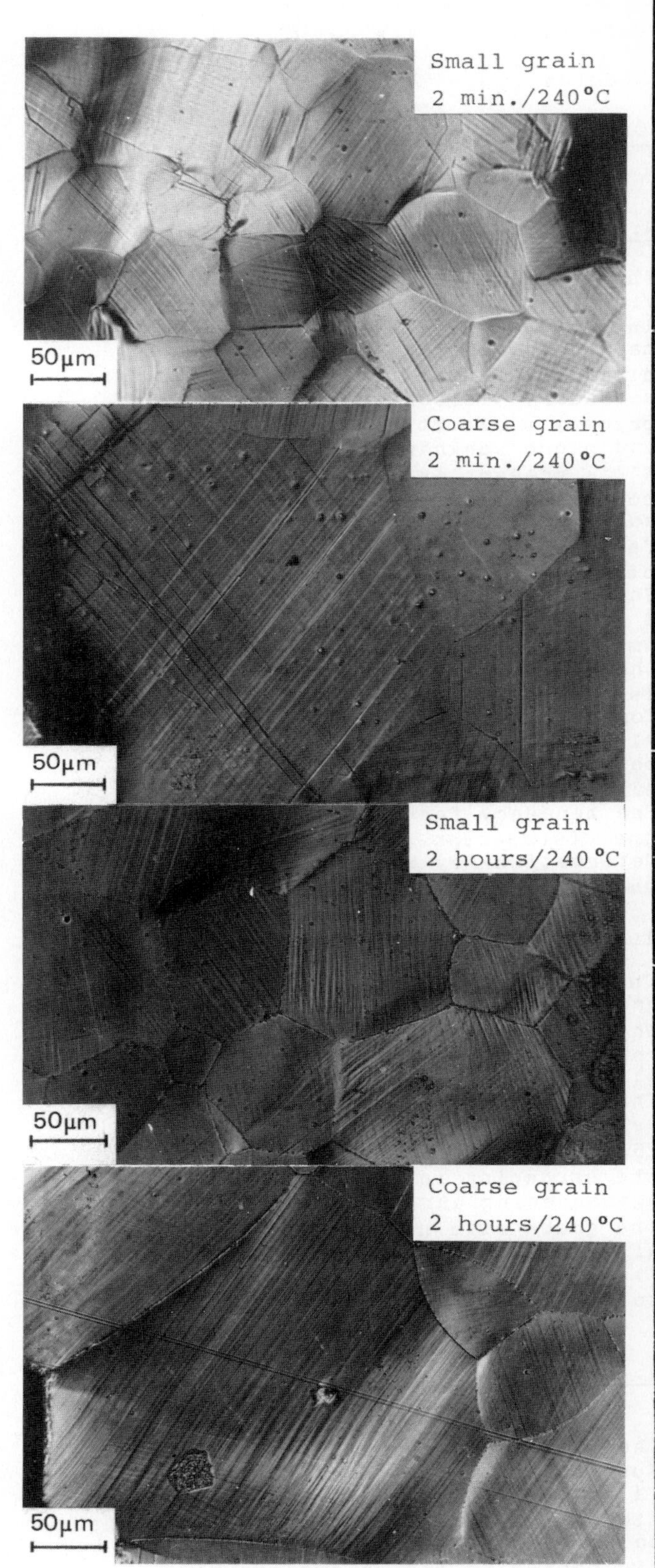

Figure 7 ▶

Light microscope pictures of the slip line distribution for tensile specimens with grain sizes 62 μm and 400 μm.

TABLE 4.1

Mechanism	equation	τ (MPa)
Coherence	4.1	8
Modulus	4.2	31
Ordering	4.3	87

4.2. Deformation hardening

The tensile tests show that the yield stress increases with increasing deformation. The deformation by rolling increases the dislocation density in the alloy. The secondary dislocations restrict the primary dislocations in their movement, and macro workhardening is observed. δ' is present in the as-quenched condition and the amount of Li in solid solution is about 9 atom %. The stacking-fault energy γ decreases as the amount of element in solid solution increases. The width of the splitted dislocation, $B = 6b/\gamma$, will increase with increasing amount of element in solid solution and the possibility for cross slip is reduced. Cross-slip is one condition for cell formation during deformation. Planar dislocation patterns should thus be expected when the amount of element in solid solution increases. The tendency for inhomogeneity in the dislocation pattern during deformation may explain the reduced ductility in the cold rolled material.

4.3. Grain size hardening

The yield stress as a function of grain size is shown in section 3.2. The factor k_f in the Hall - Petch relationship is 0.1 MPa$\sqrt{m}$ for Al + 3Li in the as-quenched condition. The increase in yield stress due to grain size refinement from 250μm to 25μm is 13 MPa. This contribution to the hardening is much less than the agehardening and deformation hardening. The grain size hardening in the Al alloys is known from the literature |14| to be low and k_f is in range 0.25 to 0.06 MPa$\sqrt{m}$.

4.4. Deformation and particle hardening

The maximum increase in yield stress due to particle hardening in Al + 3Li is 161 MPa. The increase in yield stress from deformation is proportional to the amount of deformation. The two hardening mechanisms are not linearly additive. The hardening from the precipitaties are 87 MPa after 67% deformation and before ageing. Brown and Ham |5| have shown that the sum of the flow stress contributions for two obstacles is

$$\tau = \sqrt{\tau_p^2 + \tau_d^2} \qquad (4.6)$$

a = alloy, p = particle and d = deformation

The reduced hardening potential from the particles is due to preferential coarsening of the δ' on dislocations |1|, and small precipitate free areas are formed around particle δ. The yield stress values from figure 3 inserted into eq. 4.6 gives:

$$\Delta R^{\delta'}_{po.2} = 161 \text{ MPa}$$

$$\Delta R^{\varepsilon}_{po.2} = 207 \text{ MPa}$$

$$\Delta R_{po.2} = \sqrt{161^2 + 207^2} = 262 \text{ MPa}$$

In the as-quenched condition the yield stress is 62 MPa. This gives $R_{po.2}$ = 324 MPa for the alloy which is in good agreement with the experimental value 356 MPa. The square root sum gives thus reasonable agreement with experimental values.

4.5 Grain size and fracture

Deformation and particle hardened alloys of Al + 3Li show very low ductility. In some cases the plastic deformation at fracture is less than 0.2%. Figures 4, 5 and 6 show that the grain size has a large influence on the fracture properties. Figure 8 shows that the fracture modes in aged conditions is intergranular.

4.5.1. Fracture at a critical stress.

The stress necessary for crack propogation can be related to the grain size |14|. The relationship between the fracture stress σ_f and the ultimate tensile stress is $\sigma_f = R_m (1+A_m/100)$ where A_m is the elongation at maximum load.
If σ_f is plotted against inverse square root of the grain size, the slope of the plot will be equal to $4G\gamma/k_f$; G is here the shear modulus, γ is the total work per unit area needed for continued crack growth, and k_f the slope from the Hall-Petch relationship. The values of σ_f (R_m) from figure 4, 5 and 6 give by regression γ (as-quenched) = 0.5 J/m^2, γ (underaged) = 1.0 J/m^2 and γ (overaged)= 1.6 J/m^2. In Al+Mg+Si γ is estimated to be 10 J/m^2 |14|.

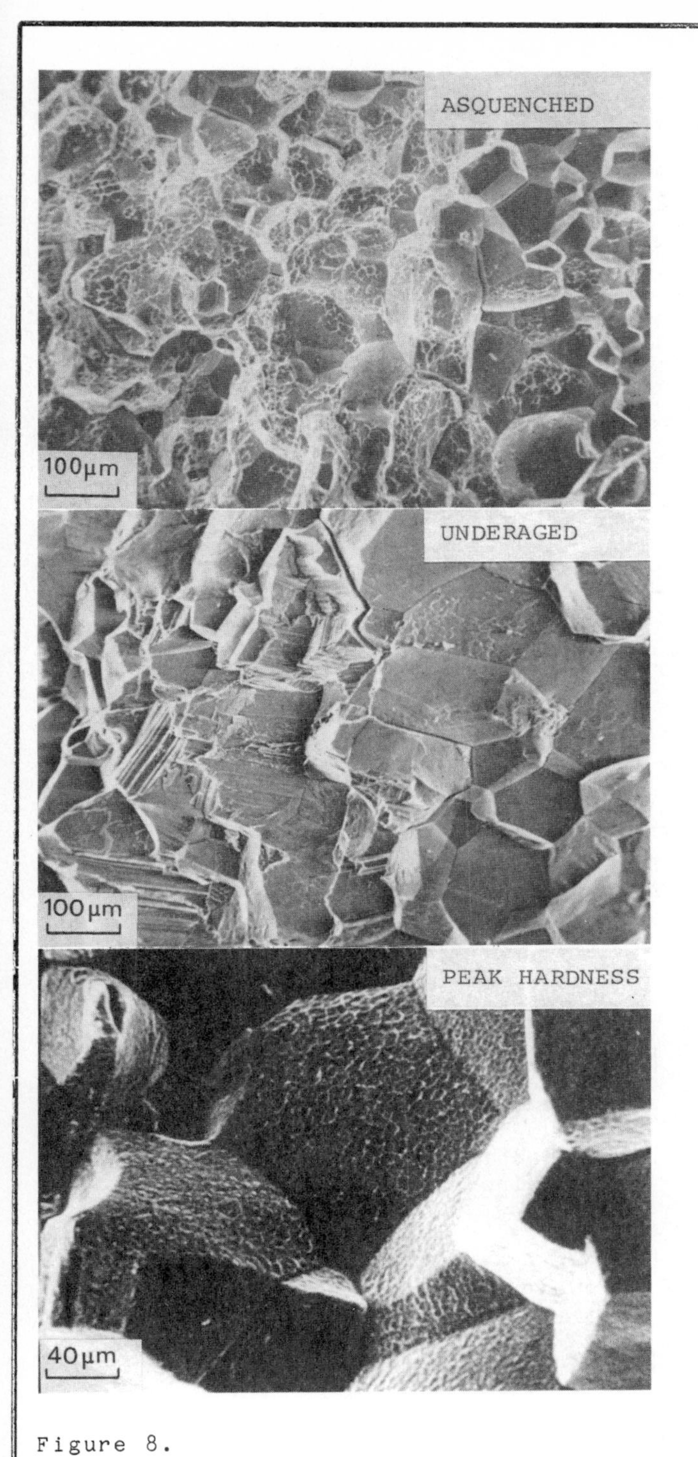

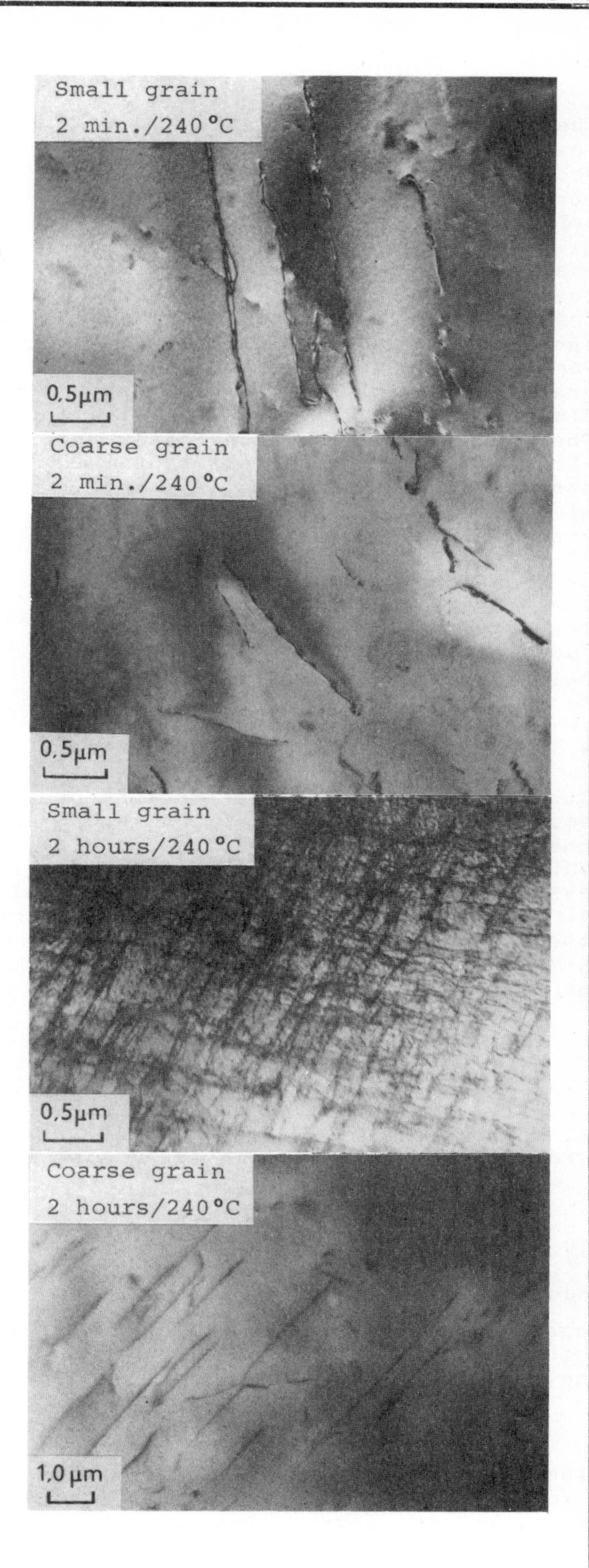

Figure 8.

Scanning electron micrographs of the fracture surfaces in three different ageing conditions.

Figure 9 ▶

TEM micrographs from tensile tested specimens.

The fracture stress of Al+Mg+Si is consequently higher than the fracture stress of Al+3Li.

The observed increase in ductility and fracture stress when the grain size decreases may be due to the reduced length of the slip bands. The work softened slip bands are long and cause a high stress at the grain boundaries in coarse grained materials (see figure 7), leading to a relatively low macro fracturestress at fracture. The steps on the fracture surfaces are due to the local work-softened slip bands. The fine grained material shows a homogeneous distribution of slip-bands in figure 7. The grain boundary precipitates, GBP, will also reduce the local cohesive stress necessary for initiation of microcracks. The precipitate free zones, PFZ, along the grain boundaries give a contribution to macro plastic deformation (figure 2 overaged) and the PFZ´s and GBP´s cause ductile intercrystalline fracture. The word ductile in this context is of course a relative characteristic.

4.5.2. Fracture at a critical strain.

Ductile intercrystalline fracture occurs because of void formation and growth at the grain boundary precipitates. The voids nucleate because of deformation in PFZ. One model for fracture at critical strain in PZF is presented in |5|, where the total plastic deformation is given as a sum of strain in matrix and strain in PFZ:

$$\varepsilon_f = \varepsilon_{PFZ} f_{PFZ} + \varepsilon_m f_m \qquad (4.7.)$$

f_{PFZ} and f_m are the volume fractions of PFZ and matrix structure. The f_{PFZ} is proportional to w/D where w is the width of the PFZ and D is the grain size.

$$\frac{\delta \varepsilon_f}{\delta 1/D} = k'w(\varepsilon_{PFZ} - \varepsilon_m) \qquad (4.8.)$$

Equation 4.8 shows that if the alloy has a hard matrix and a soft PFZ the relation-ship between ε_f, w and D will be

$$\varepsilon_f = \varepsilon_{macro} = \varepsilon_{PFZ} f_{PFZ} = k\frac{W}{D}\,\varepsilon_{PFZ} \qquad (4.9.)$$

The relative increase in ductility in overaged specimens are a result of increased PFZ width. The width of the PFZ |1| is about 0.4μm at peak hardness. Equation 4.8 shows that the fracture strain increases with decreasing grain size. If the PFZ is 0.4μm and all deformation are located in PFZ (grain size 150μm), the PFZ strain is 135%. Large strains localized in PFZ is reported (15,16) and may be the explanation why the fracture surface is dimple like.

5. Conclusions.

- The modulus and antiphase boundary model can both explain the contribution to the strengthening from the δ´ particles.
- The strength contribution from the hardening mechanisms is found to be given by the root mean square sum of the yield strength of particle- and deformation hardening.
- The grain size has a small effect on the yield stress.
- Finegrained material is more ductile then coarsegrained material.
- A small grain size reduces the effect of localized deformation which is the origin to microcracks and reduced ductility.

6. Acknowledgement.

The author is grateful to prof. Nils Ryum for valuable discussions.

7. References.

|1| O. Jensrud and N. Ryum, Mater.Sci. and Eng., 64, (1984), 229-236.

|2| B. Noble, S.J. Harris and K. Dinsdale, Met.Sci., 16. (1982), 425-430.

|3| T.H. Sanders, E.A. Ludwiczak and R.R. Sawtell, Mater.Sci, and Eng., 43, (1980), 247-260.

|4| P.M. Kelly, Inst.Met.Rev. 18, (1973), 31-36.

|5| L.M. Brown and R.K. Ham, (1971), "Strengthening Methods in Crystals", Ed. by A. Kelly and R.B. Nicholson, Applied Science Publishers Ltd. London.

|6| H. Gleither and E. Hornbogen, Phys. Status Solidi, 12, (1965), 235.

|7| M. Tamura, T. Mori and T. Nakamura, Trans. Jpn. Inst. Met., 14, (1973), 355-363.

|8| O. Jensrud, "Phasetransformations and mechanical properties of Al-Li alloys", (1982), Department of Metallurgi, University of Trondheim.

|9| B. Noble, S.J. Harris and K. Dinsdale, J.Mater.Sci., 17, (1982), 461-468.

|10| B. Noble and G.E. Thompson, Met.Sci.J., 5, (1971), 114-120.

|11| D.B. Williams and J.W. Edington, Met.Sci., 9, (1975), 529-532.

|12| M.C. Chaturvedi, D.J. Lloyd and D.W. Chung, Met.Sci., 10, (1976), 373-381.

|13| Tsyoski Yoski-Yama, Katsuhiko Hasebe and Michi-Hiko Mannami, J.Phys.Soc. Japan, 25, (1968), 908.

|14| S.D. Evensen, N. Ryum and J.D. Embury, Mat.Sci.Eng., 18, (1975), 221-229.

|15| T. Kawabata and O. Izumi, Acta.Met., 24, (1976), 817-825.

|16| E. Hornbogen and M. Gräf, Acta.Met., 25, (1977), 883-839.

□

Fundamental aspects of hardening in Al–Li and Al–Li–Cu alloys

P SAINFORT and P GUYOT

The authors are with, respectively,
Cegedur-Pechiney,
Centre de Recherches et de Développement,
BP 27 - 38340 Voreppe, France,
and Enseeg, Domaine Universitaire,
BP 75 - 38402 Saint Martin D'Heres, France

SYNOPSIS

The recent developments in the interest in Al-Li-X alloys for aerospace applications are subject to the knowledge of their fundamental strengthening mechanisms. Preliminary studies have demonstrated, for binary alloys, the prominent effect of the metastable phase δ' on the increase in the tensile strength. For ternary alloys with high lithium contents, the strengthening is due to different transition phases including δ'. The results outline the role of δ' and the contribution of each phase on the flow stress of an Al-Li-Cu alloy.

INTRODUCTION

The definition and use of new alloys require the understanding of their different mechanisms of strengthening. The present study is concerned with the yielding and tensile behaviour of binary Al-Li alloys and one ternary Al-Li-Cu alloy. The binary alloys were aged in the δ' stability domain and were studied to investigate the strengthening part of the δ' homogeneous precipitation. The ternary alloy shows a double precipitation which encloses δ' and the heterogeneous plate shape T_1 precipitation.

BINARY ALLOYS

The alloys were cast by conventional ingot metallurgy. After homogenization heat-treatment in salt bath followed by water quenching, their chemical compositions were Al-2%Li, Al-2,5%Li and Al-3%Li (weight percent). The binary Al-2,5%Li was cast under Argon and extruded into bars from which tensile specimens were prepared. Ageing treatments were selected from δ' stability domain given by COCCO et al [1] from small angle X ray scattering experiments. The Al-2%Li and Al-2,5%Li alloys were annealed at 200° C while the Al-3%Li was aged at 220° C. These isothermal treatments were performed for times varying from a few minutes to one week. Such ageing treatments resulted in three different volume fractions of precipitates [1] (0,05, 0,15 and 0,25 for the Al-2%Li, 2,5%Li and 3%Li respectively). The precipitation has been observed by conventional and high resolution transmission electron microscopy (TEM JEOL 200 CX) as well as by scanning transmission electron microscopy (STEM VG HB 501).

RESULTS

δ' coherency

The $L1_2$ crystallographic structure, reported in numerous papers, was investigated by selected area electron diffraction or by nanodiffraction with a focussed electron beam on single precipitates [2]. However, these patterns cannot be used to measure directly the lattice parameter difference between the aluminium solid solution and the precipitates. Two alternative techniques have therefore been performed :

- The ASHBY-BROWN contrast analysis [3] uses the image width of the precipitate along the diffraction vector in two beam illumination. The method uses curves established in log coordinates which are of poor precision for coherency deviation $\left|\frac{\Delta a}{a}\right|$ below 10^{-3}. The results yield a relative parameter variation of between 2.5 10^{-4} and 3.3. 10^{-3}. The previously reported values range between 8.10^{-4} and 3.10^{-3} (4, 5, 6, 7).

- An alternative and more direct method consists of imaging the lattice planes in the matrix and in the precipitates, to observe the elastic deformation along the interface. The high resolution has been performed with a tilted beam by interference of the transmitted electron beam

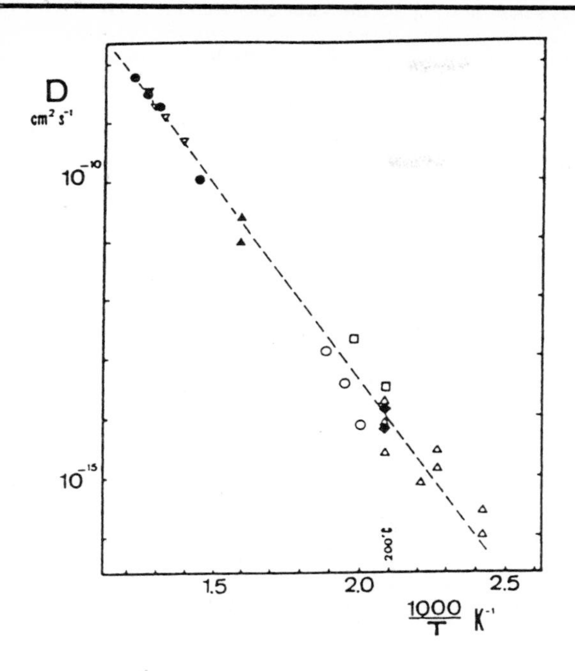

Fig. 1 Arrhenius diagram of the Lithium diffusion coefficient in aluminium (●[9], ▽[20], □[10], ▲[11], Δ[6], o[13] and ◆[12]).

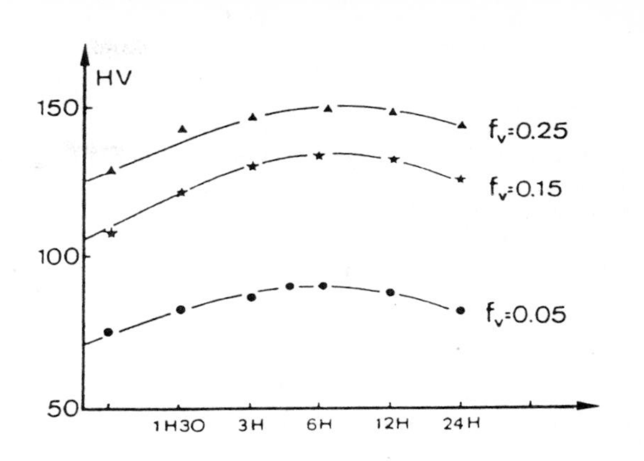

Fig. 2 Microhardness measurements versus ageing time for three different δ' precipitated volume fractions.
(f_v = 0.25 : Al-3%Li aged at 220° C,
f_v = 0.15 : Al-2,5%Li aged at 200° C
and f_v = 0.05 : Al-2%Li aged at 200° C).

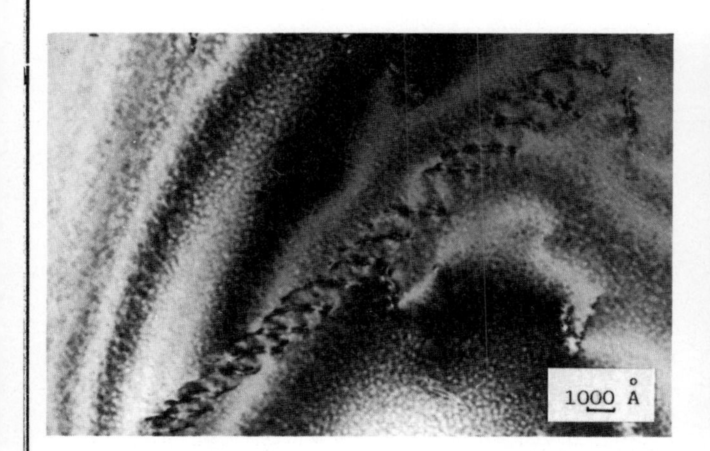

Fig. 3 Superdislocation pile up in a Al-2,5%Li aged 3 hours at 200° C (T.E.M. micrograph).

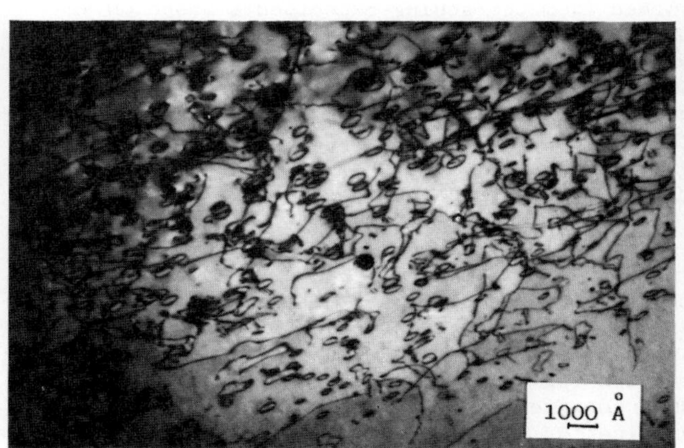

Fig. 4 Dislocation loops in a Al-2,5%Li aged 48 hours at 200° C (T.E.M. micrograph).

with the $(001)_{\delta'}$, $(002)_{\delta'}$ and $(002)_{matrix}$ beams [8]. The interplanar spacing is 2.02 Å. No deviation between the two phases was detected, even on coarsened precipitates of 0.2 µm diameter which allow a total spacing measurement from several hundreds (002) planes on each side of the interface. The maximum possible value of $\left|\frac{\Delta a}{a}\right|$ thus corresponds to the measurement uncertainty from the micrographs : $\left|\frac{\Delta a}{a}\right| \leqslant 9.10^{-4}$. This is in accordance with the mean value obtained from ASHBY-BROWN contrasts.

δ' coarsening

The kinetic of precipitate coarsening has been followed using TEM dark field observations on a δ' reflexion combined with image analysis. Histograms of the δ' size distribution were plotted for different isothermal ageing times. The results confirm the LIFSHITZ-SLYOSOV (L.S.) coarsening law : the shape of the radius distributions and the correlation between the mean diameter of each distribution and the ageing time correspond to the theoretical model ($t^{1/3}$). However, one cannot use the L.S. coarsening law to calculate directly the precipitate-matrix interfacial energy γ_S. As a matter of fact, the model was established for a precipitate volume fraction f_v close to zero. Nevertheless, the contribution of f_v on coarsening rate has been modelised by different authors and shown to increase this parameter by a factor of 2 to 3 for $f_v \simeq 25$ %. On the other hand, the knowledge of the diffusion coefficient of Li in Al severely limits the precision of the calculated value of γ_S. Figure 1 gives different experimentally measured values of D for various temperatures. At the highest temperatures, D concerns Al-Li diluted solid solutions while at medium temperatures (150-300° C) measurements have been performed from coarsening experiments based on an L.S. interpretation or on precipitate-free zones width. Our measurements, at ageing temperature T = 200° C, have been obtained from this latter method and from lithium concentration profiles through the precipitate free zones along the grain boundaries. The chemical analysis was obtained by electron energy loss spectroscopy (EELS) with the variation of the volume plasmon energy [2]. The various results at 200° C are spread over two orders of magnitude, defining the possible accuracy of the calculated γ_S value. Only a maximum evaluation of γ_S is of interest and has been calculated using no correction for f_v and the smallest reported diffusion coefficient of lithium in Al. This suggests a value for γ_S < 30 mJ/m², which is considered as reasonable for the case of a $L1_2$ structure in a fcc matrix. This may be compared to previously reported values of between 24 mJ/m² and 500 mJ/m² (4, 5, 6, 13).

Strengthening mechanisms

The structural strengthening was followed by microhardness measurements after isothermal ageing treatments. Figure 2 exhibits two domains - increasing and then decreasing hardness - with respect to ageing time, for different volume fractions f_v of precipitates. In each case, the dislocation structure was observed by TEM on polycrystalline samples having been plastically deformed by 2 %. The general structure is presented in micrographs 3 and 4. During the first stage - hardness increasing with ageing time - dislocations are organized in superdislocation pile-ups (micrograph 3). This morphology characterizes a strongly localized deformation with a shearing process of the microstructure and antiphase boundary (APB) creation inside the precipitates. The specific APB energy γ_A was measured from the splitting of each pair resolved in their glide plane, giving a value of $\gamma_A \in [130, 175]$ mJ/m². TAMURA et al [4] have measured, by the same method, γ_A = 195 mJ/m². Over a critical precipitate radius, which corresponds to the maximum hardness, the by-passing of the δ' microstructure becomes turbulent and leads to a decreasing hardness with the increase in δ' mean diameter. The dislocation distribution is then homogeneous, distinguished by numerous Orowan and double cross-slip formed loops (micrograph 4).

DISCUSSION

The strengthening mechanisms of the binary alloys may be analysed from these experimental results. The system is hardened by the simultaneous presence of two phases : the solid solution and the δ' precipitates which are characterized respectively by the residual lithium contents and by the volume fraction and mean radius of the precipitate distribution. The dominant part in strengthening has been demonstrated earlier (14,15) to be due to precipitation. Our measurements allow a quantitative but not exhaustive analysis of the relative contributions of different elementary mechanisms which may govern the strengthening. Three mechanisms have been considered :

1. Chemical strengthening due to the increase in δ'/α interfacial area during the precipitate shearing by the matrix dislocations. This hardening is governed by the interface energy γ_S.

2. Coherency strain strengthening induced by the lattice parameter difference between the precipitates and the solid solution.

3. Order effect which gives rise to the APB creation inside the precipitates consequently to their shearing. This mechanism is followed by the Orowan by-passing when the precipitate mean diameter exceeds a critical size.

For a given precipitated volume fraction f_v, their theoretical contribution on the alloy flow stress may be plotted versus the precipitates mean diameter distribution. The curves in figure 5 show these contributions using simple hardening models (16,17,18) for f_v = 0.25 and for present experimental values of γ_S (30 mJ/m²), $\left|\frac{\delta a}{a}\right|$ (9.10^{-4}) and γ_A (150 mJ/m²). The dominant effect of the order and of the APB creation

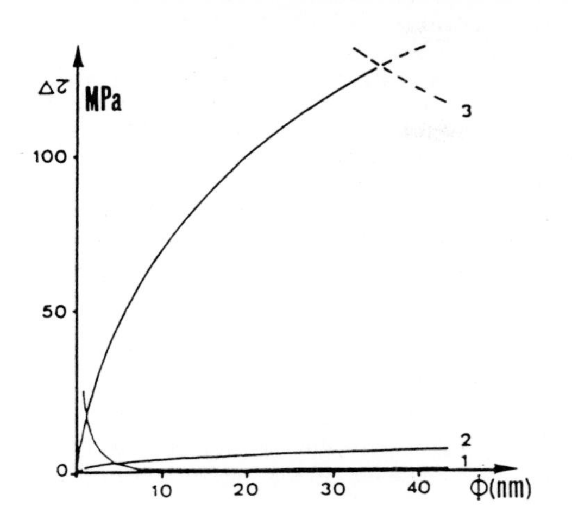

Fig. 5 Theoretical contribution of three elementary precipitation hardening mechanisms : -1- Chemical effect [16], -2- Coherency strains [17] and -3- Order effect with APB creation and OROWAN by-passing [18]

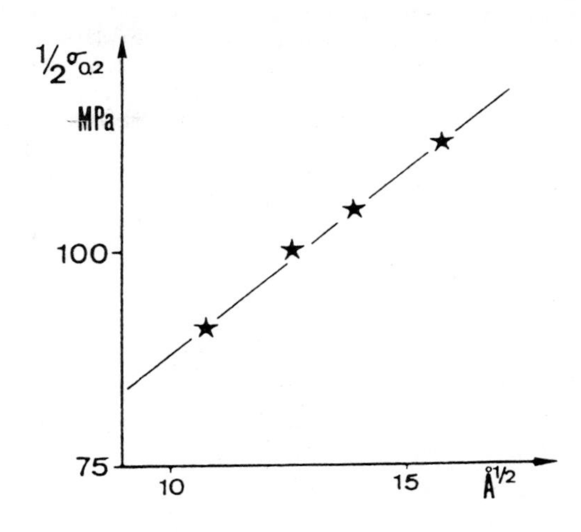

Fig. 6 Al-2,5%Li annealed at 200° C : Yield Strength versus the mean δ' diameter square root. Hardening by APB creation with γ_A = 82 mJ/m^2

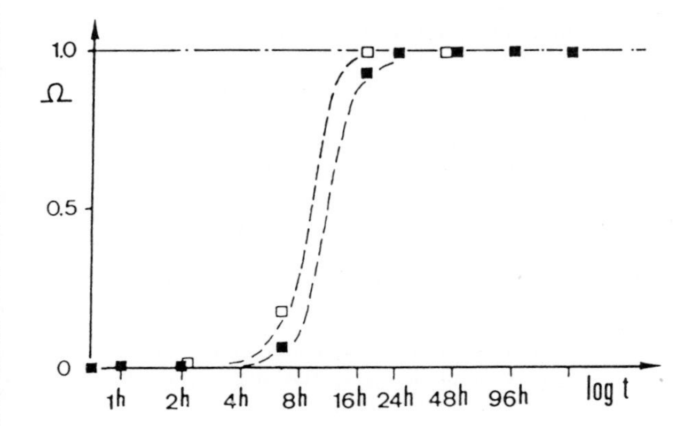

Fig. 7 Evolution of the matrix invaded volume fraction with T_1 plates. Al-2,7%Li-1,9%Cu aged at 190° C.
□ stretched 2 % before ageing,
■ unstretched before ageing.

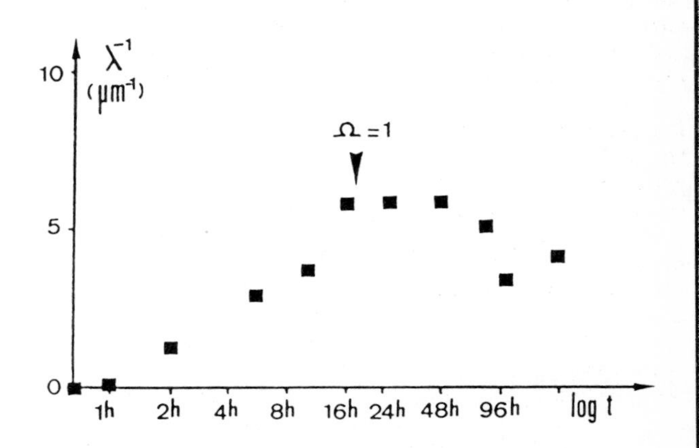

Fig. 8 Al-2,7%Li-1,9%Cu aged at 190° C. Evolution of the mean spacing between plates with a same epitaxy versus the ageing time.

can be seen. In this case, the model is based on a single unsplit matrix dislocation shearing spherical precipitates. The model has been used to calculate the critical mean radius which defines the transition between shearing and, Orowan by-passing. The experimental value is deduced from the mean size of the δ' distribution at the maximum hardness (figure 2) or from the size of the smallest Orowan loops observed by TEM on plastically deformed alloys. An excellent agreement is found between the calculated [18] and the experimental value : for $f_V = 0.25$, a transition diameter of 35 nm was determined.

These results may be finally confirmed by direct conventional yield strength measurements on an Al-2,5%Li aged at 200° C ($f_V \sim 15$ %). For coarsening treatments leading to the δ' shearing, ($t \leq 12$ h at 200° C), a linear dependence between the yield strength and the square root of the δ' mean diameter was observed. The calculation of the specific APB energy, using the slope of the straight line (Fig. 6) and the previous hardening model [18] yield a new value of $\gamma_A = 85$ mJ/m² . This result is significantly smaller than those measured from dislocation splitting width. Over 12 h at 200° C, the precipitation is by-passed by the matrix dislocations. However, the intergranular precipitation of the stable phase δ and the simultaneous formation of precipitate-free zones give rise to poor ductility. Despite a more homogeneous deformation due to the by-passing process, the conventional yield strength (0.2 % elongation) is no longer available. The decrease in the macroscopic flow stress with δ' coarsening has not been experimentally observed.

TERNARY ALLOY

The ternary alloy Al-2,7%Li-1,9%Cu (wt %) has been studied using similar investigations. The alloy was homogenized, water quenched and then aged at 190° C for various times. The δ' precipitation appears immediately while the T_1 precipitation is only observed after 30 minutes at 190° C.

RESULTS

δ' precipitation

In comparison with the binary alloys, our investigations have not demonstrated differences in the δ' behaviours in the ternary alloy. Using X ray energy dispersive spectroscopy in STEM, no copper has been detected inside the δ' precipitates while the matrix copper concentration was estimated, after ageing, to 0.5 at % [2] . The relative parameter variation between δ' and the solid solution seems to be this when observed in binary alloys, within our measurement uncertainties. In a similar manner, the superdislocation splitting widths lead to a similar specific APB energy. On the other hand, the calculation of the interfacial energy γ_S can no longer be obtained from the δ' coarsening law . As a matter of fact, the diffusion coefficient of lithium in an aluminium-copper solution is unknown. In addition, the coarsening interactions between δ' and the T_1 plates modify the coarsening rates of both precipitations. However, no apparent differences were noted between the δ' coarsening rate in binary and ternary alloys. It can thus be assumed that γ_S is similar in all cases.

Hence, the principal parameters which govern the δ' structural strengthening seem to be the same in binary and ternary alloys.

T_1 precipitation

The precipitation has been characterized to define the most important parameters which could play a part in the ternary alloy hardening. The heterogeneous T_1 precipitation associated with dislocations has often been described by earlier workers [19] . With isothermal ageing treatments, the growth leads then to matrix domains "invaded" by this precipitation. The matrix volume fraction Ω invaded by the plates may be followed by TEM observations. Figure 7 shows this evolution. Ω becomes equal to one when the alloy is completely netted by the plates, into cells containing the δ' precipitates.

Using dark field micrographs on a T_1 reflexion, the mean distance λ between plates of the same epitaxy with respect to the matrix ((111), $(1\bar{1}1)$, $(11\bar{1})$ and $(\bar{1}11)$) was measured. Figure 8 reports λ^{-1}. The minimum spacing between plates corresponds to Ω = 1. The cells are thus increasing in size by a coarsening process.

Strengthening Mechanisms

The dislocation structure was observed in 2 % tensile strained samples. Concerning the interaction with the δ' precipitates, shearing and then by-passing were always observed but a strong decrease in the critical transition diameter of the precipitates was noted despite a value of γ_A similar to the binary alloys. The by-passing process was apparent after 4 hours at 190° C when the δ' diameter was only 18 nm ($f_V \sim 0.25$).

The detailed interaction of the dislocations with the T_1 plates is difficult to analyse due to their low thickness. However, the plates seem to act as unshearable obstacles which are by-passed during the deformation.

The microhardness measurements versus ageing time exhibit two regions : a continuous hardness increase up to a maximum followed by a plateau of hardness (Fig. 9). These results are also obtained from yield strength measurements (Fig. 10). The transition between shearing and by-passing of the δ' precipitation is no longer observed on the curves (between 3 and 4 hours at 190° C).

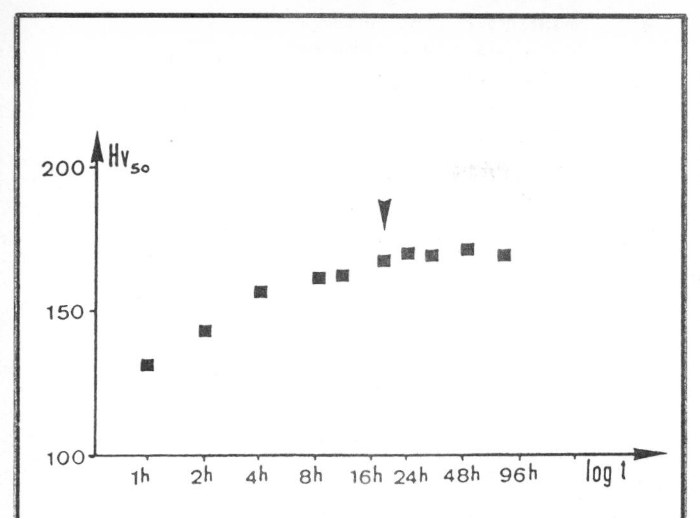

Fig. 9 Al-2,7%Li-1,9%Cu stretched 2 % before ageing at 190° C. Microhardness measurements for various ageing times.

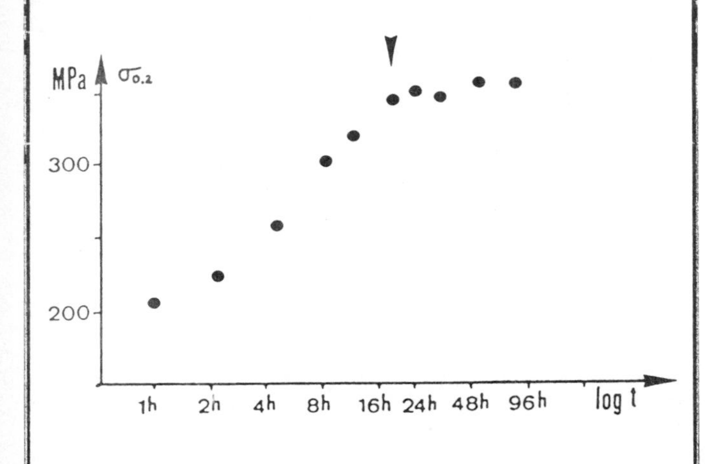

Fig. 10 Al-2,7%Li-1,9%Cu stretched 2 % before ageing at 190° C. Yield strength evolution with ageing time.

DISCUSSION

The increase in the hardness and the yield strength is in fact correlated with the progressive precipitation of T_1. The maximum hardening is reached when the matrix is completely invaded ($\Omega = 1$).

The earlier by-passing process of δ' in the ternary alloy may be explained supposing a smaller stacking fault energy of the ternary solid solution which induces a wider splitting of the dislocations. The δ' precipitate shearing by the leading partial introduces in the $L1_2$ lattice a complex order fault. The high specific energy of this fault may act against the shearing process, leading to a more favourable by-passing. Our assumption may be supported by the heterogeneous T_1 precipitation on dislocations. This point, described by [19], supposes copper segregation on matrix dislocations following a Suzuki effect. However, we have not been able to observe the splitting into partials using TEM weak beam techniques.

In the ternary alloy, the copper has a double effect : its presence promotes the by-passing of δ', leading to a homogeneous deformation, and gives rise to the T_1 precipitation which governs the observed increase in yield strength. However, the heterogeneous precipitation of T_1 introduces deformation heterogeneities. The plastic deformation being effectively localized when $\Omega < 1$ and despite the δ' presence, in regions of the matrix free of T_1 plates. This localization induces a decrease in the ductility when Ω increases and results in the observation on intergranular fracture surfaces of numerous steps at grain boundaries.

CONCLUSION

In conclusion, these results show the differences between the structural hardening mechanisms in the binary alloys and a ternary alloy. The dominant part of the order inside the δ' precipitates is demonstrated for the case of binary alloys. The strengthening is then governed by the volume fraction of precipitates and their mean diameter. The maximum strength is obtained for the diameter which leads to the transition between δ' shearing and δ' by-passing mechanisms by the glide dislocations. In ternary alloys, this transition appears at a smaller δ' size. This fact may be due to a smaller matrix stacking fault energy. The δ' strengthening is then less efficient, the evolution of the yield strength being governed by the progressive precipitation of T_1. The maximum is reached for a complete invading of the matrix by the plates. The unshearable T_1 precipitates localize the glide in the matrix regions where the plates have still not appeared. This deformation channelling reduces the ductility and overcomes the favourable effect of δ' by-passing, which would otherwise homogenize the deformation.

ACKNOWLEDGMENTS

The authors would like to thank the National Association for Technical Research and Aluminium Pechiney for providing a grant for P.S. as well as for supplying the alloys.

REFERENCES

1 COCCO G., FAGHERAZZI G. and SCHIFFINI L. J. Appl. Cryst., 10, (1977), 325-327.

2 SAINFORT P. and GUYOT P. Phil. Mag. A, 1985, Vol. 51, N° 4, 575-588.

3 ASHBY M.F. and BROWN L.M. Phil. Mag., 1963, Vol. 8, 1083 and 1649.

4 TAMURA M., MORI T. and NAKAMURA T. J. Jap. Inst. Met., Vol. 34, 1970, 919.

5 NOBLE B. and THOMPSON G.E. Met. Sci. Journal, Vol. 5, 1971, 114.

6 WILLIAMS D.B. and EDINGTON J.W. Met. Sci., Vol. 9, 1975, 529.

7 BAUMANN S.F. and WILLIAMS D.B. 2nd Int. Al-Li Conference, Eds Starke and Sanders, Warrendale P.A. (1983).

8 SAINFORT P. and GUYOT P. Al-Li III, this conference, 1985.

9 COSTAS L.P. U.S. Atomic Energy Commission Rep. DP 813, 1963.

10 WILLIAMS D.B. Ph. D Thesis, Cambridge, 1974

11 WILLIAMS D.B. and EDINGTON J.W. Phil. Mag., Vol. 30, 1974, 1147.

12 SAINFORT P. Thesis, Grenoble, 1985.

13 JENSRUD O. and RYUM N. Mat. Sci. Eng., Vol. 64, 1984, 229.

14 NOBLE B., HARRIS S.J. and DINSDALE K. Met. Sci., Vol. 16, 1982, 425.

15 NOBLE B., HARRIS S.J. and DINSDALE K. J. Mat. Sci., Vol. 17, 1982, 461.

16 KELLY A. and NICHOLSON R.B. Progress in Mat. Sci., Vol. 10, 1963, 381.

17 BROWN L.M. and HAM R.K. "Strengthening Methods in Crystals" Eds KELLY and NICHOLSON, Elsevier publishing company ltd, LONDON, (1971).

18 GUYOT P. Phil. Mag., Vol. 24, 1971, 989.

19 NOBLE B. and THOMPSON G.E. Met. Sci. J., Vol. 6, 1972, 167.

Plastic deformation of Al–Li single crystals

Y MIURA, A MATSUI,
M FURUKAWA and M NEMOTO

YM and MN are in the Department of Metallurgy, Kyushu University, Fukuoka, Japan.
AM is with Mitsubishi Heavy Ind. Co., Japan.
MF is in the Department of Technology, Fukuoka University of Education, Munakata, Japan

Synopsis

Deformation behavior of age-hardenable Al-Li single crystals containing 2mass%Li was studied as a function of aging time. Emphasis was on the nature and distribution of deformation-induced dislocations.

Observations by transmission electron microscopy revealed random arrays of dislocations in the quenched condition, but rows of long straight screw dislocations in pairs were observed in the under-aged condition, suggesting a strong characteristic interaction between screw dislocations and the ordered-precipitates. In the peak-aged or over-aged condition, dislocation loops surrounding the precipitates were observed.

Discussion on the basis of the experimental results revealed that the order hardening and the Kear-Wilsdorf hardening in the δ'-$L1_2$ ordered precipitates play important roles for the strengthening of the alloy.

Introduction

Aluminum-lithium alloys are age-hardenable by the precipitation of spherical coherent Al_3Li with $L1_2$ ordered structure(1). Much attention has been paid on the development and application of Al-Li alloys, because of their high strength and high elastic modulus(2). The most serious drawback of the alloys is the much reduced ductility, especially in age-hardened conditions, in comparison with the conventional aluminum base alloys. This poor ductility and toughness has been attributed to the intense coplaner slips caused by superdislocations cutting through the δ'-$L1_2$ precipitates, and/or to grain boundary embrittlement by the segregation of trace elements or impurities(3-5). Attempts have been made to improve ductility by the addition of the third elements, which results in the dispersion of precipitations of other phases than Al_3Li and consequently reduces slip coplanarity(6).

Similarity is expected , in the deformation characteristics, to Ni-base superalloys, where deformation mechanisms are strongly influenced by the presence of spherical coherent $L1_2$-ordered precipitates. Previous experiments on poly-crystals(4)(7-9) or single crystals(10) of Al-Li alloys revealed mechanical properties characteristic of $L1_2$ ordered structure. It seems, however, that an agreement has not been attained on the strengthening mechanisms and the deformation characteristics of Al-Li alloys. Basing upon the TEM observations, Noble et al.(4) attributed the strength of Al-Li alloys to the order hardening or a combination of order hardening and modulus hardening. On the other hand, Furukawa et al.(8) suggested that main contribution comes from order hardening and the friction stress in the δ'-particles.

The purpose of the present investigation is to determine more explicitly the nature of the deformation-induced dislocations in the age-hardenable Al-Li alloys and to discuss strengthening mechanisms in the whole aging conditions, by the experiments on the alloy single crystals.

Experimental procedure

An ingot of Al-Li binary alloy was prepared by melting 99.99%Al and 99.8%Li in an argon atmosphere and by casting into a steel mold. The ingot was rolled to a 3mm thick sheet and cut by a fine cutting machine to 3mmx3mmx80mm rods, composition of which as revealed by chemical analysis was 3.12mass%Li, 0.0073mass%Fe, 0.029mass%Si and 0.0045mass%Mg. Single crystals were grown by the modified Bridgman method in an argon atmosphere with an alumina soft mold. A considerable amount of Li was lost in melting.

The Li content of the single crystals were about 2mass%Li. The grown single crystals were then solution treated at 823K for 1.08×10^4s in glass capsules filled with argon gas, followed by quenching into iced water. Compressive specimens of 3mmx3mmx5mm were cut out and they were again sealed in capsules with argon and aged at 473K for various periods of time up to 1.2×10^6s. The specimens thus aged to different aging stages were deformed in compression at room temperature to 1-4% strain with a nominal strain rate of $3.3 \times 10^{-4} s^{-1}$. In the present paper, the results on the specimens of three different compression axes, A, B and C, are reported. Schmid factors for (111) and (010) slips, and the Schmid factor ratios are shown in Table 1.

Table 1. Schmid factors and Schmid factor ratios

Specimen	Orientation	Schmid factor		N
		(111) [$\bar{1}$01]	(010) [$\bar{1}$01]	
A	7° - [$\bar{1}$11]	0.38	0.47	1.24
B, C	10° - [$\bar{1}$23]	0.49	0.31	0.63

Surface slip traces were observed by optical interference microscopy. Interactions between precipitates and dislocations were examined by transmission electron microscopy(TEM). Observations by TEM were most extensively made of the specimen A. Out of the deformed crystals thin plates of 0.3mm thick were cut parallel to the primary (111) slip plane and foils for TEM were prepared with a twin jet electropolishing apparatus. Composition of the electrolyte was 20%$HClO_4$ and 80%C_2H_5OH. An electron microscope JEM-1000D was used.

Results

1. Growth of δ'-precipitates by aging.

Dark field electron microscopy revealed a fine dispersion of δ'-phase in the quenched condition. Precipitates of amorphous shape become spherical as they grow, to a size of 60nm in diameter at 6×10^5s. These observed precipitation characteristics are consistent with the previous observations(11)(12). Variations of size and volume fraction of δ' with aging time are shown in Fig.1. The volume fraction was evaluated from the measurements of density and size of δ', and of foil thickness by a contamination-spot-separation method(13). The value 0.11 at near saturation precipitation well agree with the estimation from the δ'-solvus at 473K for 2mass%Li(14). Coarsening of δ' at the later stage of aging apparently shows a linear relation between the diameter(d) and cube root of aging time(t), being consistent with the previous studies(1)(15).

2. Critical resolved shear stress (CRSS).

Serrations were observed in the compressive stress-strain curves for the under-aged specimens A and B. CRSS of (111) slip is shown in Fig.2 as a function of time of aging at 473K. CRSS increases with aging time from 10MPa in the quenched condition to the maximum 70MPa in the peak-aged condition, followed by a slow drop in the over-aged condition. Orientation dependence is not apparently observed. Orientation and temperature dependences of CRSS are not discussed here, but they are being studied and results will be published elsewhere. These observed variations of CRSS with aging time are in accordance with those by Tamura et al.(10) shown in Fig.2 with a broken line. Their values are higher in all the aging conditions corresponding to the higher content of Li.

3. Slip lines.

Slip lines for different aging conditions observed by an interference microscope are shown in Photo.1. The crystals of specimen A were compressed to 4% strain. Trace analysis revealed that all the operating slips were {111} type and slip on {010} planes were not detected. Slip lines in quenched or under-aged conditions (a and b in Photo.1) are straight and coarse, being consistent with the occurrence of serration in the stress-strain curves. But they become finer and shorter in the peak-aged condition (c) and wavy in the over-aged condition (d).

4. Deformation-induced dislocations revealed by TEM.

Dislocation configurations on the primary (111) plane were examined with a dark field weak beam technique using 220 or 224 reflections. As shown in Photo.2, dislocations introduced by the deformation after quenching are randomly distributed. They are in general smoothly curved and some of them form networks. Dislocations lie often in pairs. In the lightly aged condition ($t=6 \times 10^2$s), long straight dislocations tend to lie parallel to the Burgers vector (Photo.3). Contrast experiments have revealed that these dislocations are screw in character. In the lightly deformed crystal in the under-aged condition ($t=6 \times 10^3$s, 1% strain), long straight screw dislocations are observed in pairs as shown in Photo.4. By the further deformation, the pair distance increases and the pairs often break into single dislocations (Photo.5, 4% strain). Photo.6 a and b represent 220 and 110 dark field images of the crystal aged to near peak-strength. Rows of screw dislocations in pairs are no more observed, and instead small dislocation loops, the size of which is comparable to δ'-precipitates, and also heavily curved dislocations are observed. In the over-aged condition, dislocation loops surrounding the δ'-precipitates are observed, as well as heavily curved dislocations connecting the loops as shown in Photo.7 (a and b are the 220 and 110 dark field images, respectively, taken of exactly the same area of the foil). It should be noted that the size of the loops is nearly equal to that of the corresponding δ'-precipitates. There is no evidence of the standing off of the dislocation loops against the δ'-particles due to the modulus effect or the strain field effects.

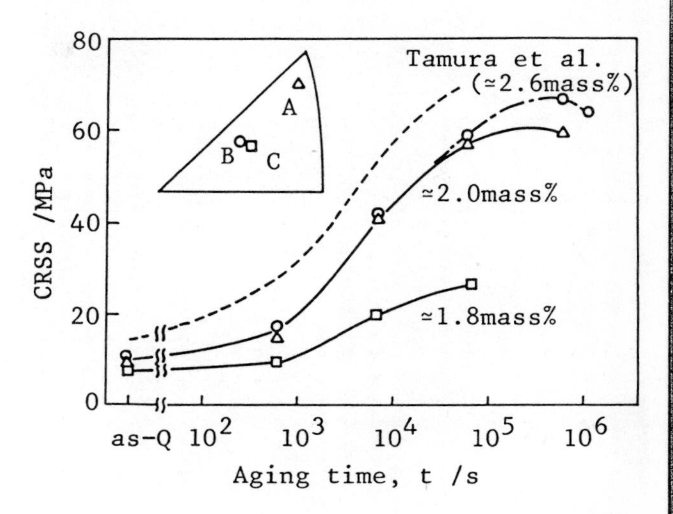

Fig. 2 Critical resolved shear stress as a function of aging time.

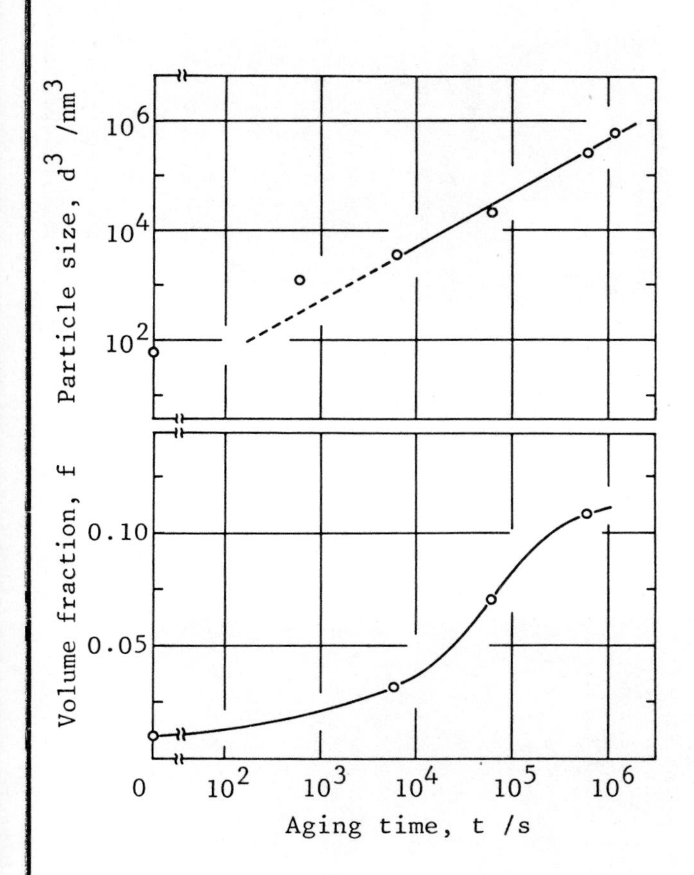

Fig. 1 Size and volume fraction of δ'-particles as a function of aging time.

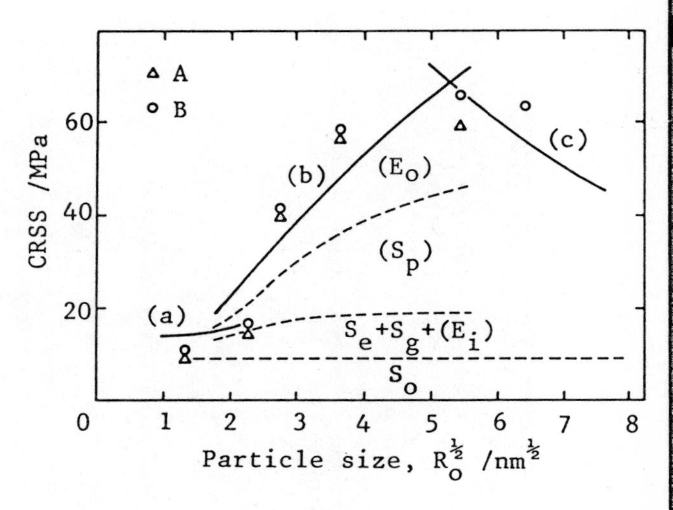

Fig. 3 Calculated curves for CRSS as a function of the particle size with the experimental plots.

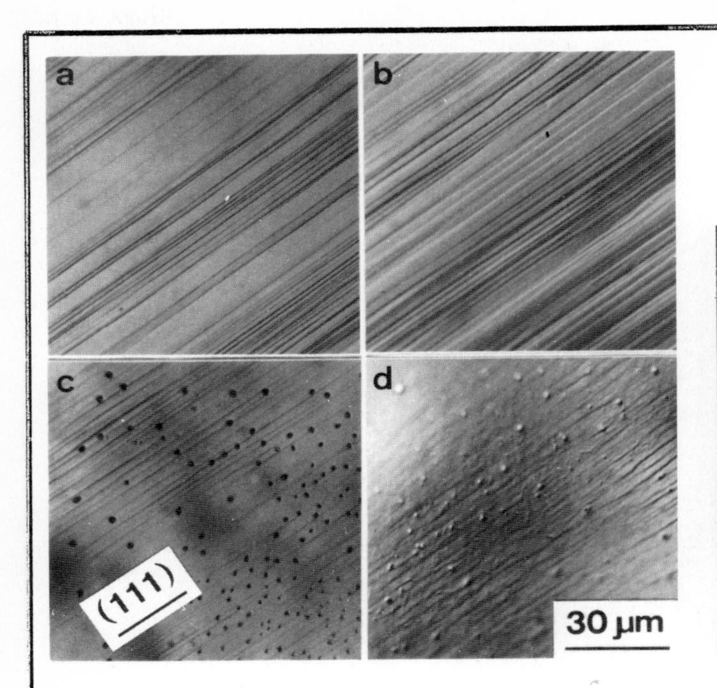

Photo. 1 Variation of slip line morphology with aging time (t). a:t=0s, b:$t=6x10^3s$, c:$t=6x10^4s$, d:$t=6x10^5s$. (deformed 4%)

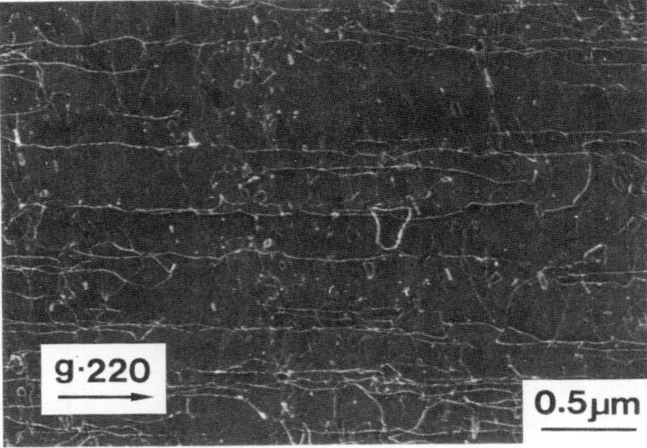

Photo. 3 220 dark field image showing long screw dislocations. ($t=6x10^2s$, deformed 4%)

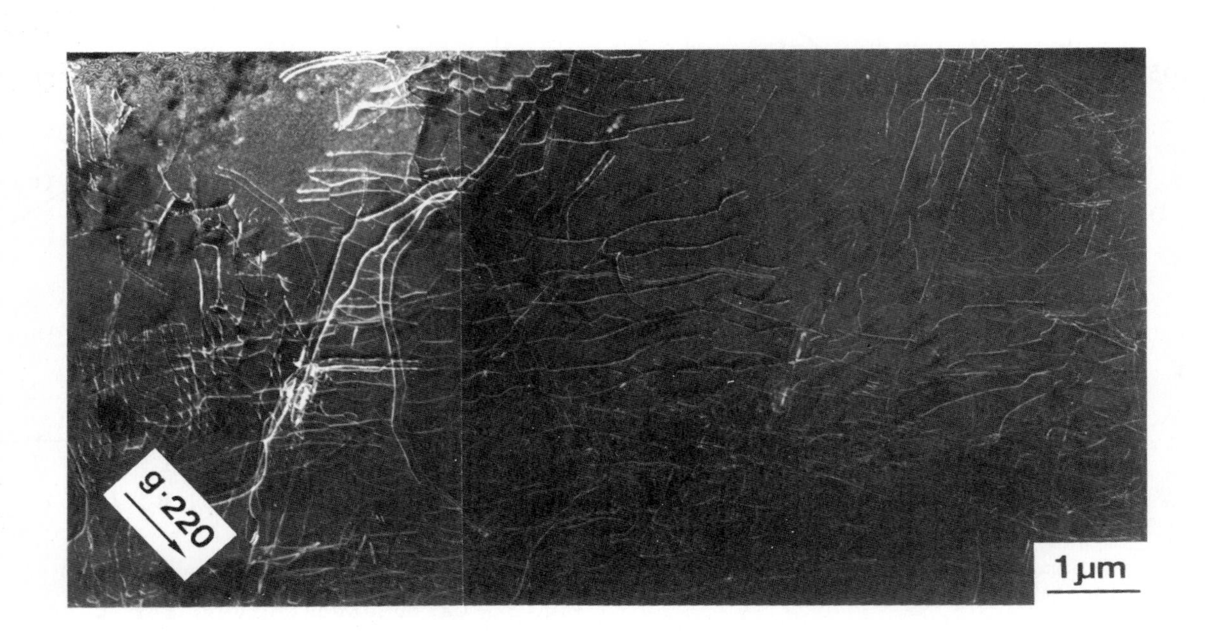

Photo. 2 220 dark field image showing smoothly curved dislocations, randomly distributed. (t=0s, deformed 4%)

Discussion

1. Interaction between moving dislocations and δ'-precipitates.

The results shown in the preceeding section indicate that, in the present Al-Li single crystals, dislocations interact with Al_3Li-precipitates in a way typical of ordinary age-hardening alloys, where dislocations cut through precipitates in the under-aged condition and bypass the precipitates in the over-aged condition. The observed arrays of straight screw dislocations in pairs in the under-aged condition imply that the resistance due to the APB energy of δ'-precipitates is considerably high and that a strong interaction exists between moving screw dislocations and Al_3Li precipitates, as in the ordinary $L1_2$ ordered single phase alloys.

Experimental works on the mechanical properties characteristic of $L1_2$ structure have been extensively carried out on Cu_3Au(16)(17), Ni_3Al(18)(19), Ni_3Ga(20)(21) and etc. It has been well understood that the characteristic positive temperature dependence of the strength of $L1_2$ ordered phases originates from the Kear-Wilsdorf sessile locking mechanism(16)(17). Long straight screw dislocations have been observed in some $L1_2$ ordered alloys deformed by the $\{111\}$ slip. The present results strongly suggest that the Kear-Wilsdorf mechanism should be the origin of the frictional stress against the dislocation motion in the δ'-particles discussed by Furukawa et al.(8).

The Kear-Wilsdorf process in Al-Li binary alloys has been supported by the observed positive temperature dependence of strength in polycrystals(9) and single crystals(10), although no direct evidence has been obtained.

The increase in the observed pair distance or break-away of dislocation pairs with increasing strain can be attributed first to the decrease in the cross section of the precipitates for dislocations to cut through, secondly to the increase in jog density of screw dislocations by a long distance motion, and thirdly to the increase in long range internal stress due to the increased dislocation density. Dislocations cut through the precipitates as long as the cutting stress remains smaller than the bypassing stress(Orowan stress). The present experiment clearly shows that the transient stage from cutting to bypassing corresponds to the peak-aging, where a maximum of CRSS is reached. In Photo.7 a good correlation is found between dislocation loops and precipitates, proving that Orowan's bypass process is prevailing in this aging condition.

2. Evaluation of CRSS as a function of aging condition.

Strength of the present alloy single crystals is dicussed in reference to the theory of precipitation hardening by coherent spherical ordered phase. Interaction of dislocation with δ'-precipitates as discussed in the previous section is tentatively classified into three cases. In quenched or lightly aged conditions, straight dislocations in pairs cut through finely dispersed precipitates (case a). With increasing aging time, precipitates grow and the leading dislocation tends to bow-out between precipitates so as line tension to assist in overcoming the increased resistance of precipitates. The trailing dislocation travels straight repairing the disorder (APB) created by the leading dislocation (case b). In the over-aged condition dislocations are not paired any more and they bypass precipitates leaving dislocation loops surrounding the precipitates(case c). These models for particle-cutting are quite similar to those studied by Gleiter and Hornbogen(22)(23) and Brown and Ham(24)(25).

An attempt is made to formulate critical shear stress for each aging condition (case a-c), considering various resistances to moving dislocations.

When dislocations cut through coherent spherical precipitates, the following factors should be considered, as well as the friction stress, S_o, of the matrix solid solution :

1) stress due to coherency strain, S_e,
2) stress due to modulus effect, S_g,
3) interface energy of δ'/matrix, E_i,
4) APB energy of δ', E_o,
5) friction stress in the precipitate, S_p.

By considering the force balance on the leading and trailing dislocations and by simply adding each contribution, the following expressions for the applied stress, S, are obtained to cut through the particles.

(case a)

$$S = S_o + S_e + S_g + E_i f/R_s + S_p f \qquad (1)$$

where f is the volume fraction of particles ($=2R_s/L$), R_s is the average radius of particles intersected by a slip plane($=\sqrt{2/3}\,R_o$), R_o is the radius of particles and L is the linear particle spacing.

(case b)

$$S = S_o + S_e + S_g + E_i(A + f)/2R_s + E_o(A - f)/2b + S_p(A + f)/2 \qquad (2)$$

The fraction of dislocation line cutting the particles is given(26) as $A=2R_s/L_1=(4E_o fR_s/\pi T)^{1/2}$, where L_1 is the effective particle spacing(27) for the leading dislocation and expressed as $L_1=(2TL^2/Sb)^{1/3}$, T is the dislocation line tension, and b is the Burgers vector.

When dislocations bypass the particles in the over-aged condition, the following equation(28) can be applied.

(case c)

$$S = S_o + 0.85(G_m b/2\pi L_o)\,\Phi\ln(R_s/2b) \qquad (3)$$

$$(L_o > 2R_s)$$

where G_m is shear modulus of matrix, L_o is Orowan spacing and $\Phi = 1/2\cdot\{1+(1+\nu)^{-1}\}$ and ν is Poisson's ratio.

Calculated curves of S vs $R_o^{1/2}$ are shown in

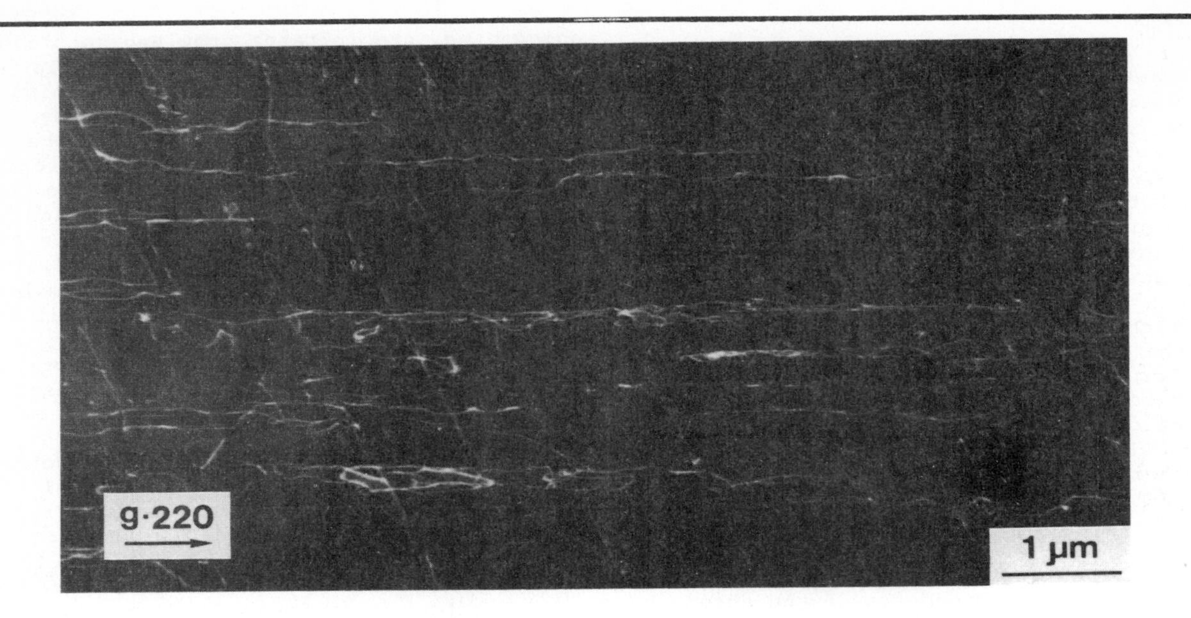

Photo. 4 220 dark field image showing long straight screw dislocations in pairs. ($t=6x10^3 s$, deformed 1%)

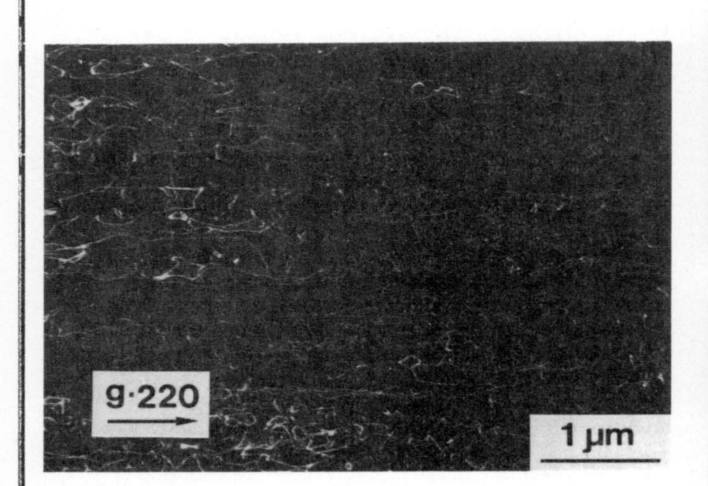

Photo. 5 220 dark field image showing the increased pair distances. ($t=6x10^3 s$, deformed 4%)

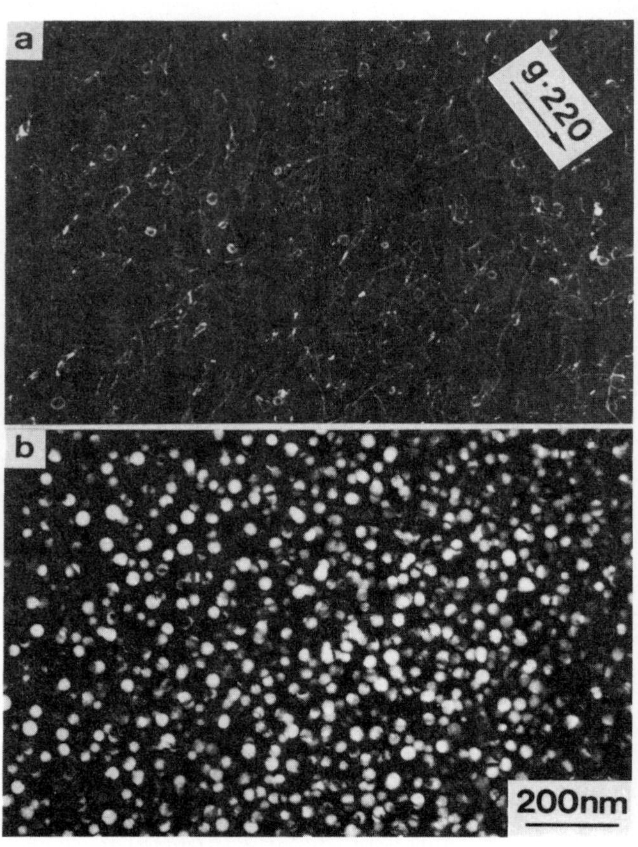

Photo. 6 220 dark field image(a) and 110 dark field image (b). ($t=6x10^4 s$, deformed 4%)

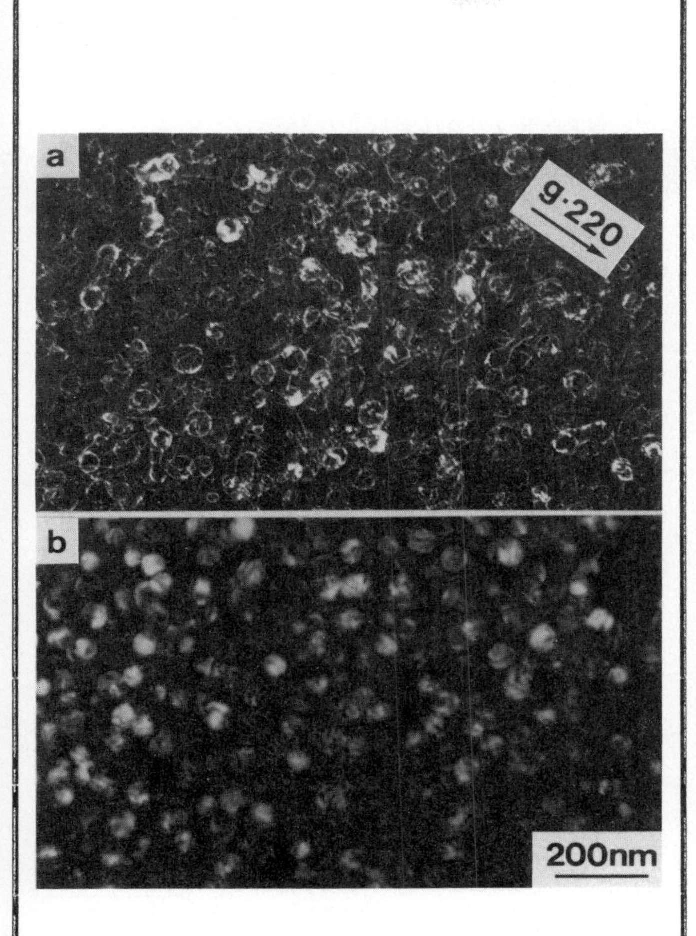

Photo. 7 220 dark field image (a) and 110 dark field image (b).
($t=6x10^5$s, deformed 4%)

Fig. 3 by solid lines, (a),(b),(c), for the three aging stages, being in good agreement with the experimental plots. Contributions from the different hardening mechanisms in the particle cutting stage are shown by dotted lines. With increasing particle size, contributions from the friction stress(S_p) and from the APB energy (E_o) of the δ'-particles increase, and in the peak aging condition, each of them amounts to roughly 30% of the total stress. While, the contributions of coherency strain hardening (S_e), modulus hardening (S_g) and hardening due to the interface energy (E_i) are relatively important in the early stage of aging, but the sum of the three increases only slightly with the growth of the particles, reaching around 15% of the total in the peak-aging condition.

The parameters, S_o, S_e, S_g, E_i, E_o, S_p, are independent of aging time and particle size, and they are estimated as follows : S_o=9MPa (the measured CRSS of Al-1.8mass%Li single crystal as quenched). S_e=4.5MPa in the peak-aged condition (since lattice mismatch e between δ'/matrix is very small, for example e=-0.09%(29), the calculation with the equation by Gerold and Harberkorn(30) gives a small value for S_e). The calculation using the equation by Weeks et al.(31) and the difference in modulus, ΔG=6.7GPa(2), yields a small value for S_g, being less than several per cent of the total stress in the peak-aged condition. This is in contrast to the reported large contribution of modulus hardening by Noble et al.(4), who used the equation developed by Kelly(32). As was already pointed out in the preceeding section, the morphology of dislocation loops surrounding the δ'-particles in the over-aged condition suggests that the contribution from the modulus effect should not be large. E_i is estimated to be 180mJ/m^2 from the data shown in Fig.1, assuming the growth of particle obeys Wagner's equation(33). E_o is estimated to be 53mJ/m^2 using Flinn's equation(34) and critical temperature T_c=548K(12). S_p=130MPa as estimated using equation (1) and the values of S, f, R_s for the linear cutting stage.

The above comparison of the calculated curves with the experimental data suggests that the strength of the δ'-precipitation hardened Al-Li alloy mainly comes from the order hardening and also from the Kear-Wilsdorf hardening in the δ'-$L1_2$ ordered precipitates.

Conclusions

1. The CRSS of (111)[$\bar{1}$01] slip of an Al-2mass%Li alloy increases with increasing time of aging at 473K and decreases after reaching a maximum of 60MPa at about 10^5s.

2. Slip lines are long, straight and coarse in the under-aged condition, being consistent with the occurrence of serrations in the stress-strain curves. In the peak-aged condition, slip lines are shorter and finer, and they become wavy in the over-aged condition.

3. Observations by TEM revealed that deformation-induced dislocations are smooth and randomly distributed, often in pairs. Rows of

long straight screw dislocations in pairs were observed in the under-aged condition, which indicate the presence of a strong interaction between $L1_2$-ordered precipitates and screw dislocations. In the peak-aged condition, besides single wavy dislocations, dislocation loops of the size corresponding to the precipitates were observed, suggesting the occurrence of Orowan bypassing, as well as cutting precipitates. In the over-aged condition, dislocations bypass the precipitates leaving Orowan loops around them.

4. The stress for superdislocations to cut through $L1_2$-precipitates is well understood in terms of the theory of hardening by the precipitation of coherent spherical ordered phase. Main contribution to CRSS in the cutting stage comes from the friction stress due to the Kear-Wilsdorf sessile locking and the APB energy of the Al_3Li precipitates.

Acknowledgment

The present work was partly supported by the Grant-in Aid for Scientific Research from the Ministry of Education, Science and Culture and also by the Research Fund from the Light Metal Educational Foundation, Inc. The Technical Research Laboratory, Sumitomo Light Metal Ind., LTD. is acknowledged for providing with the metals and alloys used in the present study.

References

(1) B.Noble and G.E.Thompson : Met. Sci. J., 5(1971),114.
(2) B.Noble, S.J.Harris and K.Dinsdale : J. Mat. Sci., 17(1982),461.
(3) T.H.Sanders and E.A.Starke : Acta Met., 30(1982),927.
(4) B.Noble, S.J.Harris and K.Dinsdale : Met. Sci., 16(1982),425.
(5) T.H.Sanders : Aluminium-Lithium Alloys, Met. Soc. AIME, (1981),p.63.
(6) P.J.Gregson and H.M.Flower : Acta Met., 33(1985),527.
(7) M.Furukawa, Y.Miura and M.Nemoto : Trans. JIM, 26(1985),225.
(8) M.Furukawa, Y.Miura and M.Nemoto : Trans. JIM, 26(1985),230.
(9) M.Furukawa, Y.Miura and M.Nemoto : Trans. JIM, 26(1985),414.
(10) M.Tamura, T.Mori and T.Nakamura : Trans. JIM, 14(1973),355.
(11) J.M.Silcox : J. Inst. Metals, 88(1959-60),357.
(12) D.B.Williams and J.W.Edington : Met. Sci., 9(1975),529.
(13) G.W.Lorimer, G.Cliff and J.N.Clark : Developments in Electron Microscopy and Analysis, Academic Press, London, (1976),p.153.
(14) G.Cocco, G.Fagherazzi and L.Schiffini : J. Appl. Cryst., 10(1977),325.
(15) M.Tamura, T.Mori and T.Nakamura : J. Jap. Inst. Metals, 34(1970),919.
(16) B.H.Kear and H.G.F.Wilsdorf : Trans. Met. Soc. AIME, 224(1962),382.
(17) B.H.Kear : Acta Met., 12(1964),555.
(18) P.H.Thornton, R.G.Davies and T.L.Johnston : Met. Trans., 1(1970),207.
(19) C.Lall, S.Chin and D.P.Pope : Met. Trans., 10A(1979),1323.
(20) S.Takeuchi and E.Kuramoto : J. Phys. Soc., Japan, 31(1971),1282.
(21) S.Takeuchi and E.Kuramoto : Acta Met., 21(1973),415.
(22) H.Gleiter and E.Hornbogen : Phys. Status Solidi, 12(1965),235.
(23) H.Gleiter and E.Hornbogen : Phys. Status Solidi, 12(1965),251.
(24) R.K.Ham : Ordered Alloys, Structural Applications and Physical Metallurgy, Claitors, Baton Rouge, Louisiana, (1970), p.365.
(25) L.M.Brown and R.K.Ham : Strengthening Methods in Crystals, Elsevier, Amsterdam,(1971),p.9.
(26) N.S.Stoloff : Superalloys, John Wiley and Sons, New York,(1972),P.79.
(27) J.Friedel : Dislocations, Pergamon Press, Oxford,(1964),p.371.
(28) M.F.Ashby : Physics of Strength and Plasticity, Cambridge, Mass.,(1966),p.113.
(29) S.F.Baumann and D.B.Williams : Aluminium-Lithium Alloys II, Met. Soc. AIME,(1984),p.17.
(30) V.Gerold and H.Harberkorn : Phys. Status Solidi, 16(1966),675.
(31) R.W.Weeks, S.R.Pati, M.F.Ashby and P.Barrand : Acta Met., 17(1969),1403.
(32) P.M.Kelly : Int. Metall. Rev., 18(1973),31.
(33) C.Wagner : Z.Elektrochem., 65(1961),581.
(34) P.A.Flinn : Trans. Met. Soc. AIME, 218(1960),145.

Elastic constants of Al–Li solid solutions and δ′ precipitates

W MÜLLER, E BUBECK and V GEROLD

The authors are with the
Institut für Metallkunde, Universität Stuttgart,
and Max-Planck-Institut für Metallforschung,
Institut für Werkstoffwissenschaften,
7000 Stuttgart 1, FRG

SYNOPSIS

The elastic constants of Al-Li single crystals have been measured for compositions up to 9.6at%Li (2.7wt%) in a temperature range from room temperature to 160°C. For the alloys containing the metastable δ' phase the variation of the constants during aging was measured in-situ. The results were correlated with X-ray small angle scattering experiments which characterize the precipitation state. From these experiments the elastic constants of the δ' precipitate were estimated.

INTRODUCTION

The elastic properties of Al-Li based alloys are of great technological interest because the addition of Li decreases the density and increases the elastic constants of aluminium alloys. The purpose of the present paper is to explain the single crystal elastic constants in the single phase and metastable two-phase region for Al-Li compositions up to 10at%Li and at various temperatures up to 160°C.

EXPERIMENTAL PROCEDURE

Several binary Al-Li alloys with Li contents up to 9.6 at% (ca. 2.7wt%) were prepared from high purity aluminium (99.9999) and Li (99.9). To avoid contamination of the lithium a hollow cylinder was made out of the aluminium and the lithium was filled into it. This was performed in a glove box containing nitrogen-free argon. The crucible was closed by an aluminium cover which could be screwed tightly on the cylinder. After melting and casting in a vacuum, single crystals were grown also in a vacuum by a Bridgman method using boron nitride crucibles. The crystals were annealed in quartz capsules filled with argon for 6 h at 580°C and then quenched in water.

The single crystal rods had a diameter of 15 mm. They were cut into cylinders of 8 mm diameter and a length of 10 mm by spark erosion. The two flat surfaces were parallel to a {110} plane within an accuracy of 2°. The elastic constants were measured by an ultrasonic pulse echo method in a temperature range from 21 to 160°C. The absolute accuracy of the measurement was within $\pm 2\cdot 10^{-3}$ whereas the relative accuracy was of the order of 5 to $8\cdot 10^{-4}$. The experimental technique allowed the measurement of all three elastic constants from a single specimen by measuring the velocity of ultrasonic waves in three different modes. The longitudinal mode results in the constant C_ℓ, the transversal [001] mode gives the constant C and the transversal [110] mode the constant C'. All three constants are related to the constants c_{ij} (Voigt notation)

$$C_\ell = (c_{11}+c_{12}+2c_{44})/2 \qquad (1)$$

$$C = c_{44} \qquad (2)$$

$$C' = (c_{11}-c_{12})/2 \qquad (3)$$

The anisotropy constant A is defined as A = C/C'. In the following only the constants C_j are discussed where C_j stands for all three constants C_ℓ, C and C'. Out of C_j the bulk modulus can be calculated. For nearly isotropic materials the quasi-isotropic constants E and G [1] can also be determined:

$$K = C_\ell - C - C'/3 \qquad (4)$$

$$G = (2C+2C')/5 \qquad (5)$$

$$E = 9KG/(3K+G) \qquad (6)$$

The alloys containing more than 5at%Li are positioned in a metastable miscibility gap where the ordered δ' phase precipitates out of the solid solution. Figure 1 shows the phase diagram according to Cocco et al. [2]. The single crystal samples of such alloys have been homogenized at 580°C in a salt bath and quenched into water immediately before an ultrasonic experiment. To protect the

Table 1. Li content and elastic constants of the depleted matrix in equilibrium with the δ' phase. A = C/C'.

Ageing temp. °C	Li conc. matrix at%	C GPa	C' GPa	C_ℓ GPa	A
21	5.0	30.7	26.6	113.2	1.15
80	5.3	29.8	25.6	110.9	1.16
120	5.6	29.2	25.1	109.4	1.16
160	6.1	28.7	24.6	107.9	1.17
250	7.8	(27.8)	(23.9)	(104.5)	(1.16)
temp. coefficient in 10^{-6} K^{-1}		-414	-433	-335	

Table 2. Li content and estimated elastic constants of the δ' phase. A = C/C'.

Ageing temp. °C	Li conc. δ' phase at%	C GPa	C' GPa	C_ℓ GPa	A
21	25	(42.8±1.3)	(43.2±4.0)	(123.2±5.4)	1.01
80	24.4	37.7±0.8	40.4±3.1	114.5±4.1	0.94
120	24.0	33.8±0.3	40.0±3.7	105.2±2.3	0.85
160	23.1	30.6±0.1	37.0±2.2	101.3±1.9	0.83
250	21.0	(22.5)	(33.7)	(85.5)	(0.67)
temp. coefficient in 10^{-6} K^{-1}		-2070	-960	-1330	

specimens they were wrapped in aluminium foil. Thereafter, the sample was mounted for ultrasonic measurements which were conducted at a constant ageing temperature for ageing times up to 5 000 min. During this time one elastic constant could be measured continuously. Subsequently the whole procedure was repeated for the measurement of the next constant.

In order to describe the precipitates, small angle X-ray scattering (SAXS) has been employed using MoKα radiation and flat specimens having a thickness of 0.6 mm. These experiments were conducted at the ageing temperature. The intensity was recorded using a position sensitive detector. The analysis gave the volume fraction f and the mean particle radius R of the precipitates.

THE ELASTIC CONSTANTS OF SOLID SOLUTIONS

The elastic constants of the solid solutions measured at room temperature (21°C) for compositions up to 4at%Li are shown in Fig. 2. All three elastic constants increase with increasing Li content. Particularly, the constant C', which increases drastically. These experimental data were converted to the elastic moduli K, E and G using Eqs. (4) to (6) and are plotted in Fig. 3. Both moduli E and G again increase with Li content whereas the bulk modulus decreases slightly. The dashed curve represents results of a static Young's modulus determination by Noble et al. [3]. Both sets of data agree with another quite well for higher solute concentrations. The static Young's modulus however deviates to lower values for the more dilute alloys and for pure aluminium. This deviation is probably caused by microplasticity which results in a lower effective modulus.

For comparison the elastic constants of Al-Zn solid solutions measured by Tempus [4] have been included as dashed lines. The slope of the constant C' is much lower in this case; for the other two constants it has even changed to negative values.

The temperature dependence of the constants C of Al-Li has been measured in the range from 21 to 120°C and found to be independent of the alloy composition. The following values dC_j/dT have been obtained for C_ℓ, C and C' (in MPa/K): -39; -18; -19, respectively.

THE ELASTIC CONSTANTS FOR ALLOYS CONTAINING THE δ' PHASE

The miscibility gap shown in Fig. 1 is in accord with own results from SAXS. Mainly two alloys have been investigated in detail. They contain 8.5 and 9.6 at%Li. Depending on the composition and the heat treatment the volume fraction f of the precipitates increased more or less rapidly until an equilibrium value was reached during isothermal ageing. During this time and thereafter a continuous growth of the δ' particles was observed. As an example, Fig. 4 shows the change of the particle radius R as a function of annealing time for the Al-9.6at%Li alloy at three different temperatures. In this case the volume fraction f was nearly constant for any given ageing temperature for ageing times above 10 min. The variation of the elastic constants C_j at 120°C is shown in Fig. 5. Obviously, the coarsening process leads to an increase in the constants C_ℓ and C' and a decrease in C. From other ageing experiments at room temperature it was found that increasing the volume fraction leads to an increase of all three constants C_j.

The elastic constants C_j of these two phase alloys are compared of those of the precipitate (C_j^P) and those of the depleted matrix (C_j^M). The latter are well known from the limit of the miscibility gap and the concentration and temperature dependence of C_j. They are listed in Table 1. In addition the dispersion of the precipitate, i. e., the total amount of particle-matrix interface, also plays a role.

In order to get an estimate for C_j^P for the precipitates the same procedure has been adopted as in the case of Al-Zn where reliable results could be obtained [5]. At first the influence of the interface has to be eliminated. The following equation

$$C_j = C_j^o + 3f\,\Delta C_j\,d/R \qquad (7)$$

approximately describes the contribution of the interface through the second term. The quantity 3f d/R is the volume fraction of an interface layer of thickness d, and ΔC_j describes the difference in its elastic constant component to the average $(C_j^A+C_j^P)/2$ of those of the matrix and precipitate. Thus ΔC_j can be either positive or negative. The first term C_j^o describes the contribution of both the matrix and the precipitate. Two limiting cases can be given known as Voigt and Reuss averaging respectively:

$$C_j^o = (1-f)C_j^A + fC_j^B \qquad (8)$$

$$(C_j^o)^{-1} = (1-f)(C_j^A)^{-1} + f\,(C_j^B)^{-1} \qquad (9)$$

As an example, Fig. 6 shows the application of Eq. (7) for Al-9.6at%Li at 120°C. The results of SAXS (Fig. 4) and $C_j(t)$ (Fig. 5) are combined and C_j has been plotted as a function of 1/R. Straight lines are obtained which intercept the ordinate at C_j^o (circles). These values were used and plotted as a function of the volume fraction according to Eqs. (8) and (9). Again an example is given in Fig. 7. From the slopes of the straight lines, values for the C_j^P of the δ' phase are obtained. From both Voigt and Reuss averaging procedures the arithmetic mean value has been adopted to be the best estimate for the elastic constants of the precipitate. They are listed in Table 2 together with the deviations defined by the averaging procedures. More details [6] will be published elsewhere.

The temperature of dependence of the elastic constants C and C' for both the depleted matrix and the precipitate are shown in Fig. 8. As can be seen from this figure and from Table 2 the temperature dependence of the elastic constants of the precipitates is extremely large. The constant C^P has a strong dependence giving a slope of $-2070\cdot10^{-6}K^{-1}$ whereas typical values range between 100 and $1000\cdot10^{-6}K^{-1}$. This large temperature dependence comes from the fact that the slopes of the straight lines in Fig. 7 increase with decreasing temperature indicating that at 80°C the elastic constant C^P of the δ' phase is considerably larger as compared to

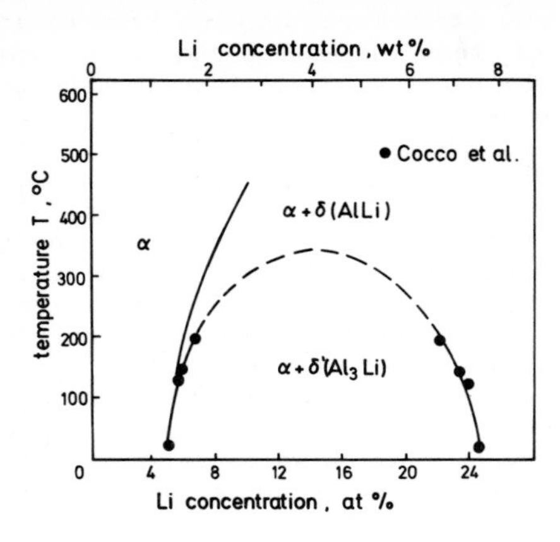

Fig. 1. Phase diagram of Al-Li including the metastable miscibility gap for δ' precipitates. After Cocco et al. [2].

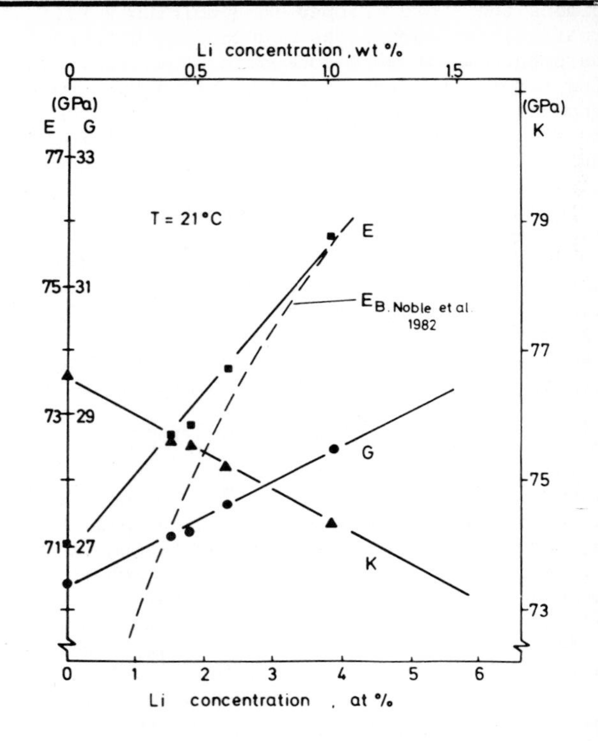

Fig. 3. Elastic moduli K, G and E calculated from data in Fig. 1. Experimental results by Noble et al. [3] are included.

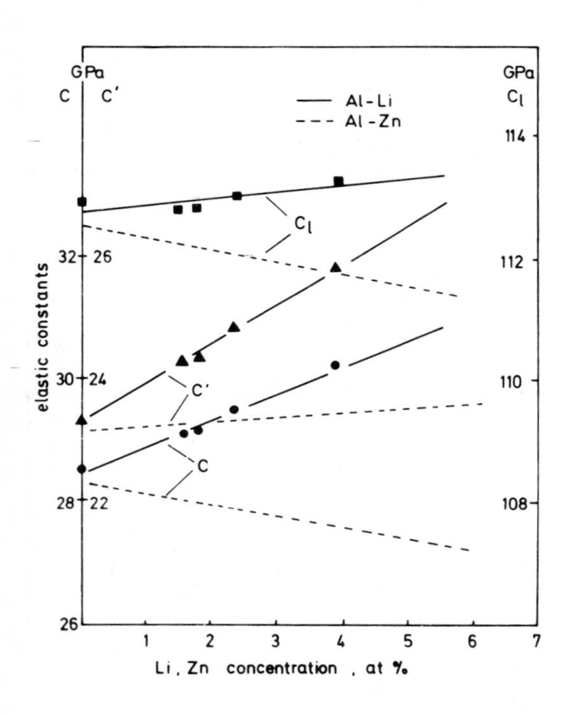

Fig. 2. Elastic constants C_j of Al-Li solid solutions at 21°C. Data for Al-Zn [4] have been included.

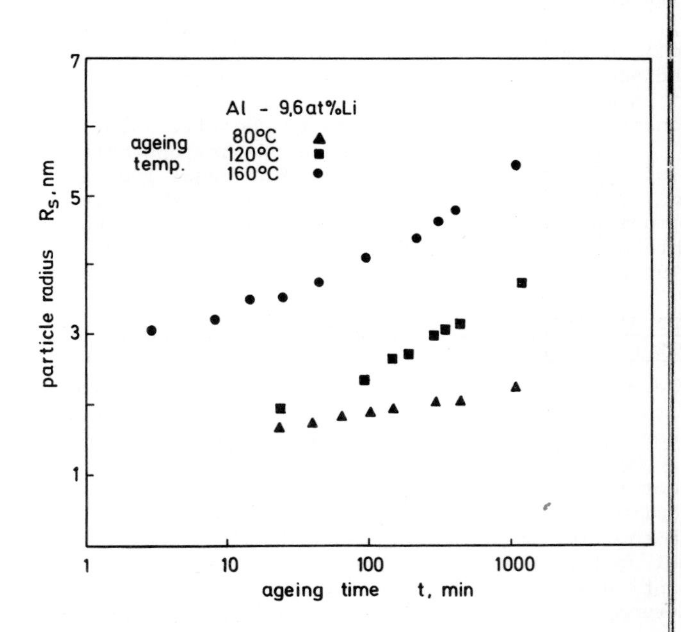

Fig. 4. Particle radius R of the δ' phase as function of annealing time for indicated ageing temperatures. Al-9.6at%Li.

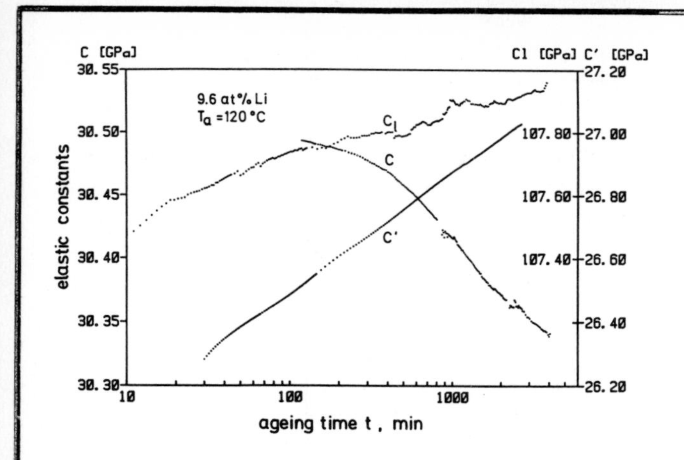

Fig. 5. Elastic constants C_j as function of the ageing time at 120°C. Al-9.6at%Li.

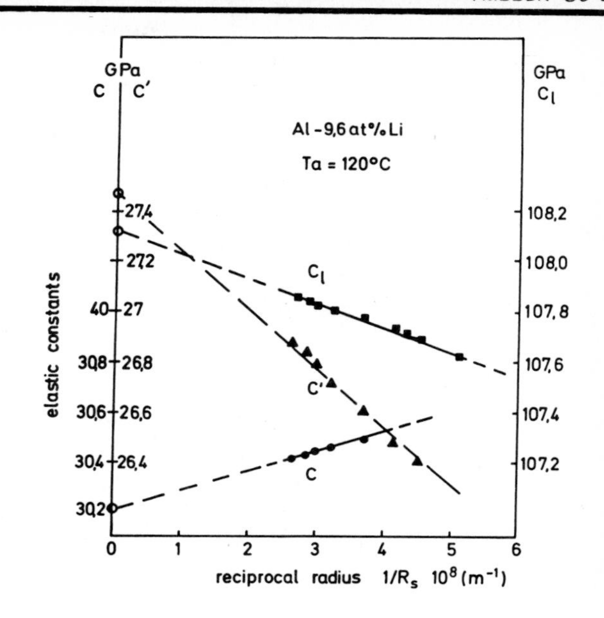

Fig. 6. Elastic constants C_j as function of 1/R for Al-9.6at%Li aged at 120°C. Evaluation of C_j^O.

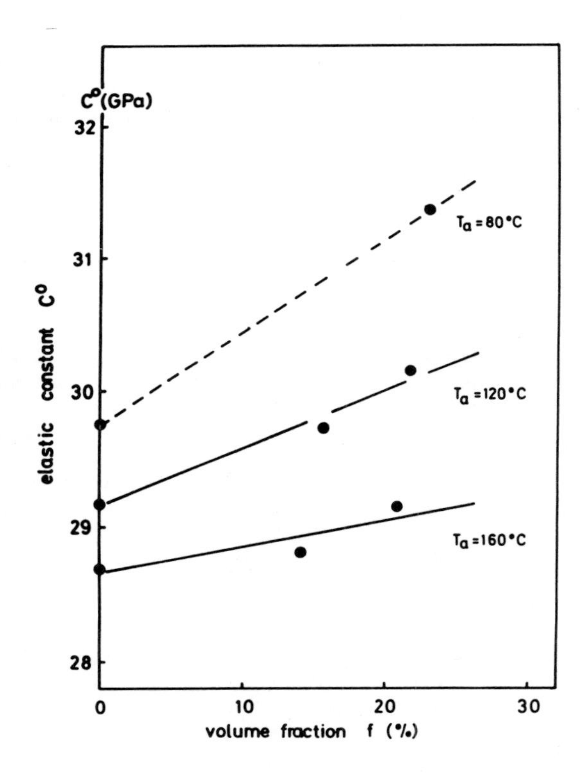

Fig. 7. The elastic constant C^O as function of the volume fraction f of the δ' phase. For f = 0 the corresponding data for the depleted matrix (Table 1) are plotted.

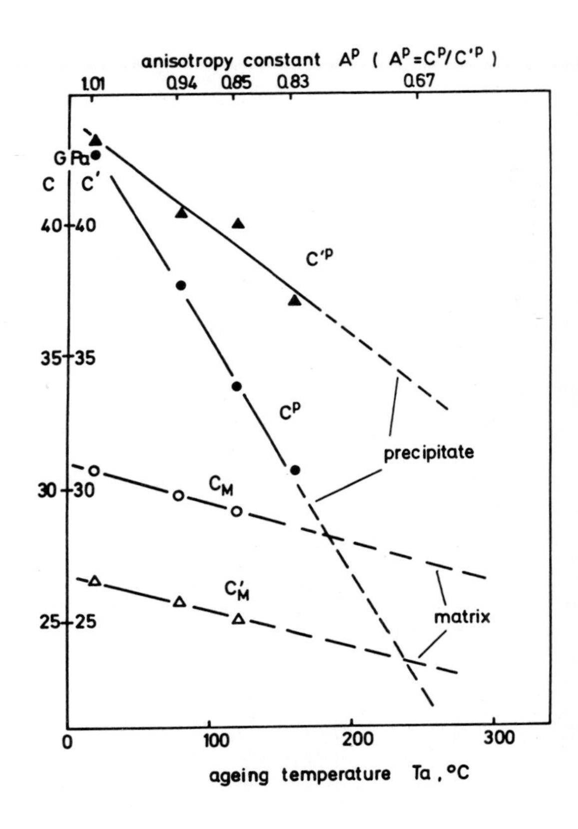

Fig. 8. The temperature dependence of the elastic constants C and C' of the precipitate (P) and the depleted matrix (M).

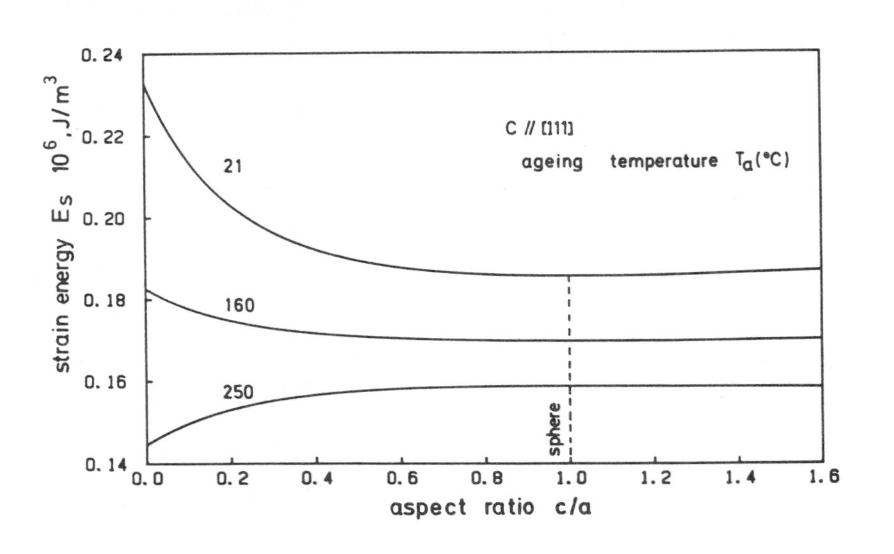

Fig. 9. Specific elastic coherency strain energy of the δ' phase as function of the aspect ratio for rotational ellipsoid particles. A stress free misfit δ = -0.13% has been assumed.

that of the depleted matrix. At 160°C there exist only minor differences.

DISCUSSION

Based on these calculations the following results are obtained: At room temperature the elastic shear constants C and C' of the δ' phase are considerably higher than those of the depleted matrix. The δ' phase is nearly isotropic (A = 1). Taking the elastic data from the solid solution, Fig. 2 and extrapolating linearly to a Li content of 25% results in C = 39.7; C' = 39.6 and C_ℓ = 115.3 GPa. These values are relatively close to the calculated values. The constants C_ℓ do not differ much for both phases. From this it can be concluded that a smaller bulk modulus for the δ' phase exists.

With increasing temperature, because of the different temperature dependencies for C and C', the anisotropy constant A becomes increasingly smaller than unity. This was not foreseen from extrapolation of the corresponding solid solution data to the high Li concentration of the δ' phase. Unfortunately, no other check of the validity of this estimate is possible. For example, it is well known from theory [7] that an elastically anisotropic coherent precipitate in a nearly isotropic matrix should show deviations from its spherical shape because of the resulting coherency strain field. In a similar analysis of the elastic properties in Al-Zn alloys containing G.P. zones it could be demonstrated that the well-known deviation from spherical shape could be totally explained from the estimated elastic anisotropy of the zones [5]. In that case the anisotropy constant was found to be about A = 0.2. In addition, a large misfit δ = -1.8% occurred leading to a relatively large strain energy. This showed a steep decrease if the particle deviated from a sphere and formed an oblate rotational ellipsoid with rotation axis parallel to <111>. For particles of a radius larger than about 2.5 nm this decrease was larger than the increase in the total surface energy resulting in stable c/a ratios depending on the particle size.

In the present case the expected elastic anisotropy of the precipitate (Fig. 8) would be large enough only for ageing temperatures above 200°C. This is shown in Fig. 9 where a decrease in the energy is expected for a temperature of 250°C. These results were obtained using the equations given by Lee et al. [7] and the elastic constants listed in Tables 1 and 2. For a stress-free misfit the value δ = -0.13% has been adopted.

In order to simplify the calculations the elastic constants of the matrix were converted to isotropic constants. Unfortunately, because of the small misfit parameter a deviation from a spherical shape can be expected only for particle radii larger than 500 nm. For this calculation an interfacial energy of 25 mJ/m^2 was assumed. Such large particles have not been observed so far. Thus, no proof of the proposed temperature dependence of the elastic properties of the δ' phase is possible.

REFERENCES

1. E. Kröner, Statistical Continuum Mechanics, Springer Berlin, 1971.
2. G. Cocco, G. Fagherazzi and L. Schiffini, J. Appl. Cryst. 10, 325 (1977).
3. B. Noble, S.J. Harris and K. Dinsdale, J. Mater. Sci. 17, 461 (1982).
4. G. Tempus, Dissertation Univ. Stuttgart, 1982.
5. G. Tempus, W. Siebke and V. Gerold, Proc. 5th Risø Intern. Symp. on Metallurgy and Materials Science, Roskilde/Denmark 1984, p. 525.
6. W. Müller, Dissertation University of Stuttgart, 1985.
7. J.K. Lee, D.M. Barnett and H.I. Aaronson, Metall. Trans. 8A, 963 (1977).

□

Influence of δ′ phase coalescence on Young's modulus in an Al–2.5wt%Li alloy

F BROUSSAUD and M THOMAS

The authors are in the Division des Matériaux of ONERA (Office National d'Etudes et de Recherches Aérospatiales) France

SYNOPSIS

A high-precision pulse-echo apparatus has been developed to measure variations of the ultrasonic velocity as a function of isothermal annealing of an Al- 2.5 wt% Li alloy. The method has already been shown to be particularly sensitive for detecting small changes in specific Young's modulus (1). Ageings were performed at temperatures below the metastable solvus line of the δ′ phase, i. e. between 190 and 250°C. Microhardness measurements were carried out together with the Young's modulus measurements, hardness being identified as very sensitive to the size and distribution of second phase particles. Coarsening behaviour was monitored by Transmission Electron Microscopy (TEM) observations.

INTRODUCTION

In spite of the high stiffness for which, in particular, Al-Li alloys are of major interest, very few attempts have been made to clear the reasons for such an unexpected modulus behaviour. Particularly important was the determination of the relative role of α and δ′ in achieving the large increase in Young's modulus with respect to pure aluminium. As a first approximation a binary age-hardened Al-Li alloy can be described as randomly distributed spherical inclusions of δ′ embedded in a matrix. Thus, the modulus of this "composite" material can be written as a function of the respective volume fractions of the matrix and δ′ precipitates and of their intrinsic modulus, according to the law of mixtures. The present work tries to get some insight on the validity of this approach by studying the effect of δ′ coalescence, i. e. δ′ coarsening at a constant volume fraction, on Young's modulus E. Since the lattice parameters of the precipitates and the matrix are close together, the δ′ phase remains coherent up to a size of about 0.3 μm (2). This offers a good opportunity to test whether the change in elastic constraints around the precipitates during coalescence have an effect on Young's modulus or not, by monitoring the annealing time over a wide range.

EXPERIMENTAL PROCEDURE

A series of Al-2.5 wt% Li (9 at% Li) test-bars were extracted from plate material delivered by Pechiney Voreppe Research Laboratories. The Li content was determined by means of emission spectrometry. Several batches of samples (80 cm long, 3.5 mm diameter) were annealed in a vertical furnace at 550°C for 30 minutes under a pressure of 3 x 10^{-5} Torr to prevent oxidation (Fig.1). After water quenching, the rods were transferred rapidly into a freezing container (-18°C) so as to prevent any precipitation. They were subsequently isothermally annealed in salt baths. The rods were oversized to avoid any detrimental effect that could arise from a possible lithium depletion in the vicinity of the surface. So, the outer shell was removed, reducing the diameter from 3.5 to 2.5 mm.

Thin foil specimens for electron microscopy were prepared by standard metallographic techniques. The 3 mm diameter discs were polished with grit papers to

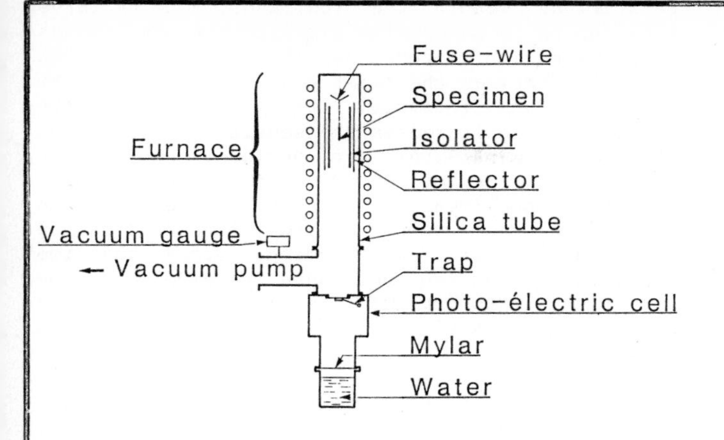

Figure 1 Water Quenching Apparatus

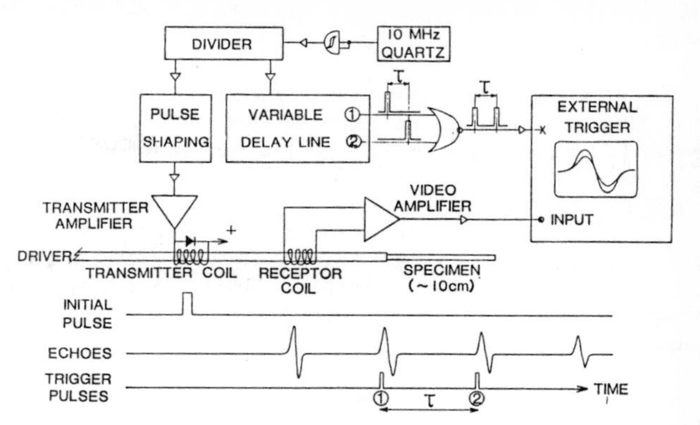

Figure 2 Ultrasonic Pulse-Echo Setup

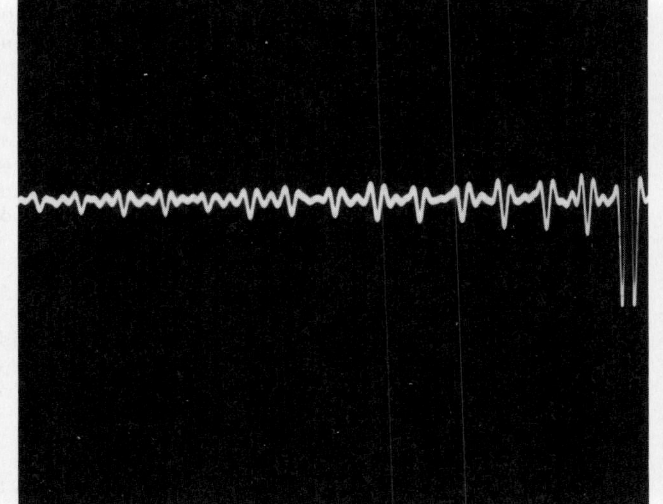

Figure 3 Pulse-Echo Pattern

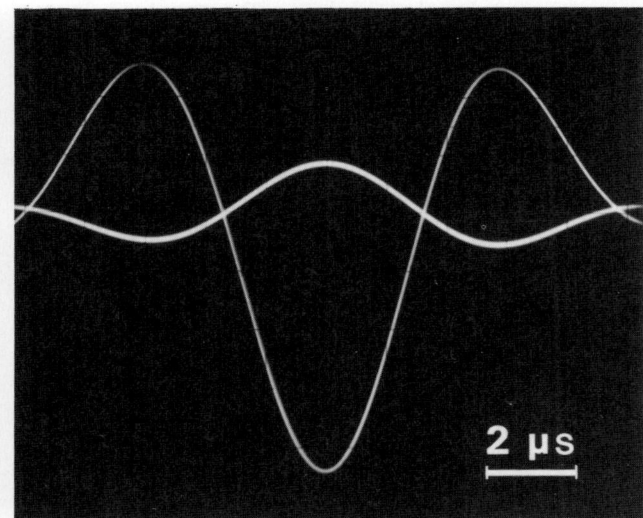

Figure 4 Overlapped Echoes

approximately 0.15 mm thickness and then thinned in a 10% perchloric acid in ethanol solution at -40°C.

X-ray measurements, using monochromatic Cu $K\alpha$ radiation, were carried out to look for any second phase particles in the as-quenched state and to detect texture effects.

Vickers microhardness measurements have been performed using a load of 200 g and making an average of five prints. Before loading, the samples were carefully polished with abrasive paper and 0.3 μm diamond past.

Ultrasonic velocity was measured using the pulse-echo overlap technique. A schematic setup of the equipment is shown in Figure 2. Basically, an ultrasonic wave train is generated by a pulse generator of 160 kHz and is propagated into the specimen through a magnetostrictive driver (Metglas 2826A) bonded to it. Part of the pulse is reflected back from the junction between the driver and the specimen and part of it from the free end of the sample. The principle of measurement is based upon the determination of the time spacing between these two reflected pulses. The propagation velocity is given by the ratio of twice the specimen length by the time spacing quoted above corresponding to a delay time for one round trip in the sample. Overlapping of the signals of interest on the oscilloscope is achieved by a system of variable trigger points operating with an electronic time base of sensitivity 0.1 μs. The specimen length is subsequently measured by a travelling microscope. Young's modulus E is given by $E=\rho V^2$ where ρ is the density of the alloy (ρ=2.51 g/cm^3). The latter has been taken as the average between a value deduced from lattice parameter calculations and Archimedian measurements found in the literature (3).

DATA EVALUATION

Caution must be exercised in preparing the samples and correct calibration must be applied to ensure maximum precision. These aspects are now considered in some detail.

In our case, the driver's width is much smaller than the wavelength, so ultrasonic waves are propagated exclusively in the longitudinal mode.

Pulse intensity was maximized and therefore the accuracy of the measurements increased by adjusting the sample length to an n-fold multiple of wavelength so as to have a resonance for the frequency of 160 kHz.

The intensity ratio between reflected and transmitted pulses at the interface of two materials is a function of their acoustic impedances, which depends on the geometry of the materials. So, the specimen dimensions were selectively chosen in order to obtain a high reflection coefficient promoting multiple reflections within the rod. Thus, echo patterns with up to 15 successive signals were obtained as illustrated in Figure 3. The delay time was measured between the first echo and several other echoes. For instance, triggering from the eleventh echo increases the sensitivity by making the measurement over ten periods. Due to the lower impedance of the driver ribbon as compared to the specimens, a phase reversal upon reflection will occur at the free end of the specimen and at the junction. Since the two distinct pulses do not have the same amplitude and sign, matching of the crests of each echo instead of their leading edge was selected (Fig.4).

Other experimental difficulties arose from the quality of the bonding film. It was necessary to determine the efficiency of various adhesives. Eventually, Cyanolit cement was tested as the most successful one. Experimental studies have been carried out to detect any transit time error due to the ultrasonic wave propagating through the bonding film adhesive. No shift in the velocity measurement has been found by successively shortening a specimen bonded to the driver, so no correction was taken into account.

The specimens were trimmed with care so as to define plane and parallel end faces to eliminate wedge effects, leading errors of 5 parts in 10^5. Some minute scratches on the surface of the rods lower the signal-to-noise ratio and generate local distortions of pulses, yielding errors in delay time up to ± 0.2 μs. These errors were minimized by means of 20 readings with varying distances of the specimen from the receptor.

Young's modulus measurements were conducted at room temperature. In order to avoid any possibility of error from temperature fluctuations, the modulus data were normalized to 20°C. Besides, the X-ray analysis revealed no texture.

Finally, the absolute accuracy of Young's modulus values was ± 0.1% with a point-to-point sensitivity of 1 in 10^4, due to the various difficulties encountered.

RESULTS

In order to magnify the effects of δ' precipitation on the variations of the Young's modulus, the range of ageing conditions was chosen as wide as possible, i. e. from 190°C to 250°C together with ageing times exceeding one month. Information concerning the volume fraction and the size of δ' precipitates has been collected by means of electron microscopy and microhardness. Dark field TEM images reveal the presence of spherical particles exhibiting a bright contrast relative to the matrix and corresponding to δ' precipitates. Micrographs (Fig. 5) show the increase in particle size over ageing time. The average particle size has been

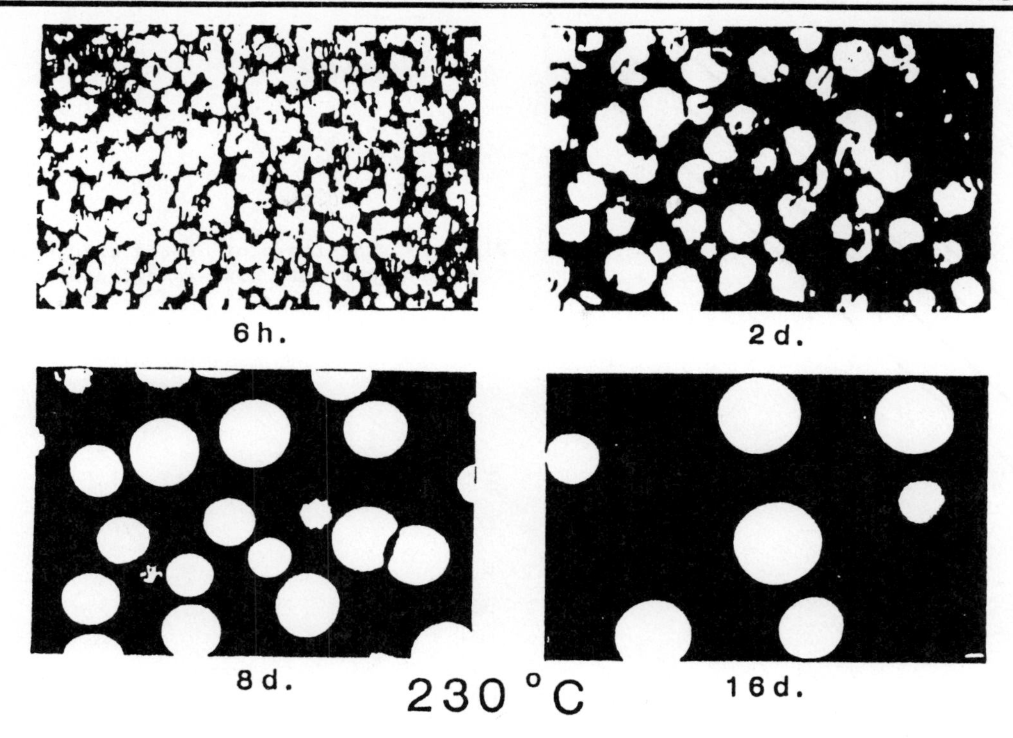

Figure 5 Increasing δ′ particle size as a function of ageing time

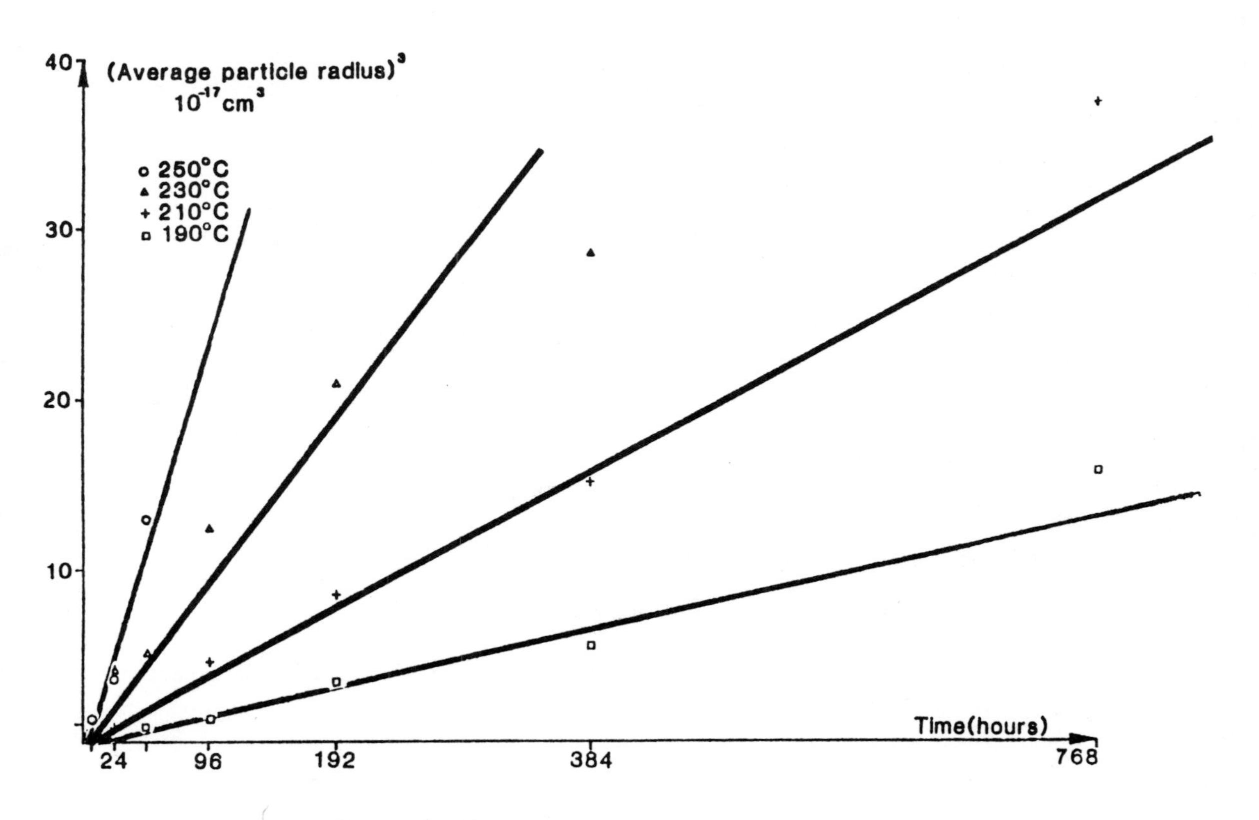

Figure 6 Change in average particle size as a function of ageing time

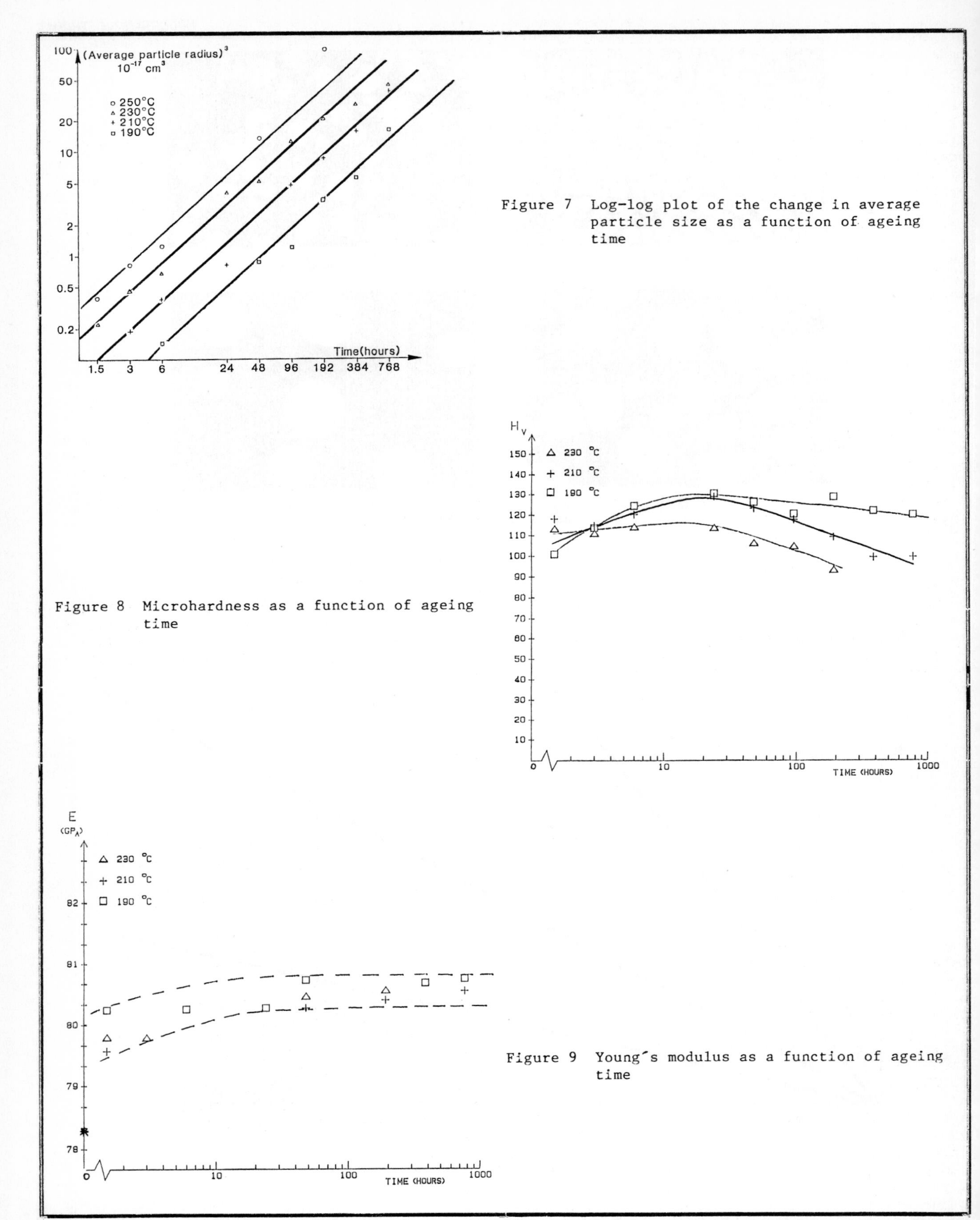

Figure 7 Log-log plot of the change in average particle size as a function of ageing time

Figure 8 Microhardness as a function of ageing time

Figure 9 Young's modulus as a function of ageing time

determined using quantitative metallography. The data shown in Figure 6 represent the change of average particle diameter of the δ´ precipitates with the ageing time for four isothermal temperatures. A log-log plot has been used due to the quick increase of the δ´ particle size with ageing time (Fig.7). The present data confirm previous investigations (4->9) showing that the Lifshitz-Slyozov-Wagner (LSW) theory can be applied to Al-Li alloys. This theory predicts a linear relationship between the cube of the mean particle radius and the ageing time. Due to the fact that some equilibrium phase particles have been detected at the higher temperature, i. e. 250°C, it was decided to carry on with the three lower temperatures.

Microhardness measurements as a function of ageing time ranging from 1.5 hour to 32 days are shown in Figure 8. Two significant features can be noted: Firstly, after an increase, microhardness tends to decrease with ageing time. This may indicate a change in the deformation mechanism from shearing of the particles for short ageing time to Orowan by-passing of the particles for long ageing time. Secondly, microhardness values are slightly higher with lower temperatures. Such changes can be directly related to the slight differences in volume fraction of the δ´ precipitates.

Young´s modulus of a two phase alloy is described in terms of the elastic moduli and volume fractions of the constituting phases. In the literature, this is generally described by two different expressions:

- The first one corresponds to the linear law of mixtures assuming an uniform strain in both phases:

$$E = f_{\delta'} E_{\delta'} + f_{\alpha} E_{\alpha} \qquad (1)$$

where $f_{\delta'}$ and f_{α} are the volume fractions of the two phases.

- The second one asssumes an uniform stress in both phases:

$$1/E = f_{\delta'}/E_{\delta'} + f_{\alpha}/E_{\alpha} \qquad (2)$$

Since the δ´ precipitates remain coherent, matching of the planes in coherence will result in equal strain on both sides of the δ´/ α interface. Therefore, the law of mixtures is more appropriate to describe Young´s modulus behaviour. An attempt was made to evaluate Young´s modulus of the three aged Al 2.5 wt% Li alloys from Equation (1). The maximum volume fractions were calculated according to solvus line determinations of the δ´ phase for the three different ageing temperatures (5,10). At 190, 210, 230°C, the volume fractions are respectively 15.7%, 14.2% and 13%. Now, replacing the values of the intrinsic modulus of the solid solution E_{α} and of the δ´ phase $E_{\delta'}$ from Noble et al. (3), yields an overall modulus almost equal for the three volume fractions: 81.4, 81.4 and 81.3 GPa. Young´s modulus experiments are in good agreement with these calculations: In Figure 9, where the quantity ρV^2 is plotted against ageing time for each isothermal temperature , all the curves are very close to each other. Scattering of the data with errors of up to o.3% may be due to a slight inhomogeneity either in the quenching rate or in composition from one sample to another. Most noteworthy is the significant difference between the Young´s modulus of an as-quenched alloy (78.2 GPa) and aged alloys. Such changes can be directly attributed to copious δ´ precipitation. On the other hand, the modulus variation is nearly minute after a few hours. Hence, the change in local lattice constraints at the δ´/α interface which occur during coarsening, should have no detectable effect on E.

CONCLUSIONS

As a result of this work, it can be stated that the coalescence of δ´ particles obeys the Lifshitz-Slyozov-Wagner law. Microhardness measurements were found to be sensitive to the volume fraction and to the size of δ´ precipitates whereas Young´s modulus are almost independant of the size and distribution of these precipitates. The coarsening process at constant volume fraction cannot really be evidenced by these modulus measurements due to the small amplitude of the phenomenon.

ACKNOWLEDGEMENTS

The authors are highly thankful to Dr Meyer of Pechiney Voreppe Research Laboratories for providing the Al-Li alloy ingots. They are also indebted to Dr Peltier of Aerospatiale for image analysis.

REFERENCES

1- M. THOMAS, G. LAPASSET and R.W. CAHN, Mém. Sci. Rev. Mét. 9 (1984) 438
2- D.B. WILLIAMS, Ph Thesis, University of Cambridge (1974)
3- B. NOBLE, S.J. HARRIS and K. DINSDALE, J. Mater. Sci. 17 (1982) 461
4- M. TAMURA, T. MORI and T. NAKAMURA, J. Jap. Inst. Met. 34 (1970) 919
5- B. NOBLE and G.E. THOMPSON, Met. Sci. J. 5 (1971) 114
6- D.B. WILLIAMS and J.W. EDINGTON, Met. Sci. 9 (1975) 529
7- D.B. WILLIAMS and J.W. EDINGTON, Acta Met. 24 (1976) 323
8- A.L. BEREZINA, L.N. TROFIMOVA, K.V. CHUISTOV, Phys. Met. Metall. 55 (1983) 111
9- O. JENSRUD and N. RYUM, Mater. Sci. Eng. 64 (1984) 229
10- S. CERESARA, G. COCCO, G. FAGHERAZZI and L. SCHIFFINI, Phil. Mag. 35 (1977) 373

Effect of precipitate type on elastic properties of Al–Li–Cu and Al–Li–Cu–Mg alloys

E AGYEKUM, W RUCH, E A STARKE Jr, S C JHA and T H SANDERS Jr

EA and WR are in the Department of Materials Science at the University of Virginia, USA.
EAS is in the School of Engineering and Applied Science at the University of Virginia, USA.
SCJ and THS are in the Department of Metallurgy at Purdue University, USA

SYNOPSIS

Lithium additions are known to significantly increase the elastic modulus of aluminum alloys when present in solid solution or as Al_3Li. However, there has been no systematic fundamental study of the effect of lithium on the elastic properties of the complex Al-Li-Cu and Al-Li-Cu-Mg alloys currently being considered for commercial utilization. A series of Al-Li-Cu and Al-Li-Cu-Mg alloys were processed to contain various amounts of lithium-containing precipitates, e.g., Al_3Li, and Al_2CuLi. The elastic moduli were determined by the pulse echo technique. The results were correlated with the type and volume fraction of precipitates. The differences in elastic properties among the various alloys and heat treatments will be described with respect to the individual contributions of the different precipitates.

INTRODUCTION

Binary aluminum-lithium alloys show low density and high stiffness but are not attractive to commercial applications mainly due to strength and ductility problems. Good overall mechanical properties are found in ternary and quaternary Al-Li alloys where copper and/or magnesium are also present and which still show low density and high elastic modulus compared to lithium free aluminum alloys. The elastic modulus of such complex alloys is composed of individual contributions from the matrix and various types of precipitates. In addition to Al_3Li, the other major strengthening precipitates that form in the ternary and quaternary alloys are the Al_2CuLi and Al_2CuMg phases.

The Young's modulus of a multi-phase alloy results from the individual contributions of each phase. In precipitation hardenable Al-Li alloys, the largest phase is the solid solution or matrix phase. The other phases present, such as constituent particles, dispersoids and strengthening precipitates exist very often in relatively low volume fractions. In order to influence the elastic properties of the alloy, the moduli of these phases must be very different from the modulus of the matrix. Some recent work on Al-Li binary and Al-Li-Mg ternary alloys has resulted in estimations of the effect of Li on the modulus of Al when it is in solid solution (1,2), and the elastic modulus of Al_3Li (1,2,3). There has been no reported estimation of the elastic modulus of other phases containing Li.

This paper describes some of our preliminary results on a study of the elastic behavior of complex Al-Li-X alloys. Although the quantitative analysis is incomplete, these early results do suggest some trends and the relative importance of the manner in which Li is present in the alloy, i.e., the type of Li-containing phases present.

EXPERIMENTAL PROCEDURES

The alloys, manufactured by ingot technology, have been received from Reynolds Aluminum Company, Richmond, Virginia, in form of cross-rolled plates with a thickness of 12 mm. After solution heat treatment and cold water quenching, stretching for 2% within an hour took place. Aging was performed at 163 °C for 16, 50 and 150 hours. The ternary Al-Li-Cu alloy has been aged at 150 °C for 36 hours. More details about the investigated alloys are listed in Table 1.

Densities have been determined using the Archimedes principle, where the samples were immersed in air and CCl_4.

Table 1. Investigated alloys.

ALLOY	Li	Cu	Mg	Zr	Cu/Mg	DENSITY (g/ml) CALC.	DENSITY (g/ml) MEAS.
	w% / at%	w% / at%	w% / at%	w% / at%			
1	2.63 / 9.59	1.86 / 0.74	1.48 / 1.54	0.12 / 0.03	1.3	2.52	2.52
2	2.31 / 8.46	0.91 / 0.36	1.30 / 1.36	0.11 / 0.03	0.7	2.53	2.54
3	1.97 / 7.32	1.85 / 0.75	1.47 / 1.56	0.13 / 0.04	1.3	2.57	2.57
4	1.94 / 7.22	1.91 / 0.78	0.90 / 0.96	0.12 / 0.03	2.1	2.58	2.58
5	2.26 / 8.37	2.54 / 1.03	–	0.11 / 0.03	–	2.58	2.59

Table 2. Results for alloys aged for 50 hrs. at 163°C.

ALLOY	Li (at%)	δ′ (vol%)	Li^R (at%)	$E_{meas.}$ (GPa)	$f_{\delta'} \cdot E_{\delta'}$ (GPa)	ΔE (GPa)	$f_{ss} \cdot E_{ss}$ (GPa)
1	"9.6"	–	–	82.9	–	–	–
2	8.46	9	0.7	81.9	9.5	72.4	70.7
3	7.32	5	0.5	81.0	5.3	75.7	73.8
4	7.22	5	0.4	80.7	5.3	75.4	73.8
5*	8.37	2.5	2.3	83.6	2.7	80.9	75.8

* 36 hrs at **150** °C

$\Delta E = E_{meas} - f_{\delta'} \cdot E_{\delta'}$

Table 3. Results for alloys aged for 150 hrs. at 163°C.

ALLOY	Li (at%)	δ′ (vol%)	Li^R (at%)	$E_{meas.}$ (GPa)	$f_{\delta'} \cdot E_{\delta'}$ (GPa)	ΔE (GPa)	$f_{ss} \cdot E_{ss}$ (GPa)
1	"9.6"	–	–	82.4	–	–	–
2	8.46	4.3	1.9	79.0	4.6	74.4	74.4
3	7.32	2.7	1.1	76.6	2.9	73.7	75.6
4	7.22	3.6	0.82	79.7	3.8	75.9	74.9

$\Delta E = E_{meas} - f_{\delta'} \cdot E_{\delta'}$

Microstructural examinations were performed with optical, scanning and transmission electron microscopes. Number fractions of second phases were measured from TEM negatives with a point grid and converted into volume fractions. The foil thicknesses have been determined by convergent beam electron diffraction patterns.

The pulse echo technique was used to determine the propagation rates of elastic ultrasonic waves (10 MHz) in longitudinal and transverse modes which allowed the calculation of the elastic moduli. Testing direction was the direction of the last rolling step. The reproducibility of this method is about plus/minus 1%.

RESULTS AND DISCUSSION

Table 1 shows the compositions of the investigated alloys and a comparison of calculated and measured densities. The calculation has been performed using the following equation from Peel et al. (4):

$$D = 2.70 + 0.024Cu - 0.01Mg - 0.079Li \quad (1)$$

where D is the density in g/ml and the atomic symbols represent the weight percent of the individual elements. In the original equation from (Peel) the density of pure aluminum was assumed to be 2.71 g/ml, but our density measurements yielded consistently a value of 2.70 for pure aluminum. For that reason the latter value was chosen in equation (1). Also the presence of zirconium has not been considered which may result in a slight underestimation of the density. The agreement between calculated and measured values is extremely good and well within experimental error.

Figure 1 shows the Young's moduli plotted versus the lithium concentration in atomic percent for the different alloys aged 16 hrs. at 163 $^{\circ}$C. Also included is the magnesium-free ternary alloy 5, represented by a filled in circle which has been heat treated 36 hrs. at 150 $^{\circ}$C. For the lower lithium containing quaternary alloys (star symbols) an increase in modulus with lithium levels takes place, as expected. If a linear regression analysis is applied to the data points from these lower lithium alloys a slope of 2 GPa per at.% lithium, and an intercept of 64 GPa, representing the modulus for a lithium free alloy, result. These numbers are reasonable for these alloys; however, alloys 1 and 5 do not follow this trend. The Al-Li-Cu alloy shows a much higher value and the quaternary alloy with the highest lithium content shows a value far too low. Because crystallographic texture was found to be only weak in these cross-rolled alloys, the cause for this behavior must lie in the composition, i.e., the phases present in these materials. In the Al-Li-Cu alloy which has been aged for 36 hrs. at 150 $^{\circ}$C mainly T_1 (Al_2CuLi) is present in addition to delta prime. T_1 seems to have a large beneficial effect on modulus as we will discuss later. The quaternary 9.6 at.% Li alloy shows a value which would be expected for a lithium content of about 7.6 at.% according to this diagram. Optical microscopy clarified this apparently anomalous behavior. Figure 2 is an optical micrograph of this alloy which shows many large particles which were found to be copper and aluminum rich using X-ray energy dispersive analysis. Obviously, the solubility limits were exceeded with this composition and, therefore, large scale precipitation of insoluble primary phases occurred during casting. The exact composition of the particles and/or whether they contained lithium could not be determined. The low atomic number of lithium does not allow its detection with X-ray energy dispersive analysis. Nonetheless, the incorporation of a substantial part of the total lithium content in these large particles seems reasonable and may explain the relatively low elastic modulus of this alloy. The other alloys contained only the typical small amounts of constituent particles.

The shear moduli show the same trend as the Young's moduli as a function of lithium content. Compared to pure aluminum they show roughly the same percent increase. The poissons ratio always stayed close to 0.30.

Figure 3a, b, c and d represent the change in elastic modulus with aging time for the investigated alloys. All materials show an increase in E in the early stages of aging, i.e., from 16 to 50 hours. The maximum difference in modulus during aging is displayed by alloy 2 (2.31 wt%Li) with about 3 GPa, i.e., roughly 4%. No measurements in the solution heat treated states were performed. A decrease in Young's modulus going from 50 to 160 hours was observed for all alloys, too. The results from transmission electron microscopy, performed on the 50 and 150 hr. treatments so far, will be used to explain this behavior. The 16 hour heat treatments are still under investigation.

The initial increase in modulus with aging is attributed to the precipitation of delta prime. At 50 and 150 hrs. aging the only strengthening precipitate present in alloy 1 and 2 (high lithium contents) was delta prime. Alloy 3 which had the same copper and magnesium levels but a lower lithium content also contained S´´, a precursor of the S´ phase (Al_2CuMg), as determined from electron diffraction patterns. In alloy 4 with the lowest Li content and the highest Cu to Mg ratio, S´´ was also found to be present. The S´´ phase is not expected to influence the modulus of elasticity to a great extent, since its effect is negligible in 2XXX alloys.

Table 2 lists the results of the quantitative transmission electron microscopy investigations for the 50 hr. aging treatments and the measured elastic moduli. In order to get some more quantitative information, the amount of lithium in solid solution was assumed to be 5.5 at.%, a value which has been used in binary alloys by Noble et al. (1) and which should represent an upper limit in the quaternary alloys. The measured volume fractions of delta

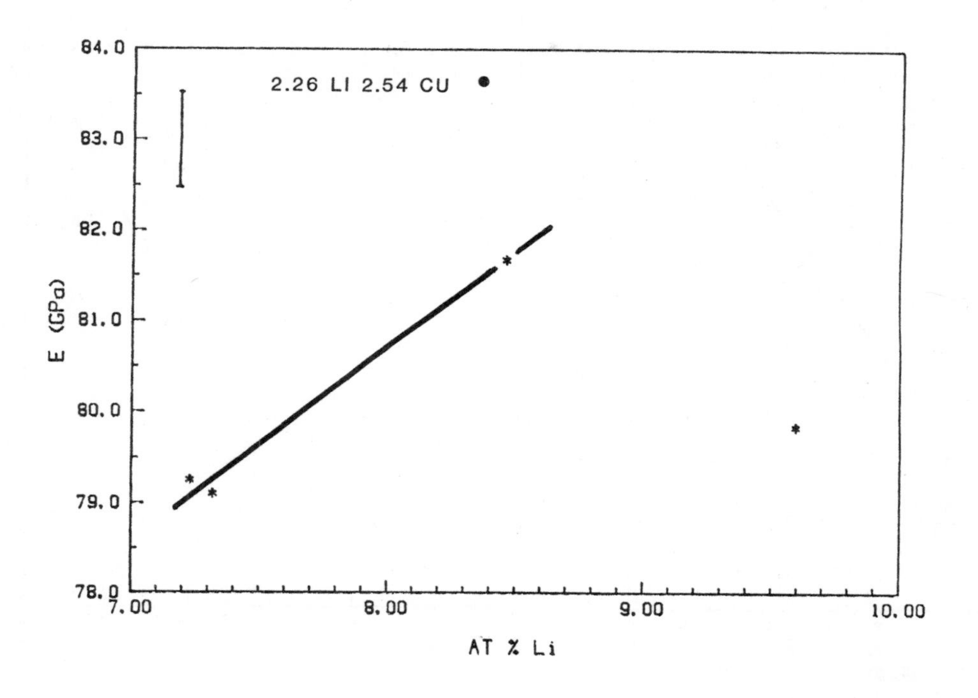

Figure 1. Young's modulus as a function of lithium concentration.

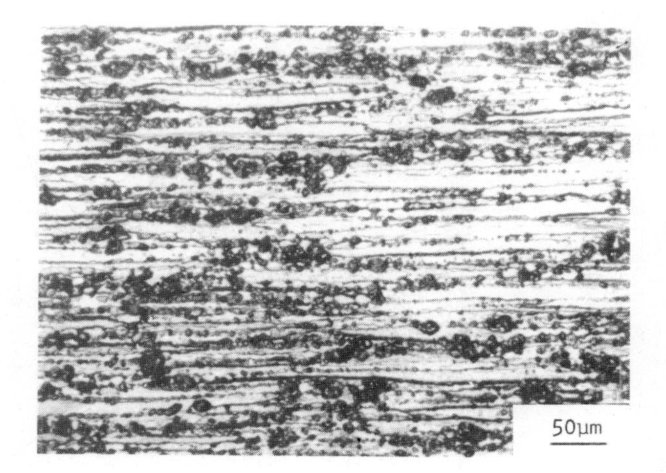

Figure 2. Optical micrograph of alloy 1 showing large particles.

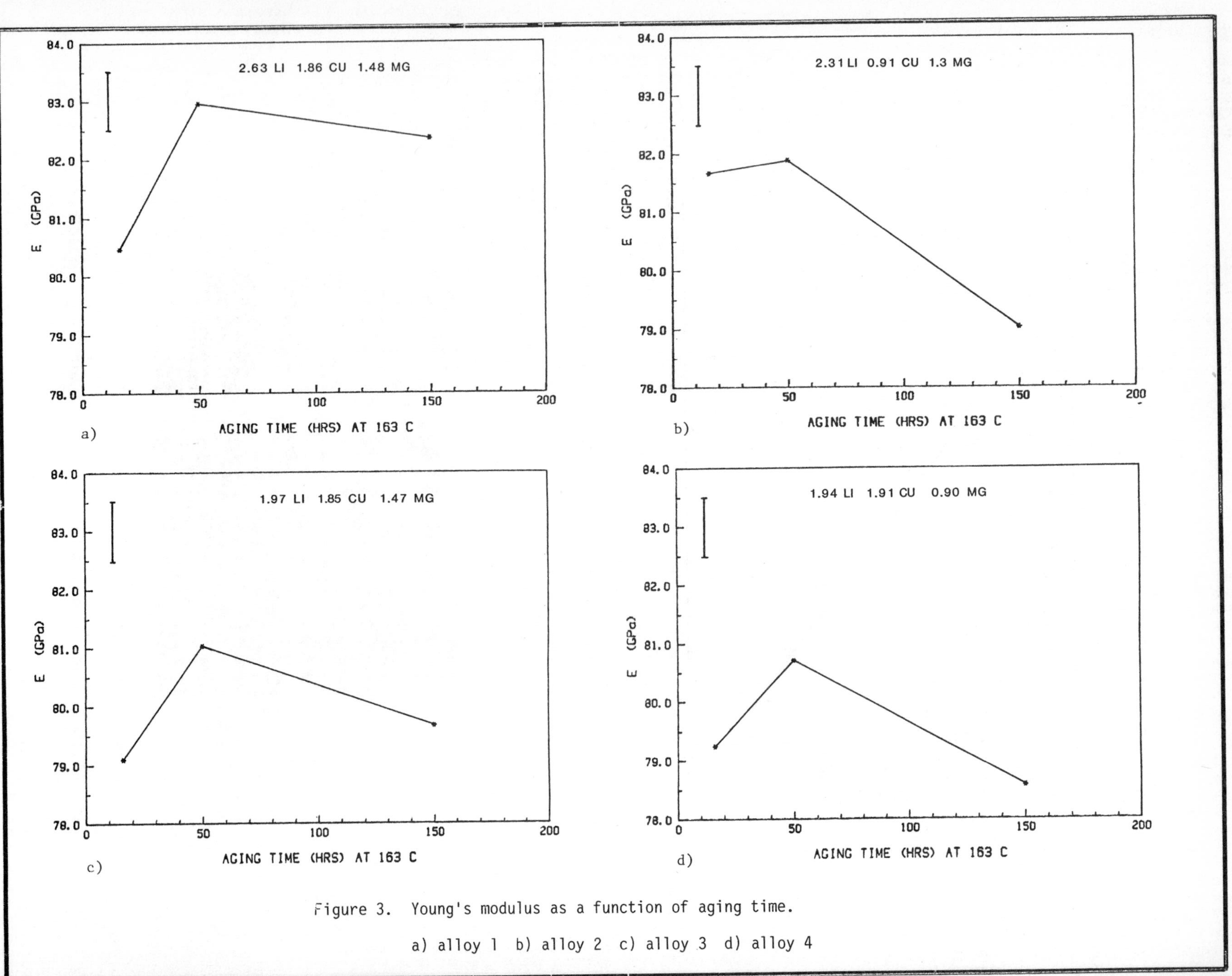

Figure 3. Young's modulus as a function of aging time.

a) alloy 1 b) alloy 2 c) alloy 3 d) alloy 4

prime have been used to calculate the amount of lithium in delta prime, and, after subtracting 5.5 at.% for solid solution, the remaining lithium content in other phases, called Li^R in Table 2, was estimated. Measurements of the S´´ volume fractions in alloys 3 and 4 have yet to be performed. In alloy 1, the delta prime particles were too small to allow quantitative measurements.

The bulk modulus of an alloy can be calculated by determining the contribution of solid solution and the contribution of precipitates. The Young's modulus of the solid solution (E_{SS}) in binary aluminum-lithium alloys has been determined by Muller and Gerold (2) from experimental measurements to be:

$$E_{SS} = 71 + 1.22Li \qquad (2)$$

where E_{SS} is the solid solution modulus in GPa and Li is the atomic concentration of lithium in solid solution. Muller and Gerold also determined a value of 106 GPa for the modulus of delta prime, and the value was selected for our calculations. The way the bulk modulus of a multi-phase material can be composed of the individual contributions is:

$$E_b^m = \sum_{i=1}^{n} (f_i \cdot E_i^m) \qquad (3)$$

where E_b is the bulk modulus, f_i the volume fraction of the phase i, and E_i the modulus of the individual phase i. The exponent m equals 1 for the linear rule of mixtures, also called Voigt (5) averaging (upper bound). There the uniform strain in both phases is assumed. For the Reuss (6) method of averaging the same stress is assumed in both phases and m equals -1 (lower bound). It turns out that in our case the difference between the bounds is small and therefore the linear rule of mixtures was applied.

The contribution of delta prime to the bulk modulus was calculated using the following equation:

$$\delta'_{CONTR} = f_{\delta'} \cdot E_{\delta'} \qquad (4)$$

where $f_{\delta'}$ is the measured volume fraction of delta prime and $E_{\delta'}$ = 106 GPa according to (2). These values are also listed in Table 2. The difference between the delta prime contribution and the measured modulus is also listed in Table 2. This difference is the contribution of solid solution and other phases. The last column in Table 2 shows the calculated contribution of the solid solution to the bulk modulus under the assumption that delta prime and the solid solution are the only two phases present. It is interesting to note that the solid solution contribution, which is believed to be overestimated anyway, cannot make up the difference to the measured value. If one would consider the amounts of Cu and Mg in solid solution, both of which lower the modulus, the resulting difference will even be higher. It has yet to be determined whether or not the lithium not in delta prime and not in solid solution is incorporated in one or several phases which have a positive influence on the modulus. Further work is necessary to identify these phases. The largest contribution of precipitation other than δ´ to the modulus was observed for the magnesium free ternary alloy 5. In this case we can assume the T_1 phase to be the major phase besides δ´ and to be of great influence. If one assumes that all copper is tied up in T_1, the resulting volume fraction for Al_2CuLi is about 4% using a density of 3.1 g/ml for T_1. Applying the linear rule of mixtures according to equation (3) and assuming the presence of only solid solution, T_1 and δ´, results in a Young's modulus of roughly 170 GPa for the T_1 phase. This value is the result of a fairly crude calculation but it stresses the point that besides delta prime there are other lithium containing phases which can contribute quite significantly to the modulus of elasticity.

In Table 3 the same analysis as in Table 2 for the quaternary alloys has been performed for the 150 hr. heat treatments. It is fair to assume that in going from 50 to 150 hours the amount of lithium in solid solution stays essentially constant, i.e., the alloys overage. Table 3 shows a decrease in delta prime volume fractions in all alloys compared to the 50 hr. treatment. It becomes obvious that the decrease in the delta prime contribution cannot alone explain the changes in measured moduli in all alloys. In alloy 2 the decrease in modulus due to a reduction in δ´ volume fraction matches the reduction of the measured value. The lithium leaving the δ´ particles apparently does not have a great influence on modulus. In alloy 4 the decrease in bulk modulus is less than the decrease in delta prime contribution where the opposite is true for alloy 3. In alloy 4, the lithium leaving delta prime apparently still makes a positive contribution to the overall modulus compared to alloy 3 where it makes a negative contribution. In all alloys an increase in the number of grain boundary precipitates was observed when going from 50 to 150 hours. Identification of these precipitates should aid in our understanding of the observed behavior and this analysis is currently being made.

CONCLUSIONS

- In Al-Li-Cu alloys, the T_1 phase contributes positively to the modulus of elasticity.
- Exceeding the solubility limit ties up lithium in large equilibrium particles which has a negative effect on Young's modulus.
- In Al-Li-Cu-Mg alloy, the modulus of elasticity goes through a maximum during aging.
- In Al-Li-Cu-Mg alloys the changes in modulus during aging may not be explained by changes in ´ volume fractions alone.

ACKNOWLEDGEMENTS

This work was sponsored by the Office of Naval Research, Contract No. N00014-85-K-0526, Dr. Bruce A. MacDonald, Program Manager.

REFERENCES

1. B. Noble, S.J. Harris and K. Dinsdale, J. Mat. Sci., **17** (1982), p. 461.

2. W. Müller, V. Gerold, to be published.

3. M.E. Fine, Acta Met., **15** (1981), p. 523.

4. C.J. Peel, B. Evans, C.A. Baker, D.A. Bennett, P.J. Gregson, and H.M. Flower, Aluminum-Lithium II, ed. T.H. Sanders, Jr., and E.A. Starke, Jr., TMS-AIME, Warrendale, PA, 1984, p. 363.

5. Voigt, W., Lehrbuch der Kristallphysik, B.G. Teubner, Verlag, Stuttgart, 1966.

6. Reuss, A., Z. angew. Math. Mech., **9** (1929), p. 49.

Microstructural evolution in two Al–Li–Cu alloys

J C HUANG and A J ARDELL

The authors are with the Department of Materials Science and Engineering at the University of California, Los Angeles, CA 90024, USA

SYNOPSIS

The microstructures of two alloys, Al-2.30Li-2.85Cu-0.12Zr (the 2-3 alloy) and Al-2.90Li-1.00Cu-0.12Zr (the 3-1 alloy), aged at 160°C and 190°C, were investigated as a function of time as part of a study of the relationship between microstructure and mechanical behavior. The δ' and T_1 phases are predominant throughout the entire aging process. Small amounts of θ' precipitates are also found in the underaged and peak-aged 2-3 alloy. A coating of the δ' phase on the θ' precipitates is frequently seen. Previous difficulties concerning identification of the T_1 and T_1' phases are discussed and an attempt is made to rationalize the observations. The evidence favors the unique existence of T_1. The δ' precipitates grow by diffusion-controlled Ostwald ripening. The δ' particle size distributions are nearly symmetrical, contrary to the predictions of the classical LSW theory but corroborating the results of other investigators. The activation energies for δ' coarsening are 127 and 112 kJ/mole for the 2-3 and 3-1 alloys, respectively. These values are well within the range of activation energies reported for the diffusion of Li and δ' coarsening in binary Al-Li alloys (110-140 kJ/mole). The volume fraction of the δ' phase is independent of aging time in the 2-3 alloy but increases somewhat during aging of the 3-1 alloy. No significant volume fraction dependence of the δ' coarsening rate was found. The diameters of the hexagonal plate-shaped T_1 particles increase linearly with the cube root of the aging time, but Ostwald ripening is unlikely to be the growth mechanism since the volume fraction increases significantly with aging time. The δ' PFZ increases linearly with the square root of aging time in both alloys, concurrent with the growth of δ particles on high-angle grain boundaries. Analysis of the kinetics yields diffusion coefficients for Li in the two alloys which are in reasonable agreement with previous measurements.

INTRODUCTION

The addition of several wt. % Cu to binary Al-Li alloys has a significant beneficial impact on the strength of these materials, largely because of its influence on the metastable and stable phase equilibria and the related sequence of precipitation reactions. As a consequence, Cu has been an essential elemental ingredient of all previous commercial Al-Li base alloys, and will continue to be one in future alloys.

Despite the importance of the role Cu plays in the physical metallurgy of Al-Li-Cu alloys, basic research addressing the effect of Cu has been relatively limited. Hardy and Silcock[1] studied the phase equilibria at several temperatures in the Al-rich corner of the Al-Li-Cu phase diagram and established the regions of stability of the equilibrium phases T_1, T_2, T_B, θ and δ. Several investigators[2-6] have studied the precipitation sequence(s) in Al-Li-Cu alloys of various compositions. The existence of metastable G.P. zones, θ'', θ' and δ' precipitates at various aging times in these alloys has been convincingly demonstrated, but according to Noble and Thompson[3], spherical δ' and hexagonal plate-shaped T_1 particles are the predominant precipitates during the early stages of aging of alloys containing at least 2% Li and less than 3% Cu. There have been no studies that we are aware of dealing with the effect of Cu on the kinetics of growth of the important metastable precipitates. In addition, new evidence[7] has been presented that another metastable phase, T_1', is an important microstructural constituent of aged Al-Li-Cu alloys.

As part of an investigation of the relationship between microstructure and properties

Table I

Compositions of the Ternary Alloys (wt. %)

Alloy	Li	Cu	Zr	Fe	Si	Ni	Cr	Ti
2-3	2.30	2.85	0.12	0.05	0.04	0.03	0.01	0.01
3-1	2.90	0.99	0.12	0.05	0.07	0.02	—	0.01

Table II

Measured Values of the Volume Fraction of δ', $f_{\delta'}$, at Selected Aging Times

Alloy	Aging Temp. (°C)	t(h)	$f_{\delta'} \times 10^{-2}$
2-3	160	8	4.88
		20	5.05
		26	5.69
	190	2	4.14
		4	4.12
		6	4.45
		132	4.15
3-1	160	10	7.01
		30	8.03
		48	7.86
	190	4	6.36
		6	6.79
		10	6.59
		132	9.25

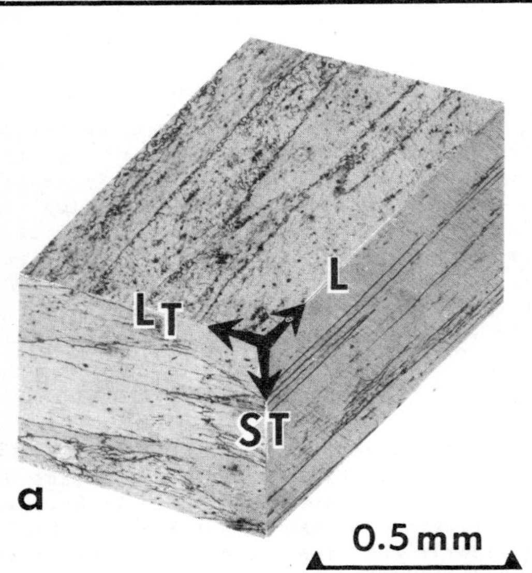

1 Optical micrographs of the grain structures observed in (a) the 2-3 alloy and (b) the 3-1 alloy.

2 Optical micrograph of the 2-3 alloy illustrating recrystallization observed in the L-LT section.

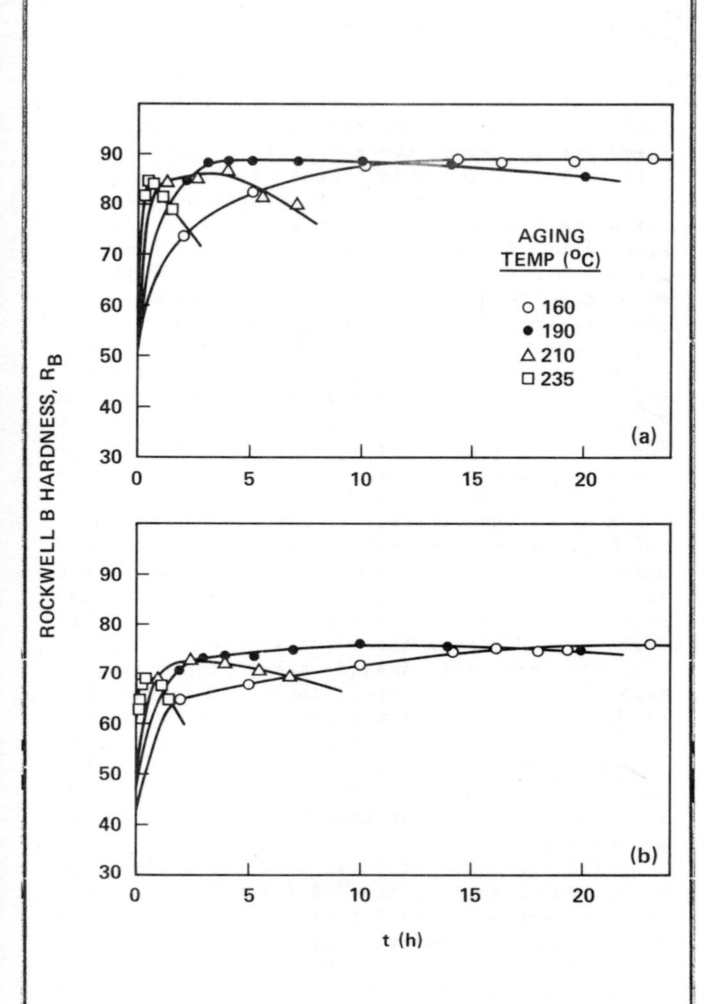

3 Rockwell B (R_B) hardness as a function of aging time for (a) the 2-3 alloy and (b) the 3-1 alloy.

of two ternary Al-Li-Cu alloys, one containing nominally 2%Li-3%Cu (the 2-3 alloy) and the other 3%Li-1%Cu (the 3-1 alloy), we have undertaken a more detailed study than hitherto of the evolution of the precipitate microstructures in this system. In this paper we report and discuss the results obtained to date, which include characterization of the precipitate dispersions with respect to the identification of the various important precipitate phases, and their kinetics of growth.

EXPERIMENTAL PROCEDURES

The two alloys studied have the exact compositions listed in Table I. The alloys were provided by Alcoa in the form of plate which had been solution treated at 538°C for 1 h and stretched approximately 1%. The grain structures in the L-LT, L-ST and LT-ST sections were determined from samples metallographically polished and etched using either Keller's reagent (the 3-1 alloy) or a modified Keller's reagent (the 2-3 alloy). The grain structures are shown in Fig. 1 and can be seen to be quite different. The 2-3 alloy has large elongated grains (Fig. 1a) measuring 3 x 1 x 0.2 mm., whereas the 3-1 alloy has much smaller elongated grains with average dimensions 1 x 0.2 x 0.05 mm. The microstructure of the 2-3 alloy also revealed evidence of considerable recrystallization (Fig. 2), with recrystallized grain sizes ranging from 10 to 25 µm. The microstructural differences between the 3-1 and 2-3 alloys are undoubtedly due to differences in the processing history, but we were not provided with those details.

The black dots visible in Figs. 1 and 2 are possibly large particles of Al_3Zr. These are located primarily at the grain boundaries, although significant quantities were observed in the grain interiors of the 2-3 alloy. For reasons that are also most likely related to its processing the addition of Zr was not very effective in preventing recrystallization of the 2-3 alloy.

Samples in the form of cubes were cut from the as-received plates for evaluating the Rockwell B hardness, R_B, as a function of aging time and temperature. At least six hardness readings were taken for each condition. The samples were aged either at 160°C in an oil bath or at 190, 210 or 235°C in a salt bath. Although several kinds of precipitates were expected, and observed, in these alloys the aging curves for both exhibited only single peaks, as shown in Fig. 3. Since the aging response at 210 and 235°C is fairly fast, the temperatures of 160 and 190°C were selected for the majority of the studies.

All the thin foil specimens for transmission electron microscopy (TEM) were prepared from aged tensile specimens machined from the as-received plate and deformed 1 to 2%. Slices were taken from the 4 mm dia. gage sections using an Ageitron electric spark-discharge machine. Electrochemical polishing was done in a solution of 30% nitric acid-70% methanol cooled to -30°C at a DC voltage of 20 to 22 V.

The thin foil specimens were examined using a JEOL 100CX TEMSCAN electron microscope operating at an accelerating voltage of 100keV.

The diameters of the spherical δ' precipitates were measured from dark-field micrographs taken using δ' superlattice reflections. The δ' particle size distributions were determined with the aid of a Zeiss particle size analyzer. In these alloys the δ' phase also coats Al_3Zr and θ' precipitates; such δ' particles were excluded from the measurements.

The thickness, h, of the thin foils was measured using the convergent beam method,[8] which proved highly reliable for foils satisfying 100 < h < 250 nm. For these measurements the foil normals were restricted to lie within ±10° of the direction of the electron beam. The volume fraction of the δ' phase, $f_{\delta'}$, was calculated using the formula given by Underwood.[9] For small values of the average particle radius ⟨r⟩, both the overlapping and truncation effects were taken into account, and the formula used was

$$f_{\delta'} = \frac{-4\langle r\rangle \ln(1 - \mathscr{A})}{2\langle r\rangle + 3h}, \qquad (1)$$

where $\mathscr{A}$ is the area fraction measured from the dark-field TEM photographs. For larger particles (in very overaged samples) where the effect of particle overlap is unimportant, equation (1) is replaced by

$$f_{\delta'} = 4\langle r\rangle\mathscr{A}/(4\langle r\rangle + 3h). \qquad (2)$$

RESULTS AND DISCUSSION

General Observations

It is evident in Fig. 3 that the 2-3 alloy ages much more rapidly than the 3-1 alloy, and that the hardness of the 2-3 alloy is always 15 to 20% higher than that of the 3-1 alloy. The addition of Cu clearly accelerates the aging response and is most likely responsible for the greater hardness of the 2-3 alloy. Both of these observations were confirmed by subsequent tensile testing. The strengthening effect of Cu in the 2-3 alloy more than compensates for whatever reduction in strength might be attributable to its far larger grain size. The maximum hardness of both alloys was obtained in samples aged at 160°C. Its value (R_B = 91) in the 2-3 alloy is fairly high and comparable to those of quaternary Al-Li-Cu-Mg alloys. That of the 3-1 alloy (R_B = 78) is, however, moderate by comparison with many other Al-Li-Cu alloys.

According to the ternary phase diagram at 350°C determined by Hardy and Silcock,[1] the 2-3 alloy is expected to contain the T_1 and T_2 phases, while the 3-1 alloy should contain the δ and T_2 phases at equilibrium. The phase diagram at 200°C and below has not been determined. While the presence of the T_2 phase at lower aging temperatures might be expected, we found no evidence for it in either alloy at either aging temperature. This is probably due to its complex crystal structure which renders nucleation difficult.[2] The important phases in both alloys are δ' and T_1 in the matrix, and δ at grain boundaries (Al_3Zr is, of course, present as well). The θ' phase is also found in the underaged 2-3 alloy. In what follows we first discuss the evidence for the existence of the phases identified, and later the kinetics of growth of the δ' and T_1 particles.

Phase Identification

The presence of the θ' phase in the 2-3 alloy was most apparent on examination of ⟨100⟩ diffraction patterns. In the initial stages of this investigation the as-received material was cold-rolled into thin sheet for subsequent heat-treatment. Solution treatment or aging of these samples in a salt bath resulted in severe loss of Li at the surface, producing microstructures containing θ' and T_1, but no δ'. Figure 4 shows the ⟨100⟩ diffraction patterns from a sample containing only θ' and T_1 (Fig. 4a) and, by comparison, one from the bulk 2-3 alloy from which no Li was lost (Fig. 4b). The reflections from all the phases are drawn schematically in Figs. 4c and d. While the θ' reflections in Fig. 4b are weak, dark-field images confirm its presence. Examples are shown in Fig. 5, in which Figs. 5b, c and d are dark-field images taken using the reflections labelled A, B and C in Fig. 4d, respectively. Reflection A belongs uniquely to δ', reflection B uniquely to θ' and reflection C to both θ' and δ'. In Fig. 5b the arrows indicate θ' particles coated by δ'. The dark gaps between the δ' coating are θ' particles that are not imaged using reflection A (Fig. 5b was printed specifically to reveal this gap clearly, which is why most of the spherical δ' particles are not in strong contrast). These same θ' particles are visible in Fig. 5c, taken using reflection B. Observations and conclusions similar to these have been made by Tosten and Howell.[10]

The presence of θ' is associated with the higher Cu content of the 2-3 alloy, since it was never observed in the 3-1 alloy. Also, θ' was found only in samples aged up to the peak-strength condition. On prolonged aging it dissolved because its stability is lower than that of the T_1 phase.

Plate-shaped precipitates of the T_1 phase were observed in both alloys at all aging times. The concentration of this phase was greater in the 2-3 alloy than in the 3-1 alloy, reflecting the larger Cu content of the former. These particles grew both within the grains and at low-angle grain boundaries. In the grain interiors the plate-shaped particles were distributed equally among all four variants on {111}. On low-angle boundaries only one variant was observed.

There has been some confusion regarding the existence of the T_1' and T_1 phases. It has been suggested that the T_1' phase is a precursor of the T_1 phase,[7,11] and that T_1' continuously transforms to T_1 as aging proceeds.[12] The evidence for this is based on the changes that occur in ⟨112⟩ diffraction patterns because in the ⟨100⟩, ⟨110⟩ and ⟨111⟩ diffraction patterns

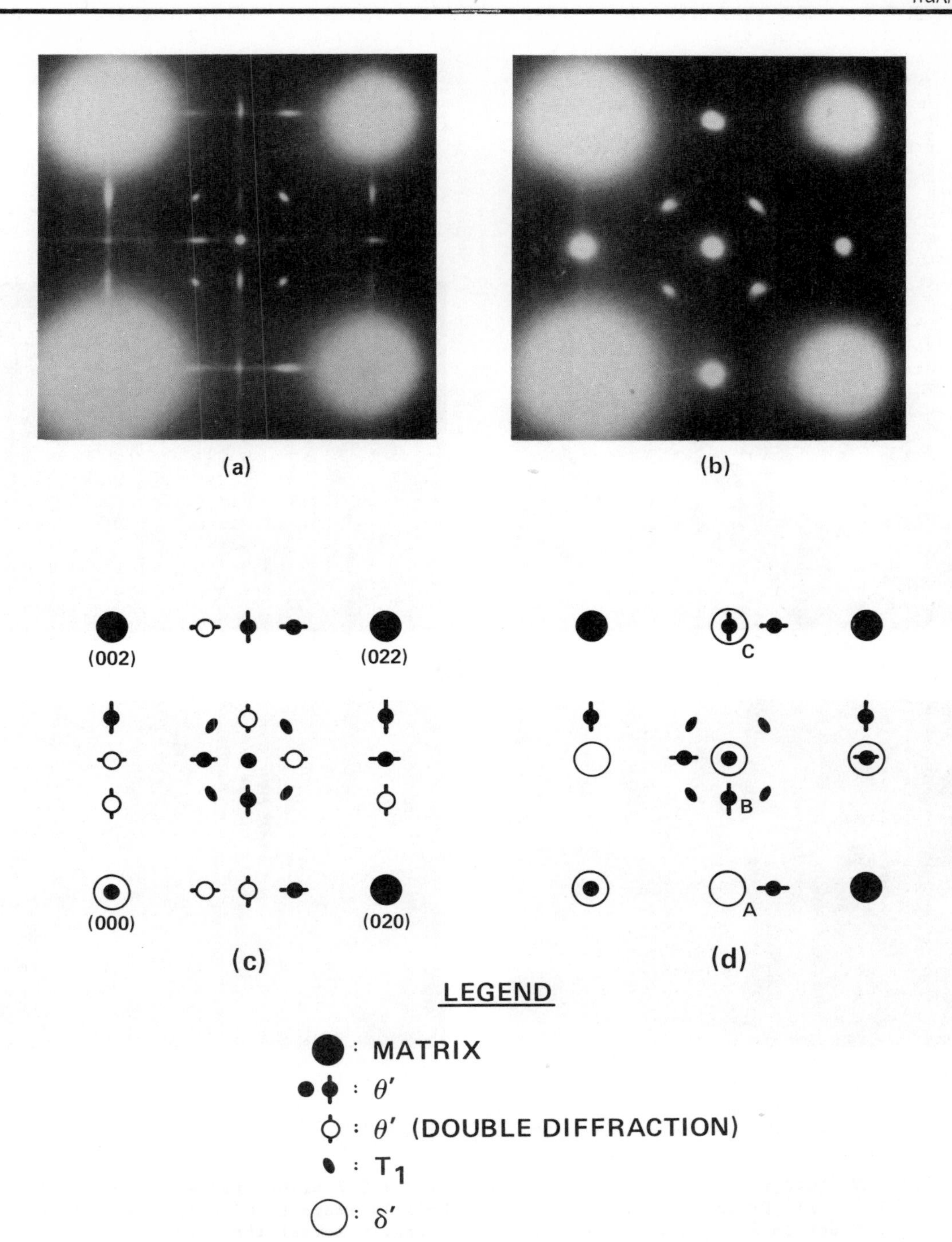

4 <100> diffraction patterns taken from: (a) a thin sheet sample of the 2-3 alloy aged at 190°C for 6 h and suffering loss of Li at the surface; (b) a sample taken from the gage section of a 2-3 alloy tensile specimen aged at 190°C for 4 h. The patterns are indexed in (c) and (d). The θ′ reflections due to double diffraction are highlighted in (c), whereas the θ′ and δ′ reflections are highlighted in (d). The reflections labelled A, B and C were used for dark-field images seen in Fig. 5.

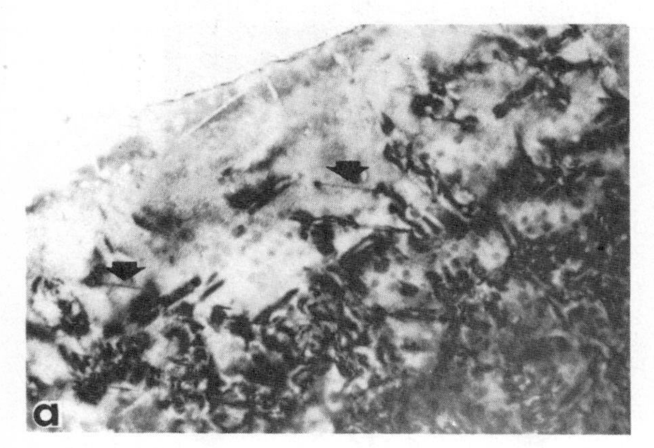

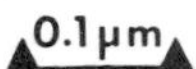

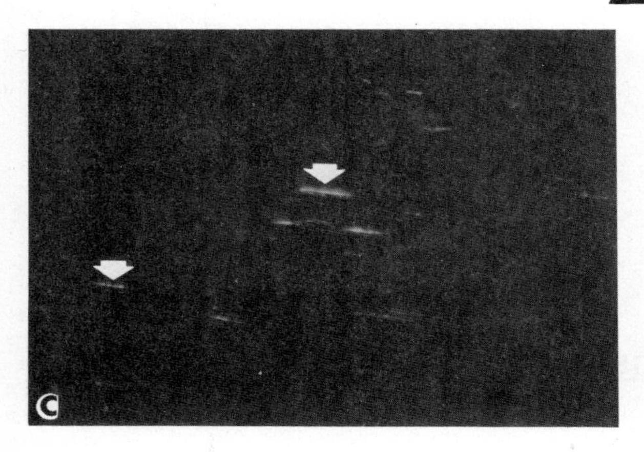

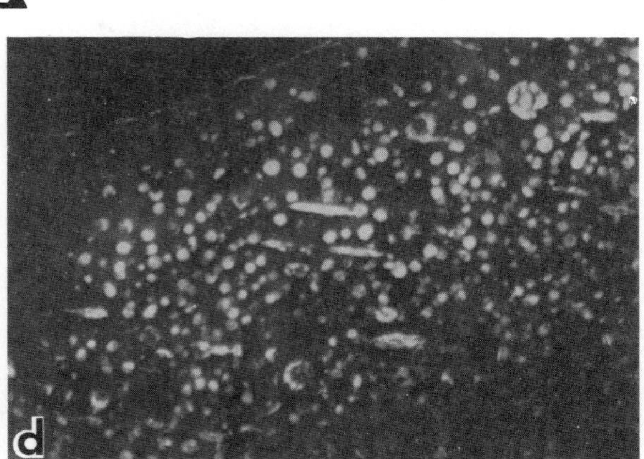

5 TEM photographs of the microstructure of the 2-3 alloy aged at 160°C for 18 h. (a) Bright-field image. (b) Dark-field image of δ′ only, taken using reflection A in Fig. 4d and printed to clearly reveal the black gap between the δ′ phase at the precipitates indicated by the arrow (these are also seen in (a) and (c)). (c) Dark-field image of θ′ only, taken using reflection B in Fig. 4d. (d) Dark-field image of θ′ and δ′, taken using reflection C in Fig. 4d. The gap at the particles in (b) indicated by the arrows are θ′ particles (out of contrast) flanked by the δ′ coating.

the T_1 and T_1' phases are expected to have the same reflections.[12] The relevant patterns are depicted schematically in Fig. 6. In the early stages (Fig. 6a) the pattern consists of intense reflections at positions of the type $\frac{1}{3}(2\bar{2}0)$, $\frac{1}{3}(3\bar{1}\bar{1})$ and $\frac{1}{3}(40\bar{2})$, etc., together with continuous streaks along $[11\bar{1}]$. As aging proceeds (Fig. 6b) the streaks break up into spots centered at integral multiples of $\frac{1}{4}(111)$ etc., while the other reflections become less intense. Eventually the streaked reflections along $[11\bar{1}]$ sharpen and become distinct spots, while the reflections at $\frac{1}{3}(311)$ and $\frac{1}{3}(402)$, etc., disappear, producing the ⟨112⟩ type patterns reported by Noble and Thompson[3] and illustrated in Fig. 6c. Dark-field photographs taken using the reflections labelled D, E, F, and G in Fig. 6b are shown in Fig. 7.

The controversy regarding T_1 and T_1' has arisen because the spots in Fig. 6a are usually so intense that it is reasonable to conclude that they cannot have originated from T_1 and must therefore be due to a new metastable phase, T_1'.[7,12] While T_1 and T_1' have different crystal structures, they are supposed to have the same morphology and habit plane, i.e. hexagonal shaped platelets parallel to {111}. Now, all the edge-on particles visible in bright-field (Fig. 7a) become visible using any of the T_1 reflections (streaked or not) in the [112] diffraction pattern (e.g. Fig. 7e). We must then argue that the edge-on particles in Fig. 7a are all T_1, while the inclined variants are all T_1', since as can be seen in Figs. 7b, c and d, these are all visible in dark-field images taken using reflections that are supposed to belong uniquely to T_1'. The logical conclusion to be reached from these observations is that in any foil oriented [112] the edge-on particles on $(11\bar{1})$ will always be T_1 while the inclined particles on the other $\{11\bar{1}\}$ planes will always be T_1'. It seems to us that this is a hopelessly unrealistic scenario.

The most sensible explanation for the observations regarding the ⟨112⟩ diffraction patterns is that the reflections depicted in Fig. 6a and visible in Figs. 6d and e are due to streaking of so-called T_1 reflections centered at reciprocal lattice points in the first Laue zone. These can be expected to be quite intense and extended in reciprocal space in the early stages of aging when the T_1 particles are very thin, consistent with the continuous streaks along ⟨111⟩. As aging proceeds and the T_1 particles become thicker, the length of the streaks diminishes (cf. Figs. 6d and 6e). Eventually the diffracted intensity is entirely localized at the reciprocal lattice points of the T_1 structure, producing the pattern in Fig. 6c.

If this suggestion is correct, the positions of the so-called T_1' reflections in the [112] diffraction pattern should shift noticeably when the sample is tilted. As demonstrated in Fig. 8, this is precisely what is observed. We are therefore convinced that the existence of the T_1' phase is doubtful, and that only T_1 nucleates and grows in our ternary alloys.

Kinetics of Precipitate Growth

Plots of $\langle r \rangle^3$ against the aging time, t, for the spherical δ′ precipitates in both alloys are shown in Fig. 9. The data show linear behavior at both aging temperatures, suggesting that the particles grow by diffusion-controlled Ostwald ripening. The curves extrapolate to $\langle r \rangle \simeq 0$ at $t = 0$, indicating that coarsening starts at the onset of aging, similar to what is found in binary Al-Li alloys.[13,14] The rate constants at both temperatures are nearly the same in the two alloys, implying that the kinetics of δ′ precipitate growth are independent of the Cu content of the sample. Also by implication, the much faster aging response of the 2-3 alloy is independent of the δ′ coarsening rate.

Measurements of $f_{\delta'}$ for all aging conditions have not yet been completed because its variability from one thin foil specimen to another, or even within the same specimen, necessitates the examination of several samples. The results of those measurements in which we have considerable confidence are shown in Table II, in which the reported values of $f_{\delta'}$ were obtained on application of equations (1) or (2). It is seen that $f_{\delta'}$ is relatively constant with aging time. Only in the overaged 3-1 alloy is there an indication that $f_{\delta'}$ increases with increasing t.

Comparisons of the coarsening rate constants, K, and activation energies, Q, with those measured for δ′ coarsening in binary alloys are shown in Fig. 10.[13-18] Q varies from 110 to 140 kJ/mole and appears to increase with increasing Li content in the binary alloys, as seen in Fig. 10a. This trend is not evident in our two ternary alloys, but the values of Q for the 2-3 and 3-1 alloys lie within the range of experimental data measured on the binary Al-Li alloys. The values of Q are also consistent with those for diffusion of Li in Al-Li alloys. This suggests that Cu is not involved in the growth of the δ′ phase. In Fig. 10b the rate constants at 200°C are plotted against Li content, with some of the data being extrapolated or interpolated from published results. Even though there is considerable scatter, a positive trend is observed, indicating that there is indeed some dependence of the coarsening rate on $f_{\delta'}$. This trend is not noticeable on examination solely of our data, however.

Histograms of the δ′ particle size in the underaged alloys aged at both 160 and 190°C were constructed from measurements of 200 to 700 particles. Figure 11 shows a typical example. All of them appear to be nearly symmetrical, corroborating the results of Kulwicki and Sanders.[14] According to them, the width of the distribution at half the value of the maximum frequency is linearly related to $\langle r \rangle$. Our data are consistent with this behavior, as shown in Fig. 12. The simple relationship between the width of the distribution and the average particle radius is valid in all the alloys aged at all temperatures. This provides an easy way to estimate the δ′ size distribution function for any Al-Li

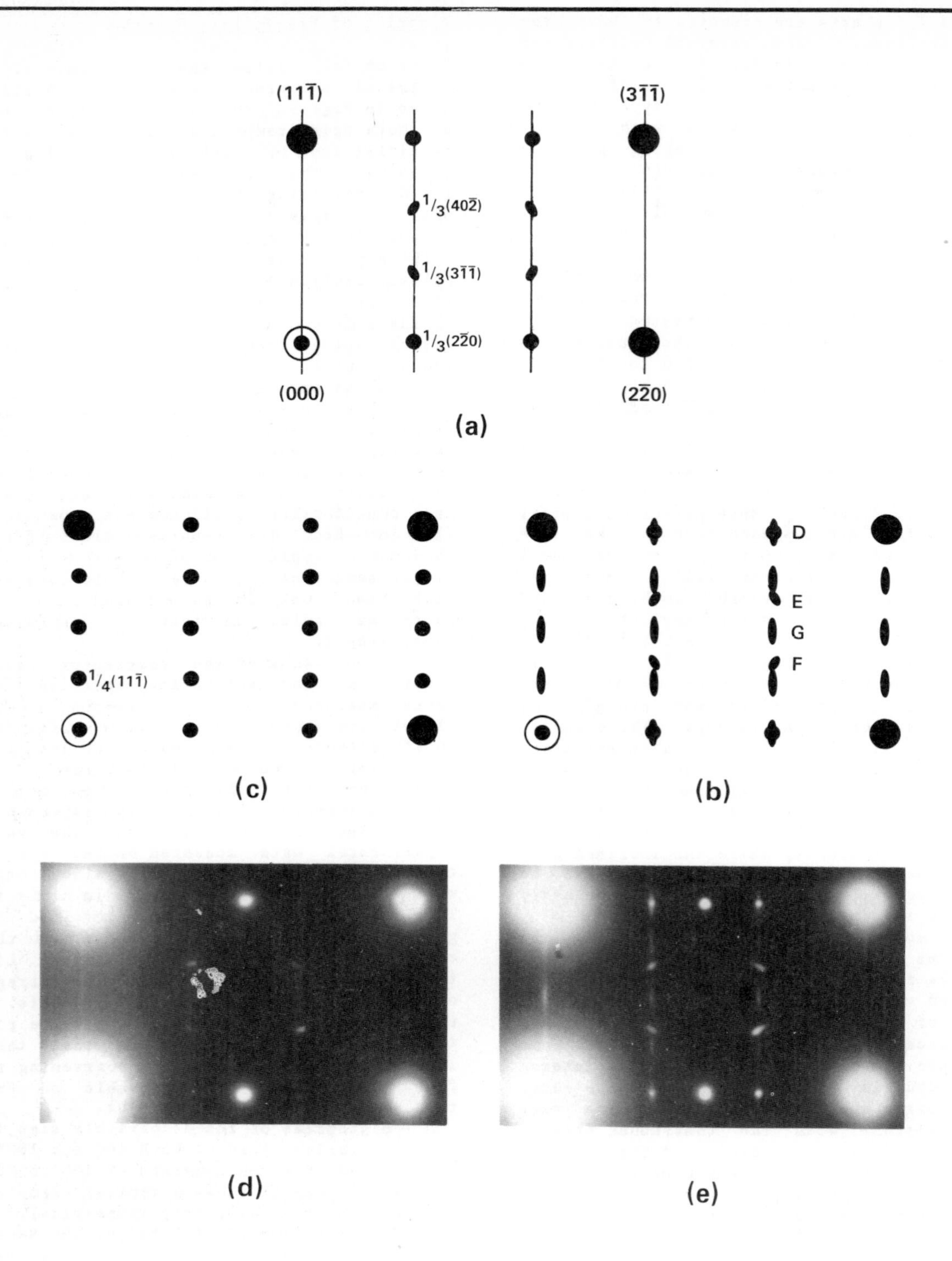

6 Illustrating several stages of evolution of the spots and streaks observed in ⟨112⟩ diffraction patterns. The stages represented are (a) early, (b) intermediate and (c) near equilibrium. The diffraction patterns shown are of the 2-3 alloy aged at (d) 160°C for 14 h and (e) 190°C for 132 h. Note the continuous streaking in (d) compared to that in (e). The reflections labelled D, E, F and G were used for dark-field images seen in Fig. 7.

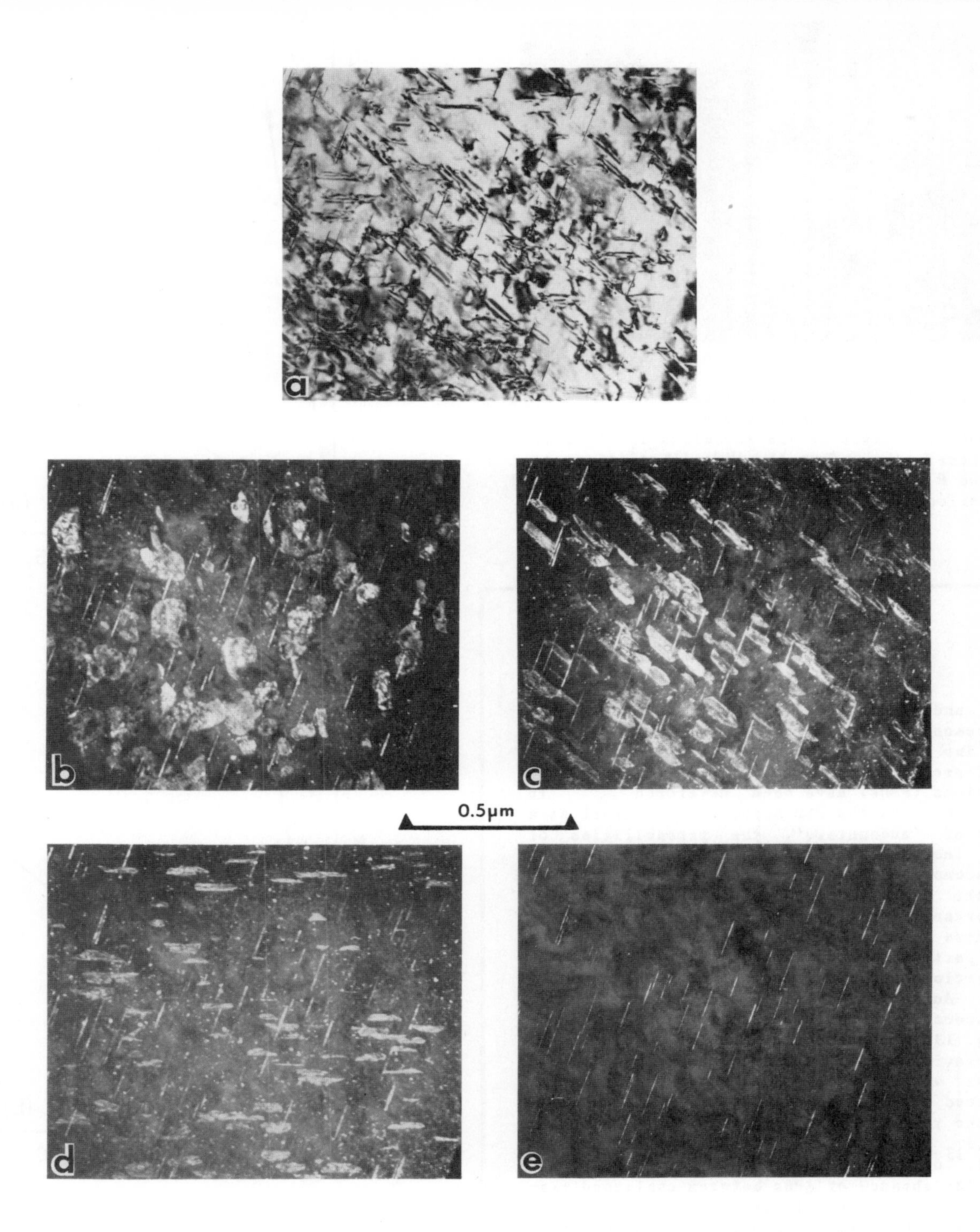

7 TEM photographs of the microstructure of the 2-3 alloy aged at 190°C for 2 h. (a) Bright-field image. The dark-field images in (b) through (e) were taken using reflections D through G, respectively, in Fig. 6. the edge-on variant is visible in (c) and (d) because intensity from the streaks cannot be excluded from the objective aperture.

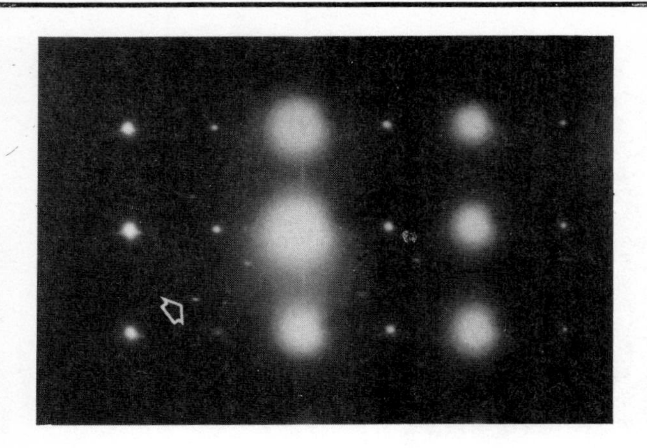

8 The [112] diffraction pattern from a sample of the 2-3 alloy aged at 190°C for 132 h. Note the shift of the so-called T_1' "reflections" away from the "ideal" positions at E and F in Fig. 6b. One of the displaced reflections is indicated by the arrow.

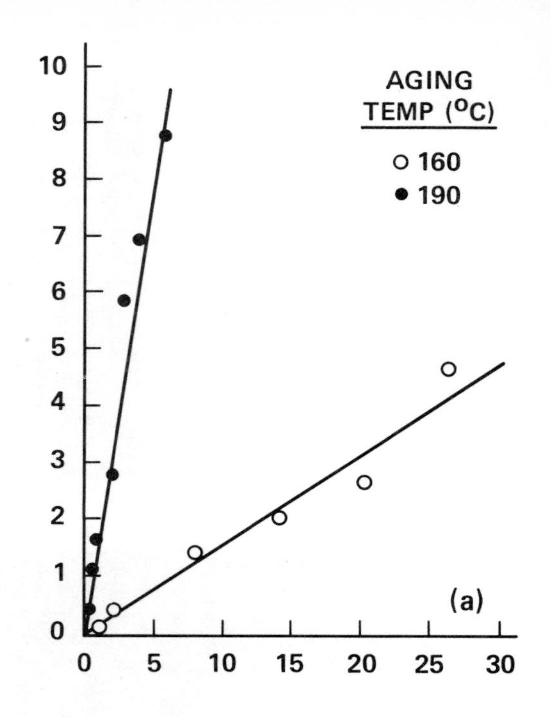

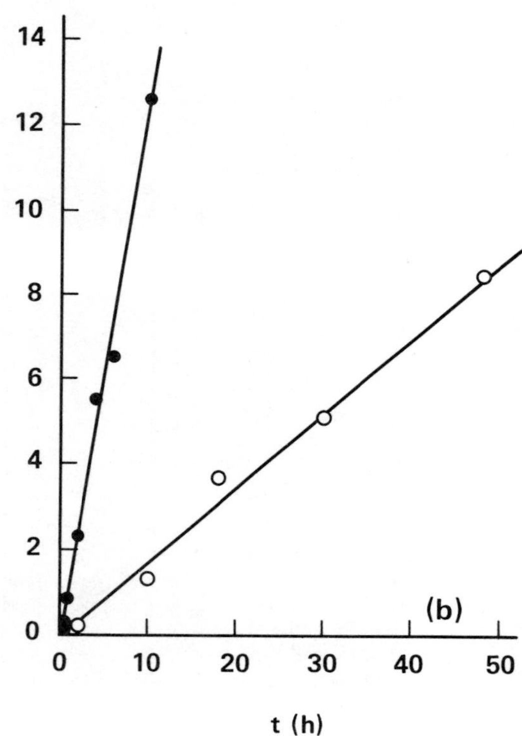

9 Plots of the cube of the average δ' particle radius, ⟨r⟩, vs. aging time, t, for (a) the 2-3 alloy and (b) the 3-1 alloy.

alloy under any aging condition, simply from a measurement of ⟨r⟩.

Two theories of particle coarsening capable of predicting symmetrical particle size distributions (PSDs) have been developed.[21,22] In these theories the PSD becomes symmetrical as a result of "encounters", the probabilities of which increase with increasing volume fraction. Coalescence is an encounter mechanism, but Kulwicki and Sanders[14] have argued that it is unimportant in the coarsening of δ' because antiphase boundaries (APBs) are never observed in δ' particles, even when the shape of a given δ' particle deviates significantly from sphericity. An example of physical contact and coalescence of δ' particles in our work is shown in Fig. 13. No APBs are visible, consistent with previous observations.[14] In fact, there are no APBs visible in published photographs of coalesced γ' particles in Ni-base alloys, even when the γ' volume fraction is quite large (an example[23] is Fig. 1b, from Chellman and Ardell, of γ' in a Ni-Cr-Al alloy).

The absence of APBs between coalesced particles suggests that those particles which must create an APB during a physical encounter tend not to coalesce. Based on the reported data on the γ' APB energy on {111} of De Hosson et al.[24] (0.14 J/m^2), Noble et al.[25] (0.16 J/m^2) and Tamura et al.[26] (0.195 J/m^2) and the interfacial energy, $\gamma_{\delta'-\alpha}$, between the α and δ' phases reported by Baumann and Williams[27] (0.014 J/m^2) the creation of one APB is probably energetically unfavorable compared to the destruction of two α-δ' interfaces (this is probably also true

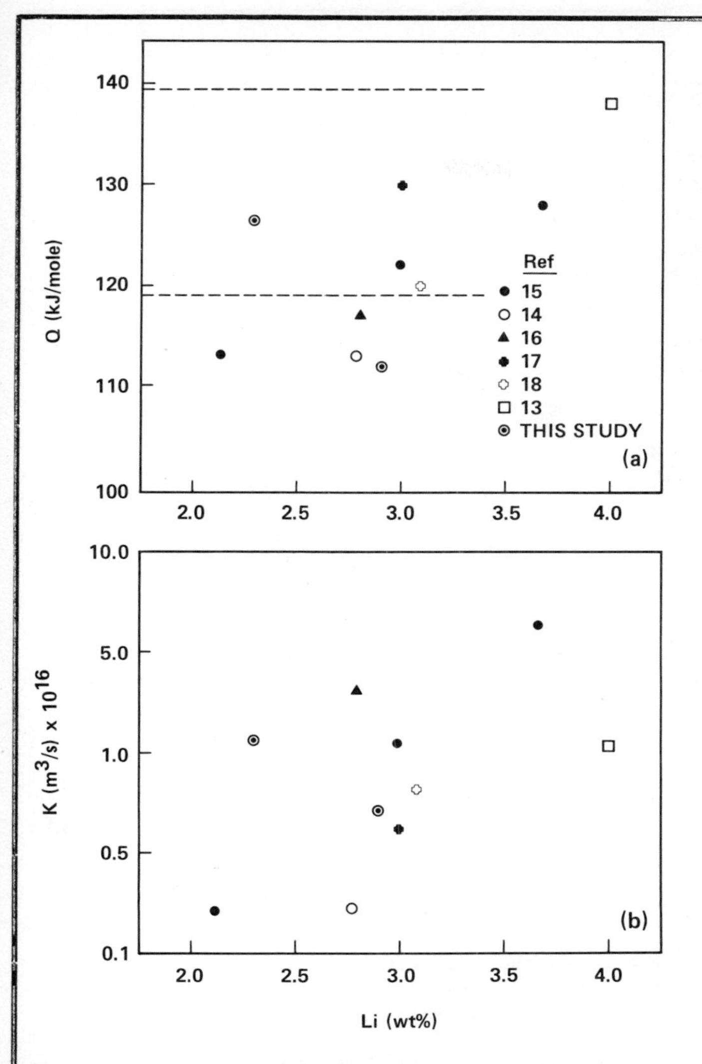

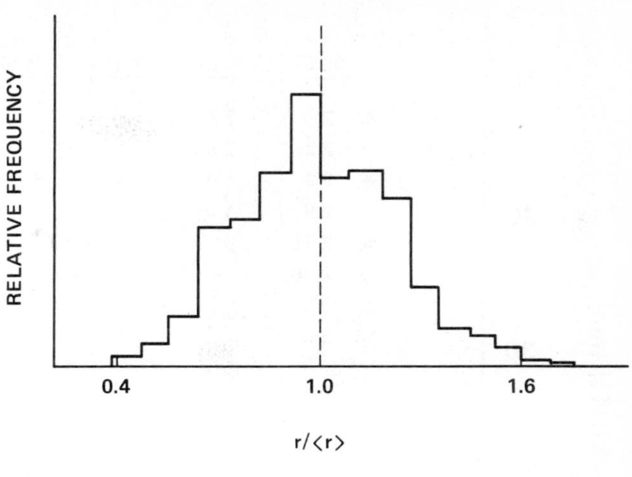

11 A representative δ' particle size distribution in both alloys. The distribution shown was determined for the 2-3 alloy aged for 20 h at 160°C. The value of ⟨r⟩ is 6.45 nm.

10 Illustrating (a) the variation of the activation energy, Q, for coarsening and (b) the rate constant, K, for coarsening at 200°C as a function of the Li content in a variety of Al-Li alloys. The horizontal lines in (a) indicate the values of Q for the diffusion of Li in Al-Li alloys.[19,20]

for the much lower energy {100} APBs). Therefore, only those particles with perfectly matching sublattices will coalesce. This is probably why the percentage of coalesced particles in Fig. 13 is typically fairly low. Takajo et al.[22] have claimed that a small amount of coalescence is enough to shift the PSD from asymmetrical (negatively skewed) to near symmetrical by assuming a linear addition rule for their two kinetic equations of coarsening and coalescence. Their theory still needs verification, but there is still no compelling reason to exclude coalescence as a possible explanation for the nearly symmetrical PSD.

It is worth pointing out that while $f_{\delta'}$ is higher in the 3-1 alloy than in the 2-3 alloy (see Table II), and higher in both alloys the lower the aging temperature (as expected), the values of $f_{\delta'}$ are much smaller than in binary Al-Li alloys containing identical concentrations of Li.

Figure 14 shows some δ phase precipitates on a high-angle grain boundary, with their associated δ' precipitate-free zones (PFZs). Coarser particles were not observed near the edge of the PFZ; these are generally found if vacancy depletion is the predominant factor involved in the formation of the PFZ. Also, the width of the PFZ, w, is a linear function of the square root of t, as shown in Fig. 15. Similar results have been found in binary Al-Li alloys.[17] This behavior also supports the idea that Li depletion near grain boundaries, concomitant with the growth of the δ phase, governs the PFZ width. Note that an incubation time is needed to develop the PFZ in the 2-3 alloy. Its

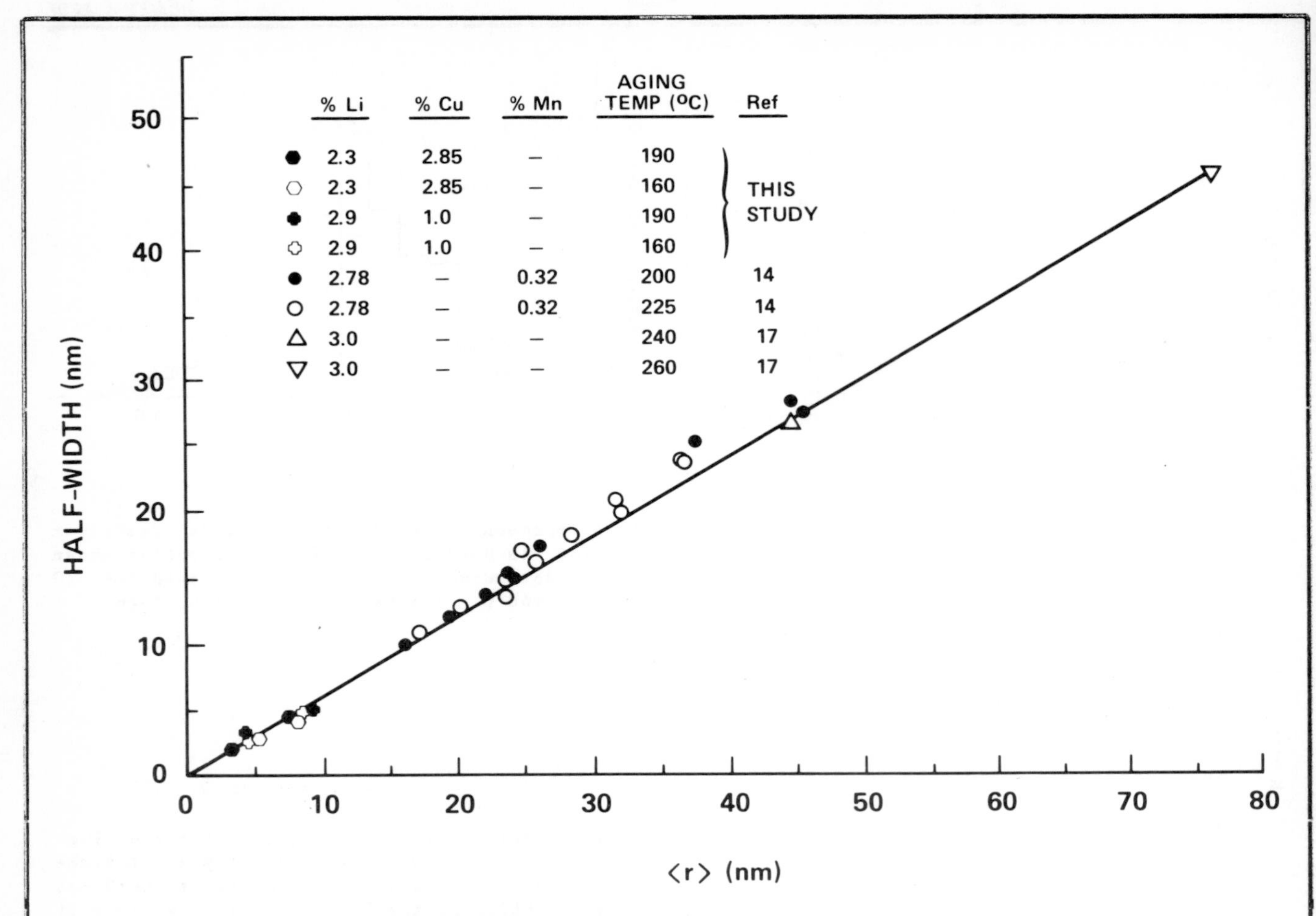

12 Illustrating the dependence of the half-width of the δ' particle size distributions on the average particle radius, ⟨r⟩, for a variety of Al-Li alloys.

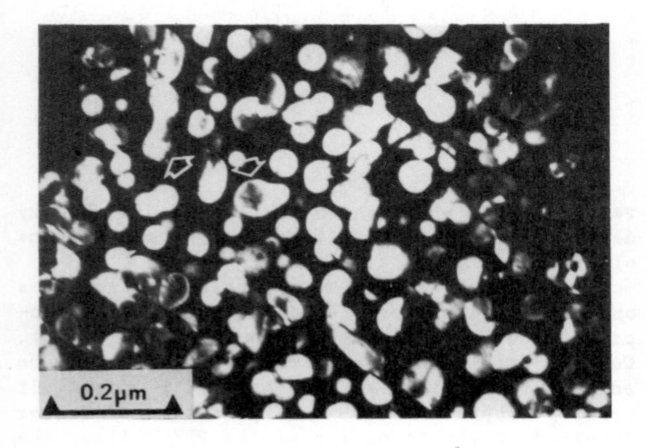

13 A dark-field TEM photograph taken using a δ' reflection in a sample of the 3-1 alloy aged at 190°C for 132 h. Coalescence at the particles indicated by the arrows is apparent.

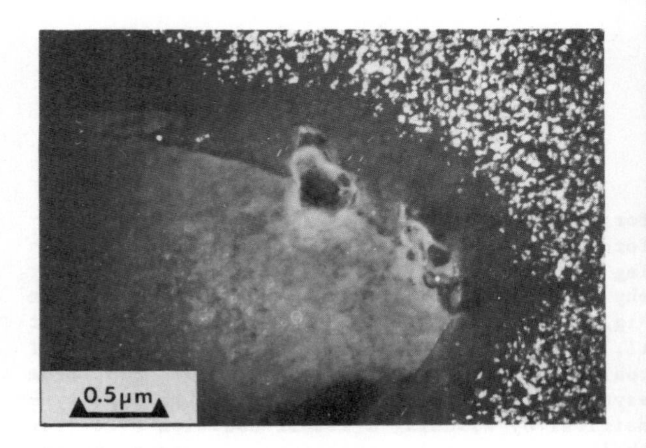

14 TEM dark-field image taken using a δ' superlattice reflection in a sample of the 2-3 alloy aged for 45 h at 190°C. The δ' PFZ is readily visible, as are δ particles at the high-angle grain boundary.

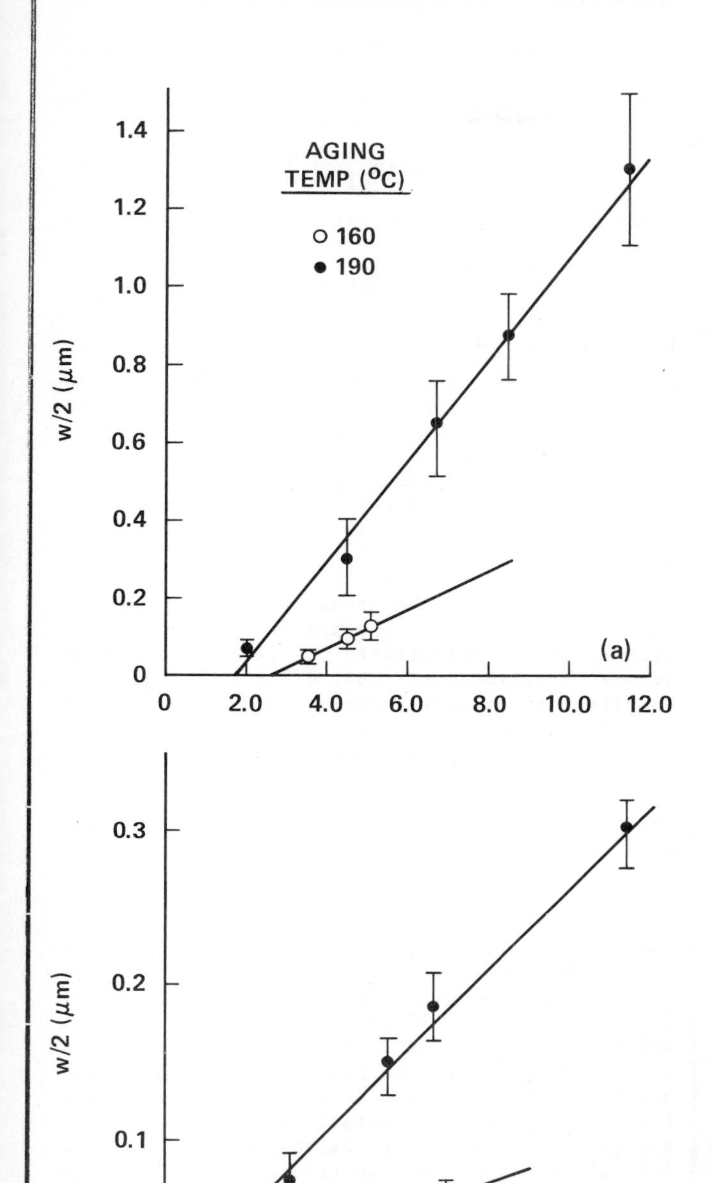

15 Illustrating the linear relationship between the PFZ width, w, and the square root of the aging time, t, for (a) the 2-3 alloy and (b) the 3-1 alloy.

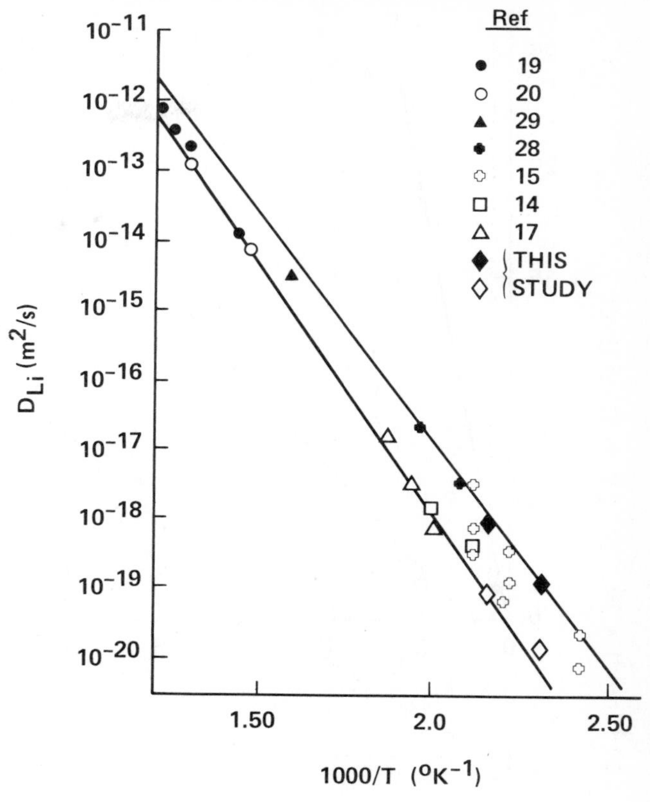

16 Arrhenius plot of various estimates and measurements of the diffusion coefficient of Li, D_{Li}, in Al-Li alloys.

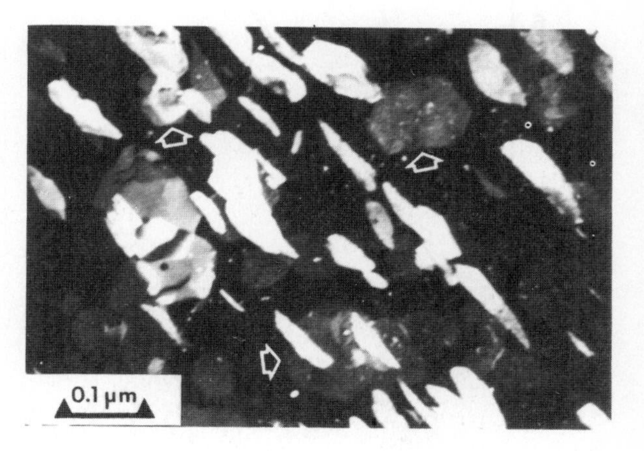

17 TEM dark-field image of inclined T_1 precipitates in the 2-3 alloy aged at 190°C for 132 h. Plate-shaped particles that are roughly hexagonal are indicated by the arrows.

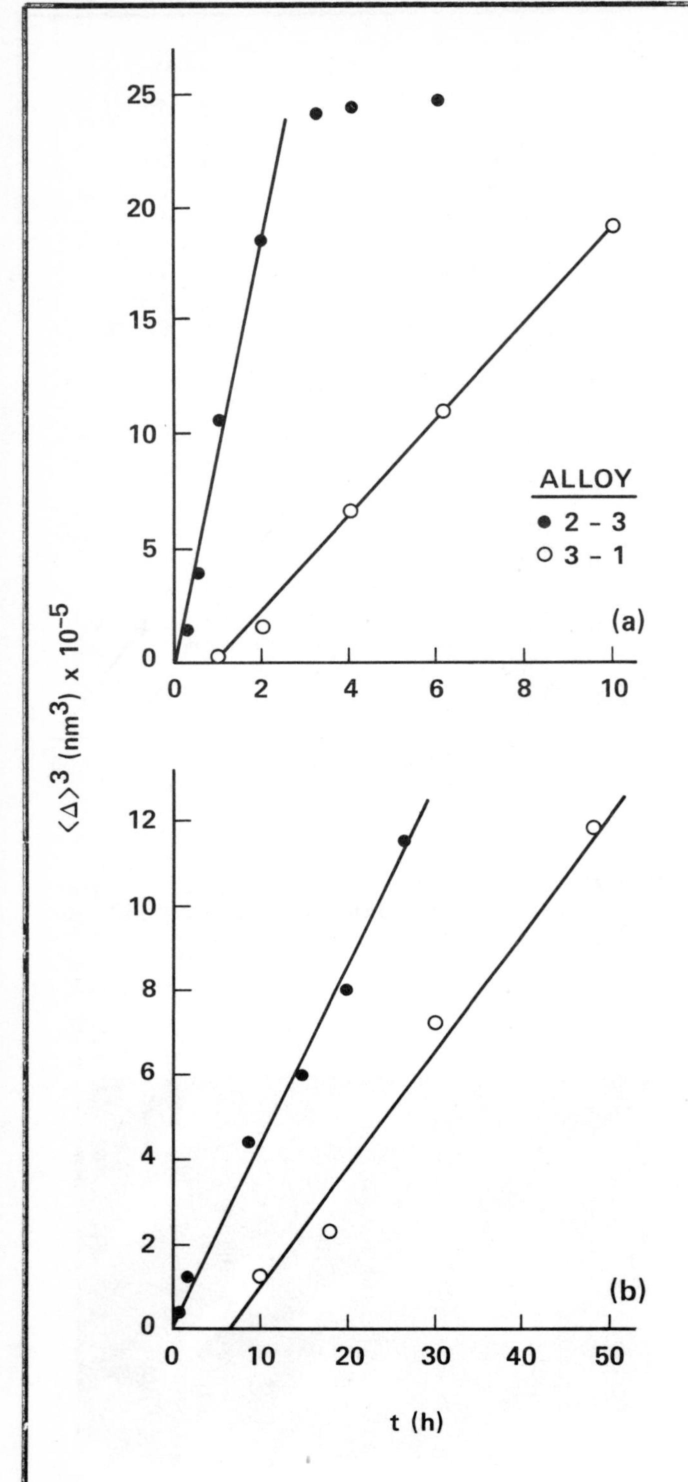

18 Plots of the cube of the average T_1 particle "diameter" $\langle\Delta\rangle$, as a function of aging time, t, for the two alloys aged at (a) 190°C and (b) 160°C.

absence at t = 0 suggests an incubation time for nucleation of the δ phase during aging of this alloy.

The kinetics of PFZ growth can be analyzed by the method of Jensrud and Ryum[17] to estimate the coefficient of diffusion of Li, D_{Li}, in the Al-rich matrix. They have shown that

$$\frac{w}{2} = 2\beta(D_{Li}t)^{1/2}, \quad (3)$$

where β is given by the solution of the equation

$$\frac{C_{\delta'} - C_\delta}{C_o - C_{\delta'}} = \pi^{1/2}\beta\exp(\beta^2)\mathrm{erf}(\beta). \quad (4)$$

In equation (4) $C_{\delta'}$ and C_δ represent the concentrations of Li in equilibrium with δ' and δ, respectively, and C_o is the Li concentration of the alloy. Due to the absence of accurate data on the phase equilibria the values of $C_{\delta'}$ and C_δ are not accurately known in the ternary alloys. It is thus not possible to obtain reliable values of D_{Li}, as has been done for the binary alloys. Nevertheless, by applying this analysis to the data in Fig. 15, tentative estimates were made by assuming that the difference $C_{\delta'} - C_\delta$ is the same as for the binary alloys, and by calculating $C_{\delta'}$ from the measured volume fractions of the δ' and T_1 phases. Despite the misgivings expressed, the values of D_{Li} for both alloys were found to lie in the same range of data on the binary alloys, as shown in Fig. 16.

Using the values of D_{Li} obtained from analysis of the data in Fig. 15, an attempt was made to estimate $\gamma_{\delta'-\alpha}$ from the formula

$$K = 8\gamma_{\delta'-\alpha}D_{Li}C_{\delta'}V_m^2/9RT, \quad (5)$$

where V_m is the molar volume of δ' and RT has its usual meaning. This analysis produces the values $\gamma_{\delta'-\alpha}$ = 0.0157 and 0.0185 J/m^2 for the 2-3 alloy and 0.132 and 0.140 J/m^2 for the 3-1 alloy aged at 160 and 190°C, respectively. The values of $\gamma_{\delta'-\alpha}$ for the 2-3 alloy are quite reasonable, since they are nearly equal to those found in the isostructural Ni-base γ/γ' alloys.[30-32] However the values of $\gamma_{\delta'-\alpha}$ for the 3-1 alloy are far too large. In our opinion none of them can be regarded with reliability until the values of $C_{\delta'}$ and C_δ are established definitively.

In binary Al-Li alloys $f_{\delta'}$ has been reported to increase substantially during aging, in some instances varying from 0.05 to 0.22 under isothermal aging conditions.[26] Similar, but not so dramatic, results have been observed in an alloy containing dilute concentrations of Cu and Mg.[33] In our alloys the increase in $f_{\delta'}$ was much smaller, varying from about 0.065 (10 h) to 0.093 (132 h) in the 3-1 alloy at 190°C (see Table II) but remaining essentially constant in the 2-3 alloy. We believe that the constancy of $f_{\delta'}$ in the more Cu-rich 2-3 alloy is partially due to the presence of the more thermodynamically stable T_1 phase, which grows at the expense of the δ' particles. This effect is smaller in the 3-1 alloy because the volume fraction of the T_1 phase is smaller than in the 2-3 alloy (these ranged from 0.01 to about 0.04 in the 2-3 alloy, and from 0.005 to 0.015 in the 3-1 alloy), hence T_1 consumes the δ' phase more slowly.

The T_1 particles were observed to be hexagonal in shape, as shown in Fig. 17, similar to the findings of Noble and Thompson.[3] The "diameters", Δ, of these plates were measured and are plotted in the form $\langle\Delta\rangle^3$ vs. t in Fig. 18. The relationship between $\langle\Delta\rangle^3$ and t is seen to be linear for most aging times. Only in the 2-3 alloy aged at 190°C (Fig. 18a) is there a deviation from this behavior, and in this case it was due to impingement of the large T_1 plates. However, we found that the number of T_1 precipitates per unit volume remained nearly constant, and that the thickness of the T_1 particles clearly increases, since the streaks sharpen with increasing aging time. This means that the volume fraction of T_1 continuously increases during aging, suggesting that Ostwald ripening is not responsible for the linear behavior observed during the early stages of T_1 growth.

Extrapolation of the curves in Fig. 17 to t = 0 indicates that the T_1 precipitates are present from the beginning of aging in the 2-3 alloy at both temperatures, while an incubation time exists for the nucleation of T_1 in the 3-1 alloy. The growth rate of T_1 in the 2-3 alloy is nearly an order of magnitude greater than that in the 3-1 alloy. We believe that this is responsible for the faster aging response of the 2-3 alloy (recalling that the kinetics of δ' growth are nearly identical in the two alloys). By implication, T_1 must contribute substantially to the strengthening of these alloys despite its small volume fraction. The contributions of the δ' and T_1 phases to the strengthening observed are currently under study.

CONCLUSIONS

1. The aging response of the 2-3 alloy is much faster than that of the 3-1 alloy. Its maximum hardness is also about 20% higher.
2. The θ' phase is observed only in the 2-3 alloy in samples aged for times up to the peak-aged condition.
3. The δ' phase grows by diffusion-controlled coarsening, with symmetrical particle size distributions. There is evidence for the coalescence of δ' precipitates. The activation energies for δ' coarsening are 127 and 112 kJ/mole for the 2-3 and 3-1 alloys, respectively. These values are consistent with Q for the diffusion of Li in Al-Li alloys. The values of $f_{\delta'}$ are nearly constant during aging of the 2-3 alloy, but exhibit a tendency to increase somewhat in the 3-1 alloy. Binary Al-Li alloys with the same Li content as the 2-3 and 3-1 alloys contain much more δ'.
4. The δ' PFZ is governed by Li solute depletion produced by the growth of δ at high angle boundaries. The PFZ width increases as $t^{1/2}$. Analysis of the kinetics of PFZ growth yield values of D_{Li} consistent with data taken from the literature.
5. The other major precipitate phase in both alloys is the T_1 phase. The existence of the T_1' phase is shown to be unlikely, since there is no need to invoke it. The T_1 particles are plate-shaped hexagons with "diameters" that increase approximately as $t^{1/3}$ during the earlier stages of aging. The volume fraction of the T_1 phase increases substantially during aging, however, suggesting that the kinetics of diametral growth are not governed by diffusion-controlled Ostwald ripening.

ACKNOWLEDGEMENTS

The authors are grateful to the Alcoa corporation for their financial support of this research. Discussions with Dr. R. J. Rioja of Alcoa have been particularly stimulating and enlightening.

REFERENCES

1. H. K. Hardy and J. M. Silcock: J. Inst. Met., 1955-56, **84**, 423.
2. J. M. Silcock: J. Inst. Met., 1959-60, **88**, 357.
3. B. Noble and G. E. Thompson: Met. Sci., 1972, **6**, 167.
4. K. Schneider and M. von Heimendahl: Z. Metallkde., 1976, **42**, 342.
5. T. V. Shchegoleva and O. F. Rybalko: Fiz. Metal. Metalloved., 1976, **42**, 322.
6. F. S. Lin, S. B. Chakrabortty and E. A. Starke, Jr.: Metall. Trans., 1982, **13A**, 401.
7. H. Suzuki, M. Kanno and H. Hayashi: J. Jpn Inst. Light Met., 1982, **32**, 88.
8. P. M. Kelly, A. Jostsons, R. G. Blake and J. G. Napier: Phys. Stat. Sol., 1975, **31**, 771.
9. E. E. Underwood: in 'Quantitative Stereology,' 1970, Reading, Mass., Addison Wesley.
10. M. H. Tosten and P. R. Howell: Private communication.
11. M. Avalos-Borja, P. P. Pizzo and L. A. Larson: 'Aluminum-Lithium Alloys II,' (ed. E. A. Starke, Jr. and T. H. Sanders, Jr.) p. 287; 1983, Warrendale, PA, AIME.
12. R. J. Rioja: private communication.
13. B. Noble and G. E. Thompson: Met. Sci., 1971, **5**, 114.
14. J. H. Kulwicki and T. H. Sanders, Jr.: 'Aluminum Lithium Alloys II,' (ed. E. A. Starke, Jr. and T. H. Sanders, Jr.) p. 31; 1983, Warrendale, PA, AIME.
15. D. B. Williams and J. W. Edington: Met. Sci., 1975, **9**, 529.
16. A. L. Berezina, L. N. Trifimova and K. V. Chuistov: Fiz. Metal. Metalloved., 1983, **55**, 543.
17. O. Jensrud and N. Ryum: Mater. Sci. Eng., 1984, **64**, 229.
18. P. L. Makin and B. Ralph: J. Mater. Sci., 1984, **19**, 3835.
19. L. P. Costas: U.S. Atomic Energy Commission Rep. DP-813, 1963.

20. C. J. Wen, W. Weppner, B. A. Boukamp and R. A. Huggins: Metall. Trans., 1980, **11B**, 131.
21. C. K. L. Davies, P. Nash and R. N. Stevens: Acta Metall., 1980, **28**, 179.
22. S. Takajo, W. A. Kaysser and G. Petzow: Acta Metall., 1984, **32**, 107.
23. D. J. Chellman and A. J. Ardell: Acta Metall., 1974, **22**, 577.
24. J. Th. M. De Hosson, A. Huis in't Veld, H. Tamler and O. Kanert: Acta Metall., 1984, **32**, 1205.
25. B. Noble, S. J. Harris and K. Dinsdale: Met. Sci., 1982, **16**, 425.
26. M. Tamura, T. Mori and T. Nakamura: J. Jpn Inst. Met., 1970, **34**, 919.
27. S. F. Baumann and D. B. Williams: Scr. Metall., 1984, **18**, 611.
28. D. B. Williams: 1974, Ph.D. Thesis, University of Cambridge.
29. D. B. Williams and J. W. Edington: Philos. Mag., 1974, **30**, 1147.
30. A. J. Ardell: Acta Metall., 1968, **16**, 611.
31. A. J. Ardell: Metall. Trans., 1970, **1**, 525.
32. P. K. Rastogi and A. J. Ardell: Acta Metall., 1971, **19**, 321.
33. P. J. Gregson and H. M. Flower: J. Mater. Sci. Lett., 1984, **3**, 829.

□

Identification of metastable phases in Al–Cu–Li alloy (2090)

R J RIOJA and E A LUDWICZAK

The authors are with the
Aluminum Company of America,
Alcoa Laboratories,
Alcoa Center,
PA 15069, USA

SYNOPSIS

Precipitation sequences for Al (rich)-Li-Cu alloys, as a function of the position of the alloy in a given phase field, are suggested. It is proposed that the metastable phases present in alloy 2090 aged to peak strength are: β', δ', T_1' and T_2'. The structures of these phases are characterized by X-ray and electron diffraction. Differential scanning calorimetry and electron microscopy revealed the typical morphologies of the strengthening phases and their regions of metastability. Since alloy 2090 is located in the $\alpha + T_2 + T_1$ phase field at aging temperatures, the T_1' and T_2' phases are considered to be precursors to the equilibrium T_1 and T_2 phases. A clustering reaction at low temperatures is documented via calorimetric analyses.

INTRODUCTION

The precipitation sequence in Al-Li-Cu alloys (on the Al-rich corner) depends on the composition of the alloy. In principle, the precipitation sequence can vary from that of the pure binary Al-Li,

$$\alpha_{ss} \dashrightarrow \alpha + \delta' \dashrightarrow \alpha + \delta \; (1)$$

to the sequence in binary Al-Cu (2), i.e.,

$$\alpha_{ss} \dashrightarrow \alpha_I + \alpha_{II} \dashrightarrow \alpha + \text{GP Zones} \dashrightarrow \alpha + \theta'' \dashrightarrow \alpha + \theta' \dashrightarrow \alpha + \theta.$$

In addition to the equilibrium phases δ and θ, the equilibrium phases T_B, T_1 and T_2 have been determined in Al (rich) -Cu-Li alloys (3); therefore, in principle, other phase separation reactions are possible. Since, as will be discussed later, there is evidence for the T_B', T_1' and T_2' phases, the following sequence of precipitation reactions is here suggested for the binaries and pseudo-binary lines:

$$\alpha_{ss} \dashrightarrow \alpha + \delta' \dashrightarrow \alpha + \delta$$

$$\alpha_{ss} \dashrightarrow \alpha + \delta' + T_2' \dashrightarrow \alpha + T_2$$

$$\alpha_{ss} \dashrightarrow \alpha + \delta' + T_1' \dashrightarrow \alpha + T_1$$

$$\alpha_{ss} \dashrightarrow \alpha + T_B' \dashrightarrow \alpha + T_B$$

$$\alpha_{ss} \dashrightarrow \alpha_I + \alpha_{II} \dashrightarrow \alpha + \text{GP Zones} \dashrightarrow \alpha + \theta'' \dashrightarrow \alpha + \theta' \dashrightarrow \alpha + \theta$$

These reactions are valid for two phase fields in the phase diagram. For three phase fields, the reactions present could be a mixture of those in adjacent fields (see Figure 1a); e.g., alloy 2090 at low temperatures is located in a three phase field, i.e., $\alpha + T_1 + T_2$, therefore, the precipitation sequence could be of the form:

$$\alpha_{ss} \dashrightarrow \alpha + \delta' + T_1' + T_2' \dashrightarrow \alpha + T_1 + T_2$$

Note that the δ' phase is present in the T_1 and T_2 reactions, but it is absent in the T_B reactions. This is based on experimental observations.

There are different interpretations in the literature with regards to phase identification of metastable phases and precipitation sequences, e.g., in pure binary alloys, although it is commonly accepted that the first phase to form is δ' ($L1_2$ structure), it has been suggested that other phases form prior to δ' (4) vis.,

$$\alpha_{ss} \dashrightarrow \alpha + \text{GP Zones} \dashrightarrow \alpha + \text{ordered GP Zones} \dashrightarrow \alpha + \delta \dashrightarrow \alpha + \delta$$

Furthermore, recent cluster variation computations of the Al-Li phase diagram predict the presence of a miscibility gap below the δ' solvus (5).

The structure of δ' is cubic, with the $L1_2$ superlattice of the Cu_3Au prototype with a = 4.01 Å and a lattice mismatch = -0.08% (1). This structure is often referred to as Al_3Li. Although δ' forms as spherical precipitates, it also forms composite precipitates wetting the surface of Al_3Zr precipitates and the coherent surface of plate-like precipitates along the traces of the (100) planes. The orientation relationship of δ' to the parent is a cube/cube orientation. In addition to the preceding morphologies of δ', it has been proposed that δ' attains a lamellae shape when formed discontinuously. The surface energy of δ' has been determined to be α = 0.014 J/m^2 via capillarity considerations. The δ phase has a cubic superlattice B32 of the NaTl prototype with a = 6.356 Å and the following orientation relationship (6, 1):

$$(100)_\delta \parallel (110)_{Al} \text{ and } [011]_\delta \parallel [1\bar{1}1]_{Al}$$

It has been suggested that two types of defects can account for the excursions in stoichiometry of this phase, namely, vacancies in the lithium sublattice and lithium antistructure atoms in the aluminum sublattice.

Suzuki, Kanno and Hayashi (7) have suggested that the T_B' and T_1' phases form in high purity Al-4%Cu-1%Li alloys. These suggestions were made on the basis of superlattice spots in diffraction patterns that could not be explained in terms of the θ' and T_1 phases. No structural analysis of T_B' and T_1' was given. However, it was suggested that T_B' and T_1' were precursors to the T_B and T_1 phases.

Schneider and Heimendahl (8) working on an Al-4.17%Cu-1.01%Li-0.55%Mn-0.23%Cd alloy concluded that the predominant metastable phase was θ' in the temperature range from room temperature to 350°C. These authors argue that incorporation of Li into the θ' structure gives rise to the T_B phase with the CaF_2 structure (a = 5.83 Å, T_B = $Al_{7.5}Cu_4$ Li). Prior to the formation of the θ' phase, the θ" phase was resolved. Extra reflections in diffraction patterns (near 1/3 (020) positions in reciprocal space) were attributed to forbidden reflections of θ' probably due to the incorporation of Cd in the θ' structure. Schneider and Heimendahl speculated that at aging times shorter than those studied, a competing process between the metastable predecessors to T_B and T_1 is involved. However, no structural discussion was presented.

Silcock (9) argued that Al-Li-Cu alloys along the pseudo-binary line Al-T_B could precipitate a structure preceding the T_B phase which could be indistinguishable from θ'. This structure could be formed if Li atoms replace Al atoms. At higher Li contents, the θ' spots developed tails into the direction of the T_B spots. Miss Silcock interpreted this as a continuous transition from θ' to T_B. This phase was denoted as θ_B'. Miss Silcock did not observe transition phases to the T_1 or T_2 phases. Noble, McLaughlin and Thompson, (10), using electrical resistivity measurements, concluded that in Al-4.5%Cu alloys with 0.4% and 1.5%Li there are two kinds of clustering phenomena prior to the formation of θ' precipitates. It was suggested that the temperature range between 0-50°C, GP zones (similar to Al-Cu alloys) consisting of Cu atoms formed. Furthermore, between 80°C-130°C GP zones consisting of Cu and Li atoms were believed to be responsible for irregular changes in resistivity.

Avalos-Borja, Pizzo and Larson (11) studying an Al-2.6%Li-1.4%Cu suggested that a precursor to the T_1 phase and δ' were the phases formed during aging, having heterogeneous and homogeneous distributions, respectively.

In summary, it has been suggested that several metastable phases form in Al-Li-Cu alloys. These phases are: δ', GP zones, θ", θ', T_B', T_1' and T_2'. The structures of T_B', T_1' and T_2' have not been resolved.

The equilibrium phases relevant to the Al rich corner of the ternary Al-Li-Cu phase diagram are: T_1, Al_2CuLi, which have an hexagonal crystal structure with a = 4.97 Å and c = 9.34 Å (3). The following orientation relationships have been proposed (12):

$$(0001)_{T_1} \parallel (111)_{Al}$$

$$(10\bar{1}0)_{T_1} \parallel (1\bar{1}0)_{Al}$$

$$(11\bar{2}0)_{T_1} \parallel (2\bar{1}\bar{1})_{Al}$$

The space group of this phase has not been determined. Although the following space groups have been proposed (3) P622, P6mm and P6/mmm. The T_2 phase, Al_5Li_3Cu, has been reported to be cubic isomorphous with $Mg_{32}(ZnAl)_{49}$, a = 13.914 Å, space group Im3 (13).

The T_B phase $Al_{7.5}LiCu_4$ is reported to be cubic (3) with the CaF_2 structure and a = 5.83 Å. The following orientation relationship has been proposed:

$$[100]_{T_B} \parallel [110]_{Al};\ [001]_{T_B} \parallel [001]_{Al}$$

The objective of this work is to document the morphology and diffraction effects associated with the strengthening phases present in alloy 2090.

EXPERIMENTAL PROCEDURE

Transmission Electron Microscopy (TEM) was performed in a Phillips 420 electron microscope operating at 120 kV. Selected Area Electron Diffraction Patterns (SAEDP) were taken in three orientations, vis., (100), (110) and (112). Standard imaging techniques were employed, i.e., bright fields (BF) and superlattice dark fields (DF). Thin foils were prepared by the double jet electropolishing technique in an electrolyte

TABLE I

LATTICE SPACINGS OF THE T_1' PHASE

Orthorhombic structure (Pt_2Mo prototype) with:

a = 2.876 A
b = 8.6 A
c = 4.06 A

d Calculated	Planes	d Measured	Relative visual Intensity
4.30	(020)	4.30	Medium
3.67	(011)	3.68	Medium
2.72	(110)	2.72	Weak
2.34	(101, 031)	2.34*	
2.15	(040)	2.15	Strong
2.06	(121)	2.06	Weak
2.03	(002, 130)	2.03*	
1.83	(022)	1.83	Weak
1.62	(112)	1.62	Weak
1.58	(141,051)	1.58	Weak
1.47	(150, 042)	not resolved	
1.43	(200, 132)	1.43*	

Ratings of the relative visual intensities were preformed among the set of reflections of T_1'. Note that the intensity of the strong T_1' line is weak compared to the lines of δ'.

* Indicates that diffuse scatter from Al lines is present at those spacings.

containing 3/4 methanol + 1/4 nitric acid at -25°C.

Differential Scanning Calorimetry (DSC) was performed in a Perkin Elmer DSCII calorimeter. Samples (45 mg) subjected to different heat treatments were heated at a heating rate of 20°C/min from room temperature to the solidus temperature. After the initial run, samples were re-run. The re-run signal was plotted as a dotted curve.

X-ray Guinier-de Wolff analysis was conducted using a quadruple focusing diffraction camera with a 57.3 mm film cylinder radius mounted on a Philips x-ray generator. CuK_{α_1} radiation was obtained through the use of a graphite crystal incident beam monochromator attached to an Amperex 1.4 kw Cu target tube. Kilovolt and milliampere levels used were 45 and 20, respectively. Exposure times of 10 hs were found to give consistent results.

Alloy 2090 has the following nominal composition:

2.7%Cu, 2.2%Li, 0.12%Zr, 0.08%SiMax, 0.12%FeMax

All percentages are given in weight percent.

EXPERIMENTAL RESULTS AND DISCUSSION

TEM analyses from aged samples reveals the presence of four metastable precipitates as shown in Figure 1b. In addition to the δ' phase, which exhibits predominantly a spherical morphology, two plate-like precipitates are resolved; one along the traces of the {100} planes, and the other along the traces of the {111} planes. Furthermore, composite precipitates consisting of an envelope of δ' on the surface of Al_3Ar (α') are also resolved. Figure 2 shows diffraction effects from these phases in three orientations, vis., (100), (110) and (211). The first phase to discuss is the one with a plate-like shape and {111} habit planes.

This phase is generally considered to be the T_1 phase (12). Diffraction patterns from this phase exhibit streaking along the {111} directions and superlattice spots at 1/3 and 2/3 of (220) matrix positions in reciprocal space. Figure 2 shows these diffraction effects. However, observed diffraction patterns in the (211) orientation exhibit the presence of diffraction spots at 1/3 and 2/3 of (131) matrix positions. The maxima in intensity at 1/3 and 2/3 (131) cannot be explained in terms of the T_1 structure. Dark field images from 1/3 (131) and 1/3 (113) spots reveal two different variants of precipitates along the {111} traces. Furthermore, DF images from spots at 1/3 (022) show a third variant. These diffraction effects and imaging results can be explained in terms of an orthorhombic structure with the symmetry of the Pt_2Mo prototype. This structure exhibits the following lattice parameters, a = 2.87 A, b = 8.6 A, c = 4.06 A and the following crystallographic orientation relationship with the matrix:

(110)m // (010)p : and [100]m // [001]p.

Figure 3 shows the Pt_2Mo structure and the (100) and (211) indexed diffraction patterns expected from the proposed crystallographic relationship between precipitates and parent.

X-ray Guinier-de Wolff analyses revealed diffraction lines consistent with the computed lattice spacings for the Pt_2Mo structure. Table I shows a comparison of computed and measured lattice spacings. Since the position of Li atoms is not known, the structure factor was estimated placing Cu atoms at (0,0,0) and (1/2,1/2,1/2) and Al atoms at (1,1/3,0), (1/2,1/6,1/2), (1/2,5/6, 1/2) and (1,2/3,0) positions.

Since the plate-like precipitates on the {111} planes have a structure different from that of the T_1 phase, these precipitates will be called hereinafter the T_1' phase. The T_1' phase was observed to precede the T_1 phase. The first indication of the T_1 phase was resolved via the development of faint <111> streaking on the (211) DP during prolonged agings. Maxima in intensity appears on the streaks at positions where the T_1 spots should be present. Figure 4 shows the transition in (211) DP from agings in which only T_1' spots are present to agings in which only T_1

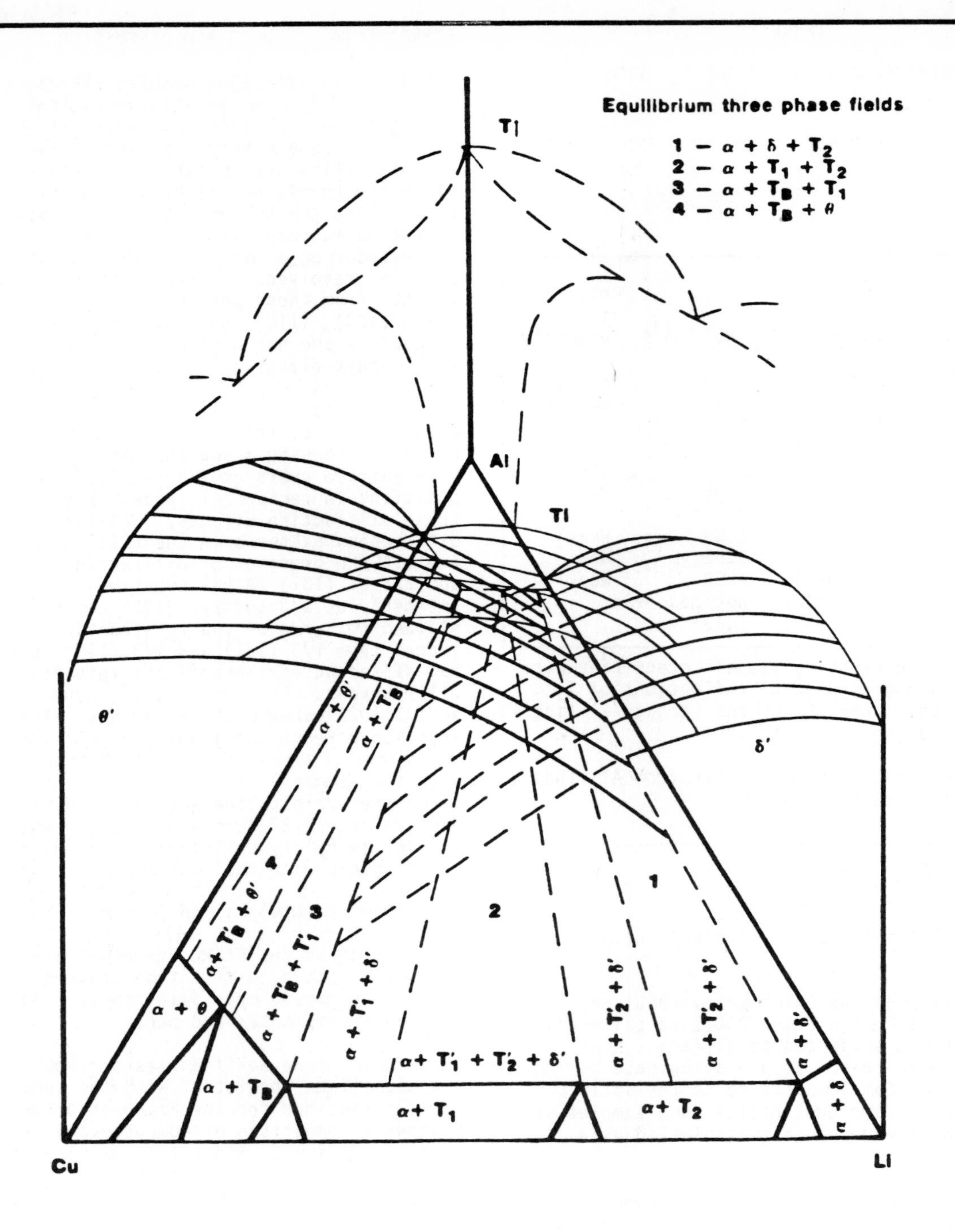

Figure 1a - Proposed regions for the formation of different metastable phases in Al-Li-Cu alloys. Solid lines inside the Al-Cu-Li triangle represent a "high temperature" isotherm. The dotted lines represent a "low temperature" isotherm. In a three-dimensional projection, the solvi for δ', T_1, and T_2' (θ'-like), are depicted as sheet boundaries terminating in certain phase fields.

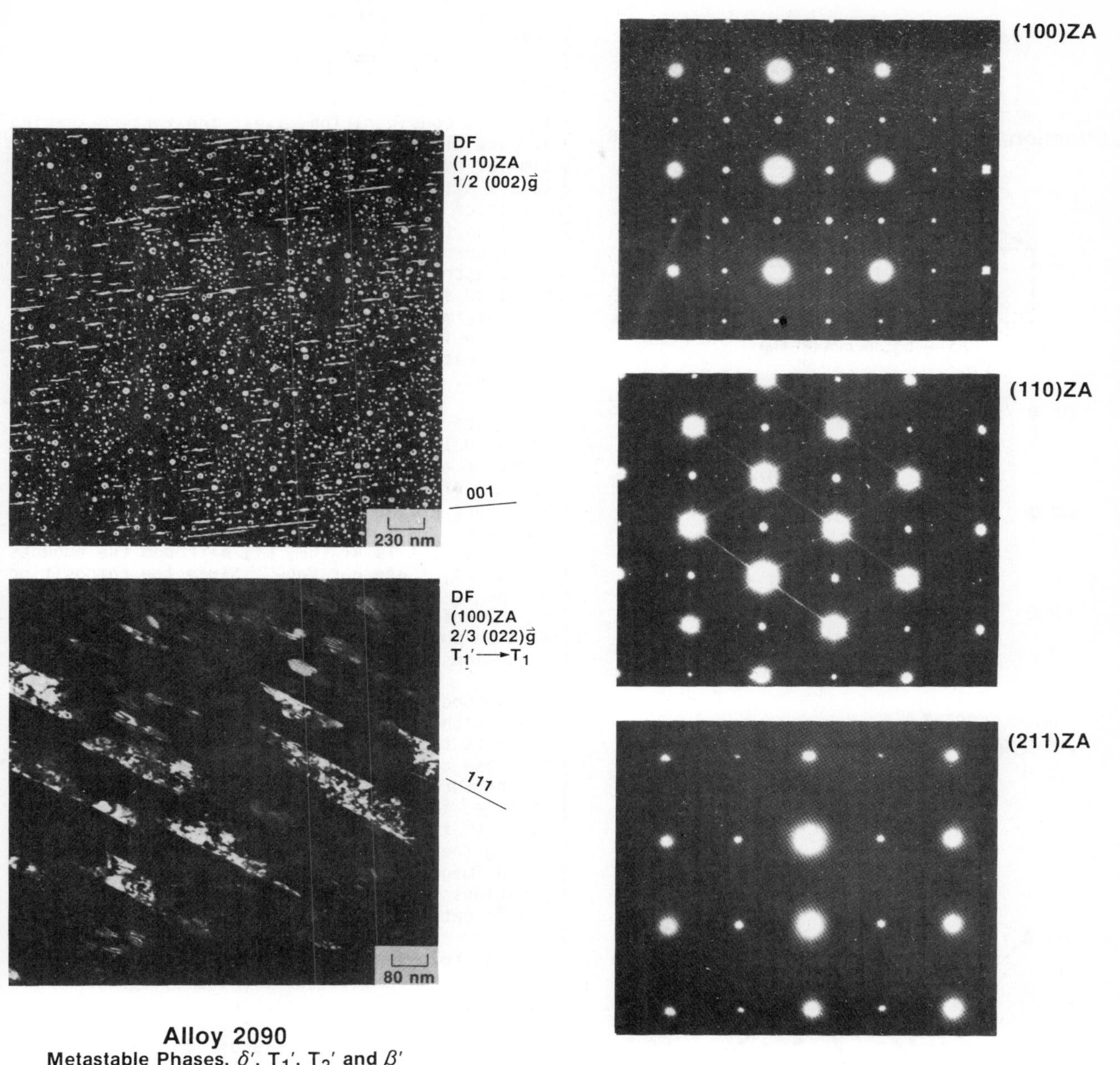

Alloy 2090, T8 Temper

Figure 1b - TEM micrograph of Alloy 2090 aged 72 hrs at 160°C
A - Note the δ' phase with three morphologies, i.e., as spherical precipitates, wetting spherical β' (Al_3Zr) precipitates and wetting the surface of plate-like T_2' precipitates.
B - Note elongated T_1 plates and small equiaxial T_1' precipitates. SADP at this aging revealed a mixture of T_1' and reflections in (211) ZA.

Figure 2 - SADP Alloy 2090 aged 24 hs at 160°C. Note <111> streaks in the (110) ZA. Also note T_1' reflections in the (211) ZA.

Diffraction From the Parent and the T_1' Phase

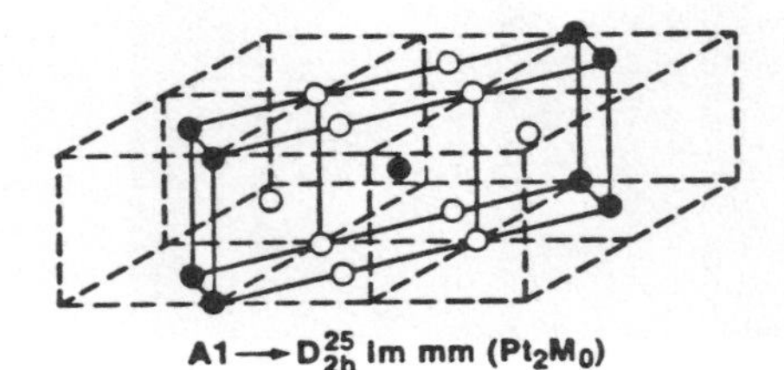

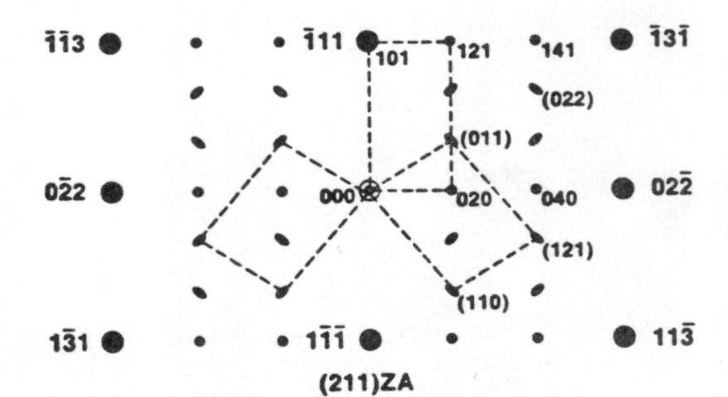

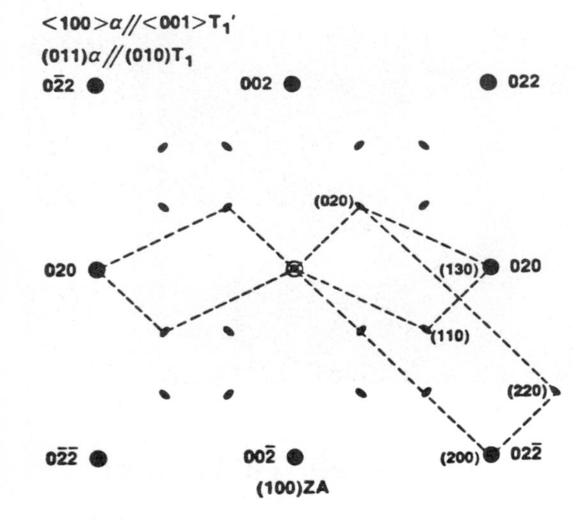

Figure 3 - Schematic diagram showing the Pt_2Mo structure oriented with the f.c.c. matrix and the resulting (100) and (211) diffraction patterns. Note that different variants can explain all superlattice reflections.

spots are observed. Dark field images from samples exhibiting a mixture of T_1' and T_1 spots revealed a transition in the morphology of precipitates from plate-like to laths. This change in shape could be due to the structural transition from T_1' to T_1. A possible reason for the {111} habit plane for the T_1' phase could be a misfit consideration, i.e., the lattice spacing of pure Al in the (111) planes (2.337 A) is similar to the spacing of the (031) T_1' planes (2.341 A). This difference in spacing implies a misfit of 0.17%. Furthermore, (111) Al is parallel to (031) T_1' under the proposed orientation relationship; however, the lattice parameter of the constrained matrix with Cu and Li in solid solution should be measured to compute an accurate misfit. This measurement was not possible in the present work due to the broadness of the matrix reflections in Guinier-de Wolff analyses.

Formation of the T_1 phase on (111) matrix planes or on (031) T_1' planes can also be explained in terms of minimum misfit arguments since the spacing of the (0004) T_1 planes is 2.333 A. The structure proposed for the T_1' phase is of the A_2B type. This implies that a possible composition for T_1' is Al_2(Cu, Li) although the amounts of Cu and Li are not know. Since the composition of the T_1 phase (Al_2CuLi) appears to be similar to that of T_1', a transition from T_1' to T_1 could also be possible without requiring large excursions in chemistry.

The second phase to discuss is the plate-like phase with {100} habit planes. This phase exhibits the morphology and habit planes of the θ' phase commonly observed in Al-Cu alloys. Furthermore, the (011) θ' reflections are readily observed in diffraction patterns. However, the (002) θ' reflections are absent in alloy 2090 during the early stages of precipitation. Dark field images obtained from 2/3 (020) matrix positions in reciprocal space fail to image these precipitates.

During prolonged agings or reversion experiments, the (002) θ' reflections eventually appear in diffraction patterns. Figure 5 shows results from imaging and diffraction analyses that illustrate the previous discussion. Note that the δ' phase generally wets the phase with the (100) habit planes. X-ray Guinier-de Wolff analyses fails to detect θ' reflections on alloy 2090 aged up to peak strength. Absence of the (002) θ' reflections at 2/3 (002) matrix positions could be due to incorporation of Li into the θ' defect structure. Since the atomic scattering factors increase in the order Li, Al, Cu, in principle, substitution of Li in the (001) and (002) planes of θ' could result in a structure factor (F(002)) for θ' equal, or close to, zero. If Li acts as a substitutional element in the structure of the θ' phase, the resulting structure would have a different chemistry than Al_2Cu, and therefore constitute a different phase. From this discussion it is possible to speculate that a θ'-like phase with Li incorporated into the θ' structure could be a precursor

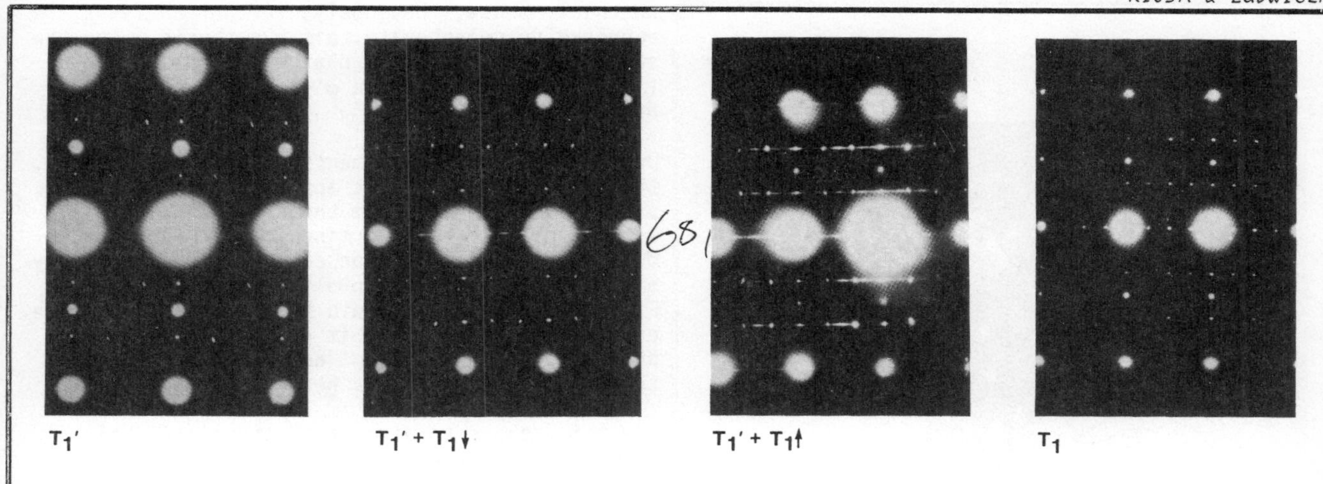

Alloy 2090
The $T_1' \longrightarrow T_1$ Transition, (211)ZA

Figure 4 - SADP showing the transition from T_1' to T_1 precipitates upon aging. Note the development of maximum in intensity on <111> streaks which eventually leads to discrete T_1 spots.

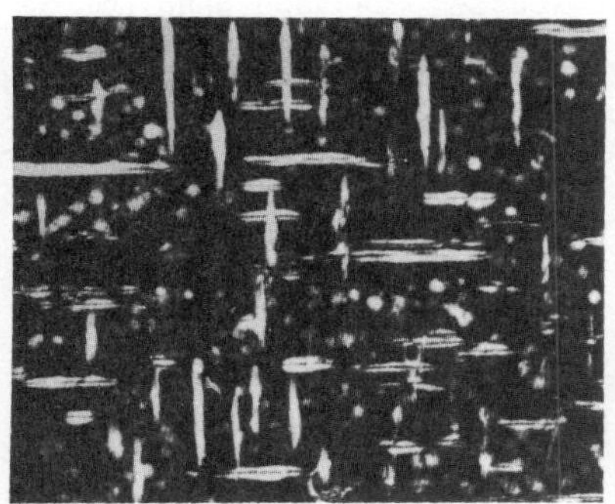

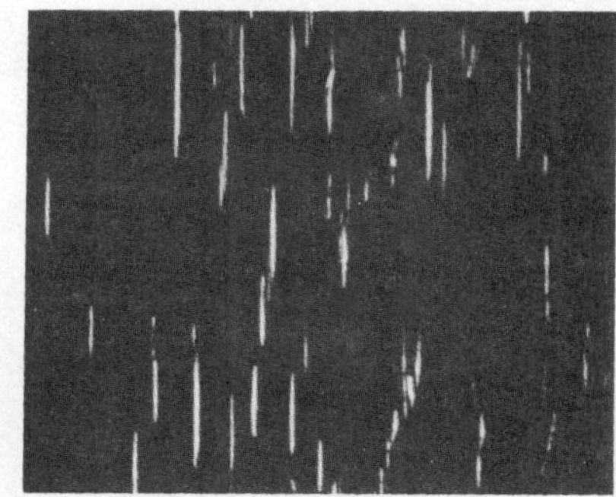

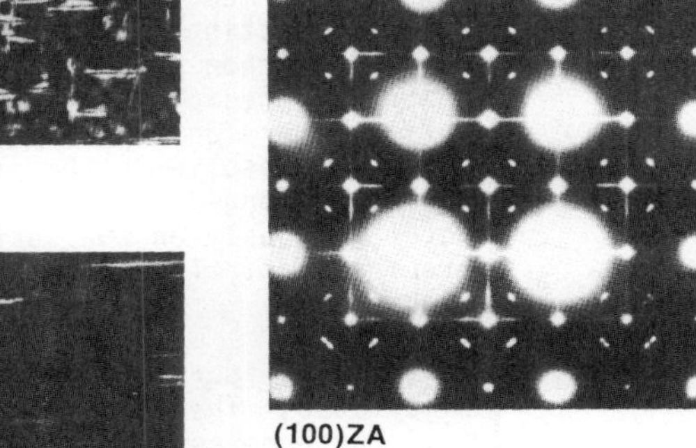

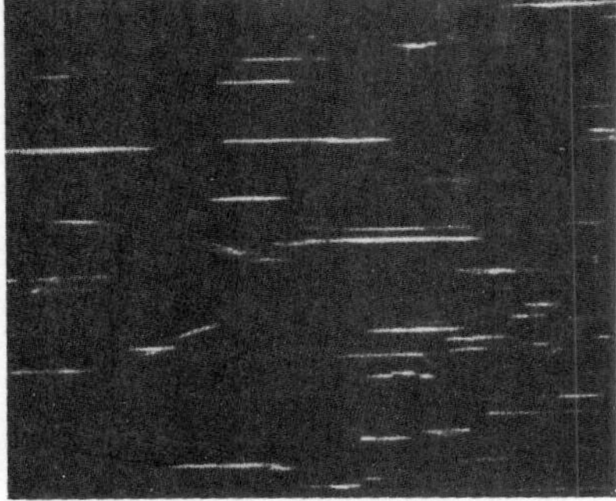

Figure 5 - TEM micrographs from alloy 2090 aged 24 hs at 160°C followed by a reversion treatment of 10 min at 200°C. Note spherical and plate-like δ' precipitates. The δ' phase with a plate-like morphology is wetting the coherent surface of the T_2' phase. Note also that the T_2' phase is not imaged with dark fields obtained at 2/3 (020) $\vec{g}$.

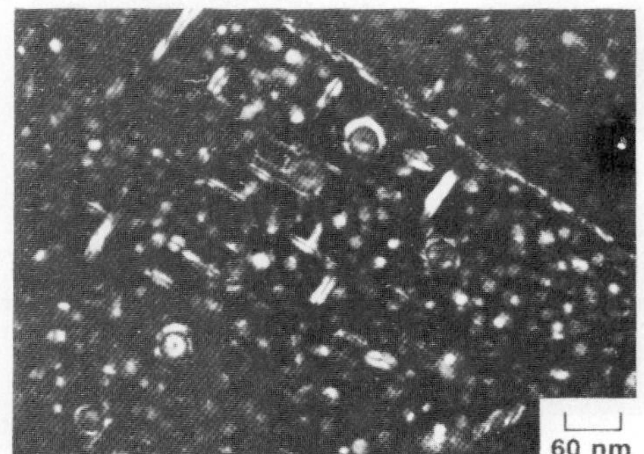

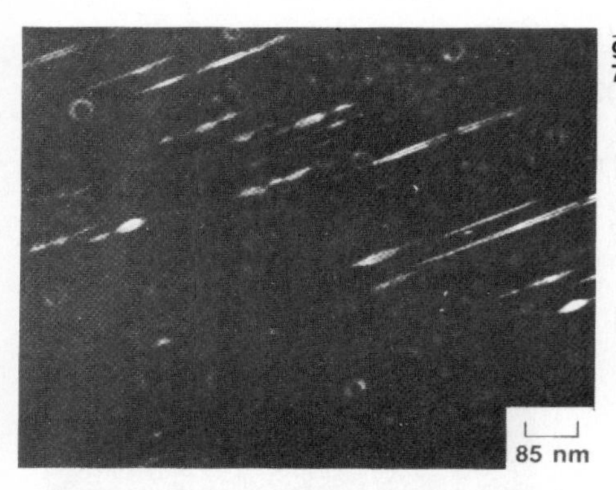

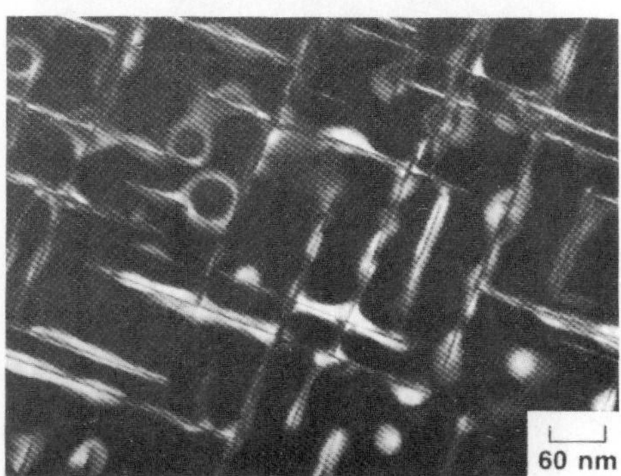

δ' Wetting on T_2'

Agings at 163°C, δ' DF

Figure 6 - TEM micrographs from alloy 2090 aged at 160°C showing the evolution of the δ' wetting on T_2' and β'. Note that coarsening of δ' takes place heterogeneously at interfaces. This phenomenon will be termed heterogeneous coarsening.

to the T_2 phase. Obviously, further analyses are required to corroborate this hypothesis. For the sake of discussion, the plate-like phase with (100) habit planes and a θ'-like structure will here be called T_2'.

During reversion experiments or prolonged agings, the (002) θ' reflections appear in diffraction patterns. However, from these observations, it is not possible to determine if this behavior is due to volume fraction or change in the chemistry effects. Future work involving the use of the Field Ion Atom Probe would help to resolve if the phase with the {100} habit planes is a ternary phase isomorphous with θ' acting as a precursor to the T_2 phase as it is here suggested.

The third phase to discuss is the δ' (Al_3Li) phase. This phase is present in aged samples from alloy 2090 with the following three different morphologies:

Spherical precipitates uniformly distributed throughout the matrix. This morphology constitutes the majority of the δ' precipitates. δ' is also present as an envelope surrounding Al_3Zr precipitates with the $L1_2$ structure. The intensity of the Al_3Zr precipitates vanishes as they are wetted by δ'. This phenomenon has previously been discussed (14) and need not be discussed here. However, note that the equilibrium Al_3Zr phase, called the β phase, has the DO_{23} structure. Therefore, in this paper, the metastable Al_3Zr phase with an $L1_2$ structure will be called the β' phase. This phase will be considered the fourth metastable phase present in alloy 2090.

Finally, as mentioned previously, δ' is also observed wetting the surface of the T_2' phase. This phenomenon gives rise to the formation of composite precipitates along the {100} planes. During reversion experiments the last morphology of δ' to dissolve is the plate-like shape, and during prolonged agings δ' is observed to coarsen heterogeneously on the surface of the T_2' precipitates. These observations are shown in Figure 6.

In summary, four metastable phases are present in alloy 2090. These phases are δ', T_1', T_2' and β'. These metastable phases are suggested to be precursors to the equilibrium T_1, T_2 and β phases.

DSC analyses from samples in the "as quenched" condition revealed the presence of several exothermic and endothermic reactions as shown in Figure 7. Endothermic reactions are considered to be due to dissolution of precipitates, whereas exothermic reactions are due to precipitation or recrystallization. To resolve the phases giving rise to the different endothermic reactions, samples were annealed 2 hs at several temperatures prior to DSC and TEM analyses.

As quenched samples exhibited diffuse $L1_2$ reflections and diffuse scatter around the matrix reflections in electron diffraction patterns.

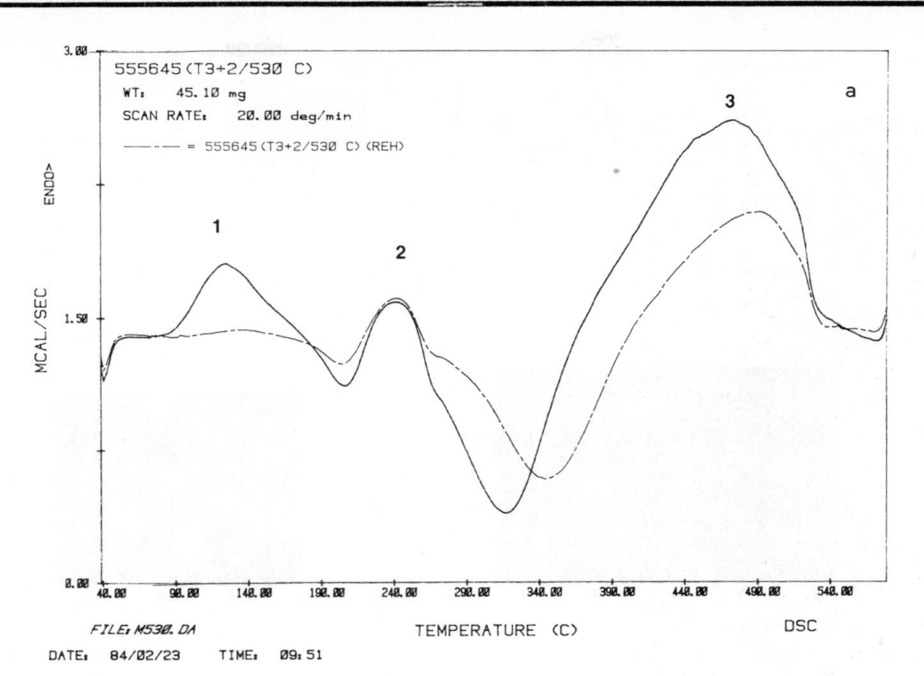

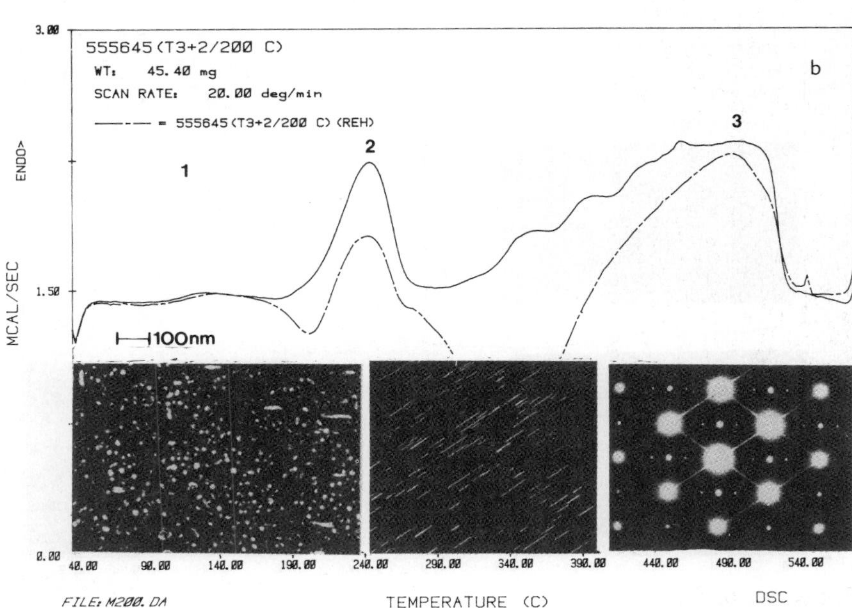

Figure 7 - DSC analyses from alloy 2090
A - Corresponds to an "as quenched" sample. Note 3 endothermic reactions. The dotted curve corresponds to a re-run of the same sample.
B - Shows the microstructure present after aging 24 hrs at 200°C, and the DSC signature resulting from this heat treatment. Note the disappearance of reaction 1, and the appearance of endothermic reactions near 360°C. The phases present prior to the DSC run are δ' and T_1'.

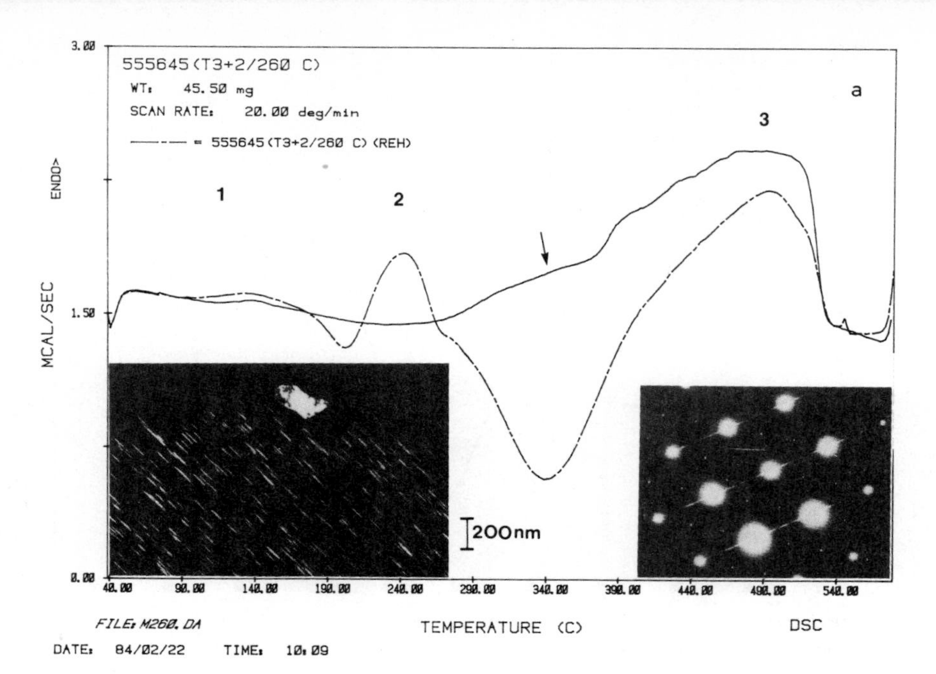

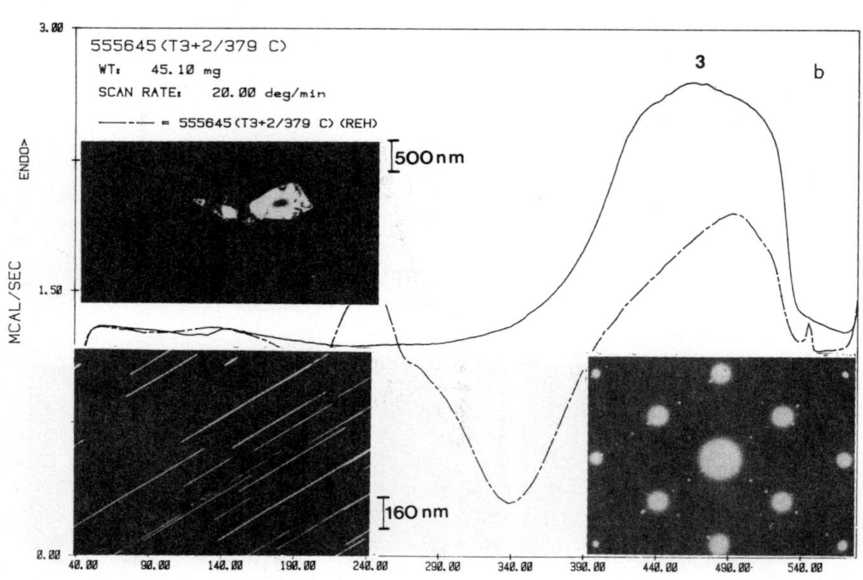

Figure 8 - DSC analyses from alloy 2090.

A - Corresponds to a sample aged 2 hrs at 260°C. The phases present prior to the DSC run are T_1' and T_2 at gram boundaries as shown in the TEM micrograph. Note the absense of reactions 1 and 2, and the presence of an endothermic reaction marked by an arrow. This reaction could be due to dissolution of the T_1' phase.

B - Shows the microstructures developed by aging 2 hrs at 379°C. The phases present are T_1 and T_2. The DSC run shows only endothermic reaction 3.

Aging treatments at temperatures above the first endothermic reaction, but below the second reaction (200°C), resulted in the growth of δ', and T_1' precipitates. Furthermore, the heat evolved by the second endothermic reaction increased and the third endotherm developed two bumps as shown in Figure 7B.

Agings above the second endotherm (260°C) resulted in the growth of the T_1' phase within the grains and the T_2 phase at the grain boundaries. Furthermore, neither the second endotherm was resolved in DSC runs nor the δ' phase was observed in TEM analyses. These results suggest that the second endothermic reaction is due to the dissolution of the δ' phase. Also, it appears that at 260°C the aging treatment is located above the δ', but below the T_1' solvi.

X-ray analyses from samples aged at 200°C and 260°C revealed greater intensity of the T_1' reflections at 260°C than at 200°C. This is consistent with the TEM observations (see Figure 8A).

Finally, agings at temperatures above the endothermic reaction (379°C) (bump in the third reaction) which develops concomitant to the appearance of T_1' precipitates, yielded microstructures containing the equilibrium T_1 and T_2 phases (Figure 8B).

Results from DSC analyses can be interpreted in the following manner:

The first endothermic reaction could be due to dissolution of solute clusters, concentration waves resulting from spinodal decomposition or small δ' precipitates formed during quenching. However, since δ' precipitates give rise to the second endotherm, a bimodal distribution of δ' would have to be invoked to explain the first endotherm. Furthermore, the Gibbs-Thompson effect would have to account for an excursion in the δ' solvi of 190°C as suggested by DSC results.

Note in Figure 7A that the samples slowly cooled do not show the first endotherm. This indicates that vacancies retained during the quench accelerate the kinetics of clustering at room temperature. The nature of this clustering reaction needs to resolved. In principle, spinodal decomposition could also be responsible for the observed clustering phenomenon. The temperature at which the maximum rate of dissolution of concentration waves occurs (133°C), would be the temperature at which the fastest growing wavelength at room temperature goes into solution. Furthermore, spherical diffuse scatter would indicate that decomposition takes place in an isotropic crystal. At this stage, all explanations seem possible, and further work is needed to resolve the phase that dissolves and gives rise to the first endothermic reaction.

The second endotherm is clearly due to dissolution of the δ' phase. Determination of the δ' solvus is complicated by the presence of different sizes and morphologies of δ'. DSC results indicate that dissolution of δ' is taking place between 200°C and 260°C. Reversion experiments for 2 hs, via hardness and TEM analyses, indicate that the δ' solvus is located near 200°C. Reversion experiments for short times yield δ' solvi which are size dependent.

The third endothermic reaction present in as quenched samples is due to dissolution of the equilibrium phases. This reaction is preceded by an exothermic reaction which is probably due to the growth of the T_1 and T_2 phases. The endothermic reaction due to dissolution of the T_1' phase in as quenched samples is hidden by the exothermic reaction previously mentioned. However, when the T_1' phase is grown during aging, the exothermic reaction disappears and an endothermic reaction near 260°C develops. This behavior indicates that the T_1' phase has a distinct solvus.

No distinct exothermic reaction could be ascribed to dissolution of the T_2' phase.

In summary, results from DSC analyses resolve the presence of a clustering reaction occurring during quenching or at room temperature. Furthermore, regions of metastability for δ' and T_1' can be defined in DSC curves from aged samples.

CONCLUSIONS

A precipitation sequence for alloy 2090 at 160°C of the following form is suggested:

$$\alpha_{ss} \dashrightarrow \alpha + \delta' + T_2' + T_1' \dashrightarrow \alpha + T_1 + T_2$$

T_1' and T_2' precipitates are plate-like with {111} and {100} habits, respectively.

δ' precipitates are observed with three morphologies which are; spherical, plate-like wetting the coherent planes of T_2' precipitates and on the surface of spherical β' (Al_3Zr) precipitates.

An orthorhombic structure for the T_1' phase is proposed. This structure is isomorphous with Pt_2Mo and it is proposed to have the $Al_2(Cu,Li)$ formula.

A structure isomorphous with θ' is envisioned to constitute the T_2' phase. It is suggested that Li is present in the θ' lattice.

The onset of dissolution of an unidentified phase is resolved at 90°C in DSC analyses. Three possibilities for the formation of this phase are discussed.

Temperature regions for the dissolution of δ' and T_1' precipitates are resolved via DSC and TEM analyses.

Acknowledgments

We express our appreciation to Mr. J. C. Casato who provided skillful assistance in the analyses of Guinier-de Wolff x-ray patterns.

REFERENCES

1. D. B. Williams and J. W. Edington, Metal Science 9, 529, (1975).

2. R. J. Rioja and D. E. Laughlin, Met. Trans., 8A, 1257, (1977).

3. H. K. Hardy and J. M. Silcock, J. Inst. Met., 84, 423, (1955-1956).

4. R. Nosato and G. Nakai, Trans. Jap. Inst. Met., 18, 679, (1977).

5. C. Sigli, J. M. Sanchez and J. M. Papazian, These Procedings, (1985).

6. B. Noble and G. E. Thompson, Met. Sci. J., 5, 114, (1971).

7. H. Suzuki, M. Kanno and N. Hayashi, Jour., Jap. Ins. Light Met., 32, 88, (1982).

8. K. Schneider and M. von Heimendahl, Z. Metallkde, 64, 342, (1973).

9. J. M. Silcock, J. Inst. Met., 88, 357, (1959-1960).

10. B. Noble, I. R. McLaughlin and G. Thompson, ActaMet., 18, 339 (1970).

11. M. Avalos-Borja, P. O. Pizzo and L. A. Larson, NASA Technical Memorandum 84385 (1983).

12. B. Noble and G. E. Thompson, Met. Sci., J., 6, 167, (1972).

13. L. E. Mondolfo, "Aluminum Alloys, Structure and Properties), Butterworths, 495, (1979).

14. F. W. Gayle and J. B. VanderSande, Scripta Met., 1, 473, (1984).

Microstructural development in Al–2%Li–3%Cu alloy

M H TOSTEN, A K VASUDÉVAN and P R HOWELL

HT and PRH are in the Department of aterials Science and Engineering at the ennsylvania State University, niversity Park, PA, USA.
KV is in the Alloy Technology Division of lcoa Laboratories, Alcoa Center, PA, USA

YNOPSIS

he development of intragranular microstructures in n Al-2% Li-3% Cu alloy, which had been aged at 90°C for times in the range of 1.25 hours to 520 ours, has been examined using transmission lectron microscopy. In the underaged and peak ged conditions (≤18 hours) the major phases resent in the matrix were δ', θ', T_1 and Al_3Zr. t was found that a tri-modal distribution of δ' eveloped, consisting of: small spherical δ', δ' hich coated the θ' and δ' which had encapsulated he Al_3Zr. In addition, evidence has been obtained hich suggests that θ' can nucleate on the Al_3Zr. t peak age (18 hours) most of the δ' was ssociated with the Al_3Zr and the θ', and little mall spherical δ' was observed.

Overaging led to the dissolution of the θ' nd, to a lesser extent, the δ'. This dissolution as accompanied by an increase in the volume raction of the T_1 phase. After 520 hours aging he δ' was associated, almost exclusively, with the l_3Zr. Intragranular precipitation of δ occurred n the overaged condition.

NTRODUCTION

nterest in Al-Li-X alloys has been extensive over he past few years (1,2) since these alloys ffer an excellent combination of properties ncluding high strength, high elastic modulus and ow density. However, Al-Li and Al-Li-X alloys can xhibit low fracture toughness and ductility. In n attempt to optimize the fracture toughness of hese alloys, a comprehensive study of the interrelation between microstructure, strengthening mechanisms, deformation behavior and fracture characteristics of a series of Al-Li-Cu-Zr and Al-Li-Cu-Mg-Zr alloys has been undertaken.

The results presented in this paper are concerned with a detailed systematic study of the microstructural characteristics of an Al-2% Li -3% Cu -0.12% Zr alloy as a function of aging time at 190°C. Specific emphasis has been placed on:

(a) determining the nature of the precipitate phases;
(b) analyzing the interaction between the various precipitate species:
(c) a quantitative analysis of the precipitate distributions in terms of: volume fraction, size and number density.

Discussion will be restricted to intragranular structures and details of the grain boundary precipitation reactions are given elsewhere (3,4).

BACKGROUND

The precipitation characteristics of Al-Li binary alloys have been documented extensively (e.g. 5,6). It has been found that low temperature aging results in a homogeneous distribution of the metastable δ' phase and after extended aging precipitation of the equilibrium δ phase occurs (5,6).

Six ternary compounds have been identified by Hardy and Silcock (10) in the Al-rich corner of the Al-Li-Cu system, the most relevant being T_1 (Al_2CuLi), T_2 (Al_6CuLi_3) and T_B ($Al_{7.5}Cu_4Li$).

Silcock (7) showed that, for an Al-Li-Cu alloy aged at 165°C, the primary phases present were θ'', θ', T_1 and δ'. Noble and Thompson (8) investigated two alloys containing 3.5% Cu, 1.5% Li and 2.5% Cu, 2.0% Li and documented the presence of T_1, GP zones, θ' and δ'. These authors showed that the T_1 phase formed on $\{111\}_\alpha$ - matrix planes with the following orientation relationship:

$$(0001)_{T_1} \parallel (111)_\alpha$$

$$(10\bar{1}0)_{T_1} \parallel (1\bar{1}0)_\alpha$$

$$(11\bar{2}0)_{T_1} \parallel (2\bar{1}\bar{1})_\alpha$$

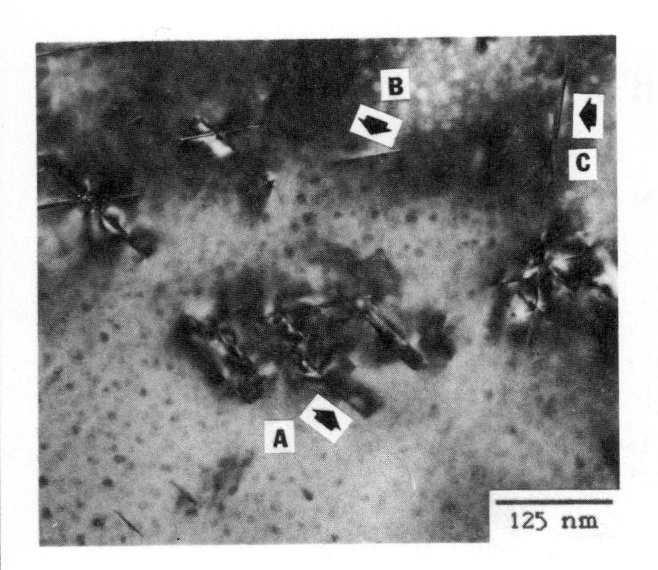

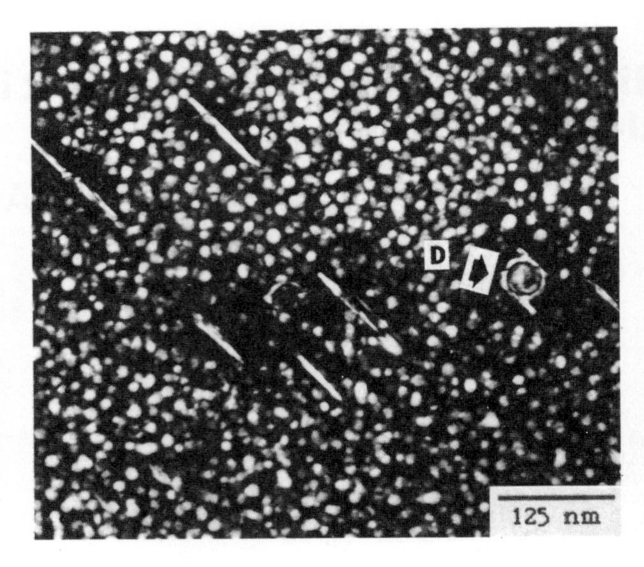

a

b

Figure 1. The microstructure developed after 1.25 hours aging; (a) bright field micrograph: (b) the corresponding δ' CDF.

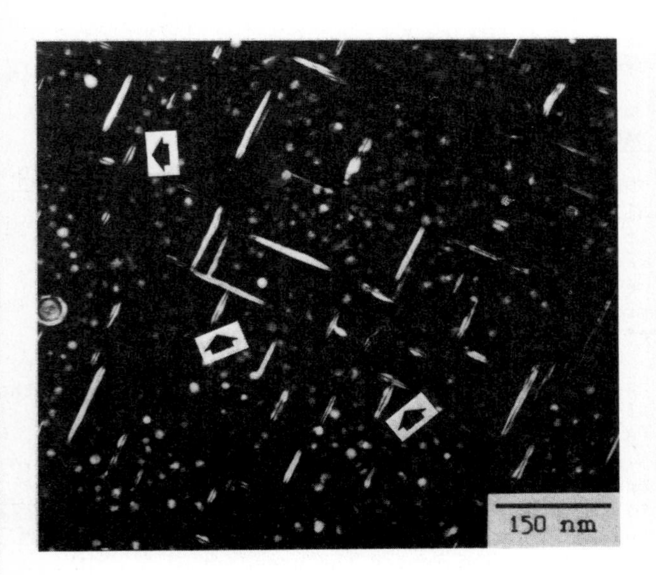

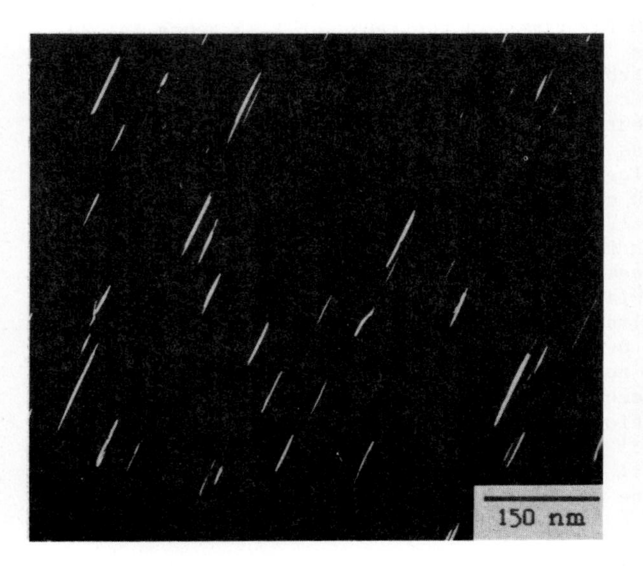

a

b

Figure 2. The microstructure developed after 2.25 hours aging: (a) δ' CDF: (b) a corresponding θ' CDF.

In addition, it was suggested that the T_1 phase could nucleate by:

(i) dissociation of a/2<110> dislocations into a/6 <112> Shockley partials, followed by copper and lithium enrichment of the intrinsic stacking fault and/or:

(ii) nucleation at α/GP zone interfaces.

Schneider and von Heimendahl (9) reported that θ' was the dominant phase in an alloy containing 4% Cu, 1% Li but that the θ' transformed to T_B at high aging temperatures.

In binary Al-Zr alloys, aging produces the metastable Al_3Zr phase (11) which can adopt a rod-like or spherical morphology. In ternary Al-Li-Zr alloys, the Al_3Zr assumes a spherical morphology for Zr contents in the range of ≃ 0.1% - 0.2% (12,13) and is rod like or spherical in ternaries containing ≃1% Zr (14).

EXPERIMENTAL

The Al-Cu-Li-Zr alloy used in this investigation was furnished by the Alloy Technology Division of Alcoa Laboratories, Alcoa Center, PA. Bulk specimens, prepared from hot rolled sheet, were solution treated at 550°C for 0.5 hours, cold water quenched, and deformed by 1% in tension prior to aging at 190°C. Aging times employed in this investigation ranged from 1.25 hours to 520 hours, with 18 hours being the peak aged condition. The composition of the alloy after heat treatment was

2.10 w/o Li, 2.90 w/o Cu, 0.12 w/o Zr, 0.06 w/o Fe, 0.03 w/o Si, 0.01 w/o Ti, balance Al.

Thin foil specimens for transmission electron microscopy (TEM) were twin jet electropolished in a 25% Nital solution cooled to - 20°C using a potential of ~12V. After thinning, the foils were rinsed for 1 min. in a steady stream of ethanol. All specimens were examined in a Phillips EM420T operating at 120 KV.

RESULTS AND DISCUSSION

The Nature of Precipitate Phases

Table 1 lists the compositions and crystal structures of the major precipitate phases encountered in the present investigation.

Table 1
The Phases Encountered in the Al-3% Cu-2% Li Alloy

Phase	Composition	Crystal Structure
δ'	Al_3Li	$L1_2$
θ'	Al_2Cu	Tetragonal
T_1	Al_2CuLi	Hexagonal
Al_3Zr	Al_3Zr	$L1_2$
δ	AlLi	Cubic

In addition to these, the iron containing phase, Al_7Cu_4Fe was also observed infrequently. Regarding Table 1, there is some uncertainty as to the exact crystal structure of the Al_2CuLi phase. For example, Rioja (15) has proposed that there is a metastable precursor to the T_1 phase, namely T_1'. It has been advanced that T_1' has the Pt_2Mo orthorhombic crystal structure and forms on the {111} α - matrix planes. However, in the present paper, the Al_2CuLi phase will be designated T_1 as originally proposed by Silcock (7) and Noble and Thompson (8).

The As-Quenched Microstructure

In the as-quenched specimen material, an inhomogeneous dispersion of spherical Al_3Zr precipitates was present. The Al_3Zr had diameters in the range of 5 nm - 40 nm. These precipitates were cube/cube related to the α-matrix and were coherent, as determined by the Ashby-Brown strain contrast obtained in strong two-beam images. Further details concerning the Al_3Zr precipitates are given in Stimpson et al (13,16). In addition to the Al_3Zr, very fine (≃ 2-4 nm diameter) δ' precipitates had nucleated during the quench from the solution treatment temperature. No evidence for θ', T_1 or δ was found.

The Microstructures of Short Term Aged Specimen Material (1.25 hours to 18 hours)

Figure 1a is a bright field micrograph ([110] foil orientation) of material that had been aged for 1.25 hours whilst figure 1b is the corresponding δ' centered dark field (CDF) image. For this foil normal one orientation variation of the θ' is upright (arrowed A) and two orientation variants of the T_1 are upright (arrowed B and C). Reference to figure 1b shows that, in addition to the small spherical δ' (average radius 6 nm), δ' is associated with the θ' and the Al_3Zr (the latter is arrowed D in figure 1b). Hence, a trimodal δ' distribution is developed in the early stages of aging and consists of:

(i) small discrete spherical precipitates of δ':

(ii) δ' which has wholly or partially encapsulated the Al_3Zr:

(iii) δ' which has coated the coherent, broad faces of the θ'.

Confirmatory evidence for (iii) above is given in the δ' and θ' CDF^s of figures 2a,b (2.5 hours aging). For figure 2, the foil normal was [001] so that two variants of the θ' are upright. Reference to figure 2a shows that two orthogonal sets of "δ' plates" are present. Closer examination shows that these plates are duplex consisting of an inner, dark imaging core together with an outer shell which is illuminated in the δ' CDF (several examples where the dark imaging core can be discerned are arrowed). Figure 2b is a θ'CDF of the same region as that shown in figure 2a. It can be seen that there is a one to one correspondence between θ' precipitates and the δ' coatings.

The initial stage in the coating of the coherent θ' precipitates can occur by one or both of the following mechanisms:

(i) impingement of the small, spherical δ' with the coherent θ' faces:

(ii) heterogeneous nucleation of δ' on the θ'/α-matrix interfaces.

In either case, complete coverage will occur by subsequent coarsening and coalescence of the δ'

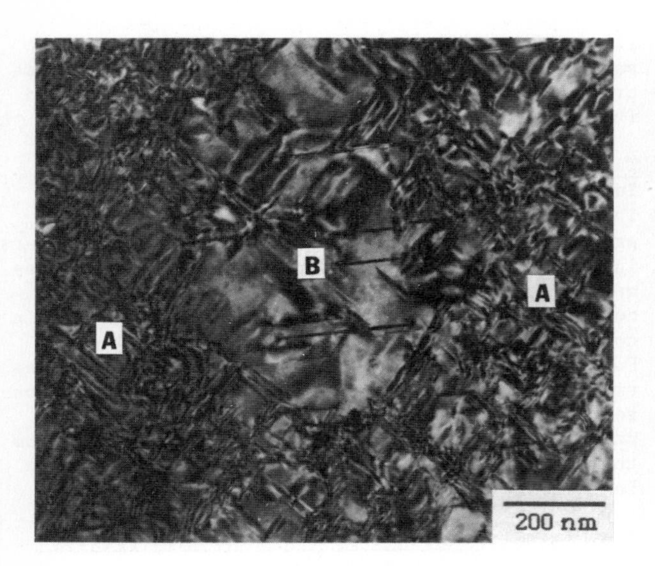

a

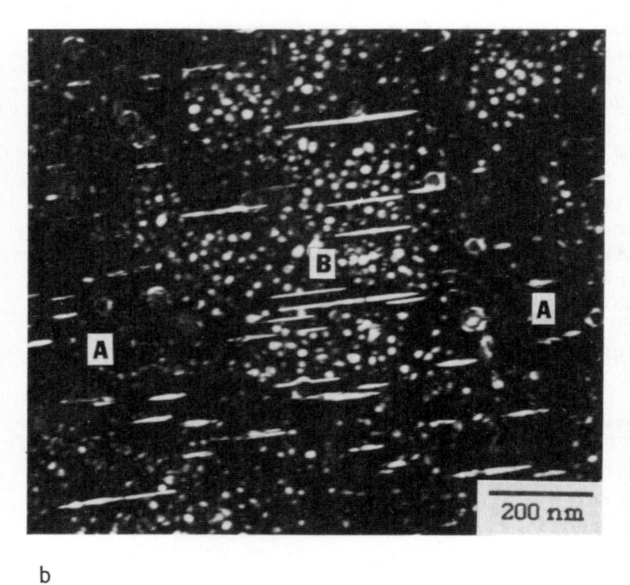

b

Figure 3. Microstructural development after 3.5 hours aging; (a) bright field image; (b) the corresponding δ' CDF.

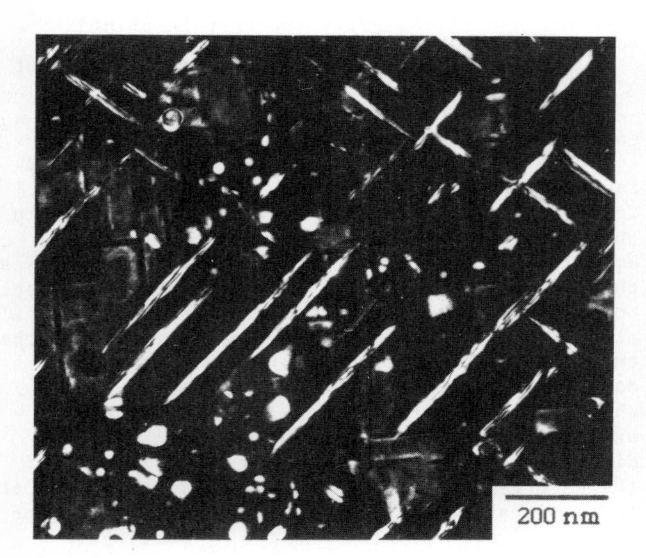

Figure 4. A δ' CDF from specimen material aged for 18 hours.

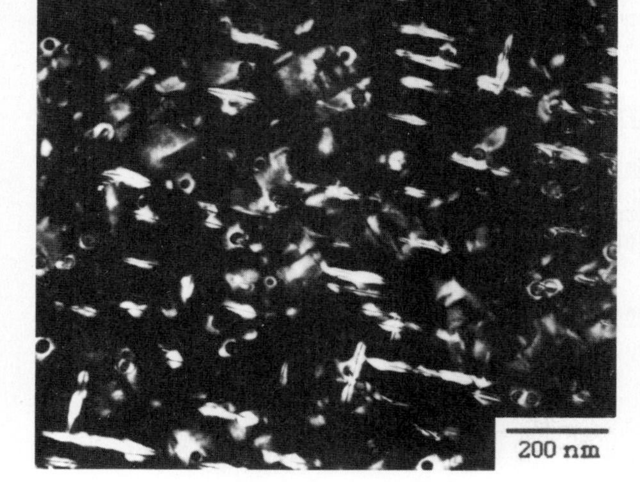

Figure 5. A δ' CDF from material aged for 70 hours.

in contact with the broad faces of the θ' plates. This will occur at the expense of the adjacent small spherical δ'. Similar arguments can be advanced for the development of the duplex δ'/Al_3Zr precipitates (13).

Returning to figure 1b, it can be seen that δ' precipitate free zones (PFZ^s) develop in the vicinity of the plate like δ'. These PFZ^s are created by diffusion of Li down the concentration gradient from the spherical δ' to the coating of δ' on the θ' (17). Similar PFZ^s were observed around the Al_3Zr although these latter PFZ^s are not so readily apparent in figure 1b. PFZ^s at Al_3Zr precipitates have also been documented in ternary Al-Li-Zr alloy (12,13).

One further point concerning figure 1b is the association between the Al_3Zr precipitate (D) and two θ' plates. Close examination of a large number of micrographs, both in the present alloy and in a 4.5% Cu-1% Li alloy has shown that this association is very frequent. Furthermore, it has been found that, in the vast majority of cases, the broad face of the θ' is in contact with the Al_3Zr (18). Hence, it can be suggested that Al_3Zr acts as a nucleation substrate, not only for δ', but also for the θ'.* Hence, there is a complex interaction between the various constituent phases in the quaternary Al-Cu-Li-Zr alloys.

Unlike the θ' and Al_3Zr phases, the T_1 precipitates do not serve as preferential nucleating/coarsening substrates for the δ' phase. Indeed there is evidence that the small spherical δ' precipitates can be cut by the advancing T_1 plates, and no coarsening of the δ' occurs in the vicinity of the T_1. It was also found that, in regions containing a large number density of the T_1 precipitates, a low number density of the small spherical δ' was present. This is not unexpected since both phases contain lithium (Table 1). Figure 3a is a bright field micrograph of specimen material that had been aged for 3.5 hours ([001] foil orientation) whilst figure 3b is the corresponding δ' CDF. It is clear from figures 3a,b that in regions containing a large number density of T_1 (labelled A), a low number density of δ' is observed and in regions which are denuded in T_1 (B) a high number density of spherical δ' is present. It is also interesting to note that in regions A, the θ' plates have consistently smaller diameters than in region B. This may be a reflection of the competition for solute, in this case copper, and/or a consequence of θ' plates impinging on the T_1.

Evidence for the growth/coarsening of the T_1 phase at the expense of the δ' has been obtained by examining low angle grain boundaries where a δ' PFZ was found to be associated with copious precipitation of the T_1 phase on the low angle boundaries (3).

Figure 3a showed that the T_1 precipitates were not uniformly distributed throughout the matrix. Reference to figure 3b shows that this is also the case for θ'. In addition to the non-uniformity in spatial distribution, the θ' did not populate the three possible crystallographic variants to the same degree and in many cases, only one θ' variant was present in a given area. For example, in figure 3b, two orthogonal, upright θ' plates might be expected whereas only one is observed.

δ, the equilibrium phase in Al-Li binary alloys, was observed in all specimens aged between 2.25 hours and 520 hours (3). Due to the high reactivity of this phase, identification using electron diffraction proved to be impossible. Instead, the presence of δ was determined by examining the oxide reaction products and the high dislocation content that is invariably associated with δ (6). In underaged to peak aged material, (2.25 to 18 hours) the δ phase was confined to the high angle boundaries. However, in overaged specimens (70 hours to 520 hours) δ was found on low angle grain boundaries as well as within the α -matrix. The significance of δ on the fracture behavior of Al-Li and Al-Li-Cu alloys is discussed by Vasudevan et al. (19).

At peak age (18 hours), the volume fraction and number density of the small spherical δ' had decreased dramatically (the volume fraction of spherical δ' was ≃8% after 1.25 hours and less than 1% after 18 hours). However, it should be noted that the overall δ' volume fraction remained essentially constant at ≃ 10%, in the range of 1.25 hours to 18 hours aging. Figure 4 is a δ' CDF of the 18 hour aged material. The majority of the δ' is associated with the θ' and, to a lesser extent, the Al_3Zr. The volume fraction of the T_1 and θ' increased with increased aging time up to peak age. Similarly, their diameters increased and after 18 hours, the average T_1 diameter was 140 nm whilst the average θ' diameter was 120 nm. Thickening kinetics of both plate-like phases were found to be inhibited and their aspect ratios increased from approximately 35:1 (1.25 hours) to approximately 75:1 (18 hours). Finally, neither the volume fraction nor the number density of the Al_3Zr were affected by aging.

The Microstructures of Overaged Specimen Material

Overaging the alloy results in the T_1 precipitate becoming the dominant matrix phase. The increase in the volume fraction of this phase occurred at the expense of the θ' and, to a lesser extent, the δ'. This effect can be seen in figure 5 ([001] foil orientation). No small spherical δ' is present and the remaining δ' is associated exclusively with the θ' and the Al_3Zr. It can also be seen that the θ' is beginning to dissolve. This may be gauged by comparing the θ' diameter after 70 hours aging (figure 5) with the θ' diameter after 18 hours aging (figure 4).

*Similar evidence has been obtained for the nucleation of T_1, on or close to the Al_3Zr/matrix interface (18).

a

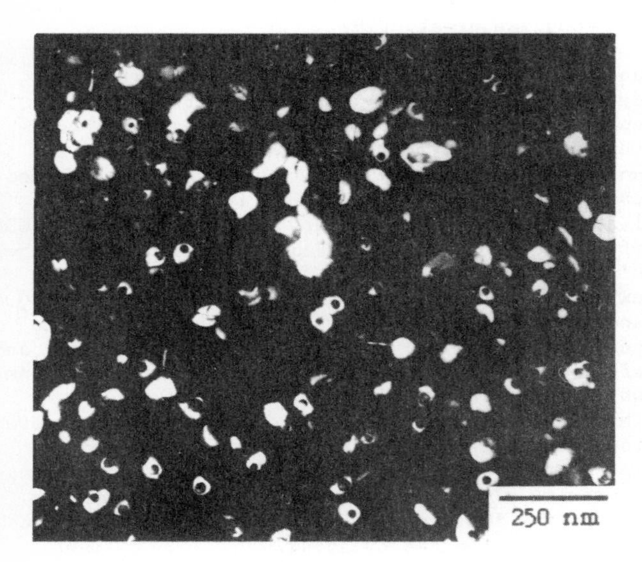

b

Figure 6. Microstructural development after 520 hours: (a) a θ' CDF; (b) the corresponding δ' CDF.

Figures 6a,b are θ' and δ' CDF^s respectively from specimen material that had been aged for 520 hours ([001] foil orientation). The θ' CDF illustrates graphically the extremely low volume fraction and number density of this phase. In addition, the average diameter of the θ' is significantly less than that found after 70 hours aging (figure 5). The almost complete elimination of the θ' has a dramatic influence on the distribution of the δ', which is now associated predominantly with the Al_3Zr* (figure 6b). The Al_3Zr in this micrograph, and in figure 5, are the dark imaging cores of the duplex precipitates. Possible reasons for the Al_3Zr appearing dark in δ' CDF images are discussed in Gayle and Vander Sande (14), Stimpson et al. (13,16).

As mentioned in the previous section, δ was observed in the matrix in the overaged condition.

SUMMARY

(i) In the as-quenched specimen material, the only precipitate phases encountered were Al_3Zr and a very fine (≃2-4 nm diameter) dispersion of δ'.
(ii) Short term aging (1.25 hours) led to the development of a trimodal δ' distribution which consisted of:
- (a) small spherical δ';
- (b) δ' which had coated the coherent θ' faces;
- (c) δ' which had encapsulated the coherent Al_3Zr precipitates.

(iii) Al_3Zr also acted as a nucleation substrate for θ'.
(iv) The θ' and T_1 phases were distributed unevenly throughout the α-matrix.
(v) δ precipitation on high angle grain boundaries occurred after 2.25 hours; on low angle grain boundaries and within the matrix in overaged material.
(vi) After 18 hours aging, the δ' was associated predominantly with the θ' and the Al_3Zr. However, after prolonged aging (520 hours), the θ' had almost wholly dissolved and the remaining δ' was associated with the Al_3Zr.
(vii) Overaging led to the T_1 phase being the dominant matrix precipitate.

ACKNOWLEDGMENTS

One of the authors (M.H.T.) is grateful to the Alloy Technology Division of Alcoa Laboratories, Alcoa Center, PA for a Grant-in-Aid.

*It should be noted that the regions shown in figures 5 and 6 contain an abnormally high number density of Al_3Zr precipitates.

REFERENCES

1. "Aluminum-Lithium Alloys", Eds. Sanders, T. H. and Starke, E. A., TMS-AIME (1981).
2. "Aluminum-Lithium Alloys II", Eds. Sanders, T. H. and Starke, E. A., TMS-AIME (1984).
3. Tosten, M. H., Vasudevan, A. K. and Howell, P. R., These Proceedings.
4. Tosten, M. H., Vasudevan, A. K. and Howell, P. R., Submitted to Mat. Sci. Eng.
5. Noble, B. E. and Thompson, G. E., Met. Sci. J., 5 (1971) p. 1147.
6. Williams, D. B. and Edington, J. W., Met. Sci., 9 (1975:) p. 299.
7. Silcock, J. M., J. Inst. Met., 88 (1959-60) p. 357.
8. Noble, B. E. and Thompson, G. E., Met. Sci. J., 6 (1972) p. 167.
9. Schneider, K. and von Heimendahl, M., J. Metallkde, 64 (1973) p. 430.
10. Hardy, H. K. and Silcock, J. M., J. Inst. Met., 84 (1955-56) p. 423.
11. Ness, E., Acta Met., 20 (1972) p. 499.
12. Makin, P. L. and Ralph, B., J. Mat. Sci., 19 (1984) p. 3835.
13. Stimpson, W., Tosten, M. H., Howell, P. R. and Williams, D. B., These Proceedings.
14. Gayle, F. W. and Vander Sande, J. B., Scripta Met, 18 (1984) p. 473.
15. Rioja, R. J., Private Communication.
16. Stimpson, W., Tosten, M. H., Howell, P. R. and Williams, D. B., Submitted to Scripta Met.
17. Tosten, M. H., Howell, P. R. and Doherty, R. D., To be submitted to Scripta Met.
18. Tosten, M. H., Galbraith, J. and Howell, P. R., Submitted to Scripta Met.
19. Vasudevan, A. K., Ludwiczak, E. A., Baumann, S. F., Howell, P. R. and Doherty, R. D., These Proceedings.

□

Grain boundary precipitation in Al–Li–Cu alloys

M H TOSTEN, A K VASUDÉVAN and P R HOWELL

MHT and PRH are in the Department of Materials Science and Engineering at the Pennsylvania State University, University Park, PA, USA.
AKV is in the Alloy Technology Division of Alcoa Laboratories, Alcoa Center, PA USA

SYNOPSIS

Grain boundary precipitation in two Al-Li-Cu alloys has been examined using transmission electron microscopy and scanning electron microscopy. In alloys aged at 190°C, the following phases are present at the boundaries: δ, T_1, θ' and Al_3Zr. For times up to, and including peak age, δ is the dominant precipitate species on high angle grain boundaries whilst T_1 and θ' are present on low angle grain boundaries (Al_3Zr is found on both types of boundary). In over aged material, δ also precipitated on low angle boundaries and eventually within the matrix. For the plate like phases (T_1 and θ'), only one crystallographic variant was found on any one planar boundary segment. In the vast majority of cases, the variant adopted was that which minimized the angle between the precipitate habit plane normal and the local boundary normal. Precipitate free zones (PFZ^s) were invariably associated with high angle grain boundaries whilst PFZ formation at low angle grain boundaries (at least up to peak age) depended on the Li:Cu ratio.

INTRODUCTION

Grain boundary precipitation is of great academic and industrial significance since this mode of precipitation can control and/or modify a wide variety of material properties. For example, precipitation of chromium rich carbides can lead to sensitization and stress corrosion cracking in a number of austenitic stainless steels and nickel-base superalloys. In addition, grain boundary precipitates can be a cause of low ductility in a variety of aluminum alloys (e.g. 1-3).

In Al-Li alloys, the dominant precipitate species on high angle grain boundaries is δ (4). Nucleation and growth of this phase leads to the development of δ' PFZ^s and it is suggested that fracture in Al-Li and Al-Li-X alloys is linked intimately to the presence of δ on the high angle grain boundaries (5).

Most modern Al-Li-X alloys contain Zr (usually ≃ 0.1 w/o). The addition of Zr to these, and other aluminum alloys, inhibits recrystallization by the formation of the metastable, coherent Al_3Zr phase (6) so that, prior to aging, the microstructure consists of the original deformed grain structure, together with sub-grain and low angle grain boundaries, the latter being formed by a recovery process. It has been noted in the literature that phases, other than δ form on low angle grain boundaries. However, to the authors' knowledge, no systematic study of precipitation on low angle grain boundaries in Al-Li-X alloys has been undertaken. Hence, as part of an overall investigation concerning the microstructures developed in a series of Al-Li-Cu alloys, the nature and distribution of grain boundary phases has been documented. In this paper, attention will be focussed on low angle grain boundaries since they have received considerably less attention than their high angle counterparts. Specifically, this paper considers the following:

(i) the nature of the precipitate phases on low angle grain boundaries:

(ii) the development of δ' PFZ^s in the vicinity of low angle boundaries;

(iii) the crystallography of low angle grain boundary precipitation.

Precipitation on high angle boundaries is also considered briefly.

EXPERIMENTAL

The alloys used in this investigation were supplied by the Alloy Technology Division, Alcoa Center, PA

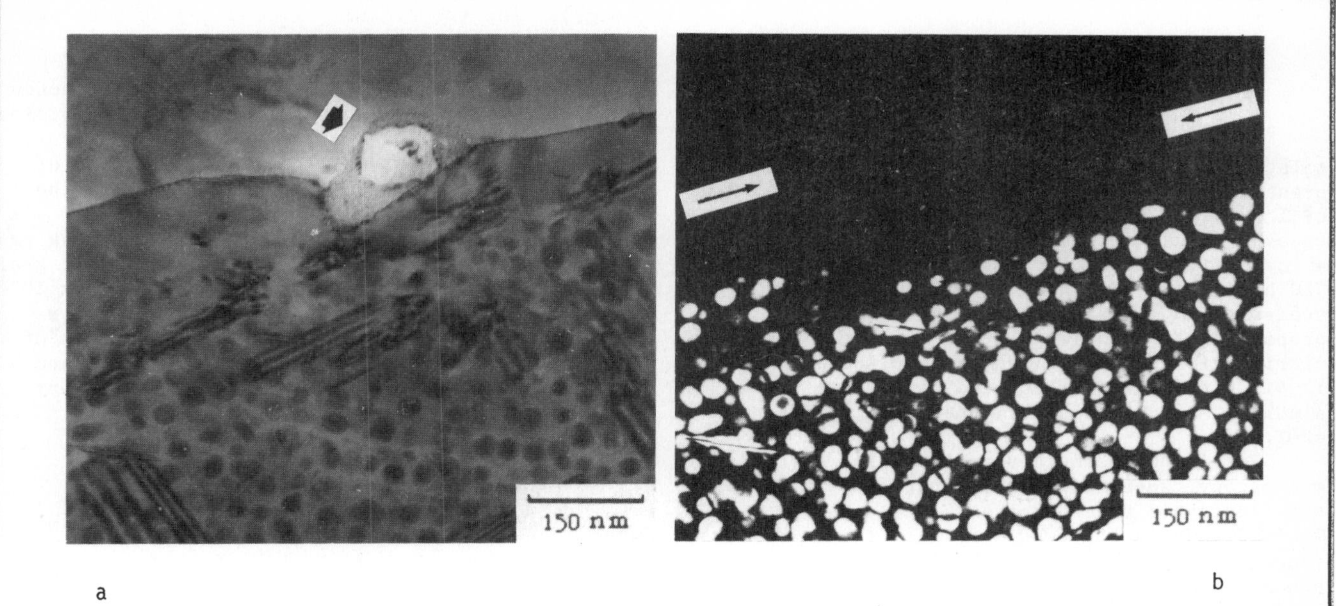

a b

Figure 1. δ precipitation on a high angle grain boundary (2.6 Li - 1 Cu alloy): (a) bright field micrograph; (b) corresponding δ' CDF image.

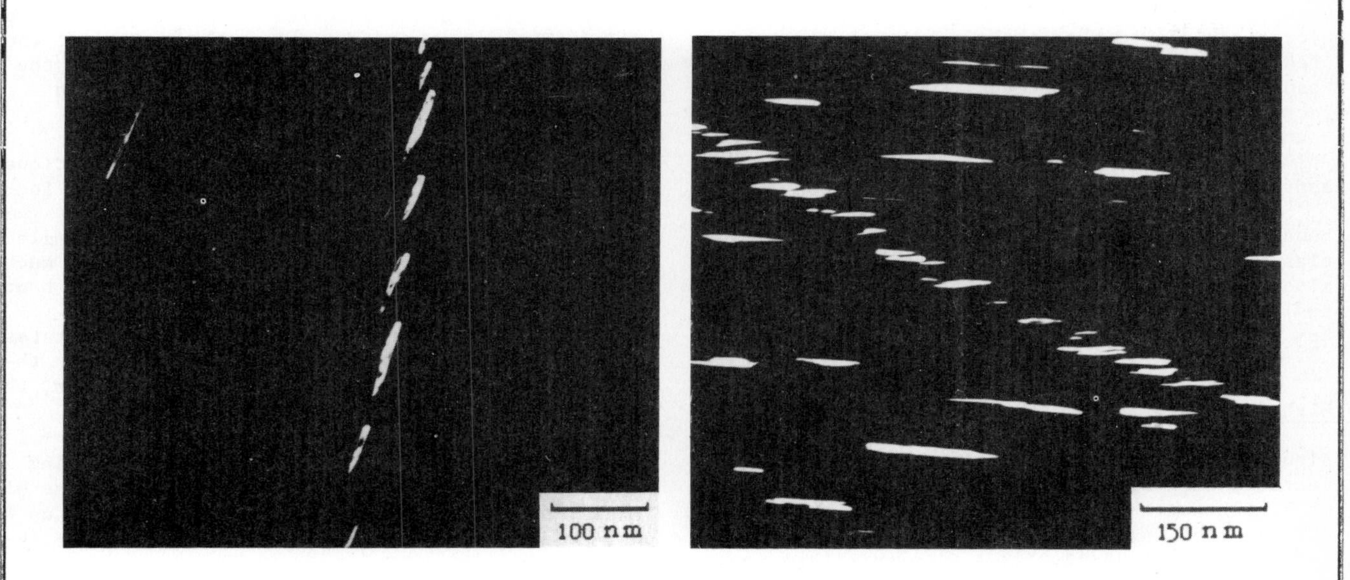

a b

Figure 2. Precipitation on low angle grain boundaries in the 2.6 Li - 1 Cu alloy; (a) θ' CDF image; (b) T_1 CDF image.

and had the following compositions (wt%):

Al - 2.6% Li - 1% Cu - 0.12% Zr

Al - 2% Li - 3% Cu - 0.12% Zr

After hot rolling, specimen material was solution treated at 550°C for 30 minutes, prestrained by 1% and aged at 190°C. For the 2.6 Li-1 Cu alloy, a single aging time of 16 hours (peak age) was employed; whilst for the 2 Li-3 Cu alloy, aging times in the range of 1.25 hours to 520 hours were investigated. Thin foils for transmission electron microscopy (TEM) were prepared in a twin-jet electropolisher using 25% nitric acid in methanol and an applied potential of 12 V. After perforation, the foils were washed in alcohol. All TEM was performed on a Philips EM 420T.

For light microscopy and scanning electron microscopy, specimens were prepared using standard metallographic procedures and etched for 10 s in Graff Sergent's followed by a 5 s etch in Keller's reagent. This two-step etch clearly delineated the δ phase. After etching, the samples were carbon coated. SEM was performed on an ISI Super III A operating at 10 kV.

The number density and volume fraction of the δ phase was determined using standard stereological techniques. Axis/angle pairs were computed using low camera length convergent beam electron diffraction (LCL-CBED) in the TEM. Details of the LCL-CBED technique are given in Tosten et al (7).

RESULTS AND DISCUSSION

Examination of the specimen material by light microscopy showed that the microstructure was predominantly unrecrystallized, consisting of large "pancaked" grains. However, some small recrystallized grains were observed to decorate the boundaries in the deformed microstructure. The volume fraction of recrystallized grains was approximately 10%*.

To facilitate further discussion, the grain boundaries as observed in the TEM will be classified as either high angle or low angle. The misorientation angle at which the low angle to high angle transition occurs is somewhat arbitrary and has been taken to be approximately 9°-10°.

High Angle Grain Boundary Precipitation

Figures 1 a,b are bright field and corresponding centered dark field (CDF) micrographs respectively of a high angle grain boundary region in the 2.6 Li - 1 Cu alloy (axis/angle pair for the boundary is [-5.59 -6.79 4.75/12.3°]) A δ' precipitate is observed in the bright field image (arrowed) whilst the δ' CDF illustrates clearly the PFZ which is associated with the boundary (the position of the boundary is denoted by the arrows in the CDF image). For the 2.6 Li - 1 Cu alloy at peak age (16 hours), it was found that δ had only precipitated at grain boundaries where the misorientation was > 9° and this value was chosen as the low angle to high angle boundary transition point.

In the 2 Li - 3 Cu alloy, precipitation of δ was not detected after 1.25 hours aging and the onset of precipitation on the high angle boundaries occurred after 2.25 hours. At peak age (18 hours), the number of δ particles per unit grain boundary area (N_A) was determined to be 4.6 x 10^{11} m^{-2} and the volume fraction of δ(V_v) was 5.6 x 10^{-3}. The number density and volume fraction of the δ continued to increase with aging time and, for example, after 70 hours aging, the relevant parameters were:

$$N_A = 7 \times 10^{11} \ m^{-2};$$

$$V_v = 1.1 \times 10^{-2},$$

It was found that the PFZ width for the 2 Li - 3 Cu alloy increased with aging time which is consistent with the findings of other investigations (e.g. 10). However, due to the large grain size of the experimental material, insufficient data have been obtained to date to determine quantitatively the increase in PFZ width with aging time.

Finally, discrete precipitates of Al_3Zr, in addition to δ, were observed infrequently on the high angle grain boundaries.

Low Angle Grain Boundary Precipitation

Figures 2 a,b are typical examples of precipitates on low angle grain boundaries in the 2.6 Li - 1 Cu alloy (These images are also representative of the 2 Li - 3 Cu alloy). In the former CDF image, the precipitates are θ' whilst the precipitates in figure 2b are T_1. The reason T_1 precipitates are illuminated in both grains is that the misorientation is small so that the T_1 reflections which originate from both grains are included in the objective aperture. In the 2.6 Li - 1 Cu alloy, T_1 was the dominant phase on the low angle grain boundaries which is a reflection of the much higher volume fraction of T_1 present after 16 hours aging. This may be gauged qualitatively by comparing the number of intragranular precipitates which are illuminated in figures 2a and b. In the 2 Li - 3 Cu alloy, T_1 still dominated. However, the incidence of θ' precipitation on low angle boundaries increased over that in the 2.6 Li - 1 Cu alloy in the underaged condition. Overaging the 2 Li - 3 Cu alloy led to dissolution of the θ' phase and after 520 hours at 190°C, virtually no θ' was evident, either on low angle boundaries or in the matrix.

Since T_1 contains lithium, it might be expected that δ' PFZ^s would form in the vicinity of low angle boundaries. This was found to be the case for the 2 Li - 3 Cu alloy as is shown in Figures 3a, b which were recorded from specimen material which had been aged for 2.25 hours. The low angle grain boundaries are arrowed on figures 3a, b and the δ' PFZ^s are clearly delineated in the δ' CDF of figure 3b. The complex contrast observed

*Details of the intragranular precipitate distributions are given in Tosten et al (8,9).

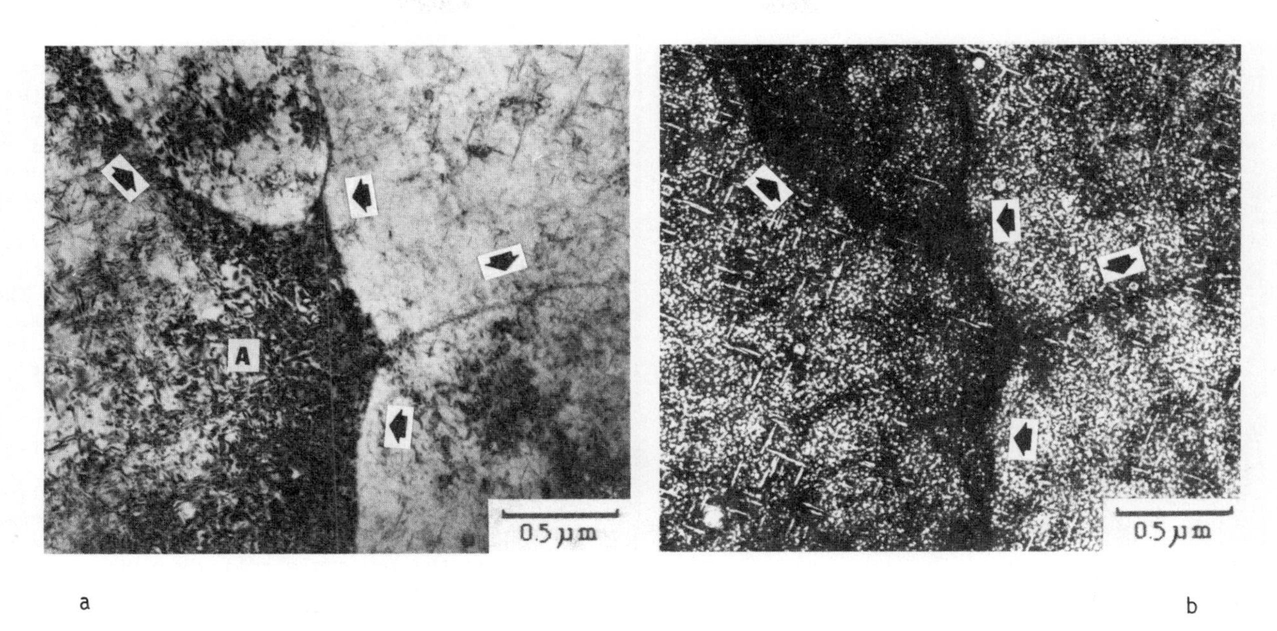

a b

Figure 3. Precipitate-free zones in the 2 Li - 3 Cu alloy; (a) bright field image; (b) corresponding δ' CDF image.

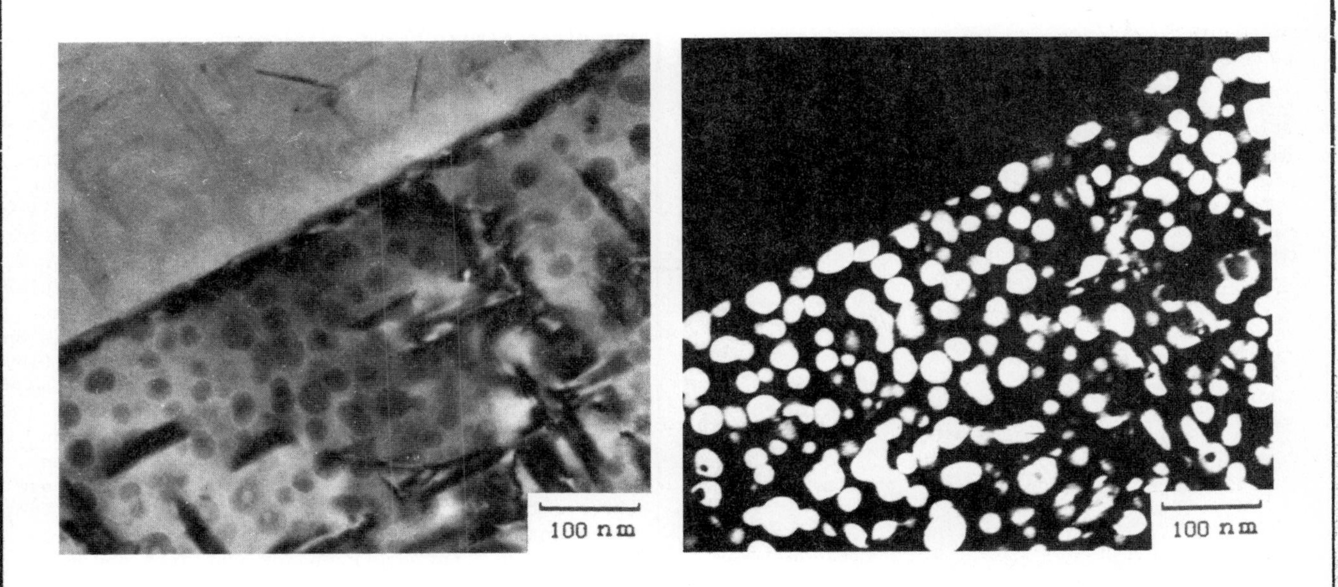

a b

Figure 4. Absence of precipitate-free zones in the 2.6 Li - 1 Cu alloy; (a) bright field image; (b) corresponding δ' dark field image.

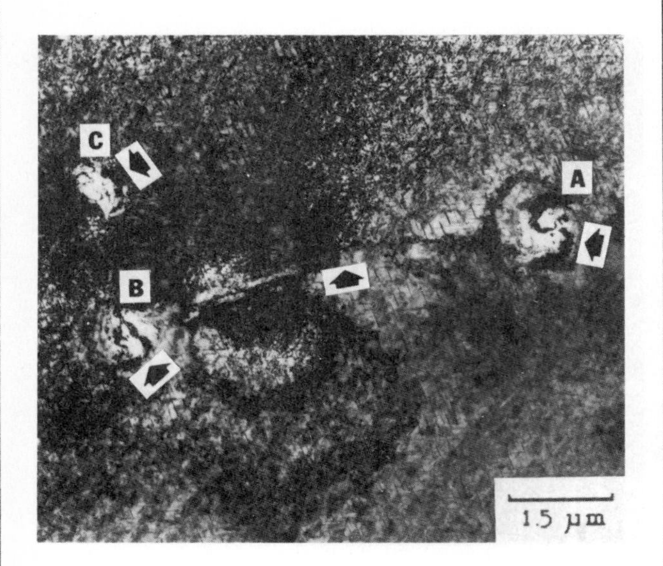

Figure 5. δ precipitation on low angle boundaries and within the matrix (2 Cu - 3 Li alloy aged at 190°C for 520 hours).

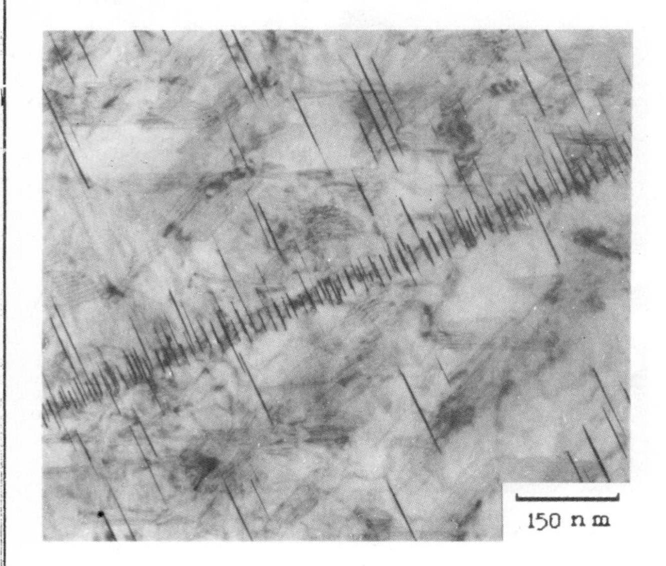

Figure 6. T_1 precipitates on a low angle boundary in the 2 Li - 3 Cu alloy after 1.25 hours aging.

in e.g. grain A in figure 3a arises from the intragranular θ' and T_1 phases. In addition the "plate-like" δ' in figure 3b is in fact a duplex precipitate consisting of an inner core of θ' which has been encapsulated by δ' (8,9).

In contrast, in the 2.6 Li - 1 Cu alloy, no PFZ^s were observed in the vicinity of low angle grain boundaries even after 16 hours aging. This is illustrated graphically in the bright field, δ' CDF images of figures 4a, b where it can be seen that the δ' is continuous right up to the boundary (which has an axis/angle pair of [353]/4°). Indeed, it was found that there was little or no interaction between the δ' and T_1 phases and the δ' was often in contact with the T_1 phase (11). The above suggests that, for a Li:Cu ratio of 2.6:1, T_1 and δ' are not competing for solute, at least for aging times ≤ 16 hours. This is obviously not the case for a Li:Cu ratio of 2:3 where preferential growth/coarsening of the T_1 on the low angle grain boundaries leads to marked δ PFZ^s after very short aging times (vid. figures 3a,b).

After extended aging times at 190°C (2 Li - 3 Cu alloy), δ began to precipitate on the low angle grain boundaries. Two δ precipitates (A, and B) are shown on the low angle grain boundary (arrowed) in figure 5. Intragranular nucleation of δ (arrowed C) is also evident in figure 5. The presence of δ, together with copious precipitation of the T_1 phase led to extensive δ' PFZ^s and after only 70 hours aging, PFZ widths of 2 μm have been observed.

As shown by a number of investigations, T_1 nucleates on the {111} matrix planes (12) whilst θ' nucleates on {001} matrix planes (e.g. 13). Hence, there are four crystallographic variants of T_1 and three crystallographic variants of θ'. However, in the vast majority of cases, it was found that, for any given planar boundary segment, only one variant of the θ' and/or T_1 was present. The single variant observed was that which minimized the angle between the low energy habit plane normal ({111} $\{0001\}T_1$ for the T_1 and {001}α {001}θ' for the θ') and the local boundary normal. Reference to figures 2a,b shows that, for both T_1 and θ', the boundary precipitate habit planes are at a shallow angle to the boundary plane. Further evidence for the nucleation of a single variant of θ' and/or T_1 on the vast majority of low angle grain boundaries is presented in Tosten et al (14). The present results are in agreement with previous findings on Al-Cu alloys by Vaughan (15), Simon and Guyot (16) and Forest and Biscondi (17). It is interesting to note that in the Al - 4% Cu alloy examined by Vaughan (15), it was found that θ' nucleated on boundaries with misorientations < 9° whilst the equilibrium θ phase nucleated on boundaries whose misorientations exceeded 9°. This is in excellent agreement with the present work where the "transition point" between low and high angle behavior was determined to be approximately 9°-10°.

Single variants of plate-like precipitates on grain boundaries have also been predicted theoretically by Lee and Aaronson (18,19) who showed that the energy barrier to nucleation would be reduced as the angle between the low energy habit plane and the local boundary plane decreased. However, the analysis of Lee and Aaronson (18,19) is only strictly applicable to high angle boundaries since low angle grain boundaries consist of arrays of discrete lattice dislocations. Precipitation on lattice dislocations has been analyzed by a variety of investigators and it is agreed generally that the maximum reduction in strain energy occurs when the precipitate adopts that variant which minimizes the angle between the direction of largest misfit and the Burgers vector (20). Considering θ', the greatest misfit occurs along [001] and it has been predicted and experimentally confirmed (21), that θ' would adopt two orientation variants on an isolated edge dislocation, and that both variants would have their [001] directions at 45° to the Burgers vector. Similarly, it can be shown that T_1 would adopt two orientation variants on an isolated dislocation. To explain this apparent discrepancy, Vaughan (15) considered that low angle grain boundaries could be classified into two groups:

(i) those where the misorientation is so small that the dislocation spacing is large compared with the critical nucleus. Nucleation would then occur as for an isolated dislocation:

(ii) those where the misorientation is sufficient that the regions where the stresses correspond to that of an isolated dislocation are small compared with the nucleus size.

It was then concluded that for group (ii) above, the maximum reduction in strain energy would occur when the angle between the habit plane and the boundary plane was minimized. This latter conclusion was substantiated by the elastic strain field calculations of Simon and Guyot (16).

The results of the present investigation lend support to Vaughan's original hypothesis in that two variants were occasionally observed on low angle boundaries of very limited misorientation. One such example is given in figure 6. In this instance, the habit plane of the single variant of T_1 which is imaged, subtends a large angle to the boundary plane. When this boundary was tilted, a second T_1 variant was imaged. The misorientation of this boundary was small enough that splitting of the diffracted spots in a selected area diffraction pattern could not be detected. The above suggests that the dislocations comprising the sub-grain boundary acted as "isolated lattice dislocations" and two T_1 variants were nucleated, rather than the single variant observed when the misorientation of the boundaries was sufficient to cause significant strain field interaction.

Finally, as shown by Tosten et al. (14), precipitation of Al_3Zr occurred on, and pinned the low angle grain boundaries.

SUMMARY

The major findings of the present investigation can be summarized as follows:

(i) δ precipitated on the high angle grain boundaries, and δ' PFZ^s were always associated with boundaries containing δ particles;

(ii) the number density and volume fraction of δ increased with aging time;

(iii) T_1 and to a lesser extent, θ' was present on low angle grain boundaries;

(iv) δ' PFZ^s invariably developed in the vicinity of low angle boundaries in the 2 Li - 3 Cu alloy. However, no PFZ^s were observed in the 2.6 Li - 1 Cu alloy at peak age

(v) in the vast majority of cases, only one variant of the T_1 and/or θ' nucleated on the grain boundaries. This is consistent with the earlier work of Vaughan (15) on an Al - 4% Cu alloy.

ACKNOWLEDGMENTS

One of the authors (M.H.T.) would like to thank the Alloy Technology Division of Alcoa Laboratories, Alcoa Center, PA for financial support in the form of a Grant-in-Aid. The experimental assistance of F. M. Adragna is gratefully acknowledged.

REFERENCES

1. Embury, J. D. and Nes, E., J. Metall, 65 (1974) p. 45.
2. Hornbogen, E. and Graf, M., Acta Met, 25 (1977) p. 883.
3. Unwin, P. N. T. and Smith, G. C., J. Inst. Met., 97 (1969) p. 299.
4. Williams, D. B. and Edington, J. W., Met Sci, 9 (1975) p. 299.
5. Vasudevan, A. K., Ludwiczak, E. A., Baumann, S. F., Howell, P. R. and Doherty, R. D., These proceedings.
6. Nes, E. and Ryum, N., Scripta Met, 5 (1971) p. 987.
7. Tosten, M. H., Thompson, S. W., Parayil, T. R. and Howell, P. R., in "Analytical Electron Microscopy - 1984", Eds. Williams, D. B. and Joy, D. C., San Francisco Press, p. 117.
8. Tosten, M. H., Vasudevan, A. K. and Howell, P. R., These proceedings.
9. Tosten, M. H., Vasudevan, A. K. and Howell, P. R., Submitted to Met Trans A.
10. Sanders, T. H., Ludwiczak, E. A. and Sawtell, R. R., Mat Sci Eng, 43 (1980) p. 247.
11. Tosten, M. H. and Howell, P. R., Unpublished research.
12. Noble, B. and Thompson, G. E., Met Sci. J, 6 (1972) p. 167.
13. Weatherly, G. C. and Nicholson, R. B., Phil Mag, 17 (1968) p. 801.
14. Tosten, M. H., Vasudevan, A. K. and Howell, P. R., Submitted to Mat Sci Eng.
15. Vaughan, D., Acta Met, 16 (1968) p. 563.
16. Simon, J. P. and Guyot, P., J. Mat Sci, 10 (1975) p. 1947.
17. Forest, B. and Biscondi, M., Met Sci, 12 (1978) p. 202.
18. Lee, J. K. and Aaronson, H. I., Acta Met, 23 (1975) p. 799.
19. Lee, J. K. and Aaronson, H. I., Acta Met, 23 (1975) p. 869.
20. Kelly, A. and Nicholson, R. B., Prog Mater Sci, 10 (1963) p. 158.
21. Silcock, J. M., Acta Met, 8 (1960) p. 589.

Evaluation of aluminium–lithium–copper–magnesium–zirconium alloy as forging material

P J DOORBAR, J B BORRADAILE and D DRIVER

The authors are in the Materials
R & D Laboratory, Rolls-Royce Ltd,
Derby, UK

SYNOPSIS

In order to exploit the low density and high stiffness of the Al-Li alloy LITAL A for aero-engine components the material has been evaluated as a forged product instead of in its more common sheet form. The forgings have the same aging sequence as sheet with homogeneous δ' matrix precipitation and slip dispersing S and T_1 heterogeneously nucleated at subgrain boundaries and deformation-induced dislocations. Typical aero-engine components had anisotropic mechanical properties due to the presence of unrecrystallised grains elongated in the working direction. The influence of grain morphology on mechanical behaviour is characterised and proposals made for minimising the anisotropy.

INTRODUCTION

LITAL A is an Al-Li-Cu-Mg-Zr alloy, (composition Table 1), with 10% lower density and 10% higher modulus than current aluminium alloys. It is currently available in sheet, plate or extruded form (1) which is often heavily worked and textured, or has been solution heat treated prior to aging to yield mechanical properties equivalent to RR58, (2618) (composition Table 1). The material is also being evaluated as forging stock in which form it would be suitable for such aero-engine components as compressor casings, stators and fan annulus fillers (Figure 1). However, forged components are likely to contain lower levels of retained work than sheet material and post solution stretching would be inappropriate. A preliminary study is therefore reported here which seeks to establish the most suitable processing and heat treatment route for forged components and to characterise the associated microstructure and mechanical properties.

EXPERIMENTAL PROCEDURE

A rectangular bar 65 x 25mm cross section was hand forged from a slice of 180mm diameter homogenised ingot. The bar was solution heat treated at 530°C for 4 hours-prior to aging trials on tensile specimens machined from the bar in the low property transverse direction.

A Spey low pressure (LP) compressor half casing (Fig 2), was then forged from the remainder of the ingot, solution heat treated for 4 hours at 530°C, warm water quenched, (to minimise distortion), and then aged for 8 hours at 200°C, this heat treatment having been selected from the earlier aging trials on the bar material to give an optimum combination of strength and ductility. Mechanical properties obtained from the casing included cold tensile results in the axial, radial and circumferential directions together with elevated temperature tensiles and room and elevated temperature fatigue testing which was restricted solely to the axial direction. Complex precision forged fan annulus fillers (Figure 3) were also produced from 55mm extruded bar and subjected to similar mechanical testing.

Representative optical metallography samples were etched in a solution of 2%HF, 10% HNO_3 in water while transmission electron microscope, (TEM) samples were prepared by a twin jetting method

Table 1 Composition of RR58 and LITAL A

Alloy	Element	Cu	Fe	Li	Mg	Ni	Si	Ti	Zr	Al
RR58	Minimum Maximum	1.8 2.7	0.9 1.4	- -	1.2 1.8	0.8 1.4	0.15 0.25	- 0.2	- -	REM
LITAL A	Minimum Maximum	1.0 1.4	- 0.3	2.3 2.6	0.5 0.9	- -	- -	- -	0.10 0.14	REM
LITAL A	Casing and forged bar	1.2	0.09	2.4	0.59	-	-	-	0.12	REM

Table 2 Room Temperature Tensile Properties

Specimen Position		UTS MPa	0.2%PS MPa	% E
Bar	Transverse	390	305	3.5
	Longitudinal	440	358	8.0
Casing	Axial	434	322	7.4
	Circumferential	451	331	7.3
	Radial	378	298	2.7
Sheet DTDXXXA T6 Spec Min		440	380	6

Table 3

Room Temperature Tensile Properties from Extruded Bar (200°C Age, 8 Hours)

Specimen Position	0.2% PS	UTS MPa	% E
Longitudinal	472	488	3
	454	508	4
Transverse	321	402	4
	314	409	4

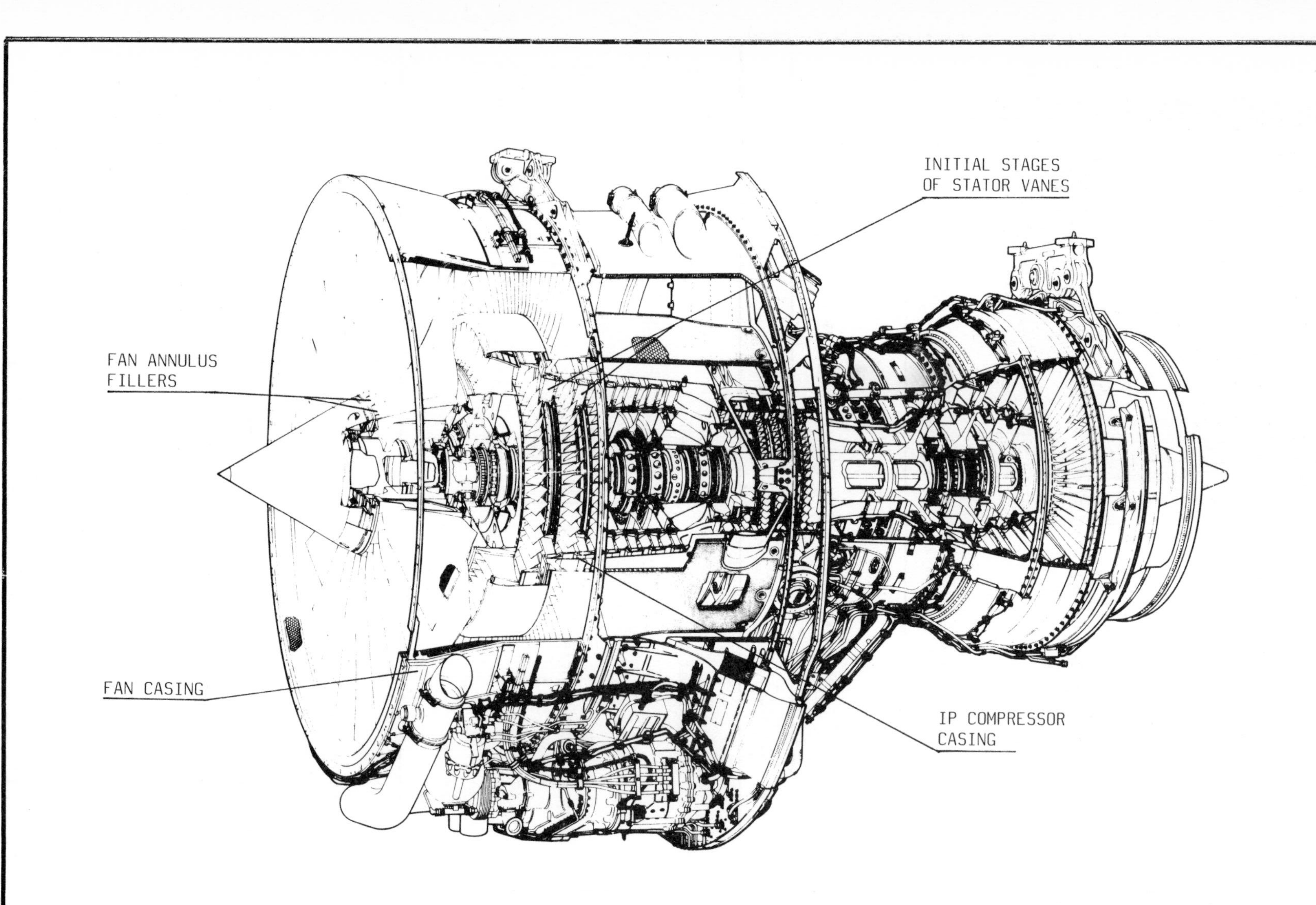

Fig 1 RB211-535 showing possible Al/Li components.

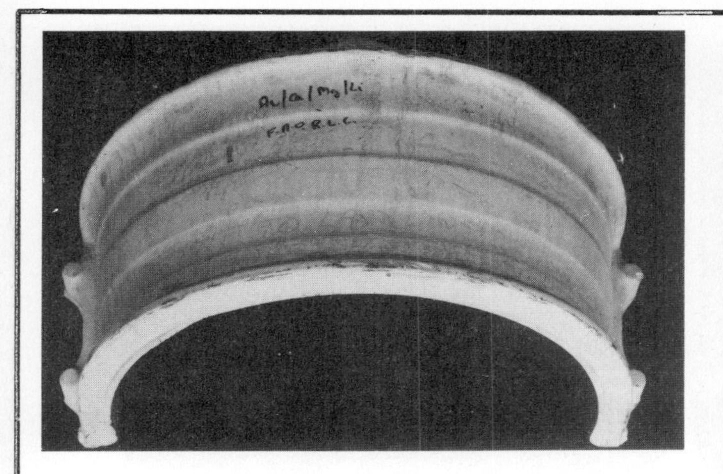

Fig 2 Rolls-Royce Spey LP compressor half casing.

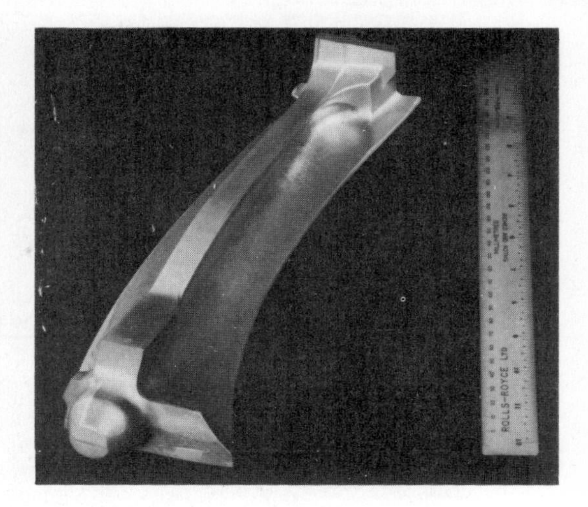

Fig 3 Rolls-Royce RB211 fan annulus filler forging.

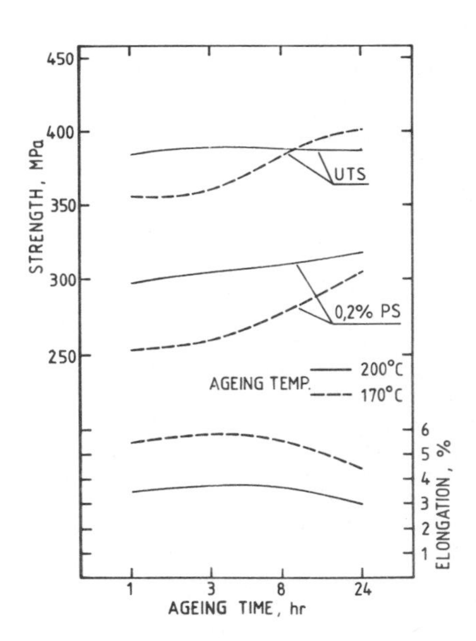

Fig 4 Effects of aging time and temperature on transverse bar tensile properties.

using 10% perchloric acid in ethanol. The TEM examination was carried out on a JEOL 200CX with fractographic characterisation undertaken using a JEOL 35CF scanning electron microscope.

RESULTS & DISCUSSION

Hand Forged Bar

Figure 4 illustrates that increased aging at 170°C improves the UTS and 0.2% PS with a corresponding reduction in ductility whereas the 200°C heat treatment produced a stronger material which was less affected by extended aging times. In consequence 8 hours at 200°C was selected as an appropriate heat treatment and additional testing was undertaken in the longitudinal direction. Table 2 summarises these results and indicates that the expected higher strength and

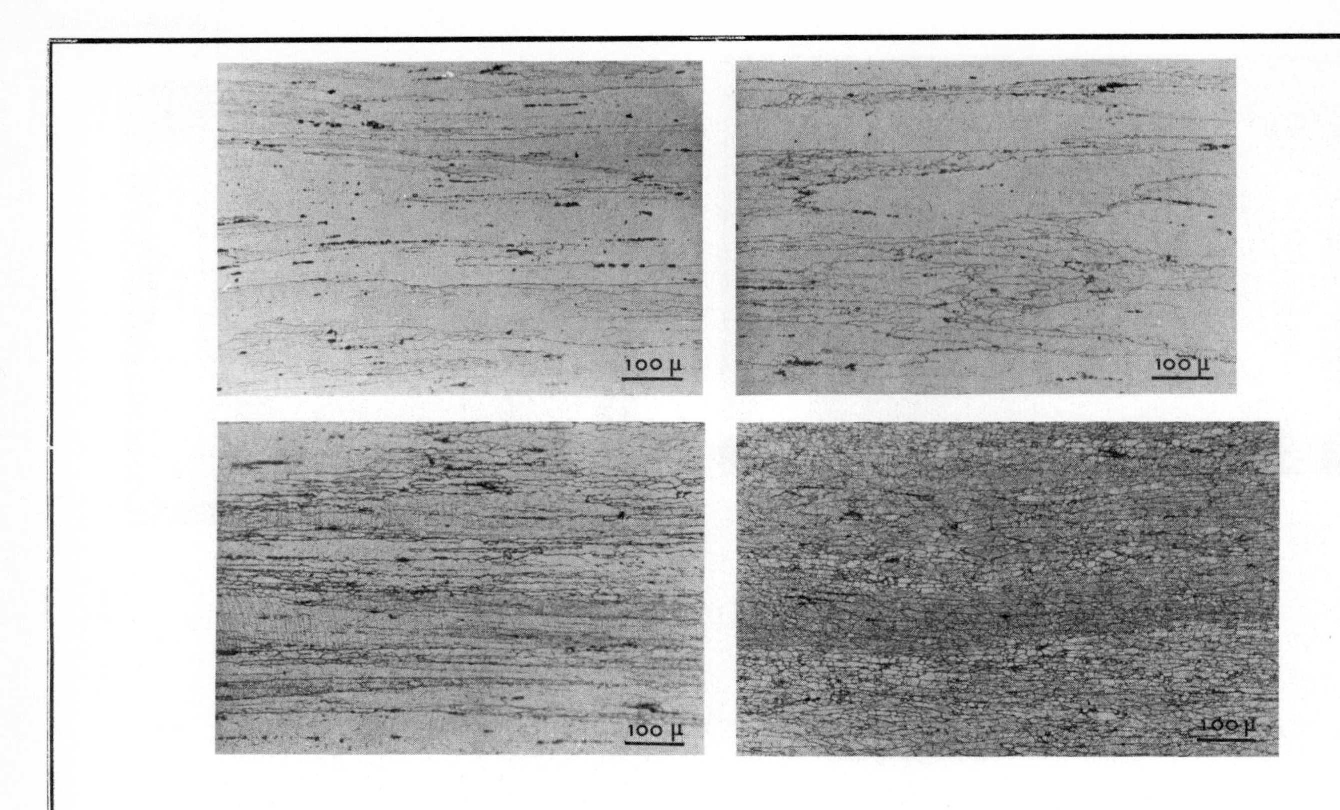

Fig 5 Optical microstructures of transverse bar material aged at 200°C for: (a) 1 hour, (b) 3 hours, (c) 8 hours (d) 24 hours.

ductility in this orientation now approaches that of the commercial sheet product (1).

Optical microscopy indicated that material aged at 200°C contained large elongated grains, (Fig 5), with occasional regions of recrystallisation. Subgrains within the unrecrystallised grains only become clearly visible after prolonged aging (cf Figures 5b and 5d). Material aged at 170°C had a similar grain morphology to the 200°C aged material but no subgrain decoration was observed within the large grains even after aging for 24 hours.

The intragranular precipitation behaviour in forged bar, compressor casing and fan annulus filler material was characterised by TEM. Since the forgings had essentially the same precipitation behaviour, description will be restricted to the bar stock. As has been reported previously (2-4) the predominant strengthening precipitate in such Al-Li alloys is the metastable ordered δ' (Al_3Li) phase which is uniformly dispersed throughout the matrix as fine coherent homogeneous precipitates interspersed with slightly coarser δ' heterogeneously nucleated at pre existing $ZrAl_3$ particles and revealed in δ' [100] dark field micrographs by a characteristic "halo", (Figures 6 and 7). Aging at 170°C and 200°C produced progressive δ' coarsening, the δ' size after 24 hours at 170°C Figure 6(d), being similar to that developed after 3 hours at 200°C, Figure 7(b). This is consistent with the similar 0.2% proof strengths at these two conditions (Fig 4). Prolonged aging times and higher temperatures led to precipitation of both S (Al_2CuMg) and T_1 (Al_2CuLi) at dislocations and subgrain boundaries, (Figure 8), copious heterogeneous precipitation being evident after 24 hours at 200°C, (Figure 9). The S and T_1 precipitates can help to disperse the intragranular planar slip thereby improving toughness (5-8) without significantly reducing strength or ductility (Fig 4). Of the two phases low volume fractions of S should be a more effective barrier to planar deformation than T_1 since the habit plane of S is not coincident with the matrix slip plane (9). In contrast T_1

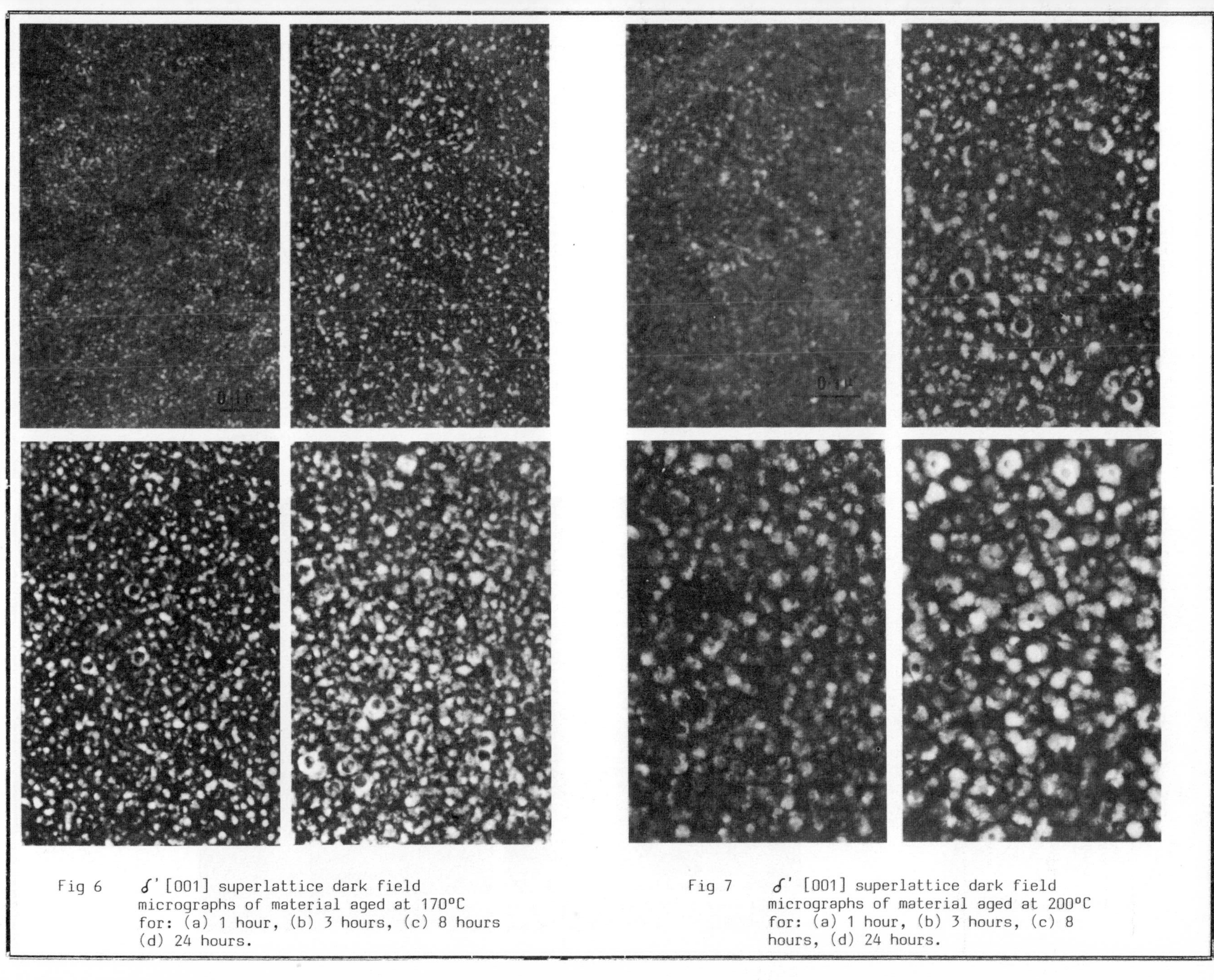

Fig 6 δ' [001] superlattice dark field micrographs of material aged at 170°C for: (a) 1 hour, (b) 3 hours, (c) 8 hours (d) 24 hours.

Fig 7 δ' [001] superlattice dark field micrographs of material aged at 200°C for: (a) 1 hour, (b) 3 hours, (c) 8 hours, (d) 24 hours.

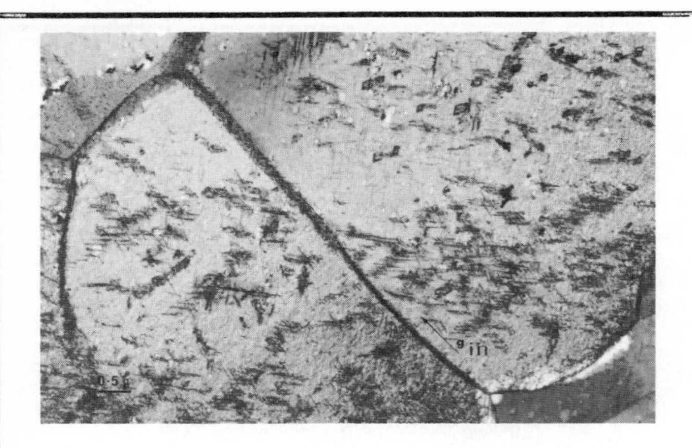

Fig 8 Bright field TEM micrograph from material aged 8 hours at 200°C showing S and T_1 formation within the grains and on subgrain boundaries.

Fig 9 Copious S and T_1 formation on aging for 24 hours at 200°C (a) Bright field (b) g, 3g weak beam image.

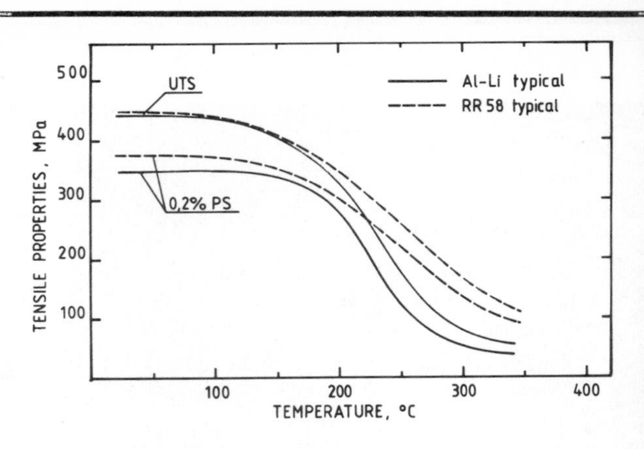

Fig 10 Temperature capability of Al/Li casing material tested in the axial direction.

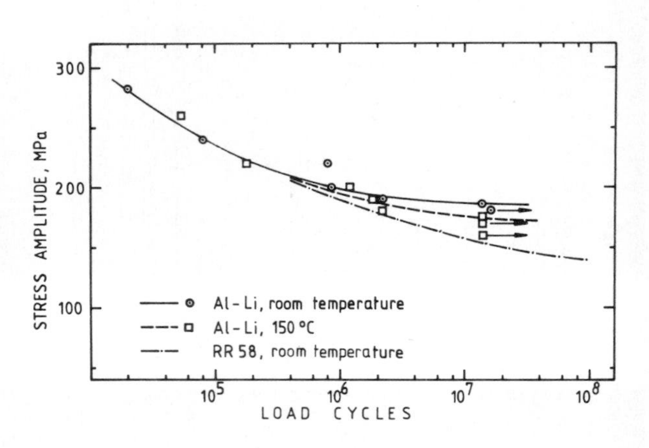

Fig 11 Rotating bend fatigue of Al/Li casing material tested in the axial direction.

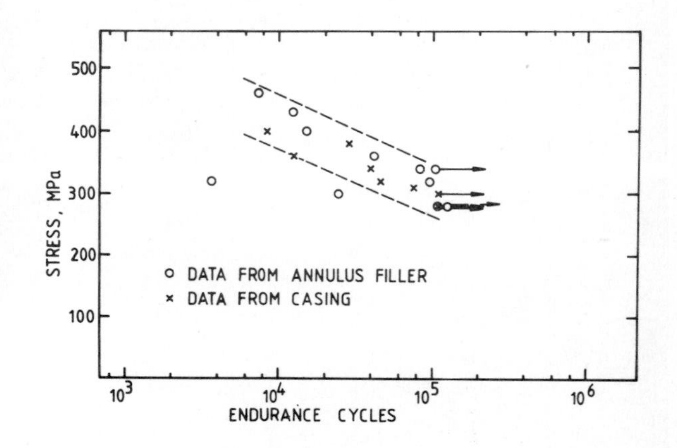

Fig 12 Room temperature low cycle fatigue data for Al/Li forgings. Dotted lines show RR58 scatterband.

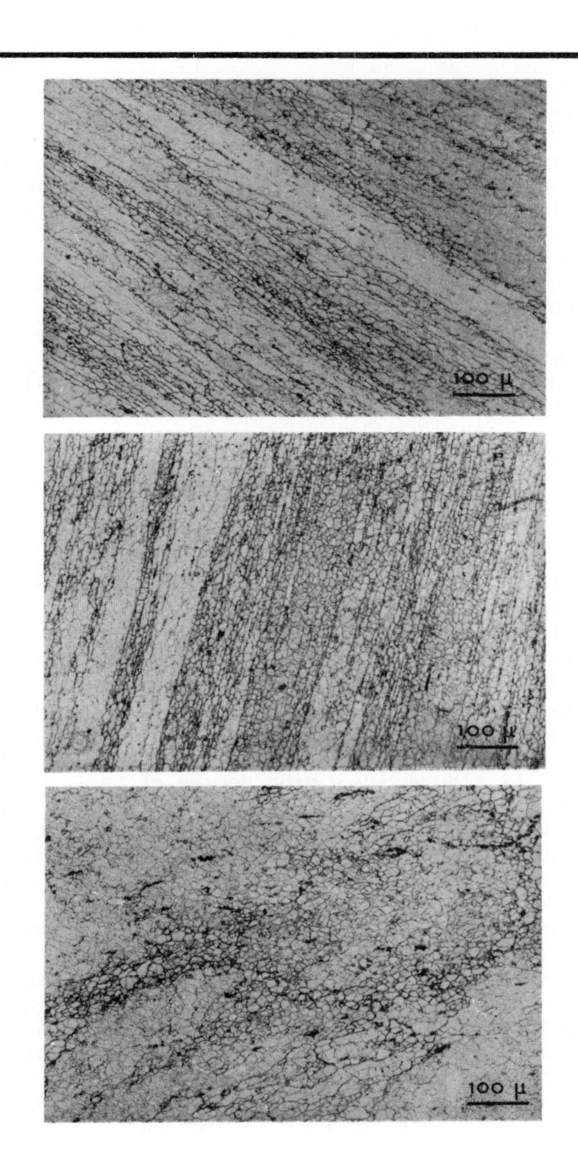

Fig 13 Optical microstructures of casing material: (a) circumferential, (b) axial, (c) radial.

Fig 14 SEM fracture surfaces of casing material: (a) circumferential, (b) radial.

has a {111} habit plane(10). The S precipitate would be favoured by higher Mg levels and therefore further work should concentrate on material with Mg target levels of 0.7 wt % or above (6) instead of the current 0.59 wt% value.

The fracture surfaces of the failed transverse tensile specimens were examined to see whether the different aging treatments produced any differences in failure modes. Irrespective of prior aging times or temperatures, all specimens had similar fracture characteristics, consisting of a mixture of intergranular failure and shear.

After 24 hours at 200°C TEM occasionally revealed PFZ's at grain boundaries, usually associated with grain boundary precipitates believed to be Al_2MgLi, (though diffraction data was inconclusive). No such precipitates or PFZ's were found at subgrain boundaries. Since it has been proposed (11.12) that the presence of PFZ's can cause strain localisation adjacent to grain boundaries leading to increased intergranular failure, the fractures of the specimens aged for 24 hours at 200°C were examined closely for evidence of increased intergranular failure characteristics. No convincing evidence was found.

Spey LP Compressor Casing

To exploit fully the higher modulus and lower density of LITAL A for aero-engine applications, it would be advantageous if the Al-Li alloy had similar temperature capabilities and fatigue characteristics to the established wrought aluminium alloy RR58. Figures 10 to 12 compare the relevant properties, from which it can be seen that in rotating bend and low cycle fatigue the LITAL A casing is at least equivalent to RR58. UTS values are also equivalent up to 150°C though 0.2% proof strength is consistently lower.

The particular advantage of RR58 is that the S phase remains stable to higher temperatures than the rapidly overaged δ' phase in Al-Li, resulting in the superior properties of RR58 above 200°C. Minor alloying additions capable of increasing S precipitation and maintaining the stability of the δ' phase to higher temperatures could extend the engineering potential of the Al-Li material. However, its current properties look sufficiently attractive to justify further evaluation, particular attention being given here to characterising and minimising the anisotropy in the forged product.

Room temperature tensile properties for the forged casing are given in Table 2 from which it can be seen that the axial and circumferential properties are significantly higher than those in the radial direction. Such low radial results can be compared with similar low properties in the short transverse direction of unrecrystallised plate material (13) and are attributable to the different grain morphology in the different orientations.

The optical micrographs of Figure 13 illustrate the grain structures in the circumferential, axial and radial directions. Large elongated unrecrystallised grains are clearly visible in the circumferential and axial sections coupled with areas of relatively fine recrystallised grains, while the radial section shows the unrecrystallised grains to have a more uniform aspect ratio. The two groups produce distinct fracture modes, (Figure 14), circumferential and axial failures occurring by transgranular shear through the unrecrystallised grains with a discontinuity generated at each intersected grain boundary, (Figure 14a), while the lower property radial sections failed by intergranular separation, (Figure 14b). The three dimensional relationship between grain morphology and fracture characteristics is illustrated schematically in Fig 15.

Previous work (7) has indicated that the anisotropi failure is compounded by texturing within the unrecrystallised regions in which adjacent grains often have less than a 3° crystallographic misorientation. Such low angle grain boundaries do not provide an effective obstacle to planar slip and the fracture passes through a number of unrecrystallised "pancake" grains without significant deviation. The present LP compressor material also shows a high degree of texture (Figure 16), which can be described as (110) [223] but with a superimposed weak cube texture, (probably from the fine recrystallised grains). In this case, however, the grain boundaries appear to be relatively effective obstacles to deformation due in part to the presence of fine recrystallised grains adjoining the unrecrystallise regions and to the intergranular precipitation of S which helps disperse the slip (5-7).Figure 18b shows a typical section through a longitudinal fan annulus filler or circumferential/axial compressor casing test piece failure in which transgranular failure is restricted to individual unrecrystallise grains with more tortuous intergranular cracking in the adjoining fine grain regions. This contrasts with characteristic shear facets of earlier work on sheet material (Figure 3 reference 7) in which the fracture traverses numerous grains.

Fan Annulus Filler

In order to investigate further the effects of grain morphology, a number of fan annulus fillers were precision forged from extruded bar with a small (ASTM 4.5) transverse grain size. Inevitably, however, the extruded bar had a "fibrous" structure in the longitudinal direction consisting of large unrecrystallised elongated grains interspersed with small recrystallised regions. Mechanical properties in the bar were markedly anisotropic, (Table 3), with longitudinal proof strengths approximately 150MPa higher than their transverse equivalents.

On forging the fan annulus fillers, the "fibrous" grain structure from the original bar was retained and partially redistributed throughout

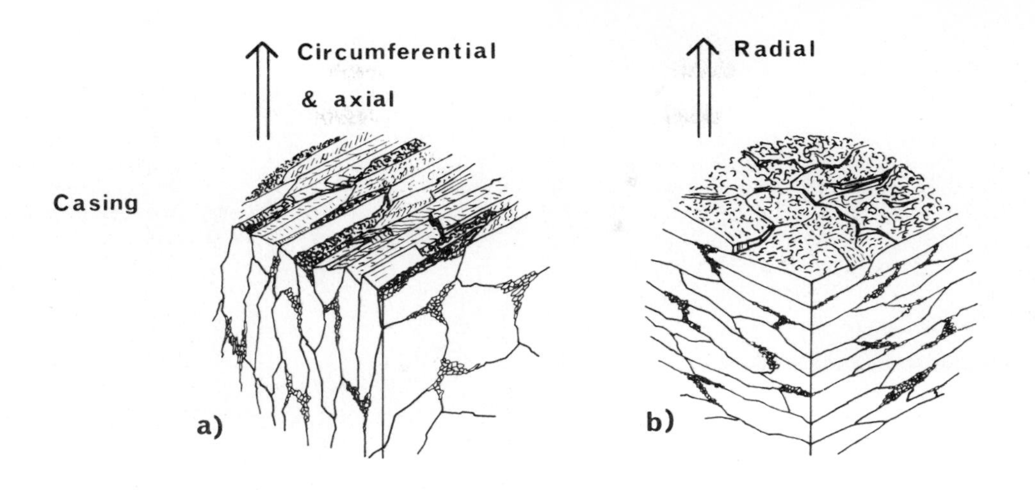

Fig 15 Schematic of casing tensile failures: (a) Largely transgranular with some intergranular fracture giving good circumferential properties, (b) Mainly intergranular fracture giving poor radial properties.

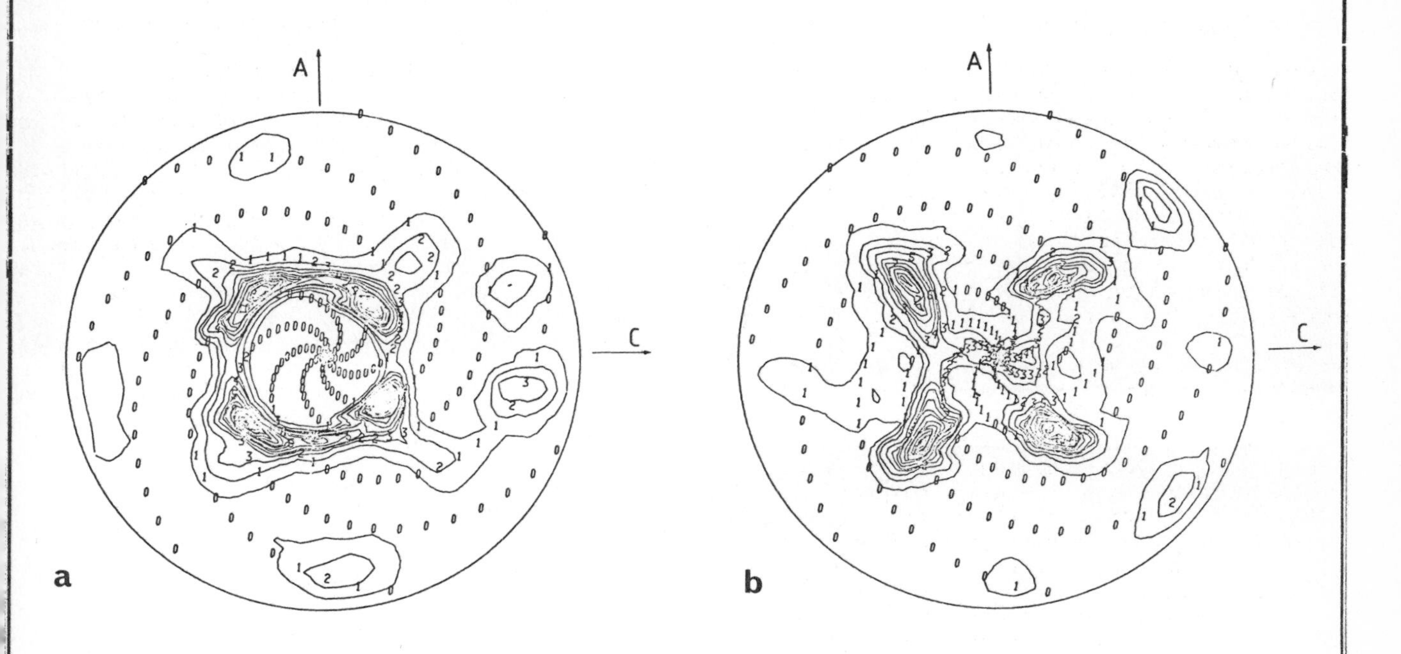

Fig 16 Pole figures from casing material: (a) (111), (b) (200). A represents the axial direction and C the circumferential.

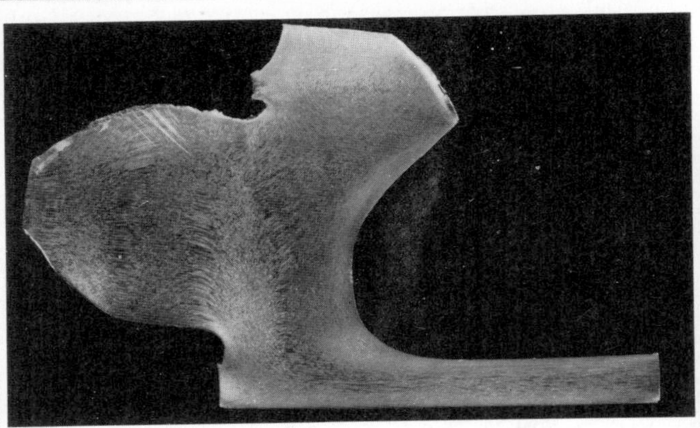

Fig 17 Grain flow in the end fixing of a fan annulus filler showing alternate longitudinal and transverse grain flow.

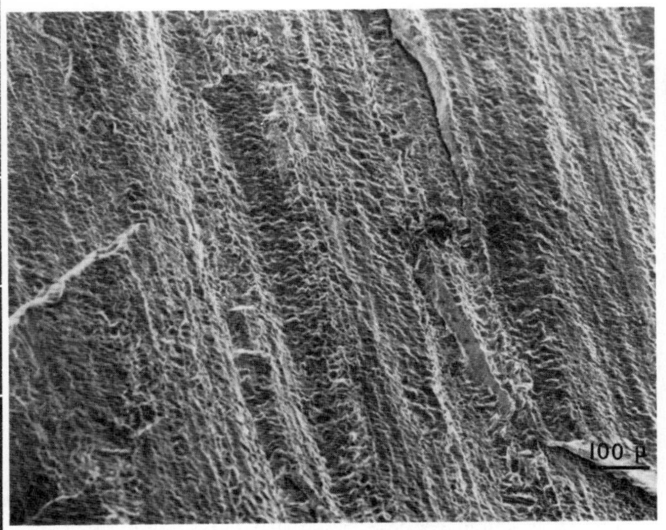

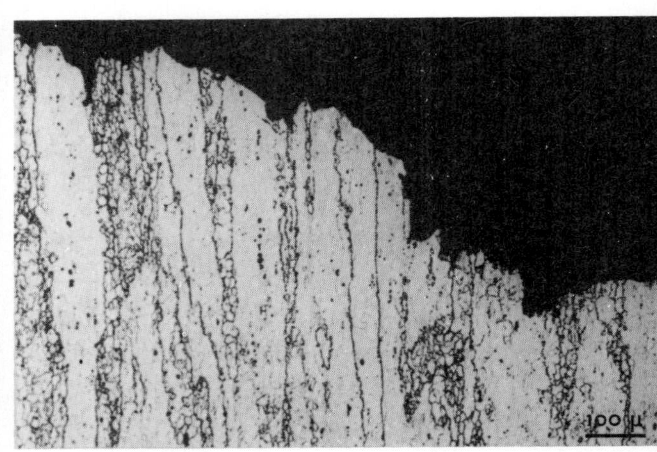

Fig 18 (a) Tensile fracture surface of longitudinal fan annulus filler specimen: (b) Optical microsection through the fracture surface.

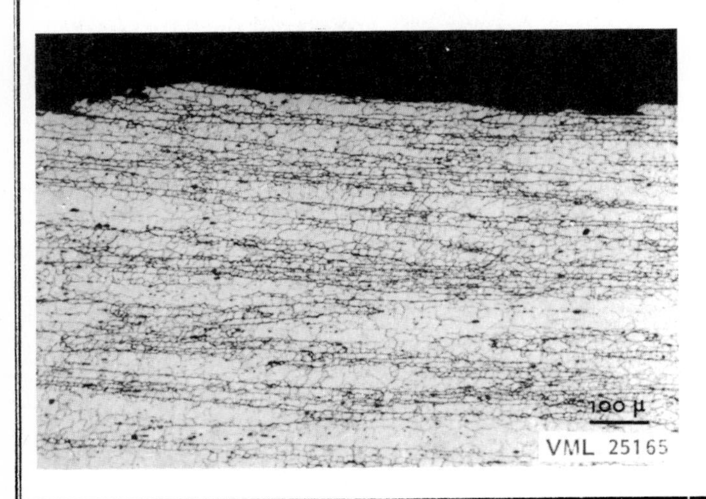

Fig 19 (a) Tensile fracture surface of transverse fan annulus filler specimen: (b) Optical microsection through the fracture surface.

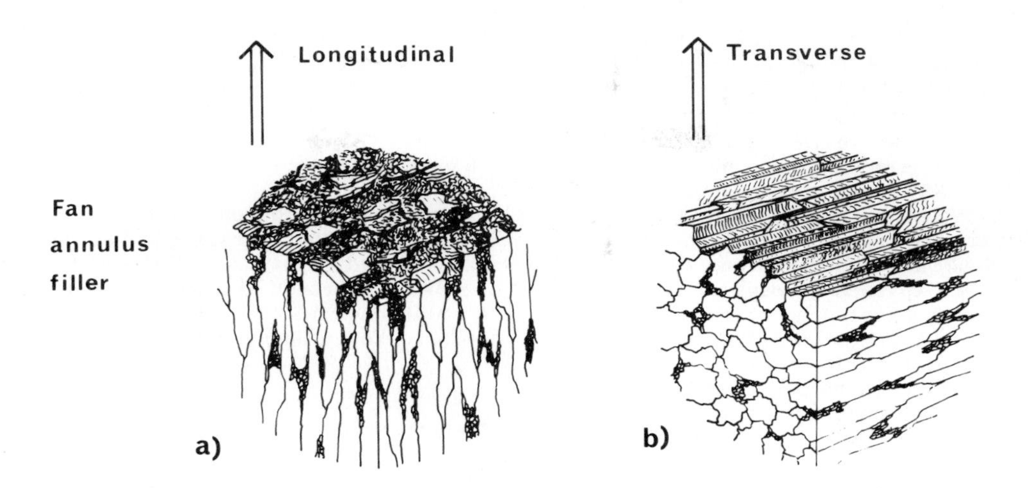

Fig 20 Schematic of fan annulus filler tensile failures: (a) A mixture of transgranular and intergranular failure giving good longitudinal properties. (b) Mainly intergranular fracture giving poor transverse (and end fixing) properties.

Fig 21 Fracture surface of good low cycle fatigue specimen from the longitudinal direction of the fan annulus filler forging.

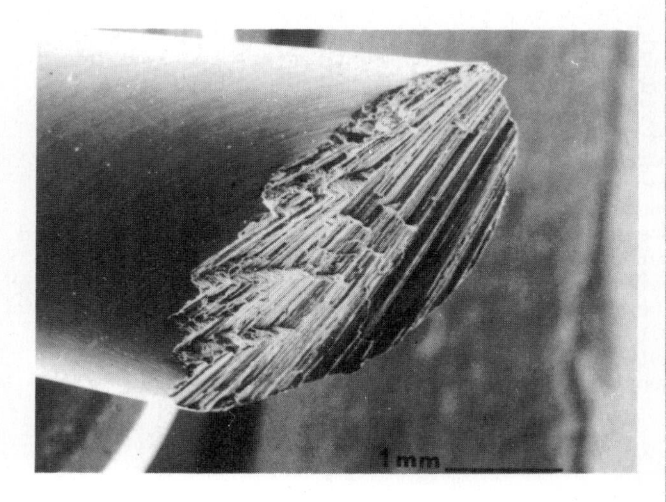

Fig 22 Fracture surface of poor low cycle fatigue specimen from the end fixing of the fan annulus filler forging.

the component. In areas of complex flow, for example at the end fixing, (Figure 17), the grain structure has "buckled" giving the appearance of alternate longitudinal and transverse grain flow with correspondingly variable property levels.

Mechanical test pieces cut from the forged fan annulus filler gave 0.2% proof strengths of around 420-490MPa and elongations up to 9% in the longitudinal direction while in regions of complex grain flow, such as the end fixing, 0.2% proof strengths of 310-360MPa were obtained with elongations of 2-5%. Optical micrographs and fractographic features from the two types of structures are shown in Figures 18 and 19. Again the grain morphology is seen to have a major influence on the mechanical properties and failure modes. In the longitudinal orientation high room temperature tensile properties are achieved as the failure is forced to propagate through unrecrystallised grains (Figure 18 and schematically in Figure 20a). In contrast the weaker transverse properties arise mainly by intergranular fracture between the elongated unrecrystallised grains (Figure 19 and schematically in Figure 20 b.

Low cycle fatigue results from the fan annulus filler (Figure 12), tend to be higher than those from the casing because most LCF specimens were taken from those regions of the fan annulus filler which had the "fibrous" longitudinal grain flow parallel to the specimen axis. A typical failed LCF specimen in this orientation is shown in Figure 21. However, two results fell below the RR58 scatter band (Figure 12). These two samples were taken from the "buckled" region of the fan annulus filler and the fibrous grain flow was disturbed such that the elongated grains were inclined at an angle to the axis of the fatigue specimen. Figure 22 shows that the resulting fracture was similarly inclined, crack propagation being accommodated mainly along the relatively weak grain boundary regions.

In order to exploit the present Al-Li alloys more fully for forged components it will be necessary to minimise the property anisotropy caused by the elongated unrecrystallised grains. One approach would be to use a fully recrystallised material as forging stock and introduce some retained work by controlled deformation to generate the appropriate shape and to provide nucleation sites for the slip dispersing S within the matrix δ' precipitation. Work is in hand to generate forged components via this route.

CONCLUSIONS

1. Forged LITAL A has the same matrix precipitation and heterogeneously nucleated S and T_1 as the sheet material.

2. The mechanical properties of the forged components are anisotropic with high properties in the direction of forging grain flow and low properties normal to grain flow.

3. High property material tends to fail in a mixed transgranular and intergranular mode while low property material fails predominantly in an intergranular mode.

4. The use of recrystallised bar stock as starting material is proposed as a means of minimising the property anisotropy.

ACKNOWLEDGEMENTS

Forging was carried out at High Duty Alloys, Redditch and their assistance is gratefully acknowledged. Thanks are also due to Dr C J Peel of RAE Farnborough for helpful discussions and to colleagues in the Materials R & D Laboratory at Rolls-Royce Derby for practical assistance.

REFERENCES

1. British Alcan Aluminium Ltd. Publ. No. PD402/3M 5.5.83. "Aluminium Lithium Alloys Development".

2. Noble B and Thompson G E. Metal Sci. J (1971) 5 114.

3. Williams D B and Edington J W. Metal Sci J (1975) 9 529.

4. Gregson P J and Flower H M. J Mat Sci Letters (1984) 3 829.

5. Gregson P J, Flower H M, Peel C J and Evans B. Inst of Met Conf March (1983) Loughborough. "Metallurgy of Light Alloys" 57

6. Peel C J, Evans B, Baker C A, Bennett D A, Gregson P J and Flower H M. Aluminium-Lithium Alloys II. Sanders T H and Starke EA (Eds) AIME - NY (1983) 363

7. Gregson P J and Flower H M Acta Met (1985) 33 527.

8. Grimes R, Cornish A J, Miller W S and Reynolds M A. Metals & Mat June (1985) 357.

9. Wilson R N and Partridge PG, Acta Met (1965) 13 1321.

10. Noble B and Thompson G E. Metal Sci J (1972) 6 167.

11. Sanders T H. Aluminium-Lithium Alloys, Sanders T H and Starke E A (Eds) AIME-NY (1981) 63.

12. Starke E A and Lin F S. Metall Trans A (1982) 13 2259.

13. Peel C J Private communication.

Coarsening of δ', T_1, S' phases and mechanical properties of two Al–Li–Cu–Mg alloys

M AHMAD and T ERICSSON

The authors are in the Department of Mechanical Engineering, Linköping Institute of Technology, S-581 83 Linköping, Sweden

The ageing response of the alloy has been determined by hardness measurements. The microstructures developed after various artificial ageing treatments were studied by transmission electron microscopy (TEM). The effect of prestretch on the precipitation of S' and T_1 was investigated. The density of S' and T_1 was increased due to the prestretch. A significant improvement in the yield strength was obtained by stretching prior to ageing. The ultimate tensile strength was also slightly improved, while the elongation to fracture was slightly reduced. In the under-aged condition deformation was localized within slip bands. The deformation was homogeneous for peak aged condition. The fracture surfaces of failed tensile specimens were studied with the help of scanning electron microscopy (SEM). The mechanical properties and fracture behaviour of the alloy has been correlated with microstructural features.

Introduction

The combination of low density and high stiffness makes the aluminium alloys containing lithium very attractive for applications in airframe and other structural components. In the binary Al-Li alloy system the δ' (Al_3Li) particles precipitate throughout the matrix and are responsible for strengthening. In the present alloy system apart from the δ', semicoherent S' (Al_2CuMg) and T_1 (Al_2CuLi) phase precipitate during artificial ageing treatment. Small amounts of plastic deformation prior to ageing are known to enhance age-hardening in a number of Al-Cu-Mg alloys (1-3). Lin et al. (4) showed that the density of T_1 precipitates was increased and as a result of this the strength was improved in the Al-Li-Cu alloy system. The objective of this study was to correlate the effect of stretching prior to ageing with the microstructure and mechanical properties of the alloy.

Experimental Procedure

A sheet of 1,6 mm thickness produced by British Alcan Aluminium was received in the as rolled condition. The composition of the alloy is given in table 1.

Solution heat treatment was carried out for 20 minutes in a salt bath at 530°C. After the solution treatment strips were quenched into water at 20°C. A 2,5% stretch was applied to the some of the specimens before artificial ageing treatment. The ageing treatments were carried out in a silicone oil bath operating at 185°C and 190°C. The ageing response of the alloy was determined by hardness measurements made on rectangular pieces cut from the sheet. After heat treatment they were ground and polished before measuring the hardness. Specimens for optical metallography were etched in a solution of 25 vol % nitric acid and 75 vol % water at 70°C.

Strips for making tensile specimens were machined both in the longitudinal and long transverse direction. The tensile specimens had a gauge length of 25 mm and a width of 6,2 mm. The tensile data presented in table 2 is based on the average of three tensile tests. An 25 mm extensometer was used to measure the elongation to fracture. Both

TABLE 1

Alloy Composition (in weight percent)

Li	Cu	Mg	Zr	Fe	Si	Ti	Al
2,55	1,97	0,78	0,13	0,06	0,08	0,02	balance

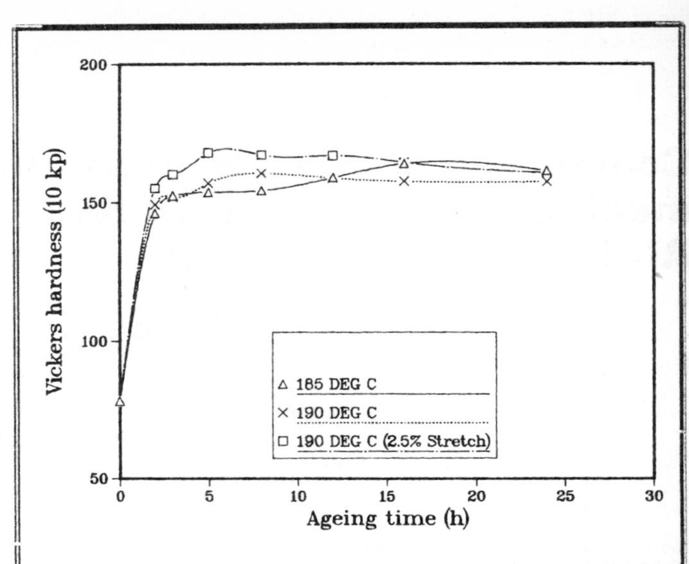

Fig.1 Hardness vs ageing time at 185°C, and 190°C with and without a 2,5% stretch prior to ageing.

TABLE 2 TENSILE PROPERTIES

Longitudinal Direction

Ageing	Yield stress at 0,2% offset (MPa)	UTS (MPa)	% elongation in 25 mm
as quenched (naturally aged 42 hours)	284	441	11,2
5h at 190°C	403	500	5,2
2,5% stretch + 5h at 190°C	487	547	4,8
8h at 190°C	421	526	5,4
2,5% stretch + 8h at 190°C	495	555	5,6
16h at 190°C	431	539	7,0
2,5% stretch + 16h at 190°C	491	538	5,3
Long Transverse Direction			
5h at 190°C	409	518	9,7
16h at 190°C	426	520	10,5

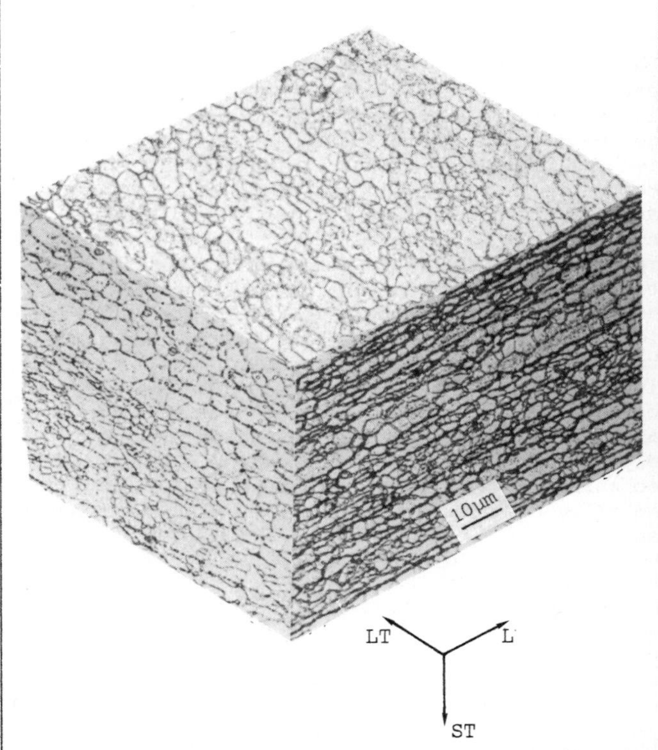

Fig.2. Optical photomicrographs showing grain structure in the solution treated and aged condition.

the prestretching treatment and tensile tests were conducted at a strain rate of $2{,}4 \cdot 10^{-4}$ per second.

Thin foils for Scanning Transmission Electron Microscopy (STEM), Philip EM 400, were prepared using the jet polishing technique with a solution of nitric acid and methanol (1:2) at $-15^{o}C$.

Fractographic study was carried out on failed tensile specimens using JEOL 25 Scanning Electron Microscope (SEM). Analysis of large inclusion particles and primary phases was carried out by Energy Dispersive Spectroscopy (EDS) coupled with SEM.

Results and Discussion

Ageing Behaviour and Microstructure

Fig. 1 shows measured hardness values versus ageing time for the ageing temperature $185^{o}C$, and $190^{o}C$ with and without a 2,5% prestretch. It is obvious that prestretching and higher ageing temperature increased the rate of ageing. Moreover, the material reaches a higher maximum hardness if it has received cold deformation before ageing. There is a slight indication that ageing at $185^{o}C$ compared to $190^{o}C$ gave higher maximum hardness but at longer ageing times.

A composite of optical photomicrographs show the grain structure in the solution treated and artificially aged condition, fig. 2. Both the TEM and optical microscopy revealed an unrecrystalized subgrain structure. Specimens aged 5 hours and 16 hours at $190^{o}C$ both with and without a prestretch were chosen for TEM investigations. A typical electron micrograph of the specimen aged 5 hours is shown in fig. 3a. The S' (Al_2CuMg) and T_1 (Al_2CuLi) phase have precipitated on a subgrain boundary and at dislocations. The density of quenching defects is low in the Al-Li-Cu-Mg alloy system. This has been attributed to the high binding energy of a lithium-vacancy couple (5,6). Both the S' and T_1 phase precipitate preferentially on quenched defects, favourably oriented subgrain boundaries and dislocations. In order to increase the number of dislocations a 2,5% stretch was given to the alloy immediately after the solution treatment. Fig. 3b is a dark field electron micrograph of the specimen which has received a 2,5% stretch before it was aged for 5 hours. Corresponding selected area diffraction (SAD) micrograph is also presented in fig. 3b, foil orientation [112]. The SAD micrograph show streaking along <110> and <210> aluminium directions caused by S' precipitates. The S' laths lying on {210} planes having common <100> growth direction can form composite sheets (7). The corrugated sheets of S' precipitates are seen in fig. 3b. Comparison of fig. 3a and 3b clearly shows that the density of S' precipitates has increased due to the prestretch. Dark field TEM micrographs in fig. 4 are from the specimens solution treated, stretched and aged for 16 hours at $190^{o}C$. Fig. 4a shows the S' laths lying along <100> Al directions, foil orientation [001]. The δ' (Al_3Li) precipitates are also visible due to the contribution of nearby δ' diffraction spot. Streaking in the <100> Al direction is caused by the S' laths. The diffraction spots of the T_1 phase are also present in the SAD micrograph, fig. 4a. One variant of T_1 and two variants of S' precipitates are seen in fig. 4b, foil orientation [112]. In addition the δ' precipitates are also seen.

The highest maximum hardness, and highest yield and tensile strength (as it will be seen later on) were obtained when the specimens have undergone cold deformation before ageing. This is explained in terms of increased nucleation sites for S' and T_1 precipitates. New dislocations which are created during cold deformation before ageing act as precipitate nuclei. Increased rate of ageing is probably a consequence of enhanced diffusion caused by the new vacancies formed during stretching treatment.

Tensile Properties

The tensile properties of the alloy are given in table 2. The stretched material reached the highest strength after 8 hours ageing. The elongation to fracture was then 5,6%. The unstretched material reached the peak strength after 16 hours ageing. The data show that a significant improvement in the yield strength, from 431 MPa to 495 MPa, can be achieved by stretching prior to ageing. The ultimate tensile strength data also showed a slight improvement from 539 MPa to 555 MPa. The elongation to fracture was reduced from 7,0% to 5,6%. The elongation to fracture was higher in the long transverse direction in comparison with the longitudinal direction. Both the yield strength and tensile strength were higher in the transverse direction for 5 hours ageing, but for 16 hours ageing the opposite was true.

In order to investigate the deformation substructure thin foils were taken just behind the fracture surface. The tensile specimens aged 5 hours showed planar slip bands, fig. 5a. The slip have propagated across several low angle boundaries. In contrast to the 5 hours ageing, the specimens aged 16 hours showed almost homogeneous deformation fig. 5b. According to Sanders and Starke (8) and Noble et. al. (9) the planar slip in the Aged Al-Li alloys is caused by the shearable δ' particles. The coherent δ' particles can be easily sheared by the moving dislocations since their slip systems are coincident with those of the matrix. Once the δ' precipitates have been sheared, the flow stress is reduced on the slip plane and further slip is encouraged on that plane. Consequently, intense planar slip occurs during deformation, leading to dislocation pile up at grain boundaries. The planar localized slip produces stress concentrations at the grain boundary and this in many cases can shear the grain boundary (8). The δ' precipitates do not become incoherent within practical ageing times. In the present alloy it is thought that the semicoherent S' precipitates should make a major contribution to achieve homogenous deformation. The T_1 precipitates may also make some contribution.

The δ' particles in the present alloy system precipitate in the early stages of ageing. With continued ageing the S' and T_1 precipitate on heterogenous nucleation sites such as quenching defects and dislocations. In the underaged condition i.e.

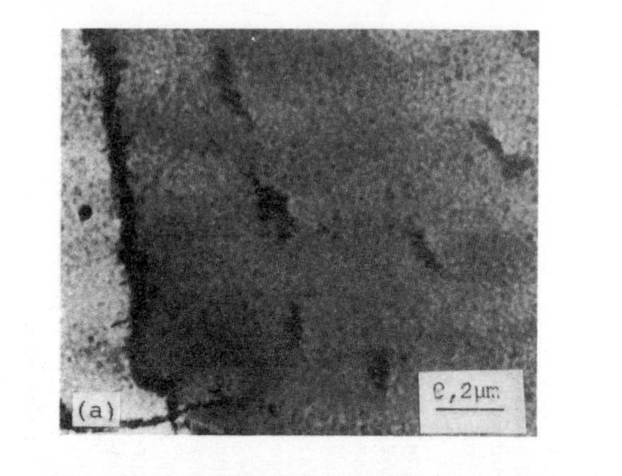

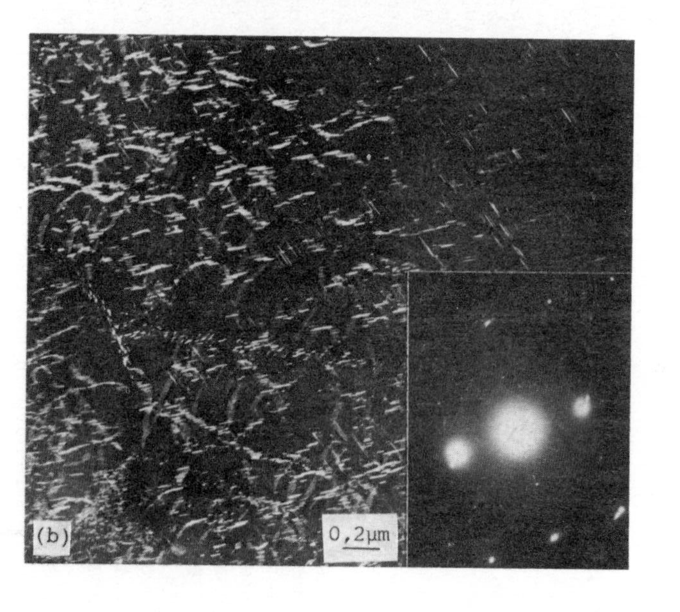

Fig.3 TEM micrographs showing the effect of stretch on the precipitation of S' and T_1, (a) bright field micrograph, 5 hours ageing, (b) dark field micrograph taken with S' streaks, 2,5% stretch + 5 hours ageing, zone axis $[112]_{Al}$.

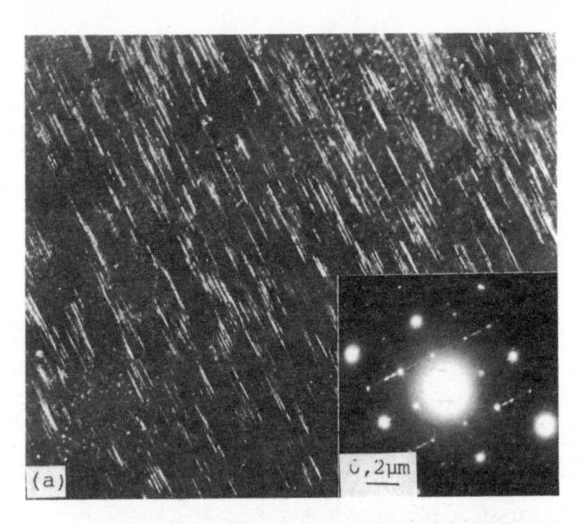

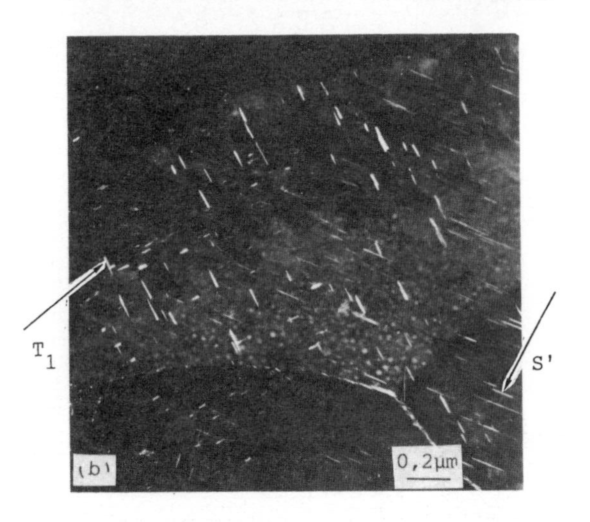

Fig.4 TEM micrographs showing distribution of S' and T_1 precipitates, 2,5% stretch + 16 hours ageing, (a) S' laths along $<100>_{Al}$ direction, zone axis $[001]_{Al}$, (b) T_1, S' and δ' precipitates, zone axis $[112]_{Al}$.

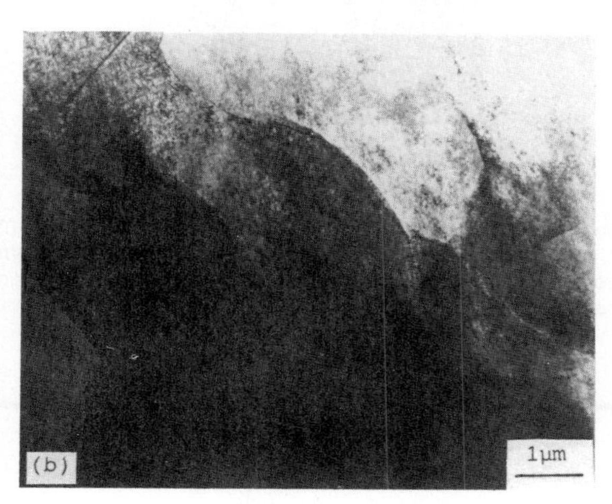

Fig.5 Deformation substructure near the fracture surface of tensile specimens, (a) planar slip in the alloy stretched and aged for 5 hours, (b) homogeneous deformation in the alloy stretched and aged for 16 hours.

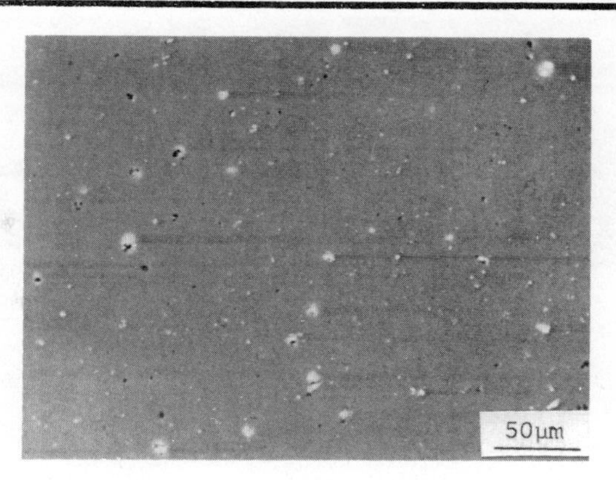

Fig.6 SEM micrograph of a polished surface, showing distribution of large constituent particles in the solution treated and aged condition.

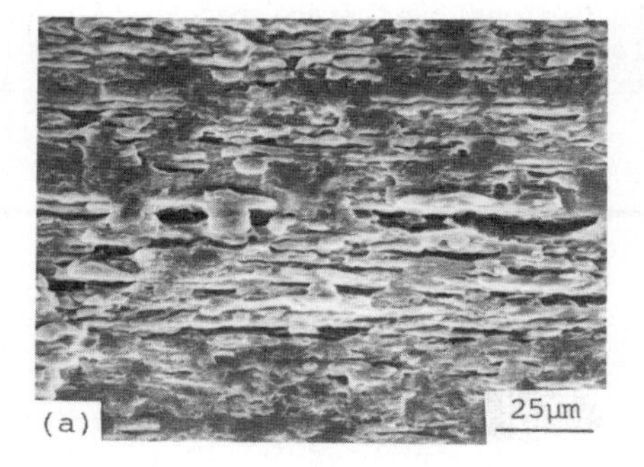

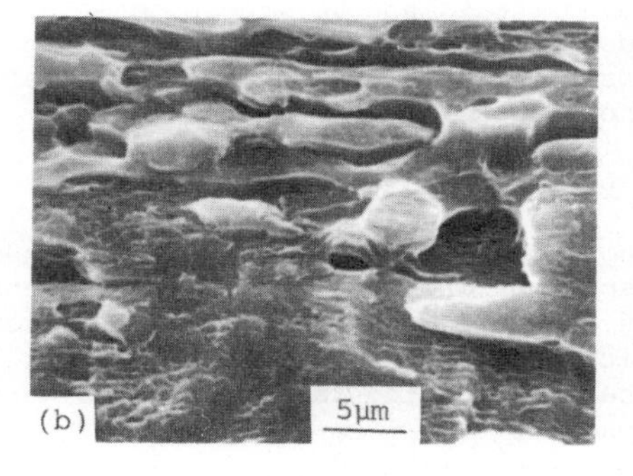

Fig.7 SEM micrographs showing secondary cracks, transgranular and intergranular fracture regions, 8 hours ageing, (a) low magnification micrograph, (b) high magnification micrograph of (a).

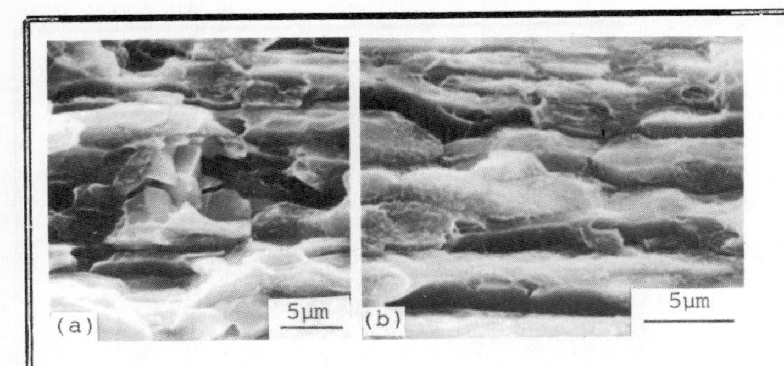

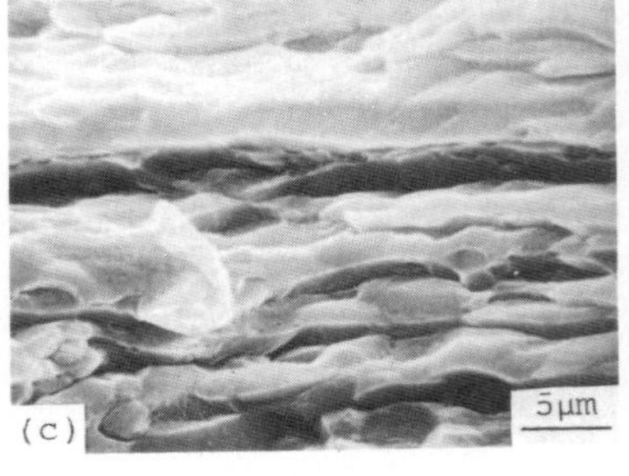
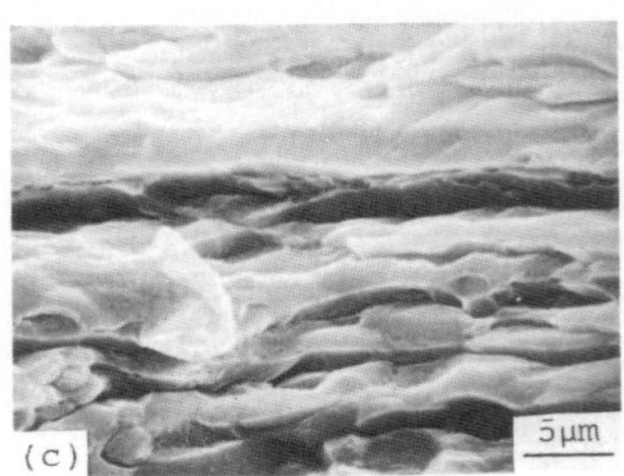

Fig.8 SEM micrographs showing intersubgranular fracture regions, (a) fractured T_2 (Al_6CuLi_3) particle, (b) subgrains covered with dimples, 2,5% stretch + 16 hours ageing, (c) subgrain fracture in 5 hours ageing condition.

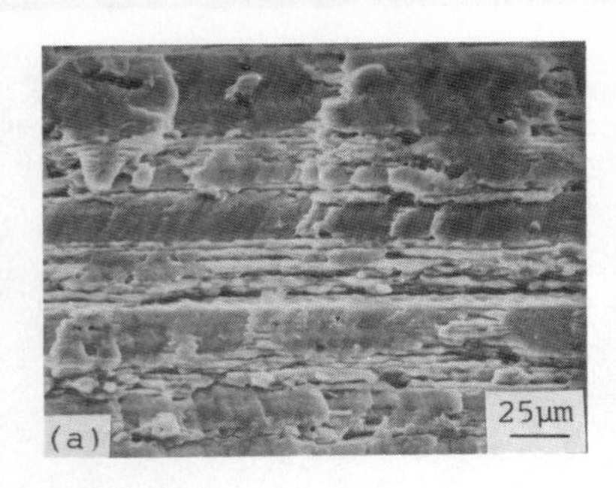

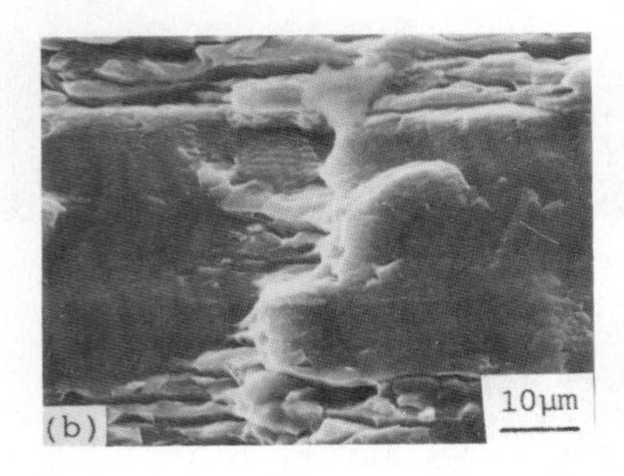

Fig.9 SEM micrographs showing slip band fracture in under-aged condition (5 hours), (a) low magnification micrograph, (b) high magnification micrograph of (a).

5 hours ageing the S' and T_1 precipitates are heterogenously distributed and their density is too low to prevent slip localization. For the 16 hours ageing the S' and T_1 precipitates are more homogenously distributed in the matrix and their density is higher and consequently deformation becomes homogenous.

Fractography

A scanning electron micrograph of a polished section of specimen in the solution treated and aged condition show the distribution of large constituent particles and inclusions, fig. 6. EDS analysis showed that most of the large particles were equilibrium phase T_2 (Al_6CuLi_3) or (Al_5CuLi_3) and few of them were Al_7Cu_2Fe. Many silicon rich particles were also found. The large T_2 particles observed on the fracture surface of failed tensile specimen were usually broken into pieces, fig. 8a.

The fracture plane was inclined approximately 45° to the stress axis for the specimen in the longitudinal direction while for the specimen in the long transverse direction the orientation was more complex. The SEM examination of fracture surfaces of failed tensile specimens showed parallel arrays of secondary cracks moving along high angle grain boundaries. The secondary cracks were found in the as quenched condition, as well as in all aged conditions. Another common observation of the aged specimens was that the fracture surfaces contained flat featureless areas and areas with intergranular type fracture (fig. 7a). The flat featureless areas showed shallow dimples at higher magnification indicating a transgranular ductile type fracture mode, fig. 7b. The high magnification SEM micrographs of the regions of intergranular fracture revealed shallow dimples on subgrain boundaries indicating a microscopic ductile type fracture mode, fig. 7b and fig. 8b. However, the specimen aged 16 hours, fig. 8b, compared to the specimen aged 5 hours, fig. 8c, showed more pronounced dimples on the subgrain boundaries. The shallow dimples formed on the subgrain boundaries may be the result of incoherent precipitates along the subgrain boundaries. Furthermore well formed ridges were observed on the fracture surface of specimen in the long transverse direction for the 5 hours ageing condition, fig. 9. The well formed ridges were absent on the fracture surface of specimen aged 16 hours. The ridges probably correspond to the coarse slip bands showed in the TEM micrograph, fig. 5a.

Conclusions

- A stretch before artificial ageing increased the rate of ageing. A microstructure study showed that the density of S' and T_1 precipitates was also increased due to the prestretch.
- A significant improvement in the yield strength from 431 MPa to 495 MPa was obtained by stretching prior to ageing. The ultimate tensile strength was then increased from 539 MPa to 555 MPa. The elongation to fracture was reduced from 7,0% to 5,6%. The improved strength was attributed to the increased nucleation sites for S' and T_1 precipitates.
- In the under-aged condition i.e. 5 hours the deformation was localized within slip bands, while for 16 hours ageing the deformation was homogeneous. The increased volume fraction of S' after 16 hours ageing is responsible for preventing slip localization.
- The ridges corresponding to the slip bands were observed on the fracture surface of specimens in the under-aged condition.
- Large secondary cracks were observed on the fracture surface of as quenched condition, as well as in all aged conditions. The fracture surfaces of aged tensile specimens contained both transgranular and intergranular fracture regions. In the peak aged and slightly overaged condition the subgrain boundaries were covered with shallow dimples indicating a microscopic ductile fracture mode.

Acknowledgments

This work has been carried out with the financial support of the National Swedish Board for Technical Development (STU) and Saab-Scania AB. Discussions with Dr. S. Johansson on various aspect of the work have been useful. The help of Dr. C. Burman with EDS analysis is appreciated.

References

1. J. T. Vietz and I. J. Polmear
   J. Inst. Metals, 94, 410 (1966)
2. D. Broek and C. Q. Bowles
   J. Inst. Metals, 99, 255 (1971)
3. A. A. Tavassoli
   Metals Science, 8, 424 (1974)
4. F. S. Lin, S. B. Chakrabortty, and E. A. Starke, Jr
   Metallurgical Transactions A, 13A, 401 (1982)
5. M. Ahmad and T. Ericsson
   Scripta Met., 19, 457 (1985)
6. P. J. Gregson and H. M. Flower
   Acta Metall., 33, 527 (1985)
7. R. N. Wilson and P. G. Partridge
   Acta Metall., 13, 1321 (1965)
8. T. H. Sanders and E. A. Starke
   Acta Metall., 30, 927 (1982)
9. B. Noble, S. J. Harris, and K. Dinsdale
   Metals Sci. J., 16, 425 (1982)

Development of properties within high-strength aluminium–lithium alloys

P J GREGSON, C J PEEL and B EVANS

PJG is with the
Department of Engineering Materials,
University of Southampton, UK.
CJP and BE are with
Materials and Structures at the
Royal Aircraft Establishment,
Farnborough, UK

SYNOPSIS

Further additions of magnesium and/or copper to Lital 'A' are shown to lead to high strengths satisfying DTD XXXB requirements; the extent of these additions is limited by the homogenisation process. Optimum properties in the unstretched sheet are associated with the coprecipitation of δ' and S phase.

INTRODUCTION

Maximum weight savings offered by the selection of the low density, high stiffness Al-Li based alloys for airframe applications can only be achieved by the replacement of all conventional aluminium alloys with the lightweight alloys, and to this end three draft property target levels have previously been identified (1). To date work has concentrated on the development of an Al-2.5% Li-1.2% Cu-0.7% Mg-0.1% Zr alloy (Lital 'A') to meet the requirements of DTD XXXA (i.e. the property levels of 2014-T651); Lital A, after different heat-treatment, may also satisfy DTD XXXC (a fatigue resistant alloy equivalent to 2024-T3) (1,2). The present research is concerned with the development of a high-strength alloy to satisfy DTD XXXB (equivalent to 7075-T6).

The attainment of adequate properties within the lightweight alloys depends strongly upon alloy chemistry, processing route, and subsequent heat-treatment in order to develop the required microstructure. The high volume fraction of the metastable δ' (Al_3Li) precipitate has been shown to be the principal strengthening agent in the binary Al-Li alloys (3), in the ternary Al-Li-Mg (4) and Al-Li-Cu alloys (5), and in the quaternary Al-Li-Cu-Mg alloys (6); the properties of the ternary and quaternary alloys are further influenced by the formation of additional phases containing copper and/or magnesium. Of these precipitates, S phase (Al_2CuMg) within the Al-Li-Cu-Mg alloys has been shown to encourage homogeneous deformation and thereby promote improvements in ductility and toughness (6); furthermore isotropic properties can be displayed by the textured sheet product. The previous work utilised a pre-ageing stretch in order to encourage widespread heterogeneous precipitation of S phase during subsequent age-hardening. Recent research however has concentrated on the properties developed in the unstretched sheet product (7,8) and the present paper describes the property changes which can be effected via variation in lithium, copper, and magnesium concentration.

EXPERIMENTAL

The composition of the alloys used in the investigation is given in Table 1, the balance being aluminium.

The alloys were either DC cast as rectangular ingots by British Alcan, or cast at the Royal Aircraft Establishment using a permanent mould and controlled cooling rate similar to the DC process (9). The material was hot rolled to 6mm and then cold rolled to 1.6mm sheet (1mm for alloy X). Alloys V and W were solution treated at 530°C for 20 minutes whilst alloys X, Y and Z were held at 540°C

TABLE I - Alloy Compositions
(weight per cent)

CODE	Li	Mg	Cu	Zr
V	2.36	0.55	1.17	0.13
W	2.54	1.02	1.37	0.12
X	2.28	0.58	1.97	0.11
Y	2.55	0.78	1.97	0.13
Z	2.8	1.00	1.75	0.14

TABLE II - Summary of tensile properties.

CODE	ALLOY COMPOSITION (Wt%)			L/T TENSILE PROPERTIES		
	Li	Mg	Cu	0.2PS	UTS	%EL
V	2.4	0.6	1.2	380	430	5
W	2.5	1.0	1.4	430	510	3(T/L: 9)
X	2.3	0.6	2.0	395	490	8
Y	2.6	0.8	2.0	420	510	5(T/L:10)
Z	2.8	1.0	1.8	450	520	3

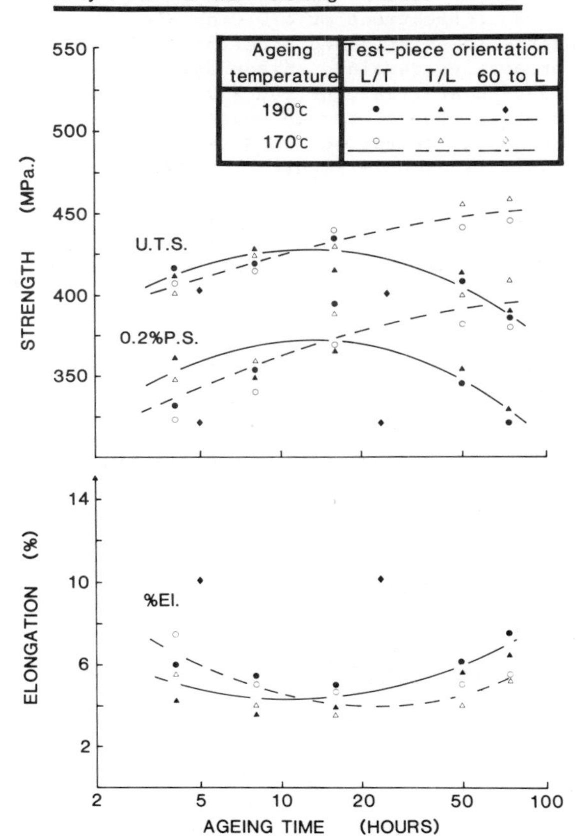

Figure 1 - Mechanical properties of alloy V after age-hardening at 170°C and 190°C.

Figure 2 - Heterogeneous precipitation of S phase in alloy V after age-hardening at 190°C for 24 hours: <110>Al orientation.

for 20 minutes before being Aquaquenched. Subsequent age-hardening was carried out at 190°C and 170°C in air-circulating ovens.

The mechanical properties were evaluated via standard tensile tests conforming to BS 4A4. Specimens for optical metallography were etched in Keller's Reagent, whilst those for Transmission Electron Microscopy (TEM) were electrolytically polished in a solution containing 30 Vol% HNO_3 and 70 Vol% CH_3OH cooled to -20°C; TEM was carried out on a JEOL 200 CX Temscan operating at 200 kV. Scanning Electron Microscopy (SEM) and X-ray microanalysis was undertaken on a Cambridge S200 SEM.

RESULTS

The tensile properties of alloy V, which had a chemical composition close to the minimum levels for the Lital A specification (7), are summarised in Figure 1. In the L/T orientation the 0.2% Proof Stress (0.2 PS) and the Ultimate Tensile Stress (UTS) reached values of 380MPa and 430MPa respectively after age-hardening for 16 hours at 190°C; after such a heat-treatment the ductility dropped to a minimum value of 5%. Similar properties were achieved by prolonged ageing at 170°C. Besides widespread precipitation of δ', TEM revealed isolated clusters of laths of S phase within the age-hardened microstructures. As previously observed (6), S precipitation was sluggish and only occurred at dislocation loops and helices, sub boundaries, and other defect sites within the alloy - Figure 2.

Alloy W, a richer alloy at the top end of the Lital A specification, displayed considerably higher strengths than V, values of 430MPa and 510MPa being recorded for 0.2 PS and UTS in the L/T orientation after ageing for 24 hours at 190°C - Figure 3; once again similar strengths were achieved after 64 hours at 170°C. Ductility showed significant variation with specimen orientation (3% El for L/T and 9% El for T/L). Optical metallography showed the textured sheet to possess an unrecrystallised grain structure in contrast to the fine recrystallised structure of alloy V - Figures 4 (a&b). TEM, however, revealed significant changes in the heat-treated microstructures compared with those of the more dilute alloy. Besides more dense S phase precipitation at defect sites, a network of homogeneously-nucleated S phase precipitates was observed; Figure 5 shows both heterogeneous and homogeneous precipitation of S phase in alloy W.

Alloys X, Y, and Z represent three alloys designed to satisfy the requirements of DTD XXXB. All had increased copper additions; alloy X, for example, was similar to alloy V but with a greater concentration of copper. Like alloy V, the age-hardening response of alloy X was rather limited in reaching a 0.2 PS of 395MPa and a UTS of 490MPa after 16 hours at 190°C - Figure 6; the ductility at about 8% El however was greater than that exhibited by alloy V, and results from L/T and T/L test pieces were very similar. After ageing at 170°C, the ductility fell to a minimum after 16 hours, and there was greater anisotropy of properties. Within this high copper alloy, heterogeneous precipitation of S phase was again evident, but after ageing at 190°C a widespread homogeneous network of S phase was observed between the heterogeneous sites.

Alloy Y had similar copper content to X, but contained increased levels of lithium and magnesium, and exhibited a good combination of properties as shown in Figure 7. After 16 hours at 190°C the 0.2% PS had increased to 420MPa and the UTS to a value of 510MPa. These high strengths were combined with an elongation to failure of 5% in the L/T orientation; test pieces cut at 60° and 90° to the rolling direction exhibited greater ductility, although the variation in properties with specimen orientation became less pronounced as ageing progressed. Heat-treatment at 170°C enabled similar strengths to be attained after 64 hours, but the properties remained anisotropic. During the 190°C heat-treatment there was again evidence of widespread homogeneous precipitation of S phase and after 48 hours there was a very dense distribution - Figure 8. Homogeneous nucleation of S phase promotes the formation of all twelve precipitate orientations, and when viewed marginally off-zone, as in Figure 8, the different variants are seen not to be precisely parallel or perpendicular to each other. At 170°C however, precipitation of the S phase was sluggish, such that after 64 hours there were indications that homogeneous precipitation had just commenced.

Alloy Z was the richest of the alloys, containing high concentrations of lithium, magnesium and copper. After ageing at 170°C or 190°C very high strengths were observed (0.2 PS of 450MPa and UTS of 520MPa after 16 hours at 190°C) - Figure 9. The corresponding ductilities however were low, especially in the L/T orientation. Optical microscopy revealed the existence of large intermetallic particles within the microstructure - Figure 10; X-ray microanalysis showed these particles to be rich in copper. Furthermore, TEM revealed an absence of any widespread precipitation of S phase during age-hardening.

DISCUSSION

For each alloy, the L/T tensile properties corresponding to the peak-aged condition after heat-treatment at 190°C, together with the concentration of the major alloying additions are given in Table II; the changes in properties associated with variation in composition

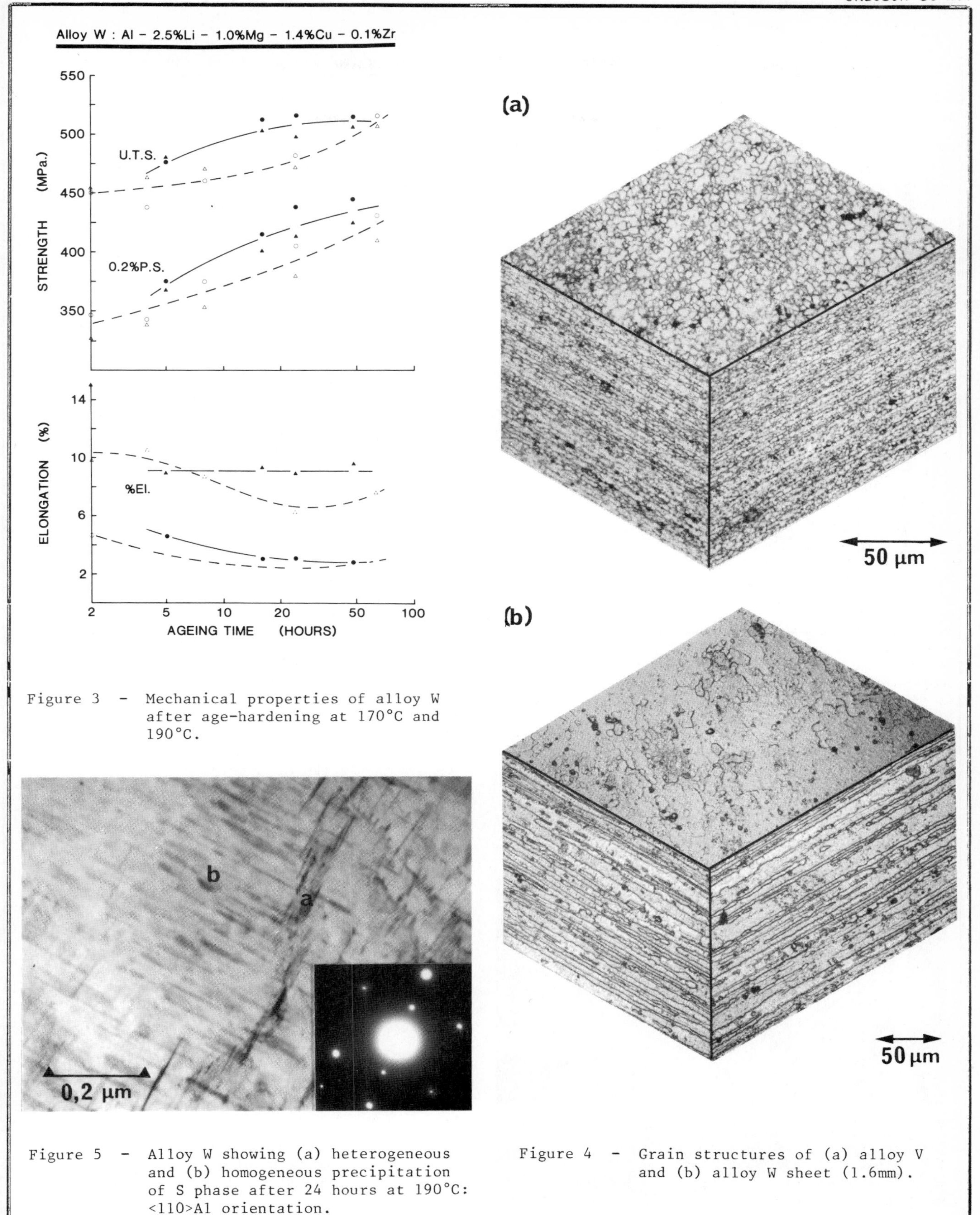

Figure 3 - Mechanical properties of alloy W after age-hardening at 170°C and 190°C.

Figure 5 - Alloy W showing (a) heterogeneous and (b) homogeneous precipitation of S phase after 24 hours at 190°C: <110>Al orientation.

Figure 4 - Grain structures of (a) alloy V and (b) alloy W sheet (1.6mm).

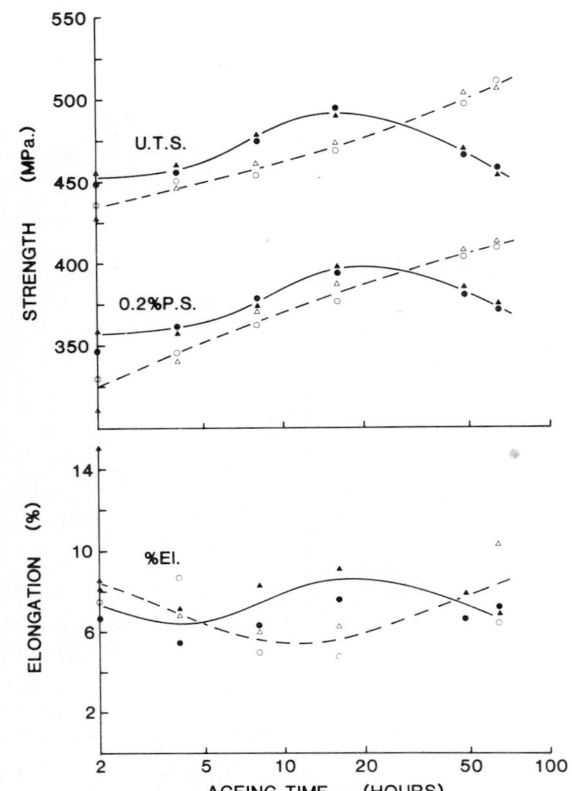

Figure 6 - Mechanical properties of alloy X after age-hardening at 170°C and 190°C.

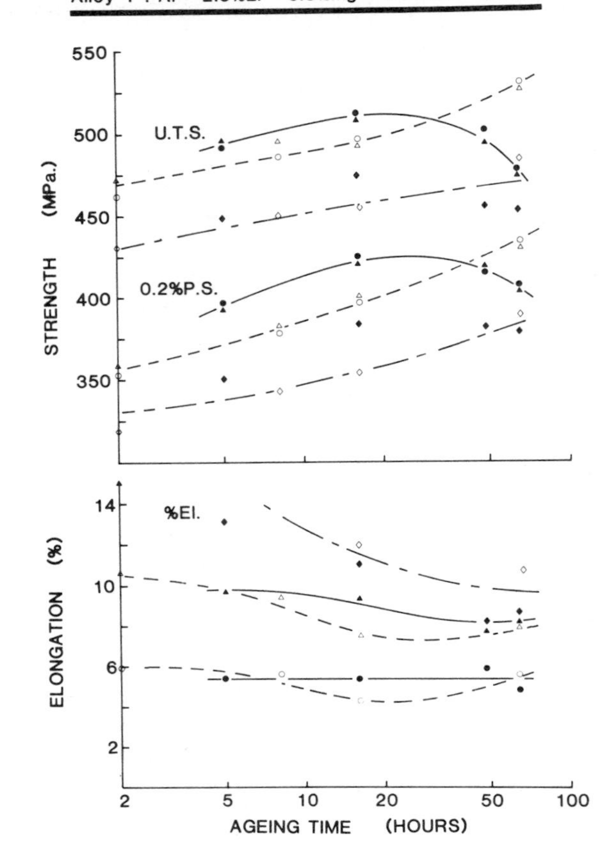

Figure 7 - Mechanical properties of alloy Y after age-hardening at 170°C and 190°C.

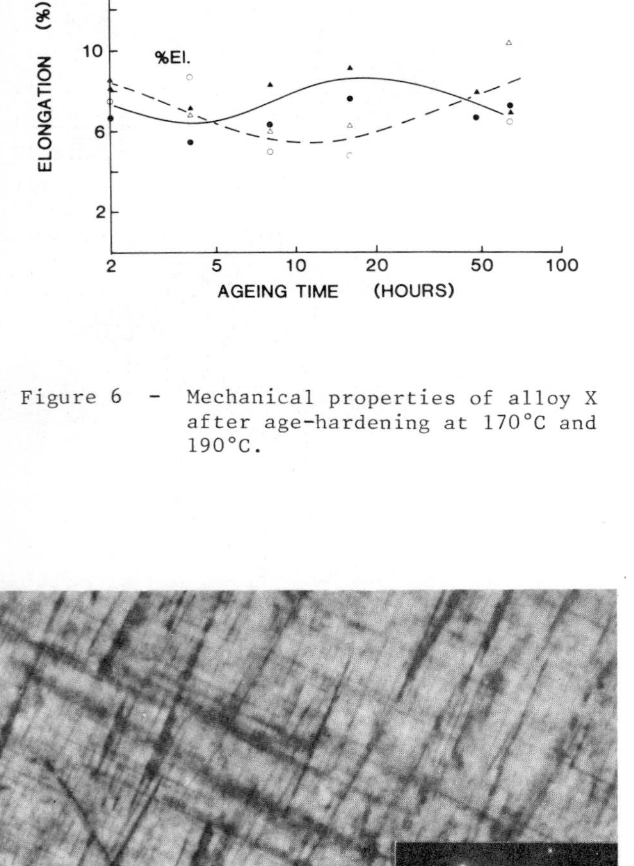

◀ Figure 8 - Homogeneous precipitation of S phase in alloy Y after 48 hours at 190°C: <110>Al orientation.

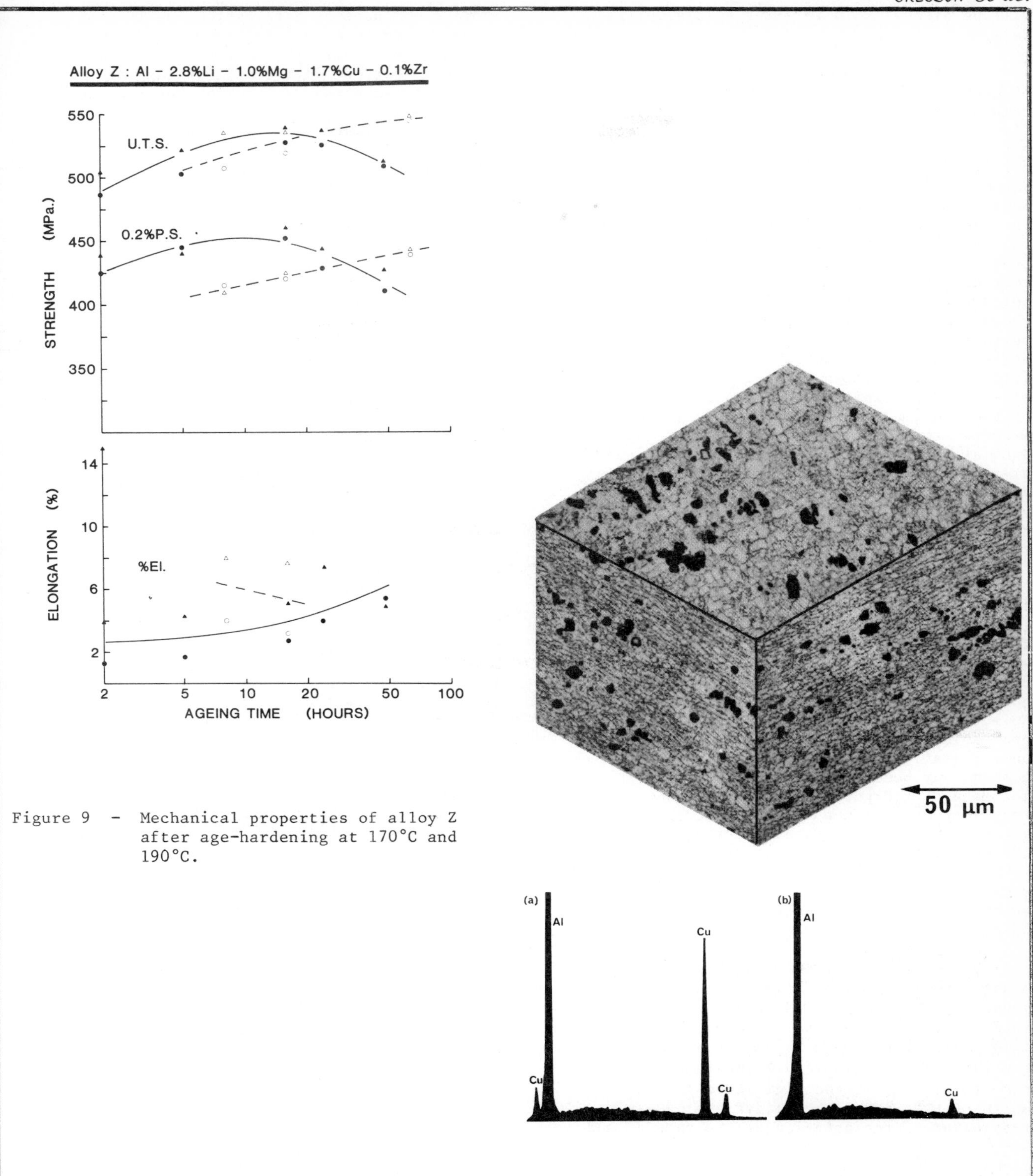

Figure 9 - Mechanical properties of alloy Z after age-hardening at 170°C and 190°C.

Figure 10 - Grain structure of alloy Z together with EDS X-ray microanalysis spectra for (a) large intermetallic particle and (b) matrix.

will be discussed in terms of this simplified data.

Alloys V and W demonstrate the wide range of properties which can be displayed by alloy compositions conforming to the Lital 'A' specification. Increased lithium content would be expected to lead to a greater volume fraction of δ', and thereby to improve strength. The magnesium additions however have a multiple influence on properties. Certainly an increase in magnesium concentration will be reflected in the solute concentration of the supersaturated solid solution (10). Furthermore, an increase in magnesium content will reduce the solid solubility of lithium in aluminium, and thereby impart additional strength by promoting further δ' precipitation; it has also been suggested that the magnesium is incorporated into the δ' particles (11). In addition, the widespread precipitation of S phase, encouraged by the high magnesium concentration, would be expected to impart an increment of strength; even at 190°C alloy W requires 48 hours heat-treatment to develop peak strength, and this delayed ageing response is consistent with precipitation of S phase which has previously been shown to be sluggish when compared with that of δ' (6).

The sparse precipitation of S phase in alloy V may help to explain the poor ductility of this low strength alloy. Conversely, despite the high strength of alloy W, the widespread precipitation of S phase might have been expected to retain reasonable L/T ductility; however the poor ductility together with the anisotropy is considered to be due to the processing route employed, with the resulting unrecrystallised grain structure and texture masking any beneficial effects of the S phase. The ease of recrystallisation may also be prone to be affected by the lithium content of the alloy. Limited recrystallisation might improve L/T ductility; a consequent drop in 0.2 PS would be expected, but such property changes would be acceptable in this Lital 'A' alloy which achieved the strengths required of DTD XXXB.

Once again the low strength of alloy X can be attributed to alloy composition; unlike magnesium, the copper additions are ineffective in increasing the 0.2 PS. However the additional precipitation of δ' and S associated with the increased lithium and magnesium additions in alloy Y enabled the high strength requirements of DTD XXXB to be satisfied. Despite relying upon precipitation of the same δ' and S phase, the ageing response of alloys X and Y can be seen to be more rapid than that of alloy W; the increased copper concentration in alloys X and Y would therefore appear to encourage earlier precipitation of S phase.

The ductility of alloys X and Y can be explained in terms of the widespread precipitation of S phase; this has been shown to promote homogeneous deformation and thereby prevent premature failure associated with coplanarity of slip (6). Previously it has been necessary to introduce a pre-ageing stretch to encourage sufficient heterogeneous precipitation of S phase to influence the deformation behaviour; the present unstretched alloys however rely upon the homogeneous precipitation of S phase and this has been the subject of a separate investigation (12).

Although still higher strengths can be achieved by further additions of lithium and magnesium (alloy Z), the ductility becomes seriously limited. By increasing the concentration of magnesium and lithium there is a danger of the alloy composition breaking into a multiple phase field containing Al_2MgLi within the quaternary phase diagram; precipitation of such a phase is reported to have a deleterious effect on ductility (10), but in alloy Z no Al_2MgLi was detected. The large copper-rich intermetallic particles present within alloy Z however would be expected to limit ductility, and these particles are thought to be due to incomplete homogenisation of the alloy. Furthermore, there is no widespread S precipitation to enhance ductility; this may be due to the increased lithium content preventing S nucleation (12) or to the intermetallic particles binding up much of the copper content. Extended homogenisation is clearly required for such a rich alloy, but this in turn could lead to excessive lithium loss (13). Higher homogenisation temperatures are limited by the relatively low solidus (546°C) of such a rich alloy.

CONCLUSIONS

It has been shown that it is possible to develop high strengths satisfying DTD XXXB within the Al-Li-Cu-Mg alloys by increasing the additions of lithium, magnesium and copper above the Lital 'A' specification. Homogeneous precipitation of S phase has been observed within the unstretched sheet product, and optimum properties are associated with the coprecipitation of δ' and S phase. Excess solute content can lead to incomplete homogenisation, and thereby to inferior properties.

REFERENCES

1. C.J. Peel, B. Evans, C.A. Baker, D.A. Bennett, P.J. Gregson and H.M. Flower, Proc. 2nd Int. Al-Li Conf, p.363, Metall. Soc. A.I.M.E. (1984).

2. R. Grimes, Al-Li Workshop, Royal Aircraft Establishment (1985).

3. B. Noble and G.E. Thompson, Metal Sci. J., 5, 114 (1971).

4. G.E. Thompson and B. Noble, J.I.M., 101, 111 (1973).

5. K.K.Sankaran and J.E. O'Neal, Proc. 2nd. Int. Al-Li Conf., p.393, Metall. Soc. A.I.M.E. (1984).

6. P.J. Gregson and H.M. Flower, Acta Metall., 33, 527 (1985).

7. W.S. Miller, Al-Li Workshop, Royal Aircraft Establishment (1985).

8. C.J. Peel, B. Evans and D.S. McDarmaid. This conference.

9. B. Evans and C.J. Peel, Technical Report 84030, Royal Aircraft Establishment.

10. K. Dinsdale, S.J. Harris and B. Noble, Proc. 1st. Int. Al-Li Conf., p.101, Metall. Soc. A.I.M.E. (1981).

11. P.J. Gregson and H.M. Flower, J. Mat. Sci. Letters, 3, 829, (1984).

12. P.J. Gregson and H.M. Flower. To be published.

13. S. Fox, H.M. Flower and D.S. McDarmaid, This conference.

□

Age hardening behavior of DTD XXXA

K WELPMANN, M PETERS
and T H SANDERS Jr

KW and MP are with the
Institut für Werkstoff-Forschung,
DFVLR, 5000 Köln 90, FRG.
THS is with the
School of Materials Engineering,
Purdue University, West Lafayette,
IN 47907, USA

Introduction

It is anticipated that Li-containing aluminium alloys will be employed to some extent in the construction of the next generation military and transport airplanes. The alloy DTD XXXA (trade name Lital A) will be one candidate alloy for this intention. This alloy has been jointly developed by the RAE and the former British Aluminium Co. and is now produced by Alcan International. It is the first higher Li content aluminium alloy which is available on a commercial scale.

Different aspects of processing, aging response and mechanical properties of this alloy have been reported in the literature [1-3]. In our investigation we studied in more detail the natural and artificial aging behavior of the alloy DTD XXXA and correlated the respective mechanical properties with microstructural features.

Experimental Procedure

The alloy was produced on a laboratory scale and supplied in the form of 1.6 mm thick sheet in the T3 condition (2 % stretch). The composition was 2.5Li-1.2Cu-0.7Mg-0.12Zr (by wt.%). When required the material was solution heat treated (SHT) in an argon furnace for 20 minutes at 520 °C and cold water quenched. Artificial aging was done in an air furnace. Age hardening was followed by hardness measurements (Vickers pyramid hardness) and longitudinal tensile tests using flat specimens with the gage dimensions 8 mm wide and 25 mm long. The strain rate for all tests was $6.7 \cdot 10^{-4}$ sec^{-1}. Testing was performed at room temperature. TEM investigations were carried out on selected aging conditions, the thin foils being prepared by conventional electrothinning techniques.

Results and Discussion

Natural Aging Behavior

Figure 1 shows the natural aging response of the alloy DTD XXXA after SHT. Hardness as well as 0.2 yield stress reveal a considerable increase after about 30 hours aging time at room temperature. At about 500 hours both curves are approaching a saturation value. Parallel to the yield stress increase a decrease of the elongation to fracture was observed.

To explain these results, it is important to know that the coherent and spherical δ' precipitates form from the supersaturated solid solution by homogeneous nucleation during or right after the water quench [4,5]. In our investigation the presence of δ' could clearly be detected by superlattice spots (less than 30 minutes after quenching), although the particles themselves were not resolved in either bright or dark field. Since there exists no transformation within the precipitation sequence of the Al-Li system (except the transformation $\delta' \rightarrow \delta$ at higher temperatures and long aging times), the strength increase during room temperature aging cannot be the result of the δ' growth but must be attributed to an additional age hardening phase which precipitates after an incubation time of about 30 hours. It is very likely that Cu and Mg bearing GPB zones are responsible for this effect [6].

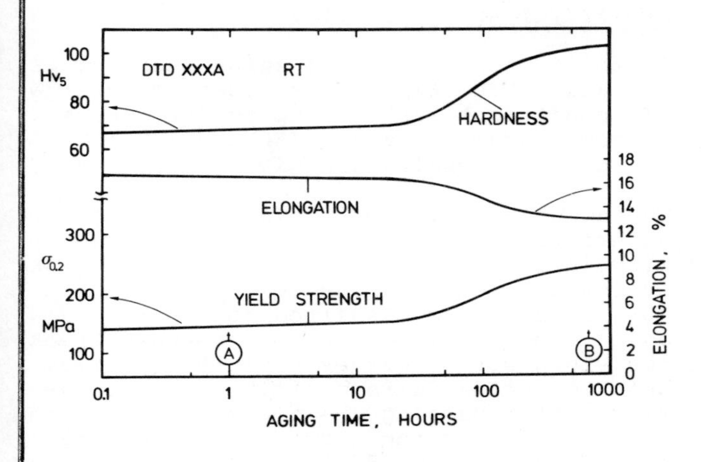

Fig. 1: Variation of hardness, yield strength and elongation to fracture with aging time at room temperature.

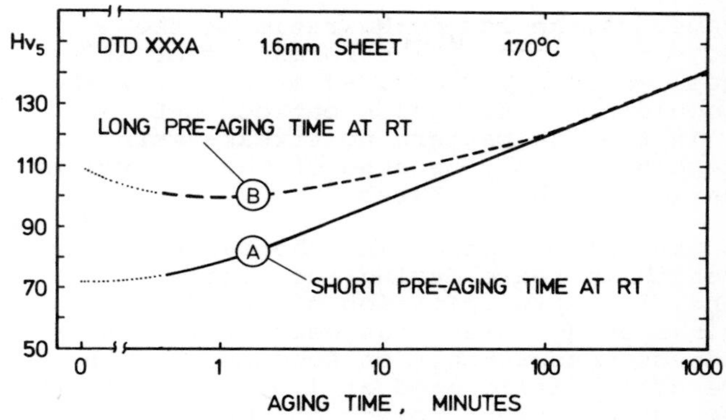

Fig. 2: Artificial aging response at 170 °C after short and long room temperature pre-aging.

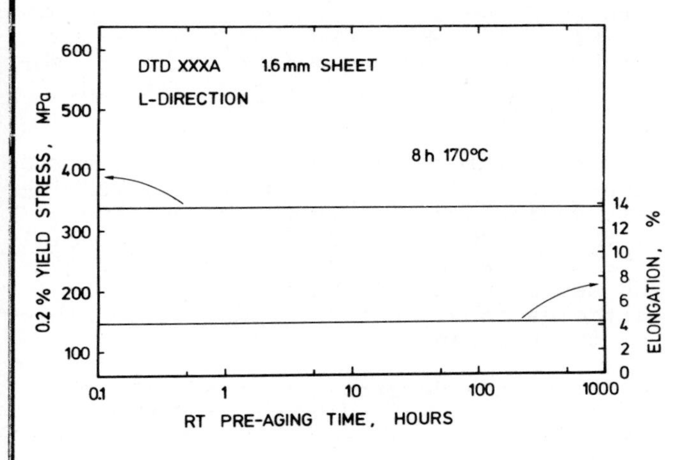

Fig. 3: Influence of room temperature pre-aging time on the tensile properties of specimens with an identical artificial aging treatment (8 hours at 170 °C).

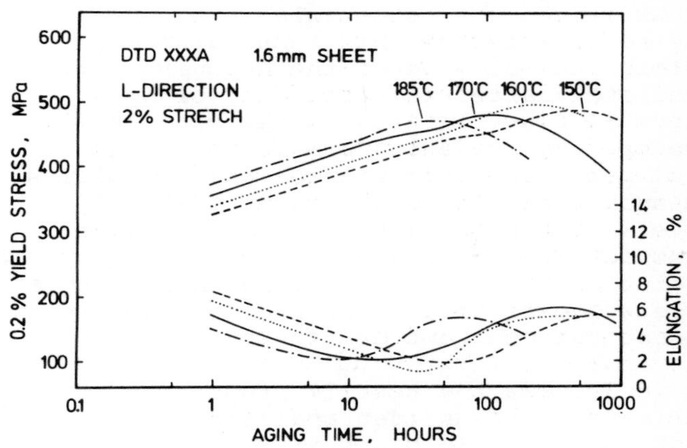

Fig. 4: Artificial aging response at different aging temperatures.

A similar age hardening curve was reported for the room temperature aging of the alloy 2024 [7], where the hardness increased in two steps. The first step was attributed to the precipitation of GPI and GPBI zones, whereas the much more pronounced second step represents the appearance of GPII and GPBII.

Because of the low Cu:Mg ratio of the alloy investigated, the formation of GP zones can be ruled out. From our TEM work we could not distinguish between GPBI and II: In the SAD pattern no streaks were detected probably because of the low volume fraction of these phases.

Our assumption concerning the cause of the strength increase during room temperature aging, i.e. precipitation of an additional age hardening phase, has been confirmed by hardness measurements on specimens which were artificially aged at 170 °C, starting from two different points of the natural aging curve (Fig. 1): A, before and B, after the hardness increase at room temperature. The results are given in Figure 2. Since at 2.5 % Li the aging temperature of 170 °C is well within the δ' metastable miscibility gap of the binary Al-Li system [8,9], a resolution of the δ' phase would not be expected. Consequently, the hardness of the material with a short natural aging interval (curve A) increases from the very beginning due to the growth of δ'. A different aging characteristic is, however, observed for the material with a long natural aging interval (curve B). Since the GPB zones are not stable at 170 °C, they re-dissolve at this temperature. This explains the decline of the hardness curve immediately after the beginning of aging (curve B, Fig. 2). After a few minutes, however, age hardening due to the growth of δ' particles predominates which explains the subsequent increase of the hardness curve. After about 1.5 hours both curves coincide. The described characteristic of resolution of an additional age hardening phase suggests that these precipitates represent the coherent GPB zones.

Figure 3 shows the tensile properties of specimens given an identical artificial aging treatment (8 hours at 170 °C) which had different natural aging intervals. The results demonstrate that all yield stress and elongation values lie on a straight line. This is in agreement with the observation that after about 1.5 hours aging at 170 °C there is no difference in hardness for the specimens with short or long natural aging intervals at room temperature (compare curves A and B in Fig. 2).

From the above the following conclusion can be drawn which is important for the commercial application of the alloy DTD XXXA: The natural aging interval has no influence on the mechanical properties achieved by subsequent artificial aging (except for short artificial aging times, i.e. before the curves A and B of Fig. 2 coincide).

Artificial Aging Behavior

To follow the artificial aging behavior of the alloy DTD XXXA the as-received material (T3 condition) was aged up to 1000 hours at temperatures between 150 and 185 °C. The tensile results are given in Fig. 4.

The aging behavior of the alloy is characteristic of an age hardenable alloy; with increasing aging temperature the peak strengths are decreased (except for the 150 °C curve) and shifted to shorter aging times. However, before reaching peak the yield stress curves exhibit a plateau which coincides with a minimum in elongation. After passing the minimum the elongation curves for the higher aging temperatures (170 °C and 185 °C) reveal a maximum which corresponds to a slightly overaged condition. For the lower aging temperatures (150 °C and 160 °C) such an elongation peak was not reached for the aging times investigated (1000 hours at 150 °C and 500 hours at 160 °C).

A characteristic feature of the aging curves is the occurrence of the plateau. Such an effect has already been found by other investigators: Yield stress curves for two different Al-Li-Cu-Mg-Zr alloys, both produced by I/M and P/M, show a comparable aging phenomenon [10], but the authors did not discuss this behavior. For the alloy DTD XXXA this effect was also reported [11].

To obtain information about the microstructural cause of the plateau effect, TEM investigations were performed on specimens of the 170 °C age hardening curve. Special attention was given to the occurrence of the semicoherent S' phase, because the δ' particles should not have been responsible for the plateau.

To follow the development of S', TEM dark field together with SAD was used because TEM bright field is not a suitable means for identifying phases with a low volume fraction in an alloy like DTD XXXA with its high δ' volume fraction. In the micrographs of Figs. 5 and 8 the zone axis is near {100}.

After 20 hours aging at 170 °C, i.e. at the onset of the plateau, S' particles could not be detected in the dark field. After 40 hours aging which is well within the plateau regime of the age hardening curve, a low density of S' was observed, as can be seen from the micrograph,

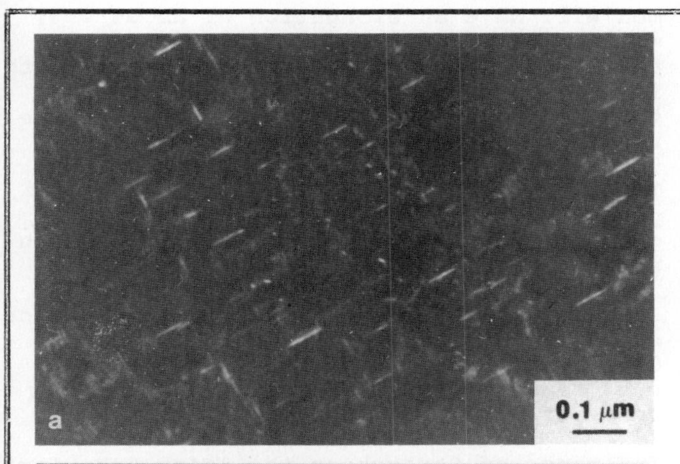

Fig. 5: TEM micrographs (dark field) of the S' phase in the plateau regime,
a) after 40 hours aging at 170 °C,
b) after 60 hours aging at 170 °C.

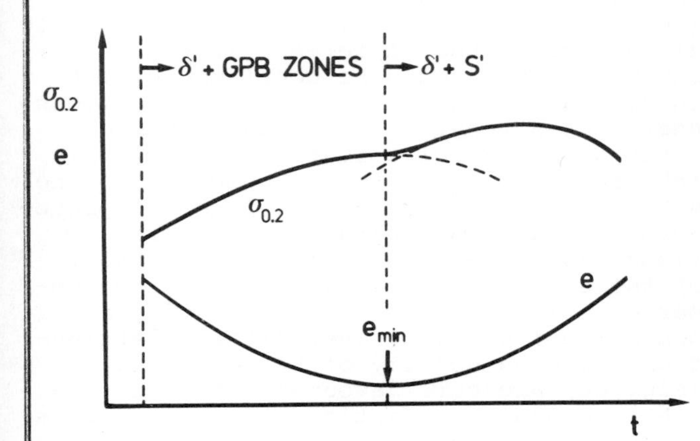

Fig. 6: Schematic explanation of the plateau effect.

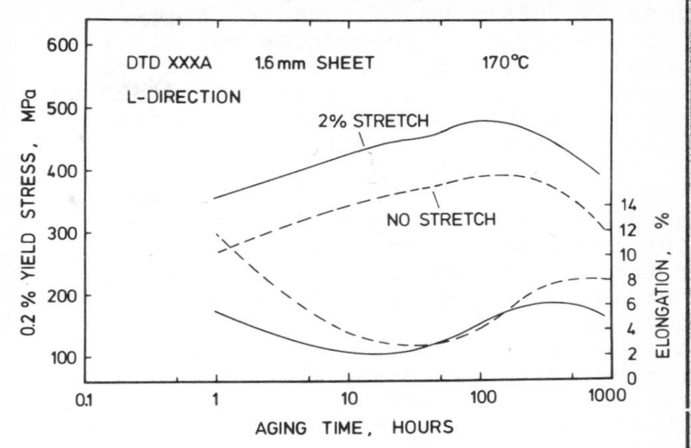

Fig. 7: Comparison of the artificial aging response of the unstretched and pre-deformed (2 % stretch) alloy conditions.

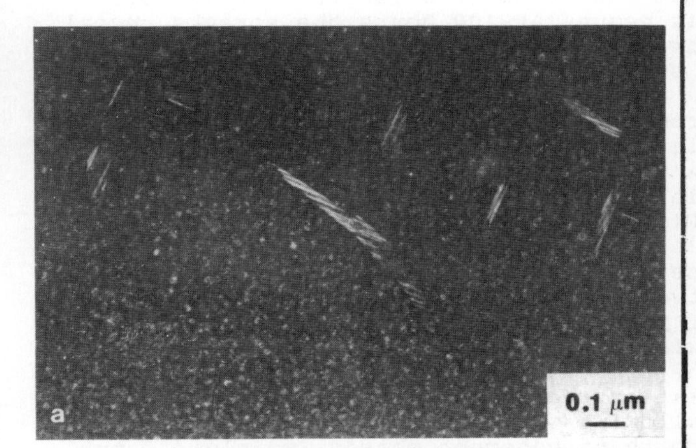

Fig. 8: TEM micrographs (dark field) of the S' phase at peak strength (100 hours at 170 °C),
a) unstretched, inhomogeneous distribution of S',
b) 2 % stretch, uniform distribution of S'.

Fig. 5a. At 60 hours aging, which is just beyond the plateau, the precipitation of S' is already more pronounced but still inhomogeneous, Fig. 5b. The S' phase does not show a homogeneous distribution before the peak strength condition (100 hours at 170 °C) (compare Fig. 8b).

The S' particles of Figs. 5 and 8 were imaged by the streaks originating from {110} superlattice spots of the δ' phase. For the 20 and 40 hours aging conditions these streaks were extremely weak but showed no discontinuities. This suggests that they were produced predominantly by the coherent GPBII phase which could, however, not be resolved by TEM bright or dark field. Since both phases, GPBII and S', have the same {100} habit plane with the fcc aluminium lattice, the low number of S' particles after 40 hours aging can be imaged by the streaks, Fig. 5a. After 60 hours aging at 170 °C the streaks begin to appear discontinuous which indicates that the transformation from GPBII to S' has begun. At 100 hours the streaks reveal the appearance which is characteristic of a semicoherent plate-like phase.

From the observed development of S' in the plateau regime it can be concluded that it is the S' phase which gives rise to the yield stress increase beyond the plateau (Fig. 4).

The schematic drawing of the 0.2 % yield stress curve in Fig. 6 gives an explanation of the plateau effect from the viewpoint of existing phases. Moreover, from this figure the shape of the elongation curve in the plateau regime can be interpreted in the following way: Before the plateau the growth of coherent phases (mainly δ', but also GPBII) gives rise to an increasing planarity of slip causing the elongation to decline. The precipitation of the semicoherent S' phase of the plateau leads to a more homogeneous distribution of slip which causes the ductility to increase.

From the tensile test data given in Fig. 4 different recommendations can be proposed concerning appropriate artificial age hardening conditions. First, of course, it is obvious that alloy conditions in the plateau regime cannot be recommended because of the very low ductility. Second, a range of underaged conditions ranking from low to medium strength apparently give a reasonable combination of yield stress and ductility [12]. Some of them have already been proposed by Alcan International. Attention should be drawn to the fact that the lower aging temperature might result in a more favourable combination of strength and ductility. For instance, a yield stress level of 375 MPa can be obtained either by 1 hour at 185 °C or by 5 hours at 150 °C. The 185 °C specimen will have an elongation of about 4 %, whereas at 150 °C 5 % elongation can be achieved. Third, the highest yield strength level (about 500 MPa) can be realized at the 160 °C aging temperature. However, the required aging time of about 200 hours is unacceptable from an economic standpoint. Thus, the higher aging temperatures at peak strength or slightly overaged will give a more favourable combination of high strength and ductility as well as cost.

It should also be mentioned that the choice of appropriate aging treatments of course depends on a number of other properties, e.g. fracture toughness, fatigue or SCC behavior.

Influence of Stretch on the Aging Behavior

The as-received T3 material was re-SHT and water quenched. From this T4 condition it was aged up to 1000 hours at 170 °C. The resultant yield stress and elongation curves are given in Fig. 7 together with the curves of the stretched alloy. The latter are identical with the 170 °C curves of Fig. 4 and have already been discussed in the previous section.

For the whole aging period, the stretched version exhibits a higher yield strength. At short aging times the higher yield stress and the lower ductility of the stretched material can mainly be explained by strain hardening due to the 2 % stretch. This could be proved by comparing the stress-strain curves of the stretched and unstretched specimens. Up to 20 hours aging at 170 °C the 0.2 % yield stress of the stretched version (T3) was nearly the same as the stress at 2 % plastic strain of the unstretched material (T4).

In the plateau and peak strength regimes a different distribution of S' additionally gives rise to the difference in yield strength. This could be confirmed by TEM investigations of the two alloy conditions near peak strength: In the unstretched material the S' distribution is extremely inhomogeneous, Fig. 8a, as compared to the stretched version, Fig. 8b. Since S' platelets nucleate predominantly on dislocations [13], the high dislocation density of the stretched material leads to the observed homogeneous distribution of this age hardening phase. The uniform distribution and high density of S' particles (Fig. 8b) are more favourable for the resistance against plastic deformation. Thus, in the peak strength regime the higher strength of the stretched material is partly due to the more uniform distribution of S', since strain hardening due to the 2 % stretch is still present.

It should also be mentioned that the plateau is less pronounced for the age hardening curve of the unstretched alloy condition. This is in agreement with the conclusion that the less favourable S' distribution of this condition gives a minor contribution to the yield stress increase compared to the stretched version.

For the alloys of the 2000 series it is well-known that plastic deformation before artificial aging has a similar influence on S' distribution and strength.

Summary and Conclusions

During natural aging of the alloy DTD XXXA a strength increase was observed after an incubation time of about 30 hours. This could be attributed to the formation and growth of GPB zones. The pre-aging time at room temperature has nearly no influence on the tensile properties obtained during artificial aging.

The artificial age hardening curves for temperatures between 150 and 185 °C reveal a plateau before reaching peak. This plateau coincides with a minimum in elongation. By TEM investigations it could be shown that the yield stress increase beyond the plateau and the corresponding increase of the elongation can be attributed to an accelerated growth of S' particles.

Stretching before artificial aging leads to a higher strength compared to an unstretched condition. This strength increase could be related to a more homogeneous distribution of the S' phase.

Acknowledgements

The authors gratefully acknowledge the experimental assistance by H. Bellingradt and J. Eschweiler. Thanks are also due to W. Zink, MBB, Bremen, for supplying the material.

References

1. W.S. Miller, A.J. Cornish, A.P. Titchener and D.A. Bennett: Development of Lithium-Containing Aluminium Alloys for the Ingot Metallurgy Production Route. Aluminum-Lithium Alloys II, T.H. Sanders, Jr. and E.A. Starke, Jr., eds., AIME, Warrendale, PA, 1984, p.335.

2. C.J. Peel, B. Evans, C.A. Baker, D.A. Bennett, P.J. Gregson and H.M. Flower: The Development and Application of Improved Aluminium-Lithium Alloys. Aluminum-Lithium Alloys II, T.H. Sanders, Jr. and E.A. Starke, Jr., eds., AIME, Warrendale, PA, 1984, p. 363.

3. C.J. Peel, B. Evans, R. Grimes and W.S. Miller: The Application of Improved Aluminium-Lithium Alloys in Aerospace Structures. 9th European Rotocraft Forum, Stresa, Italy, September 1983, Paper No. 94.

4. D.B. Williams and J.W. Edington: The Precipitation of δ' (Al_3Li) in Dilute Aluminium-Lithium Alloys. Metal Science 9 (1975), p. 529.

5. T.H. Sanders, Jr., E.A. Ludwiczak and R.R. Sawtell: The Fracture Behavior of Recrystallized Al-2.8%Li-0.3%Mn Sheet. Mat.Sci.Eng.43 (1980), p. 247.

6. J.M. Silcock: The Structural Ageing Characteristics of Al-Cu-Mg Alloys with Copper: Magnesium Weight Ratios of 7:1 and 2.2:1. J. Inst. Met. 89 (1960/61), p. 203.

7. Y. Gefen, M. Rosen and A. Rosen: Aging Phenomena in Duraluminum 2024 Studied by Resistometry and Hardness. Mat. Sci. Eng. 8 (1971), p. 181.

8. T.H. Sanders, Jr. and E.A. Starke, Jr.: Overview of the Physical Metallurgy in the Al-Li-X Systems. Aluminum-Lithium Alloys II, T.H. Sanders, Jr. and E.A. Starke, Jr., eds., AIME, Warrendale, PA, 1984, p.1.

9. K. Welpmann, M. Peters and T.H. Sanders, Jr.: Aluminium-Lithium Alloys. ALUMINIUM (English) 60 (1984), p. E641 and E709.

10. I.G. Palmer, R.E. Lewis, D.D. Crooks, E.A.Starke, Jr. and R.E. Crooks: Effect of Processing Variables on Microstructure and Properties of Two Al-Li-Cu-Mg-Zr Alloys. Aluminum-Lithium Alloys II, T.H. Sanders, Jr. and E.A. Starke, Jr., eds.,AIME, Warrendale, PA, 1984, p. 91.

11. Unpublished results. Otto Fuchs Metallwerke, Meinerzhagen, FRG.

12. M. Peters, K. Welpmann, W. Zink and T.H. Sanders, Jr.: Fatigue Behaviour of an Al-Li-Cu-Mg Alloy. Proc. 3rd Internat. Aluminium-Lithium Conference, Oxford, England, July 1985.

13. E.A. Starke, Jr. and T.H. Sanders,Jr.: New Approaches to Alloy Development in the Al-Li System. J. Metals 33 (1981), August, p. 24.

Effect of precipitation on mechanical properties of Al–Li–Cu–Mg–Zr alloy

J WHITE, W S MILLER, I G PALMER, R DAVIS and T S SAINI

The first three authors are with Alcan International Ltd, Southam Road, Banbury, Oxfordshire, UK. RD and TSS are also with Alcan International but based at Chalfont Park, Gerrards Cross, Buckinghamshire, UK

SYNOPSIS

Low ductility and poor fracture toughness in binary Al-Li alloys have been attributed to intense coplanar slip associated with the precipitation of the coherent δ' phase. The introduction of additional precipitation may disperse planar slip and hence result in an improvement in the relationship between strength, ductility and fracture toughness in more complex alloys. One such alloy system currently under development is Al-Li-Cu-Mg-Zr where S' and T_1 phases have been observed to nucleate heterogeneously on dislocations and sub-grain boundaries. The present work examines the effect of cold work prior to ageing at 190°C on the mechanical properties of commercially produced Al-Li-Cu-Mg-Zr sheet.

INTRODUCTION

Complex aluminium-lithium alloys for aerospace applications are currently under development. The main strengthening phase in these alloys (which contain > 1.4 wt% Li) is δ' (Al_3Li). This is precipitated homogeneously during the quench as spherical particles which coarsen with artificial ageing to diameters of up to 300 nm with no apparent loss of coherency[1]. During plastic deformation the coherent nature of the δ' results in shearing of the precipitates by dislocations rather than Orowan looping. As the precipitates are highly ordered, such an occurrence results in disordering within the particles which favours further slip along the same path leading to intense coplanar slip. It is believed that this results in high stress concentrations at high angle grain boundaries leading to an intergranular failure mode and consequent low ductility and fracture toughness[2].

One approach to improving the strength-ductility-fracture toughness relationship is to introduce a second phase into the matrix in order to disperse planar slip. In this respect both Al-Cu-Li-Zr and Al-Li-Cu-Mg-Zr alloys have been developed. The former primarily employs T_1 (Al_2CuLi) with some Θ" (Al_2Cu)[3] and the latter primarily S' (Al_2CuMg) with some T_1[4] as the additional precipitating phases. All of these phases precipitate heterogeneously on dislocations and thus cold work prior to ageing can produce a significant modification of the distribution of these precipitates[5].

The present work examines the effect of cold work prior to ageing at 190°C on the mechanical properties of Al-Li-Cu-Mg-Zr sheet.

EXPERIMENTAL PROCEDURE

The Al-Li-Cu-Mg-Zr alloy used in the present study was in the form of 1.6 mm sheet produced using semi-commercial sheet rolling practices. The composition of the alloy is given in Table 1. The material was solution treated at 525°C for 15 minutes and cold water quenched. Half of the material was then subjected to a 2% stretch before both sets were aged at 190°C for up to 104 hours. Tensile and fracture toughness data were obtained over the ageing regime studied.

A limited amount of tensile testing was carried out to determine the ageing response of Al-Li and Al-Li-Zr alloys produced using similar processing routes to the Al-Li-Cu-Mg-Zr alloy. The compositions of these alloys are also shown in Table 1.

Tensile tests on all alloys were carried out in the longitudinal (rolling) direction, using

Table 1: Alloy compositions (wt%).

	Li	Cu	Mg	Zr
Al-Li-Cu-Mg-Zr	2.47	1.22	0.67	0.12
Al-Li	2.7	-	-	-
Al-Li-Zr	2.5	-	-	0.15

Table 2: Mechanical properties data for Al-Li-Cu-Mg-Zr sheet aged at 190°C.

Ageing Time at 190°C (hrs)	Unstretched				2% Stretched			
	0.2% PS (MPa)	TS (MPa)	Ef (%)	Kc ($MPa\sqrt{m}$)	0.2% PS (MPa)	TS (MPa)	Ef (%)	Kc ($MPa\sqrt{m}$)
Solution treated	195	330	10	-	240	331	10	-
Naturally aged	-	-	-	-	286	355	6.0	70.4
1	315	417	5.0	65.8	375	452	3.4	68.0
2	337	446	3.8	58.8	-	-	-	-
3	341	446	3.5	57.8	398	468	2.9	62.3
5	370	462	3.1	53.5	431	498	3.0	60.2
10	380	469	3.1	53.8	447	506	3.0	52.3
16	378	461	2.9	52.4	444	510	4.5	46.7
24	380	478	4.2	51.5	440	507	5.3	47.8
64	346	448	6.8	42.5	398	467	5.7	42.2
104	329	431	6.8	41.8	390	463	5.6	39.4

test-pieces with the dimensions shown in Figure 1(a). 0.2% proof strength (PS), tensile strength (TS) and elongation to failure (E_f) were determined in each test.

Plane stress fracture toughness (K_c) testing using centre cracked panels was carried out in accordance with both the current ASTM practice for R curve determination[6] and RAE's recommendations for the measurement of R curves using centre cracked panels[7]. Tests were carried out in the L-T orientation using the specimen geometry shown in Figure 1(b).

Thin foils for TEM examination were prepared by mechanically punching 3 mm discs from the sheet and grinding down to a thickness ~ 0.4 mm. A commercial twin jet polisher (Struer's Tenupol) was used to electropolish the discs using a solution of 25% nitric acid in methanol at -15°C with an applied potential of 14V between specimen and cathode.

RESULTS AND DISCUSSION

(i) Tensile Results

The effect of a 2% stretch carried out in the solution treated condition on the ageing reponse of the Al-Li-Cu-Mg-Zr alloy is shown in Table 2. This data is presented graphically in Figure 2, where PS, TS, E_f and K_c are plotted against ageing time at 190°C. In both the stretched and unstretched conditions, the PS and TS increase rapidly on ageing to maxima between 5 and 10 hours at 190°C, thereafter the properties remain relatively constant within the ageing regime studied.

In the solution treated temper there is a relative improvement in PS but not TS in the stretched compared with unstretched material. This is simply attributable to strain hardening arising from the 2% stretch. On ageing there is a relative improvement in both PS and TS, this improvement initially increases but again reaches a maximum after 5 to 10 hours ageing. A comparison of the TEM microstructures of specimens aged for 24 hours at 190°C (Figure 3) shows that although S' and T_1 precipitates are observed on sub-grain boundaries in both conditions, these precipitates are present in a far higher density in the matrix of the stretched material. The relative increase in PS and TS in the stretched condition is thus directly attributable to the heterogeneous nucleation of S' and T_1 on dislocations, as has been shown by previous workers[5]. It is further worth noting that the greater part of this improvement in strength occurs at ageing times of less than 10 hours where the S' and T_1 precipitation has not developed to its full extent. (Figure 4).

E_f initially decreases in both stretched and unstretched material before recovering at longer ageing times. The minimum in E_f is the same in both instances but recovery appears to take place earlier in the stretched condition. The initial decrease in E_f occurs coincidentally with the rapid rise in strength reported earlier. The similar minima and difference in recovery, however, which result in an improvement in the strength-ductility relationship with stretching are not directly related to further changes in PS and TS. These observations are broadly consistent with the increase in S' and T_1 precipitation in stretched specimens which also increases markedly at ageing times in excess of 10 hours and reaches its maximum extent after ~ 24 hrs.

It has been suggested previously that this increase in S'/T_1 precipitation results in a dispersal of the planar slip commonly observed in Al-Li binary alloys. In order to examine this hypothesis thin foils for TEM examination were prepared from sheet subjected to the same thermomechanical treatments described above (with the exception of the solution treated only condition) and further deformed 5% by cold rolling. Preliminary results from these investigations show similar deformation microstructures in all samples irrespective of temper. Figure 5 shows some typical deformed microstructures, the dominant features of which are a high level of dislocation debris and regions of intense deformation. Detailed study of these regions of intense deformation show that they are regions of highly coplanar slip within which the δ' has been extensively sheared. Between these regions of intense deformation the dislocation debris is indicative of a large amount of cross-slip having taken place. The observation of similar dislocation microstructures irrespective of ageing time and prior deformation suggests that the plastic deformation of the alloy is not significantly affected by the changes in S'/T_1 precipitate density described earlier. In view of this surprising observation ageing curves and dislocation microstructures were determined in Al-Li and Al-Li-Zr alloys with similar Li and Zr levels to the more complex alloy (Figure 6 and 7).

It is readily apparent that even after long ageing times (~ 64 hours at 190°C) the Al-Li alloy does not show any recovery in ductility or drop in strength associated with overageing. The Al-Li-Zr alloy, however, shows a similar variation in tensile properties with ageing as the Al-Li-Cu-Mg-Zr alloy. Peak ageing in terms of strength occurs after ~ 10 hours at 190°C, and the minimum in ductility occurs after ~ 5 hours. This difference in behaviour between Al-Li and Al-Li-Zr may be attributable to the differences in grain structure between the two alloys. The former is recrystallised whilst the latter is unrecrystallised and hence has an additional sub-grain structure. TEM samples were prepared from regions close to the

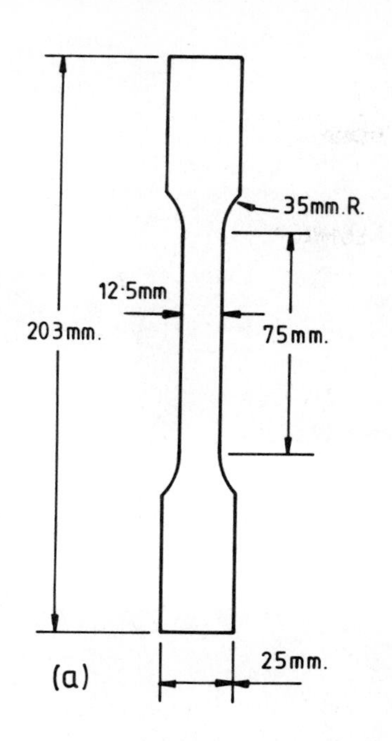

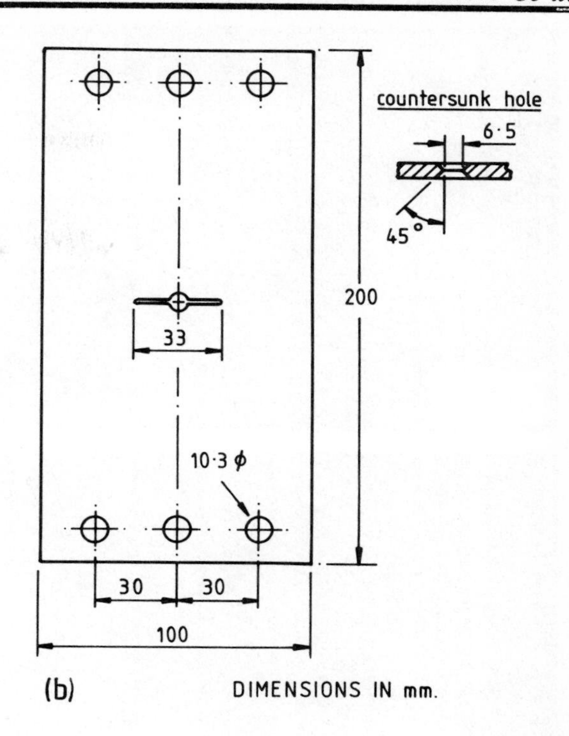

Fig.1: Specimen geometries of (a) tensile test-pieces and (b) centre cracked panels.

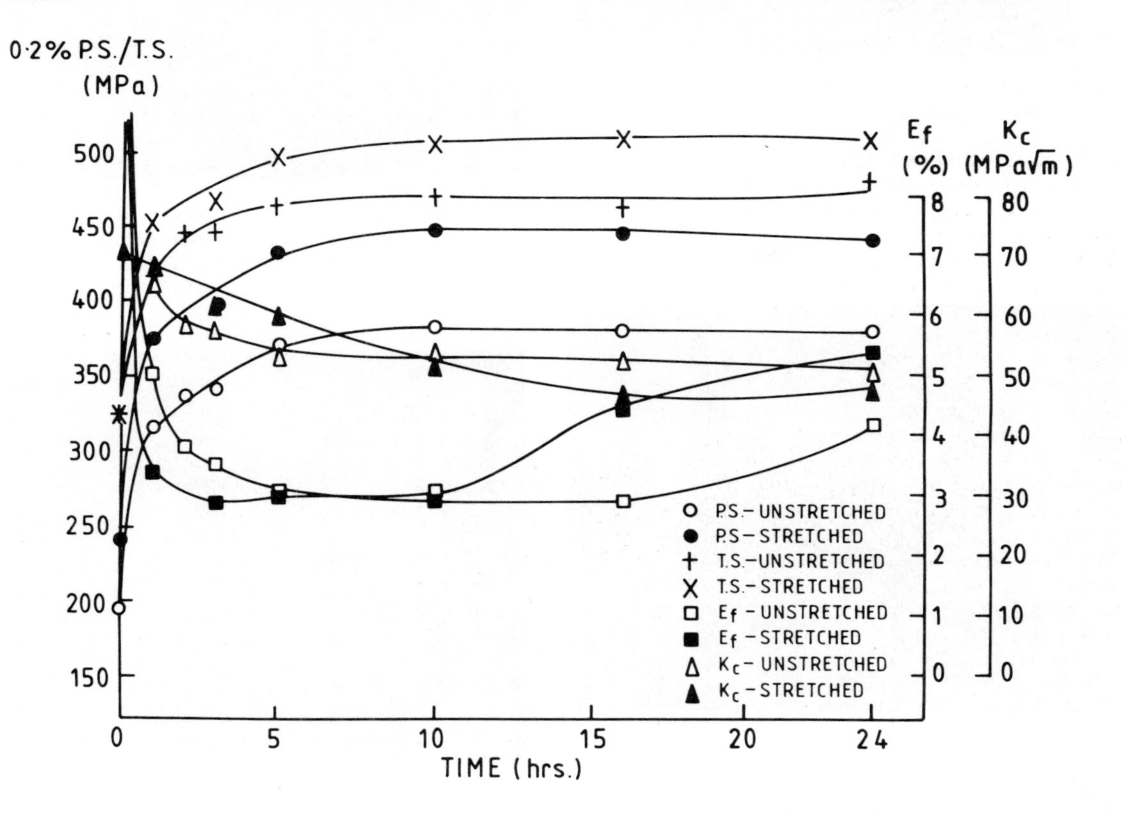

Fig.2: PS, TS, Ef and Kc plotted against ageing time at 190°C for stretched and unstretched Al-Li-Cu-Mg-Zr sheet.

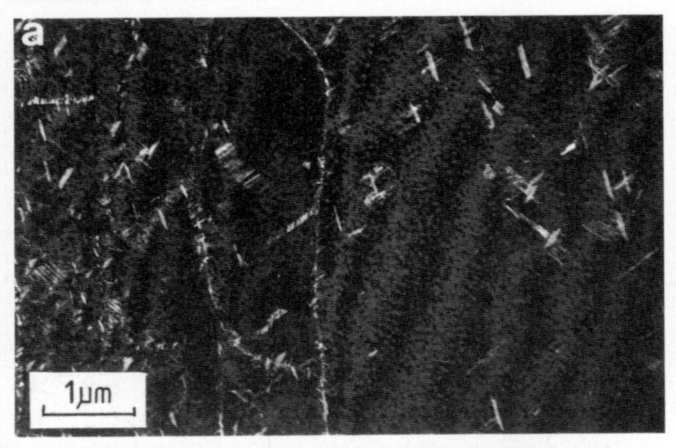

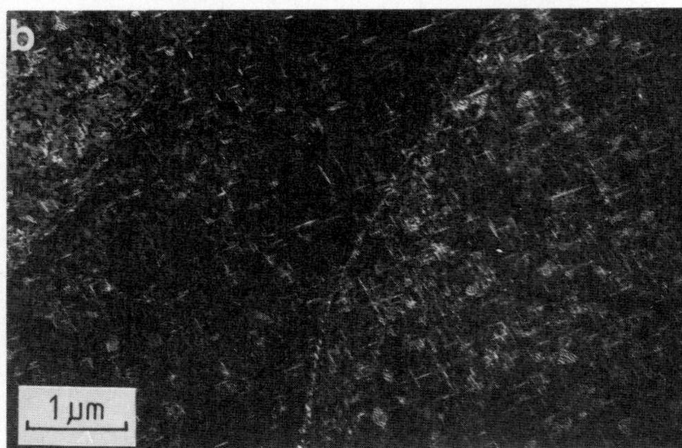

Fig.3: S' and T_1 precipitation after ageing for 24 hr at 190°C in (a) unstretched and (b) stretched Al-Li-Cu-Mg-Zr sheet. (W/B (g,3g) $11\bar{1}$ nr [112]).

Fig.4: Al-Li-Cu-Mg-Zr sheet stretched and aged 5 hr at 190°C. (W/B (g,3g) $11\bar{1}$ nr [112]).

Fig.5(a): Al-Li-Cu-Mg-Zr sheet deformed 5% in the unstretched and aged 24 hr at 190°C condition ($11\bar{1}$ nr [112]).

Fig.5(b): Al-Li-Cu-Mg-Zr sheet deformed 5% in the stretched and aged 24 hr at 190°C condition ($11\bar{1}$ nr [112]).

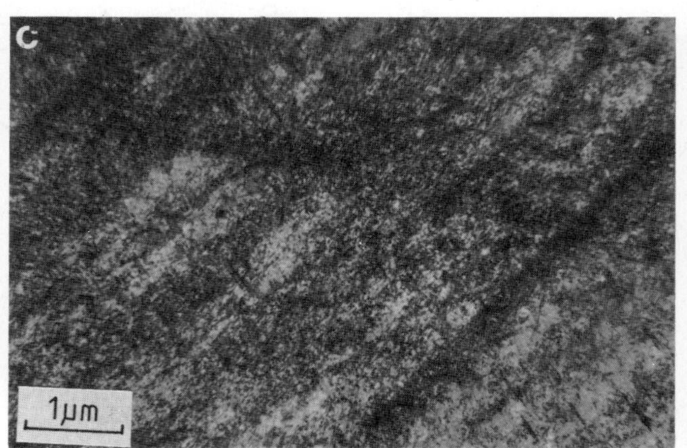

Fig.5(c): Al-Li-Cu-Mg-Zr sheet deformed 5% in the stretched and aged 24 hr at 190°C condition ($11\bar{1}$ nr [112]).

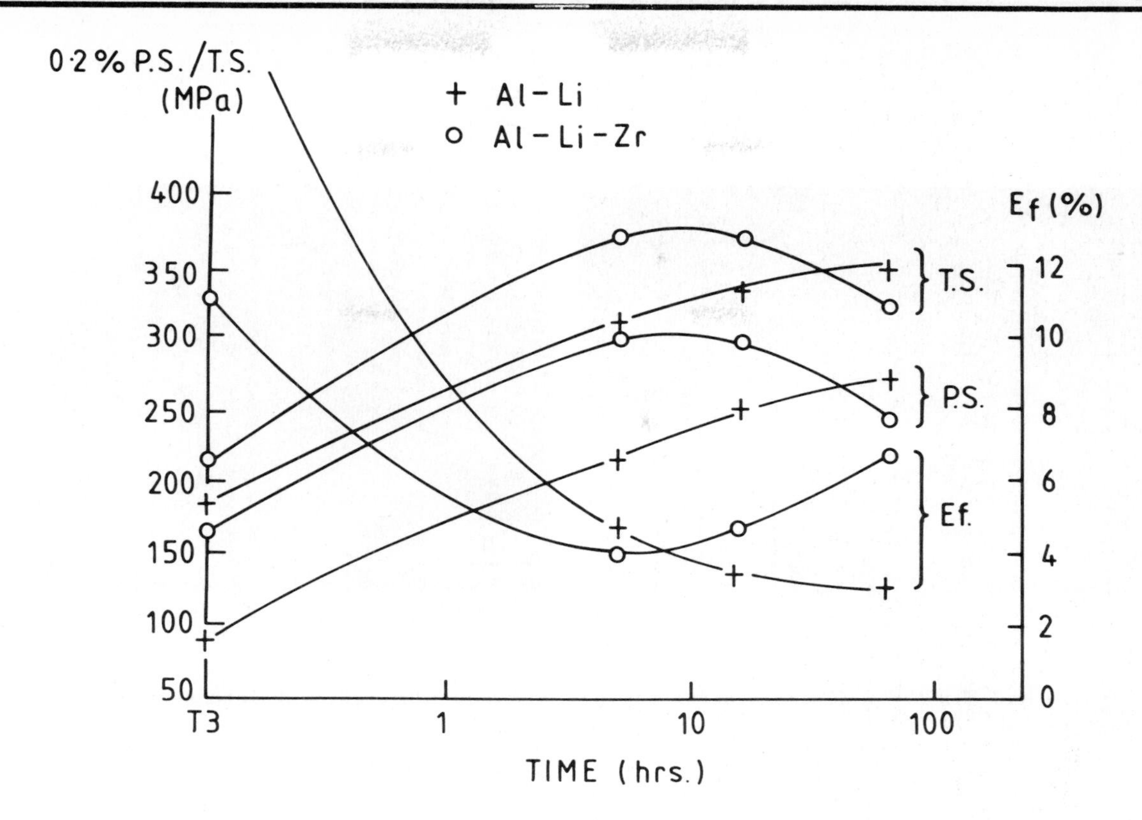

Fig.6: The variation of PS, TS and Ef with ageing at 190°C in Al-Li and Al-Li-Zr sheet.

fracture surface in tensile specimens of plate aged to 100 hours at 190°C used in a parallel investigation. The deformed microstructure of the Al-Li alloy shows well defined planar slip whilst the Al-Li-Zr shows a high level of dislocation debris in addition to slip bands (Figure 7). The close resemblance between this latter microstructure and that of the deformed complex alloy suggests similar deformation processes are occurring in both these alloys.

The TEM results presented above are preliminary and further work is currently in hand to completely characterise deformation structures in these alloys. The peak ageing response, recovery in ductility and increased cross-slip, however, in the Al-Li-Cu-Mg-Zr alloy compared with the simple Al-Li alloy are clearly largely determined by the Zr addition. Thus in order to determine whether S' precipitates are effective in dispersing slip and can therefore account for the improvement in strength-ductility relationship in stretched material, the effect of Zr on deformation must first be determined. Sufficient information is not at present available to determine this latter relationship and how it is influenced by such variables as prior deformation.

(ii) Plane Stress-Fracture Toughness Results

The fracture toughness results from the Al-Li-Cu-Mg-Zr alloy show that K_c in both stretched and unstretched material decreases with ageing to a constant minimum level (Figure 2). This minimum is lower and is attained at longer ageing times in the stretched material. This results in a relative increase in K_c for stretched compared with unstretched material in underaged tempers (< 10 hours at 190°C) but a relative decrease at longer ageing times. K_c is re-plotted against PS in Figure 8. Clearly there is an overall increase in the strength-toughness relationship in stretched material irrespective of ageing time. The improvement, however, is most marked for ageing times of

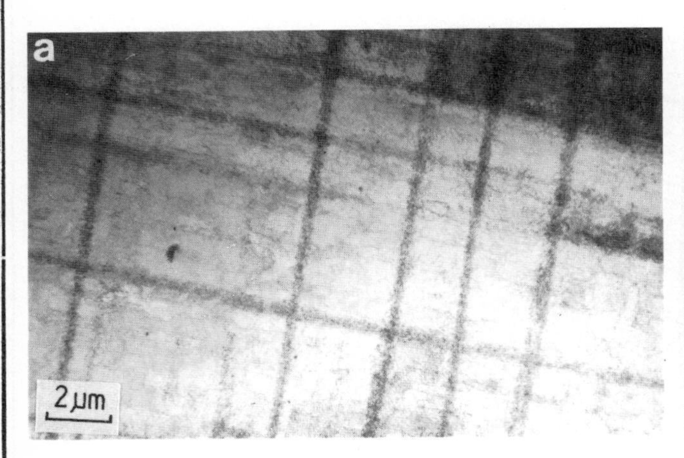

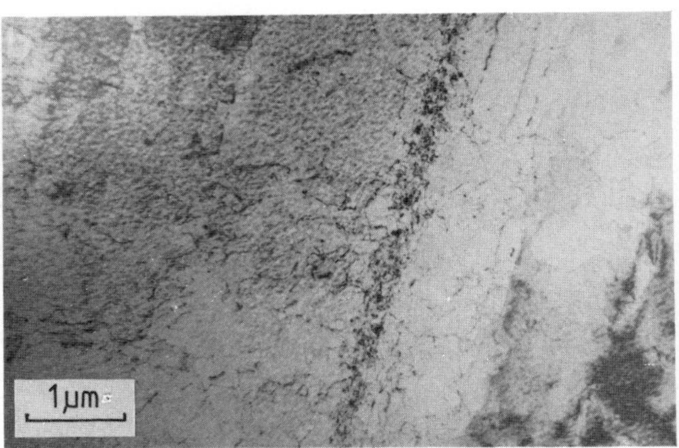

Fig.7(a and b): Al-Li deformed after ageing 100 hr at 190°C showing intense planar slip (200 nr [001]).

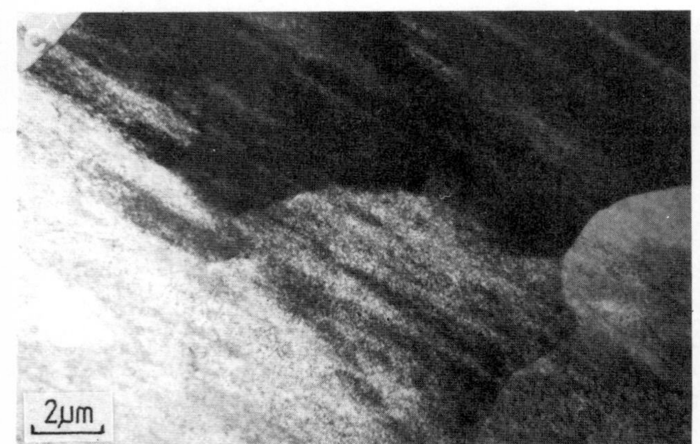

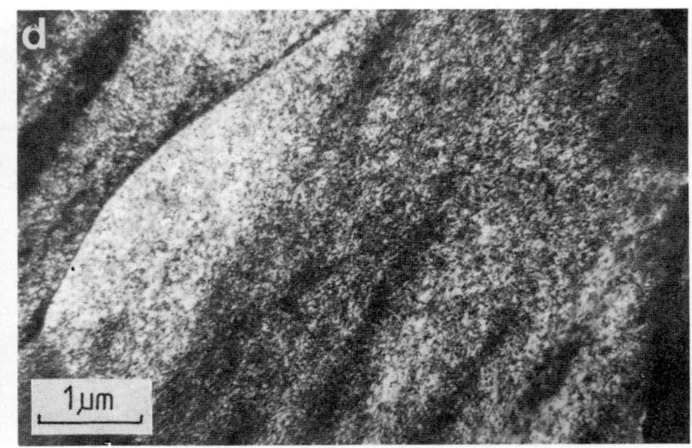

Fig.7(c and d): Al-Li-Zr deformed after ageing 100 hr at 190°C showing intense slip bands and additional dislocation debris ($11\bar{1}$ nr [112]).

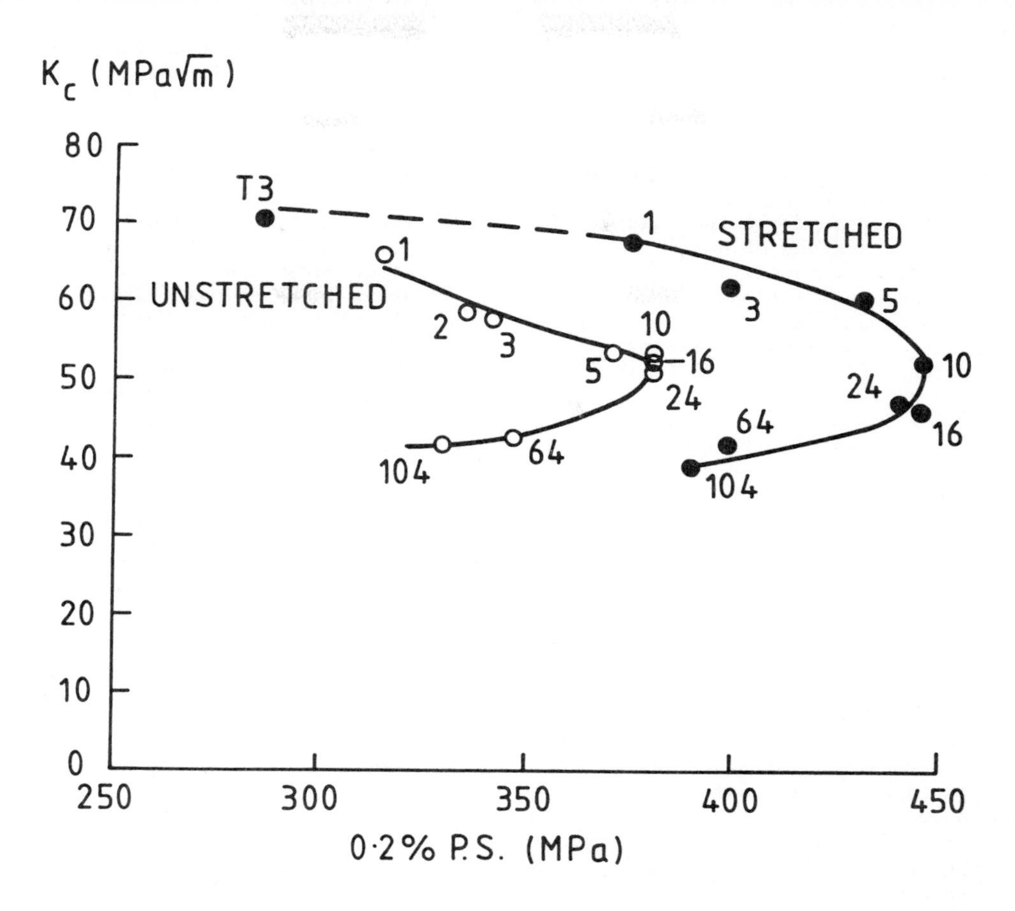

Fig.8: The relationship between fracture toughness and proof strength in stretched and unstretched Al-Li-Cu-Mg-Zr sheet aged at 190°C.

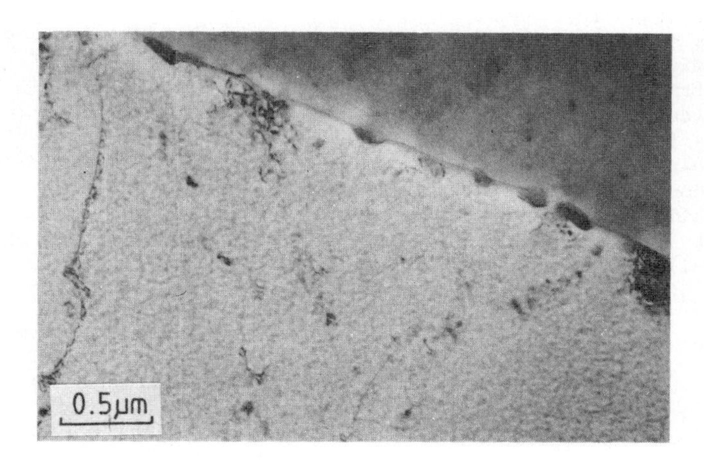

Fig.9: High angle grain boundary precipitation in Al-Li-Cu-Mg-Zr sheet aged for 5 hr at 190°C.

less than 10 hours at 190°C where both PS and K_c in stretched material are relatively increased over unstretched material. For ageing times greater than this K_c in stretched material decreases almost independently of PS and is marginally less than that in unstretched material for equivalent ageing times. These observations suggest that the factors resulting in improved fracture toughness in stretched material are most effective in underaged tempers. This would appear to be inconsistent with the previous observations of an improvement in the strength-ductility relationship at longer ageing times and suggests that the dominant factors controlling ductility and fracture toughness at ageing times of greater than 10 hours at 190°C are different.

The fracture toughness results are thus difficult to explain in terms of a dispersal of planar slip associated with increasing matrix S' precipitation as such a mechanism would be expected to increase in effectiveness with ageing time. Consequently it is worth considering alternative mechanisms that may account for the observed variation in fracture toughness in stretched and unstretched specimens. The two factors most likely to determine K_c are the deformation behaviour of the matrix and grain boundary effects. An explanation based on the former alone would be inconsistent with the increase in ductility observed at longer ageing times whilst the latter, which would require a difference in grain boundary microstructure in the underaged temper, cannot be supported by the limited observations carried out to date. It is therefore more likely that two mechanisms are operating in direct competition. The first mechanism would produce a relative increase in K_c in underaged tempers. It is suggested that this may occur as a result of S' dispersing slip more effectively during the early stages of ageing when the effect of δ' in enhancing planar slip would be less significant. The second (competing) mechanism, which would result in a decrease in fracture toughness, could be the effect of widespread grain boundary precipitation as is observed in specimens aged to more than 5 hours at 190°C (Figure 9). It is not thought that these two mechanisms would be mutually exclusive but that the former is dominant at ageing times of less than 10 hours at 190°C whilst the latter is dominant at ageing times of greater than 10 hours at 190°C.

CONCLUSIONS

(1) Improvements in the strength-ductility and strength-fracture toughness relationship associated with stretching prior to ageing have been clearly demonstrated to occur in an Al-Li-Cu-Mg-Zr alloy.

(2) The presence of a fine distribution of S'/T_1 and the development of these phases with ageing have been shown to correlate with the increase in the strength-ductility relationship with stretching. The hypothesis that this is related to a dispersal of planar-slip by the S' phase clearly requires more work to confirm or disprove the concept.

(3) The variation in tensile properties with ageing time and the deformation modes observed in the complex alloy have been shown to be mainly related to the effect of the Zr addition.

(4) The plane stress fracture toughness and ductility are clearly not interdependent for ageing times of greater than 10 hours at 190°C.

(5) The strength-fracture toughness relationship cannot be directly correlated with S'/T_1 precipitation, hence the dispersal of planar slip by S' cannot account for all observations. It has been suggested that the strength-fracture toughness relationship may be dominated by matrix precipitation effects during the initial stages of ageing (< 10 hours at 190°C) and by grain boundary effects at longer ageing times (< 10 hours at 190°C).

ACKNOWLEDGEMENTS

The authors are grateful to Alcan International Limited, for permission to publish this work.

REFERENCES

1. D.B. Williams in Aluminium-Lithium Alloys, Eds. E.A. Starke Jr., T.H. Sanders Jr., TMS-AIME, Warrendale PA, 89, (1981).

2. K. Welpmann, M. Peters and T.H. Sanders Jr., Aluminium (English) 60 (10), E641, (1984).

3. J.M. Silcock, JIM, 88, 357, (1959-60).

4. C.J. Peel, B. Evans, C.A. Baker, D.A. Bennett, P.J. Gregson and H.M. Flower in Aluminium-Lithium Alloys II, Eds. E.A. Starke Jr. and T.H. Sanders Jr., TMS-AIME, Warrendale PA, 363, (1984).

5. P.J. Gregson and H.M. Flower, Acta Met., 33, 529, 1985.

6. ASTM Standard E561/78T.

7. C. Wheeler, J.N. Eastabrook, D.P. Rooke, K.H. Schwalbe, W Setz, A.V. de Koning, RAE Tech. Report 81142, GARTEUR/TP-04 (1981).

Effect of heat treatment upon tensile strength and fracture properties of an Al–Li–Cu–Mg alloy

P J E BISCHLER and J W MARTIN

The authors are in the Department of Metallurgy and Science of Materials at the University of Oxford, UK

SYNOPSIS

The microstructure and tensile properties of an Al-Li-Cu-Mg alloy plate have been studied. The microstructure of peak-aged, stretched material was characterised, and two series of tensile tests carried out. Firstly, peak aged material was isothermally heat-treated for periods up to 1000 hr and tested at room temperature, and a second series of heat treated specimens were tested at the soaking temperature.

Mechanical properties were maintained in both series of tests up to 130°C, after which a decrease in strength was seen. This effect accompanied the progressive replacement of the δ' phase by the equilibrium δ precipitate as ageing proceeded.

A rapid rise in ductility was observed above 150°C in material tested at its soak temperature. This increased failure strain was not seen in the room temperature tests and is thus ascribed to the operation of thermally activated deformation and recovery processes.

INTRODUCTION

Al-Li-X alloys have received much attention over recent years because they offer substantial advantages in specific strength and stiffness over the present 2000 and 7000 series alloys. In the UK, research has been directed by the Royal Aircraft Establishment (RAE) with a view to developing three Al-Li-Cu-Mg alloy specifications, representing medium strength, high strength, and damage-tolerant material respectively(1). The design philosophy of this work was to develop a series of alloys that could be directly substituted for those at present used, but with a density reduction and a stiffness increase of at least 10% in each case.

Upon ageing, the δ' phase, (Al_3Li), nucleates homogeneously and grows as a fully coherent precipitate with an $L1_2$ structure. This is the major strengthening phase, and when dislocations shear precipitate particles with such an ordered structure, antiphase boundaries are generated which gives rise to paired dislocation movement. The planar slip associated with these 'superdislocations' may lead to low ductility failure, so strengthening additions of Cu and Mg are made, leading to the nucleation of S-phase which aids slip homogenisation.

Stretching the solution-treated alloy prior to ageing is found to enhance the mechanical properties over the entire ageing curve. This arises because the presence of a dislocation network promotes the S-phase nucleation, resulting in its being dispersed throughout the structure as uniform and fine particles. These improvements, including the addition of Zr as a recrystallization inhibitor and grain refiner, are discussed in detail in several recent papers (1-3).

To date, little attention has been given to the suitability of these alloys for applications at elevated temperatures. The aim of the present work was to examine the structure-property relationship of a medium-strength alloy (designated 'A'), in the form of plate designated YYYA by RAE. It may be regarded as a replacement for 2014-T651 material.

Two effects have been studied:

(a) an evaluation of the room temperature tensile properties of the alloy after isothermal soaking at various temperatures for increasing time.

Table 1.

Composition of the alloy (wt %)

Li	Cu	Mg	Zr	Fe	Si	Na	Ti	Al
2.4	1.16	0.62	0.11	0.14	0.1	0.0015	0.02	bal.

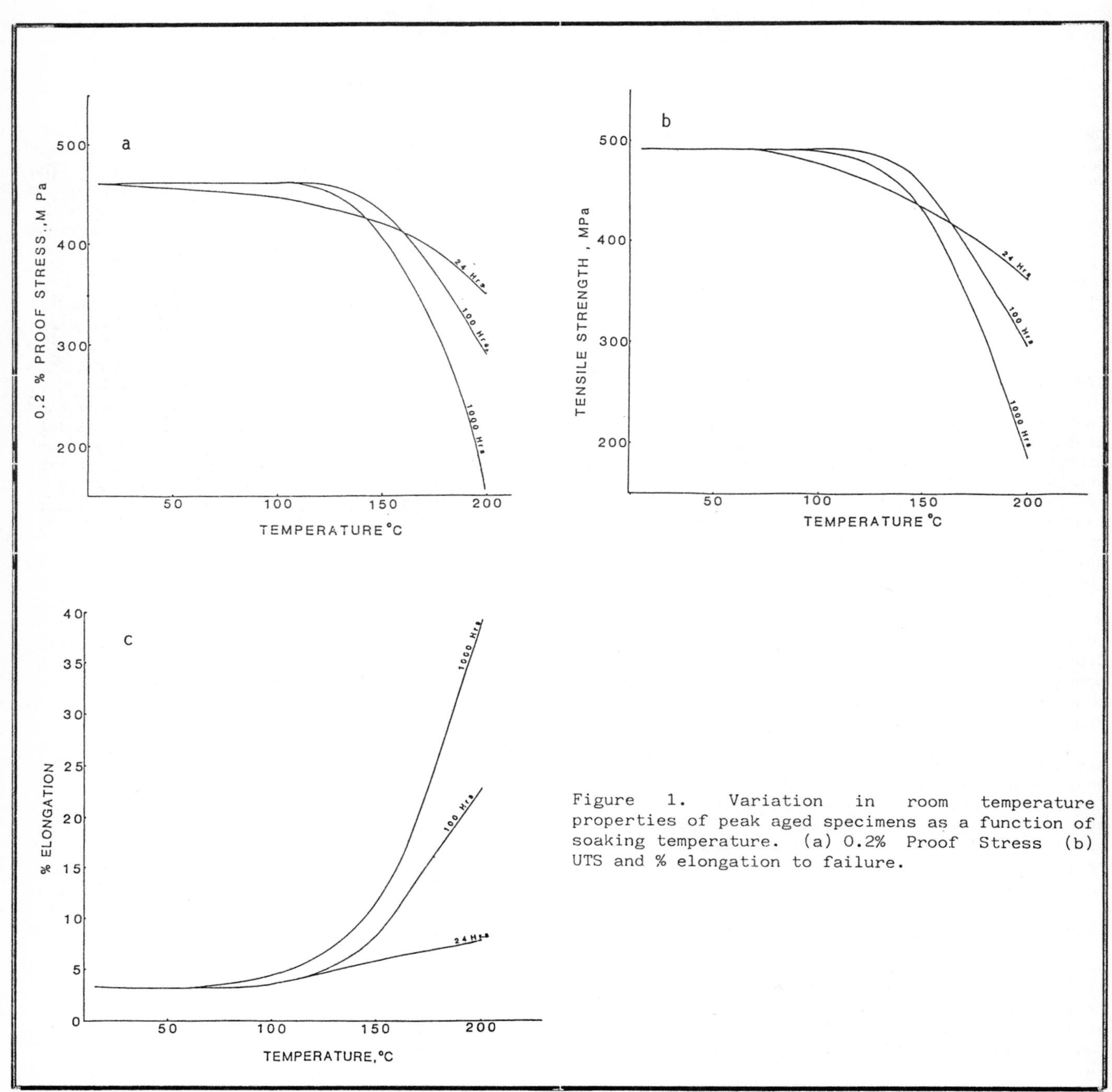

Figure 1. Variation in room temperature properties of peak aged specimens as a function of soaking temperature. (a) 0.2% Proof Stress (b) UTS and % elongation to failure.

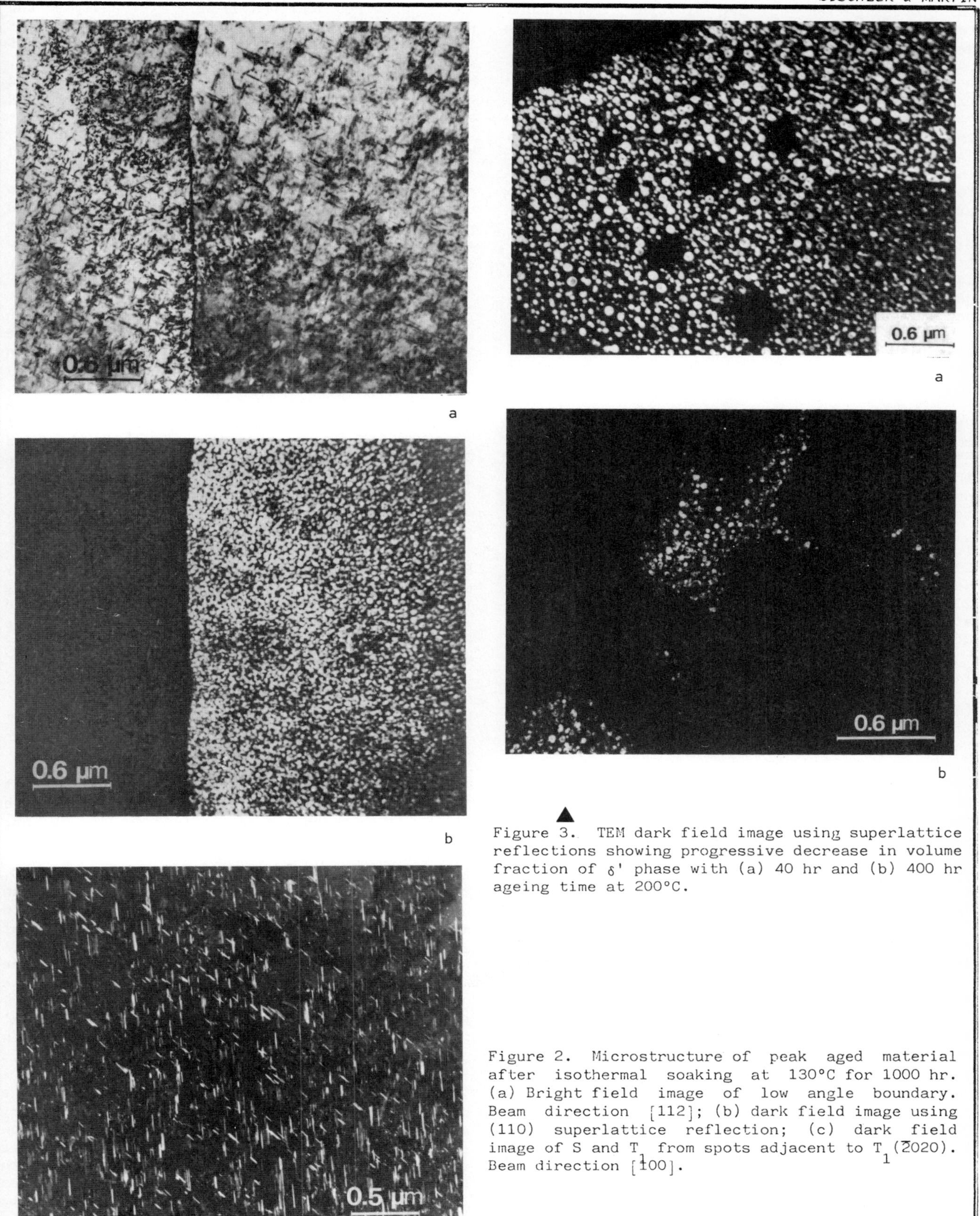

Figure 3. TEM dark field image using superlattice reflections showing progressive decrease in volume fraction of δ' phase with (a) 40 hr and (b) 400 hr ageing time at 200°C.

Figure 2. Microstructure of peak aged material after isothermal soaking at 130°C for 1000 hr. (a) Bright field image of low angle boundary. Beam direction [112]; (b) dark field image using (110) superlattice reflection; (c) dark field image of S and T_1 from spots adjacent to T_1 $(\bar{2}020)$. Beam direction [100].

(b) investigation of the elevated temperature tensile properties of the microstructures developed by the heat-treatments in (a) above.

EXPERIMENTAL

The Al-Li-Cu-Mg YYYA material was supplied by the RAE in the form of rolled plate of 25mm thickness. The plate had been solution-treated at 530°C for 16 hr, cold-water quenched and subjected to stretching of 2.5%.

Microstructure was characterised by optical and transmission electron microscopy (TEM). In the latter case, thin foils were prepared using double jet polishing techniques and a 1:3 nitric acid:methanol electrolyte.

Wave dispersive X-ray (WDX) analysis was carried out by electron-probe microanalysis on the coarse constituent particles formed during casting.

Cylindrical tensile specimens of gauge length 18mm and diameter 5mm were taken from the centre-line of the plate in order to avoid any through-thickness inhomogeneity. They were prepared from both longitudinal(L) and transverse(T) orientations.

Test-pieces were aged to peak hardness (20 hr at 190°C), and were then soaked for 24, 100 and 1000 hr in silicone oil baths at temperatures up to 200°C. Tensile tests were subsequently conducted at a strain-rate of $1x10^{-3}$ sec according to BS 3688 part 1.

Two series of tests were conducted:

(a) Peak aged specimens were soaked for the total time and air-cooled before tensile testing at room temperature (recovery tests).

(b) Peak aged specimens were soaked for the total time minus 0.75 hr, then transferred to the test machine. Here they were reheated to soak temperature over 0.5 hr and allowed a further 0.5 hr at soak temperature before testing.

Fractured specimens were examined both by SEM and by metallographic section.

RESULTS AND DISCUSSION

1. Microstructural Characterisation

The composition of the alloy is given in Table 1. The microstructure consisted of pancake-shaped grains of size 2.4mm x 0.6mm x 35 μm in the L,T and ST directions respectively .

Dislocation subgrains were present of average size 1.6 μm; these were pinned by Al_3Zr dispersoids of size approximately 40 nm. A uniform network of dislocations was present as a result of the stretch treatment.

The coarse residual phases were microanalysed and shown to be mainly iron- and silicon-bearing. The largest proportion were of Al_7Cu_2Fe in the form of stringers, and other phases identified were Al_6Fe, Mg_2Si, and Fe Si . Coarse particles containing elements from the flux and grain-refining additions were also identified.

TEM examination of the peak-aged material showed an even distribution of the δ' phase of average size 40 nm. A number of the δ' particles were observed to have formed as composites round particles of Al_3Zr. The grain boundaries (but not the subgrain boundaries) showed a narrow precipitate-free zone (PFZ).

A small amount of δ phase was present at the grain boundaries, and a uniform distribution of the S-phase (Al_2 CuMg) was present which had been nucleated on the dislocation network introduced by the stretch and also on the subgrain boundaries. A similar distribution of T_1-phase (Al_2 CuLi) was present in smaller amounts. The structures of these phases, and their orientation relationships were in agreement with those reported in the literature (4,5).

2. Recovery Tests.

Figure 1 shows the Proof Stress, UTS and ductility of L-specimens at room temperature after soaking for 24 hr, 100 hr and 1000 hr at the stated temperatures. Strength data for T-specimens have not been included, but they showed a closely similar trend.

A small rise in strength is observed after prolonged soaking at temperatures up to 130°C; this arises from an increase in the volume fraction of precipitate rejected from solution.

The TEM microstructures observed after 1000 hr at 130°C are shown in fig.2. The mean particle-size of the δ' phase has increased from 40 to 50 nm, and that of the S-phase from 120 to 150 nm.

As the soak temperature is increased above 140°C, it is seen in fig.1 that the proof stress undergoes a progressive decrease, at a rate dependent upon the soak time. After 1000 hr at 200°C, the proof stress and UTS have fallen to 40% and 60% respectively of their values in the peak-aged condition. At the same time the ductility passes through a small trough at 140°C, and then rises to a maximum of 7% after 1000 hr at 200°C. The data for T-specimens were again similar, although all values were slightly lower.

The changes in microstructure accompanying the drop in proof stress and UTS have been identified. As indicated in fig. 3, a progressive dissolution of the δ' phase occurs, accompanied by an increase in the volume fraction of the δ phase. The width of the PFZ and both grain and subgrain boundaries increases, and as soaking time increases only islands of δ' particles are observed at the

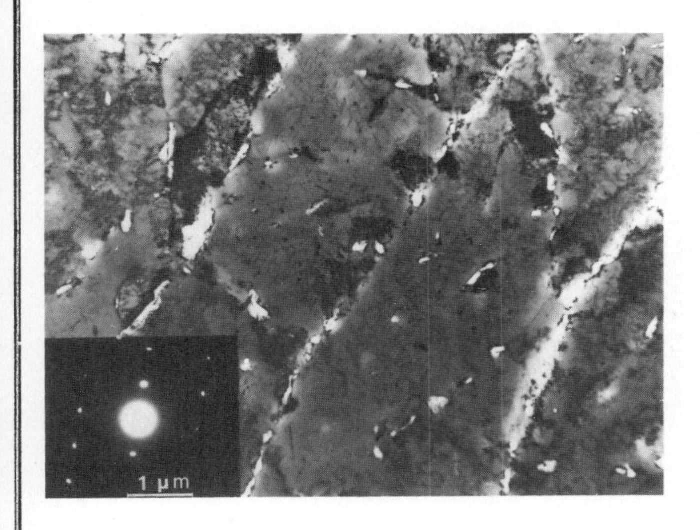

Figure 4. Bright field image of peak aged material soaked for 1000 hr at 200°C. Large particles of equilibrium phase at boundaries have dropped out during foil preparation. Beam direction [112].

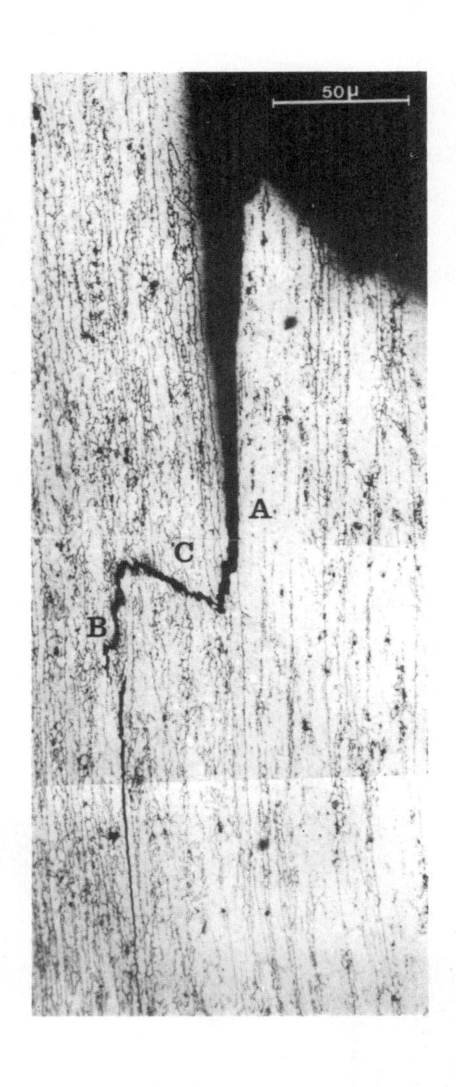

Figure 6. Section of fractured tensile specimen peak aged and soaked for 1000 hr at 130°C. Fracture shows delamination of longitudinal coplanar grain boundaries (A), intersubgranular failure (B) and transgranular failure (C).

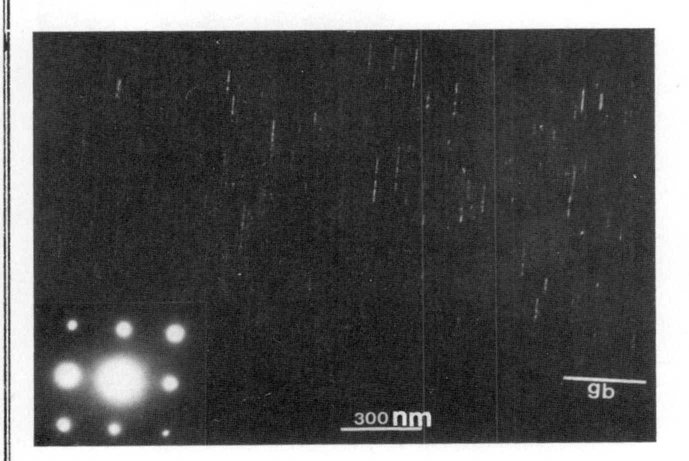

Figure 5. Dark field image of S' phase in peak aged material soaked for 1000 hr at 200°C. Beam direction [001]. A grain boundary is marked (gb).

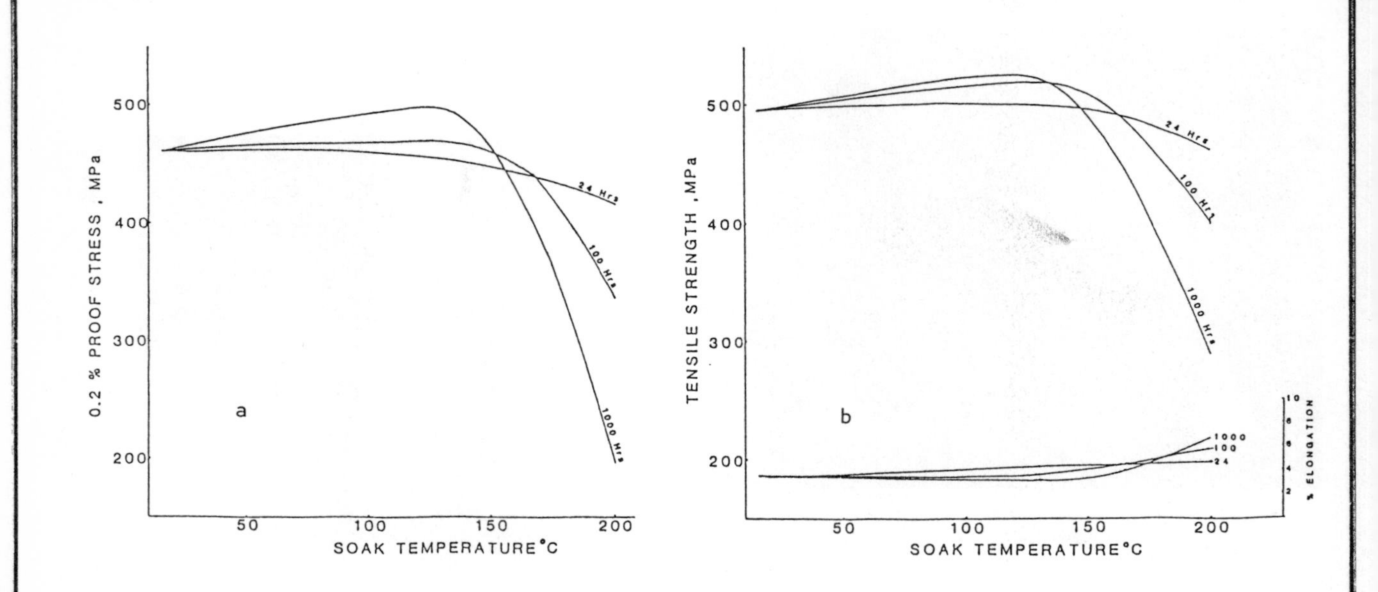

Figure 7. Variation in tensile properties of peak aged specimens with isothermal soaking temperature. (a) 0.2% Proof Stress, (b) UTS and (c) % elongation to failure of specimens tested at the temperature of soaking.

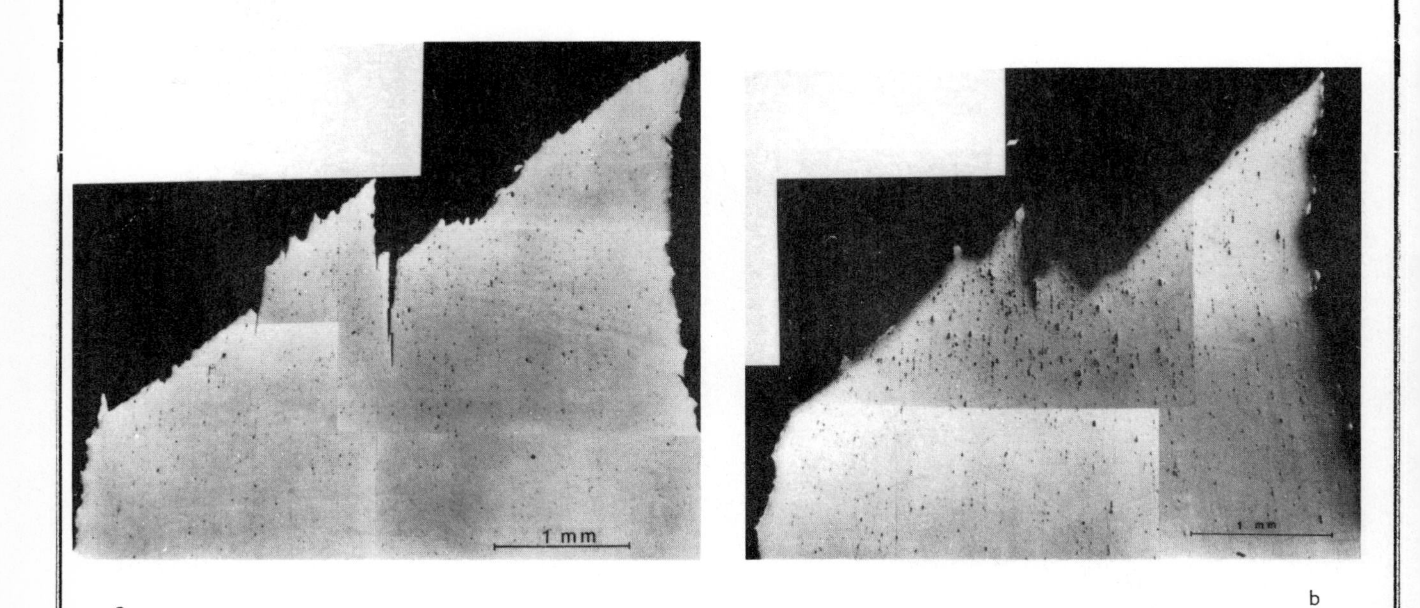

Figure 8. Longitudinal microsections of failed tensile specimens after testing at soaking temperature. (a) Peak aged and soaked for 1000 hr at 130°C and (b) Peak aged and soaked for 1000 hr at 200°C.

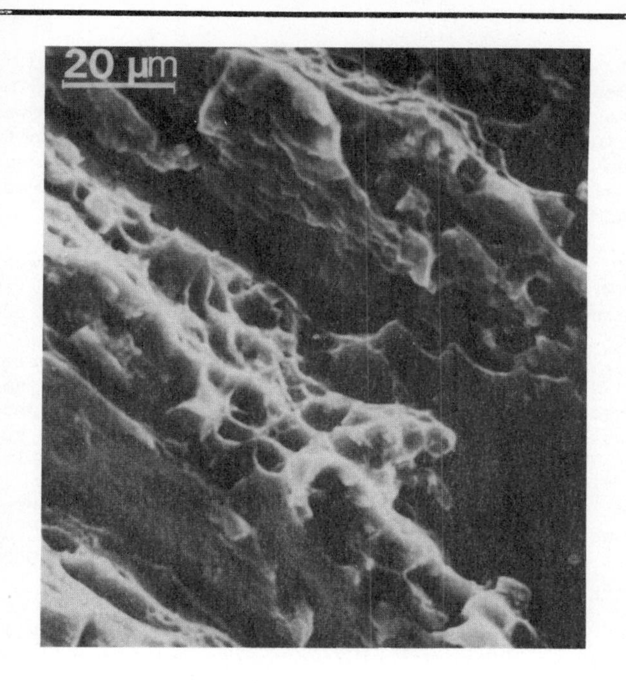

Figure 9. SEM fractograph of specimen soaked 1000 hr at 130°C, tested at soak temperature.

centres of the subgrains. After 1000 hr at 200°C, all the δ' phase had redissolved, to be replaced by δ phase in the grain interiors as well as at grain and subgrain boundaries, as shown in fig.4, although it is difficult to retain the δ phase in TEM foils.

Fig.5 shows that the S phase grows to a size of 300 nm , and a zone free of the phase is present adjacent to the boundaries. A large number of lattice dislocations were observed to be associated with the δ phase particles.

Longitudinal microsections were prepared of the fracture surfaces obtained in specimens soaked for 1000 hr at 130°C and 200°C. Macroscopically they appeared very similar, with a serrated profile. As shown in fig.6 the fracture path was identified as being made up of (a) decohesion along longitudinal coplanar grain boundaries (A) lying parallel to the stress axis , and (b) a mixture of intersubgranular failure (B), ductile rupture, and (C) transgranular shear. A common feature of room temperature fracture in this material is the continuation of longitudinal grain boundary failure, resulting in subsurface subsidiary cracks. T-specimens showed a similar behaviour, although some subsurface cracking was present in the 'L' direction, i.e. below and parallel to the fracture surface.

The inherent weakness of the longitudinal grain boundaries observed here is consistent with the observations of West and Lumsden (6) on a similar material. It has been suggested that a segregation of K to these interfaces can promote this type of intergranular failure.

The slight increase in ductility in the material soaked at 200°C was accompanied by an increased amount of cavitation associated with the coarse residual phases, and the fracture path itself exhibited less shear character and a more intersubgranular path. It is considered that the observed increase in width of the PFZ, coupled with the occurrence of the incoherent δ phase on the subgrain boundaries are likely to promote this mechanism of failure.

The quantitative metallographic analysis of the distribution of the equilibrium phases present after prolonged heat-treatment is the subject of continuing study.

3. Tests at Soak Temperature.

Fig.7 shows the Proof Stress, UTS and ductility data for L-specimens tested at the temperature of soaking. It may be seen that the trends are similar to those of the recovery tests, fig.1, but no rise in strength is seen below 140°C as in the former tests. It is however noteworthy that no significant decline in properties occurs below 140°C in the specimens which have been aged for the longer times. This presumably arises because of the increase in volume fraction of δ' precipitate leading to the elevated temperature properties being maintained.

A rapid rise in ductility takes place above 150°C, the amount of increase depending on the degree of overageing. The failure strain rises to a maximum of 40% in material held 1000 hr at 200°C.

The metallographic changes accompanying the increase in ductility may be seen in fig.8, which compares specimens held 1000 hr at 130°C and 200°C respectively. The sawtooth profile in fig.8a arises from the mixed intersubgranular fracture and ductile microvoid coalescence associated with second-phase particles. These processes may be identified from the SEM fractograph of fig.9. The smooth areas correspond to the intersubgranular fracture, and lie parallel to the stress axis. At higher magnification, fine dimples could be resolved on these surfaces. Final separation has occurred in a transgranular ductile mode associated with coarse intermetallic phases.

In fig. 8(b), the greater amount of necking in the latter specimen is apparent, as is the increased cavitation. The latter will arise because of the increased volume fraction of incoherent precipitates in the microstructure, and because of the greater strain and triaxiality of stress which have been applied . It is apparent that the high failure strain is due to the dramatic fall in Proof Stress and UTS, because of the high temperature of test.

CONCLUSIONS

1. A small rise in room temperature strength is observed after prolonged soaking at temperatures up to 130°C; this is due to further precipitation of the δ' phase from solid solution.

2. As the soak temperature is increased above 140°C, the δ' is progressively replaced by the equilibrium δ phase. This is reflected by a decline in strength, so that Proof Stress and UTS fall to 40% and 60% of their respective values at peak-ageing after the alloy is soaked for 1000 hr at 200°C.

3. When tested at the temperature of soak, no significant decline in properties occurs below 140°C, since the increase in volume fraction of δ' phase offsets any thermal softening.

4. A rapid rise in ductility occurs above 150°C in material tested at its soak temperature which is ascribed to the operation of thermally activated deformation and recovery processes.

ACKNOWLEDGEMENTS.

This work has been carried out with the support of the Procurement Executive, Ministry of Defence, and the authors are grateful to Professor Sir Peter Hirsch, FRS, for the laboratory facilities made available.

REFERENCES.

1. C.J. Peel, B. Evans, C.A. Baker, D.A. Bennett, P.J. Gregson and H.M. Flower. "The development and application of improved aluminium-lithium alloys", Proc. 2nd Int. Conf. on Aluminium-Lithium Alloys, Monterey, California,ed. T.H. Sanders and E.A. Starke. AIME, 1983.

2. P.J. Gregson and H.M. Flower. "Microstructural control of toughness in aluminium-lithium alloys". Acta Met.33 (1985) 527

3. K. Dinsdale, S.J. Harris and B. Noble. "Relationship between microstructure and mechanical properties of Al-Li-Mg alloys" Proc. 1st Int. Conf. on Aluminium-Lithium Alloys , ed. T.H. Sanders and E.A. Starke, AIME 1981, p.101.

4. R.N. Wilson and P.G. Partridge. "The nucleation and growth of S' precipitates in an Al - 2.5% Cu - 1.2% Mg alloy". Acta Met. 13 (1965) 1321

5. F.S. Lin, S.B. Chakraborthy and E.A. Starke. "Microstructure-property relationships of two Al - 3Li - 2Cu - 0.2Zr - X Cd alloys". Met.Trans 13A (1982) 401.

6. J.A. Wert and J.B. Lumsden. Scripta Met. 19 (1985) 205.

□

Elevated temperature strength of Al–Li–Cu–Mg alloys

M PRIDHAM, B NOBLE and S J HARRIS

The authors are in the Department of Metallurgy and Materials Science at the University of Nottingham, UK

SYNOPSIS

Elevated temperature tests have been carried out on an Al-2.5%Li-1.2%Cu-0.7%Mg-0.12%Zr alloy after soaking at temperature for 1h and 1000h before testing at 20-250°C. The alloy was found to maintain its strength to temperatures of 200°C (1h soak) and 160°C (1000h soak). These strength levels were superior to those of 2618A alloy for test temperatures up to 260°C (1h soak) and 180°C (1000h soak).

The aged alloy has a microstructure of small subgrains on which is precipitated $S(Al_2CuMg)$ and within the subgrains is a uniform distribution of $\delta'(Al_3Li)$. By testing alloys that did not contain copper and/or zirconium, the relative importance of subgrain boundaries, subgrain boundaries stabilised by S precipitation, and matrix δ' precipitation on the elevated temperature strength of Al-Li alloys has been assessed.

1. INTRODUCTION

An aluminium-lithium alloy that has received a great deal of attention in recent years (1) is one based on Al-2.5%Li-1.2%Cu-0.7%Mg-0.12%Zr and which has now been given the international designation 8090. Extensive work has been carried out on this alloy system to determine its tensile properties and fracture toughness (2) but relatively little is known about its behaviour at elevated temperatures. The present paper presents such data for holding times of 1h and 1000h at temperature before testing, and compares the data with that for 2618A (Al-2.5%Cu-1.5%Mg-1.0%Fe-1.0%Ni).

High temperature properties have previously been measured for Al-Li-Mg alloys (3) and rapidly solidified Al-Li-Co alloys (4). Both these studies showed that when the alloys contain zirconium as a recrystallisation inhibitor then the high temperature strength begins to fall rapidly at temperatures exceeding 75-100°C. Noble, Harlow and Harris (3) attributed this rapid decrease in strength to dynamic recovery caused by the movement of dislocations to sub-boundaries in the alloy.

The 8090 alloy examined in the present work also contains zirconium as a recrystallisation inhibitor, together with additions of copper and magnesium which act together to precipitate the S phase (Al_2CuMg). This S phase nucleates preferentially on sub-boundaries and therefore one of the objectives of the present work has been to compare the effect on high temperature strength of sub-boundaries that are precipitate-free, and sub-boundaries that are decorated with precipitates. As a further comparison, lithium containing alloys have been prepared that are free of sub-boundaries altogether, and this has been achieved by using manganese as a grain refiner instead of the more widely used zirconium.

2. EXPERIMENTAL PROCEDURE

Most of the alloys described in this work have been prepared from high purity metals, melted and cast under argon. Compositions (wt.%) are given in table 1. The cast billets were homogenised 48h at 500°C before being extruded to 10mm diameter rod. Solution treatment has been carried out at 520°C followed by ageing 24h at 195°C. Hot tensile testing was undertaken on a Mayes 100kN capacity screw driven machine operated at a strain rate of $10^{-4}sec^{-1}$. Samples were held at temperature for periods of either 1h or 1000h before testing to BS 3688.

Some tests have also been carried out on commercial 8090 rolled sheet alloy. The heat treatment and mechanical testing procedure was exactly the same as the extruded alloys.

Table 1. Compositions (wt.%) of alloys

Alloy Code	Composition (wt.%) Li	Cu	Mg	others
Al-Li-Cu-Mg-Zr, R	2.43	1.30	0.65	0.11Zr
Al-Li-Mg-Zr, R	2.45	–	0.96	0.13Zr
Al-Li-Cu-Mg-Mn, R	2.45	1.27	0.64	1.01Mn
Al-Cu-Mg-Fe-Ni, R	–	2.50	1.45	1Fe,1Ni
Al-Li-Cu-Mg-Zr, S	2.49	1.24	0.70	0.12Zr

R = extruded rod, S = rolled sheet

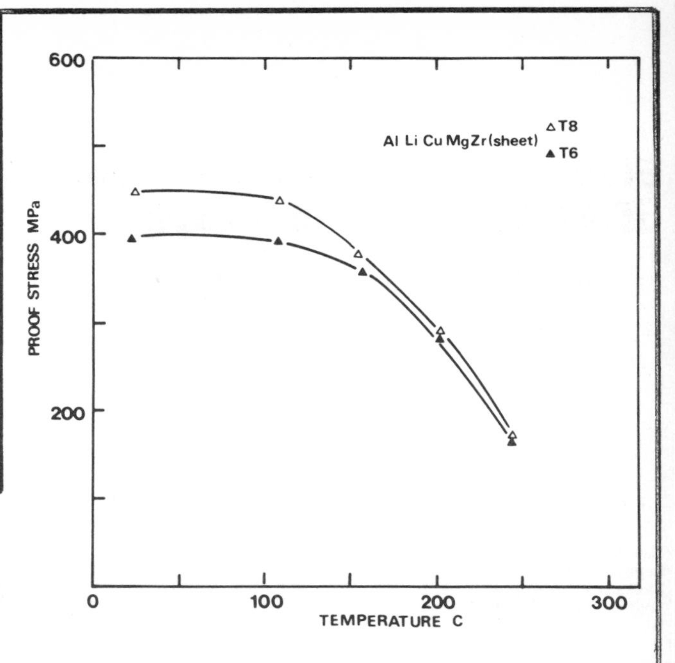

1. Short term (1h soak) elevated temperature strength of Al-Li-Cu-Mg-Zr rolled alloy sheet tested in the T6 and T8 conditions.

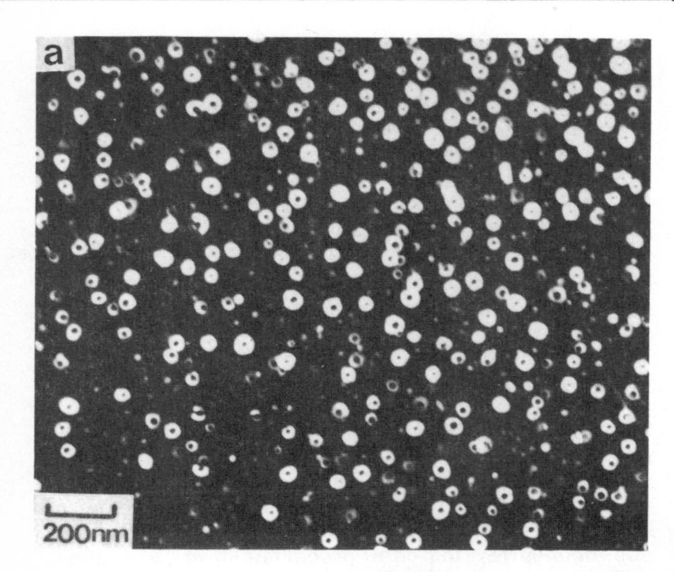

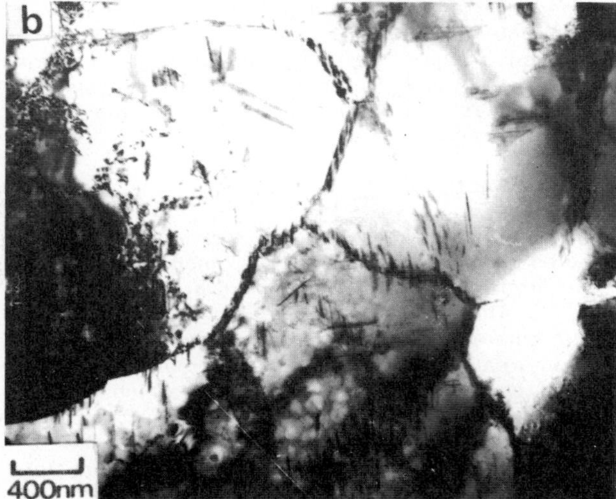

2. Precipitating phases in the Al-Li-Cu-Mg-Zr alloy (T6 condition). (a) δ'(Al_3Li) in the matrix, (b) S(Al_2CuMg) on subgrain boundaries.

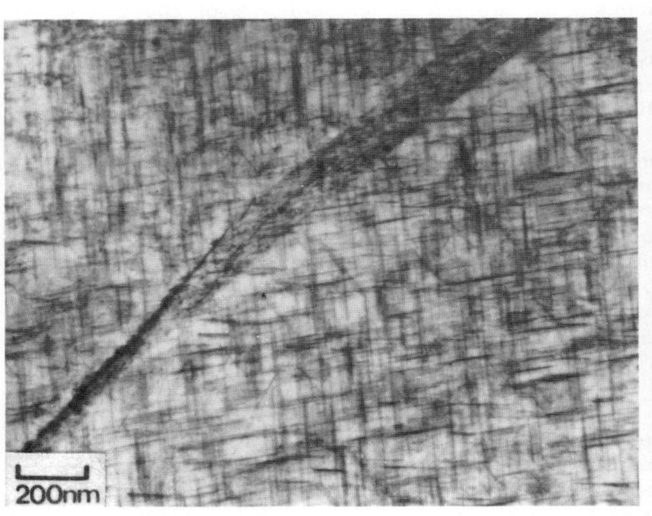

3. Precipitation of S on dislocations introduced by stretching before ageing (T8 condition).

3. RESULTS

The elevated temperature strength of 8090 alloy is presented first, the material being tested in both the T6 (solution treated and aged) and T8 (solution treated, 3% stretch followed by ageing) conditions. The contribution to the high temperature strength from sub boundaries free of precipitate and sub-boundaries decorated with precipitates is then assessed by testing Al-Li-Mg-Zr alloys. Finally, the effect of completely eliminating sub-boundaries from the microstructure is examined by testing Al-Li-Cu-Mg-Mn alloys.

3.1 Effect of stretching 8090 prior to ageing

Fig. 1 shows the short term (1h) strength of commercial Al-Li-Cu-Mg-Zr aloy sheet tested without a stretch before ageing (T6) and after a 3% extension before ageing (T8). In the T6 condition the room temperature strength is approximately 400 MPa, the strengthening phases being δ'(Al_3Li) in the matrix and S(Al_2CuMg) precipitated on sub-boundaries (fig. 2). Stretching followed by ageing (T8 temper) increases the room temperature strength to 450 MPa, the improved strength being due to the precipitation of S on dislocation lines introduced by the stretching operation (fig. 3).

The improved strength achieved by the stretch is maintained up to test temperatures of ≈100°C but thereafter the improvement decreases so that at 180°C there is little difference between the T6 and T8 conditions. It therefore appears that matrix precipitation of S is not greatly affecting the high temperature strength of these alloys. All subsequent tests have therefore been carried out with the alloys in the unstretched (T6) condition.

3.2 Sheet vs. extruded alloys

All subsequent tests to be described are on alloys extruded to 10mm bar. The different deformation texture produced by extrusion results in appreciably higher strengths and a PS/TS ratio of 0.96 compared to 0.80 for sheet alloy (fig. 4). The difference between proof stress for sheet and extrusion decreases with increasing test temperature, but even at temperatures >200°C the extruded alloy is still ≈80 MPa stronger than the rolled sheet.

3.3 Long term elevated temperature strength

The long term (1000h soak) elevated temperature strength is given in fig. 5, along with the 1h data for comparison. It can be seen that for short term testing the proof stress is maintained up to temperatures ≈200°C but after an extended (1000h) period of soaking before testing the properties start to decrease rapidly at ≈160°C.

Also included in the figure is data for 2618A alloy. The 8090 alloy has a much higher short term strength than 2618A at all temperatures up to 260°C, thereafter the strengths of the two alloys are comparable. For long term strength (1000h) 8090 is superior to 2618A at temperatures up to 180°C. Above 180°C the 2618A alloy is stronger, though on a specific strength basis the two alloys are again comparable at these elevated temperatures.

Transmission electron microscopy on alloys subjected to the 1000h soak showed the δ'(Al_3Li) particles to be resistant to coarsening at test temperatures <150°C. At higher test temperatures coarsening of δ' took place, e.g. 1000h at a test temperature of 195°C increased the δ' size from 20nm to 65nm (fig. 6). Test temperatures above 195°C caused very rapid coarsening of δ'; a 1000h soak at a test temperature of 200°C increased the δ' size from 20nm to 140nm (fig. 7). S precipitation was unchanged up to test temperatures of 195°C, but at and above this temperature the long term soak resulted in additional precipitation on sub-boundaries and on dislocations within the grains (fig. 8).

3.4 Alloys not containing the S phase

Results presented in the previous section show that Al-Li-Cu-Mg-Zr alloys perform well up to temperatures 160-200°C (depending on time of soak). This was not entirely expected since previous work (3) had indicated that when zirconium is present the resulting sub-boundaries in the microstructure encourage rapid dynamic recovery leading to significant softening of the alloy. The Al-Li-Cu-Mg-Zr alloys tested in the present work differ from those examined previously in that they contain S(Al_2CuMg) phase precipitated on the sub-boundaries. It therefore appears that the S phase is preventing the sub-boundaries from being efficient dislocation sinks and therefore delaying dynamic recovery. To investigate this further, an alloy of the 8090 type was prepared that did not contain copper. Only δ'(Al_3Li) forms in such an alloy and therefore the sub-boundaries are not decorated with precipitates (fig. 9). The elevated temperature strength of this alloy is given in fig. 10 (1h soak) and fig. 11 (1000h soak). The short term properties show an accelerated fall-off in strength with temperature in the alloy that is copper-free, and at temperatures >150°C this results in 100 MPa difference between alloy containing the S phase and alloy that is S-free. A similar trend is observed with the elongation values; these are increasing even at temperatures <100°C in the copper free alloy. In the copper-containing alloys the elongation does not start to rise until the test temperature exceeds 100°C, and even then the rate of increase with temperature is much smaller than that in the copper-free alloy.

The long term (1000h soak) properties show a more dramatic difference. The properties of the copper containing alloy do not start to fall appreciably until 160°C has been exceeded, in fact up to ≈125°C the strength actually increases. Copper-free alloys on the other hand show a rapid decrease in strength right from ambient temperatures. The curves for the two alloys converge at 175°C and thereafter there is little difference between the two alloys. These characteristics are also reflected in the elongation values, which start to increase at ≈100°C in the copper-containing alloy, but which are

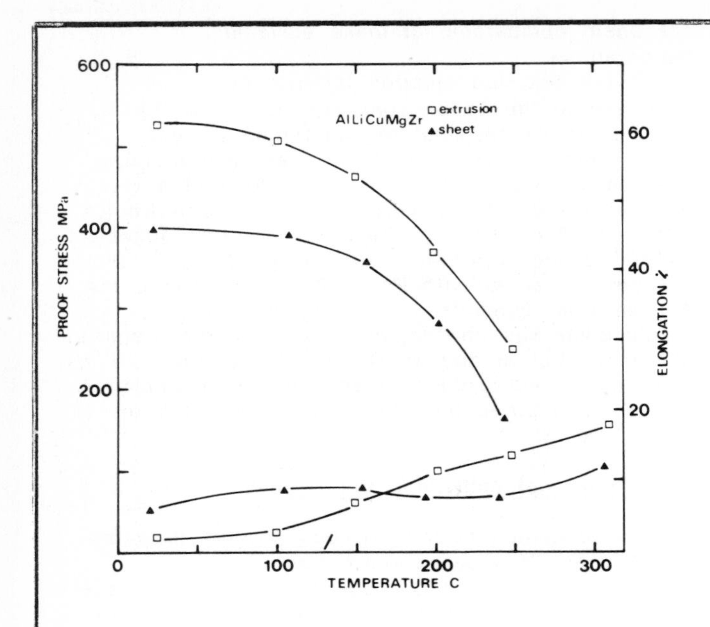

4. Short term (1h soak) elevated temperature strength and ductility of Al-Li-Cu-Mg-Zr alloy in rolled sheet and extruded rod.

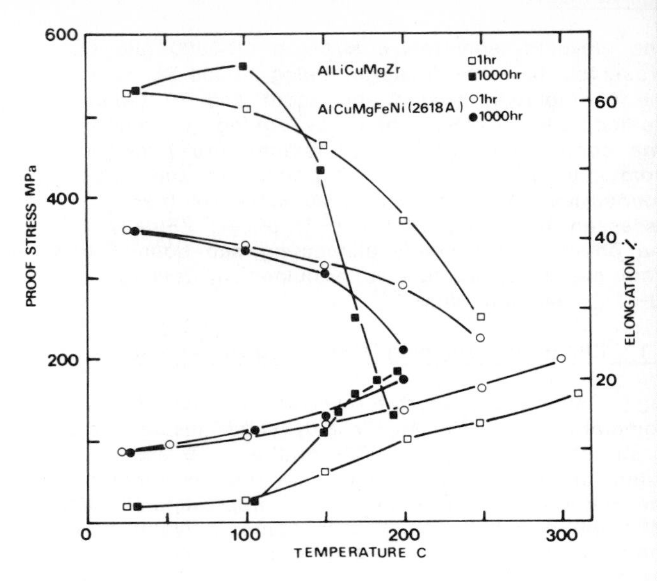

5. Short and long term (1h and 1000h soak) elevated temperature strength and ductility of extruded Al-Li-Cu-Mg-Zr. Data for 2618A alloy also shown for comparison.

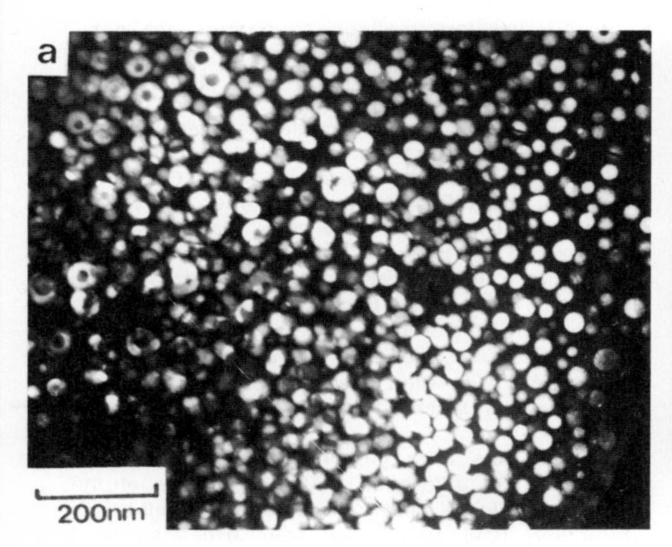

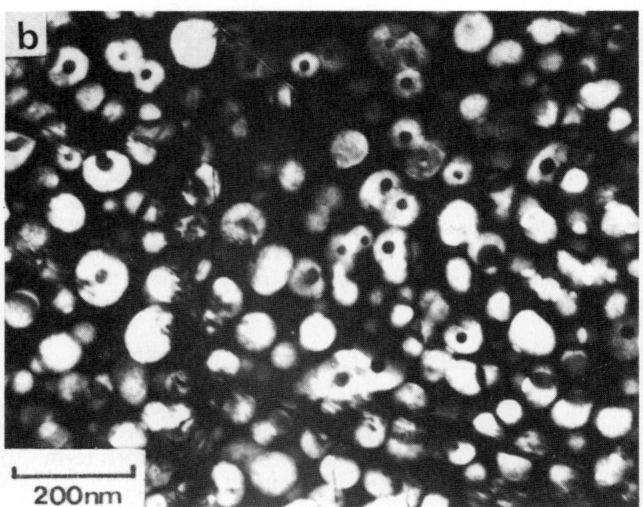

6. δ'(Al_3Li) precipitates in Al-Li-Cu-Mg-Zr alloy after ageing (a) 24h and (b) 1000h; at 195°C.

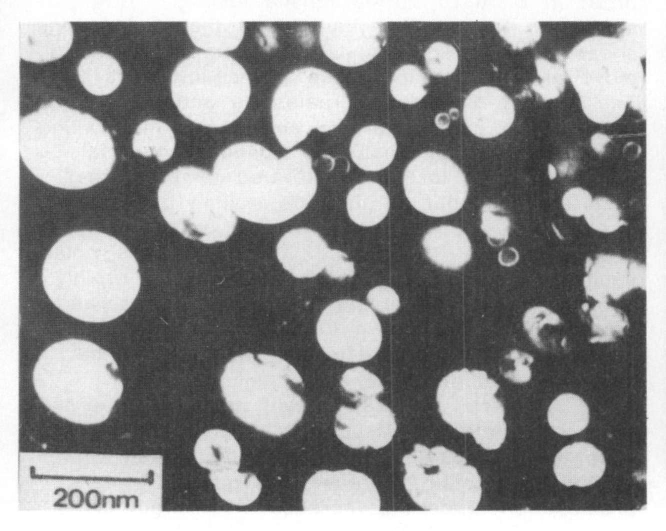

7. $\delta'(Al_3Li)$ precipitates in Al-Li-Cu-Mg-Zr alloy after 1000h at 200°C.

9. Sub-boundaries free of precipitate in aged Al-Li-Mg-Zr alloy.

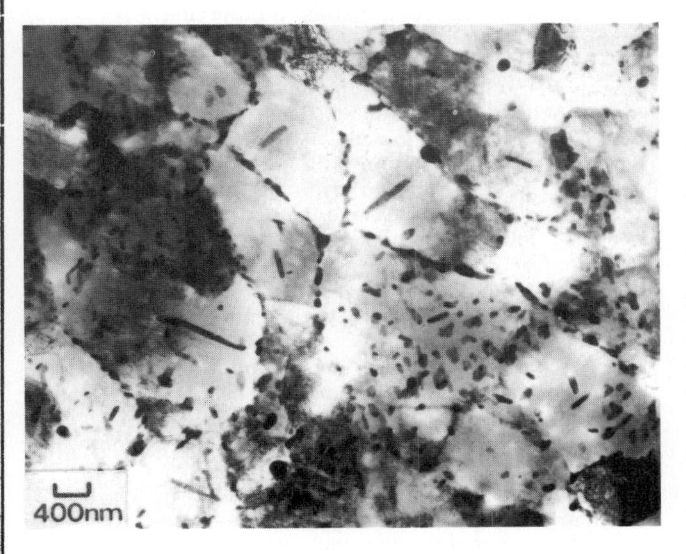

8. Coarse precipitation produced in Al-Li-Cu-Mg-Zr alloy when held at temperatures >200°C.

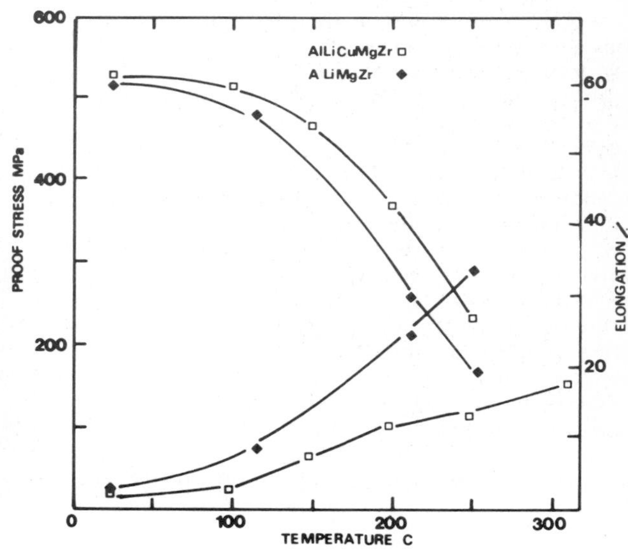

10. Short term elevated temperature strength and ductility of Al-Li-Mg-Zr, with data for Al-Li-Cu-Mg-Zr for comparison.

increasing continuously from 20°C in the copper-free alloy.

Transmission electron microscopy on the copper-free alloy showed the δ'(Al_3Li) precipitates to be approximately the same size as those in the copper-containing alloy (≈20nm) after ageing 24h at 195°C. Soaking the alloy 1000h at 195-200°C also produced approximately the same amount of coarsening of the δ' phase.

3.5 Alloys not containing sub-boundaries

Results in the previous section indicated that sub-boundaries decorated with precipitates are much more beneficial to elevated temperature strength of Al-Li alloys than are sub-boundaries free of precipitates. This now raises the question of what will be the effect on elevated temperature strength of completely eliminating sub-boundaries from the microstructure. This has been achieved using a grain refiner of manganese instead of the zirconium addition. Recrystallisation takes place in such alloys thus eliminating any subgrain structure (fig. 12).

The elevated temperature strength of the Al-Li-Cu-Mg-Mn alloy is given in fig. 13 (1h soak) and fig. 14 (1000h soak). The short term properties show the recrystallised alloy to have substantially lower strength at ambient temperatures but with increased test temperature the strength does not fall until a temperature of 230°C has been exceeded. This results in very little difference in strength between Al-Li-Cu-Mg-Mn and Al-Li-Cu-Mg-Zr alloys at test temperatures above about 230°C. At these high temperatures (>200°C) both these alloys are stronger than the Al-Li-Mg-Zr alloy that contains sub-boundaries free of precipitate.

The long term (1000h soak) mechanical properties (fig. 14) show that above 175°C the alloy free of sub-boundaries has the highest strength. The Al-Li-Cu-Mg-Zr alloy with the sub-boundaries decorated with precipitates starts to lose its strength rapidly at ≈160°C, but the Al-Li-Cu-Mg-Mn alloy without sub-boundaries does not start to significantly soften until 180°C has been exceeded. Referring back to fig. 5 shows this alloy to have approximately the same strength levels as 2618A at test temperatures above 100°C.

The elongation values of the long soak tests on the Al-Li-Cu-Mg-Mn alloy also fit in the general picture; the rapid increase in elongation does not take place until 150-160°C has been exceeded. This should be compared to a temperature of ≈100°C in the Al-Li-Cu-Mg-Zr alloy containing sub-boundaries with precipitate, and ≈20-50°C in the Al-Li-Mg-Zr alloy with sub-boundaries free of precipitate.

4. DISCUSSION

The present work shows that Al-Li-Cu-Mg-Zr (8090) alloys have good elevated temperature strength up to 200°C (short time at temperature) and 160°C (extended time at temperature). In the latter case the strength actually increases for test temperatures up to 125°C and this is probably due to a combination of further fine δ' precipitation occurring during soaking and thermal restoration of order in δ'(Al_3Li) during tensile testing. TEM studies in these alloys has shown that dislocations shear the ordered δ'(Al_3Li) particles and hence move around as superlattice dislocation pairs (5). The leading dislocation creates an antiphase domain boundary in the δ' particle and before the trailing dislocation can eliminate the boundary, thermal agitation (when testing at elevated temperatures) restores the order so that the trailing dislocation has to surmount the same obstacle as the leading dislocation (6). Sastry and O'Neal (4) have noted similar increases in strength when testing rapidly cooled Al-Li alloys at elevated temperatures, and the effect is well known in nickel based alloys containing γ'(Ni_3Al) precipitates which have the same ordered structure as δ'(Al_3Li).

Increases in strength with testing temperature were also noted in Al-Li-Cu-Mg-Mn alloys. In this case the increase in strength is due solely to thermal restoration of order and not δ' precipitation since the increase in strength is maintained to 200-210°C where, if further δ' precipitation occurred, would lead to softening (overageing) and not increased strength.

The thermal restoration of order that takes place in all the above alloys helps to maintain the high strength of the Al-Li based alloys to temperatures that compare well with those of 2618A. The short term (1h) properties of 8090 do not fall to those of 2618A until a test temperature of 260°C is reached. The long term properties (1000h) of 8090 fall to those of 2618A at a temperature of 180°C. With the Al-Li-Cu-Mg-Mn alloy the mechanical properties fall to those of 2618A at temperatures of 260°C and 200°C respectively for 1h and 1000h soak times.

The present study has also demonstrated the importance of sub-boundaries on the elevated temperature strength of Al-Li based alloys. Where sub-boundaries are present and free of any precipitates then the mechanical properties fall very rapidly with increasing test temperature, even at temperatures as low as 50°C. This agrees well with our earlier work on Al-2%Li-4%Mg-0.18%Zr alloys (3) and is due to dynamic recovery which appears to be able to take place in Al-Li alloys at relatively low test temperatures due to the retention of quenched-in vacancies by a strong Li-vacancy binding energy. The high concentration of vacancies then assists the climb of dislocations into the sub-boundaries.

Elimination of the sub-boundaries, by adding manganese instead of zirconium so that recrystallisation can take place, prevents the rapid dynamic recovery and mechanical strength is stabilised to higher test temperatures. Unfortunately, the elimination of the sub-boundaries also reduces the mechanical properties at ambient temperatures thus losing one of the major beneficial characteristics of Al-Li based alloys.

Decoration of sub boundaries with S precipitates appears to be an effective compromise between the above two effects. The high strength at ambient temperature is achieved and when the boundaries are precipitated with S phase they appear to be less effective as sinks for dislocations. This has the effect of delaying the

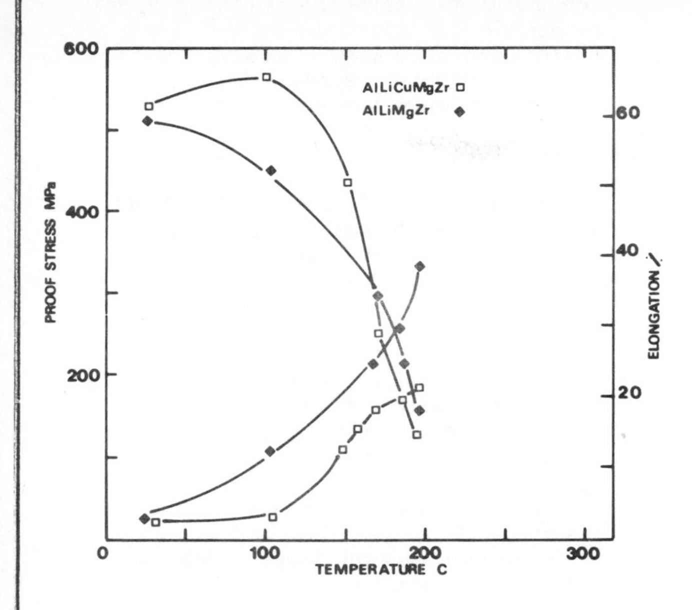

11. Long term elevated temperature strength and ductility of Al-Li-Mg-Zr, with data for Al-Li-Cu-Mg-Zr for comparison.

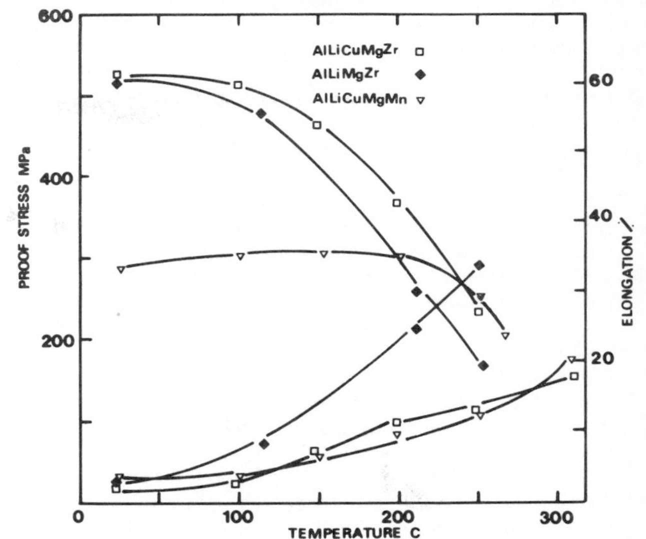

13. Short term elevated temperature strength and ductility of Al-Li-Cu-Mg-Mn, with data for Al-Li-Cu-Mg-Zr and Al-Li-Mg-Zr for comparison.

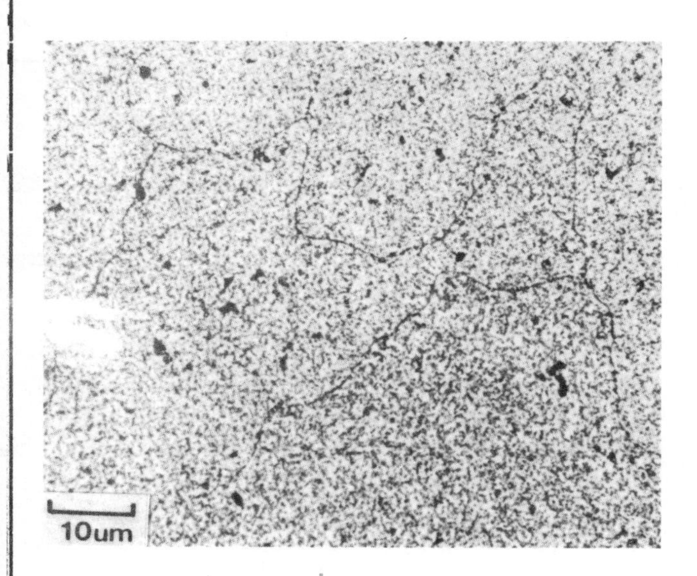

12. Recrystallised grain structure of Al-Li-Cu-Mg-Mn alloy.

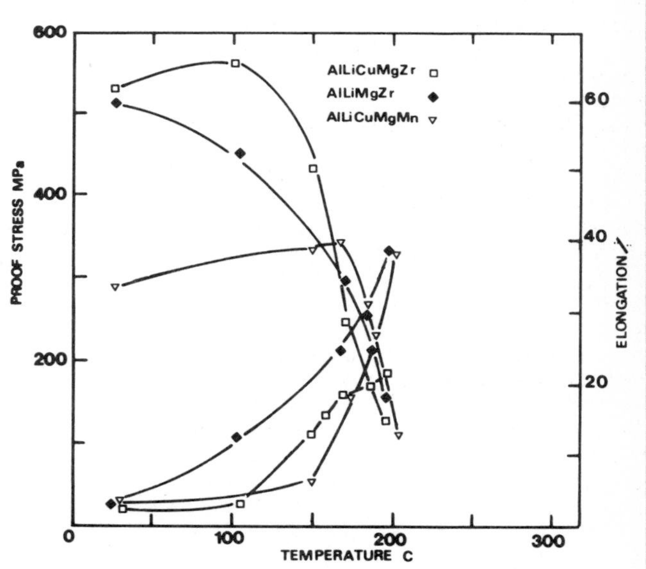

14. Long term elevated temperature strength and ductility of Al-Li-Cu-Mg-Mn, with data for Al-Li-Cu-Mg-Zr and Al-Li-Mg-Zr for comparison.

onset of rapid dynamic recovery and hence strength is maintained to relatively high temperatures.

5. CONCLUSIONS

(1) Stretching Al-Li-Cu-Mg-Zr alloys between solution treatment and ageing does not significantly improve the strength at test temperatures >150°C.
(2) Extruded Al-Li-Cu-Mg-Zr alloys have significantly higher strength than rolled sheet alloys at all test temperatures up to 250°C.
(3) Al-Li-Cu-Mg-Zr alloys maintain their strength to temperatures of 200°C (short term 1h soak) and 160°C (long term 1000h soak). The tensile properties fall to those of 2618A at test temperatures of 260°C and 180°C for short term and long term soaks respectively.
(4) The good high temperature strength of Al-Li-Cu-Mg-Zr alloys is due to the stability of $\delta'(Al_3Li)$ at temperatures up to 190°C together with the fact that sub-boundaries are heavily decorated with $S(Al_2CuMg)$. This precipitation on the sub-boundaries prevents them from acting as efficient dislocation sinks and hence delays dynamic recovery.
(5) Removal of S phase from the alloy so that sub-boundaries are not decorated with precipitates produces a drastic fall in mechanical strength, even at test temperatures as low as 50°C.
(6) Elimination of sub-boundaries altogether from the microstructure, by using manganese as a grain refiner instead of zirconium as a recrystallisation inhibitor, produces an alloy with stable mechanical properties up to test temperatures of 230°C (short term soak) and 180°C (long term soak). However, the absence of the substructure causes a significant reduction in room temperature strength.

ACKNOWLEDGEMENTS

The authors express their appreciation to SERC and Rolls Royce Ltd. for financial support, and thank Dr. D. Driver and Dr. J.B. Borradaile for many helpful discussions.

REFERENCES

(1) C.J. Peel, B. Evans, C.A. Baker, D.A. Bennett, P.J. Gregson and H.M. Flower: 'Aluminium-Lithium Alloys II' (Ed. T.H. Sanders and E.A. Starke); 1984, New York (AIME), p.363.
(2) P.J. Gregson, H.M. Flower, C.J. Peel and B. Evans: 'Metallography of Light Alloys'; 1983, London (Inst. of Metallurgists).
(3) B. Noble, K. Harlow and S.J. Harris: 'Aluminium-Lithium Alloys II' (Ed. T.H. Sanders and E.A. Starke); 1984, New York (AIME), p.65.
(4) S.M.L. Sastry and J.E. O'Neal: 'Aluminium-Lithium Alloys II' (Ed. T.H. Sanders and E.A. Starke); 1984, New York (AIME), p.79.
(5) B. Noble and G.E. Thompson, Metal Sci. J., 1971, 5, 114.
(6) A. Gysler and S. Weissman, Met. Sci. Eng., 1977, 27, 281.

Characterisation of coarse precipitates in an overaged Al–Li–Cu–Mg alloy

M D BALL and H LAGACÉ

The authors are with
Alcan International Ltd
Kingston R & D Centre,
Kingston, Ontario,
Canada

ABSTRACT

Four previously unreported phases were found to occur in Al-Li-Cu-Mg alloys aged at temperatures between 300°C and 350°C. Transmission electron microscopy and electron diffraction techniques were used to characterise these particles.

INTRODUCTION

In order to meet the varied demands of the aerospace industry, the compositions and fabrication routes for aluminium-lithium based alloys have become more diverse. To achieve specific combinations of properties alloying additions frequently include Cu, Mg and Zr.

An understanding of the precipitation reactions which occur in these alloys is important in establishing appropriate ageing reactions to optimise the performance. In the "Lital" alloy (8090) with a nominal composition of Al-2.5Li - 1.2Cu - 0.7Mg - 0.1Zr, the important precipitation reactions are found to be a combination of those which occur in binary Al-Li alloys(1) and those which occur in the ternary Al-Cu-Mg system(2). For a solution treated and quenched sample at normal ageing temperatures (e.g. at 200°C), the familiar $\delta'Al_3Li$ phase precipitates homogeneously in the form of small coherent spherical particles and the needle like S-(Al_2CuMg) particles grow heterogeneously on dislocations and grain boundaries. δ AlLi and T-(AlCuMgLi) also develop at suitable heterogeneous sites. Where zirconium is present in the alloy, it normally precipitates as small coherent Al_3Zr particles during earlier heat treatments such as homogenisation or solution heat treatments. Since the Al_3Li and Al_3Zr have the same crystal structure, the presence of Al_3Zr particles influences the precipitation of the δ' Al_3Li, leading to the formation of some complex Al_3(Zr, Li) precipitates with a zirconium rich core and a lithium rich outer shell(3,4).

Since the main strengthening mechanisms of these alloys derive from the presence of the fine δ' and S-phase precipitates, fabrication strategies are normally designed to minimise the formation of the coarser δ and T phases. Consequently, the precipitation behaviour of these alloys at higher temperatures (above 300°C) has not received much attention.

In most alloy systems, the presence of particles with sizes greater than about 1 μm can be important. For example, the recrystallisation behaviour, fracture toughness and corrosion resistance may all be influenced by the presence of coarse particles. Furthermore, they will inevitably lead to a reduction of the more desirable fine precipitate particles with a concomitant reduction in properties.

The aim of this work was to investigate the coarse precipitates which are observed in the 8090 alloy after ageing treatments in the range between 300°C and 350°C.

EXPERIMENTAL PROCEDURE

An alloy of composition Al-2.5Li - 1.2Cu-0.7Mg - 0.1Zr was cast by a modified direct chill method and hot rolled to 6 mm plate. After a solution heat treatment at 530°C, a series of samples were water quenched and then aged for one hour at temperatures between 300°C and 350°C. Specimens were prepared for transmission electron microscopy by electropolishing thin 3 mm-diameter discs in an electrolyte consisting of 30% HNO_3 in methanol at about -25°C. The specimens were always taken from the interior regions so as to

1. 2-D lattice image of a type 1 particle. The insets show a low magnification image and the corresponding diffraction pattern.

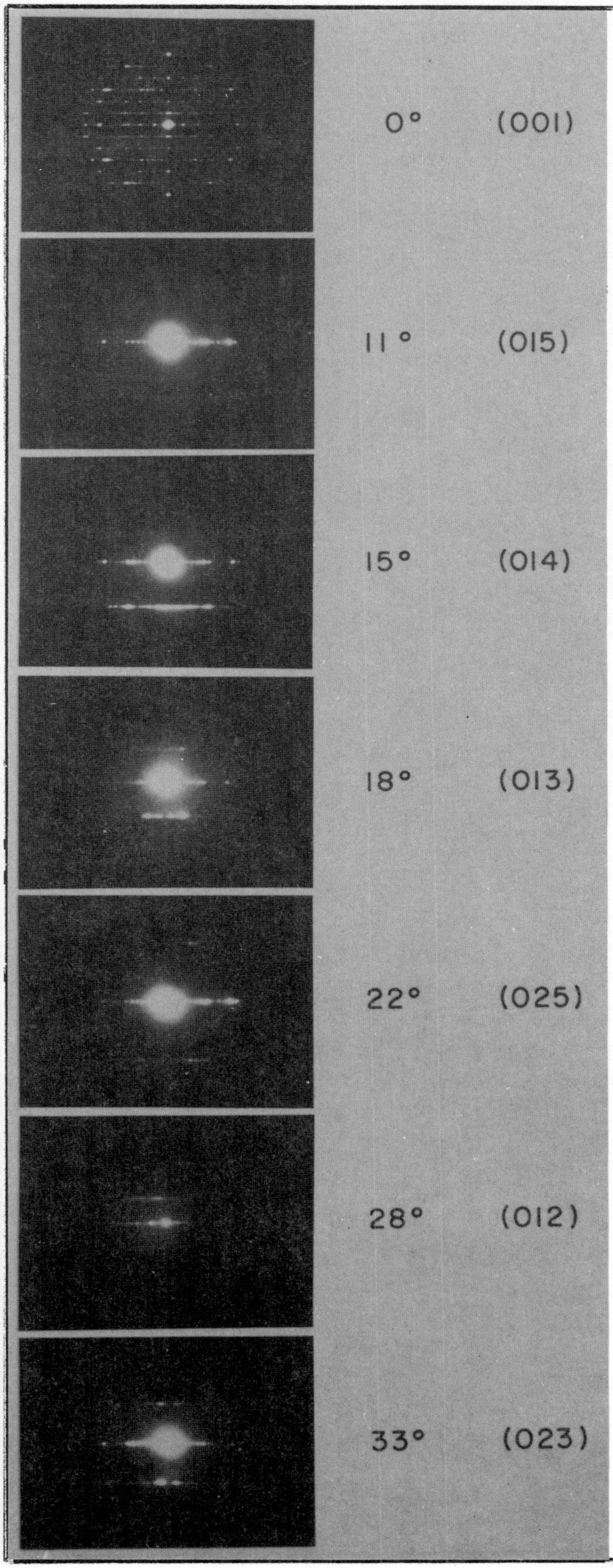

0° (001)

11° (015)

15° (014)

18° (013)

22° (025)

28° (012)

33° (023)

2. A series of electron diffraction patterns obtained from a type 1 particle by tilting systematically about the $\{h00\}$ row of spots.

3. 2-D lattice image of the type 2 precipitate. The particle and the corresponding diffraction pattern are shown in the insets.

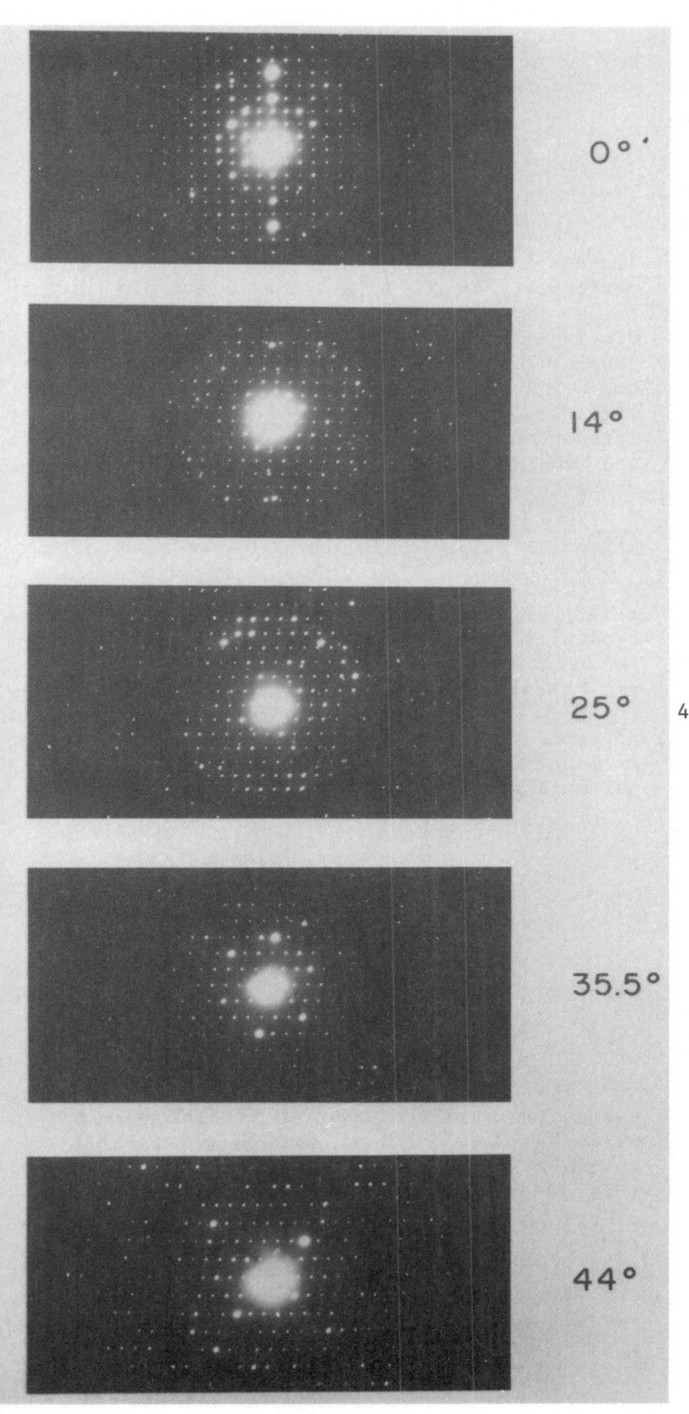

4. A systematic series of electron diffraction patterns obtained from a type 2 particle.

avoid the outer lithium depleted zone near the surface. Standard transmission electron microscopy techniques were used to characterise the various types of particles in terms of their morphology, crystallography and defect structure. Although some X-ray microanalysis and electron energy loss spectroscopy was performed, there are a number of complicating factors associated with this alloy system. The principal problems can be summarised as follows:-

(a) Lithium cannot be detected by X-ray microanalysis.

(b) Quantitative X-ray microanalysis for low levels of magnesium in aluminium alloys is prone to large errors because of the proximity of the Mg_k peak to the Al_k peak.

(c) Quantitative microanalysis in alloys containing copper is complicated by the tendency of copper to redeposit on the surface of the specimen during electropolishing. This often leads to spuriously high results for the apparent copper concentrations.

(d) Electron energy loss spectroscopy (EELS) requires very thin specimens (ideally less than about 50 nm). This cannot normally be achieved, over usefully large areas by conventional specimen preparation methods.

(e) The electron energy loss spectra obtained for intermetallics in this alloy system are difficult to interpret since the Li_K edge, the Mg_L edge and the Al_L edge are very close in energy. This complicates the background fitting procedure which is a critical part of the processing of (EELS) spectra.

Some of these problems, along with preliminary results obtained from this alloy system have been discussed by Malis[5].

In this work, the results obtained indicate that the precipitates studied all contain Mg and Cu, and probably Li.

RESULTS

Preliminary X-ray diffraction studies indicated that the δ-AlLi phase was present in significant quantities. These particles were not observed in the electropolished specimens as they invariably dissolve in the electrolyte. However a significant number of other precipitate particles were observed, ranging in size from about 0.5 μm up to about 5 μm in maximum dimension. A description of each of the four types observed is given in terms of its principal morphological and crystallographic characteristics.

TYPE 1

Figure 1 is a many beam lattice image obtained from part of the precipitate shown in the inset. The large size (up to 5 μm long) and elongated shape with well developed facets were characteristic of this phase. A series of electron diffraction patterns obtained by systematically tilting about the $\{h00\}$ row of diffraction spots is shown in Figure 2. From this diffraction data the unit cell was found to be orthorhombic with the following lattice parameters:-

$$a = 5.58 \pm 0.3 \text{ nm}$$
$$b = 1.44 \pm 0.07 \text{ nm}$$
$$c = 1.58 \pm 0.07 \text{ nm}$$

The (001) projection of this structure is imaged in Figure 1 and the large unit cell periodicity easily resolved. Several distinct types of planar defects can also be distinguished. In some cases these defects consist of planes of different structural units, such defects could accommodate significant variations in stoichiometry. Other defects may correspond to antiphase or microtwin boundaries and would not give rise to any compositional variation.

TYPE 2

The phase depicted in Figure 3 was relatively common. Examples of this type were typically in the size range between 0.5 and 1 μm in maximum dimension. No evidence of faceting was observed; the particles being generally rounded. The series of diffraction patterns depicted in Figure 4 has been indexed in terms of a monoclinic unit cell with the following parameters:-

$$a = 5.64 \pm 0.3 \text{ nm}$$
$$b = 1.40 \pm 0.05 \text{ nm}$$
$$c = 1.42 \pm 0.05 \text{ nm}$$
$$\beta = 105°$$

No obvious crystallographic defects were observed in these particles.

TYPE 3

The particle in Figure 5 is an isolated example of a third type of precipitate in this alloy system. This fan-shaped particle is multiply twinned with a high defect density. This is evident both from the heavily streaked diffraction pattern and from the lattice image. The apparent five-fold symmetry of the diffraction pattern can be explained in terms of the five twin variants. Each twin domain being rotated with respect to its neighbour by 72°.

The unit cell of this particle has not been unambiguously determined. However, the information available would be consistent with a monoclinic unit cell with

$$a = 1.07 \pm 0.05 \text{ nm}$$
$$b - \text{undetermined}$$
$$c = 0.68 \pm 0.05 \text{ nm}$$
$$\beta = 108°$$

5. Fan-shaped precipitate (Type 3). Twin boundaries and a high density of defects are apparent.

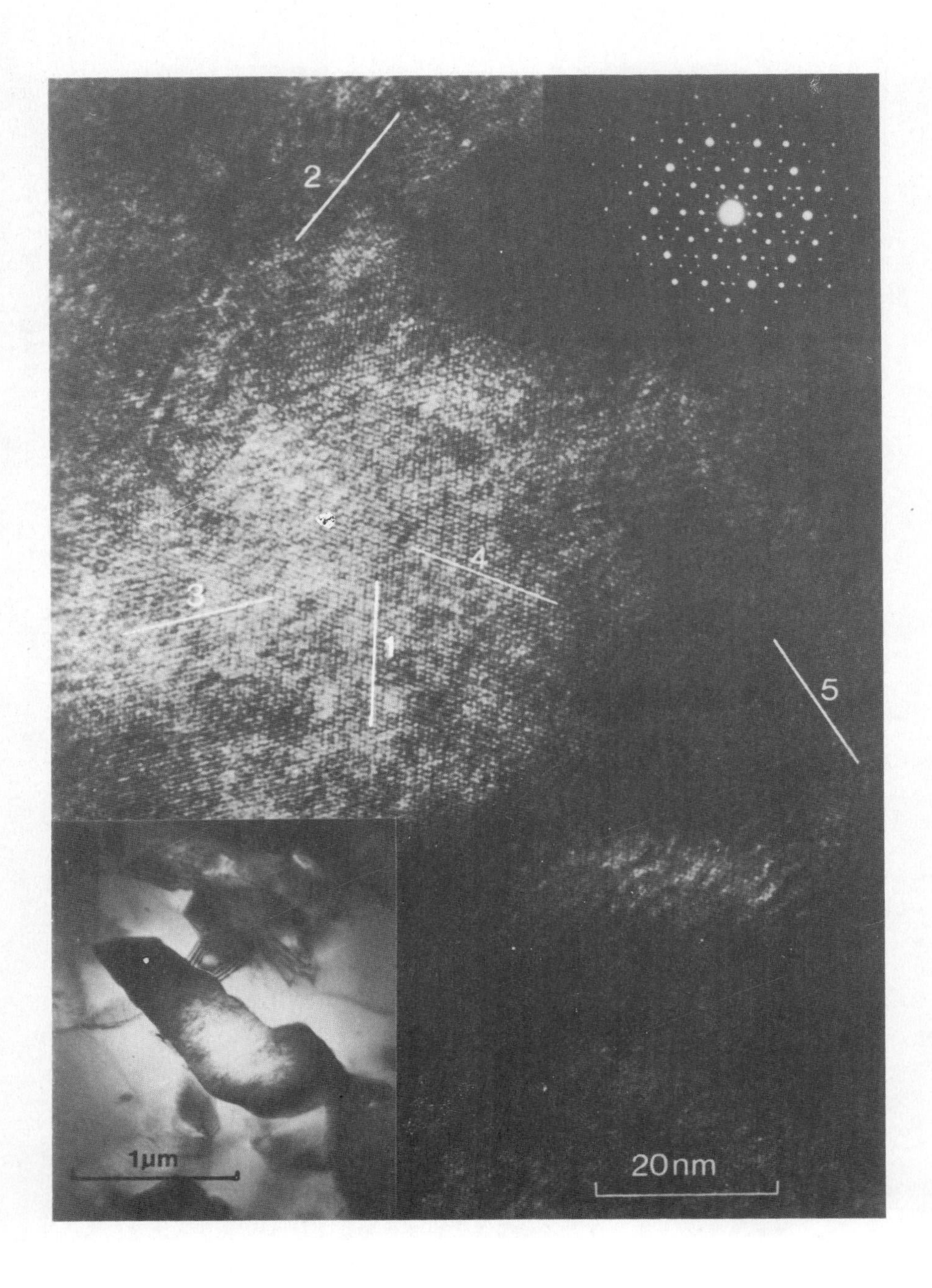

6. This 2-D lattice image of the type 4 precipitate shows five distinct lattice plane orientations (indicated). These particles always have a mottled appearance (see inset).

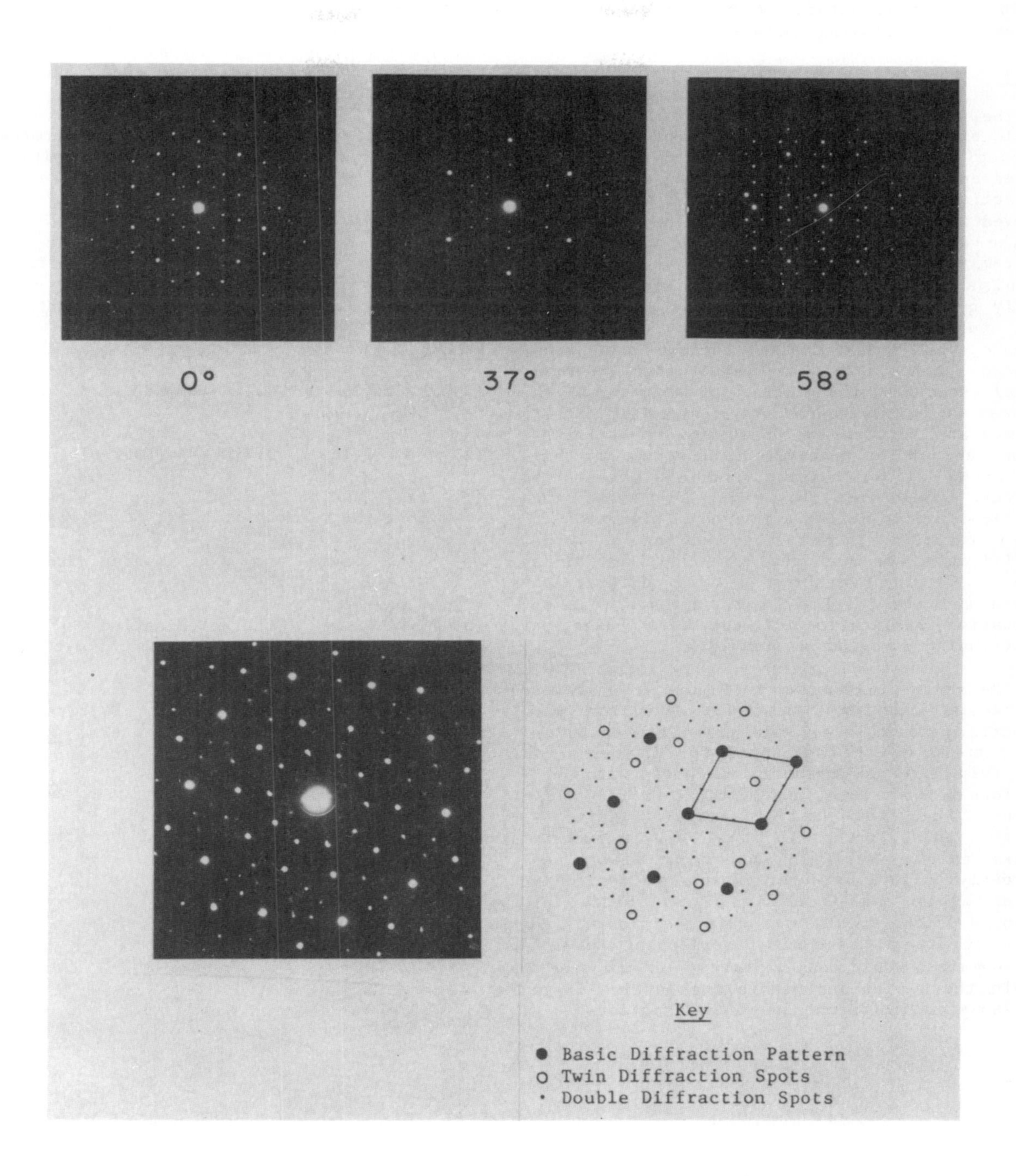

7. Diffraction patterns obtained from the type 4 precipitates. The schematic diagram indicates how the principal diffraction spots in the 5-fold pattern can be accounted for.

This particle appears to have nucleated as a number of twin related crystals which have grown by a twin plane re-entrant edge mechanism to form the fan-shaped structure. Distinct facets are apparent at the growing surfaces.

TYPE 4

The precipitate shown in Figure 6 is an example of a commonly observed species which grew to about 1 μm in size and exhibited a distinct mottled appearance under most imaging conditions. Diffraction patterns with five-fold symmetry were observed in certain orientations. These patterns and the diffraction patterns shown in Figure 7 are remarkably similar to those reported by Shechtman et al.[6,7] for particles found in rapidly quenched aluminium-manganese alloys. In order to explain their electron diffraction evidence, an icosahedral structure has been proposed. Since icosahedra cannot form a normal crystal structure, the stacking arrangement which has been suggested represents a state of matter intermediate between amorphous and crystalline. In the case of the precipitates observed in this Al-Li-Cu-Mg alloy, however, such a metastable structure would not be expected, since the particles develop during a prolonged high temperature treatment rather than during a rapid solidification as was the case for the Al-Mn phase. Careful scrutiny of the diffraction pattern and the lattice image suggests that an alternative explanation of the data based on microtwinning may also be possible.

The schematic diagram in Figure 7 indicates a possible indexing scheme for the five-fold symmetric pattern. All the principal spots in this pattern can be accounted for in terms of five twin related crystal variants with 72° rotations between them, and multiple diffraction. Evidence from the lattice image (Figure 6) supports this scheme in that the structure appears to consist of a number of domains of apparently uniform structure which are typically between 10 nm and 30 nm in size. Since the thickness of the specimen at this position was in excess of 30 nm, extensive overlapping between twin domains would be expected. This might explain the variations within the lattice image and also account for the double diffraction.

SUMMARY

Four previously unreported precipitate phases have been found to occur in our Al-Li-Cu-Mg (8090) alloy after ageing at temperatures between 300°C and 350°C. The principal morphological and structural characteristics of these phases have been described.

Further work is required to determine the detailed crystal structures and compositions of each of the phases.

ACKNOWLEDGEMENTS

The authors are grateful to Alcan International Limited for permission to publish this work and to Dr. D.J. Lloyd for many helpful discussions.

REFERENCES

(1) B. Noble and G.E. Thompson, Met. Sci. J. 5 (1971), 114.

(2) J.M. Silcock, J.I.M. 89 (1960-61), 203.

(3) P.L. Makin, D.J. Lloyd and W.M. Stobbs, Phil. Mag. 1985, Vol. 51, No. 5, L41-L47.

(4) F.W. Gayle and J.B. Vander Sande, Scr. Metall. 18 (1984), 473.

(5) T. Malis, "Characterisation of lithium distribution in aluminium alloys", (these proceedings).

(6) D. Shechtman, I. Blech, D. Gratias and J.W. Cahn, Phys. Rev. Letters, 53 (1984), 1951.

(7) D. Shechtman and I. Blech, Met. Trans., 16A (1985), 1005.

□

Effect of grain structure and texture on mechanical properties of Al–Li base alloys

I G PALMER, W S MILLER, D J LLOYD and M J BULL

IGP and WSM are with
Alcan International Limited, Southam Road,
Banbury, Oxon, OX16 7SP, UK.
DJL and MJB are with
Alcan International Limited, PO Box 8400,
Kingston, Ontario, Canada

SYNOPSIS

Al-Li alloys containing Zr often show unrecrystallised grain structures with a pronounced crystallographic texture. The influence of these factors on anisotropy of mechanical properties is discussed. The effect of various parameters on the recrystallisation behaviour of Al-Li-Cu-Mg-Zr alloys has been examined. Optimisation of these parameters has enabled fine recrystallised structures to be produced in sheet material. The texture and mechanical properties of these materials are described, with particular reference to effects on anisotropy.

INTRODUCTION

Aluminium-lithium base alloys containing zirconium often show highly elongated unrecrystallised grain structures with a pronounced crystallographic texture. These factors have a significant influence on mechanical properties and fracture behaviour. Previous work[1] has shown that the combined effects of texture and grain shape can result in anisotropy of tensile properties in rolled product. In this work it was shown that the detrimental effects of anisotropy could be overcome by using ageing treatments which resulted in copious precipitation of S phase and hence reduced planar slip. The limitations of this method are that it is only applicable to material cold worked prior to ageing and then overaged. In the present work the effects of the texture developed during rolling[2] on tensile properties are clarified for an Al-Li-Cu-Mg-Zr alloy. The factors affecting the recrystallisation characteristics of the alloys are examined together with the effect of a recrystallised structure on texture and anisotropy of mechanical properties.

The recrystallisation behaviour of Al-Li alloys containing Zr has not been extensively studied.

Lin and co-workers[3,4] examined the recrystallisation behaviour of Al-Li-Cu-Zr alloys with and without Cd additions. It was shown that recrystallisation results in a decrease in yield strength but a considerable increase in elongation[4]. In Al-Li-Cu-Mg-Zr alloys Wadsworth et al[5] used the method developed by researchers at Rockwell[6-8] involving solution treatment, overageing and warm working prior to heat treatment, to develop fine recrystallised microstructures that showed superplasticity at elevated temperatures. It was noted that small improvements in room temperature ductility were also achieved by this process. Fundamental studies on the recrystallisation and texture of Al-Li-Zr alloys are reported elsewhere[9,10], and will contribute to our understanding of these effects. Recrystallisation should result in a significant reduction in anisotropy by modification of the effect of texture on strength and ductility. This is a highly desirable feature for application of Al-Li base alloys since aerospace companies require isotropic properties in sheet material in all tempers.

EXPERIMENTAL DETAILS

Initial experiments were performed on alloys of nominal composition Al-2.5Li-1.8Cu-0.7Mg having four different Zr levels of 0.05, 0.09, 0.13 and 0.16%Zr. Subsequent experiments were

TABLE 1

Tensile Test Results of RET Material

		0.2% P.S. (MPa)	UTS (MPa)	E_U (%)	E_D (%)	E_F (%)
S.H.T.	L	219.8	344.1	9.1	1.9	11.0
S.H.T.	T	223.5	361.4	16.8	4.2	21.0
Aged	L	389.3	475.8	4.5	0	4.5
Aged	T	381.1	470.1	10.0	2.0	12.0

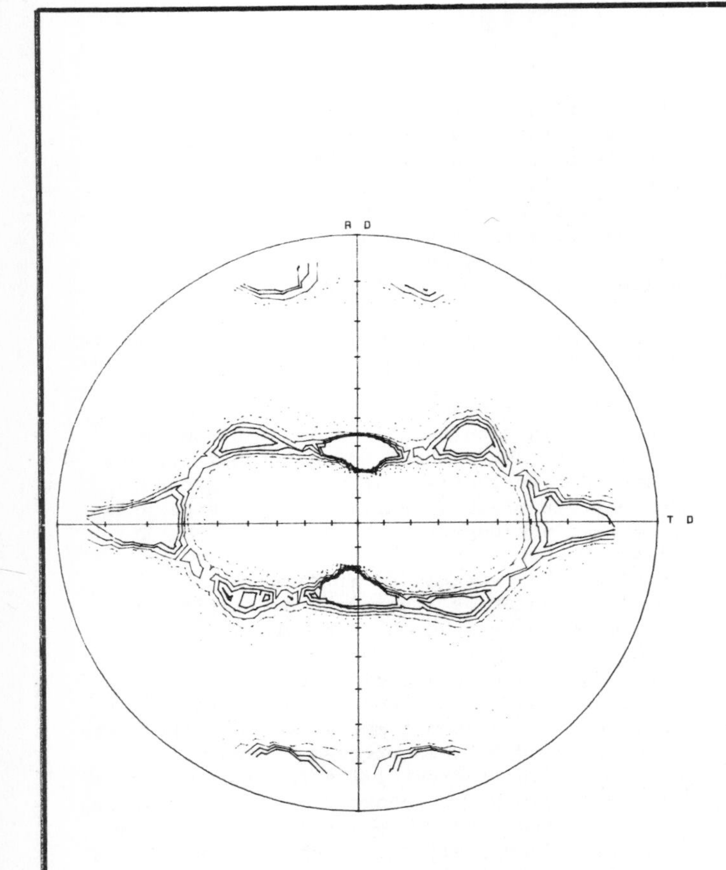

Figure 1 (111) pole figure for Al-2.58Li-1.24Cu-0.70Mg-0.12Zr sheet cold rolled from 5 mm to 1.6 mm.

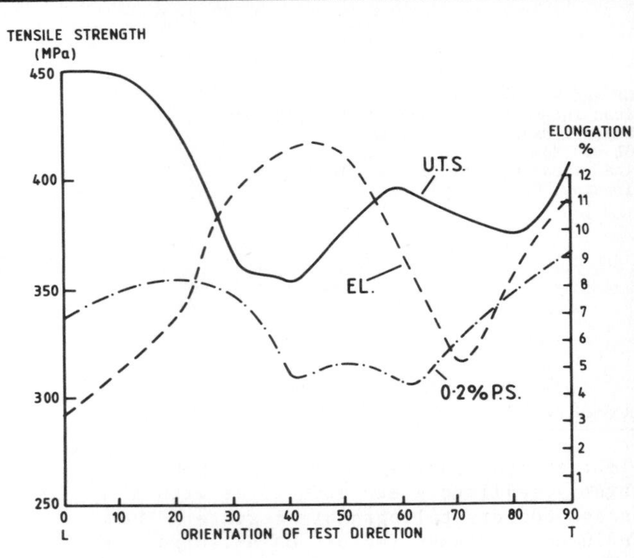

Figure 2 Anisotropy in tensile properties developed by straight rolling from 5 mm to 1.6 mm.

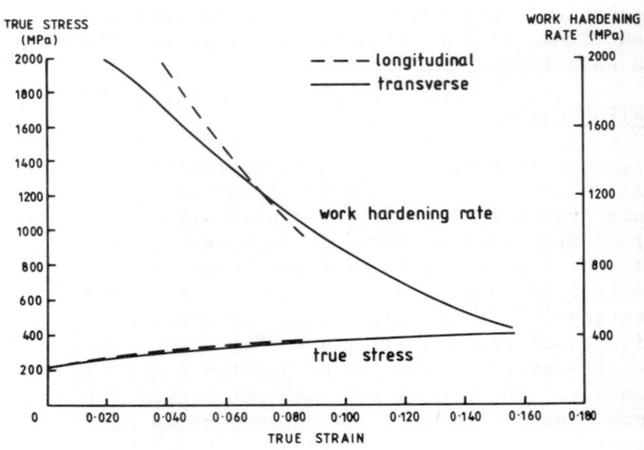

Figure 3 Stress-strain and work hardening rate-strain curves for solution treated material, longitudinal and transverse orientations.

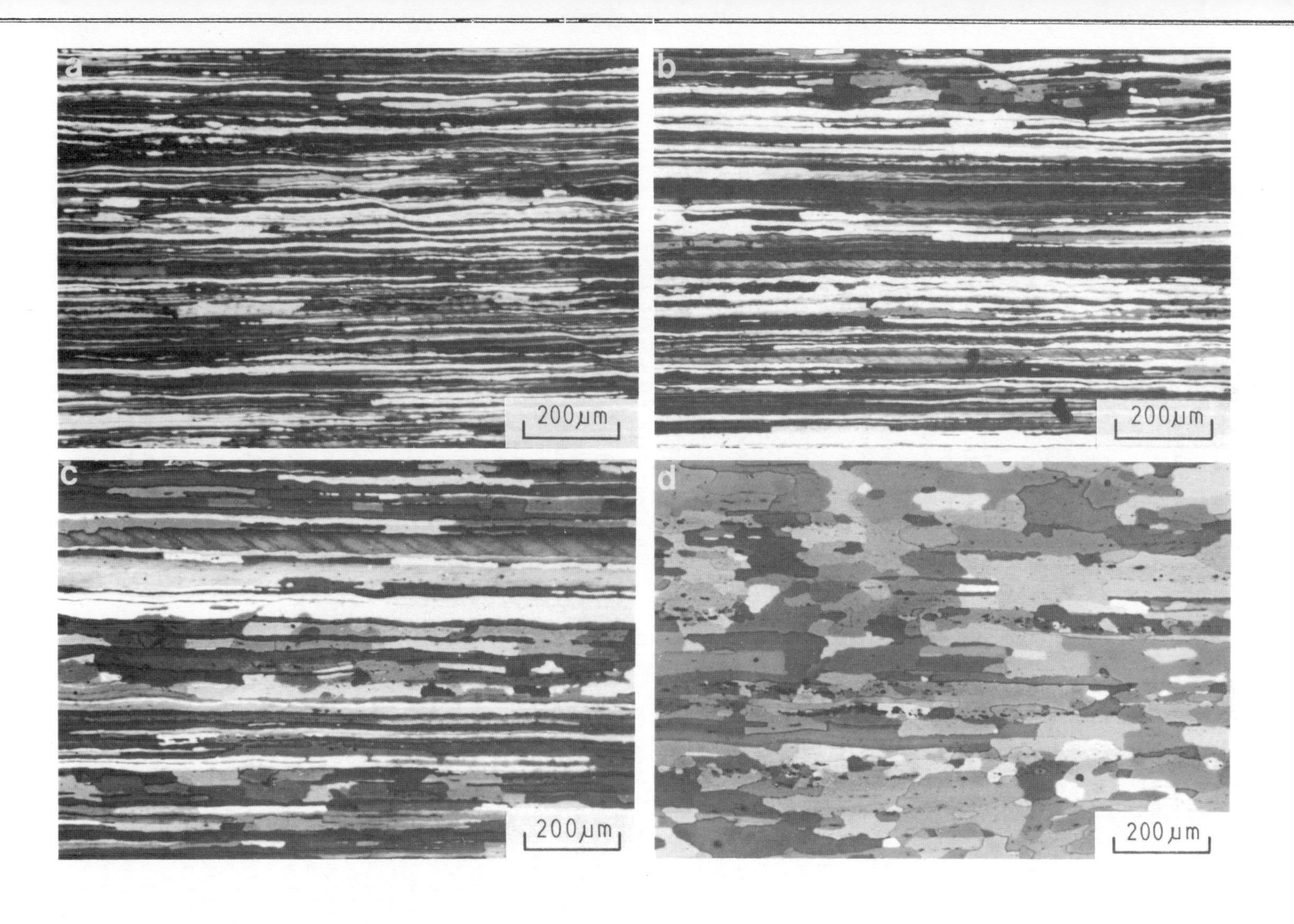

Figure 4 Influence of Zr content on grain structure (a) 0.16% Zr (b) 0.13% Zr (c) 0.09% Zr (d) 0.05% Zr.

performed using 5 mm thick hot blank starting stock of the alloy used in the texture studies[2] which had the following designation and composition:-

Alloy	Li	Cu	Mg	Zr	Fe	Si	Ti
RET	2.58	1.24	0.70	0.12	0.06	0.06	0.02

Recrystallisation heat treatments were performed either in an air recirculating furnace or in a salt bath. Ageing was carried out in air recirculating ovens. Microstructures were examined by conventional metallographic polishing and either etching in Keller's reagent, or by electropolishing and anodising.

RESULTS AND DISCUSSION

Texture and Properties of Unrecrystallised Sheet

An investigation of the development of crystallographic textures during hot and cold rolling of an Al-Li-Cu-Mg-Zr alloy and the effect of texture on mechanical properties has been conducted and reported[2]. A typical example of the texture developed in hot and cold rolled sheet is shown in Figure 1. ODF analyses were also performed and variations in planar yield strength and plastic strain ratio were predicted as a function of test angle with respect to the rolling direction. The predicted variation in yield strength is in good agreement with that observed experimentally, Figure 2. However, the variation in elongation is opposite to that observed experimentally.

Table 1 shows the anisotropy of tensile properties for 1.6 mm thick sheet cold rolled from 5 mm thick hot blank without intermediate anneals. Data are shown for both the solution treated and artificially aged condition. The strain to failure E_F is expressed as the sum of the two individual strain components E_U (uniform strain component) and E_D (localised strain component). It is apparent that both strain components are lower in the longitudinal direction and in the aged condition the L-direction has essentially zero localised strain component. The uniform strain component is controlled by geometrical considerations, provided that incipient failure mechanisms, such as voiding, are not a factor. Geometrical stability is maintained until the work hardening rate drops below the applied stress. Figure 3 shows the stress-strain and work hardening rate-strain results for solution treated longitudinal and transverse orientations. In the T-direction the two curves approach each other at 16% strain in agreement with the uniform strain value of 16.8% (true strain = 0.155), but in the L-direction the two curves are a large distance apart at the uniform strain and in fact would be far from intersecting even at the fracture strain.

From these figures it can be seen that in transverse specimens the instability developed at the maximum uniform strain is mainly a geometrical one, while the longitudinal specimen undergoes instability by some other mechanism.

The previous tensile data show that there is little difference in the yield stress between the longitudinal and transverse directions, however for the longitudinal sample the work hardening rate is initially higher and decreases more rapidly with strain. Since the geometrical instability criterion is not achieved in the longitudinal direction an alternative failure mechanism is operating, that of void nucleation and growth.

In the present alloy voids are already present, to some extent, in the form of cracked constituent particles, but they are also nucleated at particles during straining. For these alloys the calculated interfacial energy γ is low[11], leading to comparatively easy void formation at critical strains of 0.06 for aged material and 0.1 for solution heat treated material. High work hardening alloys such as these Al-Li-Cu-Mg-Zr alloys exhibit planar slip. As such, these alloys are susceptible to strain localisation and microbanding which lead to shear type failures[12 13].

Several factors are possible causes of the observed differences in behaviour between the longitudinal and transverse directions. In general these factors will be influenced by the strain path. The most favourable circumstances for premature failure (really, the most unfavourable situation) is to have the intermetallics, which exhibit low interfacial energy, aligned in the principal straining direction; coupled with a crystallographic texture such that the active slip systems produce a strong shearing component across the aligned intermetallics in the transverse direction.

The observed crystallographic texture in the solution heat-treated material displays a strong rolling texture of which the $(11\bar{2})$ $\langle 111 \rangle$ component is observed. This orientation when strained in the longitudinal direction will produce a strong shear component across the intermetallics with a slip trace of 35° in the transverse direction. This results in void linkage by localised shear between adjacent voids and premature fracture.

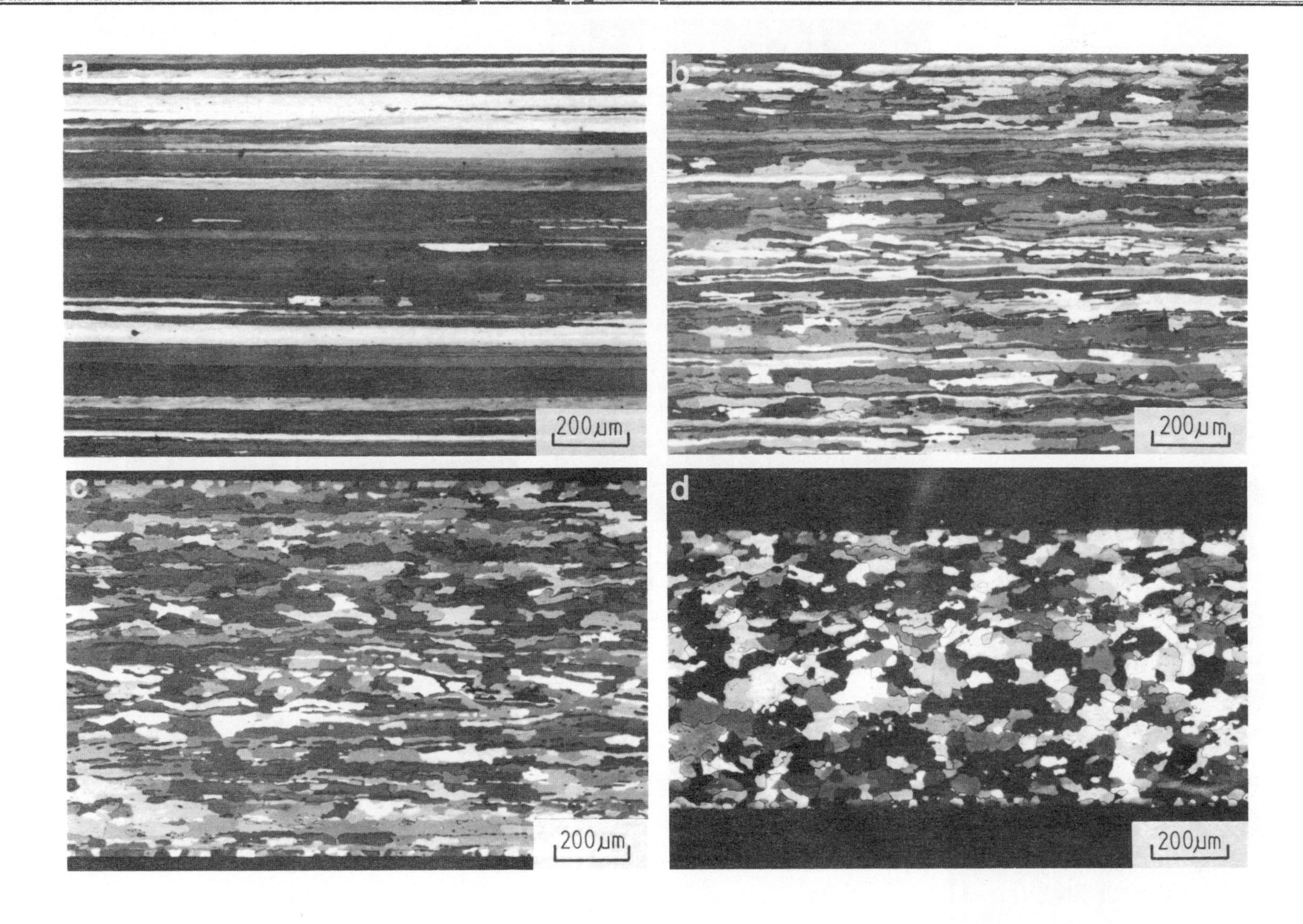

Figure 5 Influence of cold work on grain structure (a) 37.5% C.W., ε = 0.47 (b) 62% C.W., ε = 0.97 (c) 75% C.W., ε = 1.39 (d) 87.5% C.W., ε = 2.08.

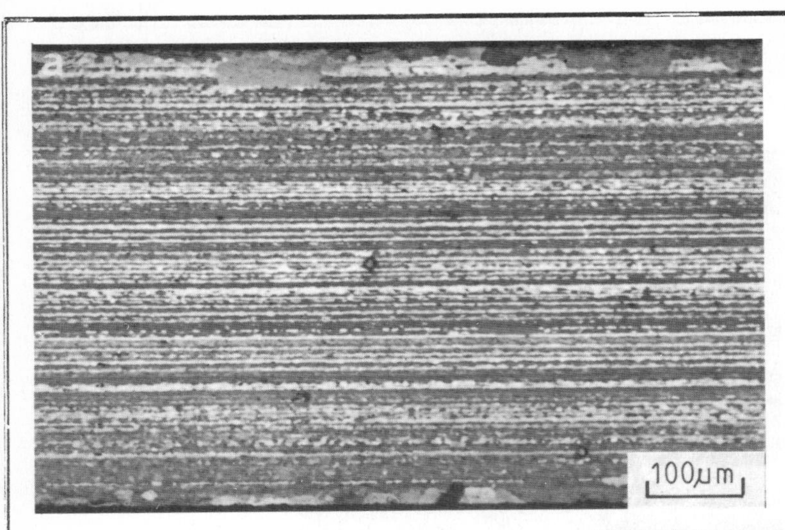

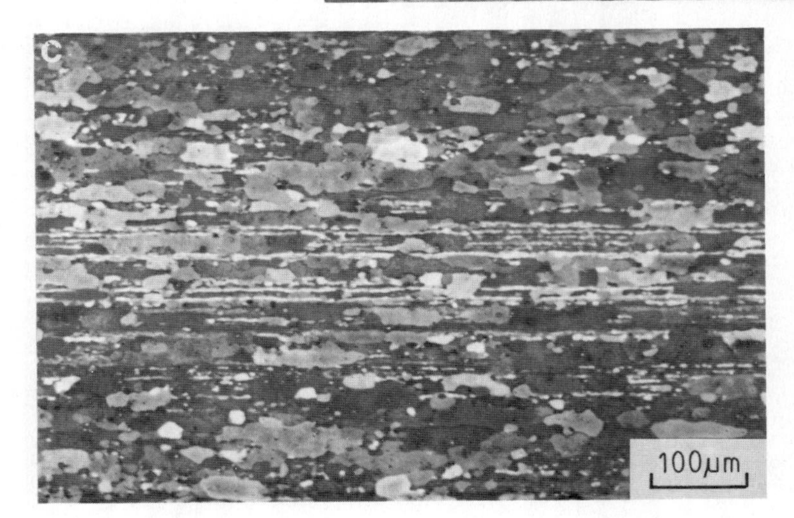

Figure 6 Influence of prior overageing on grain structure (a) 16 h, 400°C (b) 64 h, 375°C (c) 16 h, 350°C.

The scale of the microstructure in terms of grain morphology and constituent particle distribution should also be considered particularly with reference to their effects on void nucleation and growth.

Nevertheless the work shows that the anisotropy in strength and ductility can both be interpreted in terms of the texture in unrecrystallised sheet, the elongation by invoking an incipient failure mechanism.

Effect of Process Parameters on Recrystallisation

Initial experiments were performed to examine the effect of four significant variables on the microstructure of solution heat treated sheet material.

(a) Influence of Zirconium Content

Figure 4 shows the influence of zirconium level on the grain structure of a nominal Al-2.5Li-1.8Cu-0.7Mg alloy. The material was hot and cold rolled to 2.0 mm sheet and solution treated in a salt bath for 10 min. at 520 ± 2°C. In the alloy containing 0.16%Zr, Figure 4a, the material exhibits an unrecrystallised grain structure. As the zirconium level is decreased to 0.13% and then to 0.09%, Figures 4b and 4c, recrystallisation has occurred in areas of the sample. The amount of recrystallisation increases as the zirconium content decreases. In these two figures the recrystallisation occurs in the direction of working with little lateral spread giving the recrystallised grains a laminar appearance. In contrast, at the lowest zirconium level investigated, 0.05%, Figure 4d, the sheet is fully recrystallised with an almost equiaxed grain structure.

(b) Influence of Percent Cold Work

Figure 5 shows the influence of percent cold work on the grain structure of the Al-2.5Li-1.8Cu-0.7Mg alloy containing 0.09%Zr. The material was hot rolled to 4 mm, solution treated and cold rolled by various amounts as indicated in Figure 5. The samples were then solution treated in a salt bath for 10 min. at 524 ± 2°C. After cold working by 37.5% (true strain ε equal to 0.47) the alloy is essentially unrecrystallised after solution heat treatment, Figure 5a. As the amount of cold work is increased, the alloy shows increased recrystallisation and a trend for the recrystallised grains to be equiaxed. After cold working 87.5% (ε = 2.08) the grain structure is essentially equiaxed with an average grain diameter of 80 μm, Figure 5d. At the intermediate levels of cold work the grain structure shows a "zig-zag" pattern which reflects the intense planar slip formed during the cold working operation.

(c) Influence of Overageing Treatment

As indicated in the background section of this paper researchers at Rockwell[6-8] have developed techniques involving solution treatment, overageing and warm working prior to recrystallisation to develop fine recrystallised structures in 7475. Figure 6 shows the influence of three different overageing treatments prior to cold working on the recrystallisation of the Al-Li-Cu-Mg alloys. The results show that an overageing treatment similar to that used by Rockwell (16 h at 400°C) results in a grain structure showing evidence of fine recrystallised grains, but the original unrecrystallised structure predominates, Figure 6a. Reducing the overageing temperature but increasing the time at temperature (64 h at 375°C) results in partial recrystallisation with associated massive grain growth, especially near the surfaces, Figure 6b. Further reducing the overageing temperature and time (16 h at 350°C) produces more extensive recrystallisation but the grain structure is very variable, Figure 6c.

(d) Influence of Heat-up Rate

The rate of heating to the recrystallisation temperature has been shown to influence grain structure significantly. The effect of this parameter is shown in Figure 7 for the Al-2.5Li-1.8Cu-0.7Mg-0.13Zr alloy cold worked 75% prior to solution treatment. A slow heat up to the solution treatment temperature of 525°C (~ 3°C/min) results in an unrecrystallised structure, Figure 7a. Faster heat up rates (~ 100°C/min) cause recrystallisation, Figure 6d. The main growth direction of the grains is in the rolling direction resulting in a laminar grain shape. Rapid heating rates (~ 3000°C/min) result in a virtually equiaxed grain-structure, Figure 7c.

The results in this section indicate that each of the four parameters assessed has a significant influence on the grain structure of sheet after solution heat treatment. By careful optimisation of these parameters it should be possible to produce fine grain recrystallised sheet in a wide variety of sheet gauges. The following section details the structure and properties of such material.

STRUCTURE AND PROPERTIES OF RECRYSTALLISED SHEET

Fine grain recrystallised sheet was produced in RET material by optimisation of the parameters described above. The microstructure and tensile properties were examined and compared with those obtained in similar material processed by the conventional route[14]. After processing, all material was solution treated at 520°C, water quenched and aged for 16 h at 190°C, representing a peak aged condition.

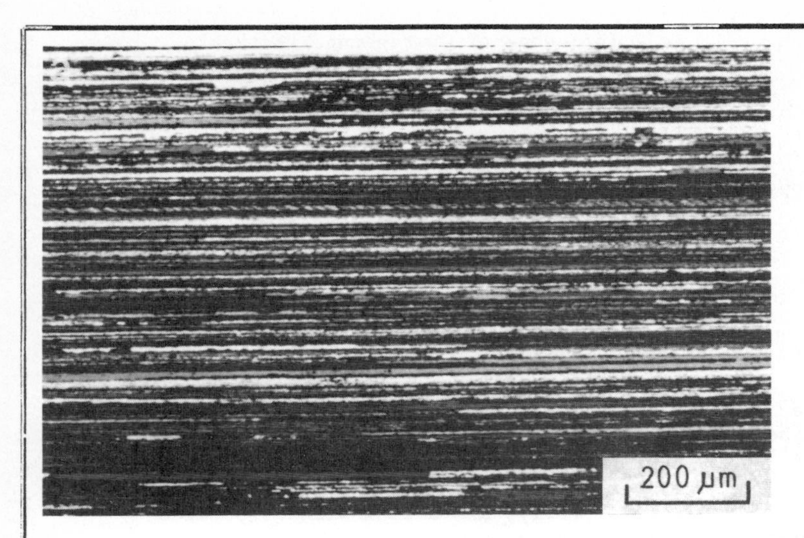

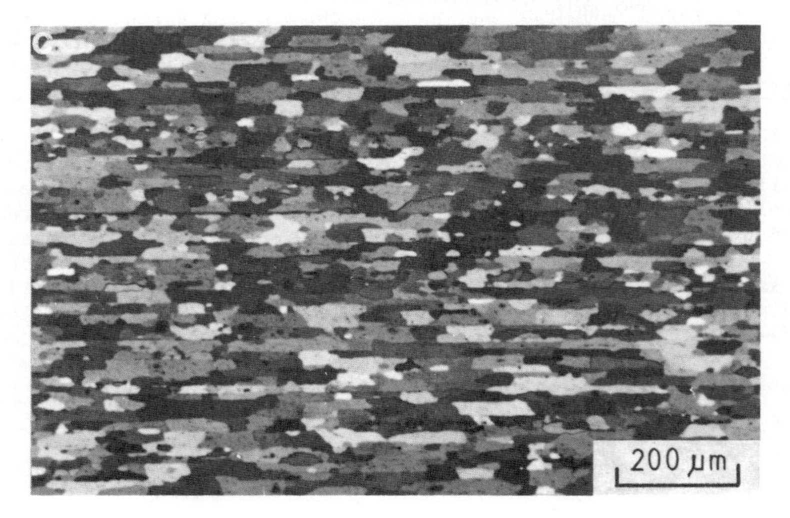

Figure 7 Influence of heat up rate on grain structure (a) 3°C/min (b) 100°C/min (c) 3000°C/min.

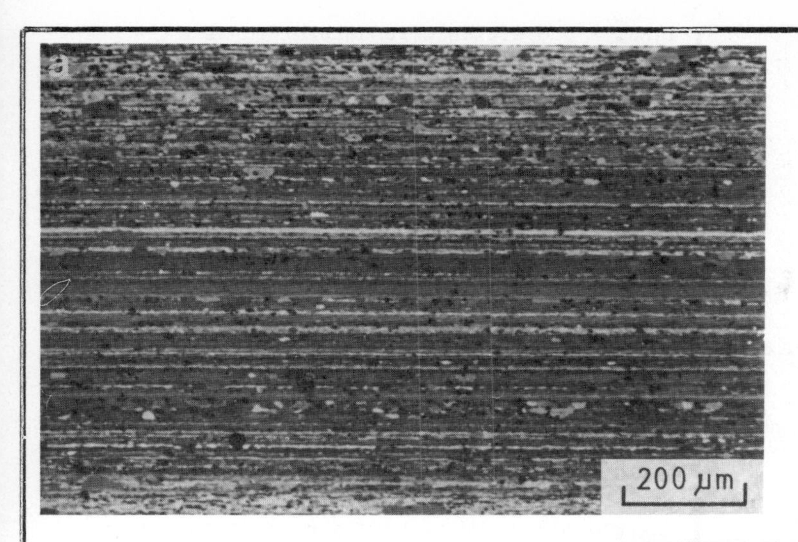

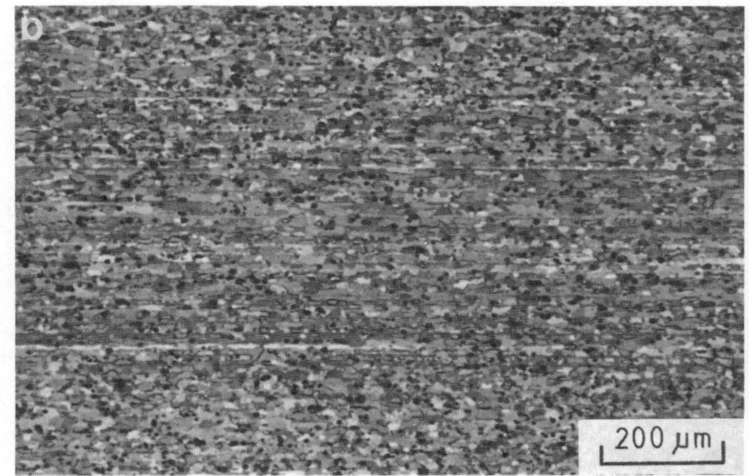

Figure 8

Microstructure of 1 mm thick sheet (a) conventionally processed (b) recrystallised.

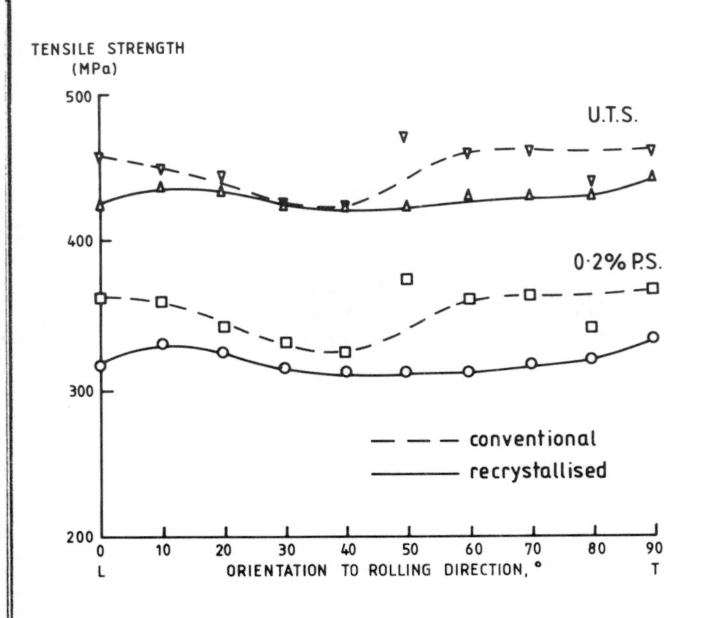

Figure 9 0.2% proof strength and ultimate strength of 1 mm thick sheet as a function of orientation.

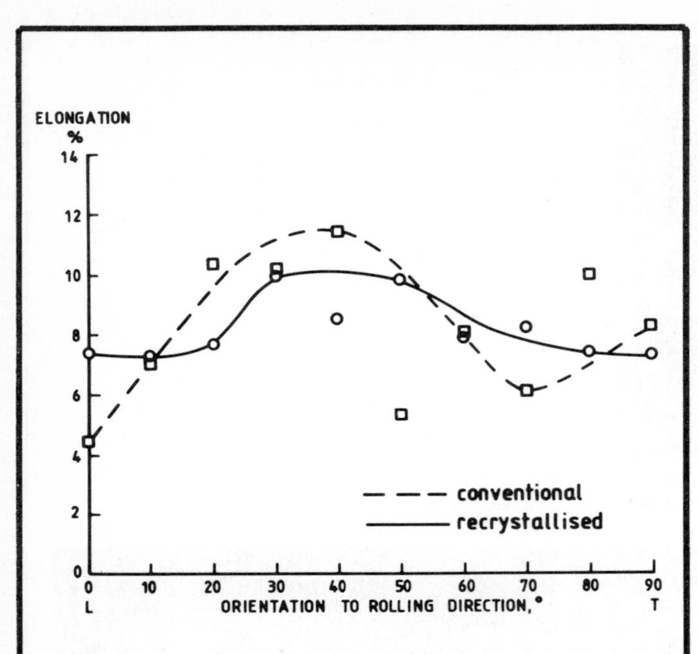

Figure 10 Elongation of 1 mm thick sheet as a function of orientation.

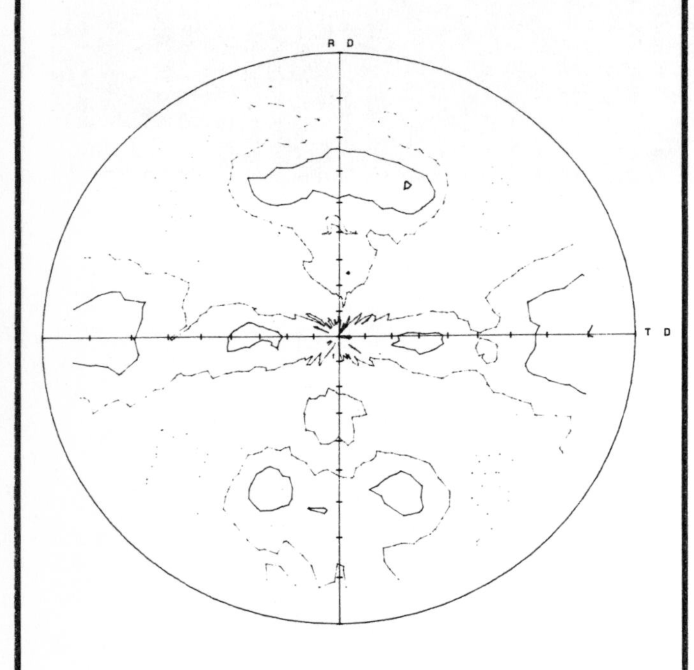

Figure 11 (111) pole figure for 0.5 mm thick recrystallised sheet material.

Tensile properties were determined as a function of orientation in the plane of the sheet by taking specimens every 10° between L and T orientations. The microstructure of 1 mm thick sheet is shown in Figure 8. The conventionally processed sheet is predominantly unrecrystallised, although areas of fine recrystallised grains are present, especially in the surface layers. The recrystallised sheet shows only a few thin bands of unrecrystallised material. The recrystallised grains are almost equiaxed and approximately 10 μm in size. The tensile properties of the two materials are shown in Figures 9 and 10. The recrystallised sheet shows slightly lower strength levels but significantly less anisotropy than the conventionally processed sheet, especially with regard to elongation values.

Texture measurements were also made on 0.5 mm thick recrystallised sheet. The (111) pole figure is shown in Figure 11 where it can be compared with the pole figure for unrecrystallised material shown in Figure 2.

The recrystallised texture is a weak (112) <110> which is a much more balanced and weaker texture about the rolling and transverse directions than is the texture of conventionally processed material. As a result it would be expected to have more isotropic properties than conventional material, and this is the case. The weak recrystallisation texture is unusual and is attributed to the rapid heat-up rate used which has resulted in the nucleation of a large number of grains having a wide spread of orientations.

The development of a recrystallised texture after the solution heat treatment is advantageous in terms of promoting an initial random selection of slip systems which delays the formation of localized slip to higher imposed strains.

CONCLUSIONS

(1) Conventional processing of Al-Li-Cu-Mg-Zr alloys gives rise to unrecrystallised grain structures with a pronounced crystallographic texture.

(2) The anisotropy in strength and ductility observed in unrecrystallised sheet material can be interpreted in terms of the texture. In the case of ductility the interpretation requires an incipient failure mechanism.

(3) Fine recrystallised grain structures can be obtained in the alloys by optimisation of the parameters controlling the recrystallisation process.

(4) The recrystallised structures exhibit almost isotropic tensile properties in the unstretched, peak aged condition.

ACKNOWLEDGEMENTS

The authors are grateful to Alcan International Limited for permission to publish this work.

REFERENCES

1. C.J. Peel, B. Evans, C.A. Baker, D.A. Bennett, P.J. Gregson and H.M. Flower, "The Development and Application of Improved Aluminium-Lithium Alloys", Aluminium-Lithium Alloys II, Ed. by T.H. Sanders and E.A. Starke, TMS-AIME, 1984, pp.363-392.

2. D.J. Lloyd and M.J. Bull, "The textures developed in an Al-Li-Cu-Mg alloy", this conference.

3. F.S. Lin, S.B. Chakrabortty and E.A. Starke "Microstructure-Property Relationships of Two Al-3Li-2Cu-XCd Alloys" Met. Trans. 1982, 13A, 401-410.

4. F.S. Lin, "The effect of grain structure on the fracture behaviour and tensile properties of an Al-Li-Cu alloy", Scripta Met. 1982, 16, 1295-1300.

5. J. Wadsworth, I.G. Palmer, D.D. Crooks and R.E. Lewis, "Superplastic behaviour of aluminium-lithium alloys", Aluminium-Lithium Alloys II, Ed. by T.H. Sanders and E.A. Starke, TMS-AIME, 1984, pp.111-135.

6. J.A. Wert, "Grain refinement and grain size control", Superplastic Forming of Structural Alloys Ed. by N.E. Paton and C.H. Hamilton, TMS-AIME, 1982, pp.69-84.

7. J.A. Wert, N.E. Paton, C.H. Hamilton and M.W. Mahoney, "Grain refinement in 7075 aluminium by thermomechanical processing", Met. Trans. 1981, 12A, 1267-1276.

8. N.E. Paton, C.H. Hamilton, J.A. Wert, M.W. Mahoney, "Characterisation of fine-grained superplastic aluminium alloys", J. Metals, 1982, 34, 21-27.

9. P.L. Makin and W.M. Stobbs, "Recrystallisation in aluminium-lithium alloys", this conference.

10. M.H. Loretto, K.M. Gatenby, R.E. Smallmall, and W.S. Miller, "Texture and Properties of Al-Li based Alloys", to be presented at Strength of Metals and Alloys, Montreal, August 1985.

11. N.J. Owen, D.J. Field and E.P. Butler, "The Initiation of Voiding at Second Phase Particles in a Quaternary Al-Li Alloy", this conference.

12. M.J. Bull and D.J. Lloyd, unpublished work.

13. J.D. Embury, A. Korbel, V.S. Raghunathan and R. Rys, "Shear Band Formation in Cold Rolled Cu-6% Al Single Crystals". Acta Met. 1984, 32, 1883-1894.

14. M.A. Reynolds, A. Gray, E. Creed, R.M. Jordan, and A.P. Titchener "Processing and Properties of Alcan Medium and High Strength Al-Li-Cu-Mg Alloys in Various Product Forms", this conference.

□

Initiation of voiding at second-phase particles in a quaternary Al–Li alloy

N J OWEN, D J FIELD and E P BUTLER

NJO is in the Department of Metallurgy and Materials Science, Imperial College, London SW7, UK.
DJF and EPB are with Alcan International Limited, Banbury, Oxon, UK

SYNOPSIS

The intermetallic phases within an as-cast and rolled Al-Li-Cu-Mg have been characterised using quantitative X-ray microanalysis and back-scattered electron imaging. An Al-Cu-Mg eutectic and an Fe/Cu rich intermetallic resembling Al_6(Fe, Cu) are the principal constituents of the as-cast alloy. During homogenisation and fabrication the eutectic dissolves but the Fe-rich phase is broken up into stringers which lie along the high angle grain boundaries. Void initiation at the particles during deformation of the wrought material has been observed and quantified. Voids nucleate following either particle cracking or particle/matrix decohesion. Evidence for particle cracking caused by coplanar slip band impingement is presented based on electron microscopical observations of conventional and in-situ deformed specimens.

INTRODUCTION

It is the generally held view that the comparatively low ductility of Al-Li alloys relative to other high strength Al alloys is associated with severe strain localisation caused by coplanar slip which is intensified by the presence of the metastable, coherent, ordered δ'(Al_3Li) phase developed on ageing[1]. Recent work utilising conventional ingot metallurgy techniques[2] has shown that the toughness and ductility of Al-Li alloys can be improved by controlled additions of Cu and Mg. Such alloys, develop S phase (Al_2CuMg) which helps to homogenise slip and inhibit strain localisation. However, ductility and fracture toughness is still below that of conventional aircraft alloys. The failure mode in Al-Li alloys is frequently intergranular. This has been rationalised in terms of either strain localisation at grain boundaries following concentrated slip, exacerbated by the presence of precipitate-free zones[3], or segregation of tramp elements such as Na, K and S[4] or Li[5] to grain boundaries. Void initiation in the later stages of deformation at second phase inclusions, particularly those lying along grain boundaries, has been identified[6] as an important subsidiary cause of intergranular failure.

It has been demonstrated[6] that quantitative back-scattered imaging (BSI) in the SEM can provide useful compositional information in Al-Li alloys, the change in back-scattered contrast being a sensitive function of Li level. Quantitative BSI and conventional EDX micro-analysis can therefore be used together to identify large second phase intermetallics. These techniques of compositional analysis, coupled with in situ deformation experiments in a high voltage electron microscope (HVEM) have been applied to study more fully void nucleation and growth at intermetallic particles in a quaternary Al-Li alloy.

EXPERIMENTAL

Material

The alloy of composition (wt %) 2.58 Li - 1.24 Cu - 0.7 Mg - 0.12 Zr - 0.06 Fe - 0.06 Si - 0.019 Ti was supplied both in as-cast and plate form by Alcan International Limited. The plate was hot rolled to 25 mm section from scalped and homogenised ingot and subsequently solution heat-treated and aged at 190°C for 16 and 64 hours to produce peak and overaged structures respectively.

Techniques

EDX microanalysis of second phase particles was conducted using a JEOL JSM-35 SEM and a JEOL 120 CX TEM. Quantitative BSI[7] was performed using an ISI DS 130 SEM with the back-scattered signals from the alloy microstructure normalised with respect to cobalt. The contrast in BSI is dependent on the mean atomic number ($\bar{z}$) of the phases present in the alloy microstructure and phases of high and low $\bar{z}$ appear in light and dark

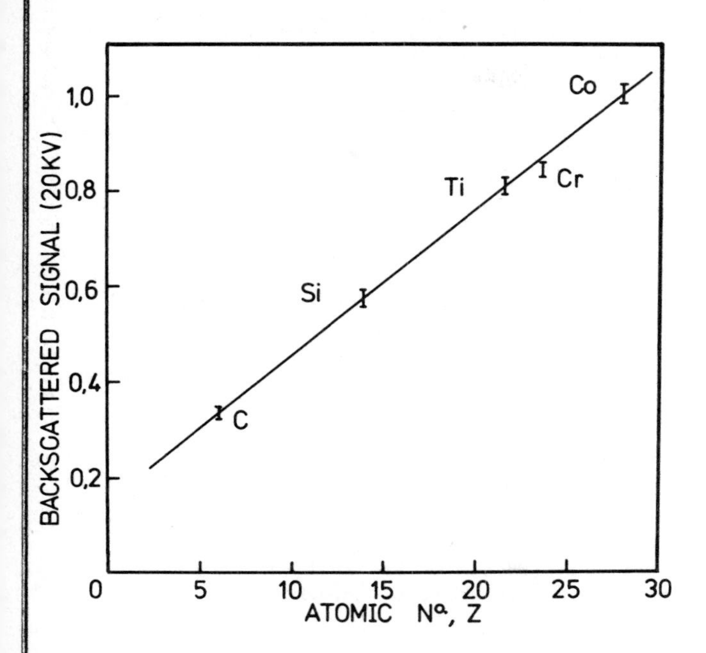

Figure 1. Calibration curve showing the relationship between back-scattered signal and atomic number (after (7)).

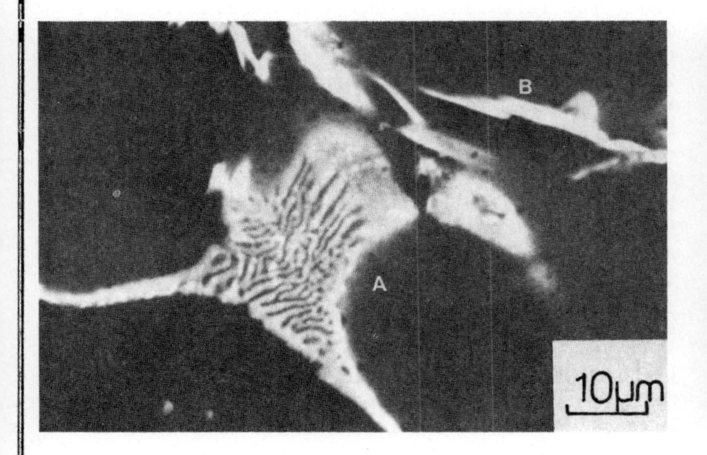

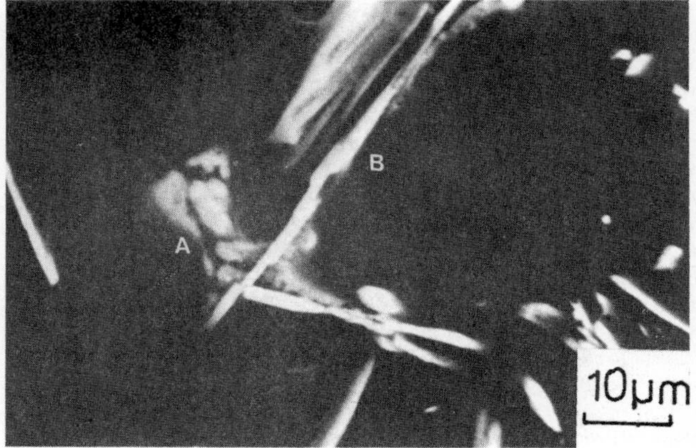

Figure 2. (a) The as-cast grain boundary structures, as revealed by secondary electron imaging: A = Al-Cu-Mg; B = Al-Fe-Cu.

(b) Back scattered imaging of as-cast material, showing the atomic number contrast differences between regions A and B.

contrast respectively. Hence compounds with two $\bar{z}$'s can readily be detected from a calibration curve such as that shown in Figure 1. The relationship between the mean atomic number and the elemental composition has been established for simple binary compounds[8] and more complex structures[9-11]. Diagrams can be constructed showing contours of equal $\bar{z}$ ("isozed") diagrams[6] to assist in the interpretation of the results. The errors in $\bar{z}$ determination are considered to be ± 0.5 atomic number[7].

In situ deformation studies were carried out in an AEl EM7 HVEM at 1MV using a Gatan hard straining stage. Both standard TEM discs (3 mm) and microtensile specimen blanks were made by spark cutting coupons mechanically thinned to 0.25 and 0.1 mm thickness respectively. The discs and centres of the microtensiles were electrolytically polished in a Struers jet-polishing machine operating at 20V using a 30% nitric/methanol solution at -20°C.

Under and overaged samples of plate were machined in the form of cylindrical notched tensile specimens as described previously[6] with the rolling direction parallel to the direction of the tensile test. The semi-circular notch in the specimen creates a hydrostatic stress which causes the voids which nucleate during straining to grow and dilate. Tensile test specimens with a known strain history were metallographically sectioned and quantitative metallography of the void population and second phase particle dimensions was carried out across the region of the minimum notch diameter using an IBAS image analysis system.

RESULTS

Phase Characterisation of As-Cast Material

Examination of the as-cast alloy microstructure by SEM (Figure 2(a,b)) revealed that the cell boundaries were covered with a eutectic phase decorated with a compound of a higher $\bar{z}$ having a needle-like morphology. The mean atomic number of the eutectic phases, the needles and the matrix were found to be 17.0, 19.2 and 12.9 respectively. X-ray microanalysis revealed that the eutectic structure had an average composition (wt %) of 60% Al, 29% Cu, 9% Mg with less than 1% Si. The high atomic number needle phase had a chemical composition of 68% Al, 23% Fe and 8% Cu. Analysis of the matrix showed that all the Fe and the majority of the Cu was concentrated in the grain boundary phases.

Phase Characterisation of Undeformed Plate Material

No Cu/Mg rich phase was detected in the plate material but the Fe-rich phase was still present, elongated along the grain boundaries. In addition, some particles containing Cu, Ti, Si and Zr were detected. The particle distribution in the solution treated alloy is illustrated in Figure 3(a). The volume fraction of grain boundary intermetallics is ~ 0.25%; this value remained approximately constant throughout ageing. The mean equivalent particle size was ~ 0.76 μm. The back-scattered contrast from the Fe-rich phase and other grain boundary phases precipitated on ageing was examined using BSI. Figure 3(b) shows grain boundary phases in an overaged sample which have both high and low $\bar{z}$'s. In addition there are zones of higher $\bar{z}$ delineating the grain boundaries. The larger phase, in light contrast, proved to be Fe/Cu rich accordingly to EDX and had an atomic number, $\bar{z}$ of 19.2 compared with the average matrix contrast corresponding to $\bar{z}$ = 12.9. The $\bar{z}$ of the smaller grain boundary phase of dark contrast was 11.7.

Characterisation of Deformed Material

Following deformation, it was observed that voiding occurred in the alloy in all three heat-treated conditions, principally at the large elongated second phase particles. Quantitative analysis of EDX spectra from bulk samples and thin foils gave virtually identical values of percentage composition for this phase to that found in the cast material; 68% Al, 24% Fe, 7% Cu. BSI confirmed that the mean atomic number of the phase was 19.2.

At low strains it was observed that the elongated particles fractured while in the later stages of straining it appeared that decohesion took place at the ends of the already cracked elongated particles and also at other smaller spherical particles. Figure 4(a) shows the voids associated with an Fe/Cu intermetallic. The particle initially fractured close to mid-point (arrrowed) and decohesion at the lower end of the intermetallic has also taken place. The slip bands running into the particle are clearly visible. Figure 4(b) shows a cracked Fe/Cu particle in a thin foil of undeformed peakaged material. Fracture of the elongated particle has taken place randomly along its length, and this was a common observation in both SEM and TEM. Once the voids formed at cracked particles, they grew laterally under the influence of the hydrostatic stress, which because of the curvature of the notch, served to dilate the voids making them barrel-shaped. Voids then coalesced with their nearest neighbours to form larger more irregular-shaped cavities. Beyond this stage, crack formation at particles could not be observed as there was little or no necking of the tensile specimens and final fracture happened catastrophically at the ultimate fracture strain. For the solution treated material this was around 18% and for the peak and overaged material it was ~ 8%. These values are low when compared with values obtained by Broek[12] for other aluminium alloys.

The degree of voiding for incremental plastic strains was evaluated on longitudinal sections of deformed specimens, and correlated with the calculated mean. The area fraction of voids is plotted against strain in Figure 5 for the three heat-treated conditions. It can be seen that the peak and overaged specimens behave similarly and undergo voiding at a lower effective strain than the solution treated material. From this figure, the critical strain for void nucleation at particles can be obtained by extrapolating the linear portion of each curve onto the strain axis[13]. This gives an effective critical

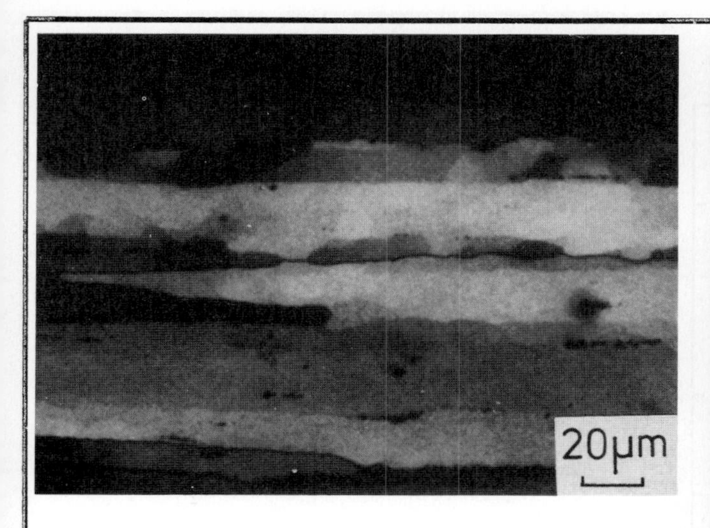

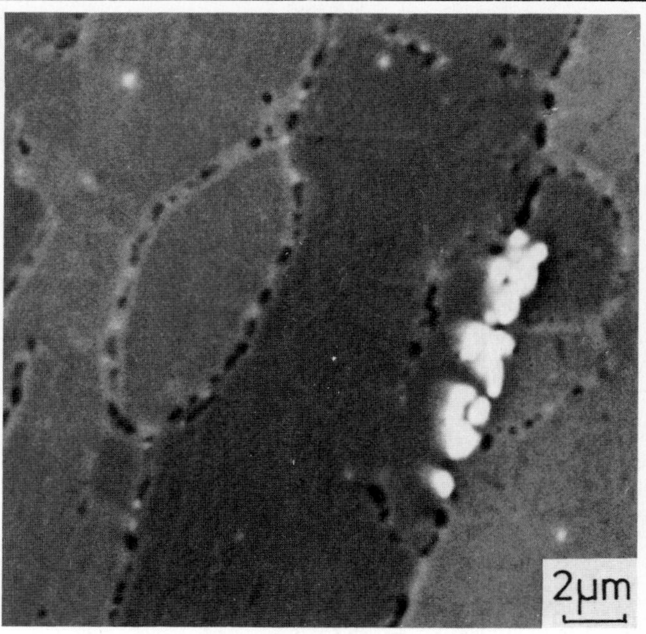

Figure 3. (a) Polarised light micrograph of polished and anodised solution treated material showing grain structure and second phase particle distribution.

(b) Back-scattered electron image of an overaged sample showing the atomic number contrast between different phases. The Fe/Cu-rich phase appears light.

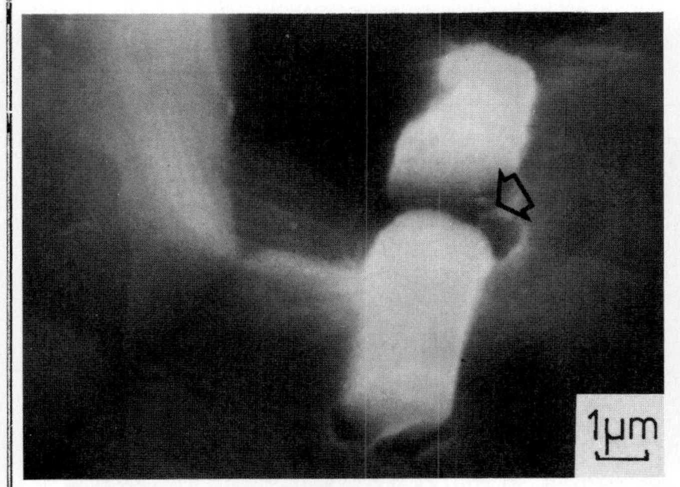

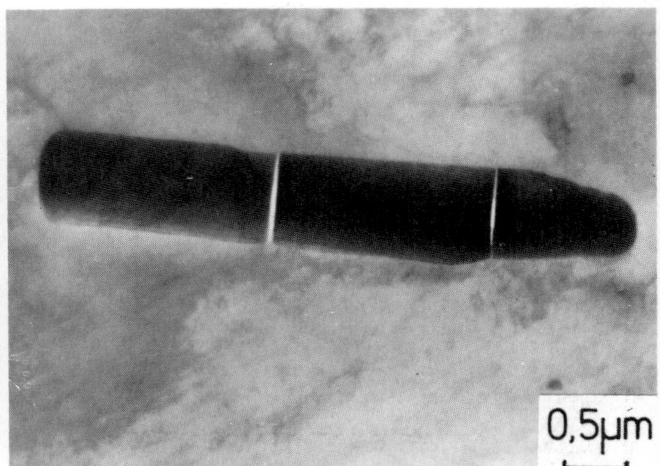

Figure 4. (a) SEM image of fracture surface of a peakaged specimen illustrating void dilation (arrowed) following particle fracture, and particle decohesion at the end of the particle. Note slip bands within matrix.

(b) TEM image of a randomly cracked Fe/Cu-rich particle in a deformed thin foil sample.

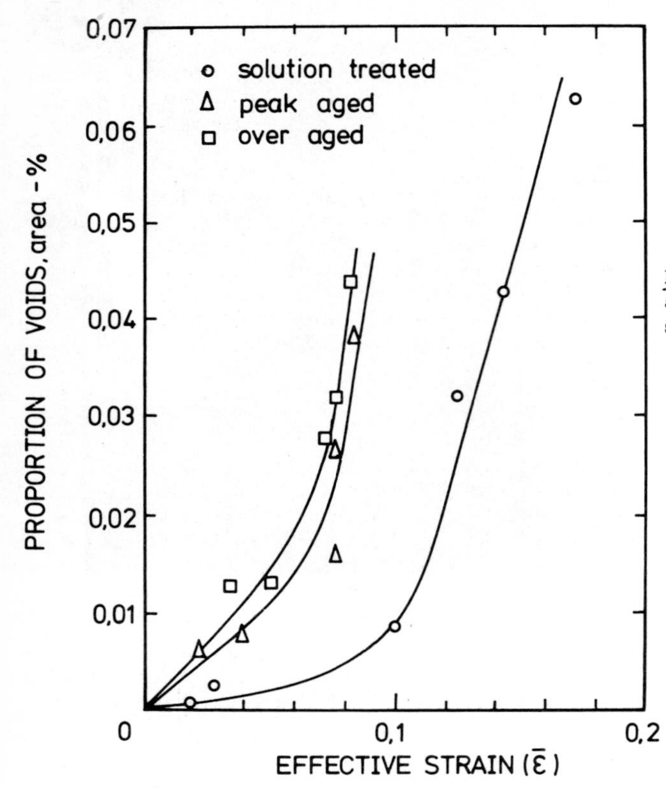

Figure 5. A graph showing the development of the area fraction of voids as a function of effective strain for solution treated and aged materials.

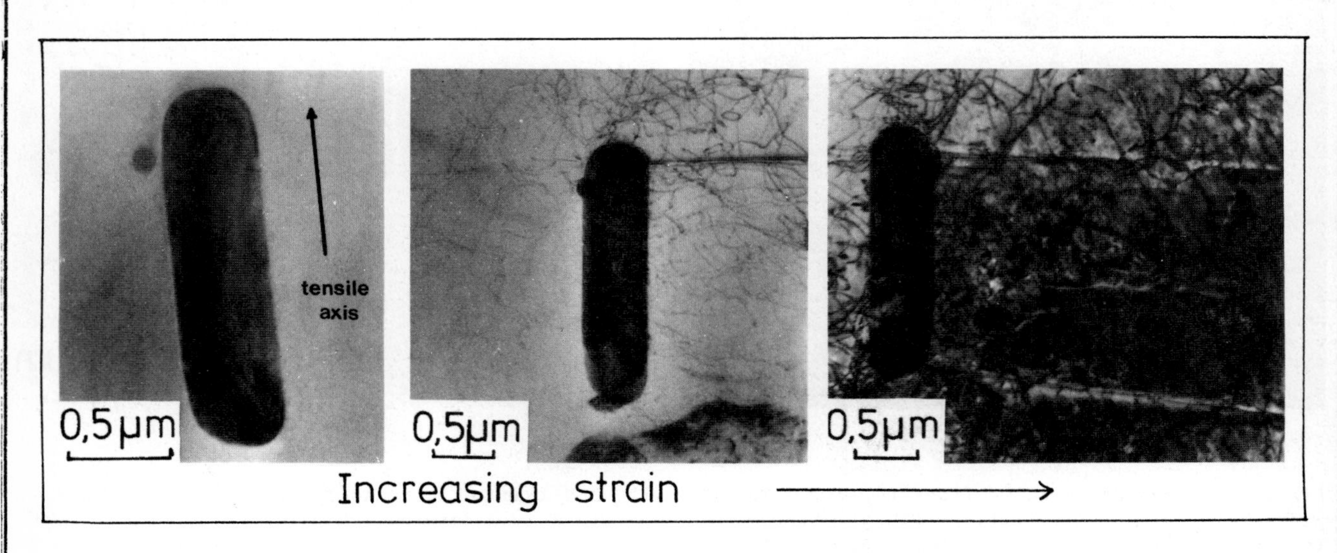

Figure 6. In-situ observations of deformation around an intergranular intermetallic showing the build-up of dislocations and slip band impingement at the particle interfaces.

nucleation strain ~ 0.1 for the solution treated material and ~ 0.06 for the aged materials.

In-Situ Straining Observations

Figure 6 shows an in-situ sequence in underaged material in the vicinity of an intergranular inclusion. It can be seen that with increasing strain, slip bands are generated and impinge upon the particle, the particle being an effective barrier to the passage of the slip bands. Further deformation to the point of void formation and particle cracking (as shown in Figure 4(b)) could not be developed via in-situ deformation due to premature foil fracture in very thin areas at points of maximum concentration. The general mode of initial deformation in the in-situ deformation experiments was, however, comparable to that observed in deformed and thinned samples.

DISCUSSION

Particle Characterisation

The eutectic phase in the as-cast material was found to have a composition of 60% Al, 29% Cu and 9% Mg and a $\bar{z}$ of 17.0. Figure 7(a) shows the computed mean atomic values of aluminium compounds as a function of Cu and Mg content. Superimposed on this "isozed" diagram is a line where the Cu/Mg ratio is 3:1, as found experimentally. This line crosses close to a number of previously reported phases, such as $Al_2Cu\ Mg$. In the Al-Cu-Mg system, it is known that the following ternary eutectic reaction occurs:

$$\text{Liquid} \rightarrow \alpha(\text{Al}) + \Theta(\text{Al}_2\text{Cu}) + \text{S}(\text{Al}_2\text{CuMg})$$

having an overall composition 64% Al, 30% Cu and 6% Mg. This eutectic composition has a calculated $\bar{z}$ of 16.9, which is close to the mean atomic number of 17 measured for the eutectic phase in the as-cast alloy. Hence, barring the interference of lithium in the cast cooling reaction, the phase is the Al-Cu-Mg eutectic, and this phase is removed by the homogenisation treatment.

The single most important intermetallic phase in the fabricated material is Fe and Cu-rich, having a composition of 68% Al, 23% Fe and 8% Cu and $\bar{z}$ of 19.2. This phase is retained through homogenisation, but fragments during rolling and is distributed as particle stringers along the high angle grain boundaries. Figure 7(b) shows the "isozed" diagram for aluminium compounds as a function of Fe and Cu content, together with previously reported phases in this ternary system. A line showing the experimentally determined Fe/Cu ratio of 3:1 is superimposed on this diagram. From the diagram the formula which is most consistent with these two independent measurement techniques is $Al_9\ Fe_3\ Cu$. The phases $Al_{23}Fe_4Cu$ and Al_7Cu_2Fe[14], which belong to a family of compounds stemming from Al_6 (Fe, Cu), lie outside the $\bar{z}$ range measured, although the Fe/Cu ratio of the former is within the experimental error of EDX. The phase Al_7Cu_2Fe has been identified by XRD in Al-Li-Cu and Al-Li-Cu-Mg alloys[15]. The morphology of the phase observed in this investigation closely resembles that of Al_6(Fe, Cu) described by Phragmén[16]. The important observation is that it is the Fe-rich phase which is associated with the nucleation of voids leading to premature cracking and reduced ductility in the Al-Li-Cu-Mg alloy examined. The findings of Feng and Lin[17] for Al-Li-Cu alloys (2020) and those of Speidel[18] for Al-Cu-Mg alloys relating to the effects of iron intermetallics on the mechanical properties of these alloys are consistent with this work.

Void Nucleation

Previous work by the present authors[6] has demonstrated that there is a direct correlation between the stress for void nucleation, and the fracture strain of the material. However, the nucleation strains of about 0.1 for the solution treated material and about 0.06 for the aged material are macroscopic values and do not indicate the nucleation conditions at individual particles. The local conditions clearly depend on such factors as particle size, shape and location[19,20] and usually a range of void sizes is observed at a given strain level with particles not nucleating voids at the same strain[20, 21]. In this investigation it was found that the average strain to nucleate a void was modified by ageing (Figure 5). As the material hardens, plastic accommodation at the particle becomes more difficult and the local stress rises to cause voids to be nucleated sooner. The early occurrence of voiding in aged materials is of importance to ductility and fracture toughness because of the strain localisation at these stress concentrators. Voids enlarge and cause microvoids to initiate in the vicinity of the highly strained regions. The microvoids then coalesce and form a sharp crack which can grow out from the voids and along the grain boundary. The cracking of particles and the formation of voids around particles thus promotes grain boundary fracture and will also have an effect on microvoid formation at a later stage.

At higher strains, decohesion of the particle/matrix interface took place at the end of the elongated already cracked particles (see Figure 4(a)) and at smaller spherical particles. This latter type of voiding is well documented and has been described in detail by Brown and Embury[22]. A void grows at either end of the particle and eventually links up with voids from neighbouring particles when the spacing between the voids equals their overall length. This mechanism of coalescence was not observed in this investigation presumably because the particles invariably lie in stringers which are well spaced apart; instead voids coalesced down the length of the grain boundaries.

Mechanism of Void Initiation

The electron metallographic results of Figures 4(a), 4(b) and 6 clearly validate the mechanism previously proposed[6] for void nucleation, where cracking is initiated ahead of blocked slip bands. This mechanism is shown schematically in Figure 8 and the sequence of events leading to fracture are presumed to be as follows:

(a) The coarse slip inherent in Al-Li alloys creates dislocation pile-ups at planar defects such as grain boundaries and large second phase particle interfaces (e.g. Figure 6),
(b) The stress at the particle interface where

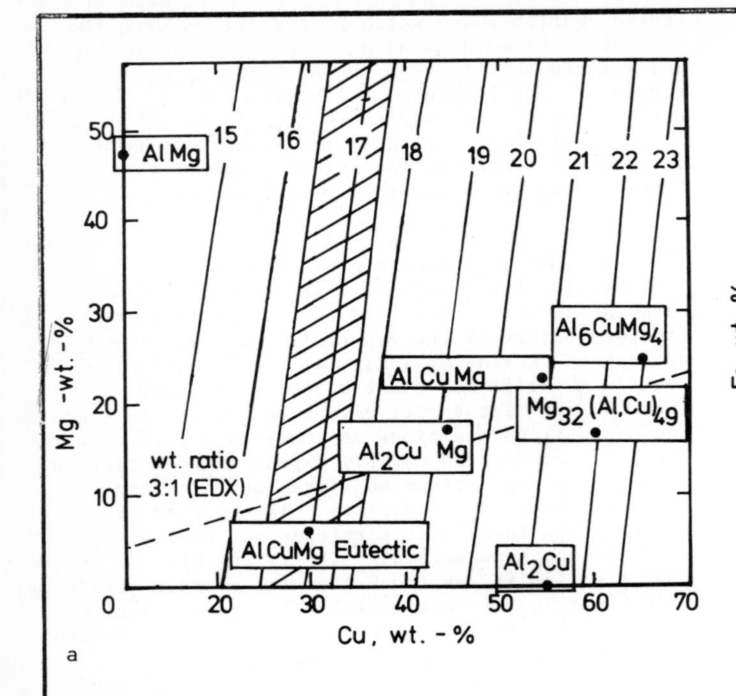

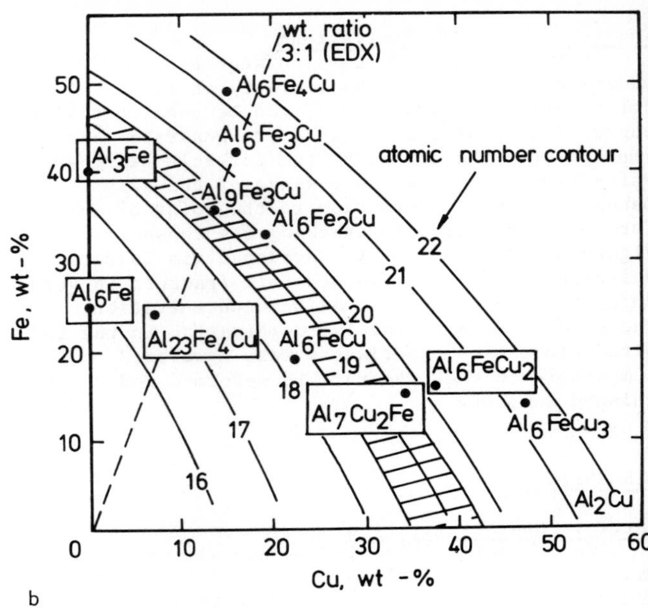

Figure 7. Schematic diagrams showing the expected variation in mean apparent atomic number as determined by BSI with (a), Cu and Mg content (b) Fe and Cu content. The EDX analysis line (dotted) and BSI determined value (hatched) meet at the expected composition of the phases. Previous reported phases in boxes.

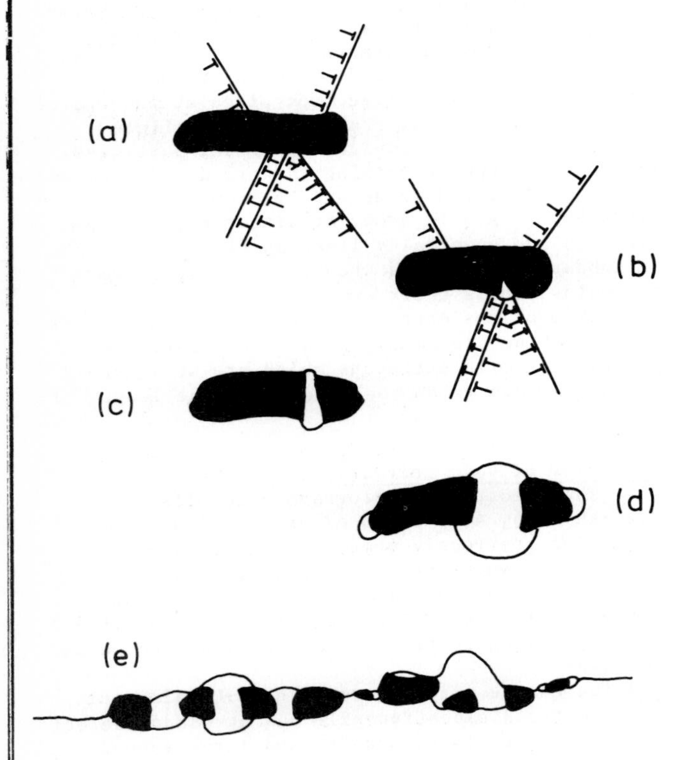

Figure 8. A proposed mechanism of fracture initiation in Al-Li-Cu-Mg alloys. At low strains, coplanar slip bands impinge (a) and fracture the second phase particles (b). Under the action of hydrostatic stresses, the voids grow (c) and decohesion occurs at the ends of already cracked particles (d). At higher strains, the voids coalesce and propagate down grain boundaries (e).

multiple pile-ups impinge exceeds the fracture strength of the particle, which fractures (e.g. Figure 4(b)),

(c) The voids grow transversely under the action of the hydrostatic stress assuming a barrel shape,

(d) As plastic deformation increases, decohesion of the particle/matrix interface occurs at the ends of the elongated particles (e.g. Figure 4(a)),

(e) Void linkage occurs within groups of particles along the grain boundary initiating cracks which propagate down the boundary.

CONCLUSIONS

1. In the as-cast material an Al-Cu-Mg eutectic and an Fe-rich intermetallic based around the composition Al_6 (Fe, Cu) have been identified at grain boundaries.

2. On homogenisation the Al-Cu-Mg eutectic dissolves but the Fe-rich phase remains, to be broken up into particle stringers along high angle grain boundaries during fabrication.

3. Void formation during deformation occurs by the fracture of the elongated Fe-rich particles.

4. The fracture strain of the second phase particles is low and the critical strain to cause voiding is about 0.06 for aged material and 0.1 for solution heat-treated material. These values are low compared to other precipitation hardened Al alloys.

5. Following particle cracking and decohesion, voids coalesce and propagate down the short transverse grain boundaries, to form isolated ligaments which finally fail by shear.

6. A proposed mechanism for fracture in which the failure of the particle is initiated by co-planar slip band impingement has been substantiated by conventional and in-situ TEM observations of deformed material.

ACKNOWLEDGEMENTS

The authors wish to thank Imperial College and Alcan International Limited for the provision of laboratory facilities. One of us (N.J.O.) wishes to acknowledge financial support from the SERC and Alcan International Limited.

REFERENCES

1. T.H. Sanders and E.S. Balmuth, Metals Progress, 1978, 113, 32.

2. C.J. Peel, B. Evans, C.A. Baker, D.A. Bennett, P.J. Gregson, H.M. Flower, Proc. 2nd Int. Conf. Al-Li Alloys (Ed. T.H. Sanders and E.A. Starke) p.363, 1984, New York, AIME.

3. T.H. Sanders and E.A. Starke, Acta Met., 1982, 30, 927.

4. A.K. Vasudevan, A.C. Miller and M.M. Kersker, Proc. 2nd Int. Conf. Al-Li Alloys, (Ed. T.H. Sanders and E.A. Starke) p.181, 1984, New York, AIME.

5. W.S. Miller, M.P. Thomas, D.J. Lloyd and D. Creber: this conference.

6. E.P. Butler, N.J. Owen and D.J. Field, Mat. Sci. & Technology 1985, 1.

7. M.D. Ball and D.G. McCartney, J. Microscopy, 1981, 124, 57.

8. L. Danguy and R. Quivy, J. Phys. Radium, 1956, 16, 320.

9. S.G. Tomlin, Proc. Phys. Soc. 1963, 82, 465.

10. H.E. Bishop, 'Optiques des Rayon X et Microanalyse', (Ed. R. Castiange, P. Deschamps and J. Philbert), p.153, 1966, Paris, Hermann.

11. S.J.B. Reed, Electron Microprobe Analysis, p.219-239, 1975, Cambridge, Cambridge University Press.

12. D. Broek, 'A Study on Ductile Fracture', National Aerospace Laboratory report - The Netherlands, 1972, NLR-TR 71021 U.

13. S.H. Goods and L.M. Brown, Acta. Met. 1979, 27, 1.

14. L.F. Mondolfo, 'Aluminium Alloys: Structure and Properties', (Butterworths, London), 1976.

15. R.J. Kar, J.W. Bohlen and G.R. Chanani, Proc. 2nd Int. Conf. Al-Li Alloys (Ed. T.H. Sanders and E.H. Starke). p.255, 1984, New York, AIME.

16. G. Phragmén, J. Inst. Metals, 1950, 77, 489.

17. W.X. Feng and F.S. Lin, J. Mat. Sci. 1984, 19, 2079.

18. M.O. Speidel, Proceeding of the 6th Int. Conf. on Light Metals, Leoben, Austria, (Aluminium-Verlag, Dusseldorf) 1975.

19. A. Brownrigg, W.A. Spitzig, O. Richmond, D. Tierlinck and J.D. Embury, Acta. Met. Embury, Acta. Met., 1983, 31, 1141.

20. G. Le Roy, D. Embury, G. Edwards and M.F. Ashby, Acta. Met. 1981, 29, 1509.

21. L.M. Brown and W.M. Stobbs, Phil. Mag., 1976, 34, 351.

22. L.M. Brown and J.D. Embury, Proc. 3rd Int. Conf. on Strength of Metals and Alloys, Cambridge, p.164, 1973.

Deformation and fracture in Al–Li base alloys

W S MILLER, M P THOMAS,
D J LLOYD and D CREBER

WSM and MPT are with
Alcan International Limited,
Southam Road, Banbury, Oxon, UK.
DJL and DC are with
Alcan International Limited,
PO Box 8400, Kingston,
Ontario, Canada

ABSTRACT

Aluminium-Lithium based alloys are of considerable interest because of their low density and high modulus. However, they have been shown to be susceptible to low ductility and poor fracture toughness. This has been attributed to a variety of factors including intense shear band formation, segregation to grain boundaries and weakened grain boundaries due to precipitation and PFZ's.

In the present paper the deformation structures observed in binary and more complex commercial alloys have been investigated. As would be expected considering the microstructure of the alloys, extensive strain localisation and shear band formation occurs in these alloys. However it will be shown that the commercial alloys are less sensitive to strain localisation than the model binary alloy systems investigated.

The stress strain behaviour has been investigated. The alloys exhibit jerky flow which is indicative of negative strain rate sensitivity and strain rate change tests show this to be the case. This is consistent with the deformation structures observed.

The effect of weakened grain boundaries due to precipitation and PFZ's has been studied by comparing the fracture characteristics of aged and unaged material. It is shown that the mode of failure is identical under the appropriate conditions.

It is concluded that segregation to grain boundaries is the major cause of the lower ductility and toughness of Al-Li alloys. This possibility has been investigated using in-situ fracture surface analysis techniques. Results are presented on grain boundary segregation and methods of reducing its influence on fracture behaviour are indicated.

INTRODUCTION

The efficiency and performance of airframe structures would greatly benefit from the use of lighter weight materials. The design advantages of the use of low density, high specific modulus alloys have been illustrated by Lewis and his co-workers[1]. The addition of lithium to aluminium gives the greatest reduction in density and increase in elastic modulus of any alloying element and the Al-Li-X system has been the subject of intensive recent studies[2,3]. The alloys have been shown to be susceptible to low ductility and poor fracture toughness[4] which restricted the use of alloys such as 2020[5]. These characteristics have been attributed to a variety of factors including intense shear band formation[6], segregation to grain boundaries of tramp elements[7], and weakened grain boundaries due to particles[8] and the formation of precipitate free zones[9]. This paper details observations of the deformation and fracture behaviour of Al-Li base alloys and discusses the results in terms of the various theories of deformation and fracture.

EXPERIMENTAL PROCEDURE

Ingots weighing 300 kg and 1000 kg were semi-continuously direct chill cast, homogenised and hot rolled to plate in the gauge range 25-50 mm. The plates were solution treated at 535°C and quenched into water at room temperature, stretched 2.5% and artificially aged at 190°C. In the results that follow a number of different batches of material were tested. However, the

<u>Figure 1</u> Three dimensional photomicrograph showing typical grain structure of Al-Li-Cu-Mg-Zr plate.

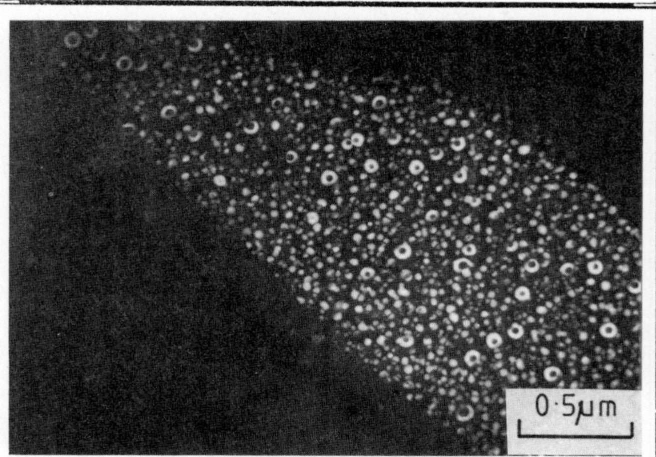

a

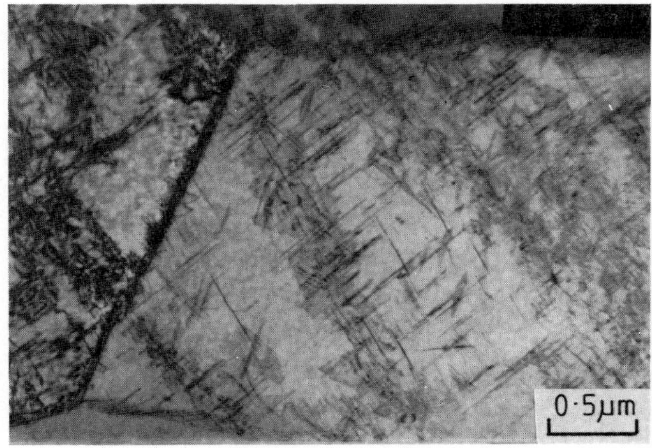

b

<u>Figure 2</u> Transmission electron micrographs showing principle microstructural features in Al-Li-Cu-Mg-Zr alloys:

(a) precipitation of δ' and formation of PFZ's at high angle grain boundaries: D/F 010 nr [001].

(b) precipitation of S and T_1 in grain interiors and at low angle grain boundaries: Multibeam Conditions nr [110].

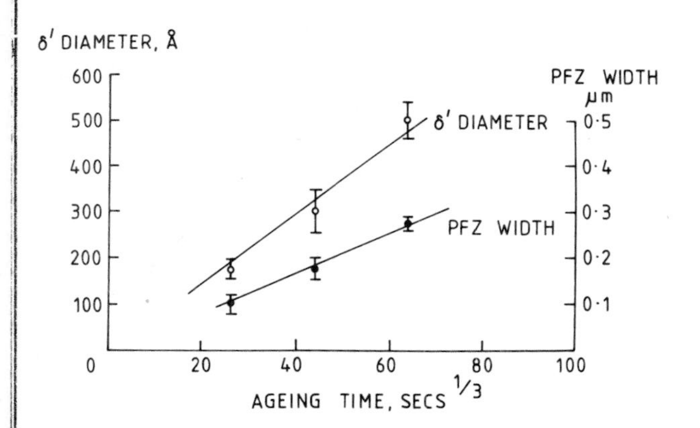

<u>Figure 3</u> Coarsening kinetics of δ' phase and the PFZ.

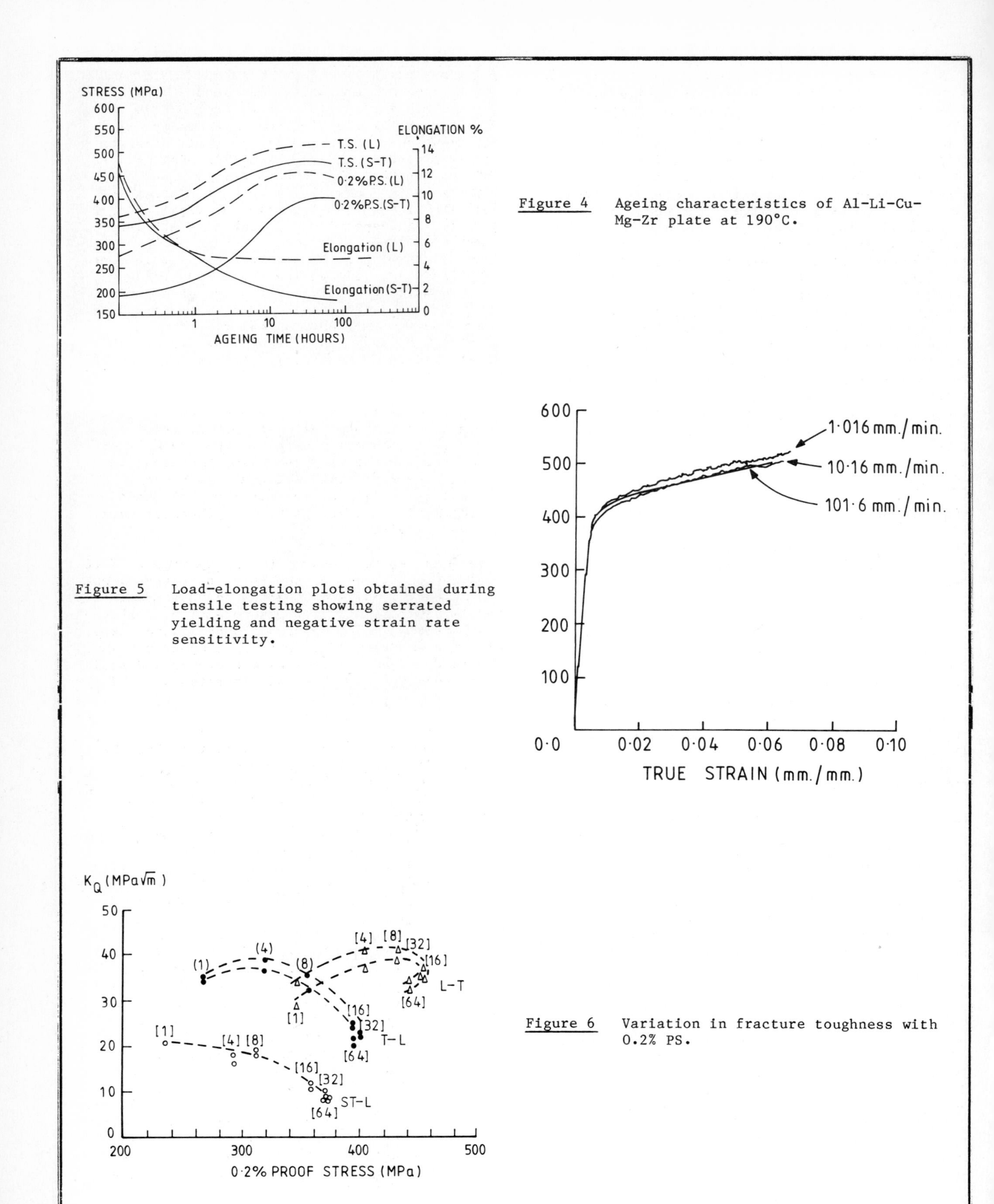

Figure 4 Ageing characteristics of Al-Li-Cu-Mg-Zr plate at 190°C.

Figure 5 Load-elongation plots obtained during tensile testing showing serrated yielding and negative strain rate sensitivity.

Figure 6 Variation in fracture toughness with 0.2% PS.

majority of the plates were based on an Al-Li-Cu-Mg-Zr alloy [10], with a nominal composition of Al-2.5Li, 1.2Cu, 0.7Mg, 0.12Zr (weight percent). Metallographic samples were prepared using standard techniques. High angle grain boundaries were revealed by microanodising in a solution of 2% HBF_4 in methanol at 20V for 120 seconds. Thin foils were prepared by electropolishing 0.2 mm x 3 mm discs in a twin-jet electropolishing device. The electrolyte was 25% HNO_3 in methanol maintained at -20°C with a potential of 15V. Deformation studies were carried out using electropolished double cantilever beams (DCB's). The stress strain behaviour was examined using round bar test pieces from the plate core; tests were carried out at a variety of strain rates. Fracture toughness measurements were carried out using compact tension and DCB specimens. Fracture surfaces were examined by optical and scanning electron microscopy. Grain boundary segregation was investigated using SIMS and Scanning Auger microscopes, the latter with in-situ fracture capability.

RESULTS

Microstructure

Microstructural evaluation was conducted to characterise the grain structure and precipitate type and distribution.

Three-dimensional optical micrographs showing the grain structure of the three principal plate axes were obtained from all plates examined. Figure 1, for an Al-2.62Li-1.14Cu-0.58Mg-0.15Zr-0.10Fe-0.03Si alloy is typical of all material produced. An unrecrystallised pancake grain structure is observed with planar grain boundaries running in the rolling plane. The unrecrystallised structure is attributed to $ZrAl_3$ precipitates inhibiting recrystallisation.

Transmission electron micrographs documenting the salient microstructural features of the Al-Li-Cu-Mg alloy are shown in Figure 2. The lithium precipitates homogeneously as Al_3Li (δ'), Figure 2a. The δ' precipitate is nucleated during the quench [11] and coarsens according to Lifshitz-Wagner kinetics [12], Figure 3. Composite precipitates exhibiting dual contrast are observed. These have been shown to be δ' surrounding an Al_3Zr core [13]. During ageing a precipitate-free zone (PFZ) is created at high angle boundaries. The PFZ also grows according to a $t^{1/3}$ law as shown in Figure 3. The Cu and Mg precipitate heterogenously during artificial ageing as S' (Al_2CuMg) and T_1 (AlCuLi), Figure 2b. The nucleation sites for these precipitates are dislocations formed during quenching, dislocations introduced during stretching and low angle sub-grain boundaries. Grain boundary precipitation of δ and complex Al-Cu-Mg phases occurs during artificial ageing [14].

Mechanical Properties

The change in longitudinal and short-transverse tensile properties with ageing at 190°C is shown in Figure 4 for the Al-Li-Cu-Mg-Zr alloy. In the solution treated temper the yield strength in the longitudinal direction (272 MPa) is ~ 80 MPa greater than that obtained in the short-transverse (193 MPa) but the tensile strengths and ductilities are virtually identical at ~ 350 MPa and 13% respectively. Peak strength is achieved after ageing for ~ 32 h at 190°C. The peak values of 0.2% PS and TS are lower in the S-T direction. Elongation to failure falls rapidly with ageing time such that after only one hour at temperature the ductility is halved. Thereafter in the longitudinal direction the ductility remains constant at ~ 6% whereas the short-transverse ductility continues to decrease to ~ 1% in the peak aged temper.

The load-elongation plots obtained during tensile testing are serrated indicative of a Portevin-Le Chatelier effect. The presence of this effect means that the alloy has negative strain rate sensitivity Figure 5, and that deformation will tend to localise into bands having little resistance to the growth of necks and early failure.

The fracture toughness of the alloy with 0.2% PS is shown in Figure 6. The results in the LT and T-L orientations were obtained using compact tension samples whilst DCB's were used in the S-L orientation. As expected toughness decreases with increasing 0.2% PS but the toughness values obtained in the S-L orientation are considerably lower than in the other two directions.

Deformation Structure

The preceding section has illustrated that the ductility of the alloys is lower than conventional high strength aerospace alloys. To follow the microstructures developed during deformation it would be preferable to examine samples tested in compression rather than tension. However, since other aspects of the deformation consider the tensile mode, the slip morphology and dislocation structure developed during tensile straining has been examined.

Tensile samples orientated in the rolling direction were electropolished and incrementally strained. In the solution treated condition the Al-Li-Cu-Mg alloy strained 3 to 4% exhibit long unusually straight slip lines extending across the diameter of the grain (Figure 7a). Strains of greater than 10% are necessary before two or more slip systems per grain are common and extensive cross-slip can occur. The formation of intense slip lines was expected since fine ordered δ' is being sheared. The slip behaviour is similar to that observed in aged binary Al-Li alloys, Figure 7b. This evidence of intense planar slip has been used to explain the low ductility in these alloys [4]. In contrast aged

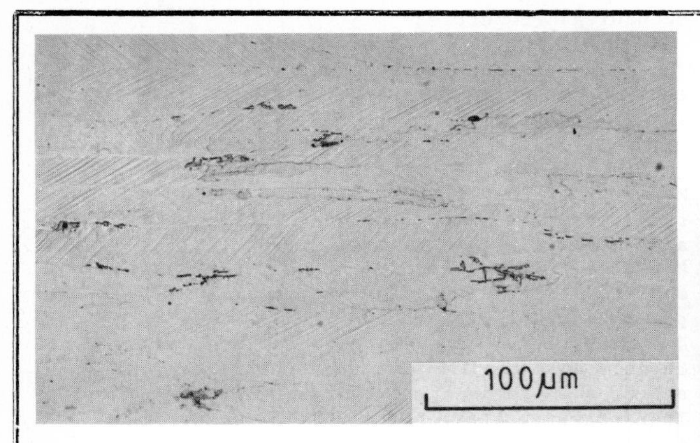

a

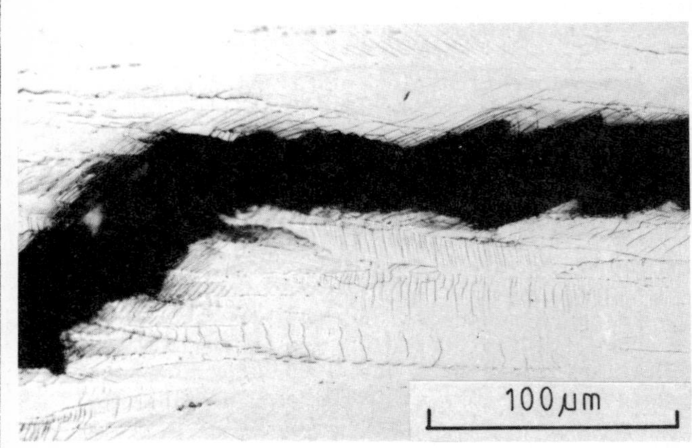

b

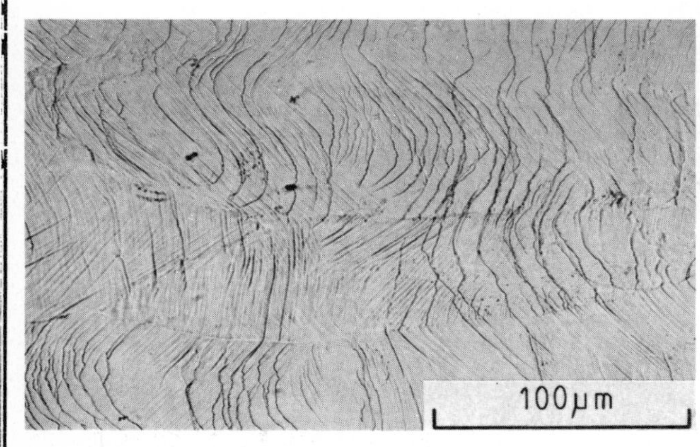

c

Figure 7 Deformation in Al-Li base alloys

(a) solution treated Al-Li-Cu-Mg-Zr alloy strained 3-4% exhibiting planar slip.

(b) planar slip in aged Al-Li-Zr alloys.

(c) extensive cross slip observed in aged Al-Li-Cu-Mg-Zr alloys.

a

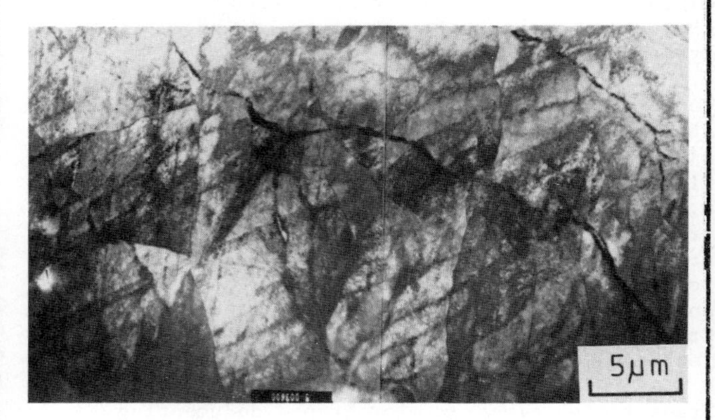

b

Figure 8 Transmission electron micrographs of the deformation in aged Al-Li-Cu-Mg-Zr alloys

(a) Low strain regions (-3%) showing diffuse cell structure formed by dislocation pinning by S phase and dislocations bands.

(b) high + strain regions (~ 6%) showing irregular intense veining superimposed on the banded structure.

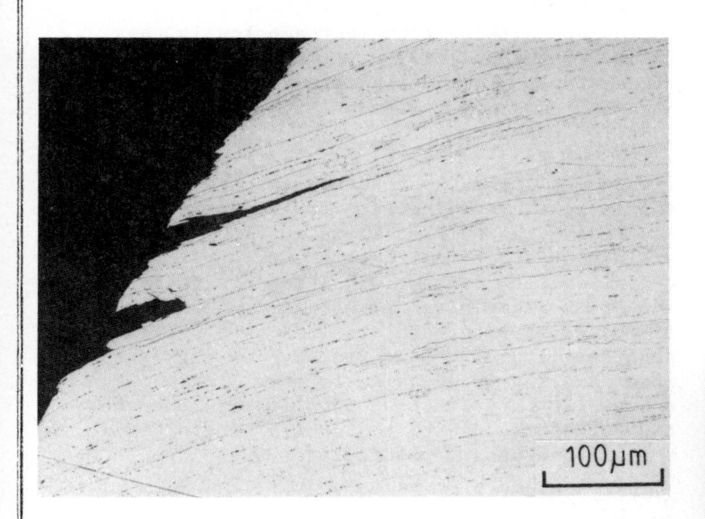

a

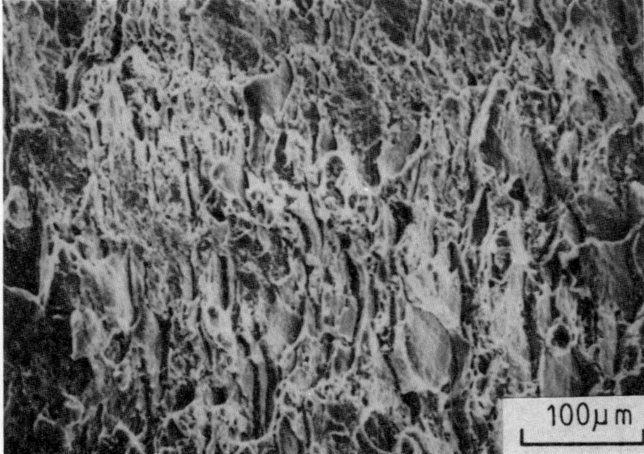

b

Figure 9 Longitudinal fracture in solution treated Al-Li-Cu-Mg-Zr plate. Tested at room temperature

(a) Optical micrograph.

(b) Scanning electron micrograph.

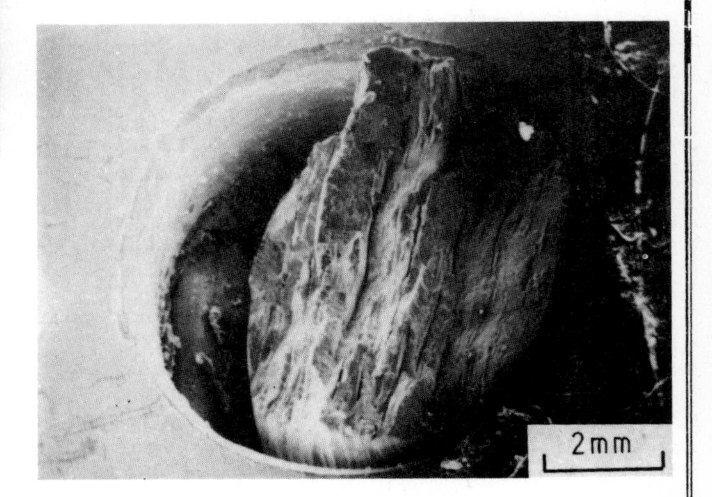

Figure 10 Longitudinal fracture in solution treated Al-Li-Cu-Mg-Zr plate. Tested at 77°K

(a) Optical micrograph.

(b) Scanning electron micrograph.

Al-Li-Cu-Mg alloys show extensive cross slip occurring at 3 to 4% strain which results in intense irregular shaped slip lines (Figure 7c) which often intersect each other. The extensive cross slip is presumably due to the dual phase precipitation system in these alloys. The laths of S phase, shown in Figure 2b, assist in dispersing slip by promoting cross slip between individual slip bands.

In order to elucidate the influence of S phase thin foils were prepared from failed peak aged samples. In the lower strain regions a diffuse cell structure is formed by dislocations being pinned by S phase. Superimposed on the cell structure are dislocation bands, Figure 8a. These bands extend across sub-grain boundaries and can shear the boundary that they intersect. At high strain regions close to the fracture surface an irregular intense banding or veining is superimposed on the banded structure (Figure 8b). It is probable that the veining is a microscopic manifestation of the irregular slip lines. One possibility is that as a procession of glide dislocations follow essentially the same convoluted glide path, debris will accumulate at particular obstacles or intersection points, building up the veined structure.

Fracture Behaviour

In solution treated material tested longitudinally the failure mode is predominantly ductile shear linkage of void sheets but there is also some interboundary failure resulting in grain boundary splitting back from the fracture surface. Extensive void formation at constituent particles occurs throughout the gauge length Figure 9a. Figure 9b is a scanning electron micrograph of the fracture surface and shows that the majority of the surface is dimpled but there is some back splitting and very smooth interboundary failure. From these observations it would appear that the failure strength of the boundary is similar to that of the matrix. To check this possibility samples were tested at 77° K. Figure 10a is an optical micrograph which illustrates the increased amount of splitting linked by shear lips in comparison with the same alloy tested at 300° K (Figure 9a). The change in fracture path is more clearly illustrated in Figure 10b which is a scanning electron micrograph of the whole of the fracture surface showing extensive splitting and also that the ductile component of the linking step is reduced. These results confirm that the boundary fracture strength is low even in the solution treated condition, when there are no particles precipitated on the boundaries. Raising the fracture strength of the matrix by testing at 77° K results in extensive interboundary failure.

All aged material shows the same overall fracture behaviour, which is one of increasing interboundary failure with increasing time. Figure 11a shows an optical micrograph of a fractured sample. The fracture surface consists of interboundary failure along boundaries and sub-boundaries parallel to the tensile axis linked by shear lips. Figure 11b is an SEM photograph showing that macroscopically the failure occurs along the maximum shear stress planes but on the microscopic scale it consists of a series of flat ledges connected by short ductile steps.

Figure 12a shows the fracture appearance in the short-transverse direction for solution treated material. The surface shows that the laminar grain boundaries have been pulled apart to reveal shallow depressions which are featureless and of the same size as sub-grains. Stringers of iron-rich intermetallics can be observed together with slip band traces. In peak aged material, the fracture surface is little changed, Figure 12b, but dimpling is found on the sub-grains due to the growth of grain and sub-grain boundary precipitates.

DISCUSSION

Fracture in these alloys appears to be according to the following mechanism; failure is predominantly along grain boundaries parallel to the rolling plane, and connected to boundaries on adjacent planes by short lengths of more ductile failure along intersecting boundaries and probably along shear bands. The lower ductility and fracture toughness observed in the short-transverse direction can thus be explained since, in this case the weak boundaries are orientated perpendicular to the applied tensile stress and are more susceptible to fracture. In addition, as a result of fabrication any constituent particles are also aligned in the plane and further reduce the load bearing capacity due to void formation. The fracture path is shown diagrammatically in Figure 13a, and can be compared with direct observations of the mode of failure obtained using electropolished DCB specimens, Figure 13b.

The tendency for grain boundary failure in these alloys has been attributed to planar slip, strain localisation in PFZ's, grain boundary embrittlement due to the precipitation of δ and T phases and impurity segregation effects.

The results in the preceding section show that a surprising amount of cross-slip occurs in the Al-Li-Cu-Mg alloys due to the presence of S phase and cross-slip is a means of dispersing slip. Furthermore in the solution treated condition strain localisation is no more extensive than in conventional high solute aluminium alloys. It may thus be concluded that whilst planar slip may tend to enhance boundary failure it is not the sole cause of the failure.

Arguments favouring strain localisation in PFZ's are not confirmed in the present experiments. In the solution treated condition no PFZ's are present and the failure mode is the same as in aged material provided the stress is high

a

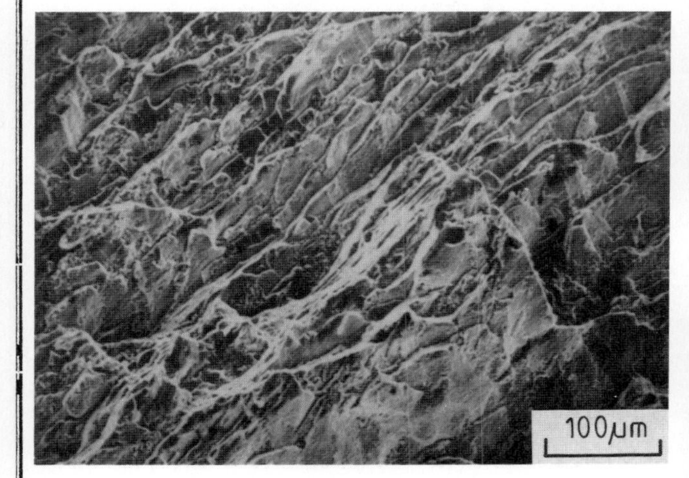

b

Figure 11 Longitudinal fracture in peak aged Al-Li-Cu-Mg-Zr plate. Tested at room temperature.

(a) Optical micrograph.

(b) Scanning electron micrograph.

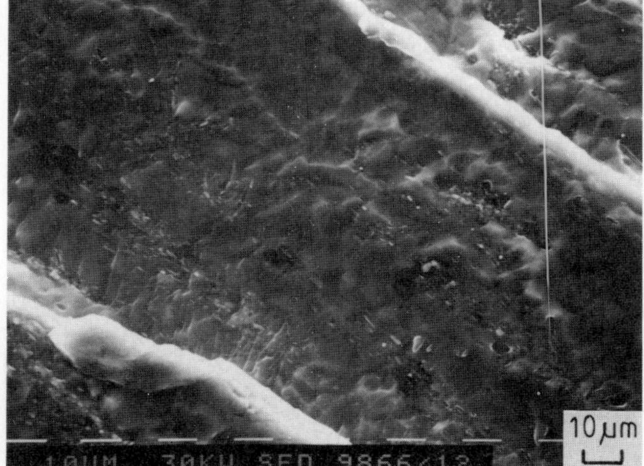

a

b

Figure 12 Scanning electron micrographs from fractures in the short-transverse direction

(a) Solution treated temper.

(b) Peak aged temper.

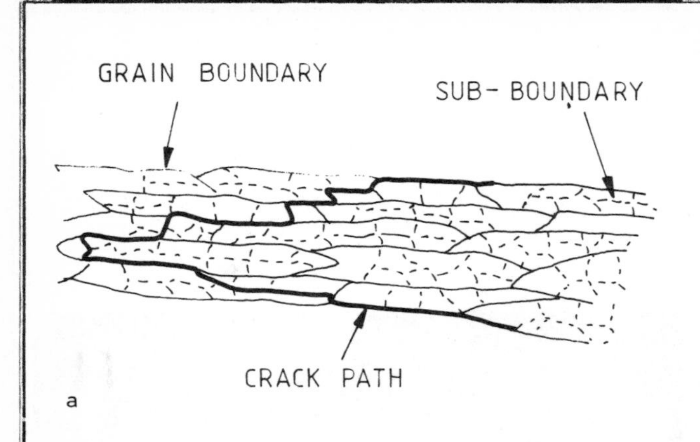

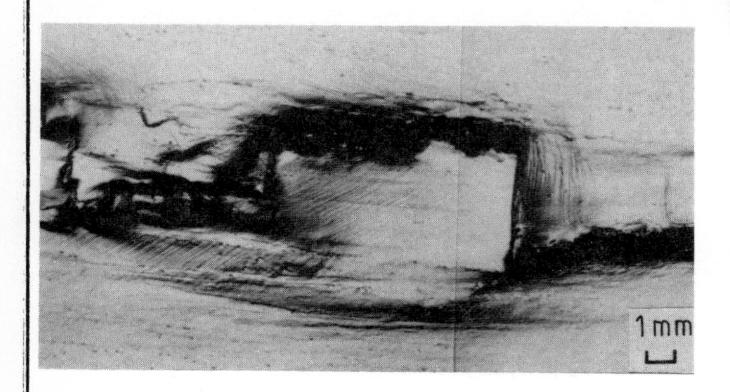

b

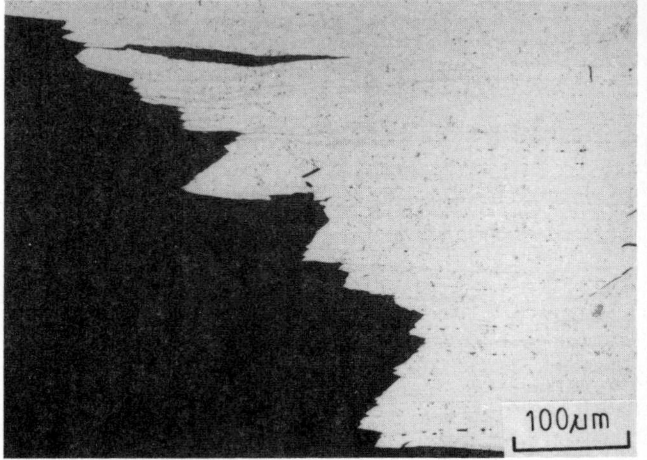

Figure 14 Longitudinal fracture in peak aged Al-Li-Cu-Mg-Zr plate with low Fe (0.01%).

Figure 13 Mode of fracture in Al-Li base alloys

(a) Theoretical model.

(b) Comparison with direct observation.

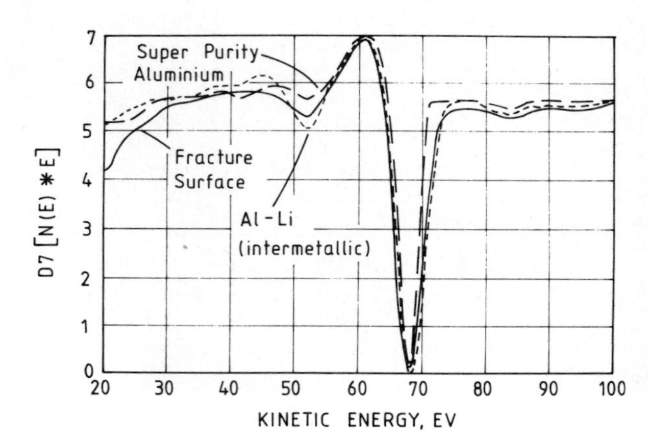

Figure 15 Lithium segregation in solution treated Al-Li-Cu-Mg-Zr plate. Auger trace from sample fractured in-situ under ultra high vacuum. Differentiated spectrum.

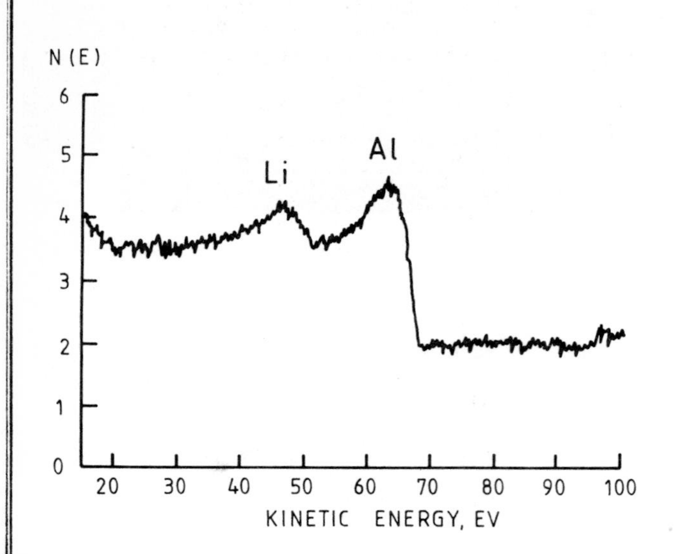

Figure 16 Lithium segregation in aged Al-Li-Cu-Mg-Zr plate. Auger trace from sample fractured in-situ under ultra high vacuum. Undifferentiated spectrum.

enough. Furthermore the tensile ductilities drop off very rapidly with ageing, particularly in the short transverse direction and certainly have no direct relationship with PFZ size. Finally extensive sub-grain fracture is also observed in these alloys but the microstructural evidence indicates that PFZ's are only formed at high angle grain boundaries.
The influence of grain boundary precipitation on the fracture behaviour of these alloys is detailed elsewhere. The fact that in the solution treated condition interboundary failure occurs under the appropriate test conditions suggests that the presence of grain boundary precipitates of δ and T phases is not the basic cause of the mode of fracture. However it has been shown that Fe/Cu-rich second phase particles increase the susceptibility to grain boundary decohesion during void fracture via void initiation and propagation in both solution treated and aged material[8].

The influence of iron-rich intermetallics in reducing toughness and ductility in aluminium alloys has been well documented[15]. In the Al-Li-Cu-Mg alloys reducing the iron level to ~ 0.01% does not alter the mode of failure, Figure 14. Thus it may be concluded that whilst the presence of coarse particles, either constituent particles aligned in the rolling direction by prior fabrication, or δ and T phase particles precipitated during ageing at grain boundaries cavitate easily (as evidenced by the extensive void formation behind the fracture surface)[8], such particles are not the underlying cause of the failure.

As a result while all the factors considered above tend to enhance boundary failure they are not the basic cause. Instead this would appear to be due to some type of embrittling segregation.

Embrittlement due to impurity elements such as Na or K has been investigated by surface analysis techniques and by impact tests.

Previous work on recrystallised binary alloys has shown a significant influence of Na content (in the range 0 to 500 ppm) on the fracture toughness[(7)]. To assess this, two alloys, one containing 10 ppm and the other 260 ppm Na in the Al-Li-Cu-Mg-Zr alloys were Charpy tested. No change in the Charpy values with Na content was found either in the solution treated or the solution treated and peak aged tempers (14.3 J/m and 5.4 J/m respectively). In addition, scanning auger microscopy (SAM) of commercial Al-Li-Cu-Mg-Zr alloy plate fractured in-situ under ultra-high vacuum showed no evidence of Na segregation on the fracture surface. Furthermore no K segregation[(16)] was detected on the fracture surface except in the neighbourhood of inclusion particles of flux. Previous work[(7)] considered coarse grain recrystallised material with a resulting low grain boundary area which could contain a high concentration of segregate even at low segregated levels. In the present alloys the grain boundary area is significantly greater resulting in lower (undetectable) segregate concentrations.

SAM examination of samples fractured in the S-T direction under ultra-high vacuum revealed evidence of Li segregation at the fractured boundaries[(17)]. Samples examined on a PHI 600 instrument in the solution heated temper show an Li peak, Figure 15, which can be separated from the overlapping Al plasmon loss peaks. Quantification of the Li segregate level was carried out using standard samples and will be fully detailed elsewhere[17]. Similar analyses on the VG Microlab were carried out on slightly overaged material. The indications were that Li segregation was still observed in regions devoid of δ phase (identified by plasmon loss peaks), Figure 16). It should be pointed out that ultra-high vacuum conditions are critical if oxidation of the freshly exposed surface is to be prevented. The fact that no other segregate species were detected (sensitively 1 a/o) suggests that if grain boundary segregation is the cause of low ductility alloys, it is not due to the presence of Na, K or Ca at the levels currently found in commercially produced material.

CONCLUSION

The fracture experiments on Al-Li-Cu-Mg plate have shown that failure is governed by the grain boundary strength. The influence of factors such as grain boundary precipitates, PFZ size and slip morphology have only secondary effects. The main cause has not been positively identified but the effects due to segregation of tramp elements such as Na, K and Ca do not occur in the alloys investigated.

In order to further improve the short transverse ductility and toughness of Al-Li base alloys several possibilities exist. Firstly optimisation of the microstructure may be employed to reduce the influence of planarity of slip[18]. Grain and sub-grain shape should be controlled to make the fracture path as tortuous as possible[19] and the harmful effects of grain boundary precipitation reduced by, for example, decreasing iron level[20].

ACKNOWLEDGEMENTS

The authors are grateful to Alcan International Limited, for permission to publish this work.

REFERENCES

1. R.E. Lewis, D. Webster, I.G. Palmer - Lockheed Palo Alto Research Laboratory Final Report, Contract F33615-77-C-5186 Technical Report No. AFML-TR-78-102 - July 1978.

2. T.H. Sanders Jr. and E.A. Starke Jr, eds. Aluminium Lithium alloys TMS-AIME Warrendale PA, 1981.

3. T.H. Sanders Jr. and E.A. Starke Jr. eds. Aluminium Lithium alloy II, TMS-AIME, Warrendale PA, 1984.

4. T.H. Sanders Jr., Naval Air Development Centre, Contract No. N62269-76-C-0271 Final Report - June 1979.

5. E.A. Balmuth, R. Schmidt in Aluminium Lithium Alloys, ed. T.H. Sanders, E.A. Starke, TMS-AIME, Warrendale PA, 1981, p.69.

6. E.A. Starke Jr., T.H. Sanders Jr., I.G. Palmer, J. of Metals - August 1981, Vol. 33, No.8, p.24.

7. A.K. Vasudevan, A.C. Miller, M.M. Kersker in Aluminium Lithium Alloys II. ed. E.A. Starke Jr., T.H. Sanders Jr., TMS-AIME, Warrendale PA, 1984, p.181.

8. E.P. Butler, N.J. Owen, D.J. Field - Mat. Sci and Tech. July 1985.

9. F.S. Lin, S.B. Chatraborthy, E.A. Starke Jr. - Acta Metall. 1982, V.30, p.927.

10. C.J. Peel, B. Evans, C. Baker, D.A. Bennett, P.J. Gregson and H.M. Flower in Aluminium-Lithium Alloys II, ed. E.A. Starke Jr.,
J.H. Sanders Jr., TMS-AIME, Warrendale PA, 1984, p.363.

11. D.B. Williams and J.W. Edington, Met. Sc. Vol.9, 1975, p.529.

12. T.H. Sanders in Aluminium-Lithium Alloys II. Ed. E.A. Starke Jr. and T.H. Sanders - TMS-AIME, Warrendale PA, 1981, p.64.

13. P.L. Makin, D.J. Lloyd and W.M. Stobbs - Scripta. Met. July 1985.

14. M.D. Ball and H. Lagace - this conference.

15. M.V. Hyatt, Aluminium Alloys in the Aircraft Industry. Proc. Symp. Turin Oct. 1976, pub. Technology Ltd. UK, p.31.

16. J.A. Wert, J.B. Lumsden, Scripta Met. 19, 2, 1985, p.205.

17. M.P. Thomas, D. Creber, D.J. Lloyd, W.S. Miller - to be published.

18. W.S. Miller, I.G. Palmer, J. White, R. Davis, T.S. Saini - This conference.

19. I.G. Palmer, W.S. Miller, D.J. Lloyd, M. Bull -This conference.

20. M.A. Reynolds, E. Creed, A. Gray - This conference.

Influence of composition and aging treatment on fracture toughness of lithium-containing aluminum alloys

S SURESH and A K VASUDÉVAN

The authors are, respectively, with the Division of Engineering, Brown University, Providence, Rhode Island 02912, USA, and the Alloy Technology Division, Alcoa Laboratories, Alcoa Center, Pennsylvania 15069, USA

SYNOPSIS

The influence of lithium and copper compositions (1.1 Li/4.5 Cu, 2.1 Li/3 Cu and 2.9 Li/1 Cu, in wt pct) and aging treatment (spanning the very underaged to the severely overaged tempers) on *plane strain* fracture toughness (K_{Ic}) are examined in alloys belonging to the Aℓ-Li-Cu-Zr system. It is found that grain boundary fracture induced by stable δ particles affects fracture toughness in the peak-aged and over-aged microstructures of alloys with high lithium content. In such cases, K_{Ic} values are not very sensitive to aging beyond the peak strength. The fracture behavior in the underaged tempers is strongly influenced by slip planarity and severe crack bifurcation promoted by metastable δ' particles.

INTRODUCTION

The improvements in mechanical properties associated with lower density and higher elastic modulus have generated considerable research interest in the aluminum-lithium alloy system (e.g., [1]). A property of great practical importance, namely, the resistance of these alloys to fracture initiation and crack growth, has not been investigated thoroughly. The design of lithium-containing aluminum alloys for potential structural applications demands a detailed understanding of the intrinsic effects of aging and microstructure on failure mechanisms. In this study, the effects of alloy composition and aging treatment on plane strain fracture initiation toughness under quasi-static loading conditions are examined in alloys belonging to the Aℓ-Li-Cu-Zr system. In particular, we have attempted to evaluate the significance of the following mechanisms in the fracture of Aℓ-Li alloys: slip planarity and crack branching promoted by the metastable δ' ($Aℓ_3Li$) precipitates and intergranular fracture induced by stable δ (AℓLi) particles. The resistance of aluminum-lithium alloys to both constant and variable amplitude fatigue crack growth is discussed in a companion paper [2].

MATERIALS AND EXPERIMENTAL METHODS

The nominal compositions (weight percent) of the three lithium-containing high purity alloys examined in this study are shown in Table I. All three materials were preheat treated at 520 °C for 8 hours and hot rolled at an initial temperature of 520 °C. The 12.74 mm thick plates of alloys A, B and C were solution-treated at 528 °C, 552 °C and 552 °C, respectively, for 1 hour, cold water quenched and stretched by 2 percent. Figures 1a and 1c show the variation of yield strength, σ_y, ultimate strength, σ_{UTS}, and elongation with aging for the three alloys. Two overaged tempers of alloy A, aged at 218 °C for 2.5 and 8 hours, were also studied. Their room temperature properties were σ_y=443 and 378 MPa, σ_{UTS}=535 and 480 MPa and Elongation = 9 and 10 percent, respectively. The elastic moduli for materials A, B and C were 76.5, 79.2 and 82.7 GPa, respectively. The microstructures of all the alloys were predominantly unrecrystallized. The subgrain size was about 2-3 μm. The presence of some recrystallized grains was observed, with the degree of recrystallization (<10%) increasing with increasing lithium content. Detailed descriptions of microstructure and experimental procedures are provided in references [3-5]. Fracture initiation experiments were made on 12.74 mm thick compact specimens machined in the T-L orientation, following ASTM standard E-399. Since the requirements for "valid" plane strain conditions cannot be met in these specimens for some of the microstructures examined, the elastic plastic fracture toughness J_{Ic} was also obtained for all the tempers. An

TABLE I

Composition (WT %) of Alloys Studied

Material	Li	Cu	Zr	Al	(Li/Cu) Ratio*
A	1.1	4.6	0.17	Bal	2.2
B	2.1	2.9	0.12	Bal	6.5
C	2.9	1.1	0.11	Bal	25.2
Fe (0.06%), Si (0.04%), Ti (0.01%), for all materials.					

*The (Li/Cu) ratio is expressed as a ratio of the atomic fractions.

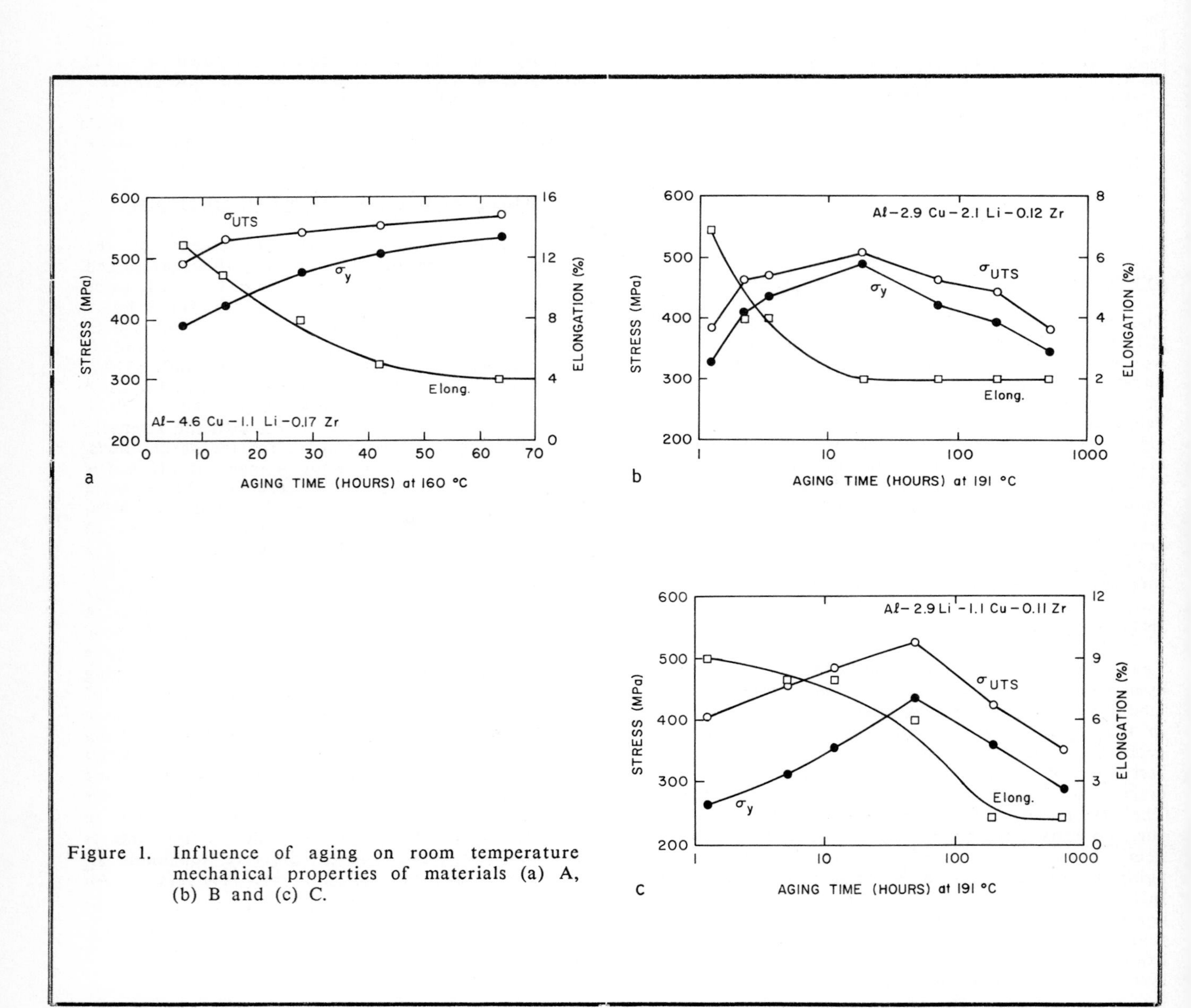

Figure 1. Influence of aging on room temperature mechanical properties of materials (a) A, (b) B and (c) C.

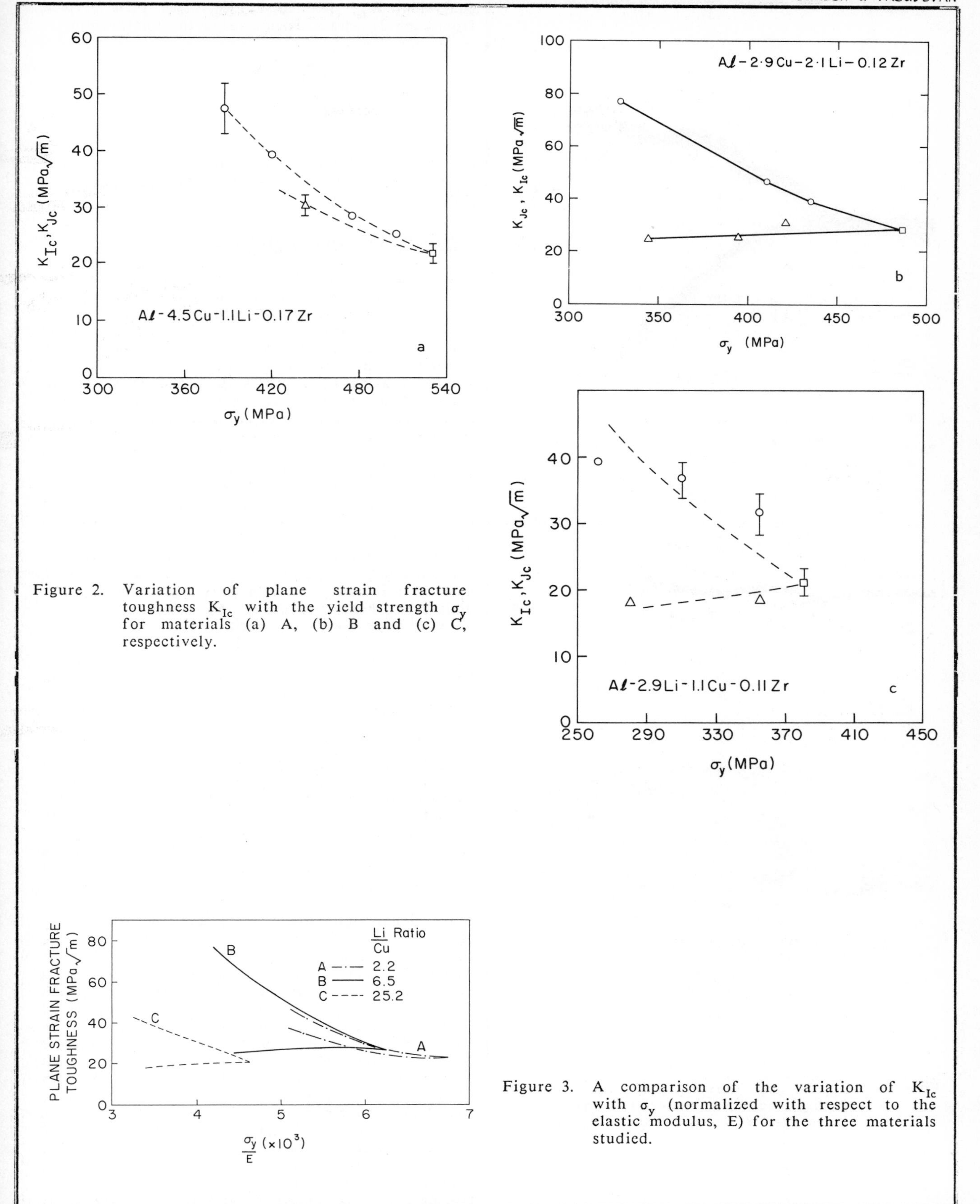

Figure 2. Variation of plane strain fracture toughness K_{Ic} with the yield strength σ_y for materials (a) A, (b) B and (c) C, respectively.

Figure 3. A comparison of the variation of K_{Ic} with σ_y (normalized with respect to the elastic modulus, E) for the three materials studied.

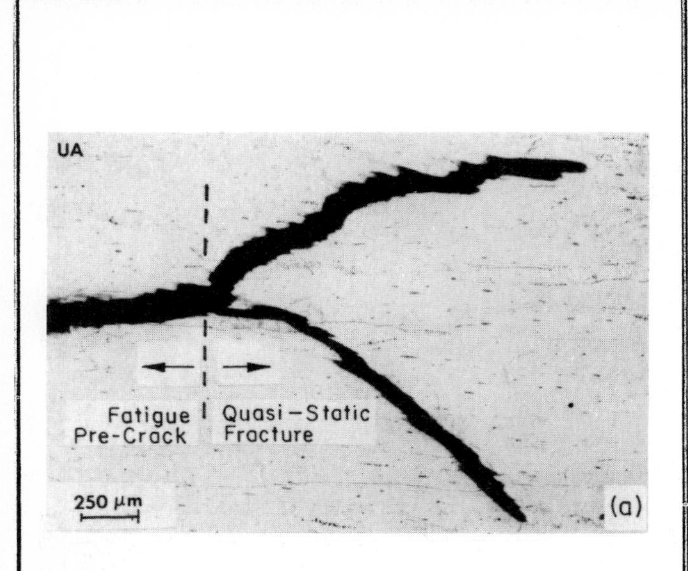

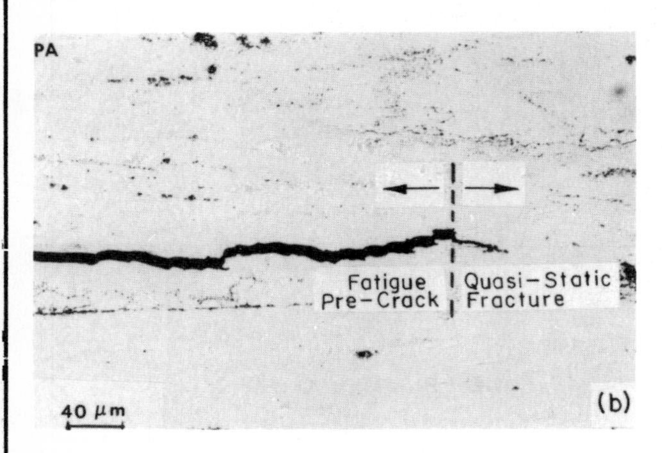

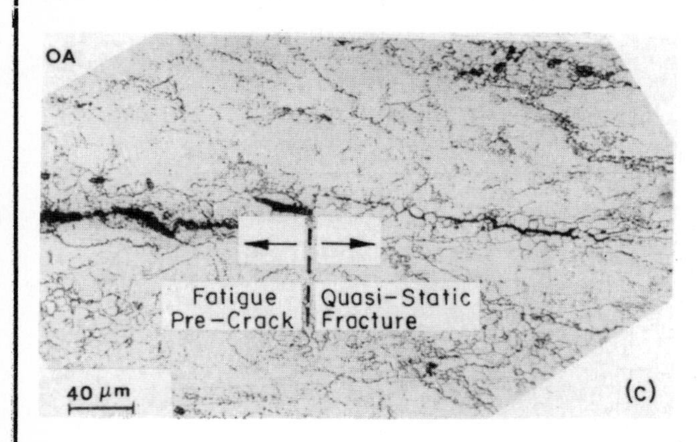

Figure 4. Optical micrographs of crack profile during fracture in material B; (a) severely under-aged, (b) peak-aged and (c) severely over-aged.

equivalent K_{Ic}, denoted as K_{Jc}, was then derived from such J_{Ic} measurements using the relationship

$$K_{Jc} = \sqrt{\frac{J_{Ic}E}{(1-\nu^2)}}$$

(see [3-5] for details). Conditions for J-dominance were ensured in accordance with ASTM guidelines (see [3,4]).

RESULTS AND DISCUSSION

The dependence of plane strain fracture toughness, K_{Ic} on yield strength σ_y is shown in Figures 2a-2c for materials A, B and C, respectively. Figure 3 shows a comparison of the variation of plane strain fracture initiation toughness K_{Ic} with the yield strength σ_y normalized with the elastic modulus E. Artificial aging of alloys A, B and C from the as-quenched to the peak strength temper results in a monotonic reduction in K_{Ic}. Aging beyond the peak strength leads to an increase in K_{Ic} for alloy A, whereas the other two materials show yield strength-independent fracture behavior in the overaged conditions. The results also indicate that increasing the lithium-content gives rise to a deleterious effect on K_{Ic}. Figures 4 and 5 show some typical examples of crack profiles observed during the fracture of alloys B and C. Alloy A exhibited a predominantly linear transgranular crack path in the underaged and peak aged tempers. The underaged heat treatments of alloys B and C developed severely branched crack paths during quasi-static fracture, with the arms of the branch extending through both low and high angle grain boundaries (Figures 4a and 5a). The peak- and over-aged tempers of these two materials, however, fractured in an intergranular mode (e.g., Figures 4b, 4c and 5b).

Conventional interpretations of the decrease in fracture toughness with increased aging to peak strength are often based on the premise that the formation of coarse slip bands leads to an enhancement in strain localization at the grain boundary and hence to a drop in toughness as the peak strength is approached [6-10]. Aging beyond the peak strength causes a dispersion of slip through the Orowan process and hence overaging of conventional aluminum alloys results in the recovery of ductility and fracture toughness. Such a behavior is also noticed in the present alloy A where a lithium content of only 1.1 wt percent is not large enough to promote substantial precipitation of ordered δ' in the underaged tempers or stable d in the overaged conditions. Alloys B and C, with larger amounts of lithium, however, exhibit fracture behavior different from those of conventional aluminum alloys without lithium or alloy A with little lithium content. In alloys B and C where considerable precipitation of δ' particles occurs, these ordered precipitates promote severe bifurcation of the crack along the bands of intense shear within the plastic zone at the crack tip in the underaged tempers. Observations of crack path and fractography (see Figure 4) and the effective stress

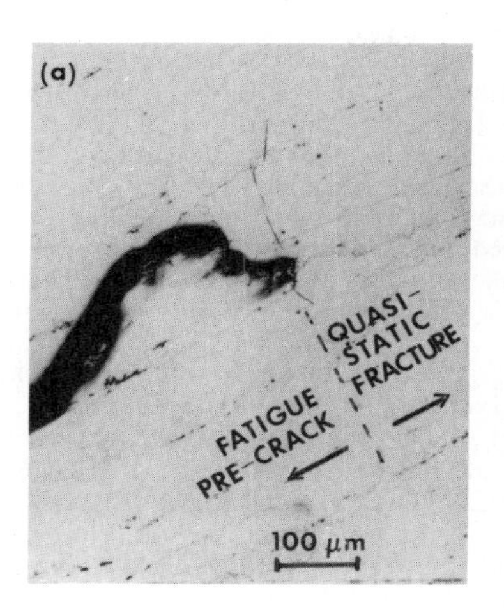

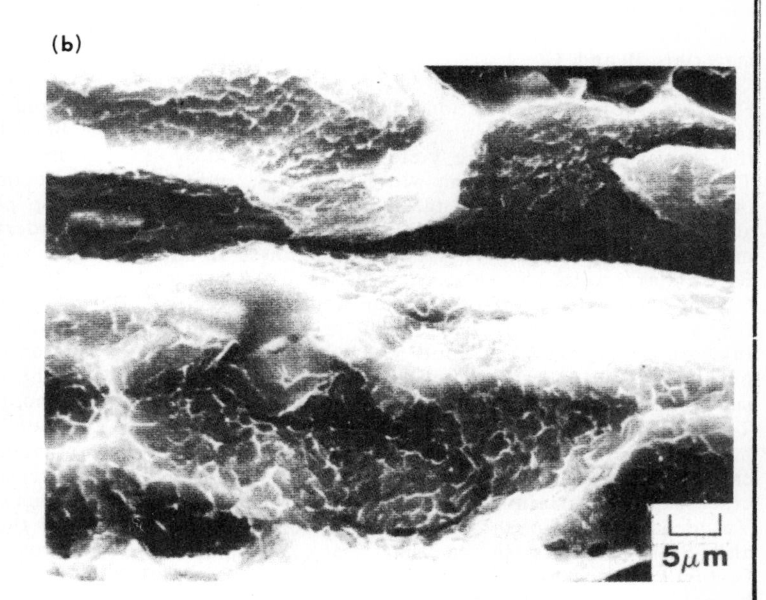

Figure 5. Fracture modes in material C; (a) crack forking in the very under-aged temper; (b) intergranular fracture in the peak-aged microstructure.

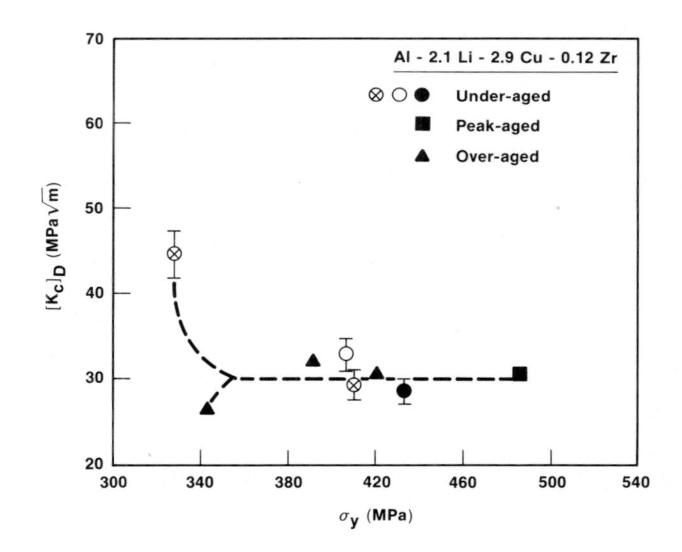

Figure 6. Variation of the effective fracture toughness (which accounts for crack branching) with the yield strength for material B.

intensity factor calculations (see [11,3]) for branched cracks, indicate that microstructurally-induced crack branching can account for a large part of the apparent increase in the nominal values of fracture toughness, K_{Ic}, in the underaged tempers of materials B and C.

Figure 6 shows the variation of an effective fracture toughness (which takes into account the branching of the crack) $[K_C]_D$, with the yield strength for alloy B. This figure, in conjunction with Figure 3, indicates that crack geometry changes during fracture can account for a substantial fraction of the microstructural effects on fracture toughness in alloy B. Similar results have also been obtained for material C [4]. We note that in some of the underaged tempers, the arms of the branched crack propagated in an intergranular mode at about 30-45 degrees from the purely Mode 1 tensile crack plane. It appears that such intergranular fracture is promoted by the T_1-phase precipitates located on low angle grain boundaries. The $[K_C]_D$ values shown in Figure 6 were calculated based on elastic solutions for branched cracks and the strain energy release rate criterion (see Vasudevan and Suresh [3]). Recent work by Suresh and Shih [12] has shown that crack tip plasticity further enhances the beneficial effect of crack branching by increasing the apparent values of fracture toughness. This result may rationalize the large fracture toughness value for the most underaged (lower strength) temper of alloy B where the use of elastic solutions for branched cracks are not strictly justifiable because of extensive crack-tip plasticity. Detailed quantitative analyses of crack branching and its dependence on the precipitation characteristics of aluminum-lithium alloys can be found elsewhere [4].

In ductile materials with high fracture toughness, stable crack growth can occur at K or J values several times greater than K_{Ic}. In such cases, microstructural effects on crack growth resistance can be far more pronounced than during crack initiation. Characterizations of crack growth resistance, for the present alloys, reveal that crack branching promoted by δ' particles improves the tearing modulus in the underaged structures of materials B and C [4]. Indeed the total beneficial effect of this mechanism on fracture resistance is felt only after the branched crack profiles are fully developed during stable crack growth. Thus, the general notion, that an increase in fracture initiation toughness (e.g., J_{Ic}) also leads to an increase in crack growth resistance (e.g., the tearing modulus), may not be fully valid if crack branching plays an important role in quasi-static fracture.

It has previously been suggested that intense slip planarity induced by ordered δ' particles, causes strain localization at the grain boundaries and a resultant loss in ductility and fracture toughness in the peak strength tempers of Aℓ-Li alloys [13-15]. This mechanism has also been considered a reason for the progressive reduction in fracture toughness with aging from the as-quenched to the peak-strength temper. Firstly, the present study indicates that stable δ particles at high angle boundaries play an important role in promoting intergranular fracture in the peak- and over-aged tempers. Furthermore, a fraction of the apparent increase in K_{Ic} in the under-aged tempers seems to arise from crack branching. Although the data presented here do not directly confirm or deny the role of slip planarity, other related results obtained in this work imply that slip planarity may play some role in influencing fracture behavior in the under-aged and peak-aged tempers. There is a reduction in fracture toughness with increasing lithium content from alloy B to C, although the degree of crack branching seemed comparable. This result, as well as slip line observations on plastically deformed tensile specimens, lend some support to the possible role of slip planarity in influencing the fracture behavior of underaged Aℓ-Li alloys. Furthermore, there appears to be a link between slip planarity and the propensity to develop a bifurcated crack during quasi-static fracture.

With aging to or beyond peak strength, there is a progressive reduction in the volume fraction of spherical δ' in the matrix and an increase in the amount of stable d located at the high angle grain boundaries. Our fractographic results clearly show that the fracture toughness in the peak-aged and over-aged heat treatments of alloys B and C is primarily dictated by intergranular fracture induced by the coalescence of voids created around δ particles. Provided there exists a sufficient amount of grain boundary δ, the fracture toughness is not significantly affected by further aging. This is evident from the results of Figure 3 which suggest that aging beyond the peak strength temper causes very little change in fracture toughness in both materials B and C. Increasing the lithium content and the volume fraction of δ (from B to C) does not lead to any appreciable drop in fracture toughness. Although δ particles located along high angle grain boundaries appeared to be the primary cause of intergranular fracture, there was also some evidence of (low angle) grain boundary failure induced by the T-type precipitates.

Such interpretations of the influence of δ particles on fracture toughness are also corroborated by recent studies involving "reversion" experiments [16]. Here microstructures "reverted" to redissolve matrix δ' (while retaining grain boundary δ) showed that the grain boundary fracture of peak-aged and over-aged tempers of Aℓ-Li alloys are severely affected by the precipitation of grain boundary δ [16].

It should be noted that all the fracture toughness results reported in this work were obtained under "valid" plane strain conditions so that comparisons can be made among different materials under conditions of a common stress state. Thus, the observed differences in fracture toughness among the various alloys are presumed to be induced solely by intrinsic microstructural effects.

CONCLUSIONS

The effects of composition and heat treatment on plane strain fracture toughness were studied in three aluminum alloys containing lithium. Artificial aging of Aℓ-Li alloys to peak strength leads to up to a three fold reduction in plane strain fracture toughness. In the alloys of higher lithium content, over-aging has little effect on fracture toughness. The quasi-static fracture of Aℓ-Li alloys appears to be governed by the competitive influence of the following mechanisms: (i) severe branching of the crack aided by the presence of metastable δ' precipitates in the under-aged microstructures; (ii) possible slip planarity and strain localization at grain boundaries in microstructures aged up to peak strength; and (iii) intergranular fracture in peak-aged and over-aged alloys induced by the coalescence of voids formed around stable δ particles located at high angle grain boundaries.

ACKNOWLEDGEMENTS

This work was supported partly by Naval Air Systems Command Grant N-00019-80-C-0569 and partly by NSF Grant NSF-ENG-8451092. Thanks are due to Mrs. Louise Gray of Brown University for her help in the preparation of this manuscript.

REFERENCES

[1] *"Aluminum-Lithium Alloys"*, Proceedings of the Second International Conference on Al-Li Alloys, edited by T. H. Sanders, Jr. and E. A. Starke, Jr., The Metallurgical Society of AIME, Warrendale, PA (1984).

[2] A. K. Vasudevan, S. Suresh, R. C. Malcolm and J. Petit, this volume.

[3] A. K. Vasudevan and S. Suresh, *Materials Science and Engineering*, Vol. 72 (1985) p. 37.

[4] S. Suresh and A. K. Vasudevan, *Metallurgical Transactions A* (1985) in press.

[5] A. K. Vasudevan, S. Suresh, M. Tosten and P. R. Howell, *Metallurgical Transactions A* (1985) in press.

[6] G. T. Hahn and A. R. Rosenfield, *Metallurgical Transactions A*, Vol. 6 (1975) p. 653.

[7] R. H. Van Stone and J. A. Psoida, *ibid*, p. 668.

[8] G. G. Garrett and J. F. Knott, *Metallurgical Transactions A*, Vol. 9 (1978) p. 1187.

[9] C. Q. Chen and J. F. Knott, *Metal Science*, Vol. 15 (1981) p. 357.

[10] R. H. Van Stone, R. H. Merchant and J. R. Low, *ASTM STP* 556 (1974) p. 93.

[11] S. Suresh, *Engineering Fracture Mechanics*, Vol. 18 (1983) p. 577.

[12] S. Suresh and C. F. Shih, *International Journal of Fracture* (1985) submitted.

[13] B. Noble, S. J. Harris and K. Dinsdale, *Metal Science*, Vol. 16 (1982) p. 425.

[14] T. H. Sanders and E. A. Starke, *Acta Metallurgica*, Vol. 30 (1982) p. 927.

[15] P. J. Gregson and H. M. Flower, *Acta Metallurgica*, Vol. 33 (1985) p. 527.

[16] A. K. Vasudevan, S. F. Bauman, E. A. Ludwiczak, R. D. Doherty and M. M. Kersker, *Materials Science and Engineering*, (1985) in press.

□

Temperature dependence of toughness in various aluminum–lithium alloys

D WEBSTER

The author is with
Kaiser Aluminum and Chemical
Corporation Center for Technology,
Pleasanton, California,
USA

SYNOPSIS

As part of a research program to understand the plastic flow and fracture behavior of Al-Li alloys, mechanical properties were measured at temperatures between 77 and 408K. Examination of fracture surfaces by SIMS was performed. High grain boundary concentrations of Na, K, In and hydrogen were observed. Both tensile elongation and toughness were found to increase with decrease of temperature. A hypothesis for the flow and fracture behavior of Al-Li alloys is presented.

INTRODUCTION

Al-Li alloys have traditionally been thought of as having lower toughness and ductility than conventional aerospace alloys. Improved manufacturing techniques in the last few years have done much to offset this problem. However, the problem still remains to some degree as shown in Figure 1 where the impact toughness (LT notch) - yield strength relationship for a semi-commercial extrusion billet is compared with commercial 7000 and 2000 series extrusions. Tensile elongation shows a similar advantage for conventional alloys. The present research program is designed to investigate this apparent embrittlement in Al-Li alloys. Since these alloys fracture along the grain boundaries in their most brittle conditions, some researchers have examined grain boundary segregation of impurities. Vasudevan[1] added up to 564 atomic ppm of sodium to Al-Li alloys and found significant embrittlement in the underaged condition but very little in the peak aged condition. Sodium was found to be concentrated in grain boundary areas. Webster[2] observed significant potassium segregation to grain boundary regions and found that the segregation was greater for lower toughness materials of similar composition. Wert and Lumsden[3] also observed potassium segregation to grain boundaries. Webster observed that the toughness of Al-Li alloys increased with decrease in temperature, a fact consistent with the theory some low melting point impurities existed as liquid regions at the grain boundaries. A more comprehensive investigation of the temperature dependent properties of Al-Li alloys, including some with added sodium and potassium, is now in progress. The initial results of this work are reported here.

EXPERIMENTAL PROCEDURE

Alloys evaluated in this work were either in the form of 58 x 14 mm extrusions from 127 mm diameter billets, 60 cm wide sheet or plate from 75 mm thick slabs or in the case of alloys with added sodium and potassium in the form of 6.4 mm plate rolled from 25 mm thick ingots. Charpy impact tests were conducted on a 32 J Manlabs impact machine. SIMS analysis on grain boundary fractures was conducted on a Physical Electronics Model 590 Scanning Auger Microprobe equipped with a 3M SIMS unit. The specimens were fractured at a pressure of 1×10^{-9} torr. Approximate concentrations of various elements using the SIMS were determined by the use of the number counts of each element and the use of the appropriate sensitivity factors.

RESULTS

(a) Temperature Effects in Standard Alloys

The toughness of most metals generally decreases as the test temperature is

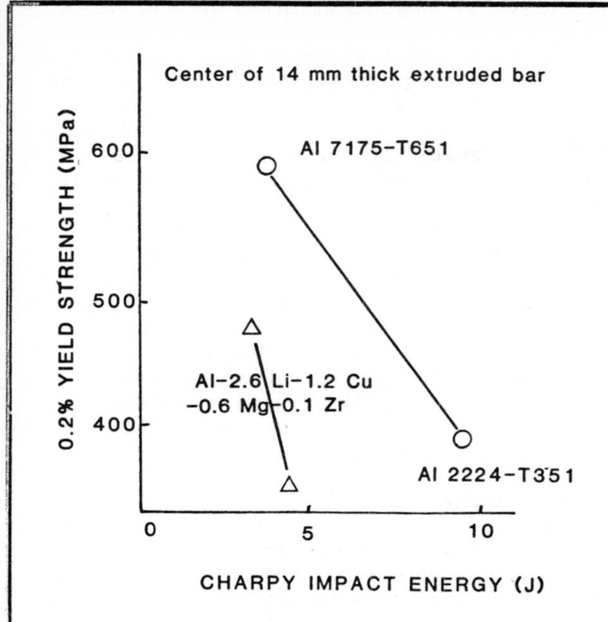

Fig. 1

Comparison of strength toughness properties of semi-commercial DC cast Al-Li with commercial Al 7175 and Al 2224.

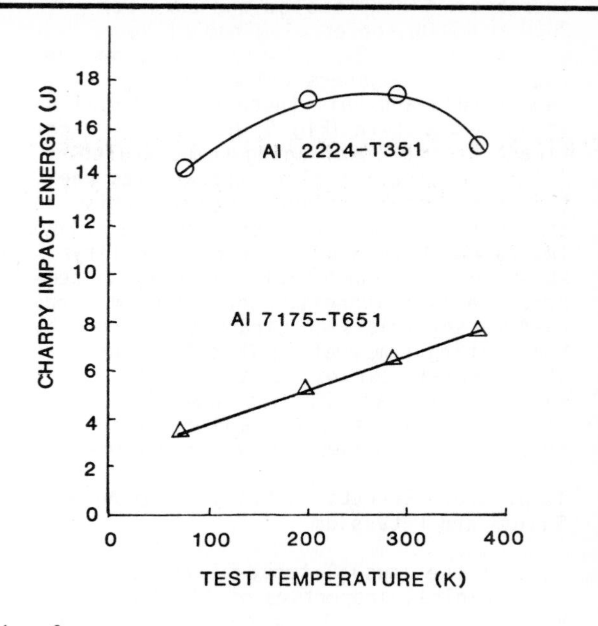

Fig. 2

Change of toughness with test temperature for Al 2224 and Al 7175.

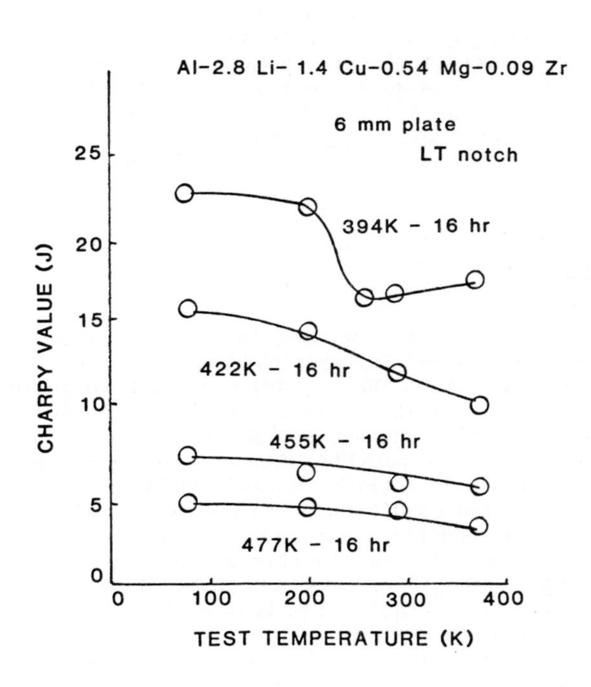

Fig. 3

Change of toughness with test temperature for various aging temperatures in Al-Li-Cu-Mg-Zr 6 mm plate.

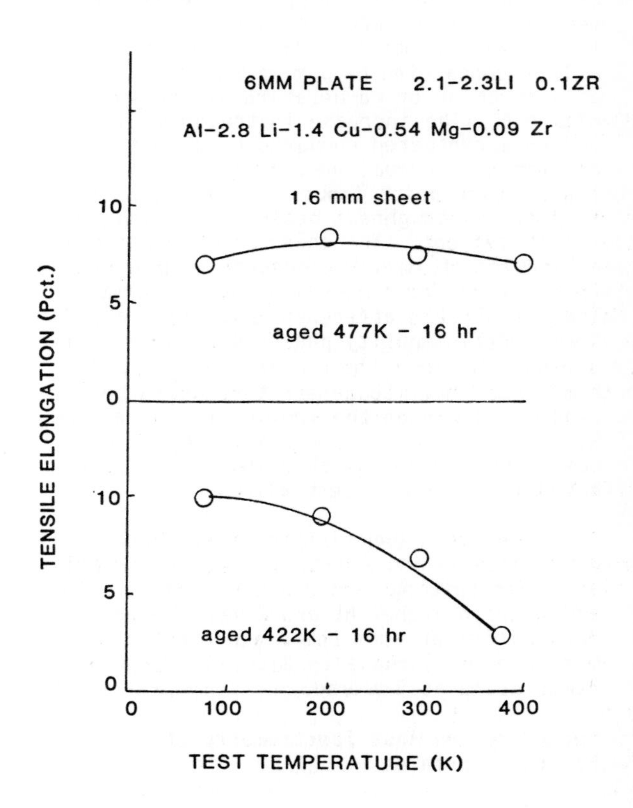

Fig. 4

Effect of test temperature and aging condition on the tensile elongation of Al-Li-Mg-Cu-Zr sheet.

decreased. This is also the case for 2000 and 7000 series aluminum alloy extrusions (Fig 2). Al-Li alloys, however, increase in toughness below room temperature and this increase varies with aging temperature (Fig 3) being more significant at the lower aging temperatures. Tensile elongation also increases as the test temperature is decreased and this effect is also unusual for aluminum alloys. The temperature dependance of ductility shows the same correlation with aging temperature as toughness, i.e., only the under aged material is significantly affected by the testing temperature (Fig 4). The stress strain curves for the sheet material aged at 422K are shown in Figure 5. Both strength and ductility increase as the temperature is lowered.

b) Temperature Effects in Alloys with Added Sodium and Potassium

If the unusual temperature dependence of mechanical properties of Al-Li alloys were due to low melting point impurities, then it should be accentuated by alloys in which additional sodium and potassium were added. Alloys were made with 2.1 Li, 0.1 Zr to which up to 434 ppm Na and 31 ppm K were added just before casting. Charpy impact values obtained from one of these alloys containing 64 ppm Na and 31 ppm K, as a function of temperature are shown in Figure 6. The increase in toughness is indeed accentuated compared to alloys containing more normal amounts (0-5 ppm) of low melting point impurities (Fig 3). The increase in toughness between 400K and 200K is 250 percent. The change in toughness as a function of impurity content is plotted in Figure 7 for three test temperatures. There is little effect of impurities at 77 K where all impurity phases should be solidified. At 295 K the alloy containing 31 ppm K has lower toughness than would be expected based on the sodium content alone. At 373 K both potassium containing alloys have lower toughness than would be expected based on sodium content alone.

Tensile elongation in these alloys with added Na and K also increases markedly at lower testing temperatures (Fig 8). The alloys with higher Na and K have lower ductilities at room temperature but show a more marked increase in ductility as the temperature is lowered.

c) Secondary Ion Mass Spectrometry of Fracture Surfaces

Specimens were fractured at a pressure of 1 x 10^{-9} and in the aging conditions investigated (near peak age) the fracture was predominantly along grain boundaries and subgrain boundaries. Elements in high concentrations in the grain boundaries of the several samples examined were H, Na, K and in some cases indium. The concentration of all these elements increased with the bulk lithium content of the alloy. The concentration of these elements decreased as the fracture surface was sputtered away. There were indications in many of the alloys of elements or compounds with atomic weights of 24 and 40. These could be magnesium and calcium, respectively, but there are reasons for believing they indicate the presence of NaH and possibly KH. For example, the peak at 24 does not change in intensity with magnesium content but does increase with hydrogen content. The approximate concentrations of the elements found at the fracture surface and at a depth of 0.7 μm below the fracture surface are shown in Figure 9.

d) Fractography

The fracture behavior of Al-Li alloys is now well documented and shows an increasing tendency for intergranular fracture as the aging temperature is increased. Grain boundary surfaces are generally smooth in the underaged conditions with isolated particles while peak aged and over aged materials show rough grain boundary surfaces which are interpreted by some workers as evidence of local plasticity. The alloys investigated in this work which contain intentional additions of Na and K show similar fracture behavior. Although this part of the investigation is only in its initial stages, it has been possible to show by energy dispersive x-ray analysis that in an alloy containing 3.1 Li and 52 ppm Na, sodium is concentrated in isolated grain boundary particles 1-5 μm in diameter.

DISCUSSION

The evidence presented in this work shows that hydrogen and at least three low melting points impurities (Na, K, In) segregate to grain boundaries in Al-Li alloys. The temperature dependence of toughness and ductility indicate that part of the metallic impurities are present at the grain boundaries in the liquid form in the underaged conditions. In the peak aged and over aged conditions the toughness of Al-Li alloys does not show significant temperature dependence (Fig 3) unless additional Na and K are added[4]. Liquid phases exist in the system Na-K-Cs down to 195 K (-78°C)[5]. It is probable that liquids exist at still lower temperatures in systems with more components.

Because the change of toughness with temperature, although significant, is not dramatic, it has been suggested that if there is a grain

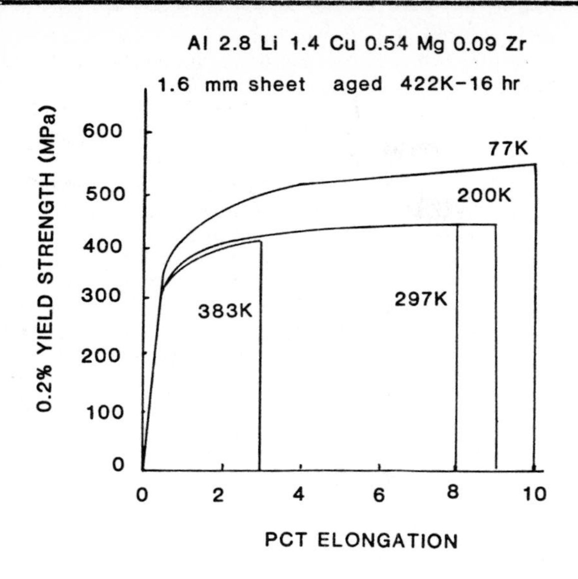

Fig. 5

Tensile stress strain curves for Al-Li-Mg-Zr alloy tested at various temperatures.

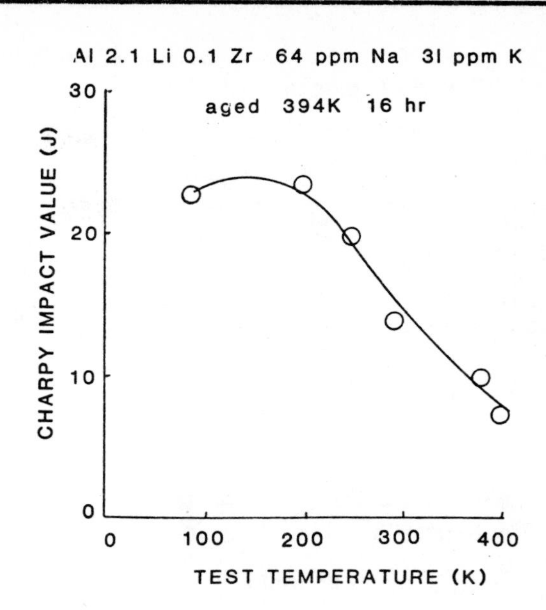

Fig. 6

Effect of test temperature on the impact toughness of Al-Li-Zr plate containing additional Na and K.

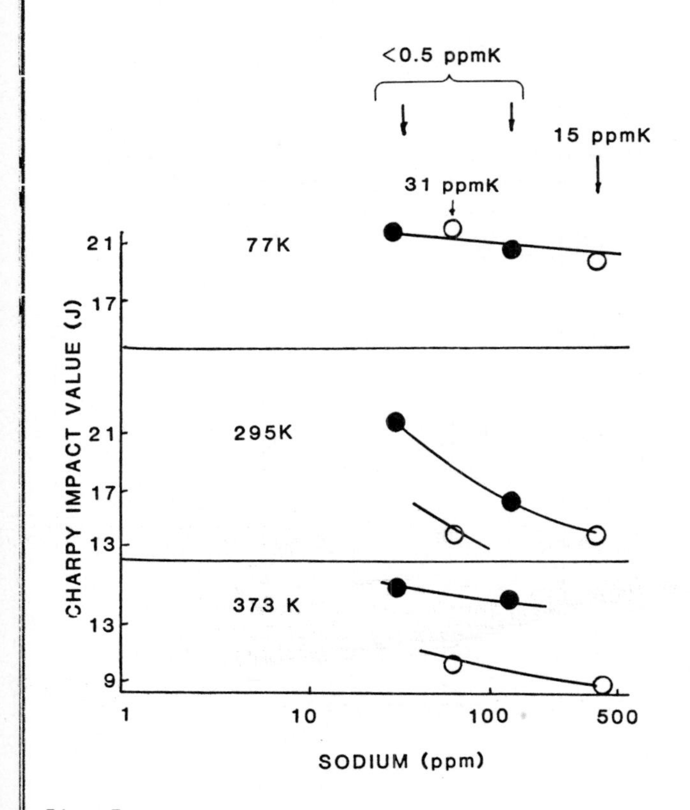

Fig. 7

Impact toughness at 3 test temperatures of 4 Al-Li-Zr alloys containing additional Na and K aged 394K 16 hours.

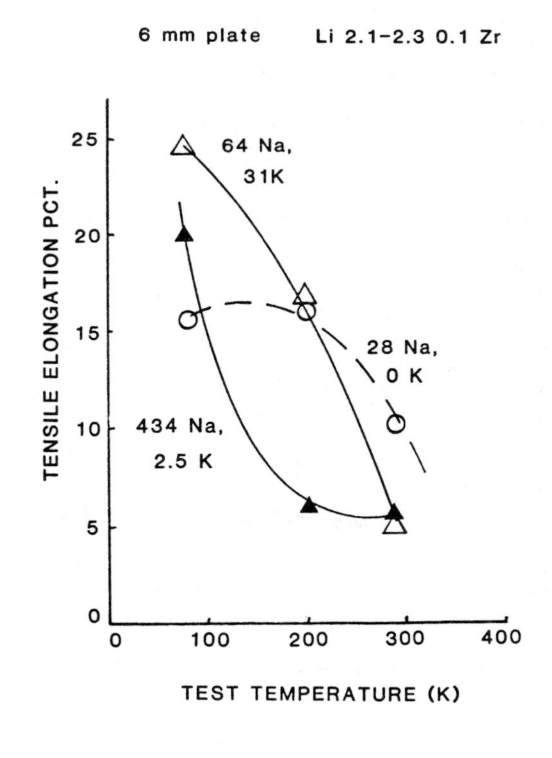

Fig. 8

Effect of test temperature on the tensile elongation of 3 Al-Li-Zr alloys containing additional Na and K aged 394K 16 hours.

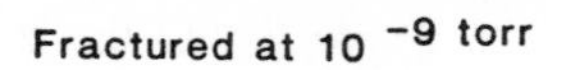

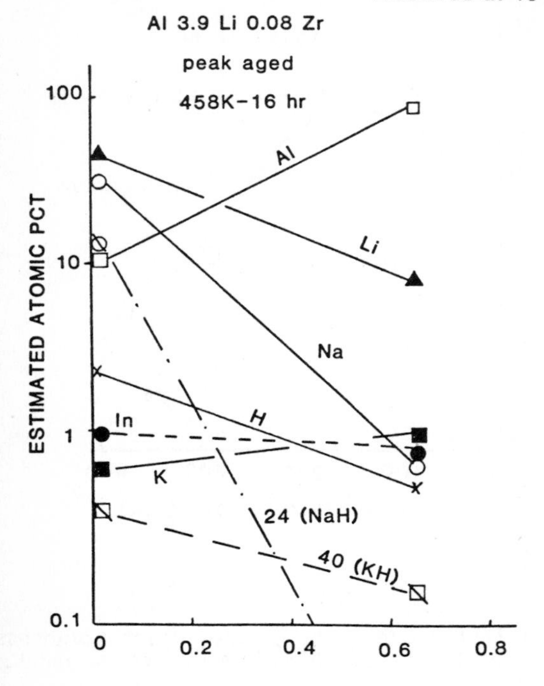

Fig. 9

Approximate grain boundary concentrations of elements at the fracture surface and 0.7 μm below the fracture surface determined by SIMS.

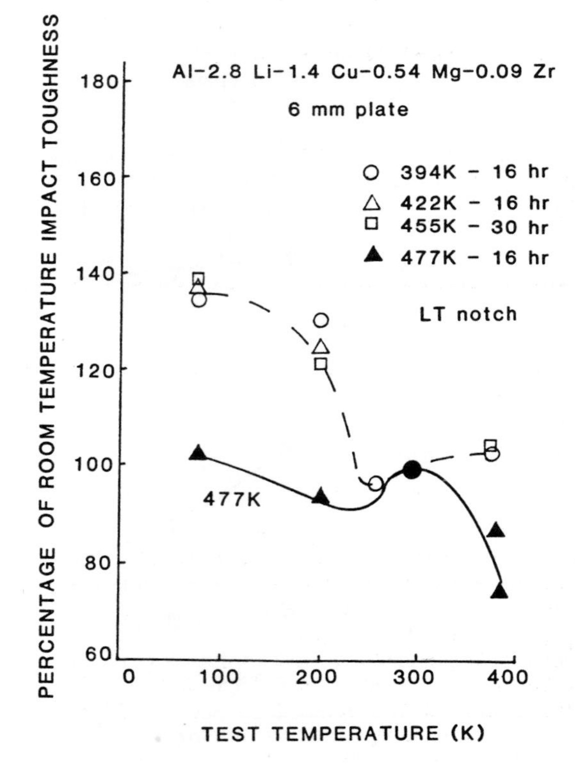

Fig. 10

Percentage change in impact toughness with test temperature for various aging temperatures for Al-Li-Cu-Mg-Zr alloy.

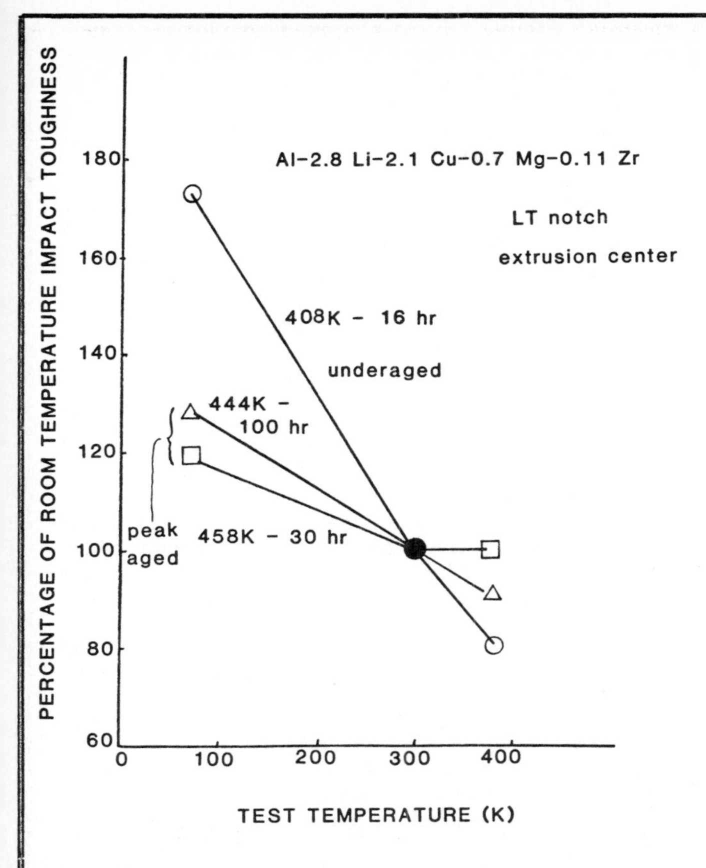

Fig. 11

Percentage change in impact toughness with test temperature for various aging temperatures for Al-Li-Cu-Mg-Zr alloy.

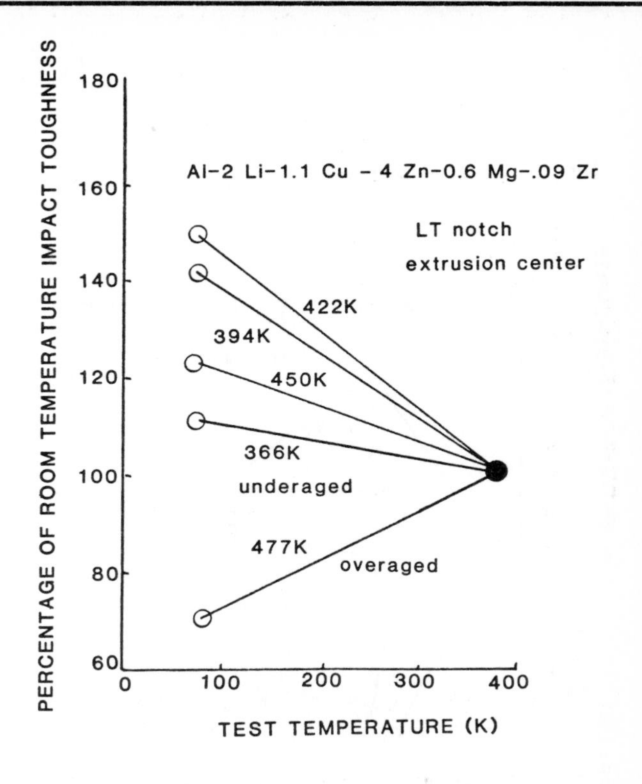

Fig. 12

Percentage change in impact toughness with test temperature for various aging temperatures for Al-Li-Cu-Zn-Mg-Zr alloy.

boundary liquid phase, it occurs as isolated regions rather than as a continuous film[2]. Depending on their size, these liquid areas could act as significant stress concentrations or merely reduce the amount of grain boundary that requires to be fractured. In either case, if their morphology is independent of aging temperature, they should exert a similar influence on toughness in all aging conditions. To examine the influence of aging temperature in more detail, the toughness of three alloys at various test temperatures is plotted as a percent of their toughness at room temperature (Figs 10, 11, 12). There is a general tendency for the temperature dependence to decrease with increase of aging temperature. Some alloys, as illustrated in Figure 12, show an increase in temperature dependence up to aging temperatures of 422 K and then a decrease as the aging temperature is increased to 477 K. This behavior suggests that the Na and K are being converted at the higher aging temperatures to compounds which are solid at the testing temperatures. In view of the high concentrations of Na, K and H in the grain boundaries and the stability of NaH and KH at low temperatures, these hydrides appear to be the most likely solid compounds forming during aging. The SIMS results are consistent with this hypothesis. The role, if any, of other stable hydrides LiH and CaH_2 is not known at this stage.

The mechanical properties of an Al-Li-Zr alloy as a function of aging temperature are

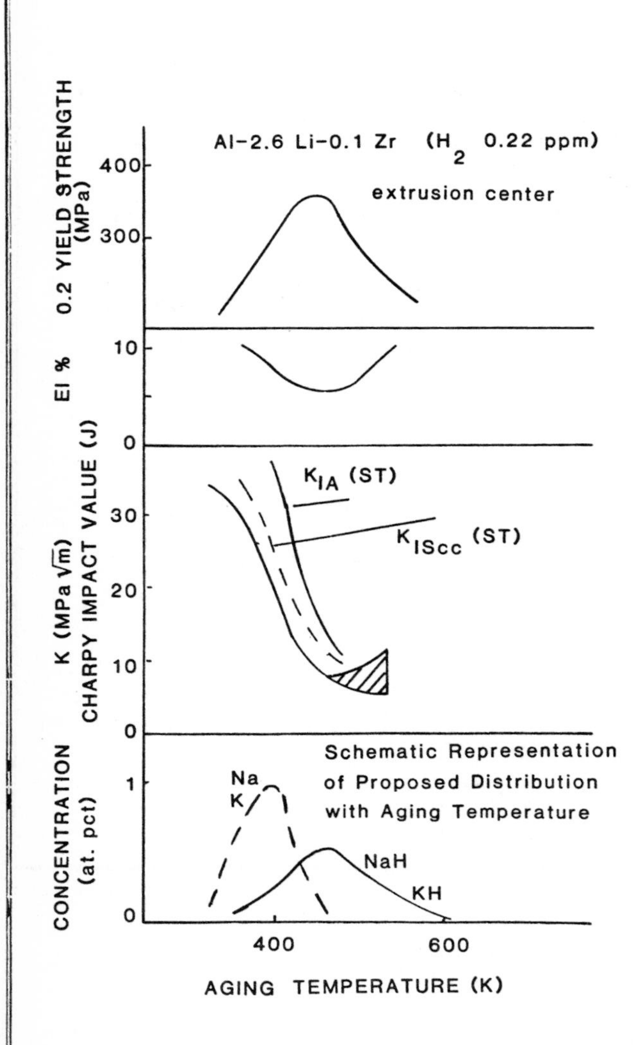

Fig. 13

Effect of aging temperature on the mechanical properties of Al-Li-Zr alloy and the proposed distribution of Na and K and their hydrides.

shown in Figure 13 and are used to illustrate the proposed influence of Na, K, NaH and KH on the mechanical properties of Al-Li alloys. The proposed distribution of Na, K and their hydrides during aging is also shown as a function of aging temperature.

In the underaged condition, the strength, ductility, toughness and short transverse stress corrosion resistance (K_{Iscc}) are high, and low melting point impurities exist as discrete particles. In this condition, mechanical properties will show marked temperature dependence with ductility and toughness increasing as the liquid phases freeze and no longer act as stress concentrations. Grain boundary fractures in the few areas where they occur are smooth and show only isolated particles which are rich in low melting point impurities. During aging at higher temperatures, the hydrogen will react with Na and K to form hydrides as thin plates covering most of the grain boundaries. Fracture will be predominantly intergranular and toughness will be low. Stress corrosion resistance will be reduced by both the reduced toughness and the fact that NaH and KH react violently with water. In peak and slightly overaged conditions the temperature dependence of ductility and toughness is low as the amount of free sodium and potassium at the boundaries is reduced. At still higher aging temperatures, where sodium and potassium are completely converted to hydrides, the temperature dependence of toughness reverses and the alloys show a normal decrease in toughness with decreasing testing temperature. At still higher aging temperatures, above the range normally considered for structural applications, NaH and KH will begin to decompose releasing free Na and K and toughness will again increase at lower testing temperatures. It is possible that between the temperature where NaH and KH dissociate and the solution treatment temperature, there will be a region where the more stable hydride LiH will exist. At the solution treatment temperature, sodium and potassium hydrides will be dissociated and Na (up to about 25 ppm) and K will be in solid solution. After quenching, these elements will again diffuse to grain boundary positions at room temperature and low aging temperatures. The tendency for Al-Li alloys to increase in toughness during overaging has been found to vary widely among alloys of nominally identical composition as indicated by the shaded area on the Charpy impact curve in Figure 13. It is thought that this effect is related to the impurity effects discussed above.

CONCLUSIONS

1. Na, K, H and sometimes In are segregated to grain boundaries in Al-Li alloys.

2. The toughness and ductility of Al-Li alloys show a marked increase as the testing temperature is decreased suggesting that K,

Na, etc. are present as liquid phases at room temperature.

3. The temperature dependence of toughness and ductility in Al-Li alloys is accentuated by the addition of Na and K.

4. The temperature dependence of mechanical properties decreases as aging temperature is increased suggesting that the liquid phases are being converted to solid compounds. Hydrides of K and Na are proposed as these compounds.

5. The toughness, stress corrosion resistance and ductility behavior of Al-Li alloys as a function of testing temperature and aging temperature can be explained by the behavior of low melting point impurities and their hydrides.

ACKNOWLEDGMENTS

The author would like to thank the following employees of Kaiser Aluminum for their contributions to this work: Ron Foreman, Wayne Erwin, Jack Go, Ray Hintze, Penny Linde, Nancy Yang and Norine Ross.

REFERENCES

1. A. K. Vasudevan, A. C. Miller and M. M. Kersker, "Proceedings of the Second International Aluminum-Lithium Conference," Published by AIME 1984, Edited by E. A. Starke, Jr. and T. H. Sanders, Jr., p. 181.

2. D. Webster, Metal Progress, V 125, No. 5, April 1984, p. 33.

3. J. A. Wert and J. B. Lumsden, Scripta Met., V 19, No. 2 1985, p. 205.

4. D. Webster, unpublished research at Kaiser Aluminum & Chemical Corporation.

5. "Multicomponent Alkali Metal Alloys", F. Tepper, J. King and J. Greer, Chem. Soc. (London) Special Publication. No. 22, 23-31 (1967) (Eng).

Mechanical properties of Al–Li–Zn–Mg alloys

S J HARRIS, B NOBLE, K DINSDALE,
C PEEL and B EVANS

SJH, BN and KD are in the
Department of Metallurgy and Materials Science
at the University of Nottingham, UK.
CP and BE are in the Materials Department,
Royal Aircraft Establishment, Farnborough, UK

SYNOPSIS

Al-Li-Zn-Mg alloys have been prepared by ingot metallurgy with Zn:Mg ratios varying from 1:1 to 5:1. The lithium concentration has been kept in the range 2-3% and recrystallisation has been inhibited by an addition of 0.15% zirconium. This has enabled alloys to be developed that precipitate two phases, δ'(Al_3Li) and a zinc rich phase. The effect of cold work before ageing and the addition of copper to the alloy on the mechanical properties has been investigated.

Proof stresses and tensile strengths have been achieved in the Al-Li-Zn-Mg alloys that compare with or exceed the target properties of Lital A (8090) and Lital B (8091).

INTRODUCTION

The advantages of reducing weight and increasing elastic modulus by making lithium additions to aluminium alloys are now well established[1,2]. Suitable alloys are available for intermediate strength applications, e.g. Lital A (8090) in sheet, plate and extruded forms which is set to compete effectively with conventional higher density materials such as 2014-T6 in aerospace applications[3]. There is still a need to develop an effective light weight replacement for the higher strength 7000 series alloys.

Problems have arisen with toughness and ductility when attempts have been made to raise the strength of lithium-containing alloys. These attempts have been based upon Al-Li-Cu-Zr or Al-Li-Mg-Cu-Zr systems. Both alloy systems promote the formation of precipitates additional to the main strengthening precipitate of δ'(Al_3Li), e.g. Al_2CuLi (T_1) in the lower magnesium or magnesium free alloys or Al_2CuMg (S) in higher magnesium-containing alloy[4]. The precipitation of S in particular helps to disperse the planarity of the slip processes which are associated with the presence of the ordered δ'. To obtain significant amounts of S type precipitation it is necessary to impose a cold stretch of ≈3% prior to the final ageing treatment[2] but this type of processing is not always convenient in all product forms. Therefore a need exists for an alloy which contains a slip dispersing precipitate which can be produced by heat treatment procedures and which may assist in strengthening and toughening the material.

This paper describes initial attempts at developing an alloy based on the Al-Li-Zn-Mg-Zr system. The philosophy of the additions to the alloy was to produce precipitation of two phases, one based on δ' and the other on η($MgZn_2$). The latter phase is that which forms in the 7000 series alloys and in such alloys is produced at relatively low ageing temperatures (<150°C), well below the temperatures (>170°C) used for precipitating δ' in lithium-containing alloys. Extruded and rolled materials have been prepared in the laboratory to examine the mechanical properties of alloys based upon the Al-Li-Zn-Mg-Zr system and these are presented in this paper. Electron microscopy has also commenced on this alloy system and preliminary results are reported.

EXPERIMENTAL

All the extruded alloys tested in this work have been prepared from high purity metals which were melted and cast under argon; the compositions are given in Table I. The cast ingots were homogenised at 490°C and hot extruded to flat strip 16mm x 3mm section After the final extrusion, the alloys were solution treated at 530°C and aged at various temperatures (T6 condition). Some of the alloys were stretched 3% after solution treatment and before ageing (T651 condition).

The sheet alloy was prepared from commercial purity metals which were melted and cast in air.

TABLE I. Chemical Compositions and Densities of Alloys

Alloy	Li	Mg	Zn	Cu	Zr	Fe	Si	Density Mg/m^3
Al-Li-Zr (extruded)	2.20	–	–	–	0.15	<0.03	<0.02	2.54
Al-Li-Mg-Zr (extruded)	2.23	2.10	–	–	0.16	<0.03	<0.02	2.51
Al-Li-Zn-Zr (extruded)	2.24	–	2.03	–	0.15	<0.03	<0.02	2.57
Al-Li-3.5Zn-1Mg-Zr (extruded)	2.35	1.01	3.68	–	0.15	<0.03	<0.02	2.58
Al-Li-3.5Zn-1Mg-Cu-Zr (extruded)	2.10	1.09	3.59	1.09	0.15	<0.03	<0.02	2.60
Al-Li-5Zn-1Mg-Zr (extruded)	2.20	1.13	5.10	–	0.19	<0.03	<0.02	2.62
Al-Li-5Zn-1Mg-Cu-Zr (extruded)	2.30	1.04	4.85	0.96	0.17	<0.03	<0.02	2.63
Al-Li-4Zn-4Mg-Zr (extruded)	2.20	4.03	4.22	–	0.20	<0.03	<0.02	2.56
Al-Li-4Zn-4Mg-Cu-Zr (extruded)	2.30	3.82	3.97	0.96	0.18	<0.03	<0.02	2.59
Al-5Zn-1Mg-Zr (extruded)	–	1.01	5.05	–	0.19	<0.03	<0.02	2.80
Al-5Zn-1Mg-Cu-Zr (extruded)	–	1.10	4.80	0.95	0.18	<0.03	<0.02	2.81
Al-4Zn-4Mg-Zr (extruded)	–	4.03	3.95	–	0.17	<0.03	<0.02	2.73
Al-4Zn-4Mg-Cu-Zr (extruded)	–	3.82	3.95	0.96	0.15	<0.03	<0.02	2.76
Al-Li-2Zn-2Mg-Cu-Zr (sheet)	2.59	1.76	2.22	0.75	0.12	0.13	0.04	2.55

TABLE II. Mechanical properties of lithium free alloys in the solution treated, double aged and T651 conditions

Alloy	0.2% Proof stress MPa			Tensile strength MPa			% Elongation		
	ST	DA	T651	ST	DA	T651	ST	DA	T651
Al-5Zn-1Mg-Zr	243	278	269	379	320	314	14.5	7.5	11.3
Al-4Zn-4Mg-Zr	297	490	338	476	544	446	17.9	11.9	14.3
Al-5Zn-1Mg-Cu-Zr	274	339	303	434	404	370	16.6	9.2	13.0
Al-4Zn-4Mg-Cu-Zr	355	402	410	527	519	517	14.6	12.2	12.8

ST = solution treated
DA = double aged
T651 = 3% stretch before ageing at 150°C

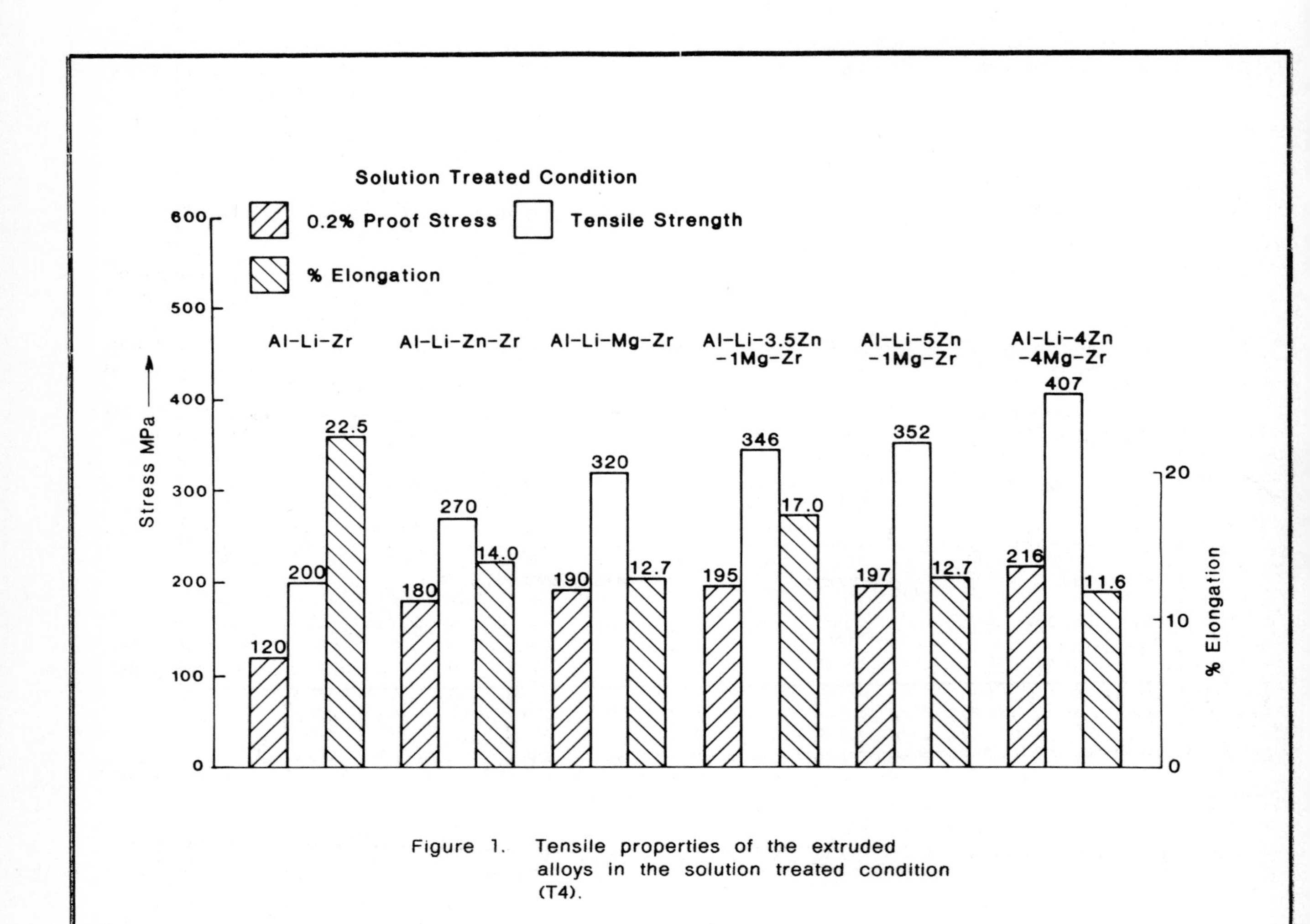

Figure 1. Tensile properties of the extruded alloys in the solution treated condition (T4).

The cast billet (≈140mm diameter) was subjected to a 24h soak at 460°C followed by a heat up to 530°C at 10°C/h and held there for 20h. The temperature was reduced to 500°C and the billet forged down to 55mm thickness. Prior to rolling the forged material was preheated for 2h at 510°C. A hot cross-roll to a thickness of 25mm was followed by a straight roll to 4mm. After solution treatment at 530°C for 15 minutes and water quenching the material was cold rolled to 1.6mm. A further solution treatment at 530°C (20 minutes) and cold water quenching preceded ageing at 170°C.

Flat tensile test samples were machined from both extruded and rolled sheet materials to a cross-sectional area of 18mm^2 and a gauge length of 25mm. Strain measurements were made during tensile testing by the prior application of strain gauges bonded on both sides of the sample. The tests were carried out in a Mayes screw driven tensile machine. Elastic modulus values were found to be in the range 78-82 GPa.

Density measurements were made on each alloy in the aged condition by weighing suitable coupons in air and in water. Sections were cut from each alloy to examine grain size and substructure in the optical microscope. All alloys had a grain size of approximately 100 x 500μm and the grains contained a well developed substructure some 2μm diameter. Transmission electron microscopy samples were prepared using standard electropolishing techniques.

RESULTS

Tensile properties of extruded alloys in the solution treated condition

The properties given in Fig. 1 are those determined after solution treatment (alloy compositions are given in Table I) and holding prior to test at ambient for periods up to 50h. The basic Al-Li-Zr alloy gives proof stress and tensile strength of 120 MPa and 200 MPa respectively. Separate additions of magnesium (2%) and zinc (2%) both promote proof stress increases of ≈65 MPa. A 2% magnesium addition increases the tensile strength by 110 MPa, a value 50 MPa in excess of that developed in the zinc-containing alloy. Elongation was reduced by ≈11% as a result of both the zinc and magnesium additions.

Combined zinc and magnesium additions to the alloy increased the proof stress above that of the Al-Li-Zr alloy by ≈75 MPa (Al-Li-5%Zn-1%Mg and Al-Li-3.5%Zn-1%Mg) and 96 MPa (Al-Li-4%Zn-4%Mg). Tensile strength levels were increased by ≈150 MPa (Al-Li-5%Zn-1%Mg and Al-Li-3.5%Zn-1%Mg) and by ≈200 MPa (Al-Li-4%Zn-4%Mg). Elongation values remained close to those achieved with the single additions. For comparison purposes lithium free alloys were also tested and these gave proof stress values which were 46 MPa (Al-5%Zn-1%Mg) and 81 MPa (Al-4%Zn-4%Mg) above the equivalent values given in lithium-containing versions (see Table II). Tensile strengths were also raised by 27 MPa (Al-5%Zn-1%Mg) and 69 MPa (Al-4%Zn-4%Mg).

Tensile properties of extruded alloys in the aged (T6) condition

Fig. 2 gives the values obtained after ageing the lithium-containing alloys for periods up to 48h at 170°C. Proof stress values in excess of 400 MPa were achieved in alloys containing Al-Li-4%Zn-4%Mg. Tensile strengths in excess of 500 MPa were measured on all Al-Li-Zn-Mg alloys whilst elongation had been reduced by the ageing treatment to values in the range 2.8-7.7%. Ageing produced an increment in proof stress of between 183 and 207 MPa in those alloys with combined additions of zinc and magnesium, see fig. 3. These values exceed the increment produced by separate additions, i.e. 126 MPa (2%Zn) and 155 MPa (2%Mg). Tensile strength increments produced by ageing lie between 126 to 201 MPa, this range including both single and combined additions of zinc and magnesium.

With lithium free alloys an ageing temperature of 170°C promoted overageing and thus low tensile properties. A two stage ageing procedure of 16h at 90°C followed by 24h at 150°C produced an increase in proof stress of 35 MPa (Al-5%Zn-1%Mg) and 193 MPa (Al-4%Zn-4%Zr) over their respective solution treated conditions, see Table III. This gave proof stresses up to 490 MPa in the Al-4%Zn-4%Mg alloy. Tensile strength values fell by 50 MPa in the Al-5%Zn-1%Mg alloy and increased by 68 MPa to 544 MPa in the Al-4%Zn-4%Mg alloy as a result of the double ageing treatment. Double ageing was also applied to the lithium-containing alloys but without promoting better properties than those achieved by a single age at 170°C.

Tensile properties of extruded alloys in the stretched and aged condition (T651)

The effect of an intermediate 3% cold stretching operation prior to final ageing for 24h at 150°C was investigated on a limited number of alloys. Fig. 3 demonstrates that in the lithium-containing alloys the cold stretch before ageing leads to the proof stress rising above 400 MPa in the Al-Li-5%Zn-1%Mg alloy and to a value of 522 MPa in the Al-Li-4%Zn-4%Mg alloy Tensile strengths which were already high in the T6 condition in these alloys were thus further increased by ≈15 MPa due to the pre-stretch. The additional treatment lowered the elongation values so that in the case of the Al-Li-4%Zn-4%Mg alloy this property had dropped below 2%.

A T651 treatment generally promoted a reduction in the strengths of the lithium free alloys, see Table II. It should be noted that the T651 treatment did not include double ageing but instead a single age of 24h at 150°C was applied after the cold stretch.

Effect of copper additions on extruded Al-Li-Zn-Mg alloys

In the solution treated condition the tensile properties of all three lithium-containing alloys were improved by the addition of 1%Cu, see fig. 4. It should be noted that the tensile tests were carried out ≈50h after quenching from the solution

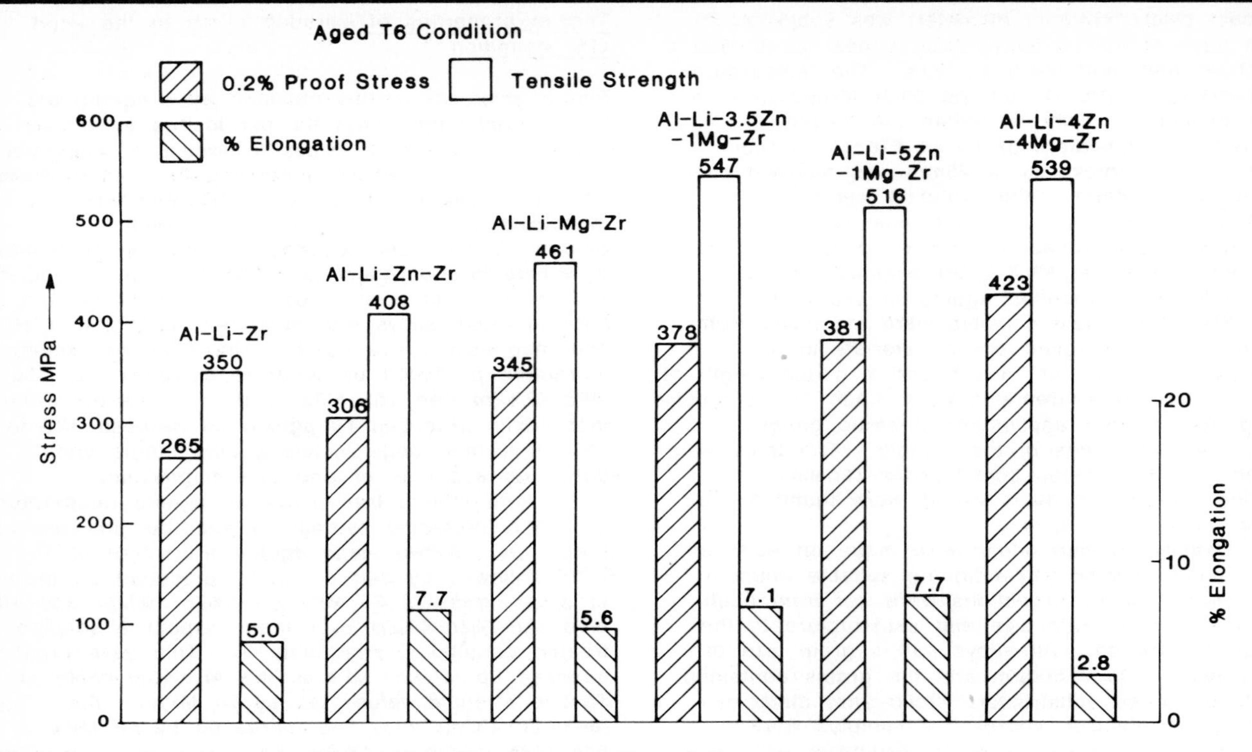

Figure 2. Tensile properties of the extruded alloys in the aged condition (T6, 170°C 48h).

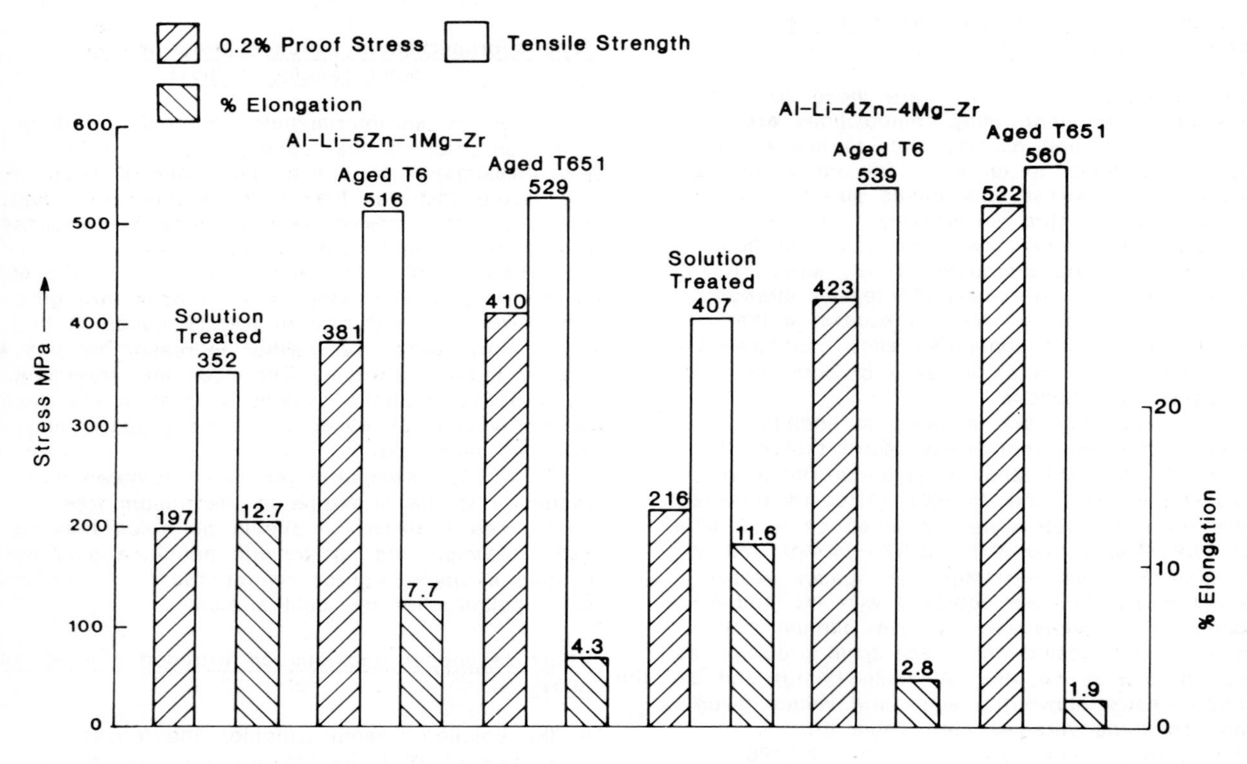

Figure 3. Effect of cold stretching prior to ageing on the tensile properties of the extruded alloys.

treatment temperature. The increments in proof stress were in the range 50 to 85 MPa whilst the tensile strength increased by 18 to 107 MPa; the lower strength increments being found in the Al-Li-4%Zn-4%Mg alloy A reduction in elongation occurred in the case of one alloy (Al-Li-3.5%Zn-1%Mg) whilst the others showed a small improvement in this property as a result of the copper addition.

Lithium free alloys also received an increase in tensile properties as a result of a copper addition in the solution treated condition, see Table II. This amounted to between 31 and 58 MPa in the case of the proof stress and ≈50 Mpa for the tensile strength.

Ageing for various times at 170°C (T6) increased the proof stress of the three copper-containing alloys to values in the region of 440 MPa, see fig. 5. This represented an ageing increment above the "solution treated" condition of between 159 and 175 MPa which was ≈20 MPa below the ageing increments achieved by the copper-free alloys. This smaller increment was always amply compensated for by the improvement of 50-85 MPa achieved in solution treated copper-containing alloys. Hence, the copper-containing alloys in the aged condition had at least a 17 MPa increase in 0.2% proof stress over the copper-free alloy. The proof stresses of the Al-Li-3.5%Zn-1%Mg and the Al-Li-5%Zn-1%Mg alloys were improved by ≈60 MPa as a result of the copper addition. Tensile strength increments from ageing the copper-containing alloys were lower than those found for proof stress, i.e. in the range 72-117 MPa. This was a significant fall from those values achieved in the copper-free alloys. However, overall tensile strengths were increased to values close to 550 MPa in the Al-Li-3.5%Zn-1%Mg and Al-Li-5%Zn-1%Mg alloys as a result of the copper additions. This again came about as a result of the improved tensile properties of the "solution treated" alloys. The Al-Li-4%Zn-4%Mg alloy showed some rather disappointing tensile strength results when copper was present and this was compounded by low elongation to fracture values. Elongation to fracture values on the Al-Li-3.5%Zn-1%Mg and Al-Li-5%Zn-1%Mg alloys were in excess of 5% and were considered to be encouraging at the measured strength levels.

Lithium-free alloys with copper additions after double ageing produced proof stress increments over the "solution treated" condition of between 47 and 65 MPa; this was much lower than that achieved in the copper-free alloy, see Table II. The resultant proof stresses in the aged (T6) copper-containing alloys showed some improvement in the Al-5%Zn-1%Mg alloy (+61 MPa) and a reduction in the case of the Al-4%Zn-4%Mg type alloy (-88 MPa). Elongation to failure values were maintained at levels in excess of 9% in all the aged copper-containing alloys.

The application of an intermediate cold stretch (3% prior to ageing at 150°C) promoted a small improvement in 0.2% proof stress of 40 MPa (Al-Li-5%Zn-1%Mg-Cu) and 58 MPa (Al-Li-4%Zn-4%Mg-Cu) above the values obtained in the T6 condition in the lithium-containing alloys (fig. 6). This was about the same order of increment achieved in the copper-free Al-5%Zn-1%Mg alloy but much less than the 100 MPa increase attained in the Al-Li-4%Zn-4%Mg alloy as a result of the 3% cold stretch. Tensile strength values remained constant in the Al-Li-5%Zn-1%Mg type alloy when the copper addition was made, but ≈50 MPa increase in this property (above the T6 condition) occurred when the prior cold stretch was applied to the Al-Li-4%Zn-4%Mg-Cu alloy.

Effect of ageing time on the T6 properties of Al-Li-Zn-Mg-Cu alloys in sheet form

With increasing times of ageing at 170°C the 0.2% proof stress properties of Al-Li-2%Zn-2%Mg-Cu-Zr increased to a maximum at 25-32h, see fig. 7. At this stage proof stress values of 415 MPa were measured on the alloys. Tensile strength followed a similar pattern to give a maximum value of 540 MPa. Elongation to fracture fell in the early stages of ageing until at the 16h stage a minimum value of 2.2% was obtained; ageing for longer periods promoted a restoration of ductility, until at 32h, where maximum strength levels were obtained, it had increased back to 4.5%.

For comparison, fig. 7 also shows strength and elongation values obtained on Lital A (8090) sheet (2.49%Li, 1.24%Cu, 0.70%Mg, 0.12%Zr, 0.09%Fe, 0.07%Si) as a function of ageing time at 170°C. It is noteworthy that the zinc-containing alloy ages more quickly at this temperature and gives strength values in excess of those measured on Lital A. Also during the early stages of ageing both alloys showed the presence of a ductility trough after ≈16-25h ageing. Compared to Lital A the zinc-containing alloy is able to maintain higher strength levels in association with slightly higher elongation to fracture values, after ageing at 170°C.

Electron microscopy

Thin films prepared from extruded Al-Li-3.5%Zn-1%Mg-Zr and Al-Li-3.5%Zn-1%Mg-Cu-Zr alloys (T6) aged for various times at 170°C have been examined under all ageing conditions. Both alloys had a subgrain structure due to the presence of zirconium in the material, and both alloys had a fine dispersion of δ'(Al_3Li) within the subgrains (fig. 8a).

In the copper-free alloy additional zinc-containing phases were present, both on the subgrain boundaries (fig. 8b) and as a dispersion within the grains (fig. 8c).

In the copper-containing alloy the subgrain structure, δ' precipitation and the zinc-containing phases were again present, but additionally precipitation was observed to occur on dislocations (fig. 8d). This precipitate has been identified as the S(Al_2CuMg) phase using electron diffraction.

DISCUSSION

Strengthening in the copper free alloys

In the solution treated condition, Fig. 1 demonstrates that single additions of zinc and magnesium to lithium-containing alloys are capable of increasing the proof stress by 35 MPa/wt.% of addition. Combined additions of zinc and magnesium do not achieve the predicted level of strengthening

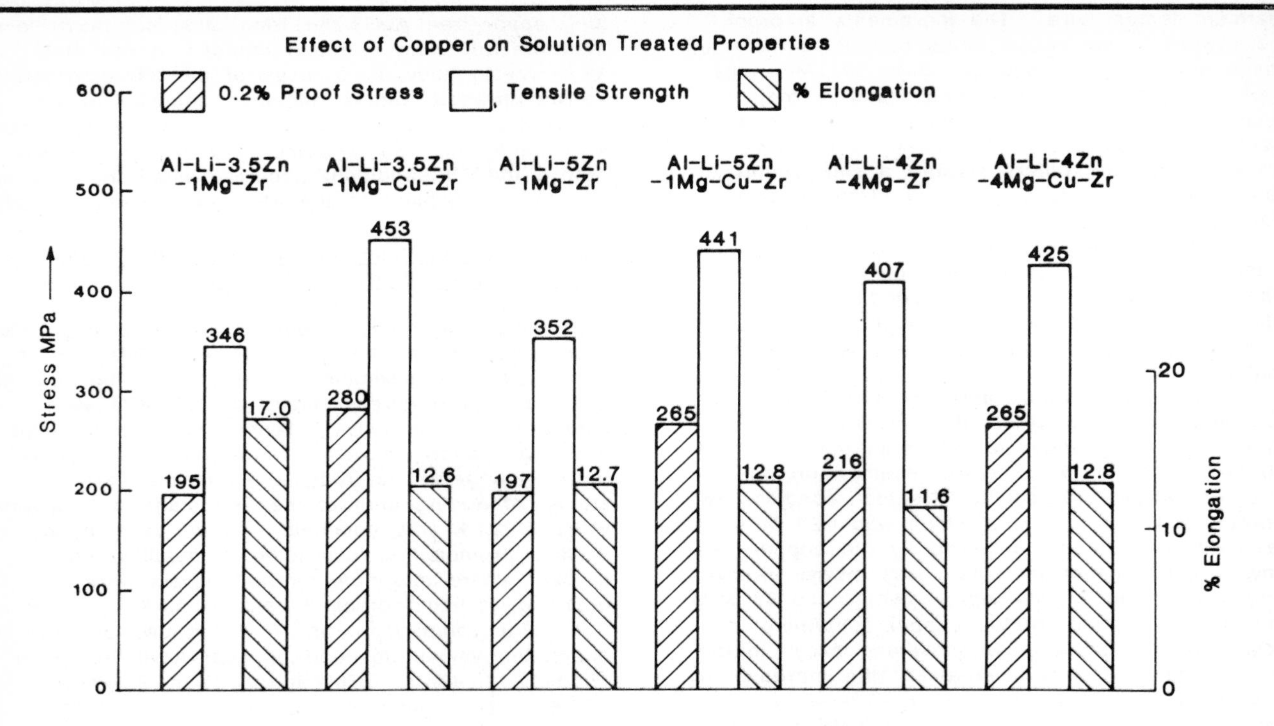

Figure 4. Tensile properties of copper-containing extruded alloys in the solution treated condition (T4).

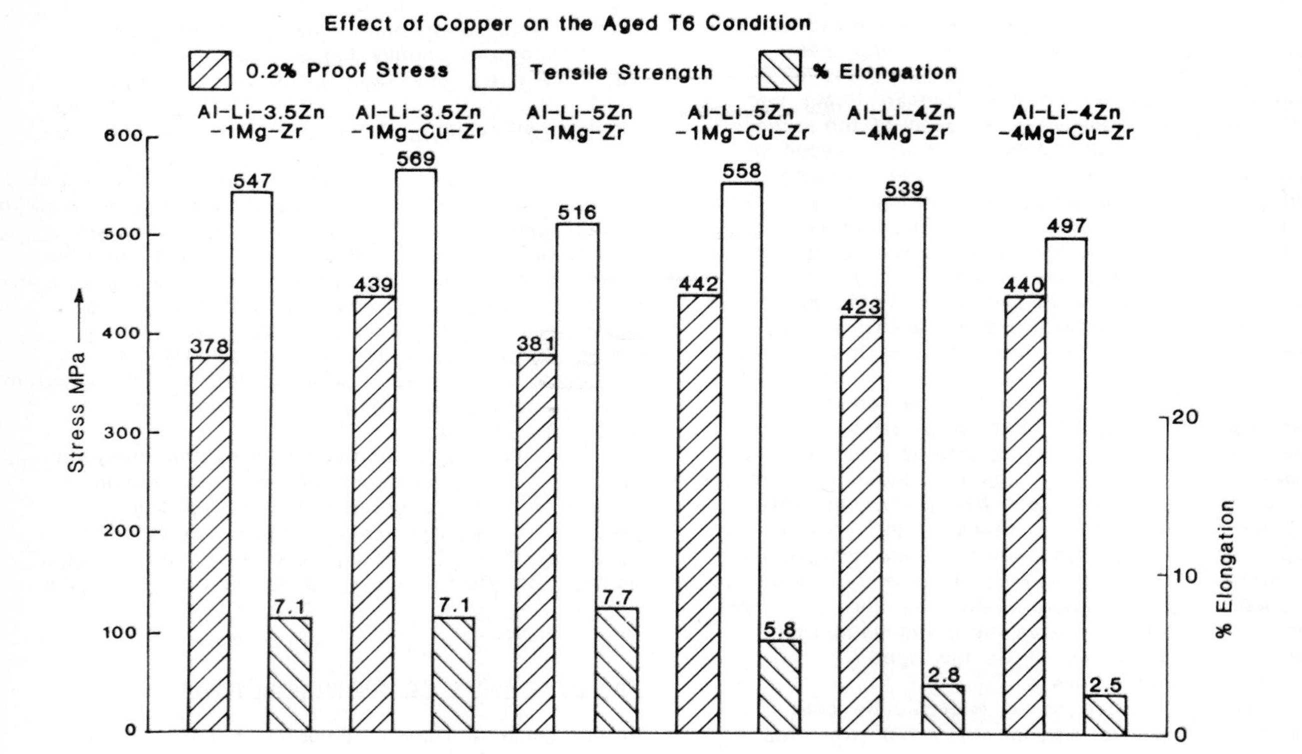

Figure 5. Tensile properties of copper-containing extruded alloys in the aged condition (T6, 170°C 48h).

e.g. the 3.5%Zn, 1%Mg containing alloy should experience an increase of proof stress of ≈150 MPa, but only one half of this increment is achieved. The reduced strengthening probably results from the interaction of zinc and magnesium in solid solution. Without lithium being present higher proof strengths are achieved, e.g. Al-5%Zn-1%Mg (243 MPa) may be compared with Al-Li-5%Zn-1%Mg (197 MPa). GP zones are produced in the lithium-free alloy which harden the alloy during natural ageing over a 50h period after solution treatment and quenching. The presence of lithium prevents or delays GP zone formation in these alloys. Tensile strengths are generally increased to a greater extent by single and combined additions of zinc and magnesium. Magnesium appears to be more potent than zinc in this respect, e.g. in the Al-Li-2%Mg and Al-Li-4%Mg-4%Zn alloys. The increases in tensile strength and the halving of associated elongation to fracture values results at least in part from the increased work hardening rate which magnesium in particular is capable of producing. It is to be noted that tensile strengths of the lithium free alloys are higher still in the solution treated condition, see Table II, than the lithium-containing versions and here zinc-magnesium interactions and zone formation may be playing a part.

An increase in proof stress of ≈200 MPa is achieved by ageing Al-Li-Zn-Mg alloys for 48h at 170°C. This increment is greater than that achieved in the alloys with single additions of magnesium and zinc by at least 30 MPa, which suggests that some kind of interaction is taking place between the zinc and magnesium, which either increases the volume fraction of δ', allows solution of elements in δ', or produces an additional precipitating phase. Electron microscopy evidence obtained to date shows that zinc-rich precipitates are formed on sub-boundaries and within the grains. Further work is being done to assess the importance of these zinc-rich phases on the mechanical properties of the alloys.

After ageing the increases in tensile strength are generally in keeping with those found for the proof stress. Elongation to fracture is halved in the majority of the Al-Li-Zn-Mg alloys as a result of the ageing treatment (see fig. 2); this is much less drastic than the reduction from 22.5% to 5% which occurs in the basic Al-Li-Zr alloy.

Ageing patterns in the lithium-free alloys showed some differences, the Al-4%Zn-4%Mg alloy was much more responsive than the Al-5%Zn-1%Mg. Also lower ageing temperatures and double ageing (90 and 150°C) was necessary to achieve the proof stress levels displayed in Table II. Tensile strength values were not changed to the same degree by the ageing treatments in these alloys as those shown for the lithium containing materials. Elongation values were maintained at higher levels in the aged lithium free alloys. Strengthening in these alloys is achieved by $\eta(MgZn_2)$ and it is clear that these precipitates do not reduce the ductility of the alloy as much as δ' in the lithium-containing alloys.

The introduction of a 3% cold stretch prior to ageing at 170°C promotes significant increases in proof stress, particularly in the Al-Li-4%Zn-4%Mg alloy where a proof stress of 522 MPa was achieved. Tensile strengths do not increase by more than 15 MPa as a result of the stretch whilst elongation values all fall to lower levels, see fig. 3.

Strengthening in copper-containing alloys

Additions of 1%Cu to the Al-Li-Mg-Zn alloys promote increases in the proof stress and tensile strength above those of the copper-free alloys in both the solution treated and aged (T6) conditions. The increases in the 'solution treated' condition have been examined further by carrying out hardness tests during natural ageing. It is clear that during a period of 50h immediately after the quench from the solution treatment temperature, the hardness increases by ≈20HV. Also similar levels of hardness increases were noted in the copper-containing lithium-free alloys after solution treating plus 50h natural ageing. It would appear that the presence of copper is capable of assisting with the natural ageing processes even in the presence of lithium.

Ageing at 170°C produces high strength levels in these Al-Li-Zn-Mg-Cu alloys. Preliminary electron microscope examination of the copper-containing alloys in the T6 condition reveals the presence of $S(Al_2CuMg)$ precipitation on dislocations, in addition to zinc-rich precipitates and fine δ'.

Tensile strengths are improved to a smaller extent as a result of copper additions, but the more important result in the Al-Li-5%Zn-1%Mg-Cu and Al-Li-3.5%Zn-1%Mg-Cu alloys is the maintenance of a high elongation to fracture value (6-8%) at high proof stress and tensile strength levels. Fig. 6 compares these properties (T6 condition) with those obtained on Lital A and Lital B in the stretched and aged condition (T651) and it is clear that in the extruded form there is no need for an intermediate stretch treatment to be given to the Al-Li-Zn-Mg-Cu alloy and improvements in ductility and toughness may be sought at high strength levels in these alloys. It is worth noting that the Al-Li-4%Zn-4%Mg-Cu alloy does not achieve as well balanced properties in the T6 condition as the other two copper containing alloys shown in fig. 5. The lower tensile strength and elongation probably result from problems which arise in connection with the solution treatment and further work is necessary in this area.

Applying an intermediate stretch to the copper-containing alloys (T651 treatment), see fig. 6, brings about a 40-50 MPa improvement in proof stress and brings the strength properties comparable with those of Lital B. However, the ductility is reduced to values below 2.5% as a result of this T651 treatment.

Ageing of sheet formed materials

Work on Lital A and Lital B has demonstrated the differences which exist between the properties of different product forms of lithium-containing alloys, e.g. sheet, plate, extrusions etc. Similar differences have been found in the Al-Li-Zn-Mg-Cu alloys studied in the present work, e.g. lower

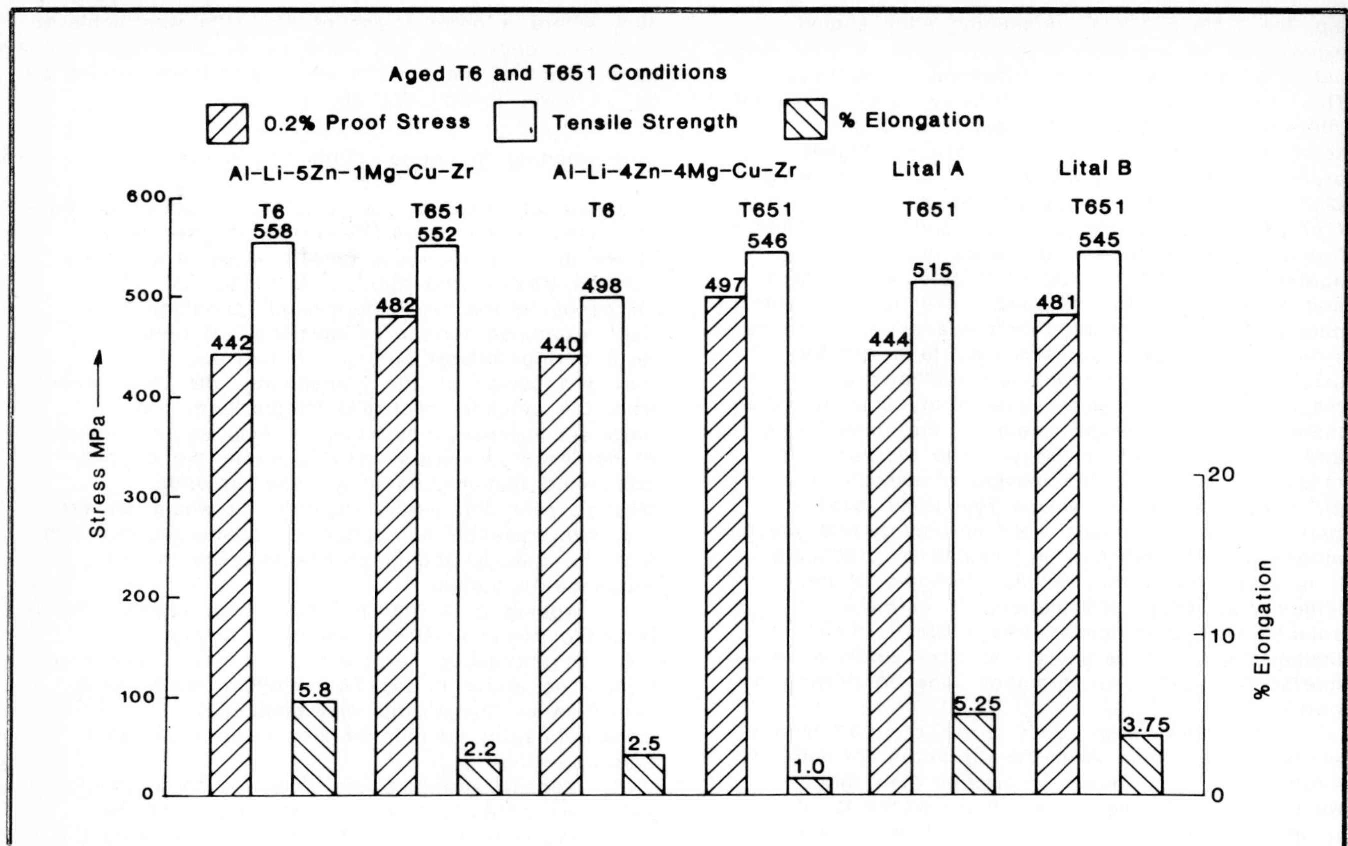

Figure 6. Effect of cold stretching prior to ageing on the tensile properties of copper-containing extruded alloys. Note that the values for the Lital alloys are target properties.

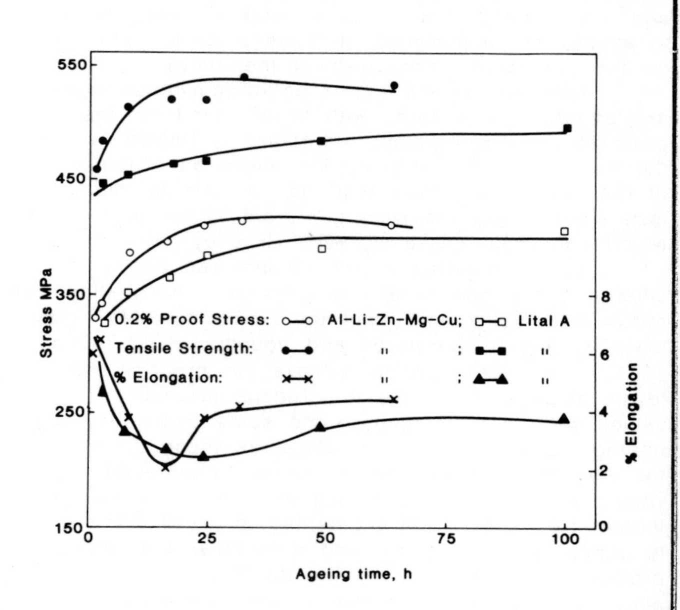

Figure 7. Effect of ageing time at 170°C on the tensile properties of sheet material.

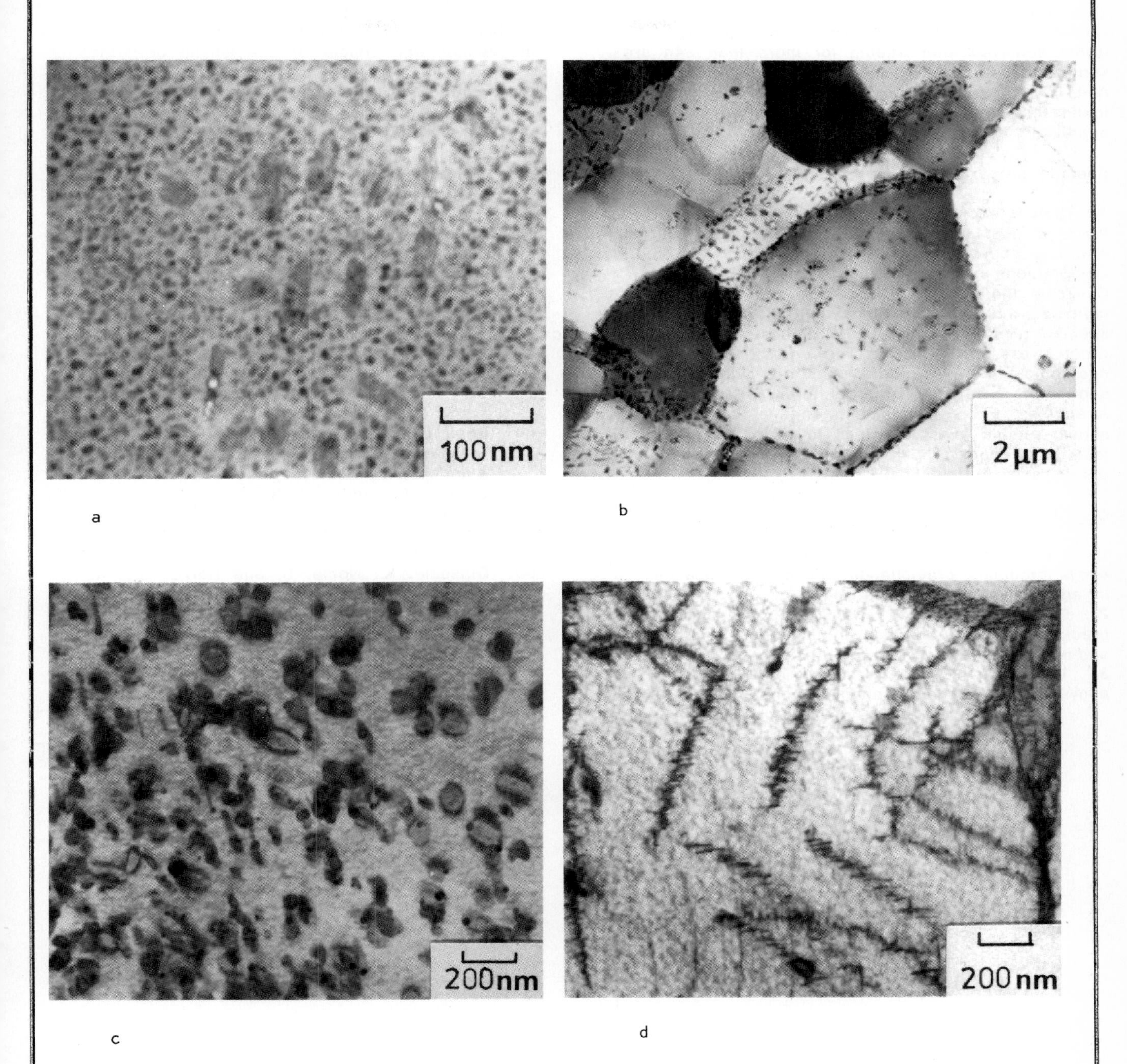

Figure 8. Transmission electron micrographs of the Al-Li-3.5%Zn-1%Mg-Zr alloy.
(a) 0%Cu, δ' precipitation.
(b) 0%Cu, precipitation of zinc-containing phase on sub boundaries.
(c) 0%Cu, precipitation of zinc-containing phase within the grains.
(d) 1%Cu, precipitation of S(Al_2CuMg) on dislocation lines.

strengths are produced in sheet material after ageing (T6) at 170°C.

Typical ageing curves for sheet material are shown in fig. 7 where the alloys are compared with results from Lital A (T6). The Al-Li-Zn-Mg-Cu alloy ages more quickly and to higher strength values than Lital A. Furthermore the ductility levels achieved after ageing for more than 24h are somewhat better than Lital A. If the results from Al-Li-Zn-Mg-Cu are compared with Lital B then it is found that similar proof and tensile strengths are achieved.

Specific properties

As Table I shows, the density of the majority of the zinc-containing alloys are close to ≈2.6 Mg/m^3, which is above the value for Lital A (2.54 Mg/m^3). Comparisons between extruded Al-Li-3.5%Zn-1%Mg-Cu-Zr in the aged (T6) condition and Lital A (T651) indicates that the zinc-containing alloy has the greater specific tensile strength (3.9% higher) and a lower specific proof stress (3.4% less). The zinc-containing sheet alloy has almost the same density as the Lital A and therefore shows improved specific proof stresses and tensile strengths. Futher refinements in alloy compositions should enable lower densities to be achieved without loss of strength properties and work is being carried out to achieve this goal[5].

CONCLUSIONS

1. The addition of zinc and magnesium to Al-Li alloys produces precipitation of zinc-rich phases in addition to the usual $\delta'(Al_3Li)$.
2. Copper additions to Al-Li-Zn-Mg produces precipitates of $S(Al_2CuMg)$ on dislocations in addition to δ' and zinc-rich precipitation.
3. The mechanical properties of Al-Li-Zn-Mg-Cu alloys in the T6 aged condition compare favourably with the target properties of Lital A and B in the T651 condition. It therefore appears possible to achieve high strength and good ductility in Al-Li-Zn-Mg-Cu alloys without a cold working operation before ageing.
4. The present paper has demonstrated the beneficial effects of combined additions of zinc, magnesium and copper on the tensile properties of Al-Li alloys. Further work is now required on other properties such as fracture toughness, fatigue and corrosion behaviour. Work is also required on the effect of zinc additions on the ease of alloy preparation and mechanical working.

ACKNOWLEDGEMENTS

The authors express their appreciation to the Procurement Executive, Ministry of Defence, for financial support of the work.

REFERENCES

(1) Quist, W.E., Narayanan, G.H. and Wingert, A.L., Aluminium-Lithium Alloys II; Sanders, T.H. and Starke, E.A. (eds), AIME, New York, 1983, 313.

(2) Peel, C.J., Evans, B., Baker, C., Bennett, D.A., Gregson, P.J. and Flower, H.M., Aluminium-Lithium Alloys II; Sanders, T.H. and Starke, E.A. (eds), AIME, New York, 1983, 363.

(3) Grimes, R., Cornish, A.J., Miller, W.S. and Reynolds, M.A., Metals & Materials, 1 (6), 1985, 357.

(4) Dinsdale, K., Noble, B. and Harris, S.J., to be published.

(5) Peel, C.J., Evans, B., Noble, B., Dinsdale, K., and Harris, S.J., U.K. Patent Application No. 8326260 (April 1984).

Concluding summary

C J PEEL

In view of the broad scope of the Conference, my concluding remarks are grouped under three main headings, subdivided as follows:

1. BASIC STUDIES	a. Metallurgy b. Microstructure
2. PROPERTIES	a. Fatigue b. Corrosion/Oxidation c. Elevated Temperature
3. PRODUCTION	a. Industrial Aspects b. Products, Processes, Joining c. Hot Forming d. RSR, Powder and Mechanical Alloying Routes

The above topics are considered to be worthy of further attention since problem areas have been highlighted at the conference. Each subheading is dealt with on the basis of positive or negative aspects in the following tables. Positive in this context means that a good understanding has already been attained. Negative describes an area where more effort or investment is required.

1. BASIC STUDIES

a. Metallurgy

Positive Aspects

i Wide range of microstructural studies
ii New techniques available; some have already been successfully applied, e.g. Auger, EELS, SAS, SIMS, LIMA, CBD, NMP
iii New phases have been identified
iv Increasing understanding of grain structure, recrystallisation and texture

Negative Aspects

i Analysis for important trace elements, e.g. Na, K at the levels found in alloys is difficult
ii Knowledge of phase diagrams, in particular Al–Li–Cu–Mg, is inadequate

b. Microstructure

Positive Aspects

i The requirement for duplex precipitation, e.g. δ' + S is widely accepted
ii The interaction of planar slip, texture and precipitation is better understood

Negative Aspects

i The suppression of second phase precipitation by lithium has an adverse effect on T6 properties but can be an advantage, i.e. slower natural ageing
ii The comparative performance of recrystallised and unrecrystallised requires more understanding
iii Complex interactions between Li, Zr and TiB_2 have been found with effects on both grain structure and properties. More research is required

2. PROPERTIES

a. Fatigue

Positive Aspects

i Fatigue performance appears to be better than current alloys
ii The reasons for the better properties seem to be understood
iii The effect of heat treatment, particularly ageing time/temperature, has been studied

Negative Aspects

i The choice of test parameters needs to be widened to include complex loading, different environments and multi-axial loading
ii The quality of metal surfaces may be important but more effort is required

b. Corrosion/Oxidation

Positive Aspects

i Solute depletion effects (Li, Mg loss during heat treatment) are better understood
ii Improved corrosion performance may be obtained by modified heat treatments
iii Metal quality may be important, particularly in the case of stress corrosion
iv Evidence of the beginnings of an understanding of corrosion and stress corrosion performance

Negative Aspects

i Heat treatments will need to be controlled to minimise lithium loss
ii Corrosion resistance may prove to be the critical factor governing the application of aluminium–lithium alloys
iii The stress corrosion performance of ingot-route alloys needs to be improved

c. Elevated Temperature

Positive Aspects

i Elevated temperature tensile properties look promising
ii High performance light alloys appear possible, perhaps exploiting powder or mechanical alloying routes

Negative Aspects

i More data is needed, particularly on creep resistance
ii Good high temperature performance may be incompatible with superplasticity
iii Effect of elevated and low temperature exposure on grain boundary embrittlement to be resolved

3. PRODUCTION

a. Industrial

Positive Aspects

i Intense interest by all major suppliers
ii Large scale production is starting
iii Three alloy composition ranges have been registered: 2090, 8090, 8091
iv Property targets defined by different users and producers are roughly similar

Negative Aspects

i Cost of the finished product may become a critical factor determining whether alloys are used. The following will make a major contribution:

cost of lithium
scrap recovery
ease of manufacture
utilisation (buy:fly ratio)
requirement for clad material
ii Competition from alternative materials; CFRP may be more cost effective
iii Number of alloys needs to be reduced

b. Products, Processes and Joining

Positive Aspects

i Salt bath heat treatment of Al–Li alloys does not introduce any new risks
ii Cold coining of forgings leads to improved properties
iii Welding may be possible for selected alloys
iv Adhesive bonding would not seem to be a problem

Negative Aspects

i Special alloys may be required for castings and forgings but these have yet to be developed
ii The quality of anodised surfaces needs to be investigated

c. Hot Forming

Positive Aspects

i Most Al–Li alloys exhibit superplastic properties
ii Back pressure can be used to suppress cavitation
iii Planar boundaries can be disrupted during deformation

Negative Aspect

i Property data for post-formed material needs to be generated

d. RSR, Powder and Mechanical Alloying

Positive Aspects

i Potential for greater weight savings in the long term
ii Mechanically alloyed material looks attractive especially for isotropic forgings

Negative Aspects

i Lithium additions might cause additional problems in powder manufacture
ii Currently there is a lack of samples large enough for critical testing

APPENDIX

Analysis of powder metallurgy and related techniques for the production of aluminum–lithium alloys

W E QUIST, G H NARAYANAN, A L WINGERT and T M F RONALD

WEQ, GHN and ALW are with the Boeing Materials Technology Group of the Boeing Commercial Airplane Company, Seattle, Washington, USA. TMFR is the focal point for Structural Metals Development at the Air Force Materials Laboratory, Dayton, Ohio, USA

SYNOPSIS

The unique characteristics of three diverse powder making techniques are compared as they apply to the production of aluminum-lithium (Al-Li) powder and products. These techniques were considered to be state-of-the-art examples of the major methods now used for aluminum powder production and included the Pratt and Whitney atomization technique, the Novamet mechanically alloying technique, and the Allied Corporation planer flow casting technique (melt spinning). Comparisons are made with respect to powder characteristics, contamination problems, and microstructural, mechanical and fracture properties of the powder products.

INTRODUCTION

Powder metallurgical (P/M) approaches to the development of aluminum-lithium (Al-Li) alloys have been extensively studied for nearly a decade.(1-7) However, a review of all production methods and their current state of development indicates that ingot metallurgical (I/M) approaches that have been concurrently developed are at present better suited to supply material in all the needed sizes and forms than are the P/M approaches. Therefore, the question arises as to why time and money are being spent on the continued development of P/M techniques for aluminum-lithium alloys. The answer lies in the fact that P/M techniques offer greater compositional and microstructural flexibility with regard to alloy design strategies and therefore, may have an important future place in the aerospace industry which demands new materials with increasingly stringent property requirements. Even if P/M eventually captures only a small part of the Al-Li market, it could nevertheless represent an important segment. It is interesting to take note that during the first International Conference on Aluminum-Lithium alloys at Stone Mountain, Georgia in 1980, most of the attention was focused on Al-Li alloy development via the P/M approach.(7) Obviously, the situation has changed greatly since then.(9,10)

The second question to be answered is: what is the potential gain from the use of P/M techniques for aluminum-lithium alloy products? The answer to this query actually has two parts; the first deals with microstructural aspects and the second with engineering properties. With regard to microstructure, several of the improvements that can be achieved through the use of P/M approaches are shown in Table 1. They include the ability to incorporate lithium contents in excess of the currently recognized practical maximum of 2.7 wt.% for I/M alloys. In addition, the P/M approaches allow the utilization of several unique strengthening mechanisms, such as dispersion and substructure hardening and Petch hardening in some cases. Also, increased amounts or different combinations of alloying elements can be utilized and explored. Finally, reduction in the size, and amount of large undesirable second-phase particles, and a reduction in the amount and extent of alloying element segregation can be expected.(4,6,7)

As a consequence of the microstructural improvements of the type described above, P/M alloys can be expected to exhibit certain engineering property advantages over their I/M ounterparts. In the case of P/M Al-Li alloys, these property improvements may include one or more of the following: (1) a greater density reduction over that achieved by I/M methods, (2) improved corrosion resistance, a property that is inherently good in most P/M materials, (3) reduced anisotropy and (4) improvements in the strength and fracture toughness tradeoffs.

The P/M techniques that have been employed to date to produce aluminum-lithium alloys can be divided into three general categories as shown in Table 2. The first covers the various types of atomization techniques that may utilize air, flue gas, and inert gas environments. The second category is mechanical alloying which is a solid state alloying technique. The final category includes various types of chill methods, including melt spinning, melt extraction and splat cooling.

When P/M techniques are employed for Al-Li alloys it is necessary to follow a number of general guidelines

Table I. Rationale for use of P/M approaches to Al-Li alloy

I. Compositional/microstructural considerations

- Incorporate lithium in excess of the upper limit for I/M alloys (>2.7 weight %)
- Incorporate unique strengthening approaches
- Incorporate unique amounts and combinations of alloying elements (i.e., Li, Cr, Mn, Ge, Cu, Mg, and Zr)
- Reduce size and amount of second phase intermetallic particles
- Reduce segregation of alloying elements

II. Engineering property improvements

- Reduce density beyond limits of I/M technology
- Improve corrosion resistance
- Reduce anisotropy of microstructure and properties
- Improve strength/fracture toughness tradeoffs

Table II. P/M approaches available for Al-Li alloy production

I. Recent approaches

Method	Cover gas	Particle morphology	Cooling rate (approx.) (°C/sec)
Atomization ● Air-atomized ● Flue gas-atomized ● Inert gas-atomized	 Air Flue gas Ar, He	 Irregular Irregular - spherical Spherical	 10^3-10^4 10^3-10^4 10^3-10^5
Mechanically alloyed	Air, Ar, He	Irregular	N/A- Powder produced at room temperature
Chill methods ● Melt - spinning ● Melt - extraction ● Splat cooling	 Air Air Air, Ar, He	 Ribbon Wire Flakes	 10^6 10^5-10^6 10^5-10^6

II. General requirements for above approaches

1. Protective requirements for above approaches
   - Melting
   - Powder production
   - Handling
2. Cooling rates of 10^5-10^6 °C/sec (for RST methods)
   - Retain high solute supersaturation
   - Refine microstructures
3. Mechanical working to break up surface oxides
4. Low processing temperatures to preserve benefits of RST

Table III. Best engineering properties achieved to date for P/M Al-Li alloys evaluated in present program

Method	Nominal composition, weight %	Longitudinal tensile properties				Notch-yield ratio	Density g/cm³ (lb/in³)
		UTS, MP (ksi)	TYS, MPa (ksi)	Elongation, %	Modulus (E) GPa (msi)		
● Rotary atomization (P&WA)	Al-3.7Li-2.4Cu-1.6Mg-0.2Zr	473 (68.6)	386 (56.0)	3.5	82.7 (12.0)	0.6	2.41 (0.087)
● Mechanically alloyed (NOVAMET)	Al-1.75Li-4Mg-0.6O-0.7C	532 (77.1)	504 (73.1)	9.5	79.3 (11.5)	1.25	2.55 (0.0922)
● Pulverized melt-spun ribbon (ALLIED)	Al-3.2Li-2.1Cu 1.0Mg-0.45Zr	525 (76.1)	479 (69.5)	6.0	79.3 (11.5)	—	2.46 (0.089)

Table IV. Summary evaluation of three P/M approaches used for Al-Li alloy production in present program

	ROTARY ATOMIZATION	MECHANICALLY ALLOYED	MELT-SPUN RIBBON
CONCERNS	1. Inadequate cooling rates 2. Contamination • Heavy metals • Other powders • Organic compounds 3. Microstructural instability	1. Partially understood strengthening mechanisms 2. Iron contamination 3. Microstructural inhomogeneities 4. High as-received strength	1. Microstructural instability during billet scale-up 2. Contamination • Heavy metals (Fe, Si, Zr) • Organic compounds 3. Strength/toughness relationships
ASSESSMENT	1. Contamination problems • Probably solvable 2. Scale-up to commercial sizes is feasible 3. Cooling rates *not* adequate for alloys explored	1. Scale-up to commercial requirements is practical 2. Property benefits over current I/M alloys 3. All concerns appear solvable	1. Contamination problems from melt appear solvable 2. Contamination problems from processing are probably solvable 3. Major hurdle: Scale-up without microstructural deterioration
PAYOFF	1. Require improvements in cooling rate/microstructure	1. Improved corrosion resistance 2. Improved transverse properties 3. Improved density reduction	1. Substantial density reductions improvement 2. Improved corrosion resistance? 3. Improved transverse properties

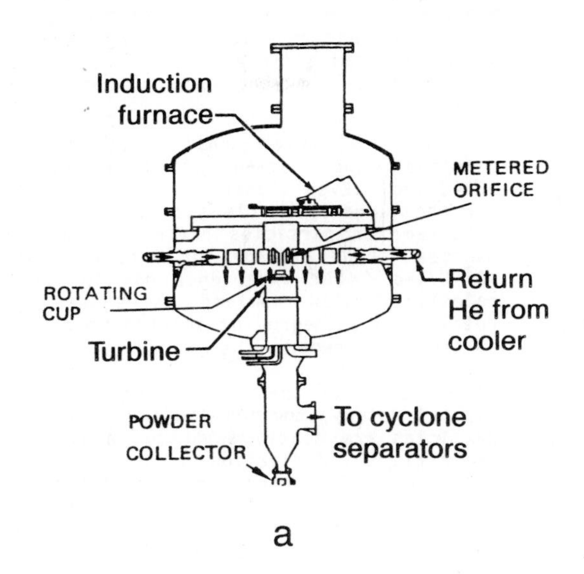

a

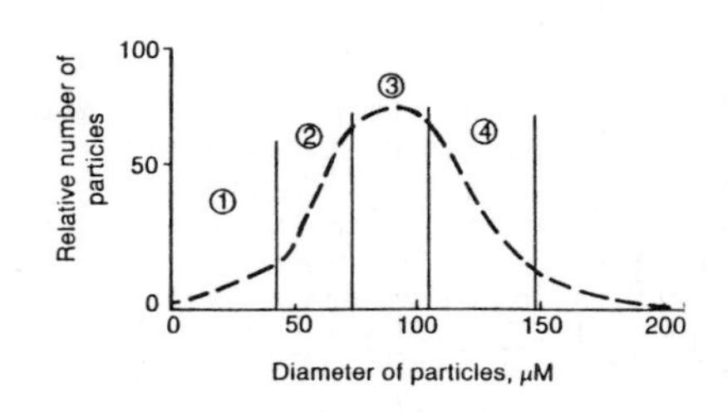

Alloy Composition
(Al—3.4Li—10 Mg—0.05Zr)

Zone	Particle size, μM	Cooling rate, °C/sec
1	-44	$9x10^4$
2	44-74	$3x10^4$
3	74-105	$1.5x10^4$
4	105-149	$8x10^3$

b

1. (a) Schematic diagram of Pratt & Whitney Aircraft's rotary atomization process (b) Typical particle size distribution and cooling rates obtained for P & WA Al-Li alloy powders

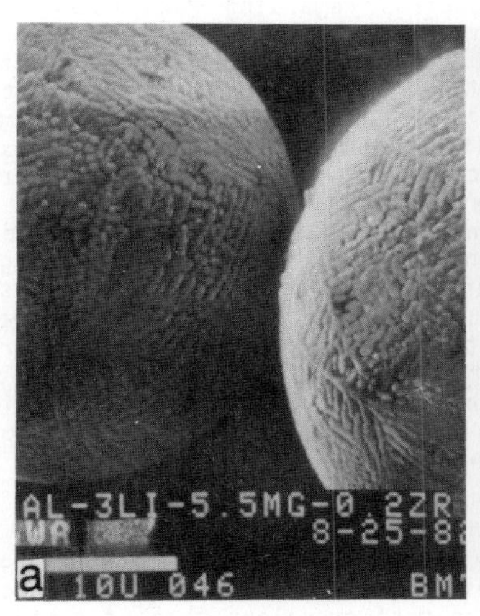

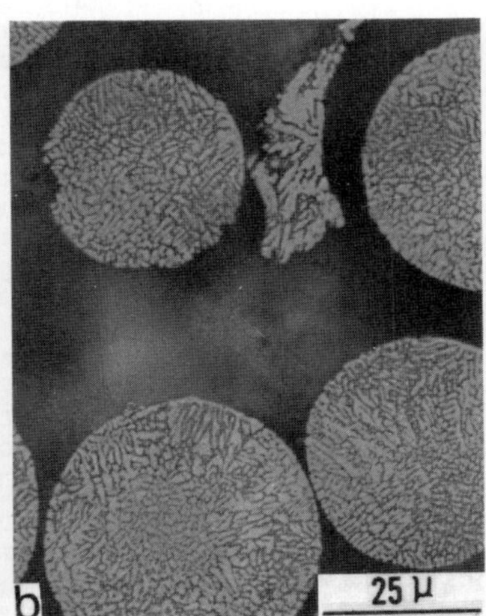

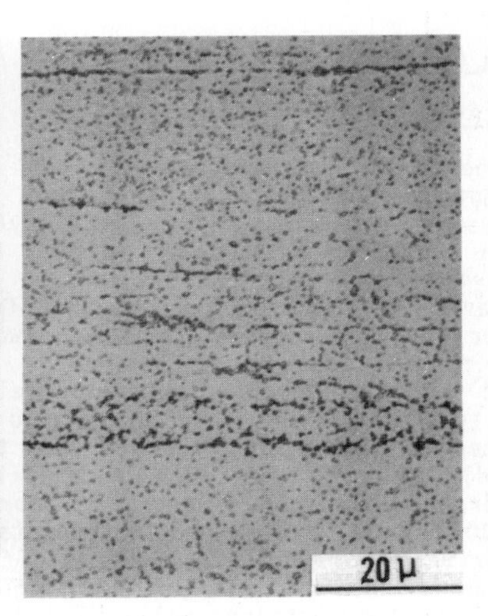

2. Some general characteristics of rotary atomized Al-Li Alloys (Al-3Li-5.5Mg-0.15Zr)

(a) powder morphology - as received
(b) typical powder microstructure - as received
(c) typical product microstructure - extrusion product; soln-treat, 547°C/30 min

to help achieve optimum properties. See Table 2. These constraints pertain to both powder production and to the subsequent consolidation and fabrication. For example, environmental protection is often required during melting, powder production and handling, in order to prevent contamination by oxygen, water vapor and other possible reactants. For processes involving rapid solidification, cooling rates of the order of 10^5 to 10^6°C/sec are thought to be necessary to produce the desired degree of refinement in the microstructure and to minimize segregation of alloying elements. With respect to processing, all Al-Li powder particles have a tenacious oxide film at the surface which requires that a critical amount of mechanical working be performed in order to break them up and insure a high degree of interparticle bonding. Another important constraint that applies to the consolidation of rapidly solidified powders is that the times and temperatures employed for vacuum degassing and hot pressing of the powders into billets be such that the primary benefit of rapid solidification is not lost - namely, the destruction of a refined microstructure due to excessive coarsening of second phase particles. Therefore, it is advisable that the lowest possible processing and fabrication temperatures be utilized consistent with equipment and commercial realities.

EXPERIMENTAL

Three unique powder production methods, each representing one of the three generic P/M approaches discussed previously and showing the best possible chance for success for the production of Al-Li alloys were selected for evaluation in this study. The three state-of-the-art techniques chosen were: (1) Pratt & Whitney Aircraft's (P&WA) inert gas rotary atomization method (11,12), (2) Novamet's (Inco's) mechanical alloying technique (13,14), and (3) Allied's planar flow casting technique (a form of melt-spinning)(15).

RESULTS AND DISCUSSION

Rotary Atomization

The general configuration of the P&WA powder-making equipment is shown in schematic form in Figure 1. The system consists of a large cylindrical two chamber container with alloy melting being carried out in the upper part and atomization in the lower, both having a flowing helium gas environment. The diameter of the rigs used for Al-Li powder production ranged from 7 feet up to 15 feet. The alloy of desired composition is induction melted under a helium gas cover and poured into a tundish and delivered through a small orifice onto the surface of a rapidly spinning shallow cup approximately 3" in diameter. Cup rotation speeds are normally between 20 to 30,000 rpm. Molten metal droplets are spun off the perimeter of the cup and solidification takes place while the droplets are in flight and before they impact the container wall. The powder falls to the bottom of the apparatus and is collected together with fines that are recovered during recirculation and purification of the helium gas. Twelve Al-Li compositions have been produced for the present investigation using the (P&WA) rotary atomization technique. The alloys that have been produced are shown in Figure 1, and include those of the Al-Li-Mg-Zr, Al-Li-Cu-Mg-Zr and Al-Li-Cu-X-Zr systems. The compositions studied in these alloys are unique only in that the lithium contents are in the range of 3-4 wt. percent. Typical powder size distributions are shown in figure 1 where it will be noted that the mean particle diameter for these powder runs ranged from 90-100 microns. It has been possible to shift the mean particle size towards lower values, for example 50 microns, by changing certain powder production parameters. Cooling rates achieved by particles from different size fractions as determined from dendrite arm spacing measurements are also shown in figure 1.

It will be noted that even for the smallest particles, the cooling rates achieved were less than 10^5°C/sec. Even slower cooling rates were estimated for the larger particle sizes. For example, the rate measured for particles having a mean diameter of 90 microns (about average for this run) was only 1.5 x 10^4°C/sec. Our experience has indicated that these relatively slow cooling rates are inadequate to retain a high degree of solute supersaturation in the as-produced powder.

Figure 2 shows the morphology and microstructure of typical rotary atomized powder particles. The powder particles are nearly spherical and show well defined dendritic solidification patterns at their surfaces. Polished and etched cross-sections through these particles exhibited predominantly a dendritic microstructure with an appreciable amount of divorced eutectic in the interdendritic regions. This type of starting powder microstructure generally has been found to be unsuitable to yield final fabricated products with acceptable microstructures, even when the consolidation and fabrication steps have been optimized to minimize high temperature exposures of the powder. On occasion, a cellular solidification structure would also be noted in some particles. Nevertheless, for properly processed powder having a suitable chemistry, (i.e. a low degree of supersaturation) a fine, homogeneous, and quite isotropic microstructure can be obtained in the final product, as shown in figure 2.

Two primary concerns regarding P&WA rotary atomized powder have surfaced during this investigation. They fall into the areas of microstructural degradation and contamination. (See figure 3.) The first aspect refers to the less-than-desirable microstructure of the as-solidified powders which show excessive divorced eutectic in the interdendritic regions, particularly with solute rich compositions. This condition often progressively worsens as the powder is put through the processing steps of degassing, compaction, and extrusion. Following solution treatment and aging the excess solute appears as coarse intermetallic particles distributed throughout the microstructure as noted in figure 3(a). As would be expected, the presence of these particles is very detrimental to the ductility and fracture toughness of the resulting products.

The second concern pertains to the contamination of the powder by various types of foreign particles which find their way into the final microstructure as inclusions. The most prevalent type of contamination encountered arose from cross contamination by iron and nickel base particles produced during previous production runs. Typical examples are shown in figure 3(b)-1, and -2. Reaction zones between the foreign particle and the aluminum matrix and the presence of voids at the particle/matrix interface are sometimes observed. This form of contamination can be minimized, or eliminated, when dedicated powder production facilities become available for Al-Li alloys. However, even under these conditions some amount of cross contamination from an Al-Li alloy of different composition may still persist if the same

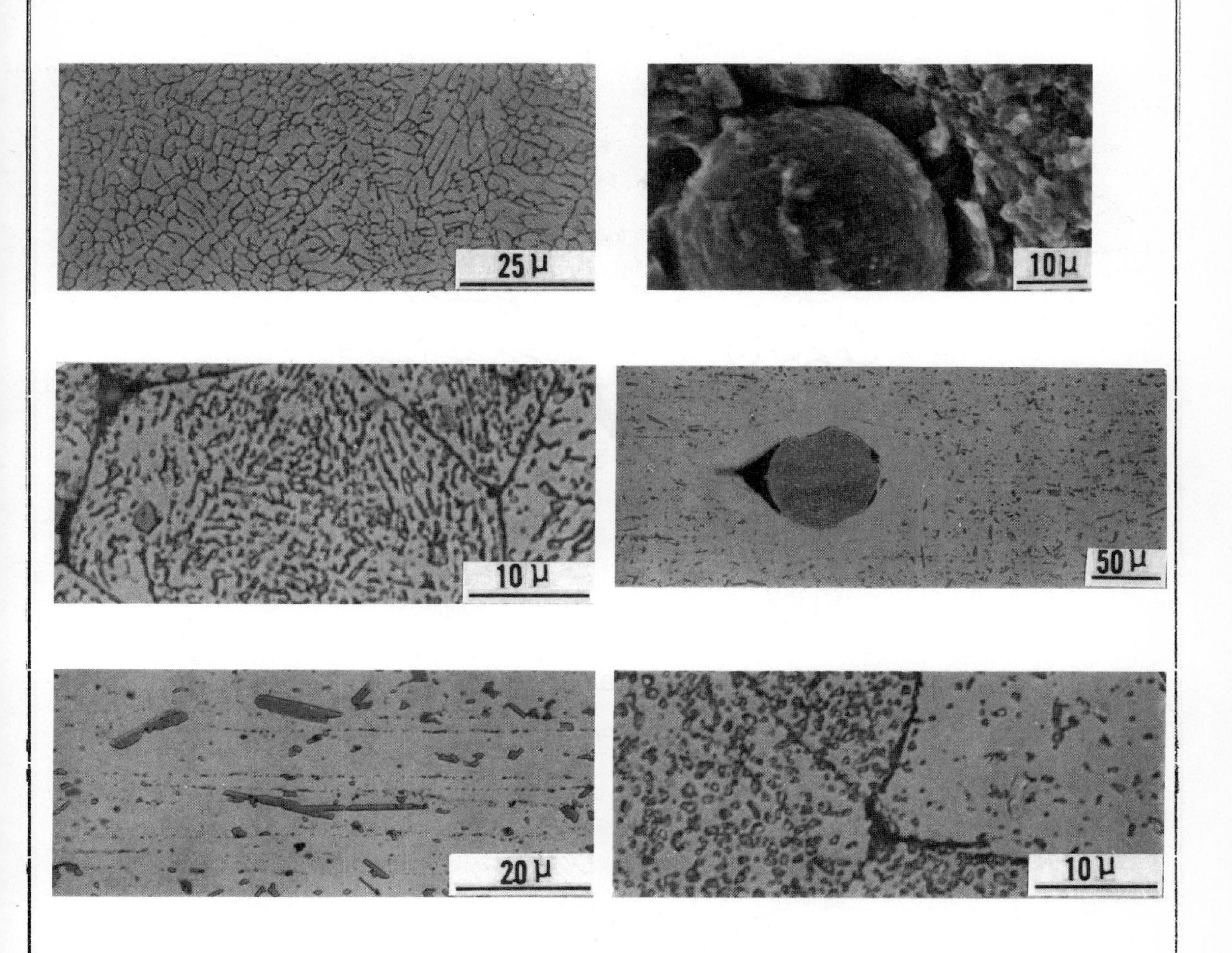

3. Typical examples of microstructural concerns regarding Al-Li P/M alloys produced by rotary atomization technique

(a) microstructural degradation (Al-3.7Li-1.6Mg-2.4Cu-0.21Zr) during processing of powder: top, powder particle; middle, 344°C/4hr, compacted billet; bottom, 344°C/4hr, extruded, 545°C/30 min

(b) foreign particle and powder contamination problems (typical for all alloys): top, Fe and Si rich particle; middle, Ni bearing particle with reaction zone; bottom, powder cross-contamination

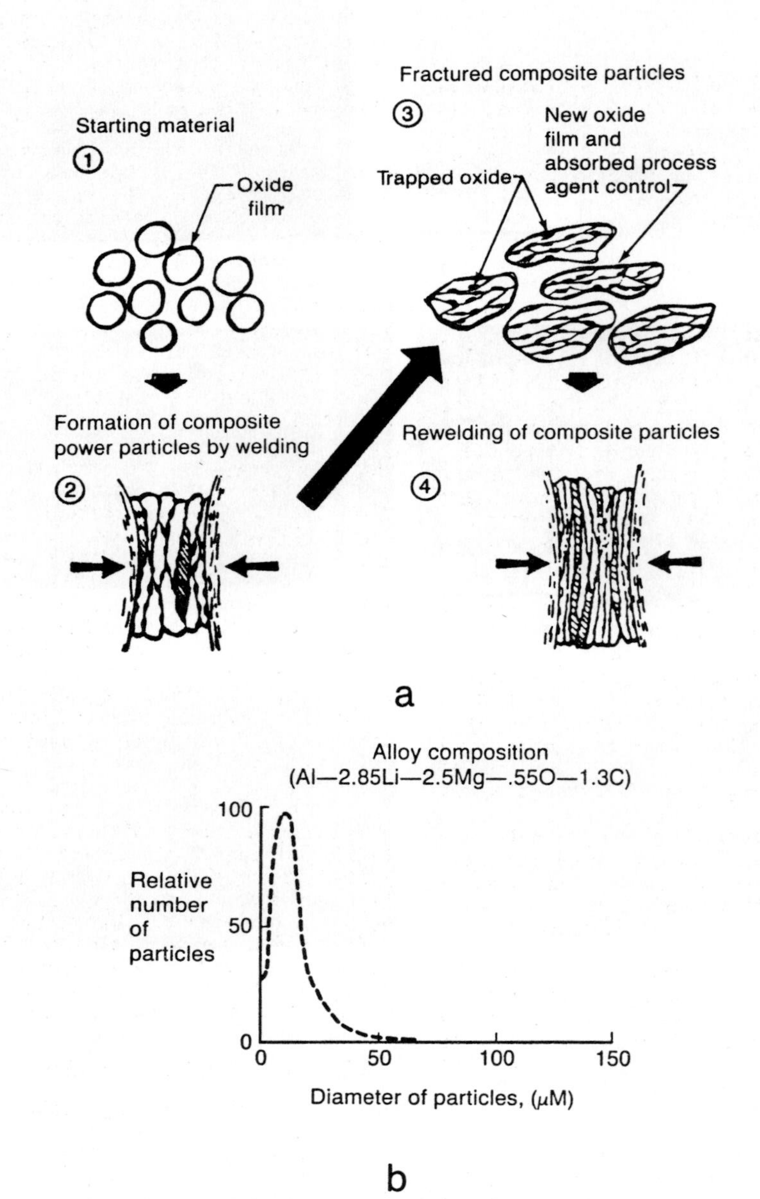

4. (a) schematic diagram of mechanical alloying (MA) process
(b) particle size distribution for MA Al-Li alloy powders

rig is used to produce various Al-Li compositions. See figure 3(b)-3. This occurs because of the difficulty in purging powder production rigs of powder from previous runs owing to the complicated and extensive nature of the ducting found in many of the atomization rigs currently in operation.

Mechanical Alloying

The process of mechanical alloying (MA) which is essentially a solid state alloying technique, is shown schematically in figure 4(a). In this process, a blend of elemental powders of the alloy constituents are normally used as starting stock and are attrited in a high-energy ball mill. During the milling operation the elemental powder particles are flattened and cold welded into composite particles, which subsequently become fragmented due to the extreme coldwork. The process of coldwelding and fracturing is repeated numerous times until a submicron homogeneity is achieved in the particles. Oxygen and carbon are an integral part of the alloy composition in this system. During the ball milling operation, the naturally occurring oxides at the particle surface are broken up and entrapped within the composite particles as an oxide dispersion. Carbon is added as an organic reagent to the powder charge which also serves as a process control agent to prevent excessive cold welding and agglomoration. The carbon reagent reacts with aluminum to produce carbide (Al_4C_3) dispersoids.

Two alloy systems have been explored using the MA approach in the present investigation - the Al-Li-Cu-Mg and the Al-Li-Mg systems - both containing substantial amounts of carbon and oxygen. The latter system has shown most promise so far. In this system, the alloys that are being actively developed have a lithium content in the range of 1.5 to 1.75 wt.%, and a magnesium content of about 4 wt.%. The carbon and oxygen range from 0.5% to over 1.0% for each element. The density of Al-Li-Mg type alloys is relatively low as no heavy alloying elements are included in the composition. Indeed, even with the modest lithium contents noted above the density reductions are very competitive with ingot produced Al-Li alloys that are currently being developed.

A typical size distribution for mechanically alloyed powder is shown in figure 4(b) where it will be noted that the average particle diameter is extremely small, in the range of 10 to 15 microns. This is substantially lower than the 50 to 100 microns average particle diameter range obtained with the rotary atomization technique.

The morphology of the mechanically alloyed powder is shown in figure 5(a). The particles are of irregular shape with complex surface contours. In cross-section the larger particles clearly show evidence that they are conglomerates of smaller particles, with numerous internal fissures and cavities along the apparent interparticle boundaries. See figure 5(b). Consolidation and extrusion of the M/A powder yields a product with homogeneous and extremely fine grained microstructure. See figure 5(c).

The strengthening mechanisms involved in mechanically alloyed products are quite unique compared to both Al-Li ingot metallurgy and other Al-Li powder metallurgy techniques. First, the MA process affords an opportunity to achieve high strengths with little or no reliance on precipitation hardening. Significant strenghtening is obtained from the extremely fine grain size (Petch hardening) and from the very high dislocation density which is developed during the ball milling operation, and which is both very uniform and very stable. In addition, the very fine dispersion of oxide and carbide particles present throughout the microstructure, provide additional dispersion hardening. Consequently, it is possible to design high strength Al-Li alloys that do not rely heavily on δ' precipitates and this is beneficial in that strain localization effects caused by δ' and the attendant adverse influence on fracture characteristics are minimized.

The uniqueness of the strengthening mechanisms present in this system are both an asset and a drawback, however. The primary drawback is that the details of these mechanisms are not sufficiently understood to tailor the alloy properties precisely. Another problem is that the as-consolidated product is considerably stronger than I/M or other P/M counterparts, which makes subsequent fabrication steps somewhat more difficult. This also means that forming operations with these alloys are more difficult because they cannot be put into a ductile heat treatment condition and then subsequently reheat treated to a high strength condition.

There are two types of problems that were encountered during the early stages of MA Al-Li alloy development. First, contamination of the mechanically alloyed powder by iron bearing intermetallic particles has been a serious and persistant problem (see figures 6(a) and (b)), but shows promise of being corrected. Figure 6(b) shows an example of this type of contamination as it influences the tensile fracture of an MA alloy. At present these iron bearing intermetallics are believed to originate, at least in part, from the attriting process. Magnetic separation, as well as other techniques, have recently been successfully employed to remove these contaminates.

The second problem has to do with inadequately processed powder particles that were dispersed throughout the "fully attrited" powders, and which gave rise to microstructural inhomogeneities in the final product. The large particle shown in figure 6(c) is believed to be Al-Li master alloy, as it contains a relatively high amount of lithium and no magnesium. In the final product, these master alloy particles show up as elongated white streaks in the microstructure as shown in figure 6(d).

Planar Flow Casting (Melt Spinning)

The planar flow casting technique pioneered by the Allied Corp. for the production of Al-Li powder is shown schematically in figure 7(a). Alloy melt of the desired composition is prepared in an environmentally-controlled crucible and is delivered through orifices of carefully controlled design onto a rapidly spinning chilled wheel. A great deal of attention has been devoted to understand and optimize the rapid solidification process in order to obtain the desired microstructures. The resulting product is a rapidly solidified thin ribbon which is spun off the wheel and collected. The ribbon product is then comminuted into a rather coarse powder having a typical size-distribution as shown in figure 7(b). As will be noted, the mean particle diameter is nearly 200 microns, considerably larger than those obtained with the powder production methods described earlier. Only two alloy systems have been investigated to date using this method, the Al-Li-Mg-Zr and the Al-Li-Mg-Cu-Zr systems. The compositions

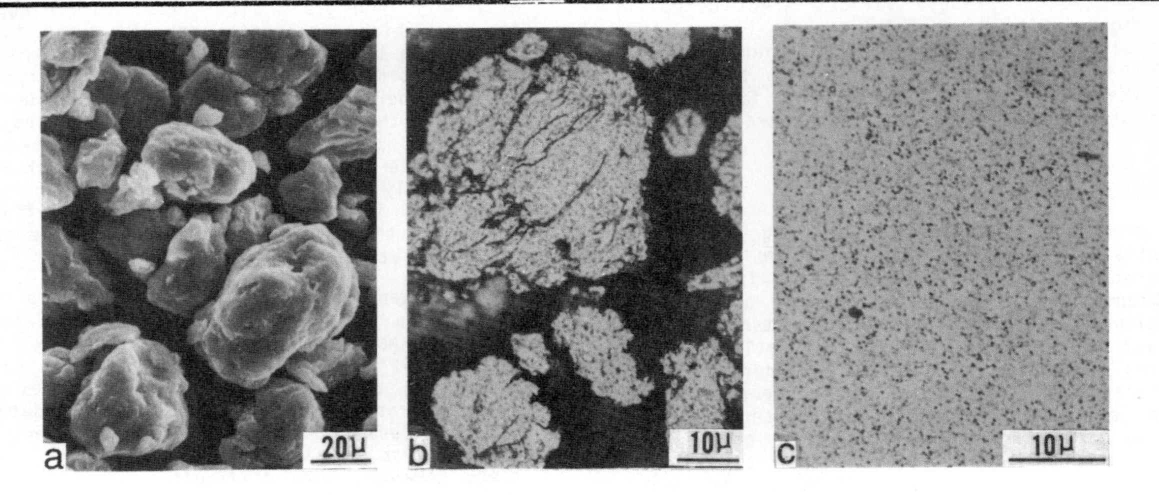

5. Some general characteristics of mechanically alloyed Al-Li alloys (Al-1.7Li-4.0Mg-0.5-1.0C)

(a) powder morphology - as received
(b) powder microstructure - as received
(c) typical product microstructure - extruded product solution treated 493°C/1 hr

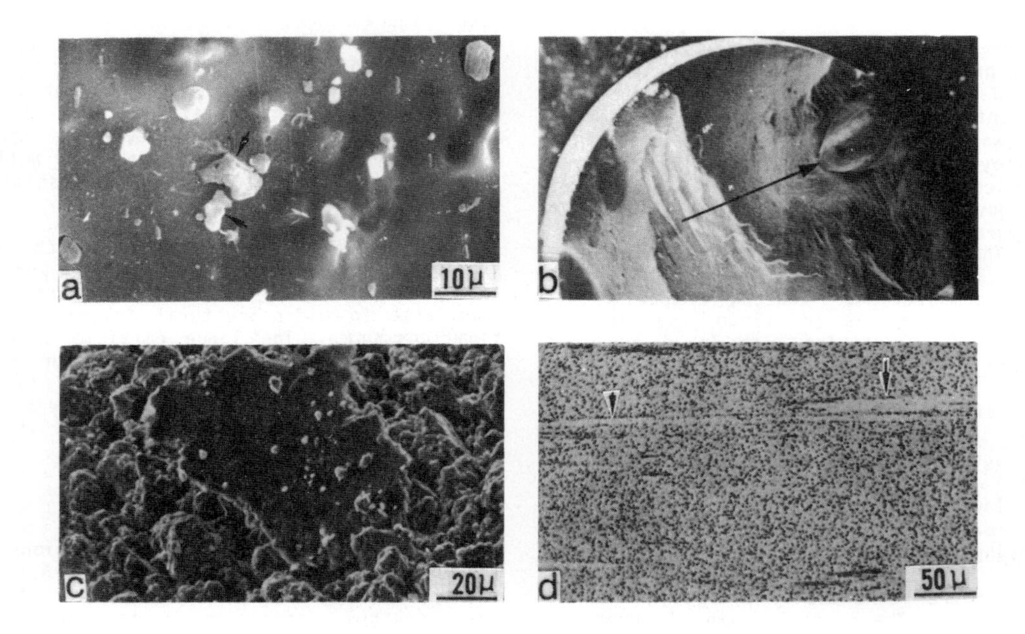

6. Typical examples of microstructural concerns regarding MA Al-Li alloys

(a) foreign particle contamination - Al, Si, Mg, Fe particles (SEM)
(b) foreign particle contamination - iron containing particle
(c) indications of inadequately alloyed master alloy particles - unalloyed constituent in powder
(d) microstructural inhomogeneities in final product - stringers in extruded product

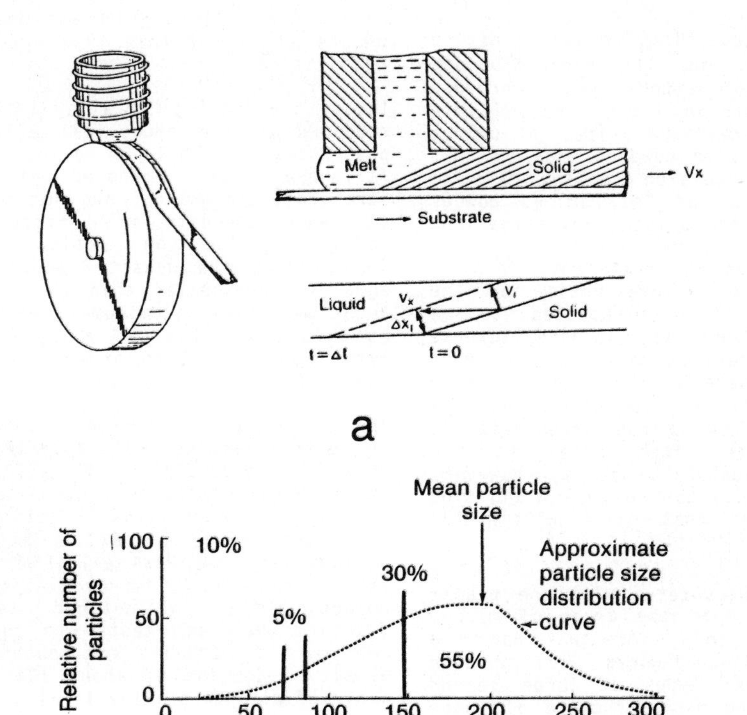

7. (a) schematic diagram of allied planar flow casting method
(b) typical particle size distribution obtained for Al-Li alloys after ribbon pulverization

studied have been relatively standard except that the lithium contents have been in the 3-4% range, and the zirconium contents are commonly about 0.5%. A quartiary Al-Li-Mg-Cu-Zr alloy has shown the most promise.

The morphology of the comminuted planar cast particles is shown in figure 8(a). Particles have a flattened ellipsoid shape with curled edges indicating that some deformation has occurred during the commutating process. In cross-sections they display a very fine microstructure, often having a mixture of type A and type B structures, the substrate side being predominantly type A. See figure 8(b).

Several additional items concerning the unique nature of the melt spin powder and its microstructure deserve comment. First, the methods used to produce the ribbon and powder result in a high cooling rate that is independent of particle size, a unique situation that occurs only with powders made by this method. Secondly, the calculated cooling rate is about 10^6C°/sec, the highest of the various powder techniques explored during the present investigation. Thirdly, the relatively large particle sizes associated with this method translate into a relatively small surface to volume ratio for the individual particles. This is expected to be beneficial, in that it tends to minimize adverse effects originating from surface oxides. The most attractive alloys made to date by the planar casting method show a homogeneous and refined microstructural condition, and are free from coarse inter-metallic particles. Streaking is often noted in the microstructure of consolidated products, however, which is considered to be due to the type A and type B solidification structures that exist within the powder particles. See figure 8(c).

Problems that have been encountered with the planar casting method again fall into the areas of micro-structural degradation and foreign particle contamination. For example, figures 9(a) through 9(c) show several different types of large second phase particles found in the microstructure of these alloys. The high zirconium content in these alloys have occasionally resulted in stringers of Zr-rich particles, as shown in figure 9(c).

Even though the as-produced powder often exhibits a highly refined microstructure, severe breakdown of this microstructure has been noted for some alloys made by the melt spinning process as noted in figure 9(d). The example shown is for a fully consolidated, extruded, and heat treated alloy where the excess solute has precipitated as a high volume fraction of coarse intermetallic particles. However, the most recent Al-Li alloy development activities by The Allied Corp. have shown that improved chemical compositions avoid such undesirable coarsening of second phase constituents and retain a refined microstructure after fabrication and final heat treatment.

Mechanical Properties of P/M Al-Li Alloys

The most attractive mechanical properties developed to date in alloys produced by each of the three P/M technques discussed herein are shown in Table III. The engineering property goals of primary interest in this program are those that emulate 7075-T73 forging material. Typical yield and ultimate strength goals for this alloy and temper are in the order of 455 MPa (65 ksi) and 525 MPa (75 ksi) respectively.

Among the alloys produced by P&WA atomization technique none developed yield and ultimate strength values sufficient to meet the 7075-T73 goals. This was also true of elongation and notch-yield strength values. See Table III. The data shown is for a rather agressive alloy composition with respect to Li content and possessed a density 14% lower than that of alloy 7075. There is certainly a possibility that these properties could be improved somewhat if a more modest density reduction were allowed through adjustment of the alloy chemistry. Nevertheless, the desired combination of properties has not yet been obtained and further process development is required before alloys of this type are suitable for commercialization.

The most promising P/M alloy investigated to date was produced by the mechanical alloying process. The properties of that MA alloy are shown in Table III, where it will be noted that ultimate and yield strengths, elongation, and notched yield ratio values all were adequate with respect to the 7075-T73 forging application goals previously outlined. Importantly, the density reduction achieved is also equal to, or better than that currently feasible by ingot metallurgy aluminum-lithium alloys. Therefore, at this time, this MA alloy appears very promising with respect to both other P/M and I/M Al-Li alloy competitors for the above application.

The properties of the most promising alloy made by the planar casting (melt spinning) technique are also shown in Table III. It will be noted that the strength values are very adequate with respect to the previously stated 7075-T73 goals. In addition, the elongation value of 6% is much better than that exhibited by the P&WA alloy having a lower strength (but somewhat improved density reduction.) Unfortunately, no notched tensile, or similar fracture toughness tests are currently available to confirm the ductility and toughness properties. The density reduction of about 12% compared to 7075-T73 achieved with this alloy is also very attractive.

The competitive position of the three P/M methods evaluated in this study in relation to current I/M Al-Li alloys is summarized in Table IV. This comparison covers the following three aspects: concerns, assessments, and payoffs. As the major concerns regarding each process have already been discussed in some detail, the following discussion will concentrate on the latter two items.

With respect to the P&WA rotary atomization method, it is believed that the various types of contamination problems previously discussed are all solvable. In addition the process is very amenable to scaleup to commercial quantities of aluminum although this will be more difficult than with superalloys which do not require such large volumes. However, in its present state of development the P&WA process does not appear to be suitable for the production of supersaturated Al-Li alloys of the type investigated here, owing to inadequate microstructural refinement and property combinations that have been achieved. In the long run, it is possible that innovative compositions or improvements in technique may brighten future prospects.

The current prospects for near term commercialization of the P/M Al-Li alloys made by the mechanically alloyed technique of Novamet appear quite bright.

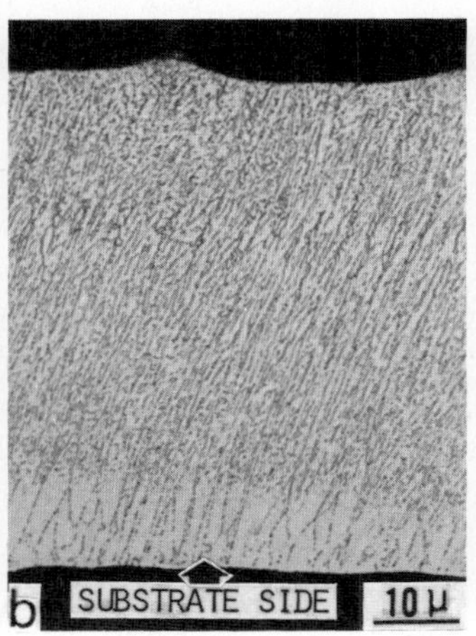

8. Some general characteristics of Al-Li alloys produced by planar flow casting method

(a) powder morphology - as received
(b) powder microstructure - as received
(c) typical microstructure of final product - as extruded

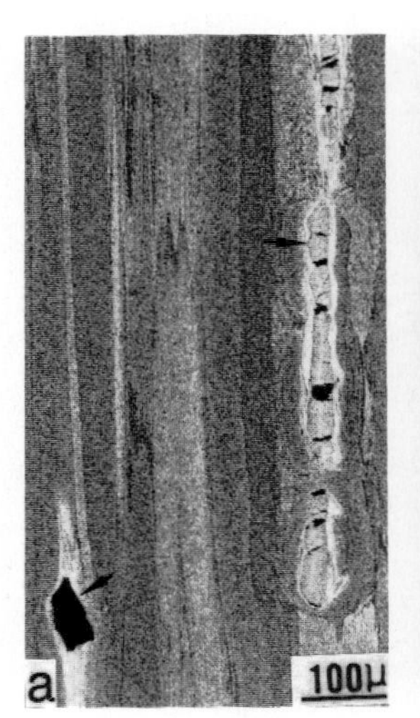

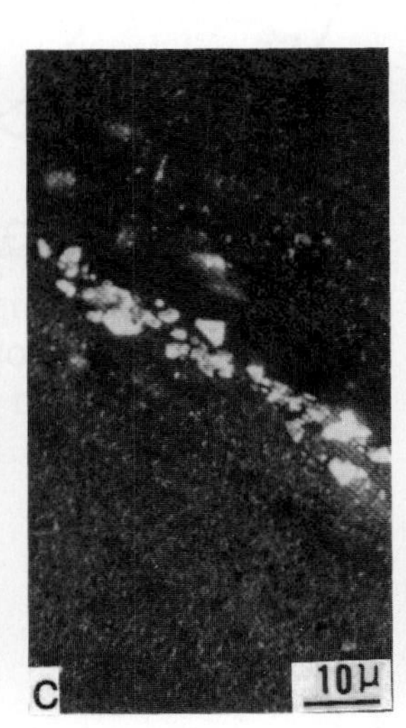

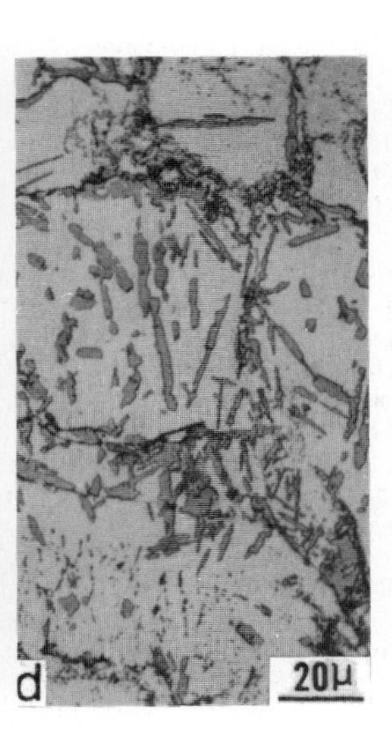

9. Examples of microstructural concerns regarding P/M Al-Li alloys produced by melt spinning method

(a) foreign particle contamination in an Al-2.7Li-5.3Mg-2Zr alloy - stringer inclusions (not identified)
(b) foreign particle contamination in an Al-2.7Li-5.3Mg-2Zr alloy - silicon-rich inclusion (SEM)
(c) foreign particle contamination in an Al-2.7Li-5.3Mg-2Zr alloy - zirconium-rich inclusions (SEM)
(d) microstructural degradation due to excessive particle coarsening - as-compacted billet

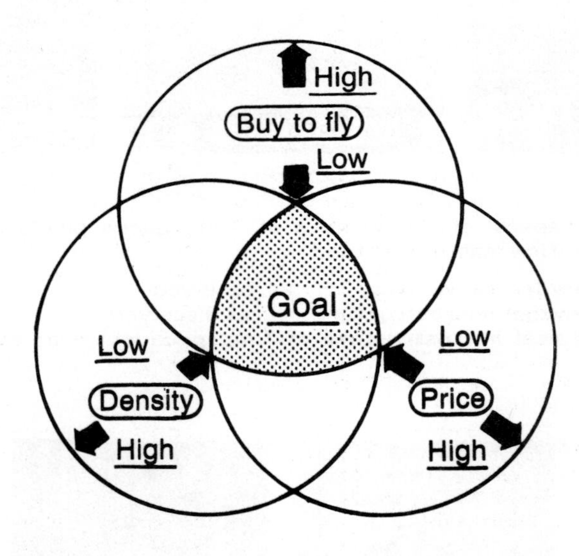

Goal

Develop an improved Al-Li aerospace alloy that is both fabricable and cost effective

10. Manufacturing ease, engineering properties, and material costs form an interlocking system that determines the acceptability of a material for aerospace use

The mechanical property benefits that are required for commercial attractiveness have been demonstrated. The payoffs that can be expected from this technique include a modest increase in density reduction over current I/M aluminum-lithium alloys, more isotropic properties, and improved corrosion and stress corrosion cracking resistance at given strength levels. Scale-up to commercial billet sizes has already been accomplished, as 11-inch diameter billets are currently being produced and further scale-up to 19-inch billet diameters is now being undertaken. It is also believed that contamination problems previously encountered are being satisfactorily resolved by employing magnetic separators and perhaps other means.

For alloys made by the planar flow casting technique of The Allied Corp., the current assessment is also positive. Contamination problems that have been encountered both during alloy melting and the comminution process, are believed to be solvable by having dedicated facilities exclusively available for Al-Li and by the use of improved techniques. One major concern at present is the degree of success that can be achieved when scaling to billet sizes that are commercially viable while avoiding degradation of microstructure and properties. Billet sizes having a minimum of 8 to 11 inches in diameter are required. Finally, a satisfactory mix of strength and fracture properties must be demonstrated and verified by more extensive fracture testing. If these aspects are satisfactorily dealt with, the planar casting technique promises alloys that provide substantial density reduction improvements as compared to current I/M Al-Li alloys. Improved corrosion resistance and more isotropic mechanical properties over current I/M Al-Li alloys are additional benefits that would also be expected.

SUMMARY AND CONCLUSION

A number of the observations and findings of this investigation are particularly noteworthy. They may be summarized as follows:

1. Three P/M techniques, each representing a generic P/M approach, have been carefully assessed with regard to their overall applicability for the productcion of advanced Al-Li alloys. They include the rotary atomization technique of Pratt and Whitney Aircraft, the mechanical alloying technique of Novamet and the planar flow casting or melt-spinning technique of the Allied Corp.

2. The data indicates that the mechanical alloying and planar flow casting techniques offer near term promise for achieving property improvements over current I/M aluminum-lithium alloys.

3. In their present state of development, alloys made by the P&WA rotary atomization technique have not demonstrated the desired microstructural refinement and engineering property combinations required to successfully compete with other P/M or I/M Al-Li alloys.

4. The current generation of MA Al-Li alloys offered by Novamet provide an 0-2% additional density reduction over current I/M Al-Li alloys, with more isotropic mechanical properties and improved corrosion and SCC resistance at given strength levels.

5. The alloys presently produced by Allied using the planar flow casting technique provide an 3-5% additional density reduction over current Al-Li alloys in conjunction with more isotropic properties and improved corrosion resistance. However, a satisfactory mix of strength and toughness must be demonstrated before their commercial viability can be fully assessed.

6. P/M alloys produced via the rapid solidification approach, are somewhat susceptible to scale-up problems that may lead to microstructural and property degradation.

7. It is envisioned that Al-Li alloys produced by P/M techniques would be used primarily for small and intermediate sized forgings and for relatively small extrusion and plate applications where the total weight of product does not exceed several hundred pounds.

8. The economics of the various Al-Li P/M approaches will be a critical aspect of commercial acceptance. If the most important considerations contributing to commercial success of Al-Li P/M products are represented by three interlocking spheres as shown in figure 10, the following observations may be made. One sphere represents the ease and efficiency of fabricating alloys into parts; that is identified as fly-to-buy ratio. The second sphere represents the attractiveness of the engineering properties and for this example is entitled density. The third sphere represents the price of the material. If these spheres fit together in proper prospective, an area of overlap or opportunity (goal) is developed as is noted in figure 10. The better the relationship between the three spheres the larger and more significant the area of opportunity. However, if one or more of the sphears move apart the area of opportunity decreases, and indeed, if the moves are too adverse, the area of opportunity will totally disappear. The challenge is to keep all three items, inclucing price, in the proper relationship. Only in that way will commercial success be achieved.

ACKNOWLEDGEMENTS

Portions of this work were supported under Air Force Contract No. F33615-81-C-5053 in cooperation with the Northrop Corporation, the University of Virginia, and the Kaiser Aluminum and Chemical Corporation. Alloys and powders were produced by Pratt & Whitney Aircraft, Novamet, and the Allied Corporation. T. H. Sanders, Jr., of Purdue University, was a consultant to the contract throughout the investigation. Each of these contributors is thanked for their valuable contributions.

REFERENCES

1. R. E. Lewis, D. Webster, and I. G. Palmer, "A Feasibility Study for Development of Structural Aluminum Alloys From Rapidly Solidified Powders for Aerospace Applications," AFML-TR-78-102 (July 1978).

2. R. E. Lewis, et al., "Development of Advanced Aluminum Alloys From Rapidly Solidified Powders for Aerospace Application," interim technical reports on DARPA Contract F33615-78-C-5203, Order 3575: (a). LMSC-D674504 (March 1979); (b). LMSC-D678772 (September 1979); (c). LMSC-D686125

(March 1980); (d). LMSC-D770654 (June 1980) (e). LMSC-D777825 (September 1980).

3. R. E. Lewis, et al., "Development of Advanced Aluminum Alloys From Rapidly Solidified Powders for Aerospace Application," Phase I, final technical report on DARPA/AFWAL Contract F33615-78-C-5203 (1983).

4. E. A. Starke, Jr., T. H. Sanders, Jr., and I. G. Palmer, "New Approaches to Alloy Development in the Al-Li System", J. of Metals, vol. 33, no. 8, pp. 24-33 (1981).

5. G. H. Narayanan, W. E. Quist, B. L. Wilson, and A. L. Wingert, "Low Density Aluminum Alloy Development", first interim reports; AFWAL Contract F33615-81-C-5053; (a) D6-51411, (August 1982); (b) D6-51411-1 (April 1983); (c) D6-51411-2 (Nov. 1983); (d) D6-5411-3 (May 1984).

6. W. E. Quist, G. H. Narayanan and A. L. Wingert, "Aluminum-Lithium Alloys for Aircraft Strutcure: An Overview" Aluminum-Lithium Alloys II, E. A. Starke, Jr. and T. H. Sanders, Jr., Editors, TMS-AIME, Warrendale, PA, 1984, page 313.

7. W. E. Quist and G. H. Narayanan, "Powder Metallurgy P/M Aluminum Alloys for Aerospace Use", SAMPE, Anaheim, CA, April 1983.

8. Aluminum-Lithium Alloys, Proceedings of the First International Aluminum-Lithium Conference, T. H. Sanders, Jr. and E. A. Starke, Jr., Eds., The Metallurgical Society of AIME, 1981.

9. Aluminum-Lithium Alloys II, Proceedings of the Second International Aluminum-Lithium Conference, T. H. Sanders, Jr. and E. A. Starke, Jr., Eds., The Metallurgical Society of AIME, 1984.

10. Third International Aluminum-Lithium Conference, Oxford, England, July 8-11, 1985.

11. Amorphous and Metastable Microcrystalline Rapidly Solidified Alloys, Status and Potential, National Materials Advisory Board, Report No. NMAB-358, May 1980, page 60.

12. Cox, A. R., et al., Superalloys-Metallurgy and Manufacture, p. 45, AIME, New York, 1976.

13. G. H. Narayanan, B. L. Wilson, and W. E. Quist, "P/M Aluminum-Lithium Alloys by the Mechanical Alloying Process", Aluminum-Lithium Alloys II, E.A. Starke, Jr. and T. H. Sanders, Jr., Editors, TMS-AIME, Warrendale, PA, 1984, page 517.

14. P. S. Gilman and S. J. Donachie, "The Microstructure and Properties of Al-4Mg-Li Alloys Prepared by Mechanical Alloying", Aluminum-Lithium Alloys II, E. A. Starke, Jr. and T. H. Sanders, Jr., Editors, TMS-AIME, Warrendale, PA, 1984, page 507.

15. N. J. Kim, D. J. Skinner and C. M. Adam, "Planar Flow Casting of RSP Al-High Li Alloys", Presented at WESTEC '85, Los Angeles, California, March 19, 1985.

INDEX OF AUTHORS